Lecture Notes in Computer Science   12359

More information about this series at http://www.springer.com/series/7412

Foreword

Hosting the European Conference on Computer Vision (ECCV 2020) was certainly an exciting journey. From the 2016 plan to hold it at the Edinburgh International Conference Centre (hosting 1,800 delegates) to the 2018 plan to hold it at Glasgow's Scottish Exhibition Centre (up to 6,000 delegates), we finally ended with moving online because of the COVID-19 outbreak. While possibly having fewer delegates than expected because of the online format, ECCV 2020 still had over 3,100 registered participants.

Although online, the conference delivered most of the activities expected at a face-to-face conference: peer-reviewed papers, industrial exhibitors, demonstrations, and messaging between delegates. In addition to the main technical sessions, the conference included a strong program of satellite events with 16 tutorials and 44 workshops.

Furthermore, the online conference format enabled new conference features. Every paper had an associated teaser video and a longer full presentation video. Along with the papers and slides from the videos, all these materials were available the week before the conference. This allowed delegates to become familiar with the paper content and be ready for the live interaction with the authors during the conference week. The live event consisted of brief presentations by the oral and spotlight authors and industrial sponsors. Question and answer sessions for all papers were timed to occur twice so delegates from around the world had convenient access to the authors.

As with ECCV 2018, authors' draft versions of the papers appeared online with open access, now on both the Computer Vision Foundation (CVF) and the European Computer Vision Association (ECVA) websites. An archival publication arrangement was put in place with the cooperation of Springer. SpringerLink hosts the final version of the papers with further improvements, such as activating reference links and supplementary materials. These two approaches benefit all potential readers: a version available freely for all researchers, and an authoritative and citable version with additional benefits for SpringerLink subscribers. We thank Alfred Hofmann and Aliaksandr Birukou from Springer for helping to negotiate this agreement, which we expect will continue for future versions of ECCV.

August 2020

Vittorio Ferrari
Bob Fisher
Cordelia Schmid
Emanuele Trucco

Preface

Welcome to the proceedings of the European Conference on Computer Vision (ECCV 2020). This is a unique edition of ECCV in many ways. Due to the COVID-19 pandemic, this is the first time the conference was held online, in a virtual format. This was also the first time the conference relied exclusively on the Open Review platform to manage the review process. Despite these challenges ECCV is thriving. The conference received 5,150 valid paper submissions, of which 1,360 were accepted for publication (27%) and, of those, 160 were presented as spotlights (3%) and 104 as orals (2%). This amounts to more than twice the number of submissions to ECCV 2018 (2,439). Furthermore, CVPR, the largest conference on computer vision, received 5,850 submissions this year, meaning that ECCV is now 87% the size of CVPR in terms of submissions. By comparison, in 2018 the size of ECCV was only 73% of CVPR.

The review model was similar to previous editions of ECCV; in particular, it was double blind in the sense that the authors did not know the name of the reviewers and vice versa. Furthermore, each conference submission was held confidentially, and was only publicly revealed if and once accepted for publication. Each paper received at least three reviews, totalling more than 15,000 reviews. Handling the review process at this scale was a significant challenge. In order to ensure that each submission received as fair and high-quality reviews as possible, we recruited 2,830 reviewers (a 130% increase with reference to 2018) and 207 area chairs (a 60% increase). The area chairs were selected based on their technical expertise and reputation, largely among people that served as area chair in previous top computer vision and machine learning conferences (ECCV, ICCV, CVPR, NeurIPS, etc.). Reviewers were similarly invited from previous conferences. We also encouraged experienced area chairs to suggest additional chairs and reviewers in the initial phase of recruiting.

Despite doubling the number of submissions, the reviewer load was slightly reduced from 2018, from a maximum of 8 papers down to 7 (with some reviewers offering to handle 6 papers plus an emergency review). The area chair load increased slightly, from 18 papers on average to 22 papers on average.

Conflicts of interest between authors, area chairs, and reviewers were handled largely automatically by the Open Review platform via their curated list of user profiles. Many authors submitting to ECCV already had a profile in Open Review. We set a paper registration deadline one week before the paper submission deadline in order to encourage all missing authors to register and create their Open Review profiles well on time (in practice, we allowed authors to create/change papers arbitrarily until the submission deadline). Except for minor issues with users creating duplicate profiles, this allowed us to easily and quickly identify institutional conflicts, and avoid them, while matching papers to area chairs and reviewers.

Papers were matched to area chairs based on: an affinity score computed by the Open Review platform, which is based on paper titles and abstracts, and an affinity score computed by the Toronto Paper Matching System (TPMS), which is based on the paper's full text, the area chair bids for individual papers, load balancing, and conflict avoidance. Open Review provides the program chairs a convenient web interface to experiment with different configurations of the matching algorithm. The chosen configuration resulted in about 50% of the assigned papers to be highly ranked by the area chair bids, and 50% to be ranked in the middle, with very few low bids assigned.

Assignments to reviewers were similar, with two differences. First, there was a maximum of 7 papers assigned to each reviewer. Second, area chairs recommended up to seven reviewers per paper, providing another highly-weighed term to the affinity scores used for matching.

The assignment of papers to area chairs was smooth. However, it was more difficult to find suitable reviewers for all papers. Having a ratio of 5.6 papers per reviewer with a maximum load of 7 (due to emergency reviewer commitment), which did not allow for much wiggle room in order to also satisfy conflict and expertise constraints. We received some complaints from reviewers who did not feel qualified to review specific papers and we reassigned them wherever possible. However, the large scale of the conference, the many constraints, and the fact that a large fraction of such complaints arrived very late in the review process made this process very difficult and not all complaints could be addressed.

Reviewers had six weeks to complete their assignments. Possibly due to COVID-19 or the fact that the NeurIPS deadline was moved closer to the review deadline, a record 30% of the reviews were still missing after the deadline. By comparison, ECCV 2018 experienced only 10% missing reviews at this stage of the process. In the subsequent week, area chairs chased the missing reviews intensely, found replacement reviewers in their own team, and managed to reach 10% missing reviews. Eventually, we could provide almost all reviews (more than 99.9%) with a delay of only a couple of days on the initial schedule by a significant use of emergency reviews. If this trend is confirmed, it might be a major challenge to run a smooth review process in future editions of ECCV. The community must reconsider prioritization of the time spent on paper writing (the number of submissions increased a lot despite COVID-19) and time spent on paper reviewing (the number of reviews delivered in time decreased a lot presumably due to COVID-19 or NeurIPS deadline). With this imbalance the peer-review system that ensures the quality of our top conferences may break soon.

Reviewers submitted their reviews independently. In the reviews, they had the opportunity to ask questions to the authors to be addressed in the rebuttal. However, reviewers were told not to request any significant new experiment. Using the Open Review interface, authors could provide an answer to each individual review, but were also allowed to cross-reference reviews and responses in their answers. Rather than PDF files, we allowed the use of formatted text for the rebuttal. The rebuttal and initial reviews were then made visible to all reviewers and the primary area chair for a given paper. The area chair encouraged and moderated the reviewer discussion. During the discussions, reviewers were invited to reach a consensus and possibly adjust their ratings as a result of the discussion and of the evidence in the rebuttal.

After the discussion period ended, most reviewers entered a final rating and recommendation, although in many cases this did not differ from their initial recommendation. Based on the updated reviews and discussion, the primary area chair then made a preliminary decision to accept or reject the paper and wrote a justification for it (meta-review). Except for cases where the outcome of this process was absolutely clear (as indicated by the three reviewers and primary area chairs all recommending clear rejection), the decision was then examined and potentially challenged by a secondary area chair. This led to further discussion and overturning a small number of preliminary decisions. Needless to say, there was no in-person area chair meeting, which would have been impossible due to COVID-19.

Area chairs were invited to observe the consensus of the reviewers whenever possible and use extreme caution in overturning a clear consensus to accept or reject a paper. If an area chair still decided to do so, she/he was asked to clearly justify it in the meta-review and to explicitly obtain the agreement of the secondary area chair. In practice, very few papers were rejected after being confidently accepted by the reviewers.

This was the first time Open Review was used as the main platform to run ECCV. In 2018, the program chairs used CMT3 for the user-facing interface and Open Review internally, for matching and conflict resolution. Since it is clearly preferable to only use a single platform, this year we switched to using Open Review in full. The experience was largely positive. The platform is highly-configurable, scalable, and open source. Being written in Python, it is easy to write scripts to extract data programmatically. The paper matching and conflict resolution algorithms and interfaces are top-notch, also due to the excellent author profiles in the platform. Naturally, there were a few kinks along the way due to the fact that the ECCV Open Review configuration was created from scratch for this event and it differs in substantial ways from many other Open Review conferences. However, the Open Review development and support team did a fantastic job in helping us to get the configuration right and to address issues in a timely manner as they unavoidably occurred. We cannot thank them enough for the tremendous effort they put into this project.

Finally, we would like to thank everyone involved in making ECCV 2020 possible in these very strange and difficult times. This starts with our authors, followed by the area chairs and reviewers, who ran the review process at an unprecedented scale. The whole Open Review team (and in particular Melisa Bok, Mohit Unyal, Carlos Mondragon Chapa, and Celeste Martinez Gomez) worked incredibly hard for the entire duration of the process. We would also like to thank René Vidal for contributing to the adoption of Open Review. Our thanks also go to Laurent Charling for TPMS and to the program chairs of ICML, ICLR, and NeurIPS for cross checking double submissions. We thank the website chair, Giovanni Farinella, and the CPI team (in particular Ashley Cook, Miriam Verdon, Nicola McGrane, and Sharon Kerr) for promptly adding material to the website as needed in the various phases of the process. Finally, we thank the publication chairs, Albert Ali Salah, Hamdi Dibeklioglu, Metehan Doyran, Henry Howard-Jenkins, Victor Prisacariu, Siyu Tang, and Gul Varol, who managed to compile these substantial proceedings in an exceedingly compressed schedule. We express our thanks to the ECVA team, in particular Kristina Scherbaum for allowing open access of the proceedings. We thank Alfred Hofmann from Springer who again serve as the publisher. Finally, we thank the other chairs of ECCV 2020, including in particular the general chairs for very useful feedback with the handling of the program.

August 2020   Andrea Vedaldi
                                                 Horst Bischof
                                                 Thomas Brox
                                                 Jan-Michael Frahm

Organization

General Chairs

Vittorio Ferrari	Google Research, Switzerland
Bob Fisher	University of Edinburgh, UK
Cordelia Schmid	Google and Inria, France
Emanuele Trucco	University of Dundee, UK

Program Chairs

Andrea Vedaldi	University of Oxford, UK
Horst Bischof	Graz University of Technology, Austria
Thomas Brox	University of Freiburg, Germany
Jan-Michael Frahm	University of North Carolina, USA

Industrial Liaison Chairs

Jim Ashe	University of Edinburgh, UK
Helmut Grabner	Zurich University of Applied Sciences, Switzerland
Diane Larlus	NAVER LABS Europe, France
Cristian Novotny	University of Edinburgh, UK

Local Arrangement Chairs

Yvan Petillot	Heriot-Watt University, UK
Paul Siebert	University of Glasgow, UK

Academic Demonstration Chair

Thomas Mensink	Google Research and University of Amsterdam, The Netherlands

Poster Chair

Stephen Mckenna	University of Dundee, UK

Technology Chair

Gerardo Aragon Camarasa	University of Glasgow, UK

Tutorial Chairs

Carlo Colombo   University of Florence, Italy
Sotirios Tsaftaris   University of Edinburgh, UK

Publication Chairs

Albert Ali Salah   Utrecht University, The Netherlands
Hamdi Dibeklioglu   Bilkent University, Turkey
Metehan Doyran   Utrecht University, The Netherlands
Henry Howard-Jenkins   University of Oxford, UK
Victor Adrian Prisacariu   University of Oxford, UK
Siyu Tang   ETH Zurich, Switzerland
Gul Varol   University of Oxford, UK

Website Chair

Giovanni Maria Farinella   University of Catania, Italy

Workshops Chairs

Adrien Bartoli   University of Clermont Auvergne, France
Andrea Fusiello   University of Udine, Italy

Area Chairs

Lourdes Agapito   University College London, UK
Zeynep Akata   University of Tübingen, Germany
Karteek Alahari   Inria, France
Antonis Argyros   University of Crete, Greece
Hossein Azizpour   KTH Royal Institute of Technology, Sweden
Joao P. Barreto   Universidade de Coimbra, Portugal
Alexander C. Berg   University of North Carolina at Chapel Hill, USA
Matthew B. Blaschko   KU Leuven, Belgium
Lubomir D. Bourdev   WaveOne, Inc., USA
Edmond Boyer   Inria, France
Yuri Boykov   University of Waterloo, Canada
Gabriel Brostow   University College London, UK
Michael S. Brown   National University of Singapore, Singapore
Jianfei Cai   Monash University, Australia
Barbara Caputo   Politecnico di Torino, Italy
Ayan Chakrabarti   Washington University, St. Louis, USA
Tat-Jen Cham   Nanyang Technological University, Singapore
Manmohan Chandraker   University of California, San Diego, USA
Rama Chellappa   Johns Hopkins University, USA
Liang-Chieh Chen   Google, USA

Yung-Yu Chuang	National Taiwan University, Taiwan
Ondrej Chum	Czech Technical University in Prague, Czech Republic
Brian Clipp	Kitware, USA
John Collomosse	University of Surrey and Adobe Research, UK
Jason J. Corso	University of Michigan, USA
David J. Crandall	Indiana University, USA
Daniel Cremers	University of California, Los Angeles, USA
Fabio Cuzzolin	Oxford Brookes University, UK
Jifeng Dai	SenseTime, SAR China
Kostas Daniilidis	University of Pennsylvania, USA
Andrew Davison	Imperial College London, UK
Alessio Del Bue	Fondazione Istituto Italiano di Tecnologia, Italy
Jia Deng	Princeton University, USA
Alexey Dosovitskiy	Google, Germany
Matthijs Douze	Facebook, France
Enrique Dunn	Stevens Institute of Technology, USA
Irfan Essa	Georgia Institute of Technology and Google, USA
Giovanni Maria Farinella	University of Catania, Italy
Ryan Farrell	Brigham Young University, USA
Paolo Favaro	University of Bern, Switzerland
Rogerio Feris	International Business Machines, USA
Cornelia Fermuller	University of Maryland, College Park, USA
David J. Fleet	Vector Institute, Canada
Friedrich Fraundorfer	DLR, Austria
Mario Fritz	CISPA Helmholtz Center for Information Security, Germany
Pascal Fua	EPFL (Swiss Federal Institute of Technology Lausanne), Switzerland
Yasutaka Furukawa	Simon Fraser University, Canada
Li Fuxin	Oregon State University, USA
Efstratios Gavves	University of Amsterdam, The Netherlands
Peter Vincent Gehler	Amazon, USA
Theo Gevers	University of Amsterdam, The Netherlands
Ross Girshick	Facebook AI Research, USA
Boqing Gong	Google, USA
Stephen Gould	Australian National University, Australia
Jinwei Gu	SenseTime Research, USA
Abhinav Gupta	Facebook, USA
Bohyung Han	Seoul National University, South Korea
Bharath Hariharan	Cornell University, USA
Tal Hassner	Facebook AI Research, USA
Xuming He	Australian National University, Australia
Joao F. Henriques	University of Oxford, UK
Adrian Hilton	University of Surrey, UK
Minh Hoai	Stony Brooks, State University of New York, USA
Derek Hoiem	University of Illinois Urbana-Champaign, USA

Timothy Hospedales	University of Edinburgh and Samsung, UK
Gang Hua	Wormpex AI Research, USA
Slobodan Ilic	Siemens AG, Germany
Hiroshi Ishikawa	Waseda University, Japan
Jiaya Jia	The Chinese University of Hong Kong, SAR China
Hailin Jin	Adobe Research, USA
Justin Johnson	University of Michigan, USA
Frederic Jurie	University of Caen Normandie, France
Fredrik Kahl	Chalmers University, Sweden
Sing Bing Kang	Zillow, USA
Gunhee Kim	Seoul National University, South Korea
Junmo Kim	Korea Advanced Institute of Science and Technology, South Korea
Tae-Kyun Kim	Imperial College London, UK
Ron Kimmel	Technion-Israel Institute of Technology, Israel
Alexander Kirillov	Facebook AI Research, USA
Kris Kitani	Carnegie Mellon University, USA
Iasonas Kokkinos	Ariel AI, UK
Vladlen Koltun	Intel Labs, USA
Nikos Komodakis	Ecole des Ponts ParisTech, France
Piotr Koniusz	Australian National University, Australia
M. Pawan Kumar	University of Oxford, UK
Kyros Kutulakos	University of Toronto, Canada
Christoph Lampert	IST Austria, Austria
Ivan Laptev	Inria, France
Diane Larlus	NAVER LABS Europe, France
Laura Leal-Taixe	Technical University Munich, Germany
Honglak Lee	Google and University of Michigan, USA
Joon-Young Lee	Adobe Research, USA
Kyoung Mu Lee	Seoul National University, South Korea
Seungyong Lee	POSTECH, South Korea
Yong Jae Lee	University of California, Davis, USA
Bastian Leibe	RWTH Aachen University, Germany
Victor Lempitsky	Samsung, Russia
Ales Leonardis	University of Birmingham, UK
Marius Leordeanu	Institute of Mathematics of the Romanian Academy, Romania
Vincent Lepetit	ENPC ParisTech, France
Hongdong Li	The Australian National University, Australia
Xi Li	Zhejiang University, China
Yin Li	University of Wisconsin-Madison, USA
Zicheng Liao	Zhejiang University, China
Jongwoo Lim	Hanyang University, South Korea
Stephen Lin	Microsoft Research Asia, China
Yen-Yu Lin	National Chiao Tung University, Taiwan, China
Zhe Lin	Adobe Research, USA

Haibin Ling	Stony Brooks, State University of New York, USA
Jiaying Liu	Peking University, China
Ming-Yu Liu	NVIDIA, USA
Si Liu	Beihang University, China
Xiaoming Liu	Michigan State University, USA
Huchuan Lu	Dalian University of Technology, China
Simon Lucey	Carnegie Mellon University, USA
Jiebo Luo	University of Rochester, USA
Julien Mairal	Inria, France
Michael Maire	University of Chicago, USA
Subhransu Maji	University of Massachusetts, Amherst, USA
Yasushi Makihara	Osaka University, Japan
Jiri Matas	Czech Technical University in Prague, Czech Republic
Yasuyuki Matsushita	Osaka University, Japan
Philippos Mordohai	Stevens Institute of Technology, USA
Vittorio Murino	University of Verona, Italy
Naila Murray	NAVER LABS Europe, France
Hajime Nagahara	Osaka University, Japan
P. J. Narayanan	International Institute of Information Technology (IIIT), Hyderabad, India
Nassir Navab	Technical University of Munich, Germany
Natalia Neverova	Facebook AI Research, France
Matthias Niessner	Technical University of Munich, Germany
Jean-Marc Odobez	Idiap Research Institute and Swiss Federal Institute of Technology Lausanne, Switzerland
Francesca Odone	Università di Genova, Italy
Takeshi Oishi	The University of Tokyo, Tokyo Institute of Technology, Japan
Vicente Ordonez	University of Virginia, USA
Manohar Paluri	Facebook AI Research, USA
Maja Pantic	Imperial College London, UK
In Kyu Park	Inha University, South Korea
Ioannis Patras	Queen Mary University of London, UK
Patrick Perez	Valeo, France
Bryan A. Plummer	Boston University, USA
Thomas Pock	Graz University of Technology, Austria
Marc Pollefeys	ETH Zurich and Microsoft MR & AI Zurich Lab, Switzerland
Jean Ponce	Inria, France
Gerard Pons-Moll	MPII, Saarland Informatics Campus, Germany
Jordi Pont-Tuset	Google, Switzerland
James Matthew Rehg	Georgia Institute of Technology, USA
Ian Reid	University of Adelaide, Australia
Olaf Ronneberger	DeepMind London, UK
Stefan Roth	TU Darmstadt, Germany
Bryan Russell	Adobe Research, USA

Kwang Moo Yi   University of Victoria, Canada
Zhaozheng Yin   Stony Brook, State University of New York, USA
Chang D. Yoo   Korea Advanced Institute of Science and Technology,
                                  South Korea
Shaodi You   University of Amsterdam, The Netherlands
Jingyi Yu   ShanghaiTech University, China
Stella Yu   University of California, Berkeley, and ICSI, USA
Stefanos Zafeiriou   Imperial College London, UK
Hongbin Zha   Peking University, China
Tianzhu Zhang   University of Science and Technology of China, China
Liang Zheng   Australian National University, Australia
Todd E. Zickler   Harvard University, USA
Andrew Zisserman   University of Oxford, UK

Technical Program Committee

Sathyanarayanan
  N. Aakur
Wael Abd Almgaeed
Abdelrahman
  Abdelhamed
Abdullah Abuolaim
Supreeth Achar
Hanno Ackermann
Ehsan Adeli
Triantafyllos Afouras
Sameer Agarwal
Aishwarya Agrawal
Harsh Agrawal
Pulkit Agrawal
Antonio Agudo
Eirikur Agustsson
Karim Ahmed
Byeongjoo Ahn
Unaiza Ahsan
Thalaiyasingam Ajanthan
Kenan E. Ak
Emre Akbas
Naveed Akhtar
Derya Akkaynak
Yagiz Aksoy
Ziad Al-Halah
Xavier Alameda-Pineda
Jean-Baptiste Alayrac

Samuel Albanie
Shadi Albarqouni
Cenek Albl
Hassan Abu Alhaija
Daniel Aliaga
Mohammad
  S. Aliakbarian
Rahaf Aljundi
Thiemo Alldieck
Jon Almazan
Jose M. Alvarez
Senjian An
Saket Anand
Codruta Ancuti
Cosmin Ancuti
Peter Anderson
Juan Andrade-Cetto
Alexander Andreopoulos
Misha Andriluka
Dragomir Anguelov
Rushil Anirudh
Michel Antunes
Oisin Mac Aodha
Srikar Appalaraju
Relja Arandjelovic
Nikita Araslanov
Andre Araujo
Helder Araujo

Pablo Arbelaez
Shervin Ardeshir
Sercan O. Arik
Anil Armagan
Anurag Arnab
Chetan Arora
Federica Arrigoni
Mathieu Aubry
Shai Avidan
Angelica I. Aviles-Rivero
Yannis Avrithis
Ismail Ben Ayed
Shekoofeh Azizi
Ioan Andrei Bârsan
Artem Babenko
Deepak Babu Sam
Seung-Hwan Baek
Seungryul Baek
Andrew D. Bagdanov
Shai Bagon
Yuval Bahat
Junjie Bai
Song Bai
Xiang Bai
Yalong Bai
Yancheng Bai
Peter Bajcsy
Slawomir Bak

Mahsa Baktashmotlagh
Kavita Bala
Yogesh Balaji
Guha Balakrishnan
V. N. Balasubramanian
Federico Baldassarre
Vassileios Balntas
Shurjo Banerjee
Aayush Bansal
Ankan Bansal
Jianmin Bao
Linchao Bao
Wenbo Bao
Yingze Bao
Akash Bapat
Md Jawadul Hasan Bappy
Fabien Baradel
Lorenzo Baraldi
Daniel Barath
Adrian Barbu
Kobus Barnard
Nick Barnes
Francisco Barranco
Jonathan T. Barron
Arslan Basharat
Chaim Baskin
Anil S. Baslamisli
Jorge Batista
Kayhan Batmanghelich
Konstantinos Batsos
David Bau
Luis Baumela
Christoph Baur
Eduardo
   Bayro-Corrochano
Paul Beardsley
Jan Bednavr'ik
Oscar Beijbom
Philippe Bekaert
Esube Bekele
Vasileios Belagiannis
Ohad Ben-Shahar
Abhijit Bendale
Róger Bermúdez-Chacón
Maxim Berman
Jesus Bermudez-cameo

Florian Bernard
Stefano Berretti
Marcelo Bertalmio
Gedas Bertasius
Cigdem Beyan
Lucas Beyer
Vijayakumar Bhagavatula
Arjun Nitin Bhagoji
Apratim Bhattacharyya
Binod Bhattarai
Sai Bi
Jia-Wang Bian
Simone Bianco
Adel Bibi
Tolga Birdal
Tom Bishop
Soma Biswas
Mårten Björkman
Volker Blanz
Vishnu Boddeti
Navaneeth Bodla
Simion-Vlad Bogolin
Xavier Boix
Piotr Bojanowski
Timo Bolkart
Guido Borghi
Larbi Boubchir
Guillaume Bourmaud
Adrien Bousseau
Thierry Bouwmans
Richard Bowden
Hakan Boyraz
Mathieu Brédif
Samarth Brahmbhatt
Steve Branson
Nikolas Brasch
Biagio Brattoli
Ernesto Brau
Toby P. Breckon
Francois Bremond
Jesus Briales
Sofia Broomé
Marcus A. Brubaker
Luc Brun
Silvia Bucci
Shyamal Buch

Pradeep Buddharaju
Uta Buechler
Mai Bui
Tu Bui
Adrian Bulat
Giedrius T. Burachas
Elena Burceanu
Xavier P. Burgos-Artizzu
Kaylee Burns
Andrei Bursuc
Benjamin Busam
Wonmin Byeon
Zoya Bylinskii
Sergi Caelles
Jianrui Cai
Minjie Cai
Yujun Cai
Zhaowei Cai
Zhipeng Cai
Juan C. Caicedo
Simone Calderara
Necati Cihan Camgoz
Dylan Campbell
Octavia Camps
Jiale Cao
Kaidi Cao
Liangliang Cao
Xiangyong Cao
Xiaochun Cao
Yang Cao
Yu Cao
Yue Cao
Zhangjie Cao
Luca Carlone
Mathilde Caron
Dan Casas
Thomas J. Cashman
Umberto Castellani
Lluis Castrejon
Jacopo Cavazza
Fabio Cermelli
Hakan Cevikalp
Menglei Chai
Ishani Chakraborty
Rudrasis Chakraborty
Antoni B. Chan

Kwok-Ping Chan
Siddhartha Chandra
Sharat Chandran
Arjun Chandrasekaran
Angel X. Chang
Che-Han Chang
Hong Chang
Hyun Sung Chang
Hyung Jin Chang
Jianlong Chang
Ju Yong Chang
Ming-Ching Chang
Simyung Chang
Xiaojun Chang
Yu-Wei Chao
Devendra S. Chaplot
Arslan Chaudhry
Rizwan A. Chaudhry
Can Chen
Chang Chen
Chao Chen
Chen Chen
Chu-Song Chen
Dapeng Chen
Dong Chen
Dongdong Chen
Guanying Chen
Hongge Chen
Hsin-yi Chen
Huaijin Chen
Hwann-Tzong Chen
Jianbo Chen
Jianhui Chen
Jiansheng Chen
Jiaxin Chen
Jie Chen
Jun-Cheng Chen
Kan Chen
Kevin Chen
Lin Chen
Long Chen
Min-Hung Chen
Qifeng Chen
Shi Chen
Shixing Chen
Tianshui Chen

Weifeng Chen
Weikai Chen
Xi Chen
Xiaohan Chen
Xiaozhi Chen
Xilin Chen
Xingyu Chen
Xinlei Chen
Xinyun Chen
Yi-Ting Chen
Yilun Chen
Ying-Cong Chen
Yinpeng Chen
Yiran Chen
Yu Chen
Yu-Sheng Chen
Yuhua Chen
Yun-Chun Chen
Yunpeng Chen
Yuntao Chen
Zhuoyuan Chen
Zitian Chen
Anchieh Cheng
Bowen Cheng
Erkang Cheng
Gong Cheng
Guangliang Cheng
Jingchun Cheng
Jun Cheng
Li Cheng
Ming-Ming Cheng
Yu Cheng
Ziang Cheng
Anoop Cherian
Dmitry Chetverikov
Ngai-man Cheung
William Cheung
Ajad Chhatkuli
Naoki Chiba
Benjamin Chidester
Han-pang Chiu
Mang Tik Chiu
Wei-Chen Chiu
Donghyeon Cho
Hojin Cho
Minsu Cho

Nam Ik Cho
Tim Cho
Tae Eun Choe
Chiho Choi
Edward Choi
Inchang Choi
Jinsoo Choi
Jonghyun Choi
Jongwon Choi
Yukyung Choi
Hisham Cholakkal
Eunji Chong
Jaegul Choo
Christopher Choy
Hang Chu
Peng Chu
Wen-Sheng Chu
Albert Chung
Joon Son Chung
Hai Ci
Safa Cicek
Ramazan G. Cinbis
Arridhana Ciptadi
Javier Civera
James J. Clark
Ronald Clark
Felipe Codevilla
Michael Cogswell
Andrea Cohen
Maxwell D. Collins
Carlo Colombo
Yang Cong
Adria R. Continente
Marcella Cornia
John Richard Corring
Darren Cosker
Dragos Costea
Garrison W. Cottrell
Florent Couzinie-Devy
Marco Cristani
Ioana Croitoru
James L. Crowley
Jiequan Cui
Zhaopeng Cui
Ross Cutler
Antonio D'Innocente

Rozenn Dahyot
Bo Dai
Dengxin Dai
Hang Dai
Longquan Dai
Shuyang Dai
Xiyang Dai
Yuchao Dai
Adrian V. Dalca
Dima Damen
Bharath B. Damodaran
Kristin Dana
Martin Danelljan
Zheng Dang
Zachary Alan Daniels
Donald G. Dansereau
Abhishek Das
Samyak Datta
Achal Dave
Titas De
Rodrigo de Bem
Teo de Campos
Raoul de Charette
Shalini De Mello
Joseph DeGol
Herve Delingette
Haowen Deng
Jiankang Deng
Weijian Deng
Zhiwei Deng
Joachim Denzler
Konstantinos G. Derpanis
Aditya Deshpande
Frederic Devernay
Somdip Dey
Arturo Deza
Abhinav Dhall
Helisa Dhamo
Vikas Dhiman
Fillipe Dias Moreira
  de Souza
Ali Diba
Ferran Diego
Guiguang Ding
Henghui Ding
Jian Ding

Mingyu Ding
Xinghao Ding
Zhengming Ding
Robert DiPietro
Cosimo Distante
Ajay Divakaran
Mandar Dixit
Abdelaziz Djelouah
Thanh-Toan Do
Jose Dolz
Bo Dong
Chao Dong
Jiangxin Dong
Weiming Dong
Weisheng Dong
Xingping Dong
Xuanyi Dong
Yinpeng Dong
Gianfranco Doretto
Hazel Doughty
Hassen Drira
Bertram Drost
Dawei Du
Ye Duan
Yueqi Duan
Abhimanyu Dubey
Anastasia Dubrovina
Stefan Duffner
Chi Nhan Duong
Thibaut Durand
Zoran Duric
Iulia Duta
Debidatta Dwibedi
Benjamin Eckart
Marc Eder
Marzieh Edraki
Alexei A. Efros
Kiana Ehsani
Hazm Kemal Ekenel
James H. Elder
Mohamed Elgharib
Shireen Elhabian
Ehsan Elhamifar
Mohamed Elhoseiny
Ian Endres
N. Benjamin Erichson

Jan Ernst
Sergio Escalera
Francisco Escolano
Victor Escorcia
Carlos Esteves
Francisco J. Estrada
Bin Fan
Chenyou Fan
Deng-Ping Fan
Haoqi Fan
Hehe Fan
Heng Fan
Kai Fan
Lijie Fan
Linxi Fan
Quanfu Fan
Shaojing Fan
Xiaochuan Fan
Xin Fan
Yuchen Fan
Sean Fanello
Hao-Shu Fang
Haoyang Fang
Kuan Fang
Yi Fang
Yuming Fang
Azade Farshad
Alireza Fathi
Raanan Fattal
Joao Fayad
Xiaohan Fei
Christoph Feichtenhofer
Michael Felsberg
Chen Feng
Jiashi Feng
Junyi Feng
Mengyang Feng
Qianli Feng
Zhenhua Feng
Michele Fenzi
Andras Ferencz
Martin Fergie
Basura Fernando
Ethan Fetaya
Michael Firman
John W. Fisher

Matthew Fisher
Boris Flach
Corneliu Florea
Wolfgang Foerstner
David Fofi
Gian Luca Foresti
Per-Erik Forssen
David Fouhey
Katerina Fragkiadaki
Victor Fragoso
Jean-Sébastien Franco
Ohad Fried
Iuri Frosio
Cheng-Yang Fu
Huazhu Fu
Jianlong Fu
Jingjing Fu
Xueyang Fu
Yanwei Fu
Ying Fu
Yun Fu
Olac Fuentes
Kent Fujiwara
Takuya Funatomi
Christopher Funk
Thomas Funkhouser
Antonino Furnari
Ryo Furukawa
Erik Gärtner
Raghudeep Gadde
Matheus Gadelha
Vandit Gajjar
Trevor Gale
Juergen Gall
Mathias Gallardo
Guillermo Gallego
Orazio Gallo
Chuang Gan
Zhe Gan
Madan Ravi Ganesh
Aditya Ganeshan
Siddha Ganju
Bin-Bin Gao
Changxin Gao
Feng Gao
Hongchang Gao

Jin Gao
Jiyang Gao
Junbin Gao
Katelyn Gao
Lin Gao
Mingfei Gao
Ruiqi Gao
Ruohan Gao
Shenghua Gao
Yuan Gao
Yue Gao
Noa Garcia
Alberto Garcia-Garcia
Guillermo
    Garcia-Hernando
Jacob R. Gardner
Animesh Garg
Kshitiz Garg
Rahul Garg
Ravi Garg
Philip N. Garner
Kirill Gavrilyuk
Paul Gay
Shiming Ge
Weifeng Ge
Baris Gecer
Xin Geng
Kyle Genova
Stamatios Georgoulis
Bernard Ghanem
Michael Gharbi
Kamran Ghasedi
Golnaz Ghiasi
Arnab Ghosh
Partha Ghosh
Silvio Giancola
Andrew Gilbert
Rohit Girdhar
Xavier Giro-i-Nieto
Thomas Gittings
Ioannis Gkioulekas
Clement Godard
Vaibhava Goel
Bastian Goldluecke
Lluis Gomez
Nuno Gonçalves

Dong Gong
Ke Gong
Mingming Gong
Abel Gonzalez-Garcia
Ariel Gordon
Daniel Gordon
Paulo Gotardo
Venu Madhav Govindu
Ankit Goyal
Priya Goyal
Raghav Goyal
Benjamin Graham
Douglas Gray
Brent A. Griffin
Etienne Grossmann
David Gu
Jiayuan Gu
Jiuxiang Gu
Lin Gu
Qiao Gu
Shuhang Gu
Jose J. Guerrero
Paul Guerrero
Jie Gui
Jean-Yves Guillemaut
Riza Alp Guler
Erhan Gundogdu
Fatma Guney
Guodong Guo
Kaiwen Guo
Qi Guo
Sheng Guo
Shi Guo
Tiantong Guo
Xiaojie Guo
Yijie Guo
Yiluan Guo
Yuanfang Guo
Yulan Guo
Agrim Gupta
Ankush Gupta
Mohit Gupta
Saurabh Gupta
Tanmay Gupta
Danna Gurari
Abner Guzman-Rivera

JunYoung Gwak
Michael Gygli
Jung-Woo Ha
Simon Hadfield
Isma Hadji
Bjoern Haefner
Taeyoung Hahn
Levente Hajder
Peter Hall
Emanuela Haller
Stefan Haller
Bumsub Ham
Abdullah Hamdi
Dongyoon Han
Hu Han
Jungong Han
Junwei Han
Kai Han
Tian Han
Xiaoguang Han
Xintong Han
Yahong Han
Ankur Handa
Zekun Hao
Albert Haque
Tatsuya Harada
Mehrtash Harandi
Adam W. Harley
Mahmudul Hasan
Atsushi Hashimoto
Ali Hatamizadeh
Munawar Hayat
Dongliang He
Jingrui He
Junfeng He
Kaiming He
Kun He
Lei He
Pan He
Ran He
Shengfeng He
Tong He
Weipeng He
Xuming He
Yang He
Yihui He

Zhihai He
Chinmay Hegde
Janne Heikkila
Mattias P. Heinrich
Stéphane Herbin
Alexander Hermans
Luis Herranz
John R. Hershey
Aaron Hertzmann
Roei Herzig
Anders Heyden
Steven Hickson
Otmar Hilliges
Tomas Hodan
Judy Hoffman
Michael Hofmann
Yannick Hold-Geoffroy
Namdar Homayounfar
Sina Honari
Richang Hong
Seunghoon Hong
Xiaopeng Hong
Yi Hong
Hidekata Hontani
Anthony Hoogs
Yedid Hoshen
Mir Rayat Imtiaz Hossain
Junhui Hou
Le Hou
Lu Hou
Tingbo Hou
Wei-Lin Hsiao
Cheng-Chun Hsu
Gee-Sern Jison Hsu
Kuang-jui Hsu
Changbo Hu
Di Hu
Guosheng Hu
Han Hu
Hao Hu
Hexiang Hu
Hou-Ning Hu
Jie Hu
Junlin Hu
Nan Hu
Ping Hu

Ronghang Hu
Xiaowei Hu
Yinlin Hu
Yuan-Ting Hu
Zhe Hu
Binh-Son Hua
Yang Hua
Bingyao Huang
Di Huang
Dong Huang
Fay Huang
Haibin Huang
Haozhi Huang
Heng Huang
Huaibo Huang
Jia-Bin Huang
Jing Huang
Jingwei Huang
Kaizhu Huang
Lei Huang
Qiangui Huang
Qiaoying Huang
Qingqiu Huang
Qixing Huang
Shaoli Huang
Sheng Huang
Siyuan Huang
Weilin Huang
Wenbing Huang
Xiangru Huang
Xun Huang
Yan Huang
Yifei Huang
Yue Huang
Zhiwu Huang
Zilong Huang
Minyoung Huh
Zhuo Hui
Matthias B. Hullin
Martin Humenberger
Wei-Chih Hung
Zhouyuan Huo
Junhwa Hur
Noureldien Hussein
Jyh-Jing Hwang
Seong Jae Hwang

Sung Ju Hwang
Ichiro Ide
Ivo Ihrke
Daiki Ikami
Satoshi Ikehata
Nazli Ikizler-Cinbis
Sunghoon Im
Yani Ioannou
Radu Tudor Ionescu
Umar Iqbal
Go Irie
Ahmet Iscen
Md Amirul Islam
Vamsi Ithapu
Nathan Jacobs
Arpit Jain
Himalaya Jain
Suyog Jain
Stuart James
Won-Dong Jang
Yunseok Jang
Ronnachai Jaroensri
Dinesh Jayaraman
Sadeep Jayasumana
Suren Jayasuriya
Herve Jegou
Simon Jenni
Hae-Gon Jeon
Yunho Jeon
Koteswar R. Jerripothula
Hueihan Jhuang
I-hong Jhuo
Dinghuang Ji
Hui Ji
Jingwei Ji
Pan Ji
Yanli Ji
Baoxiong Jia
Kui Jia
Xu Jia
Chiyu Max Jiang
Haiyong Jiang
Hao Jiang
Huaizu Jiang
Huajie Jiang
Ke Jiang

Lai Jiang
Li Jiang
Lu Jiang
Ming Jiang
Peng Jiang
Shuqiang Jiang
Wei Jiang
Xudong Jiang
Zhuolin Jiang
Jianbo Jiao
Zequn Jie
Dakai Jin
Kyong Hwan Jin
Lianwen Jin
SouYoung Jin
Xiaojie Jin
Xin Jin
Nebojsa Jojic
Alexis Joly
Michael Jeffrey Jones
Hanbyul Joo
Jungseock Joo
Kyungdon Joo
Ajjen Joshi
Shantanu H. Joshi
Da-Cheng Juan
Marco Körner
Kevin Köser
Asim Kadav
Christine Kaeser-Chen
Kushal Kafle
Dagmar Kainmueller
Ioannis A. Kakadiaris
Zdenek Kalal
Nima Kalantari
Yannis Kalantidis
Mahdi M. Kalayeh
Anmol Kalia
Sinan Kalkan
Vicky Kalogeiton
Ashwin Kalyan
Joni-kristian Kamarainen
Gerda Kamberova
Chandra Kambhamettu
Martin Kampel
Meina Kan

Christopher Kanan
Kenichi Kanatani
Angjoo Kanazawa
Atsushi Kanehira
Takuhiro Kaneko
Asako Kanezaki
Bingyi Kang
Di Kang
Sunghun Kang
Zhao Kang
Vadim Kantorov
Abhishek Kar
Amlan Kar
Theofanis Karaletsos
Leonid Karlinsky
Kevin Karsch
Angelos Katharopoulos
Isinsu Katircioglu
Hiroharu Kato
Zoltan Kato
Dotan Kaufman
Jan Kautz
Rei Kawakami
Qiuhong Ke
Wadim Kehl
Petr Kellnhofer
Aniruddha Kembhavi
Cem Keskin
Margret Keuper
Daniel Keysers
Ashkan Khakzar
Fahad Khan
Naeemullah Khan
Salman Khan
Siddhesh Khandelwal
Rawal Khirodkar
Anna Khoreva
Tejas Khot
Parmeshwar Khurd
Hadi Kiapour
Joe Kileel
Chanho Kim
Dahun Kim
Edward Kim
Eunwoo Kim
Han-ul Kim

Hansung Kim
Heewon Kim
Hyo Jin Kim
Hyunwoo J. Kim
Jinkyu Kim
Jiwon Kim
Jongmin Kim
Junsik Kim
Junyeong Kim
Min H. Kim
Namil Kim
Pyojin Kim
Seon Joo Kim
Seong Tae Kim
Seungryong Kim
Sungwoong Kim
Tae Hyun Kim
Vladimir Kim
Won Hwa Kim
Yonghyun Kim
Benjamin Kimia
Akisato Kimura
Pieter-Jan Kindermans
Zsolt Kira
Itaru Kitahara
Hedvig Kjellstrom
Jan Knopp
Takumi Kobayashi
Erich Kobler
Parker Koch
Reinhard Koch
Elyor Kodirov
Amir Kolaman
Nicholas Kolkin
Dimitrios Kollias
Stefanos Kollias
Soheil Kolouri
Adams Wai-Kin Kong
Naejin Kong
Shu Kong
Tao Kong
Yu Kong
Yoshinori Konishi
Daniil Kononenko
Theodora Kontogianni
Simon Korman

Adam Kortylewski
Jana Kosecka
Jean Kossaifi
Satwik Kottur
Rigas Kouskouridas
Adriana Kovashka
Rama Kovvuri
Adarsh Kowdle
Jedrzej Kozerawski
Mateusz Kozinski
Philipp Kraehenbuehl
Gregory Kramida
Josip Krapac
Dmitry Kravchenko
Ranjay Krishna
Pavel Krsek
Alexander Krull
Jakob Kruse
Hiroyuki Kubo
Hilde Kuehne
Jason Kuen
Andreas Kuhn
Arjan Kuijper
Zuzana Kukelova
Ajay Kumar
Amit Kumar
Avinash Kumar
Suryansh Kumar
Vijay Kumar
Kaustav Kundu
Weicheng Kuo
Nojun Kwak
Suha Kwak
Junseok Kwon
Nikolaos Kyriazis
Zorah Lähner
Ankit Laddha
Florent Lafarge
Jean Lahoud
Kevin Lai
Shang-Hong Lai
Wei-Sheng Lai
Yu-Kun Lai
Iro Laina
Antony Lam
John Wheatley Lambert

Xiangyuan lan
Xu Lan
Charis Lanaras
Georg Langs
Oswald Lanz
Dong Lao
Yizhen Lao
Agata Lapedriza
Gustav Larsson
Viktor Larsson
Katrin Lasinger
Christoph Lassner
Longin Jan Latecki
Stéphane Lathuilière
Rynson Lau
Hei Law
Justin Lazarow
Svetlana Lazebnik
Hieu Le
Huu Le
Ngan Hoang Le
Trung-Nghia Le
Vuong Le
Colin Lea
Erik Learned-Miller
Chen-Yu Lee
Gim Hee Lee
Hsin-Ying Lee
Hyungtae Lee
Jae-Han Lee
Jimmy Addison Lee
Joonseok Lee
Kibok Lee
Kuang-Huei Lee
Kwonjoon Lee
Minsik Lee
Sang-chul Lee
Seungkyu Lee
Soochan Lee
Stefan Lee
Taehee Lee
Andreas Lehrmann
Jie Lei
Peng Lei
Matthew Joseph Leotta
Wee Kheng Leow

Gil Levi
Evgeny Levinkov
Aviad Levis
Jose Lezama
Ang Li
Bin Li
Bing Li
Boyi Li
Changsheng Li
Chao Li
Chen Li
Cheng Li
Chenglong Li
Chi Li
Chun-Guang Li
Chun-Liang Li
Chunyuan Li
Dong Li
Guanbin Li
Hao Li
Haoxiang Li
Hongsheng Li
Hongyang Li
Houqiang Li
Huibin Li
Jia Li
Jianan Li
Jianguo Li
Junnan Li
Junxuan Li
Kai Li
Ke Li
Kejie Li
Kunpeng Li
Lerenhan Li
Li Erran Li
Mengtian Li
Mu Li
Peihua Li
Peiyi Li
Ping Li
Qi Li
Qing Li
Ruiyu Li
Ruoteng Li
Shaozi Li

Sheng Li
Shiwei Li
Shuang Li
Siyang Li
Stan Z. Li
Tianye Li
Wei Li
Weixin Li
Wen Li
Wenbo Li
Xiaomeng Li
Xin Li
Xiu Li
Xuelong Li
Xueting Li
Yan Li
Yandong Li
Yanghao Li
Yehao Li
Yi Li
Yijun Li
Yikang LI
Yining Li
Yongjie Li
Yu Li
Yu-Jhe Li
Yunpeng Li
Yunsheng Li
Yunzhu Li
Zhe Li
Zhen Li
Zhengqi Li
Zhenyang Li
Zhuwen Li
Dongze Lian
Xiaochen Lian
Zhouhui Lian
Chen Liang
Jie Liang
Ming Liang
Paul Pu Liang
Pengpeng Liang
Shu Liang
Wei Liang
Jing Liao
Minghui Liao

Renjie Liao
Shengcai Liao
Shuai Liao
Yiyi Liao
Ser-Nam Lim
Chen-Hsuan Lin
Chung-Ching Lin
Dahua Lin
Ji Lin
Kevin Lin
Tianwei Lin
Tsung-Yi Lin
Tsung-Yu Lin
Wei-An Lin
Weiyao Lin
Yen-Chen Lin
Yuewei Lin
David B. Lindell
Drew Linsley
Krzysztof Lis
Roee Litman
Jim Little
An-An Liu
Bo Liu
Buyu Liu
Chao Liu
Chen Liu
Cheng-lin Liu
Chenxi Liu
Dong Liu
Feng Liu
Guilin Liu
Haomiao Liu
Heshan Liu
Hong Liu
Ji Liu
Jingen Liu
Jun Liu
Lanlan Liu
Li Liu
Liu Liu
Mengyuan Liu
Miaomiao Liu
Nian Liu
Ping Liu
Risheng Liu

Sheng Liu
Shu Liu
Shuaicheng Liu
Sifei Liu
Siqi Liu
Siying Liu
Songtao Liu
Ting Liu
Tongliang Liu
Tyng-Luh Liu
Wanquan Liu
Wei Liu
Weiyang Liu
Weizhe Liu
Wenyu Liu
Wu Liu
Xialei Liu
Xianglong Liu
Xiaodong Liu
Xiaofeng Liu
Xihui Liu
Xingyu Liu
Xinwang Liu
Xuanqing Liu
Xuebo Liu
Yang Liu
Yaojie Liu
Yebin Liu
Yen-Cheng Liu
Yiming Liu
Yu Liu
Yu-Shen Liu
Yufan Liu
Yun Liu
Zheng Liu
Zhijian Liu
Zhuang Liu
Zichuan Liu
Ziwei Liu
Zongyi Liu
Stephan Liwicki
Liliana Lo Presti
Chengjiang Long
Fuchen Long
Mingsheng Long
Xiang Long

Yang Long
Charles T. Loop
Antonio Lopez
Roberto J. Lopez-Sastre
Javier Lorenzo-Navarro
Manolis Lourakis
Boyu Lu
Canyi Lu
Feng Lu
Guoyu Lu
Hongtao Lu
Jiajun Lu
Jiasen Lu
Jiwen Lu
Kaiyue Lu
Le Lu
Shao-Ping Lu
Shijian Lu
Xiankai Lu
Xin Lu
Yao Lu
Yiping Lu
Yongxi Lu
Yongyi Lu
Zhiwu Lu
Fujun Luan
Benjamin E. Lundell
Hao Luo
Jian-Hao Luo
Ruotian Luo
Weixin Luo
Wenhan Luo
Wenjie Luo
Yan Luo
Zelun Luo
Zixin Luo
Khoa Luu
Zhaoyang Lv
Pengyuan Lyu
Thomas Möllenhoff
Matthias Müller
Bingpeng Ma
Chih-Yao Ma
Chongyang Ma
Huimin Ma
Jiayi Ma

K. T. Ma
Ke Ma
Lin Ma
Liqian Ma
Shugao Ma
Wei-Chiu Ma
Xiaojian Ma
Xingjun Ma
Zhanyu Ma
Zheng Ma
Radek Jakob Mackowiak
Ludovic Magerand
Shweta Mahajan
Siddharth Mahendran
Long Mai
Ameesh Makadia
Oscar Mendez Maldonado
Mateusz Malinowski
Yury Malkov
Arun Mallya
Dipu Manandhar
Massimiliano Mancini
Fabian Manhardt
Kevis-kokitsi Maninis
Varun Manjunatha
Junhua Mao
Xudong Mao
Alina Marcu
Edgar Margffoy-Tuay
Dmitrii Marin
Manuel J. Marin-Jimenez
Kenneth Marino
Niki Martinel
Julieta Martinez
Jonathan Masci
Tomohiro Mashita
Iacopo Masi
David Masip
Daniela Massiceti
Stefan Mathe
Yusuke Matsui
Tetsu Matsukawa
Iain A. Matthews
Kevin James Matzen
Bruce Allen Maxwell
Stephen Maybank

Helmut Mayer
Amir Mazaheri
David McAllester
Steven McDonagh
Stephen J. Mckenna
Roey Mechrez
Prakhar Mehrotra
Christopher Mei
Xue Mei
Paulo R. S. Mendonca
Lili Meng
Zibo Meng
Thomas Mensink
Bjoern Menze
Michele Merler
Kourosh Meshgi
Pascal Mettes
Christopher Metzler
Liang Mi
Qiguang Miao
Xin Miao
Tomer Michaeli
Frank Michel
Antoine Miech
Krystian Mikolajczyk
Peyman Milanfar
Ben Mildenhall
Gregor Miller
Fausto Milletari
Dongbo Min
Kyle Min
Pedro Miraldo
Dmytro Mishkin
Anand Mishra
Ashish Mishra
Ishan Misra
Niluthpol C. Mithun
Kaushik Mitra
Niloy Mitra
Anton Mitrokhin
Ikuhisa Mitsugami
Anurag Mittal
Kaichun Mo
Zhipeng Mo
Davide Modolo
Michael Moeller

Pritish Mohapatra
Pavlo Molchanov
Davide Moltisanti
Pascal Monasse
Mathew Monfort
Aron Monszpart
Sean Moran
Vlad I. Morariu
Francesc Moreno-Noguer
Pietro Morerio
Stylianos Moschoglou
Yael Moses
Roozbeh Mottaghi
Pierre Moulon
Arsalan Mousavian
Yadong Mu
Yasuhiro Mukaigawa
Lopamudra Mukherjee
Yusuke Mukuta
Ravi Teja Mullapudi
Mario Enrique Munich
Zachary Murez
Ana C. Murillo
J. Krishna Murthy
Damien Muselet
Armin Mustafa
Siva Karthik Mustikovela
Carlo Dal Mutto
Moin Nabi
Varun K. Nagaraja
Tushar Nagarajan
Arsha Nagrani
Seungjun Nah
Nikhil Naik
Yoshikatsu Nakajima
Yuta Nakashima
Atsushi Nakazawa
Seonghyeon Nam
Vinay P. Namboodiri
Medhini Narasimhan
Srinivasa Narasimhan
Sanath Narayan
Erickson Rangel
  Nascimento
Jacinto Nascimento
Tayyab Naseer

Lakshmanan Nataraj
Neda Nategh
Nelson Isao Nauata
Fernando Navarro
Shah Nawaz
Lukas Neumann
Ram Nevatia
Alejandro Newell
Shawn Newsam
Joe Yue-Hei Ng
Trung Thanh Ngo
Duc Thanh Nguyen
Lam M. Nguyen
Phuc Xuan Nguyen
Thuong Nguyen Canh
Mihalis Nicolaou
Andrei Liviu Nicolicioiu
Xuecheng Nie
Michael Niemeyer
Simon Niklaus
Christophoros Nikou
David Nilsson
Jifeng Ning
Yuval Nirkin
Li Niu
Yuzhen Niu
Zhenxing Niu
Shohei Nobuhara
Nicoletta Noceti
Hyeonwoo Noh
Junhyug Noh
Mehdi Noroozi
Sotiris Nousias
Valsamis Ntouskos
Matthew O'Toole
Peter Ochs
Ferda Ofli
Seong Joon Oh
Seoung Wug Oh
Iason Oikonomidis
Utkarsh Ojha
Takahiro Okabe
Takayuki Okatani
Fumio Okura
Aude Oliva
Kyle Olszewski

Björn Ommer

Mohamed Omran

Elisabeta Oneata

Michael Opitz

Jose Oramas

Tribhuvanesh Orekondy

Shaul Oron

Sergio Orts-Escolano

Ivan Oseledets

Aljosa Osep

Magnus Oskarsson

Anton Osokin

Martin R. Oswald

Wanli Ouyang

Andrew Owens

Mete Ozay

Mustafa Ozuysal

Eduardo Pérez-Pellitero

Gautam Pai

Dipan Kumar Pal

P. H. Pamplona Savarese

Jinshan Pan

Junting Pan

Xingang Pan

Yingwei Pan

Yannis Panagakis

Rameswar Panda

Guan Pang

Jiahao Pang

Jiangmiao Pang

Tianyu Pang

Sharath Pankanti

Nicolas Papadakis

Dim Papadopoulos

George Papandreou

Toufiq Parag

Shaifali Parashar

Sarah Parisot

Eunhyeok Park

Hyun Soo Park

Jaesik Park

Min-Gyu Park

Taesung Park

Alvaro Parra

C. Alejandro Parraga

Despoina Paschalidou

Nikolaos Passalis

Vishal Patel

Viorica Patraucean

Badri Narayana Patro

Danda Pani Paudel

Sujoy Paul

Georgios Pavlakos

Ioannis Pavlidis

Vladimir Pavlovic

Nick Pears

Kim Steenstrup Pedersen

Selen Pehlivan

Shmuel Peleg

Chao Peng

Houwen Peng

Wen-Hsiao Peng

Xi Peng

Xiaojiang Peng

Xingchao Peng

Yuxin Peng

Federico Perazzi

Juan Camilo Perez

Vishwanath Peri

Federico Pernici

Luca Del Pero

Florent Perronnin

Stavros Petridis

Henning Petzka

Patrick Peursum

Michael Pfeiffer

Hanspeter Pfister

Roman Pflugfelder

Minh Tri Pham

Yongri Piao

David Picard

Tomasz Pieciak

A. J. Piergiovanni

Andrea Pilzer

Pedro O. Pinheiro

Silvia Laura Pintea

Lerrel Pinto

Axel Pinz

Robinson Piramuthu

Fiora Pirri

Leonid Pishchulin

Francesco Pittaluga

Daniel Pizarro

Tobias Plötz

Mirco Planamente

Matteo Poggi

Moacir A. Ponti

Parita Pooj

Fatih Porikli

Horst Possegger

Omid Poursaeed

Ameya Prabhu

Viraj Uday Prabhu

Dilip Prasad

Brian L. Price

True Price

Maria Priisalu

Veronique Prinet

Victor Adrian Prisacariu

Jan Prokaj

Sergey Prokudin

Nicolas Pugeault

Xavier Puig

Albert Pumarola

Pulak Purkait

Senthil Purushwalkam

Charles R. Qi

Hang Qi

Haozhi Qi

Lu Qi

Mengshi Qi

Siyuan Qi

Xiaojuan Qi

Yuankai Qi

Shengju Qian

Xuelin Qian

Siyuan Qiao

Yu Qiao

Jie Qin

Qiang Qiu

Weichao Qiu

Zhaofan Qiu

Kha Gia Quach

Yuhui Quan

Yvain Queau

Julian Quiroga

Faisal Qureshi

Mahdi Rad

Filip Radenovic
Petia Radeva
Venkatesh
  B. Radhakrishnan
Ilija Radosavovic
Noha Radwan
Rahul Raguram
Tanzila Rahman
Amit Raj
Ajit Rajwade
Kandan Ramakrishnan
Santhosh
  K. Ramakrishnan
Srikumar Ramalingam
Ravi Ramamoorthi
Vasili Ramanishka
Ramprasaath R. Selvaraju
Francois Rameau
Visvanathan Ramesh
Santu Rana
Rene Ranftl
Anand Rangarajan
Anurag Ranjan
Viresh Ranjan
Yongming Rao
Carolina Raposo
Vivek Rathod
Sathya N. Ravi
Avinash Ravichandran
Tammy Riklin Raviv
Daniel Rebain
Sylvestre-Alvise Rebuffi
N. Dinesh Reddy
Timo Rehfeld
Paolo Remagnino
Konstantinos Rematas
Edoardo Remelli
Dongwei Ren
Haibing Ren
Jian Ren
Jimmy Ren
Mengye Ren
Weihong Ren
Wenqi Ren
Zhile Ren
Zhongzheng Ren

Zhou Ren
Vijay Rengarajan
Md A. Reza
Farzaneh Rezaeianaran
Hamed R. Tavakoli
Nicholas Rhinehart
Helge Rhodin
Elisa Ricci
Alexander Richard
Eitan Richardson
Elad Richardson
Christian Richardt
Stephan Richter
Gernot Riegler
Daniel Ritchie
Tobias Ritschel
Samuel Rivera
Yong Man Ro
Richard Roberts
Joseph Robinson
Ignacio Rocco
Mrigank Rochan
Emanuele Rodolà
Mikel D. Rodriguez
Giorgio Roffo
Grégory Rogez
Gemma Roig
Javier Romero
Xuejian Rong
Yu Rong
Amir Rosenfeld
Bodo Rosenhahn
Guy Rosman
Arun Ross
Paolo Rota
Peter M. Roth
Anastasios Roussos
Anirban Roy
Sebastien Roy
Aruni RoyChowdhury
Artem Rozantsev
Ognjen Rudovic
Daniel Rueckert
Adria Ruiz
Javier Ruiz-del-solar
Christian Rupprecht

Chris Russell
Dan Ruta
Jongbin Ryu
Ömer Sümer
Alexandre Sablayrolles
Faraz Saeedan
Ryusuke Sagawa
Christos Sagonas
Tonmoy Saikia
Hideo Saito
Kuniaki Saito
Shunsuke Saito
Shunta Saito
Ken Sakurada
Joaquin Salas
Fatemeh Sadat Saleh
Mahdi Saleh
Pouya Samangouei
Leo Sampaio
  Ferraz Ribeiro
Artsiom Olegovich
  Sanakoyeu
Enrique Sanchez
Patsorn Sangkloy
Anush Sankaran
Aswin Sankaranarayanan
Swami Sankaranarayanan
Rodrigo Santa Cruz
Amartya Sanyal
Archana Sapkota
Nikolaos Sarafianos
Jun Sato
Shin'ichi Satoh
Hosnieh Sattar
Arman Savran
Manolis Savva
Alexander Sax
Hanno Scharr
Simone Schaub-Meyer
Konrad Schindler
Dmitrij Schlesinger
Uwe Schmidt
Dirk Schnieders
Björn Schuller
Samuel Schulter
Idan Schwartz

William Robson Schwartz
Alex Schwing
Sinisa Segvic
Lorenzo Seidenari
Pradeep Sen
Ozan Sener
Soumyadip Sengupta
Arda Senocak
Mojtaba Seyedhosseini
Shishir Shah
Shital Shah
Sohil Atul Shah
Tamar Rott Shaham
Huasong Shan
Qi Shan
Shiguang Shan
Jing Shao
Roman Shapovalov
Gaurav Sharma
Vivek Sharma
Viktoriia Sharmanska
Dongyu She
Sumit Shekhar
Evan Shelhamer
Chengyao Shen
Chunhua Shen
Falong Shen
Jie Shen
Li Shen
Liyue Shen
Shuhan Shen
Tianwei Shen
Wei Shen
William B. Shen
Yantao Shen
Ying Shen
Yiru Shen
Yujun Shen
Yuming Shen
Zhiqiang Shen
Ziyi Shen
Lu Sheng
Yu Sheng
Rakshith Shetty
Baoguang Shi
Guangming Shi

Hailin Shi
Miaojing Shi
Yemin Shi
Zhenmei Shi
Zhiyuan Shi
Kevin Jonathan Shih
Shiliang Shiliang
Hyunjung Shim
Atsushi Shimada
Nobutaka Shimada
Daeyun Shin
Young Min Shin
Koichi Shinoda
Konstantin Shmelkov
Michael Zheng Shou
Abhinav Shrivastava
Tianmin Shu
Zhixin Shu
Hong-Han Shuai
Pushkar Shukla
Christian Siagian
Mennatullah M. Siam
Kaleem Siddiqi
Karan Sikka
Jae-Young Sim
Christian Simon
Martin Simonovsky
Dheeraj Singaraju
Bharat Singh
Gurkirt Singh
Krishna Kumar Singh
Maneesh Kumar Singh
Richa Singh
Saurabh Singh
Suriya Singh
Vikas Singh
Sudipta N. Sinha
Vincent Sitzmann
Josef Sivic
Gregory Slabaugh
Miroslava Slavcheva
Ron Slossberg
Brandon Smith
Kevin Smith
Vladimir Smutny
Noah Snavely

Roger
   D. Soberanis-Mukul
Kihyuk Sohn
Francesco Solera
Eric Sommerlade
Sanghyun Son
Byung Cheol Song
Chunfeng Song
Dongjin Song
Jiaming Song
Jie Song
Jifei Song
Jingkuan Song
Mingli Song
Shiyu Song
Shuran Song
Xiao Song
Yafei Song
Yale Song
Yang Song
Yi-Zhe Song
Yibing Song
Humberto Sossa
Cesar de Souza
Adrian Spurr
Srinath Sridhar
Suraj Srinivas
Pratul P. Srinivasan
Anuj Srivastava
Tania Stathaki
Christopher Stauffer
Simon Stent
Rainer Stiefelhagen
Pierre Stock
Julian Straub
Jonathan C. Stroud
Joerg Stueckler
Jan Stuehmer
David Stutz
Chi Su
Hang Su
Jong-Chyi Su
Shuochen Su
Yu-Chuan Su
Ramanathan Subramanian
Yusuke Sugano

Masanori Suganuma
Yumin Suh
Mohammed Suhail
Yao Sui
Heung-Il Suk
Josephine Sullivan
Baochen Sun
Chen Sun
Chong Sun
Deqing Sun
Jin Sun
Liang Sun
Lin Sun
Qianru Sun
Shao-Hua Sun
Shuyang Sun
Weiwei Sun
Wenxiu Sun
Xiaoshuai Sun
Xiaoxiao Sun
Xingyuan Sun
Yifan Sun
Zhun Sun
Sabine Susstrunk
David Suter
Supasorn Suwajanakorn
Tomas Svoboda
Eran Swears
Paul Swoboda
Attila Szabo
Richard Szeliski
Duy-Nguyen Ta
Andrea Tagliasacchi
Yuichi Taguchi
Ying Tai
Keita Takahashi
Kouske Takahashi
Jun Takamatsu
Hugues Talbot
Toru Tamaki
Chaowei Tan
Fuwen Tan
Mingkui Tan
Mingxing Tan
Qingyang Tan
Robby T. Tan

Xiaoyang Tan
Kenichiro Tanaka
Masayuki Tanaka
Chang Tang
Chengzhou Tang
Danhang Tang
Ming Tang
Peng Tang
Qingming Tang
Wei Tang
Xu Tang
Yansong Tang
Youbao Tang
Yuxing Tang
Zhiqiang Tang
Tatsunori Taniai
Junli Tao
Xin Tao
Makarand Tapaswi
Jean-Philippe Tarel
Lyne Tchapmi
Zachary Teed
Bugra Tekin
Damien Teney
Ayush Tewari
Christian Theobalt
Christopher Thomas
Diego Thomas
Jim Thomas
Rajat Mani Thomas
Xinmei Tian
Yapeng Tian
Yingli Tian
Yonglong Tian
Zhi Tian
Zhuotao Tian
Kinh Tieu
Joseph Tighe
Massimo Tistarelli
Matthew Toews
Carl Toft
Pavel Tokmakov
Federico Tombari
Chetan Tonde
Yan Tong
Alessio Tonioni

Andrea Torsello
Fabio Tosi
Du Tran
Luan Tran
Ngoc-Trung Tran
Quan Hung Tran
Truyen Tran
Rudolph Triebel
Martin Trimmel
Shashank Tripathi
Subarna Tripathi
Leonardo Trujillo
Eduard Trulls
Tomasz Trzcinski
Sam Tsai
Yi-Hsuan Tsai
Hung-Yu Tseng
Stavros Tsogkas
Aggeliki Tsoli
Devis Tuia
Shubham Tulsiani
Sergey Tulyakov
Frederick Tung
Tony Tung
Daniyar Turmukhambetov
Ambrish Tyagi
Radim Tylecek
Christos Tzelepis
Georgios Tzimiropoulos
Dimitrios Tzionas
Seiichi Uchida
Norimichi Ukita
Dmitry Ulyanov
Martin Urschler
Yoshitaka Ushiku
Ben Usman
Alexander Vakhitov
Julien P. C. Valentin
Jack Valmadre
Ernest Valveny
Joost van de Weijer
Jan van Gemert
Koen Van Leemput
Gul Varol
Sebastiano Vascon
M. Alex O. Vasilescu

Subeesh Vasu
Mayank Vatsa
David Vazquez
Javier Vazquez-Corral
Ashok Veeraraghavan
Erik Velasco-Salido
Raviteja Vemulapalli
Jonathan Ventura
Manisha Verma
Roberto Vezzani
Ruben Villegas
Minh Vo
MinhDuc Vo
Nam Vo
Michele Volpi
Riccardo Volpi
Carl Vondrick
Konstantinos Vougioukas
Tuan-Hung Vu
Sven Wachsmuth
Neal Wadhwa
Catherine Wah
Jacob C. Walker
Thomas S. A. Wallis
Chengde Wan
Jun Wan
Liang Wan
Renjie Wan
Baoyuan Wang
Boyu Wang
Cheng Wang
Chu Wang
Chuan Wang
Chunyu Wang
Dequan Wang
Di Wang
Dilin Wang
Dong Wang
Fang Wang
Guanzhi Wang
Guoyin Wang
Hanzi Wang
Hao Wang
He Wang
Heng Wang
Hongcheng Wang

Hongxing Wang
Hua Wang
Jian Wang
Jingbo Wang
Jinglu Wang
Jingya Wang
Jinjun Wang
Jinqiao Wang
Jue Wang
Ke Wang
Keze Wang
Le Wang
Lei Wang
Lezi Wang
Li Wang
Liang Wang
Lijun Wang
Limin Wang
Linwei Wang
Lizhi Wang
Mengjiao Wang
Mingzhe Wang
Minsi Wang
Naiyan Wang
Nannan Wang
Ning Wang
Oliver Wang
Pei Wang
Peng Wang
Pichao Wang
Qi Wang
Qian Wang
Qiaosong Wang
Qifei Wang
Qilong Wang
Qing Wang
Qingzhong Wang
Quan Wang
Rui Wang
Ruiping Wang
Ruixing Wang
Shangfei Wang
Shenlong Wang
Shiyao Wang
Shuhui Wang
Song Wang

Tao Wang
Tianlu Wang
Tiantian Wang
Ting-chun Wang
Tingwu Wang
Wei Wang
Weiyue Wang
Wenguan Wang
Wenlin Wang
Wenqi Wang
Xiang Wang
Xiaobo Wang
Xiaofang Wang
Xiaoling Wang
Xiaolong Wang
Xiaosong Wang
Xiaoyu Wang
Xin Eric Wang
Xinchao Wang
Xinggang Wang
Xintao Wang
Yali Wang
Yan Wang
Yang Wang
Yangang Wang
Yaxing Wang
Yi Wang
Yida Wang
Yilin Wang
Yiming Wang
Yisen Wang
Yongtao Wang
Yu-Xiong Wang
Yue Wang
Yujiang Wang
Yunbo Wang
Yunhe Wang
Zengmao Wang
Zhangyang Wang
Zhaowen Wang
Zhe Wang
Zhecan Wang
Zheng Wang
Zhixiang Wang
Zilei Wang
Jianqiao Wangni

Anne S. Wannenwetsch
Jan Dirk Wegner
Scott Wehrwein
Donglai Wei
Kaixuan Wei
Longhui Wei
Pengxu Wei
Ping Wei
Qi Wei
Shih-En Wei
Xing Wei
Yunchao Wei
Zijun Wei
Jerod Weinman
Michael Weinmann
Philippe Weinzaepfel
Yair Weiss
Bihan Wen
Longyin Wen
Wei Wen
Junwu Weng
Tsui-Wei Weng
Xinshuo Weng
Eric Wengrowski
Tomas Werner
Gordon Wetzstein
Tobias Weyand
Patrick Wieschollek
Maggie Wigness
Erik Wijmans
Richard Wildes
Olivia Wiles
Chris Williams
Williem Williem
Kyle Wilson
Calden Wloka
Nicolai Wojke
Christian Wolf
Yongkang Wong
Sanghyun Woo
Scott Workman
Baoyuan Wu
Bichen Wu
Chao-Yuan Wu
Huikai Wu
Jiajun Wu

Jialin Wu
Jiaxiang Wu
Jiqing Wu
Jonathan Wu
Lifang Wu
Qi Wu
Qiang Wu
Ruizheng Wu
Shangzhe Wu
Shun-Cheng Wu
Tianfu Wu
Wayne Wu
Wenxuan Wu
Xiao Wu
Xiaohe Wu
Xinxiao Wu
Yang Wu
Yi Wu
Yiming Wu
Ying Nian Wu
Yue Wu
Zheng Wu
Zhenyu Wu
Zhirong Wu
Zuxuan Wu
Stefanie Wuhrer
Jonas Wulff
Changqun Xia
Fangting Xia
Fei Xia
Gui-Song Xia
Lu Xia
Xide Xia
Yin Xia
Yingce Xia
Yongqin Xian
Lei Xiang
Shiming Xiang
Bin Xiao
Fanyi Xiao
Guobao Xiao
Huaxin Xiao
Taihong Xiao
Tete Xiao
Tong Xiao
Wang Xiao

Yang Xiao
Cihang Xie
Guosen Xie
Jianwen Xie
Lingxi Xie
Sirui Xie
Weidi Xie
Wenxuan Xie
Xiaohua Xie
Fuyong Xing
Jun Xing
Junliang Xing
Bo Xiong
Peixi Xiong
Yu Xiong
Yuanjun Xiong
Zhiwei Xiong
Chang Xu
Chenliang Xu
Dan Xu
Danfei Xu
Hang Xu
Hongteng Xu
Huijuan Xu
Jingwei Xu
Jun Xu
Kai Xu
Mengmeng Xu
Mingze Xu
Qianqian Xu
Ran Xu
Weijian Xu
Xiangyu Xu
Xiaogang Xu
Xing Xu
Xun Xu
Yanyu Xu
Yichao Xu
Yong Xu
Yongchao Xu
Yuanlu Xu
Zenglin Xu
Zheng Xu
Chuhui Xue
Jia Xue
Nan Xue

Tianfan Xue
Xiangyang Xue
Abhay Yadav
Yasushi Yagi
I. Zeki Yalniz
Kota Yamaguchi
Toshihiko Yamasaki
Takayoshi Yamashita
Junchi Yan
Ke Yan
Qingan Yan
Sijie Yan
Xinchen Yan
Yan Yan
Yichao Yan
Zhicheng Yan
Keiji Yanai
Bin Yang
Ceyuan Yang
Dawei Yang
Dong Yang
Fan Yang
Guandao Yang
Guorun Yang
Haichuan Yang
Hao Yang
Jianwei Yang
Jiaolong Yang
Jie Yang
Jing Yang
Kaiyu Yang
Linjie Yang
Meng Yang
Michael Ying Yang
Nan Yang
Shuai Yang
Shuo Yang
Tianyu Yang
Tien-Ju Yang
Tsun-Yi Yang
Wei Yang
Wenhan Yang
Xiao Yang
Xiaodong Yang
Xin Yang
Yan Yang

Yanchao Yang
Yee Hong Yang
Yezhou Yang
Zhenheng Yang
Anbang Yao
Angela Yao
Cong Yao
Jian Yao
Li Yao
Ting Yao
Yao Yao
Zhewei Yao
Chengxi Ye
Jianbo Ye
Keren Ye
Linwei Ye
Mang Ye
Mao Ye
Qi Ye
Qixiang Ye
Mei-Chen Yeh
Raymond Yeh
Yu-Ying Yeh
Sai-Kit Yeung
Serena Yeung
Kwang Moo Yi
Li Yi
Renjiao Yi
Alper Yilmaz
Junho Yim
Lijun Yin
Weidong Yin
Xi Yin
Zhichao Yin
Tatsuya Yokota
Ryo Yonetani
Donggeun Yoo
Jae Shin Yoon
Ju Hong Yoon
Sung-eui Yoon
Laurent Younes
Changqian Yu
Fisher Yu
Gang Yu
Jiahui Yu
Kaicheng Yu

Ke Yu
Lequan Yu
Ning Yu
Qian Yu
Ronald Yu
Ruichi Yu
Shoou-I Yu
Tao Yu
Tianshu Yu
Xiang Yu
Xin Yu
Xiyu Yu
Youngjae Yu
Yu Yu
Zhiding Yu
Chunfeng Yuan
Ganzhao Yuan
Jinwei Yuan
Lu Yuan
Quan Yuan
Shanxin Yuan
Tongtong Yuan
Wenjia Yuan
Ye Yuan
Yuan Yuan
Yuhui Yuan
Huanjing Yue
Xiangyu Yue
Ersin Yumer
Sergey Zagoruyko
Egor Zakharov
Amir Zamir
Andrei Zanfir
Mihai Zanfir
Pablo Zegers
Bernhard Zeisl
John S. Zelek
Niclas Zeller
Huayi Zeng
Jiabei Zeng
Wenjun Zeng
Yu Zeng
Xiaohua Zhai
Fangneng Zhan
Huangying Zhan
Kun Zhan

Xiaohang Zhan
Baochang Zhang
Bowen Zhang
Cecilia Zhang
Changqing Zhang
Chao Zhang
Chengquan Zhang
Chi Zhang
Chongyang Zhang
Dingwen Zhang
Dong Zhang
Feihu Zhang
Hang Zhang
Hanwang Zhang
Hao Zhang
He Zhang
Hongguang Zhang
Hua Zhang
Ji Zhang
Jianguo Zhang
Jianming Zhang
Jiawei Zhang
Jie Zhang
Jing Zhang
Juyong Zhang
Kai Zhang
Kaipeng Zhang
Ke Zhang
Le Zhang
Lei Zhang
Li Zhang
Lihe Zhang
Linguang Zhang
Lu Zhang
Mi Zhang
Mingda Zhang
Peng Zhang
Pingping Zhang
Qian Zhang
Qilin Zhang
Quanshi Zhang
Richard Zhang
Rui Zhang
Runze Zhang
Shengping Zhang
Shifeng Zhang

Shuai Zhang
Songyang Zhang
Tao Zhang
Ting Zhang
Tong Zhang
Wayne Zhang
Wei Zhang
Weizhong Zhang
Wenwei Zhang
Xiangyu Zhang
Xiaolin Zhang
Xiaopeng Zhang
Xiaoqin Zhang
Xiuming Zhang
Ya Zhang
Yang Zhang
Yimin Zhang
Yinda Zhang
Ying Zhang
Yongfei Zhang
Yu Zhang
Yulun Zhang
Yunhua Zhang
Yuting Zhang
Zhanpeng Zhang
Zhao Zhang
Zhaoxiang Zhang
Zhen Zhang
Zheng Zhang
Zhifei Zhang
Zhijin Zhang
Zhishuai Zhang
Ziming Zhang
Bo Zhao
Chen Zhao
Fang Zhao
Haiyu Zhao
Han Zhao
Hang Zhao
Hengshuang Zhao
Jian Zhao
Kai Zhao
Liang Zhao
Long Zhao
Qian Zhao
Qibin Zhao

Qijun Zhao
Rui Zhao
Shenglin Zhao
Sicheng Zhao
Tianyi Zhao
Wenda Zhao
Xiangyun Zhao
Xin Zhao
Yang Zhao
Yue Zhao
Zhichen Zhao
Zijing Zhao
Xiantong Zhen
Chuanxia Zheng
Feng Zheng
Haiyong Zheng
Jia Zheng
Kang Zheng
Shuai Kyle Zheng
Wei-Shi Zheng
Yinqiang Zheng
Zerong Zheng
Zhedong Zheng
Zilong Zheng
Bineng Zhong
Fangwei Zhong
Guangyu Zhong
Yiran Zhong
Yujie Zhong
Zhun Zhong
Chunluan Zhou
Huiyu Zhou
Jiahuan Zhou
Jun Zhou
Lei Zhou
Luowei Zhou
Luping Zhou
Mo Zhou
Ning Zhou
Pan Zhou
Peng Zhou
Qianyi Zhou
S. Kevin Zhou
Sanping Zhou
Wengang Zhou
Xingyi Zhou

Yanzhao Zhou
Yi Zhou
Yin Zhou
Yipin Zhou
Yuyin Zhou
Zihan Zhou
Alex Zihao Zhu
Chenchen Zhu
Feng Zhu
Guangming Zhu
Ji Zhu
Jun-Yan Zhu
Lei Zhu
Linchao Zhu
Rui Zhu
Shizhan Zhu
Tyler Lixuan Zhu

Wei Zhu
Xiangyu Zhu
Xinge Zhu
Xizhou Zhu
Yanjun Zhu
Yi Zhu
Yixin Zhu
Yizhe Zhu
Yousong Zhu
Zhe Zhu
Zhen Zhu
Zheng Zhu
Zhenyao Zhu
Zhihui Zhu
Zhuotun Zhu
Bingbing Zhuang
Wei Zhuo

Christian Zimmermann
Karel Zimmermann
Larry Zitnick
Mohammadreza
   Zolfaghari
Maria Zontak
Daniel Zoran
Changqing Zou
Chuhang Zou
Danping Zou
Qi Zou
Yang Zou
Yuliang Zou
Georgios Zoumpourlis
Wangmeng Zuo
Xinxin Zuo

Additional Reviewers

Victoria Fernandez
   Abrevaya
Maya Aghaei
Allam Allam
Christine
   Allen-Blanchette
Nicolas Aziere
Assia Benbihi
Neha Bhargava
Bharat Lal Bhatnagar
Joanna Bitton
Judy Borowski
Amine Bourki
Romain Brégier
Tali Brayer
Sebastian Bujwid
Andrea Burns
Yun-Hao Cao
Yuning Chai
Xiaojun Chang
Bo Chen
Shuo Chen
Zhixiang Chen
Junsuk Choe
Hung-Kuo Chu

Jonathan P. Crall
Kenan Dai
Lucas Deecke
Karan Desai
Prithviraj Dhar
Jing Dong
Wei Dong
Turan Kaan Elgin
Francis Engelmann
Erik Englesson
Fartash Faghri
Zicong Fan
Yang Fu
Risheek Garrepalli
Yifan Ge
Marco Godi
Helmut Grabner
Shuxuan Guo
Jianfeng He
Zhezhi He
Samitha Herath
Chih-Hui Ho
Yicong Hong
Vincent Tao Hu
Julio Hurtado

Jaedong Hwang
Andrey Ignatov
Muhammad
   Abdullah Jamal
Saumya Jetley
Meiguang Jin
Jeff Johnson
Minsoo Kang
Saeed Khorram
Mohammad Rami Koujan
Nilesh Kulkarni
Sudhakar Kumawat
Abdelhak Lemkhenter
Alexander Levine
Jiachen Li
Jing Li
Jun Li
Yi Li
Liang Liao
Ruochen Liao
Tzu-Heng Lin
Phillip Lippe
Bao-di Liu
Bo Liu
Fangchen Liu

Hanxiao Liu
Hongyu Liu
Huidong Liu
Miao Liu
Xinxin Liu
Yongfei Liu
Yu-Lun Liu
Amir Livne
Tiange Luo
Wei Ma
Xiaoxuan Ma
Ioannis Marras
Georg Martius
Effrosyni Mavroudi
Tim Meinhardt
Givi Meishvili
Meng Meng
Zihang Meng
Zhongqi Miao
Gyeongsik Moon
Khoi Nguyen
Yung-Kyun Noh
Antonio Norelli
Jaeyoo Park
Alexander Pashevich
Mandela Patrick
Mary Phuong
Bingqiao Qian
Yu Qiao
Zhen Qiao
Sai Saketh Rambhatla
Aniket Roy
Amelie Royer
Parikshit Vishwas
  Sakurikar
Mark Sandler
Mert Bülent Sarıyıldız
Tanner Schmidt
Anshul B. Shah

Ketul Shah
Rajvi Shah
Hengcan Shi
Xiangxi Shi
Yujiao Shi
William A. P. Smith
Guoxian Song
Robin Strudel
Abby Stylianou
Xinwei Sun
Reuben Tan
Qingyi Tao
Kedar S. Tatwawadi
Anh Tuan Tran
Son Dinh Tran
Eleni Triantafillou
Aristeidis Tsitiridis
Md Zasim Uddin
Andrea Vedaldi
Evangelos Ververas
Vidit Vidit
Paul Voigtlaender
Bo Wan
Huanyu Wang
Huiyu Wang
Junqiu Wang
Pengxiao Wang
Tai Wang
Xinyao Wang
Tomoki Watanabe
Mark Weber
Xi Wei
Botong Wu
James Wu
Jiamin Wu
Rujie Wu
Yu Wu
Rongchang Xie
Wei Xiong

Yunyang Xiong
An Xu
Chi Xu
Yinghao Xu
Fei Xue
Tingyun Yan
Zike Yan
Chao Yang
Heran Yang
Ren Yang
Wenfei Yang
Xu Yang
Rajeev Yasarla
Shaokai Ye
Yufei Ye
Kun Yi
Haichao Yu
Hanchao Yu
Ruixuan Yu
Liangzhe Yuan
Chen-Lin Zhang
Fandong Zhang
Tianyi Zhang
Yang Zhang
Yiyi Zhang
Yongshun Zhang
Yu Zhang
Zhiwei Zhang
Jiaojiao Zhao
Yipu Zhao
Xingjian Zhen
Haizhong Zheng
Tiancheng Zhi
Chengju Zhou
Hao Zhou
Hao Zhu
Alexander Zimin

Contents – Part XIV

SipMask: Spatial Information Preservation for Fast Image and Video Instance Segmentation

Jiale Cao[1], Rao Muhammad Anwer[2,3], Hisham Cholakkal[2,3], Fahad Shahbaz Khan[2,3], Yanwei Pang[1(✉)], and Ling Shao[2,3]

[1] Tianjin University, Tianjin, China
{connor,pyw}@tju.edu.cn
[2] Mohamed bin Zayed University of Artificial Intelligence, Abu Dhabi, UAE
{rao.anwer,hisham.cholakkal,fahad.khan,ling.shao}@mbzuai.ac.ae
[3] Inception Institute of Artificial Intelligence, Abu Dhabi, UAE

Abstract. Single-stage instance segmentation approaches have recently gained popularity due to their speed and simplicity, but are still lagging behind in accuracy, compared to two-stage methods. We propose a fast single-stage instance segmentation method, called SipMask, that preserves instance-specific spatial information by separating mask prediction of an instance to different sub-regions of a detected bounding-box. Our main contribution is a novel light-weight spatial preservation (SP) module that generates a separate set of spatial coefficients for each sub-region within a bounding-box, leading to improved mask predictions. It also enables accurate delineation of spatially adjacent instances. Further, we introduce a mask alignment weighting loss and a feature alignment scheme to better correlate mask prediction with object detection. On COCO `test-dev`, our SipMask outperforms the existing single-stage methods. Compared to the state-of-the-art single-stage TensorMask, SipMask obtains an absolute gain of 1.0% (mask AP), while providing a four-fold speedup. In terms of real-time capabilities, SipMask outperforms YOLACT with an absolute gain of 3.0% (mask AP) under similar settings, while operating at comparable speed on a Titan Xp. We also evaluate our SipMask for real-time video instance segmentation, achieving promising results on YouTube-VIS dataset. The source code is available at https://github.com/JialeCao001/SipMask.

Keywords: Instance segmentation · Real-time · Spatial preservation

1 Introduction

Instance segmentation aims to classify each pixel in an image into an object category. Different from semantic segmentation [6,10,32,34,39], instance segmentation also differentiates multiple object instances. Modern instance segmentation methods typically adapt object detection frameworks, where bounding-box

© Springer Nature Switzerland AG 2020
A. Vedaldi et al. (Eds.): ECCV 2020, LNCS 12359, pp. 1–18, 2020.
https://doi.org/10.1007/978-3-030-58568-6_1

Fig. 1. Instance segmentation examples using YOLACT [2] (top) and our approach (bottom). YOLACT struggles to accurately delineate spatially adjacent instances. Our approach with novel spatial coefficients addresses this issue (marked by white dotted region) by preserving spatial information in bounding-box. The spatial coefficients split mask prediction into multiple sub-mask predictions, leading to improved mask quality.

detection is first performed, followed by segmentation inside each of detected bounding-boxes. Instance segmentation approaches can generally be divided into two-stage [8,17,21,23,31] and single-stage [2,13,36,37,42,47] methods, based on the underlying detection framework. Two-stage methods typically generate multiple object proposals in the first stage. In the second stage, they perform feature pooling operations on each proposal, followed by box regression, classification, and mask prediction. Different from two-stage methods, single-stage approaches do not require proposal generation or pooling operations and employ dense predictions of bounding-boxes and instance masks. Although two-stage methods dominate accuracy, they are generally slow, which restricts their usability in real-time applications.

As discussed above, most single-stage methods are inferior in accuracy, compared to their two-stage counterparts. A notable exception is the single-stage TensorMask [11], which achieves comparable accuracy to two-stage methods. However, TensorMask achieves this accuracy at the cost of reduced speed. In fact, TensorMask [11] is slower than several two-stage methods, including Mask R-CNN [21]. Recently, YOLACT [2] has shown to achieve an optimal tradeoff between speed and accuracy. On the COCO benchmark [29], the single-stage YOLACT operates at real-time (33 frames per second), while obtaining competitive accuracy. YOLACT achieves real-time speed mainly by avoiding proposal generation and feature pooling head networks that are commonly employed in two-stage methods. While operating at real-time, YOLACT still lags behind modern two-stage methods (*e.g.*, Mask R-CNN [21]), in terms of accuracy.

In this work, we argue that one of the key reasons behind sub-optimal accuracy of YOLACT is the loss of spatial information within an object (bounding-box). We attribute this loss of spatial information due to the utilization of a *single* set of object-aware coefficients to predict the whole mask of an object. As a result, it struggles to accurately delineate spatially adjacent object instances (Fig. 1). To address this issue, we introduce an approach that comprises a novel computationally efficient spatial preservation (SP) module to preserve spatial information in a bounding-box. Our SP module predicts object-aware *spatial* coefficients that splits mask prediction into multiple sub-mask predictions, thereby enabling improved delineation of spatially adjacent objects (Fig. 1).

Contributions: We propose a fast anchor-free single-stage instance segmentation approach, called SipMask, with the following contributions.

- We propose a novel light-weight spatial preservation (SP) module that preserves the spatial information within a bounding-box. Our SP module generates a separate set of *spatial* coefficients for each bounding-box sub-region, enabling improved delineation of spatially adjacent objects.
- We introduce two strategies to better correlate mask prediction with object detection. First, we propose a mask alignment weighting loss that assigns higher weights to the mask prediction errors occurring at accurately detected boxes. Second, a feature alignment scheme is introduced to improve the feature representation for both box classification and spatial coefficients.
- Comprehensive experiments are performed on COCO benchmark [29]. Our single-scale inference model based on ResNet101-FPN backbone outperforms state-of-the-art single-stage TensorMask [11] in terms of *both* mask accuracy (absolute gain of 1.0% on COCO test-dev) and speed (four-fold speedup). Compared with real-time YOLACT [2], our SipMask provides an absolute gain of 3.0% on COCO test-dev, while operating at comparable speed.
- The proposed SipMask can be extended to single-stage video instance segmentation by adding a fully-convolutional branch for tracking instances across video frames. On YouTube-VIS dataset [48], our single-stage approach achieves favourable performance while operating at real-time (30 fps).

2   Related Work

Deep learning has achieved great success in a variety of computer vision tasks [12,20,24,25,35,43–45,52,53]. Existing instance segmentation methods either follow bottom-up [1,19,26,30,33] or top-down [2,8,21,31,36] paradigms. Modern instance segmentation approaches typically follow top-down paradigm where the bounding-boxes are first detected and second segmented. The top-down approaches are divided into two-stage [8,17,21,23,31] and single-stage [2,13,36, 42,47] methods. Among these two-stage methods, Mask R-CNN [21] employs a proposal generation network (RPN) and utilizes RoIAlign feature pooling strategy (Fig. 2(a)) to obtain a fixed-sized features of each proposal. The pooled features are used for box detection and mask prediction. A position sensitive feature pooling strategy, PSRoI [15] (Fig. 2(b)), is proposed in FCIS [27]. PANet [31] proposes an adaptive feature pooling that allows each proposal to access information from multiple layers of FPN. MS R-CNN [23] introduces an additional branch

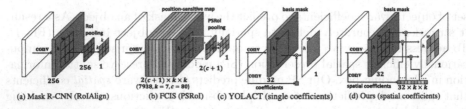

(a) Mask R-CNN (RoIAlign)   (b) FCIS (PSRoI)   (c) YOLACT (single coefficients)   (d) Ours (spatial coefficients)

Fig. 2. On the left (a and b), feature pooling strategies employed in Mask R-CNN [21] and FCIS [27] resize the feature map to a fixed resolution. Instead, both YOLACT [2] (c) and our approach (d) do not utilize any pooling operation and obtain mask prediction by a simple linear combination of basis mask and coefficient. Mask R-CNN is computationally expensive (*conv* and *deconv* operations after RoIAlign), whereas FCIS is memory demanding due to large number of channels in position-sensitive maps. Both YOLACT and our approach reduce the computational and memory complexity. However, YOLACT uses a single set of coefficients for a detected box, thereby ignoring the spatial information within a box. Our approach preserves the spatial information of an instance by using separate set of spatial coefficients for $k \times k$ sub-regions within a box.

to predict mask quality (mask-IoU). MS R-CNN performs a mask confidence rescoring without improving mask quality. In contrast, our mask alignment loss aims to improve mask quality at accurate detections.

Different to two-stage methods, single-stage approaches [2,13,42,47] typically aim at faster inference speed by avoiding proposal generation and feature pooling strategies. However, most single-stage approaches are generally inferior in accuracy compared to their two-stage counterparts. Recently, YOLACT [2] obtains an optimal tradeoff between accuracy and speed by predicting a dictionary of category-independent maps (basis masks) for an image and a single set of instance-specific coefficients. Despite its real-time capabilities, YOLACT achieves inferior accuracy compared to two-stage methods. Different to YOLACT, which has a single set of coefficients for each bounding-box (Fig. 2(c)), our novel SP module aims at preserving spatial information within a bounding-box. The SP module generates multiple sets of spatial coefficients that splits mask prediction into different sub-regions in a bounding-box (Fig. 2(d)). Further, SP module contains a feature alignment scheme that improves feature representation by aligning the predicted instance mask with detected bounding-box. Our SP module is different to feature pooling strategies, such as PSRoI [27] in several ways. Instead of pooling features into a fixed size ($k \times k$), we perform a simple linear combination between spatial coefficients and basis masks without any feature resizing operation. This preservation of feature resolution is especially suitable for large objects. PSRoI pooling (Fig. 2(b)) generates feature maps of $2(c+1) \times k \times k$ channels, where k is the pooled feature size and c is the number of classes. In practice, such a pooling operation is memory expensive (7938 channels for $k = 7$ and $c = 80$). Instead, our design is memory efficient since the basis masks are of only 32 channels for whole image and the spatial coefficients are a 32 dimensional vector for each sub-region of a bounding-box (Fig. 2(d)). Further, compared to contemporary work [7] using RoIpool based feature maps, our

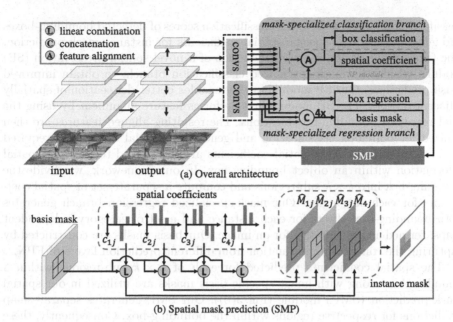

Fig. 3. (a) Overall architecture of our SipMask comprising fully convolutional mask-specialized classification (Sect. 3.1) and regression (Sect. 3.2) branches. The focus of our design is the introduction of a novel spatial preservation (SP) module in the mask-specialized classification branch. The SP module performs two-tasks: feature alignment and spatial coefficients generation. In our approach, a separate set of spatial coefficients are generated for each predicted bounding-box. These spatial coefficients are designed to preserve the spatial information within an object instance, thereby enabling improved delineation of spatially adjacent objects. The mask-specialized regression branch predicts both bounding-box offsets and a set of category-independent basis masks. The basis masks are generated by capturing contextual information from different prediction layers of FPN. (b) Both the basis masks and spatial coefficients along with predicted bounding-box locations are then input to our spatial mask prediction (SMP) module (Sect. 3.3) for predicting the final instance mask.

approach utilizes fewer coefficients on original basis mask. Moreover, our SipMask can be adapted for real-time single-stage video instance segmentation.

3 Method

Overall Architecture: Figure 3(a) shows the overall architecture of our single-stage anchor-free method, SipMask, named for its instance-specific spatial information preservation characteristic. Our architecture is built on FCOS detection method [40], due to its flexible anchor-free design. In the proposed architecture, we replace the standard classification and regression in FCOS with our mask-specialized regression and classification branches. Both mask-specialized classification and regression branches are fully convolutional. Our mask-specialized classification branch predicts the classification scores of detected bounding-boxes and generates instance-specific spatial coefficients for instance mask prediction. The focus of our design is the introduction of a novel spatial preservation (SP) module, within the mask-specialized classification branch, to obtain improved mask predictions. Our SP module further enables better delineation of spatially adjacent objects. The SP module first performs feature alignment by using the final regressed bounding-box locations. The resulting aligned features are then utilized for both box classification and generating spatial coefficients required for mask prediction. The spatial coefficients are introduced to preserve spatial information within an object bounding-box. In our framework, we divide the bounding-box into $k \times k$ sub-regions and compute a separate set of spatial coefficients for each sub-region. Our mask-specialized regression branch generates both bounding-box offsets for each instance and a set of category-independent maps, termed as basis masks, for an image. Our basis masks are constructed by capturing the contextual information from different prediction layers of FPN.

The spatial coefficients predicted for each of $k \times k$ sub-regions within a bounding-box along with image-specific basis masks are utilized in our spatial mask prediction (SMP) module (Fig. 3(b)). Our SMP generates separate map predictions for respective regions within the bounding-box. Consequently, these separate map predictions are combined to obtain final instance mask prediction.

3.1 Spatial Preservation Module

Besides box classification, our mask-specialized classification branch comprises a novel spatial preservation (SP) module. Our SP module performs two tasks: spatial coefficients generation and feature alignment. The spatial coefficients are introduced to improve mask prediction by preserving spatial information within a bounding-box. Our feature alignment scheme aims at improving the feature representation for both box classification and spatial coefficients generation.

Spatial Coefficients Generation: As discussed earlier, the recently introduced YOLACT [2] utilizes a single set of coefficients to predict the whole mask of an object, leading to the loss of spatial information within a bounding-box. To address this issue, we propose a simple but effective approach that splits mask prediction into multiple sub-mask predictions. We divide the spatial regions within a predicted bounding-box into $k \times k$ sub-regions. Instead of predicting a *single set of coefficients* for the whole bounding-box j, we predict a *separate set of spatial coefficients* $c_{ij} \in R^m$ for each of its sub-region i. Figure 3(b) shows an example where a bounding-box is divided into 2×2 sub-regions (four quadrants, *i.e.*, top-left, top-right, bottom-left and bottom-right). In practice, we observe that $k = 2$ provides an optimal tradeoff between speed and accuracy. Note that our spatial coefficients utilize improved features obtained though a feature alignment operation described next.

Feature Alignment Scheme: Generally, convolutional layer operates on a rectangular grid (*e.g.*, 3×3 kernel). Thus, the extracted features for classification and coefficients generation may fail to align with the features of regressed bounding-box. Our feature alignment scheme addresses this issue by aligning the features with regressed box location, resulting in an improved feature representation. For feature alignment, we introduce a deformable convolutional layer [5,16,51] in our mask-specialized classification branch. The input to the deformable convolutional layer are the regression offsets to left, right, top, and bottom corners of ground-truth bounding-box obtained from mask-specialized regression branch (Sect. 3.2). These offsets are utilized to estimate the kernel offset Δp_r that augments the regular sampling grid G in the deformable convolution operator, resulting in an aligned feature $y(p_0)$ at position p_0, as follows:

$$y(p_0) = \sum_{i \in G} w_r \cdot x(p_0 + p_r + \Delta p_r), \qquad (1)$$

where x is the input feature, and p_r is the original position of convolutional weight w_r in G. Different to [50,51] that aim to learn accurate geometric localization, our approach aims to generate better features for box classification and coefficient generation. Next, we describe mask-specialized regression branch.

3.2 Mask-Specialized Regression Branch

Our mask-specialized regression branch performs box regression and generates a set of category-independent basis masks for an image. Note that YOLACT utilizes a single FPN prediction layer to generate the basis masks. Instead, the basis masks in our SipMask are generated by exploiting the multi-layer information from different prediction layers of FPN. The incorporation of multi-layer information helps to obtain a continuous mask (especially on large objects) and remove background clutter. Further, it helps in scenarios, such as partial occlusion and large-scale variation. Here, objects of various sizes are predicted at different prediction layers of the FPN (*i.e.*, $P3 - P7$). To capture multi-layer information, the features from the $P3 - P5$ layers of the FPN are utilized to generate basis masks. Note that $P6$ and $P7$ are excluded for basis mask generation to reduce the computational cost. The outputs from $P4$ and $P5$ are first upsampled to the resolution of $P3$ using bilinear interpolation. The resulting features from all three prediction layers ($P3 - P5$) are concatenated, followed by a 3×3 convolution to generate feature maps with m channels. Finally, these feature maps are upsampled four times by using bilinear interpolation, resulting in m basis masks, each having a spatial resolution of $h \times w$. Both the spatial coefficients (Sect. 3.1) and basis masks are utilized in our spatial mask prediction (SMP) module for final instance mask prediction.

3.3 Spatial Mask Prediction Module

Given an input image, our spatial mask prediction (SMP) module takes the predicted bounding-boxes, basis masks and spatial coefficients as inputs and predicts the final instance mask. Let $B \in R^{h \times w \times m}$ represent m predicted basis masks for the whole image, p be the number of predicted boxes, and C_i be a $m \times p$

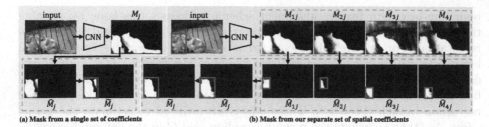

(a) Mask from a single set of coefficients   **(b) Mask from our separate set of spatial coefficients**

Fig. 4. A visual comparison between mask generation using (a) a single set of coefficients, as in YOLACT and (b) our SipMask. For simplicity, only one detected 'cat' instance and its corresponding mask generation procedure is shown here. A linear combination of single set of coefficients and basis masks leads to one map M_j. Then, the map M_j is pruned followed by thresholding to produce the final mask $\tilde{M}_j$. Instead, our SipMask generates a separate set of spatial coefficients for each sub-region (quadrant for $k = 2$) within a bounding-box. As a result, a separate set of spatial map M_{ij} is obtained for each quadrant i in the bounding-box j. Afterwards, these spatial maps are first pruned and then integrated (a simple addition) followed by thresholding to obtain final mask $\tilde{M}_j$. Our SipMask is able to reduce the influence of the adjacent object ('cat') instance, resulting in improved mask prediction.

matrix that indicates the spatial coefficients at the ith sub-region (quadrant for $k = 2$) of all p predicted bounding-boxes. Note that the column j of C_i (*i.e.*, $c_{ij} \in R^m$) indicates the spatial coefficients for the bounding-box j (Sect. 3.1). We perform a simple matrix multiplication between C_i and B to obtain p maps corresponding to the ith quadrant of all bounding-boxes as follows.

$$M_i = \sigma(B \times C_i) \quad \forall i \in [1, 4], \tag{2}$$

where σ is sigmoid normalization and $M_i \in R^{h \times w \times p}$ are the maps generated for the ith quadrant of all p bounding-boxes. Figure 4(b) shows the procedure to obtain final mask of an instance j. Let $M_{ij} \in R^{h \times w}$ be the map generated for the ith quadrant of a bounding-box j. Then, the response values of M_{ij} outside the ith quadrant of the box j are set as zero for generating a pruned map $\hat{M}_{ij}$. To obtain the instance map $\hat{M}_j$ of a bounding-box j, we perform a simple addition of its pruned maps obtained from all four quadrants, *i.e.*, $\hat{M}_j = \sum_{i=1}^{4} \hat{M}_{ij}$. Finally, the instance map at the predicted bounding-box region is binarized with a fixed threshold to obtain final mask $\tilde{M}_j$ of instance j.

Figure 4 shows a visual comparison of a single set of coefficients based mask prediction, as in YOLACT, with our separate set of spatial coefficients (for each sub-region) based mask prediction. The top-left pixels of an adjacent 'cat' instance are appearing inside the top-right quadrant of the detected 'cat' instance bounding-box (in red). In Fig. 4(a), a linear combination of a single set of instance-specific coefficients and image-level basis masks is used to obtain a map M_j. The response values of the map M_j outside the box j are assigned with zero to produce a pruned mask $\hat{M}_j$, followed by thresholding to obtain the final mask $\tilde{M}_j$. Instead, our SipMask (Fig. 4(b)) generates a separate set of instance-specific spatial coefficients for each sub-region i within a bounding-box j. By separating the mask predictions to different sub-regions of a box, our SipMask reduces the influence of adjacent (overlapping) object instance in final mask prediction.

3.4 Loss Function

The overall loss function of our framework contains loss terms corresponding to bounding-box detection (classification and regression) and mask generation. For box classification L_{cls} and box regression L_{reg}, we utilize focal loss and IoU loss, respectively, as in [40]. For mask generation, we introduce a novel mask alignment weighting loss L_{mask} that better correlate mask predictions with high quality bounding-box detections. Different to YOLACT that utilizes a standard pixel-wise binary cross entropy (BCE) loss during training, our L_{mask} improves the BCE loss with a mask alignment weighting scheme that assigns higher weights to the masks $\tilde{M}_j$ obtained from high quality bounding-box detections.

Mask Alignment Weighting: In our mask alignment weighting, we first compute the overlap o_j between a predicted bounding-box j and the corresponding ground-truth. The weighting factor α_j is then obtained by multiplying the overlap o_j and the classification score s_j of the bounding-box j. Here, a higher α_j indicates good quality bounding-box detections. Consequently, α_j is used to weight the mask loss l^j of the instance j, leading to $L_{mask} = \frac{1}{N}\sum_j l^j \times \alpha_j$. Here, N is the number of bounding-boxes. Our weighting strategy encourages the network to predict a high quality instance mask for a high quality bounding-box detections. The proposed mask alignment weighting loss L_{mask} is utilized along with loss terms corresponding to bounding-box detection (classification and regression) in our overall loss function: $L = L_{reg} + L_{cls} + L_{mask}$.

3.5 Single-Stage Video Instance Segmentation

In addition to still image instance segmentation, we investigate our single-stage SipMask for the problem of real-time video instance segmentation. In video instance segmentation, the aim is to simultaneously detect, segment, and track instances in videos.

To perform real-time single-stage video instance segmentation, we simply extend our SipMask by introducing an additional fully-convolutional branch in parallel to mask-specialized classification and regression branches for instance tracking. The fully-convolutional branch consists of two convolutional layers. After that, the output feature maps of different layers in this branch are fused to obtain the tracking feature maps, similar to basis mask generation in our mask-specialized regression branch. Different from the state-of-the-art Mask-Track R-CNN [48] that utilizes RoIAlign and fully-connected operations, our SipMask extracts a tracking feature vector from the tracking feature maps at the bounding-box center to represent each instance. The metric for matching the instances between different frames is similar to MaskTrack R-CNN. Our SipMask is very simple, efficient and achieves favourable performance for video instance segmentation (Sect. 4.4).

4   Experiments

4.1   Dataset and Implementation Details

Dataset: We conduct experiments on COCO dataset [29], where the `trainval` set has about $115k$ images, the `minival` set has $5k$ images, and the `test-dev` set has about $20k$ images. We perform training on `trainval` set and present state-of-the-art comparison on `test-dev` set and the ablations on `minival` set.

Implementation Details: We adopt ResNet [22] (ResNet50/ResNet101) with FPN pre-trained on ImageNet [38] as the backbone. Our method is trained eight GPUs with SGD for optimization. During training, the initial learning rate is set to 0.01. When conducting ablation study, we use a $1\times$ training scheme at single scale to reduce training time. For a fair comparison with the state-of-the-art single-stage methods [2,11], we follow the $6\times$, multi-scale training scheme. During inference we select top 100 bounding-boxes with highest classification scores, after NMS. For these bounding-boxes, a simple linear combination between the predicted spatial coefficients and basis masks are used to obtain instance masks.

4.2   State-of-the-art Comparison

Here, we compare our method with some two-stage [4,8,9,14,18,21,23,27,31] and single-stage [2,11,46,54] methods on COCO `test-dev` set. Table 1 shows the comparison in terms of both speed and accuracy. Most existing methods use a larger input image size, typically $\sim 1333 \times 800$ (except YOLACT [2], which operates on input size of 550×550). Among existing two-stage methods, Mask R-CNN [21] and PANet [31] achieve overall mask AP scores of 35.7 and 36.6, respectively. The recently introduced MS R-CNN [21] and HTC [8] obtain mask AP scores of 38.3 and 39.7, respectively. Note that HTC achieves this improved accuracy at the cost of a significant reduction in speed. Further, most two-stage approaches require more than 100 milliseconds (ms) to process an image.

In case of single-stage methods, PolarMask [46] obtains a mask AP of 30.4. RDSNet [42] achieves a mask AP score of 36.4. Among these single-stage methods, TensorMask [11] obtains the best results with a mask AP score of 37.1. Our SipMask under similar settings (input size and backbone) outperforms TensorMask with an absolute gain of 1.0%, while obtaining a four-fold speedup. In particular, our SipMask achieves an absolute gain of 2.7% on the large objects, compared to TensorMask.

In terms of fast instance segmentation and real-time capabilities, we compare our SipMask with YOLACT [2] when using two different backbone models (ResNet50/ResNet101 FPN). Compared to YOLACT, our SipMask achieves an absolute gain of 3.0% without any significant reduction in speed (YOLACT: 30 ms vs. SipMask: 32 ms). A recent variant of YOLACT, called YOLACT++ [3], utilizes a deformable backbone (ResNet101-Deform [55] with interval 3) and a mask scoring strategy. For a fair comparison, we also integrate the same two ingredients in our SipMask, called as SipMask++. When using a similar input size and same backbone, our SipMask++ achieves improved mask accuracy while

Table 1. State-of-the-art instance segmentation comparison in terms of accuracy (mask AP) and speed (inference time) on COCO `test-dev` set. All results are based on single-scale test and speeds are reported on a single Titan Xp GPU (except TensorMask and RDSNet that are reported on Tesla V100). When using the same large input size ($\sim 1333 \times 800$) and backbone, our SipMask outperforms all existing single-stage methods in terms of accuracy. Further, our SipMask obtains a four-fold speedup over the TensorMask. When using a similar small input size ($\sim 550 \times 550$), our SipMask++ achieves superior performance while operating at comparable speed, compared to the YOLACT++. In terms of real-time capabilities, our SipMask consistently improves the mask accuracy without any significant reduction in speed, compared to the YOLACT.

Method	Backbone	Input size	Time	AP	AP@0.5	AP@0.75	AP_s	AP_m	AP_l
Two-Stage									
MNC [14]	ResNet101-C4	$\sim 1333 \times 800$	-	24.6	44.3	24.8	4.7	25.9	43.6
FCIS [27]	ResNet101-C5	$\sim 1333 \times 800$	152	29.2	49.5	-	7.1	31.3	50.0
RetinaMask [18]	ResNet101-FPN	$\sim 1333 \times 800$	167	34.7	55.4	36.9	14.3	36.7	50.5
MaskLab [9]	ResNet101	$\sim 1333 \times 800$	-	35.4	57.4	37.4	16.9	38.3	49.2
Mask R-CNN [21]	ResNet101-FPN	$\sim 1333 \times 800$	116	35.7	58.0	37.8	15.5	38.1	52.4
Mask R-CNN* [21]	ResNet101-FPN	$\sim 1333 \times 800$	116	38.3	61.2	40.8	18.2	40.6	54.1
PANet [31]	ResNet50-FPN	$\sim 1333 \times 800$	212	36.6	58.0	39.3	16.3	38.1	53.1
MS R-CNN [23]	ResNet101-FPN	$\sim 1333 \times 800$	117	38.3	58.8	41.5	17.8	40.4	**54.4**
HTC [8]	ResNet101-FPN	$\sim 1333 \times 800$	417	39.7	**61.8**	43.1	21.0	42.2	53.5
D2Det [4]	ResNet101-FPN	$\sim 1333 \times 800$	168	**40.2**	61.5	**43.7**	**21.7**	**43.0**	54.0
Single-Stage: Large input size									
PolarMask [46]	ResNet101-FPN	$\sim 1333 \times 800$	-	30.4	51.9	31.0	13.4	32.4	42.8
RDSNet [42]	ResNet101-FPN	$\sim 1333 \times 800$	113	36.4	57.9	39.0	16.4	39.5	51.6
TensorMask [11]	ResNet101-FPN	$\sim 1333 \times 800$	380	37.1	59.3	39.4	17.1	39.1	51.6
Our SipMask	ResNet101-FPN	$\sim 1333 \times 800$	89	**38.1**	**60.2**	**40.8**	**17.8**	**40.8**	**54.3**
Single-Stage: Small input size									
YOLACT++ [3]	ResNet101-Deform	550×550	**37**	34.6	53.8	36.9	**11.9**	36.8	55.1
Our SipMask++	ResNet101-Deform	544×544	**37**	**35.4**	**55.6**	**37.6**	11.2	**38.3**	**56.8**
Real-Time									
YOLACT [2]	ResNet50-FPN	550×550	**22**	28.2	46.6	29.2	9.2	29.3	44.8
Our SipMask	ResNet50-FPN	544×544	24	31.2	51.9	32.3	9.2	33.6	49.8
YOLACT [2]	ResNet101-FPN	550×550	30	29.8	48.5	31.2	**9.9**	31.3	47.7
Our SipMask	ResNet101-FPN	544×544	32	**32.8**	**53.4**	**34.3**	9.3	**35.6**	**54.0**

operating at the same speed, compared to YOLACT++. Figure 5 shows example instance segmentation results of our SipMask on COCO `test-dev`.

4.3 Ablation Study

We perform an ablation study on COCO `minival` set with ResNet50-FPN backbone [28]. First, we show the impact of progressively integrating our different components: spatial preservation (SP) module (Sect. 3.1), contextual basis masks (CBM) obtained by integrating context information from different FPN prediction layers (Sect. 3.2), and mask alignment weighting loss (WL) (Sect. 3.4), to the baseline. Note that our baseline is similar to YOLACT, obtaining the basis masks by using only high-resolution FPN layer ($P3$) and using a single set of coefficients for mask prediction. The results are presented in Table 2. The

Fig. 5. Qualitative results on COCO `test-dev` [29] (corresponding to our 38.1 mask AP). Each color represents different object instances in an image. Our SipMask generates high quality instance segmentation masks in challenging scenarios. (Color figure online)

Table 2. Impact of progressively integrating (from left to right) different components into the baseline. All our components (SP, CBM and WL) contribute towards achieving improved mask AP.

Baseline	SP	CBM	WL	AP
✓				31.2
✓	✓			33.4
✓	✓	✓		33.8
✓	✓	✓	✓	34.3

Table 3. Impact of integrating different components individually into the baseline. Our spatial coefficients (SC) obtains the most improvement in accuracy.

Baseline	SC	FA	CBM	WL	AP
✓					31.2
✓	✓				32.9
✓		✓			31.7
✓			✓		31.9
✓				✓	32.0

baseline achieves a mask AP of 31.2. All our components (SP, CBM and WL) contribute towards achieving improved performance (mask accuracy). In particular, the most improvement in mask accuracy, over the baseline, comes from our SP module. Our final SipMask integrating all contributions obtains an absolute gain of 3.1% in terms of mask AP, compared to the baseline. We also evaluate the impact of adding our different components individually to the baseline. The results are shown in Table 3. Among these components, the spatial coefficients provides the most improvement in accuracy over the baseline. It is worth mentioning that both the spatial coefficients and feature alignment constitute our spatial preservation (SP) module. These results suggest that each of our components individually contributes towards improving the final performance.

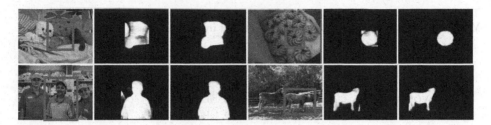

Fig. 6. Qualitative results highlighting the spatial delineation capabilities of our spatial preservation (SP) module. Input image with a detected bounding-box (red) is shown in column 1 and 4. Mask prediction obtained by the baseline that is based on a single set of coefficients is shown in column 2 and 5. Mask prediction obtained by our approach that is based on a separate set of spatial coefficients in a bounding-box is shown in column 3 and 6. Compared to the baseline, our approach is able to better delineate spatially adjacent object instances, leading to improved mask predictions. (Color figure online)

Table 4. The effect of varying the number of sub-regions to compute spatial coefficients. A separate set of spatial coefficients are generated for each sub-region.

	1×1	1×2	2×1	2×2	3×3	4×4
AP	31.2	32.2	32.1	32.9	33.1	33.1

Table 5. The effect of classification (class confidences) and localization (ground-truth overlap) scores on our mask alignment weighting loss (cls. + loc.).

	Baseline	Only cls	Only loc	cls.+loc
AP	31.2	31.8	31.7	32.0

Figure 6 shows example results highlighting the spatial delineation capabilities of our spatial preservation (SP) module. We show the input image with the detected bounding-box (red) together with the mask prediction based on a single set of coefficients (baseline) and our mask prediction based on a separate set of spatial coefficients. Our approach is able to provide improved delineation of spatially adjacent instances, leading to superior mask predictions.

As discussed in Sect. 3.1, our SP module generates a separate set of spatial coefficients for each sub-region within a bounding-box. Here, we perform a study by varying the number of sub-regions to obtain spatial coefficients. Table 4 shows that a large gain in performance is obtained going from 1×1 to 2×2. We also observe that the performance tends to marginally increase by further increasing the number of sub-regions. In practice, we found 2×2 to provide an optimal tradeoff between speed and accuracy. As discussed earlier (Sect. 3.4), our mask alignment weighting loss re-weights the pixel-level BCE loss using both classification (class scores) and localization (overlap with the ground-truth) information. Here, we analyze the effect of classification (only cls.) and localization (only loc.) on our mask alignment weighting loss in Table 5. It shows that both the classification and localization are useful to re-weight the BCE loss for improved mask prediction.

Table 6. Comparison with state-of-the-art video instance segmentation methods on YouTube-VIS validation set. Results are reported in terms of mask accuracy and recall.

Method	Category	AP	AP@0.5	AP@0.75	AR@1	AR@10
OSMN [49]	Mask propagation	23.4	36.5	25.7	28.9	31.1
FEELVOS [41]	Mask propagation	26.9	42.0	29.7	29.9	33.4
OSMN [49]	Track-by-detect	27.5	45.1	29.1	28.6	31.1
MaskTrack R-CNN [48]	Track-by-detect	30.3	51.1	32.6	31.0	35.5
Our SipMask	Track-by-detect	32.5	53.0	33.3	33.5	38.9
Our SipMask *ms-train*	Track-by-detect	**33.7**	**54.1**	**35.8**	**35.4**	**40.1**

Fig. 7. Qualitative results on example frames of different videos from Youtube-VIS validation set [48]. The object with same predicted identity has same color.

4.4 Video Instance Segmentation Results

In addition to instance segmentation, we present the effectiveness of our Sip-Mask, with the proposed modifications described in Sect. 3.5, for real-time video instance segmentation. We conduct experiments on the recently introduced large-scale YouTube-VIS dataset [48]. The YouTube-VIS dataset contains 2883 videos, 4883 objects, $131k$ instance masks, and 40 object categories. Table 6 shows the state-of-the-art comparison on the YouTube-VIS validation set. When using the same input size (640 × 360) and backbone (ResNet50 FPN), our SipMask outperforms the state-of-the-art MaskTrack R-CNN [48] with an absolute gain of 2.2% in terms of mask accuracy (AP). Further, our SipMask achieves impressive mask accuracy while operating at real-time (30 fps) on a Titan Xp. Figure 7 shows video instance segmentation results on example frames from the validation set.

5 Conclusion

We introduce a fast single-stage instance segmentation method, SipMask, that aims at preserving spatial information within a bounding-box. A novel lightweight spatial preservation (SP) module is designed to produce a separate set of spatial coefficients by splitting mask prediction of an object into different sub-regions. To better correlate mask prediction with object detection, a feature alignment scheme and a mask alignment weighting loss are further proposed. We also show that our SipMask is easily extended for real-time video instance segmentation. Our comprehensive experiments on COCO dataset show the effectiveness of the proposed contributions, leading to state-of-the-art single-stage instance segmentation performance. With the same instance segmentation framework and just changing the input resolution (544 × 544), our SipMask operates at real-time on a single Titan Xp with a mask accuracy of 32.8 on COCO `test-dev`.

This work was supported by National Key R&D Program (2018AAA0102800) and National Natural Science Foundation (61906131, 61632018) of China.

Author contributions. Jiale Cao, Rao Muhammad Anwer

References

1. Arnab, A., Torr, P.H.: Pixelwise instance segmentation with a dynamically instantiated network. In: Proceedings of the IEEE Conference on Computer Vision and Pattern Recognition (2017)
2. Bolya, D., Zhou, C., Xiao, F., Lee, Y.J.: Yolact: real-time instance segmentation. In: Proceedings of the IEEE International Conference on Computer Vision (2019)
3. Bolya, D., Zhou, C., Xiao, F., Lee, Y.J.: Yolact++: better real-time instance segmentation. arXiv:1912.06218 (2020)
4. Cao, J., Cholakkal, H., Anwer, R.M., Khan, F.S., Pang, Y., Shao, L.: D2det: towards high quality object detection and instance segmentation. In: Proceedings of the IEEE Conference on Computer Vision and Pattern Recognition (2020)
5. Cao, J., Pang, Y., Han, J., Li, X.: Hierarchical shot detector. In: Proceedings of the IEEE International Conference on Computer Vision (2019)
6. Cao, J., Pang, Y., Li, X.: Triply supervised decoder networks for joint detection and segmentation. In: Proceedings of the IEEE Conference on Computer Vision and Pattern Recognition (2019)
7. Chen, H., Sun, K., Tian, Z., Shen, C., Huang, Y., Yan, Y.: Blendmask: top-down meets bottom-up for instance segmentation. In: Proceedings of the IEEE Conference on Computer Vision and Pattern Recognition (2020)
8. Chen, K., et al.: Hybrid task cascade for instance segmentation. In: Proceedings of the IEEE Conference on Computer Vision and Pattern Recognition (2019)
9. Chen, L.C., Hermans, A., Papandreou, G., Schroff, F., Wang, P., Adam, H.: Masklab: instance segmentation by refining object detection with semantic and direction features. In: Proceedings of the IEEE Conference on Computer Vision and Pattern Recognition (2018)
10. Chen, L.C., Papandreou, G., Kokkinos, I., Murphy, K., Yuille, A.L.: Deeplab: semantic image segmentation with deep convolutional nets, atrous convolution, and fully connected CRFs. IEEE Trans. Pattern Anal. Mach. Intell. **40**(4), 834–848 (2017)
11. Chen, X., Girshick, R., He, K., Dollár, P.: Tensormask: a foundation for dense object segmentation. In: Proceedings of the IEEE International Conference Computer Vision (2019)

12. Cholakkal, H., Sun, G., Khan, F.S., Shao, L.: Object counting and instance segmentation with image-level supervision. In: Proceedings of the IEEE Conference on Computer Vision and Pattern Recognition (2019)
13. Dai, J., He, K., Li, Y., Ren, S., Sun, J.: Instance-sensitive fully convolutional networks. In: Leibe, B., Matas, J., Sebe, N., Welling, M. (eds.) ECCV 2016. LNCS, vol. 9910, pp. 534–549. Springer, Cham (2016). https://doi.org/10.1007/978-3-319-46466-4_32
14. Dai, J., He, K., Sun, J.: Instance-aware semantic segmentation via multi-task network cascades. In: Proceedings of the IEEE Conference on Computer Vision and Pattern Recognition (2016)
15. Dai, J., Li, Y., He, K., Sun, J.: R-FCN: object detection via region-based fully convolutional networks. In: Proceedings of the Advances in Neural Information Processing Systems (2016)
16. Dai, J., et al.: Deformable convolutional networks. In: Proceedings of the IEEE International Conference on Computer Vision (2017)
17. Fang, H.S., Sun, J., Wang, R., Gou, M., Li, Y.L., Lu, C.: Instaboost: boosting instance segmentation via probability map guided copy-pasting. In: Proceedings of the IEEE International Conference on Computer Vision (2019)
18. Fu, C.Y., Shvets, M., Berg, A.C.: Retinamask: learning to predict masks improves state-of-the-art single-shot detection for free. arXiv:1901.03353 (2019)
19. Gao, N., et al.: SSAP: single-shot instance segmentation with affinity pyramid. In: Proceedings of the IEEE International Conference on Computer Vision (2019)
20. Girshick, R., Donahue, J., Darrell, T., Malik, J.: Rich feature hierarchies for accurate object detection and semantic segmentation. In: Proceedings of the IEEE International Conference on Computer Vision and Pattern Recognition (2014)
21. He, K., Gkioxari, G., Dollár, P., Girshick, R.: Mask R-CNN. In: Proceedings of the IEEE International Conference on Computer Vision (2017)
22. He, K., Zhang, X., Ren, S., Sun, J.: Deep residual learning for image recognition. In: Proceedings of the IEEE International Conference on Computer Vision (2016)
23. Huang, Z., Huang, L., Gong, Y., Huang, C., Wang, X.: Mask scoring R-CNN. In: Proceedings of the IEEE Conference on Computer Vision and Pattern Recognition (2019)
24. Jiang, X., et al.: Density-aware multi-task learning for crowd counting. IEEE Trans. Multimedia (2020)
25. Khan, F.S., Xu, J., van de Weijer, J., Bagdanov, A., Anwer, R.M., Lopez, A.: Recognizing actions through action-specific person detection. IEEE Trans. Image Process. **24**(11), 4422–4432 (2015)
26. Kirillov, A., Levinkov, E., Andres, B., Savchynskyy, B., Rother, C.: Instancecut: from edges to instances with multicut. In: Proceedings of the IEEE Conference on Computer Vision and Pattern Recognition (2017)
27. Li, Y., Qi, H., Dai, J., Ji, X., Wei, Y.: Fully convolutional instance-aware semantic segmentation. In: Proceedings of the IEEE Conference on Computer Vision and Pattern Recognition (2017)
28. Lin, T.Y., Dollár, P., Girshick, R., He, K., Hariharan, B., Belongie, S.: Feature pyramid networks for object detection. In: Proceedings of the IEEE Conference on Computer Vision and Pattern Recognition (2017)
29. Lin, T.-Y., et al.: Microsoft COCO: common objects in context. In: Fleet, D., Pajdla, T., Schiele, B., Tuytelaars, T. (eds.) ECCV 2014. LNCS, vol. 8693, pp. 740–755. Springer, Cham (2014). https://doi.org/10.1007/978-3-319-10602-1_48

30. Liu, S., Jia, J., Fidler, S., Urtasun, R.: SGN: sequential grouping networks for instance segmentation. In: Proceedings of the IEEE International Conference on Computer Vision (2017)
31. Liu, S., Qi, L., Qin, H., Shi, J., Jia, J.: Path aggregation network for instance segmentation. In: Proceedings of the IEEE Conference on Computer Vision and Pattern Recognition (2018)
32. Long, J., Shelhamer, E., Darrell, T.: Fully convolutional networks for semantic segmentation. In: Proceedings of the IEEE Conference on Computer Vision and Pattern Recognition (2015)
33. Neven, D., Brabandere, B.D., Proesmans, M., Gool, L.V.: Instance segmentation by jointly optimizing spatial embeddings and clustering bandwidth. In: Proceedings of the IEEE Conference on Computer Vision and Pattern Recognition (2019)
34. Pang, Y., Li, Y., Shen, J., Shao, L.: Towards bridging semantic gap to improve semantic segmentation. In: Proceedings of the IEEE International Conference on Computer Vision (2019)
35. Pang, Y., Xie, J., Khan, M.H., Anwer, R.M., Khan, F.S., Shao, L.: Mask-guided attention network for occluded pedestrian detection. In: Proceedings of the IEEE International Conference on Computer Vision (2019)
36. Peng, S., Jiang, W., Pi, H., Li, X., Bao, H., Zhou, X.: Deep snake for real-time instance segmentation. In: Proceedings of the IEEE Conference on Computer Vision and Pattern Recognition (2020)
37. Pinheiro, P.O., Lin, T.-Y., Collobert, R., Dollár, P.: Learning to refine object segments. In: Leibe, B., Matas, J., Sebe, N., Welling, M. (eds.) ECCV 2016. LNCS, vol. 9905, pp. 75–91. Springer, Cham (2016). https://doi.org/10.1007/978-3-319-46448-0_5
38. Russakovsky, O., et al.: Imagenet large scale visual recognition challenge. Int. J. Comput. Vis. (2015)
39. Sun, G., Wang, B., Dai, J., Gool, L.V.: Mining cross-image semantics for weakly supervised semantic segmentation. In: ECCV 2020. Springer, Cham (2020)
40. Tian, Z., Shen, C., Chen, H., He, T.: FCOS: fully convolutional one-stage object detection. In: Proceedings of the IEEE International Conference on Computer Vision (2019)
41. Voigtlaender, P., Chai, Y., Schroff, F., Adam, H., Leibe, B., Chen, L.C.: Feelvos: fast end-to-end embedding learning for video object segmentation. In: Proceedings of the IEEE Conference on Computer Vision and Pattern Recognition (2019)
42. Wang, S., Gong, Y., Xing, J., Huang, L., Huang, C., Hu, W.: RDSNet: a new deep architecture for reciprocal object detection and instance segmentation. In: Proceedings of the AAAI Conference on Artificial Intelligence (2020)
43. Wang, T., Anwer, R.M., Cholakkal, H., Khan, F.S., Pang, Y., Shao, L.: Learning rich features at high-speed for single-shot object detection. In: Proceedings of the IEEE International Conference on Computer Vision (2019)
44. Wang, T., Yang, T., Danelljan, M., Khan, F.S., Zhang, X., Sun, J.: Learning human-object interaction detection using interaction points. In: Proceedings of the IEEE Conference on Computer Vision and Pattern Recognition (2020)
45. Wu, J., Zhou, C., Yang, M., Zhang, Q., Li, Y., Yuan, J.: Temporal-context enhanced detection of heavily occluded pedestrians. In: Proceedings of the IEEE Conference on Computer Vision and Pattern Recognition (2020)
46. Xie, E., et al.: Polarmask: single shot instance segmentation with polar representation. arXiv:1909.13226 (2019)

47. Xu, W., Wang, H., Qi, F., Lu, C.: Explicit shape encoding for real-time instance segmentation. In: Proceedings of the IEEE International Conference on Computer Vision (2019)
48. Yang, L., Fan, Y., Xu, N.: Video instance segmentation. In: Proceedings of the IEEE International Conference on Computer Vision (2019)
49. Yang, L., Wang, Y., Xiong, X., Yang, J., Katsaggelos, A.K.: Efficient video object segmentation via network modulation. In: Proceedings of the IEEE Conference on Computer Vision and Pattern Recognition (2018)
50. Yang, Z., Liu, S., Hu, H., Wang, L., Lin, S.: Reppoints: point set representation for object detection. In: Proceedings of the IEEE International Conference on Computer Vision (2019)
51. Yang, Z., et al.: Reppoints: point set representation for object detection. In: ECCV 2020. Springer, Cham (2020)
52. Ye, M., Shen, J., Lin, G., Xiang, T., Shao, L., Hoi, S.C.H.: Deep learning for person re-identification: a survey and outlook. arXiv:2001.04193 (2020)
53. Ye, M., Zhang, X., Yuen, P.C., Chang, S.F.: Unsupervised embedding learning via invariant and spreading instance feature. In: Proceedings of the IEEE Conference on Computer Vision and Pattern Recognition (2019)
54. Zhou, X., Zhuo, J., Krahenbuhl, P.: Bottom-up object detection by grouping extreme and center points. In: Proceedings of the IEEE Conference on Computer Vision and Pattern Recognition (2019)
55. Zhu, X., Hu, H., Lin, S., Dai, J.: Deformable convnets v2: more deformable, better results. In: Proceedings of the IEEE Conference on Computer Vision and Pattern Recognition (2019)

SemanticAdv: Generating Adversarial Examples via Attribute-Conditioned Image Editing

Haonan Qiu[1(✉)], Chaowei Xiao[2], Lei Yang[3], Xinchen Yan[2,4], Honglak Lee[2], and Bo Li[5]

[1] The Chinese University of Hong Kong, Shenzhen, China
haonanqiu@link.cuhk.edu.cn
[2] University of Michigan, Ann Arbor, USA
[3] The Chinese University of Hong Kong, Hong Kong, China
[4] Uber ATG, Pittsburgh, USA
[5] UIUC, Champaign, USA

Abstract. Recent studies have shown that DNNs are vulnerable to adversarial examples which are manipulated instances targeting to mislead DNNs to make incorrect predictions. Currently, most such adversarial examples try to guarantee "subtle perturbation" by limiting the L_p norm of the perturbation. In this paper, we propose *SemanticAdv* to generate a new type of *semantically realistic* adversarial examples via attribute-conditioned image editing. Compared to existing methods, our *SemanticAdv* enables fine-grained analysis and evaluation of DNNs with input variations in the attribute space. We conduct comprehensive experiments to show that our adversarial examples not only exhibit semantically meaningful appearances but also achieve high targeted attack success rates under both whitebox and blackbox settings. Moreover, we show that the existing pixel-based and attribute-based defense methods fail to defend against *SemanticAdv*. We demonstrate the applicability of *SemanticAdv* on both face recognition and general street-view images to show its generalization. We believe that our work can shed light on further understanding about vulnerabilities of DNNs as well as novel defense approaches. Our implementation is available at https://github.com/AI-secure/SemanticAdv.

1   Introduction

Deep neural networks (DNNs) have demonstrated great successes in various vision tasks [13,26,38,43,59,60,63,80,83]. Meanwhile, several studies have

---

H. Qiu, C. Xiao, and L. Yang—Contributed equally.

Electronic supplementary material The online version of this chapter (https://doi.org/10.1007/978-3-030-58568-6_2) contains supplementary material, which is available to authorized users.

© Springer Nature Switzerland AG 2020
A. Vedaldi et al. (Eds.): ECCV 2020, LNCS 12359, pp. 19–37, 2020.
https://doi.org/10.1007/978-3-030-58568-6_2

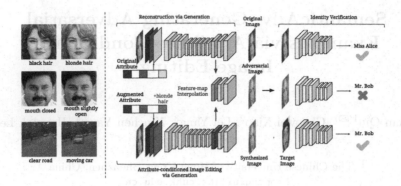

Fig. 1. Pipeline of *SemanticAdv* Left: Each row shows a pair of images differ in only one semantic aspect. One of them is sampled from the ground-truth dataset, while the other one is adversarial example created by our conditional image generator. Right: Overview of the proposed attribute-conditioned *SemanticAdv* against the face identity verification model

revealed the vulnerablity of DNNs against input variations [10–12,21,24,47,52, 64,66,70–74,85]. For example, carefully crafted L_p bounded perturbations added to the pristine input images can introduce arbitrary prediction errors during testing time. While being visually imperceptible, L_p bounded adversarial attacks have certain limitations as they only capture the variations in the raw pixel space and cannot guarantee the semantic realism for the generated instances. Recent works [33,69,75] have shown the limitations of the L_p bounded perturbation (e.g., cannot handle variations in lighting conditions). Therefore, understanding the failure modes of deep neural networks beyond raw pixel variations including semantic perturbations requires further understanding and exploration.

In this work, we focus on studying how DNNs respond towards semantically meaningful perturbations in the visual attribute space. In the visual recognition literature, visual attributes [22,39,53] are properties observable in images that have human-designated properties (e.g., *black hair* and *blonde hair*). As illustrated in Fig. 1(left), given an input image with known attributes, we would like to craft semantically meaningful (attribute-conditioned) adversarial examples via image editing along a single attribute or a subset of attributes while keeping the rest unchanged. Compared to traditional L_p bounded adversarial perturbations or semantic perturbations on global color and texture [5], such attribute-based image editing enables the users to conduct a fine-grained analysis and evaluation of the DNN models through removing one or a set of visual aspects or adding one object into the scene. We believe our attribute-conditioned image editing is a natural way of introducing semantic perturbations, and it preserves clear interpretability as: wearing a new pair of glasses or having the hair dyed with a different color.

To facilitate the generation of semantic adversarial perturbations along a single attribute dimension, we take advantage of the disentangled representation in deep image generative models [3,6,14,31,34,55,78,82]. Such disentangled representation allows us to explore the variations for a specific semantic factor while keeping the other factors unchanged. As illustrated in Fig. 1(right), we first leverage an attribute-conditioned image editing model [14] to construct a new instance which is very similar to the source except one semantic aspect (the source image is given as input). Given such pair of images, we synthesize the adversarial example by interpolating between the pair of images in the *feature-map space*. As the interpolation is constrained by the image pairs, the appearance of the resulting semantic adversarial example resembles both of them.

To validate the effectiveness of our proposed *SemanticAdv* by attribute-conditioned image editing, we consider two real-world tasks, including face verification and landmark detection. We conduct both qualitative and quantitative evaluations on CelebA dataset [42]. The results show that our *SemanticAdv* not only achieves high targeted attack success rate and also preserves the semantic meaning of the corresponding input images. To further demonstrate the applicability of our *SemanticAdv* beyond face domain, we extend the framework to generate adversarial street-view images. We treat semantic layouts as input attributes and use the layout-conditioned image editing model [27] pretrained on Cityscape dataset [16]. Our results show that a well-trained semantic segmentation model can be successfully attacked to neglect the pedestrian if we insert another object by the side using our image editing model. In addition, we show that existing adversarial training-based defense method is less effective against our attack method, which motivates further defense strategies against such semantic adversarial examples.

Our contributions are summarized as follows: (1) We propose a novel method *SemanticAdv* to generate semantically meaningful adversarial examples via attribute-conditioned image editing based on **feature-space** interpolation. Compared to existing adversarial attacks, our method enables fine-grained attribute analysis as well as further evaluation of vulnerabilities for DNN models. Such semantic adversarial examples also provide explainable analysis for different attributes in terms of their robustness and editing flexibility. (2) We conduct extensive experiments and show that the proposed feature-space interpolation strategy can generate high quality attribute-conditioned adversarial examples more effectively than the simple attribute-space interpolation. Additionally, our *SemanticAdv* exhibits high attack **transferability** as well as 67.7% query-free **black-box attack** success rate on a real-world face verification platform. (3) We empirically show that, compared to L_p attacks, the existing per-pixel based as well as attribute-based defense methods fail to defend against our *SemanticAdv*, which indicates that such semantic adversarial examples identify certain unexplored vulnerable landscape of DNNs. (4) To demonstrate the applicability and generalization of *SemanticAdv* beyond the face recognition domain, we extend the framework to generate adversarial street-view images that fool semantic segmentation models effectively.

2   Related Work

Semantic Image Editing. Semantic image synthesis and manipulation is a popular research topic in machine learning, graphics and vision. Thanks to recent advances in deep generative models [23,36,51] and the empirical analysis of deep classification networks [38,60,63], past few years have witnessed tremendous breakthroughs towards high-fidelity pure image generation [6,34,55], attribute-to-image generation [14,78], text-to-image generation [31,46,49,50,56,84], and image-to-image translation [27,29,41,68,87].

Adversarial Examples. Generating L_p bounded adversarial perturbation has been extensively studied recently [12,21,24,47,52,64,70–73]. To further explore diverse adversarial attacks and potentially help inspire defense mechanisms, it is important to generate the so-called "unrestricted" adversarial examples which contain unrestricted magnitude of perturbation while still preserve perceptual realism [7]. Recently, [20,75] propose to spatially transform the image patches instead of adding pixel-wise perturbation, while such spatial transformation does not consider semantic information. Our proposed *semanticAdv* focuses on generating unrestricted perturbation with semantically meaningful patterns guided by visual attributes.

Relevant to our work, [61] proposed to synthesize adversarial examples with an unconditional generative model. [5] studied semantic transformation in only the color or texture space. Compared to these works, *semanticAdv* is able to generate adversarial examples in a controllable fashion using specific visual attributes by performing manipulation in the feature space. We further analyze the robustness of the recognition system by generating adversarial examples guided by different visual attributes. Concurrent to our work, [32] proposed to generate semantic-based attacks against a restricted binary classifier, while our attack is able to mislead the model towards arbitrary adversarial targets. They conduct the manipulation within the attribution space which is less flexible and effective than our proposed feature-space interpolation.

3   SemanticAdv

3.1   Problem Definition

Let $\mathcal{M}$ be a machine learning model trained on a dataset $\mathcal{D} = \{(\mathbf{x}, \mathbf{y})\}$ consisting of image-label pairs, where $\mathbf{x} \in \mathbb{R}^{H \times W \times D_I}$ and $\mathbf{y} \in \mathbb{R}^{D_L}$ denote the image and the ground-truth label, respectively. Here, H, W, D_I, and D_L denote the image height, image width, number of image channels, and label dimensions, respectively. For each image $\mathbf{x}$, our model $\mathcal{M}$ makes a prediction $\hat{\mathbf{y}} = \mathcal{M}(\mathbf{x}) \in \mathbb{R}^{D_L}$. Given a target image-label pair $(\mathbf{x}^{tgt}, \mathbf{y}^{tgt})$ and $\mathbf{y} \neq \mathbf{y}^{tgt}$, a *traditional attacker* aims to synthesize adversarial examples $\mathbf{x}^{adv}$ by adding pixel-wise perturbations to or spatially transforming the original image $\mathbf{x}$ such that $\mathcal{M}(\mathbf{x}^{adv}) = \mathbf{y}^{tgt}$. In this work, we consider a *semantic attacker* that generates semantically meaningful perturbation via attribute-conditioned image editing with a conditional generative model $\mathcal{G}$. Compared to the traditional attacker, the proposed attack method generates adversarial examples in a more controllable fashion by editing a single semantic aspect through attribute-conditioned image editing.

3.2   Attribute-Conditioned Image Editing

In order to produce semantically meaningful perturbations, we first introduce how to synthesize attribute-conditioned images through interpolation.

Semantic Image Editing. For simplicity, we start with the formulation where the input attribute is represented as a compact vector. This formulation can be directly extended to other input attribute formats including semantic layouts. Let $\mathbf{c} \in \mathbb{R}^{D_C}$ be an attribute representation reflecting the semantic factors (e.g., expression or hair color of a portrait image) of image $\mathbf{x}$, where D_C indicates the attribute dimension and $c_i \in \{0, 1\}$ indicates the existence of i-th attribute. We are interested in performing semantic image editing using the attribute-conditioned image generator $\mathcal{G}$. For example, given a portrait image of a girl with black hair and the new attribute blonde hair, our generator is supposed to synthesize a new image that turns the girl's hair color from black to blonde while keeping the rest of appearance unchanged. The synthesized image is denoted as $\mathbf{x}^{\text{new}} = \mathcal{G}(\mathbf{x}, \mathbf{c}^{\text{new}})$ where $\mathbf{c}^{\text{new}} \in \mathbb{R}^{D_C}$ is the new attribute. In the special case when there is no attribute change ($\mathbf{c} = \mathbf{c}^{\text{new}}$), the generator simply reconstructs the input: $\mathbf{x}' = \mathcal{G}(\mathbf{x}, \mathbf{c})$ (ideally, we hope $\mathbf{x}'$ equals to $\mathbf{x}$). As our attribute representation is disentangled and the change of attribute value is sufficiently small (e.g., we only edit a single semantic attribute), our synthesized image $\mathbf{x}^{\text{new}}$ is expected to be close to the data manifold [4,55,57]. In addition, we can generate many similar images by linearly interpolating between the image pair $\mathbf{x}$ and $\mathbf{x}^{\text{new}}$ in the attribute-space or the feature-space of the image-conditioned generator $\mathcal{G}$, which is supported by the previous work [3,55,78].

Attribute-Space Interpolation. Given a pair of attributes $\mathbf{c}$ and $\mathbf{c}^{\text{new}}$, we introduce an interpolation parameter $\alpha \in (0, 1)$ to generate the augmented attribute vector $\mathbf{c}^* \in \mathbb{R}^{D_C}$ (see Eq. 1). Given augmented attribute $\mathbf{c}^*$ and original image $\mathbf{x}$, we produce the image $\mathbf{x}^*$ by the generator $\mathcal{G}$ through attribute-space interpolation.

$$\mathbf{x}^* = \mathcal{G}(\mathbf{x}, \mathbf{c}^*)$$
$$\mathbf{c}^* = \alpha \cdot \mathbf{c} + (1 - \alpha) \cdot \mathbf{c}^{\text{new}}, \text{where } \alpha \in [0, 1] \tag{1}$$

Feature-Map Interpolation. Alternatively, we propose to interpolate using the feature map produced by the generator $\mathcal{G} = \mathcal{G}_{\text{dec}} \circ \mathcal{G}_{\text{enc}}$. Here, $\mathcal{G}_{\text{enc}}$ is the encoder module that takes the image as input and outputs the feature map. Similarly, $\mathcal{G}_{\text{dec}}$ is the decoder module that takes the feature map as input and outputs the synthesized image. Let $\mathbf{f}^* = \mathcal{G}_{\text{enc}}(\mathbf{x}, \mathbf{c}) \in \mathbb{R}^{H_F \times W_F \times C_F}$ be the feature map of an intermediate layer in the generator, where H_F, W_F and C_F indicate the height, width, and number of channels in the feature map.

$$\mathbf{x}^* = \mathcal{G}_{\text{dec}}(\mathbf{f}^*)$$

$$\mathbf{f}^* = \boldsymbol{\beta} \odot \mathcal{G}_{\text{enc}}(\mathbf{x}, \mathbf{c}) + (\mathbf{1} - \boldsymbol{\beta}) \odot \mathcal{G}_{\text{enc}}(\mathbf{x}, \mathbf{c}^{\text{new}}) \tag{2}$$

Compared to the attribute-space interpolation which is parameterized by a scalar α, we parameterize feature-map interpolation by a tensor $\boldsymbol{\beta} \in \mathbb{R}^{H_F \times W_F \times C_F}$ ($\beta_{h,w,k} \in [0,1]$, where $1 \leq h \leq H_F$, $1 \leq w \leq W_F$, and $1 \leq k \leq C_F$) with the same shape as the feature map. Compared to linear interpolation over attribute-space, such design introduces more flexibility for adversarial attacks. Empirical results in Sect. 4.2 show such design is critical to maintain both attack success and good perceptual quality at the same time.

3.3   Generating Semantically Meaningful Adversarial Examples

Existing work obtains the adversarial image $\mathbf{x}^{\text{adv}}$ by adding perturbations or transforming the input image $\mathbf{x}$ directly. In contrast, our semantic attack method requires additional attribute-conditioned image generator $\mathcal{G}$ during the adversarial image generation through interpolation. As we see in Eq. (3), the first term of our objective function is the adversarial metric, the second term is a smoothness constraint to guarantee the perceptual quality, and λ is used to control the balance between the two terms. The adversarial metric is minimized once the model $\mathcal{M}$ has been successfully attacked towards the target image-label pair $(\mathbf{x}^{\text{tgt}}, \mathbf{y}^{\text{tgt}})$. For identify verification, $\mathbf{y}^{\text{tgt}}$ is the identity representation of the target image. For structured prediction tasks in our paper, $\mathbf{y}^{\text{tgt}}$ either represents certain coordinates (landmark detection) or semantic label maps (semantic segmentation).

$$\mathbf{x}^{\text{adv}} = \text{argmin}_{\mathbf{x}^*} \mathcal{L}(\mathbf{x}^*)$$

$$\mathcal{L}(\mathbf{x}^*) = \mathcal{L}_{\text{adv}}(\mathbf{x}^*; \mathcal{M}, \mathbf{y}^{\text{tgt}}) + \lambda \cdot \mathcal{L}_{\text{smooth}}(\mathbf{x}^*) \tag{3}$$

Identity Verification. In the identity verification task, two images are considered to be the same identity if the corresponding identity embeddings from the verification model $\mathcal{M}$ are reasonably close.

$$\mathcal{L}_{\text{adv}}(\mathbf{x}^*; \mathcal{M}, \mathbf{y}^{\text{tgt}}) = \max\{\kappa, \Phi_{\mathcal{M}}^{\text{id}}(\mathbf{x}^*, \mathbf{x}^{\text{tgt}})\} \tag{4}$$

As we see in Eq. (4), $\Phi_{\mathcal{M}}^{\text{id}}(\cdot, \cdot)$ measures the distance between two identity embeddings from the model $\mathcal{M}$, where the normalized L_2 distance is used in our setting. In addition, we introduce the parameter κ representing the constant related to the false positive rate (FPR) threshold computed from the development set.

Structured Prediction. For structured prediction tasks such as landmark detection and semantic segmentation, we use Houdini objective proposed in [15] as our adversarial metric and select the target landmark (semantic segmentation) target as $\mathbf{y}^{\text{tgt}}$. As we see in the equation, $\Phi_{\mathcal{M}}(\cdot, \cdot)$ is a scoring function for each image-label pair and γ is the threshold. In addition, $l(\mathbf{y}^*, \mathbf{y}^{\text{tgt}})$ is task loss decided by the specific adversarial target, where $\mathbf{y}^* = \mathcal{M}(\mathbf{x}^*)$.

$$\mathcal{L}_{\text{adv}}(\mathbf{x}^*; \mathcal{M}, \mathbf{y}^{\text{tgt}}) = P_{\gamma \sim \mathcal{N}(0,1)}\left[\Phi_{\mathcal{M}}(\mathbf{x}^*, \mathbf{y}^*) - \Phi_{\mathcal{M}}(\mathbf{x}^*, \mathbf{y}^{\text{tgt}}) < \gamma\right] \cdot l(\mathbf{y}^*, \mathbf{y}^{\text{tgt}}) \quad (5)$$

Interpolation Smoothness $\mathcal{L}_{\text{smooth}}$. As the tensor to be interpolated in the feature-map space has far more parameters compared to the attribute itself, we propose to enforce a smoothness constraint on the tensor α used in feature-map interpolation. As we see in Eq. (6), the smoothness loss encourages the interpolation tensors to consist of piece-wise constant patches spatially, which has been widely used as a pixel-wise de-noising objective for natural image processing [30,45].

$$\mathcal{L}_{\text{smooth}}(\boldsymbol{\beta}) = \sum_{h=1}^{H_F-1} \sum_{w=1}^{W_F} \|\boldsymbol{\beta}_{h+1,w} - \boldsymbol{\beta}_{h,w}\|_2^2 + \sum_{h=1}^{H_F} \sum_{w=1}^{W_F-1} \|\boldsymbol{\beta}_{h,w+1} - \boldsymbol{\beta}_{h,w}\|_2^2 \quad (6)$$

4   Experiments

In the experimental section, we mainly focus on analyzing the proposed *SemanticAdv* in attacking state-of-the-art face recognition systems [28,59,62,67,79,81, 86] due to its wide applicability (e.g., identification for mobile payment) in the real world. We attack both face verification and face landmark detection by generating attribute-conditioned adversarial examples using annotations from CelebA dataset [42]. In addition, we extend our attack to urban street scenes with semantic label maps as the condition. We attack the semantic segmentation model DRN-D-22 [83] previously trained on Cityscape [16] by generating adversarial examples with dynamic objects manipulated (e.g., insert a car into the scene).

4.1   Experimental Setup

Face Identity Verification. We select `ResNet-50` and `ResNet-101` [26] trained on MS-Celeb-1M [17,25] as our face verification models. The models are trained using two different objectives, namely, `softmax` loss [62,86] and `cosine` loss [67]. For simplicity, we use the notation "R-N-S" to indicate the model with N-layer residual blocks as backbone trained using `softmax` loss, while "R-N-C" indicates the same backbone trained using `cosine` loss. The distance between face features is measured by normalized L2 distance. For R-101-S model, we decide the parameter κ based on the commonly used false positive rate (FPR) for the identity verification task [35,37]. Four different FPRs have been used: 10^{-3} (with $\kappa = 1.24$), 3×10^{-4} (with $\kappa = 1.05$), 10^{-4} (with $\kappa = 0.60$), and $< 10^{-4}$ (with $\kappa = 0.30$). Supplementary provides more details on the performance of face recognition models and their corresponding κ. To distinguish between the

FPR we used in generating adversarial examples and the other FPR used in evaluation, we introduce two notations "Generation FPR (G-FPR)" and "Test FPR (T-FPR)". For the experiment with query-free black-box API attacks, we use two online face verification services provided by Face++ [2] and AliYun [1].

Semantic Attacks on Face Images. In our experiments, we randomly sample $1,280$ distinct identities form CelebA [42] and use the StarGAN [14] for attribute-conditional image editing. In particular, we re-train our model on CelebA by aligning the face landmarks and then resizing images to resolution 112×112. We select 17 identity-preserving attributes as our analysis, as such attributes mainly reflect variations in facial expression and hair color.

In feature-map interpolation, to reduce the reconstruction error brought by the generator (e.g., $\mathbf{x} \neq \mathcal{G}(\mathbf{x}, \mathbf{c})$) in practice, we take one more step to obtain the updated feature map $\mathbf{f}' = \mathcal{G}_{\mathrm{enc}}(\mathbf{x}', \mathbf{c})$, where $\mathbf{x}' = \mathrm{argmin}_{\mathbf{x}'} \|\mathcal{G}(\mathbf{x}', \mathbf{c}) - \mathbf{x}\|$.

For each distinct identity pair $(\mathbf{x}, \mathbf{x}^{\mathrm{tgt}})$, we perform *semanticAdv* guided by each of the 17 attributes (e.g., we intentionally add or remove one specific attribute while keeping the rest unchanged). In total, for each image $\mathbf{x}$, we generate 17 adversarial images with different augmented attributes. In the experiments, we select a commonly-used pixel-wise adversarial attack method [12] (referred as CW) as our baseline. Compared to our proposed method, CW does not require visual attributes as part of the system, as it only generates one adversarial example for each instance. We refer the corresponding attack success rate as the instance-wise success rate in which the attack success rate is calculated for each instance. For each instance with 17 adversarial images using different augmented attributes, if one of the 17 produced images can attack successfully, we count the attack of this instance as one success, vice verse.

Face Landmark Detection. We select Face Alignment Network (FAN) [9] trained on 300W-LP [88] and fine-tuned on 300-W [58] for 2D landmark detection. The network is constructed by stacking Hour-Glass networks [48] with hierarchical block [8]. Given a face image as input, FAN outputs 2D heatmaps which can be subsequently leveraged to yield 68 2D landmarks.

Semantic Attacks on Street-View Images. We select DRN-D-22 [83] as our semantic segmentation model and fine-tune the model on image regions with resolution 256×256. To synthesize semantic adversarial perturbations, we consider semantic label maps as the input attribute and leverage a generative image manipulation model [27] pre-trained on CityScape [16] dataset. The details are shown in supplementary materials.

4.2   *SemanticAdv* on Face Identity Verification

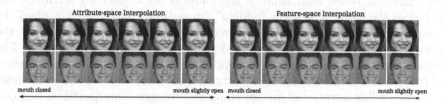

Attribute-space Interpolation   Feature-space Interpolation

mouth closed   mouth slightly open   mouth closed   mouth slightly open

Fig. 2. Qualitative comparisons between attribute-space and feature-space interpolation In our visualization, we set the interpolation parameter to be $0.0, 0.2, 0.4, 0.6, 0.8, 1.0$

Attribute-Space vs. Feature-Space Interpolation. First, we qualitatively compare the two interpolation methods and found that both attribute-space and feature-space interpolation can generate reasonably looking samples (see Fig. 2) through interpolation (these are not adversarial examples). However, we found the two interpolation methods perform differently when we optimize using the adversarial objective (Eq. 3). We measure the attack success rate of attribute-space interpolation (with G-FPR = T-FPR = 10^{-3}): 0.08% on R-101-S, 0.31% on R-101-C, and 0.16% on both R-50-S and R-50-C, which consistently fails to attack the face verification model. Compared to attribute-space interpolation, generating adversarial examples with feature-space interpolation produces much better quantitative results (see Table 1). We conjecture that this is because the high dimensional feature space can provide more manipulation freedom. This also explains one potential reason of poor samples (e.g., blurry with many noticeable artifacts) generated by the method proposed in [32]. We select $\mathbf{f}_0$, the last conv layer before up-sampling layer in the generator for feature-space interpolation due to its good performance. Note that due to the effectiveness of **feature-space** interpolation, we only use **feature-space** interpolation for *semanticAdv* in the following experiments.

Table 1. Attack success rate by selecting attribute or different layer's feature-map for interpolation on R-101-S(%) using G-FPR = T-FPR = 10^{-3}. Here, $\mathbf{f}_i$ indicates the feature-map after i-th up-sampling operation. $\mathbf{f}_{-2}$ and $\mathbf{f}_{-1}$ are the first and the second feature-maps after the last down-sampling operation, respectively. Due to the effectiveness of **feature-space** interpolation, we only use **feature-space** interpolation in the following experiments.

Interpolation/Attack Success (%)	Feature					Attribute
	$\mathbf{f}_{-2}$	$\mathbf{f}_{-1}$	$\mathbf{f}_0$	$\mathbf{f}_1$	$\mathbf{f}_2$	
$\mathbf{x}^{adv}$, G-FPR = 10^{-3}	99.38	100.00	**100.00**	100.00	99.69	0.08
$\mathbf{x}^{adv}$, G-FPR = 10^{-4}	59.53	98.44	**99.45**	97.58	73.52	0.00

Qualitative Analysis. Figure 3(top) shows the generated adversarial images and corresponding perturbations against R-101-S of *SemanticAdv* and CW respectively. The text below each figure is the name of an augmented attribute, the sign before the name represents "adding" (in red) or "removing" (in blue) the corresponding attribute from the original image. Figure 3(bottom) shows the adversarial examples with 17 augmented semantic attributes, respectively. The attribute names are shown in the bottom. The first row contains images generated by $\mathcal{G}(\mathbf{x}, \mathbf{c}^{new})$ with an augmented attribute $\mathbf{c}^{new}$ and the second row includes the corresponding adversarial images under feature-space interpolation. It shows that our *SemanticAdv* can generate examples with reasonably-looking appearance guided by the corresponding attribute. In particular, *SemanticAdv* is able to generate perturbations on the corresponding regions correlated with the augmented attribute, while the perturbations of CW have no specific pattern and are evenly distributed across the image.

To further measure the perceptual quality of the adversarial images generated by *SemanticAdv* in the most strict settings (G-FPR $< 10^{-4}$), we conduct a user study using Amazon Mechanical Turk (AMT). In total, we collect $2,620$ annotations from 77 participants. In $39.14 \pm 1.96\%$ (close to random guess 50%) of trials, the adversarial images generated by our *SemanticAdv* are selected as reasonably-looking images, while $30.27 \pm 1.96\%$ of trials by CW are selected as reasonably-looking. It indicates that *SemanticAdv* can generate more perceptually plausible adversarial examples compared with CW under the most strict setting (G-FPR $< 10^{-4}$). The corresponding images are shown in supplementary materials.

Single Attribute Analysis. One of the key advantages of our *SemanticAdv* is that we can generate adversarial perturbations in a more controllable fashion guided by the selected semantic attribute. This allows analyzing the robustness of a recognition system against different types of semantic attacks. We group the adversarial examples by augmented attributes in various settings. In Fig. 4, we present the attack success rate against two face verification models, namely, R-101-S and R-101-C, using different attributes. We highlight the bar with light blue for G-FPR $= 10^{-3}$ and blue for G-FPR $= 10^{-4}$, respectively. As shown in Fig. 4, with a larger T-FPR $= 10^{-3}$, our *SemanticAdv* can achieve almost 100% attack success rate across different attributes. With a smaller T-FPR $= 10^{-4}$, we observe that *SemanticAdv* guided by some attributes such as Mouth Slightly Open and Arched Eyebrows achieve less than 50% attack success rate, while other attributes such as Pale Skin and Eyeglasses are relatively less affected. In summary, the above experiments indicate that *SemanticAdv* guided by attributes describing the local shape (e.g., mouth, earrings) achieve a relatively lower attack success rate compared to attributes relevant to the color (e.g., hair color) or entire face region (e.g., skin). This suggests that the face verification models used in our experiments are more robustly trained in terms of detecting local shapes compared to colors. In practice, we have the flexibility to select attributes for attacking an image based on the perceptual quality and attack success rate.

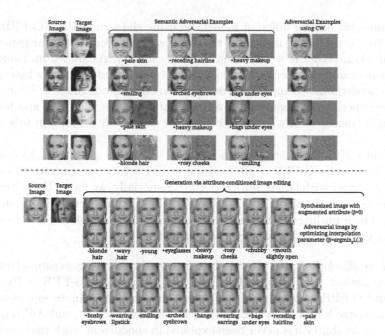

Fig. 3. Top: Qualitative comparisons between our proposed *SemanticAdv* and pixelwise adversarial examples generated by CW [12]. Along with the adversarial examples, we also provide the corresponding perturbations (residual) on the right. Perturbations generated by our *SemanticAdv* (G-FPR = 10^{-3}) are unrestricted with semantically meaningful patterns. Bottom: Qualitative analysis on single-attribute adversarial attack (G-FPR = 10^{-3}). More results are shown in the supplementary. (Color figure online)

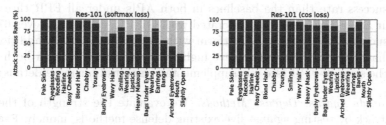

Fig. 4. Quantitative analysis on the attack success rate with different single-attribute attacks. In each figure, we show the results correspond to a larger FPR (G-FPR = T-FPR = 10^{-3}) in skyblue and the results correspond to a smaller FPR (G-FPR = T-FPR = 10^{-4}) in blue, respectively. (Color figure online)

Transferability Analysis. To generate adversarial examples under black-box setting, we analyze the transferability of *SemanticAdv* in various settings. For each model with different FPRs, we select the successfully attacked adversarial examples from Sect. 4.1 to construct our evaluation dataset and evaluate these adversarial samples across different models. Table 2(a) illustrates the transferability of *SemanticAdv* among different models by using the same FPRs (G-FPR = T-FPR = 10^{-3}). Table 2(b) illustrates the result with different FPRs for generation and evaluation (G-FPR = 10^{-4} and T-FPR = 10^{-3}). As shown in Table 2(a), adversarial examples generated against models trained with softmax loss exhibit certain transferability compared to models trained with cosine loss. We conduct the same experiment by generating adversarial examples with CW and found it has weaker transferability compared to our *SemanticAdv* (results in brackets of Table 2).

As Table 2(b) illustrates, the adversarial examples generated against the model with smaller G-FPR = 10^{-4} exhibit strong attack success rate when evaluating the model with larger T-FPR = 10^{-3}. Especially, we found the adversarial examples generated against R-101-S have the best attack performance on other models. These findings motivate the analysis of the query-free black-box API attack detailed in the following paragraph.

Query-Free Black-Box API Attack. In this experiment, we generate adversarial examples against R-101-S with G-FPR = $10^{-3}(\kappa = 1.24)$, G-FPR = $10^{-4}(\kappa = 0.60)$, and G-FPR < $10^{-4}(\kappa = 0.30)$, respectively. We evaluate our algorithm on two industry level face verification APIs, namely, Face++ and AliYun. Since attack transferability has never been explored in concurrent work that generates semantic adversarial examples, we use $\mathcal{L}_p$ bounded pixel-wise methods (CW [12], MI-FGSM[18], M-DI^2-FGSM[76]) as our baselines. We also introduce a much strong baseline by first performing attribute-conditioned image editing and running CW attack on the editted images, which we refer as and StarGAN+CW. Compared to CW, the latter two devise certain techniques to improve their transferability. We adopt the ensemble version of MI-FGSM[18] following the original paper. As shown in Table 3, our proposed *SemanticAdv* achieves a much higher attack success rate than the baselines in both APIs under all FPR thresholds (e.g., our adversarial examples generated with G-FPR < 10^{-4} achieves 67.69% attack success rate on Face++ platform with T-FPR = 10^{-3}). In addition, we found that lower G-FPR can achieve higher attack success rate on both APIs within the same T-FPR (see our supplementary material for more details).

SemanticAdv Against Defense Methods. We evaluate the strength of the proposed attack by testing against five existing defense methods, namely, Feature squeezing [77], Blurring [40], JPEG [19], AMI [65] and adversarial training [44].

Table 2. Transferability of *SemanticAdv*: cell (i, j) shows attack success rate of adversarial examples generated against j-th model and evaluate on i-th model. Results of CW are listed in brackets. Left: Results generated with G-FPR = 10^{-3} and T-FPR = 10^{-3}; Right: Results generated with G-FPR = 10^{-4} and T-FPR = 10^{-3}

$\mathcal{M}_{test}/\mathcal{M}_{opt}$	R-50-S	R-101-S	R-50-C	R-101-C	$\mathcal{M}_{test}/\mathcal{M}_{opt}$	R-50-S	R-101-S
R-50-S	1.000 (1.000)	0.108 (0.032)	0.023 (0.007)	0.018 (0.005)	R-50-S	1.000 (1.000)	0.862 (0.530)
R-101-S	0.169 (0.029)	1.000 (1.000)	0.030 (0.009)	0.032 (0.011)	R-101-S	0.874 (0.422)	1.000 (1.000)
R-50-C	0.166 (0.054)	0.202 (0.079)	1.000 (1.000)	0.048 (0.020)	R-50-C	0.693 (0.347)	0.837 (0.579)
R-101-C	0.120 (0.034)	0.236 (0.080)	0.040 (0.017)	1.000 (1.000)	R-101-C	0.617 (0.218)	0.888 (0.617)

Table 3. Quantitative analysis on query-free black-box attack. We use ResNet-101 optimized with `softmax` loss for evaluation and report the attack success rate(%) on two online face verification platforms. Note that for PGD-based attacks, we adopt MI-FGSM ($\epsilon = 8$) in [18] and M-DI^2-FGSM ($\epsilon = 8$) in [76], respectively. For CW, StarGAN+CW and *SemanticAdv*, we generate adversarial samples with G-FPR $< 10^{-4}$

API name Attacker/Metric	Face++		AliYun	
	T-FPR $= 10^{-3}$	T-FPR $= 10^{-4}$	T-FPR $= 10^{-3}$	T-FPR $= 10^{-4}$
CW [12]	37.24	20.41	18.00	9.50
StarGAN + CW	47.45	26.02	20.00	8.50
MI-FGSM [18]	53.89	30.57	29.50	17.50
M-DI^2-FGSM [76]	56.12	33.67	30.00	18.00
SemanticAdv	**67.69**	**48.21**	**36.50**	**19.50**

Figure 5 illustrates *SemanticAdv* is more robust against the pixel-wise defense methods comparing with CW. The same G-FPR and T-FPR are used for evaluation. Both *SemanticAdv* and CW achieve a high attack success rate when T-FPR $= 10^{-3}$, while *SemanticAdv* marginally outperforms CW when T-FPR goes down to 10^{-4}. While defense methods have proven to be effective against CW attacks on classifiers trained with ImageNet [38], our results indicate that these methods are still vulnerable in the face verification system with small G-FPR.

We further evaluate *SemanticAdv* on attribute-based defense method `AMI` [65] by constructing adversarial examples for the pretrained VGG-Face [54] in a black-box manner. From adversarial examples generated by R-101-S, we use `fc7` as the embedding and select the images with normalized L2 distance (to the corresponding benign images) beyond the threshold defined previously. With the benign and adversarial examples, we first extract attribute witnesses with our aligned face images and then leverage them to build a attribute-steered model.

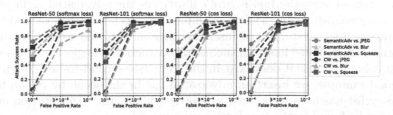

Fig. 5. Quantitative analysis on attacking several defense methods including `JPEG` [19], `Blurring` [40], and `Feature Squeezing` [77]

When misclassifying 10% benign inputs into adversarial images, it only correctly identifies 8% adversarial images from *SemanticAdv* and 12% from CW.

Moreover, we evaluate *SemanticAdv* on existing adversarial training based defense (the detailed setting is presented in supplementary materials). We find that accuracy of adversarial training based defense method is 10% against the adversarial examples generated by *SemanticAdv*, while is 46.7% against the adversarial examples generated by PGD [44]. It indicates that existing adversarial training based defense method is less effective against *SemanticAdv*, which further demonstrates that our *SemanticAdv* identifies an unexplored research area beyond previous L_p-based ones.

4.3 *SemanticAdv* on Face Landmark Detection

We evaluate the effectiveness of *SemanticAdv* on face landmark detection under two tasks, "Rotating Eyes" and "Out of Region". For the "Rotating Eyes" task, we rotate the coordinates of the eyes with 90° counter-clockwise. For the "Out of Region" task, we set a target bounding box and push all points out of the box. Figure 6 indicates that *semanticAdv* could attack landmark detection models.

Fig. 6. Qualitative results on attacking face landmark detection model

4.4 *SemanticAdv* on Street-View Semantic Segmentation

We further generate adversarial perturbations on street-view images to show the generalization of *semanticAdv*. Figure 7 illustrates the adversarial examples on semantic segmentation. In the first example, we select the leftmost pedestrian as the target object instance and insert another car into the scene to attack it. The segmentation model has been successfully attacked to neglect the pedestrian (see last column), while it does exist in the scene (see second-to-last column). In the second example, we insert an adversarial car in the scene by *SemanticAdv* and the cyclist has been recognized as a pedestrian by the segmentation model. Details can be found in the supplementary material.

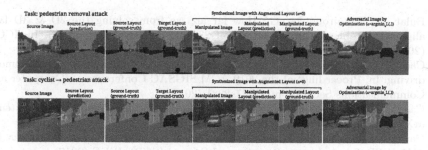

Fig. 7. Qualitative results on attacking street-view semantic segmentation model

5   Conclusions

Overall we presented a novel attack method *SemanticAdv* to generate semantically meaningful adversarial perturbations guided by single semantic attribute. Experimental evaluations demonstrate several unique properties of *semanticAdv* including attack transferability. We believe *semanticAdv* would open up new research opportunities and challenges in adversarial learning domain. For instance, how to generate semantic adversarial example in physical world and leverage semantic information to defend against such attacks.

Acknowledgments. This work was supported in part by AWS Machine Learning Research Awards, National Science Foundation under grants CNS-1422211, CNS-1616575, CNS-1739517, and NSF CAREER Award IIS-1453651.

References

1. Alibaba Cloud Computing Co., Ltd. https://help.aliyun.com/knowledge_detail/53535.html
2. Megvii Technology Co., Ltd. https://console.faceplusplus.com/documents/5679308
3. Bau, D., et al.: GAN dissection: visualizing and understanding generative adversarial networks. arXiv preprint arXiv:1811.10597 (2018)
4. Bengio, Y., Mesnil, G., Dauphin, Y., Rifai, S.: Better mixing via deep representations. In: ICML (2013)
5. Bhattad, A., Chong, M.J., Liang, K., Li, B., Forsyth, D.: Unrestricted adversarial examples via semantic manipulation. In: International Conference on Learning Representations (2020)
6. Brock, A., Donahue, J., Simonyan, K.: Large scale GAN training for high fidelity natural image synthesis. In: ICLR (2019)
7. Brown, T.B., Carlini, N., Zhang, C., Olsson, C., Christiano, P., Goodfellow, I.: Unrestricted adversarial examples. arXiv preprint arXiv:1809.08352 (2018)
8. Bulat, A., Tzimiropoulos, G.: Binarized convolutional landmark localizers for human pose estimation and face alignment with limited resources. In: Proceedings of the IEEE International Conference on Computer Vision, pp. 3706–3714 (2017)

9. Bulat, A., Tzimiropoulos, G.: How far are we from solving the 2D & 3D face alignment problem? (and a dataset of 230,000 3D facial landmarks). In: ICCV (2017)

10. Cao, Y., et al.: Adversarial sensor attack on lidar-based perception in autonomous driving. In: Proceedings of the 2019 ACM SIGSAC Conference on Computer and Communications Security, pp. 2267–2281 (2019)

11. Cao, Y., et al.: Adversarial objects against lidar-based autonomous driving systems. arXiv preprint arXiv:1907.05418 (2019)

12. Carlini, N., Wagner, D.: Towards evaluating the robustness of neural networks. In: 2017 IEEE Symposium on Security and Privacy (S&P). IEEE (2017)

13. Chen, L.C., Papandreou, G., Kokkinos, I., Murphy, K., Yuille, A.L.: Deeplab: semantic image segmentation with deep convolutional nets, atrous convolution, and fully connected CRFs. IEEE Trans. Pattern Anal. Mach. Intell. **40**(4), 834–848 (2017)

14. Choi, Y., Choi, M., Kim, M., Ha, J.W., Kim, S., Choo, J.: Stargan: unified generative adversarial networks for multi-domain image-to-image translation. In: CVPR (2018)

15. Cisse, M., Adi, Y., Neverova, N., Keshet, J.: Houdini: fooling deep structured prediction models. In: NIPS (2017)

16. Cordts, M., et al.: The cityscapes dataset for semantic urban scene understanding. In: Proceedings of the IEEE Conference on Computer Vision and Pattern Recognition, pp. 3213–3223 (2016)

17. Deng, J., Guo, J., Xue, N., Zafeiriou, S.: Arcface: additive angular margin loss for deep face recognition. In: Proceedings of the IEEE Conference on Computer Vision and Pattern Recognition, pp. 4690–4699 (2019)

18. Dong, Y., et al.: Boosting adversarial attacks with momentum. In: Proceedings of the IEEE Conference on Computer Vision and Pattern Recognition, pp. 9185–9193 (2018)

19. Dziugaite, G.K., Ghahramani, Z., Roy, D.M.: A study of the effect of JPG compression on adversarial images. arXiv preprint arXiv:1608.00853 (2016)

20. Engstrom, L., Tran, B., Tsipras, D., Schmidt, L., Madry, A.: A rotation and a translation suffice: fooling CNNs with simple transformations. arXiv preprint arXiv:1712.02779 (2017)

21. Eykholt, K., et al.: Robust physical-world attacks on deep learning visual classification. In: Proceedings of the IEEE Conference on Computer Vision and Pattern Recognition, pp. 1625–1634 (2018)

22. Farhadi, A., Endres, I., Hoiem, D., Forsyth, D.: Describing objects by their attributes. In: CVPR. IEEE (2009)

23. Goodfellow, I., et al.: Generative adversarial nets. In: NIPS (2014)

24. Goodfellow, I.J., Shlens, J., Szegedy, C.: Explaining and harnessing adversarial examples. In: ICLR (2014)

25. Guo, Y., Zhang, L., Hu, Y., He, X., Gao, J.: MS-Celeb-1M: a dataset and benchmark for large-scale face recognition. In: Leibe, B., Matas, J., Sebe, N., Welling, M. (eds.) ECCV 2016. LNCS, vol. 9907, pp. 87–102. Springer, Cham (2016). https://doi.org/10.1007/978-3-319-46487-9_6

26. He, K., Zhang, X., Ren, S., Sun, J.: Deep residual learning for image recognition. In: CVPR (2016)

27. Hong, S., Yan, X., Huang, T.S., Lee, H.: Learning hierarchical semantic image manipulation through structured representations. In: NeurIPS (2018)

28. Huang, Q., Yang, L., Huang, H., Wu, T., Lin, D.: Caption-supervised face recognition: training a state-of-the-art face model without manual annotation. In: ECCV 2020. Springer, Cham (2020)
29. Isola, P., Zhu, J.Y., Zhou, T., Efros, A.A.: Image-to-image translation with conditional adversarial networks. In: CVPR, pp. 1125–1134 (2017)
30. Johnson, J., Alahi, A., Fei-Fei, L.: Perceptual losses for real-time style transfer and super-resolution. In: Leibe, B., Matas, J., Sebe, N., Welling, M. (eds.) ECCV 2016. LNCS, vol. 9906, pp. 694–711. Springer, Cham (2016). https://doi.org/10.1007/978-3-319-46475-6_43
31. Johnson, J., Gupta, A., Fei-Fei, L.: Image generation from scene graphs. In: CVPR, pp. 1219–1228 (2018)
32. Joshi, A., Mukherjee, A., Sarkar, S., Hegde, C.: Semantic adversarial attacks: parametric transformations that fool deep classifiers. arXiv preprint arXiv:1904.08489 (2019)
33. Kang, D., Sun, Y., Hendrycks, D., Brown, T., Steinhardt, J.: Testing robustness against unforeseen adversaries. arXiv preprint arXiv:1908.08016 (2019)
34. Karras, T., Aila, T., Laine, S., Lehtinen, J.: Progressive growing of GANs for improved quality, stability, and variation. In: ICLR (2018)
35. Kemelmacher-Shlizerman, I., Seitz, S.M., Miller, D., Brossard, E.: The megaface benchmark: 1 million faces for recognition at scale. In: CVPR, pp. 4873–4882 (2016)
36. Kingma, D.P., Welling, M.: Auto-encoding variational bayes. In: ICLR (2014)
37. Klare, B.F., et al.: Pushing the frontiers of unconstrained face detection and recognition: Iarpa Janus benchmark A. In: CVPR (2015)
38. Krizhevsky, A., Sutskever, I., Hinton, G.E.: Imagenet classification with deep convolutional neural networks. In: NIPS (2012)
39. Kumar, N., Berg, A.C., Belhumeur, P.N., Nayar, S.K.: Attribute and simile classifiers for face verification. In: ICCV. IEEE (2009)
40. Li, X., Li, F.: Adversarial examples detection in deep networks with convolutional filter statistics. In: Proceedings of the IEEE International Conference on Computer Vision, pp. 5764–5772 (2017)
41. Liu, M.Y., Breuel, T., Kautz, J.: Unsupervised image-to-image translation networks. In: NIPS (2017)
42. Liu, Z., Luo, P., Wang, X., Tang, X.: Deep learning face attributes in the wild. In: ICCV (2015)
43. Long, J., Shelhamer, E., Darrell, T.: Fully convolutional networks for semantic segmentation. In: Proceedings of the IEEE Conference on Computer Vision and Pattern Recognition, pp. 3431–3440 (2015)
44. Madry, A., Makelov, A., Schmidt, L., Tsipras, D., Vladu, A.: Towards deep learning models resistant to adversarial attacks. In: ICLR (2018)
45. Mahendran, A., Vedaldi, A.: Understanding deep image representations by inverting them. In: CVPR (2015)
46. Mansimov, E., Parisotto, E., Ba, J.L., Salakhutdinov, R.: Generating images from captions with attention. In: ICLR (2015)
47. Moosavi-Dezfooli, S.M., Fawzi, A., Frossard, P.: Deepfool: a simple and accurate method to fool deep neural networks. In: Proceedings of the IEEE Conference on Computer Vision and Pattern Recognition, pp. 2574–2582 (2016)
48. Newell, A., Yang, K., Deng, J.: Stacked hourglass networks for human pose estimation. In: Leibe, B., Matas, J., Sebe, N., Welling, M. (eds.) ECCV 2016. LNCS, vol. 9912, pp. 483–499. Springer, Cham (2016). https://doi.org/10.1007/978-3-319-46484-8_29

49. Odena, A., Olah, C., Shlens, J.: Conditional image synthesis with auxiliary classifier GANs. In: ICML. JMLR (2017)
50. Van den Oord, A., Kalchbrenner, N., Espeholt, L., Vinyals, O., Graves, A., et al.: Conditional image generation with pixelcnn decoders. In: NIPS (2016)
51. Oord, A.v.d., Kalchbrenner, N., Kavukcuoglu, K.: Pixel recurrent neural networks. In: ICML (2016)
52. Papernot, N., McDaniel, P., Jha, S., Fredrikson, M., Celik, Z.B., Swami, A.: The limitations of deep learning in adversarial settings. In: 2016 IEEE European Symposium on Security and Privacy (EuroS&P) (2016)
53. Parikh, D., Grauman, K.: Relative attributes. In: ICCV. IEEE (2011)
54. Parkhi, O.M., Vedaldi, A., Zisserman, A., et al.: Deep face recognition. In: BMVC, vol. 1, p. 6 (2015)
55. Radford, A., Metz, L., Chintala, S.: Unsupervised representation learning with deep convolutional generative adversarial networks. In: ICLR (2015)
56. Reed, S., Akata, Z., Yan, X., Logeswaran, L., Schiele, B., Lee, H.: Generative adversarial text to image synthesis. In: ICML (2016)
57. Reed, S., Sohn, K., Zhang, Y., Lee, H.: Learning to disentangle factors of variation with manifold interaction. In: ICML (2014)
58. Sagonas, C., Tzimiropoulos, G., Zafeiriou, S., Pantic, M.: 300 faces in-the-wild challenge: the first facial landmark localization challenge. In: ICCV Workshop (2013)
59. Schroff, F., Kalenichenko, D., Philbin, J.: Facenet: a unified embedding for face recognition and clustering. In: Proceedings of the IEEE Conference on Computer Vision and Pattern Recognition, pp. 815–823 (2015)
60. Simonyan, K., Zisserman, A.: Very deep convolutional networks for large-scale image recognition. arXiv preprint arXiv:1409.1556 (2014)
61. Song, Y., Shu, R., Kushman, N., Ermon, S.: Constructing unrestricted adversarial examples with generative models. In: Advances in Neural Information Processing Systems, pp. 8312–8323 (2018)
62. Sun, Y., Wang, X., Tang, X.: Deep learning face representation from predicting 10,000 classes. In: CVPR (2014)
63. Szegedy, C., et al.: Going deeper with convolutions. In: CVPR (2015)
64. Szegedy, C., et al.: Intriguing properties of neural networks. arXiv preprint arXiv:1312.6199 (2013)
65. Tao, G., Ma, S., Liu, Y., Zhang, X.: Attacks meet interpretability: attribute-steered detection of adversarial samples. In: NeurIPS (2018)
66. Tong, L., Li, B., Hajaj, C., Xiao, C., Zhang, N., Vorobeychik, Y.: Improving robustness of {ML} classifiers against realizable evasion attacks using conserved features. In: 28th {USENIX} Security Symposium ({USENIX} Security 2019), pp. 285–302 (2019)
67. Wang, H., et al.: Cosface: large margin cosine loss for deep face recognition. In: CVPR (2018)
68. Wang, T.C., Liu, M.Y., Zhu, J.Y., Tao, A., Kautz, J., Catanzaro, B.: High-resolution image synthesis and semantic manipulation with conditional GANs. In: CVPR (2018)
69. Wong, E., Schmidt, F.R., Kolter, J.Z.: Wasserstein adversarial examples via projected sinkhorn iterations. In: ICML (2019)
70. Xiao, C., et al.: Advit: adversarial frames identifier based on temporal consistency in videos. In: Proceedings of the IEEE International Conference on Computer Vision, pp. 3968–3977 (2019)

71. Xiao, C., Deng, R., Li, B., Yu, F., Liu, M., Song, D.: Characterizing adversarial examples based on spatial consistency information for semantic segmentation. In: Ferrari, V., Hebert, M., Sminchisescu, C., Weiss, Y. (eds.) ECCV 2018. LNCS, vol. 11214, pp. 220–237. Springer, Cham (2018). https://doi.org/10.1007/978-3-030-01249-6_14

72. Xiao, C., Li, B., Zhu, J.Y., He, W., Liu, M., Song, D.: Generating adversarial examples with adversarial networks. In: IJCAI (2018)

73. Xiao, C., et al.: Characterizing attacks on deep reinforcement learning. arXiv preprint arXiv:1907.09470 (2019)

74. Xiao, C., Yang, D., Li, B., Deng, J., Liu, M.: MeshAdv: adversarial meshes for visual recognition. In: Proceedings of the IEEE Conference on Computer Vision and Pattern Recognition, pp. 6898–6907 (2019)

75. Xiao, C., Zhu, J.Y., Li, B., He, W., Liu, M., Song, D.: Spatially transformed adversarial examples. In: ICLR (2018)

76. Xie, C., et al.: Improving transferability of adversarial examples with input diversity. In: Proceedings of the IEEE Conference on Computer Vision and Pattern Recognition, pp. 2730–2739 (2019)

77. Xu, W., Evans, D., Qi, Y.: Feature squeezing: detecting adversarial examples in deep neural networks. arXiv preprint arXiv:1704.01155 (2017)

78. Yan, X., Yang, J., Sohn, K., Lee, H.: Attribute2Image: conditional image generation from visual attributes. In: Leibe, B., Matas, J., Sebe, N., Welling, M. (eds.) ECCV 2016. LNCS, vol. 9908, pp. 776–791. Springer, Cham (2016). https://doi.org/10.1007/978-3-319-46493-0_47

79. Yang, L., Chen, D., Zhan, X., Zhao, R., Loy, C.C., Lin, D.: Learning to cluster faces via confidence and connectivity estimation. In: Proceedings of the IEEE Conference on Computer Vision and Pattern Recognition (2020)

80. Yang, L., Huang, Q., Huang, H., Xu, L., Lin, D.: Learn to propagate reliably on noisy affinity graphs. In: ECCV 2020. Springer, Cham (2020)

81. Yang, L., Zhan, X., Chen, D., Yan, J., Loy, C.C., Lin, D.: Learning to cluster faces on an affinity graph. In: Proceedings of the IEEE Conference on Computer Vision and Pattern Recognition (CVPR) (2019)

82. Yao, S., et al.: 3D-aware scene manipulation via inverse graphics. In: Advances in Neural Information Processing Systems, pp. 1887–1898 (2018)

83. Yu, F., Koltun, V., Funkhouser, T.: Dilated residual networks. In: Computer Vision and Pattern Recognition (CVPR) (2017)

84. Zhang, H., et al.: Stackgan: text to photo-realistic image synthesis with stacked generative adversarial networks. In: ICCV (2017)

85. Zhang, H., et al.: Towards stable and efficient training of verifiably robust neural networks. In: ICLR 2020 (2019)

86. Zhang, X., Yang, L., Yan, J., Lin, D.: Accelerated training for massive classification via dynamic class selection. In: Thirty-Second AAAI Conference on Artificial Intelligence (2018)

87. Zhu, J.Y., Park, T., Isola, P., Efros, A.A.: Unpaired image-to-image translation using cycle-consistent adversarial networks. In: ICCV (2017)

88. Zhu, X., Lei, Z., Liu, X., Shi, H., Li, S.Z.: Face alignment across large poses: a 3D solution. In: CVPR (2016)

Learning with Noisy Class Labels
for Instance Segmentation

Longrong Yang, Fanman Meng, Hongliang Li[✉], Qingbo Wu,
and Qishang Cheng

School of Information and Communication Engineering,
University of Electronic Science and Technology of China, Chengdu, China
{yanglr,cqs}@std.uestc.edu.cn, {fmmeng,hlli,qbwu}@uestc.edu.cn

Abstract. Instance segmentation has achieved siginificant progress in
the presence of correctly annotated datasets. Yet, object classes in large-
scale datasets are sometimes ambiguous, which easily causes confusion.
In addition, limited experience and knowledge of annotators can also
lead to mislabeled object classes. To solve this issue, a novel method
is proposed in this paper, which uses different losses describing different roles of noisy class labels to enhance the learning. Specifically,
in instance segmentation, noisy class labels play different roles in the
foreground-background sub-task and the foreground-instance sub-task.
Hence, on the one hand, the noise-robust loss (e.g., symmetric loss) is
used to prevent incorrect gradient guidance for the foreground-instance
sub-task. On the other hand, standard cross entropy loss is used to fully
exploit correct gradient guidance for the foreground-background sub-
task. Extensive experiments conducted with three popular datasets (i.e.,
Pascal VOC, Cityscapes and COCO) have demonstrated the effectiveness of our method in a wide range of noisy class labels scenarios. Code
will be available at: github.com/longrongyang/LNCIS.

Keywords: Noisy class labels · Instance segmentation ·
Foreground-instance sub-task · Foreground-background sub-task

1 Introduction

Datasets are of crucial to instance segmentation. Large-scale datasets with clean
annotations are often required in instance segmentation. However, some classes
show similar appearance and are easily mislabeled, as shown in Fig. 1. Meanwhile, some existing papers [11,19,24] also mention that inherent ambiguity of
classes and limited experience of annotators can result in corrupted object class
labels. These mislabeled samples inevitably affect the model training. Therefore,
how to train accurate instance segmentation models in the presence of noisy class
labels is worthy to explore.

In the classification task, label noise problem has been studied for a long
time. Some existing methods apply the noise-robust loss (e.g., symmetric loss)

© Springer Nature Switzerland AG 2020
A. Vedaldi et al. (Eds.): ECCV 2020, LNCS 12359, pp. 38–53, 2020.
https://doi.org/10.1007/978-3-030-58568-6_3

(a) Image   (b) Sample without Noise   (c) Sample with Noise

Fig. 1. Example of noisy samples in the instance segmentation task. This example is selected from Cityscapes dataset [5]. In this example, object class *motocycle* is mislabeled as class *bicycle* by the annotator. We mainly discuss the noise on object class labels in this paper. Similar with methods in classification, datastes with lots of noise are produced by artificially corrupting labels.

to all samples to reduce gradients generated by noisy samples, such as [8,29,33]. These works have achieved promising results in the classification task. However, in the instance segmentation task, noisy class labels play different roles in the foreground-background sub-task (i.e., distinguishing foreground and background) and the foreground-instance sub-task (i.e., classifying different classes of foreground instances). From the perspective of the foreground-background sub-task, all class labels always provide correct guidance to the gradient update. Hence, if some key samples to the foreground-background sub-task are suppressed by the noise-robust loss in the gradient computation, the foreground-background sub-task is inevitably degenerated.

To solve this problem, we propose a novel method in this paper, which describes different roles of noisy class labels using diverse losses to enhance the learning. Firstly, some evidences provided in [1,21,32] show that models prone to fit clean and noisy samples in early and mature stages of training, respectively. Hence, in early stages of training, the classification loss remains unchanged. In mature stages of training, we observe that negative samples (i.e., samples belonging to background) are impossibly noisy and pseudo negative samples (i.e., positive samples misclassified as background) play key role in the foreground-background sub-task. Hence, cross entropy loss is applied to all negative samples and pesudo negative samples, to fully exploit correct gradient guidance provided by these samples for the foreground-background sub-task. Meanwhile, the noise-robust loss is applied to other samples for the foreground-instance sub-task. In addition, we also use loss values as the cue to detect and isolate some noisy samples, to further avoid the incorrect guidance provided by noisy class labels for the foreground-instance sub-task. This proposed method is verified on three well-known datasets, namely Pascal VOC [7], Cityscapes [5] and COCO [19]. Extensive experiments show that our method is effective in a wide range of noisy class labels scenarios.

2   Related Works

Learning with Noisy Class Labels for the Classification Task: Different methods have been proposed to train accurate classification models in the presence of noisy class labels, which can be roughly divided into three categories. The first category is to improve the quality of raw labels by modeling the noises through directed graphical models [30], conditional random fields [27], neural networks [17,28] or knowledge graphs [18]. These methods usually require a set of clean samples to estimate the noise model, which limits their applicability. Hence, the joint optimization framework [26] does unsupervised sample relabeling with own estimate of labels. Meanwhile, Label Smoothing Regularization [23,25] can alleviate over-fitting to noisy class labels by soft labels. The second category is to compensate for the incorrect guidance provided by noisy samples via modifying the loss functions. For example, Backward [22] and Forward [22] explicitly model the noise transition matrix to weight the loss of each sample. However, this method is hard to use because the noise transition matrix is not always available in practice [12]. Hence, some noise-robust loss functions are designed, such as MAE [8]. However, training a model with MAE converges slowly. To deal with this issue, Generalized Cross Entropy Loss [33] combines advantages of MAE and cross entropy loss by Box-Cox transformation [2]. Meanwhile, symmetric Cross Entropy Loss is proposed in [29], which applies the weighting sum of reverse cross entropy loss and cross entropy loss to achieve promising results in classification. These methods only require minimal intervention to existing algorithms and architectures. The third category is to introduce an auxiliary model. For example, a TeacherNet is used to provide a data-driven curriculum for a StudentNet by a learned sample weighting scheme in MentorNet [16]. To solve the accumulated error issue in MentorNet, Co-teaching [13] maintains two models simultaneously during training, with one model learning from the another model's most confident samples. Furthermore, Co-teaching+ [31] keep two networks diverged during training to prevent Co-teaching reducing to the self-training MentorNet in function.

These methods suppose that noisy class labels inevitably degenerate the model accuracy, which is suitable for the classification task, but is invalid for the instance segmentation task with multiple sub-tasks such as foreground-background and foreground-instance. It is the fact that noisy labels play different roles in the two sub-tasks, which need be treated differently.

Instance Segmentation: Some instance segmentation models have been proposed in the past few years [3,6,14,15,20]. Based on the segmentation manner, these methods can be roughly divided into two categories. The first one is driven by the success of the semantic segmentation task, which firstly predicts the class of each pixel, and then groups different instances, such as GMIS [20] and DLF [6]. The second one connects strongly with the object detection task, such as Mask R-CNN [14], Mask Scoring R-CNN [15] and HTC [3], which detects object instances firstly, and then generates masks from the bounding boxes. Among these methods, Mask R-CNN [14] selected as the reference backbone for the task of instance-level segmentation in this paper consists of four steps. The first one is to extract features of images by CNNs. The second one is to generate proposals by RPN [10]. The third one is to obtain the classification confidence and the bounding box regression. Finally, segmentation masks are generated inside of bounding boxes by the segmentation branch.

Although Mask R-CNN [14] has achieved promising results for the instance segmentation task, it is based on clean annotations. When there are noisy labels, its performance drops significantly. In contrast to the classification task, Mask R-CNN[14] has multiple sub-tasks with different roles and classification losses. From the perspective of the foreground-background sub-task, noisy class labels still provide correct guidance. Meanwhile, in instance segmentation, proposal generation and mask generation are only related with the foreground-background sub-task. Hence, the binary classification losses in RPN [10] and the segmentation branch remains unchanged. In this paper, we focus on the multi-class classification loss in the box head, which is related with the foreground-background sub-task and the foreground-instance sub-task, simultaneously.

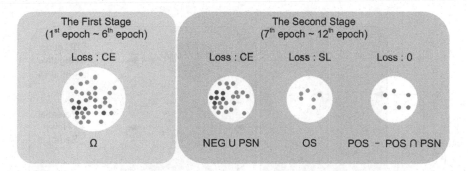

Fig. 2. Losses of different types of samples. Ω denotes the total sample space. $NEG \cup PSN$ denotes all negative samples and pseudo negative samples. $POS - POS \cap PSN$ denotes potential noisy samples classified foreground. OS denotes other samples. CE and SL denote standard cross entropy loss and symmetric loss, respectively.

3   Methodology

The multi-class classification in instance segmentation consists of the foreground-background sub-task and the foreground-instance sub-task. In general, it can be formulated as the problem to learn a classifier $f_\theta(x)$ from a set of training samples $D = (x_i, y_i)_{i=1}^N$ with $y_i \in \{0, 1, ..., K\}$. In instance segmentation, the sample x_i corresponds to an image region rather than an image. y_i is the class label of the sample x_i and can be noisy. For convenience, we assign 0 as the class label of samples belonging to background. Meanwhile, suppose the correct class label of the sample x_i is $y_{c,i}$. In this paper, we focus on the noise on object class labels and 0 is not a true object class label in datasets, so $p(y_i = 0 | y_{c,i} \neq 0) =$

$p(y_i \neq 0 | y_{c,i} = 0) = 0$. By a loss function, e.g., multi-class cross entropy loss, the foreground-background sub-task and the foreground-instance sub-task are optimized simultaneously:

$$l_{ce} = \frac{1}{N} \sum_{i=1}^{N} l_{ce,i} = -\frac{1}{N} \sum_{i=1}^{N} \sum_{k=0}^{K} q(k|x_i) log p(k|x_i) \qquad (1)$$

where $p(k|x_i)$ denotes classification confidence of each class $k \in \{0, 1, ..., K\}$ for the sample x_i, and $q(k|x_i)$ denotes the one-hot encoded label, so $q(y_i|x_i) = 1$ and $q(k|x_i) = 0$ for all $k \neq y_i$.

Cross entropy loss is sensitive to label noise. Some existing methods in classification use the noise-robust loss (e.g., symmetric loss) to replace cross entropy loss to deal with this problem. However, the noise-robust loss leads to reduced gradients of some samples, which degenerates the foreground-background sub-task in instance segmentation. To solve this problem, we describes different roles of noisy class labels using diverse losses, as shown in Fig. 2.

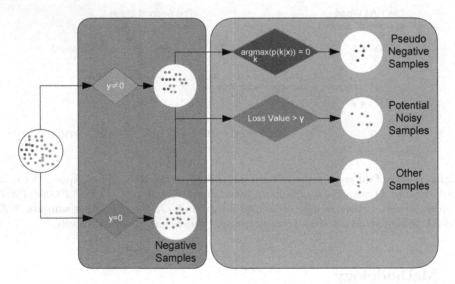

Fig. 3. Division of samples in this paper. Here, y is the class label of the sample x. γ is a hyper-parameter that can adjust. $p(k|x)$ is the confidence of x on class k. Some samples possibly belong to pseudo negative samples and potential noisy samples, simultaneously.

3.1   Division of Samples

Firstly, as shown in Fig. 3, all samples are roughly divided into two types: positive samples and negative samples. In Mask R-CNN [14], positive samples refer to the samples whose IoUs (Intersection over Union) with the corresponding ground truths are above 0.5. Hence, it is generally considered that positive samples belong to foreground (i.e., $y \neq 0$) and negative samples belong to background (i.e., $y = 0$).

Furthermore, positive samples are divided into three types: pseudo negative samples (PSN), potential noisy samples (PON) and other samples (OS). Here, we define positive samples which are misclassified as background as pseudo negative samples. Hence, it is easy to know that, for a pseudo negative samples x, $argmax_k(p(k|x)) = 0$. In addition, as shown in Fig. 3, we define samples whose loss values $l_{ce} > \gamma$ as potential noisy samples. According to our statics in Fig. 5, in instance segmentation, noisy samples usually have larger loss values than clean samples in mature stages of training. From Fig. 5, it can be seen that 88.5% of noisy samples have loss values $l_{ce} > 6.0$ while only 2.31% of clean samples have loss values $l_{ce} > 6.0$. Subjective examples are also given in Fig. 4 to explain the difference of different samples.

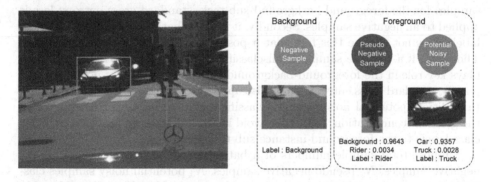

Fig. 4. Subjective examples to explain the difference of different samples. $Car : 0.9357$ denotes that the confidence of this sample is 0.9357 on class Car.

3.2 Classification Loss

In early stages of training, models favor learning clean samples, while hindering learning noisy samples [32]. As models converge to higher accuracy, noisy samples gradually contribute more to the gradient update. In mature stages of training, models prone to fit in noisy samples. Hence, in early stages of training (the first stage), the classification loss remains unchanged (i.e., cross entropy loss is applied to all samples). Suppose total sample numbers of a batch are N. Classification loss of this batch in the first stage can be described as:

$$Loss_1 = -\frac{1}{N} \sum_{i=1}^{N} \sum_{k=0}^{K} q(k|x_i) log p(k|x_i) \qquad (2)$$

where $Loss_1$ denotes the multi-class classification loss in the first stage. Suppose E_1 and E are epoch numbers of the first stage and total epoch numbers, respectively. Under different noise rates, we empirically derive the relation between E and E_1:

$$E_1 = \frac{1}{2}E, \quad s.t. \ \forall \eta \tag{3}$$

where η denotes noise rate. In general, $E = 12$ and $E_1 = 6$.

In mature stages of training (the second stage), the noise-robust loss needs be introduced. However, this loss leads to reduced gradients of some key samples to the foreground-background sub-task in instance segmentation. Hence, cross entropy loss still needs to be used, to prevent the noise-robust loss from degenerating the foreground-background sub-task. Different losses are applied to different types of samples to yield a good compromise between fast convergence and noise robustness.

Firstly, in instance segmentation, noisy class labels only exist in partial foreground regions. Hence, we think that potential noisy labels do not exist in any negative samples. To fully exploit correct gradient information provided by these samples for the foreground-background sub-task, standard cross entropy loss is applied to all negative samples. Secondly, it is clear that the noise on object class labels does not change this fact that a positive sample belongs to foreground. Therefore, if a positive sample is misclassified as background, this sample still plays key role in the foreground-background sub-task even if it is noisy. For this reason, standard cross entropy loss is also applied to all pseudo negative samples. In addition, potential noisy samples classified as foreground can be isolated in the gradient computation, to further avoid incorrect guidance provided by noisy class labels to the foreground-instance sub-task.

Suppose total sample numbers of a batch are N. Meanwhile, there are N_1 negative samples, N_2 pseudo negative samples, N_3 potential noisy samples classified foreground and N_4 other samples in this batch. $N = N_1 + N_2 + N_3 + N_4$. In mature stages of training, classification loss of this batch can be described as:

$$Loss_2 = -\frac{1}{N}[\sum_{i=1}^{N_1+N_2}\sum_{k=0}^{K} q(k|x_i)logp(k|x_i) + \sum_{m=1}^{N_3} 0 + \sum_{j=1}^{N_4}(-l_{sl,j})] \tag{4}$$

where $Loss_2$ denotes the multi-class classification loss in the second stage. $l_{sl,j}$ denotes a symmetric loss function which is robust to noise.

3.3   Reverse Cross Entropy Loss

In this paper, we select reverse cross entropy loss proposed in [29] as the symmetric loss $l_{sl,i}$. The reverse cross entropy loss is defined as:

$$l_{rce,i} = -\sum_{k=0}^{K} p(k|x_i)logq(k|x_i) \tag{5}$$

where $l_{rce,i}$ denotes the reverse cross entropy loss for the sample x_i. This kind of loss is robust to noise but takes longer time to converge [29]. Meanwhile, there is also a compromise in accuracy due to the increased difficulty to learn useful features. In this paper, the clipped replacement A of $log0$ is set to -4.

4 Theoretical Analyses

4.1 Noise Robustness

We explain the noise robustness of reverse cross entropy loss from the perspective of symmetric loss. For any classifier f, $R(f)$ or $R^{\eta}(f)$ denotes the risk of f under clean class labels or noisy class labels (noise rate η), respectively. Suppose f^* and f_{η}^* are the global minimizers of $R(f)$ and $R^{\eta}(f)$, respectively. Let $\mathcal{X} \subset R^d$ be the feature space. A symmetric loss function l is defined as:

$$\sum_{y=0}^{K} l(f(x), y) = C, \ \forall x \in \mathcal{X}, \ \forall f. \tag{6}$$

where C is a constant. In [9], it has been proven that if loss function is symmetric and noise rate $\eta < \frac{K}{K+1}$, then under symmetric label noise, for $\forall f$, $R^{\eta}(f^*) - R^{\eta}(f) = (1 - \frac{\eta(K+1)}{K})(R(f^*) - R(f)) \leqslant 0$. Hence, $f_{\eta}^* = f^*$ and l is noise robust. Moreover, it can also be proven, if $R(f^*) = 0$, that l is noise robust under asymmetric noise. Meanwhile, according to Eq. (5), we can derive:

$$\sum_{y=0}^{K} l_{rce} = -\sum_{y=0}^{K}\sum_{k=0}^{K} p(k|x)logq(k|x)$$

$$= -\sum_{y=0}^{K}(\sum_{k\neq y} p(k|x)log0 + p(y|x)log1)$$

$$= -\sum_{y=0}^{K}[1 - p(y|x)]log0 = -KA$$

where $-KA$ is a constant when class numbers K and A ($log0 = A$) are given. Hence, reverse cross entropy loss is symmetric and noise robust.

4.2 Gradients

Gradients of reverse cross entropy loss and cross entropy loss have been derived in [29]. Based on this, we can explain why models favor learning clean samples in early stages of training, while hindering learning noisy samples. This is not discussed in [29]. For brevity, we denote p_k, q_k as abbreviations for $p(k|x)$ and $q(k|x)$. We focus on a sample $\{x, y\} \subset D$. The gradient of reverse cross entropy loss with respect to the logit z_j can be derived as:

$$\frac{\partial l_{rce}}{\partial z_j} = \begin{cases} Ap_j - Ap_j^2, & q_j = q_y = 1 \\ -Ap_j p_y, & q_j = 0 \end{cases} \tag{7}$$

The gradient of cross entropy loss l_{ce} is:

$$\frac{\partial l_{ce}}{\partial z_j} = \begin{cases} p_j - 1, & q_j = q_y = 1 \\ p_j, & q_j = 0 \end{cases} \tag{8}$$

Analysis. As shown in Eq. (8), for cross entropy loss, if $q_j = 1$, samples with smaller p_j contribute more to the gradient update. Based on this, it is clear that models gradually prone to fit noisy samples as models converge to higher accuracy. To clarify this, suppose sample A and sample B belong to the same class C_1. p_{a,c_1} denotes that the classification confidence of sample A on class C_1. Meanwhile, suppose class labels of sample A and B are C_1 and C_2, respectively ($C_1 \neq C_2$). At the beginning of the training, $p_{a,c_1} \approx p_{b,c_2}$ hence their contribution to the gradient computation is approximately equal. When the training continues to later stages of training, because the accuracy of models increases, p_{a,c_1} increases and p_{b,c_2} decreases. As a result, gradients generated by noisy samples become larger and gradients generated by clean samples become smaller. When noisy samples contribute more to the gradient computation than clean samples, model begins to prone to fit noisy samples.

Secondly, the weakness of reverse cross entropy loss can also be explained from the respect of gradients. As shown in Eq. (7), if $q_j = 1$, gradients of reverse cross entropy loss are symmetric about $p_j = 0.5$. This means that, if p_j of a sample is close to 0, the gradient generated by this sample is also close to 0. This is the reason why this loss is robust to noise. However, this also leads to reduced gradients of clean samples whose p_j is close to 0 and these samples usually play key role in training.

5   Experiments

5.1   Datasets and Noise Settings

Pascal VOC Dataset: On Pascal VOC 2012 [7], the *train* subset with 5718 images and the *val* subset with 5823 images are used to train the model and evaluate the performance, respectively. There are 20 semantic classes on Pascal VOC dataset [7]. All objective results are reported following the COCO-style metrics which calculates the average AP across IoU thresholds from 0.5 to 0.95 with an interval of 0.05.

COCO Dataset: COCO dataset [19] is one of the most challenging datasets for the instance segmentation task due to the data complexity. It consists of 118,287 images for training (*train-2017*) and 20,288 images for test (*test-dev*). There are 80 semantic classes on COCO dataset [19]. We train our models on *train-2017* subset and report objective results on *test-dev* subset. COCO standard metrics are used in this paper, which keeps the same with traditional instance segmentation methods.

Cityscapes Dataset: On Cityscapes dataset [5], the fine training set which has 2975 images with fine annotations is used to train the model. The validation set has 500 images, which is used to evaluate the performance of our method. Eight semantic classes are annotated with instance masks.

Noise Settings: Noisy datasets are produced by artificially corrupting object class labels. Similar with noise settings in classification, there are mainly two types of noise in this paper: symmetric (uniform) noise and asymmetric (class-conditional) noise. If the noise is conditionally independent of correct class labels, the noise is named as symmetric or uniform noise. Class labels with symmetric noise are generated by flipping correct class labels of training samples to a different class with uniform random probability η. For class labels with asymmetric noise, similar with [22,33], flipping labels only occurs within specific classes which are easily misclassified, for example, for VOC dataset, flipping *Bird* $\rightarrow$ *Aeroplane, Diningtable* $\rightarrow$ *Chair, Bus* $\rightarrow$ *Car, Sheep* $\rightarrow$ *Horse, Bicyce* $\leftrightarrow$ *Motobike, Cat* $\leftrightarrow$ *Dog*; for Cityscapes dataset, flipping *Person* $\leftrightarrow$ *Rider, Bus* $\leftrightarrow$ *Truck, Motorcycle* $\leftrightarrow$ *Bicycle*; for COCO dataset, 80 classes are grouped into 12 super-classes based on the COCO standard, then flipping between two randomly selected sub-classes within each super-class. Super-class *Person* is not flipped because this super-class only has a sub-class.

5.2 Implementation Details

Hyper-Parameters: For COCO dataset, the overall batch size is set to 16. The initial learning rate is set to 0.02. We train for 12 epoches ($1\times$) and the learning rate reduces by a factor of 10 at 8 and 11 epoches. For Cityscapes dataset, the overall batch size is set to 2. The initial learning rate is set to 0.0025. We train for 64 epoches and the learning rate reduces by a factor of 10 at 48 epoches. For Pascal VOC dataset, the overall batch size is set to 4. The initial learning rate is set to 0.005. We train for 12 epoches and the learning rate reduces by a factor of 10 at 9 epoches. Other hyper-parameters follow the settings in MMdetection [4].

In our method, we set $\gamma = 6.0$ for all datasets. For symmertric noise, we test varying noise rates $\eta \in \{20\%, 40\%, 60\%, 80\%\}$, while for asymmetric noise, we test varying noise rates $\eta \in \{20\%, 40\%\}$.

Table 1. The results on Pascal VOC dataset. 0% denotes no artificially corrupting class labels. We report the $mAPs$ of all methods. The best results are in bold.

Methods		Symmetric noise				Asymmetric noise	
Noise rates (η)	0%	20%	40%	60%	80%	20%	40%
LSR [23]	36.7	34.5	31.9	27.0	20.3	35.8	33.2
JOL [26]	32.7	22.7	11.5	3.8	3.8	24.6	24.3
GCE [33]	38.7	35.3	32.9	26.1	15.5	36.0	32.3
SCE [29]	39.4	34.6	32.1	27.9	21.2	37.5	34.9
CE [14]	39.3	34.2	31.5	27.1	20.7	36.8	34.5
Our Method	**39.7**	**38.5**	**38.1**	**33.8**	**25.5**	**37.8**	**35.1**

Table 2. The results on COCO *test-dev* subset.

Methods		Symmetric noise				Asymmetric noise	
Noise rates (η)	0%	20%	40%	60%	80%	20%	40%
SCE [29]	32.3	31.5	29.9	28.2	22.0	32.1	31.6
CE [14]	**34.2**	31.3	29.3	27.1	21.7	31.9	31.3
Our Method	33.7	**33.1**	**31.3**	**30.8**	**26.6**	**33.3**	**33.0**

5.3   Main Results

Baselines: How to effectively train an instance segmentation model in the presence of noisy class labels is never discussed in existing papers. Hence, we mainly compare our method with some methods [23,26,29,33] proposed in the classification task as well as the standard CE loss [14]: (1) LSR [23]: training with standard cross entropy loss on soft labels; (2) JOL [26]: training with the joint optimization framework. Here, α is set to 0.8 and β is set to 1.2; (3) GCE [33]: training with generalized cross entropy loss. Here, q is set to 0.3; (4) SCE [29]: training with symmetric cross entropy loss $l_{sce} = \alpha l_{ce} + \beta l_{rce}$. Here, α is set to 1.0 and β is set to 0.1; (5) CE [14]: training with standard cross entropy loss. Based on our experiments, we select the best hyper-parameter settings for methods proposed in classification. Note we only change the multi-class classification loss in box head. Mask R-CNN [14] is used as the instance segmentation model and the backbone is ResNet-50-FPN in all methods.

Pascal VOC Dataset: The objective results are reported in Table 1. Compared with [14], methods proposed in classification bring marginal accuracy increase (below 2% under different noise rates) if directly applied to instance segmentation. However, our method can generate a substantial increase in performance by using different losses describing different roles of noisy class labels. Specifically, compared with [14], the accuracy increases 4.3%, 6.6%, 6.7% and 4.8% for 20%, 40%, 60% and 80% of symmetric label noise, respectively. Under 20% and 40%

Table 3. The results on Cityscapes dataset.

Methods	Symmetric noise					Asymmetric noise	
Noise rates (η)	0%	20%	40%	60%	80%	20%	40%
SCE [29]	30.2	26.1	22.3	12.6	9.3	28.0	18.6
CE [14]	32.5	26.0	21.0	13.3	11.6	29.8	18.9
Our Method	**32.7**	**30.8**	**29.1**	**19.1**	**15.2**	**30.9**	**21.3**

asymmetric noise, the accuracy increases by 1.0% and 0.6%, respectively. It can be seen that our method yields an ideal accuracy under different noise rates.

COCO Dataset: The objective results are reported in Table 2. Compared with [14], the accuracy increases 1.8%, 3.1%, 4.2% and 4.9% for 20%, 40%, 60% and 80% of symmetric label noise, respectively. Under 20% and 40% asymmetric noise, the accuracy increases by 1.4% and 1.7%, respectively.

Cityscapes Dataset: The objective results are reported in Table 3. Compared with [14], the accuracy increases 4.8%, 8.1%, 5.8% and 3.6% for 20%, 40%, 60% and 80% of symmetric label noise, respectively. Under 20% and 40% asymmetric noise, the accuracy increases by 1.1% and 2.4%, respectively.

5.4   Discussion

Component Ablation Study: Our component ablation study from the baseline gradually to all components incorporated is conducted on Pascal VOC dataset [7] and noise rate $\eta = 40\%$. We mainly discuss:

(i) **CE:** The baseline. The model is trained with standard cross entropy loss;

(ii) **ST:** Stage-wise training. We apply cross entropy loss and reverse cross entropy loss to all samples in early and mature stages of training, respectively.

(iii) **N & PSN:** The key contribution in this paper. Cross entropy loss is applied to all negative samples and pseudo negative samples. Meanwhile, reverse cross entropy loss is applied to other samples;

(iv) **N & PSN & PON:** Cross entropy loss is applied to all negative samples and pseudo negative samples. Meanwhile, potential noisy samples classified as foreground are isolated in the gradient computation.

(v) **ST & N:** Stage-wise training is applied. Meanwhile, cross entropy loss is applied to negative samples in mature stage of training;

(vi) **ST & N & PSN:** Stage-wise training is applied. Meanwhile, cross entropy loss is applied to all negative samples and pseudo negative samples in mature stage of training;

(vii) **ST & N & PSN & PON:** Stage-wise training is applied. Meanwhile, cross entropy loss is applied to all negative samples and pseudo negative samples in mature stage of training. In addition, in mature stage of training, potential noisy samples classified as foreground are isolated in the gradient computation.

Table 4. Component ablation study. CE denotes the baseline. ST denotes stage-wise training. N denotes applying cross entropy loss to all negative samples. PSN denotes applying cross entropy loss to all pseudo negative samples. PON denotes isolating all potential noisy samples classified as foreground.

CE	ST	N	PSN	PON	AP	AP_{50}	AP_{75}
√	-	-	-	-	31.5	57.4	31.0
	√				34.3	59.9	35.1
		√	√		37.3	63.5	39.0
		√	√	√	36.5	63.2	37.4
	√	√			37.4	64.7	39.0
	√	√	√		37.9	**64.7**	38.0
	√	√	√	√	**38.1**	64.5	**40.0**

The results of ablation study are reported in Table 4. Firstly, the key strategy in this paper (i.e., using different losses to describe different roles of noisy class labels) brings 5.8% higher accuracy than the baseline, which shows that this strategy is greatly important in instance segmentation. Secondly, stage-wise training can bring 2.8% higher accuracy than the baseline. However, if already considering special properties of negative samples and pseudo negative samples, stage-wise training can only bring about 0.6% higher accuracy. This means, the main reason that stage-wise training works in instance segmentation is factually to fully exploit correct gradient information for the foreground-background sub-task. Thirdly, using loss values as the cue to identify noisy samples (i.e., PON) brings marginal accuracy increase (about 0.2%), and stage-wise training (i.e., ST) should be applied simultaneously when PON is applied.

The Relation Between E and E_1: Stage-wise training applies cross entropy loss to all samples in early stages of training, to speed the convergence of models. In mature stages of training, different losses are applied to different samples to yield a good compromise between fast convergence and noise robustness. We conduct some experiments about different E and E_1 under noise rate $\eta = 40\%$ in Table 5. From Table 5, it can be seen that $E_1 = 0.5E$ is the best setting.

Table 5. Study about the relation between E and E_1. E_1 and E are epoch numbers of the first stage and total epoch numbers, respectively.

	AP	AP_{50}	AP_{75}
$E = 12, E_1 = 3$	37.2	63.7	39.0
$E = 12, E_1 = 6$	**38.1**	**64.5**	**40.0**
$E = 12, E_1 = 9$	34.6	61.1	34.8
$E = 18, E_1 = 6$	36.2	62.2	37.9
$E = 18, E_1 = 9$	**38.5**	**66.0**	**39.8**
$E = 18, E_1 = 12$	38.2	66.0	39.7
$E = 18, E_1 = 15$	32.5	58.8	33.3

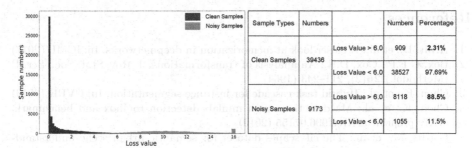

Fig. 5. Loss values l_{ce} and statistics of samples after the sixth epoch. We sample about 50000 samples.

Table 6. The settings of γ. γ controls how many samples are identified as potential noisy samples.

	AP	AP_{50}	AP_{75}
$\gamma = 4.0$	37.5	64.0	37.8
$\gamma = 5.0$	38.0	**64.9**	39.7
$\gamma = 6.0$	**38.1**	64.5	**40.0**
$\gamma = 7.0$	38.0	64.7	40.0
$\gamma = 8.0$	37.2	62.6	38.1

Hyper-Parameter Analysis: In our method, γ is a hyper-parameter that need be discussed. Positive samples whose loss values $l_{ce} > \gamma$ are named as potential noisy samples. We observe that for a sample, if $p_j = 0.01$ and $q_j = 1$, cross entropy loss $l_{ce} = 4.5052$. Meanwhile, if $p_j = 0.9$, $l_{ce} = 0.1054$. Hence, we think that γ must satisfy $\gamma \geq 5.0$ to identify noisy samples, which also fits statistics shown in Fig. 5. Several experiments are conducted about γ in Table 6. In our setting, we set $\gamma = 6.0$ for all datasets and all noise rates.

6 Conclusion

In this paper, we propose a novel method to effectively train an instance segmentation model whose performance is robust under noisy supervision. Our key strategy is to use different losses describing different roles of noisy class labels. Based on this, correct gradient information is fully exploited for the foreground-background sub-task and incorrect guidance provided by noisy samples is avoided for the foreground-instance sub-task. We have conducted sufficient experiments on three well-known datasets (i.e., Pascal VOC, Cityscapes and COCO). The results show the superiority of our method in various noisy class labels scenarios.

Acknowledgement. This work was supported in part by National Natural Science Foundation of China (No.61831005, 61525102, 61871087 and 61971095).

References

1. Arpit, D., et al.: A closer look at memorization in deep networks. In: ICML (2017)
2. Box, G.E.P., Cox, D.R.: An analysis of transformations. J. Roy. Stat. Soc.: Ser. B (Methodol.) **26**(2), 211–243 (1964)
3. Chen, K., et al.: Hybrid task cascade for instance segmentation. In: CVPR (2019)
4. Chen, K., et al.: MMdetection: open mmlab detection toolbox and benchmark. arXiv preprint arXiv:1906.07155 (2019)
5. Cordts, M., et al.: The cityscapes dataset for semantic urban scene understanding. In: Proceedings of the IEEE Conference on Computer Vision and Pattern Recognition, pp. 3213–3223 (2016)
6. De Brabandere, B., Neven, D., Van Gool, L.: Semantic instance segmentation with a discriminative loss function. arXiv preprint arXiv:1708.02551 (2017)
7. Everingham, M., Van Gool, L., Williams, C.K., Winn, J., Zisserman, A.: The pascal visual object classes (VOC) challenge. Int. J. Comput. Vision **88**(2), 303–338 (2010)
8. Ghosh, A., Kumar, H., Sastry, P.: Robust loss functions under label noise for deep neural networks. In: Thirty-First AAAI Conference on Artificial Intelligence (2017)
9. Ghosh, A., Manwani, N., Sastry, P.: Making risk minimization tolerant to label noise. Neurocomputing **160**, 93–107 (2015)
10. Girshick, R.: Fast R-CNN. In: Proceedings of the IEEE International Conference on Computer Vision, pp. 1440–1448 (2015)
11. Gygli, M., Ferrari, V.: Fast object class labelling via speech. In: Proceedings of the IEEE Conference on Computer Vision and Pattern Recognition, pp. 5365–5373 (2019)
12. Han, B., et al.: Masking: a new perspective of noisy supervision. In: Advances in Neural Information Processing Systems, pp. 5836–5846 (2018)
13. Han, B., et al.: Co-teaching: robust training of deep neural networks with extremely noisy labels. In: Advances in Neural Information Processing Systems, pp. 8527–8537 (2018)
14. He, K., Gkioxari, G., Dollár, P., Girshick, R.: Mask R-CNN. In: 2017 IEEE International Conference on Computer Vision (ICCV), pp. 2980–2988. IEEE (2017)
15. Huang, Z., Huang, L., Gong, Y., Huang, C., Wang, X.: Mask scoring R-CNN. In: Proceedings of the IEEE Conference on Computer Vision and Pattern Recognition, pp. 6409–6418 (2019)
16. Jiang, L., Zhou, Z., Leung, T., Li, L.J., Fei-Fei, L.: Mentornet: learning data-driven curriculum for very deep neural networks on corrupted labels. arXiv preprint arXiv:1712.05055 (2017)
17. Lee, K.H., He, X., Zhang, L., Yang, L.: Cleannet: transfer learning for scalable image classifier training with label noise. In: Proceedings of the IEEE Conference on Computer Vision and Pattern Recognition, pp. 5447–5456 (2018)
18. Li, Y., Yang, J., Song, Y., Cao, L., Luo, J., Li, L.J.: Learning from noisy labels with distillation. In: Proceedings of the IEEE International Conference on Computer Vision, pp. 1910–1918 (2017)
19. Lin, T.-Y., et al.: Microsoft COCO: common objects in context. In: Fleet, D., Pajdla, T., Schiele, B., Tuytelaars, T. (eds.) ECCV 2014. LNCS, vol. 8693, pp. 740–755. Springer, Cham (2014). https://doi.org/10.1007/978-3-319-10602-1_48
20. Liu, Y., et al.: Affinity derivation and graph merge for instance segmentation. In: Ferrari, V., Hebert, M., Sminchisescu, C., Weiss, Y. (eds.) ECCV 2018. LNCS, vol. 11207, pp. 708–724. Springer, Cham (2018). https://doi.org/10.1007/978-3-030-01219-9_42

21. Ma, X., et al.: Dimensionality-driven learning with noisy labels. arXiv preprint arXiv:1806.02612 (2018)
22. Patrini, G., Rozza, A., Krishna Menon, A., Nock, R., Qu, L.: Making deep neural networks robust to label noise: a loss correction approach. In: Proceedings of the IEEE Conference on Computer Vision and Pattern Recognition, pp. 1944–1952 (2017)
23. Pereyra, G., Tucker, G., Chorowski, J., Kaiser, Ł., Hinton, G.: Regularizing neural networks by penalizing confident output distributions. arXiv preprint arXiv:1701.06548 (2017)
24. Shao, S., et al.: Objects365: a large-scale, high-quality dataset for object detection. In: Proceedings of the IEEE International Conference on Computer Vision, pp. 8430–8439 (2019)
25. Szegedy, C., Vanhoucke, V., Ioffe, S., Shlens, J., Wojna, Z.: Rethinking the inception architecture for computer vision. In: Proceedings of the IEEE Conference on Computer Vision and Pattern Recognition, pp. 2818–2826 (2016)
26. Tanaka, D., Ikami, D., Yamasaki, T., Aizawa, K.: Joint optimization framework for learning with noisy labels. In: Proceedings of the IEEE Conference on Computer Vision and Pattern Recognition, pp. 5552–5560 (2018)
27. Vahdat, A.: Toward robustness against label noise in training deep discriminative neural networks. In: Advances in Neural Information Processing Systems, pp. 5596–5605 (2017)
28. Veit, A., Alldrin, N., Chechik, G., Krasin, I., Gupta, A., Belongie, S.: Learning from noisy large-scale datasets with minimal supervision. In: Proceedings of the IEEE Conference on Computer Vision and Pattern Recognition, pp. 839–847 (2017)
29. Wang, Y., Ma, X., Chen, Z., Luo, Y., Yi, J., Bailey, J.: Symmetric cross entropy for robust learning with noisy labels. In: Proceedings of the IEEE International Conference on Computer Vision, pp. 322–330 (2019)
30. Xiao, T., Xia, T., Yang, Y., Huang, C., Wang, X.: Learning from massive noisy labeled data for image classification. In: Proceedings of the IEEE Conference on Computer Vision and Pattern Recognition, pp. 2691–2699 (2015)
31. Yu, X., Han, B., Yao, J., Niu, G., Tsang, I., Sugiyama, M.: How does disagreement help generalization against label corruption? In: International Conference on Machine Learning, pp. 7164–7173 (2019)
32. Zhang, C., Bengio, S., Hardt, M., Recht, B., Vinyals, O.: Understanding deep learning requires rethinking generalization. In: ICLR (2017)
33. Zhang, Z., Sabuncu, M.: Generalized cross entropy loss for training deep neural networks with noisy labels. In: Advances in Neural Information Processing Systems, pp. 8778–8788 (2018)

Deep Image Clustering
with Category-Style Representation

Junjie Zhao[1], Donghuan Lu[2], Kai Ma[2], Yu Zhang[1(✉)], and Yefeng Zheng[2(✉)]

[1] School of Computer Science and Engineering, Southeast University, Nanjing, China
{kamij.zjj,zhang_yu}@seu.edu.cn
[2] Tencent Jarvis Lab, Shenzhen, China
{caleblu,kylekma,yefengzheng}@tencent.com

Abstract. Deep clustering which adopts deep neural networks to obtain optimal representations for clustering has been widely studied recently. In this paper, we propose a novel deep image clustering framework to learn a category-style latent representation in which the category information is disentangled from image style and can be directly used as the cluster assignment. To achieve this goal, mutual information maximization is applied to embed relevant information in the latent representation. Moreover, augmentation-invariant loss is employed to disentangle the representation into category part and style part. Last but not least, a prior distribution is imposed on the latent representation to ensure the elements of the category vector can be used as the probabilities over clusters. Comprehensive experiments demonstrate that the proposed approach outperforms state-of-the-art methods significantly on five public datasets (Project address: https://github.com/sKamiJ/DCCS).

Keywords: Image clustering · Deep learning · Unsupervised learning

1 Introduction

Clustering is a widely used technique in many fields, such as machine learning, data mining and statistical analysis. It aims to group objects 'similar' to each other into the same set and 'dissimilar' ones into different sets. Unlike supervised learning methods, clustering approaches should be oblivious to ground truth labels. Conventional methods, such as K-means [23] and spectral clustering [25], require feature extraction to convert data to a more discriminative form. Domain knowledge could be useful to determine more appropriate feature extraction strategies in some cases. But for many high-dimensional problems

---

J. Zhao and D. Lu—Equal contribution and the work was done at Tencent Jarvis Lab.

Electronic supplementary material The online version of this chapter (https://doi.org/10.1007/978-3-030-58568-6_4) contains supplementary material, which is available to authorized users.

© Springer Nature Switzerland AG 2020
A. Vedaldi et al. (Eds.): ECCV 2020, LNCS 12359, pp. 54–70, 2020.
https://doi.org/10.1007/978-3-030-58568-6_4

(*c.g.* images), manually designed feature extraction methods can easily lead to inferior performance.

Because of the powerful capability of deep neural networks to learn non-linear mapping, a lot of deep learning based clustering methods have been proposed recently. Many studies attempt to combine deep neural networks with various kinds of clustering losses [9,13,33] to learn more discriminative yet low-dimensional latent representations. To avoid trivially learning some arbitrary representations, most of those methods also minimize a reconstruction [13] or generative [24] loss as an additional regularization. However, there is no substantial connection between the discriminative ability and the generative ability of the latent representation. The aforementioned regularization turns out to be less relevant to clustering and forces the latent representation to contain unnecessary generative information, which makes the network hard to train and could also affect the clustering performance.

In this paper, instead of using a decoder/generator to minimize the reconstruction/generative loss, we use a discriminator to maximize the mutual information [14] between input images and their latent representations in order to retain discriminative information. To further reduce the effect of irrelevant information, the latent representation is divided into two parts, *i.e.*, the category (or cluster) part and the style part, where the former one contains the distinct identities of images (inter-class difference) while the latter one represents style information (intra-class difference). Specifically, we propose to use data augmentation to disentangle the category representation from style information, based on the observation [16,31] that appropriate augmentation should not change the image category.

Moreover, many deep clustering methods require additional operations [29, 33] to group the latent representation into different categories. But their distance metrics are usually predefined and may not be optimal. In this paper, we impose a prior distribution [24] on the latent representation to make the category part closer to the form of a one-hot vector, which can be directly used to represent the probability distribution of the clusters.

In summary, we propose a novel approach, **D**eep **C**lustering with **C**ategory-**S**tyle representation (DCCS) for unsupervised image clustering. The main contributions of this study are four folds:

- We propose a novel end-to-end deep clustering framework to learn a latent category-style representation whose values can be used directly for the cluster assignment.
- We show that maximizing the mutual information is enough to prevent the network from learning arbitrary representations in clustering.
- We propose to use data augmentation to disentangle the category representation (inter-class difference) from style information (intra-class difference).
- Comprehensive experiments demonstrate that the proposed DCCS approach outperforms state-of-the-art methods on five commonly used datasets, and the effectiveness of each part of the proposed method is evaluated and discussed in thorough ablation studies.

2   Related Work

In recent years, many deep learning based clustering methods have been proposed. Most approaches [9,13,36] combined autoencoder [2] with traditional clustering methods by minimizing reconstruction loss as well as clustering loss. For example, Jiang et al. [17] combined a variational autoencoder (VAE) [19] for representation learning with a Gaussian mixture model for clustering. Yang et al. [35] also adopted the Gaussian mixture model as the prior in VAE, and incorporated a stochastic graph embedding to handle data with complex spread. Although the usage of the reconstruction loss can embed the sample information into the latent space, the encoded latent representation may not be optimal for clustering.

Other than autoencoder, Generative Adversarial Network (GAN) [10] has also been employed for clustering [5]. In ClusterGAN [24], Mukherjee et al. also imposed a prior distribution on the latent representation, which was a mixture of one-hot encoded variables and continuous latent variables. Although their representations share a similar form to ours, their one-hot variables cannot be used as the cluster assignment directly due to the lack of proper disentanglement. Additionally, ClusterGAN consisted of a generator (or a decoder) to map the random variables from latent space to image space, a discriminator to ensure the generated samples close to real images and an encoder to map the images back to the latent space to match the random variables. Such a GAN model is known to be hard to train and brings irrelevant generative information to the latent representations. To reduce the complexity of the network and avoid unnecessary generative information, we directly train an encoder by matching the aggregated posterior distribution of the latent representation to the prior distribution.

To avoid the usage of additional clustering, some methods directly encoded images into latent representations whose elements can be treated as the probabilities over clusters. For example, Xu et al. [16] maximized pair-wise mutual information of the latent representations extracted from an image and its augmented version. This method achieved good performance on both image clustering and segmentation, but its batch size must be large enough (more than 700 in their experiments) so that samples from different clusters were almost equally distributed in each batch. Wu et al. [31] proposed to learn one-hot representations by exploring various correlations of the unlabeled data, such as local robustness and triple mutual information. However, their computation of mutual information required pseudo-graph to determine whether images belonged to the same category, which may not be accurate due to the unsupervised nature of clustering. To avoid this issue, we maximized the mutual information between images and their own representations instead of representations encoded from images with the same predicted category.

3   Method

As stated in the introduction, the proposed DCCS approach aims to find an appropriate encoder Q to convert the input image X into a latent representa-

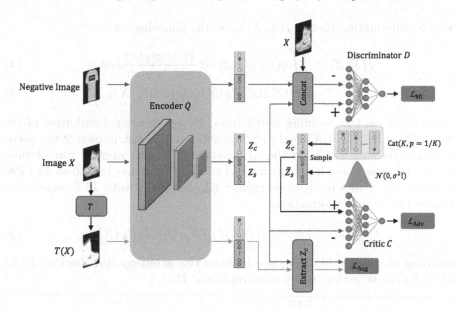

Fig. 1. The overall framework of the proposed DCCS method. The encoder Q converts the image X into a latent representation $Z = (Z_c, Z_s)$. The discriminator D maximizes the mutual information between X and Z, while the critic C imposes a prior distribution on Z to make Z_c closer to the form of a one-hot vector and constrain the range of Z_s. Z_c is also regularized to be invariant to data augmentation T

tion Z, which consists of disentangled category and style information. To be more precise, the encoded latent representation Z consists of a softmax-activated vector Z_c and a linear-activated vector Z_s, $i.e.$, $Z = (Z_c, Z_s)$, where Z_c represents the probabilities of X belonging to each class and Z_s represents the intra-class style information. To achieve this, three regularization strategies are applied to constrain the latent representation as detailed in the following three sections, and the framework is shown in Fig. 1. To clarify notations, we use upper case letters ($e.g.$ X) for random variables, lower case letters ($e.g.$ x) for their values and calligraphic letters ($e.g.$ $\mathcal{X}$) for sets. The probability distributions are denoted with upper case letters ($e.g.$ $P(X)$), and the corresponding densities are denoted with lower case letters ($e.g.$ $p(x)$).

3.1 Maximize Mutual Information

Because of the powerful capability to fit training data with complex non-linear transformations, the encoder of deep neural networks can easily map input images to arbitrary representations if without proper constraints thus losing relevant information for proceeding the target clustering task. To retain the essential information of each image and learn better discriminative latent representations, a discriminator D is introduced to maximize the mutual information $I(X, Z)$ between the input image X and its encoded latent representation Z.

Based on information theory, $I(X, Z)$ takes the following form:

$$I(X, Z) = \iint q(z|x)p_X(x) \log \frac{q(z|x)p_X(x)}{q_Z(z)p_X(x)} dxdz \tag{1}$$

$$= \text{KL}(Q(Z|X)P_X(X)\|Q_Z(Z)P_X(X)) \tag{2}$$

where $Q(Z|X)$ is the encoding distribution, P_X is the prior distribution of the images, $Q_Z = \mathbb{E}_{P_X}[Q(Z|X)]$ is the aggregated posterior distribution of the latent representation and $\text{KL}(\cdot\|\cdot)$ is the KL-divergence. In this original formulation, however, KL-divergence is unbounded and maximizing it may lead to an unstable training result. Following [14], we replace KL-divergence with JS-divergence to estimate the mutual information:

$$I^{(\text{JSD})}(X, Z) = \text{JS}(Q(Z|X)P_X(X), Q_Z(Z)P_X(X)). \tag{3}$$

According to [10,27], JS-divergence between two arbitrary distributions $P(X)$ and $Q(X)$ can be estimated by a discriminator D:

$$\text{JS}(P(X), Q(X)) = \frac{1}{2} \max_D \{\mathbb{E}_{X \sim P(X)}[\log S(D(X))] \\ + \mathbb{E}_{X \sim Q(X)}[\log(1 - S(D(X)))]\} + \log 2 \tag{4}$$

where S is the sigmoid function. Replacing $P(X)$ and $Q(X)$ with $Q(Z|X)P_X(X)$ and $Q_Z(Z)P_X(X)$, the mutual information can be maximized by:

$$\frac{1}{2} \max_{Q,D} \{\mathbb{E}_{(X,Z) \sim Q(Z|X)P_X(X)}[\log S(D(X, Z))] \\ + \mathbb{E}_{(X,Z) \sim Q_Z(Z)P_X(X)}[\log(1 - S(D(X, Z)))]\} + \log 2. \tag{5}$$

Accordingly, the mutual information loss function can be defined as:

$$\mathcal{L}_{\text{MI}} = -(\mathbb{E}_{(X,Z) \sim Q(Z|X)P_X(X)}[\log S(D(X, Z))] \\ + \mathbb{E}_{(X,Z) \sim Q_Z(Z)P_X(X)}[\log(1 - S(D(X, Z)))]) \tag{6}$$

where Q and D are jointly optimized.

With the concatenation of X and Z as input, minimizing $\mathcal{L}_{\text{MI}}$ can be interpreted as to determine whether X and Z are correlated. For discriminator D, an image X along with its representation is a positive sample while X along with a representation encoded from another image is a negative sample. As aforementioned, many deep clustering methods use the reconstruction loss or generative loss to avoid arbitrary representations. However, it allows the encoded representation to contain unnecessary generative information and makes the network, especially GAN, hard to train. The mutual information maximization only instills necessary discriminative information into the latent space and experiments in Sect. 4 confirm that it leads to better performance.

3.2   Disentangle Category-Style Information

As previously stated, we expect the latent category representation Z_c only contains the categorical cluster information while all the style information is represented by Z_s. To achieve such a disentanglement, an augmentation-invariant regularization term is introduced based on the observation that certain augmentation should not change the category of images.

Specifically, given an augmentation function T which usually includes geometric transformations (*e.g.* scaling and flipping) and photometric transformations (*e.g.* changing brightness or contrast), Z_c and Z'_c encoded from X and $T(X)$ should be identical while all the appearance differences should be represented by the style variables. Because the elements of Z_c represent the probabilities over clusters, the KL-divergence is adopted to measure the difference between $Q(Z_c|X)$ and $Q(Z_c|T(X))$. The augmentation-invariant loss function for the encoder Q can be defined as:

$$\mathcal{L}_{\text{Aug}} = \text{KL}(Q(Z_c|X)\|Q(Z_c|T(X))). \tag{7}$$

3.3   Match to Prior Distribution

There are two potential issues with the aforementioned regularization terms: the first one is that the category representation cannot be directly used as the cluster assignment, therefore additional operations are still required to determine the clustering categories; the second one is that the augmentation-invariant loss may lead to a degenerate solution, *i.e.*, assigning all images into a few clusters, or even the same cluster. In order to resolve these issues, a prior distribution P_Z is imposed on the latent representation Z.

Following [24], a categorical distribution $\tilde{Z}_c \sim \text{Cat}(K, p = 1/K)$ is imposed on Z_c, where $\tilde{Z}_c$ is a one-hot vector and K is the number of categories that the images should be clustered into. A Gaussian distribution $\tilde{Z}_s \sim \mathcal{N}(0, \sigma^2 \mathbf{I})$ (typically $\sigma = 0.1$) is imposed on Z_s to constrain the range of style variables.

As aforementioned, ClusterGAN [24] uses the prior distribution to generate random variables, applies a GAN framework to train a proper decoder and then learns an encoder to match the decoder. To reduce the complexity of the network and avoid unnecessary generative information, we directly train the encoder by matching the aggregated posterior distribution $Q_Z = \mathbb{E}_{P_X}[Q(Z|X)]$ to the prior distribution P_Z. Experiments demonstrate that such a strategy can lead to better clustering performance.

To impose the prior distribution P_Z on Z, we minimize the Wasserstein distance [1] $W(Q_Z, P_Z)$ between Q_Z and P_Z, which can be estimated by:

$$\max_{C \in \mathcal{C}} \{\mathbb{E}_{\tilde{Z} \sim P_Z}[C(\tilde{Z})] - \mathbb{E}_{Z \sim Q_Z}[C(Z)]\} \tag{8}$$

where $\mathcal{C}$ is the set of 1-Lipschitz functions. Under the optimal critic C (denoted as *discriminator* in vanilla GAN), minimizing Eq. (8) with respect to the encoder parameters also minimizes $W(Q_Z, P_Z)$:

$$\min_{Q} \max_{C \in \mathcal{C}} \{\mathbb{E}_{\tilde{Z} \sim P_Z}[C(\tilde{Z})] - \mathbb{E}_{Z \sim Q_Z}[C(Z)]\}. \tag{9}$$

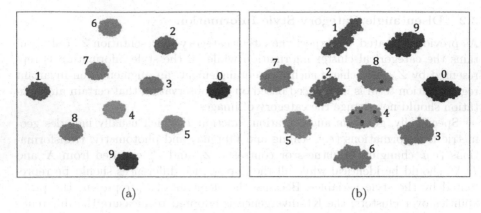

(a)   (b)

Fig. 2. The t-SNE visualization of the latent representations of MNIST dataset. The dimensions of Z_c and Z_s are set as 10 and 50, respectively. Here, (a) shows the prior representation $\tilde{Z}$ sampled from P_Z, the numbers 0–9 represent different categories, while (b) demonstrates the encoded representation Z. Each point represents a latent representation and the color refers to its ground truth label (Color figure online)

For optimization, the gradient penalty [12] is introduced to enforce the Lipschitz constraint on the critic. The adversarial loss functions for the encoder Q and the critic C can be defined as:

$$\mathcal{L}_{\text{Adv}}^Q = -\mathbb{E}_{Z \sim Q_Z}[C(Z)] \tag{10}$$

$$\mathcal{L}_{\text{Adv}}^C = \mathbb{E}_{Z \sim Q_Z}[C(Z)] - \mathbb{E}_{\tilde{Z} \sim P_Z}[C(\tilde{Z})] + \lambda \mathbb{E}_{\hat{Z} \sim P_{\hat{Z}}}[(\|\nabla_{\hat{Z}} C(\hat{Z})\|_2 - 1)^2] \tag{11}$$

where λ is the gradient penalty coefficient, $\hat{Z}$ is the latent representation sampled uniformly along the straight lines between pairs of latent representations sampled from Q_Z and P_Z, and $(\|\nabla_{\hat{Z}} C(\hat{Z})\|_2 - 1)^2$ is the one-centered gradient penalty. Q and C are optimized alternatively.

Note that the reason why we use Wasserstein distance instead of f-divergence is that Wasserstein distance is continuous everywhere and differentiable almost everywhere. Such a critic is able to provide meaningful gradients for the encoder even with an input containing discrete variables. On the other hand, the loss of the critic can be viewed as an estimation of $W(Q_Z, P_Z)$ to determine whether the clustering progress has converged or not, as shown in Sect. 4.2.

Figure 2a shows the t-SNE [22] visualization of the prior representation $\tilde{Z} = (\tilde{Z}_c, \tilde{Z}_s)$ with points being colored based on $\tilde{Z}_c$. It shows that the representations sampled from the prior distribution can be well clustered based on $\tilde{Z}_c$ while $\tilde{Z}_s$ represents the intra-class difference. After imposing the prior distribution on Z as displayed in Fig. 2b, the encoded latent representations show a similar pattern as the prior representations, therefore the cluster assignment can be easily achieved by using argmax over Z_c.

Algorithm 1. Deep Clustering with Category-Style Representation

---

Input: Dataset $\mathcal{X} = \{x^i\}_{i=1}^N$, $\theta_Q, \theta_D, \theta_C$ initial parameters of encoder Q, discriminator D and critic C, the dimensions of Z_c and Z_s, hyper-parameters σ, λ, β_{MI}, β_{Aug}, β_{Adv}, augmentation function T, the number of critic iterations per encoder iteration n_{critic}, batch size m.

1: **while** $\mathcal{L}^C$ not converged **do**
2:     **for** $t = 1, ..., n_{\text{critic}}$ **do**
3:         Sample $\{x^i\}_{i=1}^m$ from $\mathcal{X}$, $\{\tilde{z}^i\}_{i=1}^m$ from P_Z, $\{\epsilon^i\}_{i=1}^m$ from $U[0,1]$;
4:         Sample z^i from $Q(Z|X = x^i)$ for $i = 1, ..., m$;
5:         Compute $\hat{z}^i = \epsilon^i z^i + (1 - \epsilon^i) \tilde{z}^i$ for $i = 1, ..., m$;
6:         Update θ_C by minimizing $\mathcal{L}^C$ (Eq. 14);
7:     **end for**
8:     Sample $\{x^i\}_{i=1}^m$ from $\mathcal{X}$;
9:     Sample $z'^i = (z'^i_c, z'^i_s)$ from $Q(Z|X = T(x^i))$ for $i = 1, ..., m$;
10:    Sample $z^i = (z^i_c, z^i_s)$ from $Q(Z|X = x^i)$ for $i = 1, ..., m$;
11:    Sample z^j from $\{z^i\}_{i=1}^m$ for each x^i to form negative paris;
12:    Update θ_Q and θ_D by minimizing $\mathcal{L}^Q$ (Eq. 12) and $\mathcal{L}^D$ (Eq. 13);
13: **end while**
14: **for** $i = 1, ..., N$ **do**
15:    Sample $z^i = (z^i_c, z^i_s)$ from $Q(Z|X = x^i)$;
16:    Compute cluster assignment $l^i = \text{argmax}(z^i_c)$;
17: **end for**
Output: Cluster assignment $\{l^i\}_{i=1}^N$.

3.4 The Unified Model

As shown in Fig. 1, the network of DCCS consists of three parts: the encoder Q to convert images into latent representations, the discriminator D to maximize the mutual information and the critic C to impose the prior distribution. The objective functions for encoder Q, discriminator D and critic C are defined as:

$$\mathcal{L}^Q = \beta_{\text{MI}}\mathcal{L}_{\text{MI}} + \beta_{\text{Aug}}\mathcal{L}_{\text{Aug}} + \beta_{\text{Adv}}\mathcal{L}^Q_{\text{Adv}} \tag{12}$$

$$\mathcal{L}^D = \beta_{\text{MI}}\mathcal{L}_{\text{MI}} \tag{13}$$

$$\mathcal{L}^C = \beta_{\text{Adv}}\mathcal{L}^C_{\text{Adv}} \tag{14}$$

where β_{MI}, β_{Aug} and β_{Adv} are the weights used to balance each term.

As described in Algorithm 1, the parameters of Q and D are jointly updated while the parameters of C are trained separately. Note that because Q is a deterministic encoder, $i.e.$, $Q(Z|X = x) = \delta_{\mu(x)}$, where δ denotes Dirac-delta and $\mu(x)$ is a deterministic mapping function, sampling z^i from $Q(Z|X = x^i)$ is equivalent to assign z^i with $\mu(x^i)$.

4 Experiments

4.1 Experimental Settings

Datasets. We evaluate the proposed DCCS on five commonly used datasets, including MNIST [21], Fashion-MNIST [32], CIFAR-10 [20], STL-10 [6] and ImageNet-10 [4]. The statistics of these datasets are described in Table 1. As a widely adopted setting [4,31,33], the training and test sets of these datasets are jointly utilized. For STL-10, the unlabelled subset is not used. For ImageNet-10, images are selected from the ILSVRC2012 1K dataset [7] the same as in [4] and resized to 96×96 pixels. Similar to the IIC approach [16], color images are converted to grayscale to discourage clustering based on trivial color cues.

Table 1. Statistics of the datasets

Dataset	Images	Clusters	Image size
MNIST [21]	70000	10	28×28
Fashion-MNIST [32]	70000	10	28×28
CIFAR-10 [20]	60000	10	$32 \times 32 \times 3$
STL-10 [6]	13000	10	$96 \times 96 \times 3$
ImageNet-10 [4]	13000	10	$96 \times 96 \times 3$

Evaluation Metrics. Three widely used metrics are applied to evaluate the performance of the clustering methods, including unsupervised clustering accuracy (ACC), normalized mutual information (NMI), and adjusted rand index (ARI) [31]. For these metrics, a higher score implies better performance.

Implementation Details. The architectures of encoders are similar to [12,24] with a different number of layers and units being used for different sizes of images. The critic and discriminator are multi-layer perceptions. All the parameters are randomly initialized without pretraining. The Adam [18] optimizer with a learning rate of 10^{-4} and $\beta_1 = 0.5$, $\beta_2 = 0.9$ is used for optimization. The dimension of Z_s is set to 50, and the dimension of Z_c is set to the expected number of clusters. For other hyper-parameters, we set the standard deviation of prior Gaussian distribution $\sigma = 0.1$, the gradient penalty coefficient $\lambda = 10$, $\beta_{MI} = 0.5$, $\beta_{Adv} = 1$, the number of critic iterations per encoder iteration $n_{critic} = 4$, and batch size $m = 64$ for all datasets. Because β_{Aug} is related to the datasets and generally the more complex the images are, the larger β_{Aug} should be. We come up with an applicable way to set β_{Aug} by visualizing the t-SNE figure of the encoded representation Z, $i.e.$, β_{Aug} is gradually increased until the clusters visualized by t-SNE start to overlap. With this method, β_{Aug} is set to 2 for MNIST and Fashion-MNIST, and set to 4 for other datasets. The data augmentation includes

Table 2. Comparison with baseline and state-of-the-art methods on MNIST and Fashion-MNIST. The best three results of each metric are highlighted in **bold**. $\star$: Re-implemented results with the released code

Method	MNIST			Fashion-MNIST		
	ACC	NMI	ARI	ACC	NMI	ARI
K-means [30]	0.572	0.500	0.365	0.474	0.512	0.348
SC [37]	0.696	0.663	0.521	0.508	0.575	-
AC [11]	0.695	0.609	0.481	0.500	0.564	0.371
NMF [3]	0.545	0.608	0.430	0.434	0.425	-
DEC [33]	0.843	0.772	0.741	0.590*	0.601*	0.446*
JULE [34]	0.964	0.913	0.927	0.563	0.608	-
VaDE [17]	0.945	0.876	-	0.578	0.630	-
DEPICT [9]	0.965	0.917	-	0.392	0.392	-
IMSAT [15]	**0.984**	**0.956***	**0.965***	**0.736***	**0.696***	**0.609***
DAC [4]	0.978	0.935	0.949	0.615*	0.632*	0.502*
SpectralNet [29]	0.971	0.924	0.936*	0.533*	0.552*	-
ClusterGAN [24]	0.950	0.890	0.890	0.630	0.640	0.500
DLS-Clustering [8]	0.975	0.936	-	0.693	0.669	-
DualAE [36]	0.978	0.941	-	0.662	0.645	-
RTM [26]	0.968	0.933	0.932	0.710	0.685	0.578
NCSC [38]	0.941	0.861	0.875	**0.721**	**0.686**	**0.592**
IIC [16]	**0.992**	**0.978***	**0.983***	0.657*	0.637*	0.523*
DCCS (Proposed)	**0.989**	**0.970**	**0.976**	**0.756**	**0.704**	**0.623**

four commonly used approaches, *i.e.*, random cropping, random horizontal flipping, color jittering and channel shuffling (which is used on the color images before graying). For more details about network architectures, data augmentation and hyper-parameters, please refer to the supplementary materials.

4.2   Main Result

Quantitative Comparison. We first compare the proposed DCCS with several baseline methods as well as other state-of-the-art clustering approaches based on deep learning, as shown in Table 2 and Table 3. DCCS outperforms all the other methods by large margins on Fashion-MNIST, CIFAR-10, STL-10 and ImageNet-10. For the ACC metric, DCCS is 2.0%, 3.3%, 3.7% and 2.7% higher than the second best methods on these four datasets, respectively. Although for MNIST, the clustering accuracy of DCCS is slightly lower (*i.e.*, 0.3%) than IIC [16], DCCS significantly surpasses IIC on CIFAR-10 and STL-10.

Training Progress. The training progress of the proposed DCCS is monitored by minimizing the Wasserstein distance $W(Q_Z, P_Z)$, which can be estimated by the negative critic loss $-\mathcal{L}^C$. As plotted in Fig. 3, the critic loss stably converges and it correlates well with the clustering accuracy, demonstrating a robust training progress. The visualizations of the latent representations with t-SNE at three different stages are also displayed in Fig. 3. From stage A to C, the latent representations gradually cluster together while the critic loss decreases steadily.

Qualitative Analysis. Figure 4 shows images with top 10 predicted probabilities from each cluster in MNIST and ImageNet-10. Each row corresponds to a cluster and the images from left to right are sorted in a descending order based on their probabilities. In each row of Fig. 4a, the same digits are written in different ways, indicating that Z_c contains well disentangled category information. For ImageNet-10 in Fig. 4b, most objects are well clustered and the major confusion is for the airships and airplanes in the sky due to their similar shapes and backgrounds (Row 8). A possible solution is overclustering, *i.e.*, more number of clusters than expected, which requires investigation in future work.

Table 3. Comparison with baseline and state-of-the-art methods on CIFAR-10, STL-10 and ImageNet-10. The best three results of each metric are highlighted in **bold**. ⋆: Re-implemented results with the released code. †: The results are evaluated on STL-10 without using the unlabelled data subset

Method	CIFAR-10			STL-10			ImageNet-10		
	ACC	NMI	ARI	ACC	NMI	ARI	ACC	NMI	ARI
K-means [30]	0.229	0.087	0.049	0.192	0.125	0.061	0.241	0.119	0.057
SC [37]	0.247	0.103	0.085	0.159	0.098	0.048	0.274	0.151	0.076
AC [11]	0.228	0.105	0.065	0.332	0.239	0.140	0.242	0.138	0.067
NMF [3]	0.190	0.081	0.034	0.180	0.096	0.046	0.230	0.132	0.065
AE [2]	0.314	0.239	0.169	0.303	0.250	0.161	0.317	0.210	0.152
GAN [28]	0.315	0.265	0.176	0.298	0.210	0.139	0.346	0.225	0.157
VAE [19]	0.291	0.245	0.167	0.282	0.200	0.146	0.334	0.193	0.168
DEC [33]	0.301	0.257	0.161	0.359	0.276	0.186	0.381	0.282	0.203
JULE [34]	0.272	0.192	0.138	0.277	0.182	0.164	0.300	0.175	0.138
DAC [4]	0.522	0.396	0.306	0.470	0.366	0.257	**0.527**	**0.394**	**0.302**
IIC [16]	**0.617**	**0.513**⋆	**0.411**⋆	**0.499**†	**0.431**⋆†	**0.295**⋆†	-	-	-
DCCM [31]	**0.623**	0.496	0.408	0.482	0.376	0.262	**0.710**	**0.608**	**0.555**
DCCS (Proposed)	**0.656**	**0.569**	**0.469**	**0.536**	**0.490**	**0.362**	**0.737**	**0.640**	**0.560**

Table 4. Evaluation of different ways for the cluster assignment

Method	MNIST			Fashion-MNIST			CIFAR-10			STL-10		
	ACC	NMI	ARI	ACC	NMI	ARI	ACC	NMI	ARI	ACC	NMI	ARI
Argmax over Z_c	**0.9891**	0.9696	**0.9758**	**0.7564**	0.7042	**0.6225**	**0.6556**	**0.5692**	0.4685	**0.5357**	**0.4898**	**0.3617**
K-means on Z	**0.9891**	0.9694	0.9757	**0.7564**	0.7043	0.6224	0.6513	0.5588	**0.4721**	0.5337	0.4888	0.3599
K-means on Z_c	**0.9891**	0.9696	0.9757	0.7563	0.7042	0.6223	0.6513	0.5587	**0.4721**	0.5340	0.4889	0.3602
K-means on Z_s	0.5164	0.4722	0.3571	0.4981	0.4946	0.3460	0.2940	0.1192	0.0713	0.4422	0.4241	0.2658

4.3   Ablation Study

Cluster Assignment w/o K-Means. As stated in Sect. 3.3, by imposing the prior distribution, the latent category representation Z_c can be directly used as the cluster assignment. Table 4 compares the results of several ways to obtain the cluster assignment with the same encoder. We can see that using Z_c with or without K-means has similar performance, indicating that Z_c is discriminative enough to be used as the cluster assignment directly. Additional experiments on each part of Z show that applying K-means on Z_c can yield similar performance as on Z, while the performance of applying K-means on Z_s is much worse. It demonstrates that the categorical cluster information and the style information are well disentangled as expected.

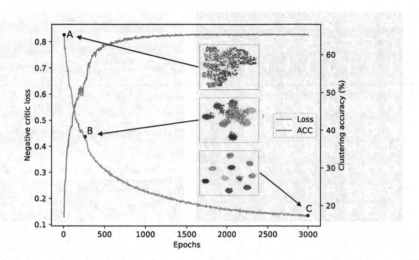

Fig. 3. Training curves of the negative critic loss and the clustering accuracy on CIFAR-10. The t-SNE visualizations of the latent representations Z for different stages are also displayed. The color of the points in the t-SNE visualizations refers to the ground truth category (Color figure online)

Table 5. Ablation study of DCCS on Fashion-MNIST and CIFAR-10

Method	Loss			Fashion-MNIST			CIFAR-10		
	$\mathcal{L}_{Adv}$	$\mathcal{L}_{MI}$	$\mathcal{L}_{Aug}$	ACC	NMI	ARI	ACC	NMI	ARI
M1	✓			0.618	0.551	0.435	0.213	0.076	0.040
M2	✓	✓		0.692	0.621	0.532	0.225	0.085	0.048
M3	✓		✓	0.725	0.694	0.605	0.645	0.557	0.463
M4	✓	✓	✓	**0.756**	**0.704**	**0.623**	**0.656**	**0.569**	**0.469**

Ablation Study on the Losses. The effectiveness of the losses is evaluated in Table 5. M1 is the baseline method, *i.e.*, the only constraint applied to the network is the prior distribution. This constraint is always necessary to ensure that the category representation can be directly used as the cluster assignment. By adding the mutual information maximization in M2 or the category-style information disentanglement in M3, the clustering performance achieves significant gains. The results of M4 demonstrate that combining all three losses can further improve the clustering performance. Note that large improvement with data augmentation for CIFAR-10 is due to that the images in CIFAR-10 have considerable intra-class variability, therefore disentangling the category-style information can improve the clustering performance by a large margin.

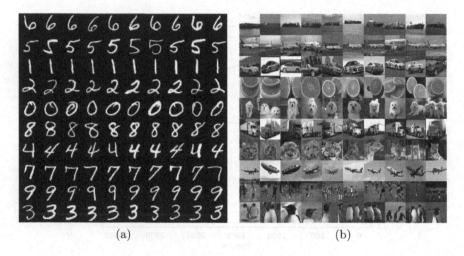

(a)   (b)

Fig. 4. Clustering images from MNIST (a) and ImageNet-10 (b). Each row contains the images with the highest probability to belong to the respective cluster

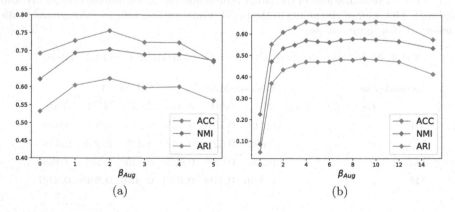

(a)   (b)

Fig. 5. The impact of β_{Aug} on Fashion-MNIST (a) and CIFAR-10 (b)

Impact of β_{Aug}. The clustering performance with different β_{Aug}, which is the weight of the data augmentation loss in Eq. (12), is displayed in Fig. 5. For Fashion-MNIST, the performance drops when β_{Aug} is either too small or too large because a small β_{Aug} cannot disentangle the style information enough, and a large β_{Aug} may lead the clusters to overlap. For CIFAR-10, the clustering performance is relatively stable with large β_{Aug}. As previously stated, the biggest β_{Aug} without overlapping clusters in the t-SNE visualization of the encoded representation Z is selected (the visualization of t-SNE can be found in the supplementary materials).

Impact of Z_s. As shown in Fig. 6, varying the dimension of Z_s from 10 to 70 does not affect the clustering performance much. However, when the dimension of Z_s is 0, *i.e.*, the latent representation only contains the category part, the performance drops a lot, demonstrating the necessity of the style representation. The reason is that for the mutual information maximization, the category representation alone is not enough to describe the difference among images belonging to the same cluster.

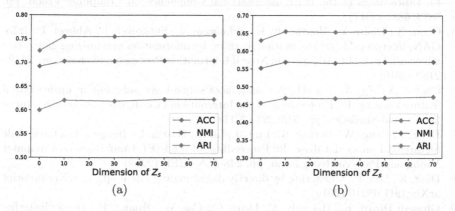

(a)   (b)

Fig. 6. The impact of Z_s on Fashion-MNIST (a) and CIFAR-10 (b)

5 Conclusions

In this work, we proposed a novel unsupervised deep image clustering framework with three regularization strategies. First, mutual information maximization was applied to retain essential information and avoid arbitrary representations. Furthermore, data augmentation was employed to disentangle the category representation from style information. Finally, a prior distribution was imposed to prevent degenerate solutions and avoid the usage of additional clustering so that the category representation could be used directly as the cluster assignment. Ablation studies demonstrated the effectiveness of each component and the extensive comparison experiments on five datasets showed that the proposed approach outperformed other state-of-the-art methods.

Acknowledgements. This work was supported by National Key Research and Development Program of China (No. 2018AAA0100100), National Natural Science Foundation of China (61702095), the Key Area Research and Development Program of Guangdong Province, China (No. 2018B010111001), National Key Research and Development Project (2018YFC2000702), Science and Technology Program of Shenzhen, China (No. ZDSYS201802021814180) and the Big Data Computing Center of Southeast University.

References

1. Arjovsky, M., Chintala, S., Bottou, L.: Wasserstein GAN. arXiv preprint arXiv:1701.07875 (2017)
2. Bengio, Y., Lamblin, P., Popovici, D., Larochelle, H.: Greedy layer-wise training of deep networks. In: Advances in Neural Information Processing Systems, pp. 153–160 (2007)
3. Cai, D., He, X., Wang, X., Bao, H., Han, J.: Locality preserving nonnegative matrix factorization. In: International Joint Conference on Artificial Intelligence (2009)
4. Chang, J., Wang, L., Meng, G., Xiang, S., Pan, C.: Deep adaptive image clustering. In: Proceedings of the IEEE International Conference on Computer Vision, pp. 5879–5887 (2017)
5. Chen, X., Duan, Y., Houthooft, R., Schulman, J., Sutskever, I., Abbeel, P.: InfoGAN: interpretable representation learning by information maximizing generative adversarial nets. In: Advances in Neural Information Processing Systems, pp. 2172–2180 (2016)
6. Coates, A., Ng, A., Lee, H.: An analysis of single-layer networks in unsupervised feature learning. In: Proceedings of the International Conference on Artificial Intelligence and Statistics, pp. 215–223 (2011)
7. Deng, J., Dong, W., Socher, R., Li, L.J., Li, K., Fei-Fei, L.: ImageNet: a large-scale hierarchical image database. In: Proceedings of the IEEE Conference on Computer Vision and Pattern Recognition, pp. 248–255. IEEE (2009)
8. Ding, F., Luo, F.: Clustering by directly disentangling latent space. arXiv preprint arXiv:1911.05210 (2019)
9. Ghasedi Dizaji, K., Herandi, A., Deng, C., Cai, W., Huang, H.: Deep clustering via joint convolutional autoencoder embedding and relative entropy minimization. In: Proceedings of the IEEE International Conference on Computer Vision, pp. 5736–5745 (2017)
10. Goodfellow, I., et al.: Generative adversarial nets. In: Advances in Neural Information Processing Systems, pp. 2672–2680 (2014)
11. Gowda, K.C., Krishna, G.: Agglomerative clustering using the concept of mutual nearest neighbourhood. Pattern Recogn. **10**(2), 105–112 (1978)
12. Gulrajani, I., Ahmed, F., Arjovsky, M., Dumoulin, V., Courville, A.C.: Improved training of Wasserstein GANs. In: Advances in Neural Information Processing Systems, pp. 5767–5777 (2017)
13. Guo, X., Liu, X., Zhu, E., Yin, J.: Deep clustering with convolutional autoencoders. In: Liu, D., Xie, S., Li, Y., Zhao, D., El-Alfy, E.S. (eds.) ICONIP 2017. LNCS, vol. 10635, pp. 373–382. Springer, Cham (2017). https://doi.org/10.1007/978-3-319-70096-0_39
14. Hjelm, R.D., et al.: Learning deep representations by mutual information estimation and maximization. arXiv preprint arXiv:1808.06670 (2018)

15. Hu, W., Miyato, T., Tokui, S., Matsumoto, E., Sugiyama, M.: Learning discrete representations via information maximizing self-augmented training. In: Proceedings of the International Conference on Machine Learning, pp. 1558–1567 (2017)
16. Ji, X., Henriques, J.F., Vedaldi, A.: Invariant information clustering for unsupervised image classification and segmentation. In: Proceedings of the IEEE International Conference on Computer Vision, pp. 9865–9874 (2019)
17. Jiang, Z., Zheng, Y., Tan, H., Tang, B., Zhou, H.: Variational deep embedding: an unsupervised and generative approach to clustering. arXiv preprint arXiv:1611.05148 (2016)
18. Kingma, D.P., Ba, J.: Adam: a method for stochastic optimization. arXiv preprint arXiv:1412.6980 (2014)
19. Kingma, D.P., Welling, M.: Auto-encoding variational Bayes. arXiv preprint arXiv:1312.6114 (2013)
20. Krizhevsky, A., Hinton, G., et al.: Learning multiple layers of features from tiny images. Technical report (2009)
21. LeCun, Y., Bottou, L., Bengio, Y., Haffner, P.: Gradient-based learning applied to document recognition. Proc. IEEE **86**(11), 2278–2324 (1998)
22. van der Maaten, L., Hinton, G.: Visualizing data using t-SNE. J. Mach. Learn. Res. **9**, 2579–2605 (2008)
23. MacQueen, J., et al.: Some methods for classification and analysis of multivariate observations. In: Proceedings of the Berkeley Symposium on Mathematical Statistics and Probability, Oakland, CA, USA, vol. 1, pp. 281–297 (1967)
24. Mukherjee, S., Asnani, H., Lin, E., Kannan, S.: ClusterGAN: latent space clustering in generative adversarial networks. In: Proceedings of the AAAI Conference on Artificial Intelligence, vol. 33, pp. 4610–4617 (2019)
25. Ng, A.Y., Jordan, M.I., Weiss, Y.: On spectral clustering: analysis and an algorithm. In: Advances in Neural Information Processing Systems, pp. 849–856 (2002)
26. Nina, O., Moody, J., Milligan, C.: A decoder-free approach for unsupervised clustering and manifold learning with random triplet mining. In: Proceedings of the Geometry Meets Deep Learning Workshop in IEEE International Conference on Computer Vision (2019)
27. Nowozin, S., Cseke, B., Tomioka, R.: f-GAN: training generative neural samplers using variational divergence minimization. In: Advances in Neural Information Processing Systems, pp. 271–279 (2016)
28. Radford, A., Metz, L., Chintala, S.: Unsupervised representation learning with deep convolutional generative adversarial networks. arXiv preprint arXiv:1511.06434 (2015)
29. Shaham, U., Stanton, K., Li, H., Nadler, B., Basri, R., Kluger, Y.: SpectralNet: spectral clustering using deep neural networks. arXiv preprint arXiv:1801.01587 (2018)
30. Wang, J., Wang, J., Song, J., Xu, X.S., Shen, H.T., Li, S.: Optimized Cartesian K-means. IEEE Trans. Knowl. Data Eng. **27**(1), 180–192 (2014)
31. Wu, J., et al.: Deep comprehensive correlation mining for image clustering. In: Proceedings of the IEEE International Conference on Computer Vision, pp. 8150–8159 (2019)
32. Xiao, H., Rasul, K., Vollgraf, R.: Fashion-MNIST: a novel image dataset for benchmarking machine learning algorithms. arXiv preprint arXiv:1708.07747 (2017)
33. Xie, J., Girshick, R., Farhadi, A.: Unsupervised deep embedding for clustering analysis. In: Proceedings of the International Conference on Machine Learning, pp. 478–487 (2016)

34. Yang, J., Parikh, D., Batra, D.: Joint unsupervised learning of deep representations and image clusters. In: Proceedings of the IEEE Conference on Computer Vision and Pattern Recognition, pp. 5147–5156 (2016)
35. Yang, L., Cheung, N.M., Li, J., Fang, J.: Deep clustering by Gaussian mixture variational autoencoders with graph embedding. In: Proceedings of the IEEE International Conference on Computer Vision, pp. 6440–6449 (2019)
36. Yang, X., Deng, C., Zheng, F., Yan, J., Liu, W.: Deep spectral clustering using dual autoencoder network. In: Proceedings of the IEEE Conference on Computer Vision and Pattern Recognition, pp. 4066–4075 (2019)
37. Zelnik-Manor, L., Perona, P.: Self-tuning spectral clustering. In: Advances in Neural Information Processing Systems, pp. 1601–1608 (2005)
38. Zhang, T., Ji, P., Harandi, M., Huang, W., Li, H.: Neural collaborative subspace clustering. arXiv preprint arXiv:1904.10596 (2019)

Self-supervised Motion Representation via Scattering Local Motion Cues

Yuan Tian, Zhaohui Che, Wenbo Bao, Guangtao Zhai[✉], and Zhiyong Gao

Shanghai Jiao Tong University, Shanghai, China
{ee_tianyuan,chezhaohui,baowenbo,zhaiguangtao,zhiyong.gao}@sjtu.edu.cn

Abstract. Motion representation is key to many computer vision problems but has never been well studied in the literature. Existing works usually rely on the optical flow estimation to assist other tasks such as action recognition, frame prediction, video segmentation, etc. In this paper, we leverage the massive unlabeled video data to learn an accurate explicit motion representation that aligns well with the semantic distribution of the moving objects. Our method subsumes a coarse-to-fine paradigm, which first decodes the low-resolution motion maps from the rich spatial-temporal features of the video, then adaptively upsamples the low-resolution maps to the full-resolution by considering the semantic cues. To achieve this, we propose a novel context guided motion upsampling layer that leverages the spatial context of video objects to learn the upsampling parameters in an efficient way. We prove the effectiveness of our proposed motion representation method on downstream video understanding tasks, *e.g.*, action recognition task. Experimental results show that our method performs favorably against state-of-the-art methods.

Keywords: Motion representation · Self-supervised learning · Action recognition

1 Introduction

Motion serves as an essential part of video semantic information, and has led to great breakthroughs in numerous tasks such as action recognition [17,21,62], video prediction [33,46], video segmentation [66,71], to name a few. Existing literature typically represents motions in the form of 2-dimensional optical flow vectors. However, optical flow estimation algorithms usually suffer from expensive computational cost or inaccurate estimates [15,24]. More seriously, recent deep learning based approaches rely on human-labeled ground-truths that are labor-consuming [4], or computer-generated synthetic samples [24,25] that may cause domain gap with natural realistic scenes. Therefore, there exists an urgent demand for unsupervised learning of motion representations.

Amounts of tasks rely on accurate motion representations. Action recognition methods [8,21,50,55,59] usually take motion modalities, *e.g.*, optical flow stream as the additional input besides RGB frames to further improve the performance. Many works in video generation [2,32,34,35,46,48] learn to predict

A. Vedaldi et al. (Eds.): ECCV 2020, LNCS 12359, pp. 71–89, 2020.
https://doi.org/10.1007/978-3-030-58568-6_5

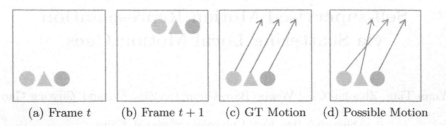

(a) Frame t (b) Frame $t + 1$ (c) GT Motion   (d) Possible Motion

Fig. 1. Schematic diagram of motion. The circle and the triangle shapes represent two different semantic groups, while the pixels of different brightness are marked with different colors. (a) and (b) are two consecutive frames. (c) is the ground-truth motion along the time step. (d) is one possible but wrong solution by only considering the brightness consistency rule.

the motions of the objects in the frame as an intermediate step. Video segmentation works [10,39,66,71] highly depend on accurate motion information to find prominent objects from frames. However, there are few works focusing on the learning of motion representation. Sun et al. [54] proposed a motion representation in feature space and shared a similar formula definition with optical flow. However, the training of their networks involves large-scale category labeled video datasets. Choutas et al. [11] proposed the PoTion method that aggregates human joints' heatmap in each frame to the pose motion representation, which is limited to the category of human motions.

Motivation: The basic self-supervised paradigm of learning motion representation is (1) first predicting a per-pixel transformation from a pair of consecutive frames and (2) then minimizing the photometric loss, $e.g.$, ℓ_1 loss between the transformed second frame and the ground-truth. So, what's the main obstacle for learning the accurate motion representation beneficial to down-stream tasks? **We argue that the current works ignore the correlation between the local motion and the high-level semantic constraints.** As shown in Fig. 1, by only considering the brightness consistency rule, one possible motion solution is (d), which is less semantic and is harmful to downstream tasks.

To tackle the above mentioned problem, we propose a coarse-to-fine motion network to extract motion maps of both high accuracy and semantics from the input video in a self-supervised manner. In the coarse stage, the network decodes the low-resolution motion maps from the video features. In the refined stage, the network upsamples the motion maps from the previous stage to high-resolution. Moreover, to make the upsampling operation learnable, the motion maps are interpolated by our proposed Context Guided Motion Upsampling Layer (CGMUL) instead of the traditional bilinear upsampling. CGMUL is carefully designed to exploit the local motion-semantics correlation in feature space for producing the full-scale motion features and aggregate the features into high-resolution motion maps in an efficient way.

To fully utilize the long-term temporal semantics in videos, our method takes video clips instead of frame pairs as input and adopt the off-the-shelf 3D CNNs, *e.g.*, C3D [56], 3DResNet [21] or SlowFast Network [17] as the video feature extractor. This reduces the semantic gap between our learned motion representations and other video understanding tasks based on these 3D CNNs. Additionally, our learning process can regularize the backbone 3D CNNs without increasing the computation cost at inference time in two ways, (1) improving the performances of other tasks in a multi-task fashion and (2) serving as a pre-training method for the backbone network.

Our contributions are summarized from the following aspects:

First, we further restrict the search space of self-supervised motion representation learning by leveraging the motion-semantics correlations in local regions. The resulting representations are accurate, of high semantics and beneficial to downstream video understanding tasks.

Second, we propose a Context Guided Motion Upsampling Layer (CGMUL) to learn the motion map upsampling parameters by exploiting the correlation between the semantic features of spatial contexts and local motions.

Third, we show that our method reaches a new state-of-the-art performance on action recognition task. Moreover, the motion representation ability of our method is competitive to other recent optical flow methods, *e.g.*, FlowNet2.0 [24].

2   Related Work

Motion Representation Learning. The most common motion representation is optical flow. Numerous works [24,25] attempt to produce flow map from coupled frames by CNN in an efficient way. However, most of their training datasets are synthetic and thus they perform poorly on real word scenes. Recently, other motion representations have been proposed. TSN [62] leverages the RGB difference between consecutive frames. OFF [54] proposes an optical flow alike motion representation in feature space. PoTion [11] temporally aggregates the human joints' heatmap in each frame to a clip-level representation with fixed dimension. In contrast, our method is self-supervised and learns more general motion representations for both articulated objects and dynamic textures.

Dynamic Filter Networks. DFN [3] first proposes to generate the variable filters dynamically conditioned on the input data. DCN [12] can also produce position-specific filters. PAC [52] proposes a pixel-adaptive convolution operation, in which the convolution filter's weights are multiplied with a spatially varying kernel. Unlike them, we produce the dynamic motion filters directly from the video features for individual spatial position.

Video Prediction. Video prediction relies on the motion cues in feature space or primal space to synthesize the future frame from past frames. BeyondMSE [36] adopts a cGAN model to make use of temporal motion information in videos implicitly. Recent works take advantage of motion cues embodied in videos by flow consistency [33,42], the retrospective cycle nature of video [29], the static and dynamic structure variation [69], etc. However, their motion generators usually adopt the frame-level spatial feature extractor. Some works such as SDC [46] rely on optical flow map and dynamic kernel simultaneously to synthesize the future frames. In comparison, our method makes use of the rich spatial-temporal features from long-term videos without leveraging optical flow maps.

Action Recognition. Recently, convolutional networks are widely adopted in many works [19,50,56,61] and have achieved great performance. Typically, two-stream networks [19,50,61] learn motion features based on extra optical flow stream separately. C3D network [56] adopts 3D convolution layers to directly capture both appearance and motion features from raw frames volume. Recent deep 3D CNN based networks [8,17,21] such as 3D-RestNet [21] have been trained successfully with the promising results upon the large scale video datasets. Our work is built upon the 3D CNNs and surpass their performance.

3   Proposed Method

In this section, we first provide an overview of our motion representation algorithm. We then introduce the proposed context guided motion upsampling layer, which plays a critical role in learning the accurate full-resolution motion maps. Finally, we demonstrate the design of all the sub-modules and clarify the implementation details of the proposed model.

3.1   Overview

Different from the previous motion representation methods that only take two consecutive frames as input, we feed a video clip consisting of T frames into the network to craft T motion maps simultaneously, where the first $T - 1$ motion maps reflect the motion representations between every consecutive frame pair, while the last one is a prediction of the possible motion $w.r.t.$ the next unknown future frame. Our method shares the video's spatial-temporal features with other tasks, $e.g.$, action recognition and benefits them in a multi-task paradigm. Moreover, the learned motion maps can serve as another input modality to further improve the performances of these downstream tasks.

3.2   Context Guided Motion Upsampling Layer

Motion Map: We first give a principled definition to the motion map in our method. Given input video $\mathbf{X} \in \mathbb{R}^{t \times w \times h \times c}$, the motion maps are composed of a series of local filters of size $k \times k$, each of which models the localized motion cues around the center pixel, where t, w, h, c and k denote video temporal length, video frame width, video frame height, the number of video frame channels and the constant parameter indicating the maximum displacement between consecutive frames. Let us denote the motion map by $\mathbf{M}_t \in \mathbb{R}^{k \times k \times w \times h}$, which describes the motions between $\mathbf{X}_t$ and $\mathbf{X}_{t+1}$. These three tensors are related by the pixel-adaptive convolution [52] operation, which can be precisely formulated as:

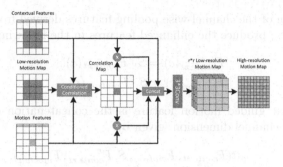

Fig. 2. Context Guided Motion Upsampling Layer. The layer exploits the correlation between the contextual features and the motion features to estimate the higher resolution motion maps. "x" and "Concat" denote the channel-wise multiplication and concatenation operations respectively.

$$\tilde{\mathbf{X}}_{t+1,x,y,c_i} = \sum_{m=-r}^{r} \sum_{n=-r}^{r} \mathbf{M}_t \left(m+r, n+r, x, y \right) \times \mathbf{X}_{t,c_i} \left(x-m, y-n \right). \quad (1)$$

where $r = \frac{k-1}{2}$, c_i denotes the color channel. Each filter of the motion map is adaptive to a single pixel of $\mathbf{X}_t$ while shared across the color channels. Compared to optical flow, this definition can represent the motions in sub-pixel level and synthesize novel pixel values to address subtle brightness changes of the moving pixels, which is common in natural scenes.

Learn to Upsample Motion Maps: To learn the non-linear interpolating parameters for upsampling the motion map, we propose the Context Guided Motion Upsampling Layer (CGMUL) to estimate the high-resolution (HR) motion maps from the low-resolution (LR) motion maps, as shown in Fig. 2. The estimation process is guided by the semantic context in the local regions.

Precisely, we denote the contextual feature, the LR motion map, and the motion feature as $F_{context} \in R^{C \times \hat{w} \times \hat{h}}$, $M_{LR} \in R^{k \times k \times \hat{w} \times \hat{h}}$ and $F_{motion} \in R^{C \times \hat{w} \times \hat{h}}$, where $\hat{w} = \frac{w}{r}$, $\hat{h} = \frac{h}{r}$ and r is the upsampling scale.

We first compute the correlational similarity map $S \in R^{\hat{w} \times \hat{h}}$ conditioned by M_{LR} between $F_{context}$ and F_{motion}:

$$S(x,y) = \sum_{c=0}^{C} \sum_{m=-r}^{r} \sum_{n=-r}^{r} (M_{LR}(m+r, n+r, x, y) \times$$
$$F_{motion}(c,x,y) \times F_{context}(c, x-m, y-n)) \times \frac{1}{C}, \quad (2)$$

where $r = \frac{k-1}{2}$ and $\frac{1}{C}$ are used for normalization. The similarity describes the relationship between three inputs explicitly.

Recent studies [31] proposed to enhance the feature's discrimination by dot-producting the channel-wise pooling features. By analogy, Eq. 2 can be viewed as the soft fusion of the channel-wise pooling features derived from $F_{context}$ and F_{motion}. Thus, we produce the enhanced features in the following way:

$$F'_{context}(c) = S \cdot F_{context}(c),$$
$$F'_{motion}(c) = S \cdot F_{motion}(c), \tag{3}$$

The final context guided motion feature is the concatenation of the features above along the channel dimension, given by

$$F = cat(F_{context}, F_{motion}, S, F'_{context}, F'_{motion}), \tag{4}$$

We perform a learnable 3×3 convolution on F to produce feature maps $F' \subset R^{(r \cdot r \cdot \hat{k} \cdot \hat{k}) \times \hat{w} \times \hat{h}}$. Finally, we utilize the periodic shuffling operator [49] on the feature maps above to get the HR motion map $M_{HR} \in R^{(\hat{k} \cdot \hat{k}) \times (r\hat{w}) \times (r\hat{h})}$:

$$M_{HR}(c, x, y) = F'_{C \cdot r \cdot \text{ mod } (y,r)+C \cdot \text{ mod } (x,r)+c, \lfloor x/r \rfloor, \lfloor y/r \rfloor}, \tag{5}$$

where $C = \hat{k} \cdot \hat{k}$ and c denotes the channel index. Noting that $\hat{k}$ is bigger than k , for the motion filters in motion maps of higher resolution require wider receptive field.

3.3 Context Guided Motion Network

As shown in Fig. 3, the proposed Context Guided Motion Network (CGM-Net) consists of the following submodules: the video encoder, the LR motion decoder, the context extractor, and the motion upsampler. We adopt the proposed context guided motion upsampling layer to upsample the LR motion map to the HR motion map in a learnable way. We illustrate every component in detail as follows.

Video Encoder. This module extracts compact video features $\mathcal{F}_v$ from the input video clip **X**, which mainly consists of a series of 3D convolution operations. Notably, the proposed method is compatible with most recent off-the-shelf 3D CNNs [8,17,56]. In our experiment, due to the space limitation, we only report the performance when using two landmark 3D CNNs (*i.e.* 3D-ResNet [21] and SlowFast network [17]) as feature extractors to derive spatial-temporal features.

LR Motion Decoder. This module reconstructs the LR motion features $\mathcal{F}_{LR}$ from video features $\mathcal{F}_v$ by deconvolution operations. To facilitate the network convergence, we replace all deconvolution operations in the network with a bilinear upsample operation followed by a convolution operation with the kernel size of 3 and the stride of 1 as suggested by the previous research [70].

Context Extractor. This module extracts semantic contextual information from each frame of the input video. We utilize the response of the $conv3\_x$ layer from ResNet-18 [23] as the contextual features and remove the max-pooling layer between the $conv\_1$ and $conv2\_x$ to maintain a high spatial resolution of the contextual features.

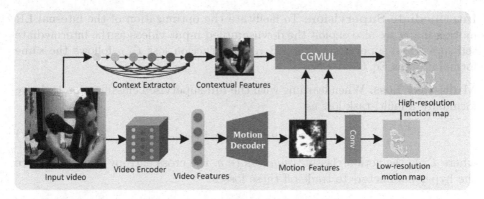

Fig. 3. Architecture of the proposed context guided motion network. CGMUL denotes the Context Guided Motion Upsampling Layer, as illustrated in Fig. 2. Given the input video, we extract the rich spatial-temporal features from the entire video clip, and extract a group of contextual features from every single frame, respectively. We first decode the low-resolution motion maps from spatial-temporal features directly following an encoder-decoder paradigm. We then adopt the proposed CGMUL to upsample the motion maps adaptively following the semantic guide of the contextual features. The final output high-resolution motion map is of both high accuracy and semantics.

3.4   Enhancing Action Recognition

After obtaining the motion maps, we feed them into a light-weight action CNN to boost the video action recognition task, because our motion maps capture more semantics than the vanilla RGB images. Concretely, the action CNN utilizes six convolution layers and one fully-connected layer to predict the action category of the input video. It is worth mentioning that we can also perform classification by adding one fully-connected layer after the backbone network directly. We fuse the prediction scores of these two methods to boost action recognition in the test.

3.5   Training Strategy

Self-supervised Learning. When learning motion representations, CGM-Net aims to (1) reconstruct all input frames and (2) predict the next future frame after the input clip simultaneously. The output frames are computed as $\tilde{\mathbf{X}}_{t+1} = \mathbf{M}_t \otimes \mathbf{X}_t$ as defined in Eq. 1, where $\mathbf{M}_t$ is the predicted motion maps. We train the network by optimizing the following reconstruction loss:

$$\mathcal{L}_{HR} = \sum_{t=0}^{T} \rho\left(\tilde{\mathbf{X}}_{t+1} - \mathbf{X}_{t+1} \right), \qquad (6)$$

where $\rho(x) = \sqrt{x^2 + \epsilon^2}$ is the Charbonnier penalty function [9]. We set the constant ϵ to 0.000001.

Intermediate Supervision: To facilitate the optimization of the internal LR motion maps, we also exploit the downsampled input videos as the intermediate self-supervised supervision. The LR rescontruction loss $\mathcal{L}_{LR}$ follows the same formulation as Eq. 6.

Multi-task Loss: When learning with the full-supervised classification task, we formulate a multi-task loss as

$$\mathcal{L} = \mathcal{L}_{HR} + \lambda_1 \mathcal{L}_{LR} + \lambda_2 \mathcal{L}_c. \tag{7}$$

where $\mathcal{L}_c$ is the action classification loss (*e.g.* the cross entropy), λ_1 and λ_2 are the hyper-parameters to trade-off these losses.

4   Experimental Results

We first carry out comparisons between our method and other recent methods regarding the motion representation ability. Then, we show that our method can facilitate the action recognition performance of the very recent 3D CNNs to achieve the new state-of-the-arts while keeping efficient. Finally, we conduct extensive ablation studies to verify the every aspects of the proposed method.

4.1   Comparison with Other Motion Representation Method

To solve the occlusion and color noise problems in the natural scenes, our method synthesizes novel pixels not in the previous frame. Therefore, we compare the motion representation errors on the natural scene dataset, *i.e.*, the UCF101 dataset [51]. We compare our method with other methods in terms of (1) motion estimation and (2) 1-step frame prediction.

Dataset. UCF-101 is a widely-used video benchmark including 101 human action classes. We choose 20 videos with neat backgrounds and obvious motions, named **UCF-Flow**, to compare the performance of motion estimation. We select 101 videos of different actions, named **UCF-Pred**, to compare the performance of frame synthesis on the video prediction task.

Implementation Details. For our method, we split the videos into clips of 16 frames and discard the too short clips. For other optical flow methods, we compute the optical flow for every two consecutive frames. We apply the ℓ_1 error between the warped second image and the ground-truth second image instead of End-Point-Error (EPE) to measure the motion representation error, for our motion map can't be transformed to an optical flow map losslessly. All the images are normalized to the range $[-1, 1]$ before computing the error. In the training process, we set the λ_1 and λ_2 of Eq. 7 as 1.0 and 0 (we do not leverage the video ground-truth label in this part) respectively. We adopt Adam optimizer [26] with a start learning rate as 0.001 and reduce the learning rate every 50 epochs.

Motion Estimation Results. As shown in Table 1 (left), our method substantially outperforms the best optical-flow methods by a large margin in terms of

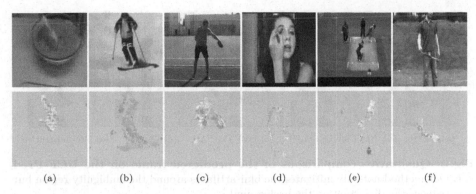

(a)   (b)   (c)   (d)   (e)   (f)

Fig. 4. The visualization of our high-resolution motion maps on UCF-Flow set. The predominant motions are consistent with the semantics of the objects. Our method is robust to human action and natural object's motion. For example, in (a) and (d), the motion of objects (*i.e.* spoon and human body) are predicted accurately. In (e), our motion map shows excellent performance in the multi-objects scenario.

Table 1. *left*: Comparison of motion estimation methods on UCF-Flow. The top part shows the performance of the current best traditional optical flow estimation methods. The middle part shows the results of the CNN based methods. *right*: Comparison of video prediction methods on UCF-Pred.

Method	$\ell 1$ Err	Train
DIS-Fast [27]	0.055	No
Deepflow [65]	0.058	No
TV-L1 [44]	0.037	No
Flownet2.0 [24]	0.057	Yes
PWC-Net [53]	0.049	Yes
TV-Net-50 [16]	0.040	Yes
Ours	**0.018**	Yes

Method	PSNR	SSIM
BeyondMSE [36]	32	0.92
ContextVP [5]	34.9	0.92
MCnet+RES [58]	31	0.91
EpicFlow [47]	31.6	0.93
DVF [35]	33.4	0.94
Ours (112px)	**35.0**	**0.96**
Ours (256px)	**36.3**	**0.96**

ℓ_1 **Error.** The explanation for the obvious improvement upon the optical flow based methods is that the environment illuminations change constantly and most objects are not rigid in natural scenes. Our motion map can synthesize new pixels around the moving objects. To further prove our method represents the motions precisely and robustly, we show the visualization of the motion map for diverse human actions on UCF101 in Fig. 4.

1-Step Frame Prediction Results. For quantitative evaluation, we utilize the SSIM and PSNR [63] as the evaluation metrics. The higher SSIM and PSNR, the better prediction performance. Table 1 (right) describes the quantitative evaluation results of the state-of-the-art methods and the proposed method on UCF101. Our method achieves the best results in terms of SSIM and PSNR. Moreover, even with lower input resolution of 112×112, the performance of our method

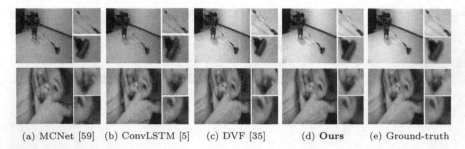

(a) MCNet [59]   (b) ConvLSTM [5]   (c) DVF [35]   (d) **Ours**   (e) Ground-truth

Fig. 5. Qualitative comparisons of the predicted frame on the UCF-Pred set. Our method not only mitigates the blur artifacts around the ambiguity region but also reduces the distortion on the background.

also keeps stable and competitive. This demonstrates our motion representations are learned from either spatially localized textures or global semantic information that are not sensitive to video resolution.

Figure 5 depicts a few results of the preceding methods, where the frames predicted by our method obtain better perceptual quality [37,38,73], even our method is not equipped with the perceptual loss. MCNet [58] and ConvLSTM [5] cause ghosts in the regions around the blob objects, e.g., the mop head, because they encode both the motion and content features into their models. DVF [35] shows unexpected artifacts in background regions around the motion objects because it doesn't synthesize novel pixels around the motion regions. However, the motion filters of our motion map are adaptive to the position and thus only acts on the activities related regions in a pixel synthesis way.

4.2 Action Recognition

Datasets. We evaluated our method on three large-scale general human action and activity datasets, including Kinetics [8], UCF101 [51], and HMDB51 [28]. We follow the original training/test splits and protocols provided by the datasets. We report the mean average accuracy over the three splits for HMDB51 and UCF101. For Kinetics, we report the performance on the validation set.

Kinetics. Kinetics is a challenging human action recognition dataset. We evaluate our method on Kinetics-400 and Kinetics-600. We report top-1 and top-5 classification accuracy (%) on Kinetics.

UCF101. UCF-101 includes 13,320 action instances from 101 human action classes. To evaluate the action recognition performance, we first train it on the Kinetics-400 dataset and then fine-tune on UCF101. For this dataset, we also study the effectiveness of our method as a pre-training strategy compared with other self-supervised pre-training methods.

HMDB51. HMDB51 includes 6,766 videos from 51 human action classes. On this dataset, we conduct all action recognition experiments mentioned in UCF101.

Table 2. *left*: Comparison of self-supervised action representation methods. The baseline methods in first group are without self-supervised pretraining. *right*: Evaluation of training on different sets of Kinetics-400. ResNext-101 and STC-ResNext101 are abbreviated as R101 and S-R101 respectively. * indicates that the corresponding method uses extra unlabeled data.

Method	UCF101	HMDB51
3D-R101 [21]	56.2	30.3
S-R101 [13]	56.7	30.8
Shuffle and learn [40]	50.9	19.8
OPN-RGB [30]	71.8	36.7
Order Prediction [68]	72.4	30.9
Odd-One-Out [20]	60.3	32.5
AOT-RGB* [64]	86.5	-
ActionFlowNet* [41]	83.9	56.4
DynamoNet(3D-R101)* [14]	87.3	58.6
DynamoNet(S-R101)* [14]	**88.1**	**59.9**
Ours(3D-R101)*	**88.1**	59.0

Method	Data	Top1
3D-R101 [21]	half	53.9
3D-R101 [21]	full	65.1
S-R101 [13]	half	55.4
S-R101 [13]	full	66.2
St-Net(R101) [22]	half	56.7
St-Net(R101) [22]	full	71.38
DynamoNet(S-R101) [14]	half	63.6
DynamoNet(S-R101) [14]	full	67.67
Ours(3D-R101)	half	**69.8**
Ours(3D-R101)	full	**76.2**

Implementation Details. We train the network with only motion representation branch (λ_2 is set to 0) as the pre-training step on 500K unlabeled video clips from YouTube8M [1] dataset. We first resize the video frames to 128px when smaller and then randomly perform 5 crops (and flips) of size 112×112 as the main network input size. When using the SlowFast [17] as backbone network, we follow the same input size as them. We adopt Adam optimizer [26] with an initial learning rate as 0.001 and batch size of 64 to train the model. In our experiments, we use the different versions of 3D-ResNet/ResNeXt as the backbone networks. Empirically, we obtain the best results when setting the motion map upsampling scale factor = 4, $\lambda_1 = 1$ and $\lambda_2 = 10$. We use the PyTorch [43] framework for the implementation and all the experiments are conducted on sixteen 2080 Ti NVIDIA GPUs.

Self-supervised Action Representation. Since motion is an important cue in action recognition, we argue the learned motion representation implied in the backbone 3D CNN can be adopted as a good initial representation for the action recognition task. Our network is firstly trained on unlabeled video clips to learn motion representation. Then, we fine-tune the full network carefully, with all losses in Eq. 7 activated.

In Table 2 (left), we observe that our method performs better in comparison to state-of-the-art self-supervised methods [14,20,30,40,41,64,68] on UCF101 and HMDB51. The performance gap between our method pretrained by unlabeled data and the 3D-ResNet101 trained on Kinetics-400 (shown in Table 5) is largely reduced to **0.8%** on UCF101. DynamoNet with the STC-ResNeXt101 indeed outperforms our method with 3D-ResNeXt101 by 0.9% on HMDB51 because STC-ResNeXt101 has a stronger ability to capture spatial-temporal correlations compared to vanilla 3D-ResNeXt101.

Table 3. Performance comparisons of our method with other state-of-the-art 3D CNNs on Kinetics-400 dataset. Y500K indicates the subset of Youtube8M.

Method	Flow	Backbone	Pretrain	Top1	Top5
3D-ResNet18 [21]	✗	✗	✗	54.2	78.1
C3D [56]	✗	✗	Sports1m	55.6	-
3D-ResNet50 [21]	✗	✗	✗	61.3	83.1
3D-ResNet101 [21]	✗	✗	✗	62.8	83.9
3D-ResNeXt101 [21]	✗	✗	✗	65.1	85.7
R(2+1)D [57]	✗	✗	✗	73.9	90.9
STC-Net [13]	✗	3D-ResNeXt101	✗	68.7	88.5
DynamoNet [14]	✗	3D-ResNeXt101	Y500K	68.2	88.1
StNet [22]		ResNet101	✗	71.4	-
DynamoNet [14]	✗	STC-ResNeXt101	Y500K	77.9	94.2
SlowFast 16×8 [17]	✗	ResNeXt101	✗	78.9	93.5
R(2+1)D Flow [57]	✓	✗	✗	67.5	87.2
I3D [8]	✓	✗	✗	71.6	90.0
R(2+1)D [57]	✓	✗	✗	73.9	90.9
Two-Stream I3D [8]	✓	BN-Inception	ImageNet	75.7	92.0
S3D-G [67]	✓	✗	ImageNet	77.2	93.0
Ours	✗	3D-ResNet50	Y500K	70.1	90.2
Ours	✗	3D-ResNeXt101	Y500K	**76.2**	**92.3**
Ours	✗	SlowFast16×8	Y500K	**80.8**	**94.5**

Table 4. Performance comparisons of our method with other state-of-the-art 3D CNNs on Kinetics-600 dataset. Y500K indicates the subset of Youtube8M.

Method	Backbone	Pretrain	Top1	Top5
P3D [45]	ResNet152	ImageNet	71.3	-
I3D [7]	BN-Inception	✗	71.9	90.1
TSN [62]	IRv2	ImageNet	76.2	-
StNet [22]	IRv2	ImageNet	79.0	-
SlowFast 16×8 [17]	ResNeXt101	✗	81.1	95.1
Ours	3D-ResNet50	Y500K	76.2	90.7
Ours	3D-ResNeXt101	Y500K	80.2	94.0
Ours	SlowFast16×8	Y500K	**81.9**	**95.1**

Table 2 (right) shows the self-supervised pre-training backbone network based on our method can alleviate the need for labeled data and achieves the best results with datasets of different sizes. Moreover, the performance of our pipeline trained with half data is competitive with other state-of-the-art methods (*e.g.*, St-Net) trained with full data.

Table 5. Performance comparisons of our method with other state-of-the-art methods on UCF101 and HMDB51. The number inside the brackets indicates the frame number of the input clip. [†] and [*] indicate the backbone network is 3D-ResNeXt101 or STC-ResNeXt101 respectively.

UCF101		HMDB51	
Method	Top1	Method	Top1
DT+MVSM [6]	83.5	DT+MVSM [6]	55.9
iDT+FV [59]	85.9	iDT+FV [59]	57.2
C3D [56]	82.3	C3D [56]	56.8
Two Stream [50]	88.6	Two Stream [50]	-
TDD+FV [60]	90.3	TDD+FV [60]	63.2
RGB+Flow-TSN [62]	94.0	RGB+Flow-TSN [62]	68.5
ST-ResNet [18]	93.5	ST-ResNet [18]	66.4
TSN [62]	94.2	TSN [62]	69.5
3D-ResNet101 [21]	88.9	3D-ResNet101 [21]	61.7
3D-ResNeXt101 [21]	90.7	3D-ResNeXt101 [21]	63.8
DynamoNet (16)[†] [14]	91.6	DynamoNet (16)[†] [14]	66.2
DynamoNet (32)[†] [14]	93.1	DynamoNet (32)[†] [14]	68.5
DynamoNet (64)[*] [14]	94.2	DynamoNet (64)[*] [14]	**77.9**
Ours (32)[†]	**94.1**	**Ours (32)[†]**	**69.8**

Comparison with the State-of-the-Art. Table 3 presents results on Kinetics-400 for our method. With 3D-ResNeXt101 backbone, our method outperforms DynamoNet, which also ensembles the motion representations in a self-supervised way, with large margins: **8.0%** and **4.2%** improvements in terms of Top1 and Top5 accuracies respectively. This indicates the superiority of our semantic guided motion maps, compared with DynamoNet [14] directly adopting the spatially shared motion kernel weights. Interestingly, we find that our method based on 3D-ResNet50 outperforms the vanilla 3D-ResNet101 obviously, by **7.3%** and **6.3%** improvements in terms of Top1 and Top5 accuracies. As shown in Table 4, our method with SlowFast backbone also achieves the best performances. We also compare our method with the other most recent 3D CNNs taking inputs RGB and optical flow modalities and verify that our method outperforms the best of them by **3.6%** while saving the inference cost *w.r.t.* the computation of optical flow maps. Table 5 demonstrates the state-of-the-art performances achieved by our method compared with the very recent methods on

UCF101 and HMDB51 datasets. DynamoNet [14] outperforms our method on
HMDB51 with more input frames (64 vs. 32), because it has been verified [13,14]
that the number of input frames has a strong impact on the final performance,
and the more input frames, the better performance.

4.3   Ablation Study

In this part, to facilitate the training process, we adopt the 3D-ResNet18 as the
backbone network.

Learnable vs. Unlearnable Upsampling Methods. We first emphasize the
superiority of our learnable motion upsampling method compared with the traditional methods: (1) nearest neighbour interpolation and (2) bilinear interpolation. For traditional methods, we upsample each channel of the motion maps
and exaggerate each motion filter with zero holes following the similar expanding
method as dilation convolution kernels [72]. As shown in Table 7 and Table 7,
our method substantially outperforms the traditional baselines in both motion
representation and action recognition. The traditional motion upsampling methods result in coarse output motion maps whereas our method hallucinates the
motion details thanks to the static contexts and the motion prior learned from
massive videos. It's also interesting to notice from Table 7 that the LR motion
map also benefits the action recognition task obviously by **3.8%** despite the
motions in this scale are imperceptible, which indicates the advantage of our
motion representation in sub-pixel level.

Impact of Different HR/LR Motion Map Scale Factors. As shown in
Fig. 6 (left), the motion estimation performance decreases as the scale factor
increases. Besides, when the scale factor < 8, the performance drop is moderate.
The trend of Fig. 6 (right) is quite different. When the scale factor $= 1$, we
got the worse performance because the motion maps are only decoded from the
video features without considering the motion-semantics correlation. When the
scale factor is quite large, $e.g.$, 16, the deficiency of the motion details causes the
performance drop. The scale factor of 4 produces the best performance result
that surpasses the baseline by **6.2%**. Therefore, in all experiments in our paper,
we select the scale factor as **4** as a good trade-off between the accuracy and the
semantics of the motion map if not specified otherwise.

Computation Cost Analysis. We list the performance and the computation
cost of each pipeline above in Table 8. The pipeline only adopting the features
from the 3D-ResNet18 backbone CNN outperforms the corresponding baseline
by **2.1% without any extra inference-time computation cost**. When fused
with the results from the LR motion map, our method outperforms the baseline
3D-ResNet18 by **3.8%**. More importantly, despite using a shallower backbone
($i.e.$, 3D-ResNet18), our method outperforms the stronger baseline 3D-ResNet34
by **0.5%**, demonstrating the lower inference-time computation cost and the better performance. The pipeline fusing the results from both LR and HR motion
map shows a superior performance - **90.6%**.

Table 6. Comparison of different motion map upsampling methods.

Method	$\ell 1$ Err
Nearest	0.042
Bilinear	0.037
Ours	**0.024**

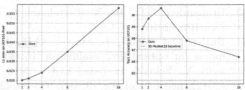

Fig. 6. *left*: The motion representation performances. Lower $\ell 1$ error indicates better motion estimation. *right*: The action recognition performances. Higher Top1 accuracy indicates better performance.

Table 7. Comparison of different motion maps for action recognition.

Method	Top1
Backbone	84.4%
+LR	**88.2%**
+LR+HR (nearest)	88.2%
+LR+HR (bilinear)	88.2%
+LR+HR (Ours)	**90.6%**

Table 8. Comparison of different pipelines on UCF101. † indicates the result is averaged with the prediction of backbone CNN.

Inference pipeline	Top1	Parameters	GFLOPs
Baseline (3D-ResNet18)	84.4	33.2M	19.3
Baseline (3D-ResNet34)	87.7	63.5M	36.7
Backbone CNN (3D-ResNet18)	**86.5**	**33.2M**	**19.3**
LR (3D-ResNet18)†	**88.2**	**45.73M**	**30.3**
HR (3D-ResNet18)†	89.4	48.12M	155.7
LR+HR (3D-ResNet18)†	**90.6**	**53.63M**	156.01

5 Conclusion

In this paper, we propose a context guided motion network, which integrates a novel context guided motion upsampling layer, in order to learn the semantic motion representation in a self-supervised manner. The learned motion representation is versatile and can be applied to boost the performance of various video-related tasks, *e.g.*, frame prediction and video recognition. We experimentally verified the superiority of the proposed method from various perspectives, showing the state-of-the-art performances over several popular video-related tasks.

Acknowledgement. This work was supported by National Natural Science Foundation of China (61831015, U1908210).

References

1. Abu-El-Haija, S., et al.: Youtube-8m: A large-scale video classification benchmark. arXiv (2016)
2. Bao, W., Lai, W.S., Ma, C., Zhang, X., Gao, Z., Yang, M.H.: Depth-aware video frame interpolation. In: CVPR (2019)

3. Brabandere, B.D., Jia, X., Tuytelaars, T., Gool, L.V.: Dynamic filter networks. In: NeurIPS (2016)
4. Butler, D.J., Wulff, J., Stanley, G.B., Black, M.J.: A naturalistic open source movie for optical flow evaluation. In: Fitzgibbon, A., Lazebnik, S., Perona, P., Sato, Y., Schmid, C. (eds.) ECCV 2012. LNCS, vol. 7577. Springer, Heidelberg (2012). https://doi.org/10.1007/978-3-642-33783-3_44
5. Byeon, W., Wang, Q., Srivastava, R.K., Koumoutsakos, P.: ContextVP: fully context-aware video prediction. In: Ferrari, V., Hebert, M., Sminchisescu, C., Weiss, Y. (eds.) ECCV 2018. LNCS, vol. 11220. Springer, Cham (2018). https://doi.org/10.1007/978-3-030-01270-0_46
6. Cai, Z., Wang, L., Peng, X., Qiao, Y.: Multi-view super vector for action recognition. In: CVPR (2014)
7. Carreira, J., Noland, E., Banki-Horvath, A., Hillier, C., Zisserman, A.: A short note about kinetics-600. arXiv (2018)
8. Carreira, J., Zisserman, A.: Quo vadis, action recognition? A new model and the kinetics dataset. In: CVPR (2017)
9. Charbonnier, P., Blanc-Feraud, L., Aubert, G., Barlaud, M.: Two deterministic half-quadratic regularization algorithms for computed imaging. In: ICIP (1994)
10. Che, Z., Borji, A., Zhai, G., Min, X., Guo, G., Le Callet, P.: How is gaze influenced by image transformations? Dataset and model. TIP **29**, 2287–2300 (2019)
11. Choutas, V., Weinzaepfel, P., Revaud, J., Schmid, C.: PoTion: pose motion representation for action recognition. In: CVPR (2018)
12. Dai, J., et al.: Deformable convolutional networks. In: ICCV (2017)
13. Diba, A., et al.: Spatio-temporal channel correlation networks for action classification. In: Ferrari, V., Hebert, M., Sminchisescu, C., Weiss, Y. (eds.) ECCV 2018. LNCS, vol. 11208. Springer, Cham (2018). https://doi.org/10.1007/978-3-030-01225-0_18
14. Diba, A., Sharma, V., Gool, L.V., Stiefelhagen, R.: DynamoNet: Dynamic action and motion network. arXiv (2019)
15. Dosovitskiy, A., et al.: FlowNet: learning optical flow with convolutional networks. In: ICCV (2015)
16. Fan, L., Huang, W., Gan, C., Ermon, S., Gong, B., Huang, J.: End-to-end learning of motion representation for video understanding. In: CVPR (2018)
17. Feichtenhofer, C., Fan, H., Malik, J., He, K.: SlowFast networks for video recognition. In: ICCV (2019)
18. Feichtenhofer, C., Pinz, A., Wildes, R.P.: Spatiotemporal residual networks for video action recognition. In: NeurIPS (2016)
19. Feichtenhofer, C., Pinz, A., Zisserman, A.: Convolutional two-stream network fusion for video action recognition. In: CVPR (2016)
20. Fernando, B., Bilen, H., Gavves, E., Gould, S.: Self-supervised video representation learning with odd-one-out networks. In: CVPR (2017)
21. Hara, K., Kataoka, H., Satoh, Y.: Learning spatio-temporal features with 3D residual networks for action recognition. In: ICCVW (2017)
22. He, D., et al.: StNet: local and global spatial-temporal modeling for action recognition. In: AAAI (2019)
23. He, K., Zhang, X., Ren, S., Sun, J.: Deep residual learning for image recognition. In: CVPR (2016)
24. Ilg, E., Mayer, N., Saikia, T., Keuper, M., Dosovitskiy, A., Brox, T.: FlowNet 2.0: evolution of optical flow estimation with deep networks. In: CVPR (2017)

25. Ilg, E., Saikia, T., Keuper, M., Brox, T.: Occlusions, motion and depth boundaries with a generic network for disparity, optical flow or scene flow estimation. In: Ferrari, V., Hebert, M., Sminchisescu, C., Weiss, Y. (eds.) ECCV 2018. LNCS, vol. 11216. Springer, Cham (2018). https://doi.org/10.1007/978-3-030-01258-8_38

26. Kingma, D.P., Ba, J.L.: Adam: a method for stochastic optimization. In: ICLR (2015)

27. Kroeger, T., Timofte, R., Dai, D., Van Gool, L.: Fast optical flow using dense inverse search. In: Leibe, B., Matas, J., Sebe, N., Welling, M. (eds.) ECCV 2016. LNCS, vol. 9908. Springer, Cham (2016). https://doi.org/10.1007/978-3-319-46493-0_29

28. Kuehne, H., Jhuang, H., Garrote, E., Poggio, T., Serre, T.: HMDB: a large video database for human motion recognition. In: ICCV (2011)

29. Kwon, Y.H., Park, M.G.: Predicting future frames using retrospective cycle GAN. In: CVPR (2019)

30. Lee, H.Y., Huang, J.B., Singh, M., Yang, M.H.: Unsupervised representation learning by sorting sequences. In: ICCV (2017)

31. Li, X., Hu, X., Yang, J.: Spatial group-wise enhance: Improving semantic feature learning in convolutional networks. arXiv (2019)

32. Li, Y., Fang, C., Yang, J., Wang, Z., Lu, X., Yang, M.H.: Flow-grounded spatial-temporal video prediction from still images. In: Ferrari, V., Hebert, M., Sminchisescu, C., Weiss, Y. (eds.) ECCV 2018. LNCS, vol. 11213. Springer, Cham (2018). https://doi.org/10.1007/978-3-030-01240-3_37

33. Liang, X., Lee, L., Dai, W., Xing, E.P.: Dual motion gan for future-flow embedded video prediction. In: ICCV (2017)

34. Liu, W., Luo, W., Lian, D., Gao, S.: Future frame prediction for anomaly detection - a new baseline. In: CVPR (2018)

35. Liu, Z., Yeh, R.A., Tang, X., Liu, Y., Agarwala, A.: Video frame synthesis using deep voxel flow. In: ICCV (2017)

36. Mathieu, M., Couprie, C., LeCun, Y.: Deep multi-scale video prediction beyond mean square error. In: ICLR (2016)

37. Min, X., Gu, K., Zhai, G., Liu, J., Yang, X., Chen, C.W.: Blind quality assessment based on pseudo-reference image. TMM **20**, 2049–2062 (2017)

38. Min, X., Zhai, G., Gu, K., Yang, X., Guan, X.: Objective quality evaluation of dehazed images. IEEE Trans. Intell. Transp. Syst. **20**, 2879–2892 (2018)

39. Min, X., Zhai, G., Zhou, J., Zhang, X.P., Yang, X., Guan, X.: A multimodal saliency model for videos with high audio-visual correspondence. TIP **29**, 3805–3819 (2020)

40. Misra, I., Zitnick, C.L., Hebert, M.: Shuffle and learn: unsupervised learning using temporal order verification. In: Leibe, B., Matas, J., Sebe, N., Welling, M. (eds.) ECCV 2016. LNCS, vol. 9905. Springer, Cham (2016). https://doi.org/10.1007/978-3-319-46448-0_32

41. Ng, J.Y.H., Choi, J., Neumann, J., Davis, L.S.: ActionFlowNet: learning motion representation for action recognition. In: WACV (2018)

42. Pan, J., et al.: Video generation from single semantic label map. In: CVPR (2019)

43. Paszke, A., et al.: Automatic differentiation in PyTorch (2017)

44. Pérez, J.S., Meinhardt-Llopis, E., Facciolo, G.: Tv-l1 optical flow estimation. Image Process. On Line **3**, 137–150 (2013)

45. Qiu, Z., Yao, T., Mei, T.: Learning spatio-temporal representation with pseudo-3D residual networks. In: ICCV (2017)

46. Reda, F.A., et al.: SDC-Net: video prediction using spatially-displaced convolution. In: Ferrari, V., Hebert, M., Sminchisescu, C., Weiss, Y. (eds.) ECCV 2018. LNCS, vol. 11211. Springer, Cham (2018). https://doi.org/10.1007/978-3-030-01234-2_44
47. Revaud, J., Weinzaepfel, P., Harchaoui, Z., Schmid, C.: EpicFlow: edge-preserving interpolation of correspondences for optical flow. In: CVPR (2015)
48. Shen, W., Bao, W., Zhai, G., Chen, L., Min, X., Gao, Z.: Blurry video frame interpolation. In: CVPR (2020)
49. Shi, W., et al.: Real-time single image and video super-resolution using an efficient sub-pixel convolutional neural network. In: CVPR (2016)
50. Simonyan, K., Zisserman, A.: Two-stream convolutional networks for action recognition in videos. In: NeurIPS (2014)
51. Soomro, K., Zamir, A.R., Shah, M.: UCF101: A dataset of 101 human actions classes from videos in the wild. arXiv (2012)
52. Su, H., Jampani, V., Sun, D., Gallo, O., Learned-Miller, E., Kautz, J.: Pixel-adaptive convolutional neural networks. arXiv (2019)
53. Sun, D., Yang, X., Liu, M.Y., Kautz, J.: PWC-Net: CNNs for optical flow using pyramid, warping, and cost volume. In: CVPR (2018)
54. Sun, S., Kuang, Z., Sheng, L., Ouyang, W., Zhang, W.: Optical flow guided feature: a fast and robust motion representation for video action recognition. In: CVPR (2018)
55. Tian, Y., Min, X., Zhai, G., Gao, Z.: Video-based early ASD detection via temporal pyramid networks. In: ICME (2019)
56. Tran, D., Bourdev, L., Fergus, R., Torresani, L., Paluri, M.: Learning spatiotemporal features with 3D convolutional networks. In: ICCV (2015)
57. Tran, D., Wang, H., Torresani, L., Ray, J., LeCun, Y., Paluri, M.: A closer look at spatiotemporal convolutions for action recognition. In: CVPR (2018)
58. Villegas, R., Yang, J., Hong, S., Lin, X., Lee, H.: Decomposing motion and content for natural video sequence prediction. arXiv (2017)
59. Wang, H., Schmid, C.: Action recognition with improved trajectories. In: ICCV (2013)
60. Wang, L., Qiao, Y., Tang, X.: Action recognition with trajectory-pooled deep-convolutional descriptors. In: CVPR (2015)
61. Wang, L., Xiong, Y., Wang, Z., Qiao, Y.: Towards good practices for very deep two-stream convnets. arXiv (2015)
62. Wang, L., et al.: Temporal segment networks: towards good practices for deep action recognition. In: Leibe, B., Matas, J., Sebe, N., Welling, M. (eds.) ECCV 2016. LNCS, vol. 9912. Springer, Cham (2016). https://doi.org/10.1007/978-3-319-46484-8_2
63. Wang, Z., Bovik, A.C., Sheikh, H.R., Simoncelli, E.P.: Image quality assessment: from error visibility to structural similarity. TIP 13, 600–612 (2004)
64. Wei, D., Lim, J., Zisserman, A., Freeman, W.T.: Learning and using the arrow of time. In: CVPR (2018)
65. Weinzaepfel, P., Revaud, J., Harchaoui, Z., Schmid, C.: DeepFlow: large displacement optical flow with deep matching. In: ICCV (2013)
66. Xiao, H., Feng, J., Lin, G., Liu, Y., Zhang, M.: MoNet: deep motion exploitation for video object segmentation. In: CVPR (2018)
67. Xie, S., Sun, C., Huang, J., Tu, Z., Murphy, K.: Rethinking spatiotemporal feature learning: speed-accuracy trade-offs in video classification. In: Ferrari, V., Hebert, M., Sminchisescu, C., Weiss, Y. (eds.) ECCV 2018. LNCS, vol. 11219. Springer, Cham (2018). https://doi.org/10.1007/978-3-030-01267-0_19

68. Xu, D., Xiao, J., Zhao, Z., Shao, J., Xie, D., Zhuang, Y.: Self-supervised spatiotemporal learning via video clip order prediction. In: CVPR (2019)
69. Xu, J., Ni, B., Li, Z., Cheng, S., Yang, X.: Structure preserving video prediction. In: CVPR (2018)
70. Xu, L., Ren, J.S., Liu, C., Jia, J.: Deep convolutional neural network for image deconvolution. In: NeurIPS (2014)
71. Xu, X., Cheong, L.F., Li, Z.: Motion segmentation by exploiting complementary geometric models. In: CVPR (2018)
72. Yu, F., Koltun, V.: Multi-scale context aggregation by dilated convolutions. arXiv (2015)
73. Zhai, G., Min, X.: Perceptual image quality assessment: a survey. Sci. China Inf. Sci. **63**, 211301 (2020). https://doi.org/10.1007/s11432-019-2757-1

Improving Monocular Depth Estimation by Leveraging Structural Awareness and Complementary Datasets

Tian Chen, Shijie An, Yuan Zhang$^{(\boxtimes)}$, Chongyang Ma, Huayan Wang, Xiaoyan Guo, and Wen Zheng

Y-tech, Kuaishou Technology, Beijing, China
zhang.yuan09@gmail.com

Abstract. Monocular depth estimation plays a crucial role in 3D recognition and understanding. One key limitation of existing approaches lies in their lack of structural information exploitation, which leads to inaccurate spatial layout, discontinuous surface, and ambiguous boundaries. In this paper, we tackle this problem in three aspects. First, to exploit the spatial relationship of visual features, we propose a structure-aware neural network with spatial attention blocks. These blocks guide the network attention to global structures or local details across different feature layers. Second, we introduce a global focal relative loss for uniform point pairs to enhance spatial constraint in the prediction, and explicitly increase the penalty on errors in depth-wise discontinuous regions, which helps preserve the sharpness of estimation results. Finally, based on analysis of failure cases for prior methods, we collect a new Hard Case (HC) Depth dataset of challenging scenes, such as special lighting conditions, dynamic objects, and tilted camera angles. The new dataset is leveraged by an informed learning curriculum that mixes training examples incrementally to handle diverse data distributions. Experimental results show that our method outperforms state-of-the-art approaches by a large margin in terms of both prediction accuracy on NYUDv2 dataset and generalization performance on unseen datasets.

1 Introduction

Recovering 3D information from 2D images is one of the most fundamental tasks in computer vision with many practical usage scenarios, such as object localization, scene understanding, and augmented reality. Effective depth estimation for a single image is usually desirable or even required when no additional signal (e.g., camera motion and depth sensor) is available. However, monocular depth

---

T. Chen and S. An–joint first authors.

Electronic supplementary material The online version of this chapter (https://doi.org/10.1007/978-3-030-58568-6_6) contains supplementary material, which is available to authorized users.

© Springer Nature Switzerland AG 2020
A. Vedaldi et al. (Eds.): ECCV 2020, LNCS 12359, pp. 90–108, 2020.
https://doi.org/10.1007/978-3-030-58568-6_6

estimation (MDE) is well known to be ill-posed due to the many-to-one mapping from 3D to 2D. To address this inherent ambiguity, one possibility is to leverage auxiliary prior information, such as texture cues, object sizes and locations, as well as occlusive and perspective clues [21,27,40].

More recently, advances in deep convolutional neural network (CNN) have demonstrated superior performance for MDE by capturing these priors implicitly and learning from large-scale dataset [9,10,19,28,36,52]. CNNs often formulate MDE as classification or regression from pixels values without explicitly accounting for global structure. That leads to loss of precision in many cases. To this end, we focus on improving structure awareness in monocular depth estimation.

Specifically, we propose a new network module, named *spatial attention block*, which extracts features via blending cross-channel information. We sequentially adopt this module at different scales in the decoding stage (as shown in Fig. 2a) to generate spatial attention maps which correspond to different levels of detail. We also add a novel loss term, named *global focal relative loss* (GFRL), to ensure sampled point pairs are ordered correctly in depth. Although existing methods attempt to improve the visual consistency between predicted depth and the RGB input, they typically lack the ability to boost performance in border areas, which leads to a large portion of quantitative error and inaccurate qualitative details. We demonstrate that simply assigning larger weights to edge areas in the loss function can address this issue effectively.

Furthermore, MDE through CNNs usually cannot generalize well to unseen scenarios [8]. We find six types of common failure cases as shown in Fig. 1 and note that the primary reason for these failures is the lack of training data, even if we train our network on five commonly used MDE datasets combined. To this end, we collect a new dataset, named *HC Depth Dataset*, to better cover these difficult cases. We also show that an incremental dataset mixing strategy inspired by curriculum learning can improve the convergence of training when we use data following diverse distributions.

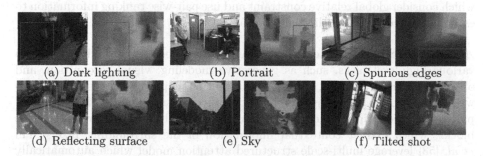

(a) Dark lighting   (b) Portrait   (c) Spurious edges

(d) Reflecting surface   (e) Sky   (f) Tilted shot

Fig. 1. Six typical hard cases for existing monocular depth estimation methods. (a), (c), and (e) show results of Alhashim *et al.* [1], while (b), (d), and (f) are based on Fu *et al.* [11]. Red boxes highlight inaccurate regions in the results. (Color figure online)

To sum up, our main contributions include:

- A novel spatial attention block in the network architecture.
- A new loss term (GFRL) and an edge-aware consistency scheme.
- A new MDE dataset featuring hard cases that are missing or insufficient in existing datasets, and a data mixing strategy for network training.

2 Related Work

Monocular Depth Estimation. Depth estimation from 2D images is an essential step for 3D reconstruction, recognition, and understanding. Early methods for depth estimation are dominated by geometry-based algorithms which build feature correspondences between input images and reconstruct 3D points via triangulation [17,23]. Recently CNN-based approaches for pixel-wise depth prediction [9,24,52] present promising results from a single RGB input, based on supervision with ground-truth training data collected from depth sensors such as LiDAR and Microsoft Kinect camera. By leveraging multi-level contextual and structural information from neural network, depth estimation has achieved very encouraging results [12,25,28,54]. The major limitation of this kind of methods is that repeated pooling operations in deep feature extractors quickly decrease the spatial resolution of feature maps. To incorporate long-range cues which are lost in downsampling operations, a variety of approaches adopt skip connections to fuse low-level depth maps in encoder layers with high-level ones in decoder layers [11,25,52].

Instead of solely estimating depth, several recent multi-task techniques [9,19,36] predict depth map together with other information from a single image. These methods have shown that the depth, normal, and class label information can be jointly and consistently transformed with each other in local areas. However, most of these approaches only consider local geometric properties, while ignoring global constraints on the spatial layout and the relationship between individual objects. The most relevant prior methods to ours are weakly-supervised approaches which consider global relative constraint and use pair-wise ranking information to estimate and compare depth values [5,6,51].

Attention Mechanism. Attention mechanisms has been successfully applied to various high-level tasks, such as generative modeling, visual recognition, and object detection [18,47,55]. In addition, attention maps are very useful in pixel-wise tasks. NLNet [49] adopts self-attention mechanism to model the pixel-level pairwise relationship. CCNet [20] accelerates NLNet by stacking two criss-cross blocks, which extract contextual information of the surrounding pixels. Yin et al. [53] leverage multi-scale structured attention model which automatically regulates information transferred between corresponding features.

Cross-Dataset Knowledge Transfer. A model trained on one specific dataset generally does not perform well on others due to dataset bias [45]. For MDE, solving different cases, e.g. indoor, outdoor, and wild scenes, usually requires explicitly training on diverse datasets [6,13,29,31]. When training on mixed

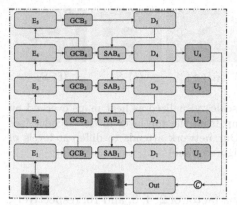

(a) Our network architecture. E_i/D_i denotes the i-th encoder/decoder. SAB_i module couples a high-level feature from D_{i+1} with a low-layer feature from GCB_i. ⓒ is the concatenate operation for the output of upsampling blocks $\{U_i\}$.

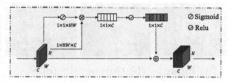

(b) Global context block (GCB) [4]. The feature maps are shown as the shape of their tensors, e.g., $H \times W \times C$. $\otimes$ denotes matrix multiplication and $\oplus$ denotes element-wise sum.

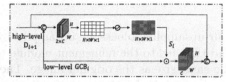

(c) Our spatial attention block (SAB). S_i represents the attention map of SAB_i and $\odot$ denotes element-wise dot product.

Fig. 2. Illustration of our network architecture and the proposed SAB.

datasets, curriculum learning [3] is needed to avoid the local minimum problem by training the model on *easier* datasets first. Moreover, when the datasets are imbalanced, resampling [15,16,35] is often performed to reshape the data distribution.

3 Our Method

3.1 Network Architecture

We illustrate our network architecture in Fig. 2a. Based on a U-shaped network with an encoder-decoder architecture [33], we add skip connections [48] from encoders to decoders for multi-level feature maps. We observe that encoder mainly extracts semantic features, while decoder pays more attention to spatial information. A light-weight attention based global context block (GCB) [4] (Fig. 2b) is sequentially applied to each residual block in the encoding stage to recalibrate channel-wise features. The recalibrated vectors with global context are then combined with high-level features as input for our spatial attention blocks (SAB) through skip connections. Then, the recalibrated features and the high-level features are fused to make the network focus on either the global spatial information or the detail structure at different stages. From a spatial point of view, the channel attention modules are applied to emphasize semantic information *globally*, while the spatial attention modules focus on where to emphasize or suppress *locally*. With skip connections, these two types of attention blocks build a 3D attention map to guide feature selection. The output features of the

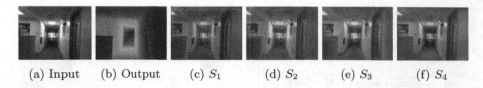

(a) Input   (b) Output   (c) S_1 (d) S_2 (e) S_3 (f) S_4

Fig. 3. Spatial attention maps at different scales visualized as heat maps overlaid on the input image. S_i is the ith attention map from the top layer in the decoder (see Fig. 2c). Red color indicates larger value. Our network attends to different levels of structural detail at different scales. (Color figure online)

four upsampling blocks are upsampled by a factor of 2, 4, 8, and 16, respectively, and then fed to the refinement module to obtain the final depth map which has the same size as the original image. The main purpose of these multi-scale output layers is to fuse information at multiple scales together. The low resolution output retains information with finer global layout, while the high resolution output is used to restore details lost after the downsampling operations.

Spatial Attention Block. Different from channel-attention mechanism which selects semantic feature in network derivation, our SAB is designed to optimize geometric spatial layout in pixel-wise regression tasks. In SAB, we perform a squeeze operation on concatenate features via 1×1 convolution to aggregate spatial context across their channel dimensions. Then we activate local attention to get a 2D attention map which encodes pixel-wise depth information over all spatial locations. The low-level features are multiplied by this 2D attention map for subsequent fusion to deliver the spatial context from a higher-level layer. Therefore, our SAB generates attention maps with richer spatial information to recalibrate the semantic features from GCB. The SAB shown in Fig. 2(c) can be formulated as

$$\mathrm{D}_i = f\Big(\sigma\big(W_1 * f(\mathrm{D}_{i+1}, \mathrm{GCB}_i)\big) \odot \mathrm{GCB}_i, \mathrm{D}_{i+1}\Big), \tag{1}$$

where f is a fusion function (e.g. element-wise sum, element-wise dot product, or concatenation), while $*$ denotes 1×1 or 3×3 convolution and $\odot$ denotes element-wise dot product. Since a depth maps has a wide range of positive values, we use ReLU as the activation function $\sigma(x)$.

As shown in Fig. 3, attention feature maps obtained using our SAB help the network focus on the specific information of relative depth across different levels. Specifically, the attention map S_4 contains low-level features which depict the semantic hierarchy and capture the overall near-and-far structure in 3D space. The closer a spatial attention feature map is to the top layer S_1, the more local details are focused for the prediction output.

3.2   Network Training

The loss function to train our network contains four terms, i.e., Berhu loss $\mathcal{L}_B$, scale-invariant gradient loss $\mathcal{L}_g$, normal loss $\mathcal{L}_n$, and global focal relative loss $\mathcal{L}_r$. We describe each loss term in detail as follows.

BerHu Loss. The BerHu loss refers to the reversed Huber penalty [58] of depth residuals, which provides a reasonable trade-off between L1 norm and L2 norm in regression tasks when the errors present a heavy-tailed distribution [28,39]. Therefore, the BerHu loss $\mathcal{L}_B$ is commonly used as the basic error metric for MDE and is defined as:

$$\mathcal{L}_B = \sum_{i,j} |d_{i,j} - \hat{d}_{i,j}|_b, \quad |x|_b = \begin{cases} |x| & |x| \leq c, \\ \frac{x^2 + c^2}{2c} & |x| > c \end{cases}, \tag{2}$$

where $d_{i,j}$ and $\hat{d}_{i,j}$ are the ground-truth and predicted depth values at the pixel location (i, j), respectively. We set c to be $0.2 \max_p(|\hat{d}_p - d_p|)$, where $\{p\}$ indicate all the pixels in one batch.

Scale-Invariant Gradient Loss. We use scale-invariant gradient loss [46] to emphasize depth discontinuities at object boundaries and to improve smoothness in homogeneous regions. To cover gradients at different scales in our model, we use 5 different spacings $\{s = 1, 2, 4, 8, 16\}$ for this loss term $\mathcal{L}_g$:

$$\mathcal{L}_g = \sum_s \sum_{i,j} |\mathbf{g}_s(i,j) - \hat{\mathbf{g}}_s(i,j)|^2,$$
$$\mathbf{g}_s(i,j) = \left(\frac{d_{i+s,j} - d_{i,j}}{|d_{i+s,j} + d_{i,j}|}, \frac{d_{i,j+s} - d_{i,j}}{|d_{i,j+s} + d_{i,j}|} \right)^\top, \tag{3}$$

Normal Loss. To deal with small-scale structures and to further improve high-frequency details in the predicted depth, we also use a normal loss term $\mathcal{L}_n$:

$$\mathcal{L}_n = \sum_{i,j} \left(1 - \frac{\langle \mathbf{n}_{i,j}, \hat{\mathbf{n}}_{i,j} \rangle}{\sqrt{\langle \mathbf{n}_{i,j}, \mathbf{n}_{i,j} \rangle} \cdot \sqrt{\langle \hat{\mathbf{n}}_{i,j}, \hat{\mathbf{n}}_{i,j} \rangle}} \right), \tag{4}$$

in which $\langle \cdot, \cdot \rangle$ denotes the inner product of two vectors. The surface normal is denoted as $\mathbf{n}_{i,j} = [-\nabla_x, -\nabla_y, 1]^\top$, where ∇_x and ∇_y are gradient vectors along the x and y-axis in the depth map, respectively.

Global Focal Relative Loss. Relative loss (RL) [5,6,30,51] is used to make depth-wise ordinal relations between sample pairs predicted by the network consistent with the ground-truth. Inspired by the focal loss defined in Lin *et al.* [32], we propose an improved version of relative loss, named *global focal relative loss* (GFRL), to put more weight on sample pairs of incorrect ordinal relationships in the prediction. To ensure uniform selection of point pairs, we subdivide the image into 16×16 blocks of the same size and randomly sample one point from each block. Each point is compared with all the other points from the same image when training the network. Our loss term of n pairs is formally defined as $\mathcal{L}_r = \sum_k^n \mathcal{L}_{r,k}$. The k-th pair loss term $\mathcal{L}_{r,k}$ is

$$
\mathcal{L}_{r,k} = \begin{cases} w_k^\gamma \log \left(1 + \exp \left(- r_k \left(d_{1,k} - d_{2,k} \right) \right) \right), & r_k \neq 0 \\ \left(d_{1,k} - d_{2,k} \right)^2, & r_k = 0 \end{cases} \tag{5}
$$
$$
w_k = 1 - 1/\left(1 + \exp \left(- r_k \left(d_{1,k} - d_{2,k} \right) \right) \right),
$$

where r_k is the ground-truth ordinal relationship and is set to -1, 0, or 1, if the first point has a smaller, equal, or a larger depth value compared to the second point. The equality holds if and only if the depth difference ratio is smaller than a threshold of 0.02.

In Eq. 5, our key idea is to introduce a modulating factor w_k^γ to scale the relative loss. Intuitively, when a pair of pixels have incorrect ordinal relationship in the prediction, the loss is unaffected since w_k is close to 1. If the depth ordinal relationship is correct and the depth difference is large enough, the weight w_k on this pair will go to 0. The parameter γ smoothly adjusts the magnitude of weight reduction on easy point pairs. When $\gamma = 0$, our GFRL is equivalent to RL [5]. As γ increases, the impact of the modulating factor becomes larger and we set $\gamma = 2$ in our experiments. It turns out our GFRL outperforms RL under various evaluation metrics (see Sect. 5.2).

Total Loss. We train our network in a sequential fashion for better convergence by using different combinations of loss terms in different stages [19]. Our total loss function $\mathcal{L}_{total}$ is defined as:

$$
\mathcal{L}_{total} = \lambda_1 \mathcal{L}_B^{\text{I–III}} + \lambda_2 \mathcal{L}_g^{\text{II–III}} + \lambda_3 \mathcal{L}_n^{\text{III}} + \lambda_4 \mathcal{L}_r^{\text{III}}, \tag{6}
$$

where $\{\lambda_i\}$ are the weights of different loss terms, and the superscripts I–III denote the stages of using the corresponding terms. BerHu loss is the basic term being used to start the training. After convergence, we first add gradient loss for smooth surface and sharp edges. To further improve the details of predicted depth and refine the spatial structure, we add normal loss and global focal relative loss in the final stage.

Edge-Aware Consistency. Depth discontinuities typically arise at the boundaries of occluding objects in a scene. Most existing MDE methods cannot recover these edges accurately and tend to generate distorted and blurry object boundaries [19]. Besides, we observe that in these areas, the average prediction error is about 4 times larger than that of other areas.

Based on these observations, we introduce an edge-aware consistency scheme to preserve sharp discontinuities in depth prediction results. Specifically, we first use Canny edge detector [2] to extract edges for the ground-truth depth map, and then dilate these edges with a kernel of 5 to get a boundary mask. We multiply the loss $\mathcal{L}_{ij}$ at the pixel p_{ij} by a weight of 5 if p_{ij} is an edge pixel in the boundary mask. Our edge-aware consistency scheme can be considered as a hard example mining method [41] to explicitly increase the penalty on prediction errors in boundary regions.

4 Datasets

4.1 HC Depth Dataset

In recent years, several RGBD datasets have been proposed to provide collections of images with associated depth maps. In Table 1, we list five open source RGBD datasets and summarize their properties such as types of content, annotation methods, and number of images. Among them, NYUDv2 [34], ScanNet [7], CAD [37], and URFall [26] are captured from real indoor scenes using Microsoft Kinect [56], while SUNCG [42] is a synthetic dataset collected by rendering manually created 3D virtual scenes. In addition, CAD and URFall contain videos of humans performing activities in indoor environments. These datasets offer a large number of annotated depth images and are widely used to train models for an MDE task. However, each of these RGBD datasets primarily focuses on only one type of scenes and may not cover enough challenging cases. In Fig. 1, we identify and summarize six types of typical failure cases for two state-of-the-art MDE methods [1,11] trained on the NYUDv2 dataset.

Table 1. Properties of different monocular depth estimation datasets. The last column denotes types of hard cases (see Fig. 1) included in these datasets.

ID	Datasets	Type	Annotation	Images	Hard cases
1	NYU	Indoor	Kinect	108,644	a,b,c
2	ScanNet	Indoor	Kinect	99,571	a,c,f
3	SUNCG	Indoor	Synthetic	454,078	-
4	CAD	Portrait	Kinect	145,155	b
5	URFall	Portrait	Kinect	10,764	b
6	HC Depth	Mixed	Kinect & RealSense	120,060	a,b,c,d,e,f

To complement existing RGBD datasets and provide sufficient coverage on challenging examples for the MDE task, we design and acquire a new dataset, named *HC Depth dataset*, which contains all the six categories of hard cases shown in Fig. 1. Specifically, we collect 24660 images using Microsoft Kinect [56], which provides dense and accurate depth maps for the corresponding RGB images. Due to the limited effective distance range of Kinect, these images are mainly about indoor scenes of portraits. We also collect 95400 images of both indoor and outdoor scenes using Intel RealSense [22], which is capable of measuring larger depth range in medium precision. In *sky* cases, we assign a predefined maximum depth value to sky regions based on semantic segmentation [57]. We also perform surface smoothing and completion for all the cases using the toolbox proposed by Silberman *et al.* [34]. We show several typical examples of our HC Depth dataset in the supplementary materials.

4.2   Incremental Dataset Mixing Strategy

Training on aforementioned datasets together poses a challenge due to the different distributions of depth data in various scenes. Motivated by curriculum learning [3] in global optimization of non-convex functions, we propose an incremental dataset mixing strategy to accelerate the convergence of network training and improve the generalization performance of trained models.

Curriculum learning is related to boosting algorithms, in which difficult examples are gradually added during the training process. In our case of MDE, we divide all the training examples into four main categories based on the content and difficulty, i.e., indoor (I), synthetic (S), portrait (PT), and hard cases (HC). First, we train our model on datasets with similar distributions (e.g., I + S) until convergence. Then we add remaining datasets (e.g., PT or HC) one by one and build a new sampler for each batch to ensure a balanced sampling from these imbalanced datasets. Specifically, we count the number of images k_i contained in each dataset, and $K = \sum_i k_i$ is the total number of training images. The probability of sampling an image from the i-th dataset is proportional to K/k_i to effectively balance different datasets.

5   Experiments

5.1   Experimental Setup

To validate each algorithm component, we first train a baseline model $\mathcal{M}_b$ without any module introduced in Sect. 3. We denote the model trained with edge-aware consistency as $\mathcal{M}_e$ and the model trained with both edge-aware consistency and our spatial attention blocks as $\mathcal{M}_{e,SAB}$. The model obtained with all the components is denoted as $\mathcal{M}_{full}$, which is essentially $\mathcal{M}_{e,SAB}$ trained with the addition of GFRL and $\mathcal{M}_{e,\mathcal{L}_r}$ trained with the addition of SAB. We also compare with the option to train $\mathcal{M}_{e,SAB}$ with an additional relative loss $\mathcal{L}_r$ described in Chen *et al.* [5]. We show additional comparisons with variations of existing attention modules, e.g., Spatial Excitation Block (sSE) [38] and Convolutional Block Attention Module (CBAM) [50]. Finally, we train the full model on different combinations of datasets to evaluate the effect of adding more training data.

We implement our method using PyTorch and train the models on four NVIDIA GPUs with 12 GB memory. We initialize all the ResNet-101 blocks with pretrained ImageNet weights and randomly initialize other layers. We use RMSProp [44] with a learning rate of 10^{-4} for all the layers and reduce the rate by 10% after every 50 epochs. β_1, β_2 and weight decay are set to 0.9, 0.99, and 0.0001 respectively. The batch size is set to be 48. The weights $\{\lambda_i\}$ of the loss terms described in Sect. 3.2 are set to 1, 1, 1, and 0.5 respectively. We pretrain

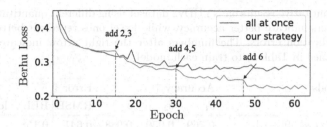

Fig. 4. BerHu loss curves of different training strategies. The red curve denotes the loss curve with all datasets added from the beginning, while the green curve represents the loss curve based on our incremental dataset mixing strategy. The blue dot lines indicate the epochs when new datasets are included. The dataset IDs are defined in Table 1. (Color figure online)

the baseline model $\mathcal{M}_b$ for 40 epochs on NYUDv2 dataset. The total number of trainable parameters for our full model is about 167M. When training on multiple datasets, we first use our multi-stage algorithm (Sect. 3.2) on the first dataset until convergence of the total loss function, and then include more data via our incremental dataset mixing strategy (Sect. 4.2).

We compare our method with several state-of-the-art MDE algorithms both quantitatively and qualitatively using the following commonly used metrics [10]:

- Average relative error (REL): $\frac{1}{n} \sum_p^n \left| d_p - \hat{d}_p \right| / \hat{d}_p$.

- Root mean squared error (RMSE): $\sqrt{\frac{1}{n} \sum_p^n \left(d_p - \hat{d}_p \right)^2}$.

- Average log10 error: $\frac{1}{n} \sum_p^n \left| \log_{10} (d_p) - \log_{10} \left(\hat{d}_p \right) \right|$.

- Threshold accuracy: $\{\delta_i\}$ are ratios of pixels $\{d_p\}$ s.t. $\max \left(\frac{d_p}{\hat{d}_p}, \frac{\hat{d}_p}{d_p} \right) < thr_i$ for $thr_i = 1.25, 1.25^2,$ and 1.25^3.

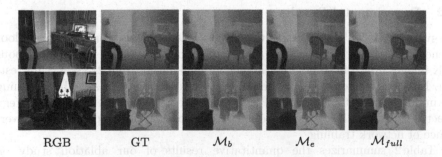

RGB GT $\mathcal{M}_b$ $\mathcal{M}_e$ $\mathcal{M}_{full}$

Fig. 5. Evaluation results of our edge-aware consistency module. From left to right: input images, ground-truth depth, results of our baseline model $\mathcal{M}_b$, the model $\mathcal{M}_e$ with our edge-aware consistency module, and our full model $\mathcal{M}_{full}$.

Table 2. Quantitative results on NYUDv2 dataset using different quantitative metrics. Higher numbers indicate better accuracy, while lower ones represent better results in terms of predication errors. The numbers after the right arrow indicate the IDs of datasets (defined in Table 1) to train each model.

Methods	Accuracy ↑			Error ↓		
	δ_1	δ_2	δ_3	RMSE	REL	log10
Eigen *et al.* [9] → 1	0.769	0.950	0.988	0.641	0.158	-
Laina *et al.* [28] → 1	0.811	0.953	0.988	0.573	0.127	0.055
Qi *et al.* [36] → 1	0.834	0.960	0.990	0.569	0.128	0.057
Hao *et al.* [14] → 1	0.841	0.966	0.991	0.555	0.127	0.053
Hu *et al.* [19] → 1	0.866	0.975	0.993	0.530	0.115	0.050
Fu *et al.* [11] → 1	0.828	0.965	0.992	0.509	0.115	0.051
Alhashim *et al.* [1] → 1	0.846	0.974	0.994	0.465	0.123	0.053
Yin *et al.* [54] → 1	0.875	0.976	0.994	0.416	0.108	0.048
$\mathcal{M}_b$ → 1	0.856	0.974	0.994	0.430	0.120	0.051
$\mathcal{M}_e$ → 1	0.860	0.974	0.994	0.426	0.118	0.050
$\mathcal{M}_{e,SAB}$ → 1	0.864	0.971	0.993	0.417	0.113	0.049
$\mathcal{M}_{full}$ → 1	0.876	0.979	0.995	0.407	0.109	0.047
$\mathcal{M}_{e,SAB} + \mathcal{L}_r$ [5] → 1	0.864	0.971	0.993	0.418	0.113	0.049
$\mathcal{M}_{e,\mathcal{L}_r}$+ sSE [38] → 1	0.857	0.968	0.992	0.433	0.117	0.051
$\mathcal{M}_{e,\mathcal{L}_r}$+ CBAM [50] → 1	0.860	0.968	0.992	0.432	0.117	0.050
$\mathcal{M}_b$ → 1,6	0.868	0.976	0.995	0.420	0.115	0.049
$\mathcal{M}_b$ → 1–6	0.874	0.978	0.995	0.414	0.111	0.048
$\mathcal{M}_{full}$ → 1–3	0.888	0.982	**0.996**	0.391	0.104	0.044
$\mathcal{M}_{full}$ → 1–5	0.888	0.981	**0.996**	0.390	0.103	0.044
$\mathcal{M}_{full}$ → 1,6	0.885	0.979	0.994	0.401	0.104	0.045
$\mathcal{M}_{full}$ → 1–6	**0.899**	**0.983**	**0.996**	**0.376**	**0.098**	**0.042**

5.2   Experimental Results

Results on NYUDv2 Dataset. The NYUDv2 dataset contains 464 indoor scenes. We follow the same train/test split as previous work, i.e., to use about 50K images from 249 scenes for training and 694 images from 215 scenes for testing. In Fig. 4, we compare the BerHu loss curves on the test set when training on multiple datasets *without* and *with* our incremental dataset mixing strategy (Sect. 4.2), which illustrates that our strategy considerably improves the convergence of network training.

Table 2 summarizes the quantitative results of our ablation study on NYUDv2 dataset together with the numbers reported in previous work. Our baseline model $\mathcal{M}_b$ uses ResNet101 as the backbone and couples with GCB modules, combining several widely used loss terms (Berhu loss, scale-invariant gradient loss, and normal loss) of previous work. As can be seen from the table,

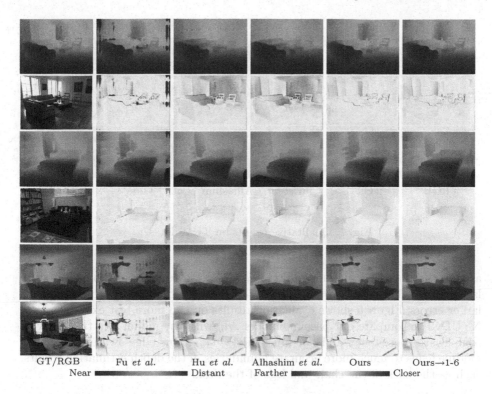

GT/RGB   Fu *et al.*   Hu *et al.*   Alhashim *et al.*   Ours   Ours→1-6
Near ▬▬▬▬▬▬▬▬▬ Distant   Farther ▬▬▬▬▬▬▬▬▬ Closer

Fig. 6. Qualitative results on four test cases from NYUDv2 Dataset. In each example, the first row shows the ground-truth depth map, results of three prior methods [1, 11, 19], and the results of our full model trained solely on NYUDv2 and on the combination of all the six datasets listed in Table 1, respectively. The second row of each example shows the input image and the error maps of the corresponding results in the first row.

all of our algorithm components can improve depth estimation results noticeably, including edge-aware consistency, SAB, and GFRL. Furthermore, adding our HC Depth dataset with our dataset mixing strategy (Sect. 4.2) significantly improves the model performance (see $\mathcal{M}_b \to 1,6$ and $\mathcal{M}_{full} \to 1,6$). By comparing $\mathcal{M}_{full}$ with $\mathcal{M}_{e,SAB} + \mathcal{L}_r$ [5], we can conclude that GFRL leads to better results than the alternative relative loss [5]. We also show that our SAB ($\mathcal{M}_{full}$) brings notable improvement over Spatial Excitation Block ($\mathcal{M}_{e,\mathcal{L}_r} +$ sSE [38]) and Convolutional Block Attention Module ($\mathcal{M}_{e,\mathcal{L}_r} +$ CBAM [50]). Finally, training our full model $\mathcal{M}_{full}$ on multiple datasets can achieve the best results and outperforms training solely on NYUDv2 dataset by a large margin.

Figure 6 presents qualitative results on three test cases to compare our full model with three prior methods [1, 11, 19]. The first example shows that our results have more reasonable spatial layout. In the second example, our model provides more accurate estimation than other methods in the regions of color boundaries. In the third example, our results preserve details in the region

RGB   GT   Fu *et al.*   Alhashim *et al.*   Ours   Ours→1-6

Fig. 7. Qualitative results on TUM dataset.

around the chandelier. Although Fu *et al.* [11] achieves structural patterns similar to our method, their results contain disordered patterns and thus have much larger errors. Figure 5 shows evaluation results on two more test examples from NYUDv2 to compare our baseline model $\mathcal{M}_b$ with the model $\mathcal{M}_e$ trained with edge-aware consistency and our full model $\mathcal{M}_{full}$. Our edge-aware consistency module leads to much more accurate boundaries in the depth estimation results, such as the back of the chair and legs of the desk.

Table 3. Quantitative results of generalization test on TUM dataset. The experimental configuration is the same as Table 2.

Methods	Accuracy ↑			Error ↓		
	δ_1	δ_2	δ_3	RMSE	REL	log10
Hu et al. [19] → 1	0.577	0.842	0.932	1.154	0.216	0.111
Fu et al. [11] → 1	0.598	0.855	0.934	1.145	0.209	0.110
Alhashim et al. [1] → 1	0.567	0.847	0.920	1.250	0.224	0.115
$\mathcal{M}_{full}$ →1	0.606	0.888	0.947	1.109	0.212	0.100
$\mathcal{M}_{full}$ → 1–3	0.665	0.903	0.955	1.031	0.194	0.091
$\mathcal{M}_{full}$ → 1–5	0.710	0.913	0.952	1.013	**0.175**	0.084
$\mathcal{M}_{full}$ →1,6	0.689	0.906	0.950	1.029	0.189	0.086
$\mathcal{M}_{full}$ → 1–6	**0.735**	**0.927**	**0.959**	**0.912**	0.177	**0.081**

Results on TUM Dataset. We use the open source benchmark TUM [43] to evaluate the generalization performance of different methods in the setting of zero-shot cross-dataset inference. The test set of TUM consists of 1815 high quality images taken in a factory including pipes, computer desks and people,

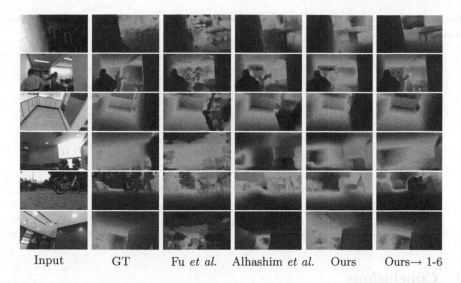

| Input | GT | Fu *et al.* | Alhashim *et al.* | Ours | Ours→ 1-6 |

Fig. 8. Qualitative results on our HC Depth dataset. From top to bottom, the examples correspond to the six types of hard cases showed in Fig. 1.

which are never seen when we train our models. Table 3 demonstrates that our full model $\mathcal{M}_{full}$ trained solely on NYUDv2 dataset outperforms previous methods. Furthermore, training the full model by adding HC Depth dataset or more datasets using our dataset mixing strategy can significantly improve the generalization performance. As shown in the qualitative results in Fig. 7, our method retains sharp edges of plants in the first example and leads to more accurate spatial layout in the second and third examples.

Results on HC Depth Dataset. To test the performance of different models on hard cases, we use the test split of our HC Depth dataset which contains 328 examples. Table 4 summarizes the corresponding quantitative results, from which we can obtain consistent findings with Table 3. Figure 8 illustrates qualitative results of six examples from the test set, which correspond to the six types of hard cases in Fig. 1. As shown in Fig. 8 and Table 4, our method based on NYUDv2 dataset already provides more faithful prediction compared to Alhashim *et al.* [1] and the predicted depth can be further improved considerably by adding our HC Depth dataset or using all the datasets through our dataset mixing strategy.

Table 4. Quantitative results on the HC Depth dataset. The experimental configuration is the same as Table 2.

Methods	Accuracy ↑			Error ↓		
	δ_1	δ_2	δ_3	RMSE	REL	log10
Hu et al. [19] → 1	0.531	0.783	0.898	1.276	0.285	0.128
Fu et al. [11] → 1	0.477	0.755	0.866	1.356	0.2962	0.145
Alhashim et al. [1] → 1	0.551	0.819	0.918	1.137	0.257	0.118
$\mathcal{M}_{full}$ →1	0.600	0.843	0.930	1.070	0.249	0.109
$\mathcal{M}_{full}$ → 1–3	0.610	0.837	0.921	1.072	0.244	0.108
$\mathcal{M}_{full}$ → 1–5	0.633	0.845	0.924	1.065	0.237	0.107
$\mathcal{M}_{full}$ →1,6	0.825	0.895	0.961	0.715	0.190	0.087
$\mathcal{M}_{full}$ → 1–6	**0.879**	**0.965**	**0.988**	**0.566**	**0.113**	**0.048**

6 Conclusions

In this paper we put together a series of coherent efforts to improve the structural awareness in monocular depth estimation, with the effectiveness and necessity of each component thoroughly verified. We introduce a novel encoder-decoder architecture using the spatial attention mechanism, and boost the network performance by proposing a global focal relative loss and an edge-aware consistency module. We further collect a dataset of hard cases for the task of depth estimation and leverage a data mixing strategy based on curriculum learning for effective network training. We validate each component of our method via comprehensive ablation studies and demonstrate substantial advances over state-of-the-art approaches on benchmark datasets. Our experimental results show that truly generic models for monocular depth estimation require not only innovations in network architecture and training algorithm, but also sufficient data for various scenarios.

Our source code, pretrained models, and the HC Depth dataset will be released to encourage follow-up research. In the future, we plan to capture more diverse scenes and further expand our HC Depth dataset. We would also like to deploy our models on mobile devices for several applications such as augmented reality.

Acknowledgements. We would like to thank the anonymous reviewers for their valuable comments, Jiwen Liu for help on preparing our dataset, and Miao Xuan for help on paper proofreading.

References

1. Alhashim, I., Wonka, P.: High Quality Monocular Depth Estimation via Transfer Learning. arXiv preprint arXiv:1812.11941 (2018)
2. Bao, P., Zhang, L., Wu, X.: Canny edge detection enhancement by scale multiplication. IEEE Trans. Pattern Anal. Mach. Intell. **27**(9), 1485–1490 (2005)
3. Bengio, Y., Louradour, J., Collobert, R., Weston, J.: Curriculum learning. In: Proceedings of the 26th Annual International Conference on Machine Learning, pp. 41–48. ACM (2009)
4. Cao, Y., Xu, J., Lin, S., Wei, F., Hu, H.: GCNet: Non-local networks meet squeeze-excitation networks and beyond. arXiv preprint arXiv:1904.11492 (2019)
5. Chen, W., Fu, Z., Yang, D., Deng, J.: Single-image depth perception in the wild. In: Advances in Neural Information Processing Systems, pp. 730–738 (2016)
6. Chen, W., Qian, S., Deng, J.: Learning single-image depth from videos using quality assessment networks. In: Proceedings of the IEEE Conference on Computer Vision and Pattern Recognition, pp. 5604–5613 (2019)
7. Dai, A., Chang, A.X., Savva, M., Halber, M., Funkhouser, T., Nießner, M.: ScanNet: richly-annotated 3D reconstructions of indoor scenes. In: Proceedings of the IEEE Conference on Computer Vision and Pattern Recognition, pp. 5828–5839 (2017)
8. van Dijk, T., de Croon, G.: How do neural networks see depth in single images? In: Proceedings of the IEEE International Conference on Computer Vision, pp. 2183–2191 (2019)
9. Eigen, D., Fergus, R.: Predicting depth, surface normals and semantic labels with a common multi-scale convolutional architecture. In: Proceedings of the IEEE International Conference on Computer Vision, pp. 2650–2658 (2015)
10. Eigen, D., Puhrsch, C., Fergus, R.: Depth map prediction from a single image using a multi-scale deep network. In: Advances in Neural Information Processing Systems, pp. 2366–2374 (2014)
11. Fu, H., Gong, M., Wang, C., Batmanghelich, K., Tao, D.: Deep ordinal regression network for monocular depth estimation. In: Proceedings of the IEEE Conference on Computer Vision and Pattern Recognition, pp. 2002–2011 (2018)
12. Garg, R., B.G., V.K., Carneiro, G., Reid, I.: Unsupervised CNN for single view depth estimation: geometry to the rescue. In: Leibe, B., Matas, J., Sebe, N., Welling, M. (eds.) ECCV 2016. LNCS, vol. 9912, pp. 740–756. Springer, Cham (2016). https://doi.org/10.1007/978-3-319-46484-8_45
13. Gordon, A., Li, H., Jonschkowski, R., Angelova, A.: Depth from videos in the wild: unsupervised monocular depth learning from unknown cameras. arXiv preprint arXiv:1904.04998 (2019)
14. Hao, Z., Li, Y., You, S., Lu, F.: Detail preserving depth estimation from a single image using attention guided networks. In: 2018 International Conference on 3D Vision (3DV), pp. 304–313. IEEE (2018)
15. He, H., Garcia, E.A.: Learning from imbalanced data. IEEE Trans. Knowl. Data Eng. **21**(9), 1263–1284 (2009)
16. He, H., Ma, Y.: Imbalanced Learning: Foundations, Algorithms, and Applications. Wiley, Hoboken (2013)
17. Hirschmuller, H.: Stereo processing by semiglobal matching and mutual information. IEEE Trans. Pattern Anal. Mach. Intell. **30**(2), 328–341 (2008)
18. Hu, H., Gu, J., Zhang, Z., Dai, J., Wei, Y.: Relation networks for object detection. In: Proceedings of the IEEE Conference on Computer Vision and Pattern Recognition, pp. 3588–3597 (2018)

19. Hu, J., Ozay, M., Zhang, Y., Okatani, T.: Revisiting single image depth estimation: toward higher resolution maps with accurate object boundaries. In: WACV, pp. 1043–1051 (2019)
20. Huang, Z., Wang, X., Huang, L., Huang, C., Wei, Y., Liu, W.: CCNet: criss-cross attention for semantic segmentation. In: Proceedings of the IEEE International Conference on Computer Vision, pp. 603–612 (2019)
21. Karsch, K., Liu, C., Kang, S.B.: Depth transfer: depth extraction from video using non-parametric sampling. IEEE Trans. Pattern Anal. Mach. Intell. **36**(11), 2144–2158 (2014)
22. Keselman, L., Iselin Woodfill, J., Grunnet-Jepsen, A., Bhowmik, A.: Intel RealSense stereoscopic depth cameras. In: Proceedings of the IEEE Conference on Computer Vision and Pattern Recognition Workshops, pp. 1–10 (2017)
23. Khamis, S., Fanello, S., Rhemann, C., Kowdle, A., Valentin, J., Izadi, S.: StereoNet: guided hierarchical refinement for real-time edge-aware depth prediction. In: Ferrari, V., Hebert, M., Sminchisescu, C., Weiss, Y. (eds.) ECCV 2018. LNCS, vol. 11219, pp. 596–613. Springer, Cham (2018). https://doi.org/10.1007/978-3-030-01267-0_35
24. Kong, S., Fowlkes, C.: Pixel-wise attentional gating for scene parsing. In: 2019 IEEE Winter Conference on Applications of Computer Vision (WACV), pp. 1024–1033. IEEE (2019)
25. Kuznietsov, Y., Stuckler, J., Leibe, B.: Semi-supervised deep learning for monocular depth map prediction. In: Proceedings of the IEEE Conference on Computer Vision and Pattern Recognition, pp. 6647–6655 (2017)
26. Kwolek, B., Kepski, M.: Human fall detection on embedded platform using depth maps and wireless accelerometer. Comput. Meth. Programs Biomed. **117**(3), 489–501 (2014)
27. Ladicky, L., Shi, J., Pollefeys, M.: Pulling things out of perspective. In: Proceedings of the IEEE Conference on Computer Vision and Pattern Recognition, pp. 89–96 (2014)
28. Laina, I., Rupprecht, C., Belagiannis, V., Tombari, F., Navab, N.: Deeper depth prediction with fully convolutional residual networks. In: 2016 4th International Conference on 3D Vision (3DV), pp. 239–248. IEEE (2016)
29. Lasinger, K., Ranftl, R., Schindler, K., Koltun, V.: Towards Robust Monocular Depth Estimation: Mixing Datasets for Zero-Shot Cross-Dataset Transfer. arXiv preprint arXiv:1907.01341 (2019)
30. Lee, J.H., Kim, C.S.: Monocular depth estimation using relative depth maps. In: Proceedings of the IEEE Conference on Computer Vision and Pattern Recognition, pp. 9729–9738 (2019)
31. Li, Z., et al.: Learning the depths of moving people by watching frozen people. In: Proceedings of the IEEE Conference on Computer Vision and Pattern Recognition, pp. 4521–4530 (2019)
32. Lin, T.Y., Goyal, P., Girshick, R., He, K., Dollár, P.: Focal loss for dense object detection. In: Proceedings of the IEEE International Conference on Computer Vision, pp. 2980–2988 (2017)
33. Mayer, N., et al.: A large dataset to train convolutional networks for disparity, optical flow, and scene flow estimation. In: Proceedings of the IEEE Conference on Computer Vision and Pattern Recognition, pp. 4040–4048 (2016)
34. Silberman, N., Hoiem, D., Kohli, P., Fergus, R.: Indoor segmentation and support inference from RGBD images. In: Fitzgibbon, A., Lazebnik, S., Perona, P., Sato, Y., Schmid, C. (eds.) ECCV 2012. LNCS, vol. 7576, pp. 746–760. Springer, Heidelberg (2012). https://doi.org/10.1007/978-3-642-33715-4_54

35. Oquab, M., Bottou, L., Laptev, I., Sivic, J.: Learning and transferring mid-level image representations using convolutional neural networks. In: Proceedings of the IEEE Conference on Computer Vision and Pattern Recognition, pp. 1717–1724 (2014)

36. Qi, X., Liao, R., Liu, Z., Urtasun, R., Jia, J.: GeoNet: geometric neural network for joint depth and surface normal estimation. In: Proceedings of the IEEE Conference on Computer Vision and Pattern Recognition, pp. 283–291 (2018)

37. Robot Learning Lab at Cornell University: Cornell Activity Datasets: CAD-60 & CAD-120 (2019). http://pr.cs.cornell.edu/humanactivities/data.php

38. Roy, A.G., Navab, N., Wachinger, C.: Concurrent spatial and channel 'squeeze & excitation' in fully convolutional networks. In: Frangi, A.F., Schnabel, J.A., Davatzikos, C., Alberola-López, C., Fichtinger, G. (eds.) MICCAI 2018. LNCS, vol. 11070, pp. 421–429. Springer, Cham (2018). https://doi.org/10.1007/978-3-030-00928-1_48

39. Roy, A., Todorovic, S.: Monocular depth estimation using neural regression forest. In: Proceedings of the IEEE Conference on Computer Vision and Pattern Recognition, pp. 5506–5514 (2016)

40. Saxena, A., Chung, S.H., Ng, A.Y.: Learning depth from single monocular images. In: Advances in Neural Information Processing Systems, pp. 1161–1168 (2006)

41. Shrivastava, A., Gupta, A., Girshick, R.: Training region-based object detectors with online hard example mining. In: Proceedings of the IEEE Conference on Computer Vision and Pattern Recognition, pp. 761–769 (2016)

42. Song, S., Yu, F., Zeng, A., Chang, A.X., Savva, M., Funkhouser, T.: Semantic scene completion from a single depth image. Proceedings of 30th IEEE Conference on Computer Vision and Pattern Recognition (2017)

43. Sturm, J., Engelhard, N., Endres, F., Burgard, W., Cremers, D.: A benchmark for the evaluation of RGB-D SLAM systems. In: 2012 IEEE/RSJ International Conference on Intelligent Robots and Systems, pp. 573–580. IEEE (2012)

44. Tieleman, T., Hinton, G.: Lecture 6.5-rmsprop: divide the gradient by a running average of its recent magnitude. COURSERA Neural Netw. Mach. Learn. 4(2), 26–31 (2012)

45. Torralba, A., Efros, A.A., et al.: Unbiased look at dataset bias. In: Proceedings of the IEEE Conference on Computer Vision and Pattern Recognition, pp. 1521–1528 (2011)

46. Ummenhofer, B., et al.: DeMoN: depth and motion network for learning monocular stereo. In: Proceedings of the IEEE Conference on Computer Vision and Pattern Recognition, pp. 5038–5047 (2017)

47. Wang, F., et al.: Residual attention network for image classification. In: Proceedings of the IEEE Conference on Computer Vision and Pattern Recognition, pp. 3156–3164 (2017)

48. Wang, P., Shen, X., Lin, Z., Cohen, S., Price, B., Yuille, A.L.: Towards unified depth and semantic prediction from a single image. In: Proceedings of the IEEE Conference on Computer Vision and Pattern Recognition, pp. 2800–2809 (2015)

49. Wang, X., Girshick, R., Gupta, A., He, K.: Non-local neural networks. In: Proceedings of the IEEE Conference on Computer Vision and Pattern Recognition, pp. 7794–7803 (2018)

50. Woo, S., Park, J., Lee, J.-Y., Kweon, I.S.: CBAM: convolutional block attention module. In: Ferrari, V., Hebert, M., Sminchisescu, C., Weiss, Y. (eds.) ECCV 2018. LNCS, vol. 11211, pp. 3–19. Springer, Cham (2018). https://doi.org/10.1007/978-3-030-01234-2_1

51. Xian, K., et al.: Monocular relative depth perception with web stereo data supervision. In: Proceedings of the IEEE Conference on Computer Vision and Pattern Recognition, pp. 311–320 (2018)
52. Xu, D., Ricci, E., Ouyang, W., Wang, X., Sebe, N.: Multi-scale continuous CRFs as sequential deep networks for monocular depth estimation. In: Proceedings of the IEEE Conference on Computer Vision and Pattern Recognition, pp. 5354–5362 (2017)
53. Xu, D., Wang, W., Tang, H., Liu, H., Sebe, N., Ricci, E.: Structured attention guided convolutional neural fields for monocular depth estimation. In: Proceedings of the IEEE Conference on Computer Vision and Pattern Recognition, pp. 3917–3925 (2018)
54. Yin, W., Liu, Y., Shen, C., Yan, Y.: Enforcing geometric constraints of virtual normal for depth prediction. In: Proceedings of the IEEE International Conference on Computer Vision, pp. 5684–5693 (2019)
55. Zhang, H., Goodfellow, I., Metaxas, D., Odena, A.: Self-attention generative adversarial networks. arXiv preprint arXiv:1805.08318 (2018)
56. Zhang, Z.: Microsoft kinect sensor and its effect. IEEE Multimedia 19(2), 4–10 (2012)
57. Zhou, B., et al.: Semantic understanding of scenes through the ADE20K dataset. Int. J. Comput. Vis. 127, 302–321 (2018)
58. Zwald, L., Lambert-Lacroix, S.: The Berhu penalty and the grouped effect. arXiv preprint arXiv:1207.6868 (2012)

BMBC: Bilateral Motion Estimation with Bilateral Cost Volume for Video Interpolation

Junheum Park[1]([⊠])[iD], Keunsoo Ko[1][iD], Chul Lee[2][iD], and Chang-Su Kim[1][iD]

[1] School of Electrical Engineering, Korea University, Seoul, Korea
{jhpark,ksko}@mcl.korea.ac.kr, changsukim@korea.ac.kr
[2] Department of Multimedia Engineering, Dongguk University, Seoul, Korea
chullee@dongguk.edu

Abstract. Video interpolation increases the temporal resolution of a video sequence by synthesizing intermediate frames between two consecutive frames. We propose a novel deep-learning-based video interpolation algorithm based on bilateral motion estimation. First, we develop the bilateral motion network with the bilateral cost volume to estimate bilateral motions accurately. Then, we approximate bi-directional motions to predict a different kind of bilateral motions. We then warp the two input frames using the estimated bilateral motions. Next, we develop the dynamic filter generation network to yield dynamic blending filters. Finally, we combine the warped frames using the dynamic blending filters to generate intermediate frames. Experimental results show that the proposed algorithm outperforms the state-of-the-art video interpolation algorithms on several benchmark datasets. The source codes and pre-trained models are available at https://github.com/JunHeum/BMBC.

Keywords: Video interpolation · Bilateral motion · Bilateral cost volume

1 Introduction

A low temporal resolution causes aliasing, yields abrupt motion artifacts, and degrades the video quality. In other words, the temporal resolution is an important factor affecting video quality. To enhance temporal resolutions, many video interpolation algorithms [2–4,12,14,16–18,20–22] have been proposed, which synthesize intermediate frames between two actual frames. These algorithms are widely used in applications, including visual quality enhancement [32], video compression [7], slow-motion video generation [14], and view synthesis [6]. However, video interpolation is challenging due to diverse factors, such as large and

Electronic supplementary material The online version of this chapter (https://doi.org/10.1007/978-3-030-58568-6_7) contains supplementary material, which is available to authorized users.

nonlinear motions, occlusions, and variations in lighting conditions. Especially, to generate a high-quality intermediate frame, it is important to estimate motions or optical flow vectors accurately.

Recently, with the advance of deep-learning-based optical flow methods [5, 10,25,30], flow-based video interpolation algorithms [2,3,14] have been developed, yielding reliable interpolation results. Niklaus et al. [20] generated intermediate frames based on the forward warping. However, the forward warping may cause interpolation artifacts because of the hole and overlapped pixel problems. To overcome this, other approaches leverage the backward warping. To use the backward warping, intermediate motions should be obtained. Various video interpolation algorithms [2–4,14,16,20,32] based on the bilateral motion estimation approximate these intermediate motions from optical flows between two input frames. However, this approximation may degrade video interpolation results.

In this work, we propose a novel video interpolation network, which consists of the bilateral motion network and the dynamic filter generation network. First, we predict six bilateral motions: two from the bilateral motion network and the other four through optical flow approximation. In the bilateral motion network, we develop the bilateral cost volume to facilitate the matching process. Second, we extract context maps to exploit rich contextual information. We then warp the two input frames and the corresponding context maps using the six bilateral motions, resulting in six pairs of warped frame and context map. Next, these pairs are used to generate dynamic blending filters. Finally, the six warped frames are superposed by the blending filters to generate an intermediate frame. Experimental results demonstrate that the proposed algorithm outperforms the state-of-the-art video interpolation algorithms [2,3,18,22,32] meaningfully on various benchmark datasets.

This work has the following major contributions:

- We develop a novel deep-learning-based video interpolation algorithm based on the bilateral motion estimation.
- We propose the bilateral motion network with the bilateral cost volume to estimate intermediate motions accurately.
- The proposed algorithm performs better than the state-of-the-art algorithms on various benchmark datasets.

2   Related Work

2.1   Deep-Learning-Based Video Interpolation

The objective of video interpolation is to enhance a low temporal resolution by synthesizing intermediate frames between two actual frames. With the great success of CNNs in various image processing and computer vision tasks, many deep-learning-based video interpolation techniques have been developed. Long et al. [18] developed a CNN, which takes a pair of frames as input and then directly generates an intermediate frame. However, their algorithm yields severe blurring since it does not use a motion model. In [19], PhaseNet was proposed using the phase-based motion representation. Although it yields robust results to lightning changes or motion blur, it may fail to faithfully reconstruct detailed texture. In [21,22], Niklaus et al. proposed kernel-based methods that estimate an adaptive convolutional kernel for each pixel. The kernel-based methods produce reasonable results, but they cannot handle motions larger than a kernel's size.

To exploit motion information explicitly, flow-based algorithms have been developed. Niklaus and Liu [20] generated an intermediate frame from two consecutive frames using the forward warping. However, the forward warping suffers from holes and overlapped pixels. Therefore, most flow-based algorithms are based on backward warping. In order to use backward warping, intermediate motions (*i.e.* motion vectors of intermediate frames) should be estimated. Jiang et al. [14] estimated optical flows and performed bilateral motion approximation to predict intermediate motions from the optical flows. Bao et al. [3] approximated intermediate motions based on the flow projection. However, large errors may occur when two flows are projected onto the same pixel. In [2], Bao et al. proposed an advanced projection method using the depth information. However, the resultant intermediate motions are sensitive to the depth estimation performance. To summarize, although the backward warping yields reasonable video interpolation results, its performance degrades severely when intermediate motions are unreliable or erroneous. To solve this problem, we propose the bilateral motion network to estimate intermediate motions directly.

2.2 Cost Volume

A cost volume records similarity scores between two data. For example, in pixelwise matching between two images, the similarity is computed between each pixel pair: one in a reference image and the other in a target image. Then, for each reference pixel, the target pixel with the highest similarity score becomes the matched pixel. The cost volume facilitates this matching process. Thus, the optical flow estimation techniques in [5,10,30,31] are implemented using cost volumes. In [5,10,31], a cost volume is computed using various features of two video frames, and optical flow is estimated using the similarity information in the cost volume through a CNN. Sun et al. [30] proposed a partial cost volume to significantly reduce the memory requirement while improving the motion estimation accuracy based on a reduced search region. In this work, we develop a novel cost volume, called bilateral cost volume, which is different from the conventional volumes in that its reference is an intermediate frame to be interpolated, instead of one of the two input frames.

3   Proposed Algorithm

Figure 1 is an overview of the proposed algorithm that takes successive frames I_0 and I_1 as input and synthesizes an intermediate frame I_t at $t \in (0,1)$ as output. First, we estimate two 'bilateral' motions $V_{t \to 0}$ and $V_{0 \to t}$ between the input

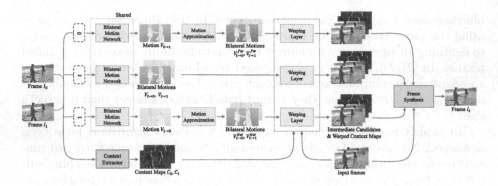

Fig. 1. An overview of the proposed video interpolation algorithm.

frames. Second, we estimate 'bi-directional' motions $V_{0\to1}$ and $V_{1\to0}$ between I_0 and I_1 and then use these motions to approximate four further bilateral motions. Third, the pixel-wise context maps C_0 and C_1 are extracted from I_0 and I_1. Then, the input frames and corresponding context maps are warped using the six bilateral motions. Note that, since the warped frames become multiple candidates of the intermediate frame, we refer to each warped frame as an intermediate candidate. The dynamic filter network then takes the input frames, and the intermediate candidates with the corresponding warped context maps to generate the dynamic filters for aggregating the intermediate candidates. Finally, the intermediate frame I_t is synthesized by applying the blending filters to the intermediate candidates.

3.1   Bilateral Motion Estimation

Given the two input frames I_0 and I_1, the goal is to predict the intermediate frame I_t using motion information. However, it is impossible to directly estimate the intermediate motion between the intermediate frame I_t and one of the input frames I_0 or I_1 because there is no image information of I_t. To address this issue, we assume linear motion between successive frames. Specifically, we attempt to estimate the backward and forward motion vectors $V_{t\to0}(\mathbf{x})$ and $V_{t\to1}(\mathbf{x})$ at $\mathbf{x}$, respectively, where $\mathbf{x}$ is a pixel location in I_t. Based on the linear assumption, we have $V_{t\to0}(\mathbf{x}) = -\frac{t}{1-t} \times V_{t\to1}(\mathbf{x})$.

We develop a CNN to estimate bilateral motions $V_{t\to0}$ and $V_{t\to1}$ using I_0 and I_1. To this end, we adopt an optical flow network, PWC-Net [30], and extend it for the bilateral motion estimation. Figure 2 shows the key components of the modified PWC-Net. Let us describe each component subsequently.

Warping Layer: The original PWC-Net uses the previous frame I_0 as a reference and the following frame I_1 as a target. On the other hand, the bilateral motion estimation uses the intermediate frame I_t as a reference, and the input frames I_0 and I_1 as the target. Thus, whereas the original PWC-Net warps

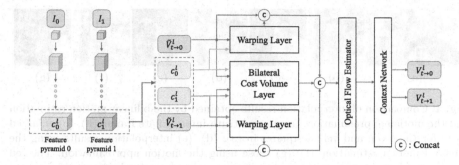

Fig. 2. The architecture of the bilateral motion network: The feature maps c_0^l and c_1^l of the previous and following frames I_0 and I_1 at the lth level and the up-sampled motion fields $\widetilde{V}_{t\rightarrow0}^l$ and $\widetilde{V}_{t\rightarrow1}^l$ estimated at the $(l-1)$th level are fed into the CNN to generate the motion fields $V_{t\rightarrow0}^l$ and $V_{t\rightarrow1}^l$ at the lth level.

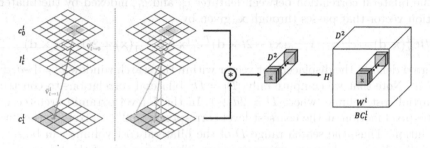

Fig. 3. Illustration of the bilateral cost volume layer for a specific time t.

the feature c_1^l of I_1 toward the feature c_0^l of I_0, we warp both features c_0^l and c_1^l toward the intermediate frame, leading to $c_{0\rightarrow t}^l$ and $c_{1\rightarrow t}^l$, respectively. We employ the spatial transformer networks [11] to achieve the warping. Specifically, a target feature map c_{tgt} is warped into a reference feature map c_{ref} using a motion vector field by

$$c_{\text{ref}}^w(\mathbf{x}) = c_{\text{tgt}}\big(\mathbf{x} + V_{\text{ref}\rightarrow\text{tgt}}(\mathbf{x})\big) \tag{1}$$

where $V_{\text{ref}\rightarrow\text{tgt}}$ is the motion vector field from the reference to the target.

Bilateral Cost Volume Layer: A cost volume has been used to store the matching costs associating with a pixel in a reference frame with its corresponding pixels in a single target frame [5,9,30,31]. However, in the bilateral motion estimation, because the reference frame does not exist and should be predicted from two target frames, the conventional cost volume cannot be used. Thus, we develop a new cost volume for the bilateral motion estimation, which we refer to as the bilateral cost volume.

Figure 3 illustrates the proposed bilateral cost volume generation that takes the features c_0^l and c_1^l of the two input frames and the up-sampled bilateral

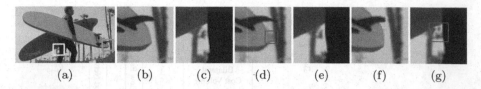

|(a)|(b)|(c)|(d)|(e)|(f)|(g)|

Fig. 4. Comparison of interpolation results obtained by the bilinear motion estimation and the motion approximation: (a) ground-truth intermediate frame; (b), (c) enlarged parts for the green and yellow squares in (a); (d), (e) interpolation results using the bilateral motion estimation; (f), (g) those using the motion approximation. The red squares in (d) and (g) contain visual artifacts caused by motion inaccuracies.

motion fields $\widetilde{V}_{t\to0}^l$ and $\widetilde{V}_{t\to1}^l$ estimated at the $(l-1)$th level. Let $\mathbf{x}$ denote a pixel location in the intermediate frame I_t^l. Then, we define the matching cost as the bilateral correlation between features c_0^l and c_1^l, indexed by the bilateral motion vector that passes through $\mathbf{x}$, given by

$$BC_t^l(\mathbf{x},\mathbf{d}) = c_0^l(\mathbf{x} + \widetilde{V}_{t\to0}^l(\mathbf{x}) - 2t \times \mathbf{d})^T c_1^l(\mathbf{x} + \widetilde{V}_{t\to1}^l(\mathbf{x}) + 2(1-t) \times \mathbf{d}) \quad (2)$$

where $\mathbf{d}$ denotes the displacement vector within the search window $\mathcal{D} = [-d,d] \times [-d,d]$. Note that we compute only $|\mathcal{D}| = D^2$ bilateral correlations to construct a partial cost volume, where $D = 2d + 1$. In the L-level pyramid architecture, a one-pixel motion at the coarsest level corresponds to 2^{L-1} pixels at the finest resolution. Thus, the search range D of the bilateral cost volume can be set to a small value to reduce the memory usage. The dimension of the bilateral cost volume at the lth level is $D^2 \times H^l \times W^l$, where H^l and W^l denote the height and width of the lth level features, respectively. Also, the up-sampled bilateral motions $\widetilde{V}_{t\to0}^l$ and $\widetilde{V}_{t\to1}^l$ are set to zero at the coarsest level.

Most conventional video interpolation algorithms generate a single intermediate frame at the middle of two input frames, i.e. $t = 0.5$. Thus, they cannot yield output videos with arbitrary frame rates. A few recent algorithms [2,14] attempt to interpolate intermediate frames at arbitrary time instances $t \in (0,1)$. However, because their approaches are based on the approximation, as the time instance gets far from either of the input frames, the quality of the interpolated frame gets worse. On the other hand, the proposed algorithm takes into account the time instance $t \in [0,1]$ during the computation of the bilateral cost volume in (2). Also, after we train the bilateral motion network with the bilateral cost volume, we can use the shared weights to estimate the bilateral motions at an arbitrary $t \in [0,1]$. In the extreme cases $t = 0$ or $t = 1$, the bilateral cost volume becomes identical to the conventional cost volume in [5,10,24,30,31], which is used to estimate the bi-directional motions $V_{0\to1}$ and $V_{1\to0}$ between input frames.

3.2 Motion Approximation

Although the proposed bilateral motion network effectively estimates motion fields $V_{t\to0}$ and $V_{t\to1}$ from the intermediate frame at t to the previous and fol-

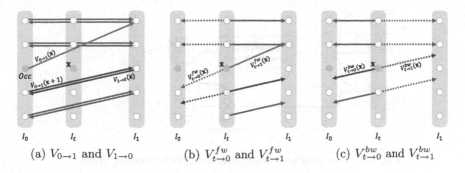

(a) $V_{0 \to 1}$ and $V_{1 \to 0}$ (b) $V_{t \to 0}^{fw}$ and $V_{t \to 1}^{fw}$ (c) $V_{t \to 0}^{bw}$ and $V_{t \to 1}^{bw}$

Fig. 5. Motion approximation: bi-directional motions in (a) are used to approximate forward bilateral motions in (b) and backward bilateral motions in (c).

lowing frames, it may fail to find accurate motions, especially at occluded regions. For example, Fig. 4(d) and (e) show that the interpolated regions, reconstructed by the bilateral motion estimation, contain visual artifacts. To address this issue and improve the quality of an interpolated frame, in addition to the bilateral motion estimation, we develop an approximation scheme to predict a different kind of bilateral motions $V_{t \to 0}$ and $V_{t \to 1}$ using the bi-directional motions $V_{0 \to 1}$ and $V_{1 \to 0}$ between the two input frames.

Figure 5 illustrates this motion approximation, in which each column represents a frame at a time instance and a dot corresponds to a pixel in the frame. In particular, in Fig. 5(a), an occluded pixel in I_0 is depicted by a green dot. To complement the inaccuracy of the bilateral motion at pixel $\mathbf{x}$ in I_t, we use two bi-directional motions $V_{0 \to 1}(\mathbf{x})$ and $V_{1 \to 0}(\mathbf{x})$, which are depicted by green and red lines, respectively. We approximate two forward bilateral motions $V_{t \to 1}^{fw}$ and $V_{t \to 0}^{fw}$ in Fig. 5(b) using $V_{0 \to 1}$. Specifically, for pixel $\mathbf{x}$ in I_t, depicted by an orange dot, we approximate a motion vector $V_{t \to 1}^{fw}(\mathbf{x})$ by scaling $V_{0 \to 1}(\mathbf{x})$ with a factor $(1 - t)$, assuming that the motion vector field is locally smooth. Since the bilateral motion estimation is based on the assumption that a motion trajectory between consecutive frames is linear, two approximate motions $V_{t \to 1}^{fw}(\mathbf{x})$ and $V_{t \to 0}^{fw}(\mathbf{x})$ should be symmetric with respect to $\mathbf{x}$ in I_t. Thus, we obtain an additional approximate vector $V_{t \to 0}^{fw}(\mathbf{x})$ by reversing the direction of the vector $V_{t \to 1}^{fw}(\mathbf{x})$. In other words, we approximate the forward bilateral motions by

$$V_{t \to 1}^{fw}(\mathbf{x}) = (1 - t) \times V_{0 \to 1}(\mathbf{x}), \tag{3}$$

$$V_{t \to 0}^{fw}(\mathbf{x}) = (-t) \times V_{0 \to 1}(\mathbf{x}). \tag{4}$$

Similarly, we approximate the backward bilateral motions by

$$V_{t \to 0}^{bw}(\mathbf{x}) = t \times V_{1 \to 0}(\mathbf{x}), \tag{5}$$

$$V_{t \to 1}^{bw}(\mathbf{x}) = -(1 - t) \times V_{1 \to 0}(\mathbf{x}), \tag{6}$$

as illustrated in Fig. 5(c). Note that Jiang et al. [14] also used these Eqs. (3)–(6), but derived only two motion candidates: $V_{t \to 1}(\mathbf{x})$ by combining (3) and (6) and

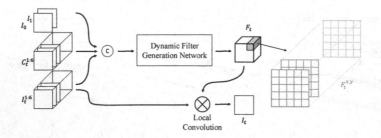

Fig. 6. Frame synthesis using dynamic local blending filters.

$V_{t\to 0}(\mathbf{x})$ by combining (4) and (5). Thus, if an approximated motion in (3)–(6) is unreliable, the combined one is also degraded. In contrast, we use all four candidates in (3)–(6) directly to choose reliable motions in Sect. 3.3.

Figure 4 shows that, whereas the bilateral motion estimation provides visual artifacts in (d), the motion approximation provides results without noticeable artifacts in (f). On the other hand, the bilateral motion estimation is more effective than the motion approximation in the cases of (e) and (g). Thus, the two schemes are complementary to each other.

3.3   Frame Synthesis

We interpolate an intermediate frame by combining the six intermediate candidates, which are warped by the warping layers in Fig. 1. If we consider only color information, rich contextual information in the input frames may be lost during the synthesis [16,20,27], degrading the interpolation performance. Hence, as in [2,3,20], we further exploit contextual information in the input frames, called context maps. Specifically, we extract the output of the conv1 layer of ResNet-18 [8] as a context map, which is done by the context extractor in Fig. 1.

By warping the two input frames and the corresponding context maps, we obtain six pairs of a warped frame and its context map: two pairs are reconstructed using the bilateral motion estimation, and four pairs using the motion approximation. Figure 1 shows these six pairs. Since these six warped pairs have different characteristics, they are used as complementary candidates of the intermediate frame. Recent video interpolation algorithms employ synthesis neural networks, which take warped frames as input and yield final interpolation results or residuals to refine pixel-wise blended results [2,3,20]. However, these synthesis networks may cause artifacts if motions are inaccurately estimated. To alleviate these artifacts, instead, we develop a dynamic filter network [13] that takes the aforementioned six pairs of candidates as input and outputs local blending filters, which are then used to process the warped frames to yield the intermediate frame. These local blending filters compensate for motion inaccuracies, by considering spatiotemporal neighboring pixels in the stack of warped frames. The frame synthesis layer performs this synthesis in Fig. 1.

Dynamic Local Blending Filters: Figure 6 shows the proposed synthesis network using dynamic local blending filters. The coefficients of the filters are learned from the images and contextual information through a dynamic blending filter network [13]. We employ the residual dense network [34] as the backbone for the filter generation. In Fig. 6, the generation network takes the input frames I_0 and I_1 and the intermediate candidates $\{I_t^{1:6}\}$ with the corresponding context maps $\{C_t^{1:6}\}$ as input. Then, for each pixel $\mathbf{x} = (x, y)$, we generate six blending filters to fuse the six intermediate candidates, given by

$$F_t^{x,y} \in \mathbb{R}^{5 \times 5 \times 6}. \tag{7}$$

For each $\mathbf{x}$, the sum of all coefficients in the six filters are normalized to 1.

Then, the intermediate frame is synthesized via the dynamic local convolution. More specifically, the intermediate frame is obtained by filtering the intermediate candidates, given by

$$I_t(x, y) = \sum_{c=1}^{6} \sum_{i=-2}^{2} \sum_{j=-2}^{2} F_t^{x,y}(i, j, c) I_t^c(x + i, y + j). \tag{8}$$

3.4   Training

The proposed algorithm includes two neural networks: the bilateral motion network and the dynamic filter generation network. We found that separate training of these two networks is more efficient than the end-to-end training in training time and memory space. Thus, we first train the bilateral motion network. Then, after fixing it, we train the dynamic filter generation network.

Bilateral Motion Network: To train the proposed bilateral motion network, we define the bilateral loss $\mathcal{L}_b$ as

$$\mathcal{L}_b = \mathcal{L}_p + \mathcal{L}_s \tag{9}$$

where $\mathcal{L}_p$ and $\mathcal{L}_s$ are the photometric loss [26,33] and the smoothness loss [17].

For the photometric loss, we compute the sum of differences between a ground-truth frame I_t^l and two warped frames $I_{0 \to t}^l$ and $I_{1 \to t}^l$ using the bilateral motion fields $V_{t \to 0}^l$ and $V_{t \to 1}^l$, respectively, at all pyramid levels,

$$\mathcal{L}_p = \sum_{l=1}^{L} \alpha_l \left[\sum_{\mathbf{x}} \rho(I_{0 \to t}^l(\mathbf{x}) - I_t^l(\mathbf{x})) + \rho(I_{1 \to t}^l(\mathbf{x}) - I_t^l(\mathbf{x})) \right] \tag{10}$$

where $\rho(x) = \sqrt{x^2 + \epsilon^2}$ is the Charbonnier function [23]. The parameters α_l and ϵ are set to 0.01×2^l and 10^{-6}, respectively. Also, we compute the smoothness loss to constrain neighboring pixels to have similar motions, given by

$$\mathcal{L}_s = \|\nabla V_{t \to 0}\|_1 + \|\nabla V_{t \to 1}\|_1. \tag{11}$$

We use the Adam optimizer [15] with a learning rate of $\eta = 10^{-4}$ and shrink it via $\eta \leftarrow 0.5\eta$ at every 0.5M iterations. We use a batch size of 4 for 2.5M iterations and augment the training dataset by randomly cropping 256×256 patches with random flipping and rotations.

Dynamic Filter Generation Network: We define the dynamic filter loss $\mathcal{L}_d$ as the Charbonnier loss between I_t and its synthesized version $\hat{I}_t$, given by

$$\mathcal{L}_d = \sum_{\mathbf{x}} \rho(\hat{I}_t(\mathbf{x}) - I_t(\mathbf{x})). \tag{12}$$

Similarly to the bilateral motion network, we use the Adam optimizer with $\eta = 10^{-4}$ and shrink it via $\eta \leftarrow 0.5\eta$ at 0.5M, 0.75M, and 1M iterations. We use a batch size of 4 for 1.25M iterations. Also, we use the same augmentation technique as that for the bilateral motion network.

Datasets: We use the Vimeo90K dataset [32] to train the proposed networks. The training set in Vimeo90K is composed of 51,312 triplets with a resolution of 448×256. We train the bilateral motion network with $t = 0.5$ at the first 1M iterations and then with $t \in \{0, 0.5, 1\}$ for fine-tuning. Next, we train the dynamic filter generation network with $t = 0.5$. However, notice that both networks are capable of handling any $t \in (0, 1)$ using the bilateral cost volume in (2).

4   Experimental Results

We evaluate the performances of the proposed video interpolation algorithm on the Middlebury [1], Vimeo90K [32], UCF101 [28], and Adobe240-fps [29] datasets. We compare the proposed algorithm with state-of-the-art algorithms. Then, we conduct ablation studies to analyze the contributions of the proposed bilateral motion network and dynamic filter generation network.

4.1   Datasets

Middlebury: The Middlebury benchmark [1], the most commonly used benchmark for video interpolation, provides two sets: Other and Evaluation. 'Other' contains the ground-truth for fine-tuning, while 'Evaluation' provides two frames selected from each of 8 sequences for evaluation.

Vimeo90K: The test set in Vimeo90K [32] contains 3,782 triplets of spatial resolution 256×448. It is not used to train the model.

UCF101: The UCF101 dataset [28] contains human action videos of resolution 256×256. Liu et al. [17] constructed the test set by selecting 379 triplets.

Adobe240-fps: Adobe240-fps [29] consists of high frame-rate videos. To assess the interpolation performance, we selected a test set of 254 sequences, each of which consists of nine frames.

Table 1. Quantitative comparisons on the Middlebury Evaluation set. For each metric, the numbers in red and <u>blue</u> denote the best and the second best results, respectively.

	[22]	[32]	[14]	[20]	[3]	[2]	Ours
Mequon	2.52	2.54	2.51	2.24	2.47	<u>2.38</u>	2.30
Schefflera	3.56	3.70	3.66	2.96	3.49	3.28	<u>3.07</u>
Urban	4.17	3.43	2.91	4.32	4.63	3.32	<u>3.17</u>
Teddy	5.41	5.05	5.05	4.21	4.94	4.65	<u>4.24</u>
Backyard	10.2	9.84	9.56	9.59	8.91	<u>7.88</u>	7.79
Basketball	5.47	5.34	5.37	5.22	<u>4.70</u>	4.73	4.08
Dumptruck	6.88	6.88	6.69	7.02	6.46	<u>6.36</u>	5.63
Evergreen	6.63	7.14	6.73	6.66	6.35	<u>6.26</u>	5.55
Average	5.61	5.49	5.31	5.28	5.24	<u>4.86</u>	4.48

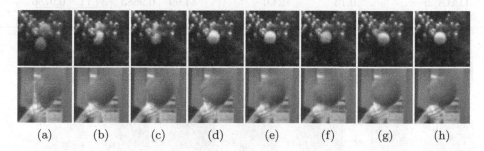

  (a)   (b)   (c)   (d)   (e)   (f)   (g)   (h)

Fig. 7. Visual comparison on the Middlebury Evaluation set. (a) Input, (b) SepConv-L_1 [22], (c) ToFlow [32], (d) SuperSlomo [14], (e) CtxSyn [20], (f) MEMC-Net* [3], (g) DAIN [2], and (h) BMBC (Ours).

4.2   Comparison with the State-of-the-Arts

We assess the interpolation performances of the proposed algorithm in comparison with the conventional video interpolation algorithms: MIND [18], DVF [17], SpyNet [25], SepConv [22], CtxSyn [20], ToFlow [32], SuperSloMo [14], MEMC-Net [3], CyclicGen [16], and DAIN [2]. For SpyNet, we generated intermediate frames using the Baker et al.'s algorithm [1].

Table 1 shows the comparisons on the Middlebury Evaluation set [1], which are also available on the Middlebury website. We compare the average interpolation error (IE). A lower IE indicates better performance. The proposed algorithm outperforms all the state-of-the-art algorithms in terms of average IE score. Figure 7 visually compares interpolation results. SepConv-L_1 [22], ToFlow [32] SuperSlomo [14], CtxSyn [20], MEMC-Net [3], and DAIN [2] yield blurring artifacts around the balls, losing texture details. On the contrary, the proposed algorithm reconstructs the clear shapes of the balls, preserving the details faithfully.

Table 2. Quantitative comparisons on the UCF101 and Vimeo90K datasets.

	Runtime (seconds)	#Parameters (million)	UCF101[17] PSNR	UCF101[17] SSIM	Vimeo90K [32] PSNR	Vimeo90K [32] SSIM
SpyNet[25]	0.11	1.20	33.67	0.9633	31.95	0.9601
MIND[18]	0.01	7.60	33.93	0.9661	33.50	0.9429
DVF[17]	0.47	1.60	34.12	0.9631	31.54	0.9462
ToFlow[32]	0.43	1.07	34.58	0.9667	33.73	0.9682
SepConv-L_f[22]	0.20	21.6	34.69	0.9655	33.45	0.9674
SepConv-L_1[22]	0.20	21.6	34.78	0.9669	33.79	0.9702
MEMC-Net[3]	0.12	70.3	34.96	0.9682	34.29	0.9739
CyclicGen[16]	0.09	3.04	<u>35.11</u>	<u>0.9684</u>	32.09	0.9490
CyclicGen_large[16]	-	19.8	34.69	0.9658	31.46	0.9395
DAIN[2]	0.13	24.0	34.99	0.9683	<u>34.71</u>	<u>0.9756</u>
BMBC (Ours)	0.77	11.0	35.15	0.9689	35.01	0.9764

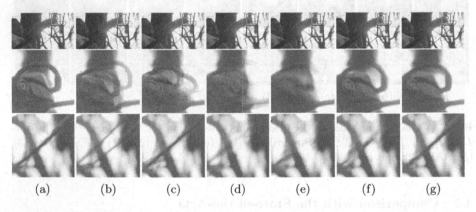

(a)   (b)   (c)   (d)   (e)   (f)   (g)

Fig. 8. Visual comparison on the Viemo90K test set. (a) Ground-truth, (b) ToFlow [32], (c) SepConv-L_f [22], (d) CyclicGen [16], (e) MEMC-Net* [3], (f) DAIN [2], and (g) BMBC (Ours).

In Table 2, we provide quantitative comparisons on the UCF101 [28] and Vimeo90K [32] datasets. We compute the average PSNR and SSIM scores. The proposed algorithm outperforms the conventional algorithms by significant margins. Especially, the proposed algorithm provides 0.3 dB higher PSNR than DAIN [2] on Vimeo90K. Figure 8 compares interpolation results qualitatively. Because the cable moves rapidly and the background branches make the motion estimation difficult, all the conventional algorithms fail to reconstruct the cable properly. In contrast, the proposed algorithm faithfully interpolates the intermediate frame, providing fine details.

Table 3. Quantitative comparisons on the Adobe240-fps dataset for ×2, ×4, and ×8 frame interpolation.

	×2		×4		×8	
	PSNR	SSIM	PSNR	SSIM	PSNR	SSIM
ToFlow[32]	28.51	0.8731	29.20	0.8807	28.93	0.8812
SepConv-L_f[22]	29.14	0.8784	29.75	0.8907	30.07	0.8956
SepConv-L_1[22]	29.31	0.8815	<u>29.91</u>	<u>0.8935</u>	30.23	<u>0.8985</u>
CyclicGen[16]	<u>29.39</u>	0.8787	29.72	0.8889	30.18	0.8972
CyclicGen_large[16]	28.90	0.8682	29.70	0.8866	<u>30.24</u>	0.8955
DAIN[2]	29.35	<u>0.8820</u>	29.73	0.8925	30.03	0.8983
BMBC (Ours)	29.49	0.8832	30.18	0.8964	30.60	0.9029

The proposed algorithm can interpolate an intermediate frame at any time instance $t \in (0, 1)$. To demonstrate this capability, we assess the ×2, ×4, and ×8 frame interpolation performance on the Adobe240-fps dataset [29]. Because the conventional algorithms in Table 3, except for DAIN [2], can generate only intermediate frames at $t = 0.5$, we recursively apply those algorithms to interpolate intermediate frames at other t's. Table 3 shows that the proposed algorithm outperforms all the state-of-the-art algorithms. As the frame rate increases, the performance gain of the proposed algorithm against conventional algorithms gets larger.

4.3   Model Analysis

We conduct ablation studies to analyze the contributions of the three key components in the proposed algorithm: bilateral cost volume, intermediate motion approximation, and dynamic filter generation network. By comparing various combinations of intermediate candidates, we analyze the efficacy of the bilateral cost volume and the intermediate motion approximation jointly.

Intermediate Candidates: To analyze the effectiveness of the bilateral motion estimation and the intermediate motion approximation, we train the proposed networks to synthesize intermediate frames using the following combinations:

- Appx4: Four intermediate candidates, obtained using approximated bilateral motions in (3)–(6), are combined.
- BM: Two intermediate candidates, obtained using bilateral motions, are combined.
- BM+Appx2: In addition to BM, two more candidates obtained using approximated bilateral motions $V_{t\to0}^{fw}$ in (4) and $V_{t\to1}^{bw}$ in (6) are used.
- BM+Appx4: Six intermediate candidates are used as well (proposed model).

Table 4 compares these models quantitatively. First, Appx4 shows the worst performance, while it is still comparable to the state-of-the-art algorithms. Second, BM provides better performance than Appx4 as well as the state-of-the-art

Table 4. PSNR comparison of combination of the intermediate candidates.

Intermediate candidates	UCF101 [28] PSNR	Vimeo90K [32] PSNR
Appx4	34.99	34.72
BM	35.12	34.93
BM+Appx2	35.14	34.95
BM+Appx4	35.15	35.01

Table 5. Analysis of the dynamic filter generation network. In all settings, the six warped frames are input to the network.

Kernel size	Input to filter generation network		UCF101 [28]	Vimeo90K [32]
	Input frames	Context maps	PSNR	PSNR
5 × 5			34.98	34.81
3 × 3	✓		35.09	34.90
5 × 5	✓		35.08	34.96
7 × 7	✓		35.02	34.98
5 × 5	✓	✓	35.15	35.01

algorithms, which confirms the superiority of the proposed BM to the approximation. Third, we can achieve even higher interpolation performance with more intermediate candidates obtained through the motion approximation.

Dynamic Blending Filters: We analyze the optimal kernel size and the input to the dynamic filter generation network. Table 5 compares the PSNR performances of different settings. First, the kernel size has insignificant impacts, although the computational complexity is proportional to the kernel size. Next, when additional information (input frames and context maps) is fed into the dynamic filter generation network, the interpolation performance is improved. More specifically, using input frames improves PSNRs by 0.10 and 0.15 dB on the UCF101 and Vimeo90K datasets. Also, using context maps further improves the performances by 0.07 and 0.05 dB. This is because the input frames and context maps help restore geometric structure and exploit rich contextual information.

5   Conclusions

We developed a deep-learning-based video interpolation algorithm based on the bilateral motion estimation, which consists of the bilateral motion network and the dynamic filter generation network. In the bilateral motion network, we developed the bilateral cost volume to estimate accurate bilateral motions. In the dynamic filter generation network, we warped the two input frames using the estimated bilateral motions and fed them to learn filter coefficients. Finally, we synthesized the intermediate frame by superposing the warped frames with the generated blending filters. Experimental results showed that the proposed algorithm outperforms the state-of-the-art video interpolation algorithms on four benchmark datasets.

Acknowledgements. This work was supported in part by the Agency for Defense Development (ADD) and Defense Acquisition Program Administration (DAPA) of Korea under grant UC160016FD and in part by the National Research Foundation of Korea (NRF) grant funded by the Korea Government (MSIP) (No. NRF-2018R1A2B3003896 and No. NRF-2019R1A2C4069806).

References

1. Baker, S., Scharstein, D., Lewis, J., Roth, S., Black, M.J., Szeliski, R.: A database and evaluation methodology for optical flow. Int. J. Comput. Vis. **92**(1), 1–31 (2011). http://vision.middlebury.edu/flow/eval/
2. Bao, W., Lai, W.S., Ma, C., Zhang, X., Gao, Z., Yang, M.H.: Depth-aware video frame interpolation. In: Proceedings of the IEEE CVPR, pp. 3703–3712 (June 2019)
3. Bao, W., Lai, W.S., Zhang, X., Gao, Z., Yang, M.H.: MEMC-Net: motion estimation and motion compensation driven neural network for video interpolation and enhancement. IEEE Trans. Pattern Anal. Mach. Intell. (2019)
4. Choi, B.D., Han, J.W., Kim, C.S., Ko, S.J.: Motion-compensated frame interpolation using bilateral motion estimation and adaptive overlapped block motion compensation. IEEE Trans. Circuits Syst. Video Technol. **17**(4), 407–416 (2007)
5. Dosovitskiy, A., et al.: FlowNet: learning optical flow with convolutional networks. In: Proceedings of the IEEE ICCV, pp. 2758–2766 (December 2015)
6. Flynn, J., Neulander, I., Philbin, J., Snavely, N.: DeepStereo: learning to predict new views from the world's imagery. In: Proceedings of the IEEE CVPR, pp. 5515–5524 (June 2016)
7. Lu, G., Zhang, X., Chen, L., Gao, Z.: Novel integration of frame rate up conversion and HEVC coding based on rate-distortion optimization. IEEE Trans. Image Process. **27**(2), 678–691 (2018)
8. He, K., Zhang, X., Ren, S., Sun, J.: Deep residual learning for image recognition. In: Proceedings of the IEEE CVPR, pp. 770–778 (June 2016)
9. Hosni, A., Rhemann, C., Bleyer, M., Rother, C., Gelautz, M.: Fast cost-volume filtering for visual correspondence and beyond. IEEE Trans. Pattern Anal. Mach. Intell. **35**(2), 504–511 (2013)
10. Ilg, E., Mayer, N., Saikia, T., Keuper, M., Dosovitskiy, A., Brox, T.: FlowNet 2.0: evolution of optical flow estimation with deep networks. In: Proceedings of the IEEE CVPR, pp. 2462–2470 (July 2017)
11. Jaderberg, M., Simonyan, K., Zisserman, A., Kavukcuoglu, K.: Spatial transformer networks. In: Proceedings of the NIPS, pp. 2017–2025 (2015)
12. Jeong, S.G., Lee, C., Kim, C.S.: Motion-compensated frame interpolation based on multihypothesis motion estimation and texture optimization. IEEE Trans. Image Process. **22**(11), 4497–4509 (2013)

13. Jia, X., De Brabandere, B., Tuytelaars, T., Gool, L.V.: Dynamic filter networks. In: NIPS, pp. 667–675 (2019)
14. Jiang, H., Sun, D., Jampani, V., Yang, M.H., Learned-Miller, E., Kautz, J.: Super SloMo: high quality estimation of multiple intermediate frames for video interpolation. In: Proceedings of the IEEE CVPR, pp. 9000–9008 (June 2018)
15. Kingma, D.P., Ba, J.: Adam: a method for stochastic optimization. In: Proceedings of the ICLR (May 2015)
16. Liu, Y.L., Liao, Y.T., Lin, Y.Y., Chuang, Y.Y.: Deep video frame interpolation using cyclic frame generation. In: Proceedings of the AAAI (January 2019)
17. Liu, Z., Yeh, R.A., Tang, X., Liu, Y., Agarwala, A.: Video frame synthesis using deep voxel flow. In: Proceedings of the IEEE ICCV, pp. 4463–4471 (October 2017)
18. Long, G., Kneip, L., Alvarez, J.M., Li, H., Zhang, X., Yu, Q.: Learning image matching by simply watching video. In: Leibe, B., Matas, J., Sebe, N., Welling, M. (eds.) ECCV 2016. LNCS, vol. 9910, pp. 434–450. Springer, Cham (2016). https://doi.org/10.1007/978-3-319-46466-4_26
19. Meyer, S., Djelouah, A., McWilliams, B., Sorkine-Hornung, A., Gross, M., Schroers, C.: PhaseNet for video frame interpolation. In: Proceedings of the IEEE CVPR, pp. 498–507 (June 2018)
20. Niklaus, S., Liu, F.: Context-aware synthesis for video frame interpolation. In: Proceedings of the IEEE CVPR, pp. 1701–1710 (June 2018)
21. Niklaus, S., Mai, L., Liu, F.: Video frame interpolation via adaptive convolution. In: Proceedings of the IEEE CVPR, pp. 670–679 (July 2017)
22. Niklaus, S., Mai, L., Liu, F.: Video frame interpolation via adaptive separable convolution. In: Proceedings of the IEEE ICCV, pp. 261–270 (October 2017)
23. Charbonnier, P., Blanc-Feraud, L., Aubert, G., Barlaud, M.: Two deterministic half-quadratic regularization algorithms for computed imaging. In: Proceedings of the IEEE ICIP, pp. 168–172 (November 1994)
24. Qifeng Chen, V.K.: Full flow: optical flow estimation by global optimization over regular grids. In: Proceedings of the IEEE CVPR, pp. 4706–4714 (June 2016)
25. Ranjan, A., Black, M.J.: Optical flow estimation using a spatial pyramid network. In: Proceedings of the IEEE CVPR, pp. 4161–4170 (July 2017)
26. Ren, Z., Yan, J., Ni, B., Liu, B., Yang, X., Zha, H.: Unsupervised deep learning for optical flow estimation. In: Proceedings of the AAAI (February 2017)
27. Ronneberger, O., Fischer, P., Brox, T.: U-Net: convolutional networks for biomedical image segmentation. In: Navab, N., Hornegger, J., Wells, W.M., Frangi, A.F. (eds.) MICCAI 2015. LNCS, vol. 9351, pp. 234–241. Springer, Cham (2015). https://doi.org/10.1007/978-3-319-24574-4_28
28. Soomro, K., Zamir, A.R., Shah, M.: UCF101: a dataset of 101 human actions classes from videos in the wild. arXiv preprint arXiv:1212.0402 (2012)
29. Su, S., Delbracio, M., Wang, J., Sapiro, G., Heidrich, W., Wang, O.: Deep video deblurring for hand-held cameras. In: Proceedings of the IEEE CVPR, pp. 1279–1288 (June 2017)
30. Sun, D., Yang, X., Liu, M.Y., Kautz, J.: PWC-Net: CNNs for optical flow using pyramid, warping, and cost volume. In: Proceedings of the IEEE CVPR, pp. 8934–8943 (June 2018)
31. Xu, J., Ranftl, R., Koltun, V.: Accurate optical flow via direct cost volume processing. In: Proceedings of the IEEE CVPR, pp. 1289–1297 (June 2017)
32. Xue, T., Chen, B., Wu, J., Wei, D., Freeman, W.T.: Video enhancement with task-oriented flow. IJCV **127**(8), 1106–1125 (2019)

33. Yu, J.J., Harley, A.W., Derpanis, K.G.: Back to basics: unsupervised learning of optical flow via brightness constancy and motion smoothness. In: Hua, G., Jégou, H. (eds.) ECCV 2016. LNCS, vol. 9915, pp. 3–10. Springer, Cham (2016). https://doi.org/10.1007/978-3-319-49409-8_1

34. Zhang, Y., Tian, Y., Kong, Y., Zhong, B., Fu, Y.: Residual dense network for image super-resolution. In: Proceedings of the IEEE CVPR, pp. 2472–2481 (June 2018)

Hard Negative Examples are Hard, but Useful

Hong Xuan[1]($\boxtimes$) (iD), Abby Stylianou[2], Xiaotong Liu[1], and Robert Pless[1]

[1] The George Washington University, Washington DC 20052, USA
{xuanhong,liuxiaotong2017,pless}@gwu.edu
[2] Saint Louis University, St. Louis, MO 63103, USA
abby.stylianou@slu.edu

Abstract. Triplet loss is an extremely common approach to distance metric learning. Representations of images from the same class are optimized to be mapped closer together in an embedding space than representations of images from different classes. Much work on triplet losses focuses on selecting the most useful triplets of images to consider, with strategies that select dissimilar examples from the same class or similar examples from different classes. The consensus of previous research is that optimizing with the *hardest* negative examples leads to bad training behavior. That's a problem – these hardest negatives are literally the cases where the distance metric fails to capture semantic similarity. In this paper, we characterize the space of triplets and derive why hard negatives make triplet loss training fail. We offer a simple fix to the loss function and show that, with this fix, optimizing with hard negative examples becomes feasible. This leads to more generalizable features, and image retrieval results that outperform state of the art for datasets with high intra-class variance. Code is available at: https://github.com/littleredxh/HardNegative.git

Keywords: Hard negative · Deep metric learning · Triplet loss

1 Introduction

Deep metric learning optimizes an embedding function that maps semantically similar images to relatively nearby locations and maps semantically dissimilar images to distant locations. A number of approaches have been proposed for this problem [3,8,11,14–16,23]. One common way to learn the mapping is to define a loss function based on triplets of images: an anchor image, a positive image from the same class, and a negative image from a different class. The loss penalizes

---

Electronic supplementary material The online version of this chapter (https://doi.org/10.1007/978-3-030-58568-6_8) contains supplementary material, which is available to authorized users.

© Springer Nature Switzerland AG 2020
A. Vedaldi et al. (Eds.): ECCV 2020, LNCS 12359, pp. 126–142, 2020.
https://doi.org/10.1007/978-3-030-58568-6_8

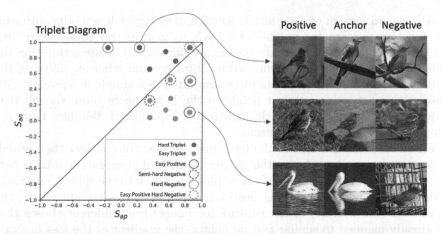

Fig. 1. The triplet diagram plots a triplet as a dot defined by the anchor-positive similarity S_{ap} on the x-axis and the anchor-negative similarity S_{an} on the y-axis. Dots below the diagonal correspond to triplets that are "correct", in the sense that the same class example is closer than the different class example. Triplets above the diagonal of the diagram are candidates for the hard negative triplets. They are important because they indicate locations where the semantic mapping is not yet correct. However, previous works have typically avoided these triplets because of optimization challenges. (Color figure online)

cases where the anchor is mapped closer to the negative image than it is to the positive image.

In practice, the performance of triplet loss is highly dependent on the triplet selection strategy. A large number of triplets are possible, but for a large part of the optimization, most triplet candidates already have the anchor much closer to the positive than the negative, so they are redundant. Triplet mining refers to the process of finding useful triplets.

Inspiration comes from Neil DeGrasse Tyson, the famous American Astrophysicist and science educator who says, while encouraging students, *"In whatever you choose to do, do it because it is hard, not because it is easy"*. Directly mapping this to our case suggests hard negative mining, where triplets include an anchor image where the positive image from the same class is less similar than the negative image from a different class.

Optimizing for hard negative triplets is consistent with the actual use of the network in image retrieval (in fact, hard negative triplets are essentially errors in the trained image mappings), and considering challenging combinations of images has proven critical in triplet based distance metric learning [3,5,7, 14,19]. But challenges in optimizing with the *hardest* negative examples are widely reported in work on deep metric learning for face recognition, people re-identification and fine-grained visual recognition tasks. A variety of work shows that optimizing with the hardest negative examples for deep metric learning leads to bad local minima in the early phase of the optimization [1,2,14,16,20,24,25].

A standard version of deep metric learning uses triplet loss as the optimization function to learn the weights of a CNN to map images to a feature vector. Very commonly, these feature vectors are normalized before computing the similarity because this makes comparison intuitive and efficient, allowing the similarity between feature vectors to be computed as a simple dot-product. We consider this network to project points to the hypersphere (even though that projection only happens during the similarity computation). We show there are two problems in this implementation.

First, when the gradient of the loss function does not consider the normalization to a hypersphere during the gradient backward propagation, a large part of the gradient is lost when points are re-projected back to the sphere, especially in the cases of triplets including nearby points. Second, when optimizing the parameters (the weights) of the network for images from different classes that are already mapped to similar feature points, the gradient of the loss function may actually pull these points together instead of separating them (the opposite of the desired behavior).

We give a systematic derivation showing when and where these challenging triplets arise and diagram the sets of triplets where standard gradient descent leads to bad local minima, and do a simple modification to the triplet loss function to avoid bad optimization outcomes.

Briefly, our main contributions are to:

- introduce the triplet diagram as a visualization to help systematically characterize triplet selection strategies,
- understand optimization failures through analysis of the triplet diagram,
- propose a simple modification to a standard loss function to fix bad optimization behavior with hard negative examples, and
- demonstrate this modification improves current state of the art results on datasets with high intra-class variance.

2   Background

Triplet loss approaches penalize the relative similarities of three examples – two from the same class, and a third from a different class. There has been significant effort in the deep metric learning literature to understand the most effective sampling of informative triplets during training. Including challenging examples from different classes (ones that are similar to the anchor image) is an important technique to speed up the convergence rate, and improve the clustering performance. Currently, many works are devoted to finding such challenging examples within datasets. Hierarchical triplet loss (HTL) [3] seeks informative triplets based on a pre-defined hierarchy of which classes may be similar. There are also stochastic approaches [19] that sample triplets judged to be informative based on approximate class signatures that can be efficiently updated during training.

However, in practice, current approaches cannot focus on the *hardest* negative examples, as they lead to bad local minima early on in training as reported in [1,2,14,16,20,24,25]. The avoid this, authors have developed alternative approaches, such as semi-hard triplet mining [14], which focuses on triplets with negative examples that are *almost* as close to the anchor as positive examples. Easy positive mining [24] selects only the closest anchor-positive pairs and ensures that they are closer than nearby negative examples.

Avoiding triplets with hard negative examples remedies the problem that the optimization often fails for these triplets. But hard negative examples are important. The hardest negative examples are literally the cases where the distance metric fails to capture semantic similarity, and would return nearest neighbors of the incorrect class. Interesting datasets like CUB [21] and CAR [9] which focus on birds and cars, respectively, have high intra-class variance – often similar to or even larger than the inter-class variance. For example, two images of the same species in different lighting and different viewpoints may look quite different. And two images of different bird species on similar branches in front of similar backgrounds may look quite similar. These hard negative examples are the most important examples for the network to learn discriminative features, and approaches that avoid these examples because of optimization challenges may never achieve optimal performance.

There has been other attention on ensure that the embedding is more spread out. A non-parametric approach [22] treats each image as a distinct class of its own, and trains a classifier to distinguish between individual images in order to spread feature points across the whole embedding space. In [26], the authors proposed a spread out regularization to let local feature descriptors fully utilize the expressive power of the space. The easy positive approach [24] only optimizes examples that are similar, leading to more spread out features and feature representations that seem to generalize better to unseen data.

The next section introduces a diagram to systematically organize these triplet selection approaches, and to explore why the hardest negative examples lead to bad local minima.

3   Triplet Diagram

Triplet loss is trained with triplets of images, (x_a, x_p, x_n), where x_a is an anchor image, x_p is a positive image of the same class as the anchor, and x_n is a negative image of a different class. We consider a convolution neural network, $f(\cdot)$, that embeds the images on a unit hypersphere, $(\mathbf{f}(\mathbf{x_a}), \mathbf{f}(\mathbf{x_p}), \mathbf{f}(\mathbf{x_n}))$. We use $(\mathbf{f_a}, \mathbf{f_p}, \mathbf{f_n})$ to simplify the representation of the normalized feature vectors. When embedded on a hypersphere, the cosine similarity is a convenient metric to measure the similarity between anchor-positive pair $S_{ap} = \mathbf{f_a^T f_p}$ and anchor-negative pair $S_{an} = \mathbf{f_a^T f_n}$, and this similarity is bounded in the range $[-1, 1]$.

The triplet diagram is an approach to characterizing a given set of triplets. Figure 1 represents each triplet as a 2D dot (S_{ap}, S_{an}), describing how similar the positive and negative examples are to the anchor. This diagram is useful because the location on the diagram describes important features of the triplet:

- **Hard triplets:** Triplets that are not in the correct configuration, where the anchor-positive similarity is less than the anchor-negative similarity (dots above the $S_{an} = S_{ap}$ diagonal). Dots representing triplets in the wrong configuration are drawn in red. Triplets that are not hard triplets we call **Easy Triplets**, and are drawn in blue.

- **Hard negative mining:** A triplet selection strategy that seeks hard triplets, by selecting for an anchor, the most similar negative example. They are on the top of the diagram. We circle these red dots with a blue ring and call them **hard negative triplets** in the following discussion.
- **Semi-hard negative mining** [14]: A triplet selection strategy that selects, for an anchor, the most similar negative example which is less similar than the corresponding positive example. In all cases, they are under $S_{an} = S_{ap}$ diagonal. We circle these blue dots with a red dashed ring.
- **Easy positive mining** [24]: A triplet selection strategy that selects, for an anchor, the most similar positive example. They tend to be on the right side of the diagram because the anchor-positive similarity tends to be close to 1. We circle these blue dots with a red ring.
- **Easy positive, Hard negative mining** [24]: A related triplet selection strategy that selects, for an anchor, the most similar positive example and most similar negative example. The pink dot surrounded by a blue dashed circle represents one such example.

4   Why Some Triplets are Hard to Optimize

The triplet diagram offers the ability to understand when the gradient-based optimization of the network parameters is effective and when it fails. The triplets are used to train a network whose loss function encourages the anchor to be more similar to its positive example (drawn from the same class) than to its negative example (drawn from a different class). While there are several possible choices, we consider NCA [4] as the loss function:

$$L(S_{ap}, S_{an}) = -log\frac{\exp{(S_{ap})}}{\exp{(S_{ap})} + \exp{(S_{an})}} \tag{1}$$

All of the following derivations can also be done for the margin-based triplet loss formulation used in [14]. We use the NCA-based of triplet loss because the following gradient derivation is clear and simple. Analysis of the margin-based loss is similar and is derived in the Appendix.

The gradient of this NCA-based triplet loss $L(S_{ap}, S_{an})$ can be decomposed into two parts: a single gradient with respect to feature vectors $\mathbf{f_a}$, $\mathbf{f_p}$, $\mathbf{f_n}$:

$$\Delta L = (\frac{\partial L}{\partial S_{ap}}\frac{\partial S_{ap}}{\partial \mathbf{f_a}} + \frac{\partial L}{\partial S_{an}}\frac{\partial S_{an}}{\partial \mathbf{f_a}})\Delta\mathbf{f_a} + \frac{\partial L}{\partial S_{ap}}\frac{\partial S_{ap}}{\partial \mathbf{f_p}}\Delta\mathbf{f_p} + \frac{\partial L}{\partial S_{an}}\frac{\partial S_{an}}{\partial \mathbf{f_n}}\Delta\mathbf{f_n} \tag{2}$$

and subsequently being clear that these feature vectors respond to changes in the model parameters (the CNN network weights), θ:

$$\Delta L = (\frac{\partial L}{\partial S_{ap}}\frac{\partial S_{ap}}{\partial \mathbf{f_a}} + \frac{\partial L}{\partial S_{an}}\frac{\partial S_{an}}{\partial \mathbf{f_a}})\frac{\partial \mathbf{f_a}}{\partial \theta}\Delta\theta + \frac{\partial L}{\partial S_{ap}}\frac{\partial S_{ap}}{\partial \mathbf{f_p}}\frac{\partial \mathbf{f_p}}{\partial \theta}\Delta\theta + \frac{\partial L}{\partial S_{an}}\frac{\partial S_{an}}{\partial \mathbf{f_n}}\frac{\partial \mathbf{f_n}}{\partial \theta}\Delta\theta$$

(3)

The gradient optimization only affects the feature embedding through variations in θ, but we first highlight problems with hypersphere embedding assuming that the optimization *could* directly affect the embedding locations without considering the gradient effect caused by θ. To do this, we derive the loss gradient, $\mathbf{g_a}$, $\mathbf{g_p}$, $\mathbf{g_n}$, with respect to the feature vectors, $\mathbf{f_a}$, $\mathbf{f_p}$, $\mathbf{f_n}$, and use this gradient to update the feature locations where the error should decrease:

$$\mathbf{f_p}^{new} = \mathbf{f_p} - \alpha\mathbf{g_p} = \mathbf{f_p} - \alpha\frac{\partial L}{\partial \mathbf{f_p}} = \mathbf{f_p} + \beta\mathbf{f_a}$$

(4)

$$\mathbf{f_n}^{new} = \mathbf{f_n} - \alpha\mathbf{g_n} = \mathbf{f_n} - \alpha\frac{\partial L}{\partial \mathbf{f_n}} = \mathbf{f_n} - \beta\mathbf{f_a}$$

(5)

$$\mathbf{f_a}^{new} = \mathbf{f_a} - \alpha\mathbf{g_a} = \mathbf{f_a} - \alpha\frac{\partial L}{\partial \mathbf{f_a}} = \mathbf{f_a} - \beta\mathbf{f_n} + \beta\mathbf{f_p}$$

(6)

where $\beta = \alpha\frac{\exp(S_{an})}{\exp(S_{ap})+\exp(S_{an})}$ and α is the learning rate.

This gradient update has a clear geometric meaning: the positive point $\mathbf{f_p}$ is encouraged to move along the direction of the vector $\mathbf{f_a}$; the negative point $\mathbf{f_n}$ is encouraged to move along the opposite direction of the vector $\mathbf{f_a}$; the anchor point $\mathbf{f_a}$ is encouraged to move along the direction of the sum of $\mathbf{f_p}$ and $-\mathbf{f_n}$. All of these are weighted by the same weighting factor β. Then we can get the new anchor-positive similarity and anchor-negative similarity (the complete derivation is given in the Appendix):

$$S_{ap}^{new} = (1+\beta^2)S_{ap} + 2\beta - \beta S_{pn} - \beta^2 S_{an}$$

(7)

$$S_{an}^{new} = (1+\beta^2)S_{an} - 2\beta + \beta S_{pn} - \beta^2 S_{ap}$$

(8)

The first problem is these gradients, $\mathbf{g_a}$, $\mathbf{g_p}$, $\mathbf{g_n}$, have components that move them off the sphere; computing the cosine similarity requires that we compute the norm of $\mathbf{f_a}^{new}$, $\mathbf{f_p}^{new}$ and $\mathbf{f_n}^{new}$ (the derivation for these is shown in Appendix). Given the norm of the updated feature vector, we can calculate the similarity change after the gradient update:

$$\Delta S_{ap} = \frac{S_{ap}^{new}}{\|\mathbf{f_a}^{new}\|\|\mathbf{f_p}^{new}\|} - S_{ap}$$

(9)

$$\Delta S_{an} = \frac{S_{an}^{new}}{\|\mathbf{f_a}^{new}\|\|\mathbf{f_n}^{new}\|} - S_{an}$$

(10)

Figure 2(left column) shows calculations of the change in the anchor-positive similarity and the change in the anchor-negative similarity. There is an area along the right side of the ΔS_{ap} plot (top row, left column) highlighting locations where the anchor and positive are not strongly pulled together. There is also a region along the top side of the ΔS_{an} plot (bottom row, left column) highlighting locations where the anchor and negative can not be strongly separated. This

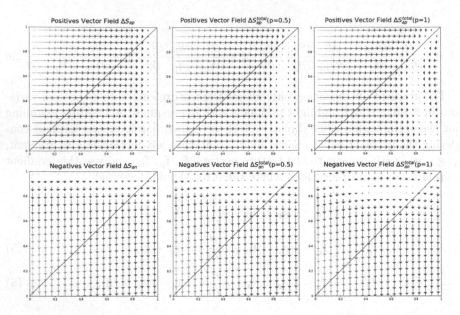

Fig. 2. Numerical simulation of how the optimization changes triplets, with 0 entanglement (left), some entanglement (middle) and complete entanglement (right). The top row shows effects on anchor-positive similarity the bottom row shows effects on anchor-negative similarity. The scale of the arrows indicates the gradient strength. The top region of the bottom-middle and bottom-right plots highlight that the hard negative triplets regions are not well optimized with standard triplet loss.

behavior arises because the gradient is pushing the feature off the hypersphere and therefore, after normalization, the effect is lost when anchor-positive pairs or anchor-negative pairs are close to each other.

The second problem is that the optimization can only control the feature vectors based on the network parameters, θ. Changes to θ are likely to affect **nearby** points in similar ways. For example, if there is a hard negative triplet, as defined in Sect. 3, where the anchor is very close to a negative example, then changing θ to move the anchor closer to the positive example is likely to pull the negative example along with it. We call this effect "entanglement" and propose a simple model to capture its effect on how the gradient update affects the similarities.

We use a scalar, p, and a similarity related factor $q = S_{ap}S_{an}$, to quantify this entanglement effect. When all three examples in a triplet are nearby to each other, both S_{ap} and S_{an} will be large, and therefore q will increase the entanglement effect; when either the positive or the negative example is far away from the anchor, one of S_{ap} and S_{an} will be small and q will reduce the entanglement effect.

The total similarity changes with entanglement will be modeled as follows:

$$\Delta S_{ap}^{total} = \Delta S_{ap} + pq\Delta S_{an} \tag{11}$$

$$\Delta S_{an}^{total} = \Delta S_{an} + pq\Delta S_{ap} \tag{12}$$

Figure 2(middle and right column) shows vector fields on the diagram where S_{ap} and S_{an} will move based on the gradient of their loss function. It highlights the region along right side of the plots where that anchor and positive examples become less similar ($\Delta S_{ap}^{total} < 0$), and the region along top side of the plots where that anchor and negative examples become more similar ($\Delta S_{an}^{total} > 0$) for different parameters of the entanglement.

When the entanglement increases, the problem gets worse; more anchor-negative pairs are in a region where they are pushed to be more similar, and more anchor-positive pairs are in a region where they are pushed to be less similar. The anchor-positive behavior is less problematic because the effect stops while the triplet is still in a good configuration (with the positive closer to the anchor than the negative), while the anchor-negative has not limit and pushes the anchor and negative to be completely similar.

The plots predict the potential movement for triplets on the triplet diagram. We will verify this prediction in the Sect. 6.

Local Minima Caused by Hard Negative Triplets. In Fig. 2, the top region indicates that hard negative triplets with very high anchor-negative similarity get pushed towards (1, 1). Because, in that region, S_{an} will move upward to 1 and S_{ap} will move right to 1. The result of the motion is that a network cannot effectively separate the anchor-negative pairs and instead pushes all features together. This problem was described in [1,2,14,16,20,24,25] as bad local minima of the optimization.

When will Hard Triplets Appear. During triplet loss training, a mini-batch of images is samples random examples from numerous classes. This means that for every image in a batch, there are many possible negative examples, but a smaller number of possible positive examples. In datasets with low intra-class variance and high inter-class variance, an anchor image is less likely to be more similar to its hardest negative example than its random positive example, resulting in more easy triplets.

However, in datasets with relatively higher intra-class variance and lower inter-class variance, an anchor image is more likely to be more similar to its hardest negative example than its random positive example, and form hard triplets. Even after several epochs of training, it's difficult to cluster instances from same class with extremely high intra-class variance tightly.

5 Modification to Triplet Loss

Our solution for the challenge with hard negative triplets is to decouple them into anchor-positive pairs and anchor-negative pairs, and ignore the anchor-positive pairs, and introduce a contrastive loss that penalizes the anchor-negative similarity. We call this Selectively Contrastive Triplet loss L_{SC}, and define this as follows:

$$L_{SC}(S_{ap}, S_{an}) = \begin{cases} \lambda S_{an} & \text{if } S_{an} > S_{ap} \\ L(S_{ap}, S_{an}) & \text{others} \end{cases} \tag{13}$$

In most triplet loss training, anchor-positive pairs from the same class will be always pulled to be tightly clustered. With our new loss function, the anchor-positive pairs in triplets will not be updated, resulting in less tight clusters for a class of instances (we discuss later how this results in more generalizable features that are less over-fit to the training data). The network can then 'focus' on directly pushing apart the hard negative examples.

We denote triplet loss with a Hard Negative mining strategy (HN), triplet loss trained with Semi-Hard Negative mining strategy (SHN), and our Selectively Contrastive Triplet loss with hard negative mining strategy (SCT) in the following discussion.

Figure 3 shows four examples of triplets from the CUB200 (CUB) [21] and CAR196 (CAR) [9] datasets at the very start of training, and Fig. 4 shows four examples of triplets at the end of training. The CUB dataset consists of various classes of birds, while the CAR196 dataset consists of different classes of cars. In both of the example triplet figures, the left column shows a positive example, the second column shows the anchor image, and then we show the hard negative example selected with SCT and SHN approach.

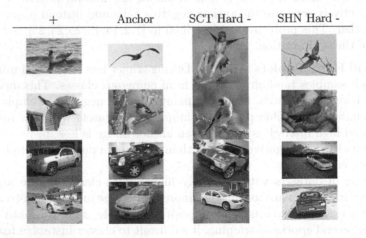

Fig. 3. Example triplets from the CAR and CUB datasets at the start of training. The positive example is randomly selected from a batch, and we show the hard negative example selected by SCT and SHN approach.

At the beginning of training (Fig. 3), both the positive and negative examples appear somewhat random, with little semantic similarity. This is consistent with its initialization from a pretrained model trained on ImageNet, which contains classes such as birds and cars – images of birds all produce feature vectors that point in generally the same direction in the embedding space, and likewise for images of cars.

Figure 4 shows that the model trained with SCT approach has truly hard negative examples – ones that even as humans are difficult to distinguish. The negative examples in the model trained with SHN approach, on the other hand, remain quite random. This may be because when the network was initialized, these anchor-negative pairs were accidentally very similar (very hard negatives) and were never included in the semi-hard negative (SHN) optimization.

6 Experiments and Results

We run a set of experiments on the CUB200 (CUB) [21], CAR196 (CAR) [9], Stanford Online Products (SOP) [16], In-shop Cloth (In-shop) [10] and Hotels-50K(Hotel) [18] datasets. All tests are run on the PyTorch platform [12], using ResNet50 [6] architectures, pre-trained on ILSVRC 2012-CLS data [13]. Training images are re-sized to 256 by 256 pixels. We adopt a standard data augmentation scheme (random horizontal flip and random crops padded by 10 pixels on each side). For pre-processing, we normalize the images using the channel means and standard deviations. All networks are trained using stochastic gradient descent (SGD) with momentum 0. The batch size is 128 for CUB and CAR, 512 for SOP, In-shop and Hotel50k. In a batch of images, each class contains 2 examples and all classes are randomly selected from the training data. Empirically, we set $\lambda = 1$ for CUB and CAR, $\lambda = 0.1$ for SOP, In-shop and Hotel50k.

+	Anchor	SCT Hard -	SHN Hard -

Fig. 4. Example triplets from the CAR and CUB datasets at the end of training. The positive example is randomly selected from a batch, and we show the hard negative example selected by SCT and SHN approach.

We calculate Recall@K as the measurement for retrieval quality. On the CUB and CAR datasets, both the query set and gallery set refer to the testing set. During the query process, the top-K retrieved images exclude the query image itself. In the Hotels-50K dataset, the training set is used as the gallery for all query images in the test set, as per the protocol described by the authors in [18].

6.1 Hard Negative Triplets During Training

Figure 5 helps to visualize what happens with hard negative triplets as the network trains using the triplet diagram described in Sect. 3. We show the triplet diagram over several iterations, for the HN approach (top), SHN approach (middle), and the SCT approach introduced in this paper (bottom).

In the HN approach (top row), most of the movements of hard negative triplets coincide with the movement prediction of the vector field in the Fig. 2 – all of the triplets are pushed towards the bad minima at the location (1, 1).

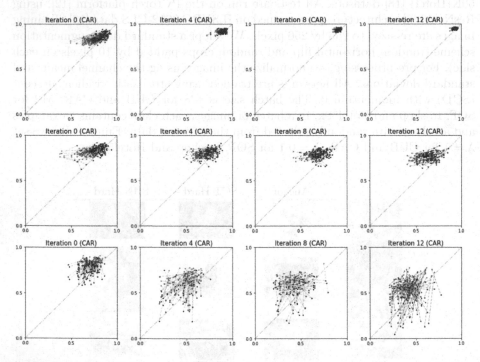

Fig. 5. Hard negative triplets of a batch in training iterations 0, 4, 8, 12. 1st row: Triplet loss with hard negative mining (HN); 2nd row: Triplet loss with semi hard negative mining (SHN). Although the hard negative triplets are not selected for training, their position may still change as the network weights are updated; 3rd row: Selectively Contrastive Triplet loss with hard negative mining (SCT). In each case we show where a set of triplets move before an after the iteration, with the starting triplet location shown in red and the ending location in blue. (Color figure online)

During the training of SHN approach (middle row), it can avoid this local minima problem, but the approach does not do a good job of separating the hard negative pairs. The motion from the red starting point to the blue point after the gradient update is small, and the points are not being pushed below the diagonal line (where the positive example is closer than the negative example).

SCT approach (bottom) does not have any of these problems, because the hard negative examples are more effectively separated early on in the optimization, and the blue points after the gradient update are being pushed towards or below the diagonal line.

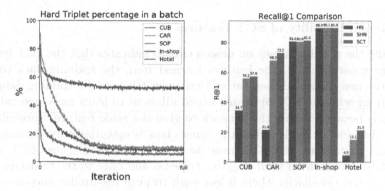

Fig. 6. Left: the percentage of hard triplets in a batch for SOP, In-shop Cloth and Hotels-50K datasets. Right: Recall@1 performance comparison between HN, SHN and SCT approaches.

In Fig. 6, we display the percentage of hard triplets as defined in Sect. 3 in a batch of each training iteration on CUB, CAR, SOP, In-shop Cloth and Hotels-50k datasets (left), and compare the Recall@1 performance of HN, SHN and SCT approaches (right). In the initial training phase, if a high ratio of hard triplets appear in a batch such as CUB, CAR and Hotels-50K dataset, the HN approach converges to the local minima seen in Fig. 5.

We find the improvement is related to the percentage of hard triplets when it drops to a stable level. At this stage, there is few hard triplets in In-shop Cloth dataset, and a small portion of hard triplets in CUB, CAR and SOP datasets, a large portion of hard triplets in Hotels-50K dataset. In Fig. 6, the model trained with SCT approach improves R@1 accuracy relatively small improvement on CUB, CAR, SOP and In-shop datasets but large improvement on Hotels-50K datasets with respect to the model trained with the SHN approach in Table 1, we show the new state-of-the-art result on Hotels-50K dataset, and tables of the other datasets are shown in Appendix (this data is visualized in Fig. 6(right)).

Table 1. Retrieval performance on the Hotels-50K dataset. All methods are trained with Resnet-50 and embedding size is 256.

Method	Hotel instance			Hotel chain		
	R@1	R@10	R@100	R@1	R@3	R@5
BATCH-ALL [18]256	8.1	17.6	34.8	42.5	56.4	62.8
Easy Positive [24]256	16.3	30.5	49.9	-	-	-
SHN256	15.1	27.2	44.9	44.9	57.5	63.0
SCT256	**21.5**	**34.9**	**51.8**	**50.3**	**61.0**	**65.9**

6.2   Generalizability of SCT Features

Improving the recall accuracy on unseen classes indicates that the SCT features are more generalizable – the features learned from the training data transfer well to the new unseen classes, rather than overfitting on the training data. The intuition for why the SCT approach would allow us to learn more generalizable features is because forcing the network to give the same feature representation to very different examples from the same class is essentially memorizing that class, and that is not likely to translate to new classes. Because SCT uses a contrastive approach on hard negative triplets, and only works to decrease the anchor-negative similarity, there is less work to push dis-similar anchor-positive

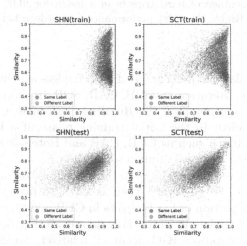

Fig. 7. We train a network on the CAR dataset with the SHN and SCT approach for 80 epochs. Testing data comes from object classes not seen in training. We make a triplet for every image in the training and testing data set, based on its easiest positive (most similar same class image) and hardest negative (most similar different class image), and plot these on the triplet diagram. We see the SHN (left) have a more similar anchor-positive than SCT (right) on the training data, but the SCT distribution of anchor-positive similarities is greater on images from unseen testing classes, indicating improved generalization performance.

pairs together. This effectively results in training data being more spread out in embedding space which previous works have suggested leads to generalizable features [22,24].

We observe this spread out embedding property in the triplet diagrams seen in Fig. 7. On training data, a network trained with SCT approach has anchor-positive pairs that are more spread out than a network trained with SHN approach (this is visible in the greater variability of where points are along the x-axis), because the SCT approach sometimes removes the gradient that pulls anchor-positive pairs together. However, the triplet diagrams on test set show that in new classes the triplets have similar distributions, with SCT creating features that are overall slightly higher anchor-positive similarity.

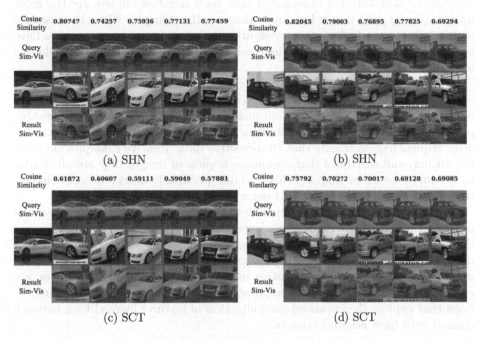

Fig. 8. The above figures show two query images from the CAR dataset (middle left in each set of images), and the top five results returned by our model trained with hard negative examples. The Similarity Visualization approach from [17] show what makes the query image similar to these result images (blue regions contribute more to similarity than red regions). In all figures, the visualization of what makes the query image look like the result image is on top, and the visualization of what makes the result image look like the query image is on the bottom. The top two visualizations (a) and (b) show the visualization obtained from the network trained with SHN approach, while the bottom two visualizations (c) and (d) show the visualization obtained from the network trained with SCT approach. The heatmaps in the SCT approach visualizations are significantly more concentrated on individual features, as opposed to being more diffuse over the entire car. This suggests the SCT approach learns specific semantic features rather than overfitting and memorizing entire vehicles. (Color figure online)

A different qualitative visualization in Fig. 8, shows the embedding similarity visualization from [17], which highlights the regions of one image that make it look similar to another image. In the top set of figures from the SHN approach, the blue regions that indicate similarity are diffuse, spreading over the entire car, while in the bottom visualization from the SCT approach, the blue regions are focused on specific features (like the headlights). These specific features are more likely to generalize to new, unseen data, while the features that represent the entire car are less likely to generalize well.

7 Discussion

Substantial literature has highlighted that hard negative triplets are the most informative for deep metric learning – they are the examples where the distance metric fails to accurately capture semantic similarity. But most approaches have avoided directly optimizing these hard negative triplets, and reported challenges in optimizing with them.

This paper introduces the triplet diagram as a way to characterize the effect of different triplet selection strategies. We use the triplet diagram to explore the behavior of the gradient descent optimization, and how it changes the anchor-positive and anchor-negative similarities within triplet. We find that hard negative triplets have gradients that (incorrectly) force negative examples closer to the anchor, and situations that encourage triplets of images that are all similar to become even more similar. This explains previously observed bad behavior when trying to optimize with hard negative triplets.

We suggest a simple modification to the desired gradients, and derive a loss function that yields those gradients. Experimentally we show that this improves the convergence for triplets with hard negative mining strategy. With this modification, we no longer observe challenges in optimization leading to bad local minima and show that hard-negative mining gives results that exceed or are competitive with state of the art approaches. We additionally provide visualizations that explore the improved generalization of features learned by a network trained with hard negative triplets.

Acknowledgements. This research was partially funded by the Department of Energy, ARPA-E award #DE-AR0000594, and NIJ award 2018-75-CX-0038. Work was partially completed while the first author was an intern with Microsoft Bing Multimedia team.

References

1. Faghri, F., Fleet, D.J., Kiros, J.R., Fidler, S.: Vse++: improving visual-semantic embeddings with hard negatives. In: Proceedings of the British Machine Vision Conference (BMVC) (2018)
2. Ge, J., Gao, G., Liu, Z.: Visual-textual association with hardest and semi-hard negative pairs mining for person search. arXiv preprint arXiv:1912.03083 (2019)

3. Ge, W., Huang, W., Dong, D., Scott, M.R.: Deep metric learning with hierarchical triplet loss. In: Ferrari, V., Hebert, M., Sminchisescu, C., Weiss, Y. (eds.) ECCV 2018. LNCS, vol. 11210, pp. 272–288. Springer, Cham (2018). https://doi.org/10.1007/978-3-030-01231-1_17

4. Goldberger, J., Hinton, G.E., Roweis, S.T., Salakhutdinov, R.R.: Neighbourhood components analysis. In: Saul, L.K., Weiss, Y., Bottou, L. (eds.) Advances in Neural Information Processing Systems, vol. 17, pp. 513–520. MIT Press (2005)

5. Harwood, B., Kumar, B., Carneiro, G., Reid, I., Drummond, T., et al.: Smart mining for deep metric learning. In: Proceedings of the IEEE International Conference on Computer Vision, pp. 2821–2829 (2017)

6. He, K., Zhang, X., Ren, S., Sun, J.: Deep residual learning for image recognition. In: Proceedings of the IEEE Conference on Computer Vision and Pattern Recognition (CVPR) (June 2016)

7. Hermans, A., Beyer, L., Leibe, B.: In Defense of the Triplet Loss for Person Re-Identification. arXiv preprint arXiv:1703.07737 (2017)

8. Kim, W., Goyal, B., Chawla, K., Lee, J., Kwon, K.: Attention-based ensemble for deep metric learning. In: Ferrari, V., Hebert, M., Sminchisescu, C., Weiss, Y. (eds.) ECCV 2018. LNCS, vol. 11205, pp. 760–777. Springer, Cham (2018). https://doi.org/10.1007/978-3-030-01246-5_45

9. Krause, J., Stark, M., Deng, J., Fei-Fei, L.: 3D object representations for fine-grained categorization. In: 4th International IEEE Workshop on 3D Representation and Recognition, 3dRR-13, Sydney, Australia (2013)

10. Liu, Z., Luo, P., Qiu, S., Wang, X., Tang, X.: DeepFashion: powering robust clothes recognition and retrieval with rich annotations. In: Proceedings of IEEE Conference on Computer Vision and Pattern Recognition (CVPR) (June 2016)

11. Movshovitz-Attias, Y., Toshev, A., Leung, T.K., Ioffe, S., Singh, S.: No fuss distance metric learning using proxies. In: Proceedings of the International Conference on Computer Vision (ICCV) (October 2017)

12. Paszke, A., et al.: Automatic differentiation in PyTorch. In: NIPS-W (2017)

13. Russakovsky, O., et al.: ImageNet large scale visual recognition challenge. Int. J. Comput. Vis. (IJCV) **115**(3), 211–252 (2015)

14. Schroff, F., Kalenichenko, D., Philbin, J.: FaceNet: a unified embedding for face recognition and clustering. In: Proceedings of the IEEE Conference on Computer Vision and Pattern Recognition (CVPR) (June 2015)

15. Sohn, K.: Improved deep metric learning with multi-class n-pair loss objective. In: Advances in Neural Information Processing Systems, pp. 1857–1865 (2016)

16. Song, H.O., Xiang, Y., Jegelka, S., Savarese, S.: Deep metric learning via lifted structured feature embedding. In: Proceedings of the IEEE Conference on Computer Vision and Pattern Recognition (CVPR) (2016)

17. Stylianou, A., Souvenir, R., Pless, R.: Visualizing deep similarity networks. In: IEEE Winter Conference on Applications of Computer Vision (WACV) (January 2019)

18. Stylianou, A., Xuan, H., Shende, M., Brandt, J., Souvenir, R., Pless, R.: Hotels-50k: a global hotel recognition dataset. In: AAAI Conference on Artificial Intelligence (2019)

19. Suh, Y., Han, B., Kim, W., Lee, K.M.: Stochastic class-based hard example mining for deep metric learning. In: The IEEE Conference on Computer Vision and Pattern Recognition (CVPR) (June 2019)

20. Wang, C., Zhang, X., Lan, X.: How to train triplet networks with 100k identities? In: Proceedings of the IEEE International Conference on Computer Vision Workshops, pp. 1907–1915 (2017)

21. Welinder, P., et al.: Caltech-UCSD Birds 200. Technical report CNS-TR-2010-001, California Institute of Technology (2010)
22. Wu, Z., Xiong, Y., Yu, S.X., Lin, D.: Unsupervised feature learning via non-parametric instance discrimination. In: The IEEE Conference on Computer Vision and Pattern Recognition (CVPR) (June 2018)
23. Xuan, H., Souvenir, R., Pless, R.: Deep randomized ensembles for metric learning. In: Ferrari, V., Hebert, M., Sminchisescu, C., Weiss, Y. (eds.) ECCV 2018. LNCS, vol. 11220, pp. 751–762. Springer, Cham (2018). https://doi.org/10.1007/978-3-030-01270-0_44
24. Xuan, H., Stylianou, A., Pless, R.: Improved embeddings with easy positive triplet mining. In: The IEEE Winter Conference on Applications of Computer Vision (WACV) (March 2020)
25. Yu, B., Liu, T., Gong, M., Ding, C., Tao, D.: Correcting the triplet selection bias for triplet loss. In: Ferrari, V., Hebert, M., Sminchisescu, C., Weiss, Y. (eds.) ECCV 2018. LNCS, vol. 11210, pp. 71–86. Springer, Cham (2018). https://doi.org/10.1007/978-3-030-01231-1_5
26. Zhang, X., Yu, F.X., Kumar, S., Chang, S.F.: Learning spread-out local feature descriptors. In: The IEEE International Conference on Computer Vision (ICCV) (October 2017)

ReActNet: Towards Precise Binary Neural Network with Generalized Activation Functions

Zechun Liu[1,2], Zhiqiang Shen[2(✉)], Marios Savvides[2], and Kwang-Ting Cheng[1]

[1] Hong Kong University of Science and Technology, Hong Kong, China
zliubq@connect.ust.hk, timcheng@ust.hk
[2] Carnegie Mellon University, Pittsburgh, USA
{zhiqians,marioss}@andrew.cmu.edu

Abstract. In this paper, we propose several ideas for enhancing a binary network to close its accuracy gap from real-valued networks without incurring any additional computational cost. We first construct a baseline network by modifying and binarizing a compact real-valued network with parameter-free shortcuts, bypassing all the intermediate convolutional layers including the downsampling layers. This baseline network strikes a good trade-off between accuracy and efficiency, achieving superior performance than most of existing binary networks at approximately half of the computational cost. Through extensive experiments and analysis, we observed that the performance of binary networks is sensitive to activation distribution variations. Based on this important observation, we propose to generalize the traditional Sign and PReLU functions, denoted as RSign and RPReLU for the respective generalized functions, to enable explicit learning of the distribution reshape and shift at near-zero extra cost. Lastly, we adopt a distributional loss to further enforce the binary network to learn similar output distributions as those of a real-valued network. We show that after incorporating all these ideas, the proposed ReActNet outperforms all the state-of-the-arts by a large margin. Specifically, it outperforms Real-to-Binary Net and MeliusNet29 by 4.0% and 3.6% respectively for the top-1 accuracy and also reduces the gap to its real-valued counterpart to within 3.0% top-1 accuracy on ImageNet dataset. Code and models are available at: https://github.com/liuzechun/ReActNet.

1 Introduction

The 1-bit convolutional neural network (1-bit CNN, also known as binary neural network) [7,30], of which both weights and activations are binary, has been recognized as one of the most promising neural network compression methods for deploying models onto the resource-limited devices. It enjoys 32× memory compression ratio, and up to 58× practical computational reduction on CPU, as

---

Z. Liu—Work done while visiting CMU.

© Springer Nature Switzerland AG 2020
A. Vedaldi et al. (Eds.): ECCV 2020, LNCS 12359, pp. 143–159, 2020.
https://doi.org/10.1007/978-3-030-58568-6_9

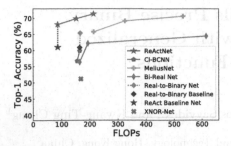

Methods	OPs ($\times 10^8$)	Top-1 Acc(%)
Real-to-Binary Baseline [3]	1.63	60.9
Our ReAct Baseline Net	**0.87**	**61.1**
XNORNet [30]	1.67	51.2
Bi-RealNet [23]	1.63	56.4
Real-to-Binary [3]	1.65	65.4
MeliusNet29 [2]	2.14	65.8
Our ReActNet-A	**0.87**	**69.4**
MeliusNet59 [2]	5.32	70.7
Our ReActNet-C	**2.14**	**71.4**

Fig. 1. Computational Cost vs. ImageNet Accuracy. Proposed ReActNets significantly outperform other binary neural networks. In particular, ReActNet-C achieves state-of-the-art result with 71.4% top-1 accuracy but being 2.5× more efficient than MeliusNet59. ReActNet-A exceeds Real-to-Binary Net and MeliusNet29 by 4.0% and 3.6% top-1 accuracy, respectively, and with more than 1.9× computational reduction. Details are described in Sect. 5.2.

demonstrated in [30]. Moreover, with its pure logical computation (*i.e.*, XNOR operations between binary weights and binary activations), 1-bit CNN is both highly energy-efficient for embedded devices [8,40], and possesses the potential of being directly deployed on next generation memristor-based hardware [17].

Despite these attractive characteristics of 1-bit CNN, the severe accuracy degradation prevents it from being broadly deployed. For example, a representative binary network, XNOR-Net [30] only achieves 51.2% accuracy on the ImageNet classification dataset, leaving a ~18% accuracy gap from the real-valued ResNet-18. Some preeminent binary networks [8,37] show good performance on small datasets such as CIFAR10 and MNIST, but still encounter severe accuracy drop when applied to a large dataset such as ImageNet.

In this study, our motivation is to further close the performance gap between binary neural networks and real-valued networks on the challenging large-scale datasets. We start with designing a high-performance baseline network. Inspired by the recent advances in real-valued compact neural network design, we choose MobileNetV1 [15] structure as our binarization backbone, which we believe is of greater practical value than binarizing non-compact models. Following the insights highlighted in [23], we adopt blocks with identity shortcuts which bypass 1-bit vanilla convolutions to replace the convolutions in MobileNetV1. Moreover, we propose to use a concatenation of two of such blocks to handle the channel number mismatch in the downsampling layers, as shown in Fig. 2(a). This baseline network design not only helps avoid real-valued convolutions in shortcuts, which effectively reduces the computation to near half of that needed in prevalent binary neural networks [3,23,30], but also achieves a high top-1 accuracy of 61.1% on ImageNet.

To further enhance the accuracy, we investigate another aspect which has not been studied in previous binarization or quantization works: activation distribution reshaping and shifting via non-linearity function design. We observed that the overall activation value distribution affects the feature representation, and this effect will be exaggerated by the activation binarization. A small distribution value shift near zero will cause the binarized feature map to have a disparate appearance and in turn will influence the final accuracy. This observation will be elaborated in Sect. 4.2. Enlightened by this observation, we propose a new generalization of Sign function and PReLU function to explicitly shift and reshape the activation distribution, denoted as ReAct-Sign (RSign) and ReAct-PReLU (RPReLU) respectively. These activation functions adaptively learn the parameters for distributional reshaping, which enhance the accuracy of the baseline network by ∼7% with negligible extra computational cost.

Furthermore, we propose a distributional loss to enforce the output distribution similarity between the binary and real-valued networks, which further boosts the accuracy by ∼1%. After integrating all these ideas, the proposed network, dubbed as ReActNet, achieves 69.4% top-1 accuracy on ImageNet with only 87M OPs, surpassing all previously published works on binary networks and reduce the accuracy gap from its real-valued counterpart to only 3.0%, shown in Fig. 1.

We summarize our contributions as follows: (1) We design a baseline binary network by modifying MobileNetV1, whose performance already surpasses most of the previously published work on binary networks while incurring only half of the computational cost. (2) We propose a simple channel-wise reshaping and shifting operation on the activation distribution, which helps binary convolutions spare the computational power in adjusting the distribution to learn more representative features. (3) We further adopt a distributional loss between binary and real-valued network outputs, replacing the original loss, which facilitates the binary network to mimic the distribution of a real-valued network. (4) We demonstrate that our proposed ReActNet, achieves 69.4% top-1 accuracy on ImageNet, for the first time, exceeding the benchmarking ResNet-level accuracy (69.3%). This result also outperforms the state-of-the-art binary network [3] by 4.0% top-1 accuracy while incurring only half the OPs[1].

2 Related Work

There have been extensive studies on neural network compression and acceleration, including quantization [39,43,46], pruning [9,12,22,24], knowledge distillation [6,14,33] and compact network design [15,25,32,41]. A comprehensive survey can be found in [35]. The proposed method falls into the category of quantization, specifically the extreme case of quantizing both weights and activations to only 1-bit, which is so-called network binarization or 1-bit CNNs.

Neural network binarization originates from EBP [34] and BNN [7], which establish an end-to-end gradient back-propagation framework for training the discrete binary weights and activations. As an initial attempt, BNN [7] demonstrated its success on small classification datasets including CIFAR10 [16] and MNIST [27], but encountered severe accuracy drop on a larger dataset such as

---

[1] OPs is a sum of binary OPs and floating-point OPs, i.e., OPs = BOPs/64 + FLOPs.

ImageNet [31], only achieving 42.2% top-1 accuracy compared to 69.3% of the real-valued version of the ResNet-18.

Many follow-up studies focused on enhancing the accuracy. XNOR-Net [30], which proposed real-valued scaling factors to multiply with each of binary weight kernels, has become a representative binarization method and enhanced the top-1 accuracy to 51.2%, narrowing the gap to the real-valued ResNet-18 to ~18%. Based on the XNOR-Net design, Bi-Real Net [23] proposed to add shortcuts to propagate real-values along the feature maps, which further boost the top-1 accuracy to 56.4%.

Several recent studies attempted to improve the binary network performance via expanding the channel width [26], increasing the network depth [21] or using multiple binary weight bases [19]. Despite improvement to the final accuracy, the additional computational cost offsets the BNN's high compression advantage.

For network compression, the real-valued network design used as the starting point for binarization should be compact. Therefore, we chose MobileNetV1 as the backbone network for development of our baseline binary network, which combined with several improvements in implementation achieves ~2× further reduction in the computational cost compared to XNOR-Net and Bi-Real Net, and a top-1 accuracy of 61.1%, as shown in Fig. 1.

In addition to architectural design [2,23,28], studies on 1-bit CNNs expand from training algorithms [1,3,36,46], binary optimizer design [13], regularization loss design [8,29], to better approximation of binary weights and activations [11, 30,37]. Different from these studies, this paper focuses on a new aspect that is seldom investigated before but surprisingly crucial for 1-bit CNNs' accuracy, *i.e.* activation distribution reshaping and shifting. For this aspect, we propose novel *ReAct* operations, which are further combined with a proposed distributional loss. These enhancements improve the accuracy to 69.4%, further shrinking the accuracy gap to its real-valued counterpart to only 3.0%. The baseline network design and ReAct operations, as well as the proposed loss function are detailed in Sect. 4.

3   Revisit: 1-Bit Convolution

In a 1-bit convolutional layer, both weights and activations are binarized to -1 and +1, such that the computationally heavy operations of floating-point matrix multiplication can be replaced by light-weighted bitwise XNOR operations and popcount operations [4], as:

$$\mathcal{X}_b * \mathcal{W}_b = \text{popcount}(\text{XNOR}(\mathcal{X}_b, \mathcal{W}_b)), \tag{1}$$

where $\mathcal{W}_b$ and $\mathcal{X}_b$ indicate the matrices of binary weights and binary activations. Specifically, weights and activations are binarized through a sign function:

$$x_b = \text{Sign}(x_r) = \begin{cases} +1, \text{ if } x_r > 0 \\ -1, \text{ if } x_r \leq 0 \end{cases}, \quad w_b = \begin{cases} +\frac{||\mathcal{W}_r||_{l1}}{n}, \text{ if } w_r > 0 \\ -\frac{||\mathcal{W}_r||_{l1}}{n}, \text{ if } w_r \leq 0 \end{cases} \tag{2}$$

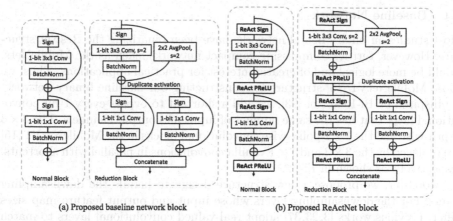

(a) Proposed baseline network block   (b) Proposed ReActNet block

Fig. 2. The proposed baseline network modified from MobileNetV1 [15], which replaces the original (3×3 depth-wise and 1×1 point-wise) convolutional pairs by the proposed blocks. (a) The baseline's configuration in terms of channel and layer numbers is identical to that of MobileNetV1. If the input and output channel numbers are equal in a dw-pw-conv pair in the original network, a normal block is used, otherwise a reduction block is adopted. For the reduction block, we duplicate the input activation and concatenate the outputs to increase the channel number. As a result, all 1-bit convolutions have the same input and output channel numbers and are bypassed by identity shortcuts. (b) In the proposed ReActNet block, ReAct-Sign and ReAct-PReLU are added to the baseline network.

The subscripts b and r denote binary and real-valued, respectively. The weight binarization method is inherited from [30], of which, $\frac{\|W_r\|_{\ell 1}}{n}$ is the average of absolute weight values, used as a scaling factor to minimize the difference between binary and real-valued weights. XNOR-Net [30] also applied similar real-valued scaling factor to binary activations. Note that with introduction of the proposed ReAct operations, to be described in Sect. 4.2, this scaling factor for activations becomes unnecessary and can be eliminated.

4   Methodology

In this section, we first introduce our proposed baseline network in Sect. 4.1. Then we analyze how the variation in activation distribution affects the feature quality and in turn influences the final performance. Based on this analysis, we introduce ReActNet which explicitly reshapes and shifts the activation distribution using ReAct-PReLU and ReAct-Sign functions described in Sect. 4.2 and matches the outputs via a distributional loss defined between binary and real-valued networks detailed in Sect. 4.3.

148   Z. Liu et al.

4.1 Baseline Network

Most studies on binary neural networks have been binarizing the ResNet structure. However, further compressing compact networks, such as the MobileNets, would be more logical and of greater interest for practical applications. Thus, we chose MobileNetV1 [15] structure for constructing our baseline binary network.

Inspired by Bi-Real Net [23], we add a shortcut to bypass every 1-bit convolutional layer that has the same number of input and output channels. The 3×3 depth-wise and the 1×1 point-wise convolutional blocks in the MobileNetV1 [15] are replaced by the 3×3 and 1×1 vanilla convolutions in parallel with shortcuts, respectively, as shown in Fig. 2.

Moreover, we propose a new structure design to handle the downsampling layers. For the downsampling layers whose input and output feature map sizes differ, previous works [3,23,37] adopt real-valued convolutional layers to match their dimension and to make sure the real-valued feature map propagating along the shortcut will not be "cut off" by the activation binarization. However, this strategy increases the computational cost. Instead, our proposal is to make sure that all convolutional layers have the same input and output dimensions so that we can safely binarize them and use a simple identity shortcut for activation propagation without additional real-valued matrix multiplications.

As shown in Fig. 2(a), we duplicate input channels and concatenate two blocks with the same inputs to address the channel number difference and also use average pooling in the shortcut to match spatial downsampling. All layers in our baseline network are binarized, except the first input convolutional layer and the output fully-connect layer. Such a structure is hardware friendly.

4.2 ReActNet

The intrinsic property of an image classification neural network is to learn a mapping from input images to the output logits. A logical deduction is that a good performing binary neural network should learn similar logits distribution as a real-valued network. However, the discrete values of variables limit binary neural networks from learning as rich distributional representations as real-valued ones. To address it, XNOR-Net [30] proposed to calculate analytical real-valued scaling factors and multiply them with the activations. Its follow-up works [4,38] further proposed to learn these factors through back-propagation.

In contrast to these previous works, this paper focuses on a different aspect: the activation distribution. We observed that small variations to activation distributions can greatly affect the semantic feature representations in 1-bit CNNs, which in turn will influence the final performance. However, 1-bit CNNs have limited capacity to learn appropriate activation distributions. To address this dilemma, we introduce generalized activation functions with learnable coefficients to increase the flexibility of 1-bit CNNs for learning semantically-optimized distributions.

Distribution Matters in 1-bit CNNs. The importance of distribution has not been investigated much in training a real-valued network, because with weights

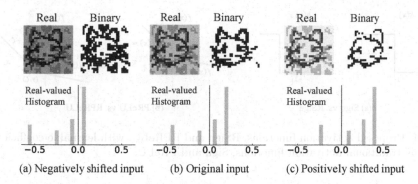

(a) Negatively shifted input   (b) Original input   (c) Positively shifted input

Fig. 3. An illustration of how distribution shift affects feature learning in binary neural networks. An ill-shifted distribution will introduce (a) too much background noise or (c) too few useful features, which harms feature learning.

and activations being continuous real values, reshaping or moving distributions would be effortless.

However, for 1-bit CNNs, learning distribution is both crucial and difficult. Because the activations in a binary convolution can only choose values from $\{-1, +1\}$, making a small distributional shift in the input real-valued feature map before the sign function can possibly result in a completely different output binary activations, which will directly affect the informativeness in the feature and significantly impact the final accuracy. For illustration, we plot the output binary feature maps of real-valued inputs with the original (Fig. 3(b)), positively-shifted (Fig. 3(a)), and negatively-shifted (Fig. 3(c)) activation distributions. Real-valued feature maps are robust to the shifts with which the legibility of semantic information will pretty much be maintained, while binary feature maps are sensitive to these shifts as illustrated in Fig. 3(a) and Fig. 3(c).

Explicit Distribution Reshape and Shift via Generalized Activation Functions. Based on the aforementioned observation, we propose a simple yet effective operation to explicitly reshape and shift the activation distributions, dubbed as ReAct, which generalizes the traditional Sign and PReLU functions to ReAct-Sign (abbreviated as RSign) and ReAct-PReLU (abbreviated as RPReLU) respectively.

Definition
Essentially, RSign is defined as a sign function with channel-wisely learnable thresholds:

$$x_i^b = h(x_i^r) = \begin{cases} +1, & \text{if } x_i^r > \alpha_i \\ -1, & \text{if } x_i^r \le \alpha_i \end{cases}. \tag{3}$$

Here, x_i^r is real-valued input of the RSign function h on the ith channel, x_i^b is the binary output and α_i is a learnable coefficient controlling the threshold. The subscript i in α_i indicates that the threshold can vary for different channels.

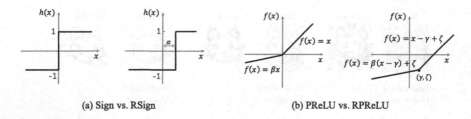

(a) Sign vs. RSign   (b) PReLU vs. RPReLU

Fig. 4. Proposed activation functions, RSign and RPReLU, with learnable coefficients and the traditional activation functions, Sign and PReLU.

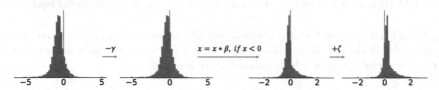

Fig. 5. An explanation of how proposed RPReLU operates. It first moves the input distribution by $-\gamma$, then reshapes the negative part by multiplying it with β and lastly moves the output distribution by ζ.

The superscripts b and r refer to binary and real values. Figure 4(a) shows the shapes of RSign and Sign.

Similarly, RPReLU is defined as

$$f(x_i) = \begin{cases} x_i - \gamma_i + \zeta_i, & \text{if } x_i > \gamma_i \\ \beta_i(x_i - \gamma_i) + \zeta_i, & \text{if } x_i \le \gamma_i \end{cases}, \quad (4)$$

where x_i is the input of the RPReLU function f on the ith channel, γ_i and ζ_i are learnable shifts for moving the distribution, and β_i is a learnable coefficient controlling the slope of the negative part. All the coefficients are allowed to be different across channels. Figure 4(b) compares the shapes of RPReLU and PReLU.

Intrinsically, RSign is learning the best channel-wise threshold (α) for binarizing the input feature map, or equivalently, shifting the input distribution to obtain the best distribution for taking a sign. From the latter angle, RPReLU can be easily interpreted as γ shifts the input distribution, finding a best point to use β to "fold" the distribution, then ζ shifts the output distribution, as illustrated in Fig. 5. These learned coefficients automatically adjust activation distributions for obtaining good binary features, which enhances the 1-bit CNNs' performance. With the introduction of these functions, the aforementioned difficulty in distributional learning can be greatly alleviated, and the 1-bit convolutions can effectively focus on learning more meaningful patterns. We will show later in the result section that this enhancement can boost the baseline network's top-1 accuracy substantially.

The number of extra parameters introduced by RSign and RPReLU is only $4 \times number\ of\ channels$ in the network, which is negligible considering the large size of the weight matrices. The computational overhead approximates a typical non-linear layer, which is also trivial compared to the computational intensive convolutional operations.

Optimization

Parameters in RSign and RPReLU can be optimized end-to-end with other parameters in the network. The gradient of α_i in RSign can be simply derived by the chain rule as:

$$\frac{\partial \mathcal{L}}{\partial \alpha_i} = \sum_{x_i^r} \frac{\partial \mathcal{L}}{\partial h(x_i^r)} \frac{\partial h(x_i^r)}{\partial \alpha_i}, \tag{5}$$

where $\mathcal{L}$ represents the loss function and $\frac{\partial \mathcal{L}}{\partial h(x_i^r)}$ denotes the gradients from deeper layers. The summation $\sum_{x_i^r}$ is applied to all entries in the ith channel. The derivative $\frac{\partial h(x_i^r)}{\partial \alpha_i}$ can be easily computed as

$$\frac{\partial h(x_i^r)}{\partial \alpha_i} = -1 \tag{6}$$

Similarly, for each parameter in RPReLU, the gradients are computed with the following formula:

$$\frac{\partial f(x_i)}{\partial \beta_i} = \mathbf{I}_{\{x_i \leq \gamma_i\}} \cdot (x - \gamma_i), \tag{7}$$

$$\frac{\partial f(x_i)}{\partial \gamma_i} = -\mathbf{I}_{\{x_i \leq \gamma_i\}} \cdot \beta_i - \mathbf{I}_{\{x_i > \gamma_i\}}, \tag{8}$$

$$\frac{\partial f(x_i)}{\partial \zeta_i} = 1. \tag{9}$$

Here, $\mathbf{I}$ denotes the indicator function. $\mathbf{I}_{\{\cdot\}} = 1$ when the inequation inside $\{\}$ holds, otherwise $\mathbf{I}_{\{\cdot\}} = 0$.

4.3   Distributional Loss

Based on the insight that if the binary neural networks can learn similar distributions as real-valued networks, the performance can be enhanced, we use a distributional loss to enforce this similarity, formulated as:

$$\mathcal{L}_{Distribution} = -\frac{1}{n} \sum_c \sum_{i=1}^n p_c^{\mathcal{R}_\theta}(X_i) \log(\frac{p_c^{\mathcal{B}_\theta}(X_i)}{p_c^{\mathcal{R}_\theta}(X_i)}), \tag{10}$$

where the distributional loss $\mathcal{L}_{Distribution}$ is defined as the KL divergence between the softmax output p_c of a real-valued network $\mathcal{R}_\theta$ and a binary network $\mathcal{B}_\theta$. The subscript c denotes classes and n is the batch size.

Different from prior work [46] that needs to match the outputs from every intermediate block, or further using multi-step progressive structural transition [3], we found that our distributional loss, while much simpler, can yield competitive results. Moreover, without block-wise constraints, our approach enjoys the flexibility in choosing the real-valued network without the requirement of architecture similarity between real and binary networks.

5 Experiments

To investigate the performance of the proposed methods, we conduct experiments on ImageNet dataset. We first introduce the dataset and training strategy in Sect. 5.1, followed by comparison between the proposed networks and state-of-the-arts in terms of both accuracy and computational cost in Sect. 5.2. We then analyze the effects of the distributional loss, concatenated downsampling layer and the RSign and the RPReLU in detail in the ablation study described in Sect. 5.3. Visualization results on how RSign and RPReLU help binary network capture the fine-grained underlying distribution are presented in Sect. 5.4.

5.1 Experimental Settings

Dataset. The experiments are carried out on the ILSVRC12 ImageNet classification dataset [31], which is more challenging than small datasets such as CIFAR [16] and MNIST [27]. In our experiments, we use the classic data augmentation method described in [15].

Training Strategy. We followed the standard binarization method in [23] and adopted the two-step training strategy as [3]. In the first step, we train a network with binary activations and real-valued weights from scratch. In the second step, we inherit the weights from the first step as the initial value and fine-tune the network with weights and activations both being binary. For both steps, Adam optimizer with a linear learning rate decay scheduler is used, and the initial learning rate is set to 5e-4. We train it for 600k iterations with batch size being 256. The weight decay is set to 1e-5 for the first step and 0 for the second step.

Distributional Loss. In both steps, we use proposed distributional loss as the objective function for optimization, replacing the original cross-entropy loss between the binary network output and the label.

OPs Calculation. We follow the calculation method in [3], we count the binary operations (BOPs) and floating point operations (FLOPs) separately. The total operations (OPs) is calculated by OPs = BOPs/64 + FLOPs, following [23,30].

5.2 Comparison with State-of-the-Art

We compare ReActNet with state-of-the-art quantization and binarization methods. Table 1 shows that ReActNet-A already outperforms all the quantizing methods in the left part, and also archives 4.0% higher accuracy than the state-of-the-art Real-to-Binary Network [3] with only approximately half of the OPs. Moreover, in contrast to [3] which computes channel re-scaling for each block with real-valued fully-connected layers, ReActNet-A has pure 1-bit convolutions except the first and the last layers, which is more hardware-friendly.

Table 1. Comparison of the top-1 accuracy with state-of-the-art methods. The left part presents quantization methods applied on ResNet-18 structure and the right part are binarization methods with varied structures (ResNet-18 if not specified). Quantization methods include weight quantization (upper left block), low-bit weight and activation quantization (middle left block) and the weight and activation binarization with the expanded network capacity (lower left block), where the number times (1/1) indicates the multiplicative factor. (W/A) represents the number of bits used in weight or activation quantization.

Methods	Bitwidth (W/A)	Acc(%) Top-1	Binary methods	BOPs ($\times 10^9$)	FLOPs ($\times 10^8$)	OPs ($\times 10^8$)	Acc(%) Top-1
BWN [7]	1/32	60.8	BNNs [7]	1.70	1.20	1.47	42.2
TWN [18]	2/32	61.8	CI-BCNN [37]	–	–	1.63	59.9
INQ [42]	2/32	66.0	Binary MobileNet [28]	–	–	1.54	60.9
TTQ [44]	2/32	66.6	PCNN [11]	–	–	1.63	57.3
SYQ [10]	1/2	55.4	XNOR-Net [30]	1.70	1.41	1.67	51.2
HWGQ [5]	1/2	59.6	Trained Bin [38]	–	–	–	54.2
LQ-Nets [39]	1/2	62.6	Bi-RealNet-18 [23]	1.68	1.39	1.63	56.4
DoReFa-Net [43]	1/4	59.2	Bi-RealNet-34 [23]	3.53	1.39	1.93	62.2
Ensemble BNN [45]	(1/1) × 6	61.1	Bi-RealNet-152 [21]	10.7	4.48	6.15	64.5
Circulant CNN [20]	(1/1) × 4	61.4	Real-to-Binary Net [3]	1.68	1.56	1.83	65.4
Structured BNN [47]	(1/1) × 4	64.2	MeliusNet29 [2]	5.47	1.29	2.14	65.8
Structured BNN* [47]	(1/1) × 4	66.3	MeliusNet42 [2]	9.69	1.74	3.25	69.2
ABC-Net [19]	(1/1) × 5	65.0	MeliusNet59 [2]	18.3	2.45	5.32	70.7
Our ReActNet-A	1/1	–	–	**4.82**	**0.12**	**0.87**	**69.4**
Our ReActNet-B	1/1	–	–	**4.69**	**0.44**	**1.63**	**70.1**
Our ReActNet-C	1/1	–	–	**4.69**	**1.40**	**2.14**	**71.4**

To make further comparison with previous approaches that use real-valued convolution to enhance binary network's accuracy [2,3,23], we constructed ReActNet-B and ReActNet-C, which replace the 1-bit 1×1 convolution with real-valued 1 × 1 convolution in the downsampling layers, as shown in Fig. 6(c). ReActNet-B defines the real-valued convolutions to be group convolutions with 4 groups, while ReActNet-C uses full real-valued convolution. We show that ReActNet-B achieves 13.7% higher accuracy than Bi-RealNet-18 with the same number of OPs and ReActNet-C outperforms MeliusNet59 by 0.7% with less than half of the OPs.

Moreover, we applied the ReAct operations to Bi-RealNet-18, and obtained 65.5% Top-1 accuracy, increasing the accuracy of Bi-RealNet-18 by 9.1% without changing the network structure.

Considering the challenges in previous attempts to enhance 1-bit CNNs' performance, the accuracy leap achieved by ReActNets is significant. It requires an ingenious use of binary networks' special property to effectively utilize every precious bit and strike a delicate balance between binary and real-valued information. For example, ReActNet-A, with 69.4% top-1 accuracy at 87M OPs, outperforms the real-valued 0.5× MobileNetV1 by 5.7% greater accuracy at 41.6% fewer OPs. These results demonstrate the potential of 1-bit CNNs and the effectiveness of our ReActNet design.

5.3 Ablation Study

We conduct ablation studies to analyze the individual effect of the following proposed techniques:

Table 2. The effects of different components in ReActNet on the final accuracy. († denotes the network not using the concatenated blocks, but directly binarizing the downsampling layers instead. * indicates not using the proposed distributional loss during training.)

Network	Top-1 Acc(%)
Baseline network † *	58.2
Baseline network †	59.6
Proposed baseline network *	61.1
Proposed baseline network	62.5
Proposed baseline network + PReLU	65.5
Proposed baseline network + RSign	66.1
Proposed baseline network + RPReLU	67.4
ReActNet-A (RSign and RPReLU)	69.4
Corresponding real-valued network	72.4

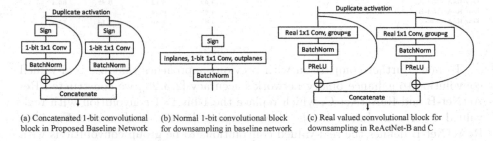

(a) Concatenated 1-bit convolutional block in Proposed Baseline Network

(b) Normal 1-bit convolutional block for downsampling in baseline network

(c) Real valued convolutional block for downsampling in ReActNet-B and C

Fig. 6. Variations in the downsampling layer design.

Block Duplication and Concatenation. Real-valued shortcuts are crucial for binary neural network accuracy [23]. However, the input channels of the down-sampling layers are twice the output channels, which violates the requirement for adding the shortcuts that demands an equal number of input and output channels. In the proposed baseline network, we duplicate the downsampling blocks and concatenate the outputs (Fig. 6(a)), enabling the use of identity shortcuts to bypass 1-bit convolutions in the downsampling layers. This idea alone results in a 2.9% accuracy enhancement compared to the network without concatenation (Fig. 6(b)). The enhancement can be observed by comparing the $2nd$ and $4th$ rows of Table 2. With the proposed downsampling layer design, our baseline network achieves both high accuracy and high compression ratio. Because it no longer requires real-valued matrix multiplications in the downsampling layers, the computational cost is greatly reduced. As a result, even without using the distributional loss in training, our proposed baseline network has already surpassed the Strong Baseline in [3] by 0.1% for top-1 accuracy at only half of the

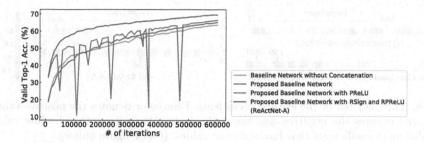

Fig. 7. Comparing validation accuracy curves between baseline networks and ReAct-Net. Using proposed RSign and RPReLU (red curve) achieves the higher accuracy and is more robust than using Sign and PReLU (green curve). (Color figure online)

OPs. With this strong performance, this simple baseline network serves well as a new baseline for future studies on compact binary neural networks.

Distributional Loss. The results in the first section of Table 2 also validate that the distributional loss designed for matching the output distribution between binary and real-valued neural networks is effective for enhancing the performance of propose baseline network, improving the accuracy by 1.4%, which is achieved independent of the network architecture design.

ReAct Operations. The introduction of RSign and RPReLU improves the accuracy by 4.9% and 3.6% respectively over the proposed baseline network, as shown in the second section of Table 2. By adding both RSign and RPReLU, ReActNet-A achieves 6.9% higher accuracy than the baseline, narrowing the accuracy gap to the corresponding real-valued network to within 3.0%. Compared to merely using the Sign and PReLU, the use of the generalized activation functions, RSign and RPReLU, with simple learnable parameters boost the accuracy by 3.9%, which is very significant for the ImageNet classification task. As shown in Fig. 7, the validation curve of the network using original Sign + PReLU oscillates vigorously, which is suspected to be triggered by the slope coefficient β in PReLU changing its sign which in turn affects the later layers with an avalanche effect. This also indirectly confirms our assumption that 1-bit CNNs are vulnerable to distributional changing. In comparison, the proposed RSign and RPReLU functions are effective for stabilizing training in addition to improving the accuracy.

5.4 Visualization

To help gain better insights, we visualize the learned coefficients as well as the intermediate activation distributions.

Learned Coefficients. For clarity, we present the learned coefficients of each layer in form of the color bar in Fig. 8. Compared to the binary network using traditional PReLU whose learned slopes β are positive only (Fig. 8(a)), ReActNet using RPReLU learns both positive and negative slopes (Fig. 8(c)), which are

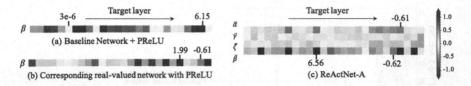

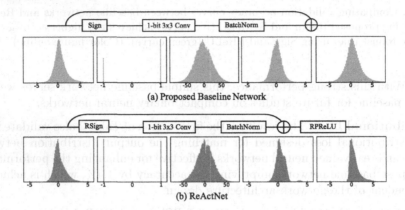

Fig. 8. The color bar of the learned coefficients. Blue color denotes the positive values while red denotes the negative, and the darkness in color reflects the absolute value. We also mark coefficients that have extreme values. (Color figure online)

Fig. 9. Histogram of the activation distribution

closer to the distributions of PReLU coefficients in a real-valued network we trained (Fig. 8(b)). The learned distribution shifting coefficients also have large absolute values as shown in Rows 1–3 of Fig. 8(c), indicating the necessity of their explicit shift for high-performance 1-bit CNNs.

Activation Distribution. In Fig. 9, we show the histograms of activation distributions inside the trained baseline network and ReActNet. Compared to the baseline network without RSign and RPReLU, ReActNet's distributions are more enriched and subtle, as shown in the forth sub-figure in Fig. 9(b). Also, in ReActNet, the distribution of -1 and +1 after the sign function is more balanced, as illustrated in the second sub-figure in Fig. 9(b), suggesting better utilization of black and white pixels in representing the binary features.

6   Conclusions

In this paper, we present several new ideas to optimize a 1-bit CNN for higher accuracy. We first design parameter-free shortcuts based on MobileNetV1 to propagate real-valued feature maps in both normal convolutional layers as well as the downsampling layers. This yields a baseline binary network with 61.1% top-1 accuracy at only 87M OPs for the ImageNet dataset. Then, based on our observation that 1-bit CNNs' performance is highly sensitive to distributional variations, we propose ReAct-Sign and ReAct-PReLU to enable shift and reshape the distributions in a learnable fashion and demonstrate their dramatical enhancements on the top-1 accuracy. We also propose to incorporate a distributional loss, which is defined between the outputs of the binary network and the real-valued reference network, to replace the original cross-entropy loss for training. With contributions jointly achieved by these ideas, the proposed ReActNet achieves 69.4% top-1 accuracy on ImageNet, which is just 3% shy of its real-valued counterpart while at substantially lower computational cost.

Acknowledgement. The authors would like to acknowledge HKSAR RGC's funding support under grant GRF-16207917.

References

1. Alizadeh, M., Fernández-Marqués, J., Lane, N.D., Gal, Y.: An empirical study of binary neural networks' optimisation (2018)
2. Bethge, J., Bartz, C., Yang, H., Chen, Y., Meinel, C.: Meliusnet: can binary neural networks achieve mobilenet-level accuracy? arXiv preprint arXiv:2001.05936 (2020)
3. Martinez, B., Yang, J., Bulat, A., Tzimiropoulos, G.: Training binary neural networks with real-to-binary convolutions. In: International Conference on Learning Representations (2020)
4. Bulat, A., Tzimiropoulos, G.: XNOR-NET++: improved binary neural networks. In: British Machine Vision Conference (2019)
5. Cai, Z., He, X., Sun, J., Vasconcelos, N.: Deep learning with low precision by half-wave gaussian quantization. In: Proceedings of the IEEE Conference on Computer Vision and Pattern Recognition, pp. 5918–5926 (2017)
6. Chen, G., Choi, W., Yu, X., Han, T., Chandraker, M.: Learning efficient object detection models with knowledge distillation. In: Advances in Neural Information Processing Systems, pp. 742–751 (2017)
7. Courbariaux, M., Hubara, I., Soudry, D., El-Yaniv, R., Bengio, Y.: Binarized neural networks: training deep neural networks with weights and activations constrained to + 1 or -1. arXiv preprint arXiv:1602.02830 (2016)
8. Ding, R., Chin, T.W., Liu, Z., Marculescu, D.: Regularizing activation distribution for training binarized deep networks. In: Proceedings of the IEEE Conference on Computer Vision and Pattern Recognition, pp. 11408–11417 (2019)
9. Ding, X., Zhou, X., Guo, Y., Han, J., Liu, J., et al.: Global sparse momentum SGD for pruning very deep neural networks. In: Advances in Neural Information Processing Systems, pp. 6379–6391 (2019)
10. Faraone, J., Fraser, N., Blott, M., Leong, P.H.: SYG: learning symmetric quantization for efficient deep neural networks. In: Proceedings of the IEEE Conference on Computer Vision and Pattern Recognition, pp. 4300–4309 (2018)
11. Gu, J., et al.: Projection convolutional neural networks for 1-bit CNNs via discrete back propagation. In: Proceedings of the AAAI Conference on Artificial Intelligence, vol. 33, pp. 8344–8351 (2019)
12. He, Y., Zhang, X., Sun, J.: Channel pruning for accelerating very deep neural networks. In: Proceedings of the IEEE International Conference on Computer Vision, pp. 1389–1397 (2017)

13. Helwegen, K., Widdicombe, J., Geiger, L., Liu, Z., Cheng, K.T., Nusselder, R.: Latent weights do not exist: rethinking binarized neural network optimization. In: Advances in Neural Information Processing Systems, pp. 7531–7542 (2019)
14. Hinton, G., Vinyals, O., Dean, J.: Distilling the knowledge in a neural network. arXiv preprint arXiv:1503.02531 (2015)
15. Howard, A.G., et al.: Mobilenets: efficient convolutional neural networks for mobile vision applications. arXiv preprint arXiv:1704.04861 (2017)
16. Krizhevsky, A., Hinton, G.: Learning multiple layers of features from tiny images. Technical report, Citeseer (2009)
17. Li, B., Shan, Y., Hu, M., Wang, Y., Chen, Y., Yang, H.: Memristor-based approximated computation. In: Proceedings of the 2013 International Symposium on Low Power Electronics and Design, pp. 242–247. IEEE Press (2013)
18. Li, F., Zhang, B., Liu, B.: Ternary weight networks. arXiv preprint arXiv:1605.04711 (2016)
19. Lin, X., Zhao, C., Pan, W.: Towards accurate binary convolutional neural network. In: Advances in Neural Information Processing Systems, pp. 345–353 (2017)
20. Liu, C., et al.: Circulant binary convolutional networks: enhancing the performance of 1-bit DCNNs with circulant back propagation. In: Proceedings of the IEEE Conference on Computer Vision and Pattern Recognition, pp. 2691–2699 (2019)
21. Liu, Z., Luo, W., Wu, B., Yang, X., Liu, W., Cheng, K.T.: Bi-real net: binarizing deep network towards real-network performance. Int. J. Comput. Vis. **128**, 1–18 (2018)
22. Liu, Z., et al.: Metapruning: meta learning for automatic neural network channel pruning. In: Proceedings of the IEEE International Conference on Computer Vision, pp. 3296–3305 (2019)
23. Liu, Z., Wu, B., Luo, W., Yang, X., Liu, W., Cheng, K.-T.: Bi-real net: enhancing the performance of 1-bit CNNs with improved representational capability and advanced training algorithm. In: Ferrari, V., Hebert, M., Sminchisescu, C., Weiss, Y. (eds.) ECCV 2018. LNCS, vol. 11219, pp. 747–763. Springer, Cham (2018). https://doi.org/10.1007/978-3-030-01267-0_44
24. Liu, Z., Li, J., Shen, Z., Huang, G., Yan, S., Zhang, C.: Learning efficient convolutional networks through network slimming. In: Proceedings of the IEEE International Conference on Computer Vision, pp. 2736–2744 (2017)
25. Ma, N., Zhang, X., Zheng, H.-T., Sun, J.: ShuffleNet V2: practical guidelines for efficient CNN architecture design. In: Ferrari, V., Hebert, M., Sminchisescu, C., Weiss, Y. (eds.) Computer Vision – ECCV 2018. LNCS, vol. 11218, pp. 122–138. Springer, Cham (2018). https://doi.org/10.1007/978-3-030-01264-9_8
26. Mishra, A., Nurvitadhi, E., Cook, J.J., Marr, D.: WRPN: wide reduced-precision networks. arXiv preprint arXiv:1709.01134 (2017)
27. Netzer, Y., Wang, T., Coates, A., Bissacco, A., Wu, B., Ng, A.Y.: Reading digits in natural images with unsupervised feature learning. In: NIPS Workshop on Deep Learning and Unsupervised Feature Learning, vol. 2011, p. 5 (2011)
28. Phan, H., Liu, Z., Huynh, D., Savvides, M., Cheng, K.T., Shen, Z.: Binarizing mobilenet via evolution-based searching. In: Proceedings of the IEEE/CVF Conference on Computer Vision and Pattern Recognition, pp. 13420–13429 (2020)
29. Qin, H., et al.: Forward and backward information retention for accurate binary neural networks. In: Proceedings of the IEEE/CVF Conference on Computer Vision and Pattern Recognition, pp. 2250–2259 (2020)

30. Rastegari, M., Ordonez, V., Redmon, J., Farhadi, A.: XNOR-Net: imageNet classi-fication using binary convolutional neural networks. In: Leibe, B., Matas, J., Sebe, N., Welling, M. (eds.) ECCV 2016. LNCS, vol. 9908, pp. 525–542. Springer, Cham (2016). https://doi.org/10.1007/978-3-319-46493-0_32
31. Russakovsky, O., et al.: Imagenet large scale visual recognition challenge. Int. J. Comput. Vis. **115**(3), 211–252 (2015)
32. Sandler, M., Howard, A., Zhu, M., Zhmoginov, A., Chen, L.C.: Mobilenetv 2: inverted residuals and linear bottlenecks. In: Proceedings of the IEEE Conference on Computer Vision and Pattern Recognition, pp. 4510–4520 (2018)
33. Shen, Z., He, Z., Xue, X.: Meal: multi-model ensemble via adversarial learning. In: Proceedings of the AAAI Conference on Artificial Intelligence, vol. 33, pp. 4886–4893 (2019)
34. Soudry, D., Hubara, I., Meir, R.: Expectation backpropagation: parameter-free training of multilayer neural networks with continuous or discrete weights. In: Advances in Neural Information Processing Systems, pp. 963–971 (2014)
35. Sze, V., Chen, Y.H., Yang, T.J., Emer, J.S.: Efficient processing of deep neural networks: a tutorial and survey. Proc. IEEE **105**(12), 2295–2329 (2017)
36. Tang, W., Hua, G., Wang, L.: How to train a compact binary neural network with high accuracy? In: Thirty-First AAAI Conference on Artificial Intelligence (2017)
37. Wang, Z., Lu, J., Tao, C., Zhou, J., Tian, Q.: Learning channel-wise interactions for binary convolutional neural networks. In: The IEEE Conference on Computer Vision and Pattern Recognition (CVPR), June 2019
38. Xu, Z., Cheung, R.C.: Accurate and compact convolutional neural networks with-trained binarization. In: British Machine Vision Conference (2019)
39. Zhang, D., Yang, J., Ye, D., Hua, G.: LQ-Nets: learned quantization for highly accurate and compact deep neural networks. In: Ferrari, V., Hebert, M., Sminchis-escu, C., Weiss, Y. (eds.) ECCV 2018. LNCS, vol. 11212, pp. 373–390. Springer, Cham (2018). https://doi.org/10.1007/978-3-030-01237-3_23
40. Zhang, J., Pan, Y., Yao, T., Zhao, H., Mei, T.: dabnn: a super fast inference framework for binary neural networks on arm devices. In: Proceedings of the 27th ACM International Conference on Multimedia, pp. 2272–2275 (2019)
41. Zhang, X., Zhou, X., Lin, M., Sun, J.: Shufflenet: an extremely efficient convolu-tional neural network for mobile devices. In: Proceedings of the IEEE Conference on Computer Vision and Pattern Recognition, pp. 6848–6856 (2018)
42. Zhou, A., Yao, A., Guo, Y., Xu, L., Chen, Y.: Incremental network quantization: towards lossless CNNs with low-precision weights. arXiv preprint arXiv:1702.03044 (2017)
43. Zhou, S., Wu, Y., Ni, Z., Zhou, X., Wen, H., Zou, Y.: Dorefa-net: training low bitwidth convolutional neural networks with low bitwidth gradients. arXiv preprint arXiv:1606.06160 (2016)
44. Zhu, C., Han, S., Mao, H., Dally, W.J.: Trained ternary quantization. arXiv preprint arXiv:1612.01064 (2016)
45. Zhu, S., Dong, X., Su, H.: Binary ensemble neural network: more bits per network or more networks per bit? In: Proceedings of the IEEE Conference on Computer Vision and Pattern Recognition, pp. 4923–4932 (2019)
46. Zhuang, B., Shen, C., Tan, M., Liu, L., Reid, I.: Towards effective low-bitwidth con-volutional neural networks. In: Proceedings of the IEEE Conference on Computer Vision and Pattern Recognition, pp. 7920–7928 (2018)
47. Zhuang, B., Shen, C., Tan, M., Liu, L., Reid, I.: Structured binary neural networks for accurate image classification and semantic segmentation. In: Proceedings of the IEEE Conference on Computer Vision and Pattern Recognition, pp. 413–422 (2019)

Video Object Detection via Object-Level Temporal Aggregation

Chun-Han Yao[1]($\boxtimes$), Chen Fang[2], Xiaohui Shen[3], Yangyue Wan[3],
and Ming-Hsuan Yang[1,4]

[1] UC Merced, Merced, USA
{cyao6,mhyang}@ucmerced.edu
[2] Tencent, Shenzhen, China
fangchen1988@gmail.com
[3] ByteDance Research, Beijing, China
{shenxiaohui,wanyangyue}@bytedance.com
[4] Google Research, Menlo Park, USA

Abstract. While single-image object detectors can be naively applied to videos in a frame-by-frame fashion, the prediction is often temporally inconsistent. Moreover, the computation can be redundant since neighboring frames are inherently similar to each other. In this work we propose to improve video object detection via temporal aggregation. Specifically, a detection model is applied on sparse keyframes to handle new objects, occlusions, and rapid motions. We then use real-time trackers to exploit temporal cues and track the detected objects in the remaining frames, which enhances efficiency and temporal coherence. Object status at the bounding-box level is propagated across frames and updated by our aggregation modules. For keyframe scheduling, we propose adaptive policies using reinforcement learning and simple heuristics. The proposed framework achieves the state-of-the-art performance on the Imagenet VID 2015 dataset while running real-time on CPU. Extensive experiments are done to show the effectiveness of our training strategies and justify the model designs.

Keywords: Video object detection · Object tracking · Temporal aggregation · Keyframe scheduling

1 Introduction

As mobile applications prevail nowadays, increasing attention has been drawn to deploying vision models on mobile devices. With the limited computation resource on mobile or embedded platforms, it is crucial to find an appropriate tradeoff between model accuracy, processing time, and memory usage. Real-time

Electronic supplementary material The online version of this chapter (https://doi.org/10.1007/978-3-030-58568-6_10) contains supplementary material, which is available to authorized users.

© Springer Nature Switzerland AG 2020
A. Vedaldi et al. (Eds.): ECCV 2020, LNCS 12359, pp. 160–177, 2020.
https://doi.org/10.1007/978-3-030-58568-6_10

inference on videos, particularly, requires heavy computation within a glimpse of an eye. This paper focuses on object detection for videos, a fundamental task of numerous downstream applications. The performance of single-image detection has been improved considerably by the advance of deep convolutional neural networks (CNNs). However, the best way to exploit temporal information in videos remains unclear. Early methods [8,15] extend single-image detectors to videos by offline linking the per-frame predictions. The flow-guided approaches [35,39,40] calculate motion field to propagate or aggregate feature maps temporally. They are shown to be effective but computationally expensive. Recent methods [20,21,38] propagate temporal information through memory modules. By inserting recurrent units between the convolutional layers, the network can extract features based on its memory of previous frames. While they achieve the state-of-the-art performance on mobile devices, we argue that aggregation at a higher level is more efficient and generic than feature-level aggregation. In particular, we propose to aggregate information at the object/bounding-box level, i.e., the output of the detection models. First, the dimensionality of object status is reduced compared to CNN feature maps. Second, it disentangles the binding between temporal aggregation and feature extraction. When adapting to a new feature extractor or distinct temporal dynamics, memory-guided feature aggregation requires model re-training as a whole. On the contrary, box predictions can be easily aggregated regardless of the semantic meaning of feature maps.

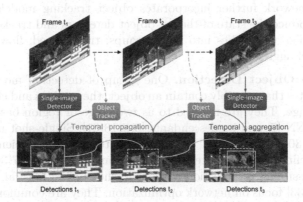

Fig. 1. Illustration of the proposed framework. Given a video sequence, we sparsely apply the detection model on certain keyframes. The object information is temporally propagated by the trackers and aggregated when the next detection is triggered.

To exploit temporal cues for object-level aggregation, tracking models come as an intuitive choice. The Detect and Track (D&T) approach [5] performs object detection and tracking with a jointly trained network, and yet its heavy computation is not applicable in real-time scenarios. For acceleration, we propose temporal aggregation modules to integrate detection and tracking information at the object level. As illustrated in Fig. 1, our framework applies detection on sparse keyframes and propagates the object status by real-time trackers. Since detection is applied sparsely, the selection of keyframes becomes crucial. We investigate the pros and cons of detection and tracking models, then propose adaptive keyframe scheduling via reinforcement learning (RL) and simple heuristics. Our experimental results demonstrate that the keyframe policies can generalize to various detection rates and produce a significant performance gain compared to fixed intervals. With the proposed aggregation modules and keyframe schedulers, we show the possibility to achieve competitive speed-accuracy tradeoffs with object-level temporal aggregation.

The contributions of this paper are as follows:

- We propose temporal aggregation modules to integrate object detection and tracking models at the object/bounding-box level.
- We present adaptive keyframe schedulers using simple heuristics and RL training with diverse video sequences.
- We conduct extensive experiments to compare our framework with the state-of-the-art methods in terms of CPU inference speed and accuracy on the Imagenet VID 2015 dataset [32].

2   Related Work

The majority of existing methods are built upon single-image object detectors. Our framework further incorporates object tracking models. Therefore, we briefly introduce the state-of-the-art object detectors and trackers, then categorize the video approaches into three groups: track-based, flow-guided, and memory-guided aggregation.

Single-Image Object Detection. One group of detection models proposes region candidates that possibly contain an object, then refines and classifies them in a second stage. They are referred to as two-stage or region-based detectors, including R-CNN [7] and its descendants [6,9,18,31]. Single-shot methods such as YOLO [29,30] and SSD [22] are proposed to improve efficiency based on a set of pre-defined anchor boxes. Bottom-up approaches [4,17,36,37] further explore the possibility of detection without anchor boxes. Several methods like Mobilenet [12,33] focus on network optimization. They are commonly combined with other detection models for acceleration on mobile devices. Since most two-stage detectors hardly run in real-time, we adopt Darknet53 [28] + YOLOv3-SPP [30] and Resnet18 [10] + CenterNet [36] as our single-image baselines, which represent the single-shot and bottom-up models, respectively.

Object Tracking. Object tracking is a common vision task which aims to trace objects throughout a video sequence. Given the initial bounding box as an object template, a tracking model can estimate the current location by correlation or CNN filtering. Most existing methods assume temporal smoothness of object appearance between frames. However, deformations, occlusions, and large motions can all pose a threat to the assumption, thus making it difficult to perform association and update object status. Fortunately, our scenario does not include long-term tracking and multi-object association. The association problem is also naturally avoided since we concern about the object class and not their identity. Considering the speed requirement, we choose the Kernelized Correlation Filter (KCF) [11] and Fully-convolutional Siamese Network (SiamFC) [1] as the trackers in our experiments.

Track-Based Aggregation. As an intuitive extension from single-images to videos, one can associate the per-frame detected boxes and refine the predictions accordingly. The Seq-NMS method [8] links the bounding boxes of an object across frames, then reassigns the maximal or average confidence score to the boxes along the track. The TCN approach [15] associates the boxes via optical flow and tracking algorithms before confidence reassignment. They can both provide a 3–5% mAP gain to most single-frame baselines, but are considered offline post-processing methods. The D&T framework [5] proposes to learn detection and tracking simultaneously. Benefiting from multi-task training, it produces a considerable accuracy gain on both tasks, whereas the expensive computation limits its application on mobile devices.

Flow-Guided Aggregation. Another group of approaches aggregates temporal information at the feature-map level via optical flow. The DFF method [40] applies expensive feature extraction on certain keyframes, then propagates the feature maps to the remaining frames. Zhu *et al.* [39] warp and average the feature maps of neighboring frames with flow-guided weighting. Wang *et al.* [35] leverage optical flow for pixel-level and instance-level calibration to boost the detection accuracy. To enable real-time inference on mobile devices, Zhu *et al.* [38] further speed up feature extraction and propagation by the Gated Recurrent Units. In spite of the acceleration efforts, most flow-guided approaches are substantially slower than other mobile-focused models. Moreover, the performance depends on the flow estimator which requires densely labeled data for training.

Memory-Guided Aggregation. Recent improvement of mobile-focused models relies on memory modules. Liu *et al.* [20] insert LSTM units between convolutional layers (ConvLSTM) to propagate temporal cues and refine feature maps. A following work by Liu *et al.* [21] proposes to balance between a small and a large feature extractor. It interleaves the extracted features by the ConvLSTM layers before the detection head. The feature extractors are trained along with the memory modules for temporal aggregation. Despite the merit of joint optimization, the binding demands intensive training when adapting to unseen domains with distinct temporal properties. We adopt a similar memory module for temporal aggregation in our framework, except that the information is aggregated at the object level. With a comparable accuracy, our method is shown to be more efficient and model-agnostic.

Adaptive Key-Frame Scheduling. Most existing methods use a fixed keyframe interval to balance between heavy and lightweight computations. Several adaptive policies are proposed for object tracking [13,34] and video segmentation [24]. For object detection, Chen *et al.* [2] select keyframes by offline measuring the difficulty of temporal propagation. Luo *et al.* [23] train a Siamese

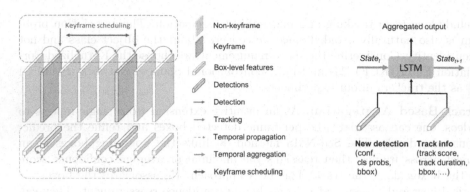

(a) Framework overview. Our aggregation module updates the object-level features and the adaptive scheduler adjusts the keyframe interval online.

(b) Memory module for temporal aggregation.

Fig. 2. The proposed framework and aggregation module (best viewed in color).

network to switch between detection and tracking models based on image-level features. Liu *et al.* [21] propose an RL policy for model interleaving, which results in marginal performance gain. Mao *et al.* [25] and Lao *et al.* [16] focus on minimizing the delay in detecting new objects. In this work we emphasize the importance of keyframe selection and propose two adaptive schedulers: a heuristic policy and a lightweight RL model.

3   Proposed Approach

Ideally, a detection model is able to capture the presence or absence of objects. The prediction of each frame is generated independently, and thus is robust from rapid motions or scene changes. Nonetheless, we often observe temporal inconsistency in the detection results, especially when an object is slightly deformed or partially occluded. The tracking models, on the other hand, excel at exploiting temporal cues to trace an object given its initial appearance. With a more lightweight computation, they can provide accurate and stable estimations of the object location. To leverage the advantages of detection and tracking models, we propose a framework which integrates them at the object level. Specifically, detection is applied on keyframes to capture new objects and initialize trackers. The trackers update object location and propagate the box features across frames, which facilitates object association and accelerates detection. Our temporal aggregation helps stabilize the confidence scores and class probabilities by utilizing past predictions. To ensure that detection is triggered when tracking fails, we propose adaptive schedulers to determine the keyframe interval based on object status. An overview of our framework is shown in Fig. 2a.

3.1 Temporal Aggregation

A typical detection model produces a confidence/objectness score c, class probabilities p, and box coordinates b for each object. Object tracking models can provide an estimate of the box coordinates $\tilde{b}$ and a tracking score s indicating how well the current box matches the object template or how confident the estimation is. The information combined with tracking duration d are exploited as features. The selection of object-level features makes our aggregation more model-agnostic, as it applies to any object detector and tracker that yield the above output. For each feature x in (c, p, b, s), we denote the value propagated by the trackers as $\tilde{x}$, and the aggregated output as x'.

At each keyframe, we initialize a tracker for each new object and terminate the old ones for occluded or disappeared objects. The tracked objects are associated with the current detection by thresholding the intersection-over-union (IOU) ratio. We adopt the convention in the existing methods [2,8] and set the IOU threshold as 0.5 in our experiments. Once associated, the features of the existing objects are aggregated with the new detection. As shown in Fig. 2b, we employ a memory module that takes the tracked information and new detection as inputs, updates the internal state, and produces an aggregated output. Our memory module is similar to the speed-optimized Bottleneck-LSTM proposed by Liu et al. [21], except that we replace convolution operations by two fully-connected layers since our inputs are object-level features instead of convolutional feature maps. We show the detail LSTM architecture in the supplementary material. Unlike the prior memory-guided methods, our aggregation module has a low-dimensional input and is only applied at sparse keyframes, thus it entails a minimal inference time.

In addition to the learning-based aggregation, we propose a simple yet effective heuristic module to mitigate the inconsistency of three predictions: 1) Box coordinates, 2) class probabilities, and 3) Confidence score. First, to produce accurate and temporally smooth box coordinates, we refine the detected boxes b_t by the tracked boxes $\tilde{b}_t$. The coordinates are averaged based on the detection confidence and tracking score. We weight higher on the detected boxes when the new detection is more confident than tracking, and vice versa. The refinement can be expressed as:

$$b'_t = \frac{\tilde{s}_t \tilde{c}_t \tilde{b}_t + c_t b_t}{\tilde{s}_t \tilde{c}_t + c_t}, \tag{1}$$

where $\tilde{c}_t$ and $\tilde{s}_t$ denote the confidence and minimal tracking score propagated from the previous keyframe to time t. Second, we aggregate the class probabilities of the previous keyframes to prevent inconsistent class predictions. The probabilities are re-weighted by the confidence scores and aggregated by cumulative average:

$$p'_t = \frac{c_t p_t + \gamma \sum_{i < t, i \in K} c_i p_i}{c_t + \gamma \sum_{i < t, i \in K} c_i}, \tag{2}$$

where K is the keyframe set, and γ is the weight decay for previous predictions. Finally, the confidence score is stabilized by track-based reassignment. We keep track of the confidence scores in the keyframes and update the current confidence by the temporal maximal:

$$c'_t = \max(\tilde{c}_t, c_t). \tag{3}$$

It aims to pick up the low-confidence detection between high-confidence frames. The final detection result for frame t is given by (b'_t, p'_t, c'_t).

3.2 Adaptive Keyframe Scheduling

Fixing the keyframe interval is a naive way to switch between lightweight and heavy and computations. However, it ignores temporal variations and fails to adapt to certain events where one model is preferred over another. If a video sequence is rather smooth, we prefer tracking over detection since it costs minimal computation. On the contrary, frequent detection is required when the object dynamics are beyond the capability of trackers, $e.g.$, large motions, deformations, and occlusions. More importantly, the selection of keyframes determines the object templates for tracker initialization, which affects tracking performance significantly. In our experiments, we observe that selecting the first frame where an object just appeared does not lead to optimal tracking results. Instead, a few frames after the first appearance typically contain more faithful and complete templates to initialize object trackers. We also find that tracking score as well as other object-level features are useful and convenient indications of the tracking and detection quality. Adding more image-level features only provides marginal information for keyframe selection.

Inspired by the above observations, we propose an adaptive policy to adjust the keyframe interval D online. The task can be formulated as an RL problem with the following state space, action space, and reward function. The state vector S is constructed from the same object-level features as leveraged in temporal aggregation. For each feature x, we encode the temporal minimal (x_{min}), maximal (x_{max}), average (x_{mean}), variance (x_{var}), and difference ($x_t - x_{t-1}$) in S to facilitate RL training. Given the state of current frame, the agent learns to predict an action a from the action space:

$$a_t = \begin{cases} 0 \text{ (trigger detection)}: & D_t = 1 \\ 1 \text{ (shorten interval)}: & D_t = \max(1, \alpha D_{t-1}), \\ 2 \text{ (fix interval)}: & D_t = D_{t-1} \end{cases} \tag{4}$$

where α is a constant multiplier. The detection model is applied if and only if the tracking duration $d_t \geq D_t$. For the instant reward R, we calculate the average IOU ratio between the predicted boxes b' and ground truth boxes $\hat{b}$ as follows:

$$R_t = \text{IOU}(b_t', \hat{b}_t). \tag{5}$$

We adopt the standard training scheme of the Asynchronous Advantage Actor-Critic (A3C) [26] model, which is shown to be effective in both continuous and discrete domains. The readers are encouraged to refer to the original paper since we omit the model details here. Considering that there is no ordering of the objects, we use PointNet [27] with 3 fully-connected layers for both the policy and value networks. The model is originally designed for point cloud data to preserve order invariance. In our work we consider each object as a point data and set a maximum number of objects according to the dataset distribution. Exceeding objects are pruned by their confidence scores.

Even though the state space and action space are low-dimensional, the diversity of video sequences makes it a challenging environment for RL training. The same action in the same state may lead to distinct rewards and next states in different videos, *i.e.*, the reward and state transition are stochastic from the perspective of the RL agent. Combining the time and accuracy constraints in the reward function is yet another critical issue. Liu *et al.* [21] add a speed reward to the accuracy reward with a specific weighting for each speed-accuracy tradeoff point. We observe that the simple weighted sum is likely to cause unstable training results. A slight difference in speed reward can easily lead to a local optimal, where the policy applies detection either most excessively or least frequently. Furthermore, the policy needs to be trained with a new reward setting whenever a new tradeoff point is desired.

We propose the following strategies to stabilize the training process and produce a more general policy. First, we mimic the concept of imitation learning by pre-training the policy network with expert supervision. Considering that exhaustive search for optimal keyframes is not feasible, a greedy algorithm is proposed to approximate an oracle policy for supervision. As described in Algorithm 1, we start from a keyframe set with fixed intervals, then iteratively replace a keyframe with the highest tracking score by a non-keyframe with the lowest tracking score. The pre-training steps using the oracle keyframes are described in the supplemental material. To train a general policy for various speed-accuracy tradeoffs, a base interval D_{base} to deviate from is randomly initialized within a range of 5 to 30 frames. The ratio between the tracking duration and base interval, d/D_{base}, is encoded in the state vector, so the agent is aware of the relative detection rate compared to the base frequency. To compensate for the varying reward distribution between diverse videos, we calculate the temporal difference of IOU scores instead of the raw values. It allows the agent to focus on policy optimization according to temporal dynamics. Finally, we add a long-term penalty at the end of an episode to constrain the policy from excessive or passive detection. The modified reward function R_t' can be expressed as:

$$
R_t' = \begin{cases} R_t - R_{t-1} + \lambda\left(\frac{1}{D_{base}} - \text{mean}(\eta)\right) & \text{if } t = \text{ end of episode} \\ R_t - R_{t-1} & \text{otherwise} \end{cases}, \quad (6)
$$

where η denotes the detection history and λ is weight of long-term penalty. We detail the training process in Algorithm 2.

To compare with the RL policy, we propose a heuristic scheduler by mapping the tracking score to the keyframe interval:

$$
D_t = \min(D_{t-1}, \mu \tilde{s}_t D_{base}), \quad (7)
$$

Algorithm 1. Greedy Keyframe Scheduling

Inputs: video length L, base keyframe interval D_{base}, number of iterations N_{iter}
Output: K - keyframe set of size $\lfloor L/D_{base} \rfloor$

1: $K \leftarrow \{k | mod(k, D_{base}) = 0, 0 \leq k < L\}$
2: $R_{max} \leftarrow$ Run the framework with K and compute mAP
3: **for** $i = 0 : N_{iter}$ **do**
4:     $k_{max} \leftarrow$ A keyframe with maximal tracking score
5:     $k_{min} \leftarrow$ A non-keyframe with minimal tracking score
6:     $R \leftarrow$ Run the framework with K $/\{k_{max}\} \cup \{k_{min}\}$ and compute mAP
7:     **if** $R > R_{max}$ **then**
8:         $K \leftarrow K$ $/\{k_{max}\} \cup \{k_{min}\}$
9:         $R_{max} \leftarrow R$
10: **return** K

Algorithm 2. RL Training for Keyframe Scheduling

Definitions: timestep t, tracking duration d, keyframe interval D, episode length L, state S, action a, reward R, detection history η
Output: keyframe policy π

1: Pre-train π with the oracle supervision
2: **repeat**
3:     Randomly initialize base detection interval D_{base}
4:     Sample video frames $\{I_0, I_1, ..., I_L\}$
5:     $t \leftarrow 0, d_0 \leftarrow 0, D_0 \leftarrow D_{base}$
6:     $S_0 \leftarrow$ Run detection on I_0 and extract object-level features
7:     $S_1 \leftarrow$ Track objects from I_0 to I_1 and propagate S_0
8:     **for** $t = 1 : L$ **do**
9:         $a_t \leftarrow$ arg max $\pi(S_t)$
10:         $D_t \leftarrow$ Update interval based on a_t and D_{t-1}
11:         $d_t \leftarrow d_{t-1} + 1$
12:         **if** $d_t \geq D_t$ **then**
13:             Run detection on I_t and update S_t
14:             Add 1 to η
15:             $D_t \leftarrow D_{base}$
16:             $d_t \leftarrow 0$
17:         **else**
18:             Add 0 to η
19:         $R_t \leftarrow$ Compute reward based on S_t and η
20:         $S_{t+1} \leftarrow$ Track objects from I_t to I_{t+1} and propagate S_t
21:         Add (S_t, a_t, R_t, S_{t+1}) to reply buffer
22:     Sample batch B from reply buffer
23:     Update π with B using A3C training
24: **until** convergence

---

where μ is a weighting parameter, $\tilde{s}_t$ is the temporally minimal tracking score since the last detection. It follows our intuition to decrease/increase the detection frequency with tracking quality. At inference time, one can easily search for the best speed-accuracy tradeoff for a specific application by setting D_{base}.

4   Performance Evaluation

4.1   Experiment Settings

We evaluate model performance on the Imagenet 2015 VID dataset [32] with 30 object classes. Following the work by Liu et al. [21], the detection models are trained on the Imagenet VID, Imagenet DET [32], and COCO [19] training sets. We sample 1 per 30 frames from the VID videos and use only the relevant classes from the DET and COCO datasets, which amount to 37K frames from VID, 147K images from DET, and 43K images from COCO. For object tracking, KCF does not require any pre-training, and we directly employ the SiamFC model trained on the GOT-10k dataset [14]. The temporal aggregation module and RL policy are trained by randomly sampling video sequences of 30 frames. We set a maximum of 10 objects per frame for PointNet since most VID videos contain less than 10 objects. The sequences where no object is detected are ignored to prevent unnecessary training. For evaluation, we calculate the mAP@0.5IOU on the VID validation set. All training and evaluation are done with an input resolution of 320×320. Based on our ablation studies, we set $\gamma = 1$ for heuristic aggregation and $\alpha = 0.5, \lambda = 1, \mu = 1$ for the keyframe schedulers.

4.2   Quantitative Comparisons

We compare our framework with the single-image baselines and previous methods in Table 1. While there are numerous related works in video processing, action recognition, frame interpolation, etc., we focus on comparing methods with similar settings: online, real-time, mobile-focused, and based on single-image detection. Existing methods evaluate model performance in terms of the raw mAP and inference time. However, we find the metrics not directly comparable since they are built upon different single-image baselines and tested on different platforms. With a better baseline, one can reach higher accuracy regardless of the temporal design for videos. Considering our focus to improve from single-image detection, we report the speed-up ratio and mAP gain (both absolute and relative values) compared to each baseline. Despite the difference in baseline models, the performance of our single-image detectors lie between the baselines of the other approaches [20,38], which makes the comparison more convincing. Combining YOLOv3-SPP and KCF at a detection interval of $D_{base} = 7$, we achieve 63.2% mAP at 31.4 fps. At $D_{base} = 15$, our framework maintains a reasonably high accuracy while having a 11.9× speed-up ratio. Note that the flow-guided approach [38] does not train the detectors on the COCO dataset, but its optical flow model requires pre-training on the Flying Chairs dataset [3].

Among all the real-time models, the memory-guided approach [21] achieves competitive accuracy based on its well-trained baseline (f_0). It can further produce a significant speed-up ratio by asynchronous inference and model quantization. Our method reaches comparable speed without any network optimization. In Fig. 3a we plot the speed-accuracy tradeoff curves, which show that the proposed framework performs favorably against the state-of-the-art methods.

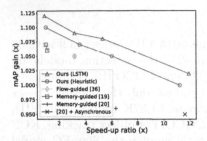

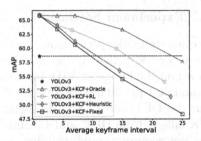

(a) Relative mAP gain versus speed-up ratio of different aggregation methods. Our approach adopts YOLOv3-SPP + KCF and RL keyframe scheduler.

(b) Speed-accuracy tradeoff curves of different keyframe policies. The data points are obtained by adjusting D_{base} and averaging the detection interval.

Fig. 3. Visualization of the quantitative comparisons.

4.3   Qualitative Comparisons

In Fig. 4 we show the qualitative results on the VID validation set. The first two videos contain an object with mild motions and occlusions, causing temporal inconsistency in the per-frame detections. The proposed aggregation effectively picks up the missed detections and corrects the erroneous classification. In the third video, the rapid motions and deformations make object tracking a tougher task. Our adaptive keyframe scheduler overcomes such difficulty by triggering more frequent detection. The importance of keyframe selection can also be observed in video 2 and 3. With a better keyframe scheduler, we can initialize the object trackers with a better object template, which results in more accurate box estimations.

4.4   Speed-Accuracy Tradeoffs

To vary how frequent the expensive feature extraction is performed, Liu *et al.* [21] train an RL policy with a specific reward weighting. On the contrary, our keyframe schedulers are generic to various detection rates as they encode the base interval D_{base} in the state vector. The average detection interval can be adjusted online simply by setting D_{base}. The mAP curves of different keyframe policies are shown in Fig. 3b. Using a fixed detection interval, the mAP drops

Table 1. Performance comparison on the Imagenet VID validation set. α is the feature extractor width multiplier described in [12], and β is the flow network width multiplier. The models of [38], [20,21], and ours are evaluated on HuaWei Mate 8 phone (*), Pixel 3 phone (†), and Intel Xeon E5 CPU, respectively. We mark the models with real-time inference speed (above 30fps) in red. All of our models adopt KCF trackers and RL keyframe scheduler.

Approach	Model	mAP	Speed (fps)	mAP gain	Speed-up ratio
Flow-guided [38]	($\alpha = 0.5$)	48.6	16.4*	-	-
	($\alpha = 0.5$) + Flow ($\beta = 0.5$)	51.2	52.6*	+2.6/×1.05	×3.2
	($\alpha = 1.0$)	58.3	4.0*	-	-
	($\alpha = 1.0$) + Flow ($\beta = 1.0$)	61.2	12.5*	+2.9/×1.05	×3.1
Memory-guided [20]	($\alpha = 0.35$)	42.0	14.4†	-	-
	($\alpha = 0.35$) + LSTM	45.1	14.6†	+3.1/×1.07	×1.0
	($\alpha = 1.4$)	60.5	3.7†	-	-
	($\alpha = 1.4$) + LSTM	**64.1**	4.1†	+3.6/×1.06	×1.1
Memory-guided [21]	Non-interleaved (f_0)	63.9	4.2†	-	-
	Interleaved + Adaptive	61.4	26.6†	−2.5/×0.96	×6.3
	Interleaved + Adaptive + Async	60.7	48.8†	−3.2/ ×0.95	×11.6
Ours	CenterNet	52.0	12.1	-	-
	CenterNet + Heuristic ($D_{base} = 7$)	54.7	57.8	+2.7/×1.05	×4.8
	CenterNet + LSTM ($D_{base} = 7$)	56.4	52.3	+4.4/×**1.08**	×4.3
	CenterNet + LSTM ($D_{base} = 15$)	53.3	95.2	+1.3/×1.03	×7.8
	YOLOv3	58.6	6.0	-	-
	YOLOv3 + Heuristic ($D_{base} = 7$)	61.3	35.8	+2.7/×1.05	×6.0
	YOLOv3 + LSTM ($D_{base} = 7$)	63.2	31.4	+4.6/×**1.08**	×5.3
	YOLOv3 + LSTM ($D_{base} = 15$)	59.9	71.5	+1.3/×1.02	×**11.9**

linearly as the interval increases. The proposed heuristic scheduler mitigates the accuracy drop by a clear margin. With the trained RL policy, we manage to extend the keyframe interval to 15 frames and still match the performance of the single-image baseline. We also plot the oracle curve obtained by running the greedy algorithm (Algorithm 1) with $N_{iter} = 20$. It demonstrates how the performance gap from the oracle is narrowed down by our RL policy.

4.5   Ablation Studies

As discussed in the previous sections, our object-level aggregation disentangles the binding between feature maps and temporal aggregation. We demonstrate the advantage by reporting the accuracy without finetuning the detection models on the VID dataset in Table 2. Unlike feature-level aggregation, our framework provides a consistent mAP gain by simply training the aggregation module on the VID dataset and not the entire feature extractor. In Table 3a we evaluate the effectiveness of LSTM-based aggregation module and the three aggregation heuristics: box coordinate refinement, class probability re-weighting, and confidence reassignment. The confidence reassignment effectively stabilizes the detection predictions. It alone contributes to 3.5% mAP gain. The aggregation of class probabilities ($\gamma = 1.0$) and box coordinates is able to produce an additional gain of 2.7%. To validate that our framework applies to arbitrary detection and tracking models, we experiment with multiple combinations of the sub-modules. The results in Table 3b show a consistent improvement by our method with either of the detection models. With a deep object tracker (SiamFC), our framework achieves higher mAP and still runs in real-time. The RL policy further produces a considerable mAP gain while entailing minimal inference time.

Table 2. Validation with/without training the detectors on the video dataset. The models are run with KCF trackers and LSTM-based aggregation at $D_{base} = 1$. The mAP scores are calculated only on the 14 relevant classes in COCO dataset.

Model	Training data	mAP	mAP gain
CenterNet	COCO	53.1	+3.7/×1.08
	VID+DET+COCO	56.8	+4.3/×1.09
YOLOv3-SPP	COCO	62.6	+6.4/×1.12
	VID+DET+COCO	66.6	+6.6/×1.11

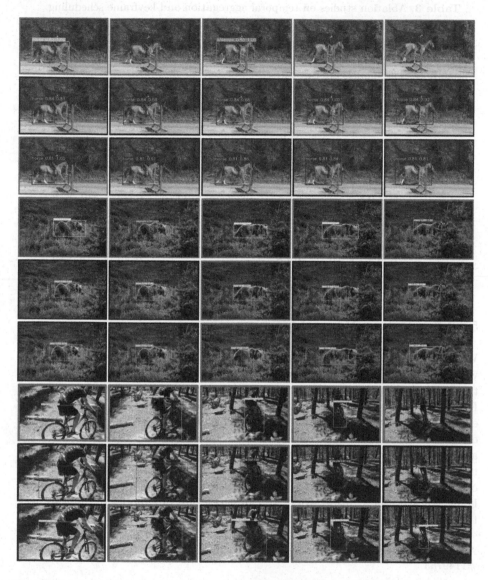

Fig. 4. Qualitative results on the Imagenet VID validation set (best viewed in color). For each video sequence, the results of per-frame detection are presented in the first row, our method (LSTM-based aggregation) with fixed keyframe interval in the second row, and adaptive scheduling (RL) in the third row. The frames where detection is applied are marked in red. Each detected object is labeled by a colored bounding box with the class name, detection confidence, and tracking score on the top. Missed detections can be observed in videos 1 and 3, and the objects are occasionally misclassified in videos 1 and 2. Our temporal aggregation effectively picks up all the missed detections and corrects the false classification. In video 3, the adaptive keyframe scheduler triggers detection more intelligently, which leads to more accurate box predictions.

Table 3. Ablation studies on temporal aggregation and keyframe scheduling.

(a) Evaluation of our heuristic and LSTM-based temporal aggregation modules. The models are run with YOLOv3-SPP + KCF at $D_{base} = 1$.

Confidence	Cls prob	Bbox	mAP
-			58.6
✓			62.1
✓	$\gamma = 0.6$		63.8
✓	$\gamma = 0.8$		64.1
✓	$\gamma = 1.0$		64.4
✓	$\gamma = 1.0$	✓	**64.8**
LSTM-based			**65.8**

(b) Evaluation on different module combinations. All are run with LSTM-based aggregation at $D_{base} = 7$.

Detector	Tracker	Keyframe scheduler	Speed (fps)	mAP
CenterNet	KCF	Fixed	**58.8**	53.2
		Heuristic	57.8	53.6
		RL	52.3	56.4
CenterNet	SiamFC	Fixed	45.9	55.3
		Heuristic	45.8	55.5
		RL	42.9	57.7
YOLOv3-SPP	KCF	Fixed	37.1	60.6
		Heuristic	35.8	61.2
		RL	31.4	63.2
YOLOv3-SPP	SiamFC	Fixed	23.8	62.0
		Heuristic	23.8	62.3
		RL	20.2	**63.7**

5 Concluding Remarks

We propose a framework to improve detection by tracking the objects across video frames. Temporal information is exploited and aggregated by real-time trackers, which makes the detection more consistent and efficient. Heuristic and RL policies are proposed for adaptive keyframe scheduling. As shown in the experimental results, our method achieves competitive speed-accuracy tradeoffs on the Imagenet VID 2015 dataset. Furthermore, the object-level aggregation alleviates the dependence on feature maps, making it more generic to arbitrary detection and tracking models. Currently, we initiate a tracker for each object, which linearly increases the inference time when multiple objects are being tracked. As a future work, we can accelerate our model by multi-object tracking and running the detection and tracking models in parallel.

Acknowledgement. This work is supported in part by the NSF CAREER Grant #1149783.

References

1. Bertinetto, L., Valmadre, J., Henriques, J.F., Vedaldi, A., Torr, P.H.S.: Fully-convolutional siamese networks for object tracking. In: Hua, G., Jégou, H. (eds.) ECCV 2016. LNCS, vol. 9914, pp. 850–865. Springer, Cham (2016). https://doi.org/10.1007/978-3-319-48881-3_56

2. Chen, K., et al.: Optimizing video object detection via a scale-time lattice. In: Proceedings of the IEEE Conference on Computer Vision and Pattern Recognition, pp. 7814–7823 (2018)
3. Dosovitskiy, A., et al.: Flownet: learning optical flow with convolutional networks. In: Proceedings of the IEEE International Conference on Computer Vision, pp. 2758–2766 (2015)
4. Duan, K., Bai, S., Xie, L., Qi, H., Huang, Q., Tian, Q.: Centernet: keypoint triplets for object detection. arXiv preprint arXiv:1904.08189 (2019)
5. Feichtenhofer, C., Pinz, A., Zisserman, A.: Detect to track and track to detect. In: Proceedings of the IEEE International Conference on Computer Vision, pp. 3038–3046 (2017)
6. Girshick, R.: Fast R-CNN. In: Proceedings of the IEEE International Conference on Computer Vision, pp. 1440–1448 (2015)
7. Girshick, R., Donahue, J., Darrell, T., Malik, J.: Rich feature hierarchies for accurate object detection and semantic segmentation. In: Proceedings of the IEEE Conference on Computer Vision and Pattern Recognition, pp. 580–587 (2014)
8. Han, W., et al.: Seq-NMS for video object detection. arXiv preprint arXiv:1602.08465 (2016)
9. He, K., Gkioxari, G., Dollár, P., Girshick, R.: Mask R-CNN. In: Proceedings of the IEEE International Conference on Computer Vision, pp. 2961–2969 (2017)
10. He, K., Zhang, X., Ren, S., Sun, J.: Deep residual learning for image recognition. In: Proceedings of the IEEE Conference on Computer Vision and Pattern Recognition, pp. 770–778 (2016)
11. Henriques, J.F., Caseiro, R., Martins, P., Batista, J.: High-speed tracking with kernelized correlation filters. IEEE Trans. Pattern Anal. Mach. Intell. **37**(3), 583–596 (2014)
12. Howard, A.G., et al.: Mobilenets: efficient convolutional neural networks for mobile vision applications. arXiv preprint arXiv:1704.04861 (2017)
13. Huang, C., Lucey, S., Ramanan, D.: Learning policies for adaptive tracking with deep feature cascades. In: Proceedings of the IEEE International Conference on Computer Vision, pp. 105–114 (2017)
14. Huang, L., Zhao, X., Huang, K.: Got-10k: a large high-diversity benchmark for generic object tracking in the wild. arXiv preprint arXiv:1810.11981 (2018)
15. Kang, K., Ouyang, W., Li, H., Wang, X.: Object detection from video tubelets with convolutional neural networks. In: Proceedings of the IEEE Conference on Computer Vision and Pattern Recognition, pp. 817–825 (2016)
16. Lao, D., Sundaramoorthi, G.: Minimum delay object detection from video. In: Proceedings of the IEEE International Conference on Computer Vision, pp. 5097–5106 (2019)
17. Law, H., Deng, J.: CornerNet: detecting objects as paired keypoints. In: Ferrari, V., Hebert, M., Sminchisescu, C., Weiss, Y. (eds.) Computer Vision – ECCV 2018. LNCS, vol. 11218, pp. 765–781. Springer, Cham (2018). https://doi.org/10.1007/978-3-030-01264-9_45
18. Li, Z., Peng, C., Yu, G., Zhang, X., Deng, Y., Sun, J.: Light-head R-CNN: in defense of two-stage object detector. arXiv preprint arXiv:1711.07264 (2017)
19. Lin, T.-Y., et al.: Microsoft COCO: common objects in context. In: Fleet, D., Pajdla, T., Schiele, B., Tuytelaars, T. (eds.) ECCV 2014. LNCS, vol. 8693, pp. 740–755. Springer, Cham (2014). https://doi.org/10.1007/978-3-319-10602-1_48
20. Liu, M., Zhu, M.: Mobile video object detection with temporally-aware feature maps. In: Proceedings of the IEEE Conference on Computer Vision and Pattern Recognition, pp. 5686–5695 (2018)

21. Liu, M., Zhu, M., White, M., Li, Y., Kalenichenko, D.: Looking fast and slow: memory-guided mobile video object detection. arXiv preprint arXiv:1903.10172 (2019)
22. Liu, W., et al.: SSD: single shot MultiBox detector. In: Leibe, B., Matas, J., Sebe, N., Welling, M. (eds.) ECCV 2016. LNCS, vol. 9905, pp. 21–37. Springer, Cham (2016). https://doi.org/10.1007/978-3-319-46448-0_2
23. Luo, H., Xie, W., Wang, X., Zeng, W.: Detect or track: towards cost-effective video object detection/tracking. In: Proceedings of the AAAI Conference on Artificial Intelligence, vol. 33, pp. 8803–8810 (2019)
24. Mahasseni, B., Todorovic, S., Fern, A.: Budget-aware deep semantic video segmentation. In: Proceedings of the IEEE Conference on Computer Vision and Pattern Recognition, pp. 1029–1038 (2017)
25. Mao, H., Yang, X., Dally, W.J.: A delay metric for video object detection: what average precision fails to tell. In: Proceedings of the IEEE International Conference on Computer Vision, pp. 573–582 (2019)
26. Mnih, V., et al.: Asynchronous methods for deep reinforcement learning. In: Proceedings of the International Conference on Machine Learning, pp. 1928–1937 (2016)
27. Qi, C.R., Su, H., Mo, K., Guibas, L.J.: Pointnet: deep learning on point sets for 3D classification and segmentation. In: Proceedings of the IEEE Conference on Computer Vision and Pattern Recognition, pp. 652–660 (2017)
28. Redmon, J.: Darknet: open source neural networks in C (2013–2016). http://pjreddie.com/darknet/
29. Redmon, J., Divvala, S., Girshick, R., Farhadi, A.: You only look once: Unified, real-time object detection. In: Proceedings of the IEEE Conference on Computer Vision and Pattern Recognition, pp. 779–788 (2016)
30. Redmon, J., Farhadi, A.: YOLOv3: an incremental improvement. arXiv preprint arXiv:1804.02767 (2018)
31. Ren, S., He, K., Girshick, R., Sun, J.: Faster R-CNN: towards real-time object detection with region proposal networks. In: Advances in Neural Information Processing Systems, pp. 91–99 (2015)
32. Russakovsky, O., et al.: Imagenet large scale visual recognition challenge. Int. J. Comput. Vis. **115**(3), 211–252 (2015). https://doi.org/10.1007/s11263-015-0816-y
33. Sandler, M., Howard, A., Zhu, M., Zhmoginov, A., Chen, L.C.: Mobilenetv 2: inverted residuals and linear bottlenecks. In: Proceedings of the IEEE Conference on Computer Vision and Pattern Recognition, pp. 4510–4520 (2018)
34. Supancic III, J., Ramanan, D.: Tracking as online decision-making: learning a policy from streaming videos with reinforcement learning. In: Proceedings of the IEEE International Conference on Computer Vision, pp. 322–331 (2017)
35. Wang, S., Zhou, Y., Yan, J., Deng, Z.: Fully motion-aware network for video object detection. In: Ferrari, V., Hebert, M., Sminchisescu, C., Weiss, Y. (eds.) ECCV 2018. LNCS, vol. 11217, pp. 557–573. Springer, Cham (2018). https://doi.org/10.1007/978-3-030-01261-8_33
36. Zhou, X., Wang, D., Krähenbühl, P.: Objects as points. In: arXiv preprint arXiv:1904.07850 (2019)
37. Zhou, X., Zhuo, J., Krahenbuhl, P.: Bottom-up object detection by grouping extreme and center points. In: Proceedings of the IEEE Conference on Computer Vision and Pattern Recognition, pp. 850–859 (2019)
38. Zhu, X., Dai, J., Zhu, X., Wei, Y., Yuan, L.: Towards high performance video object detection for mobiles. arXiv preprint arXiv:1804.05830 (2018)

39. Zhu, X., Wang, Y., Dai, J., Yuan, L., Wei, Y.: Flow-guided feature aggregation for video object detection. In: Proceedings of the IEEE International Conference on Computer Vision, pp. 408–417 (2017)
40. Zhu, X., Xiong, Y., Dai, J., Yuan, L., Wei, Y.: Deep feature flow for video recognition. In: Proceedings of the IEEE Conference on Computer Vision and Pattern Recognition, pp. 2349–2358 (2017)

Object Detection with a Unified Label Space from Multiple Datasets

Xiangyun Zhao[1]([✉]), Samuel Schulter[2], Gaurav Sharma[2], Yi-Hsuan Tsai[2], Manmohan Chandraker[2,3], and Ying Wu[1]

[1] Northwestern University, Evanston, USA
zhaoxiangyun915@gmail.com
[2] NEC Labs America, Princeton, USA
[3] UC San Diego, San Diego, USA

Abstract. Given multiple datasets with different label spaces, the goal of this work is to train a single object detector predicting over the union of all the label spaces. The practical benefits of such an object detector are obvious and significant—application-relevant categories can be picked and merged form arbitrary existing datasets. However, naïve merging of datasets is not possible in this case, due to inconsistent object annotations. Consider an object category like faces that is annotated in one dataset, but is not annotated in another dataset, although the object itself appears in the latter's images. Some categories, like face here, would thus be considered foreground in one dataset, but background in another. To address this challenge, we design a framework which works with such partial annotations, and we exploit a pseudo labeling approach that we adapt for our specific case. We propose loss functions that carefully integrate partial but correct annotations with complementary but noisy pseudo labels. Evaluation in the proposed novel setting requires full annotation on the test set. We collect the required annotations (Project page: http://www.nec-labs.com/~mas/UniDet This work was part of Xiangyun Zhao's internship at NEC Labs America.) and define a new challenging experimental setup for this task based on existing public datasets. We show improved performances compared to competitive baselines and appropriate adaptations of existing work.

1   Introduction

Object detection has made tremendous progress in recent years to become a powerful computer vision tool [10,23,25]. This is driven by the availability of large-scale datasets with bounding box annotations. However, obtaining such data is costly and time-consuming. While multiple publicly available datasets

This work was part of Xiangyun Zhao's internship at NEC Labs America.

Electronic supplementary material The online version of this chapter (https://doi.org/10.1007/978-3-030-58568-6_11) contains supplementary material, which is available to authorized users.

© Springer Nature Switzerland AG 2020
A. Vedaldi et al. (Eds.): ECCV 2020, LNCS 12359, pp. 178–193, 2020.
https://doi.org/10.1007/978-3-030-58568-6_11

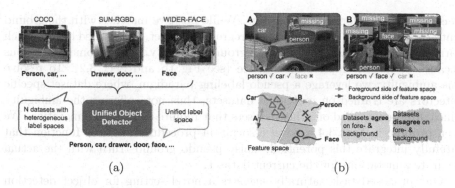

(a)   (b)

Fig. 1. **(a)** We train a single object detector from multiple datasets with heterogeneous label spaces. In contrast to prior work [31], our model unifies the label spaces of all datasets. **(b)** Illustration of the ambiguity of background in object detection when training from multiple datasets with different label spaces. Here, only "person" is consistent wrt. both datasets but "car" and "face" are missing in the other one, respectively. Naïve combination of the datasets leads to wrong training signals

with annotations for various categories already exist, their label spaces are mostly different, which makes a naïve combination of the data impossible [9,15,18]. But such a unification could be crucial for applications that require categories not labeled in some of the datasets, to avoid a costly annotation process. We address this problem and *propose to unify heterogeneous label spaces across multiple datasets*. We show how to train a single object detector capable of detecting over the union of all training label spaces on a given test image (see Fig. 1a).

Unifying different label spaces is not straightforward for object detection. A key challenge is the *ambiguity in the definition of the "background" category*. Recall that image regions in object detection datasets may contain objects without annotation because the object's category is not part of the label space. These image regions are then considered part of the background. However, their object category may be annotated in a different dataset. For instance, the label space of the COCO dataset [18] has 80 categories but it does not contain "human faces", which is present in many of the COCO images, and also annotated in other datasets like [34]. A pair of images illustrate the point in Fig. 1b. Hence, images and annotations of different datasets can not be trivially combined while preserving the annotation consistency wrt. the union of all their label spaces.

A straightforward attempt to handle the ambiguity of the label space is exhaustive manual annotation, which is very costly in terms of both time and money. Recent works have tried to address a similar problem of expanding the label space of object detectors [22,26,31], which we discuss in Sect. 2. However, none of them can truly unify multiple label spaces and are, thus, not applicable in our setting. In contrast, *we propose to handle the ambiguity directly and unify the label spaces in a single detector*. At each training iteration, we sample images from a single dataset along with the corresponding ground truth. Predicted bounding boxes of the model need to be associated with ground truth in order to compute a loss for object detectors. While positive matches with the ground truth can be assigned a category and a regression target, unmatched boxes, which normally would be assigned the background category, now become ambiguous because of the incomplete label space (see Fig. 1b and Sect. 3.2). To resolve this ambiguity, we leverage a pseudo labeling paradigm, where dataset-specific detectors are trained on individual datasets, thus not suffering the ambiguity, which are then applied on other datasets that require additional annotation. We then propose in Sect. 3.4 a novel association procedure and loss function that carefully integrate this potentially noisy pseudo ground truth with the actual accurate annotations for the current dataset.

Our proposed task naturally suggests a novel setting for object detection (Sect. 3.5): Given N datasets with heterogeneous label spaces, train a detector capable of predicting over the union of all label spaces, which is evaluated on a test set equipped with annotations for all categories. For this purpose, we choose four existing and challenging datasets (COCO [18], SUN-RGBD [27], VOC [6] and LISA-Signs [19]), mix their respective validation/test sets and collect novel annotations for missing categories. Our results in Sect. 4 show that the proposed training algorithm successfully unifies the label spaces and outperforms competitive baselines based on prior work on this practically relevant and challenging novel task.

2   Related Work

The ultimate goal of our work is to build an object detector that expands its label space beyond what is annotated in a single available training dataset. We propose an algorithmic approach to address this problem, which we contrast to related prior art and alternative approaches in this section.

Manual Annotation: The most obvious attempt to expand the label space of an object detector is by manually annotating the desired categories. New datasets are routinely proposed [9,15,16], but these come at high cost in terms of both time and money. Attempts to reduce these costs have been proposed either by interactive human-in-the-loop approaches [24,35] or by relying on cheaper forms of annotations and developing corresponding algorithms [1,20,32]. In the former attempt, one still needs to revisit every single image of the existing dataset for any new category added. The latter attempt is more promising [1,14,29,30]. However, it cannot compete with fully-supervised approaches so far. In contrast, our work shows how to combine multiple datasets with bounding box annotations for heterogeneous sets of categories.

Object Detection with Bounding Box and Image-Level Annotations: Several works try to leverage the combination of object detection datasets and image classification datasets [22,28,33]. Detection datasets typically annotate coarse-grained categories with bounding boxes [6,18], while classification datasets exist with fine-grained category annotation but only on the image level [5]. The main assumption in these works is that there are certain semantic and visual relationships between the categories with bounding box annotation and the ones with only image-level annotation. Mostly, the fine-grained categories are a sub-category of the coarse-grained categories. However, this assumption is a limiting factor because categories without clear visual or semantic relations are hard to combine, e.g., "person" and "car". Our proposed approach assumes bounding box-annotated datasets as input, but has the capability to combine any categories regardless of their visual or semantic relationship.

Universal Representations: The recently proposed work by Wang et al. [31] on universal object detection has a similar goal as our work. Inspired by a recent trend towards universal representations [2,13,21], Wang et al. propose to train a single detector from multiple datasets annotated with bounding boxes in a multi-task setup, which we review in more detail in Sect. 3.1. However, the categories in their design are not actually unified, but kept separate. At test time, the object detector knows from which dataset each test image comes from and thus makes predictions in the corresponding label space. In contrast, we propose to truly unify the label spaces from all training datasets with a more efficient network design. Moreover, our experimental setup generalizes the one from [31] as it requests from the detector to predict all categories (unified label space) for any given test image. This is a more challenging and realistic setup since there is no need to constrain each test image to the label space of a single training dataset.

Domain Adaptation: Only a few methods have been proposed on domain adaptation for object detection [3,11,12], but they are certainly related as training from multiple datasets is naturally confronted with domain gaps. While this is an interesting direction to explore, it is orthogonal to our contributions because none of these works address the problem of mismatched label spaces across datasets.

Learning from Partial Annotation: The most relevant work in this context is from Cour et al. [4], who propose loss functions for linear SVMs. Others have tried to use pseudo labeling to address this problem [7]. Most of these works, however, focus on plain classification problems, while we adapt some of these concepts to our specific task for object detection with deep networks in Sect. 3.

3   Training with Heterogeneous Label Spaces

3.1   Preliminaries

Notation: We work with N datasets, $D_1, D_2, \ldots, D_N$ and corresponding label spaces $L_1, L_2, \ldots, L_N$, each L_i being a set of categories specific to dataset i. In general, we do not constrain the label spaces to be equal, i.e., $L_i \neq L_j$ for $i \neq j$ and we allow common categories, i.e., $L_i \cap L_j \neq \emptyset$ for $i \neq j$, where $\emptyset$ is the empty set. For example, the label spaces of COCO, VOC and KITTI are different but all of them contain the category "person". We also include the special category "background", b_i, in every dataset D_i making the complete label space for D_i: $L_i \cup b_i$. However, the definition of individual "background" categories b_i is different.

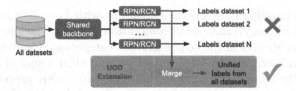

Fig. 2. Design of "universal" object detection [31] and our extension to unify the label spaces (UOD+Merge)

Thus, merging the dataset-specific background is the main challenge we address below (see also Sect. 1 and Fig. 1b).

Dataset D_i contains M_i images $I_{i,j}$, with $j = 1, \ldots, M_i$, and their corresponding sets of ground truth annotations $G_{i,j} = \{g_{i,j}^k = (x_1, y_1, x_2, y_2, c)^k, k = 1, \ldots, |G_{i,j}|\}$. Each ground truth annotation corresponds to an object present in the image and contains its bounding box coordinates, in image space, and category label $c \in L_i$. Any region of an image $I_{i,j}$ which is not covered by any bounding box in $G_{i,j}$ is considered to be in the background category b_i for the respective dataset.

Object Detection Framework: We use Faster-RCNN [23] together with the Feature Pyramid Network (FPN) [17] as our base object detector. Faster-RCNN takes as input an image and extracts convolutional features with a "backbone" network. It then uses a region proposal network (RPN) which predicts a set of bounding boxes that describe potential (category-agnostic) objects. Further, it has a region classification network (RCN) which classifies each proposal into the defined label space and also refines the localization. See [17,23] for details.

Universal Object Detection Baseline: Wang et al. [31] recently proposed a "universal" object detector (UOD) that is also trained from multiple datasets. This is done by equipping the model with multiple RPN/RCN heads, one for each dataset, and train the model similar to a multi-task setup where each dataset can be seen as one task. While this bypasses the ambiguity in the label spaces, at test-time, UOD needs to know what dataset each test image comes from and can thus activate the corresponding RPN/RCN head. This is not directly applicable in our more realistic setting, because we request to detect all categories from the training datasets on any test image. However, we introduce a simple extension to alleviate this problem, UOD+merge, which acts as a baseline to our proposed approach. It runs all RPN/RCN detector heads on a test image and merges semantically equal categories via an additional non-maximum suppression step (see Fig. 2).

3.2 Unifying Label Spaces with a Single Detector

The main goal of this work is to train a single detector with the union set of all label spaces $L_\cup = L_1 \cup L_2 \ldots \cup L_N$ such that given a new image I, all categories can be detected. This unified label space implicitly also defines a new and unique "background" category $b_\cup$, which is different to all other b_i. The intuitive benefit of a single detector over per-dataset detector heads, as in our UOD+merge baseline, is lower computational costs. The potentially bigger benefit, though, comes from better model parameters which are updated with data from all datasets, although having incomplete annotations.

The corresponding neural network architecture is simple: We use a single backbone and a single detector head with the number of categories equal to $|L_\cup| + 1$, where $+1$ is for the background category $b_\cup$. We then follow the standard training procedure of object detectors [23] by taking an image as input, predicting bounding boxes[1], and assigning them to ground truth annotations to define a loss. An association happens if predicted bounding boxes and ground truth sufficiently overlap in terms of intersection-over-union (IoU). Successfully associated bounding boxes get assigned the category and regression target of the corresponding ground truth box. The set of all remaining bounding boxes, which we denote as $\bar{D}$, would be assigned the background category (without any regression target) in a standard detection setup where ground truth is complete. However, in the proposed setting, where the ground truth label space L_i of dataset i is incomplete wrt. the unified label space $L_\cup$, the remaining bounding boxes $\bar{D}$ cannot simply be assigned the background class $b_\cup$ because it could also belong to a category from another dataset $L_\cup \setminus L_i$. To illustrate this case again, consider an object of any category $c \notin L_1$, which will be treated as part of the background b_1, even though it may be present in images of D_1. However, this category may be annotated in another dataset D_2 and, hence, not be part of background $b_\cup$. As an example, "human faces" are not part of the label space of the COCO dataset [18], but are certainly present in COCO images and other datasets exist with such annotations, like WiderFace [34].

3.3   A Loss Function to Deal with the Ambiguous Label Spaces

The problem we have to address for training a unified detector is the ambiguity of predicted bounding boxes $\bar{D}$ that are not associated with any ground truth of the given image from dataset i. These bounding boxes either belong to a category not in the label space of the current image, i.e., $L_\cup \setminus L_i$, or truly are part of the unified background $b_\cup$.

This problem can be thought of as learning with partial annotation [4], where a given sample can belong to a subset of the actual label space but the actual category is unknown. The only constraint on the true label of the ambiguous detections $\bar{D}$, without additional information, is that it *does not* belong to any label in L_i, because these categories were annotated for the given image $I_{i,j}$. Thus, the underlying ground truth category belongs to any of $L_* = (L_\cup \setminus L_i) \cup b_\cup$. As suggested in [4], we use this fact to design a classification loss function for ambiguous detections $\bar{D}$ as

---

[1]   While predicted boxes can either be anchors (RPN) or proposals (RCN), we only apply our approach on RCN. We did not observe gains when applied on RPN.

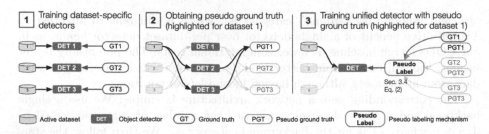

Fig. 3. Overview of our pseudo labeling approach for training a single object detector with a unified label space from multiple datasets: **(1)** We train dataset-specific detectors which do not suffer any label ambiguity. Although we draw three separate detectors, we share the weights of their backbones to encourage better performance across the different datasets. **(2)** We generate pseudo ground truth for each dataset that completes the given ground truth with bounding boxes for categories not present in the current label space. Thus, images from dataset i are put through detectors trained on other datasets $j \neq i$. The figure highlights the generation of pseudo ground truth for the first dataset. **(3)** We train a unified detector from all datasets on a unified label space $L_\cup$. The figure illustrates one training iteration with an image from the first dataset. The pseudo labeling module generates the training signal for the detector given true ground truth (GT) and pseudo ground truth (PGT) from that dataset

$$\mathcal{L}_{\text{sum}}^{-}(\mathbf{p}, L_*) = -\log \left(\sum_{c \in L_*} p_c \right), \tag{1}$$

where $\mathbf{p} = [p_1, p_2, \ldots] \in \mathbb{R}^{|L_\cup|+1}$ is the probability distribution for a predicted bounding box over the unified label space $L_\cup$ and background, with $\sum_{c=1}^{|L_\cup|+1} p_c = 1$. The loss function (1) essentially is a cross-entropy loss on the sum of the ambiguous categories and can be interpreted as merging all ambiguities into one category. An alternative is to replace the sum inside the log of (1) with a max [4]. An extension to (1) would be to add a minimum-entropy regularization to encourage selectivity instead of spreading probability mass over multiple categories.

While this sounds intuitive and we also use this loss as a baseline in our experiments, it has two flaws: First, one category, the background $b_\cup$, is never among the certainly correct classes but always ambiguous, which may lead to problems during learning. Second, this loss only considers the ambiguity but *does not resolve it*, which can be done with pseudo ground truth as described next.

3.4 Resolving the Label Space Ambiguity with Pseudo Labeling

Although the loss function (1) in the previous section encodes that one of the ambiguous categories in L_* is correct, it does not use any prior on the categories. In this section, we propose to adopt a pseudo-labeling approach to impose such a prior by estimating missing annotations. Figure 3 gives an overview.

Dataset-Specific Detectors: These are detectors trained on a single dataset D_i and have no ambiguity in their label space. We use them to augment incomplete annotations on images from different datasets D_j with different label spaces L_j ($j \neq i$). We use N dataset-specific detectors for each of the N datasets, which can be trained in different ways. The obvious option is to independently train N detectors, where each of them has its separate feature extraction backbone. Another option is to use a UOD [31] like network where a shared backbone is followed by N separate detector heads.

Since dataset-specific detectors are trained on one dataset D_j but applied to another one D_i to augment the annotations, the domain gap between the datasets should be considered. Among the two training options above, we expect the latter to be relatively better since part of the network is shared across domains. Another option for training dataset-specific detectors could be to leverage domain adaptation techniques, e.g. [3,11,12], which we expect to boost performance, but are orthogonal to our contributions and we leave this for future work.

Finally, we note that dataset-specific detectors are only used at train time, but are not required at inference time for our proposed unified object detector.

Training with Pseudo Labels: During training of our unified detector, each mini-batch contains data only from a single dataset, which enables efficient gradient computation. Suppose we have an image $I_{i,j}$ of dataset D_i with label space L_i, its ground truth for categories in L_i are available but those in $L_* = (L_\cup \setminus L_i) \cup b_\cup$ are not. To remedy this, we run all dataset-specific detectors, hence covering L_*, to obtain a set of bounding boxes $\hat{G}_{*,j}$ of label space L_* for image j, in which we consider these to be the pseudo ground truth. Each pseudo ground truth box $\hat{g}^k_{*,j}$ has a detection score $S_{\det}(\hat{g}^k_{*,j})$ associated with it. In the next paragraph, we describe a loss function that leverages this pseudo ground truth as a strong prior to resolve the label space ambiguity.

Loss Function with Pseudo Labels: Pseudo ground truth needs to be used carefully during training because it contains noise owing to (a) the domain gap between datasets and (b) the errors from the detectors. We propose a robust loss function and matching strategy as follows. Given a set of unmatched detections $\bar{D}$ and all pseudo ground truth boxes $\hat{G}_{*,j}$, we first compute the IoU similarity $s_{l,k}$ between $\bar{d}_l \in \bar{D}$ and $\hat{g}^k_{*,j} \in \hat{G}_{*,j}$ for all l and k. In contrast to standard object detectors, where the closest ground truth for each anchor/proposal is chosen, we keep all pseudo ground truth boxes with sufficiently high IoU similarity (i.e., $s_{l,k} > \tau$). We keep multiple matches with pseudo ground truth to counter the uncertainty of pseudo labeling and average out the potential noise. Suppose for each box $\bar{d}_l$ (not matched to $G_{i,j}$), we have a set of matched pseudo ground truth boxes $\hat{G}^m_{*,j} = \{\hat{g}^k_{*,j} | s_{l,k} > \tau \wedge S_{\det}(\hat{g}^k_{*,j}) > \kappa_{\mathrm{bg}}\}$, where $S_{\det}(\cdot)$ is the detection score. The threshold κ_{bg} defines a minimum score for a detection to be considered as pseudo ground truth class, and anything below it is considered as background $b_\cup$. If $\hat{G}^m_{*,j}$ is empty, we assume the ground truth label for $\bar{d}_l$ is "background" and use a standard cross-entropy loss. Otherwise, we employ the following loss on the predicted class distribution p_l of $\bar{d}_l$:

$$\mathcal{L}_{\mathrm{P}}(\mathbf{p}_l) = \frac{1}{Z} \sum_k \Gamma(\mathcal{S}_{\mathrm{det}}(\hat{g}^k_{*,j})) \cdot CE(\mathbf{p}_l, \mathrm{c}(\hat{g}^k_{*,j})) \,, \tag{2}$$

where $\hat{g}^k_{*,j} \in \hat{\mathsf{G}}^m_{*,j}$ and $\mathrm{c}(\hat{g}^k_{*,j})$ is the category of $\hat{g}^k_{*,j}$. The loss is the sum over the matched pseudo ground truth boxes, weighted by $\Gamma(\mathcal{S}_{\mathrm{det}}(\hat{g}^k_{*,j})) : \mathbb{R} \to \mathbb{R}$, which decides the importance of the pseudo ground truth. It is normalized by $Z = \max\left(\sum_k \Gamma(\mathcal{S}_{\mathrm{det}}(\hat{g}^k_{*,j})), \epsilon\right)$, where ϵ is a small constant to prevent division-by-zero in case all $\hat{g}^k_{*,j}$ have weight 0. There could be many different ways to define such a weighting function $\Gamma(\cdot)$, and we analyze various choices in our experiments. One important aspect we find to be crucial for the success of pseudo labeling is to ignore uncertain detections and not mark them as pseudo labels. Soft weighting by the score itself, i.e. $\Gamma(x) = x$ as the identity function, is one possible instantiation of such a weighting function. Another possibility is a hard threshold, i.e. $\Gamma(x; \kappa_{\mathrm{ignore}}) = 1$ if $x > \kappa_{\mathrm{ignore}}$, 0 otherwise. In the second case, the boxes with scores between κ_{bg} and κ_{ignore} are ignored and do not contribute to the loss at all, analogous to common detectors like Faster-RCNN [23].

Note that (2) only defines the classification loss based on pseudo ground truth. We can also add the corresponding regression loss as in standard detectors like [8,23], which we evaluate in our experiments.

3.5   Evaluating a Unified Object Detector

Standard object detection evaluation has typically been limited to a single dataset [10,23]. We propose a more challenging and arguably more realistic experimental setup, where we take N standard datasets with their corresponding train/val splits and *different* label spaces. A unified detector takes the training sets of these N datasets as input and, at inference time, it must be able to predict the union of all training categories on any unseen test image.

For the final evaluation, we mix the validation/test sets from each dataset together into a single larger set of images. The trained model *does not* know from which dataset the images come from, which differs from the setting in [31]. The detector *predicts all categories in all of the images*. Such a setup requires additional annotations for evaluating the test performance. Hence, we collected the annotations for the missing categories of the unified label space for all our test images. Please see the supplementary material for details.

4   Experiments

We now show results on the experimental setup proposed in Sect. 3.5, with various datasets. We evaluate the unified object detectors from Sect. 3 in this novel setting, analyze their different variants, and compare with competitive baselines.

Datasets: Recall that our goal is to train from N datasets with different label spaces, while the evaluation requires predicting over the union of all training categories on any given image. We use publicly available datasets directly for

Table 1. Different combinations of datasets that we use in our experimental setup. For each setting, a detector needs to be trained from all datasets predicting the union of their respective label spaces

Setting	Datasets
A	VOC [6] + SUN-RGBD [27] + LISA-Signs [19]
B	VOC [6] + COCO [18] (w/o VOC categories) + SUN-RGBD [27]
Ablation	VOC [6] + COCO [18] (w/o VOC categories)

training. Table 1 summarizes our dataset choices. We define several specific combinations of datasets that we believe cover a diverse set of category combinations, and one additional setup, that we use for an ablation study. Details for each setting are given in the corresponding sections below. Note that we do not include the recently proposed setting from Wang et al. [31] with 11 datasets for two main reasons: First, the combinatorial problem of annotating missing categories in 11 datasets renders it costly. Second, several datasets in the setup of [31] are completely dissimilar, in a sense that object categories are disjoint from the rest, e.g., medical images vs. usual RGB images, which defeats the purpose of what we want to demonstrate in this paper.

Implementation Details: We implement all our models with PyTorch[2]. The backbone is ResNet-50 with FPN [17], which is pretrained from ImageNet [5]. For each experiment, the network is trained for 50,000 iterations with a learning rate of 0.002. For our unified detector with pseudo labeling, we initialize the backbone with the weights from the ImageNet pretrained model. For the weighting function in (2), our default choice is the hard-thresholding described in Sect. 3.4. We follow the standard detection evaluation as in [31] and report $mAP^{0.5}$.

4.1 Ablation Study

To better understand the importance of the individual components and variations in our proposed method, we first conduct an analysis in a controlled setting.

Experimental Setup: We use a special setup where we intentionally remove certain categories, from one dataset, that are available in another dataset. Specifically, we use VOC [6] (2007) and COCO [18] (2017), with (annotations of) VOC categories removed from COCO. In this way, no additional annotation is required for evaluation and we can even monitor pseudo labeling performance on the training set. We evaluate on the combination of both datasets as described in Sect. 3.5 (MIX), but also on the individual ones (COCO and VOC), as well as specifically on the 20 VOC categories on the COCO dataset (V-on-C).

Baselines: We compare with the following baselines:

---

[2] We use the following code base as the starting point for our implementation: https://github.com/facebookresearch/maskrcnn-benchmark.

188   X. Zhao et al.

Table 2. Results of our ablation study on the four different evaluation sets. Please see text for more details on the training sets and the baselines.

Method	VOC	COCO	V-on-C	MIX
(a) Individual detectors + Merge	**81.4**	46.1	*62.0*	46.9
(b) UOD [31] + Merge	81.0	45.6	61.5	46.1
(c) Unify w/o Pseudo-Labeling	79.5	42.6	36.5	43.7
(d) Partial-loss sum [4] (1)	78.9	43.4	39.2	44.1
(e) Partial-loss sum + minimum-entropy	79.1	42.9	38.4	43.9
(f) Partial-loss max [4]	79.9	43.6	36.1	44.6
(g) Pseudo-Labeling (soft-thresh)	80.5	49.6	61.6	51.6
(h) Pseudo-Labeling (w/o regression)	80.9	*50.1*	61.9	*52.0*
(i) Pseudo-Labeling	*81.3*	**50.3**	**62.2**	**52.2**

- **Individual detectors:** The most apparent baseline is to train individual detectors on each of the datasets and run all of them on a new test image. As with our UOD+merge baseline, we apply a merging step as described in Sect. 3.1. Besides the obvious drawback of high runtime, the domain gap between datasets can lead to issues.
- **Universal detectors (UOD):** Our proposed extension of [31], UOD+ Merge, described in Sect. 3.1 and Fig. 2.
- **Unified detector ignoring label space ambiguity:** This baseline ("Unify w/o Pseudo-Labeling" in the tables) trains a single detector head with the unified label space $L_\cup$ but treats all unmatched detections $\bar{D}$ as part of the unified background $b_\cup$. It thus ignores the ambiguity in the label spaces.
- **Partial annotation losses:** This baseline implements the unified detector head and tries to resolve the label space ambiguity with the loss function (1) designed for partial annotations [4]. As mentioned in Sect. 3.3, we evaluate alternatives to (1): (i) We add a minimum entropy regularization to (1) (sum + ent) and (ii) we replace the sum in (1) with a max function as suggested in [4].
- **Pseudo labeling:** We analyze different variants of the pseudo-labeling strategy. Instead of hard-thresholding in the weighting function $\Gamma(\cdot)$ of the loss (2), which is the default, we also evaluate soft-thresholding (soft-thresh) as described in Sect. 3.4 (paragraph "Loss function with pseudo labels"). Given the inherent uncertainty in pseudo ground truth, we additionally analyze its impact on classification and regression losses by adding a variant where we remove the regression loss for every pseudo ground truth match (w/o regression).

Results: Table 2 compares our final pseudo labeling model (Pseudo-Labeling) with various baselines. We make several observations:

First and most importantly, we can observe a clear improvement of pseudo labeling (i) over the most relevant baseline, UOD+Merge (b). A big improvement

(46.1 to 52.2) can be observed on the MIX setting, which is arguably the most relevant one since it requests to detect the union of all training label spaces on test images from all domains. We identified a significantly lower number of false positives of pseudo labeling (i) as one of the reasons for this big improvement. The baselines (a–b) suffer more from the domain gap because all (a) or some parts (b) of the detector are trained only for a single dataset. The difference between having individual detectors (a) versus a shared backbone (b) seems negligible, but (b) obviously has a better runtime. Overall, the proposed pseudo-labeling (i) is still the most efficient one in terms of number of operations.

Second, methods that ignore the label space ambiguity (c) clearly underperform compared to other methods, particularly on the sets V-on-C and MIX, where it is important to have categories from one dataset transferred to another.

Third, the loss functions for problems with partial annnotation (d-f) improve only marginally over ignoring the label space ambiguity (c). This is expected to some degree, as we pointed out a potential flaw in Sect. 3.3, which is distinct to the task of object detection. The background category stays ambiguous for all examples while other categories are part of the certain classes at least every N samples, N being the number of datasets.

Finally, comparing different variants of pseudo labeling itself, i.e., (g–i), hard-thresholding (i) outperforms soft-thresholding (g) slightly. Also, using a regression loss on the pseudo ground truth (i) improves compared to (h).

Evaluation of Pseudo Labels: Our ablation study setting allows us to evaluate the performance of pseudo labeling itself, because ground truth is available. Specifically, we can evaluate the VOC detector on the COCO train set. Using all pseudo detections, we get precision and recall of 0.29 and 0.7, respectively. However, restricting the evaluation to only confident pseudo detections, we get precision and recall of 0.79 and 0.4, demonstrating the importance of thresholding (or weighing) pseudo detections based on confidence as we do in our loss (2).

4.2   Comparing Pseudo Labeling with an Upper Bound

In this section, we compare our pseudo labeling approach with an upper bound that has access to the full annotations. We split the COCO train set into two even halves, where each half sees 50% of the images and ignores 40 (different) categories. We train a standard detector on the original COCO train set and our pseudo detector and UOD on the splitted data set. The standard detector (upper bound) obtains 50.2 AP50, our pseudo detector 48.4 and UOD+merge 45.7. With domain gap absent for this experiment, we see that our pseudo labeling achieves competitive results compared to the upper bound.

4.3   Main Results

In this section, we present our main results where we compare our proposed unified detectors, (e) and (i) in Table 2, with the two most relevant baselines, UOD [31] + Merge (b) and Partial-anno loss based on the loss from [4] (d).

We first provide details on the two different settings we chose for evaluation, summarized in Table 1 before we show quantitative and qualitative results. Note that for each of the 4 datasets in use, we collect annotations for 500 images with the unified label space.

Setting A combines 3 different object detection datasets from different domains: general (VOC [6]), indoor scenes (SUN-RGBD [27]) and driving scenes (LISA-Signs [19]). The datasets have 20, 18 and 4 categories, respectively, with a few overlapping ones in the label spaces. The unified label space has 38 categories in total (4 overlapping). For the three datasets, we have a total of 1500 images for evaluation. Note that the images from different datasets are mixed for evaluation.

Setting B has three main intentions compared to setting A: First, we increase the label space with the addition of COCO [18]. Second, we increase the number of categories that need to be transferred between different datasets by removing the VOC categories from COCO. Third, we use a more focused set of scenes by using the combination of general scenes (VOC and COCO) and indoor scenes (SUN-RGBD). The datasets have 20, 60 and 18 categories, respectively. The unified label space contains 91 categories. Again, for the three datasets, we have a total of 1500 images, which are mixed together for evaluation.

Table 3. (a) Results on the two main settings as defined in Table 1. (b) Performance of pseudo labeling for overlapping classes. All numbers are AP50 [18]

Method \ Setting	A	B
UOD [31] + Merge	59.3	43.7
Partial-anno loss [4]	58.3	41.2
(Ours) Min-Entropy loss	58.7	42.9
(Ours) Pseudo-Labeling	61.1	48.5

(a)

Method \ Class	chair	sofa	tv	bed	toilet	sink
Individ (VOC)	68.6	74.8	81	-	-	-
Unified (VOC)	62.3	76.3	77.6	-	-	-
Individ (COCO)	-	-	-	55.0	66.6	51.5
Unified (COCO)	-	-	-	56.7	68.9	53.7
Individ (SUN)	75.3	68.2	85.1	78.1	67.2	78.0
Unified (SUN)	75.2	64.5	85.5	78.6	66.5	81

(a)

Quantitative Results: Table 3a provides the results on the two settings described above. We compare our proposed pseudo labeling method with three competitive baselines: UOD [31] + Merge, partial-annotation-loss [4] and minimum-entropy-loss. In both settings (A and B from Table 1), we can observe the same trend. Pseudo labeling significantly outperforms all baselines, e.g. 61.1 vs. 59.3, 58.3 and 58.7, respectively, on setting A, which again verifies the effectiveness of leveraging dataset-specific detectors as prior for training a unified detector. On setting B, we also observe an improvement of +4.8 AP50 points with pseudo labeling over the most relevant baseline UOD+Merge. The improvements are mostly attributed to the reduced number of false positives compared with UOD+Merge. Similar to the ablation study in Sect. 4.1, we see that only using loss functions designed for dealing with partial annotations is not enough to resolve the label space ambiguity for the task of object detection.

Fig. 4. Some qualitative results of our proposed unified object detector

Performance of Pseudo Labeling for Overlapping Classes: Here, we specifically analyze the performance of pseudo labeling for overlapping classes among different datasets. For setting B, we have: VOC and SUN-RGBD share "chair", "sofa" and "tv-monitor"; COCO and SUN-RGBD share "bed", "toilet" and "sink". To analyze the difference between individual detectors and our unified detector (w/ pseudo labeling), we test on each dataset separately. The results in Table 3b show that some categories improve, likely due to increased training data via pseudo labeling, while others get worse, eventually due to domain gap. However, we highlight that our proposed unified detector obtains much better results when tested on the mix of all data sets (with a complete label space) compared to individual detectors that may generate false positives on images from other data sets (MIX column in Table 2).

Qualitative Results: Finally, we show a few qualitative results for the settings A and B in Fig. 4. Most notably, we can observe in all examples, the unified detector is able to successfully predict categories that are originally not in the training dataset of the respective test image. For instance, the second image in the top row in Fig. 4 is from the LISA-Signs dataset, which does not annotate the category "person". This category is successfully transferred from other datasets.

5   Conclusions

We introduce a novel setting for object detection, where multiple datasets with heterogeneous label spaces are given for training and the task is to train an object detector that can predict over the union of all categories present in the training label spaces. This setting is challenging and more realistic than standard object detection settings [23], as well as the recent "universal" setting [31], since there can be both (i) unannotated objects of interest in training images, as well as, (ii) the original dataset of the test image is unknown. We propose training algorithms that leverage (i) loss functions for partially annotated ground truth and (ii) pseudo labeling techniques, to build a single detector with a unified label space. We also collect new annotations on test image to enable evaluation on the different methods in this novel setting.

Acknowledgements. This work was supported in part by National Science Foundation grant IIS-1619078, IIS-1815561.

References

1. Bilen, H., Vedaldi, A.: Weakly supervised deep detection networks. In: CVPR (2016)
2. Bilen, H., Vedaldi, A.: Universal representations: the missing link between faces, text, planktons, and cat breeds. arXiv:1701.07275 (2017)
3. Chen, Y., Li, W., Sakaridis, C., Dai, D., Van Gool, L.: Domain adaptive faster R-CNN for object detection in the wild. In: CVPR (2018)
4. Cour, T., Sapp, B., Taskar, B.: Learning from partial labels. JMLR **12**, 1501–1536 (2011)
5. Deng, J., Dong, W., Socher, R., Li, L.J., Li, K., Fei-Fei, L.: ImageNet: a large-scale hierarchical image database. In: CVPR (2009)
6. Everingham, M., Gool, L.V., Williams, C.K.I., Winn., J., Zisserman, A.: The pascal visual object classes (VOC) challenge. IJCV **88**(2), 303–338 (2010)
7. Feng, L., An, B.: Partial label learning with self-guided retraining. In: AAAI (2019)
8. Girshick, R.: Fast R-CNN. In: ICCV (2015)
9. Gupta, A., Dollar, P., Girshick, R.: LVIS: a dataset for large vocabulary instance segmentation. In: CVPR (2019)
10. He, K., Gkioxari, G., Dollár, P., Girshick, R.: Mask R-CNN. In: ICCV (2017)
11. Hsu, H.K., et al.: Progressive domain adaptation for object detection. In: WACV (2020)
12. Inoue, N., Furuta, R., Yamasaki, T., Aizawa, K.: Cross-domain weakly-supervised object detection through progressive domain adaptation. In: CVPR (2018)
13. Kalluri, T., Varma, G., Chandraker, M., Jawahar, C.: Universal semi-supervised semantic segmentation. In: ICCV (2019)
14. Kantorov, V., Oquab, M., Cho, M., Laptev, I.: ContextLocNet: context-aware deep network models for weakly supervised localization. In: Leibe, B., Matas, J., Sebe, N., Welling, M. (eds.) ECCV 2016. LNCS, vol. 9909, pp. 350–365. Springer, Cham (2016). https://doi.org/10.1007/978-3-319-46454-1_22
15. Kuznetsova, A., et al.: The open images dataset v4: unified image classification, object detection, and visual relationship detection at scale. arXiv:1811.00982 (2018)
16. Lambert, J., Liu, Z., Sener, O., Hays, J., Koltun, V.: MSeg: a composite dataset for multi-domain semantic segmentation. In: CVPR (2020)
17. Lin, T.Y., Dollár, P., Girshick, R., He, K., Hariharan, B., Belongie, S.: Feature pyramid networks for object detection. In: CVPR (2017)

18. Lin, T.-Y., et al.: Microsoft COCO: common objects in context. In: Fleet, D., Pajdla, T., Schiele, B., Tuytelaars, T. (eds.) ECCV 2014. LNCS, vol. 8693, pp. 740–755. Springer, Cham (2014). https://doi.org/10.1007/978-3-319-10602-1_48
19. Møgelmose, A., Trivedi, M.M., Moeslund, T.B.: Vision based traffic sign detection and analysis for intelligent driver assistance systems: perspectives and survey. IEEE Trans. Intell. Transp. Syst. **13**(4), 1484–1497 (2012)
20. Papadopoulos, D.P., Uijlings, J.R.R., Keller, F., Ferrari, V.: Training object class detectors with click supervision. In: CVPR (2017)
21. Rebuffi, S.A., Bilen, H., Vedaldi, A.: Learning multiple visual domains with residual adapters. In: NIPS (2017)
22. Redmon, J., Farhadi, A.: YOLO9000: better, faster, stronger. In: CVPR (2017)
23. Ren, S., He, K., Girshick, R., Sun, J.: Faster R-CNN: towards real-time object detection with region proposal networks. In: NIPS (2015)
24. Russakovsky, O., Li, L.J., Fei-Fei, L.: Best of both worlds: human-machine collaboration for object annotation. In: CVPR (2015)
25. Singh, B., Davis, L.S.: An analysis of scale invariance in object detection - SNIP. In: CVPR (2018)
26. Singh, B., Li, H., Sharma, A., Davis, L.S.: R-FCN-3000 at 30fps: decoupling detection and classification. In: CVPR (2018)
27. Song, S., Lichtenberg, S.P., Xiao, J.: SUN RGB-D: a RGB-D scene understanding benchmark suite. In: CVPR (2015)
28. Uijlings, J., Popov, S., Ferrari, V.: Revisiting knowledge transfer for training object class detectors. In: CVPR (2018)
29. Wan, F., Liu, C., Ke, W., Ji, X., Jiao, J., Ye, Q.: C-MIL: continuation multiple instance learning for weakly supervised object detection. In: CVPR (2019)
30. Wan, F., Wei, P., Jiao, J., Han, Z., Ye, Q.: Min-entropy latent model for weakly supervised object detection. In: CVPR (2018)
31. Wang, X., Cai, Z., Gao, D., Vasconcelos, N.: Towards universal object detection by domain attention. In: CVPR (2019)
32. Wang, Z., Acuna, D., Ling, H., Kar, A., Fidler, S.: Object instance annotation with deep extreme level set evolution. In: CVPR (2019)
33. Yang, H., Wu, H., Chen, H.: Detecting 11K classes: large scale object detection without fine-grained bounding boxes. In: ICCV (2019)
34. Yang, S., Luo, P., Loy, C.C., Tang, X.: WIDER FACE: a face detection benchmark. In: CVPR (2016)
35. Yao, A., Gall, J., Leistner, C., Gool, L.V.: Interactive object detection. In: CVPR (2012)

Lift, Splat, Shoot: Encoding Images from Arbitrary Camera Rigs by Implicitly Unprojecting to 3D

Jonah Philion[1,2,3]($\boxtimes$) and Sanja Fidler[1,2,3]

[1] NVIDIA, Santa Clara, USA
jphilion@nvidia.com
[2] University of Toronto, Toronto, Canada
[3] Vector Institute, Chennai, India

Abstract. The goal of perception for autonomous vehicles is to extract semantic representations from multiple sensors and fuse these representations into a single "bird's-eye-view" coordinate frame for consumption by motion planning. We propose a new end-to-end architecture that directly extracts a bird's-eye-view representation of a scene given image data from an arbitrary number of cameras. The core idea behind our approach is to "lift" each image individually into a frustum of features for each camera, then "splat" all frustums into a rasterized bird's-eye-view grid. By training on the entire camera rig, we provide evidence that our model is able to learn not only how to represent images but how to fuse predictions from all cameras into a single cohesive representation of the scene while being robust to calibration error. On standard bird's-eye-view tasks such as object segmentation and map segmentation, our model outperforms all baselines and prior work. In pursuit of the goal of learning dense representations for motion planning, we show that the representations inferred by our model enable interpretable end-to-end motion planning by "shooting" template trajectories into a bird's-eye-view cost map output by our network. We benchmark our approach against models that use oracle depth from lidar. Project page with code: https://nv-tlabs.github.io/lift-splat-shoot.

1 Introduction

Computer vision algorithms generally take as input an image and output either a prediction that is coordinate-frame agnostic – such as in classification [16,17,19, 30] – or a prediction in the same coordinate frame as the input image – such as in object detection, semantic segmentation, or panoptic segmentation [1,7,15,36].

This paradigm does not match the setting for perception in self-driving out-of-the-box. In self-driving, multiple sensors are given as input, each with a

Electronic supplementary material The online version of this chapter (https://doi.org/10.1007/978-3-030-58568-6_12) contains supplementary material, which is available to authorized users.

A. Vedaldi et al. (Eds.): ECCV 2020, LNCS 12359, pp. 194–210, 2020.
https://doi.org/10.1007/978-3-030-58568-6_12

Fig. 1. We propose a model that, given multi-view camera data (left), infers semantics directly in the bird's-eye-view (BEV) coordinate frame (right). We show vehicle segmentation (blue), drivable area (orange), and lane segmentation (green). These BEV predictions are then projected back onto input images (dots on the left). (Color figure online)

different coordinate frame, and perception models are ultimately tasked with producing predictions in a new coordinate frame – the frame of the ego car – for consumption by the downstream planner, as shown in Fig. 2.

There are many simple, practical strategies for extending the single-image paradigm to the multi-view setting. For instance, for the problem of 3D object detection from n cameras, one can apply a single-image detector to all input images individually, then rotate and translate each detection into the ego frame according to the intrinsics and extrinsics of the camera in which the object was detected. This extension of the single-view paradigm to the multi-view setting has three valuable symmetries baked into it:

1. **Translation equivariance** – If pixel coordinates within an image are all shifted, the output will shift by the same amount. Fully convolutional single-image object detectors roughly have this property and the multi-view extension inherits this property from them [6,11].
2. **Permutation invariance** – the final output does not depend on a specific ordering of the n cameras.
3. **Ego-frame isometry equivariance** – the same objects will be detected in a given image no matter where the camera that captured the image was located relative to the ego car. An equivalent way to state this property is that the definition of the ego-frame can be rotated/translated and the output will rotate/translate with it.

The downside of the simple approach above is that using post-processed detections from the single-image detector prevents one from differentiating from predictions made in the ego frame all the way back to the sensor inputs. As a result, the model cannot learn in a data-driven way what the best way is to fuse information across cameras. It also means backpropagation cannot be used to automatically improve the perception system using feedback from the downstream planner.

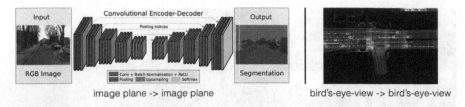

image plane -> image plane   bird's-eye-view -> bird's-eye-view

Fig. 2. (left, from SegNet [1]) Traditionally, computer vision tasks such as semantic segmentation involve making predictions in the same coordinate frame as the input image. (right, from Neural Motion Planner [41]) In contrast, planning for self-driving generally operates in the bird's-eye-view frame. Our model directly makes predictions in a given bird's-eye-view frame for end-to-end planning from multi-view images.

We propose a model named "Lift-Splat" that preserves the 3 symmetries identified above by design while also being end-to-end differentiable. In Sect. 3, we explain how our model "lifts" images into 3D by generating a frustum-shaped point cloud of contextual features, "splats" all frustums onto a reference plane as is convenient for the downstream task of motion planning. In Sect. 3.3, we propose a method for "shooting" proposal trajectories into this reference plane for interpretable end-to-end motion planning. In Sect. 4, we identify implementation details for training lift-splat models efficiently on full camera rigs. We present empirical evidence in Sect. 5 that our model learns an effective mechanism for fusing information from a distribution of possible inputs.

2   Related Work

Our approach for learning cohesive representations from image data from multiple cameras builds on recent work in sensor fusion and monocular object detection. Large scale multi-modal datasets from Nutonomy [2], Lyft [13], Waymo [35], and Argo [3], have recently made full representation learning of the entire 360° scene local to the ego vehicle conditioned exclusively on camera input a possibility. We explore that possibility with our Lift-Splat architecture.

2.1   Monocular Object Detection

Monocular object detectors are defined by how they model the transformation from the image plane to a given 3-dimensional reference frame. A standard technique is to apply a mature 2D object detector in the image plane and then train a second network to regress 2D boxes into 3D boxes [12,26,27,31]. The current state-of-the-art 3D object detector on the nuScenes benchmark [31] uses an architecture that trains a standard 2d detector to also predict depth using a loss that seeks to disentangle error due to incorrect depth from error due to incorrect bounding boxes. These approaches achieve great performance on 3D object detection benchmarks because detection in the image plane factors out the fundamental cloud of ambiguity that shrouds monocular depth prediction.

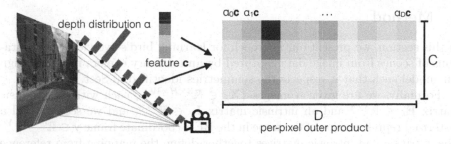

Fig. 3. We visualize the "lift" step of our model. For each pixel, we predict a categorical distribution over depth $\alpha \in \triangle^{D-1}$ (left) and a context vector $\mathbf{c} \in \mathbb{R}^{C}$ (top left). Features at each point along the ray are determined by the outer product of α and $\mathbf{c}$ (right).

An approach with recent empirical success is to separately train one network to do monocular depth prediction and another to do bird's-eye-view detection separately [39,40]. These approaches go by the name of "pseudolidar". The intuitive reason for the empirical success of pseudolidar is that pseudolidar enables training a bird's-eye-view network that operates in the coordinate frame where the detections are ultimately evaluated and where, relative to the image plane, euclidean distance is more meaningful.

A third category of monocular object detectors uses 3-dimensional object primitives that acquire features based on their projection onto all available cameras. Mono3D [4] achieved state of the art monocular object detection on KITTI by generating 3-dimensional proposals on a ground plane that are scored by projecting onto available images. Orthographic Feature Transform [29] builds on Mono3D by projecting a fixed cube of voxels onto images to collect features and then training a second "BEV" CNN to detect in 3D conditioned on the features in the voxels. A potential performance bottleneck of these models that our model addresses is that a pixel contributes the same feature to every voxel independent of the depth of the object at that pixel.

2.2 Inference in the Bird's-Eye-View Frame

Models that use extrinsics and intrinsics in order to perform inference directly in the bird's-eye-view frame have received a large amount of interest recently. MonoLayout [21] performs bird's-eye-view inference from a single image and uses an adversarial loss to encourage the model to inpaint plausible hidden objects. In concurrent work, Pyramid Occupancy Networks [28] proposes a transformer architecture that converts image representations into bird's-eye-view representations. FISHING Net [9] - also concurrent work - proposes a multi-view architecture that both segments objects in the current timestep and performs future prediction. We show that our model outperforms prior work empirically in Sect. 5. These architectures, as well as ours, use data structures similar to "multi-plane" images from the machine learning graphics community [20,32,34,38].

3   Method

In this section, we present our approach for learning bird's-eye-view representations of scenes from image data captured by an arbitrary camera rig. We design our model such that it respects the symmetries identified in Sect. 1.

Formally, we are given n images $\{\mathbf{X}_k \in \mathbb{R}^{3 \times H \times W}\}_n$ each with an extrinsic matrix $\mathbf{E}_k \in \mathbb{R}^{3 \times 4}$ and an intrinsic matrix $\mathbf{I}_k \in \mathbb{R}^{3 \times 3}$, and we seek to find a rasterized representation of the scene in the BEV coordinate frame $\mathbf{y} \in \mathbb{R}^{C \times X \times Y}$. The extrinsic and intrinsic matrices together define the mapping from reference coordinates (x, y, z) to local pixel coordinates (h, w, d) for each of the n cameras. We do not require access to any depth sensor during training or testing.

3.1   Lift: Latent Depth Distribution

The first stage of our model operates on each image in the camera rig in isolation. The purpose of this stage is to "lift" each image from a local 2-dimensional coordinate system to a 3-dimensional frame that is shared across all cameras.

The challenge of monocular sensor fusion is that we require depth to transform into reference frame coordinates but the "depth" associated to each pixel is inherently ambiguous. Our proposed solution is to generate representations at all possible depths for each pixel.

Let $\mathbf{X} \in \mathbb{R}^{3 \times H \times W}$ be an image with extrinsics $\mathbf{E}$ and intrinsics $\mathbf{I}$, and let p be a pixel in the image with image coordinates (h, w). We associate $|D|$ points $\{(h, w, d) \in \mathbb{R}^3 \mid d \in D\}$ to each pixel where D is a set of discrete depths, for instance defined by $\{d_0 + \Delta, ..., d_0 + |D|\Delta\}$. Note that there are no learnable parameters in this transformation. We simply create a large point cloud for a given image of size $D \cdot H \cdot W$. This structure is equivalent to what the multi-view synthesis community [32,38] has called a multi-plane image except in our case the features in each plane are abstract vectors instead of (r, g, b, α) values.

The context vector for each point in the point cloud is parameterized to match a notion of attention and discrete depth inference. At pixel p, the network predicts a context $\mathbf{c} \in \mathbb{R}^C$ and a distribution over depth $\alpha \in \triangle^{|D|-1}$ for every pixel. The feature $\mathbf{c}_d \in \mathbb{R}^C$ associated to point p_d is then defined as the context vector for pixel p scaled by α_d:

$$\mathbf{c}_d = \alpha_d \mathbf{c}. \tag{1}$$

Note that if our network were to predict a one-hot vector for α, context at the point p_d would be non-zero exclusively for a single depth d^* as in pseudolidar [39]. If the network predicts a uniform distribution over depth, the network would predict the same representation for each point p_d assigned to pixel p independent of depth as in OFT [29]. Our network is therefore in theory capable of choosing between placing context from the image in a specific location of the bird's-eye-view representation versus spreading the context across the entire ray of space, for instance if the depth is ambiguous.

In summary, ideally, we would like to generate a function $g_c : (x, y, z) \in \mathbb{R}^3 \rightarrow c \in \mathbb{R}^C$ for each image that can be queried at any spatial location and return a context vector. To take advantage of discrete convolutions, we choose to discretize space. For cameras, the volume of space visible to the camera corresponds to a frustum. A visual is provided in Fig. 3.

3.2   Splat: Pillar Pooling

We follow the pointpillars [18] architecture to convert the large point cloud output by the "lift" step. "Pillars" are voxels with infinite height. We assign every point to its nearest pillar and perform sum pooling to create a $C \times H \times W$ tensor that can be processed by a standard CNN for bird's-eye-view inference. The overall lift-splat architecture is outlined in Fig. 4.

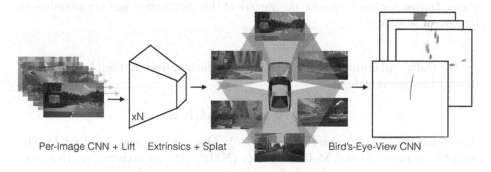

Per-Image CNN + Lift   Extrinsics + Splat   Bird's-Eye-View CNN

Fig. 4. Lift-Splat-Shoot Outline Our model takes as input n images (left) and their corresponding extrinsic and intrinsic parameters. In the "lift" step, a frustum-shaped point cloud is generated for each individual image (center-left). The extrinsics and intrinsics are then used to splat each frustum onto the bird's-eye-view plane (center-right). Finally, a bird's-eye-view CNN processes the bird's-eye-view representation for BEV semantic segmentation or planning (right).

Just as OFT [29] uses integral images to speed up their pooling step, we apply an analogous technique to speed up sum pooling. Efficiency is crucial for training our model given the size of the point clouds generated. Instead of padding each pillar then performing sum pooling, we avoid padding by using packing and leveraging a "cumsum trick" for sum pooling. This operation has an analytic gradient that can be calculated efficiently to speed up autograd as explained in Subsect. 4.2.

3.3   Shoot: Motion Planning

Key aspect of our Lift-Splat model is that it enables end-to-end cost map learning for motion planning from camera-only input. At test time, planning using the inferred cost map can be achieved by "shooting" different trajectories, scoring their cost, then acting according to lowest cost trajectory [25]. In Sect. 5.6, we probe the ability of our model to enable end-to-end interpretable motion planning and compare its performance to lidar-based end-to-end neural motion planners.

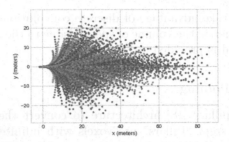

Fig. 5. We visualize the 1 K trajectory templates that we "shoot" onto our cost map during training and testing. During training, the cost of each template trajectory is computed and interpreted as a 1K-dimensional Boltzman distribution over the templates. During testing, we choose the argmax of this distribution and act according to the chosen template.

We frame "planning" as predicting a distribution over K template trajectories for the ego vehicle

$$\mathcal{T} = \{\tau_i\}_K = \{\{x_j, y_j, t_j\}_T\}_K$$

conditioned on sensor observations $p(\tau|o)$. Our approach is inspired by the recently proposed Neural Motion Planner (NMP) [41], an architecture that conditions on point clouds and high-definition maps to generate a cost-volume that can be used to score proposed trajectories.

Instead of the hard-margin loss proposed in NMP, we frame planning as classification over a set of K template trajectories. To leverage the cost-volume nature of the planning problem, we enforce the distribution over K template trajectories to take the following form

$$p(\tau_i|o) = \frac{\exp\left(-\sum\limits_{x_i, y_i \in \tau_i} c_o(x_i, y_i)\right)}{\sum\limits_{\tau \in \mathcal{T}} \exp\left(-\sum\limits_{x_i, y_i \in \tau} c_o(x_i, y_i)\right)} \tag{2}$$

where $c_o(x, y)$ is defined by indexing into the cost map predicted given observations o at location x, y and can therefore be trained end-to-end from data by optimizing for the log probability of expert trajectories. For labels, given a ground-truth trajectory, we compute the nearest neighbor in L2 distance to the template trajectories $\mathcal{T}$ then train with the cross entropy loss. This definition of $p(\tau_i|o)$ enables us to learn an interpretable spatial cost function without defining a hard-margin loss as in NMP [41].

In practice, we determine the set of template trajectories by running K-Means on a large number of expert trajectories. The set of template trajectories used for "shooting" onto the cost map that we use in our experiments are visualized in Fig. 5.

4 Implementation

4.1 Architecture Details

The neural architecture of our model is similar to OFT [29]. As in OFT, our model has two large network backbones. One of the backbones operates on each image individually in order to featurize the point cloud generated from each image. The other backbone operates on the point cloud once it is splatted into pillars in the reference frame. The two networks are joined by our lift-splat layer as defined in Sect. 3 and visualize in Fig. 4.

For the network that operates on each image in isolation, we leverage layers from an EfficientNet-B0 [37] pretrained on Imagenet [30] in all experiments for all models including baselines. EfficientNets are network architectures found by exhaustive architecture search in a resource limited regime with depth, width, and resolution scaled up proportionally. We find that they enable higher performance relative to ResNet-18/34/50 [8] across all models with a minor inconvenience of requiring more optimization steps to converge.

For our bird's-eye-view network, we use a combination of ResNet blocks similar to PointPillars [18]. Specifically, after a convolution with kernel 7 and stride 2 followed by batchnorm [10] and ReLU [22], we pass through the first 3 metalayers of ResNet-18 to get 3 bird's-eye-view representations at different resolutions x_1, x_2, x_3. We then upsample x_3 by a scale factor of 4, concatenate with x_1, apply a resnet block, and finally upsample by 2 to return to the resolution of the original input bird's-eye-view pseudo image. We count 14.3M trainable parameters in our final network.

There are several hyper-parameters that determine the "resolution" of our model. First, there is the size of the input images $H \times W$. In all experiments below, we resize and crop input images to size 128×352 and adjust extrinsics and intrinsics accordingly. Another important hyperparameter of network is the size the resolution of the bird's-eye-view grid $X \times Y$. In our experiments, we set bins in both x and y from -50 m to 50 meters with cells of size 0.5 m $\times$ 0.5 m. The resultant grid is therefore 200×200. Finally, there's the choice of D that determines the resolution of depth predicted by the network. We restrict D between 4.0 m and 45.0 m spaced by 1.0 meters. With these hyper-parameters and architectural design choices, the forward pass of the model runs at 35 hz on a Titan V GPU.

4.2 Frustum Pooling Cumulative Sum Trick

Training efficiency is critical for learning from data from an entire sensor rig. We choose sum pooling across pillars in Sect. 3 as opposed to max pooling because our "cumulative sum" trick saves us from excessive memory usage due to padding. The "cumulative sum trick" is the observation that sum pooling can be performed by sorting all points according to bin id, performing a cumulative sum over all features, then subtracting the cumulative sum values at the boundaries of the bin sections. Instead of relying on autograd to backprop through all three steps, the analytic gradient for the module as a whole can be derived, speeding up training by 2x. We call the layer "Frustum Pooling" because it handles converting the frustums produced by n images into a fixed dimensional $C \times H \times W$ tensor independent of the number of cameras n. Code can be found on our project page.

5 Experiments and Results

We use the nuScenes [2] and Lyft Level 5 [13] datasets to evaluate our approach. nuScenes is a large dataset of point cloud data and image data from 1k scenes, each of 20 s in length. The camera rig in both datasets is comprised of 6 cameras which roughly point in the forward, front-left, front-right, back-left, back-right, and back directions. In all datasets, there is a small overlap between the fields-of-view of the cameras. The extrinsic and intrinsic parameters of the cameras shift throughout both datasets. Since our model conditions on the camera calibration, it is able to handle these shifts.

We define two object-based segmentation tasks and two map-based tasks. For the object segmentation tasks, we obtain ground truth bird's-eye-view targets by projecting 3D bounding boxes into the bird's-eye-view plane. Car segmentation on nuScenes refers to all bounding boxes of class `vehicle.car` and vehicle segmentation on nuScenes refers to all bounding boxes of meta-category `vehicle`. Car segmentation on Lyft refers to all bounding boxes of class `car` and vehicle segmentation on nuScenes refers to all bounding boxes with class $\in$ { `car, truck, other_vehicle, bus, bicycle` }. For mapping, we use transform map layers from the nuScenes map into the ego frame using the provided 6 DOF localization and rasterize.

For all object segmentation tasks, we train with binary cross entropy with positive weight 1.0. For the lane segmentation, we set positive weight to 5.0 and for road segmentation we use positive weight 1.0 [24]. In all cases, we train for 300k steps using Adam [14] with learning rate $1e - 3$ and weight decay $1e - 7$. We use the PyTorch framework [23].

The Lyft dataset does not come with a canonical train/val split. We separate 48 of the Lyft scenes for validation to get a validation set of roughly the same size as nuScenes (6048 samples for Lyft, 6019 samples for nuScenes).

5.1 Description of Baselines

Unlike vanilla CNNs, our model comes equipped with 3-dimensional structure at initialization. We show that this structure is crucial for good performance by comparing against a CNN composed of standard modules. We follow an architecture similar to MonoLayout [21] which also trains a CNN to output bird's-eye-view labels from images only but does not leverage inductive bias in designing the architecture and trains on single cameras only. The architecture has an EfficientNet-B0 backbone that extracts features independently across all images. We concatenate the representations and perform bilinear interpolation to

Table 1. Segment. IOU in BEV frame

	nuScenes		Lyft	
	Car	Vehicles	Car	Vehicles
CNN	22.78	24.25	30.71	31.91
Frozen Encoder	25.51	26.83	35.28	32.42
OFT	29.72	30.05	39.48	40.43
Lift-Splat (Us)	**32.06**	**32.07**	**43.09**	**44.64**
PON* [28]	24.7	-	-	-
FISHING* [9]	-	30.0	-	56.0

Table 2. Map IOU in BEV frame

	Drivable Area	Lane Boundary
CNN	68.96	16.51
Frozen Encoder	61.62	16.95
OFT	71.69	18.07
Lift-Splat (Us)	**72.94**	**19.96**
PON* [28]	60.4	-

upsample into a $\mathbb{R}^{X \times Y}$ tensor as is output by our model. We design the network such that it has roughly the same number of parameters as our model. The weak performance of this baseline demonstrates how important it is to explicitly bake symmetry 3 from Sect. 1 into the model in the multi-view setting.

To show that our model is predicting a useful implicit depth, we compare against our model where the weights of the pretrained CNN are frozen as well as to OFT [29]. We outperform these baselines on all tasks, as shown in Tables 1 and 2. We also outperform concurrent work that benchmarks on the same segmentation tasks [9,28]. As a result, the architecture is learning both an effective depth distribution as well as effective contextual representations for the downstream task.

5.2 Segmentation

We demonstrate that our Lift-Splat model is able to learn semantic 3D representations given supervision in the bird's-eye-view frame. Results on the object segmentation tasks are shown in Table 1, while results on the map segmentation tasks are in Table 2. On all benchmarks, we outperform our baselines. We believe the extent of these gains in performance from implicitly unprojecting into 3D are substantial, especially for object segmentation. We also include reported IOU scores for two concurrent works [9,28] although both of these papers use different definitions of the bird's-eye-view grid and a different validation split for the Lyft dataset so a true comparison is not yet possible.

5.3 Robustness

Because the bird's-eye-view CNN learns from data how to fuse information across cameras, we can train the model to be robust to simple noise models that occur in self-driving such as extrinsics being biased or cameras dying. In Fig. 6, we verify that by dropping cameras during training, our model handles dropped cameras at better at test time. In fact, the best performing model when all 6 cameras are present is the model that is trained with 1 camera being randomly dropped from every sample during training. We reason that sensor dropout forces the model to learn the correlation between images on different cameras, similar to other

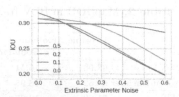
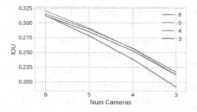

(a) Test Time Extrinsic Noise   (b) Test Time Camera Dropout

Fig. 6. We show that it is possible to train our network such that it is resilient to common sources of sensor error. On the left, we show that by training with a large amount of noise in the extrinsics (blue), the network becomes more robust to extrinsic noise at test time. On the right, we show that randomly dropping cameras from each batch during training (red) increases robustness to sensor dropout at test time. (Color figure online)

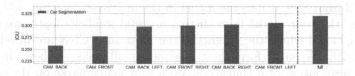

Fig. 7. We measure intersection-over-union of car segmentation when each of the cameras is missing. The backwards camera on the nuScenes camera rig has a wider field of view so it is intuitive that losing this camera causes the biggest decrease in performance relative to performance given the full camera rig (labeled "full" on the right).

Table 3. We train on images from only 4 of the 6 cameras in the nuScenes dataset. We then evaluate with the new cameras (1_{bl} corresponds to the "back left" camera and 1_{fl} corresponds to the "front left" camera) and find that the performance of the model strictly increases as we add more sensors unseen during training.

	IOU
4	26.53
$4 + 1_{fl}$	27.35
$4 + 1_{bl}$	27.27
$4 + 1_{bl} + 1_{fl}$	**27.94**

variants of dropout [5,33]. We show on the left of Fig. 6 that training the model with noisy extrinsics can lead to better test-time performance. For low amounts of noise at test-time, models that are trained without any noise in the extrinsics perform the best because the BEV CNN can trust the location of the splats with more confidence. For high amounts of extrinsic noise, our model sustains its good performance.

In Fig. 7, we measure the "importance" of each camera for the performance of car segmentation on nuScenes. Note that losing cameras on nuScenes implies

Fig. 8. For a single time stamp, we remove each of the cameras and visualize how the loss the cameras effects the prediction of the network. Region covered by the missing camera becomes fuzzier in every case. When the front camera is removed (top middle), the network extrapolates the lane and drivable area in front of the ego and extrapolates the body of a car for which only a corner can be seen in the top right camera.

Table 4. We train the model on nuScenes then evaluate it on Lyft. The Lyft cameras are entirely different from the nuScenes cameras but the model succeeds in generalizing far better than the baselines. Note that our model has widened the gap from the standard benchmark in Tables 1 and 2.

	Lyft Car	Lyft Vehicle
CNN	7.00	8.06
Frozen Encoder	15.08	15.82
OFT	16.25	16.27
Lift-Splat (Us)	**21.35**	**22.59**

that certain regions of the region local to the car have no sensor measurements and as a result performance strictly upper bounded by performance with the full sensor rig. Qualitative examples in which the network inpaints due to missing cameras are shown in Fig. 8. In this way, we measure the importance of each camera, suggesting where sensor redundancy is more important for safety.

5.4 Zero-Shot Camera Rig Transfer

We now probe the generalization capabilities of Lift-Splat. In our first experiment, we measure performance of our model when only trained on images from a subset of cameras from the nuScenes camera rig but at test time has access to images from the remaining two cameras. In Table 3, we show that the performance of our model for car segmentation improves when additional cameras are available at test time without any retraining.

We take the above experiment a step farther and probe how well our model generalizes to the Lyft camera rig if it was only trained on nuScenes data. Qualitative results of the transfer are shown in Fig. 9 and the benchmark against the generalization of our baselines is shown in Table 4.

5.5 Benchmarking Against Oracle Depth

We benchmark our model against the pointpillars [18] architecture which uses ground truth depth from LIDAR point clouds. As shown in Table 5, across all tasks, our architecture performs slightly worse than pointpillars trained with a single scan of LIDAR. However, at least on drivable area segmentation, we note that we approach the performance of LIDAR. In the world in general, not all lanes are visible in a lidar scan. We would like to measure performance in a wider range of environments in the future.

To gain insight into how our model differs from LIDAR, we plot how performance of car segmentation varies with two control variates: distance to the ego vehicle and weather conditions. We determine the weather of a scene from the description string that accompanies every scene token in the nuScenes dataset. The results are shown in Fig. 10. We find that performance of our model is much worse than pointpillars on scenes that occur at night as expected. We also find that both models experience roughly linear performance decrease with increased depth.

5.6 Motion Planning

Finally, we evaluate the capability of our model to perform planning by training the representation output by Lift-Splat to be a cost function. The trajectories that we generate are 5 s long spaced by 0.25 s. To acquire templates, we fit K-Means for $K = 1000$ to all ego trajectories in the training set of nuScenes.

Fig. 9. We qualitatively show how our model performs given an entirely new camera rig at test time. Road segmentation is shown in orange, lane segmentation is shown in green, and vehicle segmentation is shown in blue. (Color figure online)

Table 5. When compared to models that use oracle depth from lidar, there is still room for improvement. Video inference from camera rigs is likely necessary to acquire the depth estimates necessary to surpass lidar.

	Drivable Area	Lane Boundary	nuScenes		Lyft	
			Car	Vehicle	Car	Vehicle
Oracle Depth (1 scan)	74.91	25.12	40.26	44.48	74.96	76.16
Oracle Depth (>1 scan)	76.96	26.80	45.36	49.51	75.42	76.49
Lift-Splat (Us)	70.81	19.58	32.06	32.07	43.09	44.64

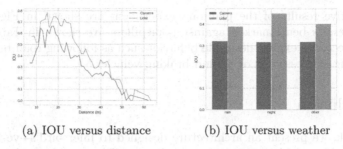

(a) IOU versus distance   (b) IOU versus weather

Fig. 10. We compare how our model's performance varies over depth and weather. As expected, our model drops in performance relative to pointpillars at nighttime.

Fig. 11. We display the top 10 ranked trajectories out of the 1k templates. Video sequences are provided on our project page. Our model predicts bimodal distributions and curves from observations from a single timestamp. Our model does not have access to the speed of the car so it is compelling that the model predicts low-speed trajectories near crosswalks and brake lights.

Table 6. Since planning is framed as classification among a set of 1 K template trajectories, we measure top-5, top-10, and top-20 accuracy. We find that our model is still lagging behind lidar-based approaches in generalization. Qualitative examples of the trajectories output by our model are shown in Fig. 11.

	Top 5	Top 10	Top 20
Lidar (1 scan)	19.27	28.88	41.93
Lidar (10 scans)	24.99	35.39	49.84
Lift-Splat (Us)	15.52	19.94	27.99

At test time, we measure how well the network is able to predict the template that is closest to the ground truth trajectory under the L2 norm. This task is an important experiment for self-driving because the ground truth targets for this experiment are orders of magnitude less expensive to acquire than ground truth 3D bounding boxes. This task is also important for benchmarking the performance of camera-based approaches versus lidar-based approaches because although the ceiling for 3D object detection from camera-only is certainly upper bounded by lidar-only, the optimal planner using camera-only should in principle upper bound the performance of an optimal planner trained from lidar-only.

Qualitative results of the planning experiment are shown in Fig. 11. The empirical results benchmarked against pointpillars are shown in Table 6. The output trajectories exhibit desirable behavior such as following road boundaries and stopping at crosswalks or behind braking vehicles.

6   Conclusion

In this work, we present an architecture designed to infer bird's-eye-view representations from arbitrary camera rigs. Our model outperforms baselines on a suite of benchmark segmentation tasks designed to probe the model's ability to represent semantics in the bird's-eye-view frame without any access to ground truth depth data at training or test time. We present methods for training our model that make the network robust to simple models of calibration noise. Lastly, we show that the model enables end-to-end motion planning that follows the trajectory shooting paradigm. In order to meet and possibly surpass the performance of similar networks that exclusively use ground truth depth data from pointclouds, future work will need to condition on multiple time steps of images instead of a single time step as we consider in this work.

References

1. Badrinarayanan, V., Kendall, A., Cipolla, R.: SegNet: a deep convolutional encoder-decoder architecture for image segmentation. CoRR arXiv:abs/1511.00561 (2015). http://arxiv.org/abs/1511.00561
2. Caesar, H., et al.: nuScenes: a multimodal dataset for autonomous driving. CoRR arXiv:abs/1903.11027 (2019). http://arxiv.org/abs/1903.11027
3. Chang, M.F., et al.: Argoverse: 3D tracking and forecasting with rich maps. In: IEEE/CVF Conference on Computer Vision and Pattern Recognition (CVPR) (2019)
4. Chen, X., Kundu, K., Zhang, Z., Ma, H., Fidler, S., Urtasun, R.: Monocular 3D object detection for autonomous driving. In: IEEE/CVF Conference on Computer Vision and Pattern Recognition (CVPR), pp. 2147–2156 (2016)
5. Ghiasi, G., Lin, T., Le, Q.V.: DropBlock: a regularization method for convolutional networks. CoRR arXiv:abs/1810.12890 (2018). http://arxiv.org/abs/1810.12890
6. Goodfellow, I., Bengio, Y., Courville, A.: Deep Learning. MIT Press (2016). http://www.deeplearningbook.org
7. He, K., Gkioxari, G., Dollár, P., Girshick, R.B.: Mask R-CNN. CoRR arXiv:abs/1703.06870 (2017). http://arxiv.org/abs/1703.06870
8. He, K., Zhang, X., Ren, S., Sun, J.: Deep residual learning for image recognition. CoRR arXiv:abs/1512.03385 (2015). http://arxiv.org/abs/1512.03385
9. Hendy, N., et al.: Fishing net: future inference of semantic heatmaps in grids (2020)
10. Ioffe, S., Szegedy, C.: Batch normalization: accelerating deep network training by reducing internal covariate shift. CoRR arXiv:abs/1502.03167 (2015). http://arxiv.org/abs/1502.03167
11. Kayhan, O.S., Gemert, J.C.v.: On translation invariance in CNNS: convolutional layers can exploit absolute spatial location. In: Proceedings of the IEEE/CVF Conference on Computer Vision and Pattern Recognition (CVPR) (2020)

12. Kehl, W., Manhardt, F., Tombari, F., Ilic, S., Navab, N.: SSD-6D: making rgb-based 3D detection and 6d pose estimation great again. CoRR arXiv:abs/1711.10006 (2017)
13. Kesten, R., et al.: Lyft level 5 AV dataset 2019 (2019). https://level5.lyft.com/dataset/
14. Kingma, D.P., Ba, J.: Adam: a method for stochastic optimization. CoRR arXiv:abs/1412.6980 (2014)
15. Kirillov, A., He, K., Girshick, R.B., Rother, C., Dollár, P.: Panoptic segmentation. CoRR arXiv:abs/1801.00868 (2018). http://arxiv.org/abs/1801.00868
16. Krizhevsky, A.: Learning multiple layers of features from tiny images (2009)
17. Krizhevsky, A., Sutskever, I., Hinton, G.E.: ImageNet classification with deep convolutional neural networks. In: Pereira, F., Burges, C.J.C., Bottou, L., Weinberger, K.Q. (eds.) Advances in Neural Information Processing Systems, vol. 25, pp. 1097–1105. Curran Associates, Inc. (2012). http://papers.nips.cc/paper/4824-imagenet-classification-with-deep-convolutional-neural-networks.pdf
18. Lang, A.H., Vora, S., Caesar, H., Zhou, L., Yang, J., Beijbom, O.: PointPillars: fast encoders for object detection from point clouds. CoRR arXiv:abs/1812.05784 (2018)
19. Lecun, Y., Bottou, L., Bengio, Y., Haffner, P.: Gradient-based learning applied to document recognition. In: Proceedings of the IEEE, pp. 2278–2324 (1998)
20. Lombardi, S., Simon, T., Saragih, J., Schwartz, G., Lehrmann, A., Sheikh, Y.: Neural volumes. ACM Trans. Graph. **38**(4), 1–14 (2019). https://doi.org/10.1145/3306346.3323020
21. Mani, K., Daga, S., Garg, S., Shankar, N.S., Jatavallabhula, K.M., Krishna, K.M.: MonoLayout: amodal scene layout from a single image. arXiv:abs/2002.08394 (2020)
22. Nair, V., Hinton, G.E.: Rectified linear units improve restricted Boltzmann machines. In: ICML (2010)
23. Paszke, A., et al.: PyTorch: an imperative style, high-performance deep learning library. In: NeurIPS (2019)
24. Philion, J.: Fastdraw: addressing the long tail of lane detection by adapting a sequential prediction network. In: Proceedings of the IEEE/CVF Conference on Computer Vision and Pattern Recognition (CVPR) (2019)
25. Philion, J., Kar, A., Fidler, S.: Learning to evaluate perception models using planner-centric metrics. In: Proceedings of the IEEE/CVF Conference on Computer Vision and Pattern Recognition (CVPR) (2020)
26. Poirson, P., Ammirato, P., Fu, C., Liu, W., Kosecka, J., Berg, A.C.: Fast single shot detection and pose estimation. CoRR arXiv:abs/1609.05590 (2016)
27. Qin, Z., Wang, J., Lu, Y.: MonoGRNet: a geometric reasoning network for monocular 3D object localization. Proc. AAAI Conf. Artif. Intell. **33**, 8851–8858 (2019). https://doi.org/10.1609/aaai.v33i01.33018851
28. Roddick, T., Cipolla, R.: Predicting semantic map representations from images using pyramid occupancy networks. In: Proceedings of the IEEE/CVF Conference on Computer Vision and Pattern Recognition (CVPR) (2020)
29. Roddick, T., Kendall, A., Cipolla, R.: Orthographic feature transform for monocular 3D object detection. CoRR arXiv:abs/1811.08188 (2018)
30. Russakovsky, O., et al.: ImageNet large scale visual recognition challenge (2014)
31. Simonelli, A., Bulò, S.R., Porzi, L., López-Antequera, M., Kontschieder, P.: Disentangling monocular 3D object detection. CoRR arXiv:abs/1905.12365 (2019)
32. Srinivasan, P.P., Mildenhall, B., Tancik, M., Barron, J.T., Tucker, R., Snavely, N.: Lighthouse: predicting lighting volumes for spatially-coherent illumination (2020)

33. Srivastava, N., Hinton, G., Krizhevsky, A., Sutskever, I., Salakhutdinov, R.: Dropout: a simple way to prevent neural networks from overfitting. J. Mach. Learn. Res. **15**, 1929–1958 (2014)
34. Su, H., et al.: SplatNet: sparse lattice networks for point cloud processing. CoRR arXiv:abs/1802.08275 (2018). http://arxiv.org/abs/1802.08275
35. Sun, P., et al.: Scalability in perception for autonomous driving: Waymo open dataset (2019)
36. Takikawa, T., Acuna, D., Jampani, V., Fidler, S.: Gated-SCNN: gated shape CNNs for semantic segmentation. In: Proceedings of the IEEE/CVF International Conference on Computer Vision (ICCV) (2019)
37. Tan, M., Le, Q.V.: EfficientNet: rethinking model scaling for convolutional neural networks. CoRR arXiv:abs/1905.11946 (2019). http://arxiv.org/abs/1905.11946
38. Tucker, R., Snavely, N.: Single-view view synthesis with multiplane images (2020)
39. Wang, Y., Chao, W., Garg, D., Hariharan, B., Campbell, M., Weinberger, K.Q.: Pseudo-LiDAR from visual depth estimation: bridging the gap in 3D object detection for autonomous driving. CoRR arXiv:abs/1812.07179 (2018)
40. You, Y., et al.: Pseudo-LiDAR++: accurate depth for 3D object detection in autonomous driving. CoRR arXiv:abs/1906.06310 (2019)
41. Zeng, W., et al.: End-to-end interpretable neural motion planner. In: IEEE/CVF Conference on Computer Vision and Pattern Recognition (CVPR), pp. 8652–8661 (2019)

Comprehensive Image Captioning
via Scene Graph Decomposition

Yiwu Zhong[1]([✉]), Liwei Wang[2], Jianshu Chen[2], Dong Yu[2], and Yin Li[1]

[1] University of Wisconsin-Madison, Madison, USA
{yzhong52,yin.li}@wisc.edu
[2] Tencent AI Lab, Bellevue, USA
{liweiwang,jianshuchen,dyu}@tencent.com

Abstract. We address the challenging problem of image captioning by revisiting the representation of image scene graph. At the core of our method lies the decomposition of a scene graph into a set of sub-graphs, with each sub-graph capturing a semantic component of the input image. We design a deep model to select important sub-graphs, and to decode each selected sub-graph into a single target sentence. By using sub-graphs, our model is able to attend to different components of the image. Our method thus accounts for accurate, diverse, grounded and controllable captioning at the same time. We present extensive experiments to demonstrate the benefits of our comprehensive captioning model. Our method establishes new state-of-the-art results in caption diversity, grounding, and controllability, and compares favourably to latest methods in caption quality. Our project website can be found at http://pages.cs.wisc.edu/~yiwuzhong/Sub-GC.html.

Keywords: Image captioning · Scene graph · Graph neural networks

1 Introduction

It is an old saying that "A picture is worth a thousand words". Complex and sometimes multiple ideas can be conveyed by a single image. Consider the example in Fig. 1. The image can be described by "A boy is flying a kite" when pointing to the boy and the kite, or depicted as "A ship is sailing on the river" when attending to the boat and the river. Instead, when presented with regions of the bike and the street, the description can be "A bike parked on the street". Humans demonstrate remarkable ability to summarize multiple ideas associated with different scene components of the same image. More interestingly, we can easily explain our descriptions by linking sentence tokens back to image regions.

Despite recent progress in image captioning, most of current approaches are optimized for caption quality. These methods tend to produce generic sentences that are minorly reworded from those in the training set, and to "look" at regions

---

Work partially done while Yiwu Zhong was an intern at Tencent AI Lab, Bellevue.

© Springer Nature Switzerland AG 2020
A. Vedaldi et al. (Eds.): ECCV 2020, LNCS 12359, pp. 211–229, 2020.
https://doi.org/10.1007/978-3-030-58568-6_13

(a) A young boy is flying   (b) A cruise ship is   (c) A bike is parked on
a kite.   sailing on the river.   the street.

Fig. 1. An example image with multiple scene components with each described by a distinct caption. *How can we design a model that can learn to identify and describe different components of an input image?*

that are irrelevant to the output sentence [10,45]. Several recent efforts seek to address these issues, leading to models designed for individual tasks including diverse [11,54], grounded [46,66] and controllable captioning [7,34]. However, no previous method exists that can address diversity, grounding and controllability at the same time—an ability seemingly effortless for we humans.

We believe the key to bridge the gap is a semantic representation that can better link image regions to sentence descriptions. To this end, we propose to revisit the idea of image captioning using scene graph—a knowledge graph that encodes objects and their relationships. Our core idea is that such a graph can be decomposed into a set of sub-graphs, with each sub-graph as a candidate scene component that might be described by a unique sentence. Our goal is thus to design a model that can identify meaningful sub-graphs and decode their corresponding descriptions. A major advantage of this design is that diversity and controllability are naturally enabled by selecting multiple distinct sub-graphs to decode and by specifying a set of sub-graphs for sentence generation.

Specifically, our method takes a scene graph extracted from an image as input. This graph consists of nodes as objects (nouns) and edges as the relations between pairs of objects (predicates). Each node or edge comes with its text and visual features. Our method first constructs a set of overlapping sub-graphs from the full graph. We develop a graph neural network that learns to select meaningful sub-graphs best described by one of the human annotated sentences. Each of the selected sub-graphs is further decoded into its corresponding sentence. This decoding process incorporates an attention mechanism on the sub-graph nodes when generating each token. Our model thus supports backtracking of generated sentence tokens into scene graph nodes and its image regions. Consequently, our method provides the *first comprehensive model for generating accurate, diverse, and controllable captions that are grounded into image regions.*

Our model is evaluated on MS-COCO Caption [6] and Flickr30K Entities [41] datasets. We benchmark the performance of our model on caption quality, diversity, grounding and controllability. Our results suggest that (1) top-ranked captions from our model achieve a good balance between quality and diversity, outperforming state-of-the-art methods designed for diverse captioning in both quality and diversity metrics and performing on par with latest methods optimized for caption quality in quality metrics; (2) our model is able to link the decoded tokens back into the image regions, thus demonstrating strong results for caption grounding; and (3) our model enables controllable captioning via the selection of sub-graphs, improving state-of-the-art results on controllability. We believe our work provides an important step for image captioning.

2   Related Work

There has recently been substantial interest in image captioning. We briefly review relevant work on conventional image captioning, caption grounding, diverse and controllable captioning, and discuss related work on scene graph generation.

Conventional Image Captioning. Major progress has been made in image captioning [17]. An encoder-decoder model is often considered, where Convolutional Neural Networks (CNNs) are used to extract global image features, and Recurrent Neural Networks (RNNs) are used to decode the features into sentences [13,20,31,33,44,52,56,63]. Object information has recently been shown important for captioning [53,62]. Object features from an object detector can be combined with encoder-decoder models to generate high quality captions [2].

Several recent works have explored objects and their relationships, encoded in the form of scene graphs, for image captioning [60,61]. The most relevant work is [61]. Their GCN-LSTM model used a graph convolutional network (GCN) [23] to integrate semantic information in a scene graph. And a sentence is further decoded using features aggregated over the full scene graph. Similar to [61], we also use a GCN for an input scene graph. However, our method learns to select sub-graphs within the scene graph, and to decode sentences from ranked sub-graphs instead of the full scene graph. This design allows our model to produce diverse and controllable sentences that are previously infeasible [2,60,61].

Grounded Captioning. A major challenge of image captioning is that recent deep models might not focus on the same image regions as a human would when generating each word, leading to undesirable behaviors, e.g., object hallucination [10,45]. Several recent work [2,19,36,46,56,59,66] has been developed to address the problem of grounded captioning—the generation of captions and the alignment between the generated words and image regions. Our method follows the weakly supervised setting for grounded captioning, where we assume that only the image-sentence pairs are known. Our key innovation is to use a sub-graph on an image scene graph for sentence generation, thus constraining the grounding within the sub-graph.

Our work is also relevant to recent work on generating text descriptions of local image regions, also known as dense captioning [19,21,59]. Both our work and dense captioning methods can create localized captions. The key difference is that our method aims to generate sentence descriptions of scene components that spans multiple image regions, while dense captioning methods focused on generating phrase descriptions for local regions [19,59] or pairs of local regions [21].

Diverse and Controllable Captioning. The generation of diverse and controllable image descriptions has also received considerable attention. Several approaches have been proposed for diverse captioning [3,8,11,27,47,51,54]. Wang et al. [54] proposed a variational auto-encoder that can decode multiple diverse sentences from samples drawn from a latent space of image features. This idea was further extended by [3], where every word has its own latent space. Moreover, Deshpande et al. [11] proposed to generate various sentences controlled by part-of-speech tags. There is a few recent work on controllable captioning. Lu et al. [34] proposed to fill a generated sentence template with the concepts from an object detector. Cornia et al. [7] selected object regions using grounding annotations and then predicted textual chunks to generate diverse and controllable sentences. Similar to [7], we address diversity and controllability within the same model. Different from [7], our model is trained using only image-sentence pairs and can provide additional capacity of caption grounding.

Scene Graph Generation. Image scene graph generation has received considerable attention, partially driven by large-scale scene graph datasets [25]. Most existing methods [9,28,29,32,55,58,64,65] start from candidate object regions given by an object detector and seek to infer the object categories and their relationships. By using a previous approach [64] to extract image scene graphs, we explore the decomposition of scene graphs into sub-graphs for generating accurate, diverse, and controllable captions. Similar graph partitioning problems have been previously considered in vision for image segmentation [15,18] and visual tracking [48,49], but has not been explored for image captioning.

3   Method

Given an input image I, we assume an image scene graph $G = (V, E)$ can be extracted from I, where V represents the set of nodes corresponding to the detected objects (nouns) in I, and E represents the set of edges corresponding to the relationships between pairs of objects (predicates). Our goal is to generate a set of sentences $C = \{C_j\}$ to describe different components of I using the scene graph G. To this end, we propose to make use of the sub-graphs $\{G_i^s = (V_i^s, E_i^s)\}$ from G, where $V_i^s \subseteq V$ and $E_i^s \subseteq E$. Our method seeks to model the joint probability $P(S_{ij} = (G, G_i^s, C_j)|I)$, where $P(S_{ij}|I) = 1$ indicates that the sub-graph G_i^s can be used to decode the sentence C_j. Otherwise, $P(S_{ij}|I) = 0$. We further assume that $P(S_{ij}|I)$ can be decomposed into three parts, given by

$$P(S_{ij}|I) = P(G|I)P(G_i^s|G, I)P(C_j|G_i^s, G, I). \tag{1}$$

Intuitively, $P(G|I)$ extracts scene graph G from an input image I. $P(G_i^s|G, I)$ decomposes the full graph G into a diverse set of sub-graphs $\{G_i^s\}$ and selects important sub-graphs for sentence generation. Finally, $P(C_j|G_i^s, G, I)$ decodes a selected sub-graph G_i^s into its corresponding sentence C_j, and also associates

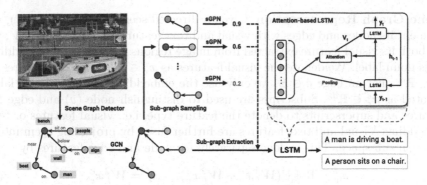

Fig. 2. Overview of our method. Our method takes a scene graph extracted from an input image, and decomposes the graph into a set of sub-graphs. We design a sub-graph proposal network (sGPN) that learns to identify meaningful sub-graphs, which are further decoded by an attention-based LSTM for generating sentences and grounding sentence tokens into sub-graph nodes (image regions). By leveraging sub-graphs, our model enables accurate, diverse, grounded and controllable image captioning.

the tokens in C_j to the nodes V_i^s of the sub-graph G_i^s (the image regions in I). Figure 2 illustrates our method. We now present details of our model.

3.1   Scene Graph Detection and Decomposition

Our method first extracts scene graph G from image I ($P(G|I)$) using MotifNet [64]. MotifNet builds LSTMs on top of object detector outputs [43] and produces a scene graph $G = (V, E)$ with nodes V for common objects (nouns) and edges E for relationship between pairs of objects (predicates), such as "holding", "behind" or "made of". Note that G is a directed graph, i.e., an edge must start from a subject noun or end at an object noun. Therefore, the graph G is defined by a collection of subject-predicate-object triplets, e.g., kid playing ball.

We further samples sub-graphs $\{G_i^s\}$ from the scene graph G by using neighbor sampling [24]. Specifically, we randomly select a set of seed nodes $\{S_i\}$ on the graph. The immediate neighbors of the seed nodes with the edges in-between define a sampled sub-graph. Formally, the sets of sub-graph nodes and edges are $V_i^s = S_i \cup \{N(v)|v \in S_i\}$ and $E_i^s = \{(v, u)|v \in S_i, u \in N(v)\}$ respectively, where $N(v)$ denotes the immediate neighbors of node v. Identical sub-graphs are removed to obtain the final set of sub-graphs $\{G_i^s = (V_i^s, E_i^s)\}$, which covers potential scene components in the input image I.

3.2   Sub-graph Proposal Network

Our next step is to identify meaningful sub-graphs that are likely to capture major scene components in the image ($P(G_i^s|G, I)$). Specifically, our model first combines visual and text features on the scene graph G, followed by an integration of contextual information within G using a graph convolutional network, and finally a score function learned to rank sub-graphs G_i^s.

Scene Graph Representation. Given a directed scene graph $G = (V, E)$, we augment its nodes and edges with visual and text features. For a node $v \in V$, we use both its visual feature extracted from image regions and the word embedding of its noun label. We denote the visual features as $\boldsymbol{x}_v^v \in \mathbb{R}^{d_v}$ and text features as $\boldsymbol{x}_v^e \in \mathbb{R}^{d_e}$. For an edge $e \in E$, we only use the embedding of its predicate label denoted as $\boldsymbol{x}_e^e \in \mathbb{R}^{d_e}$. Subscripts are used to distinguish node (v) and edge (e) features and superscripts to denote the feature type, i.e., visual features or text embedding. Visual and text features are further fused by projecting them into a common sub-space. This is done separately for node and edge features by

$$\boldsymbol{x}_v^f = \text{ReLU}(\boldsymbol{W}_f^1 \boldsymbol{x}_v^v + \boldsymbol{W}_f^2 \boldsymbol{x}_v^e), \quad \boldsymbol{x}_e^f = \boldsymbol{W}_f^3 \boldsymbol{x}_e^e, \tag{2}$$

where $\boldsymbol{W}_f^1 \in \mathbb{R}^{d_f \times d_v}$, $\boldsymbol{W}_f^2 \in \mathbb{R}^{d_f \times d_e}$ and $\boldsymbol{W}_f^3 \in \mathbb{R}^{d_f \times d_e}$ are learned projections.

Graph Convolutional Network (GCN). After feature fusion and projection, we further model the context between objects and their relationships using a GCN. The GCN aggregates information from the neighborhood within the graph and updates node and edge features. With an proper ordering of the nodes and edges, we denote the feature matrix for nodes and edges as $\boldsymbol{X}_v^f = [\boldsymbol{x}_v^f] \in \mathbb{R}^{d_f \times |V|}$ and $\boldsymbol{X}_e^f = [\boldsymbol{x}_e^f] \in \mathbb{R}^{d_f \times |E|}$, respectively. The update rule of a single graph convolution is thus given by

$$\begin{aligned}
\hat{\boldsymbol{X}}_v^f &= \boldsymbol{X}_v^f + \text{ReLU}(\boldsymbol{W}_{ps}\boldsymbol{X}_e^f \boldsymbol{A}_{ps}) + \text{ReLU}(\boldsymbol{W}_{po}\boldsymbol{X}_e^f \boldsymbol{A}_{po}), \\
\hat{\boldsymbol{X}}_e^f &= \boldsymbol{X}_e^f + \text{ReLU}(\boldsymbol{W}_{sp}\boldsymbol{X}_v^f \boldsymbol{A}_{sp}) + \text{ReLU}(\boldsymbol{W}_{op}\boldsymbol{X}_v^f \boldsymbol{A}_{op}),
\end{aligned} \tag{3}$$

where $\boldsymbol{W}_{ps}, \boldsymbol{W}_{po}, \boldsymbol{W}_{sp}, \boldsymbol{W}_{op} \in \mathbb{R}^{d_f \times d_f}$ are learnable parameters that link subject or object features (nouns) with predicate features. For example, $\boldsymbol{W}_{ps}$ connects between predicate features and subject features. $\boldsymbol{A}_{ps}, \boldsymbol{A}_{po} \in \mathbb{R}^{|E| \times |V|}$ are the normalized adjacency matrix (defined by G) between predicates and subjects, and between predicates and objects, respectively. For instance, a non-zero element in $\boldsymbol{A}_{ps}$ suggests a link between a predicate and a subject on the scene graph G. Similarly, $\boldsymbol{A}_{sp}, \boldsymbol{A}_{op} \in \mathbb{R}^{|V| \times |E|}$ are the normalized adjacency matrix between subjects and predicates, and between objects and predicates. $\boldsymbol{A}_{ps} \neq \boldsymbol{A}_{sp}^T$ due to the normalization of adjacency matrix.

Our GCN stacks several graph convolutions, and produces an output scene graph with updated node and edge features. We only keep the final node features ($\boldsymbol{X}_v^u = [\boldsymbol{x}_v^u], v \in V$) for subsequent sub-graph ranking, as the predicate information has been integrated using GCN.

Sub-graph Score Function. With the updated scene graph and the set of sampled sub-graphs, our model learns a score function to select meaningful sub-graphs for generating sentence descriptions. For each sub-graph, we index its node features as $\boldsymbol{X}_i^s = [\boldsymbol{x}_v^u], v \in V_i^s$ and construct a score function

$$s_i = \sigma(f(\Phi(\boldsymbol{X}_i^s))), \tag{4}$$

where $\Phi(\cdot)$ is a sub-graph readout function [57] that concatenates the max-pooled and mean-pooled node features on the sub-graph. $f(\cdot)$ is a score function realized by a two-layer multilayer perceptron (MLP). And $\sigma(\cdot)$ is a sigmoid function that normalizes the output score into the range of $[0, 1]$.

Learning the Score Function. The key challenge of learning the score function f is the training labels. Our goal is to rank the sampled sub-graphs and select the best ones to generate captions. Thus, we propose to use ground-truth captions provided by human annotators to guide the learning. A sub-graph with most of the nodes matched to one of the ground-truth sentences should be selected. To this end, we recast the learning of the score function f as training a binary classifier to distinguish between "good" (positive) and "bad" (negative) sub-graphs. Importantly, we design a matching score between a ground-truth sentence and a sampled sub-graph to generate the target binary labels.

Specifically, given a sentence C_j and a scene graph G, we extract a reference sub-graph on G by finding the nodes on the graph G that also appears in the sentence C_j and including their immediate neighbor nodes. This is done by extracting nouns from the sentence C_j using a part-of-speech tag parser [5], and matching the nouns to the nodes on G using LCH score [26] derived from WordNet [38]. This matching process is given by $\mathcal{M}(C_j, G)$. We further compute the node Intersection over Union (IoU) score between the reference sub-graph $\mathcal{M}(C_j, G)$ and each of the sampled sub-graph G_I^s by

$$IoU(G_i^s, C_j) = \frac{|G_i^s \cap \mathcal{M}(C_j, G)|}{|G_i^s \cup \mathcal{M}(C_j, G)|}, \tag{5}$$

where $\cap$ and $\cup$ are the intersection and union operation over sets of sub-graph nodes, respectively. The node IoU provides a matching score between the reference sentence C_j and the sub-graph G_i^s and is used to determine our training labels. We only consider a sub-graph as positive for training if its IoU with any of the target sentences is higher than a pre-defined threshold (0.75).

Training Strategy. A major issue in training is that we have many negative sub-graphs and only a few positive ones. To address this issue, a mini-batch of sub-graphs is randomly sampled to train our sGPN, where positive to negative ratio is kept as 1:1. If a ground-truth sentence does not match any positive sub-graph, we use the reference sub-graph from $\mathcal{M}(C_j, G)$ as its positive sub-graph.

3.3 Decoding Sentences from Sub-graphs

Our final step is to generate a target sentence using features from any selected single sub-graph ($P(C_j|G_i^s, G, I)$). We modify the attention-based LSTM [2] for sub-graph decoding, as shown in Fig. 2 (top right). Specifically, the model couples an attention LSTM and a language LSTM. The attention LSTM assigns each sub-graph node an importance score, further used by the language LSTM to generate the tokens. Specifically, at each time step t, the attention LSTM is given by $\boldsymbol{h}_t^A = LSTM_{Att}([\boldsymbol{h}_{t-1}^L, \boldsymbol{e}_t, \boldsymbol{x}_i^s])$, where $\boldsymbol{h}_{t-1}^L$ is the hidden state of the language LSTM at time $t-1$. $\boldsymbol{e}_t$ is the word embedding of the input token at time t and $\boldsymbol{x}_i^s$ is the sub-graph feature. Instead of averaging all region features as [2,61], our model uses the input sub-graph feature, given by $x_i^s = g(\Phi(X_i^s))$, where $g(\cdot)$ is a two-layer MLP, $\Phi(\cdot)$ is the same graph readout unit in Eq. 4.

Based on hidden states h_t^A and the node features $X_i^s = [x_v^u]$ in the sub-graph, an attention weight $a_{v,t}$ at time t for node v is computed by $a_{v,t} = w_a^T \tanh(W_v x_v^u + W_h h_t^A)$ with learnable weights W_v, W_h and w_a. A softmax function is further used to normalize a_t into α_t defined on all sub-graph nodes at time t. We use α_t to backtrack image regions associated with a decoded token for caption grounding. Finally, the hidden state of the attention LSTM h_t^A and the attention re-weighted sub-graph feature $\mathcal{V}_t = \sum_v \alpha_{v,t} x_v^u$ are used as the input of the language LSTM—a standard LSTM that decodes the next word.

3.4   Training and Inference

We summarize the training and inference schemes of our model.

Loss Functions. Our sub-graph captioning model has three parts: $P(G|I)$, $P(G_i^s|G,I)$, $P(C_j|G_i^s,G,I)$, where the scene graph generation ($P(G|I)$) is trained independently on Visual Genome [25]. For training, we combine two loss functions for $P(G_i^s|G,I)$ and $P(C_j|G_i^s,G,I)$. Concretely, we use a binary cross-entropy loss for the sub-graph proposal network ($P(G_i^s|G,I)$), and a multi-way cross-entropy loss for the attention-based LSTM model to decode the sentences ($P(C_j|G_i^s,G,I)$). The coefficient between the two losses is set to 1.

Inference. During inference, our model extracts the scene graph, samples sub-graphs and evaluates their sGPN scores. *Greedy Non-Maximal Suppression (NMS)* is further used to filter out sub-graphs that largely overlap with others, and to keep sub-graphs with high sGPN scores. The overlapping between two sub-graphs is defined by the IoU of their nodes. We find that using NMS during testing helps to remove redundant captions and to promote diversity.

After NMS, top-ranked sub-graphs are decoded using an attention-based LSTM. As shown in [35], an *optional top-K sampling* [14,42] can be applied during the decoding to further improve caption diversity. We disable top-K sampling for our experiments unless otherwise noticed. The final output is thus a set of sentences with each from a single sub-graph. By choosing which sub-graphs to decode, our model can control caption contents. Finally, we use attention weights in the LSTM to ground decoded tokens to sub-graph nodes (image regions).

4   Experiments

We now describe our implementation details and presents our results. We start with an ablation study (Sect. 4.1) for different model components. Further, we evaluate our model across several captioning tasks, including accurate and diverse captioning (Sect. 4.2), grounded captioning (Sect. 4.3) and controllable captioning (Sect. 4.4).

Implementation Details. We used Faster R-CNN [43] with ResNet-101 [16] from [2] as our object detector. Based on detection results, Motif-Net [64] was

Table 1. Ablation study on sub-graph/sentence ranking functions, the NMS thresholds and the top-K sampling during decoding. We report results for both accuracy (B4 and C) and diversity (Distinct Caption, 1/2-gram). Our model is trained on the train set of COCO caption and evaluated on the the validation set, following M-RNN split [37].

Model	Ranking function	NMS	Top-K sampling	B4	C	Distinct caption	1-gram (Best 5)	2-gram (Best 5)
Sub-GC-consensus	consensus	0.75	N/A	33.0	107.6	59.3%	0.25	0.32
Sub-GC-sGPN	sGPN	0.75	N/A	33.4	108.7	59.3%	0.28	0.37
Sub-GC	sGPN + consensus	0.75	N/A	34.3	112.9	59.3%	0.28	0.37
Sub-GC-consensus	consensus	0.55	N/A	32.5	105.6	70.5%	0.27	0.36
Sub-GC-sGPN	sGPN	0.55	N/A	33.4	108.7	70.5%	0.32	0.42
Sub-GC	sGPN + consensus	0.55	N/A	34.1	112.3	70.5%	0.32	0.42
Sub-GC-S	sGPN + consensus	0.55	T = 0.6, K = 3	31.8	108.7	96.0%	0.39	0.57
Sub-GC-S	sGPN + consensus	0.55	T = 0.6, K = 5	30.9	106.1	97.5%	0.41	0.60
Sub-GC-S	sGPN + consensus	0.55	T = 1.0, K = 3	28.4	100.7	99.2%	0.43	0.64

trained on Visual Genome [25] with 1600/20 object/predicate classes. For each image, we applied the detector and kept 36 objects and 64 triplets in scene graph. We sampled 1000 sub-graphs per image and removed duplicate ones, leading to an average of 255/274 sub-graphs per image for MS-COCO [6]/Flickr30K [41]. We used 2048D visual features for image regions and 300D GloVe [40] embeddings for node and edge labels. These features were projected into 1024D, followed by a GCN with depth of 2 for feature transform and an attention LSTM (similar to [2]) for sentence decoding. For training, we used Adam [22] with initial learning rate of 0.0005 and a mini-batch of 64 images and 256 sub-graphs. Beam search was used in decoding with beam size 2, unless otherwise noted.

4.1   Ablation Study

We first conduct an ablation study of our model components, including the ranking function, the NMS threshold and the optional top-K sampling. We now describe the experiment setup and report the ablation results.

Experiment Setup. We follow the evaluation protocol from [3,11,51,54] and report both accuracy and diversity results using the M-RNN split [37] of MS-COCO Caption dataset [6]. Specifically, this split has 118,287/4,000/1,000 images for train/val/test set, with 5 human labeled captions per image. We train the model on the train set and report the results on the *val* set. For accuracy, we report top-1 accuracy out of the top 20 output captions, using BLEU-4 [39] and CIDEr [50]. For diversity, we evaluate the percentage of distinct captions from 20 sampled output captions, and report 1/2-gram diversity of the best 5 sampled captions using a ranking function. Beam search was disabled for this ablation study. Table 1 presents our results and we now discuss our results.

Ranking function is used to rank output captions. Our sGPN provides a socre for each sub-graph and thus each caption. Our sGPN can thus be re-purposed as a ranking function. We compare sGPN with consensus re-ranking [12,37] widely used in the literature [3,11,54]. Moreover, we also experiment with applying consensus on top-scored captions (e.g., top-4) from sGPN (sGPN+consensus). Our sGPN consistently outperforms consensus re-ranking for both accuracy and diversity (+1.1 CIDEr and +12% 1-gram with NMS = 0.75). Importantly, consensus re-ranking is computational expensive, while our sGPN incurs little computational cost. Further, combining our sGPN with consensus re-ranking (sGPN + consensus) improves top-1 accuracy (+4.2 CIDEr with NMS = 0.75). sGPN + consensus produces the same diversity scores as sGPN, since only one ranking function (sGPN) is used in diversity evaluation.

NMS threshold is used during inference to eliminate similar sub-graphs (see Sect. 3.4). We evaluate two NMS thresholds (0.55 and 0.75). For all ranking functions, a lower threshold (0.55) increases diversity scores (+8%/ + 14% 1-gram for consensus/sGPN) and has comparable top-1 accuracy, expect for consensus re-ranking (−2.0 CIDEr). Note that for our sGPN, the top-1 accuracy remains the same as the top-ranked sub-graph stays unchanged.

Top-K sampling is optionally applied during caption decoding, where each token is randomly drawn from the top K candidates based on the normalized logits produced by a softmax function with temperature T. A small T favors the top candidate and a large K produces more randomness. We evaluate different combinations of K and T. Using top-K sampling decreases the top-1 accuracy yet significantly increases all diversity scores (−3.6 CIDEr yet +22% in 1-gram with $T = 0.6$, $K = 3$). The same trend was also observed in [35].

Our final model (Sub-GC) combines sGPN and consensus re-ranking for ranking captions. We set NMS threshold to 0.75 for experiments focusing on the accuracy of top-1 caption (Tables 3, 4 and 5) and 0.55 for experiments on diversity (Table 2). Top-K sampling is only enabled for additional results on diversity.

4.2 Accurate and Diverse Image Captioning

Dataset and Metric. Moving forward, we evaluate our final model for accuracy and diversity on MS-COCO caption *test* set using M-RNN split [37]. Similar to our ablation study, we report top-1 accuracy and diversity scores by selecting from a pool of top 20/100 output sentences. Top-1 accuracy scores include BLEU [39], CIDEr [50], ROUGE-L [30], METEOR [4] and SPICE [1]. And diversity scores include distinct caption, novel sentences, mutual overlap (mBLEU-4) and n-gram diversity. Beam search was disabled for a fair comparison.

Baselines. We consider several latest methods designed for diverse and accurate captioning as our baselines, including Div-BS [51], AG-CVAE [54], POS [11], POS+Joint [11] and Seq-CVAE [3]. We compare our results of Sub-GC to these baselines in Table 2. In addition, we include the results of our model with top-K sampling (Sub-GC-S), as well as human performance for references of diversity.

Diversity Results. For the majority of the diversity metrics, our model Sub-GC outperforms previous methods (+8% for novel sentences and +29%/20% for

Table 2. Diversity and top-1 accuracy results on COCO Caption dataset (M-RNN split [37]). Best-5 refers to the top-5 sentences selected by a ranking function. Note that Sub-GC and Sub-GC-S have same top-1 accuracy in terms of sample-20 and sample-100, since we have a sGPN score per sub-graph and global sorting is applied over all sampled sub-graphs. Our models outperform previous methods on both top-1 accuracy and diversity for the majority of the metrics.

Method	#	Diversity					Top-1 Accuracy							
		Distinct caption (↑)	#novel (Best 5) (↑)	mBLEU-4 (Best 5) (↓)	1-gram (Best 5) (↑)	2-gram (Best 5) (↑)	B1	B2	B3	B4	C	R	M	S
Div-BS [51]	20	100%	3106	81.3	0.20	0.26	72.9	56.2	42.4	32.0	103.2	53.6	25.5	18.4
AG-CVAE [54]		69.8%	3189	66.6	0.24	0.34	71.6	54.4	40.2	29.9	96.3	51.8	23.7	17.3
POS [11]		96.3%	3394	63.9	0.24	0.35	74.4	57.0	41.9	30.6	101.4	53.1	25.2	18.8
POS + Joint [11]		77.9%	3409	66.2	0.23	0.33	73.7	56.3	41.5	30.5	102.0	53.1	25.1	18.5
Sub-GC		71.1%	3679	67.2	0.31	0.42	**77.2**	**60.9**	**46.2**	**34.6**	**114.2**	**56.1**	**26.9**	**20.0**
Seq-CVAE [3]	20	94.0%	**4266**	52.0	0.25	0.54	73.1	55.4	40.2	28.9	100.0	52.1	24.5	17.5
Sub-GC-S		**96.2%**	4153	**36.4**	**0.39**	**0.57**	75.2	57.6	42.7	31.4	107.3	54.1	26.1	19.3
Div-BS [51]	100	100%	3421	82.4	0.20	0.25	73.4	56.9	43.0	32.5	103.4	53.8	25.5	18.7
AG-CVAE [54]		47.4%	3069	70.6	0.23	0.32	73.2	55.9	41.7	31.1	100.1	52.8	24.5	17.9
POS [11]		91.5%	3446	67.3	0.23	0.33	73.7	56.7	42.1	31.1	103.6	53.0	25.3	18.8
POS + Joint [11]		58.1%	3427	70.3	0.22	0.31	73.9	56.9	42.5	31.6	104.5	53.2	25.5	18.8
Sub-GC		65.8%	3647	69.0	0.31	0.41	**77.2**	**60.9**	**46.2**	**34.6**	**114.4**	**56.1**	**26.9**	**20.0**
Seq-CVAE [3]	100	84.2%	**4215**	64.0	0.33	0.48	74.3	56.8	41.9	30.8	104.1	53.1	24.8	17.8
Sub-GC-S		**94.6%**	4128	**37.3**	**0.39**	**0.57**	75.2	57.6	42.7	31.4	107.3	54.1	26.1	19.3
Human	5	99.8%	-	51.0	0.34	0.48	-	-	-	-	-	-	-	-

Table 3. Comparison to accuracy optimized models on COCO caption dataset using Karpathy split [20]. Our Sub-GC compares favorably to the latest methods.

Method	B1	B4	C	R	M	S
Up-Down [2]	77.2	36.2	113.5	56.4	27.0	20.3
GCN-LSTM [61]	77.3	36.8	116.3	57.0	**27.9**	**20.9**
SGAE [60]	**77.6**	**36.9**	**116.7**	**57.2**	27.7	**20.9**
Full-GC	76.7	**36.9**	114.8	56.8	**27.9**	20.8
Sub-GC	76.8	36.2	115.3	56.6	27.7	20.7
Sub-GC-oracle	90.7	59.3	166.7	71.5	40.1	30.1

1/2-gram with 20 samples), except the most recent Seq-CVAE. Upon a close inspection of Seq-CVAE model, we hypothesis that Seq-CVAE benefits from sampling tokens at each time step. It is thus meaningful to compare our model using top-K sampling (Sub-GC-S) with Seq-CVAE. Sub-GC-S outperforms Seq-CVAE in most metrics (+18%/19% for 1/2-gram with 100 samples) and remains comparable for the metric of novel sentences (within 3% difference).

Accuracy Results. We notice that the results of our sub-graph captioning models remain the same with increased number of samples. This is because our outputs follow a fixed rank from sGPN scores. Our Sub-GC outperforms all previous methods by a significant margin. Sub-GC achieves +2.6/2.1 in B4 and +11.2/9.9 in CIDEr when using 20/100 samples in comparison to previous best

Table 4. Grounded captioning results on Flickr30K Entities [41]. Our method (Sub-GC) outperforms previous weakly supervised methods.

Method	Grounding evaluation		Caption evaluation				
	F1 all	F1 loc	B1	B4	C	M	S
GVD [66]	3.88	11.70	69.2	26.9	60.1	22.1	16.1
Up-Down [2]	4.14	12.30	69.4	27.3	56.6	21.7	16.0
Cyclical [36]	4.98	13.53	69.9	27.4	61.4	22.3	16.6
Full-GC	4.90	13.08	69.8	**29.1**	**63.5**	**22.7**	**17.0**
Sub-GC	**5.98**	**16.53**	**70.7**	28.5	61.9	22.3	16.4
GVD (Sup.) [66]	7.55	22.20	69.9	27.3	62.3	22.5	16.5

Table 5. Controllable captioning results on Flickr30K Entities [41]. With weak supervision, our Sub-GC compares favorably to previous methods. With strong supervision, our Sub-GC (Sup.) achieves the best results.

Method	B1	B4	C	R	M	S	IoU
NBT [34] (Sup.)	-	8.6	53.8	31.9	13.5	17.8	49.9
SCT [7] (Sup.)	33.1	9.9	67.3	35.3	14.9	22.2	52.7
Sub-GC	33.6	9.3	57.8	32.5	14.2	18.8	50.6
Sub-GC (Sup.)	**36.2**	**11.2**	**73.7**	**35.5**	**15.9**	**22.2**	**54.1**

results. Moreover, while achieving best diversity scores, our model with top-K sampling (Sub-GC-S) also outperforms previous methods in most accuracy metrics (+0.8/0.9 in B1 and +4.1/2.8 in CIDEr when using 20/100 samples) despite its decreased accuracy from Sub-GC.

Comparison to Accuracy Optimized Captioning Models. We conduct further experiments to compare the top ranked sentence from ou Sub-GC against the results of latest captioning models optimized for accuracy, including Up-Down [2], GCN-LSTM [61] and SGAE [60]. All these previous models can only generate a single sentence, while our method (Sub-GC) can generate a set of diverse captions. As a reference, we consider a variant of our model (Full-GC) that uses a full scene graph instead of sub-graphs to decode sentences. Moreover, we include an upper bound of our model (Sub-GC-oracle) by assuming that we have an oracle ranking function, i.e., always selecting the maximum scored sentence for each metric. All results are reported on Karpathy split [20] of COCO dataset without using reinforcement learning for score optimization [44].

Our results are shown in Table 3. Our Sub-GC achieves comparable results (within 1–2 points in B4/CIDEr) to latest methods (Up-Down, GCN-LSTM and SGAE). We find that the results of our sub-graph captioning model is slightly worse than those models using the full scene graph, e.g., Full-GC, GCN-LSTM and SGAE. We argue that this minor performance gap does not diminish our contribution, as our model offers new capacity for generating diverse, controllable

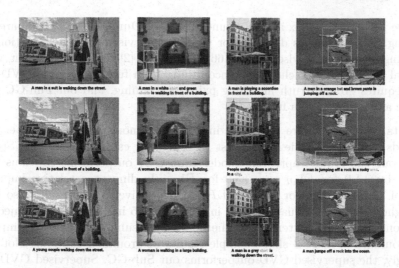

Fig. 3. Sample results of our Sub-GC on Flickr30k Entities test set. Each column shows three captions with their region groundings decoded from different sub-graphs for an input image. The first two rows are successful cases and the last row is the failure case. These sentences can describe different parts of the images. Each generated noun and its grounding bounding box are highlighted in the same color.

and grounded captions. Notably, our best case (Sub-GC-oracle) outperforms all other methods for all metrics by a very large margin (+22.4 in B4 and +50.0 in CIDEr). These results suggest that at least one high-quality caption exists among the sentences decoded from the sub-graphs. It is thus possible to generate highly accurate captions if there is a way to select this "good" sub-graph.

4.3 Grounded Image Captioning

Moreover, we evaluate our model for grounded captioning. We describe the dataset and metric, introduce our setup and baselines, and discuss our results.

Dataset and Metric. We use Flickr30k Entities [41] for grounded captioning. Flickr30k Entities has 31K images, with 5 captions for each image. The dataset also includes 275k annotated bounding boxes associated with the phrases in corresponding captions. We use the data split from [20]. To evaluate the grounding performance, we follow the protocol in GVD [66]. We report both $F1_{all}$ and $F1_{loc}$. $F1_{all}$ considers a region prediction as correct if the object word is correctly predicted and the box is correctly localized. On the other hand $F1_{loc}$ only accounts for localization quality. Moreover, we report the standard BLEU [39], CIDEr [50], METEOR [4] and SPICE [1] scores for caption quality.

Experiment Setup and Baselines. For this experiment, we only evaluate the top-ranked sentence and its grounding from our model. We select the node on the sub-graph with maximum attention weight when decoding a noun word, and use its bounding box as the grounded region. Our results are compared to a strong set of baselines designed for weakly supervised grounded captioning, including weakly supervised GVD [66], Up-Down [2] and a concurrent work Cyclical [36]. We also include reference results from fully supervised GVD [66] that requires ground-truth matching pairs for training, and our Full-GC that decode a sentence from a full graph.

Results. Our results are presented in Table 4. Among all weakly supervised methods, our model achieves the best F1 scores for caption grounding. Specifically, our sub-graph captioning model (Sub-GC) outperforms previous best results by +1.0 for $F1_{all}$ and +3.0 for $F1_{loc}$, leading to a relative improvement of 20% and 22% for $F1_{all}$ and $F1_{loc}$, respectively. Our results also have the highest captioning quality (+1.1 in B4 and +0.5 in CIDEr). We conjecture that constraining the attention to the nodes of a sub-graph helps to improve the grounding. Figure 3 shows sample results of grounded captions. Not surprisingly, the supervised GVD outperforms our Sub-GC. Supervised GVD can be considered as an upper bound for all other methods, as it uses grounding annotations for training. Comparing to our Full-GC, our Sub-GC is worse on captioning quality (−0.6 in B4 and −1.6 in CIDEr) yet has significant better performance for grounding (+1.1 in $F1_{all}$ and +3.5 in $F1_{loc}$).

4.4 Controllable Image Captioning

Finally, we report results on controllable image captioning. Again, we describe our experiments and present the results.

Dataset and Metric. Same as grounding, we consider Flickr30k Entities [41] for controllable image captioning and use the data split [20]. We follow evaluation protocol developed in [7]. Specifically, the protocol assumes that an image and a set of regions are given as input, and evaluates a decoded sentence against one or more target ground-truth sentences. These ground-truth sentences are selected from captions by matching the sentences tokens to object regions in the image. Standard captioning metrics are considered (BLEU [39], CIDEr [50], ROUGE-L [30], METEOR [4] and SPICE [1]), yet the ground-truth is different from conventional image captioning. Moreover, the IoU of the nouns between the predicted and the target sentence is also reported as [7].

Experiment Setup and Baselines. We consider (1) our Sub-GC trained with only image-sentence pairs; and (2) a supervised Sub-GC trained with ground-truth pairs of region sets and sentences as [7]. Both models follow the same inference scheme, where input controlled set of regions are converted into best matching sub-graphs for sentence decoding. However, supervised Sub-GC uses these matching during training. We compare our results to recent methods developed for controllable captioning, including NBT [34] and SCT [7]. NBT and SCT are trained with matching pairs of region sets and sentences same as our supervised Sub-GC. Results are reported without using reinforcement learning.

Results. The results are shown in Table 5. Our models demonstrate strong controllability of the output sentences. Specifically, our supervised Sub-GC outperforms previous supervised methods (NBT and SCT) by a significant margin. Comparing to previous best SCT, our results are +1.3 in B4, +6.4 in CIDEr and +1.4 in IoU. Interestingly, our vanilla model has comparable performance to previous methods, even if it is trained with only image sentence pairs. These results provide further supports to our design of using sub-graphs for image captioning.

5   Conclusion

We proposed a novel image captioning model by exploring sub-graphs of image scene graph. Our key idea is to select important sub-graphs and only decode a single target sentence from a selected sub-graph. We demonstrated that our model can generate accurate, diverse, grounded and controllable captions. Our method thus offers the first comprehensive model for image captioning. Moreover, our results established new state-of-the-art in diverse captioning, grounded captioning and controllable captioning, and compared favourably to latest method for caption quality. We hope our work can provide insights into the design of explainable and controllable models for vision and language tasks.

Acknowledgment. The work was partially developed during the first author's internship at Tencent AI Lab and further completed at UW-Madison. YZ and YL acknowledge the support by the UW VCRGE with funding from WARF.

References

1. Anderson, P., Fernando, B., Johnson, M., Gould, S.: Spice: semantic propositional image caption evaluation. In: Leibe, B., Matas, J., Sebe, N., Welling, M. (eds.) Computer Vision – ECCV 2016. Lecture Notes in Computer Science, vol. 9909, pp. 382–398. Springer, Cham (2016). https://doi.org/10.1007/978-3-319-46454-1_24
2. Anderson, P., et al.: Bottom-up and top-down attention for image captioning and visual question answering. In: Proceedings of the IEEE Conference on Computer Vision and Pattern Recognition (CVPR), pp. 6077–6086. IEEE (2018)
3. Aneja, J., Agrawal, H., Batra, D., Schwing, A.: Sequential latent spaces for modeling the intention during diverse image captioning. In: Proceedings of the IEEE International Conference on Computer Vision (ICCV), pp. 4261–4270. IEEE (2019)
4. Banerjee, S., Lavie, A.: Meteor: an automatic metric for MT evaluation with improved correlation with human judgments. In: ACL Workshop on Intrinsic and Extrinsic Evaluation Measures for Machine Translation and/or Summarization, pp. 65–72 (2005)
5. Bird, S., Loper, E.: NLTK: the natural language toolkit. In: ACL Interactive Poster and Demonstration Sessions, pp. 214–217. Association for Computational Linguistics (2004)
6. Chen, X., et al.: Microsoft coco captions: Data collection and evaluation server. arXiv preprint arXiv:1504.00325 (2015)

7. Cornia, M., Baraldi, L., Cucchiara, R.: Show, control and tell: a framework for generating controllable and grounded captions. In: Proceedings of the IEEE Conference on Computer Vision and Pattern Recognition (CVPR), pp. 8307–8316. IEEE (2019)
8. Dai, B., Fidler, S., Urtasun, R., Lin, D.: Towards diverse and natural image descriptions via a conditional GAN. In: Proceedings of the IEEE International Conference on Computer Vision (ICCV), pp. 2970–2979. IEEE (2017)
9. Dai, B., Zhang, Y., Lin, D.: Detecting visual relationships with deep relational networks. In: Proceedings of the IEEE Conference on Computer Vision and Pattern Recognition (CVPR), pp. 3076–3086. IEEE (2017)
10. Das, A., Agrawal, H., Zitnick, L., Parikh, D., Batra, D.: Human attention in visual question answering: do humans and deep networks look at the same regions? Comput. Vis. Image Underst. **163**, 90–100 (2017)
11. Deshpande, A., Aneja, J., Wang, L., Schwing, A.G., Forsyth, D.: Fast, diverse and accurate image captioning guided by part-of-speech. In: Proceedings of the IEEE Conference on Computer Vision and Pattern Recognition (CVPR), pp. 10695–10704. IEEE (2019)
12. Devlin, J., Gupta, S., Girshick, R., Mitchell, M., Zitnick, C.L.: Exploring nearest neighbor approaches for image captioning. arXiv preprint arXiv:1505.04467 (2015)
13. Donahue, J., et al.: Long-term recurrent convolutional networks for visual recognition and description. In: Proceedings of the IEEE Conference on Computer Vision and Pattern Recognition (CVPR), pp. 2625–2634. IEEE (2015)
14. Fan, A., Lewis, M., Dauphin, Y.: Hierarchical neural story generation. In: Proceedings of the 56th Annual Meeting of the Association for Computational Linguistics (ACL), pp. 889–898. Association for Computational Linguistics (2018)
15. Felzenszwalb, P.F., Huttenlocher, D.P.: Efficient graph-based image segmentation. Int. J. Comput. Vis. (IJCV) **59**(2), 167–181 (2004)
16. He, K., Zhang, X., Ren, S., Sun, J.: Deep residual learning for image recognition. In: Proceedings of the IEEE Conference on Computer Vision and Pattern Recognition (CVPR), pp. 770–778. IEEE (2016)
17. Hossain, M.Z., Sohel, F., Shiratuddin, M.F., Laga, H.: A comprehensive survey of deep learning for image captioning. ACM Comput. Surv. (CSUR) **51**(6), 1–36 (2019)
18. Shi, J., Malik, J.: Normalized cuts and image segmentation. IEEE Trans. Pattern Anal. Mach. Intell. (TPAMI) **22**(8), 888–905 (2000)
19. Johnson, J., Karpathy, A., Fei-Fei, L.: DenseCap: fully convolutional localization networks for dense captioning. In: Proceedings of the IEEE Conference on Computer Vision and Pattern Recognition (CVPR), pp. 4565–4574. IEEE (2016)
20. Karpathy, A., Fei-Fei, L.: Deep visual-semantic alignments for generating image descriptions. In: Proceedings of the IEEE Conference on Computer Vision and Pattern Recognition (CVPR), pp. 3128–3137. IEEE (2015)
21. Kim, D.J., Choi, J., Oh, T.H., Kweon, I.S.: Dense relational captioning: triple-stream networks for relationship-based captioning. In: Proceedings of the IEEE Conference on Computer Vision and Pattern Recognition (CVPR), pp. 6271–6280. IEEE (2019)
22. Kingma, D.P., Ba, J.: Adam: a method for stochastic optimization. In: International Conference on Learning Representations (ICLR) (2015)
23. Kipf, T.N., Welling, M.: Semi-supervised classification with graph convolutional networks. In: International Conference on Learning Representations (ICLR) (2016)
24. Klusowski, J.M., Wu, Y.: Counting motifs with graph sampling. In: COLT. Proceedings of Machine Learning Research, pp. 1966–2011 (2018)

25. Krishna, R., Zhu, Y., Groth, O., Johnson, J., Hata, K., Kravitz, J., Chen, S., Kalantidis, Y., Li, L.J., Shamma, D.A., et al.: Visual genome: connecting language and vision using crowdsourced dense image annotations. Int. J. Comput. Vis. (IJCV) **123**(1), 32–73 (2017)

26. Leacock, C., Miller, G.A., Chodorow, M.: Using corpus statistics and wordnet relations for sense identification. Comput. Linguist. **24**(1), 147–165 (1998)

27. Li, D., Huang, Q., He, X., Zhang, L., Sun, M.T.: Generating diverse and accurate visual captions by comparative adversarial learning. arXiv preprint arXiv:1804.00861 (2018)

28. Li, Y., Ouyang, W., Wang, X., Tang, X.: VIP-CNN: visual phrase guided convolutional neural network. In: Proceedings of the IEEE Conference on Computer Vision and Pattern Recognition (CVPR), pp. 1347–1356. IEEE (2017)

29. Li, Y., Ouyang, W., Zhou, B., Wang, K., Wang, X.: Scene graph generation from objects, phrases and region captions. In: Proceedings of the IEEE International Conference on Computer Vision (ICCV), pp. 1261–1270. IEEE (2017)

30. Lin, C.Y.: Rouge: a package for automatic evaluation of summaries. In: Text Summarization Branches Out, pp. 74–81 (2004)

31. Liu, F., Ren, X., Liu, Y., Wang, H., Sun, X.: simNet: stepwise image-topic merging network for generating detailed and comprehensive image captions. In: Proceedings of the 2018 Conference on Empirical Methods in Natural Language Processing (EMNLP), pp. 137–149. Association for Computational Linguistics (2018)

32. Lu, C., Krishna, R., Bernstein, M., Fei-Fei, L.: Visual relationship detection with language priors. In: Leibe, B., Matas, J., Sebe, N., Welling, M. (eds.) Computer Vision – ECCV 2016. Lecture Notes in Computer Science, vol. 9905, pp. 852–869. Springer, Cham (2016). https://doi.org/10.1007/978-3-319-46448-0_51

33. Lu, J., Xiong, C., Parikh, D., Socher, R.: Knowing when to look: adaptive attention via a visual sentinel for image captioning. In: Proceedings of the IEEE Conference on Computer Vision and Pattern Recognition (CVPR), pp. 375–383. IEEE (2017)

34. Lu, J., Yang, J., Batra, D., Parikh, D.: Neural baby talk. In: Proceedings of the IEEE Conference on Computer Vision and Pattern Recognition (CVPR), pp. 7219–7228. IEEE (2018)

35. Luo, R., Shakhnarovich, G.: Analysis of diversity-accuracy tradeoff in image captioning. arXiv preprint arXiv:2002.11848 (2020)

36. Ma, C.Y., Kalantidis, Y., AlRegib, G., Vajda, P., Rohrbach, M., Kira, Z.: Learning to generate grounded image captions without localization supervision. arXiv preprint arXiv:1906.00283 (2019)

37. Mao, J., Xu, W., Yang, Y., Wang, J., Huang, Z., Yuille, A.: Deep captioning with multimodal recurrent neural networks (M-RNN). In: International Conference on Learning Representations (ICLR) (2015)

38. Miller, G.A.: Wordnet: a lexical database for english. Commun. ACM **38**(11), 39–41 (1995)

39. Papineni, K., Roukos, S., Ward, T., Zhu, W.J.: BLEU: a method for automatic evaluation of machine translation. In: Proceedings of the 40th annual meeting of the Association for Computational Linguistics (ACL), pp. 311–318. Association for Computational Linguistics (2002)

40. Pennington, J., Socher, R., Manning, C.: GloVe: global vectors for word representation. In: Proceedings of the 2014 Conference on Empirical Methods in Natural Language Processing (EMNLP), pp. 1532–1543. Association for Computational Linguistics (2014)

41. Plummer, B.A., Wang, L., Cervantes, C.M., Caicedo, J.C., Hockenmaier, J., Lazeb-nik, S.: Flickr30k entities: collecting region-to-phrase correspondences for richer image-to-sentence models. In: Proceedings of the IEEE International Conference on Computer Vision (ICCV), pp. 2641–2649. IEEE (2015)
42. Radford, A., Wu, J., Child, R., Luan, D., Amodei, D., Sutskever, I.: Language models are unsupervised multitask learners. OpenAI, Technical report (2019)
43. Ren, S., He, K., Girshick, R., Sun, J.: Faster R-CNN: towards real-time object detection with region proposal networks. In: Advances in Neural Information Processing Systems (NeurIPS), pp. 91–99. Curran Associates, Inc. (2015)
44. Rennie, S.J., Marcheret, E., Mroueh, Y., Ross, J., Goel, V.: Self-critical sequence training for image captioning. In: Proceedings of the IEEE Conference on Computer Vision and Pattern Recognition (CVPR), pp. 7008–7024. IEEE (2017)
45. Rohrbach, A., Hendricks, L.A., Burns, K., Darrell, T., Saenko, K.: Object halluci-nation in image captioning. In: Proceedings of the 2018 Conference on Empirical Methods in Natural Language Processing (EMNLP), pp. 4035–4045. Association for Computational Linguistics (2018)
46. Selvaraju, R.R., et al.: Taking a hint: leveraging explanations to make vision and language models more grounded. In: Proceedings of the IEEE International Conference on Computer Vision (ICCV), pp. 2591–2600. IEEE (2019)
47. Shetty, R., Rohrbach, M., Anne Hendricks, L., Fritz, M., Schiele, B.: Speaking the same language: matching machine to human captions by adversarial training. In: Proceedings of the IEEE Conference on Computer Vision and Pattern Recognition (CVPR), pp. 4135–4144. IEEE (2017)
48. Song, J., Andres, B., Black, M.J., Hilliges, O., Tang, S.: End-to-end learning for graph decomposition. In: Proceedings of the IEEE International Conference on Computer Vision (ICCV), pp. 10093–10102. IEEE (2019)
49. Tang, S., Andres, B., Andriluka, M., Schiele, B.: Subgraph decomposition for multi-target tracking. In: Proceedings of the IEEE Conference on Computer Vision and Pattern Recognition (CVPR), pp. 5033–5041. IEEE (2015)
50. Vedantam, R., Lawrence Zitnick, C., Parikh, D.: CIDEr: consensus-based image description evaluation. In: Proceedings of the IEEE Conference on Computer Vision and Pattern Recognition (CVPR), pp. 4566–4575. IEEE (2015)
51. Vijayakumar, A.K., et al.: Diverse beam search for improved description of complex scenes. In: AAAI Conference on Artificial Intelligence (2018)
52. Vinyals, O., Toshev, A., Bengio, S., Erhan, D.: Show and tell: a neural image caption generator. In: Proceedings of the IEEE Conference on Computer Vision and Pattern Recognition (CVPR), pp. 3156–3164. IEEE (2015)
53. Wang, J., Madhyastha, P.S., Specia, L.: Object counts! bringing explicit detections back into image captioning. In: Proceedings of the 2018 Conference of the North American Chapter of the Association for Computational Linguistics (NAACL), pp. 2180–2193. Association for Computational Linguistics (2018)
54. Wang, L., Schwing, A., Lazebnik, S.: Diverse and accurate image description using a variational auto-encoder with an additive gaussian encoding space. In: Advances in Neural Information Processing Systems (NeurIPS), pp. 5756–5766. Curran Associates, Inc. (2017)
55. Xu, D., Zhu, Y., Choy, C.B., Fei-Fei, L.: Scene graph generation by iterative message passing. In: Proceedings of the IEEE Conference on Computer Vision and Pattern Recognition (CVPR), pp. 5410–5419. IEEE (2017)
56. Xu, K., et al.: Show, attend and tell: neural image caption generation with visual attention. In: International Conference on Machine Learning (ICML), pp. 2048–2057 (2015)

57. Xu, K., Hu, W., Leskovec, J., Jegelka, S.: How powerful are graph neural networks? In: International Conference on Learning Representations (ICLR) (2019)
58. Yang, J., Lu, J., Lee, S., Batra, D., Parikh, D.: Graph R-CNN for scene graph generation. In: Ferrari, V., Hebert, M., Sminchisescu, C., Weiss, Y. (eds.) Computer Vision – ECCV 2018. Lecture Notes in Computer Science, vol. 11205, pp. 690–706. Springer, Cham (2018). https://doi.org/10.1007/978-3-030-01246-5_41
59. Yang, L., Tang, K., Yang, J., Li, L.J.: Dense captioning with joint inference and visual context. In: Proceedings of the IEEE Conference on Computer Vision and Pattern Recognition (CVPR), pp. 2193–2202. IEEE (2017)
60. Yang, X., Tang, K., Zhang, H., Cai, J.: Auto-encoding scene graphs for image captioning. In: Proceedings of the IEEE Conference on Computer Vision and Pattern Recognition (CVPR), pp. 10685–10694. IEEE (2019)
61. Yao, T., Pan, Y., Li, Y., Mei, T.: Exploring visual relationship for image captioning. In: Ferrari, V., Hebert, M., Sminchisescu, C., Weiss, Y. (eds.) Computer Vision – ECCV 2018. Lecture Notes in Computer Science, vol. 11218, pp. 711–727. Springer, Cham (2018). https://doi.org/10.1007/978-3-030-01264-9_42.
62. Yin, X., Ordonez, V.: Obj2Text: generating visually descriptive language from object layouts. In: Proceedings of the 2017 Conference on Empirical Methods in Natural Language Processing (EMNLP), pp. 177–187. Association for Computational Linguistics (2017)
63. You, Q., Jin, H., Wang, Z., Fang, C., Luo, J.: Image captioning with semantic attention. In: Proceedings of the IEEE Conference on Computer Vision and Pattern Recognition (CVPR), pp. 4651–4659. IEEE (2016)
64. Zellers, R., Yatskar, M., Thomson, S., Choi, Y.: Neural motifs: scene graph parsing with global context. In: Proceedings of the IEEE Conference on Computer Vision and Pattern Recognition (CVPR), pp. 5831–5840. IEEE (2018)
65. Zhang, H., Kyaw, Z., Chang, S.F., Chua, T.S.: Visual translation embedding network for visual relation detection. In: Proceedings of the IEEE Conference on Computer Vision and Pattern Recognition (CVPR), pp. 5532–5540. IEEE (2017)
66. Zhou, L., Kalantidis, Y., Chen, X., Corso, J.J., Rohrbach, M.: Grounded video description. In: Proceedings of the IEEE Conference on Computer Vision and Pattern Recognition (CVPR), pp. 6578–6587. IEEE (2019)

Symbiotic Adversarial Learning
for Attribute-Based Person Search

Yu-Tong Cao, Jingya Wang,
and Dacheng Tao[✉]

UBTECH Sydney AI Centre, School of Computer Science, Faculty of Engineering,
The University of Sydney, Darlington, NSW 2008, Australia
ycao5602@unisydney.edu.au, {jingya.wang,dacheng.tao}@sydney.edu.au

Abstract. Attribute-based person search is in significant demand for applications where no detected query images are available, such as identifying a criminal from witness. However, the task itself is quite challenging because there is a huge modality gap between images and physical descriptions of attributes. Often, there may also be a large number of unseen categories (attribute combinations). The current state-of-the-art methods either focus on learning better cross-modal embeddings by mining only seen data, or they explicitly use generative adversarial networks (GANs) to synthesize unseen features. The former tends to produce poor embeddings due to insufficient data, while the latter does not preserve intra-class compactness during generation. In this paper, we present a symbiotic adversarial learning framework, called SAL. Two GANs sit at the base of the framework in a symbiotic learning scheme: one synthesizes features of unseen classes/categories, while the other optimizes the embedding and performs the cross-modal alignment on the common embedding space. Specifically, two different types of generative adversarial networks learn collaboratively throughout the training process and the interactions between the two mutually benefit each other. Extensive evaluations show SAL's superiority over nine state-of-the-art methods with two challenging pedestrian benchmarks, PETA and Market-1501. The code is publicly available at: https://github.com/ycao5602/SAL.

Keywords: Person search · Cross-modal retrieval · Adversarial learning

1 Introduction

The goal with "person search" is to find the same person in non-overlapping camera views at different locations. In surveillance analysis, it is a crucial tool

---

Y.-T. Cao and J. Wang are equal contribution.

Electronic supplementary material The online version of this chapter (https://doi.org/10.1007/978-3-030-58568-6_14) contains supplementary material, which is available to authorized users.

© Springer Nature Switzerland AG 2020
A. Vedaldi et al. (Eds.): ECCV 2020, LNCS 12359, pp. 230–247, 2020.
https://doi.org/10.1007/978-3-030-58568-6_14

for public safety. To date, the most common approach to person search has been to take one detected image captured from a surveillance camera and use it as a query [3,14,25,26,40,45,54,58]. However, this is not realistic in many real-world applications – for example, where a human witness has identified the criminal, but no image is available.

Attributes, such as gender, age, clothing or accessories, are more natural to us as searchable descriptions, and these can be used as soft biometric traits to search for in surveillance data [16,22,28,33]. Compared to the queries used in image-based person search, these attributes are also much easier to obtain. Further, semantic descriptions are more robust than low-level visual representations, where changes in viewpoint or diverse camera conditions can be problematic. Recently, a few studies have explored sentence-based person search [24,26]. Although this approach provides rich descriptions, unstructured text tends to introduce noise and redundancy during modeling. Attributes descriptions, on the other hand, are much cheaper to collect, and they inherently have a robust and relatively independent ability to discriminate between persons. As such, attribute descriptions have the advantage of efficiency over sentence descriptions in person search tasks.

Unfortunately, applying a cross-modal, attribute-based person search to real-world surveillance images is very challenging for several reasons. (1) There is a huge semantic gap between visual and textual modalities (i.e., attribute descriptions). Attribute descriptions are of lower dimensionality than visual data, e.g., tens versus thousands, and they are very sparse. Hence, in terms of data capacity, they are largely inferior to visual data. From performance comparisons between single-modal retrieval (image-based person search) and cross-modal retrieval (attribute-based person search) using current state-of-the-art methods, there is still a significant gap between the two, e.g., mAP 84.0% [34] vs. 24.3% [7] on Market-1501 [27,56]. (2) Most of the training and testing classes (attribute combinations) are non-overlapping, which leads to zero-shot retrieval problems – a very challenging scenario to deal with [7]. Given an attribute-style query, model aims to have more capacity to search an unseen class given a query of attributes. (3) Compared to the general zero-shot learning settings one might see in classification tasks, surveillance data typically has large intra-class variations and inter-class similarities with only a small number of available samples per class (Fig. 4), e.g., $\sim$7 samples per class in PETA [5] and $\sim$26 samples per class in Market-1501 compared with $\sim$609 samples per class in AWA2 [51]. One category (i.e., general attribute combinations that are not linked to a specific person) may include huge variations in appearance – just consider how many dark-haired, brown-eyed people you know and how different each looks. Also, the inter-class distance between visual representations from different categories can be quite small considering fine-grained attributes typically become ambiguous in low resolution with motion blur.

One general idea for addressing this problem is to use cross-modal matching to discover a common embedding space for the two modalities. Most existing methods focus on representation learning, where the goal is to find pro-

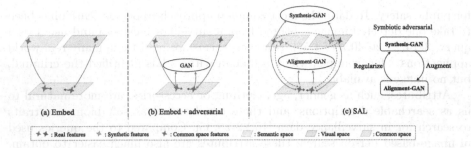

Fig. 1. Three categories of cross-modal retrieval with common space learning: (a) The embedding model that projects data from different modalities into common feature space. (b) The embedding model with common space adversarial alignment. (c) The proposed SAL that jointly considers feature-level data augmentation and cross-modal alignment by two GANs.

jections of data from different modalities and combine them in a common feature space. This way, the similarity between the two can be calculated directly [10,12,48] (Fig. 1(a)). However, these approaches typically have a weaker modality-invariant learning ability and, therefore, weaker performance in cross-modal retrieval. More recently, some progress has been made with the development of GANs [11]. GANs can better align the distributions of representations across modalities in a common space [44], plus they have been successfully applied to attribute-based person search [53]. (Fig. 1(b)). There are still some bridges to cross, however. With only a few samples, only applying cross-modal alignment to a high-level common space may mean the model fails to capture variances in the feature space. Plus, the cross-modal alignment probably will not work for unseen classes. Surveillance data has all these characteristics, so all of these problems must be overcome.

To deal with unseen classes, some recent studies on zero-shot learning have explicitly used GAN-based models to synthesize those classes [9,19,30,63]. Compared to zero-shot classification problems, our task is cross-modal retrieval, which requires learning a more complex search space with a finer granularity. As our experiments taught us, direct generation without any common space condition may reduce the intra-class compactness.

Hence, in this work, we present a fully-integrated symbiotic adversarial learning framework for cross-modal person search, called SAL. Inspired by symbiotic evolution, where two different organisms cooperate so both can specialize, we jointly explore the feature space synthesis and common space alignment with separate GANs in an integrated framework at different scales (Fig. 1(c)). The proposed SAL mainly consists of two GANs with interaction: (1) A synthesis-GAN that generates synthetic features from semantic representations to visual representations, and vice versa. The features are conditioned on common embedding information so as to preserve very fine levels of granularity. (2) An alignment-GAN that optimizes the embeddings and performs cross-modal alignment on the common embedding space. In this way, one GAN augments the data with

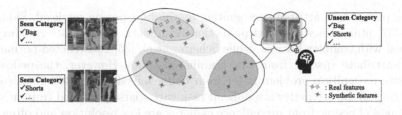

Fig. 2. An illustration of feature space data augmentation with a synthesized unseen class.

synthetic middle-level features, while the other uses those features to optimize the embedding framework. Meanwhile, a new regularization term, called common space granularity-consistency loss, forces the cross-modal generated representations to be consistent with their original representations in the high-level common space. These more reliable embeddings further boost the quality of the synthetic representations. To address zero-shot learning problems when there is no visual training data for an unseen category, we use new categories of combined attributes to synthesize visual representations for augmentation. An illustration of feature space data augmentation with a synthesized unseen class is shown in Fig. 2.

In summary, this paper makes the following main **contributions** to the literature:

- We propose a fully-integrated framework called symbiotic adversarial learning (SAL) for attribute-based person search. The framework jointly considers feature-level data augmentation and cross-modal alignment by two GANs at different scales. Plus, it handles cross-model matching, few-shot learning, and zero-shot learning problems in one unified approach.
- We introduce a symbiotic learning scheme where two different types of GANs learn collaboratively and specially throughout the training process. By generating qualified data to optimise the embeddings, SAL can better align the distributions in the common embedding space, which, in turn, yields superior cross-modal search results.
- Extensive evaluations on the PETA [5] and Market-1501 [56] datasets demonstrate the superiority of SAL at attribute-based person search over nine state-of-the-art attribute-based models.

2   Related Works

Attribute-Based Person Search. In recent years, semantic attribute recognition of pedestrians has drawn increasing attention [6,21,23,46,55] and it has been extensively exploited for image-based person search as a middle-level feature [20,27,38,39,47]. Some studies on exploit attribute-based person search for cross-modal retrieval [35,37,43,53]. Early attribute-based retrieval methods intuitively rely on attribute prediction. For example, Vaquero et al. [43] proposed a human parts based attribute recognition method for person retrieval. Siddiquie et al. [37] utilized a structured prediction framework to integrate ranking and retrieval with complex queries, while Scheirer et al. [35] constructed normalized "multi-attribute spaces" from raw classifier outputs. However, the pedestrian attribute recognition problem is far from being solved [23,29,46]. Consistently predicting all the attributes is a difficult task with sparsely-labeled training data: the images of people from surveillance cameras are low resolution and often contain motion blur. Also the same pedestrian would rarely be captured in the same pose. These "imperfect attributes" are significantly reducing the reliability of existing models. Shi et al. [36] suggested transfer learning as a way to overcome the limited availability of costly labeled surveillance data. They chose fashion images with richly labeled captions as the source domain to bridge the gap between unlabeled pedestrian surveillance images. They were able to produce semantic representations for a person search without annotated surveillance data, but the retrieval results were not as good as the supervised/semi-supervised methods, which raises a question of cost versus performance. [53] posed attribute-based person search as a cross-modality matching problem. They applied an adversarial leaning method to align the cross-modal distributions in a common space. However, their model design did not extend to the unseen class problem or few-shot learning challenges. Dong et al. [7] formulated a hierarchical matching model that fuses the similarity between global category-level embeddings and local attribute-level embeddings. Although they consider a zero-shot learning paradigm in the approach, the main disadvantage of this method is that it lacks the ability to synthesize visual representations of an unseen category. Therefore, it does not successfully handle unseen class problem for cross-modal matching.

Cross-Modal Retrieval. Our task of searching for people using attribute descriptions is closely related to studies on cross-modal retrieval as both problems require the attribute descriptions to be aligned with image data. This is a particularly relevant task to text-image embedding [26,41,44,49,52,57], where canonical correlation analysis (CCA) [12] is a common solution for aligning two modalities by maximizing their correlation. With the rapid development of deep neural networks (DNN), a deep canonical correlation analysis (DCCA) model has since been proposed based on the same insight [1]. Yan et al. [52] have subsequently extended the idea to an image-text matching method based on DCCA. Beyond these core works, a variety of cross-modal retrieval methods have been proposed for different ways of learning a common space. Most use category-level (categories) labels to learn common representations [44,53,57,57,63]. However, what might be fine-grained attribute representations often lose granularity, and semantic categories are all treated as being the same distance apart. Several more recent works are based on using a GAN [11] for cross-modal alignment in the common subspace [42,44,53,63]. The idea is intuitive since GANs have been proved to be powerful tools for distribution alignment [15,42]. Further, they have produced some impressive results in image translation [62], domain adaptation [15,42] and so on. However, when only applied to a common space, conventional adversarial learning may fail to model data variance where there are only a few samples and, again, the model design ignores the zero-shot learning challenge.

3 Symbiotic Adversarial Learning (SAL)

Problem Definition Our deep model for attribute-based person search is based on the following specifications. There are N labeled training images $\{x_i, a_i, y_i\}_{i=1}^N$ available in the training set. Each image-level attribute description $a_i = [a_{(i,1)}, \ldots, a_{(i,n_{attr})}]$ is a binary vector with n_{attr} number of attributes, where 0 and 1 indicate the absence and presence of the corresponding attribute. Images of people with the same attribute description are assigned to a unique category so as to derive a category-level label, specifically $y_i \in \{1, ..., M\}$ for M categories. The aim is to find matching pedestrian images from the gallery set $\mathcal{X}_{gallery} = \{x_j\}_{j=1}^G$ with G images given an attribute description a_q from the query set.

3.1 Multi-modal Common Space Embedding Base

One general solution for cross-modal retrieval problems is to learn a common joint subspace where samples of different modalities can be directly compared to each other. As mentioned in our literature review, most current approaches use category-level labels (categories) to generate common representations [44,53,63]. However, these representations never manage to preserve the full granularity of finely-nuanced attributes. Plus, all semantic categories are treated as being the same distance apart when, in reality, the distances between different semantic categories can vary considerably depending on the similarity of their attribute descriptions.

Thus, we propose a common space learning method that jointly considers the global category and the local fine-grained attributes. The common space embedding loss is defined as:

$$L_{embed} = L_{cat} + L_{att}. \tag{1}$$

The category loss utilizing the Softmax Cross-Entropy loss function for image embedding branch is defined as:

$$L_{cat} = -\sum_{i=1}^N \log\left(p_{cat}(x_i, y_i)\right), \tag{2}$$

where $p_{cat}(x_i, y_i)$ specifies the predicted probability on the ground truth category y_i of the training image x_i.

To make the common space more discriminative for fine-trained attributes, we use the Sigmoid Cross-Entropy loss function which considers all m attribute classes:

$$L_{att} = -\sum_{i=1}^N \sum_{j=1}^m \left(a_{(i,j)} \log\left(p_{att}^{(j)}(x_i)\right) + (1 - a_{(i,j)}) \log\left(1 - p_{att}^{(j)}(x_i)\right)\right), \tag{3}$$

where $a_{(i,j)}$ and $p_{att}^{(j)}(\boldsymbol{x}_i)$ define the ground truth label and the predicted classification probability on the j-th attribute class of the training image $\boldsymbol{x}_i$, i.e. $\boldsymbol{a}_i = [a_{(i,1)}, \cdots, a_{(i,m)}]$ and $\boldsymbol{p}_{att} = [p_{att}^{(1)}(\boldsymbol{x}_i), \cdots, p_{att}^{(m)}(\boldsymbol{x}_i)]$. Similar loss functions of Eqs. (2) and (3) are applied to the attribute embedding branch as well.

The multi-modal common space embedding can be seen as our base, named as *Embed* (Fig. 1(a)). A detailed comparison is shown in Table 2.

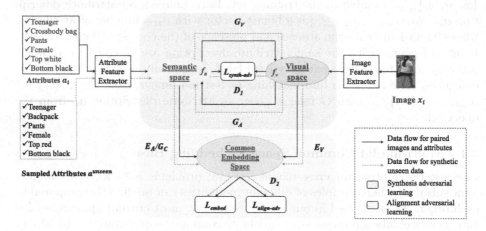

Fig. 3. An overview of the proposed SAL Framework.

3.2 Middle-Level Granularity-Consistent Cycle Generation

To address the challenge of insufficient data and bridge the modality gap between middle-level visual and semantic representations, we aim to generate synthetic features from the semantic representations to supplement the visual representations and vice versa. The synthetic features are conditioned on common embedding information so as to preserve very fine levels of granularity for retrieval. The middle-level representations are denoted as f_v for visual and f_a for attribute. More detail is provided in Fig. 3.

As paired supervision are available for seen categories during generation, conventional unsupervised cross-domain generation methods, e.g. CycleGAN [62], DiscoGAN [17] are not perfect suit. To better match joint distributions, we consider single discriminator distinguishes whether the paired data $(f_a, f_v))$ is from a real feature distribution $p(f_a, f_v)$ or not. This is inspired by Triple Generative Adversarial Networks [4]. The TripleGAN works on semi-supervised classification task with a generator, a discriminator and a classifier while our synthesis-GAN consists of two generators and a discriminator. As shown in Fig. 3, our synthesis-GAN consists of three main components: Generator G_A synthesizes semantic representations from visual representations; Generator G_V synthesizes visual representation from semantic representations, and a discriminator D_1 focused on identifying synthetic data pairs. Given that semantic representations are rather more sparse than visual representations, we add a noise vector $z \sim \mathcal{N}(0,1)$ sampled from a Gaussian distribution for G_V to get variation in visual feature generation. Thus, G_V can be regarded as a one-to-many matching from a sparse semantic space to a rich visual space. This accords with the one-to-many relationship between attributes and images. The training process is formulated as a minmax game between three players - G_V, G_A and D_1 - where, from a game theory perspective, both generators can achieve their optima. The adversarial training scheme for middle-level cross-modal feature generation can, therefore, be formulated as:

$$
\begin{aligned}
L_{\text{gan1}}(G_A, G_V, D_1) &= \mathbb{E}_{(f_a, f_v) \sim p(f_a, f_v))}[log(D_1(f_a, f_v))] \\
&+ \frac{1}{2}\mathbb{E}_{f_a \sim p(f_a)}[log(1 - D_1(f_a, \widetilde{f_v}))] + \frac{1}{2}\mathbb{E}_{f_v \sim p(f_v)}[log(1 - D_1(\widetilde{f_a}, f_v))].
\end{aligned}
\tag{4}
$$

The discriminator D_1 is fed with three types of input pairs: (1) The fake input pairs $(\widetilde{f_a}, f_v)$, where $\widetilde{f_a} = G_A(f_v)$. (2) The fake input pairs $(f_a, \widetilde{f_v})$, where $\widetilde{f_v} = G_V(f_a, z))$. (3) The real input pairs $(f_a, f_v) \sim p(f_a, f_v)$.

To constrain the generation process to only produce representations that are close to the real distribution, we assume the synthetic visual representation can be generated back to the original semantic representation, which is inspired by the CycleGAN structure [62]. Given this is a one-to-many matching problem as mentioned above, we flex the ordinary two-way cycle consistency loss to a one-way cycle consistency loss (from the semantic space to the visual space and back again):

$$
L_{\text{cyc}}(G_A, G_V) = \mathbb{E}_{f_a \sim p(f_a)}[||G_A(G_V(f_a, z)) - f_a||_2].
\tag{5}
$$

Furthermore, feature generation is conditioned on embeddings in the high-level common space, with the aim of generating more meaningful representations that preserve as much granularity as possible. This constraint is a new regularization term that we call "the common space granularity-consistency loss", formulated as follows:

$$
\begin{aligned}
L_{\text{consis}}(G_A, G_V) &= \mathbb{E}_{f_v \sim p(f_v)}[||E_A(\widetilde{f_a}) - E_V(f_v)||_2] + \mathbb{E}_{f_a \sim p(f_a)}[||E_V(\widetilde{f_v}) - E_A(f_a)||_2] \\
&+ \mathbb{E}_{(f_a, f_v) \sim p(f_a, f_v)}[||E_A(\widetilde{f_a}) - E_A(f_a)||_2] + \mathbb{E}_{(f_a, f_v) \sim p(f_a, f_v)}[||E_V(\widetilde{f_v}) - E_V(f_v)||_2],
\end{aligned}
\tag{6}
$$

where E_A and E_V stand for the common space encoders for attribute features and visual features respectively.

Lastly, the full objective of synthesis-GAN is:

$$
L_{\text{synth-adv}} = L_{\text{gan1}} + L_{\text{cyc}} + L_{\text{consis}}.
\tag{7}
$$

3.3 High-Level Common Space Alignment with Augmented Adversarial Learning

On the high-level common space which is optimized by Eq. (1), we introduce the second adversarial loss for cross-modal alignment, making the common space more modality-invariant. Here, E_A works as a synchronous common space encoder and generator. Thus, we can define E_A as G_C in this adversarial loss:

$$L_{gan2}(G_C, D_2) = \mathbb{E}_{f_v \sim p(f_v)}[log D_2(E_V(f_v))] \\ + \mathbb{E}_{f_a \sim p(f_a)}[log(1 - D_2(E_A(f_a)))]. \tag{8}$$

Benefit from the middle-level GAN that bridges the semantic and visual spaces, we can access the augmented data, $\widetilde{f}_a$ and $\widetilde{f}_v$, to optimise the common space, where $\widetilde{f}_a = G_A(f_v)$ and $\widetilde{f}_v = G_V(f_a, z)$. Thus the augmented adversarial loss is applied for cross-modal alignment by:

$$L_{aug1}(G_C, D_2) = \mathbb{E}_{f_a \sim p(f_a)}[log D_2(E_V(\widetilde{f}_v))] \\ + \mathbb{E}_{f_v \sim p(f_v)}[log(1 - D_2(E_A(\widetilde{f}_a)))]. \tag{9}$$

And can be used to optimize the common space by:

$$L_{aug2}(E_A, E_V) = L_{embed}(\widetilde{f}_a) + L_{embed}(\widetilde{f}_v). \tag{10}$$

The total augmented loss from the synthesis-GAN interaction is calculated by:

$$L_{aug} = L_{aug1} + L_{aug2}. \tag{11}$$

The final augmented adversarial loss for alignment-GAN is defined as:

$$L_{align\text{-}adv} = L_{gan2} + L_{aug}. \tag{12}$$

3.4 Symbiotic Training Scheme for SAL

Algorithm 1 summarizes the training process. As stated above, there are two GANs learn collaboratively and specially throughout the training process. The synthesis-GAN (including G_A, G_V and D_1) aims to build a bridge of semantic and visual representations in the middle-level feature space, and to synthesize features from both to each other. The granularity-consistency loss was further proposed to constrain the generation that conditions on high-level embedding information. Thus, the high-level alignment-GAN (including E_A, E_V and D_2) that optimizes the common embedding space benefits the synthesis-GAN with the better common space constrained by Eq. (6). Similarly, while the alignment-GAN attempts to optimize the embedding and perform the cross-modal alignment, the synthesis-GAN supports the alignment-GAN with more realistic and reliable augmented data pairs via Eqs. (9) and (10). Thus, the two GANs update iteratively with SAL's full objective becomes:

$$L_{SAL} = L_{embed} + L_{synth\text{-}gan} + L_{align\text{-}gan}. \tag{13}$$

Algorithm 1. Learning the SAL model.

Input: N labeled data pairs (x_i, a_i) from the training set;
Output: SAL attribute-based person search model;
for $t = 1$ **to** *max-iteration* **do**
    Sampling a batch of paired training data (x_i, a_i);
    Step 1: Multi-modal common space Embedding:
    Updating both image branch and attribute branch by common space
    embedding loss L_{embed} (Eq. (1));
    Step 2: Middle-Level Granularity-Consistent Cycle Genera-
    tion : Updating G_A, G_V and D_1 by $L_{synth\text{-}adv}$ (Eq. (7));
    Step 3: High-Level Common Space Alignment: Updating E_A,
    E_V and D_2 by $L_{align\text{-}adv}$ (Eq. (12));
    end for

Semantic Augmentation for Unseen Classes. To address the zero-shot problem, our idea is to synthesize visual representations of unseen classes from semantic representations. In contrast to conventional zero-shot settings with pre-defined category names, we rely on a myriad of attribute combinations instead. Further, this inspired us to also sample new attributes in the model design. Hence, during training, some new attribute combinations a^{unseen} are dynamically sampled in each iteration. The sampling for the binary attributes follows a Bernoulli distribution, and a 1-trial multinomial distribution for the multi-valued attributes (i.e., mutually-exclusive attributes) according to the training data probability. Once the attribute feature extraction is complete, f_a in the objective function is replaced with $[f_a, f_{a^{unseen}}]$, $f_{a^{unseen}}$ is used to generate synthetic visual feature via G_V. The synthetic feature pairs are then used for the final common space cross-modal learning.

4   Experiments

Datasets. To evaluate SAL with an attribute-based person search task, we used two widely adopted pedestrian attribute datasets[1]: (1) The ***Market-1501 attribute*** dataset [56] which the Market-1501 dataset annotated with 27 attributes for each identity [27] - see Fig. 4. The training set contains an average of ~26 labeled images in each category. During testing, 367 out of 529 (**69.4%**) were from **unseen** categories. (2) The ***PETA*** dataset [5] consists of 19,000 pedestrian images collected from 10 small-scale datasets of people. Each image is annotated with 65 attributes. Following [53], we labeled the images with categories according to their attributes, then split the dataset into a training set and a gallery set by category. In the training set, there was on average ~7 labeled images in each category, and all 200 categories in the gallery set (**100%**) were **unseen**.

Performance Metric. The two metrics used to evaluate performance were the cumulative matching characteristic (CMC) and mean Average Precision (mAP).

---

[1] The DukeMTMC dataset is not publicly available.

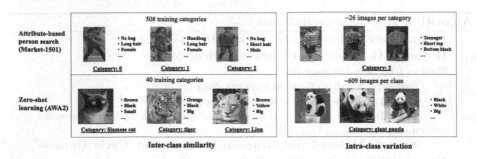

Fig. 4. Example images from the person search Market-1501 [56] dataset and compared with the general zero-shot AWA2 [51] dataset.

We followed [31] and computed the CMC on each rank k as the probe cumulative percentage of truth matches appearing at ranks $\leq k$. mAP measures the recall of multiple truth matches and was computed by first computing the area under the Precision-Recall curve for each probe, then calculating the mAP over all probes.

Table 1. Attribute-based person search performance evaluation. Best results are shown in **bold**. The second-best results are underlined.

Metric (%)		Market-1501 Attributes				PETA			
Model	Reference	mAP	rank1	rank5	rank10	mAP	rank1	rank5	rank10
DeepCCA [1]	ICML'13	17.5	30.0	50.7	58.1	11.5	14.4	20.8	26.3
DeepMAR [23]	ACPR'15	8.9	13.1	24.9	32.9	12.7	17.8	25.6	31.1
DeepCCAE [50]	ICML'15	9.7	8.1	24.0	34.6	14.5	14.2	22.1	30.0
2WayNet [8]	CVPR'17	7.8	11.3	24.4	31.5	15.4	23.7	38.5	41.9
CMCE [24]	ICCV'17	22.8	35.0	51.0	56.5	26.2	31.7	39.2	48.4
ReViSE [41]	ICCV'17	17.7	24.2	45.2	57.6	31.1	30.5	<u>57.0</u>	61.5
MMCC [9]	ECCV'18	22.2	34.9	<u>58.7</u>	<u>70.2</u>	<u>33.9</u>	33.5	<u>57.0</u>	<u>69.0</u>
AAIPR [53]	IJCAI'18	20.7	40.3	49.2	58.6	27.9	<u>39.0</u>	53.6	62.2
AIHM [7]	ICCV'19	<u>24.3</u>	<u>43.3</u>	56.7	64.5	-	-	-	-
SAL (Ours)		**29.8**	**49.0**	**68.6**	**77.5**	**41.2**	**47.0**	**66.5**	**74.0**

Implementation Details. SAL was implemented based on Torchreid [59–61] in the Pytorch framework [32] with ResNet-50 [13] as the image feature extractor. We used fully connected (FC) layers to form the attribute feature extractor (64, 128, 256, 512), the middle-level generators (256, 128, 256, 512), and the encoders (512, 256, 128). Batch normalization and a ReLU nonlinear activation followed each of the first three layers with a Tanh (activation) before the output. The discriminators were also FC layers (512, 256, 1) with batch normalization and a leaky ReLU activation following each of the first two layers, and a Sigmoid activation prior to the output. After the common space embedding, the classifiers output with 1-FC layer. **Training process:** We first pre-trained the image branch and *Embed* model as a person search baseline (Fig. 3.1) until it converges. Then we trained the full SAL model for 60 epochs with a learning rate of 0.001 for the image branch and 0.01 for the attribute branch. We chose Adam as the training optimizer [18] with a batch size of 128. During testing, we calculated the cosine similarity between the query attributes and the gallery images in the common embedding space with 128-D deep feature representations for matching. **Training time:** It took 77 min for SAL to converge (26.3M parameters) compared to 70 min for a single adversarial model Embed+adv to converge (25.0M parameters). Both training processes were run on the same platform with 2 NVIDIA 1080Ti GPUs.

4.1 Comparisons to the State-of-the-Arts

The results of the search task appear in Table 1, showing SAL's performance in comparison to 9 state-of-the-art models across 4 types of methods. These are: (I) *attribute recognition* based method: (1) DeepMAR [23]. (II) *correlation* based methods: (2) Deep Canonical Correlation Analysis (DCCA) [1]; (3) Deep Canonically Correlated Autoencoders (DCCAE) [50]; (4) 2WayNet [8] (III) *common space embedding*: (5) Cross-modality Cross-entropy (CMCE) [24]; (6) ReViSE [41]; (7) Attribute-Image Hierarchical Matching (AIHM) [7]; (IV) *adversarial learning*: (8) Multi-modal Cycle-consistent (MMCC) model [9]; (9) Adversarial Attribute-Image Person Re-ID (AAIPR) [53].

From the results, we made the following observations: (1) SAL outperformed all 9 state-of-the-art models on both datasets in terms of mAP. On the Market-1501 dataset, the improvement over the second best scores were 5.5% (29.8 v 24.3) and 5.7% (49.0 v 43.3) for rank1. On the PETA dataset, the improvement was 7.3% (41.2 v 33.9) and 8.0% (47.0 v 39.0) for rank1. This illustrates SAL's overall performance advantages in cross-modal matching for attribute-based person search. (2) Directly predicting the attributes using the existing recognition models and matching the predictions with the queries is not efficient (e.g., DeepMAR). This may be due to the relatively low prediction dimensions and the sparsity problems with attributes in semantic space. (3) Compared to the common space learning-based method (CMCE), the conventional correlation methods (e.g., DCCA, DCCAE, 2WayNet) witnessed relatively poor results. This demonstrates the power of common space learning with a common embedding space. It is worth mentioning that CMCE is specifically designed to search for people using natural language descriptions, yet SAL outperformed CMCE by 7.0% mAP/14.0% rank1 with Market-1501, and by 15.0% mAP/15.3% rank1 with PETA. (4) The adversarial model comparisons, AAIPR and MMCC, did not fare well against SAL, especially AAIPR, which utilizes single adversarial learning to align the common space. This directly demonstrates the advantages of our approach with symbiotic adversarial design compared to traditional adversarial learning methods. (5) Among the compared state-of-the-arts, MMCC, ReViSE and AIHM addressed *zero-shot problems*. Against MMCC, SAL outperformed by 7.6% mAP/14.1% rank1 on Market-1501 and 7.3% mAP/13.5% rank1 on

PETA. AIHM is the most recent state-of-the-art in this category of methods and SAL's performance improvement was 5.5% mAP/5.7% rank1 on Market-1501. This demonstrates the advantages of SAL's new regularization term, the common space granularity-consistency loss, for generating middle-level features.

Table 2. Component analysis of SAL on PETA dataset.

Metric (%)	mAP	rank1	rank5	rank10
Embed	31.3	34.0	57.0	64.5
Embed + adv	35.0	37.5	60.5	66.5
Embed + symb-adv	40.6	44.0	64.0	70.5
Embed + symb-adv + unseen(SAL)	41.2	47.0	66.5	74.0

Table 3. Effect of interactions between two GANs on PETA dataset.Effect of interactions between two GANs on PETA dataset.

Metric (%)	mAP	rank1	rank5	rank10
SAL - L_{aug}	35.4	38.0	60.0	69.0
SAL - L_{consis}	35.2	39.5	56.5	66.0
SAL (Full interaction)	41.2	47.0	66.5	74.0

4.2 Further Analysis and Discussions

To further evaluate the different components in the model, we conduct studies on the PETA dataset.

Component Analysis of SAL. Here, we compared: (1) the base embedding model (*Embed*, Fig. 1(a)), which comprises the multi-modal common space embedding base (Sect. 3.1) and is optimized by embedding loss (Eq. (1)) only; (2) the base embedding model plus single adversarial learning (*Embed + adv*, Fig. 1(b)), (3) the base embedding model plus our symbiotic adversarial learning (*Embed + symb-adv*, Fig. 1(c)), and (4) our full SAL model, which includes the attribute sampling for unseen classes. The results are shown in Table 2.

Compared to the *Embed* model, *Embed + adv* saw an improvement of 3.7% mAP/3.5% rank1, whereas *Embed + symb-adv* achieved an improvement of 9.3% mAP/10% rank1, and finally SAL (*Embed + symb-adv + unseen*) witnessed a significant improvement of 9.9% mAP/13.0% rank1. This is a clear demonstration of the benefits of jointly considering middle-level feature augmentation and high-level cross-modal alignment instead of only having common space cross-modal alignment (41.2% vs 35.0% mAP and 47.0% vs 37.5% rank1). We visualize the retrieved results in the supplementary material.

Effect of Interactions Between Two GANs. Our next test was designed to assess the influence of the symbiotic interaction, i.e., where the two GANs iteratively regularize and augment each other. Hence, we removed the interaction loss between the two GANs, which is the total augmented loss from Eq. (11) and the common space granularity-consistency loss from Eq. (6). Removing the augmented loss (SAL - L_{aug}) means the model no longer uses the augmented data. Removing the common space granularity-consistency loss (SAL - L_{consis}) means the middle-level generators are not conditioned on high-level common embedding information, which should reduce the augmented data quality. The results in Table 3 show that, without augmented data, SAL's mAP decreased by 5.8% and by 6.0% without the granularity-consistency loss as a regularizer.

Table 4. Comparing stage-wise training vs. symbiotic training scheme.

Metric (%)	mAP	rank1	rank5	rank10
SAL w/stage-wise training	35.0	41.0	58.0	65.0
SAL w/symbiotic training	41.2	47.0	66.5	74.0

Comparing Stage-Wise Training vs. Symbiotic Training Scheme. We wanted to gain a better understanding of the power of the symbiotic training scheme (Sect. 3.4). So, we replaced the iterative symbiotic training scheme (SAL w/symbiotic training) with a stage-wise training (SAL w/stage-wise training). The stage-wise training [2] breaks down the learning process into sub-tasks that are completed stage-by-stage. So during implementation, we first train the synthesis-GAN conditioned on the common embedding. Then for the next stage, we optimise the alignment-GAN on the common space with synthetic data. As shown in Table 4, there was a 6.2% drop in mAP using the stage-wise training, which endorses the merit of the symbiotic training scheme. During the symbiotic training, the synthesized augmentation data and the common space alignment iteratively boost each other's learning. A better common space constrains the data synthesis to generate better synthetic data. And, with better synthetic data for augmentation, a better common space can be learned.

5   Conclusion

In this work, we presented a novel symbiotic adversarial learning model, called SAL. SAL is an end-to-end framework for attribute-based person search. Two GANs sit at the base of the framework: one synthesis-GAN uses semantic representations to generate synthetic features for visual representations, and vice versa, while the other alignment-GAN optimizes the embeddings and performs cross-modal alignment on the common embedding space. Benefiting from the symbiotic adversarial structure, SAL is able to preserve finely-grained discriminative information and generate more reliable synthetic features to optimize the common embedding space. With the ability to synthesize visual representations of unseen classes, SAL is more robust to zero-shot retrieval scenarios, which are relatively common in real-world person search with diverse attribute descriptions. Extensive ablation studies illustrate the insights of our model designs. Further, we demonstrate the performance advantages of SAL over a wide range of state-of-the-art methods on two challenging benchmarks.

Acknowledgement. This research was supported by ARC FL-170100117, IH-180100002, LE-200100049.

Author contributions. Yu-Tong Cao, Jingya Wang : Equal contribution.

References

1. Andrew, G., Arora, R., Bilmes, J., Livescu, K.: Deep canonical correlation analysis. In: ICML (2013)
2. Barshan, E., Fieguth, P.: Stage-wise training: an improved feature learning strategy for deep models. In: Feature Extraction: Modern Questions and Challenges, pp. 49–59 (2015)
3. Chen, Y.C., Zhu, X., Zheng, W.S., Lai, J.H.: Person re-identification by camera correlation aware feature augmentation. IEEE TPAMI **40**(2), 392–408 (2018)
4. Chongxuan, L., Xu, T., Zhu, J., Zhang, B.: Triple generative adversarial nets. In: NIPS (2017)
5. Deng, Y., Luo, P., Loy, C.C., Tang, X.: Pedestrian attribute recognition at far distance. In: ACM MM. ACM (2014)
6. Deng, Y., Luo, P., Loy, C.C., Tang, X.: Learning to recognize pedestrian attribute. arXiv:1501.00901 (2015)
7. Dong, Q., Gong, S., Zhu, X.: Person search by text attribute query as zero-shot learning. In: Proceedings of the IEEE International Conference on Computer Vision, pp. 3652–3661 (2019)
8. Eisenschtat, A., Wolf, L.: Linking image and text with 2-way nets. In: CVPR (2017)
9. Felix, R., Kumar, V.B., Reid, I., Carneiro, G.: Multi-modal cycle-consistent generalized zero-shot learning. In: Ferrari, V., Hebert, M., Sminchisescu, C., Weiss, Y. (eds.) Computer Vision – ECCV 2018. Lecture Notes in Computer Science, vol. 11210, pp. 21–37. Springer, Cham (2018). https://doi.org/10.1007/978-3-030-01231-1_2
10. Feng, F., Wang, X., Li, R.: Cross-modal retrieval with correspondence autoencoder. In: ACM MM. ACM (2014)
11. Goodfellow, I., Pouget-Abadie, J., Mirza, M., Xu, B., Warde-Farley, D., Ozair, S., Courville, A., Bengio, Y.: Generative adversarial nets. In: NIPS (2014)
12. Hardoon, D.R., Szedmak, S., Shawe-Taylor, J.: Canonical correlation analysis an overview with application to learning methods. Neural Comput. **16**(12), 2639–2664 (2004)
13. He, K., Zhang, X., Ren, S., Sun, J.: Deep residual learning for image recognition. In: CVPR (2016)
14. Hermans, A., Beyer, L., Leibe, B.: In defense of the triplet loss for person re-identification. arXiv:1703.07737 (2017)
15. Hoffman, J., et al.: CyCADA: cycle-consistent adversarial domain adaptation. arXiv:1711.03213 (2017)

16. Jaha, E.S., Nixon, M.S.: Soft biometrics for subject identification using clothing attributes. In: IJCB. IEEE (2014)
17. Kim, T., Cha, M., Kim, H., Lee, J.K., Kim, J.: Learning to discover cross-domain relations with generative adversarial networks. In: Proceedings of the 34th International Conference on Machine Learning, vol. 70, pp. 1857–1865. JMLR. org (2017)
18. Kingma, D.P., Ba, J.: Adam: a method for stochastic optimization. arXiv:1412.6980 (2014)
19. Kumar Verma, V., Arora, G., Mishra, A., Rai, P.: Generalized zero-shot learning via synthesized examples. In: Proceedings of the IEEE Conference on Computer Vision and Pattern Recognition, pp. 4281–4289 (2018)
20. Layne, R., Hospedales, T.M., Gong, S.: Towards person identification and re-identification with attributes. In: Fusiello, A., Murino, V., Cucchiara, R. (eds.) Computer Vision – ECCV 2012, Workshops and Demonstrations. Lecture Notes in Computer Science, vol. 7583, pp. 402–412. Springer, Berlin, Heidelberg (2012). https://doi.org/10.1007/978-3-642-33863-2_40
21. Layne, R., Hospedales, T.M., Gong, S.: Attributes-based re-identification. In: Gong, S., Cristani, M., Yan, S., Loy, C. (eds.) Person Re-Identification. Advances in Computer Vision and Pattern Recognition, pp. 93–117. Springer, London (2014). https://doi.org/10.1007/978-1-4471-6296-4_5
22. Layne, R., Hospedales, T.M., Gong, S., Mary, Q.: Person re-identification by attributes. In: BMVC (2012)
23. Li, D., Chen, X., Huang, K.: Multi-attribute learning for pedestrian attribute recognition in surveillance scenarios. In: IAPR ACPR. IEEE (2015)
24. Li, S., Xiao, T., Li, H., Yang, W., Wang, X.: Identity-aware textual-visual matching with latent co-attention. In: ICCV (2017)
25. Li, W., Zhao, R., Xiao, T., Wang, X.: DeepREFId: deep filter pairing neural network for person re-identification. In: CVPR (2014)
26. Li, W., Zhu, X., Gong, S.: Person re-identification by deep joint learning of multi-loss classification. arXiv:1705.04724 (2017)
27. Lin, Y., Zheng, L., Zheng, Z., Wu, Y., Yang, Y.: Improving person re-identification by attribute and identity learning. arXiv:1703.07220 (2017)
28. Liu, C., Gong, S., Loy, C.C., Lin, X.: Person re-identification: what features are important? In: Fusiello, A., Murino, V., Cucchiara, R. (eds.) Computer Vision – ECCV 2012, Workshops and Demonstrations. Lecture Notes in Computer Science, vol. 7583, pp. 391–401. Springer, Berlin, Heidelberg. (2012). https://doi.org/10.1007/978-3-642-33863-2_39
29. Liu, X., et al.: HydraPlus-Net: attentive deep features for pedestrian analysis. In: ICCV (2017)
30. Long, Y., Liu, L., Shao, L., Shen, F., Ding, G., Han, J.: From zero-shot learning to conventional supervised classification: unseen visual data synthesis. In: Proceedings of the IEEE Conference on Computer Vision and Pattern Recognition, pp. 1627–1636 (2017)
31. Paisitkriangkrai, S., Shen, C., Van Den Hengel, A.: Learning to rank in person re-identification with metric ensembles. In: Proceedings of the IEEE Conference on Computer Vision and Pattern Recognition, pp. 1846–1855 (2015)
32. Paszke, A., et al.: Automatic differentiation in PyTorch. In: NIPS-W (2017)
33. Reid, D.A., Nixon, M.S., Stevenage, S.V.: Soft biometrics; human identification using comparative descriptions. IEEE TPAMI 36(6), 1216–1228 (2014)
34. Saquib Sarfraz, M., Schumann, A., Eberle, A., Stiefelhagen, R.: A pose-sensitive embedding for person re-identification with expanded cross neighborhood re-ranking. In: CVPR (2018)

35. Scheirer, W.J., Kumar, N., Belhumeur, P.N., Boult, T.E.: Multi-attribute spaces: calibration for attribute fusion and similarity search. In: CVPR. IEEE (2012)
36. Shi, Z., Hospedales, T.M., Xiang, T.: Transferring a semantic representation for person re-identification and search. In: CVPR (2015)
37. Siddiquie, B., Feris, R.S., Davis, L.S.: Image ranking and retrieval based on multi-attribute queries. In: CVPR. IEEE (2011)
38. Su, C., Yang, F., Zhang, S., Tian, Q., Davis, L.S., Gao, W.: Multi-task learning with low rank attribute embedding for person re-identification. In: ICCV (2015)
39. Su, C., Zhang, S., Xing, J., Gao, W., Tian, Q.: Deep attributes driven multi-camera person re-identification. In: Leibe, B., Matas, J., Sebe, N., Welling, M. (eds.) Computer Vision – ECCV 2016. Lecture Notes in Computer Science, vol. 9906, pp. 475–491. Springer, Cham (2016). https://doi.org/10.1007/978-3-319-46475-6_30
40. Sun, Y., Zheng, L., Deng, W., Wang, S.: SvdNet for pedestrian retrieval. In: ICCV (2017)
41. Tsai, Y.H.H., Huang, L.K., Salakhutdinov, R.: Learning robust visual-semantic embeddings. In: ICCV. IEEE (2017)
42. Tzeng, E., Hoffman, J., Saenko, K., Darrell, T.: Adversarial discriminative domain adaptation. In: CVPR (2017)
43. Vaquero, D.A., Feris, R.S., Tran, D., Brown, L., Hampapur, A., Turk, M.: Attribute-based people search in surveillance environments. In: WACV. IEEE (2009)
44. Wang, B., Yang, Y., Xu, X., Hanjalic, A., Shen, H.T.: Adversarial cross-modal retrieval. In: ACM MM. ACM (2017)
45. Wang, F., Zuo, W., Lin, L., Zhang, D., Zhang, L.: Joint learning of single-image and cross-image representations for person re-identification. In: CVPR (2016)
46. Wang, J., Zhu, X., Gong, S., Li, W.: Attribute recognition by joint recurrent learning of context and correlation. In: ICCV (2017)
47. Wang, J., Zhu, X., Gong, S., Li, W.: Transferable joint attribute-identity deep learning for unsupervised person re-identification. In: CVPR (2018)
48. Wang, K., He, R., Wang, W., Wang, L., Tan, T.: Learning coupled feature spaces for cross-modal matching. In: ICCV (2013)
49. Wang, L., Li, Y., Lazebnik, S.: Learning deep structure-preserving image-text embeddings. In: CVPR (2016)
50. Wang, W., Arora, R., Livescu, K., Bilmes, J.: On deep multi-view representation learning. In: ICML (2015)
51. Xian, Y., Lampert, C.H., Schiele, B., Akata, Z.: Zero-shot learning-a comprehensive evaluation of the good, the bad and the ugly. IEEE Trans. Pattern Anal. Mach. Intell. **41**(9), 2251–2265 (2018)
52. Yan, F., Mikolajczyk, K.: Deep correlation for matching images and text. In: CVPR (2015)
53. Yin, Z., et al.: Adversarial attribute-image person re-identification. In: IJCAI (7 2018)
54. Zhang, L., Xiang, T., Gong, S.: Learning a discriminative null space for person re-identification. In: CVPR (2016)
55. Zhao, X., Sang, L., Ding, G., Guo, Y., Jin, X.: Grouping attribute recognition for pedestrian with joint recurrent learning. In: IJCAI (2018)
56. Zheng, L., Shen, L., Tian, L., Wang, S., Wang, J., Tian, Q.: Scalable person re-identification: a benchmark. In: ICCV (2015)
57. Zheng, Z., Zheng, L., Garrett, M., Yang, Y., Shen, Y.D.: Dual-path convolutional image-text embedding with instance loss. arXiv:1711.05535 (2017)

58. Zhong, Z., Zheng, L., Cao, D., Li, S.: Re-ranking person re-identification with k-reciprocal encoding. In: CVPR (2017)
59. Zhou, K., Xiang, T.: Torchreid: A library for deep learning person re-identification in PyTorch. arXiv preprint arXiv:1910.10093 (2019)
60. Zhou, K., Yang, Y., Cavallaro, A., Xiang, T.: Learning generalisable omni-scale representations for person re-identification. arXiv preprint arXiv:1910.06827 (2019)
61. Zhou, K., Yang, Y., Cavallaro, A., Xiang, T.: Omni-scale feature learning for person re-identification. In: ICCV (2019)
62. Zhu, J.Y., Park, T., Isola, P., Efros, A.A.: Unpaired image-to-image translation using cycle-consistent adversarial networks. In: ICCV (2017)
63. Zhu, Y., Elhoseiny, M., Liu, B., Peng, X., Elgammal, A.: A generative adversarial approach for zero-shot learning from noisy texts. In: CVPR (2018)

Amplifying Key Cues
for Human-Object-Interaction Detection

Yang Liu[1]([✉]) [iD], Qingchao Chen[2] [iD], and Andrew Zisserman[1] [iD]

[1] Visual Geometry Group, University of Oxford, Oxford, UK
yangl@robots.ox.ac.uk
[2] Department of Engineering Science, University of Oxford, Oxford, UK

Abstract. Human-object interaction (HOI) detection aims to detect and recognise how people interact with the objects that surround them. This is challenging as different interaction categories are often distinguished only by very subtle visual differences in the scene. In this paper we introduce two methods to amplify key cues in the image, and also a method to combine these and other cues when considering the interaction between a human and an object. First, we introduce an encoding mechanism for representing the fine-grained spatial layout of the human and object (a subtle cue) and also semantic context (a cue, represented by text embeddings of surrounding objects). Second, we use plausible future movements of humans and objects as a cue to constrain the space of possible interactions. Third, we use a gate and memory architecture as a fusion module to combine the cues. We demonstrate that these three improvements lead to a performance which exceeds prior HOI methods across standard benchmarks by a considerable margin.

1 Introduction

Human-Object Interaction (HOI) detection—which focuses specifically on relations involving humans—requires not only to retrieve human and object locations but also to infer the relations between them. Thus, for a given image, the objective of HOI is to identify all triplets of the form *<human, verb, object>*. The ability to predict such triplets robustly is central to enabling applications in robotic manipulations [15] and surveillance event detection [1].

Driven by impressive progress on instance detection and recognition, there has been growing interest in the HOI detection problem. However, the majority of existing methods [9,30,38] first detect all human and object instances and then infer their pairwise relations using the appearance feature of the detected instances and their coarse layout (position of human and object boxes). Despite their general efficacy, the performance of prior work may still be limited by some particular design choices, which we discuss next.

Electronic supplementary material The online version of this chapter (https:// doi.org/10.1007/978-3-030-58568-6_15) contains supplementary material, which is available to authorized users.

A. Vedaldi et al. (Eds.): ECCV 2020, LNCS 12359, pp. 248–265, 2020.
https://doi.org/10.1007/978-3-030-58568-6_15

Top: <person, walk with, bicycle>   Top: <person, eat, cake>   Top: <person, catch, frisbee>
Bottom: <person, ride, bicycle>   Bottom: <person, cut, cake>   Bottom: <person, throw, frisbee>

Fig. 1. (Left): Interactions with similar spatial layouts can be resolved through *fine-grained spatial information*. (Centre): *Global and local context* encode the scene and other local objects to provide strong clues for the interaction taking place; (Right): *Plausible motion* estimation distinguishes between interactions for which dynamics play an important role.

First, although recent works have sought to introduce some fine-grained spatial configuration descriptors and context cues from the whole image into the HOI detection, the encoding mechanisms have limitations. Specifically, (1) Some approaches [14,24,38,45] use human pose to distinguish the fine-grained relation (going beyond the standard human and object coarse boxes). However, these approaches encode human pose via key-point estimation (plus post-processing), which is problematic as it loses boundary and shape information. For example in Fig. 1(left), encoding the fine-grained spatial information as key points exhibits ambiguity when distinguishing the differences between 'riding a bicycle' and 'walking with a bicycle'. We argue that the boundaries of both human parts and objects should be encoded explicitly, due to their critical ability to reveal interaction boundaries and support inference of relations. Thus we improve the encoding mechanism for the fine-grained information by leveraging the fine-grained human parsing and object semantic segmentation masks, to better capture the geometric relations between them. (2) Some approaches use the visual appearance feature from other image regions or the whole image as the auxiliary context information. However, there are a limited number of triplets in existing HOI datasets—this is insufficient to capture the full intra-class appearance variations of relationships (making it harder to generalise). We draw inspiration from classical recognition techniques (e.g. the use of context for detection [7]) and argue that the semantic categories of other objects present in the surrounding neighbourhood of the candidate instance pair (local context) and the scene category (global context) provide valuable cues for distinguishing between different interactions, but the detailed visual appearance of them is often not crucial for HOI detection. For instance, as shown in Fig. 1(middle), the surrounding neighbourhood in the 'eating a cake' category will likely comprise a spoon-like tool, whereas for the 'cutting a cake' category, it is a knife-like tool. But the colour and design of the spoon/knife do not provide useful cues when inferring its relation with the human. Instead of using visual appearance features directly, we encode categories via a semantic embedding (word2vec). This enables the model to leverage language priors to capture possible co-occurrence and affordance relations between objects and predicates. As we show through careful ablation studies in Sect. 4, these mechanisms for encoding fine-grained spatial configuration and contextual cues bring consistent improvements to HOI detection performance, highlighting the importance of studying these choices.

Second, *plausible motion*—the set of probable movements most likely to follow a static image—is not currently accounted for when detecting human object interactions. Nevertheless, humans can trivially enumerate plausible future movements (what may happen next in an image) and characterise their relative likelihood.

The inference of plausible motions brings two benefits: the first is saliency—it provides a natural attention over the key object and human body parts present in an image; the second is that it constrains the space of relations to the subset that is consistent with these dynamics. For example in Fig. 1(right), it is clear that the object and arm highlighted by motion are concerned with throwing or catching the frisbee, not concerned with eating or writing on it. Furthermore, if the estimation of the direction is also correct then that would distinguish whether the person is throwing or catching the frisbee. Note that while the plausible motion can elucidate a salient signal for human object interaction detection, it can be difficult to learn directly from the image alone. We benefit here from recent work on motion hallucination [10], that has learnt to predict local optical flow from a static image to reveal plausible future movement by identifying which regions of pixels will move (together with their velocity) in the instant following image capture. To the best of our knowledge, this work represents the first study of the utility of plausible motion as an additional cue for HOI detection.

In this paper, we aim to tackle the challenges described above with a unified framework that amplifies important cues for HOI detection (as shown in Fig. 2). We design a novel multi-expert fusion module, where different features (i.e., plausible motion, enhanced fine-grained spatial configuration and context cues) are viewed as cooperative experts to infer the human object interaction. As different cues and their relationships will have different contribution for detecting the human object interaction, we use the gate and memory mechanism to fuse the available cues sequentially, select the discriminative information and gradually generate the representation for the whole scene step by step. By doing so, the final representation is more discriminate than those from existing methods that lack a reasoning mechanism, and this leads to better HOI detection performance.

The contributions of this work are summarised as follows:

(1) We propose a mechanism for amplifying fine-grained spatial layout and contextual cues, to better capture the geometric relations and distinguish the subtle difference between relation categories.

(2) We are the first to explore the utility of the plausible motion estimation (which regions of pixels will move) as an additional cue for HOI detection.

(3) We propose a gate and memory mechanism to perform sequential fusion on these available cues to attain a more discriminative representation.

(4) Our approach achieves state-of-the-art performance on two popular HOI detection benchmarks: V-COCO and HICO-DET.

2   Related Work

Visual Relationship Detection. Visual Relationship Detection (VRD) aims to detect objects and simultaneously predict the relationship between them. This topic has attracted considerable attention, supported by the recent development of large-scale relationship datasets such as VRD [26], Visual Genome [21] and Open Images [22]. However, detecting subtle differences between visual relationship remains difficult and the task is made yet more challenging by the distribution of visual relations, which is extremely long-tailed. Several recent works have proposed various mechanisms to address this problem [5,20,23,26,42,43]. Our work focuses on one particular class of visual relationship detection problem: detecting human object interaction (HOI). HOI detection poses additional challenges over VRD: a human can also perform multiple actions with one or more objects simultaneously and the range of human actions we are interested in are typically more fine-grained and diverse than for other generic objects.

Human Object Interaction (HOI) Detection. HOI detection aims to detect and recognise how each person interacts with the objects that surround them—it provides the fundamental basis for understanding human behaviour in a complex scene. Recently, driven by the release of relevant benchmarks, HOI detection has attracted significant attention [9,12,14,24,33,38,40,41,45].

The earlier works typically focused on tackling HOI by utilizing human and object visual appearance features or by capturing their spatial relationship through their coarse layout (box locations) [9,12]. Recently, several methods [14,24] have been developed that use human pose configuration maps to distinguish fine-grained relations. Wan et al. [38] and Fang et al. [6] use human pose cues to zoom into the relevant regions of the image via attention mechanism. Zhou et al. [45] encode human pose through graph neural networks and message passing. Note however, that each of these approaches encode human pose via keypoint estimation (either draw rectangle around the key point or link the keypoint into skeleton), which removes detailed information about the boundary and shape of the human, in particular, about human parts. By contrast, we argue that the boundaries of both human parts and objects are crucial. We are the first to leverage fine-grained human parsing and object semantic segmentation masks to better capture the geometric relations between them—such cues enable discrimination between subtly different relation categories.

More recently, although some approaches [9,12,28,31,39] have sought to use contextual cues from the whole image (global context), they do so by learning a spatial attention map in pixel space based on instance visual appearance to highlight image regions, making optimisation challenging when training data is limited. By contrast, we draw inspiration from classical recognition techniques [29,34] and argue that the semantic categories of other objects present in the surrounding neighbourhood of the candidate instance pair (local context) and the scene information (global context) provide valuable cues for resolving ambiguity between different interactions, but the detailed visual appearance of them is often not crucial. Instead of using visual appearance features, we are the first to encode context information via a semantic embedding, i.e., word2vec, that enables us to leverage the language priors to capture which objects might afford or co-occur with particular predicates.

The prediction of plausible motion has received limited attention for HOI detection. Nevertheless, estimation of current movement—when coupled with an understanding of the dynamics of an interaction—provides a cue for assessing the degree to which the configuration of humans and objects is probable for that interaction. We are the first to leverage flow prediction from a static image to infer the motion most plausible for a given image (i.e., which regions of pixels will move, together with their velocity, in the instant following image capture) as an auxiliary cue for HOI detection. Our approach is related to a wide body of work on visual future prediction [8,35,36] and draws on techniques for flow prediction from static images [10,37]. Differently from prior work, our objective is to infer plausible motion as an auxiliary cue for HOI detection in static images. In this sense, our method bridges static image HOI detection with video-level action understanding by transferring a motion prior learned from videos to images.

Current approaches predominately use either late or early fusion strategy to combine multi-stream features when inferring relations, while the relationship between different streams is often overlooked. Specifically, Late fusion are adopted by [12,19,33,41], where interaction predictions are performed on each stream independently and then summed at inference time. A wide body of work, i.e., [9,14,24,30,40] use the early fusion strategy, where multi-stream features are concatenated first and then use it to predict the score (sometimes with attention mechanism as in [31,38,39,45]). In this work, we are the first to use the gate and memory mechanism to fuse the available cues for HOI detection, i.e., select the discriminative information and gradually generate the representation for the whole scene step by step.

3   Method

We now introduce our proposed Fine-grained layout-Context-Motion Network (FCMNet), for localising and recognising all human-object interaction instances in an image. We first provide a high-level overview of FCMNet in Sect. 3.1, followed by a detailed description of the model architecture in Sect. 3.2. Finally, Sect. 3.3 describes the training and inference procedures.

3.1   Overview

Our approach to human-object interaction detection comprises two main stages: (1) human and object detection and (2) interaction prediction. First, given an

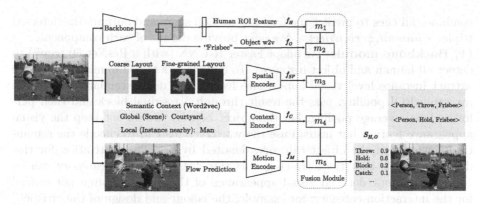

Fig. 2. The proposed **FCMNet** framework. The *backbone module* first detects all human and object boxes and encodes their representations; For each candidate pair of human-object boxes $<b_H, b_O>$, the *spatial encoder* encodes coarse and fine-grained layouts to capture the spatial configuration between them; the *context encoder* accumulates other readily available auxiliary information in the other region of the image, including local and global context; the *motion encoder* infer the likely movement of humans and objects (i.e. which regions of pixels will move together with their velocity in the instant following image capture) via flow prediction from the static image as an approximation to plausible motion. Finally, the *fusion module* combines all available knowledge about the candidate pair $<b_H, b_O>$ to predict the final interaction score for the candidate pair and outputs the detected triplet $<human, verb, object>$.

image I, we use Faster R-CNN [32] (Detectron implementation [11]) to detect all person instances $p = (p^1, p^2, ..., p^m)$ and object instances $o = (o^1, o^2, ..., o^n)$, generating a set of bounding boxes $b = (b^1, b^2, ..., b^{m+n})$ where m denotes the total number of detected person and n denotes the number of detected objects. We use $b_H \in \mathbb{R}^4$ and $b_O \in \mathbb{R}^4$ to denote the human and object bounding boxes. HOI proposals are then generated by enumerating all pairs of candidate human and object bounding boxes. Next, we process each human-object bounding box pair $<b_H, b_O>$ with FCMNet to predict an *interaction action score* $S_{H,O} \in \mathbb{R}^A$, where A represents the number of interaction classes. FCMNet encodes three image cues independently via the spatial encoder, context encoder, and motion encoder. Finally, a fusion module combines the outputs of these encoders and generates a robust HOI detection. Figure 2 shows an overview of the model.

3.2   Model Architecture

FCMNet contains the following five modules: (1) a backbone module that detects human and object boxes and encodes their representations; (2) a spatial encoder that leverages both coarse and fine-grained layout information; (3) a context encoder that accumulates auxiliary information in other regions of the image; (4) a motion encoder that predicts which regions of pixels would move in the instant following image capture via flow prediction; (5) a fusion module that combines all cues to predict the final interaction score and outputs the detected triplet $<human, verb, object>$. We next provide details of each component.

(1) Backbone module: We adopt Faster R-CNN (with a ResNet-50 trunk) to detect all human and object instances. To encode a detected human box b_H, we extract instance level visual appearance feature f_H using standard techniques: we apply ROI pooling, pass the result through a residual block and then perform global average pooling. For the object box b_O, we do not keep the visual appearance feature but instead use the word2vec instead to encode the semantic knowledge of the object category, denoted by f_O. The motivation for this design choice is that the intra-class variation for each object category can be considerable, but detailed visual appearance of the object is often not crucial for the interaction category: for example, the colour and design of the "frisbee" do not provide useful cues when inferring its relation with the human. Using the word2vec semantic embedding for the object enables the model to leverage language priors to capture which objects might afford similar predicates and therefore generalise to previously unseen or rare HOI instances [41].

(2) Spatial Encoder: The *spatial encoder* is designed to encode the spatial relationship between the human and object instance at both a coarse and fine-grained level. The encoding methods for each granularity are described next.

Coarse Layout representation: We use the two-channel binary image representation advocated by [9] to describe the interaction pattern at a coarse level. Specifically, we take the union of the human and object instance boxes as the reference box and construct two binary images (as separate channels) within it: the first channel has value 1 within the human bounding box and 0 elsewhere; the second channel has value 1 within the object box and 0 elsewhere.

Fine-Grained Layout representation: Instead of encoding the fine-grained information via key point estimation, we compute a human parse (segmentation masks for the human parts) of the detected human and a segmentation mask for the detected object instance to provide fine-grained layout information. The primary hypothesis underpinning our design is that *the shape and boundary of both the human and object can greatly help disambiguate different actions with a similar coarse spatial layout.* The reason is that these fine-grained cues can reveal the interaction boundary explicitly when inferring the relations. Moreover, fine-grained human parsing both reflects the human's pose and keeps much of the valuable information of their visual appearance. In our work, we use Mask-RCNN [16] to extract the segmentation mask of the object, and use MMAN [27] to extract the segmentation mask of all visible human parts. These two masks are stacked to form the fine-grained layout representation. Specifically, the first channel contains a set of discrete intensity values uniformly ranging from 0.05 to 0.95 to indicate different human parts; the second channel is a binary map that has a value of 1 within the object mask area. Examples of the information conveyed by each channel information can be found in Fig. 2.

We show in our ablation experiments (in Sect. 4.2) that each of the channels forming the coarse and fine-grained layout representations yield improvements in HOI detection performance. Changing the encoding mechanism of the fine-grained information outperforms the one encoding via key point estimation in the literature. In all other experiments, we stack coarse (two channels) and fine-grained (two channels) layout representations together as the input to the spatial encoder unless otherwise specified. The spatial encoder extracts the instance spatial relationship f_{SP} via a small CNN. The CNN comprises two convolutional layers (the first of which uses an SE block [17] to learn more powerful features) and is trained end-to-end.

(3) Context Encoder: The *context encoder* is designed to uncover complementary cues conveyed in other regions of the image. Instead of using visual appearance to encode the contexts information directly, we change the encoding mechanism by using semantic categories. It comprises a global context encoding, i.e., the estimated scene, and a local context encoding, namely the semantic categories of other objects present in the surrounding neighbourhood of the candidate instance pair.

Global Context: For the global context representation f_G, we use a DenseNet-161 model [18] pretrained on Places365 [44] to extract scene features. After that we encode the scene class (with largest likelihood) using word2vec. The global context embedding f_G (scene feature) is therefore a 300-dimensional vector.

Local Context: During inference of the relationship between the candidate pair containing object o^i with human h, all the other detected objects o^j where $j \neq i$ in the neighbourhood can be considered to provide local context. In particular, their semantic category and position relative to the candidate object o^i are both valuable cues for distinguishing between different interactions. For example, the objects present in the neighbourhood of an 'eating a cake' interaction will likely comprise a spoon-like tool, whereas for the 'cutting a cake' interaction, it is a knife-like tool. The distance between the knife and the cake is also important (if it is far away from the cake, it is very unlikely that the knife is being used to cut the cake).

Motivated by this observation, we first use word2vec to encode the semantic knowledge of each object o^j in the neighbourhood, and then concatenate the resulting embedding with its spatial relationship f_R (computed with respect to the candidate object o^i). Following prior work [46] on visual relationship detection, the spatial geometric relationship feature f_R between candidate object o^i and its neighbour o^j is encoded as follows:

$$ f_R = [(\frac{x_1^i - x_1^j}{x_2^j - x_1^j}), (\frac{y_1^i - y_1^j}{y_2^j - y_1^j}), log(\frac{x_2^i - x_1^i}{x_2^j - x_1^j}), log(\frac{y_2^i - y_1^i}{y_2^j - y_1^j})], \qquad (1) $$

where $(x_1^i, x_2^i, y_1^i, y_2^i)$ and $(x_1^j, x_2^j, y_1^j, y_2^j)$ are the normalised box coordinates of the candidate object o_i and its neighbour o^j. This geometric feature f_R is a measure of the scales and relative positioning of the two object entities.

To account for the variable number of objects present in the neighbourhood of an interaction, we use NetVLAD [2] to aggregate all object representations when forming the local context embedding f_L. During the end-to-end training, the NetVLAD aggregation module can learn to discriminatively select which information should be promoted (or demoted). Finally, the output of the Context Encoder, f_C, is generated by concatenating the global context f_G and local context f_L embeddings.

(4) Motion Encoder: The *Motion Encoder* aims to infer the likely movements of humans and objects in a given image, and provide cues to detect and recognise their interactions. We draw inspiration from the work of [36] which sought to learn models from video that were capable of synthesising plausible futures from a singe static image and in particular, the more recent work of [10] for static image flow prediction. In our work, we focus on the latter task of predicting local optical flow as a proxy for plausible future movements of objects. We adopt the Im2Flow model [10] to predict flow from the static image to encode plausible scene motion. The flow describes which pixel regions will move (together with their direction and velocity) in the instant following image capture. Concretely, we first predict the flow information of the image and then use the plausible motion encoder (a CNN with learnable parameters) to extract plausible motion features f_M. Qualitative examples of the predicted flow can be seen in Fig. 1 (right) and Fig. 2.

(5) Fusion Module: The *fusion module* combines the outputs of the backbone, spatial, context and motion encoders into a single feature embedding and uses it to predict a score for the interaction $s_{H,O}$ of the candidate instance pair $<b_H, b_O>$. The design of the fusion module is shown in Fig. 2. Specifically, we perform fusion by putting the sequence of available features $f^* = \{f_1, ..., f_k\} = \{f_H, f_O, f_{SP}, f_C, f_M\}$, one by one into GRUs [4].[1] The description of the whole scene gradually accumulate and update in the memory cell m_k (hidden state), where the lower index k is the number of reasoning steps. At each step k, we start with calculate the update gate z_k as:

$$z_k = \sigma_z(W_z f_k + U_z m_{k-1}) + b_z, \tag{2}$$

where W_z, U_z and b_z are weights and bias and σ_z is a sigmoid activation function. The update gate z_k analyzes the description of the whole scene at last step m_{k-1} and the current input feature f_k to decide how much the current step updates its memory cell. The new added information f_k at step k helping grow the description of the whole scene is computed as follows:

$$\hat{m}_i = \sigma_m(W_m f_k + U_m(r_k \circ m_{k-1}) + b_z), \tag{3}$$

where W_m, U_m and b_m are weights and bias and σ_m is a tanh activation function. $\circ$ is an element-wise multiplication. r_k is the reset gate that decides what content to forget based on the reasoning between the m_{k-1} and f_k, can be computed as

$$r_k = \sigma_r(W_r f_k + U_r m_{k-1}) + b_r, \tag{4}$$

where W_r, U_r and b_r are weights and bias and σ_r is a sigmoid activation function. Then the description of the whole scene m_k at the current step is a linear

---

[1] Empirically, we observe the feature ordering to GRU is not sensitive to the HOI detection performance. So we use this order in all experiments.

interpolation using the update gate z_k between the previous description m_{k-1} and the new added content $\hat{m}_k$:

$$m_k = (1 - z_k) \circ m_{k-1} + z_k \circ \hat{m}_k, \quad (5)$$

where $\circ$ is an element wise multiplication. Lastly, we take the memory cell m_k at the end of sequence f^* as the final representation to predict the relation category: the output of the fusion module is the interaction score $S_{H,O}$ for the candidate pair $<b_H, b_O>$. The proposed gate and memory mechanism allows the model to dynamically select which information should be highlighted or suppressed in the final representation, rendering it more discriminative for the final objective.

3.3 Training and Inference

Since a human can concurrently perform different actions and interact with one or more objects, e.g. "eating a cake and reading a book while sitting on the coach", HOI detection represents a multi-label classification problem in which each predicate is independent and not mutually exclusive. During training, a binary sigmoid classifier is used for each predicate that minimises the cross-entropy between the prediction score $s_{H,O}$ and the ground truth label.

During inference, all human and object instances are first detected in each image. Each human and object box pair $<b_H, b_O>$ is then passed through the network to produce a score $s_{H,O}^a$ for each predicate class $a \in 1, ..., A$, where A denotes the total number of predicate classes. The final relation score is then combined with the detection scores of the human and object instances s_H and s_O that represent the detection quality of the instance boxes b_H and b_O. The final HOI score $S_{H,O}$ for the candidate box pair $<b_H, b_O>$ is then calculated as:

$$S_{H,O} = s_{H,O} \cdot s_H \cdot s_O \quad (6)$$

4 Experiments

We first introduce the dataset used, evaluation metric and implementation details in Sect. 4.1. Next, we conduct a detailed ablation study in Sect. 4.2 to verify the effectiveness of each proposed component and present some qualitative results to demonstrate the strengths and failure cases of our approach. Finally we report our HOI detection results quantitatively and compare with state-of-the-art approaches on two benchmarks: V-COCO [13] and HICO-DET [3].

4.1 Experimental Setup

Datasets: We evaluate our proposed approach on two HOI detection benchmarks: V-COCO [13] and HICO-DET [3]. **V-COCO** contains 5400 images in the training and validation split and 4946 images in the test set. It is annotated with 26 common action category labels and the bounding boxes for human and object instances. **HICO-DET** comprises 47,776 images (38,118 images in the training set and 9,658 in the test set) with more than 150K $<human, verb, object>$ triplets. It is annotated with 600 HOI categories over 80 object classes and 117 unique verbs. HICO-DET is both larger and more diverse than V-COCO.

Evaluation Metric: We follow the standard evaluation setting in [9] and use mean average precision to measure HOI detection performance. Formally, a triplet of $<human, verb, object>$ is considered as true positive if and only if: (1) the predicted bounding box of both human and object instance overlap with the ground truth bounding box with IoU greater than 0.5, and (2) the predicted HOI matches the ground truth HOI category. For the V-COCO dataset, we evaluate mAP_{role} following [9]. For the HICO-DET dataset, mAP performance is reported for three different HOI category sets: (1) all 600 HOI categories (*Full*), (2) 138 categories with less than 10 training samples (*Rare*) and (3) the remaining 462 categories with more than 10 training samples (*Non-Rare*).

Implementation Details: Following [9], we build our approach on the Faster-RCNN [32] object detector (without FPN) with the ResNet-50 backbone to enable a fair comparison with other prior approaches. The human parse masks are obtained with a Macro-Micro Adversarial Net [27], trained on the LIP dataset [27]. Object masks are obtained with Mask-RCNN [16] pretrained on MS-COCO [25]. Global scene features are extracted using a DenseNet-161 model [18] pretrained on Places365 [44]. The object and scene semantic embeddings are obtained from Google News trained word2vec embeddings. We adopt the Im2Flow model [10] to predict flow from the static image to encode plausible motion. All pretrained models described above are kept frozen during the training in this paper. More implementation details can be found in the extended arXiv version of this paper.

4.2   Ablation Studies

In this section, we empirically investigate the sensitivity of the proposed method to different design choices. As the HICO-DET dataset is both larger and more diverse than V-COCO, we perform all ablation studies on HICO-DET. We study four aspects of FCMNet: the contribution of the network components, fine-grained spatial encodings, context features and fusion strategies. Finally, we present some qualitative examples to illustrate the challenging and failure cases.

Architectural Variants: We conduct an ablation study by examining the effectiveness of each proposed component in our network structure. We use combinations of human,object embeddings and coarse layout spatial configuration (instance boxes) as our baseline (Base). As shown in Table 1, each proposed module yields improved performance under all three HICO-DET settings. By looking at the first four rows in Table 1, we can observe that the fine-grained spatial configuration information contributes the most significant performance gain as an individual module, suggesting that fine-grained shape and boundary of the human parts and object can greatly help disambiguate different actions with

Table 1. Ablation study on different network components

Methods	Full	Rare	Non-Rare
Baseline (Base)	14.8	12.3	15.7
Base + Fine	17.6	15.4	18.3
Base + Context	16.2	14.1	16.9
Base + Motion	16.5	14.4	17.2
FCMNet (ours)	**20.4**	**17.3**	**21.6**

Table 2. Ablation study on different fine-grained information

Methods	Full	Rare	Non-Rare
Baseline (Base)	14.8	12.3	15.7
Base + HS	15.8	12.9	16.8
Base + HP	17.1	14.7	18.0
Base + OM	16.2	14.0	17.1
Base + HP + OM	**17.6**	**15.4**	**18.3**

Table 3. Ablation study on different context information

Methods	Full	Rare	Non-Rare
Baseline (Base)	14.8	12.3	15.7
Base + Local(w2v)	15.7	13.7	16.4
Base + Global(w2v)	15.1	13.0	16.2
Base + Both(visual)	15.2	12.8	15.9
Base + Both(w2v)	**16.2**	**14.1**	**16.9**

Table 4. Ablation study on different fusion strategy

Methods	Full	Rare	Non-Rare
Baseline	14.8	12.3	15.7
Late Fusion	18.1	15.3	19.0
Concatenation	17.9	14.9	20.0
Fusion (Attention)	19.7	16.6	20.1
FCMNet (ours)	**20.4**	**17.3**	**21.6**

a similar coarse spatial layout. The **FCMNet** which includes all proposed modules (human and object features, spatial encoder with coarse and fine-grained layout, context encoder, motion encoder and fusion) at the same time achieves the best performance, which verifying the effectiveness of the proposed components. More ablation studies can be found in the extended arXiv version of this paper.

Fine-Grained Spatial Configuration Encoding: We compare the performance of using different forms of fine-grained spatial information in Table 2. To enable a fair comparison, we use the same model architecture, i.e., with only human and object features and a spatial encoder. We observe that each of the investigated fine-grained spatial encoding, i.e., Human Parts (HP) and Object Mask (OM), improves performance. Since prior work ([14,38]) has shown that encoding fine-grained information via human key-point estimation is also useful to convey pose, we also compare with the Human skeleton (HS) configuration (following [14,38]) in this table. It can be seen that using human parts information outperforms human skeletons. Using both proposed fine-grained spatial encoding HP and OM concurrently outperforming the baseline by 2.8 mAP in the *Full* setting, which demonstrates the effectiveness of the proposed encoding mechanism for fine-grained spatial configuration.

Fig. 3. Samples of human object interactions detected by FCMNet. The first four illustrate correct detections; the last two shows failure cases. (Color figure online)

Table 5. Performance comparison on the V-COCO dataset. The scores are reported in mAP(role) as in the standard evaluation metric and the best score is marked in bold. Our approach sets a new state-of-the-art on this dataset.

Method	Feature Backbone	AP_{role}
InteractNet [12]	ResNet-50-FPN	40.0
BAR-CNN [19]	Inception-ResNet	41.1
GPNN [31]	ResNet-152	44.0
iHOI [40]	ResNet-50-FPN	45.8
Xu et al. [41]	ResNet-50	45.9
iCAN [9]	ResNet-50	44.7
Wang et al. [39]	ResNet-50	47.3
RPNN [45]	ResNet-50	47.5
Li et al. [24]	ResNet-50	47.8
PMFNet [38]	ResNet-50-FPN	52.0
Baseline (Ours)	ResNet-50	45.3
FCMNet (Ours)	**ResNet-50**	**53.1**

Context Features: We compare the performance of using different contextual information in Table 3. It can be seen that both local context (features of the objects in the neighbourhood) and global context (the scene) contribute to improved performance. Encoding the context via word2vec outperforms the one using the visual appearance directly.

Fusion Mechanism: We compare the performance of using different fusion mechanisms in Table 4. The proposed fusion design strongly outperforms late fusion, simple concatenation and fusion with attention mechanism, demonstrating the utility of providing the model with a gate and memory mechanism for filtering its representation of the candidate pair using all available information gradually. Nevertheless, both late fusion, concatenation, fusion with attention and proposed fusion module boost performance, verifying that the different encoders capture valuable complementary information for HOI detection.

Qualitative Visual Examples: In Fig. 3, we present qualitative examples to illustrate strengths and failure cases on the HICO-DET dataset. We highlight the detected human and object with red and blue bounding boxes respectively. The first four samples are some challenging cases where our proposed approach produce correct detection. It indicates our model can distinguish subtle visual differences between interactions (first two) and be able to detect co-occurrence relations (third and fourth). The last two show some failure cases.

4.3   Results and Comparisons

In this section, we compare our FCMNet with several existing approaches for evaluation. We use combinations of humans, objects embeddings and coarse layout spatial configuration (instance boxes) as our baseline—the final FCMNet model integrates all the proposed modules.

Table 6. Performance comparison on the HICO-DET dataset. Mean average precision (mAP) is reported for the default and Known object setting. The best score is marked in bold. Our approach sets a new state-of-the-art on this dataset.

Method	Feature Backbone	Default			Known Object		
		Full	Rare	Non-Rare	Full	Rare	Non-Rare
InteractNet [12]	ResNet-50-FPN	9.94	7.16	10.77	-	-	-
GPNN [31]	ResNet-152	13.11	9.34	14.23	-	-	-
iHOI [40]	ResNet-50-FPN	13.39	9.51	14.55	-	-	-
Xu et al. [41]	ResNet-50	14.70	13.26	15.13	-	-	-
iCAN [9]	ResNet-50	14.84	10.45	16.15	16.43	12.01	17.75
Wang et al. [39]	ResNet-50	16.24	11.16	17.75			
Li et al. [24]	ResNet-50	17.03	13.42	18.11	19.17	15.51	20.26
Gupta et al [14]	ResNet-152	17.18	12.17	18.68	-	-	-
RPNN [45]	ResNet-50	17.35	12.78	18.71	-	-	-
PMFNet [38]	ResNet-50-FPN	17.46	15.65	18.00	20.34	17.47	21.20
Baseline (Ours)	ResNet-50	14.77	12.27	15.65	16.07	13.97	16.82
FCMNet (Ours)	**ResNet-50**	**20.41**	**17.34**	**21.56**	**22.04**	**18.97**	**23.12**

For the **VCOCO** dataset, we present the quantitative results in terms of AP_{role} in Table 5. Our baseline achieves competitive performance with an AP_{role} of 45.3 (placing it above approximately half of the listed prior work). Different from those approaches, we use word2vec embeddings to represent the object rather than the visual embedding from ROI pooling, which turns out to be very effective for HOI detection in small datasets like V-COCO. Using the word2vec semantic embedding for the object representation enables us to leverage language priors to capture which objects might afford similar actions when the training data is limited. Our full FCMNet model (with all components proposed in Sect. 3) achieves 53.1 mAP, which outperforms prior approaches by a considerable margin.

For the **HICO-DET** dataset, we present quantitative results in terms of mAP in Table 6. We report results on two different settings of Default and Known Objects. Note that our baseline still performs well and surpasses nearly half of existing approaches. FCMNet improves upon our baseline by 5.64 mAP on the default setting (full split) and sets a new state-of-the-art on this dataset for both the default and Known object setting.

5   Conclusions

We have presented FCMNet, a novel framework for human object interaction detection. We illustrated the importance of the encoding mechanism for the fine-grained spatial layouts and semantic contexts, which enables to distinguish the subtle differences among interactions. We also show that the prediction of plausible motion greatly help to constrain the space of candidate interactions by considering their motion and boost performance. By combining these cues via a gate and memory mechanism, FCMNet outperforms state-of-the-art methods on standard human object interaction benchmarks by a considerable margin.

Acknowledgements. The authors gratefully acknowledge the support of the EPSRC Programme Grant Seebibyte EP/M013774/1 and EPSRC Programme Grant CALO-PUS EP/R013853/1. The authors would also like to thank Samuel Albanie and Sophia Koepke for helpful suggestions.

References

1. Adam, A., Rivlin, E., Shimshoni, I., Reinitz, D.: Robust real-time unusual event detection using multiple fixed-location monitors. IEEE Trans. Pattern Anal. Mach. Intell. **30**(3), 555–560 (2008)
2. Arandjelovic, R., Gronat, P., Torii, A., Pajdla, T., Sivic, J.: NetVLAD: CNN architecture for weakly supervised place recognition. In: Proceedings of the IEEE Conference on Computer Vision and Pattern Recognition, pp. 5297–5307 (2016)
3. Chao, Y.W., Wang, Z., He, Y., Wang, J., Deng, J.: HICO: a benchmark for recognizing human-object interactions in images. In: Proceedings of the IEEE International Conference on Computer Vision, pp. 1017–1025 (2015)
4. Chung, J., Gulcehre, C., Cho, K., Bengio, Y.: Empirical evaluation of gated recurrent neural networks on sequence modeling. arXiv preprint arXiv:1412.3555 (2014)
5. Dai, B., Zhang, Y., Lin, D.: Detecting visual relationships with deep relational networks. In: Proceedings of the IEEE Conference on Computer Vision and Pattern Recognition, pp. 3076–3086 (2017)
6. Fang, H.-S., Cao, J., Tai, Y.-W., Lu, C.: Pairwise body-part attention for recognizing human-object interactions. In: Ferrari, V., Hebert, M., Sminchisescu, C., Weiss, Y. (eds.) ECCV 2018. LNCS, vol. 11214, pp. 52–68. Springer, Cham (2018). https://doi.org/10.1007/978-3-030-01249-6_4
7. Felzenszwalb, P.F., Girshick, R.B., McAllester, D., Ramanan, D.: Object detection with discriminatively trained part-based models. IEEE Trans. Pattern Anal. Mach. Intell. **32**(9), 1627–1645 (2010)
8. Fouhey, D.F., Zitnick, C.L.: Predicting object dynamics in scenes. In: Proceedings of the IEEE Conference on Computer Vision and Pattern Recognition, pp. 2019–2026 (2014)

9. Gao, C., Zou, Y., Huang, J.B.: iCAN: instance-centric attention network for human-object interaction detection. In: British Machine Vision Conference (2018)
10. Gao, R., Xiong, B., Grauman, K.: Im2Flow: motion hallucination from static images for action recognition. In: Proceedings of the IEEE Conference on Computer Vision and Pattern Recognition, pp. 5937–5947 (2018)
11. Girshick, R., Radosavovic, I., Gkioxari, G., Dollár, P., He, K.: Detectron (2018). https://github.com/facebookresearch/detectron
12. Gkioxari, G., Girshick, R., Dollár, P., He, K.: Detecting and recognizing human-object interactions. In: Proceedings of the IEEE Conference on Computer Vision and Pattern Recognition, pp. 8359–8367 (2018)
13. Gupta, S., Malik, J.: Visual semantic role labeling. arXiv preprint arXiv:1505.04474 (2015)
14. Gupta, T., Schwing, A., Hoiem, D.: No-frills human-object interaction detection: factorization, appearance and layout encodings, and training techniques. arXiv preprint arXiv:1811.05967 (2018)
15. Hayes, B., Shah, J.A.: Interpretable models for fast activity recognition and anomaly explanation during collaborative robotics tasks. In: 2017 IEEE International Conference on Robotics and Automation (ICRA), pp. 6586–6593. IEEE (2017)
16. He, K., Gkioxari, G., Dollár, P., Girshick, R.: Mask R-CNN. In: Proceedings of the IEEE International Conference on Computer Vision, pp. 2961–2969 (2017)
17. Hu, J., Shen, L., Albanie, S., Sun, G., Wu, E.: Squeeze-and-excitation networks. IEEE Trans. Pattern Anal. Mach. Intell. **42**, 2011–2023 (2019)
18. Huang, G., Liu, Z., Van Der Maaten, L., Weinberger, K.Q.: Densely connected convolutional networks. In: Proceedings of the IEEE Conference on Computer Vision and Pattern Recognition, pp. 4700–4708 (2017)
19. Kolesnikov, A., Kuznetsova, A., Lampert, C., Ferrari, V.: Detecting visual relationships using box attention. In: Proceedings of the IEEE International Conference on Computer Vision Workshops (2019)
20. Krishna, R., Chami, I., Bernstein, M., Fei-Fei, L.: Referring relationships. In: Proceedings of the IEEE Conference on Computer Vision and Pattern Recognition, pp. 6867–6876 (2018)
21. Krishna, R., et al.: Visual genome: connecting language and vision using crowd-sourced dense image annotations. Int. J. Comput. Vis. **123**(1), 32–73 (2017)
22. Kuznetsova, A., et al.: The open images dataset v4: unified image classification, object detection, and visual relationship detection at scale. arXiv preprint arXiv:1811.00982 (2018)
23. Li, Y., Ouyang, W., Zhou, B., Wang, K., Wang, X.: Scene graph generation from objects, phrases and region captions. In: Proceedings of the IEEE International Conference on Computer Vision, pp. 1261–1270 (2017)
24. Li, Y.L., et al.: Transferable interactiveness knowledge for human-object interaction detection. In: Proceedings of the IEEE Conference on Computer Vision and Pattern Recognition, pp. 3585–3594 (2019)
25. Lin, T.-Y., et al.: Microsoft COCO: common objects in context. In: Fleet, D., Pajdla, T., Schiele, B., Tuytelaars, T. (eds.) ECCV 2014. LNCS, vol. 8693, pp. 740–755. Springer, Cham (2014). https://doi.org/10.1007/978-3-319-10602-1_48
26. Lu, C., Krishna, R., Bernstein, M., Fei-Fei, L.: Visual relationship detection with language priors. In: Leibe, B., Matas, J., Sebe, N., Welling, M. (eds.) ECCV 2016. LNCS, vol. 9905, pp. 852–869. Springer, Cham (2016). https://doi.org/10.1007/978-3-319-46448-0_51

27. Luo, Y., Zheng, Z., Zheng, L., Guan, T., Yu, J., Yang, Y.: Macro-micro adversarial network for human parsing. In: Ferrari, V., Hebert, M., Sminchisescu, C., Weiss, Y. (eds.) ECCV 2018. LNCS, vol. 11213, pp. 424–440. Springer, Cham (2018). https://doi.org/10.1007/978-3-030-01240-3_26

28. Mallya, A., Lazebnik, S.: Learning models for actions and person-object interactions with transfer to question answering. In: Leibe, B., Matas, J., Sebe, N., Welling, M. (eds.) ECCV 2016. LNCS, vol. 9905, pp. 414–428. Springer, Cham (2016). https://doi.org/10.1007/978-3-319-46448-0_25

29. Murphy, K.P., Torralba, A., Freeman, W.T.: Using the forest to see the trees: a graphical model relating features, objects, and scenes. In: Advances in Neural Information Processing Systems, pp. 1499–1506 (2004)

30. Peyre, J., Laptev, I., Schmid, C., Sivic, J.: Detecting rare visual relations using analogies. arXiv preprint arXiv:1812.05736 (2018)

31. Qi, S., Wang, W., Jia, B., Shen, J., Zhu, S.-C.: Learning human-object interactions by graph parsing neural networks. In: Ferrari, V., Hebert, M., Sminchisescu, C., Weiss, Y. (eds.) ECCV 2018. LNCS, vol. 11213, pp. 407–423. Springer, Cham (2018). https://doi.org/10.1007/978-3-030-01240-3_25

32. Ren, S., He, K., Girshick, R., Sun, J.: Faster R-CNN: towards real-time object detection with region proposal networks. In: Advances in Neural Information Processing Systems, pp. 91–99 (2015)

33. Shen, L., Yeung, S., Hoffman, J., Mori, G., Fei-Fei, L.: Scaling human-object interaction recognition through zero-shot learning. In: 2018 IEEE Winter Conference on Applications of Computer Vision (WACV), pp. 1568–1576. IEEE (2018)

34. Torralba, A., Murphy, K.P., Freeman, W.T., Rubin, M.A.: Context-based vision system for place and object recognition (2003)

35. Villegas, R., Yang, J., Hong, S., Lin, X., Lee, H.: Decomposing motion and content for natural video sequence prediction. arXiv preprint arXiv:1706.08033 (2017)

36. Vondrick, C., Pirsiavash, H., Torralba, A.: Generating videos with scene dynamics. In: Advances In Neural Information Processing Systems, pp. 613–621 (2016)

37. Walker, J., Gupta, A., Hebert, M.: Dense optical flow prediction from a static image. In: Proceedings of the IEEE International Conference on Computer Vision, pp. 2443–2451 (2015)

38. Wan, B., Zhou, D., Liu, Y., Li, R., He, X.: Pose-aware multi-level feature network for human object interaction detection. In: Proceedings of the IEEE International Conference on Computer Vision, pp. 9469–9478 (2019)

39. Wang, T., et al.: Deep contextual attention for human-object interaction detection. arXiv preprint arXiv:1910.07721 (2019)

40. Xu, B., Li, J., Wong, Y., Zhao, Q., Kankanhalli, M.S.: Interact as you intend: intention-driven human-object interaction detection. IEEE Trans. Multimed. **22**, 1423–1432 (2019)

41. Xu, B., Wong, Y., Li, J., Zhao, Q., Kankanhalli, M.S.: Learning to detect human-object interactions with knowledge. In: Proceedings of the IEEE Conference on Computer Vision and Pattern Recognition (2019)

42. Xu, D., Zhu, Y., Choy, C.B., Fei-Fei, L.: Scene graph generation by iterative message passing. In: Proceedings of the IEEE Conference on Computer Vision and Pattern Recognition, pp. 5410–5419 (2017)

43. Zhang, H., Kyaw, Z., Chang, S.F., Chua, T.S.: Visual translation embedding network for visual relation detection. In: Proceedings of the IEEE Conference on Computer Vision and Pattern Recognition, pp. 5532–5540 (2017)

44. Zhou, B., Lapedriza, A., Khosla, A., Oliva, A., Torralba, A.: Places: a 10 million image database for scene recognition. IEEE Trans. Pattern Anal. Mach. Intell. **40**(6), 1452–1464 (2017)
45. Zhou, P., Chi, M.: Relation parsing neural network for human-object interaction detection. In: Proceedings of the IEEE International Conference on Computer Vision (2019)
46. Zhuang, B., Liu, L., Shen, C., Reid, I.: Towards context-aware interaction recognition for visual relationship detection. In: Proceedings of the IEEE International Conference on Computer Vision, pp. 589–598 (2017)

Rethinking Few-Shot Image Classification: A Good Embedding is All You Need?

Yonglong Tian[1]([✉]), Yue Wang[1]([✉]), Dilip Krishnan[2], Joshua B. Tenenbaum[1], and Phillip Isola[1]

[1] MIT, Cambridge, USA
{yonglong,yuewangx}@mit.edu
[2] Google Research, Mountain View, USA

Abstract. The focus of recent meta-learning research has been on the development of learning algorithms that can quickly adapt to test time tasks with limited data and low computational cost. Few-shot learning is widely used as one of the standard benchmarks in meta-learning. In this work, we show that a simple baseline: learning a supervised or self-supervised representation on the meta-training set, followed by training a linear classifier on top of this representation, outperforms state-of-the-art few-shot learning methods. An additional boost can be achieved through the use of self-distillation. This demonstrates that using a good learned embedding model can be more effective than sophisticated meta-learning algorithms. We believe that our findings motivate a rethinking of few-shot image classification benchmarks and the associated role of meta-learning algorithms. Code: http://github.com/WangYueFt/rfs/.

1 Introduction

Few-shot learning measures a model's ability to quickly adapt to new environments and tasks. This is a challenging problem because only limited data is available to adapt the model. Recently, significant advances [11,26,28,33,41,43, 45,49,51,52,54,58] have been made to tackle this problem using the ideas of meta-learning or "learning to learn".

Meta-learning defines a family of tasks, divided into disjoint meta-training and meta-testing sets. Each task consists of limited training data, which requires fast adaptability [42] of the learner (e.g., the deep network that is fine-tuned). During meta-training/testing, the learner is trained and evaluated on a task sampled from the task distribution. The performance of the learner is evaluated by the average test accuracy across many meta-testing tasks. Methods to tackle

---

Y. Tian and Y. Wang—Equal contribution.

Electronic supplementary material The online version of this chapter (https://doi.org/10.1007/978-3-030-58568-6_16) contains supplementary material, which is available to authorized users.

A. Vedaldi et al. (Eds.): ECCV 2020, LNCS 12359, pp. 266–282, 2020.
https://doi.org/10.1007/978-3-030-58568-6_16

this problem can be cast into two main categories: optimization-based methods and metric-based methods. Optimization-based methods focus on designing algorithms that can quickly adapt to each task; while metric-based methods aim to find good metrics (usually kernel functions) to side-step the need for inner-loop optimization for each task.

Meta-learning is evaluated on a number of domains such as few-shot classification and meta-reinforcement learning. Focusing on few-shot classification tasks, a question that has been raised in recent work is whether it is the meta-learning algorithm or the learned representation that is responsible for the fast adaption to test time tasks. [37] suggested that feature reuse is main factor for fast adaptation. Recently, [9] proposed transductive fine-tuning as a strong baseline for few-shot classification; and even in a regular, inductive, few-shot setup, they showed that fine-tuning is only slightly worse than state-of-the-art algorithms. In this setting, they fine-tuned the network on the meta-testing set and *used* information from the testing data. Besides, [5] shows an improved fine-tuning model performs slightly worse than meta-learning algorithms.

In this paper, we propose an extremely simple baseline that suggests that good learned representations are more powerful for few-shot classification tasks than the current crop of complicated meta-learning algorithms. Our baseline consists of a *linear* model learned on top of a pre-trained embedding. Surprisingly, we find this outperforms *all other meta-learning algorithms* on few-shot classification tasks, often by large margins. The differences between our approach and that of [9] are: we *do not* utilize information from testing data (since we believe that inductive learning is more generally applicable to few-shot learning); and we use a fixed neural network for feature extraction, rather than fine-tuning it on the meta-testing set. The concurrent works [6,21] are inline with ours.

Our model learns representations by training a neural network on the entire meta-training set: we merge all meta-training data into a single task and a neural network is asked to perform either ordinary classification or self-supervised learning, on this combined dataset. The classification task is equivalent to the pre-training phase of TADAM [33] and LEO [41]. After training, we keep the pre-trained network up to the penultimate layer and use it as a feature extractor. During meta-testing, for each task, we fit a linear classifier on the features extracted by the pre-trained network. In contrast to [9] and [37], we *do not* fine-tune the neural network. Furthermore, we show that self-distillation on this baseline provides an additional boost.

Contributions. Our key contributions are:

- A surprisingly simple baseline for few-shot learning, which achieves the state-of-the-art. This baseline suggests that many recent meta-learning algorithms are *no better* than simply learning a good representation through a proxy task, e.g., image classification.
- Building upon the simple baseline, we use self-distillation to further improve performance. Our combined method achieves an average of 3% improvement over the previous state-of-the-art on widely used benchmarks. On the new

benchmark Meta-Dataset [50], our method outperforms previous best results
by more than 7% on average.
- Beyond supervised training, we show that representations learned with state-
of-the-art self-supervised methods achieve similar performance as fully supervised methods.

2 Related Works

Metric-Based Meta-Learning. The core idea in metric-based meta-learning is
related to nearest neighbor algorithms and kernel density estimation. Metric-
based methods embed input data into fixed dimensional vectors and use them
to design proper kernel functions. The predicted label of a query is the weighted
sum of labels over support samples. Metric-based meta-learning aims to learn
a task-dependent metric. [23] used Siamese network to encode image pairs and
predict confidence scores for each pair. Matching Networks [51] employed two
networks for query samples and support samples respectively and used an LSTM
with read-attention to encode a full context embedding of support samples.
Prototypical Networks [43] learned to encode query samples and support samples
into a shared embedding space; the metric used to classify query samples is the
distance to prototype representations of each class. Instead of using distances
of embeddings, Relation Networks [45] leveraged relational module to represent
an appropriate metric. TADAM [33] proposed metric scaling and metric task
conditioning to boost the performance of Prototypical Networks.

Optimization-Based Meta-Learning. Deep learning models are neither designed
to train with very few examples nor to converge very fast. To fix that,
optimization-based methods intend to learn with a few examples. Meta-learner
[38] exploited an LSTM to satisfy two main desiderata of few-shot learning:
quick acquisition of task-dependent knowledge and slow extraction of transferable knowledge. MAML [11] proposed a general optimization algorithm; it aims
to find a set of model parameters, such that a small number of gradient steps with
a small amount of training data from a new task will produce large improvements
on that task. In that paper, first-order MAML was also proposed, which ignored
the second-order derivatives of MAML. It achieved comparable results to complete MAML with orders of magnitude speedup. To further simplify MAML,
Reptile [32] removed re-initialization for each task, making it a more natural
choice in certain settings. LEO [41] proposed that it is beneficial to decouple
the optimization-based meta-learning algorithms from high-dimensional model
parameters. In particular, it learned a stochastic latent space from which the
high-dimensional parameters can be generated. MetaOptNet [26] replaced the
linear predictor with an SVM in the MAML framework; it incorporated a differentiable quadratic programming (QP) solver to allow end-to-end learning. For a
complete list of recent works on meta-learning, we refer readers to [55].

Towards Understanding MAML. To understand why MAML works in the first
place, many efforts have been made either through an optimization perspective

or a generalization perspective. Reptile [32] showed a variant of MAML works even without re-initialization for each task, because it tends to converge towards a solution that is close to each task's manifold of optimal solutions. In [37], the authors analyzed whether the effectiveness of MAML is due to rapid learning of each task or reusing the high quality features. It concluded that feature reuse is the dominant component in MAML's efficacy, which is reaffirmed by experiments conducted in this paper.

Meta-Learning Datasets. Over the past several years, many datasets have been proposed to test meta-learning or few-shot learning algorithms. Omniglot [24] was one of the earliest few-shot learning datasets; it contains thousands of hand-written characters from the world's alphabets, intended for one-shot "visual Turing test". In [25], the authors reported the 3-year progress for the Omniglot challenge, concluding that human-level one-shot learnability is still hard for current meta-learning algorithms. [51] introduced mini-ImageNet, which is a subset of ImageNet [8]. In [40], a large portion of ImageNet was used for few-shot learning tests. Meta-dataset [50] summarized recent datasets and tested several representative methods in a uniform fashion.

Knowledge Distillation. The idea of knowledge distillation (KD) dates back to [4]. The original idea was to compress the knowledge contained in an ensemble of models into a single smaller model. In [19], the authors generalized this idea and brought it into the deep learning framework. In KD, knowledge is transferred from the teacher model to the student model by minimizing a loss in which the target is the distribution of class probabilities induced by the teacher model. It was shown that KD has several benefits for optimization and knowledge transfer between tasks [13,14,59]. BAN [12] introduced sequential distillation, which also improved the performance of teacher models. In natural language processing (NLP), BAM [7] used BAN to distill from single-task models to a multi-task model, helping the multi-task model surpass its single-task teachers. Another two related works are [30] which provides theoretical analysis of self-distillation and CRD [47] which shows distillation improves the transferability across datasets.

3   Method

We establish preliminaries about the meta-learning problem and related algorithms in Subsect. 3.1; then we present our baseline in Subsect. 3.2; finally, we introduce how knowledge distillation helps few-shot learning in Subsect. 3.3. For ease of comparison to previous work, we use the same notation as [26].

3.1   Problem Formulation

The collection of meta-training tasks is defined as $\mathcal{T} = \{(\mathcal{D}_i^{train}, \mathcal{D}_i^{test})\}_{i=1}^{I}$, termed as meta-training set. The tuple $(\mathcal{D}_i^{train}, \mathcal{D}_i^{test})$ describes a training and a testing dataset of a task, where each dataset contains a small number of examples. Training examples $\mathcal{D}^{train} = \{(\mathbf{x}_t, y_t)\}_{t=1}^{T}$ and testing examples $\mathcal{D}^{test} = \{(\mathbf{x}_q, y_q)\}_{q=1}^{Q}$ are sampled from the same distribution.

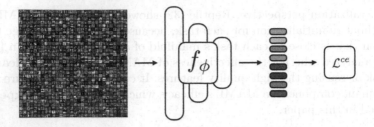

Fig. 1. In meta-training, we train on an image classification task on the merged meta-training data to learn an embedding model. This model is then re-used at meta-testing time to extract embedding for a simple linear classifier.

A base learner $\mathcal{A}$, which is given by $y_* = f_\theta(\mathbf{x}_*)$ ($*$ denotes t or q), is trained on $\mathcal{D}^{train}$ and used as a predictor on $\mathcal{D}^{test}$. Due to the high dimensionality of $\mathbf{x}_*$, the base learner $\mathcal{A}$ suffers high variance. So training examples and testing examples are mapped into a feature space by an embedding model $\boldsymbol{\Phi}_* = f_\phi(\mathbf{x}_*)$. Assume the embedding model is fixed during training the base learner on each task, then the objective of the base learner is

$$\theta = \mathcal{A}(\mathcal{D}^{train}; \phi)$$
$$= \arg\min_\theta \mathcal{L}^{base}(\mathcal{D}^{train}; \theta, \phi) + \mathcal{R}(\theta), \tag{1}$$

where $\mathcal{L}$ is the loss function and $\mathcal{R}$ is the regularization term.

The objective of the meta-learning algorithms is to learn a good embedding model, so that the average test error of the base learner on a distribution of tasks is minimized. Formally,

$$\phi = \arg\min_\phi \mathbb{E}_\mathcal{T}[\mathcal{L}^{meta}(\mathcal{D}^{test}; \theta, \phi)], \tag{2}$$

where $\theta = \mathcal{A}(\mathcal{D}^{train}; \phi)$.

Once meta-training is finished, the performance of the model is evaluated on a set of held-out tasks $\mathcal{S} = \{(\mathcal{D}_j^{train}, \mathcal{D}_j^{test})\}_{j=1}^J$, called meta-testing set. The evaluation is done over the distribution of the test tasks:

$$\mathbb{E}_\mathcal{S}[\mathcal{L}^{meta}(\mathcal{D}^{test}; \theta, \phi), \text{where } \theta = \mathcal{A}(\mathcal{D}^{train}; \phi)]. \tag{3}$$

3.2   Learning Embedding Model Through Classification

As we show in Subsect. 3.1, the goal of meta-training is to learn a transferrable embedding model f_ϕ, which generalizes to any new task. Rather than designing new meta-learning algorithms to learn the embedding model, we propose that a model pre-trained on a classification task can generate powerful embeddings for the downstream base learner. To that end, we merge tasks from meta-training set into a single task, which is given by

$$\mathcal{D}^{new} = \{(\mathbf{x}_i, y_i)\}_{k=1}^K$$
$$= \cup\{\mathcal{D}_1^{train}, \dots, \mathcal{D}_i^{train}, \dots, \mathcal{D}_I^{train}\}, \tag{4}$$

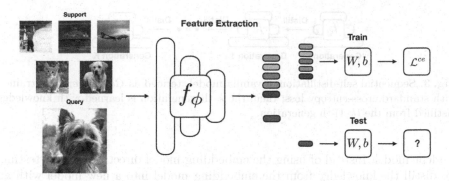

Fig. 2. We show a meta-testing case for 5-way 1-shot task: 5 support images and 1 query image are transformed into embeddings using the fixed neural network; a linear model (logistic regression (LR) in this case) is trained on 5 support embeddings; the query image is tested using the linear model.

where $\mathcal{D}_i^{train}$ is the task from $\mathcal{T}$. The embedding model is then

$$\phi = \arg\min_{\phi} \mathcal{L}^{ce}(\mathcal{D}^{new}; \phi), \tag{5}$$

and $\mathcal{L}^{ce}$ denotes the cross-entropy loss between predictions and ground-truth labels. We visualize the task in Fig. 1.

As shown in Fig. 2, for a task $(\mathcal{D}_j^{train}, \mathcal{D}_j^{test})$ sampled from meta-testing distribution, we train a base learner on $\mathcal{D}_j^{train}$. The base learner is instantiated as multivariate logistic regression. Its parameters $\theta = \{W, b\}$ include a weight term W and a bias term b, given by

$$\theta = \arg\min_{\{W,b\}} \sum_{t=1}^{T} \mathcal{L}_t^{ce}(W f_\phi(\mathbf{x}_t) + b, y_t) + \mathcal{R}(W, b). \tag{6}$$

We also evaluate other base learners such as nearest neighbor classifier with $\mathcal{L}$-2 distance and/or cosine distance in Subsect. 4.8.

In our method, the crucial difference between meta-training and meta-testing is the embedding model parameterized by ϕ is carried over from meta-training to meta-testing and kept unchanged when evaluated on tasks sampled from meta-testing set. The base learner is re-initialized for every task and trained on $\mathcal{D}^{train}$ of meta-testing task. Our method is the same with the pre-training phase of methods used in [33,41]. Unlike other methods [9,37], we *do not* fine-tune the embedding model f_ϕ during the meta-testing stage.

3.3   Sequential Self-distillation

Knowledge distillation [19] is an approach to transfer knowledge embedded in an ensemble of models to a single model, or from a larger teacher model to a smaller

Fig. 3. Sequential self-distillation: a vanilla model, termed as *Generation 0*, is trained with standard cross-entropy loss; then, the k-th generation is learned with knowledge distilled from the $(k\text{-}1)$-th generation.

student model. Instead of using the embedding model directly for meta-testing, we distill the knowledge from the embedding model into a new model with an identical architecture, training on the same merged meta-training set. The new embedding model parameterized by ϕ' is trained to minimize a weighted sum of the cross-entropy loss between the predictions and ground-truth labels and the Kullback-Leibler divergence (KL) between predictions and soft targets:

$$\phi' = \arg\min_{\phi'}(\alpha\mathcal{L}^{ce}(\mathcal{D}^{new};\phi')$$
$$+ \beta KL(f(\mathcal{D}^{new};\phi'), f(\mathcal{D}^{new};\phi))), \tag{7}$$

where usually $\beta = 1 - \alpha$.

We exploit the Born-again [12] strategy to apply KD sequentially to generate multiple generations, which is shown in Fig. 3. At each step, the embedding model of k-th generation is trained with knowledge transferred from the embedding model of $(k\text{-}1)$-th generation:

$$\phi_k = \arg\min_{\phi}(\alpha\mathcal{L}^{ce}(\mathcal{D}^{new};\phi)$$
$$+ \beta KL(f(\mathcal{D}^{new};\phi), f(\mathcal{D}^{new};\phi_{k-1}))). \tag{8}$$

We repeat the operation K times, we use ϕ_K as the embedding model to extract features for meta-testing. We analyze the sequential self-distillation in Subsect. 4.7.

4   Experiments

We conduct experiments on five widely used few-shot image recognition benchmarks: miniImageNet [51], tieredImageNet [40], CIFAR-FS [3], and FC100 [33], and Meta-Dataset [50].

4.1   Setup

Architecture. Following previous works [9,26,29,33,39], we use a ResNet12 as our backbone: the network consists of 4 residual blocks, where each has 3 convolutional layers with 3×3 kernel; a 2×2 max-pooling layer is applied after each of the first 3 blocks; and a global average-pooling layer is on top

Table 1. Comparison to prior work on miniImageNet and tieredImageNet. Average few-shot classification accuracies (%) with 95% confidence intervals on mini-ImageNet and tieredImageNet meta-test splits. Results reported with input image size of 84 × 84. a-b-c-d denotes a 4-layer convolutional network with a, b, c, and d filters in each layer. [†] results obtained by training on the union of training and validation sets.

Model	Backbone	MiniImageNet 5-way		TieredImageNet 5-way	
		1-shot	5-shot	1-shot	5-shot
MAML [11]	32-32-32-32	48.70 ± 1.84	63.11 ± 0.92	51.67 ± 1.81	70.30 ± 1.75
Matching networks [51]	64-64-64-64	43.56 ± 0.84	55.31 ± 0.73	-	-
IMP [2]	64-64-64-64	49.2 ± 0.7	64.7 ± 0.7	-	-
Prototypical networks[†] [43]	64-64-64-64	49.42 ± 0.78	68.20 ± 0.66	53.31 ± 0.89	72.69 ± 0.74
TAML [22]	64-64-64-64	51.77 ± 1.86	66.05 ± 0.85	-	-
SAML [16]	64-64-64-64	52.22 ± n/a	66.49 ± n/a	-	-
GCR [27]	64-64-64-64	53.21 ± 0.80	72.34 ± 0.64	-	-
KTN(Visual) [34]	64-64-64-64	54.61 ± 0.80	71.21 ± 0.66	-	-
PARN[57]	64-64-64-64	55.22 ± 0.84	71.55 ± 0.66	-	-
Dynamic few-shot [15]	64-64-128-128	56.20 ± 0.86	73.00 ± 0.64	-	-
Relation networks [45]	64-96-128-256	50.44 ± 0.82	65.32 ± 0.70	54.48 ± 0.93	71.32 ± 0.78
R2D2 [3]	96-192-384-512	51.2 ± 0.6	68.8 ± 0.1	-	-
SNAIL [29]	ResNet-12	55.71 ± 0.99	68.88 ± 0.92	-	-
AdaResNet [31]	ResNet-12	56.88 ± 0.62	71.94 ± 0.57	-	-
TADAM [33]	ResNet-12	58.50 ± 0.30	76.70 ± 0.30	-	-
Shot-Free [39]	ResNet-12	59.04 ± n/a	77.64 ± n/a	63.52 ± n/a	82.59 ± n/a
TEWAM [35]	ResNet-12	60.07 ± n/a	75.90 ± n/a	-	-
MTL [44]	ResNet-12	61.20 ± 1.80	75.50 ± 0.80	-	-
Variational FSL [60]	ResNet-12	61.23 ± 0.26	77.69 ± 0.17	-	-
MetaOptNet [26]	ResNet-12	62.64 ± 0.61	78.63 ± 0.46	65.99 ± 0.72	81.56 ± 0.53
Diversity w/ Cooperation [10]	ResNet-18	59.48 ± 0.65	75.62 ± 0.48	-	-
Fine-tuning [9]	WRN-28-10	57.73 ± 0.62	78.17 ± 0.49	66.58 ± 0.70	85.55 ± 0.48
LEO-trainval[†] [41]	WRN-28-10	61.76 ± 0.08	77.59 ± 0.12	66.33 ± 0.05	81.44 ± 0.09
Ours-simple	ResNet-12	62.02 ± 0.63	79.64 ± 0.44	69.74 ± 0.72	84.41 ± 0.55
Ours-distill	ResNet-12	**64.82 ± 0.60**	**82.14 ± 0.43**	**71.52 ± 0.69**	**86.03 ± 0.49**

of the fourth block to generate the feature embedding. Similar to [26], we use Dropblock as a regularizer and change the number of filters from (64,128,256,512) to (64,160,320,640). As a result, our ResNet12 is identical to that used in [26,39].

Optimization Setup. We use SGD optimizer with a momentum of 0.9 and a weight decay of $5e^{-4}$. Each batch consists of 64 samples. The learning rate is initialized as 0.05 and decayed with a factor of 0.1 by three times for all datasets, except for miniImageNet where we only decay twice as the third decay has no effect. We train 100 epochs for miniImageNet, 60 epochs for tieredImageNet,

Table 2. Comparison to prior work on CIFAR-FS and FC100. Average few-shot classification accuracies (%) with 95% confidence intervals on CIFAR-FS and FC100. a-b-c-d denotes a 4-layer convolutional network with a, b, c, and d filters in each layer.

Model	Backbone	CIFAR-FS 5-way		FC100 5-way	
		1-shot	5-shot	1-shot	5-shot
MAML [11]	32-32-32-32	58.9 ± 1.9	71.5 ± 1.0	-	-
Prototypical Networks [43]	64-64-64-64	55.5 ± 0.7	72.0 ± 0.6	35.3 ± 0.6	48.6 ± 0.6
Relation Networks [45]	64-96-128-256	55.0 ± 1.0	69.3 ± 0.8	-	-
R2D2 [3]	96-192-384-512	65.3 ± 0.2	79.4 ± 0.1	-	-
TADAM [33]	ResNet-12	-	-	40.1 ± 0.4	56.1 ± 0.4
Shot-Free [39]	ResNet-12	$69.2 \pm$ n/a	$84.7 \pm$ n/a	-	-
TEWAM [35]	ResNet-12	$70.4 \pm$ n/a	$81.3 \pm$ n/a	-	-
Prototypical Networks [43]	ResNet-12	72.2 ± 0.7	83.5 ± 0.5	37.5 ± 0.6	52.5 ± 0.6
MetaOptNet [26]	ResNet-12	72.6 ± 0.7	84.3 ± 0.5	41.1 ± 0.6	55.5 ± 0.6
Ours-simple	ResNet-12	71.5 ± 0.8	86.0 ± 0.5	42.6 ± 0.7	59.1 ± 0.6
Ours-distill	ResNet-12	$\mathbf{73.9 \pm 0.8}$	$\mathbf{86.9 \pm 0.5}$	$\mathbf{44.6 \pm 0.7}$	$\mathbf{60.9 \pm 0.6}$

and 90 epochs for both CIFAR-FS and FC100. During distillation, we use the same learning schedule and set $\alpha = \beta = 0.5$.

Data Augmentation. When training the embedding network on transformed meta-training set, we adopt random crop, color jittering, and random horizontal flip as in [26]. For meta-testing stage, we train an N-way logistic regression base classifier. We use the implementations in scikit-learn [1].

4.2   Results on ImageNet Derivatives

The miniImageNet dataset [51] is a standard benchmark for few-shot learning algorithms for recent works. It consists of 100 classes randomly sampled from the ImageNet; each class contains 600 downsampled images of size 84×84. We follow the widely-used splitting protocol proposed in [38], which uses 64 classes for meta-training, 16 classes for meta-validation, and 20 classes for meta-testing.

The tieredImageNet dataset [40] is another subset of ImageNet but has more classes (608 classes). These classes are first grouped into 34 higher-level categories, which are further divided into 20 training categories (351 classes), 6 validation categories (97 classes), and 8 testing categories (160 classes). Such construction ensures the training set is distinctive enough from the testing set and makes the problem more challenging.

Results. During meta-testing, we evaluate our method with 3 runs, where in each run the accuracy is the mean accuracy of 1000 randomly sampled tasks. We report the median of 3 runs in Table 1. Our simple baseline with ResNet-12

Table 3. Results on Meta-Dataset. Average accuracy (%) is reported with variable number of ways and shots, following the setup in [50]. We compare four variants of out method (LR, SVM, LR-distill, and SVM-distill) to the best accuracy over 7 methods in [50]. In each episode, 1000 tasks are sampled for evaluation.

Trained on ILSVRC train split					
Dataset	Best from [50]	LR (ours)	SVM (ours)	LR-distill (ours)	SVM-distill (ours)
ILSVRC	50.50	60.14	56.48	**61.48**	58.33
Omniglot	63.37	64.92	65.90	64.31	**66.77**
Aircraft	**68.69**	63.12	61.43	62.32	64.23
Birds	68.66	77.69	74.61	**79.47**	76.63
Textures	69.05	78.59	74.25	**79.28**	76.66
Quick draw	51.52	**62.48**	59.34	60.83	59.02
Fungi	39.96	47.12	41.76	**48.53**	44.51
VGG flower	87.15	**91.60**	90.32	91.00	89.66
Traffic signs	66.79	77.51	**78.94**	76.33	78.64
MSCOCO	43.74	57.00	50.81	**59.28**	54.10
Mean accuracy	60.94	68.02	65.38	**68.28**	66.86

is already comparable with the state-of-the-art MetaOptNet [26] on miniImageNet, and outperforms all previous works by at least 3% on tieredImageNet. The network trained with distillation further improves by 2–3%.

We notice that previous works [33,36,41,44] have also leveraged the standard cross-entropy pre-training on the meta-training set. In [33,41], a wide ResNet (WRN-28-10) is trained to classify all classes in the meta-training set (or combined meta-training and meta-validation set), and then frozen during the meta-training stage. [9] also conducts pre-training but the model is fine-tuned using the support images in meta-testing set, achieving 57.73 ± 0.62. We adopt the same architecture and gets 61.1 ± 0.86. So fine-tuning on small set of samples makes the performance worse. Another work [33] adopts a multi-task setting by jointly training on the standard classification task and few-shot classification (5-way) task. In another work [44], the ResNet-12 is pre-trained before mining hard tasks for the meta-training stage.

4.3 Results on CIFAR Derivatives

The CIFAR-FS dataset [3] is a derivative of the original CIFAR-100 dataset by randomly splitting 100 classes into 64, 16 and 20 classes for training, validation, and testing, respectively. The FC100 dataset [33] is also derived from CIFAR-100 dataset in a similar way to tieredImagNnet. This results in 60 classes for training, 20 classes for validation, and 20 classes for testing.

Table 4. Comparsions of embeddings from supervised pre-training and self-supervised pre-training (Moco and CMC). * the encoder of each view is 0.5× width of a normal ResNet-50.

Model	Backbone	MiniImageNet 5-way	
		1-shot	5-shot
Supervised	ResNet50	**57.56 ± 0.79**	73.81 ± 0.63
MoCo [17]	ResNet50	54.19 ± 0.93	73.04 ± 0.61
CMC [46]	ResNet50*	56.10 ± 0.89	**73.87 ± 0.65**

Table 5. Ablation study on four benchmarks with ResNet-12 as backbone network. "NN" and "LR" stand for nearest neighbour classifier and logistic regression. "$\mathcal{L}$-2" means feature normalization after which feature embeddings are on the unit sphere. "Aug" indicates that each support image is augmented into 5 samples to train the classifier. "Distill" represents the use of knowledge distillation.

NN	LR	$\mathcal{L}$-2	Aug	Distill	MiniImageNet		TieredImageNet		CIFAR-FS		FC100	
					1-shot	5-shot	1-shot	5-shot	1-shot	5-shot	1-shot	5-shot
✓					56.29	69.96	64.80	78.75	64.36	78.00	38.40	49.12
	✓				58.74	78.31	67.62	84.77	66.92	84.78	40.36	57.23
	✓	✓			61.56	79.27	69.53	85.08	71.24	85.63	42.77	58.86
	✓	✓	✓		62.02	79.64	69.74	85.23	71.45	85.95	42.59	59.13
	✓	✓	✓	✓	64.82	82.14	71.52	86.03	73.89	86.93	44.57	60.91

Results. Similar to previous experiments, we evaluate our method with 3 runs, where in each run the accuracy is the mean accuracy of 3000 randomly sampled tasks. Table 2 summarizes the results, which shows that our simple baseline is comparable to Prototypical Networks [43] and MetaOptNet [26] on CIFAR-FS dataset, and outperforms both of them on FC100 dataset. Our distillation version achieves the new state-of-the-art on both datasets. This verifies our hypothesis that a good embedding plays an important role in few-shot recognition.

4.4 Results on Meta-Dataset

Meta-Dataset [50] is a new benchmark for evaluating few-shot methods in large-scale settings. Compared to miniImageNet and tieredImageNet, Meta-Dataset provides more diverse and realistic samples.

Setup. The ILSVRC (ImageNet) subset consists of 712 classes for training, 158 classes for validation, and 130 classes for testing. We follow the setting in Meta-Dateset [50] where the embedding model is trained solely on the ILSVRC training split. We use ResNet-18 [18] as the backbone network. The input size is 128 × 128. In the pre-training stage, we use SGD optimizer with a momentum of 0.9. The learning rate is initially 0.1 and decayed by a factor of 10 for every 30 epochs. We train the model for 90 epochs in total. The batch size is 256. We use standard data augmentation, including randomly resized crop and horizontal flip. In the distillation stage, we set $\alpha = 0.5$ and $\beta = 1.0$. We perform distillation twice and use the model from the second generation for meta-testing. We do not use test-time augmentation in meta-testing. In addition to logistic regression (LR), we also provide results of linear SVM for completeness.

We select the best results from [50] for comparison – for each testing subset, we pick the best accuracy over 7 methods and 3 different architectures including 4-layer ConvNet, Wide ResNet, and ResNet-18. As shown in Table 3, our simple baselines clearly outperform the best results from [50] on 9 out of 10 testing datasets, often by a large margin. Our baseline method using LR outperforms previous best results by more than 7% on average. Also, self-distillation improves `max(LR, SVM)` in 7 out of the 10 testing subsets. Moreover, we notice empirically that logistic regression (LR) performs better than linear SVM.

4.5   Embeddings from Self-supervised Representation Learning

Using unsupervised learning [17,46,48,56] to improve the generalization of the meta-learning algorithms [53] removes the needs of data annotation. In addition to using embeddings from supervised pre-training, we also train a linear classifier on embeddings from self-supervised representation learning. Following MoCo [17] and CMC [46] (both are inspired by InstDis [56]), we train a ResNet50 [18] (without using labels) on the merged meta-training set to learn an embedding model. We compare unsupervised ResNet50 to a supervised ResNet50. From Table 4, we observe that using embeddings from self-supervised ResNet50 is only slightly worse than using embeddings from supervised ResNet50 (in 5-shot setting, the results are comparable). This observation shows the potential of self-supervised learning in the scenario of few-shot learning.

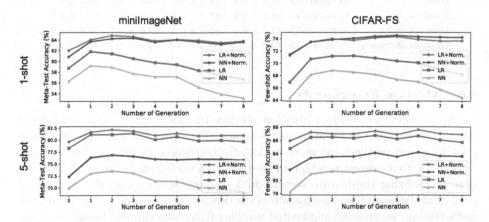

Fig. 4. Evaluation on different generations of distilled networks. The 0-th generation (or root generation) indicates the vanilla network trained with only standard cross-entropy loss. The k-th generation is trained by combining the standard classification loss and the knowledge distillation (KD) loss using the $(k\text{-}1)$-th generation as the teacher model. Logistic regression (LR) and nearest neighbours (NN) are evaluated.

Table 6. Comparisons of different backbones on miniImageNet and tieredImageNet.

Model	Backbone	MiniImageNet 5-way		TieredImageNet 5-way	
		1-shot	5-shot	1-shot	5-shot
Ours	64-64-64-64	55.25 ± 0.58	71.56 ± 0.52	56.18 ± 0.70	72.99 ± 0.55
Ours-distill	64-64-64-64	55.88 ± 0.59	71.65 ± 0.51	56.76 ± 0.68	73.21 ± 0.54
Ours-trainval	64-64-64-64	56.32 ± 0.58	72.46 ± 0.52	56.53 ± 0.68	73.15 ± 0.58
Ours-distill-trainval	64-64-64-64	**56.64 ± 0.58**	**72.85 ± 0.50**	**57.35 ± 0.70**	**73.98 ± 0.56**
Ours	ResNet-12	62.02 ± 0.63	79.64 ± 0.44	69.74 ± 0.72	84.41 ± 0.55
Ours-distill	ResNet-12	64.82 ± 0.60	82.14 ± 0.43	71.52 ± 0.69	86.03 ± 0.49
Ours-trainval	ResNet-12	63.59 ± 0.61	80.86 ± 0.47	71.12 ± 0.68	85.94 ± 0.46
Ours-distill-trainval	ResNet-12	**66.58 ± 0.65**	**83.22 ± 0.39**	**72.98 ± 0.71**	**87.46 ± 0.44**
Ours	SEResNet-12	62.29 ± 0.60	79.94 ± 0.46	70.31 ± 0.70	85.22 ± 0.50
Ours-distill	SEResNet-12	65.96 ± 0.63	82.05 ± 0.46	71.72 ± 0.69	86.54 ± 0.49
Ours-trainval	SEResNet-12	64.07 ± 0.61	80.92 ± 0.43	71.76 ± 0.66	86.27 ± 0.45
Ours-distill-trainval	SEResNet-12	**67.73 ± 0.63**	**83.35 ± 0.41**	**72.55 ± 0.69**	**86.72 ± 0.49**

Table 7. Comparisons of different backbones on CIFAR-FS and FC100.

Model	Backbone	CIFAR-FS 5-way		FC100 5-way	
		1-shot	5-shot	1-shot	5-shot
Ours	64-64-64-64	62.7 ± 0.8	78.7 ± 0.5	39.6 ± 0.6	53.5 ± 0.5
Ours-distill	64-64-64-64	63.8 ± 0.8	79.5 ± 0.5	40.3 ± 0.6	54.1 ± 0.5
Ours-trainval	64-64-64-64	63.5 ± 0.8	79.8 ± 0.5	43.2 ± 0.6	58.5 ± 0.5
Ours-distill-trainval	64-64-64-64	**64.9 ± 0.8**	**80.3 ± 0.5**	**44.6 ± 0.6**	**59.2 ± 0.5**
Ours	ResNet-12	71.5 ± 0.8	86.0 ± 0.5	42.6 ± 0.7	59.1 ± 0.6
Ours-distill	ResNet-12	73.9 ± 0.8	86.9 ± 0.5	44.6 ± 0.7	60.9 ± 0.6
Ours-trainval	ResNet-12	73.1 ± 0.8	86.7 ± 0.5	49.5 ± 0.7	66.4 ± 0.6
Ours-distill-trainval	ResNet-12	**75.4 ± 0.8**	**88.2 ± 0.5**	**51.6 ± 0.7**	**68.4 ± 0.6**
Ours	SEResNet-12	72.0 ± 0.8	86.0 ± 0.6	43.4 ± 0.6	59.1 ± 0.6
Ours-distill	SEResNet-12	74.2 ± 0.8	87.2 ± 0.5	44.9˙± 0.6	61.4 ± 0.6
Ours-trainval	SEResNet-12	73.3 ± 0.8	86.8 ± 0.5	49.9 ± 0.7	66.8 ± 0.6
Ours-distill-trainval	SEResNet-12	**75.6 ± 0.8**	**88.2 ± 0.5**	**52.0 ± 0.7**	**68.8 ± 0.6**

4.6   Ablation Experiments

In this section, we conduct ablation studies to analyze how each component affects the few-shot recognition performance. We study the following five components of our method: (a) we chose logistic regression as our base learner, and compare it to a nearest neighbour classifier with euclidean distance; (b) we find that normalizing the feature vectors onto the unit sphere, e.g., $\mathcal{L}$-2 normalization, could improve the classification of the downstream base classifier; (c) during meta-testing, we create 5 augmented samples from each support image to alleviate the data insufficiency problem, and using these augmented samples to train the linear classifier; (d) we distill the embedding network on the training set by following the sequential distillation [12] strategy.

Table 5 shows the results of our ablation studies on miniImageNet, tiered-ImageNet, CIFAR-FS, and FC100. In general, logistic regression significantly outperforms the nearest neighbour classifier, especially for the 5-shot case; $\mathcal{L}$-2 normalization consistently improves the 1-shot accuracy by 2% on all datasets; augmenting the support images leads to marginal improvement; even with all these techniques, distillation can still provide 2% extra gain.

4.7   Effects of Distillation

We use sequential self-distillation to get an embedding model, similar to the one in Born-again networks [12]. We investigate the effect of this strategy on the performance of downstream few-shot classification.

In addition to logistic regression and nearest-neighbour classifiers, we also look into a cosine similarity classifier, which is equivalent to the nearest-neighbour classifier but with normalized features (noted as "NN+Norm."). The plots of 1-shot and 5-shot results on miniImageNet and CIFAR-FS are shown in Fig. 4. The 0-th generation (or root generation) refers to the vanilla model trained with only standard cross-entropy loss, and the $(k–1)$-th generation is distilled into k-th generation. In general, few-shot recognition performance keeps getting better in the first two or three generations. After certain number of generations, the accuracy starts decreasing for logistic regression and nearest neighbour. Normalizing the features can significantly alleviate this problem. In Table 1, Table 2, and Table 5, we evalute the model of the second generation on miniImageNet, CIFAR-FS and FC100 datasets; we use the first generation on tieredImageNet. Model selection is done on the validation set.

4.8   Choice of Base Classifier

One might argue in the 1-shot case, that a linear classifier should behavior similarly to a nearest-neighbour classifier. However in Table 5 and Fig. 4, we find that logistic regression is clearly better than nearest-neighbour. We argue that this is casued by the scale of the features. After we normalize the features by the $\mathcal{L}$-2 norm, logistic regression ("LR+Norm") performs similarly to the nearest neighbour classifier ("NN+Norm."), as shown in the first row of Fig. 4. However, when increasing the size of the support set to 5, logistic regression is significantly better than nearest-neighbour even after feature normalization

4.9   Comparsions of Different Network Backbones

To further verify our assumption that the key success of few-shot learning algorithms is due to the quality of embeddings, we compare three alternatives in Table 6 and Table 7: a ConvNet with four convolutional layers (64, 64, 64, 64); a ResNet12 as in Table 1; a ResNet12 with sequeeze-and-excitation [20] modules. For each model, we have four settings: training on meta-training set; training and distilling on meta-training set; training on meta-training set and meta-validation set; training and distilling on meta-training set and meta-validation set. The results consistently improve with more data and better networks. This is inline with our hypothesis: embeddings are the most critical factor to the performance of few-shot learning/meta learning algorithms; better embeddings will lead to better few-shot testing performance (even with a simple linear classier). In addition, our ConvNet model also outperforms other few-shot learning and/or meta learning models using the same network. This verifies that in both small model regime (ConvNet) and large model regime (ResNet), few-shot learning and meta learning algorithms are *no better* than learning a good embedding model.

4.10   Multi-task vs Multi-way Classification?

We are interested in understanding whether the efficacy of our simple baseline is due to multi-task or multi-way classification. We compare to training an embedding model through *multi-task* learning: a model with shared embedding network and different classification heads is constructed, where each head is only classifying the corresponding category; then we use the embedding model to extract features as we do with our baseline model. This achieves 58.53 ± 0.8 on mini-ImageNet 5-way 1-shot case, compared to our baseline model which is 62.02 ± 0.63. So we argue that the speciality of our setting, where the few-shot classification tasks are mutually exclusive and can be merged together into a single *multi-way* classification task, makes the simple model effective.

Acknowledgement. The authors thank Hugo Larochelle and Justin Solomon for helpful discussions and feedback on this manuscript. This research was supported in part by iFlytek. This material was also based in part upon work supported by the Defense Advanced Research Projects Agency (DARPA) under Contract No. FA8750-19-C-1001.

References

1. Machine learning in python. https://scikit-learn.org/stable/
2. Allen, K., Shelhamer, E., Shin, H., Tenenbaum, J.: Infinite mixture prototypes for few-shot learning. In: ICML (2019)
3. Bertinetto, L., Henriques, J.F., Torr, P.H., Vedaldi, A.: Meta-learning with differentiable closed-form solvers. arXiv preprint arXiv:1805.08136 (2018)
4. Buciluă, C., Caruana, R., Niculescu-Mizil, A.: Model compression. In: SIGKDD (2006)
5. Chen, W.Y., Liu, Y.C., Kira, Z., Wang, Y.C., Huang, J.B.: A closer look at few-shot classification. In: ICLR (2019)
6. Chen, Y., Wang, X., Liu, Z., Xu, H., Darrell, T.: A new meta-baseline for few-shot learning. ArXiv abs/2003.04390 (2020)
7. Clark, K., Luong, M.T., Manning, C.D., Le, Q.V.: Bam! born-again multi-task networks for natural language understanding. In: ACL (2019)
8. Deng, J., Dong, W., Socher, R., Li, L.J., Li, K., Fei-Fei, L.: ImageNet: a large-scale hierarchical image database. In: CVPR09 (2009)
9. Dhillon, G.S., Chaudhari, P., Ravichandran, A., Soatto, S.: A baseline for few-shot image classification. In: ICLR (2020)
10. Dvornik, N., Schmid, C., Mairal, J.: Diversity with cooperation: ensemble methods for few-shot classification. In: ICCV (2019)

11. Finn, C., Abbeel, P., Levine, S.: Model-agnostic meta-learning for fast adaptation of deep networks. In: ICML (2017)
12. Furlanello, T., Lipton, Z.C., Tschannen, M., Itti, L., Anandkumar, A.: Born-again neural networks. In: ICML (2018)
13. Gan, C., Gong, B., Liu, K., Su, H., Guibas, L.J.: Geometry guided convolutional neural networks for self-supervised video representation learning. In: CVPR (2018)
14. Gan, C., Zhao, H., Chen, P., Cox, D., Torralba, A.: Self-supervised moving vehicle tracking with stereo sound. In: ICCV (2019)
15. Gidaris, S., Komodakis, N.: Dynamic few-shot visual learning without forgetting. In: CVPR (2018)
16. Hao, F., He, F., Cheng, J., Wang, L., Cao, J., Tao, D.: Collect and select: semantic alignment metric learning for few-shot learning. In: ICCV (2019)
17. He, K., Fan, H., Wu, Y., Xie, S., Girshick, R.B.: Momentum contrast for unsupervised visual representation learning. ArXiv abs/1911.05722 (2019)
18. He, K., Zhang, X., Ren, S., Sun, J.: Deep residual learning for image recognition. In: CVPR (2016)
19. Hinton, G., Vinyals, O., Dean, J.: Distilling the knowledge in a neural network. In: NIPS Deep Learning and Representation Learning Workshop (2015)
20. Hu, J., Shen, L., Sun, G.: Squeeze-and-excitation networks. In: CVPR (2018)
21. Huang, S., Tao, D.: All you need is a good representation: A multi-level and classifier-centric representation for few-shot learning. ArXiv abs/1911.12476 (2019)
22. Jamal, M.A., Qi, G.J.: Task agnostic meta-learning for few-shot learning. In: CVPR (2019)
23. Koch, G., Zemel, R., Salakhutdinov, R.: Siamese neural networks for one-shot image recognition. In: ICML Deep Learning Workshop (2015)
24. Lake, B.M., Salakhutdinov, R., Tenenbaum, J.B.: Human-level concept learning through probabilistic program induction. Science 350(6266), 1332–1338 (2015)
25. Lake, B.M., Salakhutdinov, R., Tenenbaum, J.B.: The Omniglot challenge: a 3-year progress report. Curr. Opin. Behav. Sci. 29, 97–104 (2019)
26. Lee, K., Maji, S., Ravichandran, A., Soatto, S.: Meta-learning with differentiable convex optimization. In: CVPR (2019)
27. Li, A., Luo, T., Xiang, T., Huang, W., Wang, L.: Few-shot learning with global class representations. In: ICCV (2019)
28. Li, H., Eigen, D., Dodge, S., Zeiler, M., Wang, X.: Finding task-relevant features for few-shot learning by category traversal. In: CVPR (2019)
29. Mishra, N., Rohaninejad, M., Chen, X., Abbeel, P.: A simple neural attentive meta-learner. arXiv preprint arXiv:1707.03141 (2017)
30. Mobahi, H., Farajtabar, M., Bartlett, P.L.: Self-distillation amplifies regularization in hilbert space. arXiv preprint arXiv:2002.05715 (2020)
31. Munkhdalai, T., Yuan, X., Mehri, S., Trischler, A.: Rapid adaptation with conditionally shifted neurons. arXiv preprint arXiv:1712.09926 (2017)
32. Nichol, A., Achiam, J., Schulman, J.: On first-order meta-learning algorithms. ArXiv abs/1803.02999 (2018)
33. Oreshkin, B., López, P.R., Lacoste, A.: Tadam: task dependent adaptive metric for improved few-shot learning. In: NIPS (2018)
34. Peng, Z., Li, Z., Zhang, J., Li, Y., Qi, G.J., Tang, J.: Few-shot image recognition with knowledge transfer. In: ICCV (2019)
35. Qiao, L., Shi, Y., Li, J., Wang, Y., Huang, T., Tian, Y.: Transductive episodic-wise adaptive metric for few-shot learning. In: ICCV (2019)
36. Qiao, S., Liu, C., Shen, W., Yuille, A.L.: Few-shot image recognition by predicting parameters from activations. In: CVPR (2018)

37. Raghu, A., Raghu, M., Bengio, S., Vinyals, O.: Rapid learning or feature reuse? towards understanding the effectiveness of maml. arXiv preprint arXiv:1909.09157 (2019)
38. Ravi, S., Larochelle, H.: Optimization as a model for few-shot learning. In: ICLR (2017)
39. Ravichandran, A., Bhotika, R., Soatto, S.: Few-shot learning with embedded class models and shot-free meta training. In: ICCV (2019)
40. Ren, M., et al.: Meta-learning for semi-supervised few-shot classification. In: ICLR (2018)
41. Rusu, A.A., Rao, D., Sygnowski, J., Vinyals, O., Pascanu, R., Osindero, S., Hadsell, R.: Meta-learning with latent embedding optimization. In: ICLR (2019)
42. Scott, T., Ridgeway, K., Mozer, M.C.: Adapted deep embeddings: a synthesis of methods for k-shot inductive transfer learning. In: NIPS (2018)
43. Snell, J., Swersky, K., Zemel, R.: Prototypical networks for few-shot learning. In: NIPS (2017)
44. Sun, Q., Liu, Y., Chua, T.S., Schiele, B.: Meta-transfer learning for few-shot learning. In: CVPR (2019)
45. Sung, F., Yang, Y., Zhang, L., Xiang, T., Torr, P.H., Hospedales, T.M.: Learning to compare: Relation network for few-shot learning. In: CVPR (2018)
46. Tian, Y., Krishnan, D., Isola, P.: Contrastive multiview coding. arXiv preprint arXiv:1906.05849 (2019)
47. Tian, Y., Krishnan, D., Isola, P.: Contrastive representation distillation. arXiv preprint arXiv:1910.10699 (2019)
48. Tian, Y., Sun, C., Poole, B., Krishnan, D., Schmid, C., Isola, P.: What makes for good views for contrastive learning? arXiv preprint arXiv:2005.10243 (2020)
49. Triantafillou, E., Zemel, R.S., Urtasun, R.: Few-shot learning through an information retrieval lens. In: NIPS (2017)
50. Triantafillou, E., et al.: Meta-dataset: a dataset of datasets for learning to learn from few examples. arXiv preprint arXiv:1903.03096 (2019)
51. Vinyals, O., Blundell, C., Lillicrap, T., kavukcuoglu, K., Wierstra, D.: Matching networks for one shot learning. In: NIPS (2016)
52. Wang, Y.X., Girshick, R.B., Hebert, M., Hariharan, B.: Low-shot learning from imaginary data. In: CVPR (2018)
53. Wang, Y.X., Hebert, M.: Learning from small sample sets by combining unsupervised meta-training with CNNs. Adv. Neural Inform. Process. Syst. 29, 244–252 (2016)
54. Wang, Y., Hebert, M.: Learning to learn: model regression networks for easy small sample learning. In: ECCV (2016)
55. Weng, L.: Meta-learning: Learning to learn fast. lilianweng.github.io/lil-log (2018). http://lilianweng.github.io/lil-log/2018/11/29/meta-learning.html
56. Wu, Z., Xiong, Y., Yu, S.X., Lin, D.: Unsupervised feature learning via nonparametric instance discrimination. In: CVPR (2018)
57. Wu, Z., Li, Y., Guo, L., Jia, K.: Parn: position-aware relation networks for few-shot learning. In: ICCV (2019)
58. Ye, H.J., Hu, H., Zhan, D.C., Sha, F.: Learning embedding adaptation for few-shot learning. CoRR abs/1812.03664 (2018)
59. Yim, J., Joo, D., Bae, J., Kim, J.: A gift from knowledge distillation: fast optimization, network minimization and transfer learning. In: CVPR (2017)
60. Zhang, J., Zhao, C., Ni, B., Xu, M., Yang, X.: Variational few-shot learning. In: ICCV (2019)

Adversarial Background-Aware Loss for Weakly-Supervised Temporal Activity Localization

Kyle Min$^{(\boxtimes)}$ and Jason J. Corso

University of Michigan, Ann Arbor, MI 48109, USA
{kylemin,jjcorso}@umich.edu

Abstract. Temporally localizing activities within untrimmed videos has been extensively studied in recent years. Despite recent advances, existing methods for weakly-supervised temporal activity localization struggle to recognize when an activity is not occurring. To address this issue, we propose a novel method named A2CL-PT. Two triplets of the feature space are considered in our approach: one triplet is used to learn discriminative features for each activity class, and the other one is used to distinguish the features where no activity occurs (i.e. background features) from activity-related features for each video. To further improve the performance, we build our network using two parallel branches which operate in an adversarial way: the first branch localizes the most salient activities of a video and the second one finds other supplementary activities from non-localized parts of the video. Extensive experiments performed on THUMOS14 and ActivityNet datasets demonstrate that our proposed method is effective. Specifically, the average mAP of IoU thresholds from 0.1 to 0.9 on the THUMOS14 dataset is significantly improved from 27.9% to 30.0%.

Keywords: A2CL-PT, Temporal activity localization, Adversarial learning, Weakly-supervised learning, Center loss with a pair of triplets

1 Introduction

The main goal of temporal activity localization is to find the start and end times of activities from untrimmed videos. Many of the previous approaches are fully supervised: they expect that ground-truth annotations for temporal boundaries of each activity are accessible during training [2,11,14,20,22,26,32]. However, collecting these frame-level activity annotations is time-consuming and difficult, leading to annotation noise. Hence, a weakly-supervised version has taken foot in the community: here, one assumes that only video-level ground-truth activity labels are available. These video-level activity annotations are

Electronic supplementary material The online version of this chapter (https://doi.org/10.1007/978-3-030-58568-6_17) contains supplementary material, which is available to authorized users.

A. Vedaldi et al. (Eds.): ECCV 2020, LNCS 12359, pp. 283–299, 2020.
https://doi.org/10.1007/978-3-030-58568-6_17

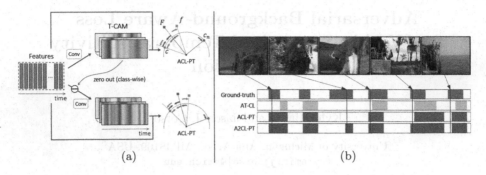

Fig. 1. (a): An illustration of the proposed A2CL-PT. F and f are aggregated video-level features where f is designed to be more attended to the background features. c is their corresponding center and c_n is the negative center. A triplet of (F, c, c_n) is used to learn discriminative features. We propose to exploit another triplet of (c, F, f) which distinguishes background features from the activity-related features. We call this method of two triplets ACL-PT. In addition, we design our network with two parallel branches so that the two separate sets of centers can be learned in an adversarial way. We call our final proposed method A2CL-PT. (b): Sample frames of a video containing *Diving* activity class from THUMOS14 dataset [5] and the corresponding results of activity localization. It is shown that our final method A2CL-PT performs the best.

much easier to collect and already exist across many datasets [6,8,15,23,31], thus weakly-supervised methods can be applied to a broader range of situations.

Current work in weakly-supervised temporal activity localization shares a common framework [9,12,16,17,19]. First, rather than using a raw video, they use a sequence of features extracted by deep networks where the features are much smaller than the raw video in size. Second, they apply a fully-connected layer to embed the pre-extracted features to the task-specific feature space. Third, they project the embedded features to the label space by applying a 1-D convolutional layer to those features. The label space has the same dimension as the number of activities, so the final output becomes a sequence of vectors that represents the classification scores for each activity over time. Each sequence of vectors is typically referred to as CAS (Class Activation Sequence) [21] or T-CAM (Temporal Class Activation Map) [17]. Finally, activities are localized by thresholding this T-CAM. T-CAM is sometimes applied with the softmax function to generate class-wise attention. This top-down attention represents the probability mass function for each activity over time.

An important component in weakly-supervised temporal activity localization is the ability to automatically determine background portions of the video where no activity is occurring. For example, BaS-Net [9] suggests using an additional suppression objective to suppress the network activations on the background portions. Nguyen et al. [18] proposes a similar objective to model the background contents. However, we argue that existing methods are not able to sufficiently distinguish background information from activities of interest for each video even though such an ability is critical to strong temporal activity localization.

To this end, we propose a novel method for the task of weakly-supervised temporal activity localization, which we call Adversarial and Angular Center Loss with a Pair of Triplets (A2CL-PT). It is illustrated in Fig. 1(a). Our key innovation is that we explicitly enable our model to capture the background region of the video while using an adversarial approach to focus on completeness of the activity learning. Our method is built on two triplets of vectors of the feature space, and one of them is designed to distinguish background portions from the activity-related parts of a video. Our method is inspired by the angular triplet-center loss (ATCL) [10] originally designed for multi-view 3D shape retrieval. Let us first describe what ATCL is and then how we develop our novel method of A2CL-PT.

In ATCL [10], a *center* is defined as a parameter vector representing the center of a cluster of feature vectors for each class. During training, the centers are updated by reducing the angular distance between the embedded features and their corresponding class centers. This groups together features that correspond to the same class and distances features from the centers of other class clusters (i.e. negative centers), making the learned feature space more useful for discriminating between classes. It follows that each training sample is a triplet of a feature vector, its center, and a negative center where the feature serves as an anchor.

Inspired by ATCL, we first formulate a loss function to learn discriminative features. ATCL cannot be directly applied to our problem because it assumes that all the features are of the same size, whereas an untrimmed video can have any number of frames. Therefore, we use a different feature representation at the video-level. We aggregate the embedded features by multiplying the top-down attention described above at each time step. The resulting video-level feature representation has the same dimension as the embedded features, so we can build a triplet whose anchor is the video-level feature vector (it is (F, c, c_n) in Fig. 1(a)). This triplet ensures that the embedded features of the same activity are grouped together and that they have high attention values at time steps when the activity occurs.

More importantly, we argue that it is possible to exploit another triplet. Let us call the features at time steps when some activity occurs *activity features*, and the ones where no activity occurs *background features*. The main idea is that the background features should be distinguished from the activity features for each video. First, we generate a new class-wise attention from T-CAM. It has higher attention values for the background features when compared to the original top-down attention. If we aggregate the embedded features with this new attention, the resulting video-level feature will be more attended to the background features than the original video-level feature is. In a discriminative feature space, the original video-level feature vector should be closer to its center than the new video-level feature vector is. This property can be achieved by using the triplet of the two different video-level feature vectors and their corresponding center where the center behaves as an anchor (it is (c, F, f) in Fig. 1(a)). The proposed triplet is novel and will be shown to be effective. Since we make use of a pair of triplets on the same feature space, we call it Angular Center Loss with a Pair of Triplets (ACL-PT).

To further improve the localization performance, we design our network to have two parallel branches which find activities in an adversarial way, also illustrated in Fig. 1(a). Using a network with a single branch may be dominated by salient activity features that are too short to localize all the activities in time. We zero out the most salient activity features localized by the first branch for each activity so that the second (adversarial) branch can find other supplementary activities from the remaining parts of the video. Here, each branch has its own set of centers which group together the features for each activity and one 1-D convolutional layer that produces T-CAM. The two adversary T-CAMs are weighted to produce the final T-CAM that is used to localize activities. We want to note that our network produces the final T-CAM with a single forward pass so it is trained in an end-to-end manner. We call our final proposed method Adversarial and Angular Center Loss with a Pair of Triplets (A2CL-PT). It is shown in Fig. 1(b) that our final method performs the best.

There are three main contributions in this paper:

- We propose a novel method using a pair of triplets. One facilitates learning discriminative features. The other one ensures that the background features are distinguishable from the activity-related features for each video.
- We build an end-to-end two-branch network by adopting an adversarial approach to localize more complete activities. Each branch comes with its own set of centers so that embedded features of the same activity can be grouped together in an adversarial way by the two branches.
- We perform extensive experiments on THUMOS14 and ActivityNet datasets and demonstrate that our method outperforms all the previous state-of-the-art approaches.

2 Related Work

Center loss (CL) [25] is recently proposed to reduce the intra-class variations of feature representations. CL learns a center for each class and penalizes the Euclidean distance between the features and their corresponding centers. Triplet-center loss (TCL) [4] shows that using a triplet of each feature vector, its corresponding center, and a nearest negative center is effective in increasing the inter-class separability. TCL enforces that each feature vector is closer to its corresponding center than to the nearest negative center by a pre-defined margin. Angular triplet-center loss (ATCL) [10] further improves TCL by using the angular distance. In ATCL, it is much easier to design a better margin because it has a clear geometric interpretation and is limited from 0 to π.

BaS-Net [9] and Nguyen et al. [18] are the leading state-of-the-art methods for weakly-supervised temporal activity localization. They take similar approaches to utilize the background portions of a video. There are other recent works without explicit usage of background information. Liu et al. [12] utilizes multi-branch

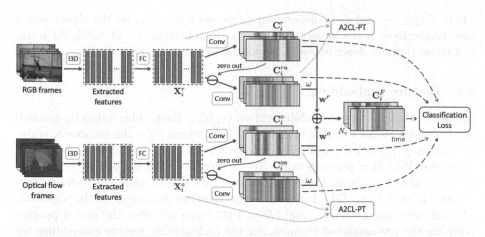

Fig. 2. An illustration of our overall architecture. It consists of two streams (RGB and optical flow), and each stream consists of two (first and adversarial) branches. Sequences of features are extracted from two input streams using pre-trained I3D networks [1]. We use two fully-connected layers with ReLU activation (FC) to compute the embedded features $\mathbf{X}_i^r, \mathbf{X}_i^o$. Next, T-CAMs $\mathbf{C}_i^r, \mathbf{C}_i^o$ are computed by applying 1-D convolutional layers (Conv). The most salient activity features localized by the first branch are zeroed out for each activity class, and the resulting features are applied with different 1-D convolutional layers (Conv) to produce $\mathbf{C}_i^{ra}, \mathbf{C}_i^{oa}$. Using the embedded features $\mathbf{X}_i^r, \mathbf{X}_i^o$ and T-CAMs $\mathbf{C}_i^r, \mathbf{C}_i^o, \mathbf{C}_i^{ra}, \mathbf{C}_i^{oa}$, we compute the term of A2CL-PT (Eq. 16). The final T-CAM $\mathbf{C}_i^F$ is computed from the four T-CAMs and these T-CAMs are used to compute the loss function for classification (Eq. 19).

network where T-CAMs of these branches differ from each other. This property is enforced by the diversity loss: the sum of the simple cosine distances between every pair of the T-CAMs. 3C-Net applies an idea of CL, but the performance is limited because CL does not consider the inter-class separability.

Using an end-to-end two-branch network that operates in an adversarial way is proposed in Adversarial Complementary Learning (ACoL) [30] for the task of weakly-supervised object localization. In ACoL, object localization maps from the first branch are used to erase the salient regions of the input feature maps for the second branch. The second branch then tries to find other complementary object areas from the remaining regions. To the best of our knowledge, we are the first to merge the idea of ACoL with center loss and to apply it to weakly-supervised temporal activity localization.

3   Method

The overview of our proposed method is illustrated in Fig. 2. The total loss function is represented as follows:

$$\mathcal{L} = \alpha \mathcal{L}_{\text{A2CL-PT}} + \mathcal{L}_{\text{CLS}} \tag{1}$$

where $\mathcal{L}_{\text{A2CL-PT}}$ and $\mathcal{L}_{\text{CLS}}$ denote our proposed loss term and the classification loss, respectively. α is a hyperparameter to control the weight of A2CL-PT term. In this section, we describe each component of our method in detail.

3.1   Feature Embedding

Let us say that we have N training videos $\{v_i\}_{i=1}^N$. Each video v_i has its ground-truth annotation for video-level label $\mathbf{y}_i \in \mathbb{R}^{N_c}$ where N_c is the number of activity classes. $\mathbf{y}_i(j) = 1$ if the activity class j is present in the video and $\mathbf{y}_i(j) = 0$ otherwise. We follow previous works [16,19] to extract the features for both RGB and optical flow streams. First, we divide v_i into non-overlapping 16-frame segments. We then apply I3D [1] pretrained on Kinetics dataset [6] to the segments. The intermediate D-dimensional ($D = 1024$) outputs after the global pooling layer are the pre-extracted features. For the task-specific feature embedding, we use two fully-connected layers with ReLU activation. As a result, sequences of the embedded features $\mathbf{X}_i^r, \mathbf{X}_i^o \in \mathbb{R}^{D \times l_i}$ are computed for RGB and optical flow stream where l_i denotes the temporal length of the features of the video v_i.

3.2   Angular Center Loss with a Pair of Triplets (ACL-PT)

For simplicity, we first look at the RGB stream. The embedded features $\mathbf{X}_i^r$ are applied with a 1-D convolutional layer. The output is T-CAM $\mathbf{C}_i^r \in \mathbb{R}^{N_c \times l_i}$ which represents the classification scores of each activity class over time. We compute class-wise attention $\mathbf{A}_i^r \in \mathbb{R}^{N_c \times l_i}$ by applying the softmax function to T-CAM:

$$\mathbf{A}_i^r(j,t) = \frac{\exp\left(\mathbf{C}_i^r(j,t)\right)}{\sum_{t'=1}^{l_i} \exp\left(\mathbf{C}_i^r(j,t')\right)} \tag{2}$$

where $j \in \{1, ..., N_c\}$ denotes each activity class and t is for each time step. Since this top-down attention represents the probability mass function of each activity over time, we can use it to aggregate the embedded features $\mathbf{X}_i^r$:

$$\mathbf{F}_i^r(j) = \sum_{t=1}^{l_i} \mathbf{A}_i^r(j,t)\mathbf{X}_i^r(t) \tag{3}$$

where $\mathbf{F}_i^r(j) \in \mathbb{R}^D$ denotes a video-level feature representation for the activity class j. Now, we can formulate a loss function that is inspired by ATCL [10] on the video-level feature representations as follows:

$$\mathcal{L}_{\text{ATCL}}^r = \frac{1}{N} \sum_{i=1}^N \sum_{j:\mathbf{y}_i(j)=1} \max\left(0, \mathcal{D}(\mathbf{F}_i^r(j), \mathbf{c}_j^r) - \mathcal{D}(\mathbf{F}_i^r(j), \mathbf{c}_{n_{i,j}^r}^r) + m_1\right) \tag{4}$$

where $\mathbf{c}_j^r \in \mathbb{R}^D$ is the center of activity class j, $n_{i,j}^r = \underset{k \neq j}{\arg\min}\, \mathcal{D}(\mathbf{F}_i^r(j), \mathbf{c}_k^r)$ is an index for the nearest negative center, and $m_1 \in [0, \pi]$ is an angular margin. It is based on the triplet of $(\mathbf{F}_i^r(j), \mathbf{c}_j^r, \mathbf{c}_{n_{i,j}^r}^r)$ that is illustrated in Fig. 1(a). Here, $\mathcal{D}(\cdot)$ represents the angular distance:

$$\mathcal{D}(\mathbf{F}_i^r(j), \mathbf{c}_j^r) = \arccos\left(\frac{\mathbf{F}_i^r(j) \cdot \mathbf{c}_j^r}{\|\mathbf{F}_i^r(j)\|_2 \|\mathbf{c}_j^r\|_2}\right) \tag{5}$$

Optimizing the loss function of Eq. 4 ensures that the video-level features of the same activity class are grouped together and that the inter-class variations of those features are maximized at the same time. As a result, the embedded features are learned to be discriminative and T-CAM will have higher values for the activity-related features.

For the next step, we exploit another triplet. We first compute a new class-wise attention $\mathbf{a}_i^r \in \mathbb{R}^{N_c \times l_i}$ from T-CAM:

$$\mathbf{a}_i^r(j, t) = \frac{\exp\left(\beta \mathbf{C}_i^r(j, t)\right)}{\sum_{t'=1}^{l_i} \exp\left(\beta \mathbf{C}_i^r(j, t')\right)} \tag{6}$$

where β is a scalar between 0 and 1. This new attention still represents the probability mass function of each activity over time, but it is supposed to have lower values for the activity features and higher values for the background features when compared to the original attention $\mathbf{A}_i^r$. Therefore, if we aggregate the embedded features $\mathbf{X}_i^r$ using $\mathbf{a}_i^r$, the resulting new video-level feature $\mathbf{f}_i^r$ should attend more strongly to the background features than $\mathbf{F}_i^r$ is. This property can be enforced by introducing a different loss function based on the new triplet of $(\mathbf{c}_j^r, \mathbf{F}_i^r(j), \mathbf{f}_i^r(j))$ that is also illustrated in Fig. 1(a):

$$\mathcal{L}_{\mathrm{NT}}^r = \frac{1}{N} \sum_{i=1}^N \sum_{j: \mathbf{y}_i(j)=1} \max\left(0, \mathcal{D}(\mathbf{F}_i^r(j), \mathbf{c}_j^r) - \mathcal{D}(\mathbf{f}_i^r(j), \mathbf{c}_j^r) + m_2\right) \tag{7}$$

where the subscript NT refers to the new triplet and $m_2 \in [0, \pi]$ is an angular margin. Optimizing this loss function makes the background features more distinguishable from the activity features. Merging the two loss functions of Eq. 4 and Eq. 7 gives us a new loss based on a pair of triplets, which we call Angular Center Loss with a Pair of Triplets (ACL-PT):

$$\mathcal{L}_{\mathrm{ACL\text{-}PT}}^r = \mathcal{L}_{\mathrm{ATCL}}^r + \gamma \mathcal{L}_{\mathrm{NT}}^r \tag{8}$$

where γ is a hyperparameter denoting the relative importance of the two losses.

Previous works on center loss [4,10,25] suggest using an averaged gradient (typically denoted as $\Delta \mathbf{c}_j^r$) to update the centers for better stability. Following this convention, the derivatives of each term of Eq. 8 with respect to the centers are averaged. For simplicity, we assume that the centers have unit length. Refer to the supplementary material for general case without such assumption. Let $\tilde{\mathcal{L}}_{\mathrm{ATCL}_{i,j}}^r$ and $\tilde{\mathcal{L}}_{\mathrm{NT}_{i,j}}^r$ be the loss terms inside the max operation of the i-th sample and of the j-th activity class as follows:

$$\tilde{\mathcal{L}}_{\mathrm{ATCL}_{i,j}}^r = \mathcal{D}(\mathbf{F}_i^r(j), \mathbf{c}_j^r) - \mathcal{D}(\mathbf{F}_i^r(j), \mathbf{c}_{n_{i,j}^r}^r) + m_1 \tag{9}$$

$$\tilde{\mathcal{L}}^r_{NT_{i,j}} = \mathcal{D}\big(\mathbf{F}^r_i(j), \mathbf{c}^r_j\big) - \mathcal{D}\big(\mathbf{f}^r_i(j), \mathbf{c}^r_j\big) + m_2 \qquad (10)$$

Next, let $\mathbf{g}^r_{1_{i,j}}$ and $\mathbf{g}^r_{2_{i,j}}$ be the derivatives of Eq. 9 with respect to $\mathbf{c}^r_j$ and $\mathbf{c}^r_{n^r_{i,j}}$, respectively; and let $\mathbf{h}^r_{i,j}$ be the derivative of Eq. 10 with respect to $\mathbf{c}^r_j$. For example, $\mathbf{g}^r_{1_{i,j}}$ is given by:

$$\mathbf{g}^r_{1_{i,j}} = -\frac{\mathbf{F}^r_i(j)}{\sin\big(\mathcal{D}\big(\mathbf{F}^r_i(j), \mathbf{c}^r_j\big)\big)\|\mathbf{F}^r_i(j)\|_2} \qquad (11)$$

Then, we can represent the averaged gradient considering the three terms:

$$\Delta\mathbf{c}^r_j = \Delta_{\mathbf{g}^r_{1_{i,j}}} + \Delta_{\mathbf{g}^r_{2_{i,j}}} + \Delta_{\mathbf{h}^r_{i,j}} \qquad (12)$$

For example, $\Delta_{\mathbf{g}^r_{1_{i,j}}}$ is computed as follows:

$$\Delta_{\mathbf{g}^r_{1_{i,j}}} = \frac{1}{N}\left(\frac{\sum_{i:\mathbf{y}_i(j)=1} \mathbf{g}^r_{1_{i,j}}\delta(\tilde{\mathcal{L}}^r_{ATCL_{i,j}} > 0)}{1 + \sum_{i:\mathbf{y}_i(j)=1}\delta(\tilde{\mathcal{L}}^r_{ATCL_{i,j}} > 0)}\right) \qquad (13)$$

Here, $\delta(condition) = 1$ if the $condition$ is true and $\delta(condition) = 0$ otherwise. Finally, the centers are updated using $\Delta\mathbf{c}^r_j$ for every iteration of the training process by a gradient descent algorithm. More details can be found in the supplementary material.

3.3 Adopting an Adversarial Approach (A2CL-PT)

We further improve the performance of the proposed ACL-PT by applying an adversarial approach inspired by ACoL [30]. For each stream, there are two parallel branches that operate in an adversarial way. The motivation is that a network with a single branch might be dominated by salient activity features that are not enough to localize all the activities in time. We zero out the most salient activity features localized by the first branch for activity class j of v_i as follows:

$$\mathbf{X}^{ra}_{i,j}(t) = \begin{cases} \mathbf{0}, & \text{if } \mathbf{C}^r_i(j,t) \in \ top\text{-}k_a \text{ elements of } \mathbf{C}^r_i(j) \\ \mathbf{X}^r_i(t), & \text{otherwise} \end{cases} \qquad (14)$$

where $\mathbf{X}^{ra}_{i,j} \in \mathbb{R}^{D\times l_i}$ denotes the input features of activity class j for the second (adversarial) branch and k_a is set to $\lfloor\frac{l_i}{s_a}\rfloor$ for a hyperparameter s_a that controls the ratio of zeroed-out features. For each activity class j, a separate 1-D convolutional layer of the adversarial branch transforms $\mathbf{X}^{ra}_{i,j}$ to the classification scores of the activity class j over time. By iterating over all the activity classes, new T-CAM $\mathbf{C}^{ra}_i \in \mathbb{R}^{N_c\times l_i}$ is computed. We argue that $\mathbf{C}^{ra}_i$ can be used to find other supplementary activities that are not localized by the first branch. By using the original features $\mathbf{X}^r_i$, new T-CAM $\mathbf{C}^{ra}_i$, and a separate set of centers $\{\mathbf{c}^{ra}_j\}^{N_c}_{j=1}$, we can compute the loss of ACL-PT for this adversarial branch $\mathcal{L}^{ra}_{ACL\text{-}PT}$ in a similar manner (Eq. 1–7). We call the sum of the losses of the two branches Adversarial and Angular Center Loss with a Pair of Triplets (A2CL-PT):

$$\mathcal{L}^r_{\text{A2CL-PT}} = \mathcal{L}^r_{\text{ACL-PT}} + \mathcal{L}^{ra}_{\text{ACL-PT}} \tag{15}$$

In addition, the losses for the optical flow stream $\mathcal{L}^o_{\text{ACL-PT}}$ and $\mathcal{L}^{oa}_{\text{ACL-PT}}$ are also computed in the same manner. As a result, the total A2CL-PT term is given by:

$$\mathcal{L}_{\text{A2CL-PT}} = \mathcal{L}^r_{\text{A2CL-PT}} + \mathcal{L}^o_{\text{A2CL-PT}} \tag{16}$$

3.4   Classification Loss

Following the previous works [12,16,19], we use the cross-entropy between the predicted pmf (probability mass function) and the ground-truth pmf of activities for classifying different activity classes in a video. We will first look at the RGB stream. For each video v_i, we compute the class-wise classification scores $\mathbf{s}^r_i \in \mathbb{R}^{N_c}$ by averaging top-k elements of $\mathbf{C}^r_i$ per activity class where k is set to $\lceil \frac{l_i}{s} \rceil$ for a hyperparameter s. Then, the softmax function is applied to compute the predicted pmf of activities $\mathbf{p}^r_i \in \mathbb{R}^{N_c}$. The ground-truth pmf $\mathbf{q}_i$ is obtained by l_1-normalizing $\mathbf{y}_i$. Then, the classification loss for the RGB stream is:

$$\mathcal{L}^r_{\text{CLS}} = \frac{1}{N} \sum_{i=1}^{N} \sum_{j=1}^{N_c} -\mathbf{q}_i(j) \log \left(\mathbf{p}^r_i(j) \right) \tag{17}$$

The classification loss for the optical flow stream $\mathcal{L}^o_{\text{CLS}}$ is computed in a similar manner. $\mathcal{L}^{ra}_{\text{CLS}}$ and $\mathcal{L}^{oa}_{\text{CLS}}$ of adversarial branches are also computed in the same way.

Finally, we compute the final T-CAM $\mathbf{C}^F_i$ from the four T-CAMs (two from the RGB stream: $\mathbf{C}^r_i, \mathbf{C}^{ra}_i$, two from the optical flow stream: $\mathbf{C}^o_i, \mathbf{C}^{oa}_i$) as follows:

$$\mathbf{C}^F_i = \mathbf{w}^r \cdot (\mathbf{C}^r_i + \omega \mathbf{C}^{ra}_i) + \mathbf{w}^o \cdot (\mathbf{C}^o_i + \omega \mathbf{C}^{oa}_i) \tag{18}$$

where $\mathbf{w}^r, \mathbf{w}^o \in \mathbb{R}^{N_c}$ are class-specific weighting parameters that are learned during training and ω is a hyperparameter for the relative importance of T-CAMs from the adversarial branch. We can then compute the classification loss for the final T-CAM $\mathcal{L}^F_{\text{CLS}}$ in the same manner. The total classification loss is given by:

$$\mathcal{L}_{\text{CLS}} = \mathcal{L}^r_{\text{CLS}} + \mathcal{L}^{ra}_{\text{CLS}} + \mathcal{L}^o_{\text{CLS}} + \mathcal{L}^{oa}_{\text{CLS}} + \mathcal{L}^F_{\text{CLS}} \tag{19}$$

3.5   Classification and Localization

During the test time, we use the final T-CAM $\mathbf{C}^F_i$ for the classification and localization of activities following the previous works [16,19]. First, we compute the class-wise classification scores $\mathbf{s}^F_i \in \mathbb{R}^{N_c}$ and the predicted pmf of activities $\mathbf{p}^F_i \in \mathbb{R}^{N_c}$ as described in Sect. 3.4. We use $\mathbf{p}^F_i$ for activity classification. For activity localization, we first find a set of possible activities that has positive classification scores, which is $\{j : \mathbf{s}^F_i(j) > 0\}$. For each activity in this set, we localize all the temporal segments that has positive T-CAM values for two or more successive time steps. Formally, a set of localized temporal segments for v_i is:

$$\{[s,e] : \forall t \in [s,e], \ \mathbf{C}_i^F(t) > 0 \text{ and } \mathbf{C}_i^F(s-1) < 0 \text{ and } \mathbf{C}_i^F(e+1) < 0\} \quad (20)$$

where $\mathbf{C}_i^F(0)$ and $\mathbf{C}_i^F(l_i+1)$ are defined to be any negative values and $e \geq s + 2$. The localized segments for each activity are non-overlapping. We assign a confidence score for each localized segment, which is the sum of the maximum T-CAM value of the segment and the classification score of it.

4   Experiments

4.1   Datasets and Evaluation

We evaluate our method on two datasets: THUMOS14 [5] and ActivityNet1.3 [3]. For the THUMOS14 dataset, the validation videos are used for training without temporal boundary annotations and the test videos are used for evaluation following the convention in the literature. This dataset is known to be challenging because each video has a number of activity instances and the duration of the videos varies widely. For the ActivityNet1.3 dataset, we use the training set for training and the validation set for evaluation. Following the standard evaluation protocol, we report mean average precision (mAP) at different intersection over union (IoU) thresholds.

4.2   Implementation Details

First, we extract RGB frames from each video at 25 fps and generate optical flow frames by using the TV-L1 algorithm [29]. Each video is then divided into non-overlapping 16-frame segments. We apply I3D networks [1] pre-trained on Kinetics dataset [6] to the segments to obtain the intermediate 1024-dimensional features after the global pooling layer. We train our network in an end-to-end manner using a single GPU (TITAN Xp).

For the THUMOS14 dataset [5], we train our network using a batch size of 32. We use the Adam optimizer [7] with learning rate 10^{-4} and weight decay 0.0005. The centers are updated using the SGD algorithm with learning rate 0.1 for the RGB stream and 0.2 for the optical flow stream. The kernel size of the 1-D convolutional layers for the T-CAMs is set to 1. We set α in Eq. 1 to 1 and γ in Eq. 8 to 0.6. For β in Eq. 6, we randomly generate a number between 0.001 and 0.1 for each training sample. We set angular margins m_1 to 2 and m_2 to 1. s_a of Eq. 14 and s for the classification loss are set to 40 and 8, respectively. Finally, ω in Eq. 18 is set to 0.6. The whole training process of 40.5k iterations takes less than 14 h.

For the ActivityNet1.3 dataset [3], it is shown from the previous works [16,19] that post-processing of the final T-CAM is required. We use an additional 1-D convolutional layer (kernel size = 13, dilation = 2) to post-process the final

Table 1. Performance comparison of A2CL-PT with state-of-the-art methods on the THUMOS14 dataset [5]. A2CL-PT significantly outperforms all the other weakly-supervised methods. † indicates an additional usage of other ground-truth annotations or independently collected data. A2CL-PT also outperforms all weakly†-supervised methods that use additional data at higher IoUs (from 0.4 to 0.9). The column AVG is for the average mAP of IoU threshold from 0.1 to 0.9.

Supervision	Method	mAP(%)@ IoU									
		0.1	0.2	0.3	0.4	0.5	0.6	0.7	0.8	0.9	AVG
Full	S-CNN [22]	47.7	43.5	36.3	28.7	19.0	10.3	5.3	-	-	-
	R-C3D [26]	54.5	51.5	44.8	35.6	28.9	-	-	-	-	-
	SSN [32]	66.0	59.4	51.9	41.0	29.8	-	-	-	-	-
	TAL-Net [2]	59.8	57.1	53.2	**48.5**	**42.8**	**33.8**	**20.8**	-	-	-
	BSN [11]	-	-	53.5	45.0	36.9	28.4	20.0	-	-	-
	GTAN [14]	**69.1**	**63.7**	**57.8**	47.2	38.8	-	-	-	-	-
Weak †	Liu et al. [12]	57.4	50.8	41.2	32.1	23.1	15.0	7.0	-	-	-
	3C-Net [16]	59.1	53.5	44.2	34.1	26.6	-	8.1	-	-	-
	Nguyen et al. [18]	64.2	59.5	**49.1**	**38.4**	**27.5**	**17.3**	8.6	3.2	0.5	29.8
	STAR [27]	**68.8**	**60.0**	48.7	34.7	23.0	-	-	-	-	-
Weak	UntrimmedNet [24]	44.4	37.7	28.2	21.1	13.7	-	-	-	-	-
	STPN [17]	52.0	44.7	35.5	25.8	16.9	9.9	4.3	1.2	0.1	21.2
	W-TALC [19]	55.2	49.6	40.1	31.1	22.8	-	7.6	-	-	-
	AutoLoc [21]	-	-	35.8	29.0	21.2	13.4	5.8	-	-	-
	CleanNet [13]	-	-	37.0	30.9	23.9	13.9	7.1	-	-	-
	MAAN [28]	59.8	50.8	41.1	30.6	20.3	12.0	6.9	2.6	0.2	24.9
	BaS-Net [9]	58.2	52.3	44.6	36.0	27.0	18.6	10.4	3.9	0.5	27.9
	A2CL-PT (Ours)	**61.2**	**56.1**	**48.1**	**39.0**	**30.1**	**19.2**	**10.6**	**4.8**	**1.0**	**30.0**

T-CAM. The kernel size of the 1-D convolutional layers for T-CAMs is set to 3. In addition, we change the batch size to 24. The learning rate for centers are 0.05 and 0.1 for the RGB and optical flow streams, respectively. We set α to 2, γ to 0.2, and ω to 0.4. The remaining hyperparameters of β, m_1, m_2, s_a, and s are the same as above. We train the network for 175k iterations.

4.3 Comparisons with the State-of-the-Art

We compare our final method A2CL-PT with other state-of-the-art approaches on the THUMOS14 dataset [5] in Table 1. Full supervision refers to training from frame-level activity annotations, whereas weak supervision indicates training only from video-level activity labels. For fair comparison, we use the symbol † to separate methods utilizing additional ground-truth annotations [16,27] or independently collected data [12,18]. The column AVG is for the average mAP of IoU thresholds from 0.1 to 0.9 with a step size of 0.1. Our method significantly outperforms other weakly-supervised methods across all metrics. Specifically, an absolute gain of 2.1% is achieved in terms of the average mAP when compared to the best previous method (BaS-Net [9]). We want to note that our method performs even better than the methods of weak† supervision at higher IoUs.

Table 2. Performance comparison on the ActivityNet1.3 dataset [3]. A2CL-PT again achieves the best performance. † indicates an additional usage of other ground-truth annotations or independently collected data. The column AVG is for the average mAP of IoU threshold from 0.5 to 0.95.

Supervision	Method	mAP(%)@ IoU										
		0.5	0.55	0.6	0.65	0.7	0.75	0.8	0.85	0.9	0.95	AVG
	Liu et al. [12]	34.0	-	-	-	-	**20.9**	-	-	-	**5.7**	21.2
Weak †	Nguyen et al. [18]	**36.4**	-	-	-	-	19.2	-	-	-	2.9	-
	STAR [27]	31.1	-	-	-	-	18.8	-	-	-	4.7	-
	STPN [17]	29.3	-	-	-	-	16.9	-	-	-	2.6	-
Weak	MAAN [28]	33.7	-	-	-	-	21.9	-	-	-	**5.5**	-
	BaS-Net [9]	34.5	-	-	-	-	**22.5**	-	-	-	4.9	22.2
	A2CL-PT (Ours)	**36.8**	33.6	30.8	27.8	24.9	22.0	18.1	14.9	10.2	5.2	**22.5**

Table 3. Performance comparison of different ablative settings on the THUMOS14 dataset [5]. The superscript + indicates that we add an adversarial branch to the baseline method. It demonstrates that both components are effective.

Method	New triplet	Adversarial	mAP(%)@ IoU					
			0.3	0.4	0.5	0.6	0.7	AVG (0.1:0.9)
ATCL			44.7	34.8	25.7	15.8	8.3	27.4
ATCL+		✓	43.7	35.1	26.3	15.7	8.3	27.2
ACL-PT	✓		46.6	37.2	28.9	18.2	10.0	29.2
A2CL-PT	✓	✓	**48.1**	**39.0**	**30.1**	**19.2**	**10.6**	**30.0**

We also evaluate A2CL-PT on the ActivityNet1.3 dataset [3]. Following the standard evaluation protocol of the dataset, we report mAP at different IoU thresholds, which are from 0.05 to 0.95. As shown in Table 2, our method again achieves the best performance.

4.4 Ablation Study and Analysis

We perform an ablation study on the THUMOS14 dataset [5]. In Table 3, we analyze the two main contributions of this work, which are the usage of the newly-suggested triplet (Eq. 7) and the adoption of adversarial approach (Eq. 15). ATCL refers to the baseline that uses only the loss term of Eq. 4. We use the superscript + to indicate the addition of adversarial branch. As described in Sect. 3.2, ACL-PT additionally uses the new triplet on top of the baseline. We can observe that our final proposed method, A2CL-PT, performs the best. It implies that both components are necessary to achieve the best performance and each of them is effective. Interestingly, adding an adversarial branch does not bring any performance gain without our new triplet. We think that although using ACL-PT increases the localization performance by learning discriminative features, it also makes the network sensitive to salient activity-related features.

We analyze the impact of two main hyperparameters in Fig. 3. The first one is α that controls the weight of A2CL-PT term (Eq. 1), and the other one is ω that is for the relative importance of T-CAMs from adversarial branches (Eq. 18). We can observe from Fig. 3(a) that positive α always brings the performance gain. It indicates that A2CL-PT is effective. As seen in Fig. 3(b), the performance is increased by using an adversarial approach when ω is less or equal to 1. If ω is greater than 1, T-CAMs of adversarial branches will play a dominant role in activity localization. Therefore, the results tell us that the adversarial branches provide mostly supplementary information.

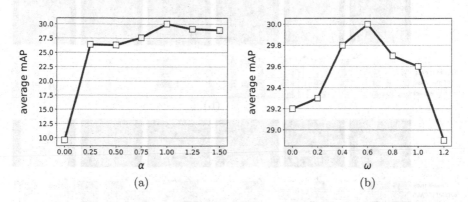

(a)   (b)

Fig. 3. We analyze the impact of two main hyperparameters α and ω. (a): Positive α always provides the performance gain, so it indicates that our method is effective. (b): If ω is too large, the performance is decreased substantially. It implies that T-CAMs of adversarial branches provide mostly supplementary information.

4.5 Qualitative Analysis

We perform a qualitative analysis to better understand our method. In Fig. 4, qualitative results of our A2CL-PT on four videos from the test set of the THUMOS14 dataset [5] are presented. (a), (b), (c), and (d) are examples of *JavelinThrow*, *HammerThrow*, *ThrowDiscus*, and *HighJump*, respectively. Detection denotes the localized activity segments. For additional comparison, we also show the results of BaS-Net [9], which is the leading state-of-the-art method. We use three different colors on the contours of sampled frames: blue, green, and red which denote true positive, false positive, and false negative, respectively. In (a), there are multiple instances of false positive. These false positives are challenging because the person in the video swings the javelin, which can be mistaken for a throw. Similar cases are observed in (b). One of the false positives includes the person drawing the line on the field, which looks similar to a *HammerThrow* activity. In (c), some false negative segments are observed. Interestingly, this is because the ground-truth annotations are wrong; that is, the *ThrowDiscus* activity is annotated but it does not actually occur in these cases. In (d), all the instances of the *HighJump* activity are successfully localized. Other than the unusual situations, our method performs well in general.

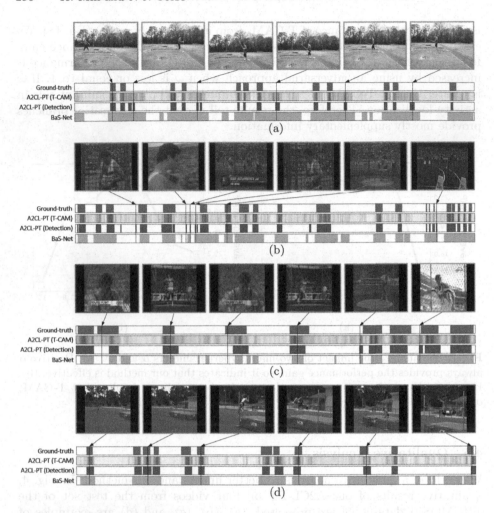

Fig. 4. Qualitative results on the THUMOS14 dataset [5]. Detection denotes the localized activity segments. The results of BaS-Net [9] are also included for additional comparison. Contours of the sampled frames have three different colors. We use blue, green, and red to indicate true positives, false positives, and false negatives, respectively. (a): An example of *JavelinThrow* activity class. The observed false positives are challenging. The person in the video swings the javelin on the frames of these false positives, which can be mistaken for a throw. (b): An example of *HammerThrow*. One of the false positives include the person who draws the line on the field. It is hard to distinguish the two activities. (c): An example of *ThrowDiscus*. Multiple false negatives are observed, which illustrates the situations where the ground-truth activity instances are wrongly annotated. (d): An example of *HighJump* without such unusual cases. It can be observed that our method performs well in general. (Color figure online)

5   Conclusion

We have presented A2CL-PT as a novel method for weakly-supervised temporal activity localization. We suggest using two triplets of vectors of the feature space to learn discriminative features and to distinguish background portions from activity-related parts of a video. We also propose to adopt an adversarial approach to localize activities more thoroughly. We perform extensive experiments to show that our method is effective. A2CL-PT outperforms all the existing state-of-the-art methods on major datasets. Ablation study demonstrates that both contributions are significant. Finally, we qualitatively analyze the effectiveness of our method in detail.

Acknowledgement. We thank Stephan Lemmer, Victoria Florence, Nathan Louis, and Christina Jung for their valuable feedback and comments. This research was, in part, supported by NIST grant 60NANB17D191.

References

1. Carreira, J., Zisserman, A.: Quo vadis, action recognition? A new model and the kinetics dataset. In: proceedings of the IEEE Conference on Computer Vision and Pattern Recognition, pp. 6299–6308 (2017)
2. Chao, Y.W., Vijayanarasimhan, S., Seybold, B., Ross, D.A., Deng, J., Sukthankar, R.: Rethinking the faster r-cnn architecture for temporal action localization. In: Proceedings of the IEEE Conference on Computer Vision and Pattern Recognition, pp. 1130–1139 (2018)
3. Caba Heilbron, F., Victor Escorcia, B.G., Niebles, J.C.: ActivityNet: a large-scale video benchmark for human activity understanding. In: Proceedings of the IEEE Conference on Computer Vision and Pattern Recognition, pp. 961–970 (2015)
4. He, X., Zhou, Y., Zhou, Z., Bai, S., Bai, X.: Triplet-center loss for multi-view 3D object retrieval. In: Proceedings of the IEEE Conference on Computer Vision and Pattern Recognition, pp. 1945–1954 (2018)
5. Jiang, Y.G., et al.: THUMOS challenge: action recognition with a large number of classes (2014). http://crcv.ucf.edu/THUMOS14/
6. Kay, W., et al.: The kinetics human action video dataset. arXiv preprint arXiv:1705.06950 (2017)
7. Kingma, D.P., Ba, J.: Adam: a method for stochastic optimization. arXiv preprint arXiv:1412.6980 (2014)
8. Kuehne, H., Jhuang, H., Garrote, E., Poggio, T., Serre, T.: HMDB: a large video database for human motion recognition. In: 2011 International Conference on Computer Vision, pp. 2556–2563. IEEE (2011)
9. Lee, P., Uh, Y., Byun, H.: Background suppression network for weakly-supervised temporal action localization. In: AAAI (2020)
10. Li, Z., Xu, C., Leng, B.: Angular triplet-center loss for multi-view 3D shape retrieval. In: Proceedings of the AAAI Conference on Artificial Intelligence, vol. 33, pp. 8682–8689 (2019)
11. Lin, T., Zhao, X., Su, H., Wang, C., Yang, M.: BSN: boundary sensitive network for temporal action proposal generation. In: Proceedings of the European Conference on Computer Vision (ECCV), pp. 3–19 (2018)

12. Liu, D., Jiang, T., Wang, Y.: Completeness modeling and context separation for weakly supervised temporal action localization. In: Proceedings of the IEEE Conference on Computer Vision and Pattern Recognition, pp. 1298–1307 (2019)
13. Liu, Z., et al.: Weakly supervised temporal action localization through contrast based evaluation networks. In: Proceedings of the IEEE International Conference on Computer Vision, pp. 3899–3908 (2019)
14. Long, F., Yao, T., Qiu, Z., Tian, X., Luo, J., Mei, T.: Gaussian temporal awareness networks for action localization. In: Proceedings of the IEEE Conference on Computer Vision and Pattern Recognition, pp. 344–353 (2019)
15. Monfort, M., et al.: Moments in time dataset: one million videos for event understanding. IEEE Trans. Pattern Anal. Mach. Intell. **42**, 1–8 (2019). https://doi.org/10.1109/TPAMI.2019.2901464
16. Narayan, S., Cholakkal, H., Khan, F.S., Shao, L.: 3C-Net: category count and center loss for weakly-supervised action localization. In: Proceedings of the IEEE International Conference on Computer Vision, pp. 8679–8687 (2019)
17. Nguyen, P., Liu, T., Prasad, G., Han, B.: Weakly supervised action localization by sparse temporal pooling network. In: Proceedings of the IEEE Conference on Computer Vision and Pattern Recognition, pp. 6752–6761 (2018)
18. Nguyen, P.X., Ramanan, D., Fowlkes, C.C.: Weakly-supervised action localization with background modeling. In: Proceedings of the IEEE International Conference on Computer Vision, pp. 5502–5511 (2019)
19. Paul, S., Roy, S., Roy-Chowdhury, A.K.: W-TALC: weakly-supervised temporal activity localization and classification. In: Proceedings of the European Conference on Computer Vision (ECCV), pp. 563–579 (2018)
20. Shou, Z., Chan, J., Zareian, A., Miyazawa, K., Chang, S.F.: CDC: convolutional-de-convolutional networks for precise temporal action localization in untrimmed videos. In: Proceedings of the IEEE Conference on Computer Vision and Pattern Recognition, pp. 5734–5743 (2017)
21. Shou, Z., Gao, H., Zhang, L., Miyazawa, K., Chang, S.F.: AutoLoc: weakly-supervised temporal action localization in untrimmed videos. In: Proceedings of the European Conference on Computer Vision (ECCV), pp. 154–171 (2018)
22. Shou, Z., Wang, D., Chang, S.F.: Temporal action localization in untrimmed videos via multi-stage CNNs. In: Proceedings of the IEEE Conference on Computer Vision and Pattern Recognition, pp. 1049–1058 (2016)
23. Soomro, K., Zamir, A.R., Shah, M.: Ucf101: a dataset of 101 human actions classes from videos in the wild. arXiv preprint arXiv:1212.0402 (2012)
24. Wang, L., Xiong, Y., Lin, D., Van Gool, L.: UntrimmedNets for weakly supervised action recognition and detection. In: Proceedings of the IEEE Conference on Computer Vision and Pattern Recognition, pp. 4325–4334 (2017)
25. Wen, Y., Zhang, K., Li, Z., Qiao, Yu.: A discriminative feature learning approach for deep face recognition. In: Leibe, B., Matas, J., Sebe, N., Welling, M. (eds.) ECCV 2016. LNCS, vol. 9911, pp. 499–515. Springer, Cham (2016). https://doi.org/10.1007/978-3-319-46478-7_31
26. Xu, H., Das, A., Saenko, K.: R-C3D: region convolutional 3D network for temporal activity detection. In: Proceedings of the IEEE International Conference on Computer Vision, pp. 5783–5792 (2017)
27. Xu, Y., et al.: Segregated temporal assembly recurrent networks for weakly supervised multiple action detection. In: Proceedings of the AAAI Conference on Artificial Intelligence, vol. 33, pp. 9070–9078 (2019)

28. Yuan, Y., Lyu, Y., Shen, X., Tsang, I.W., Yeung, D.Y.: Marginalized average attentional network for weakly-supervised learning. In: International Conference on Learning Representations (ICLR) (2019)
29. Zach, C., Pock, T., Bischof, H.: A duality based approach for realtime TV-L^1 optical flow. In: Hamprecht, F.A., Schnörr, C., Jähne, B. (eds.) DAGM 2007. LNCS, vol. 4713, pp. 214–223. Springer, Heidelberg (2007). https://doi.org/10.1007/978-3-540-74936-3_22
30. Zhang, X., Wei, Y., Feng, J., Yang, Y., Huang, T.S.: Adversarial complementary learning for weakly supervised object localization. In: Proceedings of the IEEE Conference on Computer Vision and Pattern Recognition, pp. 1325–1334 (2018)
31. Zhao, H., Torralba, A., Torresani, L., Yan, Z.: HACS: human action clips and segments dataset for recognition and temporal localization. In: Proceedings of the IEEE International Conference on Computer Vision, pp. 8668–8678 (2019)
32. Zhao, Y., Xiong, Y., Wang, L., Wu, Z., Tang, X., Lin, D.: Temporal action detection with structured segment networks. In: Proceedings of the IEEE International Conference on Computer Vision, pp. 2914–2923 (2017)

Action Localization Through Continual Predictive Learning

Sathyanarayanan Aakur[1]($\boxtimes$) (iD) and Sudeep Sarkar[2] (iD)

[1] Oklahoma State University, Stillwater, OK 74074, USA
saakur@okstate.edu
[2] University of South Florida, Tampa, FL 33620, USA
sarkar@usf.edu

Abstract. The problem of action localization involves locating the action in the video, both over time and spatially in the image. The current dominant approaches use supervised learning to solve this problem. They require large amounts of annotated training data, in the form of frame-level bounding box annotations around the region of interest. In this paper, we present a new approach based on continual learning that uses feature-level predictions for self-supervision. It does not require any training annotations in terms of frame-level bounding boxes. The approach is inspired by cognitive models of visual event perception that propose a prediction-based approach to event understanding. We use a stack of LSTMs coupled with a CNN encoder, along with novel attention mechanisms, to model the events in the video and use this model to predict high-level features for the future frames. The prediction errors are used to learn the parameters of the models continuously. This self-supervised framework is not complicated as other approaches but is very effective in learning robust visual representations for both labeling and localization. It should be noted that the approach outputs in a streaming fashion, requiring only a single pass through the video, making it amenable for real-time processing. We demonstrate this on three datasets - UCF Sports, JHMDB, and THUMOS'13 and show that the proposed approach outperforms weakly-supervised and unsupervised baselines and obtains competitive performance compared to fully supervised baselines. Finally, we show that the proposed framework can generalize to egocentric videos and achieve state-of-the-art results on the *unsupervised* gaze prediction task. Code is available on the project page(https://saakur. github.io/Projects/ActionLocalization/.).

Keywords: Action localization · Continuous learning · Self-supervision

Electronic supplementary material The online version of this chapter (https:// doi.org/10.1007/978-3-030-58568-6_18) contains supplementary material, which is available to authorized users.

A. Vedaldi et al. (Eds.): ECCV 2020, LNCS 12359, pp. 300–317, 2020.
https://doi.org/10.1007/978-3-030-58568-6_18

1 Introduction

We develop a framework for jointly learning spatial and temporal localization through continual, self-supervised learning, in a streaming fashion, requiring only *a single pass through the video*. Visual understanding tasks in computer vision have focused on the problem of recognition [1,3,23,25] and captioning [1,9,46, 47], with the underlying assumption that each input video is already localized both spatially and temporally. While there has been tremendous progress in action localization, it has primarily been driven by the dependence on large amounts of tedious, spatial-temporal annotations. In this work, we aim to tackle the problem of spatial-temporal segmentation of streaming videos in a continual, self-supervised manner, without any training annotations.

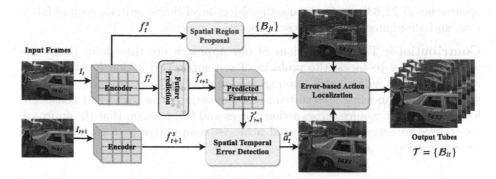

Fig. 1. The **Proposed approach** has four core components: (i) feature extraction and spatial region proposal, (ii) a future prediction framework, (iii) a spatial-temporal error detection module and (iv) the error-based action localization process.

Drawing inspiration from psychology [13,14,52], we consider the underlying mechanism for both event understanding and attention selection in humans to be the idea of *predictability*. Defined as the surprise-attention hypothesis [13], unpredictable factors such as large changes in motion, appearance, or goals of the actor have a substantial effect on the event perception and human attention. Human event perception studies [2,52] have shown that longer-term, temporal surprise have a strong correlation with event boundary detection. In contrast, short-term spatial surprise (such as those caused by motion) has a more substantial effect on human attention and localization [14]. Our approach combines both spatial and temporal surprise to formulate a computational framework to tackle the problem of self-supervised action localization in streaming videos in a continual manner.

We formulate our computational framework on the idea of spatial-temporal feature anticipation to model predictability of perceptual features. The main assumption in our framework is that expected, unpredictable features require attention and often point to the actor performing the action of interest. In contrast, predictable features can belong to background clutter and are not relevant to the action of interest. It is to be noted that unpredictability or *surprise* is not the same as *rarity*. It refers to short-term changes that aid in the completion of an overall task, which can be recurring [13]. We model the perceptual features using a hierarchical, cyclical, and recurrent framework, whose predictions are influenced by current and prior observations as well as current perceptual predictions. Hence, the predictive model's output can influence the perception of the current frame being observed. The predictions are constantly compared with the incoming observations to provide self-supervision to guide future predictions.

We leverage these characteristics to derive and quantify spatial-temporal predictability. Our framework performs continuous learning to generate *"attention maps"* that overlap with the action being performed. Using these attention maps, we leverage advances in region proposals [29,31,44,54] to localize actions in streaming videos without any supervision. Contrary to other attention-based approaches [5,28,33], we do not use the object-level characteristics such as label, role, and affordance in the proposal generation process.

Contributions: The contributions of our approach are three-fold: (i) we are among the first to tackle the problem of self-supervised action localization in *streaming videos without any training data such as labels or bounding boxes*, (ii) we show that modeling spatial-temporal prediction error can yield consistent localization performance across action classes and (iii) we show that the approach generalizes to egocentric videos and achieves competitive performance on the *unsupervised gaze prediction* task.

2   Related Work

Supervised action localization approaches tackle action localization through the simultaneous generation of bounding box proposals and labeling each bounding box with the predicted action class. Both bounding box generation and labeling are fully supervised, i.e., they require ground truth annotations of both bounding boxes and labels. Typical approachesleverage advances in object detection to include temporal information [7,16,18,36,37,40,43,50] for proposal generation. The final step typically involves the use of the Viterbi algorithm [7] to link the generated bounding boxes across time.

Weakly-supervised action localization approaches have been explored to reduce the need for extensive annotations [5,26,28,33]. They typically only require video-level labels and rely on object detection-based approaches to generate bounding box proposals. It is to be noted that weakly supervised approaches also use object-level labels and characteristics to guide the bounding box selection process. Some approaches [5] use a similarity-based tracker to connect bounding boxes across time to incorporate temporal consistency.

Unsupervised action localization approaches have not been explored to the same extent as supervised and weakly-supervised approaches. These approaches do not require any supervision - both labels or bounding boxes. The two more common approaches are to generate action proposals using (i) super-voxels [18,38] and (ii) clustering motion trajectories [45]. It should be noted that [38] also uses object characteristics to evaluate the "humanness" of each super-voxel to select bounding box proposals. Our approach falls into the class of unsupervised action localization approaches. The most closely related approaches (with respect to architecture and theme) to ours are VideoLSTM [28] and Actor Supervision [5], which use attention in the selection process for generating bounding box proposals, but require video-level labels. We, on the other hand, do not require any labels or bounding box annotations for training.

While fully supervised approaches have more precise localization and achieve better recognition, the required number of annotations is rather large. It is not amenable to an increase in the number of classes and a decrease in the number of training videos. While not requiring frame-level annotations, weakly supervised approaches have the underlying assumption that there exists a large, annotated training set that allows for effective detection of all possible actors (both human and non-human) in the set of action classes. Unsupervised approaches, such as ours, do not make any such assumptions but can result in poorer localization performance. We alleviate this to an extent by leveraging advances in region proposal mechanisms and self-learning robust representations for obtaining video-level labels.

3 Self-supervised Action Localization

In this section, we introduce our self-supervised action localization framework, as illustrated in Fig. 1. Our approach has four core components: (i) feature extraction and spatial region proposal, (ii) a self-supervised future prediction framework, (iii) a spatial-temporal error detection module, and (iv) the error-based action localization process.

3.1 Feature Extraction and Spatial Region Proposal

The first step in our approach is feature extraction and the subsequent *per-frame* region proposal generation for identifying possible areas of actions and associated objects. Considering the tremendous advances in deep learning architectures for learning robust spatial representations, we use pre-trained convolutional neural networks to extract the spatial features for each frame in the video. We use a region proposal module, based on these spatial features, to predict possible action-agnostic spatial locations. We use class-agnostic proposals (i.e., the object category is ignored, and only feature-based localizations are taken into account) for two primary reasons. First, we do not want to make any assumptions on the actor's characteristics, such as label, role, and affordance. Second, despite significant progress in object detection, there can be many missed detections, especially when the object (or actor) performs actions that can transform their physical appearance. It is to be noted that these considerations can result in a large number of region proposals that require careful and robust selection but can yield higher chances of correct localization.

3.2 Self-supervised Future Prediction

The second stage in our proposed framework is the self-supervised future prediction framework. We consider the future prediction module to be a generative model whose output is conditioned on two factors - the current observation and an *internal event model*. The current observation f_t^S is the feature-level encoding of the presently observed frame, I_t. We use the same feature encoder as the region proposal module to reduce the approach's memory footprint and complexity. The *internal event model* is a set of parameters that can effectively capture the spatial-temporal dynamics of the observed event. Formally, we define the predictor model as $P(\hat{f}_{t+1}^S | W_e, f_t^S)$, where W_e represents the internal event model and $\hat{f}_{t+1}^S$ is the predicted features at time $t + 1$. Note that features f_t^S is not a one-dimensional vector, but a tensor (of dimension $w_f \times h_f \times d_f$) representing the features at each spatial location.

We model temporal dynamics of the observed event using Long Short Term Memory Networks (LSTMs) [12]. While other approaches [21,48,49] can be used for prediction, we consider LSTMs to be more suited for the following reasons. First, we want to model the temporal dynamics across *all* frames of the observed action (or event). Second, LSTMs can allow for multiple possible futures and hence will not tend to average the outcomes of these possible futures, as can be the case with other prediction models. Third, since we work with error-based localization, using LSTMs can ensure that the learning process propagates the spatial-temporal error across time and can yield progressively better predictions, especially for actions of longer duration. Formally, we can express LSTMs as

$$i_t = \sigma(W_i x_t + W_{hi} h_{t-1} + b_i); \quad f_t = \sigma(W_f x_t + W_{hf} h_{t-1} + b_f) \qquad (1)$$

$$o_t = \sigma(W_o x_t + W_{ho} h_{t-1} + b_o); \quad g_t = \phi(W_g x_t + W_{hg} h_{t-1} + b_g) \qquad (2)$$

$$m_t = f_t \cdot m_{t-1} + i_t \cdot g_t; \quad h_t = o_t \cdot \phi(m_t) \qquad (3)$$

where x_t is the input at time t, σ is a non-linear activation function, $(\cdot)$ represents element-wise multiplication, ϕ is the hyperbolic tangent function ($tanh$) and W_k and b_k represent the trained weights and biases for each of the gates.

As opposed to [2], who also use an LSTM-based predictor and a decoder network, we use a hierarchical LSTM model (with three LSTM layers) as our event model. This modification allows us to model both spatial and temporal dependencies, since each higher-level LSTMs act as a progressive decoder framework that captures the temporal dependencies captured by the lower-level LSTMs. The first LSTM captures the spatial dependency that is propagated up the prediction stack. The updated hidden state of the first (bottom) LSTM layer (h_t^1) depends on the current observation f_t^S, the previous hidden state (h_{t-1}^1) and memory state (m_{t-1}^1). Each of the higher-level LSTMs at level l take the output of the bottom LSTM's output h_t^{l-1} and memory state m_t^{l-1} and can be defined as $(h_t^l, m_t^l) = LSTM(h_{t-1}^l, h_t^{l-1}, m_t^{l-1})$. Note this is different from a typical hierarchical LSTM model [35] in that the higher LSTMs are impacted by the output of the lower level LSTMs at current time step, as opposed to that from the previous time step. Collectively, the event model W_e is described by the learnable parameters and their respective biases from the hierarchical LSTM stack.

Hence, the top layer of the prediction stack acts as the decoder whose goal is to predict the next feature f_{t+1}^S given all previous predictions $\hat{f}_1^S, \hat{f}_2^S, \dots \hat{f}_t^S$, an event model W_e and the current observation f_t^S. We model this prediction function as a log-linear model characterized by

$$\log p(\hat{f}_{t+1}^s|h_t^l) = \sum_{n=1}^{t} f(W_e, f_t^S) + \log Z(h_t) \tag{4}$$

where h_t^l is the hidden state of the lth level LSTM at time t and $Z(h_t)$ is a normalization constant. The LSTM prediction stack acts as a generative process for anticipating future features.

The **objective function** for training the predictive stack is a weighted zero order hold between the predicted features and the actual observed features, weighted by the zero order hold difference. The prediction error at time t is given by $E(t) = \frac{1}{n_f} \sum_{i=1}^{w_f} \sum_{j=1}^{h_f} e_{ij}$, where

$$e_{ij} = \hat{m}_t(i,j) \odot \|f_{t+1}^S(i,j) - \hat{f}_{t+1}^S(i,j)\|_{\ell_2}^2 \tag{5}$$

Each feature f_t^S has dimensions $w_f \times h_f \times d_f$ and $\hat{m}_t(i,j)$ is a function that returns the zero order difference between the observed features at times t and $t+1$ at location (i,j). Note that the prediction is done at the feature level and not at the pixel level, so the spatial quantization is coarser than pixels.

3.3 Prediction Error-Based Attention Map

At the core of our approach is the idea of spatial-temporal prediction error for localizing the actions of interest in the video. It takes into account the quality of the predictions made and the relative spatial alignment of the prediction errors. The input to the error detection module is the quantity from Eq. 5. We compute a weight α_{ij} associated with each spatial location (i,j) in the predicted feature $\hat{f}_{t+1}^S$ as

$$\alpha_{ij} = \frac{\exp(e_{ij})}{\sum_{m=1}^{w_k} \sum_{n=1}^{h_k} \exp(e_{mn})} \tag{6}$$

where e_{ij} represents the weighted prediction error at location (i,j) (Eq. 5). It can be considered to be a function $a(f_t^S, h_{t-1}^l)$ of the state of the top-most LSTM and the input feature f_t^S at time t. The resulting matrix is an error-based attention map that allows us to localize the prediction error at a specific spatial location. And the average spatial error over time, $E(t)$, is used for temporal localization.

One may remark that the formulation of α_{ij} is very similar to Bahdanau attention [4]. However, there are two key differences. First, our formulation is not parametrized and does not add to the number of learnable parameters in the framework. Second, our attention map is a characterization of the difficulty in anticipating unpredictable motion. In contrast, Bahdanau attention is an effort to increase the decoder's encoding ability and does not characterize the unpredictability of the future feature. We compare the use of both types of attention in Sect. 5.4, where we see that error-based localization is more suitable for our application.

3.4   Extraction of Action Tubes

The action localization module receives a stream of bounding box proposals and an error-based attention map to select an output tube. The action localization is a selection algorithm that filters *all* region proposals from Sect. 3.1 and returns the collection of proposals that have a higher probability of action localization. We do so by assigning an energy term to each of the bounding box proposals ($\mathcal{B}_{it}$) at time t and choosing the top k bounding boxes with least energy as our final proposals. The energy of a bounding box $\mathcal{B}_{it}$ is defined as

$$E(\mathcal{B}_{it}) = w_\alpha \ \phi(\alpha_{ij}, \mathcal{B}_{it}) + w_t \delta(\mathcal{B}_{it}, \{\mathcal{B}_{j,t-1}\}) \tag{7}$$

where $\phi(\cdot)$ is a function that returns a value characteristic of the distance between the bounding box center and location of maximum error, $\delta(\cdot)$ is a function that returns the minimum spatial distance between the current bounding box and the closest bounding box from the previous time step. The constants w_α and w_t are scaling factors. Note that $\delta(\cdot)$ is introduced to enforce temporal consistency in predictions, but we find that it is optional since the LSTM prediction stack implicitly enforces the temporal consistency through its memory states. In our experiments we set $k = 10, w_\alpha = 0.75$.

3.5   Implementation Details

In our experiments, we use a VGG-16 [34] network pre-trained on ImageNet as our feature extraction network. We use the output of the last convolutional layer before the fully connected layers as our spatial features. Hence the dimensions of the spatial features are $w_f = 14, h_f = 14, d_f = 512$. These output features are then used by an SSD [29] to generate bounding box proposals. Note that we take the generated bounding box proposals without taking into account classes and associated probabilities. We use a three-layer hierarchical LSTM model with the hidden state size as 512 as our predictor module. We use the vanilla LSTM as proposed in [12]. Video level-features are obtained by max-pooling the element-wise dot-product of the hidden state of the top-most LSTM and the attention values across time. We train with the adaptive learning mechanism proposed in [2], with the initial learning rate set to be 1×10^{-8} and scaling factors Δ_t^- and Δ_t^+ as 1×10^{-2} and 1×10^{-3}, respectively. The network was trained for 1 epoch on a computer with one Titan X Pascal.

4   Experimental Setup

4.1   Data

We evaluate our approach on three publicly available datasets for evaluating the proposed approach on the action localization task.

UCF Sports [32] is an action localization dataset consisting of 10 classes of sports actions such as skating and lifting collected from sports broadcasts. It is an interesting dataset since it has a high concentration of distinct scenes and motions that make it challenging for localization and recognition. We use the splits (103 training and 47 testing videos) as defined in [26] for evaluation.

JHMDB [19] is composed of 21 action classes and 928 trimmed videos. All videos are annotated with human-joints for every frame. The ground truth bounding box for the action localization task is chosen such that the box encompasses all the joints. This dataset offers several challenges, such as increasing amounts of background clutter, high inter-class similarity, complex motion (including camera motion), and occluded objects of interest. We report all results as the average across all three splits.

THUMOS'13 [22] is a subset of the UCF-101 [39] dataset, consisting of 24 classes and 3, 207 videos. Ground truth bounding boxes are provided for each of the classes for the action localization task. It is also known as the **UCF-101-24** dataset. Following prior works [28,38], we perform our experiments and report results on the first split.

We also evaluate the proposed approach's generalization ability on egocentric videos by evaluating it on the *unsupervised gaze prediction task*. There has been evidence from cognitive psychology that there is a strong correlation between gaze points and action localization [41]. Hence, the gaze prediction task would be a reasonable measure of the generalization to action localization in egocentric videos. We evaluate the performance on the **GTEA Gaze** [6] dataset, which consists of 17 sequences of tasks performed by 14 subjects, with each sequence lasting about 4 minutes. We use the official splits for the GTEA datasets as defined in prior works [6].

4.2 Metrics and Baselines

For the **action localization** task, we follow prior works [28,38] and report the mean average precision (mAP) at various overlap thresholds, obtained by computing the Intersection Over Union (IoU) of the predicted and ground truth bounding boxes. We also evaluate the quality of bounding box proposals by measuring the average, per-frame IoU, and the bounding box *recall* at varying overlap ratios.

Since ours is an unsupervised approach, we obtain class labels by clustering the learned representations using the *k-means* algorithm. While more complicated clustering may yield better recognition results [38], the k-means approach allows us to evaluate the robustness of learned features. We evaluate our approach in two settings K_{gt} and K_{opt}, where the number of clusters is set to the number of ground truth action classes and an optimal number obtained through the elbow method [24], respectively. From our experiments, we observe that K_{opt} is three times the number of ground truth classes, which is not unreasonable and has been a working assumption in other deep learning-based clustering approaches [11]. Clusters are mapped to the ground truth clusters for evaluation using the Hungarian method, as done in prior unsupervised approaches [20,51].

We also compare against other LSTM and attention-based approaches (Sect. 5.3) to the action localization problem for evaluating the effectiveness of the proposed training protocol.

For the **gaze prediction** task, we evaluate the approaches using **Area Under the Curve** (AUC), which measures the area under the curve on saliency maps for true positive versus false-positive rates under various threshold values. We also report the **Average Angular Error** (AAE), which measures the angular distance between the predicted and ground truth gaze positions. Since our model's output is a saliency map, AUC is a more appropriate metric compared to average angular error (AAE), which requires specific locations.

5   Quantitative Evaluation

In this section, we present the quantitative evaluation of our approach on two different tasks, namely action localization, and egocentric gaze prediction. For the action localization task, we evaluate our approach on two aspects - the quality of proposals and spatial-temporal localization.

5.1   Quality of Localization Proposals

We first evaluate the quality of our localization proposals by assuming perfect class prediction. This allows us to independently assess the quality of localization performed in a self-supervised manner. We present the results of the evaluation in Table 1 and compare against fully supervised, weakly supervised, and unsupervised baselines. As can be seen, we outperform many supervised and weakly supervised baselines. APT [45] achieves a higher localization score. However, it produces, on average, 1,500 proposals per video, whereas our approach returns approximately 10 proposals. A large number of localization proposals per video can lead to higher recall and IoU but makes the localization task, i.e., action labeling per video harder and can affect the ability to generalize across domains. Also, it should be noted that our approach produces proposals in *streaming* fashion, as opposed to many of the other approaches, which produce action tubes based on motion computed across the entire video. This can make real-time action localization in streaming videos harder.

5.2   Spatial-Temporal Action Localization

We also evaluate our approach on the spatial-temporal localization task. This evaluation allows us to analyze the robustness of the self-supervised features learned through prediction. We generate video-level class labels through clustering and use the standard evaluation metrics (Sect. 4.2) to quantify the performance. The AUC curves with respect to varying overlap thresholds are presented in Fig. 2. We compare against a mix of supervised, weakly-supervised, and unsupervised baselines on all three datasets.

Table 1. Comparison with fully supervised and weakly supervised baselines on class-agnostic action localization on UCF Sports dataset. We report the average localization accuracy of each approach i.e. average IoU.

Supervision	Approach	Average
Full	STPD[42]	44.6
	Max path search [43]	**54.3**
Weak	Ma et al. [30]	44.6
	GBVS [8]	42.1
	Soomro et al. [38]	**47.7**
None	IME tublets [18]	51.5
	APT [45]	**63.7**
	Proposed approach	55.7

On the **UCF Sports** dataset (Fig. 2(a)), we outperform all baselines including several supervised baselines except for Gkioxari and Malik [7] at higher overlap thresholds ($\sigma > 0.4$) when we set number of clusters k to the number of ground truth classes. When we allow for some over-segmentation and use the *optimal* number of clusters, we outperform all baselines till $\sigma > 0.5$.

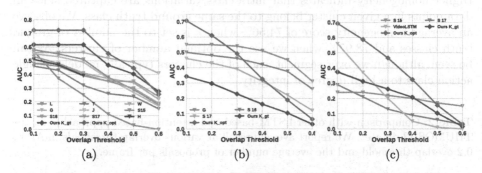

Fig. 2. AUC for the action localization tasks are shown for (a) UCF Sports, (b) JHMDB and (c) THUMOS13 datasets. We compare against baselines with varying levels of supervision such as **Lan** *et al.* [26], **Tian** *et al.* [40], **Wang** *et al.* [50], **Gkioxari** and Malik [7], **Jain** *et al.* [18], **Soomro** *et al.* [36–38], **Hou** *et al.* [16], and VideoLSTM [28].

On the **JHMDB** dataset (Fig. 2(b)), we find that our approach, while having high recall (77.8%@$\sigma = 0.5$), the large camera motion and intra-class variations have a significant impact on the classification accuracy. Hence, the mAP suffers when we set k to be the number of ground truth classes. When we set the number of clusters to the optimal number of clusters, we outperform other baselines at lower thresholds ($\sigma < 0.5$). It should be noted that the other unsupervised baseline (Soomro *et al.* [38]) uses object detection proposals from a Faster R-CNN backbone to score the *"humanness"* of a proposal. This assumption tends to make the approach biased towards human-centered action localization and affects its ability to generalize towards actions with non-human actors. On the other hand, we do not make any assumptions on the characteristics of the actor, scene, or motion dynamics.

On the **THUMOS'13** dataset (Fig. 2(c)), we achieve consistent improvements over unsupervised and weakly supervised baselines, at $k = k_{gt}$ and achieve state-of-the-art mAP scores when $k = k_{opt}$. It is interesting to note that we perform competitively (when $k = k_{gt}$) the weakly-supervised attention-based VideoLSTM [28], which uses a convLSTM for temporal modeling along with a CNN-based spatial attention mechanism. It should be noted that we have a higher recall rate ($0.47@\sigma = 0.4$ and $0.33@\sigma = 0.5$) at higher thresholds than other state-of-the-art approaches on THUMOS'13 and shows the robustness of the error-based localization approach to intra-class variation and occlusion.

Clustering Quality. Since there is a significant difference in the mAP score when we set a different number of clusters in k-means, we measured the homogeneity (or purity) of the clustering. The homogeneity score measures the "quality" of the cluster by measuring how well a cluster models a given ground-truth class. Since we allow the over-segmentation of clusters when we set k to the optimal number of clusters, this is an essential measure of feature robustness. Higher homogeneity indicates that intra-class variations are captured since all data points in a given cluster belong to the same ground truth class. We observe an average homogeneity score of 74.56% when k is set to the number of ground truth classes and 78.97% when we use the optimal number of clusters. As can be seen, although we over-segment, each of the clusters typically models a single action class to a high degree of integrity.

Table 2. Comparison with other LSTM-based and attention-based approaches on the THUMOS'13 dataset. We report average recall at various overlap thresholds, mAP at 0.2 overlap threshold and the average number of proposals per frame.

Approach	Annotations		# proposals	Average recall					mAP
	Labels	Boxes		0.1	0.2	0.3	0.4	0.5	@0.2
ALSTM [33]	✓	✗	1	0.46	0.28	0.05	0.02	-	0.06
VideoLSTM [28]	✓	✗	1	0.71	0.52	0.32	0.11	-	0.37
Actor Supervision [5]	✓	✗	~1000	**0.89**	-	-	-	**0.44**	0.46
Proposed Approach	✗	✗	~10	0.84	**0.72**	**0.58**	**0.47**	0.33	**0.59**

5.3   Comparison with Other LSTM-Based Approaches

We also compare our approach with other LSTM-based and attention-based models to highlight the importance of the proposed self-supervised learning paradigm. Since LSTM-based frameworks can have highly similar architectures, we consider different requirements and characteristics, such as the level of annotations required for training and the number of localization proposals returned per video. We compare with three approaches similar in spirit to our approach - ALSTM [33], VideoLSTM [28] and Actor Supervision [5] and summarize the results in Table 2. It can be seen that we significantly outperform VideoLSTM and ALSTM on the THUMOS'13 dataset in both recall and $mAP@\sigma = 0.2$. Actor Supervision [5] outperforms our approach on recall, but it is to be noted that the region proposals are dependent on two factors - (i) object detection-based actor proposals and (ii) a filtering mechanism that limits proposals based on ground truth action classes, which can increase the training requirements and limit generalizability. Also, note that returning a higher number of localization proposals can increase recall at the cost of generalization.

5.4 Ablative Studies

The proposed approach has three major units that affect its performance the most - (i) the region proposal module, (ii) future prediction module, and (iii) error-based action localization module. We consider and evaluate several alternatives to all three modules.

We choose selective search [44] and EdgeBox [54] as alternative region proposal methods to SSD. We use an attention-based localization method for action localization as an approximation of the ALSTM [33] to evaluate the effectiveness of using the proposed error-based localization. We also evaluate a 1-layer LSTM predictor with a fully connected decoder network to approximate [2] on the localization task. We evaluate the effect of attention-based prediction by introducing a Bahdanau [4] attention layer before prediction as an alternative to the error-based action localization module.

These ablative studies are conducted on the UCF Sports dataset. The results are plotted in Fig. 3(a). It can be seen that the use of the prediction error-based localization has a significant improvement over a trained attention-based localization approach. We can also see that the choice of region proposal methods do have some effect on the performance of the approach, with selective search and EdgeBox proposals doing slightly better at higher thresholds ($\sigma \in (0.4, 0.5)$) at the cost of inference time and additional bounding box proposals (50 compared to the 10 from SSD-based region proposal). Using SSD for generating proposals allows us to share weights across the frame encoder and region proposal tasks and hence reduce the memory and computational footprint of the approach. We also find that using attention as part of the prediction module significantly impacts the architecture's performance. It could, arguably, be attributed to the objective function, which aims to minimize the prediction error. Using attention to encode the input could impact the prediction function.

5.5   Unsupervised Egocentric Gaze Prediction

Finally, we evaluate the ability to generalize to egocentric videos by quantifying the model's performance on the unsupervised gaze prediction task. Given that we do not need any annotations or other auxiliary data, we employ the same architecture and training strategy for this task. We evaluate on the GTEA gaze dataset and compare it with other unsupervised models in Table 3. As can be seen, we obtain competitive results on the gaze prediction task, outperforming all baselines on both the AUC and AAE scores. It is to be noted that we outperform the center bias method on the AUC metric. Center bias exploits the spatial bias in egocentric images and always predicts the center of the video frame as the predicted gaze position. The AUC metric's significant improvement indicates that our approach predicts gaze fixations that are more closely aligned with the ground truth than the center bias approach. Given that the model was not designed explicitly for this task, it is a remarkable performance, especially given the performance of fully supervised baselines such as DFG [53], which achieves 10.6 and 88.3 for AUC and AAE.

Table 3. Comparison with state-of-the-art on the unsupervised egocentric gaze prediction task on the GTEA dataset.

	Itti *et al.* [17]	GBVS [10]	AWS-D [27]	Center Bias	OBDL [15]	Ours
AUC	0.747	0.769	0.770	0.789	0.801	**0.861**
AAE	18.4	15.3	18.2	**10.2**	15.6	13.6

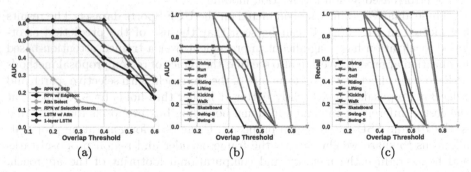

Fig. 3. Qualitative analysis of the proposed approach on UCF Sports dataset (a) ablative variations on AUC. (a) class-wise AUC, and (c) class-wise bounding box recall at different overlap thresholds.

5.6   Qualitative Evaluation

We find that our approach has a consistently high recall for the localization task across datasets and domains. We consider that an action is correctly localized if the average IoU across all frames is higher than 0.5, which indicates that most, if not all, frames in a video are correctly localized. We illustrate the recall scores and subsequent AUC scores for each class in the UCF sports dataset in Figs. 3(b) and (c). For many classes (7/10 to be specific), we have more than 80% recall at an overlap threshold of 0.5. We find, through visual inspection, that the spatial-temporal error is often correlated with the actor, but is usually not at the center of the region of interest and thus reduces the quality of the chosen proposals. We illustrate this effect in Fig. 4. The first row shows the input frame, the second shows the error-based attention, and the last row shows the final localization proposals. If more proposals are returned (as is the case with selective search and EdgeBox), we can obtain a higher recall (Fig. 3(b)) and higher mAP.

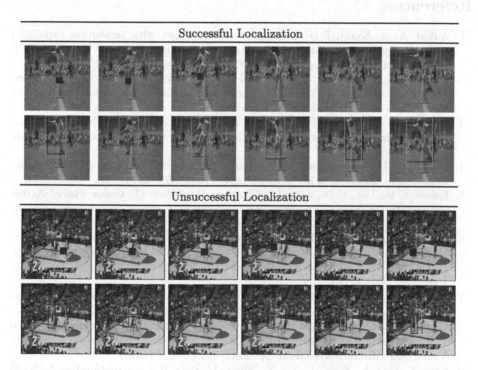

Fig. 4. Qualitative examples: we present the error-based attention location and the final prediction, for both successful and unsuccessful localizations. Green BB: Prediction, Blue BB: Ground truth (Color figure online)

6   Conclusion

In this work, we introduce a self-supervised approach to action localization, driven by spatial-temporal error localization. We show that the use of self-supervised prediction using video frames can help learn highly robust features and obtain state-of-the-art results on localization without any training annotations. We also show that the proposed framework can work with a variety of proposal generation methods without losing performance. We also show that the approach can generalize to egocentric videos without changing the training methodology or the framework and obtain competitive performance on the unsupervised gaze prediction task.

Acknowledgement. This research was supported in part by the US National Science Foundation grants CNS 1513126, IIS 1956050, and IIS 1955230.

References

1. Aakur, S., de Souza, F.D., Sarkar, S.: Going deeper with semantics: exploiting semantic contextualization for interpretation of human activity in videos. In: IEEE Winter Conference on Applications of Computer Vision (WACV). IEEE (2019)
2. Aakur, S.N., Sarkar, S.: A perceptual prediction framework for self supervised event segmentation. In: The IEEE Conference on Computer Vision and Pattern Recognition (CVPR) (June 2019)
3. Aakur, S.N., de Souza, F.D., Sarkar, S.: Towards a knowledge-based approach for generating video descriptions. In: Conference on Computer and Robot Vision (CRV). Springer (2017)
4. Bahdanau, D., Cho, K., Bengio, Y.: Neural machine translation by jointly learning to align and translate. arXiv preprint arXiv:1409.0473 (2014)
5. Escorcia, V., Dao, C.D., Jain, M., Ghanem, B., Snoek, C.: Guess where? Actor-supervision for spatiotemporal action localization. Comput. Vis. Image Underst. **192**, 102886 (2020)
6. Fathi, A., Li, Y., Rehg, J.M.: Learning to recognize daily actions using gaze. In: Fitzgibbon, A., Lazebnik, S., Perona, P., Sato, Y., Schmid, C. (eds.) ECCV 2012. LNCS, vol. 7572, pp. 314–327. Springer, Heidelberg (2012). https://doi.org/10.1007/978-3-642-33718-5_23
7. Gkioxari, G., Malik, J.: Finding action tubes. In: Proceedings of the IEEE Conference on Computer Vision and Pattern Recognition, pp. 759–768 (2015)
8. Grundmann, M., Kwatra, V., Han, M., Essa, I.: Efficient hierarchical graph-based video segmentation. In: 2010 IEEE Computer Society Conference on Computer Vision and Pattern Recognition, pp. 2141–2148. IEEE (2010)
9. Guo, Z., Gao, L., Song, J., Xu, X., Shao, J., Shen, H.T.: Attention-based LSTM with semantic consistency for videos captioning. In: ACM Conference on Multimedia (ACM MM), pp. 357–361. ACM (2016)
10. Harel, J., Koch, C., Perona, P.: Graph-based visual saliency. In: Advances in Neural Information Processing Systems, pp. 545–552 (2007)
11. Hershey, J.R., Chen, Z., Le Roux, J., Watanabe, S.: Deep clustering: discriminative embeddings for segmentation and separation. In: 2016 IEEE International Conference on Acoustics, Speech and Signal Processing (ICASSP), pp. 31–35. IEEE (2016)

12. Hochreiter, S., Schmidhuber, J.: Long short-term memory. Neural Comput. **9**(8), 1735–1780 (1997)
13. Horstmann, G., Herwig, A.: Surprise attracts the eyes and binds the gaze. Psychon. Bull. Rev. **22**(3), 743–749 (2015)
14. Horstmann, G., Herwig, A.: Novelty biases attention and gaze in a surprise trial. Atten. Percept. Psychophys. **78**(1), 69–77 (2016)
15. Hossein Khatoonabadi, S., Vasconcelos, N., Bajic, I.V., Shan, Y.: How many bits does it take for a stimulus to be salient? In: Proceedings of the IEEE Conference on Computer Vision and Pattern Recognition, pp. 5501–5510 (2015)
16. Hou, R., Chen, C., Shah, M.: Tube convolutional neural network (T-CNN) for action detection in videos. In: Proceedings of the IEEE International Conference on Computer Vision (ICCV), pp. 5822–5831 (2017)
17. Itti, L., Koch, C.: A saliency-based search mechanism for overt and covert shifts of visual attention. Vis. Res. **40**(10–12), 1489–1506 (2000)
18. Jain, M., Van Gemert, J., Jégou, H., Bouthemy, P., Snoek, C.G.: Action localization with tubelets from motion. In: Proceedings of the IEEE Conference on Computer Vision and Pattern Recognition, pp. 740–747 (2014)
19. Jhuang, H., Gall, J., Zuffi, S., Schmid, C., Black, M.J.: Towards understanding action recognition. In: Proceedings of the IEEE International Conference on Computer Vision, pp. 3192–3199 (2013)
20. Ji, X., Henriques, J.F., Vedaldi, A.: Invariant information clustering for unsupervised image classification and segmentation. In: Proceedings of the IEEE International Conference on Computer Vision, pp. 9865–9874 (2019)
21. Jia, X., De Brabandere, B., Tuytelaars, T., Gool, L.V.: Dynamic filter networks. In: Neural Information Processing Systems, pp. 667–675 (2016)
22. Jiang, Y.G., et al.: THUMOS challenge: action recognition with a large number of classes (2014)
23. Karpathy, A., Toderici, G., Shetty, S., Leung, T., Sukthankar, R., Fei-Fei, L.: Large-scale video classification with convolutional neural networks. In: IEEE Conference on Computer Vision and Pattern Recognition (CVPR), pp. 1725–1732 (2014)
24. Kodinariya, T.M., Makwana, P.R.: Review on determining number of cluster in k-means clustering. Int. J. **1**(6), 90–95 (2013)
25. Kuehne, H., Arslan, A., Serre, T.: The language of actions: recovering the syntax and semantics of goal-directed human activities. In: IEEE Conference on Computer Vision and Pattern Recognition (CVPR), pp. 780–787 (2014)
26. Lan, T., Wang, Y., Mori, G.: Discriminative figure-centric models for joint action localization and recognition. In: 2011 International Conference on Computer Vision, pp. 2003–2010. IEEE (2011)
27. Leboran, V., Garcia-Diaz, A., Fdez-Vidal, X.R., Pardo, X.M.: Dynamic whitening saliency. IEEE Trans. Pattern Anal. Mach. Intell. **39**(5), 893–907 (2016)
28. Li, Z., Gavrilyuk, K., Gavves, E., Jain, M., Snoek, C.G.: Videolstm convolves, attends and flows for action recognition. Comput. Vis. Image Underst. **166**, 41–50 (2018)
29. Liu, W., et al.: SSD: single shot multibox detector. In: Leibe, B., Matas, J., Sebe, N., Welling, M. (eds.) ECCV 2016. LNCS, vol. 9905, pp. 21–37. Springer, Cham (2016). https://doi.org/10.1007/978-3-319-46448-0_2
30. Ma, S., Zhang, J., Ikizler-Cinbis, N., Sclaroff, S.: Action recognition and localization by hierarchical space-time segments. In: Proceedings of the IEEE International Conference on Computer Vision, pp. 2744–2751 (2013)

31. Redmon, J., Divvala, S., Girshick, R., Farhadi, A.: You only look once: unified, real-time object detection. In: Proceedings of the IEEE Conference on Computer Vision and Pattern Recognition, pp. 779–788 (2016)
32. Rodriguez, M.D., Ahmed, J., Shah, M.: Action mach a spatio-temporal maximum average correlation height filter for action recognition. In: 2008 IEEE Conference on Computer Vision and Pattern Recognition, pp. 1–8. IEEE (2008)
33. Sharma, S., Kiros, R., Salakhutdinov, R.: Action recognition using visual attention. In: Neural Information Processing Systems: Time Series Workshop (2015)
34. Simonyan, K., Zisserman, A.: Very deep convolutional networks for large-scale image recognition. arXiv preprint arXiv:1409.1556 (2014)
35. Song, J., Gao, L., Guo, Z., Liu, W., Zhang, D., Shen, H.T.: Hierarchical LSTM with adjusted temporal attention for video captioning. In: Proceedings of the 26th International Joint Conference on Artificial Intelligence, pp. 2737–2743. AAAI Press (2017)
36. Soomro, K., Idrees, H., Shah, M.: Action localization in videos through context walk. In: Proceedings of the IEEE International Conference on Computer Vision, pp. 3280–3288 (2015)
37. Soomro, K., Idrees, H., Shah, M.: Predicting the where and what of actors and actions through online action localization. In: Proceedings of the IEEE Conference on Computer Vision and Pattern Recognition, pp. 2648–2657 (2016)
38. Soomro, K., Shah, M.: Unsupervised action discovery and localization in videos. In: Proceedings of the IEEE International Conference on Computer Vision, pp. 696–705 (2017)
39. Soomro, K., Zamir, A.R., Shah, M.: Ucf101: a dataset of 101 human actions classes from videos in the wild. arXiv preprint arXiv:1212.0402 (2012)
40. Tian, Y., Sukthankar, R., Shah, M.: Spatiotemporal deformable part models for action detection. In: Proceedings of the IEEE Conference on Computer Vision and Pattern Recognition, pp. 2642–2649 (2013)
41. Tipper, S.P., Lortie, C., Baylis, G.C.: Selective reaching: evidence for action-centered attention. J. Exp. Psychol.: Hum. Percept. Perform. **18**(4), 891 (1992)
42. Tran, D., Yuan, J.: Optimal spatio-temporal path discovery for video event detection. In: CVPR 2011, pp. 3321–3328. IEEE (2011)
43. Tran, D., Yuan, J.: Max-margin structured output regression for spatio-temporal action localization. In: Advances in Neural Information Processing Systems, pp. 350–358 (2012)
44. Uijlings, J.R., Van De Sande, K.E., Gevers, T., Smeulders, A.W.: Selective search for object recognition. Int. J. Comput. Vis. (IJCV) **104**(2), 154–171 (2013)
45. Van Gemert, J.C., Jain, M., Gati, E., Snoek, C.G., et al.: APT: action localization proposals from dense trajectories. In: BMVC, vol. 2, p. 4 (2015)
46. Venugopalan, S., Rohrbach, M., Donahue, J., Mooney, R., Darrell, T., Saenko, K.: Sequence to sequence-video to text. In: IEEE International Conference on Computer Vision (ICCV), pp. 4534–4542 (2015)
47. Venugopalan, S., Xu, H., Donahue, J., Rohrbach, M., Mooney, R., Saenko, K.: Translating videos to natural language using deep recurrent neural networks. arXiv preprint arXiv:1412.4729 (2014)
48. Vondrick, C., Pirsiavash, H., Torralba, A.: Anticipating visual representations from unlabeled video. In: IEEE Conference on Computer Vision and Pattern Recognition (CVPR), pp. 98–106 (2016)
49. Vondrick, C., Torralba, A.: Generating the future with adversarial transformers. In: Proceedings of the IEEE Conference on Computer Vision and Pattern Recognition, pp. 1020–1028 (2017)

50. Wang, L., Qiao, Yu., Tang, X.: Video action detection with relational dynamic-poselets. In: Fleet, D., Pajdla, T., Schiele, B., Tuytelaars, T. (eds.) ECCV 2014. LNCS, vol. 8693, pp. 565–580. Springer, Cham (2014). https://doi.org/10.1007/978-3-319-10602-1_37

51. Xie, J., Girshick, R., Farhadi, A.: Unsupervised deep embedding for clustering analysis. In: International Conference on Machine Learning (ICML), pp. 478–487 (2016)

52. Zacks, J.M., Tversky, B., Iyer, G.: Perceiving, remembering, and communicating structure in events. J. Exp. Psychol.: Gen. **130**(1), 29 (2001)

53. Zhang, M., Teck Ma, K., Hwee Lim, J., Zhao, Q., Feng, J.: Deep future gaze: gaze anticipation on egocentric videos using adversarial networks. In: Proceedings of the IEEE Conference on Computer Vision and Pattern Recognition, pp. 4372–4381 (2017)

54. Zhu, G., Porikli, F., Li, H.: Tracking randomly moving objects on edge box proposals. arXiv preprint arXiv:1507.08085 (2015)

Generative View-Correlation Adaptation for Semi-supervised Multi-view Learning

Yunyu Liu[1(✉)], Lichen Wang[1], Yue Bai[1], Can Qin[1], Zhengming Ding[2], and Yun Fu[1]

[1] Northeastern University, Boston, MA, USA
{liu.yuny,bai.yue,qin.ca}@northeastern.edu, wanglichenxj@gmail.com,
yunfu@ece.neu.edu
[2] Indiana University-Purdue University Indianapolis, Indianapolis, IN, USA
zd2@iu.edu

Abstract. Multi-view learning (MVL) explores the data extracted from multiple resources. It assumes that the complementary information between different views could be revealed to further improve the learning performance. There are two challenges. First, it is difficult to effectively combine the different view data while still fully preserve the view-specific information. Second, multi-view datasets are usually small, which means the model can be easily overfitted. To address the challenges, we propose a novel View-Correlation Adaptation (*VCA*) framework in semi-supervised fashion. A semi-supervised data augmentation me-thod is designed to generate extra features and labels based on both labeled and unlabeled samples. In addition, a cross-view adversarial training strategy is proposed to explore the structural information from one view and help the representation learning of the other view. Moreover, an effective and simple fusion network is proposed for the late fusion stage. In our model, all networks are jointly trained in an end-to-end fashion. Extensive experiments demonstrate that our approach is effective and stable compared with other state-of-the-art methods (Code is available on: https://github.com/wenwen0319/GVCA).

Keywords: Multi-view learning · Data augmentation · Semi-supervised learning

1 Introduction

Multi-view data refers to the data captured from multiple resources such as RGB, depth, and infrared in visual space [23,33,38]. MVL methods assume that the information from different views is unique and complementary. By learning from different views, MVL methods could achieve better performance. Several researches, such as [4,8,9,13,14,31], have studied MVL in supervised setting. However, the challenge is that labeling all the multi-view data can be extremely expensive. Therefore, semi-supervised learning setting is a more practical strategy since obtaining unlabeled data is easy.

© Springer Nature Switzerland AG 2020
A. Vedaldi et al. (Eds.): ECCV 2020, LNCS 12359, pp. 318–334, 2020.
https://doi.org/10.1007/978-3-030-58568-6_19

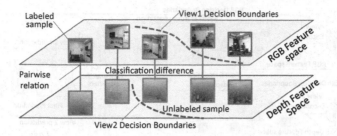

Fig. 1. The main challenge of MVL is that it is difficult to explore the latent correlations across different views due to the view heterogeneity. For instance, different view data from the same sample sometimes have the different predictions. Thus, effectively learn the cross-view information and fully explore the unlabeled data can potentially improve the learning performance.

As shown in Fig. 1, the main challenge of MVL is that different views have different data formats, feature distributions, and predicted results, namely data heterogeneity. Multi-view fusion is an efficient strategy. It mainly divided into three parts: raw-data fusion, feature-level fusion, and label-level fusion. [16] utilized feature-level and label-level fusion for learning a better representation. [10] applied GAN and directly translated one view to the other view which belongs to raw-data fusion. Although these methods achieved good performance, they considered the three fusion strategies independently which cannot fully reveal the latent relations between these steps. Semi-supervised learning [7,10,29,30,32] is an effective approach which explores the unlabeled samples to increase the learning performance. [7] proposed a new co-training method for the training process. [10] used the unlabeled data to enhance the view translation process. However, these methods combined semi-supervised setting and multi-view learning in an empirical way while lacked a unified framework to jointly solve the challenges.

In this work, we propose a View-Correlation Adaptation (*VCA*) framework in semi-supervised scenario. The framework is shown in Fig. 2. First, a specifically designed View-Correlation Adaptation method is proposed. It effectively aligns the feature distributions across different views. Second, a semi-supervised feature augmentation (*SeMix*) is proposed. It augments training samples by exploring both labeled and unlabeled samples. Third, we propose an effective and simple fusion network to learn both inter-view and intra-view high-level label correlations. By these ways, the label-level and feature-level fusion are considered simultaneously in an unified framework. This further helps our model obtain more distinctive representation for classification. Extensive experiments demonstrate that our method achieves the highest performance compared with state-of-the-art methods. Ablation studies illustrate the effectiveness of each module in the framework. The main novelties are listed below.

- A novel fusion strategy View-Correlation Adaptation (*VCA*) is deployed in both feature and label space. *VCA* makes one view learn from the other view by optimizing the decision discrepancy in an adversarial training manner.

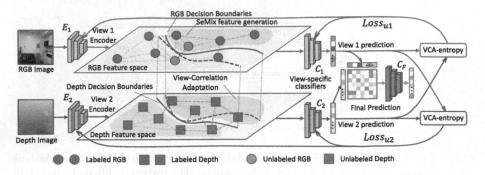

Fig. 2. Framework of our model. Multi-view samples first go through the view encoders $E_1(\cdot)$ and $E_2(\cdot)$ and get the representation features. We design *SeMix* to the feature space in order to expand the feature space and align the distribution of the labeled and unlabeled data. The representation features are sent to the view-specific classifiers $C_1(\cdot)$ and $C_2(\cdot)$. Then *VCA-entropy* is applied to learn general representations while keeping view-specific characteristics. For two single views' predictions of the unlabeled data, the one with a high entropy should learn form the other one since the entropy indicates the uncertainty of predictions. By playing an adversarial training strategy, the view encoders learn general feature distributions and the view specific classifiers keep view-specific characteristics. A cross-view fusion network $C_F(\cdot)$ is used to fuse the predictions and get a final result by capturing the inter-view and intra-view label correlation.

- A new generative *SeMix* approach is proposed to fully utilize the labeled and unlabeled data to expand the feature distributions and make the model more robust.
- An effective label-level fusion network is proposed to obtain the final classification result. This module captures both the inter-view and intra-view high-level label correlations to obtain a higher performance.

2   Related Work

2.1   Multi-view Learning

Multi-view learning explores data from different views/resources for downstream tasks (*e.g.*, classification and clustering). RGB and depth are the two commonly explored views of MVL. [4,8,13,14,28] focused on learning a better representation for actions using supervised information. [2,7,10,12,16,27] studied multi-view scene recognition. Conventional MVL methods focused on improving the multi-view features. [4] introduced a multi-view super vector to fuse the action descriptors. [13] combined optical flow with enhanced 3D motion vector fields to fuse features. [12] used contours of depth images as the local features for scene classification. [2] proposed a second-order pooling to extract the local features. Deep learning framework was proposed in the fusion procedure recently due to its potential in a wide range of applications. [27] utilized a component-aware fusion method to fuse the features extracted from deep models. [28] introduced a generative/mapping framework which learns the view-independent and view-specific representations. [16] implemented a method to learn the distinctive embedding and correlative embedding simultaneously. [10] proposed a generative method and translated one view to the other to help classification.

However, the fusion modules are separated with other modules in most methods, which degrades the potential of the methods and make the training procedure tedious to obtain good results. Our model is a union framework with both the label-level and feature-level fusion strategies. In addition, our model is a more general framework which and can be adapted to different tasks.

2.2   Semi-supervised Learning

Semi-supervised learning deploys both unlabeled and labeled data in the training procedure. It is an effective strategy for a situation where collecting data is easy while labeling the data is extremely difficult. Semi-supervised learning explores the structural information of the labeled and unlabeled data to improve the performance. The general introduction of semi-supervised learning could be found in [6]. [35] designed co-training method for semi-supervised learning. Two-learner framework is utilized and the unlabeled samples would be assigned confident labels based on the guidance of the two learners. [3] proposed a MixMatch framework which guesses low-entropy labels for data-augmented unlabeled examples. [36] proposed a multi-modal method based on Alexnet [15] to solve the missing view issues in semi-supervised fashion. [7] introduced a co-training method combined with deep neural network. [21] implemented a graph-based method to solve the multi-view semi-supervised problem. [10] trained a translation network using both labeled and unlabeled data to obtain representations.

Although high performance is achieved by the methods. They are mainly focusing on single-view setting while ignore the sophisticated cross-view relations and the extra latent connections between labeled and unlabeled samples. Our approach is specifically designed for multi-view semi-supervised learning. It jointly explores the cross-view and the cross-instance relations. Based on our experiments, the revealed knowledge considerably improves the performance and robustness of our approach.

3   Our Approach

3.1   Preliminaries and Motivation

$X_l^1 \in \mathbb{R}^{d_1 \times n_l}$ and $X_l^2 \in \mathbb{R}^{d_2 \times n_l}$ are the feature matrices of two views, and they belong to labeled samples. n_l stands for the labeled instance number. d_1, d_2 are the feature dimensions of view1 and view2, and each column is an instance. Similarly, $X_u^1 \in \mathbb{R}^{d_1 \times n_u}$ and $X_u^2 \in \mathbb{R}^{d_2 \times n_u}$ are the feature matrices of unlabeled samples obtained from view1 and view2. n_u is the unlabeled instance number. The label matrix $Y_l \in \mathbb{R}^{d_l \times n_l}$ of labeled samples are given, where d_l is the dimension of the label space. We denote the feature and label vector of the i-th instance by x_{li}^1, x_{li}^2 and y_{li}. The goal is to recover the label of the unlabeled set given X_l^1, X_l^2, X_u^1, X_u^2 and Y_l. Generally, our framework consists view-specific encoders $E_1(\cdot)$, $E_2(\cdot)$, view-specific classifiers $C_1(\cdot)$, $C_2(\cdot)$, and a fusion network $C_F(\cdot)$. The encoders encode the original data to a subspace. The classifiers obtain classification score L_1 and L_2. Then a label-level fusion mechanism $C_F(\cdot)$ is applied to derive a final result. Details are introduced in the following sections.

3.2   Semi-supervised Mixup

Limited training samples is a common challenge for general machine learning tasks and it is significant in MVL scenario. Mixup [37] is an effective data augmentation strategy. Although various modifications are proposed [3,26], the label information is always required. In this work, we propose a novel semi-supervised version Mixup approach *SeMix*. Different from existing methods, *SeMix* fully explores feature distributions of both labeled and unlabeled samples, and generates more general samples with confident labels. *SeMix* could be divided into two parts: 1) *Labeled augmentation* and 2) *Unlabeled augmentation*.

Labeled Augmentation. Given labeled samples, *SeMix* augments labeled data based on the follow steps. A random variable $\lambda' \sim Beta(\alpha, \alpha)$ is sampled where $Beta(\alpha, \alpha)$ is the Beta distribution. And the weight parameter λ is obtained via $\lambda = \max(\lambda', 1 - \lambda')$, where $\max(\cdot, \cdot)$ is the max value of the inputs. Then, the augmented representations, $\tilde{r}^i$, and the corresponding labels, $\tilde{y}$, could be obtained from the equations below:

$$
\begin{aligned}
\tilde{r}_{nm}^i &= \lambda E_i(x_{ln}^i) + (1 - \lambda)E_i(x_{lm}^i), i = \{1, 2\}, \\
\tilde{y}_{nm} &= \lambda y_{ln} + (1 - \lambda)y_{lm},
\end{aligned}
\tag{1}
$$

where x_{ln}^i and x_{lm}^i are two randomly selected labeled samples from the i-th view, y_{ln} and y_{lm} are the corresponding label vectors. The obtained $\tilde{r}_{nm}^i$ and $\tilde{y}_{nm}$ could be directly involved in the training procedure. In Eq. (1), *SeMix* augments the samples in the encoded spaces instead of the original feature space since we observed that the performance is higher and more stable.

Equation (1) is the redesigned version of Mixup for labeled samples. There are two-fold improvements compared with conventional Mixup methods [37]. First, *SeMix* achieves augmentations in the low-dimensional subspace instead of the original feature space. The representation of each sample in the subspace becomes more distinctive with lower noise. By this way, *SeMix* could effectively and accurately explore the structural knowledge and generate clear and high quality samples. The decision boundaries of the learned classifiers would become smoother [26] which improves the generalization and the model. Second, we define λ' following a Beta distribution instead of a uniform distribution and make sure λ is bigger than 0.5 by the equation $\lambda = \max(\lambda', 1 - \lambda')$. This strategy promises $E_i(x_{ln}^i)$ is in a dominate place compared with $E_i(x_{lm}^i)$. This constraint is also useful for the unlabeled data and will be discusses in the following section.

Unlabeled Augmentation. The above mentioned approach could only work when the labels of both the pairwise samples are known. To fully explore the distribution knowledge from the unlabeled samples, an effective algorithm is designed which bypass the label missing issue while still guarantee the quality of the augmented samples. Randomly select an unlabeled sample x_u^i and a labeled sample x_{lm}^i, where $i = \{1, 2\}$ means both of them are from the i-th view. Assuming the label of x_u^i is y_u'. According to Eq. (1), replacing x_{ln}^i and y_{ln} with x_u^i and y_u', we can obtain the representations and labels of x_u^i and x_{lm}^i pair as:

$$
\begin{aligned}
\tilde{r}_{um}^i &= \lambda E_i(x_u^i) + (1 - \lambda)E_i(x_{lm}^i), i = \{1, 2\}, \\
\tilde{y}_{um} &= \lambda y_u' + (1 - \lambda)y_{lm}.
\end{aligned}
\tag{2}
$$

Similarly, when x_u^i is paired with another labeled sample x_{ln}^i, we can obtain another set of representation and label as below:

$$
\begin{aligned}
\tilde{r}_{un}^i &= \lambda E_i(x_u^i) + (1 - \lambda)E_i(x_{ln}^i), i = \{1, 2\}, \\
\tilde{y}_{un} &= \lambda y_u' + (1 - \lambda)y_{ln}.
\end{aligned}
\tag{3}
$$

Since y_u' is unknown, Eq. (2) and Eq. (3) cannot be directly deployed in the training process. Therefore, we calculate the difference of the obtained labels, $\tilde{y}_{um}$ and $\tilde{y}_{un}$, and we can obtain the result as follow:

$$
\tilde{y}_{un} - \tilde{y}_{um} = (1 - \lambda)(y_{ln} - y_{lm}).
\tag{4}
$$

Therefore, Eq. (4) effectively bypass the missing label issue. When training, we randomly selected an unlabeled sample x_u^i and two labeled samples x_{lm}^i and x_{ln}^i to obtain the label difference. We also deploy the prediction differences instead of exact prediction results in the objective function which is shown below:

$$
L_{semi} = \sum_{i=1}^{2} \|(C_i(\tilde{r}_{um}^i) - C_i(\tilde{r}_{un}^i)) - (1 - \lambda)(y_{ln} - y_{lm})\|_F^2,
\tag{5}
$$

where L_{semi} is the objective value based on the label differences and we minimize L_{semi} by jointly optimizing both the view-specific encoders and classifiers (*i.e.*, $E_1(\cdot)$, $E_2(\cdot)$, $C_1(\cdot)$, and $C_2(\cdot)$). In this way, *SeMix* fully explores the distribution knowledge across the labeled and unlabeled samples which enhance the diversity and generality of the augmented samples.

We give an intuitive explanation of *SeMix* in word embedding scenario for easy understanding. For instance, we want the learned encoders and classifiers are robust which could obtain accurate representation such as *man-woman = king-queen*. This also explains why we set λ in the larger side (*i.e.*, $\max(\lambda', 1 - \lambda')$). If λ is small, the unlabeled data will be regarded as a noise since its scale is small. It is hard for the model to find useful information in the unlabeled data.

3.3   Dual-Level View-Correlation Adaptation

In MVL, classification discrepancy is that different views have different classification results. This commonly exists that because different views have their unique characteristics. It can be a valuable clue for MVL.

Here, we propose a novel way to explore and align the discrepancy in both feature space and label space, and we name this mechanism as Dual-level View-Correlation Adaptation (VCA). For the labeled data, the classification results should be the same as the ground-truth. The objective function is as follow:

$$L_{labeled} = \sum_{i=1}^{2} \|\tilde{y}^i - y\|_{F}^2,$$
(6)

where $\tilde{y}_i$ is the classification result from the i-th view and y is the ground truth. The training samples are labeled samples and the generated samples via Eq. (1).

For the unlabeled data, instead of forcibly eliminating the discrepancy of $C_1(\cdot)$ and $C_2(\cdot)$, we optimize the encoders $E_1(\cdot)$ and $E_2(\cdot)$ to minimize the discrepancy while the classifiers $C_1(\cdot)$ and $C_2(\cdot)$ to maximize the discrepancy. As shown in Fig. 2, this adversarial training strategy is the crucial module for cross-view structure learning process. We define the objective function:

$$L_{unlabeled} = W(C_1(\tilde{r}_n^1), C_2(\tilde{r}_n^2)),$$
(7)

where $W(\cdot, \cdot)$ is the Wasserstein Distance, the inputs are the classification results from two views. We assume the prediction of the unlabeled samples are noisy and uncertain, using Wasserstein distance can be more effective and stable.

Both of Eq. (6) and Eq. (7) are used to calculate the classification discrepancies. We empirically tested different evaluation setups and we found out this setup achieves the best performance.

The four networks are alternately optimized. Firstly, we use the labeled data to train $E_1(\cdot)$, $E_2(\cdot)$ and $C_1(\cdot)$, $C_2(\cdot)$. The loss can be defined as follows.

$$\min_{C_1,C_2,E_1,E_2} L_{labeled}.$$
(8)

The second step is letting the encoders minimize the discrepancy.

$$\min_{E_1,E_2} L_{labeled} + L_{unlabeled}.$$
(9)

The third step is letting the classifiers to enlarge the discrepancy.

$$\min_{C_1,C_2} L_{labeled} - L_{unlabeled}.$$
(10)

Labeled loss $L_{labeled}$ to Eq. (9) and Eq. (10) are added to stabilize the optimization. The empirical experiments also show that the performance decreases significantly without $L_{labeled}$.

By adversarial training strategy, $E_i(\cdot)$ could borrow the structural information from the other view and obtain more distinctive feature representations in their subspaces. In addition, $C_i(\cdot)$ would obtain robust classification boundaries. Overall, the adversarial manner promises: 1) the training step is more stable and 2) the model is more robust and has less possibility to be over-fitting.

Entropy-Based Modification. For Eq. (9) and Eq. (10), ideally, we want the encoders to learn a more robust representations based on the guidance of the other view. However, in some cases, this process may decrease the capacity of the encoders and make the final prediction results worse. For example, the view has a wrong result affects the view with the right result by pulling them together.

Therefore, we introduce an entropy based module to handle this challenge. Our goal is to make the view with the confident prediction to guide the other and avoid the reverse effect. To achieve this goal, the entropy is used to evaluate the confidence of a classification. The higher the entropy, the more uncertainty(lower confidence) of the classification results. Specifically, if view 1 has a higher entropy than view 2, the loss of view 1 should be larger than the loss of view 2. This will encourage view 1 learns from view 2 while view 2 tries to preserve its own knowledge. Therefore, Eq. (6) and Eq. (7) can be revised as below,

$$L^1_{unlabeled} = \frac{H(C_1(\tilde{r}^1))}{H(C_2(\tilde{r}^2))} W(C_1(\tilde{r}^1), C_2(\tilde{r}^2)), \tag{11}$$

$$L^2_{unlabeled} = \frac{H(C_2(\tilde{r}^2))}{H(C_1(\tilde{r}^1))} W(C_1(\tilde{r}^1), C_2(\tilde{r}^2)), \tag{12}$$

where $H(\cdot)$ stands for the entropy. The inputs of $H(\cdot)$ are two classification results from two views. For $W(\cdot, \cdot)$, we have $W(C_1(\tilde{r}^1), C_2(\tilde{r}^2)) = W(C_2(\tilde{r}^2), C_1(\tilde{r}^1))$. Meanwhile, Eq. (9) and Eq. (10) are changed correspondingly as follows.

$$\min_{E_i} L_{labeled} + L^i_{unlabeled}, i = \{1, 2\}, \tag{13}$$

$$\min_{C_i} L_{labeled} - L^i_{unlabeled}, i = \{1, 2\}. \tag{14}$$

3.4   Label-Level Fusion

The discrepancy still exists even after the alignment procedure. We assume different views have the unique characteristics and some categories could achieve more accurate/reliable classification results from a specific view. An intuitive example, if there are "Police" and "Doctor" views, "Police" view is good at distinguishing good/bad people, while "Doctor" view is good at distinguishing healthy/sick bodies. Conventional methods utilize naive mechanisms (e.g., mean or max pooling) which lost valuable information. Based on this assumption, we deploy a novel fusion strategy in label space. It automatically learns the prediction confidences between different views for each category. Then, the best prediction is learned as the final classification result.

In our model, we apply a cross-view fusion network $C_F(\cdot)$ to handle this challenge. For two predictions from different views $\tilde{y}_1$ and $\tilde{y}_2$, we multiply the predictions with the transpose to obtain a matrix. There are totally three matrices(i.e., $\tilde{y}_1^\top \tilde{y}_1$, $\tilde{y}_2^\top \tilde{y}_2$ and $\tilde{y}_1^\top \tilde{y}_2$) In the matrix, each element is the multiplication of the pairwise prediction scores. Then, we sum up three matrices and reshape the matrix to a vector and forward to the fusion network $C_F(\cdot)$. The framework is illustrated in Fig. 2. The objective function is shown below:

$$L_F = \|C_F(reshape(\tilde{y}_1^\top \tilde{y}_1 + \tilde{y}_2^\top \tilde{y}_2 + \tilde{y}_1^\top \tilde{y}_2)) - y\|^2_{\mathrm{F}}, \tag{15}$$

where y is the ground truth labels. By this way, $C_F(\cdot)$ would hopefully discover latent relations between different views and categories.

4   Experiments

4.1   Dataset

Three action recognition multi-view datasets are deployed for our evaluation. **Depth-included Human Action dataset (DHA)** [17] is an RGB-D multi-view dataset. It contains 23 kinds of actions performed by 21 subjects. We randomly choose 50% as the training set and the others as the testing set. We pick the RGB and depth views for the evaluation. **UWA3D Multiview Activity (UWA)** [24] is a multi-view action dataset collected by Kinect sensors. It contains 30 kinds of actions performed by 10 subjects. 50% samples are randomly extracted as the training set and the others are assigned to the testing set. Similarly, we choose the RGB videos and depth videos for the experiments. **Berkeley Multimodal Human Action Database (MHAD)** [22] is a comprehensive multimodal human action dataset which contains 11 kinds of actions. Each action is performed by 12 subjects with 5 repetitions. We choose RGB videos and depth videos for the evaluation. 50% samples are set as the training set and the others as the testing set.

4.2   Baselines

We test our approach under the scenario of multi-view (RGB-D) action recognition. We deploy some baseline methods for comparison. **Least Square Regression (LSR)** is a linear regression model. We concatenate the features from different views together as the input. LSR learns a linear mapping from the feature space to the label spaces. **Multi-layer Neural Network (NN)** is a classical multi-layer neural network. It is deployed as a classifier with the multi-view features concatenated as the input. **Support Vector Machine (SVM)** [25] attempts to explore the hyper-planes in the high-dimensional space. The features are concatenated from multiply views and the SVM module we used is implemented by LIBSVM [5]. **Action Vector of Local Aggregated Descriptor (VLAD)** [11] is an action representation method. It is able to capture local convolutional features and spatio-temporal relationship of the videos. **Auto-Weight Multiple Graph Learning (AMGL)** [20] is a graph-based multi-view classification method designed for semi-supervised learning. Optimal weights for each graph are automatically calculated. **Multi-view Learning with Adaptive Neighbours (MLAN)** [19] deploys adaptive graph to learn the local and global structure and the weight of each view. **Adaptive MUltiview SEmi-supervised model (AMUSE)** [21] is a semi-supervised model for image classification task. It learns the parameters from the graph to obtain the classification results. **Generative Multi-View Action Recognition (GMVAR)** [31] is a multi-view action recognition method. It augments samples guided by another view and applies a label-level fusion module to get the final classification.

Table 1. Baselines and performance. Classification accuracy (%) on DHA, UWA, and MHAD datasets.

Setting	Method	DHA	UWA	MHAD
RGB	LSR	65.02	67.59	96.46
	SVM [25]	66.11	69.77	96.09
	VLAD [11]	67.85	71.54	97.17
	TSN [34]	67.85	71.01	97.31
Depth	LSR	82.30	45.45	47.63
	SVM [25]	78.92	34.92	45.39
	WDMM [1]	81.05	46.58	66.41
RGB+D	LSR	77.36	68.77	97.17
	NN	86.01	73.70	96.88
	SVM [25]	83.47	72.72	96.80
	AMGL [20]	74.89	68.53	94.70
	MLAN [19]	76.13	66.64	96.46
	AMUSE [21]	78.12	70.32	97.23
	GMVAR [31]	88.72	76.28	98.94
	Ours	**89.31**	**77.08**	**98.94**

Table 2. Ablation study. Classification accuracy (%) on DHA [17] dataset.

Setting	RGB	Depth	RGB+D
TSN [34]	67.85	-	-
WDMM [1]	-	81.05	-
MLP	77.10	79.01	79.12
Mixup	68.51	81.43	81.48
SeMix	69.37	**82.73**	83.15
VCA	75.26	80.86	81.32
VCA-entropy	**80.86**	82.61	84.10
Ours complete	-	-	**89.31**

4.3 Implementation

We extract initial action features from the original visual spaces first. We deploy Temporal Segment Networks (TSN) [34] for RGB view and Weighted Depth Motion Maps (WDMM) [1] for depth view respectively. The implementation and feature extraction details are introduced below.

Temporal Segment Networks (TSN) [34] divides the video into several segments. Then it randomly samples and classifies the frames in each segment. The fusion of all the classification results(e.g. mean) is the final result. We sample 5 frames for training and 3 frames for testing. Each video is described by a concatenation of features from 3 frames.

Weighted Depth Motion Maps (WDMM) [1] is designed for human gestures recognition in depth view. It proposed a new sampling method to do a linear aggregation for spatio-temporal information from three projections views.

For LSR, NN and SVM, the original RGB features and depth features are extracted by TSN and WDMM respectively. We apply a normalization to both features since directly concatenating features from different views is not reasonable due to the view heterogeneity. Then we concatenate these two features as a single feature. We use LSR, NN and SVM to classify the single feature. For AMGL, MLAN, AMUSE and GMVAR, we consider RGB features extracted

Table 3. Classification accuracy(%) given different ratios of labeled training samples.

Dataset	Ratio	10%	20%	30%	40%	50%	60%	70%	80%	90%	100%
DHA	GMVAR [31]	44.86	62.55	69.14	73.25	77.37	80.66	84.36	84.36	86.01	88.72
	Ours	48.15	69.14	74.90	79.42	83.54	83.56	85.60	86.83	87.24	89.31
UWA	GMVAR [31]	35.18	46.64	54.15	58.57	65.61	69.17	73.91	75.49	76.28	76.28
	Ours	36.36	57.71	62.85	64.03	67.98	70.75	73.12	76.56	76.66	77.08
MHAD	GMVAR [31]	53.36	72.79	90.11	92.64	93.41	94.76	95.91	95.49	96.28	98.94
	Ours	52.30	75.83	91.17	92.23	92.93	93.11	95.05	96.82	97.88	98.94

by TSN and depth features extracted by WDMM as two views for multi-view setting. We do not include extra preprocessing steps for fair comparison.

4.4 Performance

The experimental result is shown in Table 1. From the results, we observe that our approach outperforms other state-of-the-art methods. It illustrates the effectiveness of our approach. Our model achieves similar performance in MHAD dataset compared with the best performance, we assume the classification accuracy in MHAD dataset is already high (*i.e.*, 98.94%) and is too challenging to obtain considerable improvements. For DHA and UWA datasets, our performance is slightly higher than GMVAR [31]. We assume the datasets are well-explored and hard to achieve great improvement. Meanwhile, Table 3 shows that our improvements are higher when less labeled samples (<50%) are given, which reveals the potential and advantages of our model in semi-supervised setting.

4.5 Ablation Study

To prove the effectiveness of all the components in our approach, we intentionally remove some components and evaluate the performance based on partial of our model. We also evaluated the efficiency of our approach utilizing labeled samples. The effectiveness of the generated samples are also explored. In Table 2, the performance of *TSN* and *WDMM* are listed as baselines. The baseline model only has an encoder and a classifier for each view and only optimized by Eq. (8).

Mixup and SeMix. We conduct two experiments to prove the effectiveness of *SeMix* compared with *Mixup*. The first experiment is *Mixup* with labeled data. We apply the data augmentation method from [37] with Eq. (1) as the baseline model. The second experiment is *SeMix* with both labeled and unlabeled data. We apply Eq. (1) and Eq. (5) simultaneously to the baseline model. We concatenate the obtained representations of the two views and send it to a two-layer neural network to get the RGB+D multi-view performance. The results are shown in Table 2 and Fig. 3 (a) and (b). In Table 2, we observe that both *Mixup* and *SeMix* achieve considerable improvements in single-view scenario. We can conclude that our framework does learned the extra structural knowledge from the other view's distribution and help the single view classification.

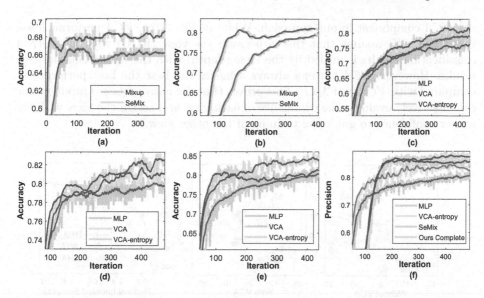

Fig. 3. Ablation study results. (a) and (b) are the performance of *Mixup* and *SeMix* in RGB view and depth view. Blue lines are the performance of *Mixup* and red lines are the performance of *SeMix*. (c), (d) and (e) are the performance of *MLP*, *VCA* and *VCA-entropy* in RGB view, depth view and after the cross-view fusion. Blue lines, green lines and red lines indicate *MLP*, *VCA* and *VCA-entropy* respectively. (f) is the performance of cross-view fusion. The whole framework is proposed to achieve a higher cross-view fusion. Blue line, yellow line, green line and red line indicate the *MLP*, *VCA-entropy*, *SeMix* and our whole framework respectively. The shadow lines are the exact performance, indicating the robustness and stability of the model. (Color figure online)

From Fig. 3 (a) and (b), we can see that *SeMix* achieves higher and more stable performance than the *Mixup* baselines. We assume *SeMix* fully utilizes the unlabeled information, which helps the classification and representation.

VCA and VCA-Entropy. We conduct several experiments to prove the effectiveness of *VCA* and the entropy-based VCA, *VCA-entropy*. *VCA* and *VCA-entropy* aims to operate the Dual-level View-Correlation Adaptation and we evaluate the performance with and without the entropy algorithm. *MLP* is the baseline model. It has an encoder and a classifier for each view. A cross-view fusion mechanism is implemented to obtain the final result. For *VCA*, we simply add the cross-view adaptation part (*i.e.* Eq. (9) and Eq. (10)) to *MLP*. For *VCA-entropy*, we change the loss functions Eq. (9) and (10) in *VCA* to Eq. (11) and (13). Our results are shown in Table 2 and Fig. 3 (c), (d) and (e). They illustrate the classification performance of view1 (RGB), view2 (Depth), and the final fusion result respectively. For each graph, there are three different lines with different colors. The blue, green and red lines indicate the results of the *MLP*, *VCA* and *VCA-entropy* respectively. We observe that in most of the cases the *VCA* performs better than *MLP*, which demonstrates the effectiveness of the *VCA* component in our approach. While in Fig. 3 (c), *VCA* performs lower than *MLP*. We assume that this is because the view (RGB) with the correct classification results is misled by the view (depth) with the wrong classification results. However, *VCA-entropy* always achieves the best the best performance compared with *VCA* and *MLP*. This proves that the entropy-based modification can effectively evaluate the confidence of each view and assign the view with the higher confidence to guide the training of the other view.

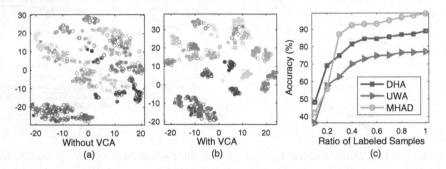

Fig. 4. (a) t-SNE Visualization of the feature extracted by TSN [34]. We use the pretrained ResNet-101 model and finetune it with raw DHA RGB features. (b) t-SNE Visualization of the output of encoder $E_1(\cdot)$. We train the whole framework for 800 epochs. (c) Performance when we use different amount of labeled data.

Complete Framework. We conduct experiments to prove that by combining these two modules, our methods can be further improved. Since our target is classification result based on two views, we show four final results of our models: 1) *MLP* with the cross-view fusion module, 2) *SeMix* with the cross-view fusion module, 3) *VCA-entropy* with cross-view fusion module, and 4) the complete model. Our results are shown in Table 2 and Fig. 3 (f). Figure 3 (f) indicates that both *SeMix* or *VCA-entropy* are helpful to improve the final performance, and by adding both *SeMix* and *VCA-entropy*, the performance achieves the best. This means our framework is able to compromise the merits of both *SeMix* and *VCA-entropy* and further improve the classification performance.

Visualization. We utilize t-SNE [18] to visualize the distribution of the RGB features and its representations after the encoder. DHA dataset is deployed for this experiment. The RGB feature is extracted by TSN [34] model. The representation is the output of the encoder in our model after training 800 epochs. The solid circles are labeled data while hollow circles are unlabeled data. Different colors stand for different actions. Figure 4 (a) is the distribution without *VCA* while Fig. 4 (b) is the distribution with *VCA*. We observe that the representations from the same class cluster better than the feature extracted by TSN. This can potentially help the model with the higher classification performance. The unlabeled data also clusters better after representation. This proves the effectiveness of our model. In summary, the results indicate the RGB encoder learns from the depth view.

Labeled Samples. We conduct the experiments to prove our model still works well even when fewer labeled samples are provided. We change the number of available labeled samples from 10% of the training samples to 100% of the training samples. The performance is shown in Fig. 4 (c). The x-axis indicates the ratio of the labeled samples to the whole training set. The y-axis is the classification accuracy. From Fig. 4 (c), we observe that the performance grows quickly in the beginning for all datasets. When the ratio is larger than 40%, the growth trend becomes slow. This indicates our semi-supervised model can achieve a comparable result with only 50% labeled data are available. The results further prove that our model can efficiently utilize the labeled sample and practically avoid the extremely expensive data labeling procedure.

Table 4. Accuracy (%) of different generate number

Dataset	0x	0.1x	0.3x	0.5x	1x	2x	3x
DHA	84.10	87.24	88.07	88.07	89.31	88.89	89.31
UWA	75.49	76.30	77.08	76.68	77.08	77.47	76.28
MHAD	98.23	98.59	98.23	98.23	98.94	98.94	97.88

Generated Samples. We evaluate the helpfulness of the generated samples. In Table 4, different amount of the generated samples are utilized for training the model. The first row indicates the ratio of the generated samples to labeled samples. Specifically, 0x indicates there are only real samples. We observe that for DHA and UWA datasets, the performance under different ratios are higher than the results without the generative samples. The performance reaches the peak around 1x and 2x. More samples could not improve the performance and we assume it reaches the limitation of *SeMix*, and the slight performance decrease could be fluctuations. For MHAD dataset, since the 0x performance is relatively high, it is hard to significantly improve the performance. While, our model still slightly outperforms the 0x when the modal generates 1x and 2x samples.

5   Conclusion

We propose a novel Generative View-Correlation Adaptation framework for semi supervised multi-view learning. A new data augmentation mechanism, *SeMix*, is proposed which utilizes both labeled and unlabeled data to generate more diverse and robust auxiliary samples. In addition, a multi-view dual-level alignment strategy is designed. The classification results from both views are used to guide the training of the encoders and classifiers where the structural information from both feature and label space are effectively explored. Moreover, a simple yet effective view-correlation fusion network is applied which reveals the latent label relations and obtains the final result. Extensive experiments are conducted which demonstrate the effectiveness of our approach. More comprehensive ablation studies illustrate that all the modules in our approach are necessary and indispensable which considerably improve the final performance.

Acknowledgement. This research is supported by the U.S. Army Research Office Award W911NF-17-1-0367.

References

1. Azad, R., Asadi-Aghbolaghi, M., Kasaei, S., Escalera, S.: Dynamic 3D hand gesture recognition by learning weighted depth motion maps. IEEE Trans. Circuits Syst. Video Technol. **29**, 1729–1740 (2018)
2. Banica, D., Sminchisescu, C.: Second-order constrained parametric proposals and sequential search-based structured prediction for semantic segmentation in RGB-D images. In: Proceedings of the IEEE Conference on Computer Vision and Pattern Recognition, pp. 3517–3526 (2015)
3. Berthelot, D., Carlini, N., Goodfellow, I., Papernot, N., Oliver, A., Raffel, C.: MixMatch: a holistic approach to semi-supervised learning. arXiv preprint arXiv:1905.02249 (2019)
4. Cai, Z., Wang, L., Peng, X., Qiao, Y.: Multi-view super vector for action recognition. In: Proceedings of the IEEE Conference on Computer Vision and Pattern Recognition, pp. 596–603 (2014)
5. Chang, C.C., Lin, C.J.: LIBSVM: a library for support vector machines. ACM Trans. Intell. Syst. Technol. **2**(3), 27 (2011)
6. Chapelle, O., Scholkopf, B., Zien, A.: Semi-supervised learning. IEEE Trans. Neural Netw. **20**(3), 542 (2009)
7. Cheng, Y., Zhao, X., Cai, R., Li, Z., Huang, K., Rui, Y., et al.: Semi-supervised multimodal deep learning for RGB-D object recognition (2016)
8. Cheng, Z., Qin, L., Ye, Y., Huang, Q., Tian, Q.: Human daily action analysis with multi-view and color-depth data. In: Fusiello, A., Murino, V., Cucchiara, R. (eds.) ECCV 2012. LNCS, vol. 7584, pp. 52–61. Springer, Heidelberg (2012). https://doi.org/10.1007/978-3-642-33868-7_6
9. Ding, Z., Shao, M., Fu, Y.: Robust multi-view representation: a unified perspective from multi-view learning to domain adaption. In: Proceedings of the International Joint Conferences on Artificial Intelligence, pp. 5434–5440 (2018)
10. Du, D., Wang, L., Wang, H., Zhao, K., Wu, G.: Translate-to-recognize networks for RGB-D scene recognition. In: Proceedings of the IEEE Conference on Computer Vision and Pattern Recognition, pp. 11836–11845 (2019)
11. Girdhar, R., Ramanan, D., Gupta, A., Sivic, J., Russell, B.: ActionVLAD: learning spatio-temporal aggregation for action classification. In: Proceedings of the IEEE Conference on Computer Vision and Pattern Recognition, vol. 2, p. 3 (2017)
12. Gupta, S., Hoffman, J., Malik, J.: Cross modal distillation for supervision transfer. In: Proceedings of the IEEE Conference on Computer Vision and Pattern Recognition, pp. 2827–2836 (2016)

13. Holte, M.B., Moeslund, T.B., Nikolaidis, N., Pitas, I.: 3D human action recognition for multi-view camera systems. In: Proceedings of the International conference on 3D Imaging, Modeling, Processing, Visualization and Transmission, pp. 342–349 (2011)

14. Ji, X., Wang, C., Li, Y.: A view-invariant action recognition based on multi-view space hidden Markov models. Int. J. Hum. Robot. **11**(01), 1450011 (2014)

15. Krizhevsky, A., Sutskever, I., Hinton, G.E.: Imagenet classification with deep convolutional neural networks. In: Proceedings of Advances in Neural Information Processing Systems, pp. 1097–1105 (2012)

16. Li, Y., Zhang, J., Cheng, Y., Huang, K., Tan, T.: DF2Net: discriminative feature learning and fusion network for RGB-D indoor scene classification. In: Proceedings of AAAI Conference on Artificial Intelligence (2018)

17. Lin, Y.C., Hu, M.C., Cheng, W.H., Hsieh, Y.H., Chen, H.M.: Human action recognition and retrieval using sole depth information. In: Proceedings of the ACM International Conference on Multimedia, pp. 1053–1056 (2012)

18. Maaten, L.V.D., Hinton, G.: Visualizing data using t-SNE. J. Mach. Learn. Res. **9**, 2579–2605 (2008)

19. Nie, F., Cai, G., Li, X.: Multi-view clustering and semi-supervised classification with adaptive neighbours. In: Proceedings of AAAI Conference on Artificial Intelligence (2017)

20. Nie, F., Li, J., Li, X., et al.: Parameter-free auto-weighted multiple graph learning: a framework for multiview clustering and semi-supervised classification. In: Proceedings of International Joint Conferences on Artificial Intelligence, pp. 1881–1887 (2016)

21. Nie, F., Tian, L., Wang, R., Li, X.: Multiview semi-supervised learning model for image classification. IEEE Trans. Knowl. Data Eng. (2019)

22. Ofli, F., Chaudhry, R., Kurillo, G., Vidal, R., Bajcsy, R.: Berkeley MHAD: a comprehensive multimodal human action database. In: IEEE Workshop on Applications of Computer Vision, pp. 53–60 (2013)

23. Pagliari, D., Pinto, L.: Calibration of Kinect for Xbox one and comparison between the two generations of Microsoft sensors. Sensors **15**, 27569–27589 (2015)

24. Rahmani, H., Mahmood, A., Huynh, D., Mian, A.: Histogram of oriented principal components for cross-view action recognition. IEEE Trans. Pattern Anal. Mach. Intell. **38**(12), 2430–2443 (2016)

25. Scholkopf, B., Smola, A.J.: Learning with Kernels: Support Vector Machines, Regularization, Optimization, and Beyond. MIT Press, Cambridge (2001)

26. Verma, V., Lamb, A., Beckham, C., Courville, A., Mitliagkis, I., Bengio, Y.: Manifold mixup: encouraging meaningful on-manifold interpolation as a regularizer. stat 1050, vol. 13 (2018)

27. Wang, A., Cai, J., Lu, J., Cham, T.J.: Modality and component aware feature fusion for RGB-D scene classification. In: Proceedings of the IEEE Conference on Computer Vision and Pattern Recognition, pp. 5995–6004 (2016)

28. Wang, D., Ouyang, W., Li, W., Xu, D.: Dividing and aggregating network for multi-view action recognition. In: Proceedings of European Conference on Computer Vision (September 2018)

29. Wang, L., Ding, Z., Fu, Y.: Learning transferable subspace for human motion segmentation. In: Proceedings of the AAAI Conference on Artificial Intelligence (2018)

30. Wang, L., Ding, Z., Fu, Y.: Low-rank transfer human motion segmentation. IEEE Trans. Image Process. **28**(2), 1023–1034 (2019)

31. Wang, L., Ding, Z., Tao, Z., Liu, Y., Fu, Y.: Generative multi-view human action recognition. In: Proceedings of the IEEE International Conference on Computer Vision, pp. 6212–6221 (2019)
32. Wang, L., Liu, Y., Qin, C., Sun, G., Fu, Y.: Dual relation semi-supervised multi-label learning. In: Proceedings of the AAAI Conference on Artificial Intelligence (2020)
33. Wang, L., Sun, B., Robinson, J., Jing, T., Fu, Y.: EV-Action: electromyography-vision multi-modal action dataset. In: Proceedings of IEEE International Conference on Automatic Face and Gesture Recognition (2020)
34. Wang, L., et al.: Temporal segment networks: towards good practices for deep action recognition. In: Proceedings of European Conference on Machine Learning, pp. 20–36 (2016)
35. Wang, W., Zhou, Z.-H.: Analyzing co-training style algorithms. In: Kok, J.N., Koronacki, J., Mantaras, R.L., Matwin, S., Mladenič, D., Skowron, A. (eds.) ECML 2007. LNCS (LNAI), vol. 4701, pp. 454–465. Springer, Heidelberg (2007). https://doi.org/10.1007/978-3-540-74958-5_42
36. Yang, Y., Zhan, D.C., Sheng, X.R., Jiang, Y.: Semi-supervised multi-modal learning with incomplete modalities. In: Proceedings of International Joint Conferences on Artificial Intelligence, pp. 2998–3004 (2018)
37. Zhang, H., Cisse, M., Dauphin, Y.N., Lopez-Paz, D.: Mixup: beyond empirical risk minimization. In: Proceedings of International Conference on Learning Representations (2018)
38. Zhang, Z.: Microsoft Kinect sensor and its effect. IEEE Multimed. **19**(2), 4–10 (2012)

READ: Reciprocal Attention Discriminator for Image-to-Video Re-identification

Minho Shim[1]([✉])[iD], Hsuan-I Ho[2][iD], Jinhyung Kim[3][iD], and Dongyoon Wee[4][iD]

[1] Seoul, South Korea
[2] Department of Computer Science, ETH Zürich, Zürich, Switzerland
hohs@student.ethz.ch
[3] School of Electrical Engineering, KAIST, Daejeon, South Korea
kkjh0723@kaist.ac.kr
[4] Clova AI, NAVER Corporation, Seongnam, South Korea
dongyoon.wee@navercorp.com

Abstract. Person re-identification (re-ID) is the problem of visually identifying a person given a database of identities. In this work, we focus on image-to-video re-ID which compares a single query image to videos in the gallery. The main challenge is the asymmetry association of an image and a video, and overcoming the difference caused by the additional temporal dimension. To this end, we propose an attention-aware discriminator architecture. The attention occurs across different modalities, and even different identities to aggregate useful spatio-temporal information for comparison. The information is effectively fused into a united feature, followed by the final prediction of a similarity score. The performance of the method is shown with image-to-video person re-identification benchmarks (DukeMTMC-VideoReID, and MARS).

Keywords: Image and video understanding · Identity retrieval · Re-identification · Attention

1 Introduction

Computer vision is all about teaching machines to identify and understand objects in images and videos. Among all, identification is to distinguish between objects within the same category. Practical applications of identification include unmanned surveillance, human-robot interaction, and so on. Specifically, person re-identification (re-ID) is the problem of identifying a person from a given set of person identities. The task involves multiple views of a moving person taken from a single or multiple cameras, so person re-ID suffers from pose variance,

Electronic supplementary material The online version of this chapter (https://doi.org/10.1007/978-3-030-58568-6_20) contains supplementary material, which is available to authorized users.

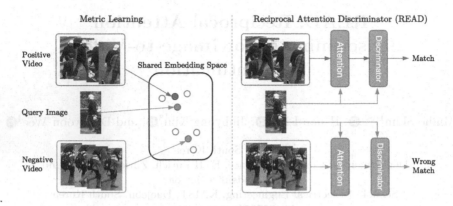

Fig. 1. Comparison between metric learning and the proposed Reciprocal Attention Discriminator in the image-to-video person re-identification setting. Metric learning projects images and videos into a shared embedding space, but those features are independent and do not refer to each other at the moment of comparison. Our Reciprocal Attention Discriminator creates a reciprocally coalesced feature using attention that occurs across images and videos.

illumination change, occlusion, and background clutter. For this reason, re-ID demands a fine-grained level of image understanding.

For example, domain knowledge such as body parts [14,38] or pose [25,26] can be used to extract fine-grained features like head, torso, and so on. To restrain using domain knowledge, attention based approaches [2,6,18,36] are proposed to adaptively find locations of interest. However, these methods do not consider the semantic relationship between objects since attention independently occurs within one object of a single identity.

To compare two images, it is natural for a human to perceive images side-by-side, just like spot-the-difference puzzle game. For instance, if a person is wearing a eye-catching furry red scarf, we can first try to find whether a person in another image is also wearing the red scarf. However, existing person re-ID methods focus on extracting one person's feature independently without consideration of the others.

Image-to-video (I2V) re-ID is a task of comparing a single image (query) to videos (gallery). The I2V re-ID task brings out another challenge for identification. Contrary to image-to-image (I2I) or video-to-video (V2V) re-ID, I2V re-ID is about connecting bridges between image representations [8,29] and video representations [1,5,30]. Existing I2V re-ID methods [32,35] suggest to project images and videos into a shared embedding space. Instead, another approach proposes to transfer representative power of video embedding network to image embedding network [7]. However, these metric learning based approaches encourage image and video information to resemble each other even though a video is innately different from an image, because of the temporal dimension (Fig. 1).

To this end, we devise the **R**eciprocal **A**ttention **D**iscriminator (READ) for image-to-video re-identification. First, the READ is designed with a reciprocal attention structure. The attention is reciprocal because it not only uses self-attention within each gallery video, but also promotes the observation to occur across between query images and gallery videos' spatio-temporal dimension. In addition, this attention mechanism efficiently aggregates the temporal dimension of videos, which naturally solves the aforementioned asymmetry problem. Compared to average pooling multiple image features across the temporal dimension, such mechanism enables more expressive power of video embedding. The module aggregates videos' spatio-temporal information with attention, and then efficiently fuses image-video feature maps into a united feature. Finally, instead of measuring similarity based on the distance between image embedding and video embedding, the READ uses discriminator in order to actually observe query and gallery at the same time to calculate similarity score.

Extensive experiments show the effectiveness of the READ on large scale I2V re-ID benchmarks: DukeMTMC-VideoReID (Duke) and MARS. In I2V re-ID benchmarks, pedestrian image sequences from multiple camera views formulate gallery, while each query is a still image.

In summary, we make following contributions:

- We propose the READ, a novel attention based discriminator, to deal with asymmetric image and video information in I2V re-ID.
- We train the READ in an end-to-end manner by designing two losses and sampling strategy.
- We demonstrate the effectiveness of our method on two benchmark databases Duke and MARS. The READ outperforms previous I2V re-ID methods on both datasets.

2   Related Work

Person re-identification (re-ID) is a branch of identity retrieval task that usually involves multiple camera views to identify a bounding box pedestrian image from existing ID images in the gallery. The field of re-ID can be divided into three major branches, image-to-image (I2I) re-ID, image-to-video (I2V) re-ID, and video-to-video (V2V) re-ID. Considering its practicality, re-ID has a large body of literature so we focus on recent advances that relate to our work. We refer readers to [42] for more comprehensive review on re-ID.

Recent I2I re-ID focuses on data-driven approaches to learn features suitable for classifying IDs and computing distance between images [3,9,24]. Triplet based loss functions have been extensively studied [3,9], as they can formulate distances between features to follow similarities among IDs, i.e. features from images with similar appearances are close to each other. Spatial-attention approaches [2,15, 20,36,39] or part-based models [11,13,25–28,38] are adopted to further guide CNNs to filter out unnecessary segments of images and concentrate on interesting parts, especially human bodies in the case of pedestrian images.

Zhang et al. [36] proposed key-value memory matching which utilizes an attention-based matching mechanism, for computing similarity between images represented by position-aware key-value memory. They showed that the attention module could attend semantically corresponding regions, e.g. body parts, bags, or shoes. As of constructing triplet samples for training, Zhang et al. [37] uses hard identity sampling and multi-stage training strategy for maximizing margins between distant identities. On the other hand, our reciprocal attention module explicitly observes each pair of a query and a gallery identity for comparison. In other words, attention is not only applied within one identity, but it also occurs across identities to determine the best spatio-temporal locations suitable for distinguishing identity.

I2V re-ID is a problem domain of re-ID comparing a single image to a sequence of images. Early works [32,35,43,44] focused on embedding images and videos into a shared feature space. Gu et al. [7] proposed a training method to solve the lack of temporal information in image by transferring knowledge from video representation network to image representation network and building a unified feature space. In addition, non-local neural network [33] is intensively embedded into recent video embedding networks for re-ID [7,18], alongside with triplet loss [9]. While the video embedding networks are able to extract self-attention features, it cannot handle asymmetry features of I2V re-ID. In contrast, our model learns correlation between a query image and a video in an end-to-end manner, thus the information asymmetry between image and video is naturally solved and extra steps of training different networks are not demanded. In this paper, we search for a method to embrace the advances made so far while wisely handling immanent problems of I2V identity retrieval task.

3   Proposed Method

In this section, we explain our new I2V re-ID framework (Fig. 2); in the order of the image and video embedding sub-networks, the Reciprocal Attention Discriminator (READ), and its training strategy. The framework measures the matching probability given two types of inputs, an image query I_q, and a video V_i sampled from the gallery G. The gallery G consists of videos for each identity, $G = \{V_i | i \in [1, 2, ..., M]\}$, where M is the number of videos in the gallery. The image and video embedding sub-networks respectively encode each input as 2D or 3D feature maps, which serve as intermediate representations for reciprocal comparison in the READ. The comparison is to observe one image and one video at the same time, and to create an attention map to aggregate spatio-temporal information from the video. The query image information and query-specific understanding of the video is then combined together for determining the final similarity score.

3.1   Image Embedding Network

Spatial information of the query image should be maintained to compare each spatial location of query against global spatio-temporal location of gallery video. Given a query image I_q, the image embedding network extracts a 2D feature map $f_I \in \mathbb{R}^{H \times W \times C}$ from *res*4 layer of ResNet-50 [8] in order to maintain spatial information [28]. The parameters are initialized with the weights pretrained on ImageNet [4].

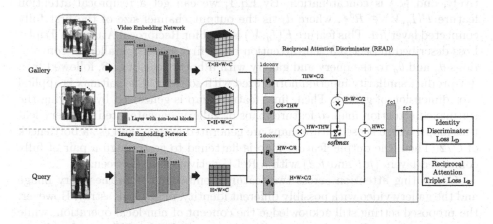

Fig. 2. Illustration of Reciprocal Attention Discriminator (READ), showing an example of comparing one query against multiple gallery videos. In the training phase, multiple queries and gallery sequences will form a minibatch and used for computing losses. ϕ and θ are 1D convolution blocks for channel size reduction. $\oplus$ denotes concatenation, and $\otimes$ denotes matrix multiplication.

3.2 Video Embedding Network

The video embedding network extracts a 3D feature map $f_{V_i} \in \mathbb{R}^{T \times H \times W \times C}$ from a video input V_i. Each video $V_i = \{I_j | j \in [1, 2, ..., L]\}$ includes images of a specific identity. The length L of the video sequences might vary within the gallery G, so we sample a fixed number of frames from videos (Sect. 3.4). In the video embedding network, we add two non-local blocks to $res3$ and three non-local blocks to $res4$ on ResNet-50 [8]. The non-local blocks enable self-attention within the video, where each location combine information from global spacetime locations. Unlike [7], we prune out $res5$ layer just like the image embedding network to keep the number of channels small enough to keep computation feasible.

3.3 Reciprocal Attention Discriminator (READ)

Given a query embedding and a video embedding, the READ measures matching probability of two different embeddings. The discriminator D consists of a reciprocal attention block, and fully connected layers. We use a spatio-temporal non-local attention block [5, 19, 22, 31, 33] to compare query feature maps against gallery feature maps. In short,

$$F(I_q, V) = fc1([softmax(\theta_g(V)^T \cdot \theta_q(I_q)) \cdot \phi_g(V), \phi_q(I_q)]), \tag{1}$$

$$D(I_q, V) = fc2(relu(F(I_q, V))), \qquad (2)$$

where θ and ϕ are 1D convolution layers as bottleneck of $C/8$ and $C/2$ respectively, and $[\cdot, \cdot]$ is concatenation. By Eq. 1, we can get a reciprocal attention feature $F(I_q, V) \in R^{d_F}$, where d_F is the output channel size of the first fully connected layer $fc1$. This feature $F(I_q, V)$ is used for Reciprocal Attention Triplet Loss described in Sect. 3.5. The attention block first applies parallel 1D convolutions θ_q and θ_g to the query and gallery with a bottleneck of $C/8$, followed by a dot-product similarity function normalized with softmax. The softmax is applied over dimension of gallery. Then, final feature map is generated by applying the softmaxed attention map to feature maps of gallery, concatenated by query feature maps, where those feature maps are generated with ϕ_g and ϕ_q, a bottleneck of $C/2$. Then, the output feature map is flattened to go through a pair of fully connected layers ($fc1$ and $fc2$) with a ReLU activation in between.

Computing attention across two features, embedded from the query image and the gallery video with possibly different identity, might look astray. However, the proposed setting still acknowledge the concept of non-local operation, while effectively solving the asymmetry between image and video. When embedding a video, the idea of the non-local operation is to enrich information of a source pixel by integrating information from pixels in global spatio-temporal locations. In the case of previous re-ID methods, the *source* and *global locations* remain within a video of a single identity. Instead, we let the READ bring global information from gallery, across different modalities and no matter the identity of the gallery matches with a given query. In the end, the network first observes a query image and a gallery video, to determine which spatio-temporal locations are valuable for comparison.

3.4 Sampling

Training the READ involves two types of sampling, 1) sampling a subset of frames from videos, and 2) sampling pairs of query and gallery. In order to sample a subset of input video frames, we utilize restricted random sampling (RRS) [14]. Each video of variable length L is divided into T splits with equal duration, and one frame is randomly sampled from each split, resulting in T images per video. The randomness of the sampling method naturally leads to data augmentation and regularization. We empirically found RRS works better than sampling sequential frames. Following the observation, RRS is used throughout all experiments.

Since our model adopts discriminator, each training batch is a combined set of query images and gallery videos. In order to make the discriminator converge, we guarantee a query in each set to have at least one positive sample in the gallery of the set. Therefore, when we sample one query image, one gallery video with the same identity is sampled, to form an image-video pair of the same ID.

Furthermore, we test the sampling method used in [7,18] to sample the same identity multiple times in one minibatch. *#samples per person* denotes the number of samples with the same identity in a minibatch. If *#samples per person* is

2 and the size of minibatch is 32, there will be 16 identity query-gallery pairs in each minibatch, and we denote this as #avgID=16.

While our I2V framework requires only a single query frame for both training and testing, the existing re-ID benchmarks are originally made up for the V2V setting. However during training, it turns out that it is beneficial to sample multiple query frames and use them as a normalization to accelerate training. We sample T_q queries per training sample using RRS, and average $D(I_q, V)$ over the number of queries per video:

$$logit_{final} = \sum_{t=1}^{T_q} \frac{logit_t}{T_q}, \tag{3}$$

then $logit_{final}$ is passed to compute the discriminator loss (Sect. 3.5). Compared to only using a single randomly sampled query frame from the training set, this strategy further improve the speed and performance.

3.5 Training Objective

The whole framework aims to minimize two training objectives, the discriminator loss and the reciprocal attention triplet loss together:

$$\mathcal{L} = \mathcal{L}_D + \mathcal{L}_R, \tag{4}$$

so the two embedding networks and the READ are jointly trained in an end-to-end manner.

Discriminator Loss. The discriminator learns to classify a given image-video pair as positive if two IDs are same or as negative otherwise. The discriminator loss $\mathcal{L}_D$ is based on the binary cross-entropy loss defined as:

$$\mathcal{L}_D = -\mathbb{E}_{V_i \sim P} \log D(I_q, V_i) - \mathbb{E}_{V_i \sim N} \log (1 - D(I_q, V_i)), \tag{5}$$

where P and N are sets of positive and negative samples from the gallery minibatch, respectively. We use $\mathcal{L}_D$ to impose the attention and embedding networks to generate features with distinctive differences when observing positive and negative pairs.

Reciprocal Attention Triplet Loss. The READ aims at constructing a manifold to classify the concatenated query-specific features. However, the features are combinations of high dimensional image/video visual information that are complex and not easily linearly separable. Extending the hard-mine triplet loss [9], we devise the Reciprocal Attention Triplet Loss (RATL) to provide extra constraints to ensure better manifold learning before the final classification:

$$\mathcal{L}_R = \frac{1}{|G|} \sum_{j=1}^{|G|} \sum_{k=1}^{|Q|} [m + \max_{p \in P_{q_k}} d(F(I_p, V_j), F(I_{q_k}, V_j)) \\ - \min_{n \in N_{q_k}} d(F(I_n, V_j), F(I_{q_k}, V_j))]_+, \tag{6}$$

where P_{q_k} and N_{q_k} are the groups of positive and negative query samples of the query I_{q_k} (and $I_{q_k} \notin P_{q_k}$), m is the margin, $d(\cdot, \cdot)$ denotes Euclidean distance and $[\cdot]_+$ denotes $\max(0, \cdot)$, and k and j are indices for query images and gallery videos in a minibatch, respectively. Step-by-step explanation about the RATL is detailed in the supplementary material.

Triplet loss is popularly used in identity retrieval tasks [16,18,21,27,37] to promote metric learning. After training, the distance in the shared embedding space becomes the criterion of similarity between two targets. On the contrary, computation of the RATL is not based on a shared embedding space, but a query-specific understanding against each gallery video. We use the RATL to encourage the reciprocal attention block to focus on specific spatio-temporal regions, where useful information is available for discriminating between gallery identities.

4 Experiments

4.1 Benchmark

We evaluate our method on video based person re-identification (re-ID) datasets: DukeMTMC-VideoReID (Duke) [34], and MARS [40]. Video frames from both datasets are cropped by the bounding boxes from a person detector. **Duke** contains 702 identities for training, 702 for testing, and 408 distractors. There are 2,196 videos for training, 2,636 videos for testing; and 6 cameras are used to capture the videos. **MARS** dataset contains 625 identities for training, and 635 identities for testing. Unlike Duke, MARS dataset's distractors do not have respective ID, so there are $625 + 635 + 1$ (distractor) $= 1,261$ identities in total. Training split of MARS contains 8,298 tracklets, test split contains 11,310 tracklets (excluding 'junk' images provided in the original dataset that do not affect retrieval accuracy), and 6 cameras are used. It is worthwhile to note that query and gallery sets could share same camera views in the test split, however for each query, his/her gallery samples from the same camera are excluded during evaluation.

Following the standard evaluation metrics for both datasets, we report the the cumulative matching curve (CMC) at top-1, top-5, top-10 accuracy and the mean average precision (mAP). Identity retrieval tasks demand high top-1 accuracy compared to general image retrieval tasks since the goal is to precisely identify whom the query is. Yet, the database contains multiple matching answers for each query, so mAP is also used [41] to reflect recall as well as precision. The prediction of the READ is an affinity score (i.e., the probability of matching) between a query and its gallery sample. Hence, the list of ranked gallery samples is sorted in a descending order of the output probabilities instead of their feature distances in the embedding space. During testing, the first frame of each query video is used as the query image following the previous I2V re-ID context [7,32]. As for the test gallery videos, we follow [18] to sample the first frame from T equally-divided chunks which would also ensure consistent evaluation results over repeated tests.

4.2   Methods to be Compared

We analyze the effectiveness of architecture by comparing with two baseline models. One baseline model is designed without reciprocal attention, i.e., only comprising of the image embedding and the non-local video embedding. Since the output of the video embedding network has a time dimension T, we average the feature over its time dimension so the size of the image and video embedding would match. Image and video features are concatenated together then go through two fully connected layers, trained with the discriminator loss and the RATL as in Sect. 3.5. The other baseline is a metric learning architecture which has been tested in [7]. We report their performance in our experiment to show the differences with the discriminator architecture.

In Sect. 3.4, various sampling related concerns are shared. Therefore, we examine how the parameter *#samples per person* in each training minibatch affects performance. However, the *#samples per person* parameter sometimes cannot be directly applied owing to database statistics. For example, Duke has a smaller number of tracklets per identity compared to MARS. To correctly show the difference, we record the average number of identities (*#avgID*) sampled in a minibatch. If there are plenty of tracklets, a minibatch of size 32 ($B = 32$) and 4 *samples per person* ($SP = 4$) will contain 8 identities in the batch. Duke does not have that many tracks, so the *#avgID* becomes 13.3 when $B = 32$ and $SP = 4$, as insufficient tracks are randomly sampled from different identities. To match the *#avgID* with Duke, we give randomness to the number of identities and set *#avgID* around 13.8 in the case of MARS.

4.3   Implementation Detail

We sample 4 frames (i.e. $T = 4$) from each video, and image height and width are resized to 256 and 128 pixels respectively. Adam [12] is adopted to optimize the parameters, with a weight decay of 5e−4 and a starting learning rate of 1e−4. The learning rate is divided by 10 after 60 or 100, and 180 or 200 epochs until it reaches 1e−6. We use a batch size of 32, and the margin of the RATL is $m = 0.3$. In addition, we apply random horizontal flip to the training input images or videos. We report the result of models with the best top-1 accuracy. Scikit-learn [23] version <0.19 is used to calculate mAP, the reason is detailed in the supplementary material.

5   Results

5.1   Ablation Study

Improvement by the READ. Table 1 shows the results of ablation experiments. Beginning from 'baseline (discriminator)', 'READ (*w/o triplet*)' contributes 8.9 and 15.3 top-1 accuracy improvements respectively on Duke and MARS. On top of 'READ (*w/o triplet*)', the RATL adds 1.7-2.2 and 1.2 top-1

Table 1. Results of ablation experiments. Results of baseline (metric) is brought from [7]. Note that three READ experiments in the middle (without triplet loss, or with/without random horizontal flip augmentation) defaults to #avgID of 32. The other experiments with specified #avgID defaults to use horizontal flip augmentation, even if the performance of not using the augmentation is slightly better, for the sake of readability. See details in Sect. 4.2 and Sect. 5.

Method		DukeMTMC-VideoReID				MARS			
		top-1	top-5	top-10	mAP	top-1	top-5	top-10	mAP
baseline (metric) [7]		67.5	–	–	65.6	67.1	–	–	55.5
baseline (discriminator)		75.2	88.9	92.9	71.7	65.0	81.0	85.7	53.2
READ (*w/o triplet*)		84.1	93.5	95.0	80.9	80.3	90.2	93.1	68.6
READ (*w/o hor. flip*)		85.8	93.0	95.7	82.0	81.5	92.1	93.8	70.4
READ (*w/ hor. flip*)		86.3	94.4	96.2	83.3	81.5	91.2	93.3	69.9
READ	*#avgID*								
	32	86.3	94.4	96.2	83.3	81.5	91.2	93.3	69.9
	16	86.0	93.7	95.3	83.4	76.6	86.9	89.6	64.6
	13.8	–	–	–	–	77.6	88.2	90.8	65.7
	13.3	84.9	94.3	96.6	82.9	–	–	–	–

accuracy depending on the database, while the horizontal flip augmentation does not seriously impact the results. Details follow about each ablation experiment.

Baseline. The difference between two baselines is the use of metric learning or a discriminator to distinguish identities. Also, the video embedding in 'baseline (discriminator)' does not have *res*5 layer compared to 'baseline (metric)', meaning a lower network capacity. The results show both baselines do not necessarily solve the issue of asymmetry. They both roughly perform pooling across the temporal dimension to match the image embedding dimension, and 'baseline (discriminator)' only concatenates those features. The READ is able to address the asymmetry of two different embeddings without dropping the temporal information.

RATL. The RATL also plays an important role for instructing reciprocal attention module, to learn where to focus on the gallery videos based on a given query. Without the RATL, performance degrades by 2.2 top-1 accuracy in Duke and 1.2 in MARS.

Augmentation. We examine the effect of random horizontal flip. The experimental results show that the random horizontal flip does not have significant influence to the performance. It implies the READ is robust to the direction of pedestrian given unflipped training data. Unless specified, all experimental results are derived from models trained with the horizontal flip augmentation.

#avgID. Applying various average numbers of identities in each minibatch (*#avgID*) displays different trends depending on the dataset. In the case of

Duke, the range of disparity between different #avgID is small and does not seem significant, since changing #avgID from 32 to 16 causes 0.3 drop of top-1 accuracy. MARS yet outputs highly variable results. There is a 3.9 top-1 accuracy gap between #avgID=32 and #avgID=16. This possibly results from the distribution difference as MARS has larger number of tracklets for each identity. Thus, this parameter should be carefully selected based on the dataset.

5.2 Comparison

Table 2 presents the results of comparison with state-of-the-art I2V re-ID methods on DukeMTMC-VideoReID (Duke) and MARS benchmark datasets. The READ shows a significant improvement over the state-of-the-art methods on both datasets. On Duke dataset, the READ improves top-1 accuracy and mAP by around 8 and 6 respectively compared to TKP [7]. On MARS dataset, the READ outperforms all models from P2SNet [32], ResNet-50+XQDA [17], to TKP [7], by a large margin of at least 5.9 top-1 accuracy and 5.3 mAP.

Table 2. Benchmark comparison with state-of-the-art I2V re-ID methods.

(a) DukeMTMC-VideoReID.

Method	top-1	top-5	top-10	mAP
TKP [7]	77.9	–	–	75.9
READ (*ours*)	**86.3**	94.4	96.2	**83.4**

(b) MARS.

Method	top-1	top-5	top-10	mAP
P2SNet [32]	55.3	72.9	78.7	–
ResNet-50+XQDA [17]	67.2	81.9	86.1	54.9
TKP [7]	75.6	87.6	90.9	65.1
READ (*ours*)	**81.5**	**92.1**	**93.8**	**70.4**

5.3 Analysis

In this section, we examine the effect of various options on the performance on DukeMTMC-VideoReID (Duke) dataset.

Normalization. Without the query sample normalization (Eq. 3) for training, only a single image is randomly sampled as a query. Table 3(a) shows the effect of the query sample normalization with various #avgID. When #avgID=32, the top-1 accuracy is dropped by 6. The gap is huge considering the network does see all training query images even without the normalization. However, there are less positive samples without normalization in each minibatch, so it could have also negatively impacted the RATL. Thus, the experiments with smaller #avgID of 16 and 13.3 and without normalization are additionally conducted. Those tests improved top-1 accuracy to 83.6 and 83.0 from 80.3; and mAP from 77.8 to 81.0

and 81.1. Combined with the results in Table 1, we can conclude that the query sample normalization greatly helps the discriminator loss $\mathcal{L}_D$, and the RATL $\mathcal{L}_R$ requires balance between the number of positive and negative samples, $|P_q|$ and $|N_q|$, respectively. Table 3(c) shows $|P_q|$ and $|N_q|$ in the perspective of the RATL.

Table 3. Experimental results of analysis. Experiments in (a)–(d) are performed with DukeMTMC-VideoReID database.

(a) Without normalization.

#avgID	top-1	top-5	top-10	mAP
32	80.3	91.2	94.6	77.8
16	83.6	92.7	95.0	81.0
13.3	83.0	93.3	95.4	81.1

(b) Peformance on variable T.

Metric	$T = 2$	$T = 4$	$T = 6$	$T = 8$
top-1	83.2	86.3	85.2	85.9
top-5	92.6	94.4	94.4	94.4
mAP	79.0	83.4	82.2	82.7

(c) #samples for RATL.

norm.	#avgID	$|P_q|$	$|N_q|$
yes	32	4	124
yes	16	8	120
no	32	1	31
no	16	2	30

(d) Direction of RATL.

Direction	top-1	top-5	top-10	mAP
query → gallery	83.0	93.0	95.0	79.9
gallery → query	86.3	94.4	96.2	83.3

Sample Length. Table 3(b) shows the experiments conducted by varying the length T of samples from the gallery videos. We test four variants of T from 2 to 8 with a stride of 2, and the results show the performances do not increase beyond $T = 4$. This result is consistent with [7]. In contrast, it is different to non-local neural networks applied to action recognition [33], where longer input video clips coherently shows better performance. This inconsistency is possibly on grounds of differences between action recognition and re-identification. In our re-ID benchmarks, gallery videos are guaranteed to contain an identity's visual information from a dedicated camera viewpoint. Hence, $T = 4$ can be the point where additional information does not further contribute.

Direction of RATL. A single operation of the RATL is described to be operated within the pool Q of query samples, in the perspective of a gallery sample $V_j \in G$ (Eq. 6 of Sect. 3.3). We analyze another case of direction, where the RATL operated in the pool of gallery samples with a query sample as its basis. The results are presented in Table 3(d). Compared to our default setting, the top-1 accuracy drops by 3.3, and mAP is degraded by 3.5. This result can be interpreted in the similar context of sample distribution. The asymmetry of I2V allows abundant sampling of query images compared to the amount of gallery video samples, thus #avgID parameter and normalization is exploited in our

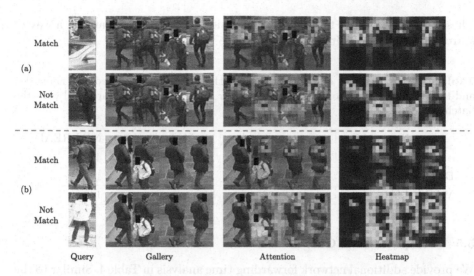

Fig. 3. Visualization of attention. Softmax normalized attention maps in accordance with the image-video pair are visualized. The attention focuses on the different spatio-temporal region of the gallery depending on the cases when the pair has same IDs (Match) and different IDs (Not Match). The detailed analysis can be found in the main text. Best viewed in color. (Color figure online)

experiments. When a gallery sample is the basis of the RATL, there are 128 query images in our default setting. On the other side, there are only 32 gallery videos when a query sample is the basis. So the *gallery* → *query* direction allows each gallery sample to observe more diverse counterpart query samples.

5.4 Visualization

We qualitatively evaluate our proposed method by visualizing the attention map on Duke database as in Fig. 3. Attention created by the READ occurs across global spatio-temporal dimension. So as to visualize the attention as 2D map, the softmaxed attention of dimension $HW \times THW$ is aggregated by averaging over HW, then is reshaped to $T \times H \times W$.

To analyze the effect of reciprocal attention with different query images, we compare the attention map generated by a matched image and a non-matched image given the same gallery video. The attention focuses on the target when the identity of image-video pair matches. On the other hand, attention often spotlights other person or background if the pair has different IDs. For instance in Fig. 3(a), the attention focuses on the upper body of the matching target, whereas different people and backgrounds are attended when IDs do not match. Similarly, in Fig. 3(b), the target is attended with an image-video pair of the same ID. On the contrary, in case the query with white jacket is chosen, a person with a white jacket in the second frame of the gallery is focused. After all, the READ tries to find the information that matches the query if unexpectedly different gallery video is given.

These results show that the proposed attention mechanism operates in a way of searching query related information from the video.

Table 4. Computational cost for forwarding the image/video embedding networks, and the reciprocal attention discriminator. Four TESLA P40 GPUs are used with the batch size of 32.

	N	M	N image embedding	M video embedding	N*M READ
–	32	32	49 ms	168 ms	4 ms
Duke	702	2,636	1 s	14 s	12 s
MARS	1,980	11,310	3 s	59 s	140 s

5.5   Computational Cost

We provide additional network forwarding time analysis in Table 4. Similar to the existing re-ID pipelines, our method is able to prefetch image/video embeddings in N+M forward passes. The remaining cost is O(NM) forward passes of the READ, which is a feasible overhead even in MARS and the cost similarly exists in other discriminator based re-ID works [36]. Also in the light of the READ, the expensive video embedding can be fully replaced by an image embedding network (x2.2 FLOPs smaller) with a marginal performance drop as discussed in the supplementary material.

Besides retrieving from a massive database (e.g. image search engine), re-ID aims at matching subjects across multiple camera views where occlusion and visual degradation might occur. The underlying motivation of this paper is an application on real-world MOT tasks, e.g. tracking person in dance videos [10], similarly described in the supplementary material. In such scenario, the scale of N*M stays feasible for real-time speed.

6   Conclusion

In this paper, we propose the Reciprocal Attention Discriminator (READ), the novel attention-based discriminator framework for I2V re-ID task along with two losses, the discriminator loss and the Reciprocal Attention Triplet Loss (RATL), for training the model. The READ can successfully integrate asymmetric information of image-video pair using non-local operation. We validate the effectiveness of our method quantitatively and qualitatively on two widely-used databases. Our method surpasses other previous arts by a wide margin. We also reported extensive ablation studies to verify the design choices.

References

1. Carreira, J., Zisserman, A.: Quo vadis, action recognition? A new model and the kinetics dataset. In: CVPR (2017)

2. Chen, B., Deng, W., Hu, J.: Mixed high-order attention network for person re-identification. In: ICCV (2019)
3. Chen, W., Chen, X., Zhang, J., Huang, K.: Beyond triplet loss: a deep quadruplet network for person re-identification. In: CVPR (2017)
4. Deng, J., Dong, W., Socher, R., Li, L.J., Li, K., Fei-Fei, L.: ImageNet: a large-scale hierarchical image database. In: CVPR (2009)
5. Feichtenhofer, C., Fan, H., Malik, J., He, K.: Slowfast networks for video recognition. In: ICCV (2019)
6. Fu, Y., Wang, X., Wei, Y., Huang, T.: STA: spatial-temporal attention for large-scale video-based person re-identification. In: AAAI (2019)
7. Gu, X., Ma, B., Chang, H., Shan, S., Chen, X.: Temporal knowledge propagation for image-to-video person re-identification. In: ICCV (2019)
8. He, K., Zhang, X., Ren, S., Sun, J.: Deep residual learning for image recognition. In: CVPR (2016)
9. Hermans, A., Beyer, L., Leibe, B.: In defense of the triplet loss for person re-identification. arXiv preprint (2017)
10. Ho, H.I., Shim, M., Wee, D.: Learning from dances: pose-invariant re-identification for multi-person tracking. In: ICASSP (2020)
11. Hou, R., Ma, B., Chang, H., Gu, X., Shan, S., Chen, X.: Interaction-and-aggregation network for person re-identification. In: CVPR (2019)
12. Kingma, D.P., Ba, J.: Adam: a method for stochastic optimization. In: ICLR (2015)
13. Li, D., Chen, X., Zhang, Z., Huang, K.: Learning deep context-aware features over body and latent parts for person re-identification. In: CVPR (2017)
14. Li, S., Bak, S., Carr, P., Wang, X.: Diversity regularized spatiotemporal attention for video-based person re-identification. In: CVPR (2018)
15. Li, W., Zhu, X., Gong, S.: Harmonious attention network for person re-identification. In: CVPR (2018)
16. Li, Y.J., Chen, Y.C., Lin, Y.Y., Du, X., Wang, Y.C.F.: Recover and identify: a generative dual model for cross-resolution person re-identification. In: ICCV (2019)
17. Liao, S., Hu, Y., Zhu, X., Li, S.Z.: Person re-identification by local maximal occurrence representation and metric learning. In: CVPR (2015)
18. Liu, C.T., Wu, C.W., Wang, Y.C.F., Chien, S.Y.: Spatially and temporally efficient non-local attention network for video-based person re-identification. In: BMVC (2019)
19. Liu, D., Wen, B., Fan, Y., Loy, C.C., Huang, T.S.: Non-local recurrent network for image restoration. In: NeurIPS (2018)
20. Liu, X., et al.: HydraPlus-Net: attentive deep features for pedestrian analysis. In: ICCV (2017)
21. Liu, Y., Yuan, Z., Zhou, W., Li, H.: Spatial and temporal mutual promotion for video-based person re-identification. In: AAAI (2019)
22. Oh, S.W., Lee, J.Y., Xu, N., Kim, S.J.: Video object segmentation using space-time memory networks. In: ICCV (2019)
23. Pedregosa, F., et al.: Scikit-learn: machine learning in Python. JMLR **12**, 2825–2830 (2011)
24. Shen, Y., Li, H., Yi, S., Chen, D., Wang, X.: Person re-identification with deep similarity-guided graph neural network. In: Ferrari, V., Hebert, M., Sminchisescu, C., Weiss, Y. (eds.) ECCV 2018. LNCS, vol. 11219, pp. 508–526. Springer, Cham (2018). https://doi.org/10.1007/978-3-030-01267-0_30
25. Su, C., Li, J., Zhang, S., Xing, J., Gao, W., Tian, Q.: Pose-driven deep convolutional model for person re-identification. In: ICCV (2017)

26. Suh, Y., Wang, J., Tang, S., Mei, T., Lee, K.M.: Part-aligned bilinear represen-tations for person re-identification. In: Ferrari, V., Hebert, M., Sminchisescu, C., Weiss, Y. (eds.) Computer Vision – ECCV 2018. LNCS, vol. 11218, pp. 418–437. Springer, Cham (2018). https://doi.org/10.1007/978-3-030-01264-9_25

27. Sun, Y., et al.: Perceive where to focus: learning visibility-aware part-level features for partial person re-identification. In: CVPR (2019)

28. Sun, Y., Zheng, L., Yang, Y., Tian, Q., Wang, S.: Beyond part models: person retrieval with refined part pooling (and a strong convolutional baseline). In: Fer-rari, V., Hebert, M., Sminchisescu, C., Weiss, Y. (eds.) ECCV 2018. LNCS, vol. 11208, pp. 501–518. Springer, Cham (2018). https://doi.org/10.1007/978-3-030-01225-0_30

29. Szegedy, C., et al.: Going deeper with convolutions. In: CVPR (2015)

30. Tran, D., Bourdev, L., Fergus, R., Torresani, L., Paluri, M.: Learning spatiotem-poral features with 3D convolutional networks. In: ICCV (2015)

31. Vaswani, A., et al.: Attention is all you need. In: NIPS (2017)

32. Wang, G., Lai, J., Xie, X.: P2SNet: can an image match a video for person re-identification in an end-to-end way? TCSVT 28(10), 2777–2787 (2017)

33. Wang, X., Girshick, R., Gupta, A., He, K.: Non-local neural networks. In: CVPR (2018)

34. Wu, Y., Lin, Y., Dong, X., Yan, Y., Ouyang, W., Yang, Y.: Exploit the unknown gradually: one-shot video-based person re-identification by stepwise learning. In: CVPR (2018)

35. Zhang, D., Wu, W., Cheng, H., Zhang, R., Dong, Z., Cai, Z.: Image-to-video person re-identification with temporally memorized similarity learning. TCSVT 28(10), 2622–2632 (2017)

36. Zhang, Y., Li, X., Zhang, Z.: Learning a key-value memory co-attention matching network for person re-identification. In: AAAI (2019)

37. Zhang, Y., Zhong, Q., Ma, L., Xie, D., Pu, S.: Learning incremental triplet margin for person re-identification. In: AAAI (2019)

38. Zhao, H., et al.: Spindle net: person re-identification with human body region guided feature decomposition and fusion. In: CVPR (2017)

39. Zhao, L., Li, X., Zhuang, Y., Wang, J.: Deeply-learned part-aligned representations for person re-identification. In: ICCV (2017)

40. Zheng, L., et al.: MARS: a video benchmark for large-scale person re-identification. In: Leibe, B., Matas, J., Sebe, N., Welling, M. (eds.) ECCV 2016. LNCS, vol. 9910, pp. 868–884. Springer, Cham (2016). https://doi.org/10.1007/978-3-319-46466-4_52

41. Zheng, L., Shen, L., Tian, L., Wang, S., Wang, J., Tian, Q.: Scalable person re-identification: a benchmark. In: ICCV (2015)

42. Zheng, L., Yang, Y., Hauptmann, A.G.: Person re-identification: past, present and future. arXiv preprint (2016)

43. Zhu, X., Jing, X.Y., Wu, F., Wang, Y., Zuo, W., Zheng, W.S.: Learning heteroge-neous dictionary pair with feature projection matrix for pedestrian video retrieval via single query image. In: AAAI (2017)

44. Zhu, X., Jing, X.Y., You, X., Zuo, W., Shan, S., Zheng, W.S.: Image to video person re-identification by learning heterogeneous dictionary pair with feature projection matrix. TIFS 13(3), 717–732 (2017)

3D Human Shape Reconstruction from a Polarization Image

Shihao Zou[1]([⊠]), Xinxin Zuo[1], Yiming Qian[2], Sen Wang[1], Chi Xu[3],
Minglun Gong[4], and Li Cheng[1]

[1] University of Alberta, Edmonton, Canada
{szou2,xzuo,sen9,lcheng5}@ualberta.ca
[2] Simon Fraser University, Burnaby, Canada
yimingq@sfu.ca
[3] School of Automation, China University of Geosciences,
Wuhan 430074, China
xuchi@cug.edu.cn
[4] University of Guelph, Guelph, Canada
minglun@uoguelph.ca

Abstract. This paper tackles the problem of estimating 3D body shape of clothed humans from single polarized 2D images, i.e. polarization images. Polarization images are known to be able to capture polarized reflected lights that preserve rich geometric cues of an object, which has motivated its recent applications in reconstructing surface normal of the objects of interest. Inspired by the recent advances in human shape estimation from single color images, in this paper, we attempt at estimating human body shapes by leveraging the geometric cues from single polarization images. A dedicated two-stage deep learning approach, SfP, is proposed: given a polarization image, stage one aims at inferring the fined-detailed body surface normal; stage two gears to reconstruct the 3D body shape of clothing details. Empirical evaluations on a synthetic dataset (SURREAL) as well as a real-world dataset (PHSPD) demonstrate the qualitative and quantitative performance of our approach in estimating human poses and shapes. This indicates polarization camera is a promising alternative to the more conventional color or depth imaging for human shape estimation. Further, normal maps inferred from polarization imaging play a significant role in accurately recovering the body shapes of clothed people.

Keywords: Human pose and shape estimation · Clothed 3D human body · Shape from polarization

Electronic supplementary material The online version of this chapter (https://doi.org/10.1007/978-3-030-58568-6_21) contains supplementary material, which is available to authorized users.

1   Introduction

Compared to the task of color-image based pose estimation [1–20] that predicts 3D joint positions of an articulated skeleton, human shapes provide much richer information of a human body in 3D and are visually more appealing. It, on the other hand, remains a challenging problem, partly owing to the relative high-dimensional space of human body shapes. The issue is somewhat alleviated by the emerging low-dimensional modelling of human shape, such as SCAPE [21] and SMPL [22], statistical models that are learned from large sets of carefully scanned 3D body shapes. Based on these low-dimensional human shape representations, a number of end-to-end deep learning methods [23–37] are subsequently developed to estimate human shapes directly from color images. The predicted human shapes, however, are usually naked and lacking in surface details, since e.g. SMPL model is learned from naked human body scans.

Volume-based techniques [38,39] are widely used in capturing surface details of a clothed human body from a single image. Due to finite computational resource, the estimated human shapes from these methods are usually of low resolution. Saito et al. [40] consider to remedy this by predicting a pixel-aligned implicit surface function that captures more detailed body surface. It however relies on a large training set of detailed 3D human bodies, and the method is still unable to handle complex poses. In the meantime, the methods of [41] and [42] aim to exploit additional geometric cues arising from color image inputs; [41] instead focuses on predicting fine depth maps, and [42] takes on the shading aspect. Unfortunately, accurate and reliable prediction of these geometric cues from a color image is yet another challenging issue - it remains unclear how much one can leverage from such cues. Motivated by these efforts and their limitations, we consider in this paper to work with a new 2D imaging modality, polarization camera, that is known at better preserving fine-scale geometric properties of 3D objects, including human shapes. The intuition comes from basic physics principle: when a light ray reflects off an object, it is polarized and conveys ample geometric cues concerning local surface details of the object, usually represented as surface normal [43,44]. It may be found to note some biological species are even able to directly perceive light polarization [45,46], which significantly facilitates their 3D sensing. Empirically, our experiments support that the surface normal maps obtained out of the input 2D polarization images could play an instrumental role in producing accurate and reliable 3D clothed human shapes.

As shown in Fig. 1, our approach, also called SfP, contains two stages. Stage 1 concentrates on predicting accurate surface normal maps from single polarization images[1] by exploiting the associated physics laws as priors. It is then fed into stage 2 in reconstructing the final clothed human shape.

Unlike existing efforts in normal map prediction [41,42], our approach predicts normal maps by explicitly incorporating the underlying physical laws of polarization imaging, which results in more *reliable* performance. To achieve

---

[1] In this paper, an polarization image has four channels with each channel corresponding to a specific polarizer degree of (0, 45, 90 and 135).

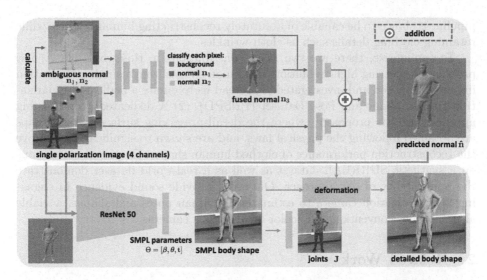

Fig. 1. Given a single polarization image, a two-stage process is executed in our approach. (1) Stage 1, in blue, estimates the surface normal from the polarization image based on the physical assumption that reflected light from an object is polarized. After calculating the two ambiguous normal maps, $(\mathbf{n}_1, \mathbf{n}_2)$, as physical priors from the polarization image (see Sect. 3.1 for details), image pixels are classified as belonging to either of the two normals or a background, thus obtaining the fused normal $\mathbf{n}_3$. Unfortunately, this normal is often noisy, thud a further step is carried out in regressing a final accurate surface normal $\hat{\mathbf{n}}$, by integrating these physical normal maps and the raw polarization image. (2) Stage 2, in orange, concatenates the polarization image and the surface normal as the input to estimate clothed body shape in two steps. The first step focuses on estimating the parameters of SMPL, a rough & naked shape model parameterized by Θ; the pose (3D joint positions) $\mathbf{J}$ is directly obtained as a by-product of the rigged shape model. The next step deforms the SMPL shape guided by the final surface normal of stage 1, to reconstruct the refined 3D human shape with clothing details. (Color figure online)

this, there are two main challenges we need to overcome, namely π-ambiguity of the azimuth angle and the possibly large noise in practical applications. To this end we introduce two ambiguous normal maps $\mathbf{n}_1$ and $\mathbf{n}_2$ (Sect. 3.1) as a physical prior, based on the assumption that the light reflected by human clothing is mostly diffused. Different from [44], each pixel is then classified into one of the three types: the two ambiguous normal maps and background. This is followed by a refinement step to deliver the final surface normal prediction of $\hat{\mathbf{n}}$, that accounts for the possibly-noisy fused normal map output owing to environmental noise and the digital quantization of the polarization camera. Based on the raw polarization image and output of stage 1, stage 2 concerns the estimation of clothed human shape. It starts from predicting a coarse SMPL shape model, which is then deformed by leveraging the geometric details from surface normal, our stage 1 output, to form the final human shape. Empirically our two-stage pipeline is shown to be capable of accurately reconstructing human shapes, while retaining clothing details such as cloth wrinkles.

To summarize, there are two main contributions in this work. (1) A new problem of inferring high-resolution 3D human shapes from a single polarization image is proposed and investigated. This lead us to curate a dedicated Polarization Human Shape and Pose Dataset (PHSPD). (2) A dedicated deep learning approach, SfP, is proposed[2], where the detail-preserving surface normal maps are obtained following the physical laws, and are shown to significantly improve the reconstruction performance of clothed human shapes. Empirical evaluations on a synthetic SURREAL dataset as well as a real-world dataset demonstrate the applicability of our approach. Our work provide sound evidence in engaging 2D polarization camera to estimate 3D human poses and shapes, a viable alternative to conventional 2D color or 3D depth cameras.

2   Related Work

Shape from polarization (SfP) focuses on the inference of shape (normally represented as *surface normal*) from the polarimetric information in the multiple channels of a polarization image, captured under linear polarizers with different angles. The main issue of SfP is angle ambiguity. Previous methods are mainly physics-based that rely on other additional information or assumptions to elucidate the possible ambiguities, such as smooth object surfaces [47], coarse depth map [43,48] and multi-view geometric constraint [49,50]. The recent work of [44] proposes to blend physical priors (ambiguous normal maps) with deep learning in uncovering the normal map. Using physical priors as part of the input, deep learning model can then be trained to account for the ambiguity and be noise-resilient. We improve upon [44] by classifying ambiguous normal and background for each pixel, and regressing the normal given the ambiguous and classified physical priors.

3D human pose estimation from single images has been extensively investigated in the past five years, centering around color or depth imaging. Many of the studies [51–57] utilize dictionary-based learning strategies. More recent efforts aim to directly regress 3D pose using deep learning techniques, including CNNs [1–3] and Graph CNNs [58,59]. In particular, several recent efforts [4–12,12–20] look into a common framework of estimating 2D pose (either 2D joint positions or heatmap), which is then lifted to 3D. Ideas from self-supervised learning [17,20] and adversarial learning [11,18] also gain attentions in e.g. predicting 3D pose under additional constraints imposed from re-projection or adversarial losses.

Human shape estimation from single images has drawn growing attentions recently, thanks to development of human shape models of SCAPE and SMPL [21,22]. These two statistical models learn low-dimensional representations of

---

[2] Our project website is https://jimmyzou.github.io/publication/2020-polarization-clothed-human-shape.

human shape from large corpus of human body scans. Together with deep learning techniques, it has since been feasible to estimate human body shapes from single color or depth images. Earlier activities focus more on optimizing the SCAPE or SMPL model parameters toward better fitting to various dedicated visual or internal representations, such as foreground silhouette [23–25] and pose [26,27]. Deep learning based approaches are more commonplace in recent efforts [28–31], which typically learn to predict the SMPL parameters by incorporating the constrains from 2/3D pose, silhouette, as well as adversarial learning losses. [32] takes the body pixel-to-surface correspondence map as proxy representation and then performs estimation of parameterized human pose and shape. In [33], optimization-based methods [26] and regression-based methods [28] are combined to form a self-improved fitting loop. Point cloud is considered as input in [60] to regress SMPL parameters. Instead of single color images, our work is based on single polarization image; rather than inferring coarse human body shape, we aim to recover high-res human shapes.

As for the estimation of clothed human shape, volume-based methods [38–40] are proposed to reconstruct textured body shapes. They unfortunately suffer from the low resolution issue of volumetric representation. Our work is closely related to [42], which combines the robustness of parametric model and the flexibility of free-form 3D deformation in a hierarchical manner. The major difference is, the clothing details of our work are provided by the reliable normal map estimated from the polarization image, whereas the network in [42] deforms depth image by employing the shading information trained on additional data, that are inherently unreliable due to the lack of ground-truth information of surface normal, albedo and environmental lighting. Our work is also related to [41] which recovers detailed human shape from a color image, by iteratively incorporating both rough depth map and estimated surface normal for improved surface details.

3   The Proposed SfP Approach

There are two main stages in our approach: (1) estimate surface normal from a single polarization image; (2) estimate human pose and shape from the estimated surface normal and the raw polarization image, followed by body shape refinement from the estimated surface normal.

3.1   Surface Normal Estimation

The reflected light from a surface mainly includes three components [50], the polarized specular reflection, the polarized diffuse reflection, and the unpolarized diffuse reflection. A polarization camera has an array of linear polarizer mounted right on top of the CMOS imager, similar to the RGB Bayer filters. During the imaging process of a polarization camera, a pixel intensity typically varies sinusoidally with the angle of the polarizer [43]. In this work, we assume that the light reflected off human clothes is dominated by polarized diffuse reflection and unpolarized diffuse reflection. For a specific polarizer angle ϕ_{pol}, the illumination intensity at a pixel with dominant diffuse reflection is

$$I(\phi_{pol}) = \frac{I_{max} + I_{min}}{2} + \frac{I_{max} - I_{min}}{2} \cos(2(\phi_{pol} - \varphi)). \tag{1}$$

Here φ is the azimuth angle of surface normal, I_{max} and I_{min} are the upper and lower bounds of the illumination intensity. I_{max} and I_{min} are mainly determined by the unpolarized diffuse reflection, and the sinusoidal variation is mainly determined by the polarized diffuse reflection. Note that there is π-ambiguity in the azimuth angle φ in Eq. (1), which means that φ and $\pi + \varphi$ will result in the same illumination intensity of the pixel. As for the zenith angle θ, it is related to the degree of polarization ρ, where

$$\rho = \frac{I_{max} - I_{min}}{I_{max} + I_{min}}. \tag{2}$$

According to [47], when diffuse reflection dominates, the degree of polarization ρ is a function of the zenith angle θ and the refractive index n,

$$\rho = \frac{(n - \frac{1}{n})^2 \sin^2 \theta}{2 + 2n^2 - (n + \frac{1}{n})^2 \sin^2 \theta + 4 \cos \theta \sqrt{n^2 - \sin^2 \theta}}. \tag{3}$$

In this paper, we assume the refractive index $n = 1.5$ since the material of human clothes is mainly cotton or nylon. With n known, the solution of θ in Eq. (3) is a close-form expression of n and ρ.

Taking into account the π-ambiguity of φ, we have two possible solutions to the surface normal for each pixel, that form the physical priors. We propose to train a network to classify each pixel into three categories: background, ambiguous normal $\mathbf{n}_1(\varphi, \theta)$ and ambiguous normal $\mathbf{n}_2(\pi + \varphi, \theta)$ with probability p_0, p_1, and p_2 respectively. Then we have the fused normal as follows,

$$\mathbf{n}_3 = (1 - p_0) \cdot \frac{p_1 \mathbf{n}_1 + p_2 \mathbf{n}_2}{\|p_1 \mathbf{n}_1 + p_2 \mathbf{n}_2\|_2}, \tag{4}$$

where $(1 - p_0)$ is a soft mask of the foreground human body. Unfortunately, due to the environmental noise and the digital quantization of camera in real-world applications, the fused normal map $\mathbf{n}_3$ is noisy and non-smooth. Thus taking the fused noisy normal as an *improved* physical prior, a denoising network is further trained to take both the polarization image and the physical priors $(\mathbf{n}_1, \mathbf{n}_2, \mathbf{n}_3)$ as input, and to produce a smoothed normal $\hat{\mathbf{n}}$. The loss function for normal estimation consists of the cross entropy (CE) loss of classification and the L1 loss of the cosine similarity,

$$L_n = \frac{1}{HW} \sum_{i=1}^{H} \sum_{j=1}^{W} \left[\lambda_c CE(y^{i,j}, p^{i,j}) + \lambda_n (1 - \langle \hat{\mathbf{n}}^{i,j}, \mathbf{n}^{i,j} \rangle) \right], \tag{5}$$

where λ_c and λ_n are the weights of each loss, $y^{i,j}$ is the label indicating which category the pixel (i, j) belongs to, and $\langle \hat{\mathbf{n}}^{i,j}, \mathbf{n}^{i,j} \rangle$ denotes the cosine similarity between the predicted and target normal vectors of pixel (i, j). Note that the category label $y^{i,j}$ is created by discriminating whether the pixel is background or which ambiguous normal has higher cosine similarity with the target normal. λ_c and λ_n is 2 and 1 respectively in our experiment.

3.2   Human Pose and Shape Estimation

To start with, the SMPL [22] representation is used for describing 3D human shapes, which is a differentiable function $\mathcal{M}(\boldsymbol{\beta}, \boldsymbol{\theta}) \in \mathbb{R}^{6,890 \times 3}$ that outputs a triangular mesh with 6,890 vertices given 82 parameters $[\boldsymbol{\beta}, \boldsymbol{\theta}]$. The shape parameter $\boldsymbol{\beta} \in \mathbb{R}^{10}$ is the linear coefficients of a PCA shape space that mainly determines individual body features such height, weight and body proportions. The PCA shape space is learned from a large dataset of body scans [22]. $\boldsymbol{\theta} \in \mathbb{R}^{72}$ is the pose parameter that mainly describes the articulated pose, consisting of one global rotation of the body and the relative rotations of 23 joints in axis-angle representation. Finally, our clothed body shape is produced by first applying shape-dependent and pose-dependent deformations to the template pose, then using forward-kinematics to articulate the body shape back to its current pose, and deforming the surface mesh by linear blend skinning. $\mathbf{J} \in \mathbb{R}^{24 \times 3}$ are the 3D joint positions that can be obtained by linear regression from the output mesh vertices.

In addition to the SMPL parameters, we also need to predict the global translation $\mathbf{t} \in \mathbb{R}^3$. Thus for the task of human pose and shape estimation, the output vector is of 85-dimension, $\hat{\Theta} = [\hat{\boldsymbol{\beta}}, \hat{\boldsymbol{\theta}}, \hat{\mathbf{t}}]$. Given $\hat{\Theta}$, we can also obtain the predicted 3D joint positions $\hat{\mathbf{J}}$. To this end, the loss function is defined as

$$L_s = \lambda_\beta \|\boldsymbol{\beta} - \hat{\boldsymbol{\beta}}\|_2^2 + \lambda_\theta \|\boldsymbol{\theta} - \hat{\boldsymbol{\theta}}\|_2^2 + \lambda_t \|\mathbf{t} - \hat{\mathbf{t}}\|_2^2 + \lambda_J \|\mathbf{J} - \hat{\mathbf{J}}\|_2^2, \qquad (6)$$

where λ_β, λ_θ, λ_t and λ_J are weights of each component in the loss function, which are fixed to 0.2, 0.5, 100, and 3, respectively.

The reconstructed SMPL human shape thus far is naked 3D shape and lacking fine surface details. Our goal is to refine this intermediate naked shape under the guidance of our smoothed surface normal estimate. It is carried out as follows. The SMPL body shape is rendered on the image plane to form a base depth map. The technique in [61] is then engaged here to obtain an optimized depth map from the predicted surface normal and the base depth map. It is carried out under three constraints: first, the predicted normal should be perpendicular to the local tangent of the optimized depth surface; second, the optimized depth should be close to the base depth; Third, a smoothness constraint is enforced on nearby pixels of the optimized depth map. This depth map is obtained as a solution of a linear least-squares system. Weights of the normal term, the depth data term, and the smoothness term are empirically set to 1.0, 0.06, and 0.55, respectively. Finally, our clothed body shape is produced by upsampling & deforming the SMPL mesh according to the Laplacian of the optimized depth map.

3.3 Polarization Human Pose and Shape Dataset

To facilitate empirical evaluation of our approach in real-world scenarios, a home-grown dataset is curated, referred as Polarization Human Shape and Pose Dataset, or PHSPD. A complete description of this PHSPD dataset is provided in [62]. In data Requisition stage, a system of four soft-synchronized cameras are engaged, consisting of a polarization camera and three Kinects V2, with each Kinect v2 having a depth and a color cameras. 12 subjects are recruited in data collection, where 9 are male and 3 are female. Each subject performs 3 different groups of actions (out of 18 different action types) 4 times, plus an addition period of free-form motion at the end of the session. Thus for each subject, there are 13 short videos (of around 1,800 frames per video with 10–15 FPS); the total number of frames for each subject amounts to 22 K. Overall, our dataset consists of 287K frames, each frame here contains a synchronized set of images - one polarization image, three color and three depth images.

The SMPL shape parameters and the 3D joint positions of a body shape are obtained from the image collection of current frame as follows. For each frame, its initial 3D pose estimation is obtained by integrating the Kinect readouts as well as the corresponding 2D joint estimation from OpenPose [63] across the depth and color sensors. Then the body shape, i.e. parameters of the SMPL model, is estimated as optimal fit to the initial pose estimate [26]. The 3D point cloud of body surface acquired from three depth cameras are now utilized in our final step, resulting in the estimation of refined body shape with clothing details [64], by iteratively minimizing the distance of SMPL shape vertex to its nearest point of the 3D point cloud. Exemplar clothed human shapes are shown in Fig. 2.

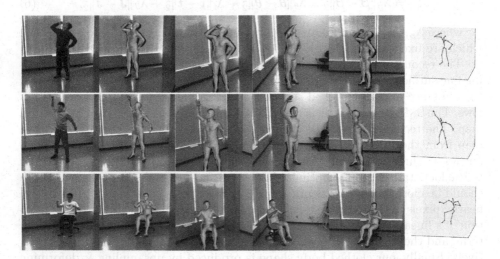

Fig. 2. Exemplar 3D poses and SMPL shapes in the real-world PHSPD dataset. We render the SMPL shape on four images (one polarization image and three-view color images) and we also show the pose in 3D space. (Color figure online)

4   Empirical Evaluations

Empirical evaluations are carried out on two major aspects. (1) For normal estimation, we report the mean angle error (MAE), which measures the angle between the target and estimated normal map, $e_{\text{angle}} = \arccos(\langle \mathbf{n}^{i,j}, \hat{\mathbf{n}}^{i,j} \rangle)$ for pixel (i, j), where $\langle \cdot, \cdot \rangle$ denotes cosine similarity. (2) For human pose and shape estimation, we report the mean per joint position error (MPJPE) and the 3D surface distance error. MPJPE is defined as the average distance between predicted and annotated joints of the test samples. In both SURREAL and PHSPD datasets, there are 24 annotated 3D joints. We also report the MPJPE for 20 joints by removing the hand and foot joints. The 3D surface error measures the distance between the predicted mesh and the ground truth mesh, by averaged distance of vertex pairs, as follows: for each vertex of the human body mesh, its closest vertex in ground truth mesh is identified to form its vertex pair; the average distance of all such vertex pairs is then computed.

For the real-world PHSPD dataset, subject 4 is chosen to form the validation set (23,786 samples); the test set contains those of subjects 7, 11, and 12 (69,283 samples); the train set has everything else (186,746 samples).

We also demonstrate the effectiveness of our SfP approach on SURREAL [29], a synthetic dataset of color images rendered from motion-captured human body shapes. Polarization images can be synthesized using color and depth images (details are in supplementary material). We choose subset "run1" and select one frame with a gap of ten frames. Finally, the train set has 245,759 samples, validation set has 14,528 samples and test set has 52,628 samples.

4.1   Evaluation of Surface Normal Estimation

In this task, our approach is compared with a recent work *Physics* [44], a traditional method *Linear* [65], and three ablation variants of our method as baselines. *Ours (color image)* uses only color image for estimating the normal map. *Ours (no physical priors)* does not incorporate the ambiguous normal maps as the physical priors and employs the polarization image as the only input. *Ours (no fused normal)* is similar to Physics [44], in which we use the two ambiguous normal maps as the only priors, discarding the fused normal maps.

Through both the quantitative results of Table 1 and the visual results of Fig. 3, it is observed that our method has consistently outperforms the state-of-the-art surface normal prediction methods [44,65] in both SURREAL and PHSPD datasets. The poor performance of [65] may be attributed to its unrealistic assumption of noise-free environment in the captured images. Let us look at the three ablation baselines of our approach: using only color images delivers relatively similar performance to that of removing physical priors when compared in PHSPD. Intuitively, it is challenging for neural networks to encode information of ambiguous normal maps (physical priors) directly from raw polarization images. Therefore, removing the physical priors results in similar performance to that of using only color images. [44] and *ours (no fused normal)* both utilize

Table 1. Comparison of surface normal estimation evaluated in MAE. The competing methods include *Linear* [65], *Physics* [44], *ours*, and three ablation variants of our method.

	SURREAL	PHSPD
Linear [65]	20.03	34.97
Physics [44]	7.45	21.45
ours (color image)	19.49	25.02
Ours (no physical priors)	13.89	24.71
Ours (no fused normal)	7.43	21.65
Ours	**7.10**	**20.75**

ambiguous normal as a physical prior, thus produce similar results. By incorporating the fused normal which discriminates the ambiguity of azimuth angle estimation, the results of our full-fledged approach surpasses those of [44].

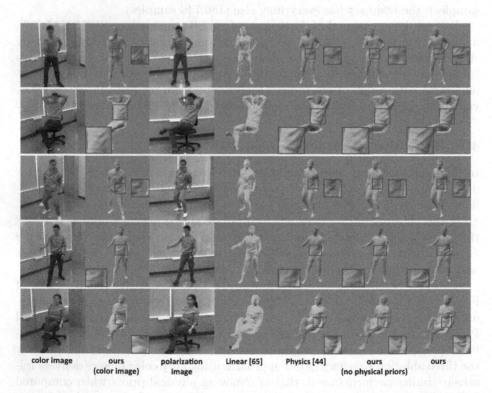

Fig. 3. Exemplar results of normal map prediction by five competing methods: [44,65], *ours (no physical priors)*, *ours (color image)*, and *ours*. Original color images and polarization images are shown in the first and third column with pixelated faces. (Color figure online)

4.2 Evaluation of Pose and Shape Estimation

The focus of this section is qualitative and quantitative evaluations on estimating poses & SMPL shapes, as well as our final estimation of clothed human shapes.

In pose estimation, it is of interest to inspect the effect of engaging surface normal maps in our SfP approach. Besides our SfP approach, the competing methods consist of *HMR* [28] and a ablation variant of SfP, *ours (w/o normal)*. The latter is obtained by engaging only the polarization image, without considering normal map estimation. Since *HMR* is trained on single color images, it is re-trained using the first three channels of a polarization image. In addition to *HMR* that works on color images, for fair comparison, *HMR* is also re-trained on the polarization images of our PHSPD dataset, as *HMR (polarization)*. From Table 2, it is observed that our method produces the lowest MPJPE values of all competing methods; the results of *ours (w/o normal)* is comparable to those of *HMR (polarization)*. The quantitative results confirm that the polarization images is capable of producing accurate estimation of human poses. Moreover, the visual results in Fig. 4 provide qualitative evidence that further performance gain is to be expected, when we have access to the normal maps. Similar observation is again obtained in Table 3, when quantitative examination is systematically carried out over w/ and w/o estimated normal map, on color and polarization images, in both datasets. Note the performance gain is particularly significant for polarization images, which may attribute to the rich geometric information encoded in the normal map representation. On color images, there is still noticeable improvement, also less significant. Our explanation is that the normal maps estimated from color images are not as reliable as those obtained from the polarization image counterparts.

Table 2. Quantitative evaluations using MPJPE evaluation metric on both SURREAL and PHSPD datasets. The unit of the error is millimeter. GT-t means the camera translation is known and Pred-t means the predicted camera translation is used to compute the joint error. We report the MPJPE results of 20/24 joints, which removes two hand and two foot joints following similar settings of previous work [28,66].

	SURREAL	PHSPD	
	GT-t	GT-t	Pred-t
HMR [28]	116.68/136.32	82.96/91.46	–
HMR (polarization)	–	77.57/88.74	97.24/106.20
Ours (w/o normal)	83.43/94.00	84.44/96.42	93.38/104.48
Ours	**67.25/75.94**	**66.32/74.46**	**74.58/81.85**

To evaluate the effectiveness of our approach on clothed human shape recovery, the state-of-the-art methods on human surface reconstruction from single color images are recruited. They are *PIFu* [40], *Depth Human* [41] and *HMD* [42]. Quantitative results are obtained in the PHSPD dataset by computing the

Table 3. Qualitative ablation study of our SfP approach (w/ vs. w/o the estimated surface normal). MPJPE is the evaluation metric with millimeter unit. Experiments are carried out on both color and polarization images of SURREAL and PHSPD datasets.

	SURREAL		PHSPD	
	ours (w/o normal)	ours	ours (w/o normal)	ours
Polarization image	83.43/94.00	67.25/75.94	84.44/96.42	66.32/74.46
Improvement	**16.18/18.06**		**18.12/21.96**	
Color image	88.53/100.32	80.70/91.51	85.67/80.34	77.72/70.07
Improvement	**7.82/8.81**		**7.95/10.27**	

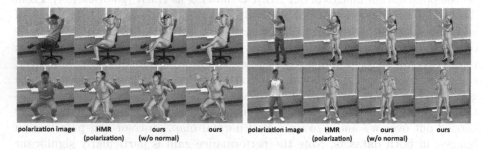

Fig. 4. Exemplar shape estimation results. The first column is polarization images. HMR (polarization) means the HMR model is retrained on polarization images of our PHSPD dataset. Ours (w/o normal) means the model is trained without the normal map as a part of the input.

3D surface error of the predicted human mesh with respect to the ground-truth mesh. Scaled rigid ICP is performed before the evaluation so as to scale and transform the predicted mesh into the same coordinates as the ground-truth surface. The results are displayed in Table 4. *PIFu* [40] performs the worst, partly as it does not take human pose into consideration when predicting the implicit surface function inside a volume. The 3D surface error from *HMD* [42] and Depth *Human* [41] are relatively small; our SfP approach achieves the best performance, which is partly due to its exploitation of the estimated normal maps. Note the comparison methods of *PIFu* [40], *Depth Human* [41] and *HMD* [42] only work with color images as input. In this experiment, for each of the polarization images used by the two SfP variants, namely *our (w/o deform)* and *ours*, the closet color image captured in the multi-camera setup of PHSPD is taken as its corresponding input to the three comparison methods.

Exemplar visual results are presented in Fig. 5, where the predicted body shapes are overlaid onto the input images. It is observed that the body shapes predicted by *PIFu* and *Depth Human* are generally well-aligned with the input image as they are actually predicting the implicit function value or depth value for each pixel of the foreground human shape. Meanwhile, it does not necessarily indicate accurate alignment of 3D surface mesh, as is evidenced in Table 4. For

Table 4. Quantitative evaluation of clothed human shape recovery performance methods in the PHSPD dataset.

	PIFu [40]	Depth Human [41]	HMD [42]	Ours (w/o deform)	Ours
3D surface error (mm)	73.13	51.02	51.71	41.10	**38.92**

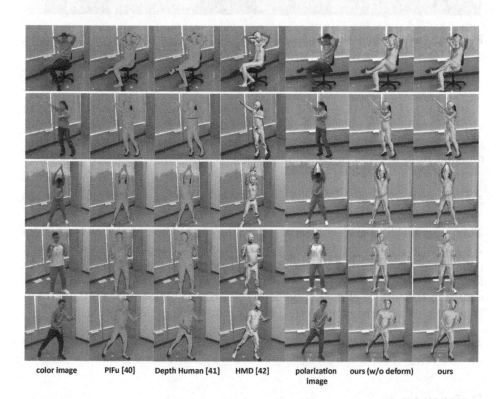

Fig. 5. Exemplar estimation results of clothed body shapes. The first and fifth column are color images and polarization images, respectively. *PIFu* [40], *Depth Human* [41] and *HMD* [42] are the results based on color input images. *Ours (w/o deformation)* and *ours* are the results with the polarization image as the input. (Color figure online)

PIFu and *Depth Human*, the exterior surfaces tend to be overly smooth. Besides, in *Depth Human*, only a partial mesh with respect to the view in the input image is produced. *HMD*, on the other hand, does not work well, as evidenced by the often error-prone surface details. This may be attributed to the less reliable shading representation, given the new environmental lighting and texture ambiguities existed in these color images. Our SfP approach is shown capable of producing reliable prediction of clothed body shapes, which again demonstrates the applicability of polarization imaging in shape estimation, as well as the benefit of engaging the surface normal maps in our approach.

Fig. 6. Exemplar estimation results of clothed body shapes, obtained on polarization images from novel test scenarios (new human subject and scene context).

Qualitative results presented in Fig. 6 showcase the robust test performance in novel settings. Note the polarization images are intentionally acquired from unseen human subjects at new geo-locations, so the background scenes are very different from those in the training images.

5   Conclusion

This paper tackles a new problem of estimating clothed human shapes from single 2D polarization images. Our work demonstrate the applicability of engaging polarization cameras as a promising alternative to the existing imaging sensors for human pose and shape estimation. Moreover, by exploiting the rich geometric details in the surface normal of the input polarization images, our SfP approach is capable of reconstructing clothed human body shapes of surface details.

Acknowledgement. This work is supported by the NSERC Discovery Grants, and the University of Alberta-Huawei Joint Innovation Collaboration grants.

References

1. Park, S., Hwang, J., Kwak, N.: 3D human pose estimation using convolutional neural networks with 2D pose information. In: Hua, G., Jégou, H. (eds.) ECCV 2016. LNCS, vol. 9915, pp. 156–169. Springer, Cham (2016). https://doi.org/10.1007/978-3-319-49409-8_15
2. Li, S., Zhang, W., Chan, A.B.: Maximum-margin structured learning with deep networks for 3D human pose estimation. In: Proceedings of the IEEE International Conference on Computer Vision, pp. 2848–2856 (2015)
3. Tekin, B., Katircioglu, I., Salzmann, M., Lepetit, V., Fua, P.: Structured prediction of 3D human pose with deep neural networks. In: British Machine Vision Conference (BMVC) (2016)
4. Tome, D., Russell, C., Agapito, L.: Lifting from the deep: convolutional 3D pose estimation from a single image. In: Proceedings of the IEEE Conference on Computer Vision and Pattern Recognition, pp. 2500–2509 (2017)
5. Martinez, J., Hossain, R., Romero, J., Little, J.J.: A simple yet effective baseline for 3D human pose estimation. In: Proceedings of the IEEE International Conference on Computer Vision, pp. 2640–2649 (2017)

6. Zhao, R., Wang, Y., Martinez, A.M.: A simple, fast and highly-accurate algorithm to recover 3D shape from 2D landmarks on a single image. IEEE Trans. Pattern Anal. Mach. Intell. **40**(12), 3059–3066 (2017)

7. Moreno-Noguer, F.: 3D human pose estimation from a single image via distance matrix regression. In: Proceedings of the IEEE Conference on Computer Vision and Pattern Recognition, pp. 2823–2832 (2017)

8. Nie, B.X., Wei, P., Zhu, S.C.: Monocular 3D human pose estimation by predicting depth on joints. In: Proceedings of the IEEE International Conference on Computer Vision, pp. 3467–3475. IEEE (2017)

9. Zhou, X., Huang, Q., Sun, X., Xue, X., Wei, Y.: Towards 3D human pose estimation in the wild: a weakly-supervised approach. In: Proceedings of the IEEE International Conference on Computer Vision, pp. 398–407 (2017)

10. Wang, M., Chen, X., Liu, W., Qian, C., Lin, L., Ma, L.: DRPose3D: depth ranking in 3D human pose estimation. In: Proceedings of the Twenty-Seventh International Joint Conference on Artificial Intelligence, pp. 978–984 (2018)

11. Yang, W., Ouyang, W., Wang, X., Ren, J., Li, H., Wang, X.: 3D human pose estimation in the wild by adversarial learning. In: Proceedings of the IEEE Conference on Computer Vision and Pattern Recognition, pp. 5255–5264 (2018)

12. Fang, H.S., Xu, Y., Wang, W., Liu, X., Zhu, S.C.: Learning pose grammar to encode human body configuration for 3D pose estimation. In: Thirty-Second AAAI Conference on Artificial Intelligence, pp. 6821–6828 (2018)

13. Pavlakos, G., Zhou, X., Daniilidis, K.: Ordinal depth supervision for 3D human pose estimation. In: Proceedings of the IEEE Conference on Computer Vision and Pattern Recognition, pp. 7307–7316 (2018)

14. Sun, X., Xiao, B., Wei, F., Liang, S., Wei, Y.: Integral human pose regression. In: Ferrari, V., Hebert, M., Sminchisescu, C., Weiss, Y. (eds.) ECCV 2018. LNCS, vol. 11210, pp. 536–553. Springer, Cham (2018). https://doi.org/10.1007/978-3-030-01231-1_33

15. Liu, J., et al.: Feature boosting network for 3D pose estimation. IEEE Trans. Pattern Anal. Mach. Intell. **42**(2), 494–501 (2020)

16. Sharma, S., Varigonda, P.T., Bindal, P., Sharma, A., Jain, A., Bangalore, S.B.: Monocular 3D human pose estimation by generation and ordinal ranking. In: Proceedings of the IEEE International Conference on Computer Vision, pp. 2325–2334 (2019)

17. Habibie, I., Xu, W., Mehta, D., Pons-Moll, G., Theobalt, C.: In the wild human pose estimation using explicit 2D features and intermediate 3D representations. In: Proceedings of the IEEE Conference on Computer Vision and Pattern Recognition, pp. 10905–10914 (2019)

18. Wandt, B., Rosenhahn, B.: RepNet: weakly supervised training of an adversarial reprojection network for 3D human pose estimation. In: Proceedings of the IEEE Conference on Computer Vision and Pattern Recognition, pp. 7782–7791 (2019)

19. Li, C., Lee, G.H.: Generating multiple hypotheses for 3D human pose estimation with mixture density network. In: Proceedings of the IEEE Conference on Computer Vision and Pattern Recognition, pp. 9887–9895 (2019)

20. Wang, K., Lin, L., Jiang, C., Qian, C., Wei, P.: 3D human pose machines with self-supervised learning. IEEE Trans. Pattern Anal. Mach. Intell. **42**, 1069–1082 (2019)

21. Anguelov, D., Srinivasan, P., Koller, D., Thrun, S., Rodgers, J., Davis, J.: Scape: shape completion and animation of people. ACM Trans. Graph. (TOG) **24**, 408–416 (2005)

22. Loper, M., Mahmood, N., Romero, J., Pons-Moll, G., Black, M.J.: SMPL: a skinned multi-person linear model. ACM Trans. Graph. (TOG) **34**(6), 248 (2015)

23. Balan, A.O., Sigal, L., Black, M.J., Davis, J.E., Haussecker, H.W.: Detailed human shape and pose from images. In: Proceedings of the IEEE Conference on Computer Vision and Pattern Recognition, pp. 1–8. IEEE (2007)

24. Dibra, E., Jain, H., Oztireli, C., Ziegler, R., Gross, M.: Human shape from silhouettes using generative HKS descriptors and cross-modal neural networks. In: Proceedings of the IEEE Conference on Computer Vision and Pattern Recognition, pp. 4826–4836 (2017)

25. Dibra, E., Jain, H., Öztireli, C., Ziegler, R., Gross, M.: HS-Nets: estimating human body shape from silhouettes with convolutional neural networks. In: Fourth International Conference on 3D Vision (3DV), pp. 108–117. IEEE (2016)

26. Bogo, F., Kanazawa, A., Lassner, C., Gehler, P., Romero, J., Black, M.J.: Keep it SMPL: automatic estimation of 3D human pose and shape from a single image. In: Leibe, B., Matas, J., Sebe, N., Welling, M. (eds.) ECCV 2016. LNCS, vol. 9909, pp. 561–578. Springer, Cham (2016). https://doi.org/10.1007/978-3-319-46454-1_34

27. Lassner, C., Romero, J., Kiefel, M., Bogo, F., Black, M.J., Gehler, P.V.: Unite the people: closing the loop between 3D and 2D human representations. In: Proceedings of the IEEE Conference on Computer Vision and Pattern Recognition, pp. 6050–6059 (2017)

28. Kanazawa, A., Black, M.J., Jacobs, D.W., Malik, J.: End-to-end recovery of human shape and pose. In: Proceedings of the IEEE Conference on Computer Vision and Pattern Recognition, pp. 7122–7131 (2018)

29. Varol, G., et al.: Learning from synthetic humans. In: Proceedings of the IEEE Conference on Computer Vision and Pattern Recognition, pp. 109–117 (2017)

30. Pavlakos, G., Zhu, L., Zhou, X., Daniilidis, K.: Learning to estimate 3D human pose and shape from a single color image. In: Proceedings of the IEEE Conference on Computer Vision and Pattern Recognition, pp. 459–468 (2018)

31. Omran, M., Lassner, C., Pons-Moll, G., Gehler, P., Schiele, B.: Neural body fitting: unifying deep learning and model based human pose and shape estimation. In: International Conference on 3D Vision (3DV), pp. 484–494. IEEE (2018)

32. Xu, Y., Zhu, S.C., Tung, T.: DenseRaC: joint 3D pose and shape estimation by dense render-and-compare. In: Proceedings of the IEEE International Conference on Computer Vision, pp. 7760–7770 (2019)

33. Kolotouros, N., Pavlakos, G., Black, M.J., Daniilidis, K.: Learning to reconstruct 3D human pose and shape via model-fitting in the loop. In: Proceedings of the IEEE International Conference on Computer Vision, pp. 2252–2261 (2019)

34. Zanfir, A., Marinoiu, E., Sminchisescu, C.: Monocular 3D pose and shape estimation of multiple people in natural scenes-the importance of multiple scene constraints. In: Proceedings of the IEEE Conference on Computer Vision and Pattern Recognition, pp. 2148–2157 (2018)

35. Sun, Y., Ye, Y., Liu, W., Gao, W., Fu, Y., Mei, T.: Human mesh recovery from monocular images via a skeleton-disentangled representation. In: Proceedings of the IEEE International Conference on Computer Vision, pp. 5349–5358 (2019)

36. Kanazawa, A., Zhang, J.Y., Felsen, P., Malik, J.: Learning 3D human dynamics from video. In: Proceedings of the IEEE Conference on Computer Vision and Pattern Recognition, pp. 5614–5623 (2019)

37. Arnab, A., Doersch, C., Zisserman, A.: Exploiting temporal context for 3D human pose estimation in the wild. In: Proceedings of the IEEE Conference on Computer Vision and Pattern Recognition, pp. 3395–3404 (2019)

38. Varol, G., et al.: BodyNet: volumetric inference of 3D human body shapes. In: Ferrari, V., Hebert, M., Sminchisescu, C., Weiss, Y. (eds.) ECCV 2018. LNCS, vol. 11211, pp. 20–38. Springer, Cham (2018). https://doi.org/10.1007/978-3-030-01234-2_2

39. Zheng, Z., Yu, T., Wei, Y., Dai, Q., Liu, Y.: DeepHuman: 3D human reconstruction from a single image. In: Proceedings of the IEEE International Conference on Computer Vision, pp. 7739–7749 (2019)

40. Saito, S., Huang, Z., Natsume, R., Morishima, S., Kanazawa, A., Li, H.: PIFu: pixel-aligned implicit function for high-resolution clothed human digitization. In: Proceedings of the IEEE International Conference on Computer Vision, pp. 2304–2314 (2019)

41. Tang, S., Tan, F., Cheng, K., Li, Z., Zhu, S., Tan, P.: A neural network for detailed human depth estimation from a single image. In: Proceedings of the IEEE International Conference on Computer Vision, pp. 7750–7759 (2019)

42. Zhu, H., Zuo, X., Wang, S., Cao, X., Yang, R.: Detailed human shape estimation from a single image by hierarchical mesh deformation. In: Proceedings of the IEEE Conference on Computer Vision and Pattern Recognition, pp. 4491–4500 (2019)

43. Yang, L., Tan, F., Li, A., Cui, Z., Furukawa, Y., Tan, P.: Polarimetric dense monocular SLAM. In: Proceedings of the IEEE Conference on Computer Vision and Pattern Recognition, pp. 3857–3866 (2018)

44. Ba, Y., Chen, R., Wang, Y., Yan, L., Shi, B., Kadambi, A.: Physics-based neural networks for shape from polarization. arXiv preprint arXiv:1903.10210 (2019)

45. Wehner, R., Müller, M.: The significance of direct sunlight and polarized skylight in the ant's celestial system of navigation. Proc. Natl. Acad. Sci. 103(33), 12575–12579 (2006)

46. Daly, I.M., et al.: Dynamic polarization vision in mantis shrimps. Nat. Commun. 7, 12140 (2016)

47. Atkinson, G.A., Hancock, E.R.: Recovery of surface orientation from diffuse polarization. IEEE Trans. Image Process. 15(6), 1653–1664 (2006)

48. Kadambi, A., Taamazyan, V., Shi, B., Raskar, R.: Depth sensing using geometrically constrained polarization normals. Int. J. Comput. Vis. 125(1–3), 34–51 (2017). https://doi.org/10.1007/s11263-017-1025-7

49. Chen, L., Zheng, Y., Subpa-asa, A., Sato, I.: Polarimetric three-view geometry. In: Ferrari, V., Hebert, M., Sminchisescu, C., Weiss, Y. (eds.) ECCV 2018. LNCS, vol. 11220, pp. 21–37. Springer, Cham (2018). https://doi.org/10.1007/978-3-030-01270-0_2

50. Cui, Z., Gu, J., Shi, B., Tan, P., Kautz, J.: Polarimetric multi-view stereo. In: Proceedings of the IEEE Conference on Computer Vision and Pattern Recognition, pp. 1558–1567 (2017)

51. Zhou, X., Zhu, M., Leonardos, S., Derpanis, K.G., Daniilidis, K.: Sparseness meets deepness: 3D human pose estimation from monocular video. In: Proceedings of the IEEE Conference on Computer Vision and Pattern Recognition, pp. 4966–4975 (2016)

52. Akhter, I., Black, M.J.: Pose-conditioned joint angle limits for 3D human pose reconstruction. In: Proceedings of the IEEE Conference on Computer Vision and Pattern Recognition, pp. 1446–1455 (2015)

53. Wang, C., Wang, Y., Lin, Z., Yuille, A.L., Gao, W.: Robust estimation of 3D human poses from a single image. In: Proceedings of the IEEE Conference on Computer Vision and Pattern Recognition, pp. 2361–2368 (2014)

54. Ramakrishna, V., Kanade, T., Sheikh, Y.: Reconstructing 3D human pose from 2D image landmarks. In: Fitzgibbon, A., Lazebnik, S., Perona, P., Sato, Y., Schmid, C. (eds.) ECCV 2012. LNCS, vol. 7575, pp. 573–586. Springer, Heidelberg (2012). https://doi.org/10.1007/978-3-642-33765-9_41

55. Zhou, X., Zhu, M., Pavlakos, G., Leonardos, S., Derpanis, K.G., Daniilidis, K.: MonoCap: monocular human motion capture using a CNN coupled with a geometric prior. IEEE Trans. Pattern Anal. Machine Intell. **41**(4), 901–914 (2019)

56. Zhou, X., Zhu, M., Leonardos, S., Daniilidis, K.: Sparse representation for 3D shape estimation: a convex relaxation approach. IEEE Trans. Pattern Anal. Mach. Intell. **39**(8), 1648–1661 (2016)

57. Chen, C.H., Ramanan, D.: 3D human pose estimation = 2D pose estimation + matching. In: Proceedings of the IEEE Conference on Computer Vision and Pattern Recognition, pp. 7035–7043 (2017)

58. Ci, H., Wang, C., Ma, X., Wang, Y.: Optimizing network structure for 3D human pose estimation. In: Proceedings of the IEEE International Conference on Computer Vision, pp. 2262–2271 (2019)

59. Cai, Y., et al.: Exploiting spatial-temporal relationships for 3D pose estimation via graph convolutional networks. In: Proceedings of the IEEE International Conference on Computer Vision, pp. 2272–2281 (2019)

60. Jiang, H., Cai, J., Zheng, J.: Skeleton-aware 3D human shape reconstruction from point clouds. In: Proceedings of the IEEE International Conference on Computer Vision, pp. 5431–5441 (2019)

61. Nehab, D., Rusinkiewicz, S., Davis, J., Ramamoorthi, R.: Efficiently combining positions and normals for precise 3D geometry. ACM Trans. Graph. (TOG) **24**(3), 536–543 (2005)

62. Zou, S., et al.: Polarization human shape and pose dataset. arXiv preprint arXiv:2004.14899 (2020)

63. Cao, Z., Martinez, G.H., Simon, T., Wei, S., Sheikh, Y.A.: OpenPose: realtime multi-person 2D pose estimation using part affinity fields. IEEE Trans. Pattern Anal. Mach. Intell. (2019)

64. Zuo, X., et al.: SparseFusion: dynamic human avatar modeling from sparse RGBD images. IEEE Trans. Multimed. (2020)

65. Smith, W.A.P., Ramamoorthi, R., Tozza, S.: Linear depth estimation from an uncalibrated, monocular polarisation image. In: Leibe, B., Matas, J., Sebe, N., Welling, M. (eds.) ECCV 2016. LNCS, vol. 9912, pp. 109–125. Springer, Cham (2016). https://doi.org/10.1007/978-3-319-46484-8_7

66. Ionescu, C., Papava, D., Olaru, V., Sminchisescu, C.: Human3.6M: large scale datasets and predictive methods for 3D human sensing in natural environments. IEEE Trans. Pattern Anal. Mach. Intell. **36**(7), 1325–1339 (2014)

The Devil Is in the Details: Self-supervised Attention for Vehicle Re-identification

Pirazh Khorramshahi[1]($\boxtimes$), Neehar Peri[1], Jun-cheng Chen[2], and Rama Chellappa[1]

[1] Center for Automation Research, UMIACS, and the Department of Electrical and Computer Engineering, University of Maryland, College Park, USA
pirazhkhorramshahi@gmail.com
[2] Research Center for Information Technology Innovation, Academia Sinica, Taipei, Taiwan

Abstract. In recent years, the research community has approached the problem of vehicle re-identification (re-id) with attention-based models, specifically focusing on regions of a vehicle containing discriminative information. These re-id methods rely on expensive key-point labels, part annotations, and additional attributes including vehicle make, model, and color. Given the large number of vehicle re-id datasets with various levels of annotations, strongly-supervised methods are unable to scale across different domains. In this paper, we present Self-supervised Attention for Vehicle Re-identification (SAVER), a novel approach to effectively learn vehicle-specific discriminative features. Through extensive experimentation, we show that SAVER improves upon the state-of-the-art on challenging VeRi, VehicleID, Vehicle-1M and VERI-Wild datasets.

Keywords: Vehicle re-identification · Self-supervised learning · Variational auto-encoder · Deep representation learning

1 Introduction

Re-identification (re-id), the task of identifying all images of a specific object ID in a gallery, has been recently revolutionized with the advancement of Deep Convolutional Neural Networks (DCNNs). This revolution is most notable in the area of person re-id. Lou *et al.* [28] recently developed a strong baseline method that supersedes state-of-the-art person re-id methods by a large margin, using an empirically derived "Bag of Tricks" to improve the discriminative capacity of DCNNs. This has created a unique opportunity for the research community to develop innovative yet simple methods to push the boundaries of object re-id.

P. Khorramshahi and N. Peri—Equally contributed to this work.

© Springer Nature Switzerland AG 2020
A. Vedaldi et al. (Eds.): ECCV 2020, LNCS 12359, pp. 369–386, 2020.
https://doi.org/10.1007/978-3-030-58568-6_22

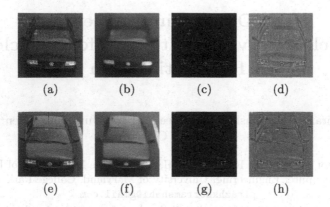

(a)   (b)   (c)   (d)

(e)   (f)   (g)   (h)

Fig. 1. Vehicle image decomposition into coarse reconstruction and residual images, left-most column (a, e): vehicle image, second column (b, f): coarse reconstruction, third column (c, g): residual, right-most column (d, h): normalized residual (for the sake of visualization). Despite having the same coarse reconstruction, both vehicles have different residuals highlighting key areas, *e.g.*, the windshield stickers, bumper design.

Specifically, vehicle re-id has great potential in intelligent transportation applications. However, the task of vehicle re-id is particularly challenging since vehicles with different identities can be of the same make, model and color. Moreover, the appearance of a vehicle varies significantly across different viewpoints. Therefore, recent DCNN-based re-id methods focus attention on discriminative regions to improve robustness to orientation and occlusion. To this end, many high performing re-id approaches rely on additional annotations for local regions that have been shown to carry identity-dependent information, *i.e.* key-points [16,17,41] and parts bounding boxes [11,46] in addition to the ID of the objects of interest. These extra annotations help DCNNs jointly learn improved global and local representations and significantly boost performance [16,48] at the cost of increased complexity. Despite providing considerable benefit, gathering costly annotations such as key-point and part locations cannot be scaled to the growing size of vehicle re-id datasets. As manufacturers change the design of their vehicles, the research community has the burdensome task of annotating new vehicle models. In an effort to re-design the vehicle re-id pipeline without the need for expensive annotations, we propose SAVER to automatically highlight salient regions in a vehicle image. These vehicle-specific salient regions carry critical details that are essential for distinguishing two visually similar vehicles. Specifically, we design a Variational Auto-Encoder (VAE) [19] to generate a vehicle image template that is free from manufacturer logos, windshield stickers, wheel patterns, and grill, bumper and head/tail light designs. By obtaining this coarse reconstruction and its pixel-wise difference from the original image, we construct **residual** image. This residual contains crucial details required for re-id, and acts as a pseudo-saliency or pseudo-attention map highlighting discriminative regions in an image. Figure 1 shows how the residual map highlights valuable fine-grained details needed for re-identification between two visually similar vehicles.

The rest of the paper is organized as follows. In Sect. 2, we briefly review recent works in vehicle re-id. The detailed architecture of each step in the proposed approach is discussed in Sect. 3. Through extensive experimentation in Sect. 4, we show the effectiveness of our approach on multiple challenging vehicle re-id benchmarks [9,22,24,27,43], obtaining state-of-the-art results. Finally, in Sect. 5 we validate our design choices.

2   Related Works

Learning robust and discriminative vehicle representations that adapt to large viewpoint variations across multiple cameras, illumination and occlusion is essential for re-id. Due to a large volume of literature, we briefly review recent works on vehicle re-identification.

With recent breakthroughs due to deep learning, we can easily learn discriminative embeddings for vehicles by feeding images from large-scale vehicle datasets, such as VehicleID, VeRi, VERI-Wild, Vehicle-1M, PKU VD1&VD2 [43], CompCars [44], and CityFlow [40], to train a DCNN that is later used as the feature extractor for re-id. However, for vehicles of the same make, model, and color, this global deep representation usually fails to discriminate between two similar-looking vehicles. To address this issue, several auxiliary features and strategies are proposed to enhance the learned global appearance representation. Cui et al. [4] fuse features from various DCNNs trained with different objectives. Suprem et al. [36] propose the use of an ensemble of re-id models for vehicle identity and attributes for robust matching. [11,16,23,41,46] propose learning enhanced representation by fusing global features with auxiliary local representations learned from prominent vehicle parts and regions, e.g., headlights, mirrors. Furthermore, Peng et al. [31] leverage an image-to-image translation model to reduce cross-camera bias for vehicle images from different cameras before learning auxiliary local representation. Zhou et al. [50] learn vehicle representation via viewpoint-aware attention. Similarly, [32,48] leverage attention guided by vehicle attribute classification, e.g., color and vehicle type, to learn attribute-based auxiliary features to enhance the global representation. Metric learning is another popular approach to make representations more discriminative. [2,3,21,47] propose various triplet losses to carefully select hard triplets across different viewpoints and vehicles to learn an improved appearance-robust representation.

Alternatively, to augment training data for more robust training, [45] adopts a graphic engine and [39,42] use generative adversarial networks (GANs) to synthesize vehicle images with diverse orientations, appearance variations, and other attributes. [14,15,25,26,29,34,38] propose methods for improving the matching performance by also making use of spatio-temporal and multi-modal information, such as visual features, license plates, inter-camera vehicle trajectories, camera locations, and time stamps.

In contrast with prior methods, SAVER benefits from self-supervised attention generation and does not assume any access to extra annotations, attributes, spatio-temporal and multi-modal information.

3   Self-supervised Attention for Vehicle Re-identification

Our proposed pipeline is composed of two modules, namely, **Self-Supervised Residual Generation, and Deep Feature Extraction**. Figure 2 presents the proposed end-to-end pipeline. The self-supervised reconstruction network is responsible for creating the overall shape and structure of a vehicle image while obfuscating discriminative details. This enables us to highlight salient regions and remove background distractors by subtracting the reconstruction from the input image. Next, we feed the convex combination (with trainable parameter α) of the residual and original input images to ResNet-50 [12] model to generate robust discriminative features. To train our deep feature extraction module, we use techniques proposed in "Bag of Tricks" [28] and adapt them for vehicle re-identification, offering a strong baseline.

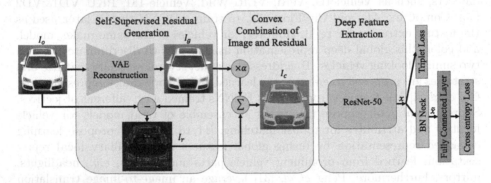

Fig. 2. Proposed SAVER pipeline. The input image is passed through the VAE-based reconstruction module to remove vehicle-specific details. Next, the reconstruction is subtracted from the input image to form the residual image containing vehicle-specific details. Later, the convex combination (with trainable parameter α) of the input and residual is calculated and passed through the re-id backbone for deep feature extraction. The entire pipeline is trained via triplet and cross entropy losses, separated via a batch normalization layer (BN Neck) proposed in [28].

3.1   Self-supervised Residual Generation

In order to generate the crude shape and structure of a vehicle while removing small-scale discriminative information, we leverage prior work in image segmentation [1] and generation [19]. Specifically, we construct a novel VAE architecture that down-samples the input image of spatial size $H \times W$ through max-pooling into a latent space of spatial size $\frac{H}{16} \times \frac{W}{16}$. Afterwards, we apply the re-parameterization trick introduced in [19] to the latent features via their mean and covariance. Next, we up-sample the latent feature map as proposed by [30] to prevent checkerboard artifacts. This step generates the reconstructed image of size $H \times W$. Figure 3 illustrates the proposed self-supervised reconstruction network.

Formally, we pre-train our reconstruction model using the mean squared error (MSE) and Kullback-Leibler (KL) divergence such that

$$\mathcal{L}_{reconstruction} = \mathcal{L}_{MSE} + \lambda \mathcal{L}_{KL} \tag{1}$$

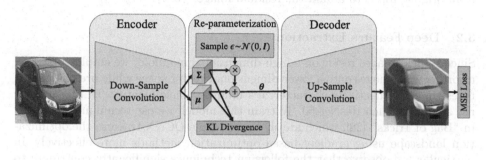

Fig. 3. Self-Supervised image reconstruction required for subsequent residual generation. The input image goes through the convolutional encoder and is mapped to 3-dimensional latent variable. Using the VAE re-parameterization trick, a sample from the standard multivariate Gaussian ϵ is drawn and scaled via mean μ and co-variance Σ of the latent variable. Lastly, θ is up-sampled with a convolutional decoder to generate the input image template with most fine grained details removed.

where

$$\mathcal{L}_{MSE} = \frac{1}{H \times W} \sum_{j=1}^{H} \sum_{k=1}^{W} |I_o(j,k) - I_g(j,k)|^2 \tag{2}$$

and

$$\mathcal{L}_{KL} = \frac{1}{2 \times (\frac{H}{16} \times \frac{W}{16})} \sum_{m=1}^{M} \left[\mu_m^2 + \sigma_m^2 - \log(\sigma_m^2) - 1 \right] \tag{3}$$

In Eq. 1, λ sets the balance between the MSE and KL objective functions. Also, I_o and I_g in Eq. 2 refer to the original and generated images respectively. Finally, in Eq. 3, M is the dimensionality of the latent features $\theta \in \mathbb{R}^M$ with mean $\mu = [\mu_1, \ldots, \mu_M]$ and covariance matrix $\Sigma = \text{diag}(\sigma_1^2, \ldots, \sigma_M^2)$, that are re-parameterized via sampling from standard multivariate Gaussian $\epsilon \sim \mathcal{N}(\mathbf{0}, I_M)$, i.e. $\theta = \mu + \Sigma^{\frac{1}{2}} \epsilon$.

We pre-train this model on the large-scale Vehicle Universe dataset, introduced in Sect. 4.2.1, prior to training our end-to-end pipeline, as described in Sect. 4. This pre-training allows the reconstruction model to generalize to vehicle images with a larger variety of make, model, color, orientation, and image quality. Hence, it captures domain invariant features that can later be fined-tuned for a particular dataset. Additionally, pre-training improves the rate of convergence for end-to-end pipeline training. It is important to note that unlike traditional VAE implementations, we use three-dimensional latent feature maps, *i.e.*, channel, height and width dimensions, rather than one-dimensional latent vectors with only channel dimension, for improving the reconstruction quality and preserve more spatial information. Moreover, we scale $\mathcal{L}_{KL}$ when calculating Eq. 1 to improve the reconstruction quality. We further explore the effect of the KL divergence scaling factor λ in Sect. 5. Once the self-supervised image reconstruction network generates the coarse image template I_g, we subtract it from original input to obtain the residual image, *i.e.* $I_r = I_o - I_g$.

3.2   Deep Feature Extraction

Since vehicle images reside on a high-dimensional manifold, we employ a DCNN to project the images onto a lower-dimensional vector space while preserving features that can effectively characterize a unique vehicle identity. To this end, we use a single-branch ResNet-50. To train this model, we use techniques proposed in "Bag of Tricks" [28], which are shown to help a DCNN traverse the optimization landscape using gradient-based optimization methods more effectively. In particular, we observe that the following techniques significantly contribute to the performance of the vehicle re-id baseline model:

1 - **Learning Rate Warm-Up:** [6] has suggested increasing the learning rate linearly in initial epochs of training to obtain improved weight initialization. This significantly contributes to the enhanced performance of our baseline.

2 - **Random Erasing Augmentation (REA):** To better handle the issue of occlusion, [13] introduced REA with the goal of encouraging a network to learn more robust representations.

3 - **Label Smoothing:** In order to alleviate the issue of over-fitting to the training data, [37] proposed smoothing the ground-truth labels.

4 - **Batch Normalization (BN) Neck:** To effectively apply both classification and triplet losses to the extracted features, a BN layer is proposed by [28]. This also significantly improves vehicle re-id performance.

The ResNet-50 feature extractor model is trained to optimize for triplet and cross entropy classification losses which are calculated as follows:

$$\mathcal{L}_{triplet} = \frac{1}{B} \sum_{i=1}^{B} \sum_{a \in b_i} \left[\gamma + \max_{p \in \mathcal{P}(a)} d(x_a, x_p) - \min_{n \in \mathcal{N}(a)} d(x_a, x_n) \right]_+ \tag{4}$$

and

$$\mathcal{L}_{classification} = -\frac{1}{N} \sum_{i=1}^{N} \log \frac{\exp(W_{c(\hat{x}_i)}^T \tilde{x}_i + b_{c(\hat{x}_i)})}{\sum_{j=1}^{C} \exp(W_j^T \tilde{x}_i + b_j)} \tag{5}$$

In Eq. 4, B, b_i, a, γ, $\mathcal{P}(a)$ and $\mathcal{N}(a)$ are the total number of batches, i^{th} batch, anchor sample, distance margin threshold, positive and negative sample sets corresponding to a given anchor respectively. Moreover, x_a, x_p, x_n represent the ResNet-50 extracted features associated with anchor, positive and negative samples. In addition, function $d(.,.)$ calculates the Euclidean distance of the two extracted features. Note that in Eq. 4, we used the batch hard triplet loss [13] to overcome the computational complexity of calculating the distances to all unique triplets of data points. Here we construct batches so that they have exactly K instances of each ID used in a particular batch, *i.e.* B is a multiple of K. In Eq. 5, $\tilde{x}_i$ and $c(\hat{x}_i)$ refer to the extracted feature for the ith image in the training set after passing through the BN Neck layer and its corresponding ground-truth class label respectively. Furthermore, W_j, b_j are the weight vector and bias associated with class j in the final classification layer. N and C represent the total number of samples and classes in the training process respectively.

3.3 End-To-End Training

After pre-training the self-supervised residual generation module, we jointly train the VAE and deep feature extractor. We compute the convex combination of input images and their respective residuals using a learnable parameter α, *i.e.* $I_c = \alpha \times I_o + (1 - \alpha) \times I_r$, allowing the feature extractor network to weight the importance of each input source. Moreover, the end-to-end training helps the entire pipeline adapt the residual generation such that it is suited for the re-id task. In summary, the loss function for end-to-end training is the following:

$$\mathcal{L}_{total} = \mathcal{L}_{triplet} + \mathcal{L}_{classification} + \eta \mathcal{L}_{reconstruction} \qquad (6)$$

In Eq. 6, the scaling factor η is empirically set to 100.

4 Experiments

In this section, we first present the different datasets on which we evaluate the proposed approach and describe how vehicle re-identification systems are evaluated in general. Next, we present implementation details for the proposed self-supervised residual generation, deep feature extraction and end-to-end training steps. Finally, we report experimental results of the proposed approach.

4.1 Vehicle Re-identification Datasets

We evaluate SAVER on six popular vehicle re-id benchmarks, including VeRi, VehicleID, VERI-Wild, Vehicle-1M and PKU VD1&VD2. Table 1 presents the statistics of these datasets in terms of the number of unique identities, images and cameras. Additionally, we highlight four additional datasets of unconstrained vehicle images, including CityFlow, CompCars, BoxCars116K [35], and Stanford-Cars [20], used in the pre-training of our self-supervised reconstruction network.

Re-id systems are commonly evaluated using the Cumulative Match Curve (CMC) and Mean Average Precision (mAP). A fixed gallery set is ranked with

Table 1. Vehicle re-id datasets statistics. ID, IM, Cam refer to number of unique identities, images and cameras respectively. Note that the evaluation set of VehicleID, VERI-Wild, Vehicle-1M, VD1 & VD2 are partitioned into small (S), medium (M) and large (L) splits.

	Split Set	VeRi	VehicleID			VERI-Wild			Vehicle-1M			VD1			VD2		
			S	M	L	S	M	L	S	M	L	S	M	L	S	M	L
Train	ID	576	13164			30671			50000			70591			39619		
	IM	37746	113346			277797			844571			422326			342608		
	Cam	20	-			173			-			-			-		
Gallery	ID	200	800	1600	2400	3000	5000	10000	1000	2000	3000	18000	131275	141757	12000	70755	79764
	IM	11579	800	1600	2400	38861	64389	128517	1000	2000	3000	104887	602032	1095649	103550	455910	805260
	Cam	19	-	-	-	146	153	161	-	-	-	-	-	-	-	-	-
Query	ID	200	800	1600	2400	3000	5000	10000	1000	2000	3000	2000	2000	2000	2000	2000	2000
	IM	1678	5693	11777	17377	3000	5000	10000	15123	30539	45069	2000	2000	2000	2000	2000	2000
	Cam	19	-	-	-	105	113	126	-	-	-	-	-	-	-	-	-

respect to the similarity score, e.g., L_2 distance, of its images and a given query image. CMC@K measures the probability of having a vehicle with the same ID as the query within the top K elements of the ranked gallery. It is a common practice to report CMC@1 and CMC@5. Similarly, mAP measures the average precision over all images in a query set.

4.2   Implementation Details

Here we discuss the implementation of both the self-supervised residual generation and deep feature extraction modules. In general, we resize all the images to $(256, 256)$ and normalize them by a mean and standard deviation of 0.5 across RGB channels before passing them through the respective networks. Moreover, similar to [17], we pre-process all images across all the experiments with the Detectron object detector [7] to minimize background noise.

4.2.1   Self-supervised Residual Generation

To pre-train the self-supervised residual generation module, we construct the large-scale Vehicle Universe dataset. We specifically consider vehicles from a variety of distributions to improve the robustness of our model. We utilize data from several sources, including CompCars, StanfordCars, BoxCars116K, CityFlow, PKU VD1&VD2, Vehicle-1M, VehicleID, VeRi and VeRi-Wild. In total, Vehicle Universe has 3706670, 1103404 and 11146 images in the train, test and validation sets respectively.

4.2.2   Deep Feature Extraction

As mentioned in Sect. 3.2, we use ResNet-50 for feature extraction. In all of our experiments learning rate starts from $3.5e-5$ and is linearly increased with the slope of $3.1e-5$ in the first ten epochs. Afterwards, it is decayed by a factor of ten every 30th epoch. In total, the end-to-end pipeline is trained for 150 epochs via Adam [18] optimizer. Furthermore, we use an initial value of $\alpha = 0.5$ for convex combination and $\gamma = 0.3$ for the triplet loss in Eq. 4.

4.3   Experimental Evaluation

In this section, we present evaluation results of the global appearance model (baseline) and global appearance model augmented with self-supervised attention (SAVER) on different re-id benchmarks discussed in Sect. 4.1.

4.3.1   VeRi

Table 2 reports the evaluation results on VeRi, a popular dataset for vehicle re-id. SAVER improves upon the strong baseline model. Most notably, SAVER gives 1.4% improvement on the mAP metric. We note that α in the convex combination of the input and residual saturates at 0.96, which means the model relies on 96% percent of the original image and 4% of the residual to construct more robust features.

4.3.2   VehicleID

Table 3 presents results of baseline and SAVER on test sets of varying sizes. Performance improvement of +1.0% in CMC@1 over the baseline model can be observed for all the test splits. To better demonstrate the discriminating capability of the proposed model, we visualize the attention map of both baseline and the proposed SAVER models on an image of this dataset using Gradient Class Activation Mapping (Grad-CAM) [33]. In Fig. 4, it is clear that SAVER is able to effectively construct attention on regions containing discriminative information such as headlights, hood and windshield stickers.

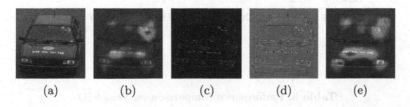

(a)   (b)   (c)   (d)   (e)

Fig. 4. Grad-CAM visualization of baseline and SAVER; (a) original image, (b) Grad-CAM visualization corresponding to the baseline model, *e.i.*, $\alpha = 1$, (c),(d) are residual and normalized residual maps (for the sake of visualization) obtain via our proposed self-supervised model respectively. (e) is the Grad-CAM visualization of proposed model, *e.i.*, $\alpha = 0.97$ in VehicleID dataset.

4.3.3   VERI-Wild

Evaluation results on the VERI-Wild dataset are presented in Table 4. Notably, our proposed residual generation model is improved upon the baseline by +2.0% and +1.0% for mAP and CMC@1 metrics on all evaluation splits respectively. The final alpha value $\alpha = 0.94$ suggests that the residual information contributes more in extracting robust features in this dataset.

4.3.4   Vehicle-1M

Table 5 reports the results of baseline and the proposed methods. Similar to VehicleID dataset, Vehicle-1M does not include fixed evaluation sets, therefore we randomly construct the evaluation splits and keep them fixed throughout the experiments. With the value of $\alpha = 0.98$ the proposed self-supervised residual generation module improves upon the baseline model in all metrics across all evaluation sets.

4.3.5   PKU VD1&2

Table 6 highlights the evaluation results on both PKU VD datasets. Similar to most re-id datasets, VD1&2 have S/M/L evaluation sets. However, due to the extreme size of these data splits, as shown in Table 1, we are only able to report numbers on the small evaluation set. The performance of SAVER is comparable to our baseline model. Moreover, the final value of $\alpha = 0.99$ indicates that baseline models is already very strong, and has almost no room for improvement. We can conclude that our performance on these data sets are saturated. Qualitatively, in Fig. 5 we show two failure cases of SAVER on these datasets. Note that how extremely similar these images are and it is nearly impossible to differentiate them based on only visual information.

Table 2. Performance comparison on VeRi

Model	mAP (%)	CMC@1 (%)	CMC@5 (%)	
Baseline	78.2	95.5	97.9	
SAVER	**79.6**	**96.4**	**98.6**	$\alpha = 0.96$

Table 3. Performance comparison on VehicleID

Model	CMC@1 (%)			CMC@5 (%)			
	S	M	L	S	M	L	
Baseline	78.4	76.0	74.1	92.5	89.1	86.4	
SAVER	**79.9**	**77.6**	**75.3**	**95.2**	**91.1**	**88.3**	$\alpha = 0.97$

4.3.6   State-of-the-Art Comparison

In this section, we present the latest state-of-the-art vehicle re-id methods and highlight the performance of the proposed SAVER model. Table 7 reports the state-of-the-art on re-id benchmarks. It can be seen that our proposed model, despite its simplicity, surpasses the most recent state-of-the-art vehicle re-id

Table 4. Performance comparison on VERI-Wild

Model	mAP (%)			CMC@1 (%)			CMC@5 (%)			
	S	M	L	S	M	L	S	M	L	
Baseline	78.5	72.8	65.0	92.9	91.3	88.1	97.3	96.8	95.0	
SAVER	**80.9**	**75.3**	**67.7**	**94.5**	**92.7**	**89.5**	**98.1**	**97.4**	**95.8**	$\alpha = 0.94$

Table 5. Performance comparison on Vehicle-1M

Model	CMC@1 (%)			CMC@5 (%)			
	S	M	L	S	M	L	
Baseline	93.6	94.9	91.7	97.9	99.1	98.0	
SAVER	**95.5**	**95.3**	**93.1**	**98.0**	**99.4**	**98.6**	$\alpha = 0.98$

Table 6. Performance comparison on VD1&VD2

Dataset	Model	mAP (%)	CMC@1 (%)	CMC@5 (%)	
VD1	Baseline	96.4	96.2	98.9	
	SAVER	**96.7**	**96.5**	**99.1**	$\alpha = 0.99$
VD2	Baseline	**96.8**	**97.9**	99.0	
	SAVER	96.7	97.8	99.0	$\alpha = 0.99$

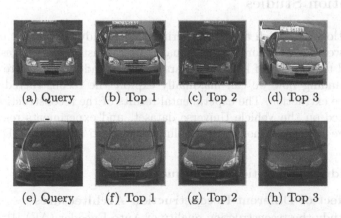

(a) Query   (b) Top 1   (c) Top 2   (d) Top 3

(e) Query   (f) Top 1   (g) Top 2   (h) Top 3

Fig. 5. Examples of SAVER failure on VD1 (sub-figures (a–d)) and VD2 (sub-figures (e–h)). The overall appearance of the query and top ranked images of the gallery are nearly identical. Visual cues such as windshield sticker placement are almost indistinguishable.

works without relying on any extra annotations or attributes. For the case of VeRi and VERI-Wild datasets, we also try the method of re-ranking suggested in [49] and achieved considerable mAP scores of 82.0 and 84.4 respectively.

Table 7. Comparison with recent state-of-the-arts methods

Method	VeRi			VehicleID					
	mAP(%)	CMC(%)		S CMC(%)		M CMC(%)		L CMC(%)	
		@1	@5	@1	@5	@1	@5	@1	@5
AAVER [16]	66.35	90.17	94.34	74.69	93.82	68.62	89.95	63.54	85.64
CCA [31]	68.05	91.71	96.90	75.51	91.14	73.60	86.46	70.08	83.20
BS [21]	67.55	90.23	96.42	78.80	**96.17**	73.41	**92.57**	69.33	**89.45**
AGNet [48]	71.59	95.61	96.56	71.15	83.78	69.23	81.41	65.74	78.28
VehicleX [45]	73.26	94.99	97.97	79.81	93.17	76.74	90.34	73.88	88.18
PRND[11]	74.3	94.3	**98.7**	78.4	92.3	75.0	88.3	74.2	86.4
Ours	**79.6**	**96.4**	98.6	**79.9**	95.2	**77.6**	91.1	**75.3**	88.3
Ours + Re-ranking	**82.0**	**96.9**	97.7						

Method	VERI-Wild									Vehicle-1M						VD1			VD2		
	S			M			L			S		M		L			CMC			CMC	
	mAP	@1	@5	mAP	@1	@5	mAP	@1	@5	@1	@5	@1	@5	@1	@5	mAP	@1	@5	mAP	@1	@5
BS[21]	70.54	84.17	95.30	62.83	78.22	93.06	51.63	69.99	88.45	-	-	-	-	-	-	87.48	-	-	84.55	-	-
AAVER[16]	62.23	75.80	92.70	53.66	68.24	88.88	41.68	58.69	81.59	-	-	-	-	-	-	-	-	-	-	-	-
TAMR [10]	-	-	-	-	-	-	-	-	-	95.95	99.24	94.27	98.86	92.91	98.30	-	-	-	-	-	-
Ours	**80.9**	**94.5**	**98.1**	**75.3**	**92.7**	**97.4**	**67.7**	**89.5**	**95.8**	95.5	98.0	**95.3**	**99.4**	**93.1**	**98.6**	**96.7**	**96.5**	**99.1**	**96.7**	**97.8**	**99.0**
Ours + Re-ranking	**84.4**	**95.3**	97.6																		

5   Ablation Studies

In this section, we design a set of experiments to study the impact of different neural network architectures on the quality of reconstructed images, and also understand the impact of key hyper-parameters. In addition, we are interested in understanding how we can maximally exploit the reconstructed images in deep feature extraction. The experimental results of the reconstruction network are evaluated on the Vehicle Universe dataset, and experiments regarding the deep feature extraction module are evaluated on VeRi and VehicleID datasets.

5.1   Residual Generation Techniques

5.1.1   Effect of Different Reconstruction Architectures

Here, we study the reconstruction quality of Auto-Encoder (AE) [1], VAE [19], and GAN [8] methods. Moreover, we study the use of Bilateral Filtering (BF) as a baseline for texture smoothing, subsequent residual generation and vehicle re-id. Figure 6 qualitatively illustrates the reconstruction of each method for a given vehicle identity. We notice that both AE and GAN models attempt to recreate

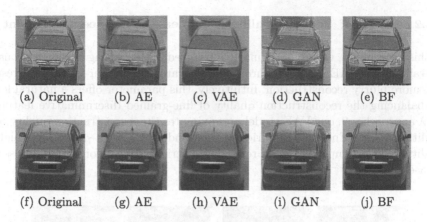

(a) Original   (b) AE   (c) VAE   (d) GAN   (e) BF

(f) Original   (g) AE   (h) VAE   (i) GAN   (j) BF

Fig. 6. Different image reconstruction methods.

fine-grained details, but often introduce additional distortions. Specifically, the GAN model generates new textures, modifies the logo and distorts the overall shape of the vehicle. As a result, GANs produce sharper images with various artifacts that diminish the quality of the residual image required by the re-id network. Also note that although bilateral filtering attempts to smooth images, it is unable to remove the critical details needed in residuals and vehicle re-id. The VAE is able to reconstruct the image by removing minute details and smoothing out textures. As a result, the VAE is able to generate the detailed residual maps needed for our proposed re-id method. Table 8 presents evaluation metrics on VeRi-766 and VehicleID for each of the generative models and bilateral filtering.

Table 8. Performance comparison of different image reconstruction methods

Method	Dataset							
	VeRi			VehicleID				
	mAP (%)	CMC (%)		S		M		L
				CMC (%)		CMC (%)		CMC (%)
		@1	@5	@1	@5	@1	@5	@1	@5
AE	79.0	96.0	98.2	79.0	93.9	76.8	90.5	74.9	87.9
VAE	**79.6**	**96.4**	**98.6**	**79.9**	**95.2**	**77.6**	**91.1**	**75.3**	**88.3**
GAN	78.3	95.6	98.1	78.5	93.0	75.6	89.1	73.4	85.7
BF	78.5	95.5	97.6	78.7	76.6	74.5	94.2	90.2	87.4

5.1.2   Effect of Scaling Kullbeck-Leibler Divergence Coefficient λ in Eq. 1

In this experiment, we are particularly interested in the scaling parameter λ used in training the VAE model. Figure 7 demonstrates how larger values of λ result in a more blurry reconstruction. Intuitively, this parameter offers a natural level for balancing the reconstruction quality of fine-grained discriminative features. As λ approaches 0, our VAE model approximates the reconstruction quality of a traditional Auto-Encoder. Empirically, we found that $\lambda = 1e-3$ produces higher quality vehicle templates, while removing discriminative information across all datasets.

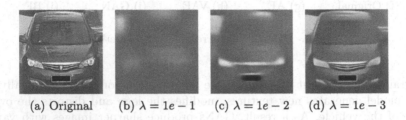

(a) Original   (b) $\lambda = 1e-1$ (c) $\lambda = 1e-2$ (d) $\lambda = 1e-3$

Fig. 7. Effect of scaling KL loss in image generation

5.2   Incorporating Residual Information

To effectively exploit complimentary information provided by the residuals, we design a set of four additional experiments on the VeRi and VehicleID datasets as follows:

A. We only feed the VAE reconstruction I_g as input to the re-id network. The purpose of this experiment is to understand how much critical information can be inferred from the VAE reconstruction.

B. We only feed the residual image I_r into the re-id pipeline. In this experiment we are interested to find out how much identity-dependent information can be extracted from only the residual image.

C. We use the residual maps to excite the actual image of the vehicle through point-wise matrix multiplication.

D. We concatenate the residual image with the original input image. Therefore, in this experiment we feed a six-channel tensor to the feature extraction module.

Table 9 presents the results of experiments A to D and highlights their performance against the baseline and SAVER models. In experiment A, the deep feature extractor is trained using the reconstructed image from the VAE. Intuitively, this method provides the lowest performance since all discriminating details are obfuscated. Interestingly, experiment B, training a deep feature extractor using only residual images, is able to perform nearly as well as our standard baseline. This reaffirms the idea that local information is essential for vehicle re-id. Experiment C performs considerably worse than the baseline model, indicating that point-wise multiplication with the sparse residual removes key information. Lastly, experiment D performs lower than our baseline. This can be attributed to the ImageNet [5] weight initialization, which is not well suited for six-channel images.

Table 9. Evaluation of different designs of employing residuals

Experiment	Dataset								
	VeRi		VehicleID						
	mAP (%)	CMC (%)	S		M		L		
			CMC (%)		CMC (%)		CMC (%)		
		@1	@5	@1	@5	@1	@5	@1	@5
A	67.5	91.4	96.4	64.2	80.6	62.9	76.3	59.4	73.5
B	77.5	94.5	98.2	77.9	92.7	74.7	89.0	73.4	86.2
C	71.4	91.9	96.4	76.3	92.6	73.3	86.8	70.7	83.5
D	75.7	94.8	98.3	78.9	93.1	75.3	89.2	73.3	86.1
Baseline	78.2	95.5	97.9	78.4	92.5	76.0	89.1	74.1	86.4
SAVER	**79.6**	**96.4**	**98.6**	**79.9**	**95.2**	**77.6**	**91.1**	**75.3**	**88.3**

6 Conclusion

In this paper we have shown the benefits of using simple, highly-scalable network architectures and training procedures to generate robust deep features for the task of vehicle re-identification. Our model highlights the importance of attending to discriminative regions without additional annotations, and outperforms existing state-of-the-art methods on benchmark datasets including VeRi, VehicleID, Vehicle-1M, and VeRi-Wild.

Acknowledgement. This research is supported in part by the Northrop Grumman Mission Systems Research in Applications for Learning Machines (REALM) initiative, and by the Office of the Director of National Intelligence (ODNI), Intelligence Advanced Research Projects Activity (IARPA), via IARPA R&D Contract No. D17PC00345. The views and conclusions contained herein are those of the authors and should not be interpreted as necessarily representing the official policies or endorsements, either expressed or implied, of ODNI, IARPA, or the U.S. Government. The U.S. Government is authorized to reproduce and distribute reprints for Governmental purposes notwithstanding any copyright annotation thereon.

References

1. Badrinarayanan, V., Kendall, A., Cipolla, R.: SegNet: a deep convolutional encoder-decoder architecture for image segmentation. IEEE Trans. Pattern Anal. Mach. Intell. **39**(12), 2481–2495 (2017)
2. Bai, Y., Lou, Y., Gao, F., Wang, S., Wu, Y., Duan, L.Y.: Group-sensitive triplet embedding for vehicle reidentification. IEEE Trans. Multimed. **20**(9), 2385–2399 (2018)
3. Chu, R., Sun, Y., Li, Y., Liu, Z., Zhang, C., Wei, Y.: Vehicle re-identification with viewpoint-aware metric learning. In: IEEE International Conference on Computer Vision (ICCV), pp. 8282–8291 (2019)
4. Cui, C., Sang, N., Gao, C., Zou, L.: Vehicle re-identification by fusing multiple deep neural networks. In: International Conference on Image Processing Theory, Tools and Applications (IPTA), pp. 1–6. IEEE (2017)
5. Deng, J., Dong, W., Socher, R., Li, L.J., Li, K., Fei-Fei, L.: ImageNet: a large-scale hierarchical image database. In: 2009 IEEE Conference on Computer Vision and Pattern Recognition, pp. 248–255. IEEE (2009)
6. Fan, X., Jiang, W., Luo, H., Fei, M.: SphereReID: deep hypersphere manifold embedding for person re-identification. J. Vis. Commun. Image Represent. **60**, 51–58 (2019)
7. Girshick, R., Radosavovic, I., Gkioxari, G., Dollár, P., He, K.: Detectron (2018). https://github.com/facebookresearch/detectron
8. Goodfellow, I., et al.: Generative adversarial nets. In: Advances in Neural Information Processing Systems, pp. 2672–2680 (2014)
9. Guo, H., Zhao, C., Liu, Z., Wang, J., Lu, H.: Learning coarse-to-fine structured feature embedding for vehicle re-identification. In: McIlraith, S.A., Weinberger, K.Q. (eds.) Proceedings of the Thirty-Second AAAI Conference on Artificial Intelligence, pp. 6853–6860. AAAI Press (2018)
10. Guo, H., Zhu, K., Tang, M., Wang, J.: Two-level attention network with multi-grain ranking loss for vehicle re-identification. IEEE Trans. Image Process. **28**(9), 4328–4338 (2019)
11. He, B., Li, J., Zhao, Y., Tian, Y.: Part-regularized near-duplicate vehicle re-identification. In: Proceedings of the IEEE Conference on Computer Vision and Pattern Recognition, pp. 3997–4005 (2019)
12. He, K., Zhang, X., Ren, S., Sun, J.: Deep residual learning for image recognition. In: 2016 IEEE Conference on Computer Vision and Pattern Recognition, CVPR 2016, pp. 770–778. IEEE Computer Society (2016)
13. Hermans, A., Beyer, L., Leibe, B.: In defense of the triplet loss for person re-identification. arXiv preprint arXiv:1703.07737 (2017)
14. Hsu, H.M., Huang, T.W., Wang, G., Cai, J., Lei, Z., Hwang, J.N.: Multi-camera tracking of vehicles based on deep features re-id and trajectory-based camera link models. In: IEEE Conference on Computer Vision and Pattern Recognition (CVPR), AI City Challenge Workshop (2019)
15. Huang, T.W., Cai, J., Yang, H., Hsu, H.M., Hwang, J.N.: Multi-view vehicle re-identification using temporal attention model and metadata re-ranking. In: IEEE Conference on Computer Vision and Pattern Recognition Workshops (CVPRW), pp. 434–442 (2019)
16. Khorramshahi, P., Kumar, A., Peri, N., Rambhatla, S.S., Chen, J.C., Chellappa, R.: A dual-path model with adaptive attention for vehicle re-identification. In: The IEEE International Conference on Computer Vision (ICCV), October 2019

17. Khorramshahi, P., Peri, N., Kumar, A., Shah, A., Chellappa, R.: Attention driven vehicle re-identification and unsupervised anomaly detection for traffic understanding. In: The IEEE Conference on Computer Vision and Pattern Recognition (CVPR) Workshops, June 2019
18. Kingma, D.P., Ba, J.: Adam: a method for stochastic optimization. arXiv preprint arXiv:1412.6980 (2014)
19. Kingma, D.P., Welling, M.: Auto-encoding variational bayes. In: 2nd International Conference on Learning Representations (2014)
20. Krause, J., Stark, M., Deng, J., Fei-Fei, L.: 3D object representations for fine-grained categorization. In: 4th International IEEE Workshop on 3D Representation and Recognition, 3DRR 2013 (2013)
21. Kumar, R., Weill, E., Aghdasi, F., Sriram, P.: Vehicle re-identification: an efficient baseline using triplet embedding. In: 2019 International Joint Conference on Neural Networks (IJCNN), pp. 1–9. IEEE (2019)
22. Liu, H., Tian, Y., Wang, Y., Pang, L., Huang, T.: Deep relative distance learning: tell the difference between similar vehicles. In: Proceedings of the IEEE Conference on Computer Vision and Pattern Recognition, pp. 2167–2175 (2016)
23. Liu, X., Zhang, S., Huang, Q., Gao, W.: RAM: a region-aware deep model for vehicle re-identification. In: IEEE International Conference on Multimedia and Expo (ICME), pp. 1–6. IEEE (2018)
24. Liu, X., Liu, W., Ma, H., Fu, H.: Large-scale vehicle re-identification in urban surveillance videos. In: IEEE International Conference on Multimedia and Expo, ICME, pp. 1–6. IEEE Computer Society (2016)
25. Liu, X., Liu, W., Mei, T., Ma, H.: A deep learning-based approach to progressive vehicle re-identification for urban surveillance. In: Leibe, B., Matas, J., Sebe, N., Welling, M. (eds.) ECCV 2016. LNCS, vol. 9906, pp. 869–884. Springer, Cham (2016). https://doi.org/10.1007/978-3-319-46475-6_53
26. Liu, X., Liu, W., Mei, T., Ma, H.: PROVID: progressive and multimodal vehicle reidentification for large-scale urban surveillance. IEEE Trans. Multimed. **20**(3), 645–658 (2017)
27. Lou, Y., Bai, Y., Liu, J., Wang, S., Duan, L.: VERI-Wild: a large dataset and a new method for vehicle re-identification in the wild. In: IEEE Conference on Computer Vision and Pattern Recognition, CVPR, pp. 3235–3243. Computer Vision Foundation/IEEE (2019)
28. Luo, H., Gu, Y., Liao, X., Lai, S., Jiang, W.: Bag of tricks and a strong baseline for deep person re-identification. In: Proceedings of the IEEE Conference on Computer Vision and Pattern Recognition Workshops (2019)
29. Lv, K., et al.: Vehicle reidentification with the location and time stamp. In: IEEE Conference on Computer Vision and Pattern Recognition Workshops (CVPRW) (2019)
30. Odena, A., Dumoulin, V., Olah, C.: Deconvolution and checkerboard artifacts. Distill **1**, e3 (2016)
31. Peng, J., Jiang, G., Chen, D., Zhao, T., Wang, H., Fu, X.: Eliminating cross-camera bias for vehicle re-identification. arXiv preprint arXiv:1912.10193 (2019)
32. Qian, J., Jiang, W., Luo, H., Yu, H.: Stripe-based and attribute-aware network: a two-branch deep model for vehicle re-identification. arXiv preprint arXiv:1910.05549 (2019)
33. Selvaraju, R.R., Cogswell, M., Das, A., Vedantam, R., Parikh, D., Batra, D.: Grad-CAM: visual explanations from deep networks via gradient-based localization. In: Proceedings of the IEEE International Conference on Computer Vision, pp. 618–626 (2017)

34. Shen, Y., Xiao, T., Li, H., Yi, S., Wang, X.: Learning deep neural networks for vehicle re-id with visual-spatio-temporal path proposals. In: IEEE International Conference on Computer Vision (ICCV), pp. 1900–1909 (2017)
35. Sochor, J., Špaňhel, J., Herout, A.: BoxCars: improving fine-grained recognition of vehicles using 3-D bounding boxes in traffic surveillance. IEEE Trans. Intell. Transp. Syst. **20**, 97–108 (2018)
36. Suprem, A., Lima, R.A., Padilha, B., Ferreira, J.E., Pu, C.: Robust, extensible, and fast: teamed classifiers for vehicle tracking in multi-camera networks. arXiv preprint arXiv:1912.04423 (2019)
37. Szegedy, C., Vanhoucke, V., Ioffe, S., Shlens, J., Wojna, Z.: Rethinking the inception architecture for computer vision. In: Proceedings of the IEEE Conference on Computer Vision and Pattern Recognition, pp. 2818–2826 (2016)
38. Tan, X., et al.: Multi-camera vehicle tracking and re-identification based on visual and spatial-temporal features. In: IEEE Conference on Computer Vision and Pattern Recognition Workshops (CVPRW), pp. 275–284 (2019)
39. Tang, Z., : PAMTRI: pose-aware multi-task learning for vehicle re-identification using highly randomized synthetic data. In: IEEE International Conference on Computer Vision (ICCV), pp. 211–220 (2019)
40. Tang, Z., et al.: CityFlow: a city-scale benchmark for multi-target multi-camera vehicle tracking and re-identification. In: IEEE Conference on Computer Vision and Pattern Recognition, CVPR, pp. 8797–8806. Computer Vision Foundation/IEEE (2019)
41. Wang, Z., et al.: Orientation invariant feature embedding and spatial temporal regularization for vehicle re-identification. In: Proceedings of the IEEE International Conference on Computer Vision, pp. 379–387 (2017)
42. Wu, F., Yan, S., Smith, J.S., Zhang, B.: Joint semi-supervised learning and re-ranking for vehicle re-identification. In: International Conference on Pattern Recognition (ICPR), pp. 278–283. IEEE (2018)
43. Yan, K., Tian, Y., Wang, Y., Zeng, W., Huang, T.: Exploiting multi-grain ranking constraints for precisely searching visually-similar vehicles. In: Proceedings of the IEEE International Conference on Computer Vision, pp. 562–570 (2017)
44. Yang, L., Luo, P., Loy, C.C., Tang, X.: A large-scale car dataset for fine-grained categorization and verification. In: IEEE Conference on Computer Vision and Pattern Recognition, CVPR, pp. 3973–3981. IEEE Computer Society (2015)
45. Yao, Y., Zheng, L., Yang, X., Naphade, M., Gedeon, T.: Simulating content consistent vehicle datasets with attribute descent. arXiv preprint arXiv:1912.08855 (2019)
46. Zhang, X., Zhang, R., Cao, J., Gong, D., You, M., Shen, C.: Part-guided attention learning for vehicle re-identification. arXiv preprint arXiv:1909.06023 (2019)
47. Zhang, Y., Liu, D., Zha, Z.J.: Improving triplet-wise training of convolutional neural network for vehicle re-identification. In: IEEE International Conference on Multimedia and Expo (ICME), pp. 1386–1391. IEEE (2017)
48. Zheng, A., Lin, X., Li, C., He, R., Tang, J.: Attributes guided feature learning for vehicle re-identification (2019)
49. Zhong, Z., Zheng, L., Cao, D., Li, S.: Re-ranking person re-identification with k-reciprocal encoding. In: Proceedings of the IEEE Conference on Computer Vision and Pattern Recognition, pp. 1318–1327 (2017)
50. Zhou, Y., Shao, L.: Viewpoint-aware attentive multi-view inference for vehicle re-identification. In: Proceedings of the IEEE Conference on Computer Vision and Pattern Recognition (CVPR), June 2018

Improving One-Stage Visual Grounding
by Recursive Sub-query Construction

Zhengyuan Yang[1]($\boxtimes$), Tianlang Chen[1], Liwei Wang[2], and Jiebo Luo[1]

[1] University of Rochester, Rochester, USA
{zyang39,tchen45,jluo}@cs.rochester.edu
[2] Tencent AI Lab, Bellevue, USA
liweiwang@tencent.com

Abstract. We improve one-stage visual grounding by addressing current limitations on grounding long and complex queries. Existing one-stage methods encode the entire language query as a single sentence embedding vector, *e.g.*, taking the embedding from BERT or the hidden state from LSTM. This single vector representation is prone to overlooking the detailed descriptions in the query. To address this query modeling deficiency, we propose a recursive sub-query construction framework, which reasons between image and query for multiple rounds and reduces the referring ambiguity step by step. We show our new one-stage method obtains $5.0\%, 4.5\%, 7.5\%, 12.8\%$ absolute improvements over the state-of-the-art one-stage approach on ReferItGame, RefCOCO, RefCOCO+, and RefCOCOg, respectively. In particular, superior performances on longer and more complex queries validates the effectiveness of our query modeling. Code is available at https://github.com/zyang-ur/ReSC.

Keywords: Visual grounding · Query modeling · Referring expressions

1 Introduction

Visual grounding aims to ground a natural language query onto a region of the image. There are mainly two threads of works in visual grounding: the two-stage approach [3,32,40,41,48,50] and one-stage approach [5,37,47]. Two-stage approaches first extract region proposals and then rank the proposals based on their similarities with the query. The recently proposed one-stage approach takes a different paradigm but soon becomes prevailing. The one-stage approach fuses visual-text features at image-level and directly predicts bounding boxes to ground the referred object. By densely sampling the possible object locations and reducing the redundant computation over region proposals, the one-stage methods [5,37,47] are simple, fast, and accurate.

Electronic supplementary material The online version of this chapter (https://doi.org/10.1007/978-3-030-58568-6_23) contains supplementary material, which is available to authorized users.

© Springer Nature Switzerland AG 2020
A. Vedaldi et al. (Eds.): ECCV 2020, LNCS 12359, pp. 387–404, 2020.
https://doi.org/10.1007/978-3-030-58568-6_23

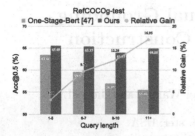
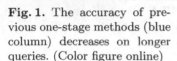

Fig. 1. The accuracy of previous one-stage methods (blue column) decreases on longer queries. (Color figure online)

Fig. 2. Previous one-stage methods' representative failure cases of (a) overlooking detailed descriptions, (b) misinterpreting the query by keywords. Blue/yellow boxes are the predicted regions [47]/ground truths. (Color figure online)

In this paper, we improve the state-of-the-art one-stage methods by addressing their weaknesses in modeling long and complex queries. The overall advantage of our method is shown in Fig. 1. Compared to the current state-of-the-art one-stage method [47], whose performance *drops dramatically on longer queries*, our approach achieves remarkably superior performance.

We analyze the limitations of current one-stage methods as follows. Existing one-stage methods [5,37,47] encode the entire query as a single embedding vector, such as directly adopting the first token's embedding ([CLS]) from BERT [8,47] or aggregating hidden states from LSTM [5,12,37,47]. The single vector is then concatenated at all spatial locations with visual features to obtain the fused features for grounding box prediction. Modeling the entire language query as a single embedding vector tends to increase representation ambiguity, such as focusing on some words, yet overlooking other important ones. Such a problem potentially causes the loss of referring information, especially on those long and complex queries. For example in Fig. 2 (a), the model seems to overlook detailed descriptions such as "sitting on the couch" or "looking tv," and grounds the wrong region with the head noun "man." As for Fig. 2 (b), the model appears to look into the wrong word "mountain" and grounds the target without full consideration of the modifier of "water." Neglecting the query modeling problem, thus, causes the performance drop on long queries for the one-stage approach.

Several two-stage visual grounding works [24,27,42,45,46,48] have studied a similar query modeling problem. The main idea of these works is to link object regions with the parsed sub-queries to have a comprehensive understanding of the referring. Among them, MattNet [48] parses the query into the subject, location, and relationship phrases, and links each phrase with the related object regions for matching score computing. NMTREE [24] parses the query with a dependency tree parser [2] and links each tree node with a visual region. DGA [46] parses the query with text self-attention and links the text with regions via dynamic graph attention. Though elegant enough, these methods are designed intuitively for two-stage methods, requiring candidate region features to be extracted at the first stage. Since the main benefit of doing one-stage visual grounding is to avoid explicitly extracting candidate region features for the sake of computational cost, the query modeling in two-stage methods cannot be directly applied to the one-stage framework [5,37,47]. Therefore, in this paper, to address the query modeling problem in a unified one-stage framework, we propose the recursive sub-query construction framework.

When presented with a referring problem such as Fig. 2 (a), humans tend to solve it by reasoning back-and-forth between the image and query for multiple rounds and recursively reduce the referring ambiguity, $i.e.$, the possible region that contains the referred object. Inspired by this, we proposed to represent the intermediate understanding of the referring in each round as the *text-conditional visual feature*, which starts as the image feature and updated after multiple rounds, ending up as the fused visual-text feature ready for box prediction. In each round, the model constructs a new sub-query as a group of words attended with scores to refine the text-conditional visual feature. Gradually, with multiple rounds, our model reduces the referring ambiguity. Such a multi-round solution is in contrast to previous one-stage approaches that try to remember the entire query and ground the region in a single round.

Our framework recursively constructs sub-queries to refine the grounding prediction. Each round faces with two core problems that facilitate recursive reasoning, namely 1) how to construct the sub-query; and 2) how to refine the text-conditional visual feature with the sub-query. We propose a sub-query learner and a sub-query modulation network to solve the above two problems, respectively. They work alternately and recursively to reduce the referring ambiguity. Using the text-conditional visual features in the last round, a final output stage predicts bounding boxes to grounding the referred object.

We benchmark our framework on the ReferItGame [17], RefCOCO [49], Ref-COCO+ [49], RefCOCOg [29] datasets, with $5.0\%, 4.5\%, 7.5\%, 12.8\%$ absolute improvements over the state-of-the-art one-stage method [47]. Meanwhile, our method runs fast at 38 FPS (26 ms). Moreover, the relative gain curve according to the query length changes in Fig. 1 shows the effectiveness of our approach in solving the aforementioned query modeling problem.

Our main contributions are:

- We improve one-stage visual grounding by addressing previous one-stage methods' limitations on grounding long and complex queries.
- We propose a recursive sub-query construction framework that recursively reduces the referring ambiguity with different constructed sub-queries.
- Our proposed method shows significantly improved results on multiple datasets and meanwhile maintains the real-time inference speed. Extensive experiments and ablations validate the effectiveness of our method.

2   Related Work

There exists two major categories of visual grounding methods: phrase localization [17,33,40] and referring expression comprehension [15,18,29,48,49]. Most previous visual grounding methods are composed of two stages. In the first stage, a number of region proposals are generated by an off-line module such as EdgeBox [53], selective search [39] or pretrained detectors [13,26,35]. In the second stage, each region is compared to the input query and outputs a similarity score. During inference, the region with the highest similarity score is output as the final prediction. Under the two-stage framework, various works explore different aspects to improve visual grounding, such as better exploiting attributes [25,27,48], object relationships [24,42,45,46], phrase co-occurrences [1,4,9], *etc.*

Recently, several works [5,20,28,37,47] propose a different paradigm of one-stage visual grounding. The primary motivation is to solve the two limitations of two-stage methods, *i.e.*, the performance cap caused by the sparsely sampled region proposals, and the slow inference speed caused by the region feature computation. Instead of explicitly extracting the features for all proposed regions, one-stage methods fuse the visual-text feature densely at all spatial locations, and directly predict bounding boxes to ground the target. Previous one-stage methods usually encode the query as a single language vector and concatenate the feature along the channel dimension of the visual feature. Despite the effectiveness of one-stage methods, modeling the language query as a single vector could lead to the loss of referring information, especially on long and complex queries. Though two-stage methods [24,27,42,45,46,48] had studied the similar problem of language query modeling, the explorations can not be directly applied to the one-stage approach given the intrinsic difference between two paradigms.

Besides, an intuitive alternative is to model the query phrase by the attention mechanism. Lin *et al.* [22] propose to extract sentence embedding with self-attention. Modeling query with attention mechanism is also explored in various vision-language tasks [19,48]. In experiments, we observe that our proposed multi-round solution performs better than the simple query attention method (*cf.* "Single/ Multi-head attention" and "Sub-query learner (ours)" in Table 3).

3   Approach

In this section, we will introduce our query modeling in a unified one-stage grounding framework. Previous one-stage grounding methods encode the language query as a C_l-dimension language feature and concatenate the text feature at all spatial locations with the visual feature $v \in R^{H \times W \times C_v}$. H, W, C_v are the height, width, and dimension of the visual feature. The visual feature and the text feature are usually mapped to the same dimension C before the concatenation. Extra convolutional layers then further refine the fused feature $f \in R^{H \times W \times 2C}$ and predict bounding boxes at each spatial location $H \times W$ to ground the target. Such single-round query modeling tends to overlook important query details and lead to incorrect predictions. The problems become increasingly severe on longer and more complex queries, as shown quantitatively in Fig. 1.

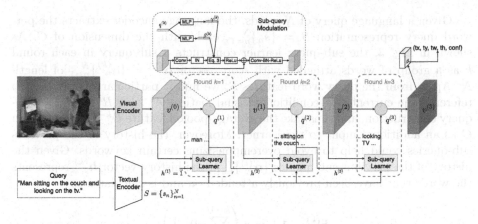

Fig. 3. The architecture for the recursive sub-query construction framework. In each round, the framework constructs a new sub-query to refine the text-conditional visual feature $v^{(k)}$ (shown in purple). $q^{(k)}$ is the feature for the constructed sub-query.

To address this problem in a unified one-stage grounding system, we propose a recursive sub-query construction framework that step by step refines the visual-text feature $v^{(k)}$ to get better prediction. As shown in Fig. 3, the initial feature $v^{(0)}$ is the image feature $v \in R^{H \times W \times C}$ encoded by the visual encoder [34]. In each round k, the framework constructs a new sub-query as a group of words attended with score vector $\alpha^{(k)}$, and obtains the sub-query embedding $q^{(k)}$ to refine the visual-text feature. The framework ends up with the refined feature $v^{(K)}$ after K rounds, and predicts bounding boxes on each spatial location of $v^{(K)}$ to ground the target. We name $v^{(k)}$ the text-conditional visual feature.

In each round, we address how to construct the sub-query and how to refine the feature $v^{(k)}$ with the sub-query embedding to better ground the target. For the first problem, we propose a sub-query learner in Sect. 3.1. The objective is to construct the sub-query that could best resolve the current referring ambiguity. We find it important to refer to the text-conditional visual feature $v^{(k)}$ in each round. For the second problem, we propose a sub-query modulation network that scales and shifts the feature $v^{(k)}$ with the sub-query feature. We introduce the details in Sect. 3.2. The sub-query learner and the modulation network operate alternately for multiple rounds, and recursively reduce the referring ambiguity. The feature $v^{(K)}$ in the last round contains the full object referring information and is used for target box prediction.

3.1 Sub-query Learner

Our method addresses the visual grounding problem as a multi-round reasoning process. In each round, the sub-query learner refers to the text-conditional visual feature $v^{(k)}$ and constructs the sub-query that could gradually reduce the referring ambiguity.

Given a language query of N words, the language encoder extracts the per-word query representation $S = \{s_n\}_{n=1}^N$, each with the dimension of C. As shown in Fig. 3, the sub-query learner constructs a sub-query in each round k as a group of words attended with score vector $\boldsymbol{\alpha}^{(k)} = \{\alpha_n^{(k)}\}_{n=1}^N$ of length N. Apart from the query word features S, we find it particularly important to reference the current text-conditional visual feature $v^{(k-1)} \in R^{H \times W \times C}$ in sub-query construction, and thus take the average-pooled feature $\bar{v}^{(k-1)}$ of dimension C as an additional input to the learner. Moreover, the history of the previous sub-queries could help to avoid overemphasizing certain keywords. Given the history of the previous sub-queries $\{\boldsymbol{\alpha}^{(i)}\}_{i=1}^{k-1}$, the history vector $\boldsymbol{h}^{(k)}$ represents the words that have been previously attended on and is calculates as

$$
\boldsymbol{h}^{(k)} = \boldsymbol{1} - \min\left(\sum_{i=1}^{k-1} \boldsymbol{\alpha}^{(i)}, \boldsymbol{1}\right),
$$

where $\boldsymbol{1}$ is the all-ones vector. Both $\boldsymbol{h}^{(k)}$ and $\boldsymbol{\alpha}^{(i)}$ are N-Dimension vectors with values range from 0 to 1. The sub-query learner takes the query word feature $\{s_n\}_{n=1}^N$, the text-conditional visual feature $\bar{v}^{(k-1)}$, and the history vector $\boldsymbol{h}^{(k)} = \{h_n^{(k)}\}_{n=1}^N$ to construct the sub-query for round k by predicting score vector $\boldsymbol{\alpha}^{(k)}$:

$$
\alpha_n^{(k)} = softmax\left[W_{a1}^{(k)} \tanh\left(W_{a0}^{(k)} h_n^{(k)}(\bar{v}^{(k-1)} \odot s_n) + b_{a0}^{(k)}\right) + b_{a1}^{(k)}\right], \qquad (1)
$$

where $\odot$ represents hadamard product, and $W_{a0}, b_{a0}, W_{a1}, b_{a1}$ are learnable parameters. We compute the softmax over the N attention scores.

To guide the multi-round reasoning, explicit regularization is imposed on the word attention scores. Intuitively, the constructed sub-queries at each round should focus on different elements of the query, and in the end, most words in the query should have been looked at. Therefore, we add two regularization terms:

$$
L_{div} = \left\|A^T A \odot (\boldsymbol{1} - I)\right\|_F^2, \quad L_{cover} = \left\|\boldsymbol{1} - \min\left(\sum_{i=1}^K \boldsymbol{\alpha}^{(i)}, \boldsymbol{1}\right)\right\|_1, \qquad (2)
$$

where matrix A is the predicted attention score matrix $A = \{\alpha_n^{(k)}\}_{n,k=1,1}^{N,K}$, $\boldsymbol{1}$ is the matrix of ones and I is an identity matrix. L_{div} avoids any words being focused on in more than one round and thus enforces the diversity. L_{cover} helps the model looks at all words in the query and thus improves the coverage.

The adopted technique of attention-based sub-query learning is related to previous compositional reasoning studies [16,46]. The major difference is that our method refers to the text-conditional visual feature $v^{(k)}$ to recursively construct the sub-query in each round. In contrast, the sub-query learning in previous studies [16,46] is purely based on the word feature $\{s_n\}_{n=1}^N$, and generates all sub-queries in prior to visual-text fusion.

3.2 Sub-query Modulation

In each round, the sub-query learner constructs a sub-query as a group of words attended by a score vector $\alpha^{(k)}$, and generates the sub-query feature $q^{(k)}$ as

$$q^{(k)} = \sum_{n=1}^{N} \alpha_n^{(k)} s_n.$$

The goal for the sub-query modulation is to refine the text-conditional visual feature $v^{(k-1)}$ with the new sub-query feature $q^{(k)}$, such that the refined feature $v^{(k)}$ performs better in grounding box prediction.

Inspired by conditional normalization on image-level tasks [7,10,31], we encode the sub-query representation $q^{(k)}$ to modulate the previous visual-text representation $v^{(k-1)}$ by scaling and shifting. To be specific, $q^{(k)}$ is projected into a scaling vectors $\gamma^{(k)}$ and a shifting vector $\beta^{(k)}$ with two MLPs respectively:

$$\gamma^{(k)} = \tanh\left(W_\gamma^{(k)} q^{(k)} + b_\gamma^{(k)}\right), \quad \beta^{(k)} = \tanh\left(W_\beta^{(k)} q^{(k)} + b_\beta^{(k)}\right).$$

The text-conditional visual feature $v^{(k)}$ is then refined from $v^{(k-1)}$ with the two modulation vectors and extra learnable parameters:

$$v^{(k)}(i,j) = f_2\left\{ ReLU\left[f_1(v^{(k-1)}(i,j)) \odot \gamma^{(k)} + \beta^{(k)}\right] + v^{(k-1)}(i,j)\right\}, \quad (3)$$

where (i,j) are the spatial coordinates, f_1, f_2 are learnable mapping layers as shown in Fig. 3. f_1 consists of a 1×1 convolution followed by an instance normalization layer. f_2 consists of a 3×3 convolution followed by a batch normalization layer and ReLU activation. The grounding module takes the text-conditional visual feature in the final round $v^{(K)}$ as input, and predicts bounding boxes to ground the referred object. With the extra referring information in each sub-query $q^{(k)}$, we expect the modulation in each round to strength the feature of the referred object, and meanwhile suppress the ones for distracting objects and the background.

Our proposed sub-query modulation in Eq. 3 has shared modulation vectors $\gamma^{(k)}, \beta^{(k)}$ over all spatial locations (i,j). One intuitive alternative is to predict different modulation vectors $\gamma^{(k)}(i,j), \beta^{(k)}(i,j)$ for each spatial location. This can be done by constructing sub-queries $\alpha^{(k)}(i,j)$ for each location with the corresponding text-conditional visual feature $v^{(k-1)}(i,j)$. Despite using different modulation parameters at different spatial locations seems more intuitive, we show that the modulation along the channel dimension achieves the same objective and meanwhile is computationally efficient (cf. "Spatial-independent sub-query" and "Sub-query learner (ours)" in Table 3).

3.3 Framework Details

Visual and Text Feature Encoder. We resize the input image to $3 \times 256 \times 256$ and use Darknet-53 [34] pretrained on COCO object detection [21] as the visual encoder. We adopt the visual feature from the 102-th convolutional layer that has a dimension $32 \times 32 \times 256$. We map the raw visual feature into the visual input $v^{(0)}$ with a 1×1 convolutional layer with batch normalization and ReLU. We set the shared dimension $C = 512$.

We encode the each word in the query as a $768D$ vector with the uncased base version of BERT [8,43]. We sum the representations for each word in the last four layers and map the features with two fully connected layers to obtain the representation $S = \{s_n\}_{n=1}^{N}$. N is the number of query words and does not include special tokens such as [CLS], [SEP] and [PAD].

Grounding Module. The grounding module takes the visual-text feature $v^{(K)}$ as input and generates object prediction at each spatial location to ground the target. We use the same two 1×1 convolutional layers as in One-Stage-BERT [47] for box prediction. There are $32 \times 32 = 1024$ spatial locations, and we predict 9 anchor boxes at each location. We follow the anchor selection steps in a previous one-stage grounding method [47] with the same anchor boxes used. For each one of the $1024 \times 9 = 9216$ anchor boxes, we predict the relative offset and confidence score. A cross-entropy loss between the softmax over all the 9216 boxes and the one-hot target center vector, a regression loss of the relative location and size offset, and the regularization in Eq. 2 are used to train the model. We use the same classification and regression losses as in One-Stage-BERT [47].

4   Experiments

4.1   Datasets

RefCOCO/RefCOCO+/RefCOCOg. RefCOCO [49], RefCOCO+ [49], and RefCOCOg [29] are three visual grounding datasets with images and referred objects selected from MSCOCO [21]. The referred objects are selected from the MSCOCO object detection annotations and belong to 80 object classes. Ref-COCO [49] has 19,994 images with 142,210 referring expressions for 50,000 object instances. RefCOCO+ has 19,992 images with 141,564 referring expressions for 49,856 object instances. RefCOCOg has 25,799 images with 95,010 referring expressions for 49,822 object instances. On RefCOCO and RefCOCO+, we follow the split [49] of train/ validation/ testA/ testB that has 120,624/ 10,834/ 5,657/ 5,095 expressions for RefCOCO and 120,191/ 10,758/ 5,726/ 4,889 expressions for RefCOCO+, respectively. Images in "testA" are of multiple people and the ones in "testB" contain all other objects. The queries in RefCOCO+ contains no absolute location words, such as "on the right" that describes the object's location in the image. On RefCOCOg, we experiment with the splits of RefCOCOg-google [29] and RefCOCOg-umd [30], and refer to the splits as the val-g, val-u, and test-u in Table 1. The queries in RefCOCOg are generally longer than those in RefCOCO and RefCOCO+: the average lengths are 3.61, 3.53, 8.43, respectively, for RefCOCO, RefCOCO+, RefCOCOg.

ReferItGame. The ReferItGame dataset [17] has 20,000 images from SAIAPR-12 [11]. We follow a cleaned version of split [4,15,36], which has 54,127, 5,842, and 60,103 referring expressions in train, validation, and test set, respectively.

Table 1. The performance comparisons (Acc@0.5%) on RefCOCO, RefCOCO+, Ref-COCOg (upper table), and ReferItGame, Flickr30K Entities (lower table). We highlight the best one-stage performance with **bold** and the best two-stage performance with underline. The *COCO-trained detector* generates ideal proposals only for images in COCO. This leads to the two-stage methods' good performance on RefCOCO, Ref-COCO+, RefCOCOg, as well as the accuracy drop on other datasets (lower table).

Method	Feature	RefCOCO			RefCOCO+			RefCOCOg			Time
		val	testA	testB	val	testA	testB	val-g	val-u	test-u	(ms)
Two-stage Methods											
MMI [29]	VGG16-Imagenet	-	64.90	54.51	-	54.03	42.81	45.85	-	-	-
Neg Bag [30]	VGG16-Imagenet	-	58.60	56.40	-	-	-	-	-	49.50	-
CMN [14]	VGG16-COCO	-	71.03	65.77	-	54.32	47.76	57.47	-	-	-
ParallelAttn [52]	VGG16-Imagenet	-	75.31	65.52	-	61.34	50.86	58.03	-	-	-
VC [51]	VGG16-COCO	-	73.33	67.44	-	58.40	53.18	62.30	-	-	-
LGRAN [42]	VGG16-Imagenet	-	76.6	66.4	-	64.0	53.4	61.78	-	-	-
SLR [50]	Res101-COCO	69.48	73.71	64.96	55.71	60.74	48.80	-	60.21	59.63	-
MAttNet [48]	Res101-COCO	76.40	80.43	69.28	64.93	70.26	56.00	-	66.67	67.01	320
DGA [46]	Res101-COCO	-	78.42	65.53	-	69.07	51.99	-	-	63.28	341
One-stage Methods											
SSG [5]	Darknet53-COCO	-	76.51	67.50	-	62.14	49.27	47.47	58.80	-	25
One-Stage-BERT [47]	Darknet53-COCO	72.05	74.81	67.59	55.72	60.37	48.54	48.14	59.03	58.70	23
One-Stage-BERT*	Darknet53-COCO	72.54	74.35	68.50	56.81	60.23	49.60	56.12	61.33	60.36	23
Ours-Base	Darknet53-COCO	76.59	78.22	**73.25**	63.23	66.64	55.53	60.96	64.87	64.87	26
Ours-Large	Darknet53-COCO	**77.63**	**80.45**	72.30	**63.59**	**68.36**	**56.81**	**63.12**	**67.30**	**67.20**	36

Method	Feature	ReferItGame test	Flickr30K Entities test	Time (ms)
Two-stage Methods				
CMN [14]	VGG16-COCO	28.33	-	-
VC [51]	VGG16-COCO	31.13	-	-
MAttNet [48]	Res101-COCO	29.04	-	320
Similarity Net [40]	Res101-COCO	34.54	60.89	184
CITE [32]	Res101-COCO	35.07	61.33	196
One-stage Methods				
SSG [5]	Darknet53-COCO	54.24	-	25
ZSGNet [37]	Res50-FPN	58.63	63.39	-
One-Stage-BERT [47]	Darknet53-COCO	59.30	68.69	23
One-Stage-BERT*	Darknet53-COCO	60.67	68.71	23
Ours-Base	Darknet53-COCO	64.33	69.04	26
Ours-Large	Darknet53-COCO	**64.60**	**69.28**	36

Flickr30K Entities. Flickr30K Entities [33] has 31,783 images with 427K referred entities. We follow the same split in previous works [32,33,40]. We note that the queries in Flickr30K Entities are mostly short noun phrases and do not well reflect the difficulty of comprehensive phrase understanding. We still benchmark our method on Flickr30K Entities and compare it with other baselines for experiments completeness.

4.2 Implementation Details

Training. Following the standard setting [34,47], we keep the aspect ratio of the input image and resize the long edge to 256. We then pad the resized image to a size of 256×256 with the mean pixel value. We follow the data augmentation in previous one-stage studies [34,47]. The RMSProp [38] optimizer, with an initial learning rate of 10^{-4} is used to train the model with a batch size of 8. The learning rate decreases by half every 10 epochs for a total of 100 epochs. We set the weight for L_{div}, L_{cov} as 1. We select $K = 3$ as the default number of rounds and defer the related ablation studies to supplementary materials.

Evaluation. We follow the same *Acc@0.5* evaluation protocol in prior works [33,36]. Given a language query, this metric is to consider the predicted region correct if its IoU is at least 0.5 with the ground truth bounding box.

4.3   Quantitative Results

Experiment Settings. Table 1 reports visual grounding results on RefCOCO, RefCOCO+, RefCOCOg (the upper table), and ReferItGame, Flickr30K Entities (the lower table). The *top part* of each table contain results of the state-of-the-art two-stage visual grounding methods [14,27,29,30,32,40,42,46,48,50–52]. The *"Feature"* column lists the backbone and pretrained dataset used for proposal feature extraction. COCO-trained Faster-RCNN [35] detector is used for region proposal generation for the experiments on RefCOCO [49], RefCOCO+ [49] and RefCOCOg [29]. We quote the two-stage methods' results on ReferItGame and Flickr30K Entities reported by SSG [5] and One-Stage-BERT [47] where Edgebox [53] is used for proposal generation.

The *bottom part* of Table 1 compares the performance of our method to other state-of-the-art one-stage methods [5,37,47]. The *"Feature"* column shows the adopted visual backbone and its pretrained dataset, if any. For a fair comparison, we modify One-Stage-BERT [47] to have the exact same training details as ours, and observe a small accuracy improvement by the modification. Specifically, we 1). encode the query as the averaged BERT word embedding instead of the BERT sentence embedding at the first token's position ([CLS]), 2). remove the feature pyramid network, and 3). follow the implementation details in Sect. 4.2. We refer to the modified version *"One-Stage-BERT*."* Other than the state-of-the-art, we design and compare to additional alternatives to our methods such as *"single/ multi-head attention query modeling," "per-word sub-query,"* etc., in Sect. 4.5 and Table 3.

We obtain our main results by the method described in Sect. 3 and refer to it as *"Ours-Base"* in Table 1. Furthermore, we observe that a larger input image size of 512 and a ConvLSTM [6,44] grounding module increase the accuracy, but meanwhile slightly slow the inference speed. We refer to the corresponding model as *"Ours-Large"* and analyze each modification in supplementary materials.

Visual Grounding Results. Our proposed method outperforms the state-of-the-art one-stage grounding methods [5,37,47] by over 5% absolute accuracy on all experimented datasets.

The two-stage methods [27,42,46,48,51,52] also show good performance on COCO-series datasets (*i.e.*, RefCOCO, RefCOCO+, and RefCOCOg) by using the COCO-trained detector [35]. For example, we notice that MAttNet [48] achieves comparable performance with our best model in RefCOCO+, though our best model obviously surpasses MAttNet on RefCOCO, RefCOCOg, and the testB of RefCOCO+. However, in the ReferItGame dataset, as listed in the lower part of Table 1, MAttNet's accuracy drops dramatically. The findings of previous one-stage work [5,47] show that two-stage visual grounding methods rely highly on the region proposals quality. Since RefCOCO/ RefCOCO+/ RefCOCOg are subsets of COCO and have shared images and objects, the COCO-trained detector generates nearly perfect region proposals on COCO-series datasets. When used in other datasets, e.g., ReferItGame and Flickr30K Entities datasets, their proposal quality and grounding accuracy drop, such as the MAttNet's degraded performance in ReferItGame. Nonetheless, our method performs stably across all datasets and, meanwhile being significantly faster.

Inference Time. The real-time inference speed is one major advantage of the one-stage visual grounding method. We conduct all the experiments on a single NVIDIA 1080TI GPU. We observe our method achieves a real-time inference speed of 26 ms. The method is more than 10 times faster than typical two-stage methods such as the MattNet [48] of 320 ms.

4.4 Performance Break-Down Studies

We show the effectiveness of our method in modeling long queries by breaking down the test set. We split the test set of ReferItGame [17], RefCOCO [49], RefCOCO+ [49], and RefCOCOg [29] each into four sub-sets based on the query lengths (we combine the testA and testB for RefCOCO and RefCOCO+). Table 2 compares our method to One-Stage-BERT [47] on the generated sub-sets. We adopt "Ours-Base" instead of "Ours-Large" for comparison because the inference speed of "Ours-Base" is more comparable with One-Stage-BERT. The first row shows the experimented dataset and the number of query words in each sub-set. The second row shows the portion of samples in each subset. We generate sub-sets that are roughly with the same size. The middle two rows compare the accuracy of our method to the state-of-the-art one-stage grounding method [47]. The last row computes the relative gain obtained by our method as (*Ours* − *Base*) /*Base*. We observe a larger relative gain of our method on longer queries. The relative gain is around 20% on the longest query sub-set. The consistent increases in the relative gain on all experimented datasets suggest the effectiveness of our recursive sub-query construction framework in modeling and grounding long queries.

4.5 Ablation Studies

In this section, we conduct ablation studies to understand our method better. We perform the study on RefCOCOg-google [29] as it has, on average, longer queries than other datasets, which can better reflect the query modeling problem.

Query Modeling. Table 3 shows the ablation studies on different query modeling choices. Specifically, we systematically study the following settings.

Table 2. The performance break-down with query lengths. The first row shows the experimented dataset and the number of query words in each sub-set.

RefCOCO	1-2	3	4-5	6+	RefCOCO+	1-2	3	4-5	6+
Percent (%)	36.22	23.87	25.60	14.30	Percent (%)	37.79	19.48	27.40	15.33
One-Stage-BERT	77.68	76.04	66.98	55.59	One-Stage-BERT	66.59	55.42	47.40	39.03
Ours-Base	79.35	79.28	72.65	66.19	Ours-Base	71.08	60.01	56.24	49.35
Relative Gain	2.15	4.26	8.46	19.07	**Relative Gain**	6.74	8.28	18.65	26.44

RefCOCOg	1-5	6-7	8-10	11+	ReferItGame	1	2	3-4	5+
Percent (%)	23.54	22.80	28.30	25.37	Percent (%)	25.78	16.76	31.53	25.93
One-Stage-BERT	63.41	59.57	56.97	55.46	One-Stage-BERT	82.33	66.66	56.64	34.89
Ours-Base	65.49	65.37	63.97	64.86	Ours-Base	82.12	69.46	61.43	46.84
Relative Gain	3.28	9.74	12.29	16.95	**Relative Gain**	-0.26	4.20	8.46	34.25

- **Average vector.** We average the BERT embedded word features $S = \{s_n\}_{n=1}^{N}$ to form a single $512D$ vector as the query representation.
- **Per word sub-query.** We consider each word as a sub-query. The per-word sub-query modeling is used by RMI [23] for referring image segmentation.
- **Single-head attention.** A set of self-attention scores of size N is learned from the word features $\{s_n\}_{n=1}^{N}$. We obtain a single query feature vector by weighted sum the BERT embedded word features. We use the same self-attention method as in Eq. 1 to obtain the attention scores [22], expect only the text feature s_n is used as the input.
- **Multi-head attention.** Self-attention scores of size $N \times K$ are learned from the word features. One unique sub-query feature vector is formed as the input to each round.
- **Spatial-independent sub-query.** We discuss one alternative to our approach in the end of Sect. 3.2. We refer to it as "Spatial-independent sub-query" as shown in the last row of Table 3.
- **Sub-query learner (ours).** Instead of jointly predicting the sub-queries for all steps, our proposed sub-query learner, as introduced in Sect. 3.1, recursively constructs the sub-query by referring to the current text-conditional visual feature $v^{(k)}$.

Our proposed sub-query learner boosts the baseline accuracy with no attention by 1.7% (*cf.* "Average vector" and "Sub-query learner (ours)"). The query attention without the visual contents shows limited improvements over the no attention baseline (*cf.* "Average vector" and "Single/ Multi-head attention"). Instead, by referring to the text-conditional visual feature in each round, our proposed sub-query learner further improve the attention baseline by 1.5% (*cf.* "Single/ Multi-head attention" and "Sub-query learner (ours)"). This shows the importance of recursive sub-query construction. Furthermore, "spatial-independent sub-query" constructs the sub-query independently at each spatial location. This alternative leads to extra computation while is not more accurate.

Sub-query Modulation. We compare with the "Concat-Conv" fusion used in One-Stage-BERT [47]. To be specific, the query feature is duplicated spatially and is concatenated with the visual and spatial features to form a

Table 3. Ablation studies on query modeling. The sub-query modulation introduced in Sect. 3.2 is used for fusion.

Query modeling	Acc@0.5
Average vector	59.24
Per word sub-query	58.36
Single-head attention	59.43
Multi-head attention	59.25
Spatial-independent sub-query	60.81
Sub-query learner (ours)	**60.96**

$512 + 512 + 8 = 1032D$ feature. One 1×1 and one 3×3 convolution layers then generate a $512D$ fused feature. In contrast, the sub-query modulation introduced in Sect. 3.2 converts the sub-query feature into scaling and shifting parameters to refine the text-conditional visual feature $v^{(k)}$. Our proposed sub-query modulation improves the accuracy by 1.8% with the similar amount of fusion parameters (*cf.* "Concat-Conv": 59.20% and "Ours": 60.96%).

(a). persons head with drill in the middle.   (b). man sitting on the couch and looking on the tv.   (c). the man in tan shirt in the back.   (d). the rail car on the other track.   (e). the water in the background below the mountain.   (f). white thing beside girl with red hair.

Fig. 4. Failure cases of One-Stage-BERT [47] (top row) that can be corrected by our method (bottom row). Blue/ yellow boxes are the predicted regions/ground truths. Constructed sub-queries and per-round visualizations are in Fig. 5. (Color figure online)

4.6   Qualitative Results

Figure 4 shows the failure cases made by a previous one-stage method [47] that can be correctly predicted by our method. We observe that previous methods appear to fail because of neglecting those detailed descriptions or modifiers (*e.g.*, Figs. 4 (a)–(d)), or attending on the wrong keyword (*e.g.*, Figs. 4 (e), (f)). In contrast, our method corrects these errors by better modeling the query.

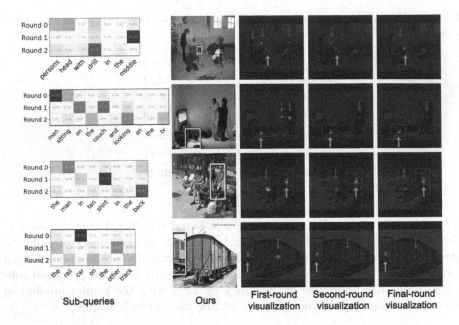

Fig. 5. Visualization of the constructed sub-queries and the intermediate text-conditional visual feature at each round. The green up arrow and the red down arrow point to the target and the major distracting object on heatmaps, respectively. Bested viewed in color. More examples and detailed analyses are in supplementary materials. (Color figure online)

More importantly, in order to understand what happened inside our model and explain why it works, we show the visualization of intermediate results of our model in Fig. 5. The left column in Fig. 5 shows the constructed sub-query in each round. The right three columns visualize the intermediate text-conditional visual feature $v^{(k)}$. For visualization, we adopt an extra output head over the feature $v^{(k)}$ in all steps and obtain the confidence score heatmaps. The confidence score indicates the probability of the object center. Therefore, the heatmap contains the peaks of object centers instead of the object contours. We highlight the referred object and the major distracting object with the green up arrow and red down arrow, respectively. We note that the intermediate prediction is just for visualization purpose, and is not in our proposed framework.

We observe that the model tends to focus on all head nouns in the first round, e.g., "man", "head", "car", etc., because such keywords are the most informative sub-query when given a raw image. Then, in the next few rounds, our method can refine the intermediate text-conditional visual representation and reduce the referring ambiguity. For example, in the first row of Fig. 5, the model first focuses on the head-noun "head". Our model refines its prediction by the constructed sub-queries "in the middle" and "with drill" in the following rounds. Accordingly, from the heatmap visualization, we observe such a disambiguation process that the refined visual-text feature step by step generates more accurate and confident predictions. In the first round with the sub-query "persons head," the model predicts four peaks in the heatmap, each centering at an appeared person. In the second round with sub-query "in the middle," the model focuses on the two person in the middle and eliminates two distracting objects. In the final round with the sub-query "with drill," the model successfully focuses on the referred person, and the heatmap values for all other distracting objects are greatly suppressed. We observe similar recursive disambiguation processes in other examples in Fig. 5 and supplementary materials.

5   Conclusions

We have proposed a recursive sub-query construction framework to address the limitation of previous one-stage methods when understanding complex queries. We recursively construct sub-queries to refine the visual-text feature for grounding box prediction. Extensive experiments and ablation studies have validated the high effectiveness of our method. Our proposed framework significantly outperforms the state-of-the-art one-stage methods by over 5% in absolute accuracy on multiple datasets while still maintaining a real-time inference speed.

Acknowledgment. This work is supported in part by NSF awards IIS-1704337, IIS-1722847, and IIS-1813709, Twitch Fellowship, as well as our corporate sponsors.

References

1. Bajaj, M., Wang, L., Sigal, L.: G3raphground: Graph-based language grounding. In: ICCV (2019)
2. Chen, D., Manning, C.D.: A fast and accurate dependency parser using neural networks. In: EMNLP, pp. 740–750 (2014)
3. Chen, K., Kovvuri, R., Gao, J., Nevatia, R.: MSRC: multimodal spatial regression with semantic context for phrase grounding. In: Proceedings of the 2017 ACM on International Conference on Multimedia Retrieval, pp. 23–31. ACM (2017)
4. Chen, K., Kovvuri, R., Nevatia, R.: Query-guided regression network with context policy for phrase grounding. In: ICCV (2017)
5. Chen, X., Ma, L., Chen, J., Jie, Z., Liu, W., Luo, J.: Real-time referring expression comprehension by single-stage grounding network. arXiv preprint arXiv:1812.03426 (2018)
6. Choy, C.B., Xu, D., Gwak, J.Y., Chen, K., Savarese, S.: 3D-R2N2: a unified approach for single and multi-view 3D object reconstruction. In: Leibe, B., Matas, J., Sebe, N., Welling, M. (eds.) ECCV 2016. LNCS, vol. 9912, pp. 628–644. Springer, Cham (2016). https://doi.org/10.1007/978-3-319-46484-8_38
7. De Vries, H., Strub, F., Mary, J., Larochelle, H., Pietquin, O., Courville, A.C.: Modulating early visual processing by language. In: Advances in Neural Information Processing Systems, pp. 6594–6604 (2017)
8. Devlin, J., Chang, M.W., Lee, K., Toutanova, K.: Bert: Pre-training of deep bidirectional transformers for language understanding. arXiv preprint arXiv:1810.04805 (2018)

9. Dogan, P., Sigal, L., Gross, M.: Neural sequential phrase grounding (seqground). In: CVPR (2019)
10. Dumoulin, V., Shlens, J., Kudlur, M.: A learned representation for artistic style. In: ICLR (2017)
11. Escalante, H.J., et al.: The segmented and annotated IAPR TC-12 benchmark. CVIU (2010)
12. Greff, K., Srivastava, R.K., Koutník, J., Steunebrink, B.R., Schmidhuber, J.: LSTM: a search space odyssey. IEEE Trans. Neural Networks Learn. Syst. **28**, 2222–2232 (2016)
13. He, K., Gkioxari, G., Dollár, P., Girshick, R.: Mask R-CNN. In: ICCV (2017)
14. Hu, R., Rohrbach, M., Andreas, J., Darrell, T., Saenko, K.: Modeling relationships in referential expressions with compositional modular networks. In: CVPR (2017)
15. Hu, R., Xu, H., Rohrbach, M., Feng, J., Saenko, K., Darrell, T.: Natural language object retrieval. In: CVPR (2016)
16. Hudson, D.A., Manning, C.D.: Compositional attention networks for machine reasoning. In: ICLR (2018)
17. Kazemzadeh, S., Ordonez, V., Matten, M., Berg, T.: Referitgame: referring to objects in photographs of natural scenes. In: EMNLP (2014)
18. Li, J., et al.: Deep attribute-preserving metric learning for natural language object retrieval. In: Proceedings of the 25th ACM International Conference on Multimedia, pp. 181–189. ACM (2017)
19. Li, S., Xiao, T., Li, H., Zhou, B., Yue, D., Wang, X.: Person search with natural language description. In: Proceedings of the IEEE Conference on Computer Vision and Pattern Recognition, pp. 1970–1979 (2017)
20. Liao, Y., et al.: A real-time cross-modality correlation filtering method for referring expression comprehension. In: Proceedings of the IEEE/CVF Conference on Computer Vision and Pattern Recognition, pp. 10880–10889 (2020)
21. Lin, T.-Y., et al.: Microsoft COCO: common objects in context. In: Fleet, D., Pajdla, T., Schiele, B., Tuytelaars, T. (eds.) ECCV 2014. LNCS, vol. 8693, pp. 740–755. Springer, Cham (2014). https://doi.org/10.1007/978-3-319-10602-1_48
22. Lin, Z., et al.: A structured self-attentive sentence embedding. In: ICLR (2017)
23. Liu, C., Lin, Z., Shen, X., Yang, J., Lu, X., Yuille, A.: Recurrent multimodal interaction for referring image segmentation. In: ICCV (2017)
24. Liu, D., Zhang, H., Wu, F., Zha, Z.J.: Learning to assemble neural module tree networks for visual grounding. In: ICCV (2019)
25. Liu, J., Wang, L., Yang, M.H.: Referring expression generation and comprehension via attributes. In: ICCV (2017)
26. Liu, W., et al.: SSD: single shot MultiBox detector. In: Leibe, B., Matas, J., Sebe, N., Welling, M. (eds.) ECCV 2016. LNCS, vol. 9905, pp. 21–37. Springer, Cham (2016). https://doi.org/10.1007/978-3-319-46448-0_2
27. Liu, X., Wang, Z., Shao, J., Wang, X., Li, H.: Improving referring expression grounding with cross-modal attention-guided erasing. In: CVPR (2019)
28. Luo, G., et al.: Multi-task collaborative network for joint referring expression comprehension and segmentation. In: Proceedings of the IEEE/CVF Conference on Computer Vision and Pattern Recognition, pp. 10034–10043 (2020)
29. Mao, J., Huang, J., Toshev, A., Camburu, O., Yuille, A.L., Murphy, K.: Generation and comprehension of unambiguous object descriptions. In: CVPR (2016)
30. Nagaraja, V.K., Morariu, V.I., Davis, L.S.: Modeling context between objects for referring expression understanding. In: Leibe, B., Matas, J., Sebe, N., Welling, M. (eds.) ECCV 2016. LNCS, vol. 9908, pp. 792–807. Springer, Cham (2016). https://doi.org/10.1007/978-3-319-46493-0_48

31. Perez, E., Strub, F., De Vries, H., Dumoulin, V., Courville, A.: Film: Visual reasoning with a general conditioning layer. In: AAAI (2018)
32. Plummer, B.A., Kordas, P., Kiapour, M.H., Zheng, S., Piramuthu, R., Lazebnik, S.: Conditional image-text embedding networks. In: Ferrari, V., Hebert, M., Sminchisescu, C., Weiss, Y. (eds.) ECCV 2018. LNCS, vol. 11216, pp. 258–274. Springer, Cham (2018). https://doi.org/10.1007/978-3-030-01258-8_16
33. Plummer, B.A., Wang, L., Cervantes, C.M., Caicedo, J.C., Hockenmaier, J., Lazebnik, S.: Flickr30k entities: collecting region-to-phrase correspondences for richer image-to-sentence models. Int. J. Comput. Vis. (2017)
34. Redmon, J., Farhadi, A.: Yolov3: an incremental improvement. arXiv preprint arXiv:1804.02767 (2018)
35. Ren, S., He, K., Girshick, R., Sun, J.: Faster R-CNN: towards real-time object detection with region proposal networks. In: Advances in Neural Information Processing Systems, pp. 91–99 (2015)
36. Rohrbach, A., Rohrbach, M., Hu, R., Darrell, T., Schiele, B.: Grounding of textual phrases in images by reconstruction. In: Leibe, B., Matas, J., Sebe, N., Welling, M. (eds.) ECCV 2016. LNCS, vol. 9905, pp. 817–834. Springer, Cham (2016). https://doi.org/10.1007/978-3-319-46448-0_49
37. Sadhu, A., Chen, K., Nevatia, R.: Zero-shot grounding of objects from natural language queries. In: ICCV (2019)
38. Tieleman, T., Hinton, G.: Lecture 6.5-RMSPROP: divide the gradient by a running average of its recent magnitude. COURSERA: Neural Networks Mach. Learn. **4**, 26–30 (2012)
39. Uijlings, J.R., Van De Sande, K.E., Gevers, T., Smeulders, A.W.: Selective search for object recognition. Int. J. Comput. Vis. **104**, 154–171 (2013)
40. Wang, L., Li, Y., Huang, J., Lazebnik, S.: Learning two-branch neural networks for image-text matching tasks. IEEE Trans. Pattern Anal. Mach. Intell. (2018)
41. Wang, L., Li, Y., Lazebnik, S.: Learning deep structure-preserving image-text embeddings. In: CVPR (2016)
42. Wang, P., Wu, Q., Cao, J., Shen, C., Gao, L., Hengel, A.V.D.: Neighbourhood watch: referring expression comprehension via language-guided graph attention networks. In: CVPR (2019)
43. Wolf, T., et al.: Huggingface's transformers: state-of-the-art natural language processing. arXiv preprint arXiv:1910.03771 (2019)
44. Xingjian, S., Chen, Z., Wang, H., Yeung, D.Y., Wong, W.K., Woo, W.C.: Convolutional LSTM network: a machine learning approach for precipitation nowcasting. In: Advances in Neural Information Processing Systems, pp. 802–810 (2015)
45. Yang, S., Li, G., Yu, Y.: Cross-modal relationship inference for grounding referring expressions. In: CVPR (2019)
46. Yang, S., Li, G., Yu, Y.: Dynamic graph attention for referring expression comprehension. In: ICCV (2019)
47. Yang, Z., Gong, B., Wang, L., Huang, W., Yu, D., Luo, J.: A fast and accurate one-stage approach to visual grounding. In: ICCV (2019)
48. Yu, L., et al.: Mattnet: Modular attention network for referring expression comprehension. In: CVPR (2018)
49. Yu, L., Poirson, P., Yang, S., Berg, A.C., Berg, T.L.: Modeling context in referring expressions. In: Leibe, B., Matas, J., Sebe, N., Welling, M. (eds.) ECCV 2016. LNCS, vol. 9906, pp. 69–85. Springer, Cham (2016). https://doi.org/10.1007/978-3-319-46475-6_5
50. Yu, L., Tan, H., Bansal, M., Berg, T.L.: A joint speaker-listener-reinforcer model for referring expressions. In: CVPR (2017)

51. Zhang, H., Niu, Y., Chang, S.F.: Grounding referring expressions in images by variational context. In: CVPR (2018)
52. Zhuang, B., Wu, Q., Shen, C., Reid, I., van den Hengel, A.: Parallel attention: a unified framework for visual object discovery through dialogs and queries. In: CVPR (2018)
53. Zitnick, C.L., Dollár, P.: Edge boxes: locating object proposals from edges. In: Fleet, D., Pajdla, T., Schiele, B., Tuytelaars, T. (eds.) ECCV 2014. LNCS, vol. 8693, pp. 391–405. Springer, Cham (2014). https://doi.org/10.1007/978-3-319-10602-1_26

Multi-level Wavelet-Based Generative Adversarial Network for Perceptual Quality Enhancement of Compressed Video

Jianyi Wang[1], Xin Deng[2], Mai Xu[1,4]($\boxtimes$), Congyong Chen[3],
and Yuhang Song[5]

[1] School of Electronic and Information Engineering,
Beihang University, Beijing, China
{iceclearwjy,maixu}@buaa.edu.cn
[2] School of Cyber Science and Technology, Beihang University, Beijing, China
cindydeng@buaa.edu.cn
[3] College of Software, Beihang University, Beijing, China
ndsffx304ccy@buaa.edu.cn
[4] Hangzhou Innovation Institute, Beihang University, Zhejiang, China
[5] Department of Computer Science, University of Oxford, Oxford, UK
yuhang.song@some.ox.ac.uk

Abstract. The past few years have witnessed fast development in video quality enhancement via deep learning. Existing methods mainly focus on enhancing the objective quality of compressed video while ignoring its perceptual quality. In this paper, we focus on enhancing the perceptual quality of compressed video. Our main observation is that enhancing the perceptual quality mostly relies on recovering high-frequency sub-bands in wavelet domain. Accordingly, we propose a novel generative adversarial network (GAN) based on multi-level wavelet packet transform (WPT) to enhance the perceptual quality of compressed video, which is called multi-level wavelet-based GAN (MW-GAN). In MW-GAN, we first apply motion compensation with a pyramid architecture to obtain temporal information. Then, we propose a wavelet reconstruction network with wavelet-dense residual blocks (WDRB) to recover the high-frequency details. In addition, the adversarial loss of MW-GAN is added via WPT to further encourage high-frequency details recovery for video frames. Experimental results demonstrate the superiority of our method.

Keywords: Video perceptual quality enhancement · Wavelet packet transform · GAN

Electronic supplementary material The online version of this chapter (https://doi.org/10.1007/978-3-030-58568-6_24) contains supplementary material, which is available to authorized users.

© Springer Nature Switzerland AG 2020
A. Vedaldi et al. (Eds.): ECCV 2020, LNCS 12359, pp. 405–421, 2020.
https://doi.org/10.1007/978-3-030-58568-6_24

1   Introduction

Nowadays, a large amount of videos are available on the Internet, such as YouTube and Facebook, which exerts huge pressure on the communication bandwidth. According to the Cisco Visual Networking Index [7], video causes 75% of Internet traffic in 2017, and this figure is predicted to reach 82% by 2022. As a result, video compression has to be applied to save the communication bandwidth [32]. However, compressed video inevitably suffers from compression artifacts, which severely degrades the quality of experience (QoE) [1,22]. Therefore, it is necessary to study on quality enhancement on compressed videos.

Recently, there have been extensive works for enhancing the quality of compressed images and videos [9,11,13,14,21,23,28,33,37,40–43]. Among them, a four-layer convolutional neural network (CNN) called AR-CNN was proposed in [9] to improve the quality of JPEG compressed images. Then, Zhang *et al.* proposed a denoising CNN (DnCNN) [43] for image denoising and JPEG image deblocking, through a residual learning strategy. Later, a multi-frame quality enhancement network (MFQE) was proposed in [12,42] to improve the objective quality of compressed video, through leveraging the information from neighboring frames. Most recently, Yang *et al.* [39] have proposed a quality-gated convolutional long short-term memory (QG-ConvLSTM) network, which takes advantage of the bi-directional recurrent structure to fully exploit the useful information in a large range of frames. Unfortunately, the existing methods mainly focus on improving the objective quality of compressed images/videos while ignoring the perceptual quality. Actually, high objective quality, i.e., peak signal-to-noise ratio (PSNR), is not always consistent with the human visual system (HVS) [20]. Besides, according to the perception-distortion tradeoff [3], improving objective quality will inevitably lead to a decrease of perceptual quality. As illustrated in Fig. 1, although the frames generated by state-of-the-art methods [12,21,34,41,42] have high PSNR values, they are not perceptually

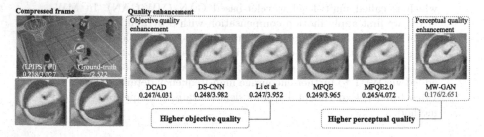

Fig. 1. Objective quality enhancement (traditional works) vs. perceptual quality enhancement (our work). State-of-the-art image [21] and video [12,34,41,42] quality enhancement methods mainly focus on objective quality enhancement, but ignore perceptual quality, which leads to low perceptual quality with high LPIPS [44] and perceptual index (PI) [2]. As the first attempt to enhance the perceptual quality of compressed video, our method achieves better perceptual quality with lower LPIPS and PI. Zoom in for best view.

photorealistic with high LPIPS value [44] and perceptual index (PI) [2], due to the lack of high-frequency details and fine textures.

In this paper, we propose a multi-level wavelet-based generative adversarial network (MW-GAN) for perceptual quality enhancement of compressed video, which recovers the high-frequency details via wavelet packet transform (WPT) [24] at multiple levels. The key insight to adopt WPT is that the high-frequency details are usually missing due to the compression and they can be regarded as the high-frequency sub-bands after WPT. Our MW-GAN has a generator and a discriminator. Specifically, the generator is composed of two main components: motion compensation and wavelet reconstruction. Motion compensation is first developed to compensate the motion between the target frame and its neighbors with a pyramid architecture following [12]. Then, the wavelet reconstruction network consisting of a bunch of wavelet-dense residual blocks (WDRB) is adopted to reconstruct the sub-bands of the target frame. Besides, WPT is also considered in the discriminator of MW-GAN, such that the generator is further encouraged to recover high-frequency details. As shown in Fig. 1, our method is able to generate photorealistic frames with sufficient texture details.

To the best of our knowledge, our work is the first attempt to enhance the perceptual quality of compressed video, using a wavelet-based GAN. The main contributions of this paper are as follows: (1) We investigate that the high-frequency sub-bands in wavelet domain is highly related to the perceptual quality of compressed video. (2) We propose a novel network architecture called MW-GAN, which learns to recover the high-frequency information in wavelet domain for perceptual quality enhancement of compressed video. (3) Extensive experiments have been conducted to demonstrate the ability of the proposed method in enhancing the perceptual quality of compressed video.

2   Related Work

Quality Enhancement of Compressed Images/Videos. In the past few years, extensive works [4,5,9,11–14,17,21,23,33,37,41–43] have been proposed to enhance the objective quality of compressed images/videos. Specifically, for compressed images, Foi *et al.* [11] applied point-wise shape-adaptiveDCT (SADCT) to reduce the blocking and ringing effects caused by JPEG compression. Then, regression tree fields (RTF) were adopted in [14] to reduce JPEG image blocking effects. Moreover, [17] and [5] utilized sparse coding to remove JPEG artifacts. Recently, deep learning has also been successfully applied for quality enhancement. Particularly, Dong *et al.* [9] proposed a four-layer AR-CNN to reduce the JPEG artifacts of images. Afterwards, $\mathbf{D}^3$ [37] and deep dual-domain convolutional network (DDCN) [13] were proposed for enhancing the quality of JPEG compressed images, utilizing the prior knowledge of JPEG compression. Later, DnCNN was proposed in [43] for multiple tasks of image restoration, including quality enhancement. Li *et al.* [21] proposed a 20-layer CNN for enhancing image quality, which achieves state-of-the-art performance on objective quality enhancement of compressed images.

For the compressed videos, Wang *et al.* [34] proposed a deep CNN-based auto decoder (DCAD) which applies 10 CNN layers to reduce the distortion of compressed video. Later, DS-CNN [41] was proposed for video quality enhancement, in which two sub-networks are used to reduce the artifacts of intra- and inter-coding frames, respectively. Besides, Yang *et al.* [42] proposed a multi-frame quality enhancement network (MFQE) to take advantage of neighbor high-quality frames. Most recently, a quality-gated convolutional long short-term memory (QG-ConvLSTM) [39] network was proposed to enhance video quality via learning the "forget" and "input" gates in the ConvLSTM [38] cell from quality-related features. Then, Guan *et al.* [12] proposed a new method for multi-frame quality enhancement called MFQE 2.0, which is an extended version of [42] with substantial improvement and achieved state-of-the-art performance. All the above methods aim at minimizing the pixel-wise loss, such as mean square error (MSE), to obtain high objective quality. However, according to [20], MSE cannot always reflect the perceptually relevant differences. Actually, minimizing MSE encourages finding pixel-wise averages of plausible solutions, leading to overly-smooth images/videos with poor perceptual quality [10,16,26].

Perception-Driven super-Resolution. To the best of our knowledge, there exists no work on perceptual quality enhancement of compressed video. The closest work to ours is the perception-driven image/video super-resolution that aims to restore perception-friendly high-resolution images/videos from low-resolution ones. Specifically, Johnson *et al.* [16] proposed to minimize perceptual loss defined in the feature space to enhance the perceptual image quality in single image super-resolution. Then, contextual loss [27] was developed to generate images with natural image statistics via using an objective that focuses on the feature distribution. SRGAN [20] was proposed to generate natural images using perceptual loss and adversarial loss. Sajjadi *et al.* [30] developed a similar method with the local texture matching loss. Later, Wang *et al.* [35] proposed spatial feature transform to effectively incorporate semantic prior in an image and improve the recovered textures. They also developed an enhanced version of SRGAN (ESRGAN) [36] with a deeper and more efficient network architecture. Most recently, TecoGAN [6] was proposed to achieve state-of-the-art performance in video super-resolution, in which a spatial-temporal discriminator and a ping-pong loss are applied. Similar to super-resolution, perceptual quality is also important for quality enhancement of compressed video. To the best of our knowledge, our MW-GAN in this paper is the first attempt in this direction.

3 Motivation for WPT

The wavelet transform allows the multi-resolution analysis of images [15], and it can decompose an image into multiple sub-bands, i.e., low- and high-frequency sub-bands. As verified in [8], the high-frequency sub-bands play an important role in enhancing the perceptual quality of a super-resolved image. Here, we further verify that the high-frequency sub-bands are also crucial for the perceptual quality enhancement of compressed video. Specifically, given a lossless

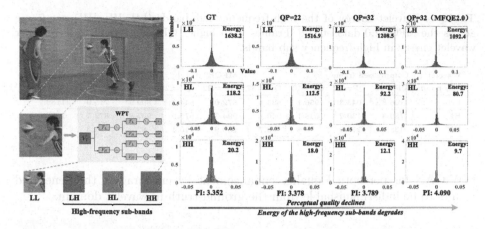

Fig. 2. Histograms of the wavelet sub-bands (zoom in for details). Compressed frames with higher QP have lower perceptual quality (i.e., higher PI) while high-frequency wavelets fade with degraded energy alongside the increase of the QP value. Besides, the state-of-the-art method [12] obtains low perceptual quality, due to the failure of enhancing high-frequency sub-bands. Similar results can be found for other state-of-the-art methods [21,34,41,42].

video frame, a series of compressed frames are obtained via HM16.5 [32] at low-delay configuration, with the quantization parameter (QP) of 22, 27, 32 and 37. The corresponding wavelet sub-bands are achieved via WPT using *haar* filter at one level. For each frame, we obtain four sub-bands called LL, LH, HL and HH. Here, LL is the low-frequency sub-band, LH, HL and HH are the sub-bands with high-frequency information at horizontal, vertical and diagonal directions, respectively. Figure 2 shows the histograms of wavelet coefficients of the LH, HL and HH sub-bands. As we can see, the compressed frames with higher QP values tend to have lower perceptual quality (i.e., higher PI) and insufficient high-frequency wavelet coefficients. Moreover, although the state-of-the-art method [12] can obtain high objective quality, it fails to recover the high-frequency sub-bands in wavelet domain, leading to low perceptual quality. In order to measure the richness of high-frequency wavelet coefficients, we further calculate the wavelet energy by summing up their squared coefficients. As shown in Table 1, the compressed frames with higher QP values tend to have lower wavelet energy in high-frequency sub-bands. All these demonstrate that the high-frequency wavelet sub-bands play an important role in enhancing the perceptual quality of compressed video.

Based on the above investigation, we propose our MW-GAN as follows. (1) Since high-frequency wavelet sub-bands are crucial for the perceptual quality, the generator of our MW-GAN directly outputs these wavelet sub-bands. (2) To restore the high-frequency wavelet sub-bands as much as possible, a wavelet-dense residual block (WDRB) is proposed in our MW-GAN to further recover the wavelet sub-bands. (3) The WPT is also applied in the discriminator of our

Table 1. Wavelet energy of the high-frequency sub-bands (LH, HL, HH) per frame across the MFQE2.0 dataset [12]. Frames with higher QP values tend to have lower wavelet energy in high-frequency sub-bands.

Wavelet Sub-bands	Compressed frame				MFQE2.0 [12]				Ground-Truth
	QP = 22	QP = 27	QP = 32	QP = 37	QP = 22	QP = 27	QP = 32	QP = 37	
LH	1106.8	1043.3	956.1	835.4	872.0	844.2	809.9	715.9	1189.7
HL	1189.3	1130.2	1048.1	940.5	994.3	973.9	937.8	872.1	1273.0
HH	138.4	115.9	93.8	70.7	130.9	110.4	92.6	65.3	176.2

MW-GAN for perceptual quality enhancement by encouraging the generated results to be indistinguishable from the ground-truth in wavelet domain.

4   The Proposed MW-GAN

The architecture of the proposed MW-GAN is shown in Fig. 3. To enhance the perceptual quality of a video frame V_t, we simultaneously train a generator G_{θ_G} parameterized by θ_G and a discriminator D_{θ_D} parameterized by θ_D in an adversarial manner. The ultimate goal is to obtain the enhanced frame $\hat{O}_t$ with high perceptual quality, similar to its ground-truth O_t. To take advantage of the temporal information in adjacent frames, the generator G_{θ_G} takes both V_t and its neighbor frames $\{V_{t\pm n}\}_{n=1}^N$ as inputs. The details about the generator G_{θ_G} are introduced in Sect. 4.1. To further encourage high perceptual quality, we propose a multi-level wavelet-based discriminator, which distinguishes the generated wavelet sub-bands from the ground-truths, as introduced in Sect. 4.2. Finally, we present the loss functions to train the MW-GAN in Sect. 4.3.

4.1   Multi-level Wavelet-Based Generator

The generator G_{θ_G} of our MW-GAN is mainly composed of two parts: motion compensation and wavelet reconstruction. Given the $2N+1$ frames as inputs, we first perform motion compensation across frames to align content across frames. After that, wavelet reconstruction is applied to reconstruct the target wavelet sub-bands using the information of sub-bands from both current and compensated frames. Finally, the output reconstructed sub-bands are used to obtain the enhanced frame through IWPT.

Motion Compensation: To take advantage of the temporal information across frames, we adopt the motion compensation network with a pyramid architecture following [42]. To obtain better performance, we further make some modifications. First, for flow estimation, it is crucial to have large-size filters in the first few layers of the network to capture intense motion. Thus, the kernel sizes of the first two convolutional layers at each pyramid level are 7×7 and 5×5, respectively, instead of 3×3 in [42]. Second, we replace the pooling operation at each pyramid level by WPT. Since WPT is invertible, this pooling operation

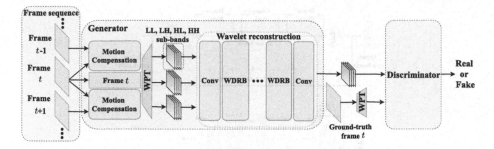

Fig. 3. Framework of our MW-GAN. Perceptual quality is enhanced via recovering wavelet sub-bands in wavelet domain. Specifically, we propose a generator with motion composition and WDRB to capture both temporal and high-frequency information. Then a multi-level wavelet-based discriminator is proposed to evaluate generated results in both pixel and wavelet domain.

is able to make all information of the original frames be preserved, meanwhile efficiently enlarging the receptive field for flow estimation.

Wavelet Reconstruction: Given the current and compensated frames, we further propose a wavelet reconstruction network to reconstruct the wavelet sub-bands $\hat{\mathbf{S}}_t$, which contains a cascade of wavelet-dense residual blocks (WDRB). Specifically, WPT is first applied to generate sub-bands as inputs. Then, several convolutional layers are applied to extract corresponding feature maps. To further capture high-frequency details and reduce the computational cost, we develop the wavelet-based residual block (WDRB) with a residual-in-residual structure, as shown in Fig. 4. Similar to [36], in our WDRB, several dense blocks are applied in the main path of the residual block to enhance the feature representation with high capacity. In addition, we adopt WPT and IWPT in the main path to learn the residual in wavelet domain. With this simple yet effective design, the receptive field is further enlarged without losing information, thus enabling the network to capture more high-frequency details. The efficiency of the WDRB is also investigated in the ablation study. The final output of G_{θ_G} are the wavelet sub-bands $\hat{\mathbf{S}}_t$, and the enhanced frame $\hat{\mathbf{O}}_t$ can be obtained by combining $\hat{\mathbf{S}}_t$ via IWPT. Assuming that $\omega^{-1}(\cdot)$ denotes the IWPT operation, the output of the generator can be summarized as follows,

$$\hat{\mathbf{S}}_t = G_{\theta_G}(\mathbf{V}_t, \{\mathbf{V}_{t\pm n}\}_{n=1}^N),$$
$$\hat{\mathbf{O}}_t = \omega^{-1}(\hat{\mathbf{S}}_t). \tag{1}$$

4.2 Multi-level Wavelet-Based Discriminator

In this section, a multi-level wavelet-based discriminator is proposed to encourage the generated results indistinguishable from the ground-truth. The structure of our discriminator is illustrated in Fig. 4. Specifically, the discriminator takes

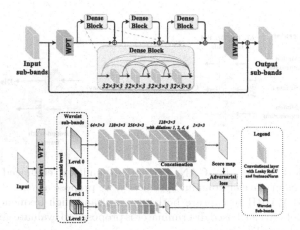

Fig. 4. Details about our framework. Top: Architecture of the proposed WDRB; Down: Architecture of the multi-level wavelet-based discriminator in MW-GAN.

the generated frames (obtained via IWPT) or the real ones (ground-truth) as input. Then, WPT is applied to both generated and real frames for obtaining their wavelet sub-bands at multiple levels. After that, wavelet sub-bands of different levels are fed into the corresponding pyramid level. Each pyramid level employs a fully convolutional architecture, in which a group of convolutional layers with different dilations are applied to extract features with various receptive fields. Subsequently, the input wavelet sub-bands at each level are mapped into a single channel of the score map with the same size as the input, which measures the similarity between the generated and real frames. Let $\omega_l(\cdot)$ denote the l-th level WPT. The output of the discriminator is a set of score maps of different sizes, which can be formulated as follows,

$$D^l_{\boldsymbol{\theta}_D}(\mathbf{O}_t) = f_l(\omega_l(\mathbf{O}_t)), \qquad (2)$$

where $f_l(\cdot)$ is the corresponding map function at the l-th level of the discriminator, L is the total number of the pyramid levels and $D^l_{\boldsymbol{\theta}_D}(\mathbf{O}_t)$ is the output score map at l-th level. Note that the $l = 0$ indicates that there is no WPT operation.

The advantage of the proposed multi-level wavelet-based discriminator is efficiency and efficacy. (1) The computational cost can be significantly reduced by adopting WPT, since the calculation of wavelet sub-bands can be regarded as forwarding through a single convolutional layer. In contrast, the traditional methods need to extract multi-scale features via a sequence of convolutional layers, thus consuming expensive computational complexity. (2) Our discriminator can distinguish the generated and real frames in both pixel ($l = 0$) and wavelet domains ($l > 0$), leading to better generative results.

4.3 Loss Functions

In this section, we propose a loss function for training the MW-GAN. Specifically, we formulate the loss of the generator $L_G(\boldsymbol{\theta}_M; \boldsymbol{\theta}_G)$ as the weighted sum of a motion loss $L_M(\boldsymbol{\theta}_M)$, a wavelet loss $L_W(\boldsymbol{\theta}_G)$ and an adversarial loss $L_{\text{Adv}}(\boldsymbol{\theta}_G)$. Then, $L_G(\boldsymbol{\theta}_M; \boldsymbol{\theta}_G)$ can be written as,

$$L_G(\boldsymbol{\theta}_M; \boldsymbol{\theta}_G) = L_W(\boldsymbol{\theta}_G) + \alpha L_M(\boldsymbol{\theta}_M) + \beta L_{\text{Adv}}(\boldsymbol{\theta}_G), \tag{3}$$

where α and β are the coefficients to balance different loss terms. In the following, we explain the different components of the loss functions.

Motion Loss: Since it is hard to obtain the ground-truth of optical flow, the motion compensation network $M_{\boldsymbol{\theta}_M}$ parameterized by $\boldsymbol{\theta}_M$ is directly trained under the supervision of the ground-truth $\mathbf{O}_t$. That is, the neighboring frames $\{\mathbf{V}_{t\pm n}\}_{n=1}^N$ are first wrapped using the optical flow estimated by $M_{\boldsymbol{\theta}_M}$; then the loss between the compensated frames $\{M_{\boldsymbol{\theta}_M}(\mathbf{V}_{t\pm n})\}_{n=1}^N$ and the ground-truth $\mathbf{O}_t$ is minimized. Here, we adopt Charbonnier penalty function [19] as the motion loss, defined by:

$$L_M(\boldsymbol{\theta}_M) = \frac{1}{2N} \sum_{n=1}^N \sqrt{\|M_{\boldsymbol{\theta}_M}(\mathbf{V}_{t\pm n}) - \mathbf{O}_t\|_F^2 + \epsilon_m^2}, \tag{4}$$

where ϵ_m is a scaling parameter and is empirically set to 1×10^{-3}.

Wavelet Loss: The wavelet loss function models how close the predicted sub-bands $\hat{\mathbf{S}}_t$ are to the ground truth $\mathbf{S}_t$. It is defined by a weighted Charbonnier penalty function in wavelet domain as follows:

$$L_W(\boldsymbol{\theta}_G) = \sqrt{\left\|\mathbf{W}^{1/2} \odot (\hat{\mathbf{S}}_t - \mathbf{S}_t)\right\|_F^2 + \epsilon_w^2}, \tag{5}$$

where $\odot$ represents dot product, $\hat{\mathbf{S}}_t = \{\hat{\mathbf{s}}_t^1, \hat{\mathbf{s}}_t^2, \ldots, \hat{\mathbf{s}}_t^{n_w}\}$ are the predicted sub-bands with the number of n_w in total and $\mathbf{S}_t$ are their corresponding ground-truths. In addition, $\mathbf{W} = \{w_1, w_2, \ldots, w_{n_w}\}$ is the weight matrix to balance the importance of each sub-band and ϵ_w is a scaling parameter set to 1×10^{-3}.

Adversarial Loss: In addition to the above two loss functions, another important loss of the GAN is the adversarial loss. Note that we follow [25] to adopt ℓ_2 loss for training. Recall that $D_{\boldsymbol{\theta}_D}^l(\mathbf{O}_t)$ represents the outputs of the discriminator in (2) at the l-th level, taking $\mathbf{O}_t$ as input. Then, the discriminator loss can be defined as:

$$L_D(\boldsymbol{\theta}_D) = \frac{1}{2L}\mathbb{E}_{\mathbf{O}_t}[\sum_{l=0}^{L-1} \|D_{\boldsymbol{\theta}_D}^l(\mathbf{O}_t) - \mathbf{1}\|_F^2] + \frac{1}{2L}\mathbb{E}_{\hat{\mathbf{O}}_t}[\sum_{l=0}^{L-1} \|D_{\boldsymbol{\theta}_D}^l(\hat{\mathbf{O}}_t)\|_F^2], \tag{6}$$

where $\mathbb{E}_{\hat{\mathbf{O}}_t}[\cdot]$ represents the average for all $\hat{O}_t$ in the mini-batch and L is the number of the pyramid levels in the discriminator. Note that $\mathbf{1} \in \mathcal{R}^{W_l \times H_l}$, where W_l and H_l are the width and height of the score map at the l-th level, respectively. Symmetrically, the adversarial loss for generator is as follow:

$$L_{\text{Adv}}(\boldsymbol{\theta}_G) = \frac{1}{2L}\mathbb{E}_{\hat{\mathbf{O}}_t}[\sum_{l=0}^{L-1} \|D_{\boldsymbol{\theta}_D}^l(\hat{\mathbf{O}}_t(\boldsymbol{\theta}_G)) - \mathbf{1}\|_F^2]. \tag{7}$$

5 Experiments

In this section, the experimental results are presented to validate the effectiveness of our MW-GAN method. Section 5.1 introduces the experimental settings. Section 5.2 compares the results between our method and other state-of-the-art methods over the test sequences of JCT-VC [29]. In Sect. 5.3, the mean opinion score (MOS) test is performed to compare the subjective quality of videos by different methods. Finally, the ablation study is presented in Sect. 5.4.

5.1 Settings

Datasets: We train the MW-GAN model using the database introduced in [12]. Following [12], except the 18 common test sequences of Joint Collaborative Team on Video Coding (JCT-VC) [29], the other 142 sequences are randomly divided into non-overlapping training set (106 sequences) and validation set (36 sequences). All 160 sequences are compressed by HM16.5 [32] under Low-Delay configuration, setting QP to 32. Note that different from existing methods [4,12, 21,33,39,41,42] that train models for each QP individually, we only train our MW-GAN under QP = 32, and we test on both QP = 32 and QP = 37. The results show the generalization ability of our method.

Parameter Settings: Here, we mainly introduce the settings and hyperparameters of our experiments. Specifically, the total number of levels L of each component in MW-GAN is set to 3 and the generator takes 3 consecutive frames as inputs for a better tradeoff between performance and efficiency. The training process is divided into two stages. We first train our model without adversarial loss (i.e., $\beta = 0$). The motion loss weight α is initialed as 10 and decayed by a factor of 10 every 5×10^4 iterations to encourage the learning of motion compensation first. The iteration number and mini-batch size are 3×10^5 and 32, respectively, and the input frames are cropped to 256×256. The Adam algorithm [18] with the step size of 2.5×10^{-4} is adopted; the learning rate is initialized as 2×10^{-4} and decayed by a factor of 2 every 1×10^5 iterations. In the second stage, the pre-trained model is employed as an initialization for the generator. The generator is trained using the loss function in (3) with $\alpha = 1 \times 10^{-2}$ and $\beta = 5 \times 10^{-3}$. The iteration number and mini-batch size are 6×10^5 and 32, respectively, and the cropped size is reduced to 128×128. The learning rate is set to 1×10^{-4} and halved every 1×10^5 iterations. Note that the above hyperparameteres are tuned over the training set. We apply the Adam algorithm [18] and alternately update the generator and discriminator network until convergence. The wavelet sub-bands are obtained by wavelet packet decomposition with *haar* filter. More details can be found in the supplementary material.

Table 2. Overall ΔLPIPS and ΔPI between enhanced and compressed frames on the test set of JCT-VC [29] at QP= 32 and QP= 37. Our MW-GAN achieves the best perceptual quality across all the test sequences.

QP	Video sequence	Li et al. [21]		DCAD [34]		DS-CNN [41]		MFQE [42]		MFQE 2.0 [12]		Ours	
		ΔLPIPS	ΔPI	ΔLPIPS	ΔPI	ΔLPIPS	ΔPI	ΔLPIPS	ΔPI	ΔLPIPS	ΔPI	ΔLPIPS	ΔPI
32 A	Traffic	0.021	0.501	0.019	0.419	0.017	0.373	0.016	0.441	0.014	0.430	−0.032	−1.123
	PeopleOnStreet	0.020	0.865	0.020	0.668	0.019	0.663	0.017	0.807	0.017	0.794	−0.020	−0.558
B	Kimono	0.038	0.479	0.034	0.403	0.033	0.332	0.034	0.440	0.036	0.443	−0.069	−1.561
	ParkScene	0.010	0.299	0.010	0.377	0.009	0.291	0.012	0.239	0.010	0.346	−0.032	−0.159
	Cactus	0.032	0.264	0.028	0.264	0.027	0.236	0.028	0.248	0.028	0.274	−0.109	−0.953
	BQTerrace	0.028	0.384	0.025	0.413	0.025	0.362	0.026	0.418	0.026	0.447	−0.099	−0.326
	BasketballDrive	0.036	0.534	0.031	0.501	0.032	0.462	0.032	0.595	0.032	0.618	−0.106	−0.921
C	RaceHorses	0.023	0.504	0.022	0.536	0.021	0.469	0.025	0.489	0.027	0.566	−0.021	−0.579
	BQMall	0.022	0.457	0.019	0.507	0.018	0.426	0.021	0.467	0.021	0.499	−0.033	−0.908
	PartyScene	0.032	0.436	0.027	0.381	0.024	0.347	0.025	0.408	0.025	0.412	−0.075	−0.765
	BasketballDrill	0.027	0.782	0.023	0.926	0.026	0.842	0.027	0.772	0.025	0.929	−0.047	−0.532
D	RaceHorses	0.019	0.628	0.018	0.657	0.017	0.581	0.021	0.657	0.021	0.685	−0.005	−0.199
	BQSquare	0.012	0.285	0.012	0.599	0.011	0.382	0.013	0.344	0.011	0.570	−0.037	−0.525
	BlowingBubbles	0.013	0.492	0.014	0.585	0.011	0.458	0.012	0.394	0.009	0.472	−0.039	−0.241
	BasketballPass	0.020	0.537	0.016	0.564	0.016	0.517	0.020	0.574	0.019	0.594	−0.021	−0.794
E	FourPeople	0.012	0.265	0.010	0.231	0.010	0.234	0.010	0.238	0.008	0.219	−0.040	−1.129
	Johnny	0.010	0.308	0.010	0.339	0.013	0.215	0.012	0.363	0.010	0.259	−0.065	−1.171
	KristenAndSara	0.016	0.278	0.015	0.316	0.016	0.360	0.015	0.338	0.014	0.352	−0.026	−1.305
	Average	0.022	0.461	0.020	0.483	0.019	0.419	0.020	0.457	0.020	0.495	−0.049	−0.764
37	Average	0.027	0.541	0.026	0.621	0.024	0.562	0.027	0.617	0.024	0.614	−0.046	−0.993

5.2   Quantatative Comparison

In this section, we evaluate the performance of our method in terms of learned perceptual image patch similarity (LPIPS) [44] and perceptual index (PI) [2], which are metrics widely used for perceptual quality evaluation. For results in terms of other evaluation metrics, please see the supplementary material. We compare our method with Li et al. [21], DS-CNN [41], DCAD [34], MFQE [42] and MFQE 2.0 [12]. Among them, Li et al. is the latest quality enhancement methods for compressed images. DCAD and DS-CNN are single-frame enhancement methods for compressed videos, while MFQE and MFQE 2.0 are state-of-the-art multi-frame quality enhancement methods for compressed videos. All these methods are trained over the same training set as ours for fair comparison.

Table 2 reports the ΔLPIPS and ΔPI results which are calculated between enhanced and compressed frames averaged over each test sequence. Note that ΔLPIPS < 0 and ΔPI < 0 indicate improvement in perceptual quality. As shown in this table, our MW-GAN outperforms all 6 compared methods in terms of perceptual quality, for all test sequences, while all the compared methods fail to enhance the perceptual quality of compressed video. Specifically, at QP = 32, the average ΔLPIPS and ΔPI in our MW-GAN are −0.049 and −0.764, respectively, while other methods [12, 21, 34, 41–43] all have positive ΔLPIPS and ΔPI values. Furthermore, our MW-GAN also gains the largest decrease of ΔLPIPS with −0.046 and ΔPI with −0.993 at QP = 37, verifying the generalization ability of our MW-GAN for perceptual quality enhancement of compressed video.

Fig. 5. Qualitative comparison on the test sequences of JCT-VC [29]. Our MW-GAN can generate much more realistic results (Zoom in for best view).

5.3   Subjective Comparison

In this section, we mainly focus on the subjective evaluation of our method. Figure 5 visualizes the enhanced video frames of *Traffic* at QP = 32, *Basketball-Pass* at QP = 32, *BQTerrace* at QP = 37 and *RaceHorses* at QP = 32 We can observe that our MW-GAN method outperforms other methods with sharper edges and more vivid details. For instance, the car in *Traffic*, the face in *BasketballPass*, the pedestrians in *BQTerrace* and the horse in *RaceHorses* can be finely restored in our MW-GAN, while the existing PSNR-oriented methods generate blurry results with low perceptual quality.

To further evaluate the subjective quality of our method, we also conduct a mean opinion score (MOS) test at QP = 37. Specifically, we asked 15 subjects to rate an integral score from 1 to 100 following [31] on the enhanced videos. The higher score indicates better perceptual quality. The subjects are required to rate the scores for compressed video sequences, raw sequences, sequences enhanced by Li *et al.* [21], MFQE [42], MFQE 2.0 [12], our MW-GAN method. Table 3 presents ΔMOS which is calculated between enhanced and compressed sequences. As can be seen, our MW-GAN method obtains the highest ΔMOS score for each class of sequences, and it obtains an average increase of 4.55 in terms of MOS, considerably better than other methods.

5.4   Ablation Study

In this section, we mainly analyze the effectiveness of WPT applied in each component of our MW-GAN.

Effectiveness of WDRB in Generator: In our generator, the multi-level WPT is achieved via adopting WDRB. Thus, it is essential to validate the effectiveness of the proposed WDRB. Here, we conduct the test with two strategies. (1) MW-GAN-RRDB: We directly replace the WDRB by the residual-in-residual dense block (RRDB) proposed in [36]. Note that the channels of the first convolutional layer before the first RRDB are increased to keep the structure of RRDB unchanged. (2) MW-GAN-CNN: We directly replace the WPT and IWPT in WDRB by average pooling and upsampling layers. Note that the

Table 3. ΔMOS calculated between enhanced and compressed sequences at QP = 37 over the test sequences of JCT-VC [29]. Our MW-GAN obtains the highest ΔMOS score for each class of sequences.

Sequence	Li et al. [21]	MFQE [42]	MFQE 2.0 [12]	Ours
A	−0.47	−0.83	−1.21	**4.74**
B	1.86	1.42	0.82	**3.95**
C	−4.76	−2.74	−5.07	**4.05**
D	−1.08	−0.30	−1.90	**5.15**
E	1.11	1.04	0.70	**4.86**
Average	−0.69	−0.28	−1.33	**4.55**

above two strategies all lead to parameter increasing. Figure 6 shows the results of the above two ablation strategies over different classes of test sequences at QP = 32. We can see from this figure that WDRB has a positive impact on the perceptual quality, leading to 0.170 and 0.126 decrease in ΔPI compared with MW-GAN-CNN and MW-GAN-RRDB, respectively. Therefore, we can conclude that WDRB plays an effective role in MW-GAN.

Effectiveness of WPT in the Discriminator: To evaluate the effectiveness of the multi-level wavelet-based discriminator, we add a baseline without WPT in the discriminator, i.e., MW-GAN-D w/o WPT. Specifically, we replace WPT in the discriminator with an average pooling layer that extracts features with the same size as wavelet sub-bands and keeps other components unchanged for a fair comparison. As Fig. 6 shows, without WPT in the discriminator, the ΔPI score increases by average 0.192 over the whole test set, which indicates the effectiveness of the proposed multi-level wavelet-based discriminator.

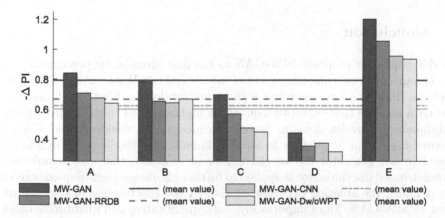

Fig. 6. Comparison between our MW-GAN and corresponding baselines on the test sequences of JCT-VC [29] according to $-\Delta$PI between enhanced and compressed sequences. Effectiveness of the proposed MW-GAN can be verified.

6   Generalization Ability

In this section, we mainly focus on the generalization ability of the proposed MW-GAN. For more details, please see our supplementary material.

Transfer to H.264. To further verify the generalization capability of our MW-GAN, we also test on the video sequences compressed by other standards. Specifically, we compress the test sequences of JCT-VC with H.264 (the JM encoder with the low-delay P configuration) at QP = 32 and QP = 37. Then, we directly test the performance of our method over the above test sequences without fine-tuning. Consequently, the average PI decrease of test sequences is 0.548 at QP = 32 and 0.527 at QP = 37, which implies the generalization ability of our MW-GAN approach across different compression standards.

Performance on Other Sequences. In addition to the common-used test sequences of JCT-VC, we also test the performance of our and other methods over the test set in [42], which is different from the test sequences of JCT-VC. Results show that our method has 0.039 LPIPS decrease and 1.058 PI decrease at QP = 32, while other methods all increase the PI value, which again indicates the superiority of our method according to the perceptual quality enhancement.

Perception-Distortion Tradeoff. Our work mainly focuses on enhancing perceptual quality of compressed video. However, according to the perception-distortion tradeoff [3], improving perceptual quality fully will inevitably lead to a decrease of PSNR. We further evaluate the objective quality of the frames generated by our method over the test sequences of JCT-VC. Results show that our method leads to 1.040 dB PSNR decrease at QP = 32 and 0.651 dB PSNR decrease at QP = 37. It is promising future work for perception-distortion tradeoff of compressed video quality enhancement, while at the current stage we mainly focus on perceptual quality enhancement (w/o considering PSNR) as the first attempt in this direction.

7   Conclusion

In this paper, we proposed MW-GAN as the first attempt for perceptual quality enhancement of compressed video, which embeds WPT in both generator and discriminator for recovering high-frequency details. First, we found from the data analysis that perceptual quality is highly related to the high-frequency sub-bands in wavelet domain. Second, we designed a wavelet-based GAN to recover high-frequency details in wavelet domain. Specifically, a WDRB is proposed in our generator for wavelet sub-band reconstruction and a multi-level wavelet-based discriminator is applied to further encourage high-frequency recovery. Finally, the ablation experiments showed the effectiveness of each component of MW-GAN. More importantly, both quantitative and qualitative results in extensive experiments verified that our MW-GAN outperforms other state-of-the-art methods according to perceptual quality enhancement of compressed video.

Acknowledgement. This work was supported by the NSFC under Project 61876013, Project 61922009, and Project 61573037.

References

1. Bampis, C.G., Li, Z., Moorthy, A.K., Katsavounidis, I., Aaron, A., Bovik, A.C.: Study of temporal effects on subjective video quality of experience. IEEE Trans. Image Process. (TIP) **26**(11), 5217–5231 (2017)
2. Blau, Y., Mechrez, R., Timofte, R., Michaeli, T., Zelnik-Manor, L.: The 2018 PIRM challenge on perceptual image super-resolution. In: Leal-Taixé, L., Roth, S. (eds.) ECCV 2018. LNCS, vol. 11133, pp. 334–355. Springer, Cham (2019). https://doi.org/10.1007/978-3-030-11021-5_21
3. Blau, Y., Michaeli, T.: The perception-distortion tradeoff. In: Proceedings of the IEEE Conference on Computer Vision and Pattern Recognition (CVPR), pp. 6228–6237 (2018)
4. Cavigelli, L., Hager, P., Benini, L.: CAS-CNN: a deep convolutional neural network for image compression artifact suppression. In: 2017 International Joint Conference on Neural Networks (IJCNN), pp. 752–759. IEEE (2017)
5. Chang, H., Ng, M.K., Zeng, T.: Reducing artifacts in jpeg decompression via a learned dictionary. IEEE Trans. Sig. Process. **62**(3), 718–728 (2013)
6. Chu, M., Xie, Y., Leal-Taixé, L., Thuerey, N.: Temporally coherent GANs for video super-resolution (tecogan). arXiv preprint arXiv:1811.09393 (2018)
7. CVNI: Cisco visual networking index: global mobile data traffic forecast update, 2016–2021 white paper. https://www.cisco.com/c/en/us/solutions/collateral/service-provider/visual-networking-index-vni/white-paper-c11-741490.html (2017)
8. Deng, X., Yang, R., Xu, M., Dragotti, P.L.: Wavelet domain style transfer for an effective perception-distortion tradeoff in single image super-resolution. In: Proceedings of the IEEE International Conference on Computer Vision (CVPR), pp. 3076–3085 (2019)
9. Dong, C., Deng, Y., Change Loy, C., Tang, X.: Compression artifacts reduction by a deep convolutional network. In: Proceedings of the IEEE International Conference on Computer Vision (ICCV), pp. 576–584 (2015)
10. Dosovitskiy, A., Brox, T.: Generating images with perceptual similarity metrics based on deep networks. In: Advances in Neural Information Processing Systems (NIPS), pp. 658–666 (2016)
11. Foi, A., Katkovnik, V., Egiazarian, K.: Pointwise shape-adaptive DCT for high-quality denoising and deblocking of grayscale and color images. IEEE Trans. Image Process. (TIP) **16**(5), 1395–1411 (2007)
12. Guan, Z., Xing, Q., Xu, M., Yang, R., Liu, T., Wang, Z.: MFQE 2.0: a new approach for multi-frame quality enhancement on compressed video. IEEE Trans. Pattern Anal. Machine Intelligence (TPAMI), p. 1 (2019)
13. Guo, J., Chao, H.: Building dual-domain representations for compression artifacts reduction. In: Leibe, B., Matas, J., Sebe, N., Welling, M. (eds.) ECCV 2016. LNCS, vol. 9905, pp. 628–644. Springer, Cham (2016). https://doi.org/10.1007/978-3-319-46448-0_38
14. Jancsary, J., Nowozin, S., Rother, C.: Loss-specific training of non-parametric image restoration models: a new state of the art. In: Fitzgibbon, A., Lazebnik, S., Perona, P., Sato, Y., Schmid, C. (eds.) ECCV 2012. LNCS, vol. 7578, pp. 112–125. Springer, Heidelberg (2012). https://doi.org/10.1007/978-3-642-33786-4_9

15. Jawerth, B., Sweldens, W.: An overview of wavelet based multiresolution analyses. SIAM Rev. **36**(3), 377–412 (1994)
16. Johnson, J., Alahi, A., Fei-Fei, L.: Perceptual losses for real-time style transfer and super-resolution. In: Leibe, B., Matas, J., Sebe, N., Welling, M. (eds.) ECCV 2016. LNCS, vol. 9906, pp. 694–711. Springer, Cham (2016). https://doi.org/10.1007/978-3-319-46475-6_43
17. Jung, C., Jiao, L., Qi, H., Sun, T.: Image deblocking via sparse representation. Sig. Process. Image Commun. **27**(6), 663–677 (2012)
18. Kingma, D.P., Ba, J.: Adam: a method for stochastic optimization (2015)
19. Lai, W.S., Huang, J.B., Ahuja, N., Yang, M.H.: Deep laplacian pyramid networks for fast and accurate super-resolution. In: Proceedings of the IEEE Conference on Computer Vision and Pattern Recognition (CVPR), pp. 624–632 (2017)
20. Ledig, C., et al.: Photo-realistic single image super-resolution using a generative adversarial network. In: Proceedings of the IEEE Conference on Computer Vision and Pattern Recognition (CVPR), pp. 4681–4690 (2017)
21. Li, K., Bare, B., Yan, B.: An efficient deep convolutional neural networks model for compressed image deblocking. In: 2017 IEEE International Conference on Multimedia and Expo (ICME), pp. 1320–1325. IEEE (2017)
22. Li, S., Xu, M., Deng, X., Wang, Z.: Weight-based r-λ rate control for perceptual HEVC coding on conversational videos. Sig. Process. Image Commun. **38**, 127–140 (2015)
23. Liew, A.C., Yan, H.: Blocking artifacts suppression in block-coded images using overcomplete wavelet representation. IEEE Trans. Circuits Syst. Video Technol. (TCSVT) **14**(4), 450–461 (2004)
24. Mallat, S.: A Wavelet Tour of Signal Processing. Elsevier, New York (1999)
25. Mao, X., et al.: Least squares generative adversarial networks. In: Proceedings of the IEEE International Conference on Computer Vision (CVPR), pp. 2794–2802 (2017)
26. Mathieu, M., Couprie, C., LeCun, Y.: Deep multi-scale video prediction beyond mean square error. arXiv preprint arXiv:1511.05440 (2015)
27. Mechrez, R., Talmi, I., Shama, F., Zelnik-Manor, L.: Maintaining natural image statistics with the contextual loss. In: Jawahar, C.V., Li, H., Mori, G., Schindler, K. (eds.) ACCV 2018. LNCS, vol. 11363, pp. 427–443. Springer, Cham (2019). https://doi.org/10.1007/978-3-030-20893-6_27
28. Meng, X., et al.: MGANET: a robust model for quality enhancement of compressed video. arXiv preprint arXiv:1811.09150 (2018)
29. Ohm, J.R., Sullivan, G.J., Schwarz, H., Tan, T.K., Wiegand, T.: Comparison of the coding efficiency of video coding standards including high efficiency video coding (HEVC). IEEE Trans. Circuits Syst. Video Technol. (TCSVT) **22**(12), 1669–1684 (2012)
30. Sajjadi, M.S., Scholkopf, B., Hirsch, M.: Enhancenet: single image super-resolution through automated texture synthesis. In: Proceedings of the IEEE International Conference on Computer Vision (ICCV), pp. 4491–4500 (2017)
31. Seshadrinathan, K., Soundararajan, R., Bovik, A.C., Cormack, L.K.: Study of subjective and objective quality assessment of video. IEEE Transactions on Image Processing (TIP) **19**(6), 1427–1441 (2010)
32. Sullivan, G.J., Ohm, J.R., Han, W.J., Wiegand, T.: Overview of the high efficiency video coding (HEVC) standard. IEEE Trans. Circuits Syst. Video Technol. (TCSVT) **22**(12), 1649–1668 (2012)

33. Tai, Y., Yang, J., Liu, X., Xu, C.: MEMNET: a persistent memory network for image restoration. In: Proceedings of the IEEE International Conference on Computer Vision (ICCV), pp. 4539–4547 (2017)

34. Wang, T., Chen, M., Chao, H.: A novel deep learning-based method of improving coding efficiency from the decoder-end for HEVC. In: 2017 Data Compression Conference (DCC), pp. 410–419. IEEE (2017)

35. Wang, X., Yu, K., Dong, C., Change Loy, C.: Recovering realistic texture in image super-resolution by deep spatial feature transform. In: Proceedings of the IEEE Conference on Computer Vision and Pattern Recognition (CVPR), pp. 606–615 (2018)

36. Wang, X., et al.: ESRGAN: enhanced super-resolution generative adversarial networks. In: Leal-Taixé, L., Roth, S. (eds.) ECCV 2018. LNCS, vol. 11133, pp. 63–79. Springer, Cham (2019). https://doi.org/10.1007/978-3-030-11021-5_5

37. Wang, Z., Liu, D., Chang, S., Ling, Q., Yang, Y., Huang, T.S.: D3: deep dual-domain based fast restoration of jpeg-compressed images. In: Proceedings of the IEEE Conference on Computer Vision and Pattern Recognition (CVPR), pp. 2764–2772 (2016)

38. Xingjian, S., Chen, Z., Wang, H., Yeung, D.Y., Wong, W.K., Woo, W.C.: Convolutional LSTM network: a machine learning approach for precipitation nowcasting. In: Advances in Neural Information Processing Systems (NIPS), pp. 802–810 (2015)

39. Yang, R., Sun, X., Xu, M., Zeng, W.: Quality-gated convolutional LSTM for enhancing compressed video. In: 2019 IEEE International Conference on Multimedia and Expo (ICME), pp. 532–537. IEEE (2019)

40. Yang, R., Xu, M., Liu, T., Wang, Z., Guan, Z.: Enhancing quality for HEVC compressed videos. IEEE Trans. Circuits Syst. Video Technol. (TCSVT) (2018)

41. Yang, R., Xu, M., Wang, Z.: Decoder-side HEVC quality enhancement with scalable convolutional neural network. In: 2017 IEEE International Conference on Multimedia and Expo (ICME), pp. 817–822. IEEE (2017)

42. Yang, R., Xu, M., Wang, Z., Li, T.: Multi-frame quality enhancement for compressed video. In: Proceedings of the IEEE Conference on Computer Vision and Pattern Recognition (CVPR), pp. 6664–6673 (2018)

43. Zhang, K., Zuo, W., Chen, Y., Meng, D., Zhang, L.: Beyond a gaussian denoiser: residual learning of deep CNN for image denoising. IEEE Trans. Image Process. (TIP) 26(7), 3142–3155 (2017)

44. Zhang, R., Isola, P., Efros, A.A., Shechtman, E., Wang, O.: The unreasonable effectiveness of deep features as a perceptual metric. In: Proceedings of the IEEE Conference on Computer Vision and Pattern Recognition (CVPR), pp. 586–595 (2018)

Example-Guided Image Synthesis Using Masked Spatial-Channel Attention and Self-supervision

Haitian Zheng$^{(\boxtimes)}$, Haofu Liao, Lele Chen, Wei Xiong, Tianlang Chen, and Jiebo Luo

University of Rochester, New York, USA
{hzheng15,hliao6,lchen63,wxiong5,tchen45,jluo}@cs.rochester.edu

Abstract. Example-guided image synthesis has recently been attempted to synthesize an image from a semantic label map and an exemplary image. In the task, the additional exemplar image provides the style guidance that controls the appearance of the synthesized output. Despite the controllability advantage, the existing models are designed on datasets with specific and roughly aligned objects. In this paper, we tackle a more challenging and general task, where the exemplar is a scene image that is semantically different from the given label map. To this end, we first propose a Masked Spatial-Channel Attention (MSCA) module which models the correspondence between two scenes via efficient decoupled attention. Next, we propose an end-to-end network for joint global and local feature alignment and synthesis. Finally, we propose a novel self-supervision task to enable training. Experiments on the large-scale and more diverse COCO-stuff dataset show significant improvements over the existing methods. Moreover, our approach provides interpretability and can be readily extended to other content manipulation tasks including style and spatial interpolation or extrapolation.

Keywords: Example-guided image synthesis · Self-supervised learning · Correspondence modeling · Efficient attention

1 Introduction

Conditional generative adversarial network (cGAN) [34] has recently made substantial progress in realistic image synthesis. In cGAN, a generator $\hat{x} = G(c, z)$ aims to output a realistic image $\hat{x}$ with a constraint implicitly encoded by c. Conversely, a discriminator $D(x, c)$ learns such a constraint from ground-truth pairs $\langle x, c \rangle$ by predicting if $\langle \hat{x}, c \rangle$ is real or generated.

The current cGAN models [20,36,43] for semantic image synthesis aim to solve the *structural consistency* constraint where the output image $\hat{x} = G(c)$ is

Electronic supplementary material The online version of this chapter (https://doi.org/10.1007/978-3-030-58568-6_25) contains supplementary material, which is available to authorized users.

A. Vedaldi et al. (Eds.): ECCV 2020, LNCS 12359, pp. 422–439, 2020.
https://doi.org/10.1007/978-3-030-58568-6_25

Structure
constraints   Style constraints

Outputs

Fig. 1. Our task aims to synthesize style-consistent images from a semantic label map (row 1, column 1) and a **structurally and semantically different** exemplar image (row 1, columns 2–7). In spite of the differences between the two scenes, our model can synthesize high-quality images of consistent styles with the reference images.

required to be aligned to a semantic label map c. However, for such a model, the style of $\hat{x}$ is inherently determined by the model and thus cannot be controlled by the user. To provide desired controllability over the generated styles, previous studies [27,41] impose additional constraints and allow more inputs to the generator: $\hat{x}_{2\rightarrow1} = G(c_1, x_2, z)$, where x_2 is an exemplar image that guides the style of c_1. However, previous studies are designed on specified datasets such as face [30,37], dancing [41] or street view [46], where the exemplar images and semantic label map usually contain similar semantics and spatial structures.

Different from the previous studies, we address a more challenging example-guided synthesis task that transfers styles *across semantically very different scenes*. As shown in Fig. 1, given a semantic label map c_1 (column 1) and an arbitrary scene image x_2 (row 1, column 2–7), the task aims to generate a new scene image $\hat{x}_{2\rightarrow1}$ (row 2) that matches the semantic structure of c_1 and the scene style of x_2. The challenge is that scene images have complex semantic structures as well as diversified scene styles, and more importantly, the inputs c_1 and x_2 can be *structurally unaligned* and *semantically different*. Therefore, a mechanism is required to better match the structures and semantics for coherent synthesis.

In this paper, we propose a novel Masked Spatial-Channel Attention (MSCA) module (Sect. 3.2) to propagate features across unstructured scenes. Our module is inspired by a recent work [7] for attention-based object recognition, but instead, we propose a new cross-attention mechanism to model the semantic correspondence for image synthesis. Moreover, our method is based on the novel design of spatial-channel decoupling that allows efficient computation. To facilitate example-guided synthesis, we further improve the module by including: i) feature masking for semantic outlier filtering, ii) multi-scaling for global-local feature processing, and iii) resolution extending for image synthesis. As a result, our module provides both clear physical meaning and interpretability for the example-guided synthesis task.

We formulate the proposed approach under a unified synthesis network for joint feature extraction, alignment and image synthesis. We achieve this by applying MSCA modules to the extracted features for multi-scale feature domain alignment. Next, we apply a recent feature normalization technique, SPADE [36] on the aligned features to allow spatially-controllable synthesis. To facilitate the learning of this network, we propose a novel self-supervision task. As opposed to [41], our scheme requires only semantically parsed images for training and does not rely on video data. We show that a model trained with this approach generalizes across different scene semantics (See Fig. 1).

Our main contributions include the following:

- A novel masked spatial-channel attention (MSCA) module to propagate features between two semantically different scenes.
- A unified example-guided synthesis network for joint feature extraction, alignment and image synthesis.
- A novel self-supervision scheme that only requires semantically annotated images for training but not at the testing (image synthesis) stage.
- Significant improvements over the existing methods on the COCO-stuff [3] dataset, as well as interpretability and easy extensions to other content manipulation tasks.

2   Related Work

Generative Adversarial Networks. Recent years have witnessed the progress of generative adversarial networks (GANs) [11] for image synthesis. A GAN model consists of a generator and a discriminator where the generator serves to produce realistic images that cannot be distinguished from the real ones by the discriminator. Recent techniques for realistic image synthesis include modified losses [1,33,38], model regularization [35], self-attention [2,48], feature normalization [24] and progressive synthesis [23].

Image-to-Image Translation (I2I). I2I translation aims to translate images from a source domain to a target domain. The initial work of Isola *et al.* [20] proposes a conditional GAN framework to learn I2I translation with paired images. Wang *et al.* [43] improve the conditional GAN for high-resolution synthesis and content manipulation. To enable I2I translation without using paired data, a few works [4,18,25,29,50] apply the cycle consistency constraint in training. Recent works on photo-realistic image synthesis take semantic label maps as inputs for image synthesis. Specifically, Wang *et al.* [43] extend the conditional GAN for high-resolution synthesis, Chen *et al.* [6] propose a cascade refinement pipeline. More recently, Park *et al.* [36] propose spatial-adaptive normalization for realistic image generation.

Example-Guided Style Transfer and Synthesis. Example guided style transfer [8,13] aims to transfer the style of an example image to a target image. Recent works [4,10,12,16,17,22,26,31,45] utilize deep neural network features to model and transfer styles. Several frameworks [18,32] perform style transfer via image domain style and content disentanglement. In addition, domain adaptation [4] applies a cycle consistency loss to perform cross-domain style transformation.

More recently, example-guided synthesis [27,41] is proposed to transfer the style of an example image to a target condition, e.g.. a semantic label map. Specifically, Lin *et al.* [27] apply dual learning to disentangle the style for guided synthesis, Wang *et al.* [41] extract style-consistent data pairs from videos for model training. In addition, Park *et al.* [36] adopt an I2I network under the auto-encoding framework for example-guided image synthesis. Different from [27,36,41], we address style alignment issue between arbitrary scenes for *region and semantic aware* style integration. Furthermore, our self-supervised learning scheme does not require video data and is a generalize and more challenging auto-encoding task.

Correspondence Matching for Synthesis. Finding correspondence is critical for many synthesis tasks. For instance, Siarohin *et al.* [39] apply the affine transformation on reference person images to improve pose-guided person image synthesis, Wang *et al.* [42] use optical flow to align frames for coherent video synthesis. However, the affine transformation and optical flow cannot adequately model the correspondences between two structurally very different scenes.

Efficient Attention Modeling. The self-attention [44,48] can capture general pair-wise correspondences. However, it is computationally intensive at high-resolution. To enable fast attention computation, GCNL [47] and CCCA [19] respectively apply Taylor series expansion and criss-cross attention to approximate self-attention. Alternatively, A^2-Nets [7] factorize self-attention to solve video classification tasks. Inspired by [7], we propose an attention-based module named MSCA. It is worth noting MSCA is based on cross-attention and feature masking for modeling image correspondence.

3   Method

The proposed approach aims to generate scene images that align with given semantic maps. Differ from conventional semantic image synthesis methods [20,36,43], our model takes an exemplary scene as an extra input to provide more controllability over the generated scene image. Unlike existing example-based approaches [27,41], our model addresses a more challenging case where the exemplary inputs are structurally and semantically unaligned with the given semantic map.

Our method takes a semantic label map c_1, a reference image x_2 and its corresponding parsed semantic label map $\tilde{c}_2$ as inputs and synthesizes an image $\hat{x}_{2\to1}$ which matches the style of x_2 and structure of x_1 using a generator G, $\hat{x}_{2\to1} = G(c_1, x_2, \tilde{c}_2)$. As shown in Fig. 2 left, the generator G consists of three parts, namely i) feature extraction ii) feature alignment and iii) image synthesis. In Sect. 3.1, we describe the first part that extracts features from inputs of both scenes. In Sect. 3.2, we propose a masked spatial-channel attention (MSCA) module to distill features and discover relations between two arbitrarily structured scenes. Unlike the affine-transformation [21] and flow-base warping [42], MSCA provides better interpretability to the scene alignment task. In Sect. 3.3, we introduce how to use the aligned features for image synthesis. Finally, in Sect. 3.4, we propose a self-supervised scheme to facilitate learning.

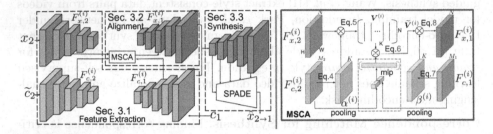

Fig. 2. Left: our generator consists of three steps, namely *feature extraction, feature alignment*, and *image synthesis*. We describe each step in its corresponding section, respectively. **Right**: The MSCA module for feature alignment (at scale i). Our module takes image feature map $F_{x,2}^{(i)}$ and segmentation feature map $F_{c,1}^{(i)}$, $F_{c,2}^{(i)}$ as inputs to output a new image feature map $F_{x,1}^{(i)}$ that is aligned to condition c_1.

3.1 Feature Extraction

Taking an image x_2 and label maps $c_1, \widetilde{c}_2$ as inputs, the feature extraction module extracts multi-scale feature maps for each input. Specifically, the feature map $F_{x,2}^{(i)}$ of image x_2 at scale i is computed by:

$$F_{x,2}^{(i)} = W_x^{(i)} * F_{\text{vgg}}^{(i)}(x_2), \quad \text{for } i \in \{0, \ldots, L\}, \tag{1}$$

where $*$ denotes the convolution operation, $F_{\text{vgg}}^{(i)}$ denotes the feature map extracted by VGG-19 [40] at scale i, and $W_x^{(i)}$ denotes a 1×1 convolutional kernel for feature compression. L is the number scales and we set $L = 4$ in this paper.

For label map c_1, its feature $F_{c,1}^{(i)}$ is computed by:

$$F_{c,1}^{(i)} = \begin{cases} \text{LReLU}(W_c^{(i)} * c_1^{(i)}) & \text{for } i = L, \\ \text{LReLU}(W_c^{(i)} * [\Uparrow (F_{c,1}^{(i+1)}), c_1^{(i)}]) & \text{otherwise,} \end{cases} \tag{2}$$

where $\Uparrow (\cdot)$ denotes $\times 2$ bilinear interpolation, $c_1^{(i)}$ denotes the resized label map, $W_c^{(i)}$ denotes a 1×1 convolutional kernel for feature extraction, and operation $[\cdot, \cdot]$ denotes channel-wise concatenation. Note that as scale i decreases from L

down to 0, the feature resolutions in Eq. 2 are progressively increased to match a finer label maps $c_1^{(i)}$.

Similarly, applying Eq. 2 with the same weights to label map $\tilde{c}_2$, we can extract its features $F_{c,2}^{(i)}$:

$$F_{c,2}^{(i)} = \begin{cases} \text{LReLU}(W_c^{(i)} * c_2^{(i)}) & \text{for } i = L \\ \text{LReLU}(W_c^{(i)} * [\Uparrow (F_{c,2}^{(i+1)}), \tilde{c}_2^{(i)}]) & \text{otherwise} \end{cases}. \qquad (3)$$

3.2 Masked Spatial-Channel Attention Module

As shown in Fig. 2 right, taking the image features $F_{x,2}^{(i)}$ and the label map features $F_{c,1}^{(i)}$, $F_{c,2}^{(i)}$ as inputs[1], the MSCA module generates a new image feature map $F_{x,1}^{(i)}$ that has the content of $F_{x,2}^{(i)}$ but is aligned with $F_{c,1}^{(i)}$. We elaborate the detailed procedures as follows:

Spatial Attention. Given feature maps $F_{x,2}^{(i)}, F_{c,2}^{(i)}$ of the exemplar scene, the module first computes a spatial attention tensor $\alpha^{(i)} \in [0,1]^{K \cdot H \cdot W}$:

$$\alpha^{(i)} = \text{softmax}_{2,3}(\phi^{(i)} * [F_{x,2}^{(i)}, F_{c,2}^{(i)}]), \qquad (4)$$

with $\phi^{(i)} \in \mathbb{R}^{(N+M_2) \cdot K}$ denoting a 1×1 convolutional filter and $\text{softmax}_{2,3}$ denoting a 2D softmax function on spatial dimensions $\{2,3\}$. The output tensor contains K attention maps of resolution $H \times W$, which serve to attend K different spatial regions on image feature $F_{x,2}^{(i)}$.

Spatial Aggregation. Then, the module aggregates K feature vectors from $F_{x,2}^{(i)}$ using the K spatial attention maps of $\alpha^{(i)}$ from Eq. 4. Specifically, a matrix dot product is performed:

$$V^{(i)} = F_{x,2}^{(i)}(\alpha^{(i)})^{\mathsf{T}}, \qquad (5)$$

with $\alpha^{(i)} \in [0,1]^{K \cdot HW}$ and $F_{x,2}^{(i)} \in \mathbb{R}^{N \cdot HW}$ denoting the reshaped versions of $\alpha^{(i)}$ and $F_{x,2}^{(i)}$, respectively. The output $V^{(i)} \in \mathbb{R}^{N \cdot K}$ stores feature vectors spatially aggregated from the K independent regions of $F_{x,2}^{(i)}$.

Feature Masking. The exemplar scene x_2 may contain irrelevant semantics to the label map c_1, and conversely, c_1 may contain semantics that are unrelated to x_2. To address this issue, we apply feature masking on the output of Eq. 5 by multiplying $V^{(i)}$ with a length-K gating vector at each row:

$$\widetilde{V}^{(i)} = (V^{(i)})^T \circ \text{mlp}([\text{gap}(F_{c,1}^{(i)}), \text{gap}(F_{c,2}^{(i)})]), \qquad (6)$$

---

[1] We assume spatial resolution at scale i being $H \times W$ and channel size of $F_{x,2}^{(i)}$, $F_{c,1}^{(i)}$, $F_{c,2}^{(i)}$ being N, M_1, M_2, respectively.

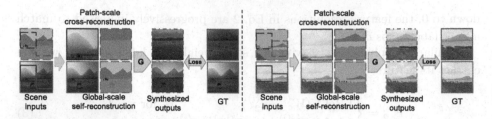

Fig. 3. Our self-supervision scheme performs cross-reconstruction at the patch scale (top row) and self-reconstruction at the global scale (bottom row). The solid, dashed and dotted bounding boxes respective represent images, semantic label maps, and synthesized outputs. Boxes with the same color are cropped from the same position.

where $\mathrm{mlp}(\cdot)$ denotes a 2-layer MLP followed by a sigmoid function, gap denotes a global average pooling layer, $\circ$ denotes broadcast element-wise multiplication, and $\widetilde{V}^{(i)}$ denotes the masked features. The design of feature masking in Eq. 6 resembles to Squeeze-and-Excitation [15]. Using the integration of global information from label maps c_1 and $\tilde{c}_2$, features are filtered.

Channel Attention. Given feature $F_{c,1}^{(i)}$ of label map c_1, a channel attention tensor $\beta^{(i)} \in [0,1]^{K \cdot H \cdot W}$ is generated as follows:

$$\beta^{(i)} = \mathrm{softmax}_1(\psi^{(i)} * F_{c,1}^{(i)}), \tag{7}$$

with $\psi^{(i)} \in \mathbb{R}^{M_1 \cdot K}$ denoting a 1×1 convolutional filter and $\mathrm{softmax}_1$ denoting a softmax function on channel dimension. The output $\beta^{(i)}$ serves to dynamically reuse features from $\widetilde{V}^{(i)}$.

Channel Aggregation. With channel attention $\beta^{(i)}$ computed in Eq. 7, feature vectors at HW spatial locations are aggregated again from $\widetilde{V}^{(i)}$ via matrix dot product:

$$F_{x,1}^{(i)} = \widetilde{V}^{(i)}(\beta^{(i)})^{\mathsf{T}}, \tag{8}$$

where $\beta^{(i)} \in \mathbb{R}^{K \cdot HW}$ denotes the reshaped version of $\beta^{(i)}$. The output $F_{x,1}^{(i)} \in \mathbb{R}^{N \cdot HW}$ represents the aggregated features at HW locations. The output feature map $F_{x,1}^{(i)}$ is generated by reshaping $F_{x,1}^{(i)}$ to size $N \times H \times W$.

Remarks. Spatial attention (Eq. 4) and aggregation (Eq. 5) attend to K independent regions from feature $F_{x,2}^{(i)}$, then store the K features into $V^{(i)}$. After feature masking, given a new label map c_1, channel attention (Eq. 4) and aggregation (Eq. 8) combine $\widetilde{V}^{(i)}$ at each location to compute an output feature map. As a result, each output location finds its correspondent regional features or ignored via feature masking. In this way, the feature of example scene is aligned. Note that when $K = 1$ and $\alpha^{(i)}$ is constant, the above operations is essentially a global average pooling. We show in the experiment that $K = 8$ is sufficient to dynamically capture visually significant scene regions for alignment.

Multi-scaling. Both global color tone and local appearances are informative for the style-constraint synthesis. Therefore, we apply MSCA modules at all scales $i \in \{0, \dots, L\}$ to generate global and local features $F_{x,1}^{(i)}$.

Fig. 4. Green box from left to right: the inputs for example-guided synthesis, i.e. target label maps, exemplar label parsing from Deeplab-v2 [5], and exemplar images. Red box from left to right: visual comparisons with cI2I [27], EGSC-IT [32], SPADE_VAE [36], four ablation models, and our full model. Blue box from left to right: the retrieved ground-truth before and after color correction [45]. Our full model generates the most style-consistent results with the exemplar images. (Color figure online)

3.3 Image Synthesis

The extracted features $F_{c,1}^{(i)}$ in Sect. 3.1 capture the semantic structure of c_1, whereas the aligned features $F_{x,1}^{(i)}$ in Sect. 3.2 capture the appearance style of the example scene. In this section, we leverage $F_{c,1}^{(i)}$ and $F_{x,1}^{(i)}$ as control signals to generate output images with desired structures and styles.

Specifically, we adopt a recent synthesis model, SPADE [36], and feed the concatenation of $F_{x,1}^{(i)}$ and $F_{c,1}^{(i)}$ to the spatially-adaptive denormalization layer of SPADE at each scale. By taking the style and structure signal as inputs, spatially-controllable image synthesis is achieved. We refer readers to appendix for more network details of the synthesis module.

3.4 Self-supervised Training

Training an example-guided synthesis model that can transfer styles across semantically different scenes is challenging. First, style-consistent scene images are hard to acquire. A previous work [41] generates style-consistent pairs from videos. However, collecting scene videos can be more labor intensive. Second, even with ground truth style-consistent pairs, the trained model is not guaranteed to generalize to a new arbitrary scene.

We propose a novel self-supervised scheme to enable style-transfer between two structurally and semantically different scenes. Our solution is motivated by the fact that the style of a scene image is stationary, meaning that patches cropped from the same scene share largely the same style. Moreover, non-overlapping patches from the same scene may contain new structures and semantic labeling, which is essential for the learned model to generalize better.

We first design a *cross-reconstruction* task at the patch scale: given patches x_p and x_q cropped from the same scene image x, the generator is asked to reconstruct x_p using x_q. Formally,

$$\hat{x}_p = G(c_p, x_q, \widetilde{c}_q). \tag{9}$$

Note that c_p and $\widetilde{c}_q$ contain different semantic labeling. Therefore, the generator are required to infer the correlation between different semantic labeling for coherent style transfer. An illustrative example is shown in Fig. 3. More details on patch sampling is included in the appendix.

The cross-reconstruction task is designed at the patch scale and may not generalize well to the global scale. In fact, the generator trained with the patch-level task alone tends to generate repetitive local textures (in Sect. 4). Therefore, we further design a *self-reconstruction* task at the global scale, which reconstructs an global image x from itself:

$$\hat{x} = G(c, x, \widetilde{c}). \tag{10}$$

Our training objective for generator G and discriminator D is formulated as:

$$\mathcal{L}(G, D) = \log D(x_p, c_p, x_q, \widetilde{c}_q) + \log(1 - D(\hat{x}_p, c_p, x_q, \widetilde{c}_q)) + \mathcal{L}_{spade}(\hat{x}_p, x_p)$$
$$+ \lambda\{\log D(x, c, x, \widetilde{c}) + \log(1 - D(\hat{x}, c, x, \widetilde{c})) + \mathcal{L}_{spade}(\hat{x}, x)\} \tag{11}$$

where $\mathcal{L}_{spade}$ refers to the VGG and GAN feature matching losses defined in [36] and λ is a parameter that controls the importance of the two self-supervised tasks. We set $\lambda = 1$ in our experiments. Our full objective for self-supervised training is:

$$G^* = \arg \min_{G} \max_{D} \mathcal{L}(G, D) \tag{12}$$

4   Experiments

Dataset. Our model is trained on the *COCO-stuff* dataset [3]. It contains densely annotated images captured from various scenes. We remove images with indoor scenes and large objects from the dataset, resulting in $34,698/499$ scene images for training/testing, respectively. The COCO-stuff dataset does not provide ground-truth for the example-guided synthesis task, i.e. two images with

Table 1. Quantitative comparisons of different methods and ablation models in terms of PSNR, LPIPS [49], Fréchet Inception Distance (FID) [14] and style loss ($\mathcal{L}_{style}$) [9]. Higher scores are better for metrics with uparrow (↑), and vice versa.

Task	Measures	cI2I [27]	EGSC-IT [41]	SPADE-VAE [36]	Ours GAP	Ours MSCA w/o att	Ours MSCA w/o fm	Ours MSCA w/o global	Ours full
retrieving	PSNR↑	9.50	12.57	15.77	15.85	11.96	16.24	15.98	**16.65**
	LPIPS↓	0.757	0.581	0.483	0.457	0.522	0.451	0.446	**0.437**
	FID↓	228.63	163.23	102.68	101.74	112.83	100.01	96.66	**91.91**
	$\mathcal{L}_{style}$ ↓	3.53e−3	1.69e−3	1.07e−3	6.40e−4	7.10e−3	7.62e−4	7.21e−4	**5.34e−4**
mirroring	PSNR↑	9.46	12.44	15.37	15.80	11.95	16.02	16.58	**17.03**
	LPIPS↓	0.759	0.602	0.477	0.438	0.510	0.437	0.421	**0.397**
	FID↓	242.73	190.01	90.99	89.15	102.41	90.52	85.92	**76.75**
	$\mathcal{L}_{style}$ ↓	4.14e−3	2.03e−3	1.67e−3	7.45e−4	7.99e−3	8.76e−4	6.69e−4	**3.96e−4**
duplicating	PSNR↑	9.46	12.45	15.43	15.80	11.94	16.02	16.62	**17.03**
	LPIPS↓	0.759	0.602	0.476	0.438	0.510	0.438	0.421	**0.397**
	FID↓	242.81	190.02	90.97	89.22	103.23	90.56	86.20	**76.64**
	$\mathcal{L}_{style}$ ↓	4.15e−3	2.03e−3	1.66e−3	7.41e−4	7.94e−3	8.81e−4	6.53e−4	**3.97e−4**

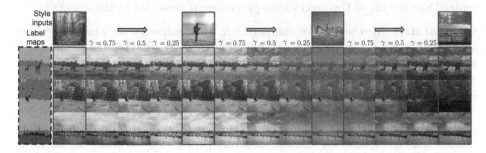

Style inputs

Label maps

$\gamma = 0.75$ $\gamma = 0.5$ $\gamma = 0.25$ $\gamma = 0.75$ $\gamma = 0.5$ $\gamma = 0.25$ $\gamma = 0.75$ $\gamma = 0.5$ $\gamma = 0.25$

Fig. 5. With a slight modification and no further training, our model can perform style interpolation between exemplar inputs. Note how our model interpolates styles for new semantics, e.g. "river" in row 3.

the exact same styles. To qualitatively evaluate the performances, we designed three tasks where the ground-truth image can be obtained: i) *duplicating task* self-reconstructs an image using itself as exemplar and its semantic label map as layout condition, ii) *mirroring task* reconstructs an image using its mirrored version as exemplar and the semantic label map as layout condition, iii) *retrieving task*: requires a model to reconstruct an ground-truth (GT) image using its semantic label map and a retrieved image from a image pool. To retrieve an image that best match GT in styles, we first select 20 candidate images from the image pool that has the greatest label histogram intersections with the GT image. Afterwards, the best-matched image is select out of candidates using SIFT Flow [28]. Finally, since the color of GT is different from the retrieved image, we apply color correction [45] on GT to eliminate color discrepancy. Examples of GT before and after color correction are shown in the blue box in Fig. 4.

Implementation Details. We use a COCO-stuff pretrained Deeplab-v2 [5] model to generate semantic label maps from exemplar images. During training, we resize images to 512×512 then crop two non-overlapping patches of size 256×256 to facilitate patch-based cross-reconstruction. After 20 epochs, we increase the patch size to 384×384 for cross-patch reconstruction in order to improve generalization to global scenes. Details of the patch sampling procedure are provided in the appendix.

For the MSCA modules from scale 0 to 4, the number of attention maps K are respectively set to $8, 16, 16, 16, 16$. The learning rate is set to 0.0002 for the generator and the discriminator. The weights of generator are updated every 5 iterations. Our synthesis model and all comparative models based on SPADE backbone are trained for 40 epochs to generate the results in the experiments.

Before training, we pretrain the spatial-channel attention with a lightweight feature decoder to improve training efficiency. Specifically, at each scale, the concatenation of $F_{x,1}^{(i)}$ and $F_{c,1}^{(i)}$ in Sect. 3.3 at each scale is fed into a 1×1 convolutional layer to reconstruct the ground-truth VGG feature at the corresponding scale. More details of the pretraining procedure is provided in the appendix.

Comparative Methods. We compare our approach with an example-guided synthesis approach: variational autoencoding SPADE (SPADE_VAE) [36] which is based on a self-reconstruction loss for training. We also trained cI2I [27], EGSC-IT [32] and SCGAN [41] on COCO-stuff dataset. cI2I and EGSC-IT are originally designed for exemplar-guided image-to-image translation. As a result, we observed that cI2I and EGSC-IT have difficulty generating images from one-hot encoded semantic label maps. However, these models can synthesize reasonable images from color-encoded semantic label maps. Finally, we note that SCGAN is not directly applicable to COCO-stuff dataset, as its positive pairs are sampled from video data. We attempted to modify SCGAN such that its positive pairs can be generated from our self-supervision task. However, we could not achieve reasonable image outputs. We speculate that the negative sampling and semantic consistency loss of SCGAN is not optimal for COCO-stuff dataset, as COCO-stuff dataset contains much larger variations for negative pairs. Finally, four ablation models are evaluated (see Ablation Study).

Quantitative Evaluation. For quantitative evaluation, we apply PSNR as the low-level metric. Furthermore, perceptual-level metrics including Perceptual Image Patch Similarity Distance (LPIPS) [49], Fréchet Inception Distance (FID) [14] and style loss ($\mathcal{L}_{style}$) of [9] are evaluated on different methods. The linearly calibrated VGG model is used to compute LPIPS distance.

Among the four competitive methods (cI2I, EGSC-IT, SPADE_VAE and ours full) in Table 1, our method clearly outperforms the remaining methods both in low-level and perceptual-level measurements, suggesting that our model can better preserve color and texture appearances. Also, we observe that without further modification, the off-the-shelf example-guided image translation approaches cannot perform well on image synthesis tasks (cI2I, EGSC-IT). It suggests that example-guided image-synthesis task can be more challenging. Finally, a simple

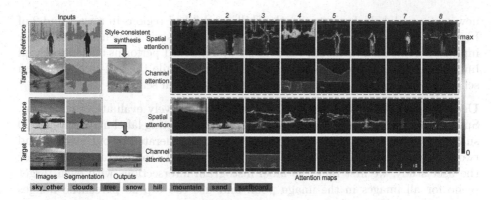

Fig. 6. Left: inputs and outputs of our model. Right: the $K = 8$ learned spatial and channel attention that attends and transfer feature between individual exemplar and target regions. By examining the semantics label maps, we observe the following transformation patterns: sky_other → clouds, tree → {tree, hill} for the 1st sample, and clouds → clouds, snow → sand, other → {surfboard,other} for the 2nd sample.

synthesis model (ours GAP) outperforms SPADE_VAE, suggesting that the self-supervised task in Sect. 3.4 is beneficial to the exampled-guided synthesis task (see Ablation Study for more details).

Qualitative Evaluation. Figure 4 qualitatively compares our approach against the remaining approach on four scenes. We observe that our full model generates more style-consistent results with the exemplar images. In comparisons, SPADE_VAE tends to generate results with low color contrast, as it lacks the mechanism and supervision to perform region-aware style transformation. In addition, the existing example-guided image-to-image approaches (cI2I, EGSC-IT) cannot generalize well to the image synthesis tasks.

Ablation Study. To evaluate the effectiveness of our design, we separately train four variants of our model: i) our GAP that replaces the MSCA module with global average pooling, ii) ours MSCA w/o att that keeps MSCA modules but replaces spatial and channel attention with one-hot label maps from source and target domains, respectively. In such a way, alignment is performed only for regions with the same semantic labeling, iii) ours MSCA w/o fm that keeps MSCA modules but removes the feature masking procedures, and iv) ours MSCA w/o global that is trained without using global-level self-reconstruction (Eq. 12) or increased patchsize.

In Table 1, our full model clearly achieves the best qualitative results. In Fig. 4, ours GAP tends to produce images with deviated colors since it averages the style features from all exemplar regions. In contrast, our model dynamically transfers appearance for individual regions. We observe that ours w/o att is less stable in training and cannot generate plausible results. We suspect that the label-level alignment generates more misaligned and noisier feature maps, thus hurting training. ours MSCA w/o fm tends to generate inconsistent colors for new semantic labels, for instance, the "hill" and "sky" regions in rows 1 and 2 of Fig. 4. In contrast, our model can eliminate the undesired influence of exemplar inputs on new semantic labels. `ours MSCA w/o global` performs reasonably well but it tends to generate repetitive local textures, while the self-reconstruction scheme helps our model generalize better at the global scale.

User Study. We conduct a user study to qualitatively evaluate our method. Specifically, we retrieve an exemplar image for each testing label map, and ask 20 subjects to choose the most style-consistent results generated by our method and two competitive baselines (`SPADE_VAE` and `ours GAP`). To generate samples for the user study, we first rank the label histogram intersections with each target scene for all images in the image pool, and use the top 20 percentile images as exemplars[2]. The subjects are given unlimited time to make their selections. For each subject, we randomly generate 100 questions from the dataset. Table 2 shows the evaluation results. First, all subjects strongly favor our results. Second, `ours GAP` is favored more than twice over `SPADE_VAE` [36], further suggesting that the proposed self-supervision scheme is effective since `ours GAP` is also trained with self-supervision.

Table 2. User preference study. The numbers indicate the percentage of user who favors the result generated by different methods. Two com

Methods	SPADE_VAE [36]	Ours GAP	Ours full
Choose rate	15.6	29.3	55.0

Effect of Attention. To understand the effect of spatial-channel attention, we visualize the learned spatial and channel attention in Fig. 6. We observe that: a) spatial attention can attend to multiple regions of the reference image. For each reference region, channel attention finds the corresponding target region. b) spatial-channel attention can detect and utilize the similarities of semantic labels to facilitate style features transfer. In the first sample of Fig. 6, attention in channels 1, 4 respectively perform transformations: `sky_other` → `clouds`, `tree` → {`tree, hill`}. In the second sample, attention in channels 1, 2, 7 respectively perform transformations: `clouds` → `clouds`, `snow` → `sand` and `other` → {`surfboard, other`}. We provide more analysis on the effect of attention in the appendix.

Interpolation. We can easily control the synthesized styles in the test stage by manipulating spatial and channel attentions. First, by manipulating the spatial attention of two exemplar inputs, our trained model can perform style interpolation between the two exemplar. The results are shown in Fig. 5. Next, by manipulating the channel attention, our trained model can perform spatial style

---

[2] This differs from the typical retrieve task that uses the top-1 image since the top 20 percentile images tend to be more semantically different from the target label maps.

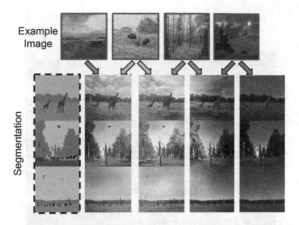

Fig. 7. With a slight modification and no further training, our model can perform spatial style interpolation. In this figure, we demonstrate a horizontal gradient style change on the output image. Please refer to Interpolation, Sect. 4 for more details.

Fig. 8. Given an exemplar patch at the center and the global semantic label map, our trained model can perform example-guided scene image extrapolation, i.e. generating style-consistent beyond-the-border images with semantic maps guidance.

interpolation. Figure 7 shows our model can interpolate between two images and generate horizontally gradient style changes. More details are included in the appendix.

Extrapolation. Given a scene patch at the center our model can achieve scene extrapolation, i.e. generating beyond-the-border image content according to the semantic map guidance. A 512×512 extrapolated image is generated by weighted combining synthesized 256×256 patches at 4 corners and 10 other random locations. As shown in Fig. 8, our model generates visually plausible extrapolated images, showing the promise of our proposed framework for guided scene panorama generation.

Style Swapping. Figure 9 shows reference-guided style swapping between six arbitrary scenes. Our model can generalize across recognizably different scenes semantics and appearances, including snow, mountain, seashore, grassland, dessert, artistic effect, and synthesize image with reasonable and consistent styles. More results and comparisons to other approaches are included in the appendix.

Fig. 9. Style-structure swapping on 6 arbitrary scenes at resolution 256 × 256. Our model can generalize across recognizably different scenes of different semantics. Note that the images along the diagonal (red boxes) are *self-reconstruction* (Color figure online)

5 Conclusion

We propose to address a challenging example-guided image synthesis task between semantically very different scenes. To propagate information between two structurally unaligned and semantically different scenes, we propose an MSCA module that leverages decoupled cross-attention for adaptive correspondence modeling. With MSCA, we propose a unified model for joint global-local alignment and image synthesis. We further propose a patch-based self-supervision scheme that enables training. Experiments on the COCO-stuff dataset show significant improvements over the existing methods. Furthermore, our approach provides interpretability and can be extended to other content manipulation tasks.

Acknowledgment. This work is supported in part by NSF awards IIS-1704337, IIS-1722847, and IIS-1813709, as well as our corporate sponsors.

References

1. Arjovsky, M., Chintala, S., Bottou, L.: Wasserstein gan. arXiv preprint arXiv:1701.07875 (2017)
2. Brock, A., Donahue, J., Simonyan, K.: Large scale GAN training for high fidelity natural image synthesis. arXiv preprint arXiv:1809.11096 (2018)

3. Caesar, H., Uijlings, J., Ferrari, V.: Coco-stuff: thing and stuff classes in context. In: Proceedings of the IEEE Conference on Computer Vision and Pattern Recognition, pp. 1209–1218 (2018)

4. Chang, H., Lu, J., Yu, F., Finkelstein, A.: Pairedcyclegan: asymmetric style transfer for applying and removing makeup. In: Proceedings of the IEEE Conference on Computer Vision and Pattern Recognition, pp. 40–48 (2018)

5. Chen, L.C., Papandreou, G., Kokkinos, I., Murphy, K., Yuille, A.L.: Deeplab: Semantic image segmentation with deep convolutional nets, Atrous convolution, and fully connected CRFs. IEEE Trans. Pattern Anal. Mach. Intell. **40**(4), 834–848 (2017)

6. Chen, Q., Koltun, V.: Photographic image synthesis with cascaded refinement networks. In: Proceedings of the IEEE International Conference on Computer Vision, pp. 1511–1520 (2017)

7. Chen, Y., Kalantidis, Y., Li, J., Yan, S., Feng, J.: $A^{\wedge}$ 2-nets: double attention networks. In: Advances in Neural Information Processing Systems, pp. 352–361 (2018)

8. Efros, A.A., Freeman, W.T.: Image quilting for texture synthesis and transfer. In: Proceedings of the 28th Annual Conference on Computer Graphics and Interactive Techniques, pp. 341–346. ACM (2001)

9. Gatys, L.A., Ecker, A.S., Bethge, M.: A neural algorithm of artistic style. arXiv preprint arXiv:1508.06576 (2015)

10. Gatys, L.A., Ecker, A.S., Bethge, M.: Image style transfer using convolutional neural networks. In: Proceedings of the IEEE Conference on Computer Vision and Pattern Recognition, pp. 2414–2423 (2016)

11. Goodfellow, I., et al.: Generative adversarial nets. In: Advances in Neural Information Processing Systems, pp. 2672–2680 (2014)

12. Gu, S., Chen, C., Liao, J., Yuan, L.: Arbitrary style transfer with deep feature reshuffle. In: Proceedings of the IEEE Conference on Computer Vision and Pattern Recognition, pp. 8222–8231 (2018)

13. Hertzmann, A., Jacobs, C.E., Oliver, N., Curless, B., Salesin, D.H.: Image analogies. In: Proceedings of the 28th Annual Conference on Computer Graphics and Interactive Techniques, pp. 327–340. ACM (2001)

14. Heusel, M., Ramsauer, H., Unterthiner, T., Nessler, B., Hochreiter, S.: GANs trained by a two time-scale update rule converge to a local Nash equilibrium. In: Advances in Neural Information Processing Systems, pp. 6626–6637 (2017)

15. Hu, J., Shen, L., Sun, G.: Squeeze-and-excitation networks. In: Proceedings of the IEEE Conference on Computer Vision and Pattern Recognition, pp. 7132–7141 (2018)

16. Huang, H., et al.: Real-time neural style transfer for videos. In: Proceedings of the IEEE Conference on Computer Vision and Pattern Recognition, pp. 783–791 (2017)

17. Huang, X., Belongie, S.: Arbitrary style transfer in real-time with adaptive instance normalization. In: Proceedings of the IEEE International Conference on Computer Vision, pp. 1501–1510 (2017)

18. Huang, X., Liu, M.-Y., Belongie, S., Kautz, J.: Multimodal unsupervised image-to-image translation. In: Ferrari, V., Hebert, M., Sminchisescu, C., Weiss, Y. (eds.) ECCV 2018. LNCS, vol. 11207, pp. 179–196. Springer, Cham (2018). https://doi.org/10.1007/978-3-030-01219-9_11

19. Huang, Z., Wang, X., Huang, L., Huang, C., Wei, Y., Liu, W.: CCNET: Criss-cross attention for semantic segmentation. In: Proceedings of the IEEE International Conference on Computer Vision, pp. 603–612 (2019)

20. Isola, P., Zhu, J.Y., Zhou, T., Efros, A.A.: Image-to-image translation with conditional adversarial networks. In: Proceedings of the IEEE Conference on Computer Vision and Pattern Recognition, pp. 1125–1134 (2017)
21. Jaderberg, M., Simonyan, K., Zisserman, A., et al.: Spatial transformer networks. In: Advances in Neural Information Processing Systems, pp. 2017–2025 (2015)
22. Johnson, J., Alahi, A., Fei-Fei, L.: Perceptual losses for real-time style transfer and super-resolution. In: Leibe, B., Matas, J., Sebe, N., Welling, M. (eds.) ECCV 2016. LNCS, vol. 9906, pp. 694–711. Springer, Cham (2016). https://doi.org/10.1007/978-3-319-46475-6_43
23. Karras, T., Aila, T., Laine, S., Lehtinen, J.: Progressive growing of GANs for improved quality, stability, and variation. arXiv preprint arXiv:1710.10196 (2017)
24. Karras, T., Laine, S., Aila, T.: A style-based generator architecture for generative adversarial networks. In: Proceedings of the IEEE Conference on Computer Vision and Pattern Recognition, pp. 4401–4410 (2019)
25. Lee, H.Y., Tseng, H.Y., Huang, J.B., Singh, M., Yang, M.H.: Diverse image-to-image translation via disentangled representations. In: Proceedings of the European Conference on Computer Vision (ECCV), pp. 35–51 (2018)
26. Liao, J., Yao, Y., Yuan, L., Hua, G., Kang, S.B.: Visual attribute transfer through deep image analogy. arXiv preprint arXiv:1705.01088 (2017)
27. Lin, J., Xia, Y., Qin, T., Chen, Z., Liu, T.Y.: Conditional image-to-image translation. In: Proceedings of the IEEE Conference on Computer Vision and Pattern Recognition, pp. 5524–5532 (2018)
28. Liu, C., Yuen, J., Torralba, A.: Sift flow: dense correspondence across scenes and its applications. IEEE Trans. Pattern Anal. Mach. Intell. 33(5), 978–994 (2010)
29. Liu, M.Y., Breuel, T., Kautz, J.: Unsupervised image-to-image translation networks. In: Advances in Neural Information Processing Systems, pp. 700–708 (2017)
30. Liu, Z., Luo, P., Wang, X., Tang, X.: Deep learning face attributes in the wild. In: Proceedings of the IEEE International Conference on Computer Vision, pp. 3730–3738 (2015)
31. Luan, F., Paris, S., Shechtman, E., Bala, K.: Deep photo style transfer. In: Proceedings of the IEEE Conference on Computer Vision and Pattern Recognition, pp. 4990–4998 (2017)
32. Ma, L., Jia, X., Georgoulis, S., Tuytelaars, T., Van Gool, L.: Exemplar guided unsupervised image-to-image translation with semantic consistency. arXiv preprint arXiv:1805.11145 (2018)
33. Mao, X., Li, Q., Xie, H., Lau, R.Y., Wang, Z., Paul Smolley, S.: Least squares generative adversarial networks. In: Proceedings of the IEEE International Conference on Computer Vision, pp. 2794–2802 (2017)
34. Mirza, M., Osindero, S.: Conditional generative adversarial nets. arXiv preprint arXiv:1411.1784 (2014)
35. Miyato, T., Kataoka, T., Koyama, M., Yoshida, Y.: Spectral normalization for generative adversarial networks. arXiv preprint arXiv:1802.05957 (2018)
36. Park, T., Liu, M.Y., Wang, T.C., Zhu, J.Y.: Semantic image synthesis with spatially-adaptive normalization. In: Proceedings of the IEEE Conference on Computer Vision and Pattern Recognition (2019)
37. Rössler, A., Cozzolino, D., Verdoliva, L., Riess, C., Thies, J., Nießner, M.: Faceforensics: a large-scale video dataset for forgery detection in human faces. arXiv preprint arXiv:1803.09179 (2018)
38. Salimans, T., Goodfellow, I., Zaremba, W., Cheung, V., Radford, A., Chen, X.: Improved techniques for training GANs. In: Advances in Neural Information Processing Systems, pp. 2234–2242 (2016)

39. Siarohin, A., Sangineto, E., Lathuilière, S., Sebe, N.: Deformable GANs for pose-based human image generation. In: Proceedings of the IEEE Conference on Computer Vision and Pattern Recognition, pp. 3408–3416 (2018)
40. Simonyan, K., Zisserman, A.: Very deep convolutional networks for large-scale image recognition. arXiv preprint arXiv:1409.1556 (2014)
41. Wang, M., et al.: Example-guided style consistent image synthesis from semantic labeling. arXiv preprint arXiv:1906.01314 (2019)
42. Wang, T.C., et al.: Video-to-video synthesis. arXiv preprint arXiv:1808.06601 (2018)
43. Wang, T.C., Liu, M.Y., Zhu, J.Y., Tao, A., Kautz, J., Catanzaro, B.: High-resolution image synthesis and semantic manipulation with conditional GANs. In: Proceedings of the IEEE Conference on Computer Vision and Pattern Recognition, pp. 8798–8807 (2018)
44. Wang, X., Girshick, R., Gupta, A., He, K.: Non-local neural networks. In: Proceedings of the IEEE Conference on Computer Vision and Pattern Recognition, pp. 7794–7803 (2018)
45. Yoo, J., Uh, Y., Chun, S., Kang, B., Ha, J.W.: Photorealistic style transfer via wavelet transforms. arXiv preprint arXiv:1903.09760 (2019)
46. Yu, F., et al.: Bdd100k: a diverse driving video database with scalable annotation tooling. arXiv preprint arXiv:1805.04687 (2018)
47. Yue, K., Sun, M., Yuan, Y., Zhou, F., Ding, E., Xu, F.: Compact generalized non-local network. In: Advances in Neural Information Processing Systems, pp. 6510–6519 (2018)
48. Zhang, H., Goodfellow, I., Metaxas, D., Odena, A.: Self-attention generative adversarial networks. arXiv preprint arXiv:1805.08318 (2018)
49. Zhang, R., Isola, P., Efros, A.A., Shechtman, E., Wang, O.: The unreasonable effectiveness of deep features as a perceptual metric. In: Proceedings of the IEEE Conference on Computer Vision and Pattern Recognition, pp. 586–595 (2018)
50. Zhu, J.Y., Park, T., Isola, P., Efros, A.A.: Unpaired image-to-image translation using cycle-consistent adversarial networks. In: Proceedings of the IEEE International Conference on Computer Vision, pp. 2223–2232 (2017)

Content-Consistent Matching for Domain Adaptive Semantic Segmentation

Guangrui Li[1]($\boxtimes$), Guoliang Kang[2], Wu Liu[3], Yunchao Wei[1], and Yi Yang[1]

[1] ReLER, Centre for Artificial Intelligence, University of Technology Sydney,
Ultimo, Australia
guangrui.li-1@student.uts.edu.au, {yunchao.wei,yi.yang}@uts.edu.au
[2] Carnegie Mellon University, Pittsburgh, USA
gkang@andrew.cmu.edu
[3] JD AI Research, Beijing, China
liuwu1@jd.com
https://github.com/Solacex/CCM

Abstract. This paper considers the adaptation of semantic segmentation from the synthetic source domain to the real target domain. Different from most previous explorations that often aim at developing adversarial-based domain alignment solutions, we tackle this challenging task from a new perspective, *i.e.*, content-consistent matching (CCM). The target of CCM is to acquire those synthetic images that share similar distribution with the real ones in the target domain, so that the domain gap can be naturally alleviated by employing the content-consistent synthetic images for training. To be specific, we facilitate the CCM from two aspects, *i.e.*, semantic layout matching and pixel-wise similarity matching. First, we use all the synthetic images from the source domain to train an initial segmentation model, which is then employed to produce coarse pixel-level labels for the unlabeled images in the target domain. With the coarse/accurate label maps for real/synthetic images, we construct their semantic layout matrixes from both horizontal and vertical directions and perform the matrixes matching to find out the synthetic images with similar semantic layout to real images. Second, we choose those predicted labels with high confidence to generate feature embeddings for all classes in the target domain, and further perform the pixel-wise matching on the mined layout-consistent synthetic images to harvest the appearance-consistent pixels. With the proposed CCM, only those content-consistent synthetic images are taken into account for learning the segmentation model, which can effectively alleviate the domain bias caused by those content-irrelevant synthetic images. Extensive experiments are conducted on two popular domain adaptation tasks, *i.e.*, GTA5→Cityscapes and SYNTHIA→Cityscapes. Our CCM yields consistent improvements over the baselines and performs favorably against previous state-of-the-arts.

Keywords: Semantic segmentation · Domain adaptation

G. Li and G. Kang—Equal contribution.

© Springer Nature Switzerland AG 2020
A. Vedaldi et al. (Eds.): ECCV 2020, LNCS 12359, pp. 440–456, 2020.
https://doi.org/10.1007/978-3-030-58568-6_26

1   Introduction

Semantic segmentation [4,5,8,19,28,30,52] plays an important role in many real-world applications. However, in practice, the off-the-shelf segmentation model trained on one scenario (source) usually cannot generalize well to the new one (target). For example, for the self-driving task, we may collect data and train a segmentation model in one city, but such a model may fail to give accurate pixel-level predictions for the scenes of another unfamiliar city. As achieving massive in-domain pixel-level annotations are expensive and sometimes impossible, practitioners usually resort to domain adaptive training to achieve satisfactory results on the target.

Generally, a domain adaptive segmentation model is established on the labeled source images and the unlabeled target images. And the training tries to utilize the knowledge learned from the source and mitigate the domain shift. Previous methods usually achieved the adapted model through the adversarial training [2,12,17,23,26,33,40,43] or by self-training [27,48,56,57]. All those methods employed the entire source domain data throughout the training process, which neglects the fact that not all the source images could contribute to the improvement of adaptation performance, especially at certain training stages. We empirically find that there usually exist "negative" source images which may even harm the adaptation. As shown in Fig. 1, compared to the positive source images, the negative ones appear quite dissimilar to the target images. Moreover, from Fig. 1, we observe that for the visually similar pair of source (i.e positive source image) and target images, the pixel-wise similarities vary a lot spatially, which implies that the pixels of a source image also should not be treated equally.

In this paper, we propose Content-Consistent Matching (CCM) to match and select the effective source information actively to facilitate the adaptation process. To be specific, we perform *Semantic Layout Matching* to select the positive source samples and *Pixel-wise Similarity Matching* to emphasize effective pixels. For the Semantic Layout Matching, we propose a novel image representation that encodes the semantic layout information. Based on such semantic layout representation, we perform clustering on the target to discover the underlying patterns in the target domain. Then the source sample is selected as the positive one if it is close enough to these patterns. Moreover, we further select the positive source pixels to mitigate the negative transfer through proposed pixel-wise similarity matching. Similar to the matching strategy in the image level, pixel-wise similarity matching selects the source pixels that share similar feature distributions with the target samples.

As the target feature evolves during training, the same source sample may contribute differently to the adaptation process, e.g. a source sample could be negative before a certain stage but positive afterward. Thus we choose to iteratively update the matching results during training to enable more effective adaptation. Specifically, we perform the CCM along with a self-training paradigm, i.e. we alternatively update the representations through self-training and the source matching results through CCM. These two parts depend on each other and cooperate to mitigate the domain shift.

(a). Negative Source Samples   (b). Target Samples   (c). Positive Source Samples   (d). Similarity Heatmap for (c)

Fig. 1. Examples of positive and negative source samples (Best viewed in color). Generally, positive source images (c) share similar layout with the target (b) while the negative source ones (a) do not. Intuitively, samples like (c) should be selected and samples like (a) should be excluded to help the adaptation. Moreover, the heatmap (d) indicates the pixel-wise similarities between target and positive source embeddings, in which red indicates higher similarity. It can be seen that the similarities vary a lot, even for the semantic-consistent pixels of the source image, which implies the pixels of positive source images should not be equally treated. Detailed information could be found in Sect. 3.2.

In a nutshell, our contributions are three-fold:

- We deal with the domain adaptive segmentation task from a new perspective, *i.e.* actively selecting positive source information for training to avoid negative transfer, which has not been investigated by previous methods.
- We propose Content-Consistent Matching (CCM), which consists of Semantic Layout Matching and Pixel-wise Similarity Matching, to select the positive source samples and their positive pixels to facilitate the adaptation process.
- Experiments on two representative benchmarks (*i.e.* GTA5 → Cityscapes and SYNTHIA → Cityscapes) demonstrate that our method performs favorably against previous methods. Ablation studies also verify the effectiveness of the key components of our framework.

2   Related Works

Unsupervised Domain Adaptation (UDA). Unsupervised domain adaptation aims to minimize the distribution discrepancy between the source domain and the target domain. To achieve this goal, some earlier [30,31,38,42] methods proposed to learn domain invariant features via directly minimizing the discrepancy of feature distribution. More recent approaches [13,15,18,22,32,55] employed adversarial training in image level. [2,14,29,41,44,46] leveraged adversarial training to learn domain-invariant representations in feature level. There are some works [2,21,37,47,49] using self-training to mitigate the domain gap via assigning labels to the most confident samples in the target domain.

Self-training. Self-training, which assigns and updates pseudo labels in an alternative style, has attracted wide attention for its simplicity and effectiveness. Self-training has been exploited in various tasks such as semi-supervised learning [20,24], domain adaptation [37,56], and noisy label learning [34,39]. Most existing UDA methods established on self-training mainly focused on how to utilizing the pseudo labels in the training while neglecting the selection of source samples.

UDA for Semantic Segmentation. Unsupervised domain adaptation for semantic segmentation is the task that applies domain adaptation at pixel-level. Many approaches [2,9,15,50,51] have been proposed. There are mainly two ways to tackle this problem, *i.e.* via adversarial training or self-training. The works that exploited adversarial training can be categorized into the feature-level adaptation and the image level adaptation. Some works [12,23,33,40,43,45] adopted adversarial training at feature level to learn domain-invariant features to reduce the discrepancy across domains. [7,17,26] applied adversarial training at the image level to make features invariant to illumination, color and other style factors. Some recent approaches adopted self-training to perform adaptation. [56] proposed to assign pseudo labels in a curriculum way and [27,48,53,57] combined self-training with other constraints to improve the quality of pseudo labels.

3   Content-Consistent Matching

We aim to train a segmentation network with parameters θ to give accurate pixel-level predictions $P(c|x,y;I^T,\theta)$ on the target T, where $c \in \{0,1,\cdots,C-1\}$ denotes the underlying categories and $x \in \{0,1,\cdots,W-1\}, y \in \{0,1,\cdots,H-1\}$ are the horizontal and vertical coordinates of a pixel in a target image I^T, respectively. The segmentation network is trained with the combination of labeled source images D^S and unlabeled target images D^T. During training, we propose to use content-consistent matching (CCM) to select positive source samples and their effective pixels. Our CCM is performed upon the self-training paradigm, *i.e.* with selected positive source samples and their effective pixels, the network is trained with ground-truth source labels and pseudo target labels to update the feature representation, and based on the updated feature representation, the set of positive source samples and pixels are reconstructed.

3.1   Semantic Layout Matching

Semantic layout means how the categories are distributed spatially in an image (*i.e.* $P(x,y|c)$). It could be an important prior during the training of segmentation models. However, the semantic layout patterns may vary a lot across domains, leading to the domain shift and degenerating the generalization. For example, it is natural that part of the source domain images is captured from a distinct perspective compared to the target. Thus the semantic layout of these source images will be quite different from the target. In this section, we propose using semantic layout matching to select the positive source samples to mitigate such domain shift.

Semantic Layout Matrix (SLM). Directly using $P(x, y|c)$ to model the semantic layout would be inefficient due to its high dimension, and ineffective because it is not robust to the inaccurate target predictions.

Following the naive Bayes assumption, we propose to decouple $P(x, y|c)$ into the horizontal one $P(x|c)$ and vertical $P(y|c)$ one, i.e.,

$$P(x, y|c) \propto P(x|c)P(y|c). \tag{1}$$

Specifically, take the vertical distribution $P(y|c)$ for an example, $P(y|c)$ can be represented as

$$P(y|c) = \frac{P(c|y)P(y)}{P(c)} = \frac{\sum_x P(c|x, y)P(x)P(y)}{\sum_x \sum_y P(c|x, y)P(x)P(y)}, \tag{2}$$

Assuming $P(x)$ and $P(y)$ are the uniform distributions, i.e. $P(x = i) = \frac{1}{W}, i \in \{0, 1, \cdots, W - 1\}$ and $P(y = j) = \frac{1}{H}, j \in \{0, 1, \cdots, H - 1\}$, then

$$P(y|c) = \frac{\sum_x P(c|x, y)}{\sum_x \sum_y P(c|x, y)}. \tag{3}$$

For the source image, suppose its ground-truth label is c', $P(c|x, y; I^S) = \begin{cases} 1 \text{ if } c = c' \\ 0 \text{ otherwise} \end{cases}$. For the target images, as we don't know the ground-truth pixel-wise labels, we adopt the probability predictions $P(c|x, y; I^T, \theta)$ of current segmentation model with parameters θ to compute Eq. (3).

Following the general practice, the images (source and target) are customized as the same size during training, and the vertical semantic layout matrix $M_v \in \mathbb{R}^{C \times H}$ can be expressed as (we omit the domain subscript for simplification)

$$M_v(\hat{c}, j) = \frac{\sum_x P(\hat{c}|x, y = j)}{\sum_x \sum_y P(\hat{c}|x, y)}. \tag{4}$$

Similarly, the horizontal semantic layout matrix $M_h \in \mathbb{R}^{C \times W}$ is

$$M_h(\hat{c}, i) = \frac{\sum_y P(\hat{c}|x = i, y)}{\sum_x \sum_y P(\hat{c}|x, y)}. \tag{5}$$

Finally, the semantic layout matrix $M \in \mathbb{R}^{C \times (H+W)}$ can be represented as

$$M = [M_h, M_v]. \tag{6}$$

Note that because the assumption in Eq. (1) may not be exactly satisfied in practice, we choose to concatenate the horizontal and vertical semantic layout matrix together rather than multiply them, which makes the training less dependent on the conditional independence assumption.

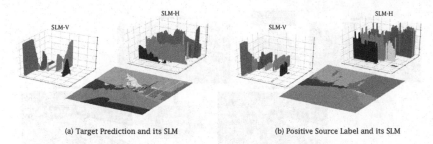

(a) Target Prediction and its SLM   (b) Positive Source Label and its SLM

Fig. 2. Illustration of SLM (Best viewed in color). Taking the class sky (annotated with sky blue) as an example, we explain how to use SLM to represent the spatial distribution. From SLM-V, we could know its vertical distribution: most of the pixels belonging to the sky are at the top. Through SLM-H, we could also found that the sky mostly lies in the middle and right of the images. (a) and (b) are pairs that share most similar spatial distributions. (Color figure online)

Matching and Selection. Based on the proposed SLM, we can encode the semantic layout of each target image. We then adopt k-means clustering to discover the underlying K patterns of target SLMs. The source image, which is close enough to these patterns, can be viewed as a positive sample. Note that because we perform single-direction selection, clustering on the source images is not needed.

Specifically, we denote the centers of the K target clusters as $\hat{M}^{T,k}, k \in \{0, 1, \cdots, K-1\}$ and compute the similarity between the source sample and each of these cluster centers through

$$Sim(M^S, \hat{M}^{T,k}) = -\sum_c D_{KL}(\hat{M}_h^{T,k}(c,:)||M_h^S(c,:)) + D_{KL}(\hat{M}_v^{T,k}(c,:)||M_v^S(c,:)),$$

(7)

where the $D_{KL}(\cdot||\cdot)$ denotes the KL divergence. And the similarity score of a source image is

$$Score(M^S) = \frac{1}{K} \sum_k Sim(M^S, \hat{M}^{T,k}).$$

(8)

Based on the ranking of above similarity scores among all the source samples, only the top ranking source samples are selected for training. In our experiment, we set the selection proportion γ_{img} as a hyper-parameter to control the number of selected source images. We will discuss this further in our experiment part. Our proposed SLM is illustrated in Fig. 2.

3.2   Pixel-Wise Similarity Matching

For a source image, it is possible that partial regions or pixels are similar to the target, while others not. That means the pixels in a source image should

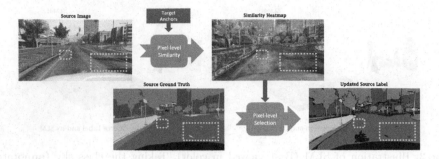

Fig. 3. Illustration of pixel-wise similarity matching. As marked by the green box, the leaves on the road are hardly spotted even by a human but annotated in the ground truth. Pixel-wise similarity matching excludes these pixels which may hinder the adaptation. The black area in the figure denotes the ignored pixels. Best viewed in color. (Color figure online)

not be equally treated during the adaptation. Thus besides selecting positive source samples, we propose to select the positive source pixels that share similar characters with the target to mitigate the domain shift further. We name such pixel-level selection as pixel-wise similarity matching, as illustrated in Fig. 3.

For a target image, based on the network's outputs $P(c|x, y; I^T, \theta)$, we could assign a pseudo label to each pixel, $i.e.$

$$L^T(x, y) = \arg\max_c P(c|x, y; I^T, \theta). \tag{9}$$

Then the pixels are classified into C groups. The pixel with low confidence prediction $P(L^T(x, y)|x, y; I^T, \theta)$ is filtered (see Sect. 3.3 for more details). And the average class distribution is calculated among each group

$$Q^T(c) = \frac{1}{|D^T|} \sum_i \frac{1}{|G_{i,c}^T|} \sum_{(\hat{x}, \hat{y}) \in G_{i,c}^T} P_i(c|\hat{x}, \hat{y}; I_i^T, \theta), c \in \{0, 1, \cdots, C-1\}, \tag{10}$$

where $G_{i,c}^T = \{(\hat{x}, \hat{y})|L_i^T(\hat{x}, \hat{y}) = c\}$, $Q^T(c) \in \mathbb{R}^C$, and the subscript i denotes the i-th target sample here, $|D^T|$ is the number of samples within target domain. By this way, it is expected that $Q^T(c)$ could describe the relationships between class c and all the other classes, based on the current predictions of the network.

From the network, we can also get the predictions for the source image, $i.e.$ $P(c|x, y; I^S, \theta)$, thus it is natural to select such source pixels $\{(\tilde{x}, \tilde{y})\}$ where $P(c|\tilde{x}, \tilde{y}; I^S, \theta)$ matches well with Q^T. We adopt KL divergence to measure the distance between each pair of $P(c|x, y; I^S, \theta)$ and $Q^T(c)$. And the matching score for a source pixel at (x, y) with ground-truth label c is computed as

$$Score(x, y) = -D_{KL}(Q^T(c)||P(c|x, y; I^S, \theta)). \tag{11}$$

We rank the source pixels within the same ground-truth class according to their similarity score and select the top ranking pixels for each class. In our experiment, we select the same proportion of pixels γ_{pix} for each class.

3.3   Active Matching with Self-training

As the target feature evolves, the effect of the same source sample on the adaptation process may be different. In this paper, we choose to update the source matching results actively throughout the adaptation process. Notice that purely training with the source data may lead to the model biased towards the source distribution, we choose to employ our matching strategy along with the self-training paradigm.

To obtain a good initialization of target predictions, we start the self-training from the segmentation network trained on all the labeled source images D^S. Then we alternatively update the network parameters θ and assign pseudo labels $L^T(x, y)$ on the target D^T according to Eq. (9).

Through pseudo labeling, the target pixels are grouped into C classes. For each class of pixels, we rank them according to the prediction confidences (*i.e.* $P(L^T(x, y)|x, y; I^T, \theta)$). Only the top ranking target pixels are selected for training, and the ratio of selection is set to r, which is shared among all the classes. To enable each target sample to have enough selected pixels, we also perform pixel ranking within each image. Then the top r pixels of a target image are also selected. The selected pixels $\{(\hat{x}, \hat{y})\}$ are assumed to have reliable pseudo labels.

The positive source samples $\hat{D}^S$ selected through our matching strategy, together with the pseudo-labeled target samples D^T, are adopted to train the network. And the network is trained with pixel-wise cross-entropy loss

$$\mathcal{L}_{ce} = \mathcal{L}_{ce}(\hat{D}^S; \theta) + \mathcal{L}_{ce}(D^T; \theta), \tag{12}$$

where

$$\mathcal{L}_{ce}(\hat{D}^S; \theta) = -\sum_i \sum_{(\tilde{x}, \tilde{y}) \in I_i^S} \log[P(L_i^S(\tilde{x}, \tilde{y})|\tilde{x}, \tilde{y}; I_i^S, \theta)], \tag{13}$$

$$\mathcal{L}_{ce}(D^T; \theta) = -\sum_i \sum_{(\hat{x}, \hat{y}) \in I_i^T} \log[P(L_i^T(\hat{x}, \hat{y})|\hat{x}, \hat{y}; I_i^T, \theta)]. \tag{14}$$

In Eq. (13) and Eq. (14), I_i^S and I_i^T are the i-th images in the $\hat{D}^S$ and D^T, respectively. Note that only the gradients coming from the positive source pixels $(\tilde{x}, \tilde{y}) \in I^S$ and target pixels $(\hat{x}, \hat{y})$ with reliable pseudo labels are backpropagated in each iteration.

3.4   Objective

Additionally, we introduce entropy regularization to regularize the adaptation

$$\mathcal{L}_{ent}(D^S; \theta) = -\sum_i \sum_c \sum_{(x, y) \in I_i^S} P(c|x, y; I_i^S, \theta) \log[P(c|x, y; I_i^S, \theta)], \tag{15}$$

$$\mathcal{L}_{ent}(D^T; \theta) = -\sum_i \sum_c \sum_{(x, y) \in I_i^T} P(c|x, y; I_i^T, \theta) \log[P(c|x, y; I_i^T, \theta)], \tag{16}$$

Algorithm 1: Content-Consistent Matching

---

1 **Input:** parameters θ; source images D^S and labels L^S, target images from D^T
2 Initialize θ with source trained segmentation model.
3 **for** *m=1 to M* **do**
4   Update target pseudo labels L^T for each $I^T \in D^T$ and select target pixels $(\hat{x}, \hat{y})$ with reliable pseudo labels;
5   Select positive source samples $\hat{D}^S$ and their positive pixels $(\tilde{x}, \tilde{y})$
6   **for** *n=1 to N* **do**
7   1) forward and compute the $\mathcal{L}$ according to Eq. (17);
8   2) back-propagating the gradients and update θ.
9   **end**
10 **end**

And the entropy regularization is imposed on all the source and target images.

In total, the objective of our training procedure is

$$\mathcal{L} = \mathcal{L}_{ce}(\hat{D}^S; \theta) + \mathcal{L}_{ce}(D^T; \theta) + \lambda(\mathcal{L}_{ent}(D^S; \theta) + \mathcal{L}_{ent}(D^T; \theta)). \qquad (17)$$

where λ is a constant indicating the strength of entropy regularization.

Our algorithm is summarized in Algorithm 1. Note that we update the target pseudo labels and perform source selection every N steps of network update. We perform selection and network update in such an asynchronous way because the network update is a relatively slower process, and this way enables more efficient and effective training.

4   Experiments

4.1   Dataset and Evaluation Metric

We evaluate our methods on two popular transfer tasks, GTA5 [35] → Cityscapes [10] and SYNTHIA [36] → Cityscapes. For the source dataset, GTA5 contains 24996 images with resolution 1914 × 1052, and SYNTHIA contains 9400 images with resolution 1280 × 760. For the target, Cityscapes contains 2975 images for training and 500 images for validation with image resolution 2048 × 1024. Following the settings in [33, 40, 43], we train the model on the source dataset (GTA5 or SYNTHIA) and the training set of Cityscapes and report the result on the validation set of Cityscapes. We only transfer on the classes shared between the source domain and the target domain. For the evaluation metric, we evaluate our methods with mean Intersection over Union (mIoU).

4.2   Implementation Detail

We start from DeepLabV2-Res101 [3,16] with the backbone pretrained on the ImageNet [11]. Then we firstly finetune the whole network on the source data and use such a source-trained network to initialize the target (adaptation) model.

Table 1. Experiment results of GTA5 → Cityscapes.

GTA5 → CityScapes	Meth.	road	side.	buil.	wall	fence	pole	light	sign	vege.	terr.	sky	pers.	rider	car	truck	bus	train	motor	bike	mIoU
Source Only	−	60.6	17.4	73.9	17.6	20.6	21.9	31.7	15.3	79.8	18.1	71.1	55.2	22.8	68.1	32.3	13.8	3.4	34.1	21.2	35.7
AdaptSeg [40]	AT	86.5	36.0	79.9	23.4	23.3	23.9	35.2	14.8	83.4	33.3	75.6	58.5	27.6	73.7	32.5	35.4	3.9	30.1	28.1	42.4
ADVENT [43]	AT	89.4	33.1	81.0	26.6	26.8	27.2	33.5	24.7	83.9	36.7	78.8	58.7	30.5	84.8	38.5	44.5	1.7	31.6	32.4	45.5
CLAN [33]	AT	87.0	27.1	79.6	27.3	23.3	28.3	35.5	24.2	83.6	27.4	74.2	58.6	28.0	76.2	33.1	36.7	6.7	31.9	31.4	43.2
DISE [1]	AT	91.5	47.5	82.5	31.3	25.6	33.0	33.7	25.8	82.7	28.8	82.7	62.4	30.8	85.2	27.7	34.5	6.4	25.2	24.4	45.4
SWD [23]	AT	92.0	46.4	82.4	24.8	24.0	35.1	33.4	34.2	83.6	30.4	80.9	56.9	21.9	82.0	24.4	28.7	6.1	25.0	33.6	44.5
SSF-DAN [12]	AT	90.3	38.9	81.7	24.8	22.9	30.5	37.0	21.2	84.8	38.8	76.9	58.8	30.7	85.7	30.6	38.1	5.9	28.3	36.9	45.4
MaxSquare [6]	−	89.4	43.0	82.1	30.5	21.3	30.3	34.7	24.0	85.3	39.4	78.2	63.0	22.9	84.6	36.4	43.0	5.5	**34.7**	33.5	46.4
MRNet [54]	−	90.5	35.0	84.6	34.3	24.0	36.8	**44.1**	42.7	84.5	33.6	82.5	**63.1**	34.4	**85.8**	32.9	38.2	2.0	27.1	41.8	48.3
PyCDA [27]	ST	90.5	36.3	84.4	32.4	**28.7**	34.6	36.4	31.5	**86.8**	37.9	78.5	62.3	21.5	85.6	27.9	34.8	18.0	22.9	**49.3**	47.4
CRST [57]	ST	91.0	55.4	80.0	33.7	21.4	37.3	32.9	24.5	85.0	34.1	80.8	57.7	24.6	84.1	27.8	30.1	26.9	26.0	42.3	47.1
CAG [48]	ST	90.4	51.6	83.8	34.2	27.8	**38.4**	25.3	**48.4**	85.4	38.2	78.1	58.6	**34.6**	84.7	21.9	42.7	**41.1**	29.3	37.2	50.2
SIM [44]	ST	90.1	44.7	**84.8**	34.3	28.7	31.6	35.0	37.6	84.7	43.3	85.3	57.0	31.5	83.8	42.6	48.5	1.9	30.4	39.0	49.2
BDL [54]	AS	91.0	44.7	84.2	34.6	27.6	30.2	36.0	36.0	85.0	**43.6**	83.0	58.6	31.6	83.3	35.3	49.7	3.3	28.8	35.6	48.5
Ours (CCM)	ST	**93.5**	**57.6**	84.6	**39.3**	24.1	25,2	35.0	17.3	85.0	40.6	**86.5**	58.7	28.7	**85.8**	**49.0**	**56.4**	5.4	31.9	43.2	**49.9**

Table 2. Experiment results of SYNTHIA → Cityscapes. The mIoU* denotes the mean IoU over classes without "*".

SYNTHIA → CityScapes	Meth.	road	side.	buil.	wall*	fence*	pole*	light	sign	vege.	sky	pers.	rider	car	bus	motor	bike	mIoU	mIoU*
Source Only	−	47.1	23.3	75.6	7.1	0.1	23.9	5.1	9.2	74.0	73.5	51.1	20.9	39.1	17.7	18.4	34.0	34.5	40.1
AdaptSeg [40]	AT	84.3	42.7	77.5	−	−	−	4.7	7.0	77.9	82.5	54.3	21.0	72.3	32.2	18.9	32.3	−	46.7
ADVENT [43]	AT	85.6	42.2	79.7	−	−	−	5.4	8.1	80.4	84.1	57.9	23.8	73.3	36.4	14.2	33.0	−	48.0
CLAN [33]	AT	81.3	37.0	80.1	−	−	−	16.1	13.7	78.2	81.5	53.4	21.2	73.0	32.9	22.6	30.7	−	47.8
SSF-DAN [12]	AT	84.6	41.7	80.8	−	−	−	11.5	14.7	80.8	85.3	57.5	21.6	82.0	36.0	19.3	34.5	−	50.0
MaxSquare [6]	−	82.9	40.7	80.3	10.2	**0.8**	25.8	12.8	18.2	82.5	82.2	53.1	18.0	79.0	31.4	10.4	35.6	41.4	48.2
CAG [48]	ST	84.7	40.8	81.7	7.8	0.0	**35.1**	13.3	22.7	84.5	77.6	**64.2**	27.8	80.9	19.7	22.7	48.3	44.5	−
pyCDA [27]	ST	75.5	30.9	**83.3**	**20.8**	0.7	32.7	**27.3**	**33.5**	**84.7**	85.0	64.1	25.4	**85.0**	45.2	21.2	32.0	**46.7**	**53.3**
SIM [44]	ST	83.0	44.0	80.3	−	−	−	17.1	15.8	80.5	81.8	59.9	**33.1**	70.2	37.3	28.5	45.8	−	52.1
BDL [26]	AS	**86.0**	**46.7**	80.3	−	−	−	14.1	11.6	79.2	81.3	54.1	27.9	73.7	42.2	25.7	45.3	−	51.4
ours (CCM)	ST	79.6	36.4	80.6	13.3	0.3	25.5	22.4	14.9	81.8	77.4	56.8	25.9	80.7	**45.3**	**29.9**	**52.0**	45.2	52.9

We choose to use Stochastic Gradient Descent (SGD) with momentum of 0.9 and weight decay of 5×10^{-4}. The learning rate decreases following the poly policy with power at 0.9. The initial learning rate is set to 7.5×10^{-5}. The M in Algorithm 1 is set to 6 and the N is set to 2 epochs, *i.e.* we train for 6 loops where each loop contains 2 epochs. For all the transfer tasks, the hyper-parameters γ_{img}, λ, r, and K are set to 0.4, 0.4, 0.1, and 10 respectively. The γ_{pix} is set to 0.9, 0.6 for GTA5 and SYNTHIA respectively.

For image preprocessing, we resize the shorter side of images to 720 and crop a patch with resolution 600×600 randomly. Besides, horizontal flip and random scale between 0.5 and 1.5 are introduced as data augmentation. For evaluation, images from Cityscapes are resized to 1024×512 as input and the mIoU is calculated on predictions upsampled to 2048×1024.

4.3 Comparison with the State-of-the-arts

We evaluate our method on two unsupervised domain adaptation tasks: GTA5 → Cityscapes and SYNTHIA → Cityscapes. The results are presented in Table 1 and Table 2, respectively. In both tables, we use "AT" and "ST" to denote approaches established on adversarial training and self-training respectively,

Table 3. Effect of different key components. The "CCM-SLM" stands for semantic layout matching, and "CCM-Fix" denotes source samples and pixels are only selected at the start of self-training. All the results are compared with our self-training baseline. Self-training means the network trained with cross-entropy loss and entropy regularization, without the source selection via CCM.

Module	GTA5→Cityscapes		SYNTHIA→Cityscapes	
	mIoU	Gain	mIoU	Gain
Self-training	48.1	-	41.2	-
CCM-Fix	48.9	+0.8	44.2	+3.0
CCM-SLM	48.8	+0.7	41.9	+0.7
CCM	49.9	+1.8	45.2	+4.0

while "AS" indicates methods utilizing both. All the models are based on DeepLabV2-Res101 backbone, except that pyCDA [27] is based on PSPNet [52] and CAG [48] is based on DeepLabV3+ [5][1]. It can be seen that our method outperforms source only baseline with a large margin, which verifies the effectiveness of our approach.

For the task GTA5 → Cityscapes, we achieve 49.9% on mIoU, comparable to previous state-of-the-art method CAG [48] (50.2%) which is trained with a much larger resolution (*i.e.* 2200 × 1100). For the task from SYNTHIA → Cityscapes, to make a fair comparison, we report the mIoU on 13 classes (excluding "Wall", "Fence", and "Pole") and 16 classes. Our method achieves 52.9% and 45.2% mIoU on 13 classes and 16 classes respectively, both of which perform favorably against previous state-of-the-arts.

Specifically, despite its simplicity, CCM outperforms previous state-of-the-art adversarial-training (denote as "AT") based method "SSF-DAN" [12] by +4.5% and +2.9% on GTA5 → Cityscapes and SYNTHIA → Cityscapes, respectively. Compared with methods established on self-training, CCM achieves comparable or even better results. For example, our method is on par with pyCDA [27], *i.e.* 49.9% (ours) vs. 47.4% (pyCDA) on GTA5 → Cityscapes and 45.2%/52.9% (mIoU and mIoU* of ours) vs. 46.7%/53.3% (mIoU and mIoU* of pyCDA) on SYNTHIA → Cityscapes. It is worth noting that pyCDA adopts AdaBN [25] to enhance its adaptation performance, which can also be employed in our framework to improve the performance further. Also, our method mainly focuses on selecting positive source information to mitigate the domain shift and help the adaptation, which is also complementary to these methods and can be combined with them to boost the adaptation performance.

4.4   Ablation Studies

Effect of Different Key Components. We verify the effect of each key component in our framework in Table 3. It can be seen that compared to the

---

[1] https://github.com/RogerZhangzz/CAG_UDA/issues/6.

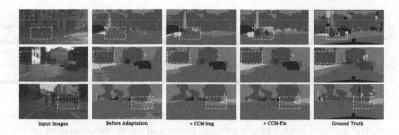

Fig. 4. Visualization of the segmentation results (GTA5 → Cityscapes). Pay attention to the dashed box to see the effect of different modules.

source-only results, self-training improves the adaptation performance apparently. Despite such a strong baseline, "CCM-SLM", which selects positive source samples through proposed SLM, improves beyond self-training by +0.7% on both tasks. Further, through combining SLM with pixel similarity matching, "CCM" improves beyond self-training by +1.8% and +4.0% for the GTA5 → Cityscapes and SYNTHIA → Cityscapes, respectively. These noticeable performance gains verify the effectiveness of the proposed matching and selection strategy.

Figure 4 gives an intuitive illustration about the effect of CCM. It can be seen that through adaptation with SLM, the pixel-level predictions have been largely improved. Further, pixel-wise similarity matching enables the adapted model to learn more details about the object and thus leads to more accurate predictions.

Compared with the "CCM-Fix" which only selects positive source samples and their positive pixels at the start of self-training and adopts them throughout the adaptation, actively update the positive source set (denoted as "CCM") achieves noticeable improvement (*i.e.* +1.0% for both tasks). This is because source samples may contribute differently to the adaptation at different training stages and the matching results should be updated as the target predictions evolve, The results imply that self-training and CCM could benefit each other and cooperate to mitigate the domain shift.

Fig. 5. Examples retrieved by semantic layout matrix(SLM). In each row, (b–e) are source images retrieved by target sample (a), where (b–d) are positive samples and (e) are negative ones.

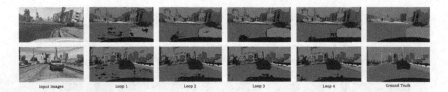

Fig. 6. Visualization of selected source pixels at different training stages. As the training goes on, the ignored source pixels become more and more concentrated on the object boundary. The black area denotes the ignored pixels during training. The results are based on task GTA5 → Cityscapes.

Visualization of Semantic Layout Matching Results. In Fig. 5, we show the source images retrieved by individual target images via semantic layout matching at the final training stage. In each row, (c–e) are source images retrieved by the target sample (a) via SLM, in which (b–d) are top positive samples and (e) are negative ones with the lowest matching scores.

It can be seen that similar layout is shared among the target samples and the retrieved source samples. For example, all positive source samples in the first row have trees on the left. Moreover, it is also obvious that negative source samples on the right-most column (e) have totally different layout. Additionally, the retrieved source samples remain reasonable variations in appearances, which will also benefit the generalization of adapted model. All of these results give an intuitive illustration why semantic layout matching can help reduce the domain shift and improve the generalization ability.

Effect of Pixel-Wise Similarity Matching. Figure 6 demonstrates the selected source pixels through pixel-wise similarity matching during the training. It can be seen that as the training goes on, the ignored pixels become more and more concentrated on the object boundary, which is reasonable and implies that the adaptation keeps improving. Moreover, at the early stage, we notice that the ignored pixels are ambiguous ones that are hard to distinguish, *e.g.* the pixels of the cracks on the road. The pixel selection through pixel-wise similarity matching enables the model to learn in a curriculum way to an extent.

Sensitivity to the Hyper-parameters. We investigate the sensitivity of our method to the hyper-parameters γ_{img}, γ_{pix}, λ, r, and show the results in Fig. 7. From Fig. 7 (a), it can be seen that trained with SLM, with the increase of γ_{img}, the mIoU firstly increases then decreases, illustrating a bell shape curve. The mIoU decreases when

Table 4. Sensitivity to K

K	5	10	15
mIoU(%)	49.8	49.9	49.7

γ_{img} is above a certain threshold, indicating that there exist negative samples harming the adaptation and it is necessary to perform source sample selection to exclude such negative samples. With SLM, the optimal mIoU achieved is 0.7%

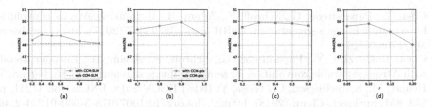

Fig. 7. (a): Performance with/without CCM-SLM under different γ_{img}. (b): Performance with/without CCM-Pix under different γ_{pix}. (c): Sensitivity analysis of λ. (d): Sensitivity analysis of r. The results shown are based on the task GTA5 $\rightarrow$ Cityscapes. The trend on another task is similar. (Color figure online)

higher than that trained without SLM. The trend of sensitivity to γ_{pix} is similar to that of γ_{img}. Our method achieves consistent improvement over baselines (the red lines) within a wide range of γ_{img} and γ_{pix}.

As illustrated in Fig. 7 (c), entropy regularization provides consistent improvement within a wide range of λ. From Fig. 7 (d), we observe that our method is also robust to r within a wide range. When r is above a certain threshold, the performance drops because more inaccurate target data is involved in training. Besides, we analyze the sensitivity of our model to K and report the results in Table 4, which further verifies the robustness of our approach.

5   Conclusion

In this paper, we propose using Content-Consistent Matching (CCM), which consists of Semantic Layout Matching and Pixel-wise Similarity Matching, to match and select positive source data to facilitate the adaptive training of the segmentation model. Our matching strategy is performed from both the image-level and the pixel-level, *i.e.* semantic layout matching selects the positive source samples, and pixel-wise similarity matching emphasizes the effective source pixels. Experiment results on two representative benchmarks demonstrate that our method performs favorably against previous state-of-the-arts.

Acknowledgement. This work is in part supported by ARC DECRA DE190101315 and ARC DP200100938.

References

1. Chang, W.L., Wang, H.P., Peng, W.H., Chiu, W.C.: All about structure: adapting structural information across domains for boosting semantic segmentation. In: IEEE CVPR, June 2019
2. Chen, C., et al.: Progressive feature alignment for unsupervised domain adaptation. In: IEEE CVPR, June 2019
3. Chen, L.C., Papandreou, G., Kokkinos, I., Murphy, K., Yuille, A.L.: DeepLab: semantic image segmentation with deep convolutional nets, Atrous convolution, and fully connected CRFs. IEEE TPAMI **40**(4), 834–848 (2017)

4. Chen, L., Papandreou, G., Schroff, F., Adam, H.: Rethinking Atrous convolution for semantic image segmentation. CoRR abs/1706.05587 (2017). http://arxiv.org/abs/1706.05587

5. Chen, L.-C., Zhu, Y., Papandreou, G., Schroff, F., Adam, H.: Encoder-decoder with Atrous separable convolution for semantic image segmentation. In: Ferrari, V., Hebert, M., Sminchisescu, C., Weiss, Y. (eds.) ECCV 2018. LNCS, vol. 11211, pp. 833–851. Springer, Cham (2018). https://doi.org/10.1007/978-3-030-01234-2_49

6. Chen, M., Xue, H., Cai, D.: Domain adaptation for semantic segmentation with maximum squares loss. In: IEEE ICCV, October 2019

7. Chen, Y.C., Lin, Y.Y., Yang, M.H., Huang, J.B.: CrDoCo: pixel-level domain transfer with cross-domain consistency. In: IEEE CVPR, June 2019

8. Cheng, B., et al.: SPGNet: Semantic prediction guidance for scene parsing. In: IEEE ICCV, pp. 5218–5228 (2019)

9. Choi, J., Kim, T., Kim, C.: Self-ensembling with GAN-based data augmentation for domain adaptation in semantic segmentation. In: IEEE ICCV, October 2019

10. Cordts, M., et al.: The cityscapes dataset for semantic urban scene understanding. In: IEEE CVPR (2016)

11. Deng, J., et al.: ImageNet: a large-scale hierarchical image database. In: CVPR (2009)

12. Du, L., et al.: SSF-DAN: separated semantic feature based domain adaptation network for semantic segmentation. In: IEEE ICCV, October 2019

13. Feng, Q., Kang, G., Fan, H., Yang, Y.: Attract or distract: exploit the margin of open set. In: Proceedings of the IEEE International Conference on Computer Vision, pp. 7990–7999 (2019)

14. Ganin, Y., Lempitsky, V.: Unsupervised domain adaptation by backpropagation. In: ICML, ICML2015, pp. 1180–1189. JMLR.org (2015)

15. Gong, R., Li, W., Chen, Y., Gool, L.V.: DLOW: domain flow for adaptation and generalization. In: IEEE CVPR, June 2019

16. He, K., Zhang, X., Ren, S., Sun, J.: Deep residual learning for image recognition. In: CVPR, pp. 770–778 (2016)

17. Hoffman, J., et al.: CyCADA: Cycle-consistent adversarial domain adaptation. arXiv preprint arXiv:1711.03213 (2017)

18. Huang, X., Liu, M.-Y., Belongie, S., Kautz, J.: Multimodal unsupervised image-to-image translation. In: Ferrari, V., Hebert, M., Sminchisescu, C., Weiss, Y. (eds.) ECCV 2018. LNCS, vol. 11207, pp. 179–196. Springer, Cham (2018). https://doi.org/10.1007/978-3-030-01219-9_11

19. Huang, Z., et al.: CCNet: Criss-cross attention for semantic segmentation. TPAMI (2020)

20. Jiang, L., Meng, D., Zhao, Q., Shan, S., Hauptmann, A.G.: Self-paced curriculum learning. In: AAAI (2015)

21. Kang, G., Jiang, L., Yang, Y., Hauptmann, A.G.: Contrastive adaptation network for unsupervised domain adaptation. In: CVPR, pp. 4893–4902 (2019)

22. Kang, G., Zheng, L., Yan, Y., Yang, Y.: Deep adversarial attention alignment for unsupervised domain adaptation: the benefit of target expectation maximization. In: Ferrari, V., Hebert, M., Sminchisescu, C., Weiss, Y. (eds.) ECCV 2018. LNCS, vol. 11215, pp. 420–436. Springer, Cham (2018). https://doi.org/10.1007/978-3-030-01252-6_25

23. Lee, C.Y., Batra, T., Baig, M.H., Ulbricht, D.: Sliced Wasserstein discrepancy for unsupervised domain adaptation. In: IEEE CVPR, June 2019

24. Lee, D.H.: Pseudo-label: the simple and efficient semi-supervised learning method for deep neural networks. In: ICML, vol. 3, p. 2 (2013)

25. Li, Y., Wang, N., Shi, J., Hou, X., Liu, J.: Adaptive batch normalization for practical domain adaptation. Pattern Recognit. **80**, 109–117 (2018). https://doi.org/10.1016/j.patcog.2018.03.005, https://doi.org/10.1016/j.patcog.2018.03.005

26. Li, Y., Yuan, L., Vasconcelos, N.: Bidirectional learning for domain adaptation of semantic segmentation. In: IEEE CVPR, June 2019

27. Lian, Q., Lv, F., Duan, L., Gong, B.: Constructing self-motivated pyramid curriculums for cross-domain semantic segmentation: a non-adversarial approach. In: IEEE ICCV, October 2019

28. Lin, G., Milan, A., Shen, C., Reid, I.: RefineNet: multi-path refinement networks for high-resolution semantic segmentation. In: IEEE CVPR, July 2017

29. Liu, X., et al.: Feature-level Frankenstein: eliminating variations for discriminative recognition. In: IEEE CVPR, June 2019

30. Long, J., Shelhamer, E., Darrell, T.: Fully convolutional networks for semantic segmentation. In: CVPR, pp. 3431–3440 (2015)

31. Long, M., Cao, Y., Wang, J., Jordan, M.I.: Learning transferable features with deep adaptation networks. In: ICML, ICML2015, p. 97–105. JMLR.org (2015)

32. Long, M., Cao, Y., Wang, J., Jordan, M.I.: Learning transferable features with deep adaptation networks. arXiv preprint arXiv:1502.02791 (2015)

33. Luo, Y., Zheng, L., Guan, T., Yu, J., Yang, Y.: Taking a closer look at domain shift: category-level adversaries for semantics consistent domain adaptation. In: IEEE CVPR (2019)

34. Reed, S., Lee, H., Anguelov, D., Szegedy, C., Erhan, D., Rabinovich, A.: Training deep neural networks on noisy labels with bootstrapping. arXiv preprint arXiv:1412.6596 (2014)

35. Richter, S.R., Vineet, V., Roth, S., Koltun, V.: Playing for data: ground truth from computer games. In: Leibe, B., Matas, J., Sebe, N., Welling, M. (eds.) ECCV 2016. LNCS, vol. 9906, pp. 102–118. Springer, Cham (2016). https://doi.org/10.1007/978-3-319-46475-6_7

36. Ros, G., Sellart, L., Materzynska, J., Vazquez, D., Lopez, A.M.: The synthia dataset: a large collection of synthetic images for semantic segmentation of urban scenes. In: IEEE CVPR, June 2016

37. Saito, K., Ushiku, Y., Harada, T.: Asymmetric tri-training for unsupervised domain adaptation. In: ICML, pp. 2988–2997. JMLR.org (2017)

38. Sun, B., Saenko, K.: Deep CORAL: correlation alignment for deep domain adaptation. In: Hua, G., Jégou, H. (eds.) ECCV 2016. LNCS, vol. 9915, pp. 443–450. Springer, Cham (2016). https://doi.org/10.1007/978-3-319-49409-8_35

39. Tanaka, D., Ikami, D., Yamasaki, T., Aizawa, K.: Joint optimization framework for learning with noisy labels. In: IEEE CVPR (2018)

40. Tsai, Y.H., Hung, W.C., Schulter, S., Sohn, K., Yang, M.H., Chandraker, M.: Learning to adapt structured output space for semantic segmentation. In: IEEE CVPR (2018)

41. Tzeng, E., Hoffman, J., Saenko, K., Darrell, T.: Adversarial discriminative domain adaptation. In: IEEE CVPR, July 2017

42. Tzeng, E., Hoffman, J., Zhang, N., Saenko, K., Darrell, T.: Deep domain confusion: maximizing for domain invariance. CoRR abs/1412.3474 (2014). http://arxiv.org/abs/1412.3474

43. Vu, T.H., Jain, H., Bucher, M., Cord, M., Pérez, P.: Advent: adversarial entropy minimization for domain adaptation in semantic segmentation. In: IEEE CVPR (2019)

44. Wang, Z., et al.: Differential treatment for stuff and things: a simple unsupervised domain adaptation method for semantic segmentation. In: IEEE CVPR, pp. 12635–12644 (2020)
45. Wu, J., et al.: Sliced Wasserstein generative models. In: IEEE CVPR, June 2019
46. Wu, Z., et al.: DCAN: dual channel-wise alignment networks for unsupervised scene adaptation. In: Ferrari, V., Hebert, M., Sminchisescu, C., Weiss, Y. (eds.) ECCV 2018. LNCS, vol. 11209, pp. 535–552. Springer, Cham (2018). https://doi.org/10.1007/978-3-030-01228-1_32
47. Xie, S., Zheng, Z., Chen, L., Chen, C.: Learning semantic representations for unsupervised domain adaptation. In: Dy, J., Krause, A. (eds.) ICML. Proceedings of Machine Learning Research, vol. 80, pp. 5423–5432. PMLR, Stockholmsmässan, Stockholm Sweden, 10–15 July 2018. http://proceedings.mlr.press/v80/xie18c.html
48. Zhang, Q., Zhang, J., Liu, W., Tao, D.: Category anchor-guided unsupervised domain adaptation for semantic segmentation. In: NeuralPS, pp. 433–443 (2019)
49. Zhang, W., Ouyang, W., Li, W., Xu, D.: Collaborative and adversarial network for unsupervised domain adaptation. In: IEEE CVPR, June 2018
50. Zhang, Y., David, P., Foroosh, H., Gong, B.: A curriculum domain adaptation approach to the semantic segmentation of urban scenes. IEEE TPAMI, p. 1 (2019). https://doi.org/10.1109/TPAMI.2019.2903401
51. Zhang, Y., David, P., Gong, B.: Curriculum domain adaptation for semantic segmentation of urban scenes. In: IEEE ICCV, p. 6, October 2017
52. Zhao, H., Shi, J., Qi, X., Wang, X., Jia, J.: Pyramid scene parsing network. In: IEEE CVPR (2017)
53. Zheng, Z., Yang, Y.: Rectifying pseudo label learning via uncertainty estimation for domain adaptive semantic segmentation. arXiv preprint arXiv:2003.03773 (2020)
54. Zheng, Z., Yang, Y.: Unsupervised scene adaptation with memory regularization in vivo. IJCAI (2020)
55. Zhu, J.Y., Park, T., Isola, P., Efros, A.A.: Unpaired image-to-image translation using cycle-consistent adversarial networks. In: IEEE ICCV (2017)
56. Zou, Y., Yu, Z., Vijaya Kumar, B.V.K., Wang, J.: Unsupervised domain adaptation for semantic segmentation via class-balanced self-training. In: Ferrari, V., Hebert, M., Sminchisescu, C., Weiss, Y. (eds.) ECCV 2018. LNCS, vol. 11207, pp. 297–313. Springer, Cham (2018). https://doi.org/10.1007/978-3-030-01219-9_18
57. Zou, Y., Yu, Z., Liu, X., Kumar, B.V., Wang, J.: Confidence regularized self-training. In: IEEE ICCV, October 2019

AE TextSpotter: Learning Visual and Linguistic Representation for Ambiguous Text Spotting

Wenhai Wang[1], Xuebo Liu[2], Xiaozhong Ji[1], Enze Xie[3], Ding Liang[2],
ZhiBo Yang[4], Tong Lu[1(✉)], Chunhua Shen[5], and Ping Luo[3]

[1] National Key Lab for Novel Software Technology,
Nanjing University, Nanjing, China
wangwenhai362@smail.nju.edu.cn, shawn_ji@163.com, lutong@nju.edu.cn
[2] SenseTime Research, Dubai, UAE
{liuxuebo,liangding}@sensetime.com
[3] The University of Hong Kong, Pokfulam, Hong Kong
xieenze@hku.hk, pluo@cs.hku.hk
[4] Alibaba-Group, Hangzhou, China
zhibo.yzb@alibaba-inc.com
[5] The University of Adelaide, Adelaide, Australia
chunhua.shen@adelaide.edu.au

Abstract. Scene text spotting aims to detect and recognize the entire word or sentence with multiple characters in natural images. It is still challenging because ambiguity often occurs when the spacing between characters is large or the characters are evenly spread in multiple rows and columns, making many visually plausible groupings of the characters (*e.g.* "BERLIN" is incorrectly detected as "BERL" and "IN" in Fig. 1(c)). Unlike previous works that merely employed visual features for text detection, this work proposes a novel text spotter, named Ambiguity Eliminating Text Spotter (AE TextSpotter), which learns both visual and linguistic features to significantly reduce ambiguity in text detection. The proposed AE TextSpotter has three important benefits. 1) The linguistic representation is learned together with the visual representation in a framework. To our knowledge, it is the first time to improve text detection by using a language model. 2) A carefully designed language module is utilized to reduce the detection confidence of incorrect text lines, making them easily pruned in the detection stage. 3) Extensive experiments show that AE TextSpotter outperforms other state-of-the-art methods by a large margin. For example, we carefully select a set of extremely ambiguous samples from the IC19-ReCTS dataset, where our approach surpasses other methods by more than 4%.

Keywords: Text spotting · Text detection · Text recognition · Text detection ambiguity

Electronic supplementary material The online version of this chapter (https://doi.org/10.1007/978-3-030-58568-6_27) contains supplementary material, which is available to authorized users.

ⓒ Springer Nature Switzerland AG 2020
A. Vedaldi et al. (Eds.): ECCV 2020, LNCS 12359, pp. 457–473, 2020.
https://doi.org/10.1007/978-3-030-58568-6_27

1   Introduction

Text analysis in unconstrained scene images like text detection and text recognition is important in many applications, such as document recognition, license plate recognition, and visual question answering based on texts. Although previous works [12,13,19,20,27,33] have acquired great success, there are still many challenges to be solved.

This work addresses one of the important challenges, which is reducing the ambiguous bounding box proposals in scene text detection. These ambiguous proposals widely occur when the spacing of the characters of a word is large or multiple text lines are juxtaposed in different rows or columns in an image. For example, as shown in Fig. 1(c)(d), the characters in an image can be grouped into multiple visually plausible words, making the detection results ambiguous and significantly hindering the accuracy of text detectors and text spotters.

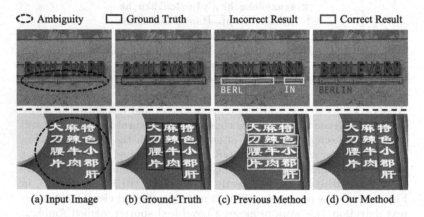

Fig. 1. The detection results of different methods in cases of ambiguity. Top row: an example of large character spacing. Bottom row: an example of juxtaposed text lines. (a) are original images. (b) are ground-truths. (c) are detection results of mask TextSpotter [13]. (d) are our detection results.

Existing text spotters [7,13,19,22] typically follow a pipeline as shown in Fig. 2(a), where a text recognition module is stacked after a text detection module, and linguistic knowledge (*e.g.* lexicon) is employed for text recognition. However, their language-based modules are isolated from the detection modules, which only detect text lines by learning visual features, leading to ambiguous proposals and hindering the accuracy of the entire systems. As shown in Fig. 1(c), these vision-based text detectors are insufficient to detect text lines correctly in ambiguous samples (*i.e.* horizontal bounding box versus vertical bounding box in this example).

In contrast, unlike existing text detectors that only utilize vision knowledge, this work incorporates linguistic knowledge into text detection by learning linguistic representation to reduce ambiguous proposals. As shown in Fig. 2(b), the re-scoring step is not only the main difference between our approach and the previous methods, but also the key step to remove ambiguity in text detection. Intuitively, without a natural language model, it is difficult to identify the correct text line in ambiguous scenarios. For example, prior art [13] detected incorrect text lines as shown in Fig. 1(c), neglecting the semantic meaning of text lines. In our approach, the knowledge of natural language can be used to reduce errors as shown in Fig. 1(d), because the text content may follow the distribution of natural language if the detected text line is correct, otherwise the text content is out of the distribution of natural language. Therefore, the linguistic representation can help refine the detection results.

This paper proposes a novel framework called Ambiguity Eliminating Text Spotter (AE TextSpotter) as shown in Fig. 2(b), to solve the ambiguity problem in text detection. The rationale of our method is to re-score the detection results by using linguistic representation, and eliminate ambiguity by filtering out low-scored text line proposals. Specifically, AE TextSpotter consists of three important components, including detection, recognition and rescoring.

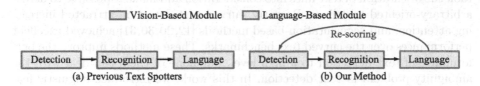

Fig. 2. Comparisons between (a) previous text spotters and (b) our AE TextSpotter. Different from previous methods [7,13,19,22] where the modules of text detection and linguistic knowledge are isolated, our method lifts the language module ahead in text detection to re-score the detection results by learning linguistic representation.

In the detection step, we build a text detection module (TDM) to provide candidates of text lines with high recall, in order to avoid missed detection. In this step, the number of candidates is very large because of the high-recall rate. In the step of recognition, we carefully design a character-based recognition module (CRM), to efficiently recognize the candidate text lines. In the re-scoring step, we propose a language module (LM) that can learn linguistic representation to re-score the candidate text lines and to eliminate ambiguity, making the text lines that correspond to natural language have higher scores than those not. The three steps above complement each other, enabling the AE TextSpotter to successfully detect and recognize text lines even in extremely challenging scenarios. As shown in Fig. 1(d), AE TextSpotter can correctly detect text lines when multiple vertical and horizontal text lines are presented.

Our main **contributions** are summarized as follows. **1)** The AE TextSpotter identifies the importance of linguistic representation for text detection, in order to solve the text detection ambiguity problem. To our knowledge, this is the first time to learn both visual features and linguistic features for text detection in a framework. **2)** We design a novel text recognition module, called the character-based recognition module (CRM), to achieve fast recognition of numerous candidate text lines. **3)** Extensive experiments on several benchmarks demonstrate the effectiveness of AE TextSpotter. For example, AE TextSpotter achieves state-of-the-art performance on both text detection and recognition on IC19-ReCTS [32]. Moreover, we construct a hard validation set for benchmarking **T**ext **D**etection **A**mbiguity, termed TDA-ReCTS, where AE TextSpotter achieves the F-measure of 81.39% and the 1-NED of 51.32%, outperforming its counterparts by 4.01% and 4.65%.

2   Related Work

Scene text detection has been a research hotspot in computer vision for a long period. Methods based on deep learning have become the mainstream of scene text detection. Tian *et al.* [27] and Liao *et al.* [15] successfully adopted the framework of object detection into text detection and achieved good performance on horizontal text detection. After that, many works [4,14,17,24,33] took the orientation of text lines into consideration and make it possible to detect arbitrary-oriented text lines. Recently, curved text detection attracted increasing attention, and segmentation-based methods [12,20,30,31] achieved excellent performances over the curved text benchmarks. These methods improve the performance of text detection to a high level, but none of them can deal with the ambiguity problem in text detection. In this work, we introduce linguistic features in the text detection module to solve the text detection ambiguity problem.

Scene text recognition targets to decode a character sequence from variable-length text images. CRNN [25] introduced CTC [8] in a text recognition model, and made the model trained end-to-end. Following this pipeline, other CTC-based methods [18,28] were proposed and achieved significant improvement in text recognition. Another direction of text recognition is attention-based encoder and decoder framework [1,2,26], which focuses on one position and predict the corresponding character at each time step. These attention-based methods have achieved high accuracy on various benchmarks. However, most existing methods work slowly when recognizing a large number of candidate text lines in an image. In this paper, we design a character-based recognition module to solve this problem.

Scene text spotting aims to detect and recognize text lines simultaneously. Li *et al.* [11] proposed a framework for horizontal text spotting, which contains a text proposal network for text detection and an attention-based method for text recognition. FOTS [19] presented a method for oriented text spotting by new RoIRotate operation, and significantly improved the performance by jointly training detection and recognition tasks. Mask TextSpotter [13] added character-level supervision to Mask R-CNN [9] to simultaneously identity text lines and characters, which is suitable for curved text spotting. Qin *et al.* [22] also developed a curved text spotter based on Mask R-CNN, and used a RoI masking operation to extract the feature for curved text recognition. However, in these methods, linguistic knowledge (*e.g.* lexicon) is only used in post-processing, so it is isolated from network training. Different from them, the linguistic representation in our method is learned together with the visual representation in a framework, making it possible to eliminate ambiguity in text detection.

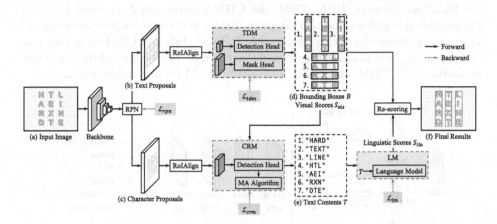

Fig. 3. The overall architecture of AE TextSpotter, which has three key modules: 1) Text detection module (TDM); 2) Character-based recognition module (CRM); 3) Language module (LM).

3 Proposed Method

3.1 Overall Architecture

Figure 3 shows the overall architecture of AE TextSpotter, which consists of two vision-based modules and one language-based module, namely, the text detection module (TDM), the character-based recognition module (CRM), and the language module (LM). Among these modules, TDM and CRM aim to detect the bounding boxes and recognize the content of candidate text lines; and LM is applied to lower the scores of incorrect text lines by utilizing linguistic features, which is the key module to remove ambiguous samples.

In the forward phase, we firstly feed the input image to the backbone network (*i.e.*, ResNet-50 [10] + FPN [16]) and Region Proposal Network (RPN) [23], to produce text line proposals and character proposals. Secondly, based on text line proposals, TDM predicts tight bounding boxes B (*i.e.*, rotated rectangles or polygons) and visual scores S_{vis} for candidate text lines (see Fig. 3(d)). Thirdly, using character proposals and bounding boxes B as input, CRM recognizes the text contents T of candidate text lines (see Fig. 3(e)). After that, we feed the text contents T to LM, to predict the linguistic scores S_{lin} for candidate text lines. Then we combine the visual scores S_{vis} and linguistic scores S_{lin} into the final scores S as Eq. 1.

$$S = \lambda S_{vis} + (1 - \lambda)S_{tex}, \tag{1}$$

where λ is used to balance S_{vis} and S_{tex}. Finally, on the basis of the final scores S, we select the final bounding boxes and text contents of text lines (see Fig. 3(f)), by using NMS and removing low-score text lines.

The training process of AE TextSpotter is divided into two stages: 1) training the backbone network, RPN, TDM, and CRM end-to-end; 2) training LM with the predicted text lines and text contents. In the first stage, we fix the weights of LM, and optimize the backbone network, RPN, TDM, and CRM by a joint loss of $\mathcal{L}_{rpn}$, $\mathcal{L}_{tdm}$, and $\mathcal{L}_{crm}$. In the second stage, we fix the weights of the backbone network, RPN, TDM, and CRM, and optimize LM by loss function $\mathcal{L}_{lm}$.

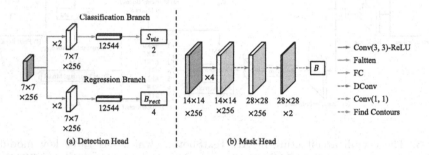

Fig. 4. The details of the detection head and the mask head in the text detection module (TDM). "Conv(3, 3)", "FC" and "DConv" represent the convolution layer with the kernel size of 3×3, fully connected layer, and deconvolution layer, respectively. "$\times 2$" denotes a stack of two consecutive convolution layers.

3.2 Text Detection Module

The text detection module (TDM) aims to detect the candidate text lines with extremely high recall. Due to the requirement of high recall, TDM needs to be implemented by a regression-based method, whose recall can be controlled by the classification score threshold and the NMS threshold. In this work, following [13, 22, 31], TDM is modified based on Mask R-CNN [9]. For each text line, it predicts an axis-aligned rectangular bounding box and the corresponding segmentation mask.

As shown in Fig. 4, there are a detection head and a mask head in TDM. Firstly, we use RoIAlign [9] to extract feature patches $(7 \times 7 \times 256)$ for text line proposals predicted by RPN. Secondly, we feed feature patches to the classification and regression branches in the detection head, to predict visual scores S_{vis} and rectangular bounding boxes B_{rect} of text lines. Then, the new feature patches $(14 \times 14 \times 256)$ are extracted for rectangular bounding boxes B_{rect}. After that, using the new feature patches as input, the mask head predicts masks of text lines, and the mask contours are tight bounding boxes B. Finally, we use a loose score threshold Thr_{score} and a loose NMS threshold Thr_{nms} to filter out redundant boxes in tight bounding boxes B, and the rest are candidate text lines.

3.3   Character-Based Recognition Module

The character-based recognition module (CRM) is applied to efficiently recognize a large number of candidate text lines produced by TDM. CRM includes a detection head and a match-assemble (MA) algorithm. The detection head runs only once for each input image to detect and classify characters; and the proposed MA algorithm is used to group characters into text line transcriptions, which is rule-based and very fast. Therefore the CRM is very suitable for recognizing numerous candidate text lines predicted by TDM.

The detection head of CRM follows a pipeline similar to the detection head of TDM (see Fig. 4 (a)). Due to the large number of character categories (about 3,600 in IC19-ReCTS), the detection head of CRM needs a better feature patch as input and a deeper classification branch. Therefore, in the detection head of CRM, the size of feature patches extracted by RoIAlign is $14 \times 14 \times 256$, and we change the number of convolutional layers in the classification branch to 4. The output of the detection head is rectangular bounding boxes and contents of characters.

After character detection and classification, we propose a match-assemble (MA) algorithm to obtain text contents. The procedure of the MA algorithm is presented in Fig. 5. For each bounding box $b \in B$ of candidate text line, we firstly **match** the predicted characters (see Fig. 5(a)) with the candidate text line b by using Eq. 2, and get the internal characters C_{in} (see Fig. 5 (c)) for the candidate text line b.

$$C_{in} = \{c \mid \frac{\text{Inter}(c, b)}{\text{Area}(c)} > Thr_{match}, c \in C\} \qquad (2)$$

Here, $\text{Inter}(\cdot)$, $\text{Area}(\cdot)$ and C are the intersection area of two boxes, the area of a box, and bounding boxes of detected characters, respectively. Secondly, we **assemble** the internal characters C_{in} according to their center positions. Specifically, we arrange the characters from left to right, when the width of the external horizontal rectangle of the candidate text line is larger than its height,

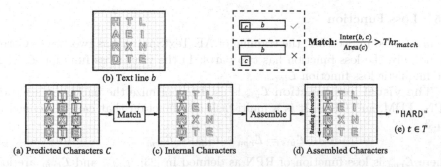

Fig. 5. The details of match-assemble (MA) algorithm. "Match" is the operation of matching character boxes with a text line boxes. "Assemble" is the operation of assembling the internal characters to a text line transcription.

or else we arrange the characters from top to down. After the arrangement, we concatenate character contents into the text content T of the candidate text line.

3.4 Language Module

The language module (LM) is a language-based module that can re-score candidate text lines according to their contents. Specifically, LM predicts a linguistic score to determine whether the text content follows the distribution of natural language.

The pipeline of LM is presented in Fig. 6. At first, we use BERT [6], a widely-used pre-trained model in NLP, to extract sentence vectors (16×768) for the input text content T. Here, 16 is the maximum length of text content, and we fill 0 for text lines whose length is less than 16. Then we feed sentence vectors to a classification network composed of BiGRU, AP, and FC, to predict the linguistic scores S_{lin} of the input text contents. Here, BiGRU, AP, and FC represent binary input gated recurrent unit [3], average pooling layer, and fully connected layer, respectively.

The label generation for LM is based on Intersection over Union (IoU) metric. Concretely, for every candidate text line, we consider the label of its text content to be positive if and only if there is a matched ground-truth whose IoU with the candidate text line is greater than a threshold IoU_{pos}.

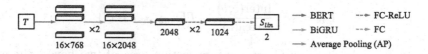

Fig. 6. The details of Language Module (LM). "BiGRU", and "FC" represent the binary input gated recurrent unit [3], and fully connected layer, respectively. "× 2" denotes a stack of two consecutive layers.

3.5 Loss Function

As mentioned in Sect. 3.1, the training of AE TextSpotter has two stages Correspondingly, the loss function has two parts: 1) the visual loss function $\mathcal{L}_{vis}$; 2) the linguistic loss function $\mathcal{L}_{lm}$.

The visual loss function $\mathcal{L}_{vis}$ is used to optimize the backbone network, RPN, TDM, and CRM. It is a multi-task loss function that can be defined as Eq. 3.

$$\mathcal{L}_{vis} = \mathcal{L}_{rpn} + \mathcal{L}_{tdm} + \mathcal{L}_{crm}, \tag{3}$$

where $\mathcal{L}_{rpn}$ is loss function of RPN as defined in [23]. $\mathcal{L}_{tdm}$ and $\mathcal{L}_{crm}$ are loss functions of TDM and CRM respectively, which can be formulated as Eq. 4 and Eq. 5.

$$\mathcal{L}_{tdm} = \mathcal{L}_{tdm}^{cls} + \mathcal{L}_{tdm}^{box} + \mathcal{L}_{tdm}^{mask}. \tag{4}$$

$$\mathcal{L}_{crm} = \mathcal{L}_{crm}^{cls} + \mathcal{L}_{crm}^{box}. \tag{5}$$

Here, $\mathcal{L}_{tdm}^{cls}$ and $\mathcal{L}_{crm}^{cls}$ are loss functions of classification branches in TDM and CRM respectively, which are implemented by cross-entropy losses. $\mathcal{L}_{tdm}^{box}$ and $\mathcal{L}_{crm}^{box}$ are loss functions of regression branches in TDM and CRM respectively, which are implemented by smooth L_1 [3]. $\mathcal{L}_{tdm}^{mask}$ is the loss function of mask head in TDM, which is the same as the mask loss in [9].

The linguistic loss function $\mathcal{L}_{lm}$ is used to optimize LM. Actually, LM only involves a task of classifying the correct and incorrect text lines in candidates, so $\mathcal{L}_{lm}$ is implemented by a cross-entropy loss.

4 Experiment

4.1 Datasets

AE TextSpotter needs character-level annotations in the training phase. In addition text detection ambiguity is relatively rare in English due to its writing habit. Therefore, we evaluate our method on a multi-language dataset with character-level annotations.

ICDAR 2019 ReCTS (IC19-ReCTS) [32] is a newly-released large-scale dataset that includes 20,000 training images and 5,000 testing images, covering multiple languages, such as Chinese, English and Arabic numerals. The images in this dataset are mainly collected in the scenes with signboards, All text lines and characters in this dataset are annotated with bounding boxes and transcripts.

TDA-ReCTS is a validation set for benchmarking text detection ambiguity, which contains 1,000 ambiguous images selected from the training set of IC19-ReCTS. To minimize the impact of subjective factors, we designed rules to pick out images with the case of large character spacing or juxtaposed text lines, and then randomly sample 1,000 images among them as the validation set.

When selecting ambiguous images, an image is regarded as a sample with large character spacing, if at least one text line in the image has large character spacing. In addition, an image is treated as a sample with juxtaposed text lines, if there is a pair of text lines aligned with the top, bottom, left or right direction, and their characters have similar scales. More details about the both rules can be found in Appendix A.1.

4.2 Implementation Details

In the backbone network, the ResNet50 [10] is initialized by the weights pretrained on ImageNet [5]. In RPN, the aspect ratios of text line anchors are set to {1/8, 1/4, 1/2, 1, 2, 4, 8} to match the extreme aspect ratios of text lines, and the aspect ratios of character anchors are set to {1/2, 1, 2}. All other settings in RPN are the same as in Mask R-CNN. In TDM, the classification branch makes a binary classification of text and non-text, and the two post-processing thresholds (i.e., Thr_{score} and Thr_{nms}) in the regression branch are set to 0.01 and 0.9, respectively. In CRM, the match threshold Thr_{match} is set to 0.3. In LM, candidate text lines are classified into correct and incorrect ones., and the training label threshold IoU_{pos} is set to 0.8. When combining the visual and linguistic scores, the balance weight λ in Eq. 1 is set to 0.7. In the post-processing, the NMS threshold is set to 0.1 and the classification score threshold is set to 0.6.

As mentioned in Sect. 3.1, the training of AE TextSpotter is divided into two phases. In the first phase, we train backbone network, RPN, TDM, and CRM with batch size 16 on 8 GPUs for 12 epochs, and the initial learning rate is set to 0.02 which is divided by 10 at 8 epoch and 11 epoch. In the second phase, we fix the weights of the backbone network, RPN, TDM, and CRM, and train LM with batch size 16 on 8 GPUs for 12 epochs, and the initial learning rate is set to 0.2 which is divided by 10 at 8 epoch and 11 epoch. We optimize all models using stochastic gradient descent (SGD) with a weight decay of 1×10^{-4} and a momentum of 0.9.

In the training phase, we ignore the blurred text regions labeled as "DO NOT CARE" in all datasets; and apply random scale, random horizontal flip, and random crop on training images.

4.3   Ablation Study

The Proportion of Ambiguous Text Lines. Text detection ambiguity is caused by large character spacing and juxtaposed text lines. To count the proportion of the two problematic cases, we use the rules mentioned in Sect. 4.1 to count the ambiguous text lines in the IC19-ReCTS training set and TDA-ReCTS. As shown in Table 1, ambiguous text lines occupy 5.97% of total text lines in the IC19-ReCTS training set, which reveals that text detection ambiguity is not rare in the natural scene. Moreover, the proportion of ambiguous text lines in TDA-ReCTS reaches 25.80%. Although the proportion in TDA-ReCTS is much larger than in the IC19-ReCTS training set, the normal text lines are still in the majority.

Table 1. The proportion of text lines with the problem of text detection ambiguity.

Type	IC19-ReCTS training set		TC-ReCTS	
	Number	Proportion	Number	Proportion
Large character spacing	1,241	1.14%	589	7.07%
Juxtaposed text lines	5,379	4.94%	1,615	19.39%
Union of two categories	6,509	5.97%	2,148	25.80%
All	108,963	-	8,327	-

The High Recall of the Text Detection Module. To verify the high recall of the text detection module (TDM), we compare the recall and the number of candidate text lines per image on TDA-ReCTS, under different visual score thresholds Thr_{score} and the NMS thresholds Thr_{nms}. As shown in Table 2, when $Thr_{score} = 0.0$ and $Thr_{nms} = 1.0$, the recall of TDM reaches the upper bound of 0.971. However, in this case, the number of candidate text lines per image is 971.5, which is too large to be recognized. When $Thr_{score} = 0.01$ and $Thr_{nms} = 0.9$, the recall of TDM is 0.969, slightly lower than the upper bound, but the number of candidate text lines per image is reduced to 171.7. If we use common post-processing thresholds (*i.e.*, $Thr_{score} = 0.5$ and $Thr_{nms} = 0.5$), the recall of TDM is only 0.852, which will result in many missed detection of text lines. Therefore, in experiments, the values of Thr_{score} and Thr_{nms} are set to 0.01 and 0.9 by default, which can achieve a high recall without too many candidate text lines.

The Effectiveness of the Character-Based Recognition Module. The character-based recognition module (CRM) aims to fast recognize numerous candidate text lines predicted by TDM. To test the speed of CRM, we compare the time cost per text line of CRM and mainstream recognizers (*e.g.* CRNN [25] and ASTER [26]). For a fair comparison, we run these methods on 1080Ti GPU and report their time cost per image on the IC19-ReCTS test set. Note that, the input text images of CRNN and ASTER are scaled to a height of 32 when testing. In AE TextSpotter, the recognition module is used to recognize the candidate text lines, whose number reaches **171.7 per image**. Table 3 shows the speed of different methods. CRNN and ASTER cost 966.7 ms and 4828.2 ms respectively to recognize **all candidate text lines** in each image. The time cost of CRM on each image is only 122.3 ms, so CRM runs much faster than CRNN and ASTER. In addition, using the same detection result predicted by AE TextSpotter, our CRM can achieve the 1-NED of 71.81 on the test set of IC19-ReCTS, which is better than CRNN and ASTER.

Table 2. The recall of TDM and the number of candidate text lines per image under different post-processing thresholds.

Thr_{score}	Thr_{nms}	Recall	Number of candidates
0	1	0.971	971.5
0.01	0.9	0.969	171.7
0.5	0.5	0.852	11.1

The Effectiveness of the Language Module. To investigate the effectiveness of the language module (LM), we compare the performance of the models with and without LM on TDA-ReCTS. For a fair comparison, we build a model without LM by removing LM from AE TextSpotter in the training phase and setting λ in Eq. 1 to 1.0 in the testing phase. As shown in Table 4, on TDA-ReCTS, the model with LM obtains the F-measure of 81.39 and the 1-NED of 51.32%, significantly surpassing the model without LM by 3.46% and 3.57%.

In addition, we analyze the quality of the detection results predicted by the models with and without LM on TDA-ReCTS. Generally, the quality of a

Table 3. The time cost per image and 1-NED of different recognizers on the IC19-ReCTS test set.

Method	Time cost	1-NED
CRNN [25]	966.7 ms	68.32
ASTER [26]	4828.2 ms	58.49
CRM (ours)	122.3 ms	71.81

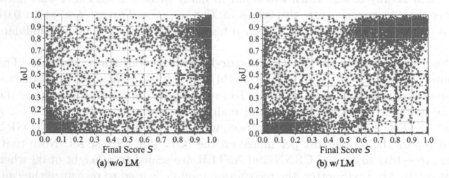

(a) w/o LM   (b) w/ LM

Fig. 7. The distributions of detection results predicted by models with and without LM. The points in **red dotted boxes** are false positives with high scores. (Color online figure)

bounding box can be measured by its IoU with the matching ground truth, and a bounding box is considered to be correct if its IoU exceeds 0.5. As shown in Fig. 7 (a), the detection results of the model without LM have many incorrect results (IoU < 0.5), even though their final scores are greater than 0.8. However, this problem is alleviated in the model with LM (see Fig. 7 (b)), where there are few detection results with high scores but low IoU. These experiment results demonstrate the power of LM.

4.4   Comparisons with State-of-the-Art Methods

To show the effectiveness of the proposed AE TextSpotter, we compare it with state-of-the-art methods on TDA-ReCTS and IC19-ReCTS test set.

On TDA-ReCTS, we compare the proposed AE TextSpotter with some representative methods. All results are obtained by using authors provided or officially released codebases, and these methods are trained on the 19,000 training images in IC19-ReCTS (except 1,000 images in TDA-ReCTS). In the testing phase, we scale the short side of images to 800, and evaluate the performance of these methods on TDA-ReCTS by the evaluation metric defined in [32]. Table 4 shows the performance of these methods. The proposed AE TextSpotter achieves the F-measure of 81.39% and the 1-NED of 51.32%, outperforming the counterparts by at least 4.01% and 4.65% respectively. Here, 1-NED denotes normalized edit distance, a recognition metric defined in [32]. Note that, the proportion of

Table 4. The single-scale results on TDA-ReCTS. "P", "R", "F" and "1-NED" mean the precision, recall, F-measure, and normalized edit distance [32], respectively.

Method	Venue	Backbone	TDA-ReCTS			
			P	R	F	1-NED
EAST [33]	CVPR'17	VGG16	70.58	61.59	65.78	-
PSENet [29]	CVPR'19	ResNet50	72.14	65.53	68.68	-
FOTS [19]	CVPR'18	ResNet50	68.20	70.00	69.08	34.87
Mask TextSpotter [13]	TPAMI	ResNet50	80.07	74.86	77.38	46.67
Detection only	-	ResNet50	79.58	74.93	77.19	-
AE TextSpotter w/o LM	-	ResNet50	79.18	76.72	77.93	47.74
AE TextSpotter	-	ResNet50	**84.78**	**78.28**	**81.39**	**51.32**

Table 5. The single-scale results on the IC19-ReCTS test set. "P", "R", "F" and "1-NED" represent the precision, recall, F-measure, and normalized edit distance, respectively. "*" denotes the methods in competition [32], which use extra datasets, multi-scale testing, and model ensemble. "800 ×" means that the short side of input images is scaled to 800.

Method	Venue	Backbone	IC19-ReCTS Test set			
			P	R	F	1-NED
Tencent-DPPR Team* [32]	Competition	-	93.49	92.49	92.99	81.45
SANHL_v1* [32]	Competition	-	91.98	93.86	92.91	81.43
HUST_VLRGROUP* [32]	Competition	-	91.87	92.36	92.12	79.38
EAST [33]	CVPR'17	VGG16	74.25	73.65	73.95	-
PSENet [29]	CVPR'19	ResNet50	87.31	83.86	85.55	-
FOTS [19]	CVPR'18	ResNet50	78.25	82.49	80.31	50.83
Mask TextSpotter [13]	TPAMI	ResNet50	89.30	88.77	89.04	67.79
Detection Only	-	ResNet50	90.72	90.73	90.72	-
AE TextSpotter w/o LM	-	ResNet50	91.54	90.78	91.16	70.95
AE TextSpotter (800×)	-	ResNet50	92.60	91.01	91.80	71.81
AE TextSpotter (1000×)	-	ResNet50	**92.28**	**91.45**	**91.86**	**73.12**

ambiguous text lines in TDA-ReCTS is only 25.80%, and we believe the improvement will be more significant if the proportion is larger.

On the IC19-ReCTS test set, we train models on the complete training set of IC19-ReCTS. As shown in Table 5, our AE TextSpotter achieves the F-measure of 91.80% and the 1-NED of 71.81%, surpassing other methods. Moreover, when we enlarge the scale of input images (short size 1000), the F-measure and 1-NED of our method can be boosted to 91.86% and 73.12%. Without extra datasets, the single-scale detection performance of our method is comparable to the methods that use extra datasets, multi-scale testing, and model ensemble (91.86% *v.s.* 92.99%). Unlike the detection task, the multi-language recognition task relies more heavily on extra datasets. Therefore, our recognition performance is lower than the methods using extra datasets, multi-scale testing, and ensemble.

(a) Large Character Spacing   (b) Juxtaposed Text Lines

Fig. 8. Visualization results of AE TextSpotter on TDA-ReCTS.

These results demonstrate that the proposed AE TextSpotter can correctly detect and recognize multi-language text lines, even in ambiguous scenarios. Some qualitative results on TDA-ReCTS are shown in Fig. 8.

4.5 Time Cost Analysis of AE TextSpotter

We analyze the time cost of all modules in AE TextSpotter. Specifically, we evaluate our methods on images in TDA-ReCTS and calculate the average time cost per image. All results are tested by PyTorch [21] with the batch size of 1 on one 1080Ti GPU in a single thread. Table 6 shows the time cost of backbone, TDM, CRM, LM, and post-processing respectively. Among them, the detection module takes up most of the execution time, because it needs to predict the mask for every candidate text line. The entire framework can run 1.3 FPS when the short size of input images are 800.

Table 6. The time cost of all modules in AE TextSpotter.

Scale	FPS	Time cost per image (ms)					
		Backbone	RPN	TDM	CRM	LM	Post-proc
800×	1.3	31.3	35.1	440.0	122.3	118.6	3.8
1000×	1.2	63.1	42.6	487.2	127.8	121.1	3.9

4.6 Discussion

As demonstrated in previous experiments, the proposed AE TextSpotter works well in most cases, including scenarios of text detection ambiguity. Nevertheless, there is still room for improvement in some parts as follows. **1)** The recognition performance of our method is not high (only 51.32% in Table 4), supressing the effectiveness of LM. **2)** TDA-ReCTS is selected by rules that only give a vague description of ambiguity in text detection. Therefore, a manually selected dataset is important for further exploring text detection ambiguity. **3)** The proposed AE TextSpotter is a baseline for removing text detection ambiguity, and it is still worthwhile to further explore a better way to combine visual and linguistic representation.

Although our AE TextSpotter works well in most cases. It still fails in some intricate images, such as text-like regions, strange character arrangement, and texts in fanc styles. More details about these failed cases can be found in Appendix A.2.

5 Conclusion and Future Work

In this paper, we first conducted a detailed analysis of the ambiguity problem in text detection, and revealed the importance of linguistic representation in solving this problem. Then we proposed a novel text spotter, termed AE TextSpotter, which introduces linguistic representation to eliminate ambiguity in text detection. For the first time, linguistic representation is utilized in scene text detection to deal with the problem of text detection ambiguity. Concretely, we propose the Language Module (LM), which can learn to re-score the text proposals by linguistic representation. LM can effectively lower the scores of incorrect text lines while improve the scores of correct proposals. Furthermore, a new validation set is proposed for benchmarking the ambiguity problem. Extensive experiments demonstrate the advantages of our method, especially in scenarios of text detection ambiguity.

In the future, we will strive to build a more efficient and fully end-to-end network for ambiguous text spotting, and plan to build a large-scale and more challenging dataset which is full of ambiguous text samples.

Acknowledgments. This work is supported by the Natural Science Foundation of China under Grant 61672273 and Grant 61832008, the Science Foundation for Distinguished Young Scholars of Jiangsu under Grant BK20160021, and Scientific Foundation of State Grid Corporation of China (Research on Ice-wind Disaster Feature Recognition and Prediction by Few-shot Machine Learning in Transmission Lines).

References

1. Cheng, Z., Bai, F., Xu, Y., Zheng, G., Pu, S., Zhou, S.: Focusing attention: towards accurate text recognition in natural images. In: Proceedings of the IEEE International Conference on Computer Vision, pp. 5076–5084 (2017)
2. Cheng, Z., Xu, Y., Bai, F., Niu, Y., Pu, S., Zhou, S.: Aon: Towards arbitrarily-oriented text recognition. In: Proceedings of the IEEE Conference on Computer Vision and Pattern Recognition, pp. 5571–5579 (2018)
3. Cho, K., et al.: Learning phrase representations using RNN encoder-decoder for statistical machine translation. arXiv preprint arXiv:1406.1078 (2014)
4. Deng, D., Liu, H., Li, X., Cai, D.: Pixellink: Detecting scene text via instance segmentation. In: Thirty-Second AAAI Conference on Artificial Intelligence (2018)
5. Deng, J., Dong, W., Socher, R., Li, L.J., Li, K., Fei-Fei, L.: Imagenet: A large-scale hierarchical image database. In: 2009 IEEE Conference on Computer Vision and Pattern Recognition, pp. 248–255. IEEE (2009)
6. Devlin, J., Chang, M.W., Lee, K., Toutanova, K.: Bert: Pre-training of deep bidirectional transformers for language understanding. arXiv preprint arXiv:1810.04805 (2018)

7. Feng, W., He, W., Yin, F., Zhang, X.Y., Liu, C.L.: Textdragon: An end-to-end framework for arbitrary shaped text spotting. In: Proceedings of the IEEE International Conference on Computer Vision, pp. 9076–9085 (2019)
8. Graves, A., Fernández, S., Gomez, F., Schmidhuber, J.: Connectionist temporal classification: labelling unsegmented sequence data with recurrent neural networks. In: Proceedings of the 23rd International Conference on Machine Learning, pp. 369–376. ACM (2006)
9. He, K., Gkioxari, G., Dollár, P., Girshick, R.: Mask r-CNN. In: Proceedings of the IEEE International Conference on Computer Vision, pp. 2961–2969 (2017)
10. He, K., Zhang, X., Ren, S., Sun, J.: Deep residual learning for image recognition. In: Proceedings of the IEEE Conference on Computer Vision and Pattern Recognition, pp. 770–778 (2016)
11. Li, H., Wang, P., Shen, C.: Towards end-to-end text spotting with convolutional recurrent neural networks. arXiv preprint arXiv:1707.03985 (2017)
12. Li, X., Wang, W., Hou, W., Liu, R.Z., Lu, T., Yang, J.: Shape robust text detection with progressive scale expansion network. arXiv preprint arXiv:1806.02559 (2018)
13. Liao, M., Lyu, P., He, M., Yao, C., Wu, W., Bai, X.: Mask textspotter: an end-to-end trainable neural network for spotting text with arbitrary shapes. IEEE Transactions on Pattern Analysis and Machine Intelligence (2019)
14. Liao, M., Shi, B., Bai, X.: Textboxes++: a single-shot oriented scene text detector. IEEE Trans. Image Process. **27**(8), 3676–3690 (2018)
15. Liao, M., Shi, B., Bai, X., Wang, X., Liu, W.: Textboxes: a fast text detector with a single deep neural network. In: AAAI, pp. 4161–4167 (2017)
16. Lin, T.Y., Dollár, P., Girshick, R., He, K., Hariharan, B., Belongie, S.: Feature pyramid networks for object detection. In: Proceedings of the IEEE Conference on Computer Vision and Pattern Recognition, pp. 2117–2125 (2017)
17. Liu, J., Liu, X., Sheng, J., Liang, D., Li, X., Liu, Q.: Pyramid mask text detector. arXiv preprint arXiv:1903.11800 (2019)
18. Liu, W., Chen, C., Wong, K.Y.K., Su, Z., Han, J.: Star-net: a spatial attention residue network for scene text recognition. In: BMVC, vol. 2, p. 7 (2016)
19. Liu, X., Liang, D., Yan, S., Chen, D., Qiao, Y., Yan, J.: Fots: fast oriented text spotting with a unified network. In: Proceedings of the IEEE Conference on Computer Vision and Pattern Recognition, pp. 5676–5685 (2018)
20. Long, S., Ruan, J., Zhang, W., He, X., Wu, W., Yao, C.: Textsnake: a flexible representation for detecting text of arbitrary shapes. In: Proceedings of the European Conference on Computer Vision (ECCV), pp. 20–36 (2018)
21. Paszke, A., et al.: Pytorch: an imperative style, high-performance deep learning library. In: Advances in Neural Information Processing Systems, pp. 8024–8035 (2019)
22. Qin, S., Bissacco, A., Raptis, M., Fujii, Y., Xiao, Y.: Towards unconstrained end-to-end text spotting. In: Proceedings of the IEEE International Conference on Computer Vision, pp. 4704–4714 (2019)
23. Ren, S., He, K., Girshick, R., Sun, J.: Faster R-CNN: Towards real-time object detection with region proposal networks. In: Advances in Neural Information Processing Systems, pp. 91–99 (2015)
24. Shi, B., Bai, X., Belongie, S.: Detecting oriented text in natural images by linking segments. arXiv preprint arXiv:1703.06520 (2017)
25. Shi, B., Bai, X., Yao, C.: An end-to-end trainable neural network for image-based sequence recognition and its application to scene text recognition. IEEE Trans. Pattern Anal. Mach. Intell. **39**(11), 2298–2304 (2016)

26. Shi, B., Yang, M., Wang, X., Lyu, P., Yao, C., Bai, X.: Aster: an attentional scene text recognizer with flexible rectification. IEEE Transactions on Pattern Analysis and Machine Intelligence (2018)
27. Tian, Z., Huang, W., He, T., He, P., Qiao, Yu.: Detecting text in natural image with connectionist text proposal network. In: Leibe, B., Matas, J., Sebe, N., Welling, M. (eds.) ECCV 2016. LNCS, vol. 9912, pp. 56–72. Springer, Cham (2016). https://doi.org/10.1007/978-3-319-46484-8_4
28. Wang, J., Hu, X.: Gated recurrent convolution neural network for OCR. In: Advances in Neural Information Processing Systems, pp. 335–344 (2017)
29. Wang, W., et al.: Shape robust text detection with progressive scale expansion network. In: Proceedings of the IEEE Conference on Computer Vision and Pattern Recognition, pp. 9336–9345 (2019)
30. Wang, W., et al.: Efficient and accurate arbitrary-shaped text detection with pixel aggregation network. In: Proceedings of the IEEE International Conference on Computer Vision (2019)
31. Xie, E., Zang, Y., Shao, S., Yu, G., Yao, C., Li, G.: Scene text detection with supervised pyramid context network. Proc. AAAI Conf. Artif. Intell. **33**, 9038–9045 (2019)
32. Zhang, R., et al.: ICDAR 2019 robust reading challenge on reading chinese text on signboard. In: 2019 International Conference on Document Analysis and Recognition (ICDAR), pp. 1577–1581. IEEE (2019)
33. Zhou, X., et al.: East: An efficient and accurate scene text detector. arXiv preprint arXiv:1704.03155 (2017)

History Repeats Itself: Human Motion Prediction via Motion Attention

Wei Mao[1]([✉]), Miaomiao Liu[1], and Mathieu Salzmann[2]

[1] Australian National University, Canberra, Australia
{wei.mao,miaomiao.liu}@anu.edu.au
[2] EPFL–CVLab and ClearSpace, Lausanne, Switzerland
mathieu.salzmann@epfl.ch

Abstract. Human motion prediction aims to forecast future human poses given a past motion. Whether based on recurrent or feed-forward neural networks, existing methods fail to model the observation that human motion tends to repeat itself, even for complex sports actions and cooking activities. Here, we introduce an attention-based feed-forward network that explicitly leverages this observation. In particular, instead of modeling frame-wise attention via pose similarity, we propose to extract *motion attention* to capture the similarity between the current motion context and the historical motion sub-sequences. Aggregating the relevant past motions and processing the result with a graph convolutional network allows us to effectively exploit motion patterns from the long-term history to predict the future poses. Our experiments on Human3.6M, AMASS and 3DPW evidence the benefits of our approach for both periodical and non-periodical actions. Thanks to our attention model, it yields state-of-the-art results on all three datasets. Our code is available at https://github.com/wei-mao-2019/HisRepItself.

Keywords: Human motion prediction · Motion attention

1 Introduction

Human motion prediction consists of forecasting the future poses of a person given a history of their previous motion. Predicting human motion can be highly beneficial for tasks such as human tracking [6], human-robot interaction [13], and human motion generation for computer graphics [14,15,26]. To tackle the problem effectively, recent approaches use deep neural networks [5,11,21] to model the temporal historical data.

Traditional methods, such as hidden Markov models [3] and Gaussian Process Dynamical Models [30], have proven effective for simple motions, such as walking and golf swings. However, they are typically outperformed by deep learning ones on more complex motions. The most common trend in modeling the

Electronic supplementary material The online version of this chapter (https://doi.org/10.1007/978-3-030-58568-6_28) contains supplementary material, which is available to authorized users.

A. Vedaldi et al. (Eds.): ECCV 2020, LNCS 12359, pp. 474–489, 2020.
https://doi.org/10.1007/978-3-030-58568-6_28

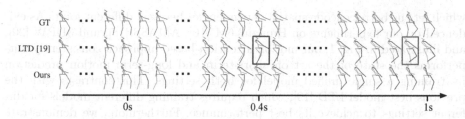

Fig. 1. Human motion prediction aims to forecast future human poses ($>0s$) given past ones. From top to bottom, we show the ground-truth pose sequence, the predictions of LTD [19] and those of our approach. Frames where LTD [19] makes larger error on arms and legs are highlighted in blue and red box respectively. Note that our results better match the ground truth than those of LTD [19]. (Color online figure)

sequential data that constitutes human motion consists of using Recurrent Neural Networks (RNNs) [5,11,21]. However, as discussed in [16], in the mid- to long-term horizon, RNNs tend to generate static poses because they struggle to keep track of long-term history. To tackle this problem, existing works [8,16] either rely on Generative Adversarial Networks (GANs), which are notoriously hard to train [1], or introduce an additional *long-term encoder* to represent information from the further past [16]. Unfortunately, such an encoder treats the entire motion history equally, thus not allowing the model to put more emphasis on some parts of the past motion that better reflect the context of the current motion.

In this paper, by contrast, we introduce an attention-based motion prediction approach that effectively exploits historical information by dynamically adapting its focus on the previous motions to the current context. Our method is motivated by the observation that humans tend to repeat their motion, not only in short periodical activities, such as walking, but also in more complex actions occurring across longer time periods, such as sports and cooking activities [25]. Therefore, we aim to find the relevant historical information to predict future motion.

To the best of our knowledge, only [28] has attempted to leverage attention for motion prediction. This, however, was achieved in a frame-wise manner, by comparing the human pose from the last observable frame with each one in the historical sequence. As such, this approach fails to reflect the motion direction and is affected by the fact that similar poses may appear in completely different motions. For instance, in most Human3.6M activities, the actor will at some point be standing with their arm resting along their body. To overcome this, we therefore propose to model *motion attention*, and thus compare the last visible sub-sequence with a history of motion sub-sequences.

To this end, inspired by [19], we represent each sub-sequence in trajectory space using the Discrete Cosine Transform (DCT). We then exploit our motion attention as weights to aggregate the entire DCT-encoded motion history into a future motion estimate. This estimate is combined with the latest observed motion, and the result acts as input to a graph convolutional network (GCN), which lets us better encode spatial dependencies between different joints.As evidenced by our experiments on Human3.6M [10], AMASS [18], and 3DPW [20], and illustrated in Fig. 1, our motion attention-based approach consistently outperforms the state of the art on short-term and long-term motion prediction by training a single unified model for both settings. This contrasts with the previous-best model LTD [19], which requires training different models for different settings to achieve its best performance. Furthermore, we demonstrate that it can effectively leverage the repetitiveness of motion in longer sequences.

Our contributions can be summarized as follows. (i) We introduce an attention-based model that exploits motions instead of static frames to better leverage historical information for motion prediction; (ii) Our motion attention allows us to train a unified model for both short-term and long-term prediction; (iii) Our approach can effectively make use of motion repetitiveness in long-term history; (iv) It yields state-of-the-art results and generalizes better than existing methods across datasets and actions.

2   Related Work

RNN-based Human Motion Prediction. RNNs have proven highly successful in sequence-to-sequence prediction tasks [12,27]. As such, they have been widely employed for human motion prediction [5,7,11,21]. For instance, Fragkiadaki et al. [5] proposed an Encoder-Recurrent-Decoder (ERD) model that incorporates a non-linear multi-layer feedforward network to encode and decode motion before and after recurrent layers. To avoid error accumulation, curriculum learning was adopted during training. In [11], Jain et al. introduced a Structural-RNN model relying on a manually-designed spatio-temporal graph to encode motion history. The fixed structure of this graph, however, restricts the flexibility of this approach at modeling long-range spatial relationships between different limbs. To improve motion estimation, Martinez et al. [21] proposed a residual-based model that predicts velocities instead of poses. Furthermore, it was shown in this work that a simple zero-velocity baseline, i.e., constantly predicting the last observed pose, led to better performance than [5,11]. While this led to better performance than the previous pose-based methods, the predictions produced by the RNN still suffer from discontinuities between the observed poses and predicted ones. To overcome this, Gui et al. proposed to adopt adversarial training to generate smooth sequences [8]. In [9], Ruiz et al. treat human motion prediction as a tensor inpainting problem and exploit a generative adversarial network for long-term prediction. While this approach further improves performance, the use of an adversarial classifier notoriously complicates training [1], making it challenging to deploy on new datasets.

Feed-Forward Methods and Long Motion History Encoding. In view of the drawbacks of RNNs, several works considered feed-forward networks as an alternative solution [4,16,19]. In particular, in [4], Butepage et al. introduced a fully-connected network to process the recent pose history, investigating different

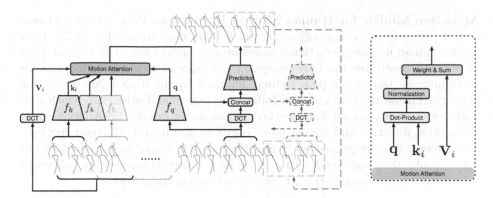

Fig. 2. Overview of our approach. The past poses are shown as blue and red skeletons and the predicted ones in green and purple. The last observed M poses are initially used as query. For every M consecutive poses in the history (key), we compute an attention score to weigh the DCT coefficients (values) of the corresponding sub-sequence. The weighted sum of such values is then concatenated with the DCT coefficients of the last observed sub-sequence to predict the future. At test time, to predict poses in the further future, we use the output of the predictor as input and predict future motion recursively (as illustrated by the dashed line).

strategies to encode temporal historical information via convolutions and exploiting the kinematic tree to encode spatial information. However, similar to [11], and as discussed in [16], the use of a fixed tree structure does not reflect the motion synchronization across different, potentially distant, human body parts. To capture such dependencies, Li *et al.* [16] built a convolutional sequence-to-sequence model processing a two-dimensional pose matrix whose columns represent the pose at every time step. This model was then used to extract a prior from long-term motion history, which, in conjunction with the more recent motion history, was used as input to an autoregressive network for future pose prediction. While more effective than the RNN-based frameworks, the manually-selected size of the convolutional window highly influences the temporal encoding.

Our work is inspired by that of Mao *et al.* [19], who showed that encoding the short-term history in frequency space using the DCT, followed by a GCN to encode spatial and temporal connections led to state-of-the-art performance for human motion prediction up to 1s. As acknowledged by Mao *et al.* [19], however, encoding long-term history in DCT yields an overly-general motion representation, leading to worse performance than using short-term history. In this paper, we overcome this drawback by introducing a *motion attention* based approach to human motion prediction. This allows us to capture the motion recurrence in the long-term history. Furthermore, in contrast to [16], whose encoding of past motions depends on the manually-defined size of the temporal convolution filters, our model dynamically adapts its history-based representation to the context of the current prediction.

Attention Models for Human Motion Prediction. While attention-based neural networks are commonly employed for machine translation [2,29], their use for human motion prediction remains largely unexplored. The work of Tang *et al.* [28] constitutes the only exception, incorporating an attention module to summarize the recent pose history, followed by an RNN-based prediction network. This work, however, uses frame-wise pose-based attention, which may lead to ambiguous motion, because static poses do not provide information about the motion direction and similar poses occur in significantly different motions. To overcome this, we propose to leverage *motion attention*. As evidenced by our experiments, this, combined with a feed-forward prediction network, allows us to outperform the state-of-the-art motion prediction frameworks.

3   Our Approach

Let us now introduce our approach to human motion prediction. Let $\mathbf{X}_{1:N} = [\mathbf{x}_1, \mathbf{x}_2, \mathbf{x}_3, \cdots, \mathbf{x}_N]$ encode the motion history, consisting of N consecutive human poses, where $\mathbf{x}_i \in \mathbb{R}^K$, with K the number of parameters describing each pose, in our case 3D coordinates or angles of human joints. Our goal is to predict the poses $\mathbf{X}_{N+1:N+T}$ for the future T time steps. To this end, we introduce a motion attention model that allows us to form a future motion estimate by aggregating the long-term temporal information from the history. We then combine this estimate with the latest observed motion and input this combination to a GCN-based feed-forward network that lets us learn the spatial and temporal dependencies in the data. Below, we discuss these two steps in detail.

3.1   Motion Attention Model

As humans tend to repeat their motion across long time periods, our goal is to discover sub-sequences in the motion history that are similar to the current sub-sequence. In this paper, we propose to achieve this via an attention model.

Following the machine translation formalism of [29], we describe our attention model as a mapping from a *query* and a set of *key-value* pairs to an output. The output is a weighted sum of *values*, where the weight, or *attention*, assigned to each value is a function of its corresponding *key* and of the *query*. Mapping to our motion attention model, the *query* corresponds to a learned representation of the last observed sub-sequence, and the *key-value* pairs are treated as a dictionary within which *keys* are learned representations for historical sub-sequences and *values* are the corresponding learned future motion representations. Our motion attention model output is defined as the aggregation of these future motion representations based on *partial motion similarity* between the latest motion sub-sequence and historical sub-sequences.

In our context, we aim to compute attention from short sequences. To this end, we first divide the motion history $\mathbf{X}_{1:N} = [\mathbf{x}_1, \mathbf{x}_2, \mathbf{x}_3, \cdots, \mathbf{x}_N]$ into $N - M - T + 1$ sub-sequences $\{\mathbf{X}_{i:i+M+T-1}\}_{i=1}^{N-M-T+1}$, each of which consists of $M + T$ consecutive human poses. By using sub-sequences of length $M + T$, we assume that the predictor, which we will introduce later, exploits the past M

frames to predict the future T frames. We then take the first M poses of each sub-sequence $\mathbf{X}_{i:i+M-1}$ to be a key, and the whole sub-sequence $\mathbf{X}_{i:i+M+T-1}$ is then the corresponding value. Furthermore, we define the query as the latest sub-sequence $\mathbf{X}_{N-M+1:N}$ with length M.

To leverage the state-of-the-art representation introduced in [19] and make the output of our attention model consistent with that of the final predictor, we map the resulting values to trajectory space using the DCT on the temporal dimension. That is, we take our final values to be the DCT coefficients $\{\mathbf{V}_i\}_{i=1}^{N-M-T+1}$, where $\mathbf{V}_i \in \mathbb{R}^{K \times (M+T)}$. Each row of $\mathbf{V}_i$ contains the DCT coefficients of one joint coordinate sequence. In practice, we can truncate some high frequencies to avoid predicting jittery motion.

As depicted by Fig. 2, the query and keys are used to compute attention scores, which then act as weights to combine the corresponding values. To this end, we first map the query and keys to vectors of the same dimension d by two functions $f_q : \mathbb{R}^{K \times M} \rightarrow \mathbb{R}^d$ and $f_k : \mathbb{R}^{K \times M} \rightarrow \mathbb{R}^d$ modeled with neural networks. This can be expressed as

$$\mathbf{q} = f_q(\mathbf{X}_{N-M+1:N}), \mathbf{k}_i = f_k(\mathbf{X}_{i:i+M-1}) \qquad (1)$$

where $\mathbf{q}, \mathbf{k}_i \in \mathbb{R}^d$, and $i \in \{1, 2, \cdots, N-M-T+1\}$. For each key, we then compute an attention score as

$$a_i = \frac{\mathbf{q}\mathbf{k}_i^T}{\sum_{i=1}^{N-M-T+1} \mathbf{q}\mathbf{k}_i^T}. \qquad (2)$$

Note that, instead of the softmax function which is commonly used in attention mechanisms, we simply normalize the attention scores by their sum, which we found to avoid the gradient vanishing problem that may occur when using a softmax. While this division only enforces the sum of the attention scores to be 1, we further restrict the outputs of f_q and f_k to be non-negative with ReLU [22] to avoid obtaining negative attention scores.

We then compute the output of the attention model as the weighed sum of values. That is,

$$\mathbf{U} = \sum_{i=1}^{N-M-T+1} a_i \mathbf{V}_i, \qquad (3)$$

where $\mathbf{U} \in \mathbb{R}^{K \times (M+T)}$. This initial estimate is then combined with the latest sub-sequence and processed by the prediction model described below to generate future poses $\hat{\mathbf{X}}_{N+1:N+T}$. At test time, to generate longer future motion, we augment the motion history with the last predictions and update the query with the latest sub-sequence in the augmented motion history, and the key-value pairs accordingly. These updated entities are then used for the next prediction step.

3.2   Prediction Model

To predict the future motion, we use the state-of-the-art motion prediction model of [19]. Specifically, as mentioned above, we use a DCT-based representation to

Table 1. Short-term prediction of 3D joint positions on H3.6M. The error is measured in millimeter. The two numbers after the method name "LTD" indicate the number of observed frames and that of future frames to predict, respectively, during training. Our approach achieves state of the art performance across all 15 actions at almost all time horizons, especially for actions with a clear repeated history, such as "Walking".

milliseconds	Walking				Eating				Smoking				Discussion			
	80	160	320	400	80	160	320	400	80	160	320	400	80	160	320	400
Res. Sup. [21]	23.2	40.9	61.0	66.1	16.8	31.5	53.5	61.7	18.9	34.7	57.5	65.4	25.7	47.8	80.0	91.3
convSeq2Seq [16]	17.7	33.5	56.3	63.6	11.0	22.4	40.7	48.4	11.6	22.8	41.3	48.9	17.1	34.5	64.8	77.6
LTD-50-25[19]	12.3	23.2	39.4	44.4	7.8	16.3	31.3	38.6	8.2	16.8	32.8	39.5	11.9	25.9	55.1	68.1
LTD-10-25[19]	12.6	23.6	39.4	44.5	7.7	15.8	30.5	37.6	8.4	16.8	32.5	39.5	12.2	25.8	53.9	66.7
LTD-10-10[19]	11.1	21.4	37.3	42.9	7.0	14.8	29.8	37.3	7.5	15.5	30.7	37.5	10.8	24.0	52.7	65.8
Ours	10.0	19.5	34.2	39.8	6.4	14.0	28.7	36.2	7.0	14.9	29.9	36.4	10.2	23.4	52.1	65.4

milliseconds	Directions				Greeting				Phoning				Posing				Purchases				Sitting			
	80	160	320	400	80	160	320	400	80	160	320	400	80	160	320	400	80	160	320	400	80	160	320	400
Res. Sup. [21]	21.6	41.3	72.1	84.1	31.2	58.4	96.3	108.8	21.1	38.9	66.0	76.4	29.3	56.1	98.3	114.3	28.7	52.4	86.9	100.7	23.8	44.7	78.0	91.2
convSeq2Seq [16]	13.5	29.0	57.6	69.7	22.0	45.0	82.0	96.0	13.5	26.6	49.9	59.9	16.9	36.7	75.7	92.9	20.3	41.8	76.5	89.9	13.5	27.0	52.0	63.1
LTD-50-25[19]	8.8	20.3	46.5	58.0	16.2	34.2	68.7	82.6	9.8	19.9	40.8	50.8	12.2	27.5	63.1	79.9	15.2	32.9	64.9	78.1	10.4	21.9	46.6	58.3
LTD-10-25[19]	9.2	20.6	46.9	58.8	16.7	33.9	67.5	81.6	10.2	20.2	40.9	50.9	12.5	27.5	62.5	79.6	15.5	32.3	63.6	77.3	10.4	21.4	45.4	57.3
LTD-10-10[19]	8.0	18.8	**43.7**	**54.9**	14.8	31.4	65.3	79.7	9.3	19.1	39.8	49.7	10.9	25.1	59.1	75.9	13.9	30.3	62.2	75.9	9.8	20.5	**44.2**	**55.9**
Ours	**7.4**	**18.4**	44.5	56.5	**13.7**	**30.1**	**63.8**	**78.1**	**8.6**	**18.3**	**39.0**	**49.2**	**10.2**	**24.2**	**58.5**	**75.8**	**13.0**	**29.2**	**60.4**	**73.9**	**9.3**	**20.1**	44.3	56.0

milliseconds	Sitting Down				Taking Photo				Waiting				Walking Dog				Walking Together				Average			
	80	160	320	400	80	160	320	400	80	160	320	400	80	160	320	400	80	160	320	400	80	160	320	400
Res. Sup. [21]	31.7	58.3	96.7	112.0	21.9	41.4	74.0	87.6	23.8	44.2	75.8	87.7	36.4	64.8	99.1	110.6	20.4	37.1	59.4	67.3	25.0	46.2	77.0	88.3
convSeq2Seq [16]	20.7	40.6	70.4	82.7	12.7	26.0	52.1	63.6	14.6	29.7	58.1	69.7	27.7	53.6	90.7	103.3	15.3	30.4	53.1	61.2	16.6	33.3	61.4	72.7
LTD-50-25[19]	17.1	34.2	63.6	76.4	9.6	20.3	43.3	54.3	10.4	22.1	47.9	59.2	22.8	44.7	77.2	88.7	10.3	21.2	39.4	46.3	12.2	25.4	50.7	61.5
LTD-10-25[19]	17.0	33.4	61.6	74.4	9.9	20.5	43.8	55.2	10.5	21.6	45.9	57.1	22.9	43.5	74.5	86.4	10.8	21.7	39.6	47.0	12.4	25.2	49.9	60.9
LTD-10-10[19]	15.6	31.4	**59.1**	71.7	8.9	18.9	41.0	51.7	9.2	19.5	**43.3**	**54.4**	20.9	40.7	73.6	86.6	9.6	19.4	36.5	44.0	11.2	23.4	47.9	58.9
Ours	**14.9**	**30.7**	**59.1**	72.0	**8.3**	**18.4**	**40.7**	**51.5**	**8.7**	**19.2**	43.4	54.9	**20.1**	**40.3**	**73.3**	**86.3**	**8.9**	**18.4**	**35.1**	**41.9**	**10.4**	**22.6**	**47.1**	**58.3**

encode the temporal information for each joint coordinate or angle and GCNs with learnable adjacency matrices to capture the spatial dependencies among these coordinates or angles.

Temporal Encoding. Given a sequence of kth joint coordinates or angles $\{x_{k,l}\}_{l=1}^{L}$ or its DCT coefficients $\{C_{k,l}\}_{l=1}^{L}$, the DCT and Inverse-DCT (IDCT) are,

$$C_{k,l} = \sqrt{\tfrac{2}{L}} \sum_{n=1}^{L} x_{k,n} \tfrac{1}{\sqrt{1+\delta_{l1}}} \cos\left(\tfrac{\pi}{L}(n-\tfrac{1}{2})(l-1)\right), \; x_{k,n} = \sqrt{\tfrac{2}{L}} \sum_{l=1}^{L} C_{k,l} \tfrac{1}{\sqrt{1+\delta_{l1}}} \cos\left(\tfrac{\pi}{L}(n-\tfrac{1}{2})(l-1)\right) \quad (4)$$

where $l \in \{1, 2, \cdots, L\}, n \in \{1, 2, \cdots, L\}$ and $\delta_{ij} = \begin{cases} 1 & \text{if } i = j \\ 0 & \text{if } i \neq j \end{cases}$.

To predict future poses $\mathbf{X}_{N+1:N+T}$, we make use of the latest sub-sequence $\mathbf{X}_{N-M+1:N}$, which is also the query in the attention model. Adopting the same padding strategy as [19], we then replicate the last observed pose $\mathbf{X}_N$ T times to generate a sequence of length $M + T$ and the DCT coefficients of this sequence are denoted as $\mathbf{D} \in \mathbb{R}^{K \times (M+T)}$. We then aim to predict DCT coefficients of the future sequence $\mathbf{X}_{N-M+1:N+T}$ given $\mathbf{D}$ and the attention model's output $\mathbf{U}$.

Spatial Encoding. To capture spatial dependencies between different joint coordinates or angles, we regard the human body as a fully-connected graph with K nodes. The input to a graph convolutional layer p is a matrix $\mathbf{H}^{(p)} \in \mathbb{R}^{K \times F}$, where each row is the F dimensional feature vector of one node. For example, for the first layer, the network takes as input the $K \times 2(M + T)$ matrix that concatenates $\mathbf{D}$ and $\mathbf{U}$. A graph convolutional layer then outputs a $K \times \hat{F}$ matrix of the form

$$\mathbf{H}^{(p+1)} = \sigma(\mathbf{A}^{(p)} \mathbf{H}^{(p)} \mathbf{W}^{(p)}), \quad (5)$$

Table 2. Long-term prediction of 3D joint positions on H3.6M. On average, our approach performs the best.

milliseconds	Walking 560	720	880	1000	Eating 560	720	880	1000	Smoking 560	720	880	1000	Discussion 560	720	880	1000
Res. Sup. [21]	71.6	72.5	76.0	79.1	74.9	85.9	93.8	98.0	78.1	88.6	96.6	102.1	109.5	122.0	128.6	131.8
convSeq2Seq [16]	72.2	77.2	80.9	82.3	61.3	72.8	81.8	87.1	60.0	69.4	77.2	81.7	98.1	112.9	123.0	129.3
LTD-50-25[19]	50.7	54.4	57.4	60.3	51.5	62.6	71.3	75.8	50.5	59.3	67.1	72.1	88.9	103.9	113.6	118.5
LTD-10-25[19]	51.8	56.2	58.9	60.9	50.0	61.1	69.6	74.1	51.3	60.8	68.7	73.6	87.6	103.2	113.1	118.6
LTD-10-10[19]	53.1	59.9	66.2	70.7	51.1	62.5	72.9	78.6	49.4	59.2	66.9	71.8	88.1	104.5	115.5	121.6
Ours	47.4	52.1	55.5	58.1	50.0	61.4	70.6	75.7	47.6	56.6	64.4	69.5	86.6	102.2	113.2	119.8

milliseconds	Directions 560	720	880	1000	Greeting 560	720	880	1000	Phoning 560	720	880	1000	Posing 560	720	880	1000	Purchases 560	720	880	1000	Sitting 560	720	880	1000
Res. Sup. [21]	101.1	114.5	124.5	129.1	126.1	138.8	150.3	153.9	94.0	107.7	119.1	126.4	140.3	159.8	173.2	183.2	122.1	137.2	148.0	154.0	113.7	130.5	144.4	152.6
convSeq2Seq [16]	86.6	99.8	109.9	115.8	116.9	130.7	142.7	147.3	77.1	92.1	105.5	114.0	122.5	148.8	171.8	187.4	111.3	129.1	143.1	151.5	82.4	98.8	112.4	120.7
LTD-50-25[19]	74.2	88.1	99.4	105.5	104.8	119.7	132.1	136.8	68.8	83.6	96.8	105.1	110.2	137.8	160.8	174.8	99.2	114.9	127.1	134.9	79.2	96.2	110.3	118.7
LTD-10-25[19]	76.1	91.0	102.8	108.8	104.3	120.9	134.6	140.2	68.7	84.0	97.2	105.1	109.9	136.8	158.3	171.7	99.4	114.9	127.9	135.9	78.5	95.7	110.0	118.8
LTD-10-10[19]	72.2	86.7	98.5	105.8	103.7	120.6	134.7	140.9	67.8	83.0	96.4	105.1	107.6	136.1	159.5	175.0	98.3	115.1	130.1	139.3	76.4	93.1	106.9	115.7
Ours	73.9	88.2	100.1	106.5	101.9	118.4	132.7	138.8	67.4	82.9	96.5	105.0	107.6	136.8	161.4	178.2	95.6	110.9	125.0	134.2	76.4	93.1	107.0	115.9

milliseconds	Sitting Down 560	720	880	1000	Taking Photo 560	720	880	1000	Waiting 560	720	880	1000	Walking Dog 560	720	880	1000	Walking Together 560	720	880	1000	Average 560	720	880	1000
Res. Sup. [21]	138.8	159.0	176.1	187.4	110.6	128.9	143.7	153.9	105.4	117.3	128.1	135.4	128.7	141.1	155.3	164.5	80.2	87.3	92.8	98.2	106.3	119.4	130.0	136.6
convSeq2Seq [16]	106.5	125.1	139.8	150.3	84.4	102.4	117.7	128.1	87.3	100.3	110.7	117.7	122.4	133.8	151.1	162.4	72.0	77.7	82.9	87.4	90.7	104.7	116.7	124.2
LTD-50-25[19]	100.2	118.2	133.1	143.8	75.3	93.5	108.4	118.8	77.2	90.6	101.1	108.3	107.8	120.3	136.3	146.4	56.0	60.3	63.1	65.7	79.6	93.6	105.2	112.4
LTD-10-25[19]	99.5	118.5	133.6	144.1	76.8	95.3	110.3	120.2	75.1	88.7	99.5	106.9	105.8	118.7	132.8	142.2	58.0	63.6	67.0	69.6	79.5	94.0	105.6	112.7
LTD-10-10[19]	96.2	115.2	130.8	142.2	72.5	90.9	105.9	116.3	73.4	88.2	99.8	107.5	109.7	122.8	139.0	150.1	55.7	61.3	66.4	69.8	78.3	93.3	106.0	114.0
Ours	97.0	116.1	132.1	143.6	72.1	90.4	105.5	115.9	74.5	89.0	100.3	108.2	108.2	120.6	135.9	146.9	52.7	57.8	62.0	64.9	77.3	91.8	104.1	112.1

where $\mathbf{A}^{(p)} \in \mathbb{R}^{K \times K}$ is the trainable adjacency matrix of layer p, representing the strength of the connectivity between nodes, $\mathbf{W}^{(p)} \in \mathbb{R}^{F \times \hat{F}}$ also encodes trainable weights but used to extract features, and $\sigma(\cdot)$ is an activation function, such as $tanh(\cdot)$. We stack several such layers to form our GCN-based predictor.

Given $\mathbf{D}$ and $\mathbf{U}$, the predictor learns a residual between the DCT coefficients $\mathbf{D}$ of the padded sequence and those of the true sequence. By applying IDCT to the predicted DCT coefficients, we obtain the coordinates or angles $\hat{\mathbf{X}}_{N-M+1:N+T}$, whose last T poses $\hat{\mathbf{X}}_{N+1:N+T}$ are predictions in the future.

3.3 Training

Let us now introduce the loss functions we use to train our model on either 3D coordinates or joint angles. For 3D joint coordinates prediction, following [19], we make use of the Mean Per Joint Position Error (MPJPE) proposed in [10]. In particular, for one training sample, this yields the loss

$$\ell = \frac{1}{J(M+T)} \sum_{t=1}^{M+T} \sum_{j=1}^{J} \|\hat{\mathbf{p}}_{t,j} - \mathbf{p}_{t,j}\|^2, \tag{6}$$

where $\hat{\mathbf{p}}_{t,j} \in \mathbb{R}^3$ represents the 3D coordinates of the jth joint of the tth human pose in $\hat{\mathbf{X}}_{N-M+1:N+T}$, and $\mathbf{p}_{t,j} \in \mathbb{R}^3$ is the corresponding ground truth.

For the angle-based representation, we use the average ℓ_1 distance between the predicted joint angles and the ground truth as loss. For one sample, this can be expressed as

$$\ell = \frac{1}{K(M+T)} \sum_{t=1}^{M+T} \sum_{k=1}^{K} |\hat{x}_{t,k} - x_{t,k}|, \tag{7}$$

where $\hat{x}_{t,k}$ is the predicted k^{th} angle of the tth pose in $\hat{\mathbf{X}}_{N-M+1:N+T}$ and $x_{t,k}$ is the corresponding ground truth.

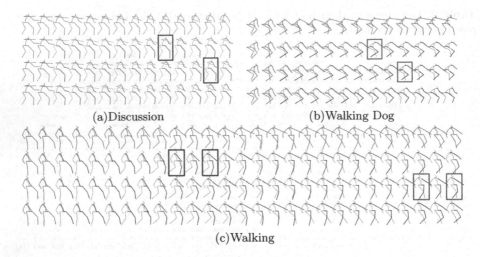

(a)Discussion   (b)Walking Dog

(c)Walking

Fig. 3. Qualitative comparison of short-term ("Discussion" and "Walking Dog") and long-term ("Walking") predictions on H3.6M. From top to bottom, we show the ground truth, and the results of LTD-10–25, LTD-10–10 and our approach on 3D positions. The ground truth is shown as blue-red skeletons, and the predictions as green-purple ones. (Color figure online)

3.4 Network Structure

As shown in Fig. 2, our complete framework consists of two modules: a motion attention model and a predictor. For the attention model, we use the same architecture for f_q and f_k. Specifically, we use a network consisting of two 1D convolutional layers, each of which is followed by a ReLU [22] activation function. In our experiments, the kernel size of these two layers is 6 and 5, respectively, to obtain a receptive field of 10 frames. The dimension of the hidden features, the query vector $\mathbf{q}$ and the key vectors $\{\mathbf{k}_i\}_{i=1}^{N-M-T+1}$ is set to 256.

For the predictor, we use the same GCN with residual structure as in [19]. It is made of 12 residual blocks, each of which contains two graph convolutional layers, with an additional initial layer to map the DCT coefficients to features and a final layer to decode the features to DCT residuals. The learnable weight matrix $\mathbf{W}$ of each layer is of size 256×256, and the size of the learnable adjacency matrix $\mathbf{A}$ depends on the dimension of one human pose. For example, for 3D coordinates, $\mathbf{A}$ is of size 66×66. Thanks to the simple structure of our attention model, the overall network remains still compact. Specifically, in our experiments, it has around 3.4 million parameters for both 3D coordinates and angles. The implementation details are included in supplementary material.

4   Experiments

Following previous works [7,16,19,21,23], we evaluate our method on Human3.6m (H3.6M) [10] and AMASS [18]. We further evaluate our method on 3DPW [20] using our model trained on AMASS to demonstrate the generalizability of our approach. Below, we discuss these datasets, the evaluation metric and the baseline methods, and present our results using joint angles and 3D coordinates.

Table 3. Short-term prediction of joint angles on H3.6M. Note that, we report results on 8 sub-sequences per action for fair comparison with MHU [28].

milliseconds	Walking 80	160	320	400	Eating 80	160	320	400	Smoking 80	160	320	400	Discussion 80	160	320	400
Res. sup. [21]	0.28	0.49	0.72	0.81	0.23	0.39	0.62	0.76	0.33	0.61	1.05	1.15	0.31	0.68	1.01	1.09
convSeq2Seq [16]	0.33	0.54	0.68	0.73	0.22	0.36	0.58	0.71	0.26	0.49	0.96	0.92	0.32	0.67	0.94	1.01
MHU [28]	0.32	0.53	0.69	0.77	-	-	-	-	-	-	-	-	0.31	0.66	0.93	1.00
LTD-10-25 [19]	0.20	0.34	0.52	0.59	0.17	0.31	0.52	0.64	0.23	0.42	0.85	0.80	0.22	0.58	0.87	0.96
LTD-10-10 [19]	**0.18**	0.31	0.49	0.56	**0.16 0.29 0.50**			0.62	**0.22 0.41 0.86 0.80**				**0.20 0.51 0.77 0.85**			
Ours	**0.18 0.30 0.46 0.51**				**0.16 0.29** 0.49 **0.60**				**0.22** 0.42 **0.86 0.80**				**0.20** 0.52 0.78 0.87			

milliseconds	Directions 80	160	320	400	Greeting 80	160	320	400	Phoning 80	160	320	400	Posing 80	160	320	400	Purchases 80	160	320	400	Sitting 80	160	320	400
Res. sup. [21]	0.26	0.47	0.72	0.84	0.75	1.17	1.74	1.83	0.23	0.43	0.69	0.82	0.36	0.71	1.22	1.48	0.51	0.97	1.07	1.16	0.41	1.05	1.49	1.63
convSeq2Seq [16]	0.39	0.60	0.80	0.91	0.51	0.82	1.21	1.38	0.59	1.13	1.51	1.65	0.29	0.60	1.12	1.37	0.63	0.91	1.19	1.29	0.39	0.61	1.02	1.18
MHU [28]	-	-	-	-	0.54	0.87	1.27	1.45	-	-	-	-	0.33	0.64	1.22	1.47	-	-	-	-	-	-	-	-
LTD-10-25 [19]	0.29	0.47	0.69	0.76	0.36	0.61	0.97	1.14	0.54	1.03	1.34	1.47	0.21	0.47	1.07	1.31	0.50	0.72	1.06	1.12	0.31	0.46	**0.79**	**0.95**
LTD-10-10 [19]	0.26	0.45	0.71	0.79	0.36	**0.60**	0.95	1.13	**0.53**	1.02	1.35	1.48	**0.19 0.44** 1.01 1.24				0.43	0.65	1.05	1.13	**0.29 0.45** 0.80 0.97			
Ours	**0.25 0.43 0.60 0.69**				**0.35 0.60 0.95** 1.14				**0.53 1.01 1.31 1.43**				**0.19** 0.46 1.09 1.35				**0.42 0.65 1.00 1.07**				**0.29** 0.47 0.83 1.01			

milliseconds	Sitting Down 80	160	320	400	Taking Photo 80	160	320	400	Waiting 80	160	320	400	Walking Dog 80	160	320	400	Walking Together 80	160	320	400	Average 80	160	320	400
Res. sup. [21]	0.39	0.81	1.40	1.62	0.24	0.51	0.90	1.05	0.28	0.53	1.02	1.14	0.56	0.91	1.26	1.40	0.31	0.58	0.87	0.91	0.36	0.67	1.02	1.15
convSeq2Seq [16]	0.41	0.78	1.16	1.31	0.23	0.49	0.88	1.06	0.30	0.62	1.09	1.30	0.59	1.00	1.32	1.44	0.27	0.52	0.71	0.74	0.38	0.68	1.01	1.13
MHU [28]	-	-	-	-	0.27	0.54	0.84	0.96	-	-	-	-	0.56	0.88	1.21	1.37	-	-	-	-	0.39	0.68	1.01	1.13
LTD-10-25 [19]	0.31	0.64	0.94	1.07	0.17	0.38	0.62	0.74	0.25	0.52	0.96	1.17	0.49	0.80	1.11	1.26	0.18	0.39	0.56	0.63	0.30	0.54	0.86	0.97
LTD-10-10 [19]	0.30	**0.61 0.90 1.00**			**0.14 0.34 0.58 0.70**				0.23	0.50	**0.91**	1.14	0.46	0.79	1.12	1.29	0.15	0.34	0.52	0.57	**0.27 0.52** 0.83 0.95			
Ours	**0.30** 0.63 0.92 1.04				0.16 0.36 **0.58 0.70**				**0.22** 0.49 0.92 1.14				**0.46 0.78 1.05 1.23**				**0.14 0.32 0.50 0.55**				**0.27 0.52 0.82 0.94**			

4.1 Datasets

Human3.6M [10] is the most widely used benchmark dataset for motion prediction. It depicts seven actors performing 15 actions. Each human pose is represented as a 32-joint skeleton. We compute the 3D coordinates of the joints by applying forward kinematics on a standard skeleton as in [19]. Following [16,19,21], we remove the global rotation, translation and constant angles or 3D coordinates of each human pose, and down-sample the motion sequences to 25 frames per second. As previous work [16,19,21], we test our method on subject 5 (S5). However, instead of testing on only 8 random sub-sequences per action, which was shown in [23] to lead to high variance, we report our results on 256 sub-sequences per action when using 3D coordinates. For fair comparison, we report our angular error on the same 8 sub-sequences used in [28]. Nonetheless, we provide the angle-based results on 256 sub-sequences per action in the supplementary material.

AMASS. The Archive of Motion Capture as Surface Shapes (AMASS) dataset [18] is a recently published human motion dataset, which unifies many mocap datasets, such as CMU, KIT and BMLrub, using a SMPL [17,24] parameterization to obtain a human mesh. SMPL represents a human by a shape vector and joint rotation angles. The shape vector, which encompasses coefficients of different human shape bases, defines the human skeleton. We obtain human poses in 3D by applying forward kinematics to one human skeleton. In AMASS,

Table 4. Long-term prediction of joint angles on H3.6M.

milliseconds	Walking				Eating				Smoking				Discussion			
	560	720	880	1000	560	720	880	1000	560	720	880	1000	560	720	880	1000
convSeq2Seq [16]	0.87	0.96	0.97	1.00	0.86	0.90	1.12	1.24	0.98	1.11	1.42	1.67	1.42	1.76	1.90	2.03
MHU [28]	1.44	1.46	-	1.44	-	-	-	-	-	-	-	-	1.37	1.66	-	1.88
LTD-10-25 [19]	0.65	0.69	0.69	0.67	0.76	0.82	1.00	1.12	0.87	0.99	1.33	1.57	1.33	1.53	1.62	1.70
LTD-10-10 [19]	0.69	0.77	0.76	0.77	0.76	0.81	1.00	1.10	0.88	1.01	1.36	1.58	1.27	1.51	1.66	1.75
Ours	0.59	0.62	0.61	0.64	0.74	0.81	1.01	1.10	0.86	1.00	1.35	1.58	1.29	1.51	1.61	1.63

milliseconds	Directions				Greeting				Phoning				Posing				Purchases				Sitting			
	560	720	880	1000	560	720	880	1000	560	720	880	1000	560	720	880	1000	560	720	880	1000	560	720	880	1000
convSeq2Seq [16]	1.00	1.18	1.41	1.44	1.73	1.75	1.92	1.90	1.66	1.81	1.93	2.05	1.95	2.26	2.49	2.63	1.68	1.65	2.13	2.50	1.31	1.43	1.66	1.72
MHU [28]	-	-	-	-	1.75	1.74	-	1.87	-	-	-	-	1.82	2.17	-	2.51	-	-	-	-	-	-	-	-
LTD-10-25 [19]	0.84	1.02	1.23	1.26	1.43	1.44	1.59	1.59	1.45	1.57	1.66	1.65	1.62	1.94	2.22	2.42	1.42	1.48	1.93	2.21	1.08	1.20	1.39	1.45
LTD-10-10 [19]	0.90	1.07	1.32	1.35	1.47	1.47	1.63	1.59	1.49	1.64	1.75	1.74	1.61	2.02	2.35	2.55	1.47	1.57	1.99	2.27	1.12	1.25	1.46	1.52
Ours	0.81	1.02	1.22	1.27	1.47	1.47	1.61	1.57	1.41	1.55	1.68	1.68	1.60	1.78	2.10	2.32	1.43	1.53	1.94	2.22	1.16	1.29	1.50	1.55

milliseconds	Sitting Down				Taking Photo				Waiting				Walking Dog				Walking Together				Average			
	560	720	880	1000	560	720	880	1000	560	720	880	1000	560	720	880	1000	560	720	880	1000	560	720	880	1000
convSeq2Seq [16]	1.45	1.70	1.85	1.98	1.09	1.18	1.27	1.32	1.68	2.02	2.33	2.45	1.73	1.85	1.99	2.04	0.82	0.89	0.95	1.29	1.35	1.50	1.69	1.82
MHU [28]	-	-	-	-	1.04	1.14	-	1.35	-	-	-	-	1.67	1.81	-	1.90	-	-	-	-	1.34	1.49	1.69	1.80
LTD-10-25 [19]	1.26	1.54	1.70	1.87	0.85	0.92	0.99	1.06	1.55	1.89	2.20	2.29	1.52	1.63	1.78	1.84	0.70	0.75	0.82	1.16	1.15	1.29	1.48	1.59
LTD-10-10 [19]	1.17	1.40	1.54	1.67	0.81	0.89	0.97	1.05	1.57	1.94	2.29	2.37	1.58	1.66	1.80	1.86	0.65	0.73	0.81	1.16	1.16	1.32	1.51	1.62
Ours	1.18	1.42	1.55	1.70	0.82	0.91	1.00	1.08	1.54	1.90	2.22	2.30	1.57	1.63	1.76	1.82	0.63	0.68	0.79	1.16	1.14	1.28	1.46	1.57

a human pose is represented by 52 joints, including 22 body joints and 30 hand joints. Since we focus on predicting human body motion, we discard the hand joints and the 4 static joints, leading to an 18-joint human pose. As for H3.6M, we down-sample the frame-rate 25 Hz. Since most sequences of the official testing split[1] of AMASS consist of transition between two irrelevant actions, such as dancing to kicking, kicking to pushing, they are not suitable to evaluate our prediction algorithms, which assume that the history is relevant to forecast the future. Therefore, instead of using this official split, we treat BMLrub[2] (522 min video sequence), as our test set as each sequence consists of one actor performing one type of action. We then split the remaining parts of AMASS into training and validation data.

3DPW. The 3D Pose in the Wild dataset (3DPW) [20] consists of challenging indoor and outdoor actions. We only evaluate our model trained on AMASS on the test set of 3DPW to show the generalization of our approach.

4.2   Evaluation Metrics and Baselines

Metrics. For the models that output 3D positions, we report the Mean Per Joint Position Error (MPJPE) [10] in millimeter, which is commonly used in human pose estimation. For those that predict angles, we follow the standard evaluation protocol [16,19,21] and report the Euclidean distance in Euler angle representation.

Baselines. We compare our approach with two RNN-based methods, Res. sup. [21] and MHU [28], and two feed-forward models, convSeq2Seq [16] and LTD [19], which constitutes the state of the art. The angular results of Res. sup. [21], convSeq2Seq [16] and MHU on H3.6M are directly taken from the

---

[1] Described at https://github.com/nghorbani/amass.
[2] Available at https://amass.is.tue.mpg.de/dataset.

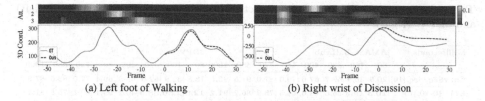

(a) Left foot of Walking   (b) Right wrist of Discussion

Fig. 4. Visualization of attention maps and joint trajectories. The x-axis denotes the frame index, with prediction starting at frame 0. The y-axis of the attention map (top) is the prediction step. Specifically, since the model is trained to predict 10 future frames, we recursively perform prediction for 3 steps to generate 30 frames. Each row of an attention map is then the attention vector when predicting the corresponding 10 future frames. For illustration purpose, we show *per-frame* attention, which represents the attention for *its motion subsequence* consisting of M-1 frames forward and T frames afterwards. (a) Predicted attention map and trajectory of the left foot's x coordinate for 'Walking', where the future motion closely resembles that between frames -45 and -10. Our model correctly attends to that very similar motion in the history. (b) Predicted attention map and trajectory of the right wrist's x coordinate for 'Discussion'. In this case, the attention model searches for the most similar motion in the history. For example, in the 1st prediction step, to predict frames 0 to 10 where a peak occurs, the model focuses on frames -30 to -20, where a similar peak pattern occurs.

respective paper. For the other results of Res. sup. [21] and convSeq2Seq [16], we adapt the code provided by the authors for H3.6M to 3D and AMASS. For LTD [19], we rely on the pre-trained models released by the authors for H3.6M, and train their model on AMASS using their official code. While Res. sup. [21], convSeq2Seq [16] and MHU [28] are all trained to generate 25 future frames, LTD [19] has 3 different models, which we refer to as LTD-50-25 [19], LTD-10–25 [19], and LTD-10-10 [19]. The two numbers after the method name indicate the number of observed past frames and that of future frames to predict, respectively, during training. For example, LTD-10–25 [19] means that the model is trained to take the past 10 frames as input to predict the future 25 frames.

4.3   Results

Following the setting of our baselines [16,19,21,28], we report results for short-term (<500 ms) and long-term (>500 ms) prediction. On H3.6M, our model is trained using the past 50 frames to predict the future 10 frames, and we produce poses further in the future by recursively applying the predictions as input to the model. On AMASS, our model is trained using the past 50 frames to predict the future 25 frames.

Human3.6M. In Tables 1 and 2, we provide the H3.6M results for short-term and long-term prediction in 3D space, respectively. Note that we outperform all the baselines on average for both short-term and long-term prediction. In particular, our method yields larger improvements on activities with a clear repeated history, such as "Walking" and "Walking Together". Nevertheless, our

Table 5. Short-term and long-term prediction of 3D joint positions on BMLrub (left) and 3DPW (right).

Milliseconds	AMASS-BMLrub								3DPW							
	80	160	320	400	560	720	880	1000	80	160	320	400	560	720	880	1000
ConvSeq2Seq [16]	20.6	36.9	59.7	67.6	79.0	87.0	91.5	93.5	18.8	32.9	52.0	58.8	69.4	77.0	83.6	87.8
LTD-10-10 [19]	**10.3**	**19.3**	36.6	44.6	61.5	75.9	86.2	91.2	**12.0**	**22.0**	**38.9**	46.2	59.1	69.1	76.5	81.1
LTD-10-25 [19]	11.0	20.7	37.8	45.3	57.2	65.7	71.3	75.2	12.6	23.2	39.7	46.6	57.9	65.8	71.5	75.5
Ours	11.3	20.7	**35.7**	**42.0**	**51.7**	**58.6**	**63.4**	**67.2**	12.6	23.1	39.0	**45.4**	**56.0**	**63.6**	**69.7**	**73.7**

approach remains competitive on the other actions. Note that we consistently outperform LTD-50-25, which is trained on the same number of past frames as our approach. This, we believe, evidences the benefits of exploiting attention on the motion history.

Let us now focus on the LTD [19] baseline, which constitutes the state of the art. Although LTD-10-10 is very competitive for short-term prediction, when it comes to generate poses in the further future, it yields higher average error, i.e., 114.0 mm at 1000 ms. By contrast, LTD-10–25 and LTD-50-25 achieve good performance at 880 ms and above, but perform worse than LTD-10-10 at other time horizons. Our approach, however, yields state-of-the-art performance for both short-term and long-term predictions. To summarize, our motion attention model improves the performance of the predictor for short-term prediction and further enables it to generate better long-term predictions. This is further evidenced by Tables 3 and 4, where we report the short-term and long-term prediction results in angle space on H3.6M, and by the qualitative comparison in Fig. 3. More qualitative results are provided in the supplementary material.

AMASS and 3DPW. The results of short-term and long-term prediction in 3D on AMASS and 3DPW are shown in Table 5. Our method consistently outperforms baseline approaches, which further evidences the benefits of our motion attention model. Since none of the methods were trained on 3DPW, these results further demonstrate that our approach generalizes better to new datasets than the baselines.

Visualisation of Attention. In Fig. 4, we visualize the attention maps computed by our motion attention model on a few sampled joints for their corresponding coordinate trajectories. In particular, we show attention maps for joints in a periodical motion ("Walking") and a non-periodical one ("Discussion"). In both cases, the attention model can find the most relevant sub-sequences in the history, which encode either a nearly identical motion (periodical action), or a similar pattern (non-periodical action).

Motion Repeats Itself in Longer-Term History. Our model, which is trained with fixed-length observations, can nonetheless exploit longer history at test time if it is available. To evaluate this and our model's ability to capture long-range motion dependencies, we manually sampled 100 sequences from the test set of H3.6M, in which similar motion occurs in the further past than that used to train our model.

Table 6. Short-term and long-term prediction of 3D positions on selected sequences where similar patterns occur in the longer history. The number after "Ours" indicates the observed frames during testing. Both methods observed 50 frames during training.

Milliseconds	80	160	320	400	560	720	880	1000
Ours-50	**10.7**	**22.4**	46.9	58.3	79.0	97.1	111.0	121.1
Ours-100	**10.7**	22.5	**46.4**	**57.5**	**77.8**	**95.1**	**107.6**	**116.9**

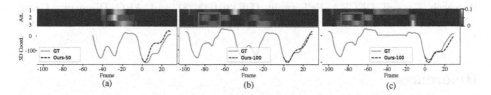

Fig. 5. Visualization of attention maps and joint coordinate trajectories for "Smoking" on H3.6M. (a) Results of our model observing 50 past frames. (b) Results of our model observing 100 frames. (c) Results obtained when replacing the motion of the past 40 frames with a constant pose. (Color figure online)

In Table 6, we compare the results of a model trained with 50 past frames and using either 50 frames (Ours-50) or 100 frames (Ours-100) at test time. Although the performance is close in the very short term (< 160 ms), the benefits of our model using longer history become obvious when it comes to further future, leading to a performance boost of 4.2 mm at 1s. In Fig. 5, we compare the attention maps and predicted joint trajectories of Ours-50 (a) and Ours-100 (b). The highlighted regions (in red box) in the attention map demonstrate that our model can capture the repeated motions in the further history if it is available during test and improve the motion prediction results.

To show the influence of further historical frames, we replace the past 40 frames with a static pose, thus removing the motion in that period, and then perform prediction with this sequence. As shown in Fig. 5 (c), attending to the similar motion between frames −80 and −60, yields a trajectory much closer to the ground truth than only attending to the past 50 frames.

5 Conclusion

In this paper, we have introduced an attention-based motion prediction approach that selectively exploits historical information according to the similarity between the current motion context and the sub-sequences in the past. This has led to a predictor equipped with a *motion attention* model that can effectively make use of historical motions, even when they are far in the past. Our approach achieves state-of-the-art performance on the commonly-used motion prediction benchmarks and on recently-published datasets. Furthermore, our experiments have demonstrated that our network generalizes to previously-unseen datasets without re-training or fine-tuning, and can handle longer history than that it was trained with to further boost performance on non-periodical motions with repeated history. In the future, we will further investigate the use of our motion attention mechanisms to discover human motion patterns in body parts level such as legs and arms to get more flexible attentions and explore new prediction frame works.

Acknowledgements. This research was supported in part by the Australia Research Council DECRA Fellowship (DE180100628) and ARC Discovery Grant (DP200102274). The authors would like to thank NVIDIA for the donated GPU (Titan V).

References

1. Arjovsky, M., Bottou, L.: Towards principled methods for training generative adversarial networks. In: ICLR (2017)
2. Bahdanau, D., Cho, K., Bengio, Y.: Neural machine translation by jointly learning to align and translate (2015)
3. Brand, M., Hertzmann, A.: Style machines. In: Proceedings of the 27th Annual Conference on Computer Graphics and Interactive Techniques, pp. 183–192. ACM Press/Addison-Wesley Publishing Company (2000)
4. Butepage, J., Black, M.J., Kragic, D., Kjellstrom, H.: Deep representation learning for human motion prediction and classification. In: CVPR (July 2017)
5. Fragkiadaki, K., Levine, S., Felsen, P., Malik, J.: Recurrent network models for human dynamics. In: ICCV. pp. 4346–4354 (2015)
6. Gong, H., Sim, J., Likhachev, M., Shi, J.: Multi-hypothesis motion planning for visual object tracking. In: ICCV, pp. 619–626. IEEE (2011)
7. Gopalakrishnan, A., Mali, A., Kifer, D., Giles, L., Ororbia, A.G.: A neural temporal model for human motion prediction. In: CVPR, pp. 12116–12125 (2019)
8. Gui, L.Y., Wang, Y.X., Liang, X., Moura, J.M.: Adversarial geometry-aware human motion prediction. In: ECCV, pp. 786–803 (2018)
9. Hernandez, A., Gall, J., Moreno-Noguer, F.: Human motion prediction via spatio-temporal inpainting. In: ICCV, pp. 7134–7143 (2019)
10. Ionescu, C., Papava, D., Olaru, V., Sminchisescu, C.: Human3.6m: large scale datasets and predictive methods for 3d human sensing in natural environments. TPAMI **36**(7), 1325–1339 (2014)
11. Jain, A., Zamir, A.R., Savarese, S., Saxena, A.: Structural-RNN: Deep learning on spatio-temporal graphs. In: CVPR, pp. 5308–5317 (2016)
12. Kiros, R., et al.: Skip-thought vectors. In: NIPS, pp. 3294–3302 (2015)
13. Koppula, H.S., Saxena, A.: Anticipating human activities for reactive robotic response. In: IROS, p. 2071. Tokyo (2013)
14. Kovar, L., Gleicher, M., Pighin, F.: Motion graphs. In: ACM SIGGRAPH 2008 classes, pp. 1–10 (2008)
15. Koppula, H.S., Saxena, A.: Anticipating human activities for reactive robotic response. In: IROS, p. 2071. Tokyo (2013)
16. Li, C., Zhang, Z., Lee, W.S., Lee, G.H.: Convolutional sequence to sequence model for human dynamics. In: CVPR, pp. 5226–5234 (2018)

17. Loper, M., Mahmood, N., Romero, J., Pons-Moll, G., Black, M.J.: SMPL: a skinned multi-person linear model. ACM Trans. Graph. (Proc. SIGGRAPH Asia) **34**(6), 248:1–248:16 (2015)
18. Mahmood, N., Ghorbani, N., Troje, N.F., Pons-Moll, G., Black, M.J.: Amass: archive of motion capture as surface shapes. In: ICCV (October 2019). https://amass.is.tue.mpg.de
19. Mao, W., Liu, M., Salzmann, M., Li, H.: Learning trajectory dependencies for human motion prediction. In: ICCV, pp. 9489–9497 (2019)
20. von Marcard, T., Henschel, R., Black, M., Rosenhahn, B., Pons-Moll, G.: Recovering accurate 3d human pose in the wild using IMUs and a moving camera. In: ECCV (September 2018)
21. Martinez, J., Black, M.J., Romero, J.: On human motion prediction using recurrent neural networks. In: CVPR (July 2017)
22. Nair, V., Hinton, G.E.: Rectified linear units improve restricted boltzmann machines. In: ICML, pp. 807–814 (2010)
23. Pavllo, D., Feichtenhofer, C., Auli, M., Grangier, D.: Modeling human motion with quaternion-based neural networks. In: IJCV, pp. 1–18 (2019)
24. Romero, J., Tzionas, D., Black, M.J.: Embodied hands: modeling and capturing hands and bodies together. ACM Trans. Graph. (Proc. SIGGRAPH Asia) **36**(6), 245 (2017)
25. Runia, T.F., Snoek, C.G., Smeulders, A.W.: Real-world repetition estimation by div, grad and curl. In: CVPR, pp. 9009–9017 (2018)
26. Sidenbladh, Hedvig., Black, Michael J., Sigal, Leonid: Implicit probabilistic models of human motion for synthesis and tracking. In: Heyden, Anders, Sparr, Gunnar, Nielsen, Mads, Johansen, Peter (eds.) ECCV 2002. LNCS, vol. 2350, pp. 784–800. Springer, Heidelberg (2002). https://doi.org/10.1007/3-540-47969-4_52
27. Sutskever, I., Martens, J., Hinton, G.E.: Generating text with recurrent neural networks. In: ICML, pp. 1017–1024 (2011)
28. Tang, Y., Ma, L., Liu, W., Zheng, W.S.: Long-term human motion prediction by modeling motion context and enhancing motion dynamics. IJCAI (July 2018). https://doi.org/10.24963/ijcai.2018/130, http://dx.doi.org/10.24963/ijcai.2018/130
29. Vaswani, A., et al.: Attention is all you need. In: NIPS, pp. 5998–6008 (2017)
30. Wang, J.M., Fleet, D.J., Hertzmann, A.: Gaussian process dynamical models for human motion. TPAMI **30**(2), 283–298 (2008)

Unsupervised Video Object Segmentation with Joint Hotspot Tracking

Lu Zhang[1], Jianming Zhang[2], Zhe Lin[2], Radomír Měch[2], Huchuan Lu[1(✉)], and You He[3]

[1] Dalian University of Technology, Dalian, China
lhchuan@dlut.edu.cn
[2] Adobe Research, Beijing, China
[3] Naval Aviation University, Yantai, China

Abstract. Object tracking is a well-studied problem in computer vision while identifying salient spots of objects in a video is a less explored direction in the literature. Video eye gaze estimation methods aim to tackle a related task but salient spots in those methods are not bounded by objects and tend to produce very scattered, unstable predictions due to the noisy ground truth data. We reformulate the problem of detecting and tracking of salient object spots as a new task called *object hotspot tracking*. In this paper, we propose to tackle this task jointly with unsupervised video object segmentation, in real-time, with a unified framework to exploit the synergy between the two. Specifically, we propose a Weighted Correlation Siamese Network (WCS-Net) which employs a Weighted Correlation Block (WCB) for encoding the pixel-wise correspondence between a template frame and the search frame. In addition, WCB takes the initial mask/hotspot as guidance to enhance the influence of salient regions for robust tracking. Our system can operate online during inference and jointly produce the object mask and hotspot tracklets at 33 FPS. Experimental results validate the effectiveness of our network design, and show the benefits of jointly solving the hotspot tracking and object segmentation problems. In particular, our method performs favorably against state-of-the-art video eye gaze models in object hotspot tracking, and outperforms existing methods on three benchmark datasets for unsupervised video object segmentation.

Keywords: Unsupervised video object segmentation · Hotspot tracking · Weighted correlation siamese network

1 Introduction

Unsupervised video object segmentation (UVOS) aims to generate the masks of the primary objects in the video sequence without any human annotation.

Electronic supplementary material The online version of this chapter (https://doi.org/10.1007/978-3-030-58568-6_29) contains supplementary material, which is available to authorized users.

© Springer Nature Switzerland AG 2020
A. Vedaldi et al. (Eds.): ECCV 2020, LNCS 12359, pp. 490–506, 2020.
https://doi.org/10.1007/978-3-030-58568-6_29

It has attracted a lot of interests due to the wide application scenarios such as autonomous driving, video surveillance and video editing.

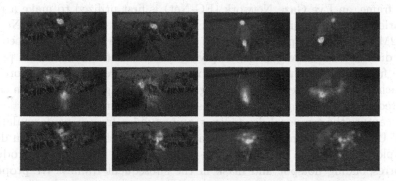

Fig. 1. The visual comparison of our object hotspot tracking results (top row) with the video eye gaze estimation results in [22] (middle row) and the eye gaze tracking ground truth in [48]. The peaks of hotspot map indicate the saliency distribution inside the object. Please zoom in for details.

However, in some applications, such as video cropping, zooming and auto-focus, tracking the full mask of a dominant object may not be sufficient. In video zooming, for example, the deformable shape of object mask may not provide a stable focal point for the zooming window. Furthermore, the mask only can't support the users to zoom in to the attractive part of the object (*e.g.,* the head of a person or the hood of a car). In such scenario, the attention distribution inside the object region can be very useful. The peak of the attention distribution can be thought of as a virtual focal region of the object, which provides a clean saliency indication on the object part (see the highlighted regions in the first row of Fig. 1). We refer to this virtual focal region of an object as an Object Hotspot. In this work, we propose a new task of Hotspot Tracking, which aims to produce clean, stable and temporally consistent object hotspot distribution along video sequence (see the first row in Fig. 1). This task is different from the video eye gaze prediction task [22,48] in which the eye gaze tracking ground truth is intrinsically unstable and noisy due to the saccadic eye movement [13,36]. However, our hotspot tracking is a clean and stable intra-object focal saliency estimation and thus a more useful evidence to facilitate the video editing applications.

The UVOS and hotspot tracking are highly related tasks. The UVOS provides a sequence of object masks, which can provide a strong spatial guidance for the hotspot detection and tracking. Meanwhile, the hotspot of an object can also help the object mask tracking, as it tends to be more robust to object deformation, occlusion and appearance change. Jointly solving both tasks can leverage the synergy between them and provides a richer supervision for the model training.

In this paper, we show how to tackle the unsupervised video object segmentation and hotspot tracking in a simple unified framework. Our model consists of two modules, Target Object Initialization (TOI) and Object Segmentation and Hotspot Tracking (OSHT). The TOI module exploits the idea of the correlation between human eye gaze and UVOS [48] and builds an efficient method for automatically determining the target object in the video. Specifically, for an initial frame, an Eye Gaze Network (EG-Net) is first utilized to make a rough estimation on the location of the target object. Then a Gaze2Mask Network (Gaze2Mask-Net) is proposed to predict the mask of the target object according to the distribution of eye gaze. By applying the automatic initialization process on the first frame, our model can perform online inference. Furthermore, this approach also allows interactive tracking, where the eye gaze distribution can be indicated by a real human via eye gaze sensors or gestures.

To capture the temporal dependency among video frames, we formulate the task of UVOS and hotspot tracking as a template matching problem. In detail, we employ the eye gaze map and mask from TOI as a template for producing the corresponding hotspot and mask in the subsequent frames. We propose a novel Weighted Correlation Siamese Network (WCS-Net) for the joint tracking of object mask and hotspot. In the WCS-Net, the Weighted Correlation Block (WCB) is exploited to calculate the cross-pixel correspondence between a template frame and the current search frame. In WCB, a weighted pooling operation is conducted between the mask/hotspot and feature of template frame to emphasize the contribution of foreground pixels. The WCB is built on multi-level side-outputs to encode the correlation features, which will be used in the parallel decoder branches to produce the track-lets of mask and hotspot across time.

Considering the difficulty in annotating hotspot using eye tracker, we design a training strategy on multi-source data. The network learns to predict the intra-object hotspots from large-scale image-level eye gaze data [19] and the temporal correlation feature representation via limited video object segmentation data [34]. Once trained, our model can operate online and produce the object mask and hotspot at a real-time speed of 33 fps. To investigate the efficacy of our proposed model, we conduct thorough experiments including overall comparison and ablation study on three benchmark datasets [12,29,34]. The results show that our proposed method performs favorable against state-of-the-arts on UVOS. Meanwhile, our model could achieve promising results on hotspot tracking with only training on image-level data.

Our contributions can be summarized as follows:

- We introduce the problem of video object hotspot tracking, which tries to consistently track the focal saliency distribution of target object. This task is beneficial for applications like video editing and camera auto-focus.
- We present a multi-task systematic design for unsupervised video object segmentation and hotspot tracking, which supports both online real-time inference and user interaction.
- We conduct experiments on three benchmark datasets (DAVIS-2016 [34], SegTrackv2 [29] and Youtube-Objects [12]) to demonstrate that our model[1] performs favorable on both tasks while being able to run at 33 FPS.

---

[1] Project Page: https://github.com/luzhangada/code-for-WCS-Net.

2   Related Work

Unsupervised VOS. Different from SVOS, unsupervised VOS (UVOS) methods aim to automatically detect the target object without any human definition. Early approaches use the low-level cues like visual saliency [18,47] or optical flow [21,23,30,31] to segment the objects. With the recent success of CNNs on segmentation tasks [7–10,16,49,50,54–57], many methods [5,31] develop various network architectures and have achieved promising performance. For example, some models [20,31,42] propose to build a bilateral network to combine the appearance feature and motion cues. In [20], two parallel streams are built to extract features from raw image and optical flow, which are further fused in the decoder for predicting the segmentation results. Except for the bilateral network, recurrent neural network (RNN) is also exploited in UVOS to capture temporal information [39,48]. However, the RNN-based methods are not robust enough to handle the long-term videos. Inspired by the success of siamese network [27,28] in video object tracking, several methods [3,30–32,53] propose to build siamese architecture to extract features for two frames to calculate their pixel-wise correspondence. [30,31] construct a seed-propagation method to obtain segmentation mask, in which the foreground/background seeds are obtained through temporal link. Yang *et al.* [53] propose an anchor diffusion module which computes the similarity of each frame with an anchor frame. The anchor-diffusion is then combined with the intra-diffusion to predict the segmentation results. In this paper, we put forward a weighted correlation siamese network for UVOS. Our model is different from the aforementioned methods on three folds. First, our WCS-Net is designed as a multi-task architecture for jointly addressing UVOS and hotspot tracking. Second, a weighted correlation block is proposed, which exploits the mask and hotspot from the template frame to highlight the influence of foreground pixel in the correlation feature calculation. Third, our correlation block is performed on multiple side-outputs to capture the rich CNN feature.

Video Eye Gaze Tracking. Video eye gaze tracking aims to record human attention in the dynamic scenario. The CNN+RNN framework is widely used in this task to encode spatio-temporal information. Wang *et al.* [46] propose to learn spatial attention from static data and utilize LSTM to predict temporal eye tracking. In [22], the spatial and temporal cues are respectively produced from objectness sub-net and motion sub-net, which are further integrated to produce eye tracking by LSTM. The current eye tracking training data are annotated with eye tracker in a free-view manner, which would incur eye gaze shifting among several salient objects in the video. This limits the existing models to apply on the video editing tasks. To address this issue, Wang *et al.* [48] construct eye gaze data on existing video object segmentation datasets, where the eye fixation is consistent with the moving object. However, due to the eye tracker annotation, the proposed eye gaze data often flickers inside the object. In this paper, we propose a problem of object hotspot tracking, which aims to produce a clean and temporally consistent track-let for the salient region inside the target object.

3   Methodology

We present a neural network model for the task of unsupervised video object segmentation and hotspot tracking. Given a video sequence $I_i \in \{I_1, ..., I_N\}$, our model aims to produce the corresponding object masks $\{M_i\}_{i=1}^N$ and hotspots $\{H_i\}_{i=1}^N$. Our model contains two sub-modules, Target Object Initialization (TOI) and Object Segmentation and Hotspot Tracking (OSHT). We formulate our task as a template matching problem, where the TOI module determines the target object of the video sequence and the OSHT module aims to predict the target object mask and its hotspot across time. In the subsequent sections, we will introduce the details of TOI, OSHT, and the training strategy, respectively.

3.1   Target Object Initialization

In the UVOS, it is very important to determine what the target object is to be segmented in a video sequence. Some recent CNN-based models [32,48] propose to incorporate visual attention to localize the target object in a video, where the attention is temporally propagated with the RNN architecture. These works have demonstrated the correlation between human attention and video object determination. Inspired by this, we build a target object initialization module to identify the target object from the guidance of human eye gaze. Different from the previous methods [32,48] where the human attention is encoded implicitly in the LSTM, we propose an Eye Gaze Network (EG-Net) to make an explicit eye gaze estimation on the initial frame and a Gaze to Mask Network (Gaze2Mask-Net) to generate the corresponding object mask. The advantages of our TOI module are two folds. First, instead of scanning multiple frames during object determination [3,30,31], using single frame could largely reduce the computational cost and meet the applications in real-time scenarios. Second, the initialization order from eye gaze to object mask makes it possible for our model to extend in the interactive applications where the eye gaze data can be easily acquired from user click or eye tracker.

Eye Gaze Network. Given a template frame I^t, the EG-Net aims to produce an eye gaze estimation E^t to indicate the location of the target object in the video. Considering the network efficiency, we build our EG-Net on the EfficientNet [40], which is a very small-size network with impressive performance. Specifically, we first exploit the encoder to extract features for the template frame I^t. Here, we use the last three level features, which are represented as $F^E = \{f_j^E\}_{j=3}^5$ (the feature size can be referred in Fig. 2). In the decoder, we stack three residual refinement blocks [6] to produce the eye gaze map in a coarse-to-fine manner, which can be represented as follows:

$$O_j^E = Conv^2(Cat(f_j^E, Up(O_{j+1}^E))) + Up(O_{j+1}^E) \tag{1}$$

where $Up()$ is the upsampling operation with stride 2. $Cat()$ is the channel-wise concatenation operation. $Conv^2()$ indicates the operation with two convolutional layers. O_j^E is the output of current residual block. Note that the term $Up(O_{j+1}^E)$

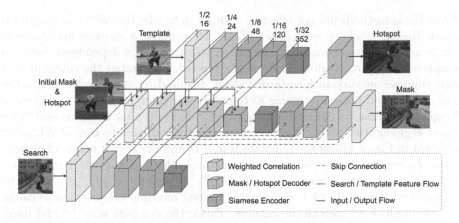

Fig. 2. The framework of Weighted Correlation Siamese Network (WCS-Net). The WCS-Net is a siamese architecture for jointly video object segmentation and hotspot tracking. Given the template frame and search frame, we first use siamese encoder to extract their multi-level features. The weighted correlation block, which takes the initial mask and eye gaze map (see Sect. 3.2) as guidance, is exploited to calculate the correspondence between template and search features. Then we build two parallel decoder branches to generate the hotspot and mask for the search frame.

is ignored when $j = 5$. We take the final output from 3rd decoder block as the eye gaze prediction E^t.

Gaze to Mask Network. With the EG-Net, we can obtain an eye gaze estimation on the template frame. The highlighted region is capable of indicating the location of the target region. In order to generate the mask of the target object, we propose a Gaze to Mask Network (Gaze2Mask-Net). Given the template frame I^t and its eye gaze map E^t, the Gaze2Mask-Net aims to segment out the object mask M^t according to the highlight region in eye gaze map. In the Gaze2Mask-Net, we also exploit EfficientNet [40] as encoder. To emphasize the guidance of eye gaze, the encoder takes the concatenation of both template frame and eye gaze map as the input. We extract five level features, which are represented as $F^{G2M} = \{f_j^{G2M}\}_{j=1}^5$. In Gaze2Mask-Net, we utilize a revised residual block, which adds the eye gaze map E^t into architecture.

$$O_j^{G2M} = Conv^2(Cat(f_j^{G2M}, Up(O_{j+1}^{G2M}), E^t)) + Up(O_{j+1}^{G2M}) \qquad (2)$$

O_j^{G2M} is the output of the revised residual block. Note that the eye gaze map E^t should be resized according to the corresponding feature resolution. We stack five residual blocks as decoder and exploit the output from the first block O_1^{G2M} as object mask M^t.

3.2 Object Segmentation and Hotspot Tracking

The previous UVOS methods [17,32,48] usually utilize LSTM or mask propagation method to capture the temporal consistency between adjacent frames.

While these methods are not effective enough to handle the object in long-term video. Recent tracking approaches [27,28,44] propose a siamese architecture, which uses the cross correlation operation to capture the dependence between template and search region. Their works have demonstrated the robustness of such siamese network in long-term videos. Inspired by this, we formulate the task of VOS and hotspot tracking as template matching problem. We propose a Weighted Correlation Siamese Network (WCS-Net) for joint unsupervised video object segmentation and hotspot tracking. The overall architecture of WCS-Net is shown in Fig. 2. Given the template frame I^t and the search region of current frame I^i, we first build a siamese encoder network to extract their multi-level features, which are represented as $F^t = \{f_j^t\}_{j=1}^5$ and $F^i = \{f_j^i\}_{j=1}^5$. Then the weighted correlation block is constructed among multiple side-outputs to calculate the multi-level correlation features. Taken the eye gaze map E^t and mask M^t from EG-Net and Gaze2Mask-Net as template guidance, we implement two parallel decoder branches to generate the corresponding hotspot H_i and mask M_i for each frame in the video sequence.

Weighted Correlation Block. We propose a weighted correlation block to calculate the correspondence between template and search features. Taken the template feature f_j^t, search feature f_j^i and the template guidance G as input, the weighted correlation block produces the corresponding correlation feature by:

$$C_j^i = W(f_j^t, R_j(G)) \star f_j^i \tag{3}$$

where $\star$ denotes the depth-wise cross correlation layer [27,44]. C_j^i is the correlation feature between template and search frames. $R_j()$ is the resize operation. $W()$ indicates the weighted pooling operation, which transfers the feature map $(h \times w \times c)$ into feature vector $(1 \times 1 \times c)$ by weighted summation with $R_j(G)$. $G = M^t$ when constructing the correlation block for the video object segmentation, or $G = H^t$ for the hotspot tracking (see Fig. 3 for more details). Compared with the original cross correlation [27,44] (formulated as $f_j^t \star f_j^i$), our weighted correlation block is more effective at our pixel-wise prediction problem. On one hand, the weighted pooling with template guidance is able to highlight the contribution of foreground and reduce the noise from background. On the other hand, the correlation between template vector and search feature would not decrease the resolution of search feature and thus helps to remain the details of target object. We conduct the comparison experiment in Sect. 4.4 to demonstrate the effectiveness of our weighted correlation on UVOS.

Mask Decoder Branch. Different from previous methods [32,48] in which only the single level feature is used to generate object mask, we exploit multi-level feature representations. Specifically, we build the weighted correlation blocks among five side-outputs to produce the multi-level correlation feature between template and search frames. Similarly with the decoder in EG-Net, we stack five residual refinement blocks to produce the object mask in a coarse to fine manner.

$$O_j^{i,M} = Conv^2(Cat(C_j^{i,M}, Up(O_{j+1}^{i,M}))) + Up(O_{j+1}^{i,M}) \tag{4}$$

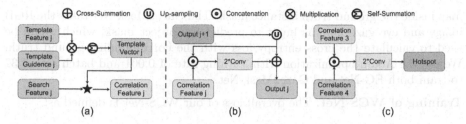

Fig. 3. The details of components in weighted correlation siamese network. (a) The details of weighted correlation block at j-th level. (b) The details of mask decoder block at j-th level. (c) The details of hotspot decoder block.

where the $O_j^{i,M}$ represents the output in the j-th decoder residual block. $C_j^{i,M}$ is the correlation feature calculated by weighted correlation block with initial mask M^t as template guidance (*i.e.*, $G = M^t$ in Eq. 3). We take the output from the 1st residual decoder block as the final object mask M_i for i-th frame. The architecture of the decoder block for mask branch is illustrated in Fig. 3 (b).

Hotspot Decoder Branch. The function of hotspot decoder is to generate the track-let of hotspots $\{H^i\}_{i=1}^N$ for the video sequence according to the initial eye gaze. We visualize the framework of hotspot decoder branch in Fig. 3 (c). In the hotspot decoder branch, the correlation features on both 3-rd and 4-th side-outputs are used for generating the hotspot map,

$$O^{i,H} = Conv^2(Cat(C_3^{i,H}, Up(C_4^{i,H}))) \tag{5}$$

where $O^{i,H}$ is the hotspot map H^i for i-th frame. $C_3^{i,H}$ and $C_4^{i,H}$ are the correlation features by Eq. 3 with G set as initial eye gaze map E^t.

3.3 Network Training

Implementation Details. The input size of EG-Net and Gaze2Mask-Net is set to 320×320. In the WCS-Net, the template and search frames fed into the siamese encoder network share the same resolution as 320×320. Similar with the object tracking methods [27,44], we take the sub-region centered on the target object as template frame. The current search region is a local crop centered on the last estimated position of the target object. We use the EfficientNetv2 [40] as the backbone of siamese encoder. For all the residual refinement blocks in the decoders of EG-Net, Gaze2Mask-Net and WCS-Net, the two convolutional layers are set with kernel size 3×3 and 1×1, respectively.

Training of EG-Net and Gaze2Mask-Net. We exploit the SALICON dataset [19] to train our EG-Net. The loss for EG-Net is formulated as the cross entropy loss between predicted eye gaze map and ground truth. For the Gaze2Mask-Net, we produce the eye gaze and object mask pair on the PASCAL VOC dataset [14]. Specifically, we utilize the EG-Net to generate the eye gaze map inside the ground truth instance mask. The Gaze2Mask-Net takes the RGB image and eye gaze map as input to predict the object mask, which would be used to calculate the cross entropy loss with the corresponding ground truth. We use the Adam optimization with learning rate of 0.001 and batch size of 32 to train both EG-Net and Gaze2Mask-Net.

Training of WCS-Net. The overall loss of our WCS-Net is defined as:

$$L = \lambda_1 L_{Hotspot} + \lambda_2 L_{Mask} \tag{6}$$

where $L_{Hotspot}$ is the cross entropy loss between predicted hotspots with ground truth. L_{Mask} is the cross entropy loss between predicted masks and ground truth mask. For hotspot detection, we generate the ground truth based on SALICON data [19]. The WCS-Net is trained in two steps. We first pre-train the WCS-Net on the static image datasets of SALICON [19] and DUTS [43]. We use the human eye gaze annotations from SALICON to generate sufficient synthetic hotspot data for training our hotspot branch. Specifically, we use a large threshold to extract the peak distribution and take the focal regions as hotspot ground truth. We randomly combine the images from both datasets. The trade-off λ_1 is set to 1 if the data comes from SALICON, and 0 vise versa. The Adam with learning rate of 0.0001 and batch size of 15 is used in this stage to train the network. At the second stage, the WCS-Net is trained on the DAVIS-2016 training set [34] and SALICON datasets [19]. The network is trained using Adam with learning rate 0.0001 and batch size 10. Similarly, we integrate the data from two benchmarks and take the same setting on the trade-offs as the first stage.

4 Experiments

4.1 Dataset and Metrics

Dataset. To evaluate the performance of our proposed model, we conduct comparison experiments on three public VOS datasets, including DAVIS-2016 [34], SegTrackv2 [29] and Youtube-Objects [12]. The DAVIS-2016 dataset [34] contains 50 high-quality videos with 3455 densely annotated frames. All the videos in DAVIS-2016 are annotated with only one foreground object and they are splitted into 30 for training and 20 for testing. The SegTrackv2 dataset [29] is another challenging VOS dataset, which has 14 video sequences with densely annotated pixel-level ground truth. The videos in SegTrackv2 have large variations in resolution, object size and occlusion. The Youtube-Objects dataset is a large-size dataset, which contains 126 video sequences of 10 object categories. The ground truth in Youtube-Objects is sparsely annotated in every 10 frames.

Metrics. To compare the performance of our model with other state-of-the-arts in UVOS, we exploit two metrics, which are mean region similarity ($\mathcal{J}$ Mean) and mean contour accuracy ($\mathcal{F}$ Mean) [35]. To evaluate our hotspot tracking results, we exploit CC, SIM, KLD and NSS [1]. Besides, we also provide the run time of our method for efficiency evaluation. All the experiments are conducted on one NVIDIA 1080Ti GPU.

Table 1. Overall comparison with state-of-the-arts on the DAVIS-2016 validation dataset. The "✓" is used to indicate whether the method contains First Frame (FF), Online Finetuning (OF) or Post-processing (PP).

Method	FF	OF	PP	$\mathcal{J}$ Mean	$\mathcal{F}$ Mean
OSVOS [2]	✓	✓	✓	79.8	80.6
PLM [37]	✓	✓	✓	70.0	62.0
SegFlow [5]	✓	✓		74.8	74.5
RGMP [51]	✓			81.5	**82.0**
TrackPart [4]	✓	✓		77.9	76.0
OSMN [52]	✓			74.0	72.9
Siammask [44]	✓			71.7	67.8
FSEG [20]		✓		70.0	65.3
LMP [41]		✓	✓	70.0	65.9
LVO [42]		✓	✓	75.9	72.1
ARP [24]			✓	76.2	70.6
PDB [39]			✓	77.2	74.5
MOA [38]		✓	✓	77.2	78.2
AGS [48]			✓	79.7	77.4
COSNet [32]			✓	80.5	79.4
AGNN [45]			✓	80.7	79.1
Ours				**82.2**	**80.7**

Table 2. Overall comparison with state-of-the-arts on SegTrackv2 dataset.

Method	KEY [26]	FST [33]	NLC [11]	FSEG [20]	MSTP [18]	Ours
$\mathcal{J}$ Mean	57.3	52.7	67.2	61.4	70.1	**72.2**

4.2 Evaluation on Unsupervised Video Object Segmentation

DAVIS 2016. We compare our proposed model with state-of-the-art approaches on DAVIS-2016 dataset [34]. In Table 1, we list comparison results with methods from both SVOS and UVOS. Besides, we provide the indicator of some operations in the existing methods, including first frame annotation (FF), online finetuning (OF) and post-processing (PP). Compared with the existing UVOS methods, our model outperforms the second-best AGNN [45] by 1.8%, 2.0% on $\mathcal{J}$ Mean and $\mathcal{F}$-Mean, respectively. Note that we do not implement any post-processing in our model as other methods. We also propose the comparison results between our model and SVOS methods. We can observe that even without providing first frame's ground truth during testing, our model can perform favorably against most SVOS approaches (without online training).

SegTrackv2. We also illustrate the comparison results on SegTrackv2 dataset in Table 2. We report the $\mathcal{J}$ Mean performance as suggested by [18,29]. As can

Table 3. Overall comparison with state-of-the-arts on Youtube-Objects dataset. We report the per-category $\mathcal{J}$ Mean and the average result. Since COSNet and AGNN use dense-CRF post-processing, we also report our method with the same post-processing, denoted as Ours*.

Method	PP	Airplane (6)	Bird (6)	Boat (15)	Car (7)	Cat (16)	Cow (20)	Dog (27)	Horse (14)	Motor (10)	Train (5)	Avg.
FST [33]		70.9	70.6	42.5	65.2	52.1	44.5	65.3	53.5	44.2	29.6	53.8
ARP [24]		73.6	56.1	57.8	33.9	30.5	41.8	36.8	44.3	48.9	39.2	46.2
PDB [39]	✓	78.0	80.0	58.9	76.5	63.0	64.1	70.1	67.6	58.3	35.2	65.4
FSEG [20]		81.7	63.8	**72.3**	74.9	68.4	68.0	69.4	60.4	62.7	62.2	68.4
AGS [48]	✓	**87.7**	76.7	72.2	78.6	**69.2**	64.6	73.3	64.4	62.1	48.2	69.7
COSNet [32]	✓	81.1	75.7	71.3	77.6	66.5	**69.8**	**76.8**	67.4	67.7	46.8	70.5
AGNN [45]	✓	81.1	75.9	70.7	78.1	67.9	69.7	77.4	67.3	68.3	47.8	70.8
Ours		81.8	81.1	67.7	79.2	64.7	65.8	73.4	68.6	**69.7**	49.2	70.5
Ours*	✓	81.8	**81.2**	67.6	**79.5**	65.8	66.2	73.4	**69.5**	69.3	**49.7**	**70.9**

Table 4. Run time comparison on the DAVIS-2016 dataset. "Ours" is the model implemented on EfficientNet [40] and "Ours†" is the model built on Resnet101 [15].

Method	Siammask [44]	OSMN [52]	RGMP [51]	AGS [48]	Ours	Ours†
Time (s)	0.02	0.14	0.13	0.60	0.03	0.04

be seen, our model significantly outperforms state-of-the-art methods and has an improvement of 3% on $\mathcal{J}$ Mean against MSTP [18].

Youtube-Objects. Table 3 lists the $\mathcal{J}$ Mean results of the state-of-the-arts on Youtube-Objects. Our model achieves the best results on the categories of bird, car, horse, motorbike and train. It is also comparable with recent UVOS methods AGS [48], COSNet [32] and AGNN [45] across all categories. For a fair comparison with methods using Dense-CRF [25], we evaluate our method with the same post-processing (named as "Ours*"), and it achieves the best performance on this dataset.

Qualitative Results. We illustrate the visual results of our model in Fig. 4. For the video object segmentation, our model can keep tracking the target object with shape deformation and occlusion and produce accurate segmentation masks with well-defined details.

Run Time. We also report the comparison on run time to verify the efficiency of our model. The run time of our model and other state-of-the-arts on DAVIS-2016 are shown in Table 4. The results illustrate that our model can not only produce accurate segmentation and eye gaze results, but achieves real-time inference speed. The Siammask [44] performs much faster than our model. However, our method does not need any ground truth indicator as [44] and we significantly outperform Siammask in accuracy with 15% and 19% improvement on $\mathcal{J}$ Mean and $\mathcal{F}$ Mean, respectively.

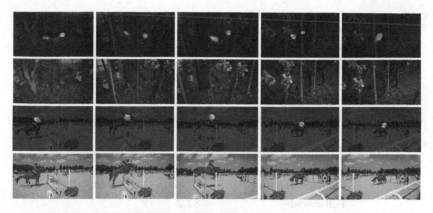

Fig. 4. Qualitative results of our proposed model on DAVIS-2016. For the two sequences, the top row is the hotspot tracking results and the bottom row lists the object segmentation. Zoom in to see the details.

4.3   Evaluation on Hotspot Tracking

Hotspot Ground Truth for Testing. As illustrated in Sect. 1, the existing video eye gaze data is noisy and unstable due to the saccade in eye movement. It can not meet the problem formulation of our hotspot tracking. To evaluate the effectiveness of our hotspot tracking, we provide the hotspot data on DAVIS-2016 validation dataset. To annotate the hotspots for the target object in DAVIS-2016, we exploit the mouse click to stimulate the human attention as [19]. The reason why we don't use the eye tracker is two folds. First, it is hard to constrain the eye gaze inside the object using eye tracker. Second, the delay between human eye gaze and fast object movement makes it difficult for users to keep tracking the salient region. Such cases would produce flicker annotations inside the object (See the third row in Fig. 1. Compared with the eye tracker, using the mouse click is more suitable for our task to produce a clean and consistent hotspot ground truth [19]. Specifically, we sample the video clips from DAVIS-2016 test set in every 10 frame for annotating. We randomly disrupt the frame order in each video sequence and ask five users to annotate. The users are required to first determine the salient part inside the object and provide consistent mouse click on that region along the video sequence.

Comparison Results. In our experiment, we exploit the existing metrics in eye gaze to evaluate the stability of our hotspot tracking [48]. The comparison results with other video eye tracking models are shown in Table 5. We can observe that our model outperforms state-of-the-arts on all metrics. The qualitative results in Fig. 4 illustrate that our model can produce clean and consistent hotspot tracking for the video objects.

Table 5. Quantitative results of hotspot tracking on the DAVIS-2016 dataset.

Method	KLD ↓	NSS ↑	CC↑	SIM↑
DeepVS [22]	3.148	2.010	0.189	0.089
AGS [48]	2.704	2.944	0.275	0.138
Ours	**2.592**	**3.174**	**0.429**	**0.333**
Ours w/o. mask branch	2.701	3.025	0.399	0.291

Table 6. Ablation study on the DAVIS-2016 validation dataset.

Model setting		$\mathcal{J}$ Mean	$\Delta\mathcal{J}$ Mean
Full model		82.2	-
Multi-level feature	45	74.5	−7.7
	345	78.0	−4.2
	2345	79.9	−2.3
w/o. weighted correlation		75.3	−6.9
w/o. weighted pooling		78.3	−3.9
w/o. hotspot branch		81.5	−0.7
Resnet101		81.7	−0.5

4.4 Ablation Study

In this section, we analyze the contribution of each component in our weighted correlation siamese network. The results in terms of $\mathcal{J}$ Mean on DAVIS-2016 dataset are shown in Table 6.

Effectiveness of Multi-level Construction. In our model, we implement the weighted correlation block on the multiple side-outputs to calculate the correspondence between template and search frames. The generated multi-level correlation features are further fed into two sub-branches to produce both hotspot and object mask. To demonstrate the effectiveness of such multi-level architecture, we gradually remove the skip connection from each feature level. The results can be referred in the rows named "Multi-level feature" in Table 6. For example, the item named "45" indicates that only features from the 4-th and 5-th encoder blocks are used to produce the correlation feature and generate the final results. The results in Table 6 verify the efficacy of our multi-level architecture.

Effectiveness of Joint Training. Our model builds a unified architecture for joint video object segmentation and hotspot tracking. Specifically, with a shared siamese encoder, we implement two parallel decoder branches for both tasks. The decoder branches are jointly trained with both object mask and hotspot annotations using Eq. 6. To investigate the efficacy of joint training strategy on video object segmentation, we remove the hotspot tracking branch and train the network for only video object segmentation. Its comparison result between full model ("w/o. hotspot tracking branch" in Table 6) demonstrate the effectiveness for such joint training method. We also implement the ablation experiment on the hotspot tracking branch. The result "Ours w/o. mask branch" in Table 5 verifies the joint training strategy on hotspot tracking task.

Effectiveness of Weighted Correlation. To verify the effect of weighted correlation block, we build a baseline network. Instead of conducting the correlation operation, we first concatenate the template and search feature in channel wise. Then the concatenated feature would be fed into the decoder block to produce the object mask. Note that the multi-level construction is also implemented in this baseline network. The result of this baseline network is illustrated in "w/o. weighted correlation" of Table 6. The comparison result with baseline verifies the efficacy of our weighted correlation block.

Effectiveness of Weighted Pooling. In our weighted correlation block, we transfer the template feature map into 1×1 feature vector via weighted pooling with the initial mask and hotspot. To demonstrate the effect of weighted pooling, we implement a model with the original correlation operation as Siammask [44]. Specifically, the size of template frame is half of the search frame, and they are fed into the siamese encoder to produce multi-level features. Instead of conduct the weighted pooling operation, we directly use the template and search features to calculate correlation features for mask generation. From the results in Table 6, we can observe that the weighted pooling is more effective compared with the original correlation operation in tracking methods [27,28,44].

Implementation using Resnet101. We implement our WCS-Net on the Resnet101 [15]. The results of $\mathcal{J}$ Mean and the run time are listed in Table 6 and Table 4, respectively. They demonstrate that our WCS-Net also works on the Resnet on both accuracy and efficiency.

5   Conclusion

In this paper, we propose a Weighted Correlation Siamese Network (WCS-Net) for joint unsupervised video object segmentation and hotspot tracking. We introduce a novel weighted correlation block (WCB) to calculate the cross-correlation between template frame and the search frame. The correlation feature from WCB is used in both sub-branches for generating the track-lets of mask and hotspots. The experimental results on three benchmarks demonstrate our proposed model outperforms existing competitors on both unsupervised video object segmentation and hotspot tracking with a significantly faster speed of 33 FPS.

Acknowledgements. The paper is supported in part by the National Key R&D Program of China under Grant No. 2018AAA0102001 and National Natural Science Foundation of China under grant No. 61725202, U1903215, 61829102, 91538201, 61771088, 61751212 and the Fundamental Research Funds for the Central Universities under Grant No. DUT19GJ201 and Dalian Innovation leader's support Plan under Grant No. 2018RD07.

References

1. Bylinskii, Z., Judd, T., Oliva, A., Torralba, A., Durand, F.: What do different evaluation metrics tell us about saliency models? arXiv preprint arXiv:1604.03605 (2016)
2. Caelles, S., Maninis, K.K., Pont-Tuset, J., Leal-Taixé, L., Cremers, D., Van Gool, L.: One-shot video object segmentation. In: CVPR (2017)
3. Chen, Y., Pont-Tuset, J., Montes, A., Van Gool, L.: Blazingly fast video object segmentation with pixel-wise metric learning. In: CVPR (2018)
4. Cheng, J., Tsai, Y.H., Hung, W.C., Wang, S., Yang, M.H.: Fast and accurate online video object segmentation via tracking parts. In: CVPR (2018)
5. Cheng, J., Tsai, Y.H., Wang, S., Yang, M.H.: SegFlow: joint learning for video object segmentation and optical flow. In: CVPR (2017)
6. Deng, Z., et al.: R^3Net: recurrent residual refinement network for saliency detection. In: IJCAI (2018)
7. Ding, H., Cohen, S., Price, B., Jiang, X.: PhraseClick: toward achieving flexible interactive segmentation by phrase and click. In: ECCV (2020)
8. Ding, H., Jiang, X., Liu, A.Q., Thalmann, N.M., Wang, G.: Boundary-aware feature propagation for scene segmentation. In: ICCV (2019)
9. Ding, H., Jiang, X., Shuai, B., Liu, A.Q., Wang, G.: Context contrasted feature and gated multi-scale aggregation for scene segmentation. In: CVPR (2018)
10. Ding, H., Jiang, X., Shuai, B., Liu, A.Q., Wang, G.: Semantic correlation promoted shape-variant context for segmentation. In: CVPR (2019)
11. Faktor, A., Irani, M.: Video segmentation by non-local consensus voting. In: BMVC (2014)
12. Ferrari, V., Schmid, C., Civera, J., Leistner, C., Prest, A.: Learning object class detectors from weakly annotated video. In: CVPR (2012)
13. Gegenfurtner, K.R.: The interaction between vision and eye movements. Perception 45(12), 1333–1357 (2016)
14. Hariharan, B., Arbeláez, P., Bourdev, L., Maji, S., Malik, J.: Semantic contours from inverse detectors. In: CVPR (2011)
15. He, K., Zhang, X., Ren, S., Sun, J.: Deep residual learning for image recognition. In: CVPR (2016)
16. Hu, P., Caba, F., Wang, O., Lin, Z., Sclaroff, S., Perazzi, F.: Temporally distributed networks for fast video semantic segmentation. In: CVPR (2020)
17. Hu, Y.T., Huang, J.B., Schwing, A.: MaskRNN: instance level video object segmentation. In: Advances in Neural Information Processing Systems (2017)
18. Hu, Y.T., Huang, J.B., Schwing, A.G.: Unsupervised video object segmentation using motion saliency-guided spatio-temporal propagation. In: ECCV (2018)
19. Huang, X., Shen, C., Boix, X., Zhao, Q.: SALICON: reducing the semantic gap in saliency prediction by adapting deep neural networks. In: ICCV (2015)
20. Jain, S.D., Xiong, B., Grauman, K.: FusionSeg: learning to combine motion and appearance for fully automatic segmentation of generic objects in videos. In: CVPR (2017)
21. Jang, W.D., Lee, C., Kim, C.S.: Primary object segmentation in videos via alternate convex optimization of foreground and background distributions. In: CVPR (2016)
22. Jiang, L., Xu, M., Liu, T., Qiao, M., Wang, Z.: DeepVS: a deep learning based video saliency prediction approach. In: ECCV (2018)

23. Keuper, M., Andres, B., Brox, T.: Motion trajectory segmentation via minimum cost multicuts. In: CVPR (2015)
24. Koh, Y.J., Kim, C.S.: Primary object segmentation in videos based on region augmentation and reduction. In: CVPR (2017)
25. Krähenbühl, P., Koltun, V.: Efficient inference in fully connected CRFs with Gaussian edge potentials. In: Advances in Neural Information Processing Systems, pp. 109–117 (2011)
26. Lee, Y.J., Kim, J., Grauman, K.: Key-segments for video object segmentation. In: ICCV (2011)
27. Li, B., Wu, W., Wang, Q., Zhang, F., Xing, J., Yan, J.: SiamRPN++: evolution of Siamese visual tracking with very deep networks. In: CVPR (2019)
28. Li, B., Yan, J., Wu, W., Zhu, Z., Hu, X.: High performance visual tracking with Siamese region proposal network. In: CVPR (2018)
29. Li, F., Kim, T., Humayun, A., Tsai, D., Rehg, J.M.: Video segmentation by tracking many figure-ground segments. In: ICCV (2013)
30. Li, S., Seybold, B., Vorobyov, A., Fathi, A., Huang, Q., Jay Kuo, C.C.: Instance embedding transfer to unsupervised video object segmentation. In: CVPR (2018)
31. Li, S., Seybold, B., Vorobyov, A., Lei, X., Jay Kuo, C.C.: Unsupervised video object segmentation with motion-based bilateral networks. In: ECCV (2018)
32. Lu, X., Wang, W., Ma, C., Shen, J., Shao, L., Porikli, F.: See more, know more: unsupervised video object segmentation with co-attention Siamese networks. In: CVPR (2019)
33. Papazoglou, A., Ferrari, V.: Fast object segmentation in unconstrained video. In: ICCV (2013)
34. Perazzi, F., Pont-Tuset, J., McWilliams, B., Van Gool, L., Gross, M., Sorkine-Hornung, A.: A benchmark dataset and evaluation methodology for video object segmentation. In: CVPR (2016)
35. Pont-Tuset, J., Perazzi, F., Caelles, S., Arbeláez, P., Sorkine-Hornung, A., Van Gool, L.: The 2017 DAVIS challenge on video object segmentation. arXiv:1704.00675 (2017)
36. Rommelse, N.N., Van der Stigchel, S., Sergeant, J.A.: A review on eye movement studies in childhood and adolescent psychiatry. Brain Cogn. **68**(3), 391–414 (2008)
37. Shin Yoon, J., Rameau, F., Kim, J., Lee, S., Shin, S., So Kweon, I.: Pixel-level matching for video object segmentation using convolutional neural networks. In: ICCV (2017)
38. Siam, M., et al.: Video object segmentation using teacher-student adaptation in a human robot interaction (HRI) setting. In: 2019 International Conference on Robotics and Automation (2019)
39. Song, H., Wang, W., Zhao, S., Shen, J., Lam, K.M.: Pyramid dilated deeper ConvLSTM for video salient object detection. In: ECCV (2018)
40. Tan, M., Le, Q.V.: EfficientNet: rethinking model scaling for convolutional neural networks. arXiv preprint arXiv:1905.11946 (2019)
41. Tokmakov, P., Alahari, K., Schmid, C.: Learning motion patterns in videos. In: CVPR (2017)
42. Tokmakov, P., Alahari, K., Schmid, C.: Learning video object segmentation with visual memory. In: CVPR (2017)
43. Wang, L., Lu, H., Wang, Y., Feng, M., Ruan, X.: Learning to detect salient objects with image-level supervision. In: CVPR (2017)
44. Wang, Q., Zhang, L., Bertinetto, L., Hu, W., Torr, P.H.: Fast online object tracking and segmentation: a unifying approach. In: CVPR (2019)

45. Wang, W., Lu, X., Shen, J., Crandall, D.J., Shao, L.: Zero-shot video object segmentation via attentive graph neural networks. In: CVPR (2019)
46. Wang, W., Shen, J., Guo, F., Cheng, M.M., Borji, A.: Revisiting video saliency: a large-scale benchmark and a new model. In: CVPR (2018)
47. Wang, W., Shen, J., Porikli, F.: Saliency-aware geodesic video object segmentation. In: CVPR (2015)
48. Wang, W., et al.: Learning unsupervised video object segmentation through visual attention. In: CVPR (2019)
49. Wei, Z., et al.: Sequence-to-segments networks for detecting segments in videos. IEEE Trans. Pattern Anal. Mach. Intell. (2019)
50. Wei, Z., et al.: Sequence-to-segment networks for segment detection. In: Advances in Neural Information Processing Systems, pp. 3507–3516 (2018)
51. Wug Oh, S., Lee, J.Y., Sunkavalli, K., Joo Kim, S.: Fast video object segmentation by reference-guided mask propagation. In: CVPR (2018)
52. Yang, L., Wang, Y., Xiong, X., Yang, J., Katsaggelos, A.K.: Efficient video object segmentation via network modulation. In: CVPR (2018)
53. Yang, Z., Wang, Q., Bertinetto, L., Hu, W., Bai, S., Torr, P.H.S.: Anchor diffusion for unsupervised video object segmentation. In: ICCV (2019)
54. Yang, Z., et al.: Predicting goal-directed human attention using inverse reinforcement learning. In: CVPR (2020)
55. Zhang, L., Dai, J., Lu, H., He, Y.: A bi-directional message passing model for salient object detection. In: CVPR (2018)
56. Zhang, L., Lin, Z., Zhang, J., Lu, H., He, Y.: Fast video object segmentation via dynamic targeting network. In: ICCV (2019)
57. Zhang, L., Zhang, J., Lin, Z., Lu, H., He, Y.: CapSal: leveraging captioning to boost semantics for salient object detection. In: CVPR (2019)

SRNet: Improving Generalization in 3D Human Pose Estimation with a Split-and-Recombine Approach

Ailing Zeng[1,2], Xiao Sun[2(✉)], Fuyang Huang[1], Minhao Liu[1], Qiang Xu[1], and Stephen Lin[2]

[1] The Chinese University of Hong Kong, Hong Kong, China
{alzeng,fyhuang,mhliu,qxu}@cse.cuhk.edu.hk
[2] Microsoft Research Asia, Beijing, China
{xias,stevelin}@microsoft.com

Abstract. Human poses that are rare or unseen in a training set are challenging for a network to predict. Similar to the long-tailed distribution problem in visual recognition, the small number of examples for such poses limits the ability of networks to model them. Interestingly, *local* pose distributions suffer less from the long-tail problem, i.e., local joint configurations within a rare pose may appear within other poses in the training set, making them less rare. We propose to take advantage of this fact for better generalization to rare and unseen poses. To be specific, our method splits the body into local regions and processes them in separate network branches, utilizing the property that a joint's position depends mainly on the joints within its local body region. Global coherence is maintained by recombining the global context from the rest of the body into each branch as a low-dimensional vector. With the reduced dimensionality of less relevant body areas, the training set distribution within network branches more closely reflects the statistics of *local* poses instead of global body poses, without sacrificing information important for joint inference. The proposed split-and-recombine approach, called *SRNet*, can be easily adapted to both single-image and temporal models, and it leads to appreciable improvements in the prediction of rare and unseen poses.

Keywords: Human pose estimation · 2D to 3D · Long-tailed distribution

1 Introduction

Human pose estimation is a longstanding computer vision problem with numerous applications, including human-computer interaction, augmented reality, and

---

A. Zeng—The work is done when Ailing Zeng is an intern at Microsoft Research Asia.

Electronic supplementary material The online version of this chapter (https://doi.org/10.1007/978-3-030-58568-6_30) contains supplementary material, which is available to authorized users.

© Springer Nature Switzerland AG 2020
A. Vedaldi et al. (Eds.): ECCV 2020, LNCS 12359, pp. 507–523, 2020.
https://doi.org/10.1007/978-3-030-58568-6_30

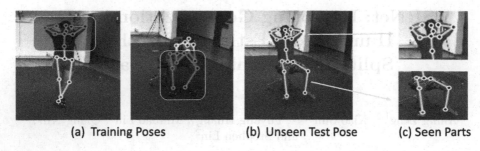

(a) Training Poses   (b) Unseen Test Pose   (c) Seen Parts

Fig. 1. An unseen test pose (b) may be decomposed into local joint configurations (c) that appear in poses that exist in the training set (a). Our method takes advantage of this property to improve estimation of rare and unseen poses.

computer animation. For predicting 3D pose, a common approach is to first estimate the positions of keypoints in the 2D image plane, and then lift these keypoints into 3D. The first step typically leverages the high performance of existing 2D human pose estimation algorithms [4,21,28]. For the second stage, a variety of techniques have been proposed, based on structural or kinematic body models [2,6,7,14,31], learned dependencies and relationships among body parts [7,22], and direct regression [18].

Besides algorithm design, an important factor in the performance of a machine learning system is its training data. A well-known issue in pose estimation is the difficulty of predicting poses that are rare or unseen in the training set. Since few examples of such poses are available for training, it is hard for the network to learn a model that can accurately infer them. Better generalization to such poses has been explored by augmenting the training set with synthetically generated images [3,12,19,27,29,32]. However, the domain gap that exists between real and synthesized images may reduce their efficacy. Different viewpoints of existing training samples have also been simulated to improve generalization to other camera positions [7], but this provides only a narrow range of pose augmentations.

In this work, we propose to address the rare/unseen pose problem through a novel utilization of the data in the original training set. Our approach is based on the observation that rare poses at the global level are composed of local joint configurations that are generally less rare in the training set. For example, a bicycling pose may be uncommon in the dataset, but the left leg configuration of this pose may resemble the left legs of other poses in the dataset, such as stair climbing and marching. Many instances of a local pose configuration may thus exist in the training data among different global poses, and they could be leveraged for learning local pose. Moreover, it is possible to reconstruct unseen poses as a combination of local joint configurations that are presented in the dataset. For example, an unseen test pose may be predicted from the upper body of one training pose and the lower body of another, as illustrated in Fig. 1.

Based on this observation, we design a network structure that splits the human body into local groups of joints that have strong inter-relationships within each group and relatively weak dependencies on joints outside the group. Each group is processed in a separate network branch. To account for the weak dependencies that exist with the rest of the body, low-dimensional global context is computed from the other branches and recombined into the branch. The dimensionality reduction of less-relevant body areas within the global context decreases their impact on the feature learning for local joints within each branch. At the same time, accounting for some degree of global context can avoid local pose estimates that are incoherent with the rest of the body. With this split-and-recombine approach, called *SRNet*, generalization performance is enhanced to effectively predict global poses that are rare or absent from the training set.

In extensive comparisons to state-of-the-art techniques, SRNet exhibits competitive performance on single-frame input and surpasses them on video input. More importantly, we show that SRNet can elevate performance considerably on rare/unseen poses. Moreover, we conduct various ablation studies to validate our approach and to examine the impact of different design choices.

2   Related Work

Extensive research has been conducted on reconstructing 3D human pose from 2D joint predictions. In the following, we briefly review methods that are closely related to our approach.

Leveraging Local Joint Relations. Many recent works take advantage of the physical connections that exist between body joints to improve feature learning and joint prediction. As these connections create strong inter-dependencies and spatial correlations between joints, they naturally serve as paths for information sharing [2,5,14,35] and for encoding kinematic [7] and anatomical [6,8,23,31] constraints.

The structure of these connections defines a locality relationship among joints. More closely connected joints have greater inter-dependency, while distantly connected joints have less and indirect dependence via the joints that lie on the paths between them. These joint relationships have been modeled hierarchically, with feature learning that starts within local joint groups and then expands to account for all the joint groups together at a global level [22]. Alternatively, these relationships have been represented in a graph structure, with feature learning and pose estimation conducted in a graph convolutional network (GCN) [2,5,35]. Within a GCN layer, dependencies are explicitly modeled between connected joints and expand in scope to more distant joints through the stacking of layers.

For both hierarchical and GCN based techniques, feature learning within a local group of joints can be heavily influenced by joints outside the group. In contrast, the proposed approach SRNet restricts the impact of non-local joints on local feature learning through dimensionality reduction of the global context, which facilitates the learning of local joint configurations without ignoring global pose coherence. By doing so, SRNet can achieve better generalization to rare and unseen poses in the training data.

510   A. Zeng et al.

Generalization in Pose Estimation. A common approach to improve generalization in pose estimation is to generate more training images through data augmentation. Approaches to augmentation have included computer graphics rendering [3,29], image-based synthesis [27], fitting 3D models to people and deforming them to generate new images [25], and changing the background, clothing and occluders in existing images [19]. These methods can produce a large amount of data for training, but the gap in realism between artificially constructed images and actual photographs may limit their effectiveness. Since our technique obtains improved training data distributions by focusing on local pose regions instead of global poses, it does not involve image manipulation or synthesis, thereby maintaining the realism of the original training set.

Other works have explored generalization to different viewpoints [7,30] and to in-the-wild scenes [8,34], which are orthogonal to our goal of robustly estimating rare and unseen poses.

Robustness to Long-tailed Distributions. In visual recognition, there exists a related problem of long-tailed training set distributions, where classes in the distribution tail have few examples. As these few examples have little impact on feature learning, the recognition performance for tail classes is often poor. Recent approaches to this problem include metric learning to enforce inter-class margins [9] and meta-learning that learns to regress many-shot model parameters from few-shot model parameters [33]. Such techniques improve the discriminability of tail classes from other classes, but they are not compatible with the problem of keypoint localization in pose estimation.

3   Method

To address the issue of rare and unseen poses, our approach is to decompose global pose estimation into a set of local pose estimation problems. For this strategy to be effective, the local problems must be defined in a manner that allows the feature learning of a local region to primarily reflect the statistical distribution of its local poses, yet account for other local regions such that the final overall pose estimate is globally coherent.

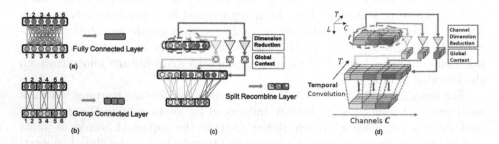

Fig. 2. Illustration of (a) a fully connected layer, (b) group connected layer, (c) our split-and-recombine layer, and (d) our convolution layer for temporal models. The four types of layers can be stacked to form different network structures as shown in Fig. 3.

In the following, we start with the most common baseline for lifting 2D joints to 3D, and then describe how to modify this baseline to follow our strategy.

Fully-Connected Network Baseline. For lifting 2D keypoints to 3D, a popular yet simple baseline is to use a fully-connected network (FCN) consisting of several layers [18]. Given the 2D keypoint detections $\mathcal{K} = \{K_i|i, ..., N\} \in \mathbb{R}^{2N}$ in the image coordinate system, the FCN estimates the 3D joint locations $\mathcal{J} = \{J_i|i, ..., N\} \in \mathbb{R}^{3N}$ in the camera coordinate system with the origin at the root joint J_0. Formally, a fully-connected network layer can be expressed as

$$\mathbf{f}^{l+1} = \Theta^l \mathbf{f}^l \tag{1}$$

where $\Theta^l \in \mathbb{R}^{D^{l+1} \times D^l}$ is the fully-connected weight matrix to be learned, and D^l is the feature dimension for the lth layer, namely $\mathbf{f}^l \in \mathbb{R}^{D^l}$. Batch normalization and ReLU activation are omitted for brevity. For the input and output layers, their feature dimensions are $D^1 = 2N$ and $D^{L+1} = 3N$, respectively.

It can be noted that in this FCN baseline, each output joint and each intermediate feature is connected to all of the input joints indiscriminately, allowing the prediction of an output joint to be overfitted to the positions of distant joints with little relevance. In addition, all the output joints share the same set of features entering the final layer and have only this single linear layer (Θ_i^L) to determine a solution particular to each joint.

Body Partitioning into Local Pose Regions. In turning global pose estimation into several local pose estimation problems, a suitable partitioning of the human body into local pose regions is needed. A local pose region should contain joints whose positions are heavily dependent on one another, but less so on joints outside the local region.

Here, we adopt the partitioning used in [22], where the body is divided into left/right arms, left/right legs, and torso. These parts have distinctive and coordinated behaviors such that the joint positions within each group are highly correlated. In contrast, joint positions between groups are significantly less related.

To accommodate this partitioning, the FCN layers are divided into groups. Formally, $\mathcal{G}_g^l$ represents the feature/joint indexes of the gth group at layer l. Specifically, for the first input layer,

$$\mathcal{G}_0^1 = [\text{joint indices of the right arm}], \tag{2}$$

and for the intermediate feature layers, we have

$$\mathcal{G}_0^l = [\text{feature indices of the right leg at the } l_{th} \text{ layer}]. \tag{3}$$

Then, a *group connected layer* for the g_{th} group is expressed as

$$\mathbf{f}^{l+1}[\mathcal{G}_g^{l+1}] = \Theta_g^l \mathbf{f}^l[\mathcal{G}_g^l]. \tag{4}$$

In a *group connected layer*, the connections between joint groups are removed. This "local connectivity" structure is commonly used in convolutional neural networks to capture spatially local patterns [13]. The features learned in a group depend on the statistical distribution of the training data for only its local pose region. In other words, this local feature is learned independently of the pose configurations of other parts, and thus it generalizes well to any global pose that includes similar local joint configurations.

However, a drawback of group connected layers is that the status of the other body parts is completely unknown when inferring the local pose. As a result, the set of local inferences may not be globally coherent, leading to low performance. There is thus a need to account for global information while largely preserving local feature independence.

SRNet: Incorporating Low-Dimensional Global Context. To address this problem, we propose to incorporate *Low-Dimensional Global Context* (LDGC) in a group connected layer. It coarsely represents information from the less relevant joints, and is brought back to the local group in a manner that limits disruption to the local pose modeling while allowing the local group to account for non-local dependencies. This split-and-recombine approach for global context can be expressed as the following modification of Eq. 4:

$$\mathbf{f}^{l+1}[\mathcal{G}_g^{l+1}] = \Theta_g^l(\mathbf{f}^l[\mathcal{G}_g^l] \circ \mathcal{M}\mathbf{f}^l[\mathcal{G}^l \setminus \mathcal{G}_g^l]) \qquad (5)$$

where $\mathbf{f}^l[\mathcal{G}^l \setminus \mathcal{G}_g^l]$ is the global context for the gth group. $\mathcal{M}$ is a mapping function that defines how the global context is represented. Special cases of the mapping function are $\mathcal{M} = Identity$, equivalent to a fully-connected layer, and $\mathcal{M} = Zero$, which is the case for a group connected layer. The mapped global context is recombined with local features by an operator $\circ$, typically concatenation for an *FCN*.

The mapping function $\mathcal{M}$ acts as a gate controlling the information passed from the non-local joints. If the gate is wide open (FCN), the local feature learning will account for the exact joint positions of other body parts, weakening the ability to model the local joint configurations of rare global poses. However, if the gate is closed (Group), the local feature learning receives no knowledge about the state of other body parts and may lose global coherence. We argue that the key to achieving high performance and strong generalization to rare poses is to learn a low-dimensional representation of the global context, namely

$$\mathcal{M} = \Gamma_g^l \in \mathbb{R}^{H \times (D^l - D_g^l)} \qquad (6)$$

where D_g^l is the feature dimension for the gth group at layer l. Γ_g^l is a weight matrix that maps the global context $\mathbf{f}^l[\mathcal{G}^l \setminus \mathcal{G}_g^l]$ of dimensions $D^l - D_g^l$ to a small number of dimensions H.

In our experiments, we empirically evaluate different design choices for the mapping function $\mathcal{M}$ and the combination operator $\circ$.

Network Structure. With our split-and-recombine approach for processing local pose regions and global context, the *SRNet* model can be illustrated as shown at the right side of Fig. 3. This network follows the structure of a group connected network, with the body joints split into separate groups according to

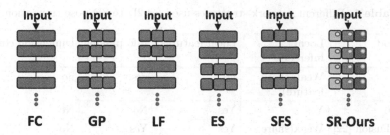

Fig. 3. Illustration of fully connected (FC), group connected (GP), late fusion (LF), early split (ES), split-fuse-split (SPS) and split-and-recombine (SR) models. The components of each layer are shown in Fig. 2.

their local pose region. Added to this network are connections to each group in a layer from the other groups in the preceding layer. As illustrated in Fig. 2 (c,d), these connections representing the global context reduce the dimensionality of the context features and recombine them with the local features in each group. These inputs are mapped into outputs of the original feature dimensions.

SRNet balances feature learning dependency between the most related local region and the less related global context. For demonstrating the efficacy of this split-and-recombine strategy, we describe several straightforward modifications to the FC network, illustrated in Fig. 3, that will serve as baselines in our experiments:

- *Late Fusion (LF)*, which postpones feature sharing among groups by cutting their connections in the earlier layers. This structure is similar to that used in [22], which first learns features within each group and fuses them later with stacked fully connected layers. Formally, the *LF* baseline is defined as

$$\mathbf{f}^{l+1} = \begin{cases} Cat\{\mathbf{f}^l[\mathcal{G}_g^l] | g = 1, ..., G\}, & \text{if } l < L_{fuse} \\ \Theta^l \mathbf{f}^l, & \text{otherwise.} \end{cases} \quad (7)$$

- *Early Split (ES)*, which aims to provide more expressive features for each group, by cutting the connections between groups in the latter layers. Formally,

$$\mathbf{f}^{l+1} = \begin{cases} \Theta^l \mathbf{f}^l, & \text{if } l < L_{split} \\ Cat\{\mathbf{f}^l[\mathcal{G}_g^l] | g = 1, ..., G\}, & \text{otherwise.} \end{cases} \quad (8)$$

- *Group Connected (GP)*, which is the standard group connected network. *Late Fusion* degenerates to this form when $L_{fuse} = L$, and *Early Split* does when $L_{split} = 1$.
- *Split-Fuse-Split (SFS)*, where the middle L_{link} layers of the network are fully-connected and the connections between joint groups are cut in the first and last layers.

The differences among FCN, these baselines, and SRNet are summarized in Table 1. The table also highlights differences of SRNet from GCN [35] and its

Table 1. Different network structures used for 2D to 3D pose estimation.

Method	Local inference	Local features	Info passing	Dimension control
FCN	Weak(share feature)	no	yes	no
Group	Yes	Yes	No	No
Late Fusion [22]	Weak(share feature)	Yes	Yes	No
Early Split	Yes	No	Yes	No
Split-Fuse-Split	Yes	Yes	Yes	No
GCN [35]	No(share weight)	Yes(overlapped group)	Yes	No
LCN [5]	Yes	Yes(overlapped group)	Yes	No
Ours (SRNet)	Yes	Yes	Yes	**Yes**

modification LCN [5], specifically in controlling the feature dependencies between local regions and global context.

SR Convolution for Temporal Models. Temporal information in 2D to 3D mapping is conventionally modeled with Recurrent Neural Networks (RNNs) [2, 14]. Recently, Pavllo et al. [23] introduce an alternative temporal convolutional model that stacks all the spatial information of a frame into the channel dimensions and replaces fully-connected operations in the spatial domain by convolutions in the temporal domain. The temporal convolution enables parallel processing of multiple frames and brings greater accuracy, efficiency, and simplicity.

As illustrated in Fig. 2 (d), our split-and-recombine modification to the fully connected layer can be easily adapted to the temporal convolution model by applying the same split-and-recombine strategy to the channels during convolution. Specifically, the group connected operations for local joint groups are replaced with corresponding temporal convolutions for local feature learning, where the features from other joint groups undergo an extra convolution for dimension reduction in channels and are concatenated back with the local joint group as global context. We call this split-and-recombine strategy in the channel dimensions during convolution the *SR convolution*.

4   Datasets and Rare-Pose Evaluation Protocols

4.1   Datasets and Evaluation Metrics

Our approach is validated on two popular benchmark datasets:

– *Human3.6M* [11] is a large benchmark widely-used for 3D human pose estimation. It consists of 3.6 million video frames from four camera viewpoints with 15 activities. Accurate 3D human joint locations are obtained from motion capture devices. Following convention, we use the mean per joint position error (MPJPE) for evaluation, as well as the Procrustes Analysis MPJPE (PA-MPJPE).

– *MPI-INF-3DHP* [19, 20] is a more challenging 3D pose dataset, containing not only constrained indoor scenes but also complex outdoor scenes. Compared to Human3.6M, it covers a greater diversity of poses and actions. For evaluation, we follow common practice [5, 31, 32, 34] by using the Percentage of Correct Keypoints (PCK) with a threshold of 150 mm and the Area Under Curve (AUC) for a range of PCK thresholds.

4.2 Evaluation Protocols

The most widely-used evaluation protocol [5, 10, 18, 23, 32] on Human3.6M uses five subjects (S1, S5, S6, S7, S8) as training data and the other two subjects (S9, S11) as testing data. We denote it as the *Subject Protocol*. However, the rare pose problem is not well examined in this protocol since each subject is asked to perform a fixed series of actions in a similar way.

To better demonstrate the generalization of our method to rare/unseen poses, we use two other protocols, introduced in [5]. The first is the *Cross Action Protocol* that trains on only one of the 15 actions in the Human3.6M dataset and tests on all actions. The second is the *Cross Dataset Protocol* that applies the model trained on Human3.6M to the test set of MPI-INF-3DHP. In addition, we propose a new protocol called *Rare Pose Protocol*.

Rare Pose Protocol. In this protocol, we use all the poses of subjects S1, S5, S6, S7, S8 for training but only use a subset of poses from subjects S9, S11 for testing. The subset is selected as the rarest poses in the testing set. To identify rare poses, we first define pose similarity (PS) as

$$PS(\mathcal{J}, \mathcal{I}) = \frac{1}{N} \sum_{i}^{N} exp(-\frac{||J_i - I_i||^2}{2\sigma_i^2}) \qquad (9)$$

where $||J_i - I_i||$ is the Euclidean distance between corresponding joints of two poses. To compute PS, we pass ΔJ_i through an unnormalized Gaussian with a standard deviation σ_i controlling falloff. This yields a pose similarity that ranges between 0 and 1. Perfectly matched poses will have $PS = 1$, and if all joints are distant by more than the standard deviation σ_i, PS will be close to 0. The occurrence of a pose $\mathcal{J}$ in a pose set $\mathbb{O}$ is defined as the average similarity between itself and all the poses in $\mathbb{O}$:

$$OCC_{\mathbb{O}}(\mathcal{J}) = \frac{1}{M} \sum_{\mathcal{I}}^{\mathbb{O}} PS(\mathcal{J}, \mathcal{I}) \qquad (10)$$

where M is the total number of poses in the pose set $\mathbb{O}$. A rare pose will have a low occurrence value with respect to other poses in the set. We select the $R\%$ of poses with the lowest occurrence values in the testing set for evaluation.

5   Experiments

5.1   Ablation Study

All our ablation studies are evaluated on the Human3.6M dataset. Both the *Subject Protocol* and our new *Rare Pose Protocol* are used for evaluation. MPJPE serves as the evaluation metric. *Please refer to the supplementary material for details on data pre-processing and training settings.*

Table 2. Comparing the *SR* network to different baseline networks under the *Subject* protocol and the *Rare Pose* protocol (with 10% and 20% of the rarest poses). MPJPE is used as the evaluation metric. The improvements of *SFS* from *FC*, and of *SR* from *SFS*, are shown as subscripts.

Protocol	GP	FC	LF	ES	SFS	Ours (SR)
Subject (100%)	62.7	<u>46.8</u>	39.8	42.0	<u>39.4</u> $\downarrow_{7.4}$	**36.6** $\downarrow_{2.8(7.1\%)}$
Rare Pose (20%)	89.1	<u>76.0</u>	60.1	62.5	<u>59.2</u> $\downarrow_{16.8}$	**48.6** $\downarrow_{10.6(17.9\%)}$
Rare Pose (10%)	98.9	<u>89.1</u>	69.9	73.5	<u>68.2</u> $\downarrow_{20.9}$	**53.7** $\downarrow_{14.5(21.3\%)}$

Effect of the SRNet. In Table 2, we compare our *SR* network with the baseline networks *FC*, *GP*, *LF*, *ES*, *SFS* described in Sect. 3. By default, the mapped dimensions H is set to 1 and the combination operator ∘ is set to multiplication.

It is found that both the *LF* and *ES* baselines are superior to the *FC* and *GP* baselines. A more detailed comparison is shown in Fig. 4, with different splitting configurations for each. In Fig. 4, both *LF* and *ES* start on the left with the *FC* network ($L_{fuse} = 0$ and $L_{split} = 6$). When the fully connected layers are gradually replaced by group connected layers, from the beginning of the network for *LF* (L_{fuse} ↑) or the end for *ES* (L_{split} ↓), testing error decreases at first. This indicates that *local feature learning* and *local inference* are each helpful. However, replacing all the fully connected layers with group layers leads to a sharp performance drop. This indicates that a local joint region still needs information from the other regions to be effective.

Combining the *LF* and *ES* baselines to create *SFS* yields further improvements from sharing the merits of both. These two simple modifications already improve upon *FC* by a large margin. Specifically, it improves upon *FC* by 7.4 mm (relative 15.8%), 16.8 mm (relative 22.1%) and 20.9 mm (relative 23.5%) on the *Subject, Rare Pose 20%* and *Rare Pose 10%* protocols, respectively.

SRNet with its use of low-dimensional global context performs significantly better than the *SFS* baseline. Specifically, it improves upon *SFS* by 2.8 mm (relative 7.1%), 10.6 mm (relative 17.9%) and 14.5 mm (relative 21.2%) on the *Subject, Rare Pose 20%* and *Rare Pose 10%* protocols, respectively. Note that the improvement is especially large on rare poses, indicating the stronger generalization ability of *SRNet* to such poses. Figure 5 breaks down the results of

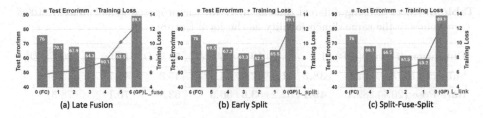

Fig. 4. Training loss and test error (MPJPE in mm) of the baseline networks (a) *LF*, (b) *ES*, and (c) *SFS* with respect to different splitting configurations. Rare Pose Protocol 20% is used for evaluation.

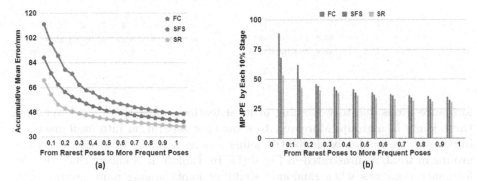

Fig. 5. Testing errors of *FC*, *SFS* and our *SR* networks on poses with different degrees of rareness. The horizontal axis represents the top $R\%$ of rarest poses. (a) Cumulative mean error. (b) Mean error at 10% intervals.

the *FC*, *SFS* and *SR* networks for different degrees of pose rareness. It can be seen that the performance improvements of SRNet increase for rarer poses.

Low-Dimensional Representation for Global Context. Table 3 presents the performance of SRNet with different dimensions H for the global context. To readily accommodate different dimensions, the combination operator ∘ is set to concatenation. When $H = 0$, *SRNet* degenerates into the *GP* baseline. It can be seen that the *GP* baseline without any global context obtains the worst performance. This shows the importance of bringing global context back to local groups. Also, it is found that a lower dimension of global context leads to better results. The best result is achieved by keeping *only a single dimension* for global context. This indicates that a low-dimensional representation of the global context can better account for the information of less relevant joints while preserving effective local feature learning.

Effect of Grouping Strategy. We compare the results of using different numbers of local groups in Table 4. The corresponding sets of joints for each group are shown in Fig. 6. More joint groups leads to greater local correlation of joints within a group. It is shown that the performance first improves with more groups, especially in the *Rare Pose* case. However, with more than five groups, the per-

Table 3. Effect of using different dimension H for the proposed global context. 100% corresponds to the same dimensionality as the local feature.

Dimension H	0(GP)	1	3	10	25% D_g^l	50% D_g^l	100% D_g^l	200% D_g^l
Subject (100%)	62.7	**38.3**	41.9	42.9	44.6	45.3	45.7	45.8
Rare Pose (20%)	89.1	**49.1**	52.2	53.6	56.5	63.8	70.1	78.5

Table 4. Mean testing error with different group numbers for *Subject* and *Rare Pose* 20% protocols.

Group Num	1(FC)	2	3	5	6	8	17
Subj.(100%)	46.8	41.4	37.7	**36.6**	41.1	46.7	91.5
Rare(20%)	76.0	55.9	51.0	**48.6**	55.6	60.6	115.3

Table 5. Mean testing error with respect to the number of shuffled groups (5 in total).

Shuffled Groups	5	4	3	2	0(Ours)
Subject (100%)	53.8	49.7	45.9	43.4	**36.6**

formance drops due to weakening of local features when there are fewer than three joints in a group. Moreover, to show that separation into *local* groups is important, we randomly shuffle the joints among a subset of groups, with five groups in total, as illustrated in Fig. 6 (f). In Table 5, it is shown that the performance decreases when randomly shuffling joints among more groups. This indicates that a strong physical relationship among joints in a group is a prerequisite for learning effective local features.

5.2   Comparison with State-of-The-Art Methods

We compare with the state of the art in three different experiment settings: cross-dataset on MPI-INF-3DHP, single-frame model on Human3.6M, and temporal model on Human3.6M. MPJPE is used as the evaluation metric on Human3.6M. *Please refer to the supplementary material for more results that use PA-MPJPE as the evaluation metric.*

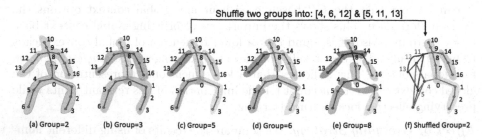

Fig. 6. Different numbers of groups and their corresponding sets of joints. Joints in the same group are shown in the same color. A random shuffle setting is also shown and evaluated in Table 5.

Cross-Dataset Results on MPI-INF-3DHP. In this setting, we follow common practice [5,16,32,34,36] by training the model on the Human3.6M training set and evaluating on the MPI-INF-3DHP test set to examine cross-dataset generalization ability. PCK and AUC are used as evaluation metrics, for which higher is better. In Table 6, our approach achieves the best cross-data generalization performance. It improves upon the state-of-the-art [5] by a large margin (4.9% on PCK and 19.3% on AUC), indicating superior generalization ability.

Table 6. Cross-dataset results on MPI-INF-3DHP. All models are trained on Human3.6M and tested on the MPI-INF-3DHP test set.

Method	Martinez [18]	Mehta [19]	Luo [16]	Biswas [1]	Yang [34]	Zhou [36]	Wang [32]	Ci [5]	Ours
Outdoor	31.2	58.8	65.7	67.4	-	72.7	-	77.3	**80.3** ↑$_{3.9\%}$
PCK	42.5	64.7	65.6	65.8	69.0	69.2	71.2	74.0	**77.6** ↑$_{4.9\%}$
AUC	17.0	31.7	33.2	31.2	32.0	32.5	33.8	36.7	**43.8** ↑$_{19.3\%}$

Table 7. *Cross Action* results compared with *Fully Connected Network* and *Locally Connected Network* on Human3.6M. Smaller values are better.

Method	Direct	Discuss	Eat	Greet	Phone	Photo	Pose	Purcha	Sit	SitD	Smoke	Wait	WalkD	Walk	WalkT	Avg.
Martinez [18]	127.3	104.6	95.0	116.1	95.5	117.4	111.4	125.0	116.9	93.6	111.0	126.0	131.0	106.3	140.5	114.5
Ci [5]	102.8	88.6	79.1	99.3	80.0	91.5	93.2	89.6	90.4	76.6	89.6	102.1	108.8	90.8	118.9	93.4
Ours	92.3	71.4	71.8	86.4	66.8	79.1	82.5	86.6	88.9	93.4	66.1	83.0	74.4	90.0	97.8	82.0

Table 8. Comparison on *single-frame 2D pose detection input* in terms of mean per-joint position error (MPJPE). Best result in bold.

Method	Luvizon [17]	Martinez [18]	Park [22]	Wang [32]	Zhao [35]	Ci [5]	Pavllo [23]	Cai [2]	Ours
Subject (100%)	64.1	62.9	58.6	58.0	57.6	52.7	51.8	50.6	**49.9**

Single-Frame Model Results on Human3.6M. In this setting, we first compare with the state-of-the-art method [5] using the *Cross Action Protocol*. Table 7 shows that our method yields an overall improvement of 11.4 mm (relative 12.2% improvement) over [5] and performs better on 93% of the actions, indicating strong generalization ability on unseen actions. We also compare our model to previous works using the standard *Subject Protocol* in Table 8 and Table 9, which use 2D keypoint detection and 2D ground truth as inputs, respectively. Under the standard protocol, our model surpasses the state-of-the-art in each case, namely [2] for 2D keypoint detection and [5] for 2D ground truth as input.

Temporal Model Results on Human3.6M. In Table 10, our approach achieves the new state-of-the-art with either 2D keypoint detection or 2D ground

Table 9. Comparison on *single-frame 2D ground truth pose input* in terms of MPJPE. With ground truth 2D pose as input, the upper bound of these methods is explored.

Method	Martinez [18]	Pham [24]	Biswas [1]	Zhao [35]	Wang [32]	Ci [5]	Ours
Subject (100%)	45.5	42.4	42.8	43.8	37.6	36.4	**33.9**

truth (with $\triangledown$) as input. Specifically, SRNet improves upon [23] from 37.2 mm to 32.0 mm (relative 14.0% improvement) with 2D ground truth input and from 46.8 mm to 44.8 mm (relative 4.3% improvement) with 2D keypoint detection input. Besides, SRNet has around one fifth parameters 3.61M of [23](16.95M) with 243 frame poses as input.

Table 10. Comparison on *Temporal Pose input* in terms of mean per-joint position error (MPJPE). Below the double line, $\triangledown$ indicates use of 2D ground truth pose as input, which is examined to explore the upper bound of these methods. Best results in bold.

Method	Direct	Discuss	Eat	Greet	Phone	Photo	Pose	Purcha	Sit	SitD	Smoke	Wait	WalkD	Walk	WalkT	Avg.
Hossain et al. [26]	48.4	50.77	57.2	55.2	63.1	72.6	53.0	51.7	66.1	80.9	59.0	57.3	62.4	46.6	49.6	58.3
Lee et al. [14]	**40.2**	49.2	47.8	52.6	50.1	75.0	50.2	43.0	55.8	73.9	54.1	55.6	58.2	43.3	43.3	52.8
Cai et al. [2]	44.6	47.4	45.6	48.8	50.8	59.0	47.2	43.9	57.9	61.9	49.7	46.6	51.3	37.1	39.4	48.8
Pavllo et al. [23]	45.2	46.7	43.3	45.6	48.1	55.1	44.6	44.3	57.3	65.8	47.1	44.0	49.0	32.8	33.9	46.8
Lin et al. [15]	42.5	**44.8**	**42.6**	44.2	48.5	57.1	**42.6**	41.4	56.5	64.5	47.4	43.0	48.1	33.0	35.1	46.6
Ours	46.6	47.1	43.9	**41.6**	**45.8**	49.6	46.5	**40.0**	**53.4**	61.1	46.1	**42.6**	43.1	**31.5**	**32.6**	**44.8**
Hossain et al. [26] $\triangledown$	35.2	40.8	37.2	37.4	43.2	44.0	38.9	35.6	42.3	44.6	39.7	39.7	40.2	32.8	35.5	39.2
Lee et al. [14] $\triangledown$	32.1	36.6	34.3	37.8	44.5	49.9	40.9	36.2	44.1	45.6	35.3	35.9	37.6	30.3	35.5	38.4
Pavllo et al. [23] $\triangledown$	-	-	-	-	-	-	-	-	-	-	-	-	-	-	-	37.2
Ours $\triangledown$	**34.8**	**32.1**	**28.5**	**30.7**	**31.4**	**36.9**	**35.6**	**30.5**	**38.9**	**40.5**	**32.5**	**31.0**	**29.9**	**22.5**	**24.5**	**32.0**

6   Conclusion

In this paper, we proposed SRNet, a split-and-recombine approach that improves generalization performance in 3D human pose estimation. The key idea is to design a network structure that splits the human body into local groups of joints and recombines a low-dimensional global context for more effective learning. Experimental results show that SRNet outperforms state-of-the-art techniques, especially for rare and unseen poses.

References

1. Biswas, S., Sinha, S., Gupta, K., Bhowmick, B.: Lifting 2D human pose to 3D: a weakly supervised approach. In: 2019 International Joint Conference on Neural Networks (IJCNN), pp. 1–9. IEEE (2019)
2. Cai, Y., et al.: Exploiting spatial-temporal relationships for 3D pose estimation via graph convolutional networks. In: Proceedings of the IEEE International Conference on Computer Vision, pp. 2272–2281 (2019)
3. Chen, W., et al.: Synthesizing training images for boosting human 3D pose estimation. In: 2016 4th International Conference on 3D Vision (3DV), pp. 479–488. IEEE (2016)
4. Chen, Y., Wang, Z., Peng, Y., Zhang, Z., Yu, G., Sun, J.: Cascaded pyramid network for multi-person pose estimation. In: Proceedings of the IEEE Conference on Computer Vision and Pattern Recognition, pp. 7103–7112 (2018)
5. Ci, H., Wang, C., Ma, X., Wang, Y.: Optimizing network structure for 3D human pose estimation. In: Proceedings of the IEEE International Conference on Computer Vision, pp. 2262–2271 (2019)
6. Dabral, R., Mundhada, A., Kusupati, U., Afaque, S., Sharma, A., Jain, A.: Learning 3D human pose from structure and motion. In: Proceedings of the European Conference on Computer Vision (ECCV), pp. 668–683 (2018)
7. Fang, H.S., Xu, Y., Wang, W., Liu, X., Zhu, S.C.: Learning pose grammar to encode human body configuration for 3D pose estimation. In: 32nd AAAI Conference on Artificial Intelligence (2018)
8. Habibie, I., Xu, W., Mehta, D., Pons-Moll, G., Theobalt, C.: In the wild human pose estimation using explicit 2D features and intermediate 3D representations. In: Proceedings of the IEEE Conference on Computer Vision and Pattern Recognition, pp. 10905–10914 (2019)
9. Huang, C., Li, Y., Loy, C.C., Tang, X.: Learning deep representation for imbalanced classification. In: Proceedings of the IEEE Conference on Computer Vision and Pattern Recognition (2016)
10. Huang, F., Zeng, A., Liu, M., Lai, Q., Xu, Q.: Deepfuse: an imu-aware network for real-time 3D human pose estimation from multi-view image. arXiv preprint arXiv:1912.04071 (2019)
11. Ionescu, C., Papava, D., Olaru, V., Sminchisescu, C.: Human3.6m large scale datasets and predictive methods for 3D human sensing in natural environments. IEEE Trans. Pattern Anal. Mach. Intell. 36(7), 1325–1339 (2014)
12. Jahangiri, E., Yuille, A.L.: Generating multiple diverse hypotheses for human 3D pose consistent with 2D joint detections. In: Proceedings of the IEEE International Conference on Computer Vision Workshops, pp. 805–814 (2017)
13. Krizhevsky, A., Sutskever, I., Hinton, G.E.: Imagenet classification with deep convolutional neural networks. In: Advances in Neural Information Processing Systems, pp. 1097–1105 (2012)
14. Lee, K., Lee, I., Lee, S.: Propagating lstm: 3D pose estimation based on joint interdependency. In: Proceedings of the European Conference on Computer Vision (ECCV), pp. 119–135 (2018)
15. Lin, J., Lee, G.H.: Trajectory space factorization for deep video-based 3D human pose estimation. arXiv preprint arXiv:1908.08289 (2019)
16. Luo, C., Chu, X., Yuille, A.: Orinet: a fully convolutional network for 3D human pose estimation. arXiv preprint arXiv:1811.04989 (2018)

17. Luvizon, D.C., Picard, D., Tabia, H.: 2D/3D pose estimation and action recognition using multitask deep learning. In: Proceedings of the IEEE Conference on Computer Vision and Pattern Recognition, pp. 5137–5146 (2018)
18. Martinez, J., Hossain, R., Romero, J., Little, J.J.: A simple yet effective baseline for 3D human pose estimation. In: Proceedings of the IEEE International Conference on Computer Vision, pp. 2640–2649 (2017)
19. Mehta, D., et al.: Monocular 3D human pose estimation in the wild using improved cnn supervision. In: 2017 International Conference on 3D Vision (3DV), pp. 506–516. IEEE (2017)
20. Mehta, D., et al.: Vnect: real-time 3D human pose estimation with a single RGB camera. ACM Trans. Graph. (TOG) **36**(4), 1–14 (2017)
21. Newell, A., Yang, K., Deng, J.: Stacked hourglass networks for human pose estimation. In: Leibe, B., Matas, J., Sebe, N., Welling, M. (eds.) ECCV 2016. LNCS, vol. 9912, pp. 483–499. Springer, Cham (2016). https://doi.org/10.1007/978-3-319-46484-8_29
22. Park, S., Kwak, N.: 3D human pose estimation with relational networks. arXiv preprint arXiv:1805.08961 (2018)
23. Pavllo, D., Feichtenhofer, C., Grangier, D., Auli, M.: 3D human pose estimation in video with temporal convolutions and semi-supervised training. In: Proceedings of the IEEE Conference on Computer Vision and Pattern Recognition, pp. 7753–7762 (2019)
24. Pham, H.H., Salmane, H., Khoudour, L., Crouzil, A., Zegers, P., Velastin, S.A.: A unified deep framework for joint 3D pose estimation and action recognition from a single RGB camera. arXiv preprint arXiv:1907.06968 (2019)
25. Pishchulin, L., Jain, A., Andriluka, M., Thorm ahlen, T., Schiele, B.: Articulated people detection and pose estimation: reshaping the future. In: Proceedings of the IEEE Conference on Computer Vision and Pattern Recognition (2012)
26. Rayat Imtiaz Hossain, M., Little, J.J.: Exploiting temporal information for 3D human pose estimation. In: Proceedings of the European Conference on Computer Vision (ECCV), pp. 68–84 (2018)
27. Rogez, G., Schmid, C.: Mocap-guided data augmentation for 3D pose estimation in the wild. In: Advances in Neural Information Processing Systems, pp. 3108–3116 (2016)
28. Sun, K., Xiao, B., Liu, D., Wang, J.: Deep high-resolution representation learning for human pose estimation. In: Proceedings of the IEEE Conference on Computer Vision and Pattern Recognition, pp. 5693–5703 (2019)
29. Varol, G., et al.: Learning from synthetic humans. In: Proceedings of the IEEE Conference on Computer Vision and Pattern Recognition, pp. 109–117 (2017)
30. Véges, M., Varga, V., Lőrincz, A.: 3D human pose estimation with siamese equivariant embedding. Neurocomputing **339**, 194–201 (2019)
31. Wandt, B., Rosenhahn, B.: Repnet: weakly supervised training of an adversarial reprojection network for 3D human pose estimation. In: Proceedings of the IEEE Conference on Computer Vision and Pattern Recognition, pp. 7782–7791 (2019)
32. Wang, L., et al.: Generalizing monocular 3D human pose estimation in the wild. arXiv preprint arXiv:1904.05512 (2019)
33. Wang, Y.X., Ramanan, D., Hebert, M.: Learning to model the tail. In: Conference on Neural Information Processing Systems (2017)
34. Yang, W., Ouyang, W., Wang, X., Ren, J., Li, H., Wang, X.: 3D human pose estimation in the wild by adversarial learning. In: Proceedings of the IEEE Conference on Computer Vision and Pattern Recognition, pp. 5255–5264 (2018)

35. Zhao, L., Peng, X., Tian, Y., Kapadia, M., Metaxas, D.N.: Semantic graph convolutional networks for 3D human pose regression. In: Proceedings of the IEEE Conference on Computer Vision and Pattern Recognition, pp. 3425–3435 (2019)
36. Zhou, X., Huang, Q., Sun, X., Xue, X., Wei, Y.: Towards 3D human pose estimation in the wild: a weakly-supervised approach. In: Proceedings of the IEEE International Conference on Computer Vision, pp. 398–407 (2017)

CAFE-GAN: Arbitrary Face Attribute Editing with Complementary Attention Feature

Jeong-gi Kwak[1], David K. Han[2], and Hanseok Ko[1(✉)]

[1] Korea University, Seoul, Korea
hsko@korea.ac.kr
[2] Army Research Laboratory, Adelphi, MD, USA

Abstract. The goal of face attribute editing is altering a facial image according to given target attributes such as hair color, mustache, gender, etc. It belongs to the image-to-image domain transfer problem with a set of attributes considered as a distinctive domain. There have been some works in multi-domain transfer problem focusing on facial attribute editing employing Generative Adversarial Network (GAN). These methods have reported some successes but they also result in unintended changes in facial regions - meaning the generator alters regions unrelated to the specified attributes. To address this unintended altering problem, we propose a novel GAN model which is designed to edit only the parts of a face pertinent to the target attributes by the concept of Complementary Attention Feature (CAFE). CAFE identifies the facial regions to be transformed by considering both target attributes as well as "complementary attributes", which we define as those attributes absent in the input facial image. In addition, we introduce a complementary feature matching to help in training the generator for utilizing the spatial information of attributes. Effectiveness of the proposed method is demonstrated by analysis and comparison study with state-of-the-art methods.

Keywords: Face attribute editing · GAN · Complementary attention feature · Complementary feature matching

1 Introduction

Since the advent of GAN by Goodfellow *et al.* [8], its application literally exploded into a variety of areas, and many variants of GAN emerged. Conditional GAN (CGAN) [25], one of the GAN variants, adds an input as a condition on how the synthetic output should be generated. An area in CGAN receiving a particular attention in the media is "Deep Fake" in which an input image

Electronic supplementary material The online version of this chapter (https://doi.org/10.1007/978-3-030-58568-6_31) contains supplementary material, which is available to authorized users.

© Springer Nature Switzerland AG 2020
A. Vedaldi et al. (Eds.): ECCV 2020, LNCS 12359, pp. 524–540, 2020.
https://doi.org/10.1007/978-3-030-58568-6_31

| Original | AttGAN | StarGAN | STGAN | CAFE-GAN (Ours) |

Fig. 1. Face editing results of AttGAN [10], StarGAN [6], STGAN [21] and our model given a target attribute *Blond hair*. While AttGAN, StarGAN and STGAN deliver blond hair, they also create unwanted changes (e.g.. halo, different hair style, etc.) in resultant images.

is transformed into a new image of different nature while key elements of the original image are retained and transposed [11,25]. GAN based style transfer is often the method of choice for achieving the domain transfer of the input to another domain in the output, such as generating a hypothetical painting of a well known artist from a photograph. CycleGAN [40] has become a popular method in image domain transfer, because it uses cycle-consistency loss from a single image input in training, and thus its training is unsupervised.

Nevertheless, single domain translation models [31,37] including CycleGAN are incapable of learning domain transfer to multiple domains. Thus, in these approaches, multiple models are required to match the number of target domains. One of the multi-domain transfer problems is manipulation of one's facial characteristics. The goal of facial attribute editing is to convert specific attributes of a face, such as wearing eyeglasses, to a face without eyeglasses. Other attributes may include local properties e.g., beard as well as global properties e.g.., face aging. Obviously this requires multiples of domain transfer models if a single domain transfer concept is to be used. The number of single domain transfer models required in such a case is a function of the attribute combination since these facial attributes are mostly independent. Even for relatively small number of attributes, single domain transfer approaches would require a significantly high number of separate models. A model such as CycleGAN, therefore, would become impractical.

To address the multi-domain transfer problem, StarGAN [6] and AttGAN [10], have been proposed by introducing a target vector of multiple attributes as an additional input. These target attribute vector based GAN models have shown some impressive images, but, they often result in unintended changes - meaning the generator alters regions unrelated to specified attributes as shown in Fig. 1. It stems from these models driven to achieve higher objective scores in classifying attributes at the expense of affecting areas unrelated to the target attribute. Some methods [31,37] have used a strategy that adds only the values of attribute-specific region to an input image, but these methods have achieved limited success. Hence, changing only pertinent regions remains as an important but challenging problem. Their limitation stems from considering only structural improvement in the generator. However, more effective approach may be possible by exploring decision making process of the discriminator.

Visual explanation [7,30,34,39], which is known effective in interpreting Convolutional Neural Network (CNN) by highlighting response areas critical in recognition, is considered here to address the problem. Our model is mainly motivated by Attention Branch Network (ABN) [7] which extends response-based visual explanation to an attention mechanism. In ABN, the attention branch takes mid-level features then extracts attention feature maps. Attention feature maps are then downsampled through a global average pooling layer (GAP) [20] and subsequently utilized as class probability. However, the problem with the response based visual explanation methods is that they can only extract attention feature maps of the attributes already present in the image. Thus, these methods are effective only in manipulations of existing attributes, such as removing beard or changing hair color.

To address this issue, we propose a method of identifying the regions of attributes that are not present in an input image, via a novel concept of Complementary Attention FEature (CAFE). With the idea of creating a complementary attribute vector, CAFE identifies the regions to be transformed according to the input attributes even when the input image lacks the specified attributes. With CAFE, our discriminator can generate and exploit the spatial attention feature maps of all attributes. We will demonstrate CAFE's effectiveness both in local as well as in global attributes in generating plausible and realistic images.

Our contributions are as follows:

- We present a novel approach for facial attribute editing designed to only edit the parts of a face pertinent to the target attributes based on the proposed concept of Complementary Attention FEature (CAFE).
- We introduce a complementary feature matching loss designed to aid in training the generator for synthesizing images with given attributes rendered accurately and in appropriate facial regions.
- We demonstrate effectiveness of CAFE in both local as well as global attribute transformation with both qualitative and quantitative results.

2   Related Work

Generative Adversarial Networks. Since Goodfellow et al. [8] proposed Generative Adversarial Network (GAN), GAN-based generative models have attracted significant attention because of their realistic output. However, the original formulation of GAN suffers from training instability and mode collapse. Numerous studies have been made to address these problems in various ways such as formulating alternative objective function [2,9,24] or developing modified architecture [12,13,29]. Several conditional methods [11,25] have extended GAN to image-to-image translation. CycleGAN [40] proposed the use of cycle consistency loss to overcome the lack of paired training data. Advancing from single domain transfer methods, StarGAN [6] can handle image translation among multiple domains. These developments enabled GANs to deliver some remarkable results in various tasks such as style transfer [3], super-resolution [16,18], and many other real-world applications [1,26,28].

Face Attribute Editing. The goal of face attribute editing is to transform the input facial image according to a given set of target attributes. Several methods have been proposed with Deep Feature Interpolation (DFI) scheme [4,5,35]. By shifting deep features of an input image with a certain direction of target attributes, a decoder takes interpolated features and outputs an edited image. Although they produce some impressive results, limitation of these methods is that they require a pre-trained encoder such as VGG network [32] besides they have a weakness in multi-attribute editing. Recently, GAN based frameworks have become the dominant form of face attribute manipulation. A slew of studies for single attribute editing [14,19,22,31,37,40] have been conducted. However these methods cannot handle manipulation of multiple attributes with a unified model. Several efforts [17,27] have been extended to an arbitrary attribute editing but they achieved limited quality of image. Several methods [6,10,38] have shown remarkable results in altering multiple attributes by taking the target attribute vector as an input of their generator or adopting additional network. STGAN [21] and RelGAN [36] further improved face editing ability by using difference between a target and a source vector to constrain in addressing selected attributes. However these methods still suffer from change of irrelevant regions. SaGAN [37] exploits spatial attention to solve the problem, but it is only effective for editing local attributes like adding mustache.

Interpreting CNN. Several researches [7,30,33,34,39] have visualized the decision making of CNN by highlighting important region. Gradient-based visual explanation methods [30,33,39] have been widely used because they are applicable to pre-trained models. Nonetheless, these methods are inappropriate for providing spatial information to our discriminator because they require back propagation to obtain attention maps and are not trainable jointly with a discriminator. In addition to gradient-based methods, several response based methods [7,39] have been proposed for visual explanation. They obtain attention map using only response of feed forward propagation. ABN [7] combines visual explanation and attention mechanism by introducing an attention branch. Therefore, we adopt ABN to guide attention features in our model. However, there is a problem when applying ABN in our discriminator because it can visualize only attributes present in the input image. We combine ABN and arbitrary face attribute editing by introducing the concept of complementary attention feature to address the difficulty of localizing facial regions when the input image doesn't contain the target attribute.

3 CAFE-GAN

This section presents our proposed CAFE-GAN, a framework to address arbitrary image editing. Figure 2 shows an overview of our model which consists of a generator G and a discriminator D. We first describe our discriminator that recognizes attribute-relevant regions by introducing the concept of complementary attention feature (CAFE), and then describe how the generator learns to reflect the spatial information by describing complementary feature matching. Finally, we present our training implementation in detail.

3.1 Discriminator

The discriminator D takes both real images and fake images modified by G as input. D consists of three main parts, i.e., D_{att}, D_{adv}, and D_{cls} as illustrated in right side of Fig. 2. Unlike other arbitrary attribute editing methods [6,10, 21], a spatial attention mechanism is applied to mid-level features f in our discriminator. D_{att} plays a major role in applying attention mechanism with a novel concept of adopting complementary feature maps. D_{att} consists of an attention branch (AB) and a complementary attention branch (CAB) and they generate k attention maps, which are the number of attributes, respectively. $M = \{M_1, ..., M_k\}$, a collection of attention maps from AB, contains important regions of attributes that exist in an input image while $M^c = \{M_1^c, ..., M_k^c\}$ from CAB contains casual region of attributes that do not exist. These attention maps are applied to mid-level features by the attention mechanism as

$$f_i' = f \cdot M_i, \tag{1}$$
$$f_i'' = f \cdot M_i^c, \tag{2}$$

where M_i and M_i^c are the i-th attention map from AB and CAB respectively and ($\cdot$) denotes element-wise multiplication.

The following paragraph describes D_{att} in detail as illustrated in Fig. 3. As explained above, we adopt Attention Branch (AB) to identify attribute-relevant

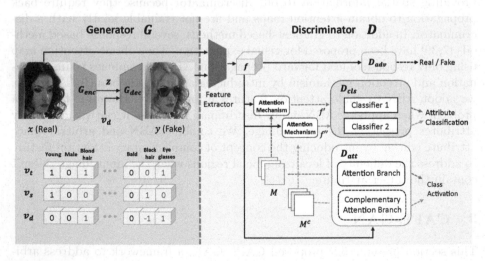

Fig. 2. Overview of our model. On the left side is the generator G which edits source image according to given target attribute vector. G consists of G_{enc} and G_{dec}. Then the discriminator D takes both source and edited image as input. D consists of feature extractor, D_{att}, D_{adv} and D_{cls}.

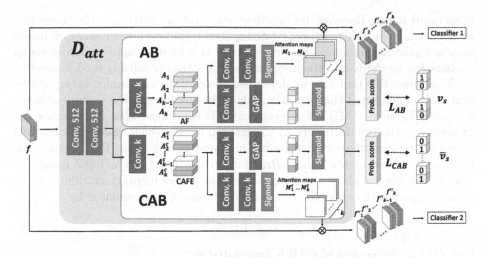

Fig. 3. Details on structure of attention branch (AB) and complementary attention branch (CAB) in D_{att}.

regions following ABN [7]. AB takes mid-level features of an input image and generate $h \times w \times k$ attention features (AF), denoted by A, with a $1 \times 1 \times k$ convolution layer. Here, k denotes the number of channels in A which is the same as the number of attributes. h and w denote the height and width of the feature map respectively. AB outputs k attention maps $M_1, ... M_k$ with $1 \times 1 \times k$ convolution layers and a sigmoid layer. It also outputs activation of each attribute class by global average pooling (GAP) [20]. The $h \times w \times k$ attention feature map A is converted to $1 \times 1 \times k$ feature map by GAP to produce probability score of each class with a sigmoid layer. The probability score is compared to label v_s with a cross-entropy loss to train D to minimize classification errors when a real image (source image) is given as an input. Therefore, attention loss of AB is defined as

$$\mathcal{L}_{D_{AB}} = -\mathbb{E}_x \sum_{i=1}^{k} \left[v_s^{(i)} \log D_{AB}^{(i)}(x) + (1 - v_s^{(i)}) \log(1 - D_{AB}^{(i)}(x)) \right], \quad (3)$$

where x is real image, $v_s^{(i)}$ denotes the i-th value of source attribute vector and $D_{AB}^{(i)}(x)$ denotes the i-th probability score that is output of AB. Therefore, the values of each channel in A are directly related to activation of the corresponding attribute. AB can extract A which represents spatial information about attributes contained in the input image. However A does not include the information about attributes that are not present in the image because A_i, the i-th channel of feature map A, does not have response if i-th attribute is not in the input image.

This aspect does not influence the classification models like ABN because it only needs to activate a channel that corresponds to the correct attribute shown in an input image. However, for handling any arbitrary attribute, the generative model needs to be able to expect related spatial region even when an input image does not possess the attribute. Hence, there is a limit to apply existing visual explanation methods into the discriminator of attribute editing model directly.

To address this problem, we developed the concept of complementary attention and implemented the idea in our algorithm by integrating Complementary Attention Branch (CAB).The concept of CAB is intuitive that it extracts complementary attention feature (CAFE), denoted by A^c, which represents causative region of attribute that are not present in image. For example, CAB detects the lower part of face about attribute *Beard* if the beard is not in an input image. To achieve this inverse class activation, we exploit complementary attribute vector $\bar{v}_s$ to compare with probability score from A^c and $\bar{v}_s$ is formulated by

$$\bar{v}_s = 1 - v_s, \tag{4}$$

hence the attention loss of CAB is formulated as

$$\mathcal{L}_{D_{CAB}} = -\mathbb{E}_x \sum_{i=1}^{k} \left[(1 - v_s^{(i)}) \log D_{CAB}^{(i)}(x) + v_s^{(i)} \log(1 - D_{CAB}^{(i)}(x)) \right], \tag{5}$$

where, $D_{CAB}^{(i)}(x)$ denotes the i-th probability score that is output of CAB. CAB is designed to generate a set of attention maps M^c for attention mechanism from A^c. Therefore A^c should contain spatial information to help D_{cls} classify attributes. In other words, A^c represents causative regions of non-existing attribute. With AB and CAB, our model extracts attention feature map about all attributes because A and A^c are complementary. In other words, for any i-th attribute, if A_i does not have response values, A_i^c has them and vice versa.

Two groups of attention maps M and M^c, outputs of AB and CAB respectively, have different activation corresponding to attributes. In other words, M is about existing attributes of the input image while M^c is about absent attributes of that. After attention mechanism, the transformed features are forwarded to two multi-attribute classifiers in D_{cls} and the classifier 1 and classifer 2 classify correct label of image with f' and f'' respectively. Each classifier outputs the probability of each attribute with cross-entropy loss. For discriminator, it learns to classify real image x with two different attention mechanism, i.e.,

$$\mathcal{L}_{D_{cls}} = -\mathbb{E}_x \sum_{n=1,2} \sum_{i=1}^{k} \left[v_s^{(i)} \log D_{cls_n}^{(i)}(x) + (1 - v_s^{(i)}) \log(1 - D_{cls_n}^{(i)}(x)) \right], \tag{6}$$

where D_{cls_1} and D_{cls_2} stand for two classifiers using collections of attention maps $M = \{M_1, ..., M_k\}$ and $M^c = \{M_1^c, ..., M_k^c\}$, respectively. Therefore, the reason CAFE can represent the spatial information of non-existent attributes is that CAB has to generate the attention maps that can help to improve performance of the classifiers while reacting to non-existent attributes by GAP.

In D, there is another branch D_{adv} distinguishes real image x and fake image y in order to guarantee visually realistic output with adversarial learning. In particular, we employ adversarial loss in WGAN-GP [9], hence the adversarial loss of D is given as

$$\mathcal{L}_{D_{adv}} = \mathbb{E}_x(D_{adv}(x)) - \mathbb{E}_y(D_{adv}(y)) - \lambda_{gp}\mathbb{E}_{\hat{x}}\left[(\|\nabla_{\hat{x}}D_{adv}(\hat{x})\|_2 - 1)^2\right], \quad (7)$$

where $\hat{x}$ is weighted sum of real and fake sample with randomly selected weight $\alpha \in [0, 1]$.

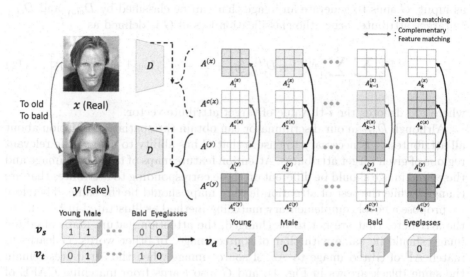

Fig. 4. Details on the proposed complementary feature matching with an example. G learns to match AF and CAFE of edited image corresponding to attribute to be changed, i.e., *To old* and *To bald*, to CAFE and AF of source image, respectively. On the contrary, G learns to match AF and CAFE of edited image to AF and CAFE of source image for the attribute not to be changed.

3.2   Generator

The generator G takes both the source image x and the target attribution label v_t as input, and then conducts a transformation of x to y, denoted by $y = G(x, v_t)$. The goal of G is to generate an image y with attributes according to v_t while maintaining the identity of x. G consists of two components: an encoder G_{enc} and a decoder G_{dec}. From the given source image x, G_{enc} encodes the image into a latent representation z. Then, the decoder generates a fake image y with the latent feature z and the target attribute vector v_t. Following [21], we compute a difference attribute vector v_d between the source and the target attribute vectors and use it as an input to our decoder.

$$v_d = v_t - v_s, \quad (8)$$

Thus, the process can be expressed as

$$z = G_{enc}(x), \tag{9}$$
$$y = G_{dec}(z, v_d). \tag{10}$$

In addition, we adopt the skip connection methodology used in STGAN [21] between G_{enc} and G_{dec} to minimize loss of fine-scale information due to down sampling. After that, D takes both the source image x and the edited image y as input. G aims to generate an image that can be classified by D_{cls_1} and D_{cls_2} as target attribute, hence the classification loss of G is defined as

$$\mathcal{L}_{G_{cls}} = -\mathbb{E}_y \sum_{n=1,2} \sum_{i=1}^k \left[v_t^{(i)} \log D_{cls_n}^{(i)}(y) + (1 - v_t^{(i)}) \log(1 - D_{cls_n}^{(i)}(y)) \right], \tag{11}$$

where $v_t^{(i)}$ denotes the i-th value of target attribute vector.

Although D_{att} in our discriminator can obtain the spatial information about all attributes, it is necessary to ensure that G has ability to change the relevant regions of given target attributes. Attention feature maps of the source image and the edited image should be different on those corresponding to attributes that are changed while the rest of attention feature maps should be the same. Therefore we propose a novel complementary matching method as illustrated in Fig. 4. For the attributes that are not to be changed, the attention feature maps of edited image should be same with that of source image. In other words, G learns to match AF of edited image to AF of source image when the attributes remain the same (black arrows in Fig. 4), and G also learns from matching CAFE of the source image. When the given target attributes are different from the source image, G learns to match AF of edited image to CAFE of source image (red arrows in Fig. 4), same with CAFE of edited image to AF of source image. Let $\{A^{(x)}, A^{c(x)}\}$ and $\{A^{(y)}, A^{c(y)}\}$ denote set of AF and CAFE from two different samples, real image x and fake image y, respectively. Complementary matching is conducted for changed attributes and thus the complementary matching loss is defined as

$$\mathcal{L}_{CM} = \mathbb{E}_{(x,y)} \sum_{i=1}^k \frac{1}{N_i} [\|A_i^{(x)} - P_i\|_1 + \|A_i^{c(x)} - Q_i\|_1],$$

$$\text{where } \{P_i, Q_i\} = \begin{cases} \{A_i^{(y)}, A_i^{c(y)}\} & \text{if } |v_d^{(i)}| = 0, \\ \{A_i^{c(y)}, A_i^{(y)}\} & \text{if } |v_d^{(i)}| = 1, \end{cases} \tag{12}$$

where k is the number of attributes and $N(i)$ denotes the number of elements in feature map. $A_i^{(x)/(y)}$ and $A_i^{c(x)/(y)}$ denote i-th channel of $A^{(x)/(y)}$ and $A^{c(x)/(y)}$ respectively and $v_d^{(i)}$ denotes i-th value of difference attribute vector v_d.

For adversarial training of GAN, we also adopt the adversarial loss to generator used in WGAN-GP [9], i.e.,

$$\mathcal{L}_{G_{adv}} = \mathbb{E}_{x,v_d}[D_{adv}(G(x, v_d))], \tag{13}$$

Although the generator can edit face attribute with $\mathcal{L}_{G_{cls}}$ and generate realistic image with $\mathcal{L}_{G_{adv}}$, it should preserve identity of image. Therefore, G should reconstruct the source image when difference attribute vector is zero. We adopt the pixel-level reconstruction loss, i.e.,

$$\mathcal{L}_{rec} = \mathbb{E}_x[\|x - G(x, \mathbf{0})\|_1], \tag{14}$$

where we use ℓ_1 loss for sharpness of reconstructed image and $\mathbf{0}$ denotes zero vector.

3.3   Model Objective

Finally, the full objective to train discriminator D is formulated as

$$\mathcal{L}_D = -\mathcal{L}_{D_{adv}} + \lambda_{att}\mathcal{L}_{D_{att}} + \lambda_{D_{cls}}\mathcal{L}_{D_{cls}}, \tag{15}$$

and that for the generator G is formulated as

$$\mathcal{L}_G = \mathcal{L}_{G_{adv}} + \lambda_{CM}\mathcal{L}_{CM} + \lambda_{G_{cls}}\mathcal{L}_{G_{cls}} + \lambda_{rec}\mathcal{L}_{rec}, \tag{16}$$

where $\lambda_{att}, \lambda_{D_{cls}}, \lambda_{CM}, \lambda_{G_{cls}}$ and λ_{rec} are hyper-parameters which control the relative importance of the terms.

4   Experiments

In this section, we first explain our experimental setup and then present qualitative and quantitative comparisons of our model with the state-of-the-art models. Finally, we demonstrate effectiveness of CAFE with visualization results and ablation study. The experiments not included in this paper can be found in supplementary material.

4.1   Experimental Setup

We use CelebFaces Attributes (CelebA) dataset [23] which consists of 202,599 facial images of celebrities. Each image is annotated with 40 binary attribute labels and cropped to 178×218. We crop each image to 170×170 and resize to 128×128. For comparison, we choose the same attributes used in the state-of-the-art models [6,10,21]. In our experiment, coefficients of the objective functions in Eq. (15) and (16) are set to $\lambda_{att} = \lambda_{D_{cls}} = \lambda_{CM} = 1, \lambda_{G_{cls}} = 10$, and $\lambda_{rec} = 100$. We adopt ADAM [15] solver with $\beta_1 = 0.5$ and $\beta_2 = 0.999$, and the learning rate is set initially to 0.0002 and set to decay to 0.0001 after 100 epochs.

4.2   Qualitative Result

First, we conduct qualitative analysis by comparing our approach with three state-of-the-art methods in arbitrary face editing, i.e., AttGAN [10], StarGAN [6] and STGAN [21]. The results are shown in Fig. 5. Each column represents different attribute manipulation and each row represents qualitative results from the methods compared. The source image is placed in the leftmost of each row and we analyze results about single attribute as well as multiple attributes. First, AttGAN [10] and StarGAN [6] perform reasonably on attributes such as *Add bangs*. However they tend to edit irrelevant regions for attributes such as *Blond hair* or *Pale skin* and they also result in blurry images for attributes such as *To bald*. STGAN [21] improves manipulating ability by modifying structure of the generator, but this model also presents unnatural images for some attributes like *To Bald* and *Add Bangs*. In addition, unwanted changes are inevitable for some attributes such as *Blond hair*. As shown in Fig. 5, our model can successfully convert local attributes like *Eyeglasses* as well as global attributes like *To female*

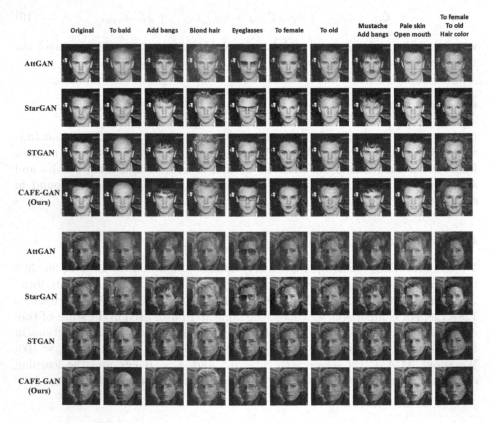

Fig. 5. Qualitative comparison with arbitrary facial attribute editing models. The row from top to down are result of AttGAN [10], StarGAN [6], STGAN [21] and our model. **Please zoom in to see more details.**

and *To old.* Last three columns represent results of multi-attribute editing. *Hair color* in the last column denotes *Black ↔ Brown hair.* It can be seen that our model delivers more natural and well-edited images than the other approaches.

We also compare our model with SaGAN [37] which adopt spatial attention concept. As shown in Fig. 6, SaGAN could not edit well in global attributes like *To male* and *To old.* Even when the target attribute is on a localized region like *Eyeglasses,* it performed poorly. However, our model calculates the spatial information in feature-level, not in pixel-level. As a result, our model shows well-edited and natural results in both local and global attributes. In addition, note that SaGAN is a single-attribute editing model, wherein it requires one model to be trained per attribute.

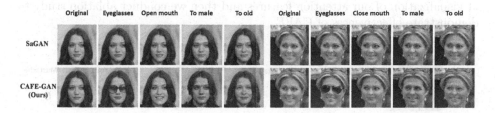

Fig. 6. Qualitative comparison with SaGAN [37] which adopt spatial attention in the generator. **Please zoom in to see more details.**

Table 1. Comparisons on the attribute classification accuracy. Numbers indicate the classification accuracy on each attribute.

	Bald	Bangs	Blond h	Musta	Gender	Pale s	Aged	Open m	Avg
AttGAN [10]	23.67	91.08	41.51	21.78	82.85	86.28	65.69	96.91	63.72
STGAN [21]	59.76	95.48	79.98	**42.10**	92.70	97.81	85.86	**98.65**	81.54
CAFE-GAN	**79.03**	**98.59**	**88.14**	40.13	**95.22**	**98.20**	**88.61**	97.15	**85.64**

4.3 Quantitative Result

We present quantitative results by comparison with two of the three models compared earlier [10,21]. In the absence of the ground-truth, the success of arbitrary attribute editing can be determined by a well trained multi-label classifier, i.e., well edited image would be classified to target domain by a well-trained attribute classifier. For fair comparison, we adopted the classifier that was used to evaluate attribute generation score for STGAN [21] and AttGAN [10]. To evaluate quantitative result, we use official code which contains network architecture and weights of parameter in well-trained model. We exclude StarGAN [6] in this section because the official code of StarGAN provides only few attributes. There are 2,000 edited images evaluated for each attribute and their source images come from the test set in CelebA dataset. We measure the classification score of each attribute of edited images and they are listed in Table 1. In Table 1, "Blond h." and "Open m." denote *Blond hair* and *Open mouth* respectively. While our model shows competitive scores on the average for many attributes, it also delivered overwhelming results compared to the other models for specific attributes such as *Bald*. For attributes such as *Mustache* and *Open mouth*, STGAN results in slightly better performance over our model.

4.4   Analysis of CAFE

This section presents analysis of the proposed method. First we show the result of visualization of our attention features and then we conduct ablation study to demonstrate effectiveness of CAFE.

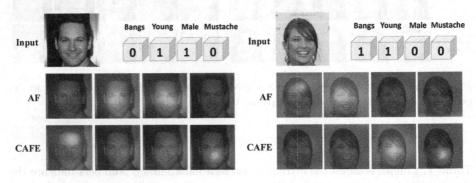

Fig. 7. Visualization results of the attention features (AF) and the complementary attention features (CAFE).

Visualization of CAFE. Figure 7 shows visualization results of AF and CAFE to examine whether they activate related region correctly. Because the man in left in Fig. 7 has no bangs, AF rarely activates but CAFE activates and highlights the related region correctly. The result on the right shows AF correctly activating the region relevant to bangs while CAFE doesn't. Since CAFE only finds complement features absent in the specified attributes, not activating the region specified in the attribute is the correct response. The remainder of the figure demonstrates that CAFE lights up the regions complementary to the given attributes accurately. For global attributes like *Young* and *Male*, both AF and CAFE respond correctly. Although AF and CAFE does not detect attribute-relevant regions at pixel level since they are considered at feature-level, they highlight the corresponding regions accurately per given attribute. As such, our model performs better on editing both global and local attributes compared with other methods.

Ablation Study. To demonstrate effectiveness of the proposed method, we evaluate performance of our model by excluding key components one by one. We compare our model with two different versions, i.e., (i) CAFE-GAN w.o. CM: excluding complementary matching loss ($\mathcal{L}_{CM}$) in training process and (ii) CAFE-GAN w.o. CAB: removing complementary attention branch (CAB) in our discriminator. As shown in Fig. 8, the generator has a difficulty to determine where to change without complementary matching loss though the discriminator can extract AF and CAFE. Some results from the model without CM show unwanted and over-emphasized changes. Excluding CAB leads to artifacts and unwanted changes in generated images because the discriminator has limited spatial information about the attributes that are contained in the input image We also measure classification accuracy of each attribute by the pre-trained classifier which was used in Sect. 4.3 and the results are listed in Table 2. In the absence of CM or CAB, the classification accuracy decreases in all attributes except *Open mouth*, and the model without CAB shows the lowest accuracy.

Fig. 8. Qualitative results of CAFE-GAN variants.

Table 2. Attribute classification accuracy of ablation study.

	Bald	Bangs	Blond h	Musta	Gender	Pale s	Aged	Open m	Avg
Ours	**79.03**	**98.59**	**88.14**	**40.13**	**95.22**	**98.20**	**88.61**	97.15	**85.64**
w.o. CM	61.68	97.46	87.01	39.78	85.93	92.38	86.39	**97.56**	81.02
w.o. CAB	32.43	91.45	70.41	36.13	81.93	92.22	65.95	94.72	70.66

5   Conclusion

We introduced the CAFE-GAN based on the complementary attention features and attention mechanism for facial attribute editing. The CAFE controls facial editing to occur only on parts of the facial image pertinent to specified target attributes by exploiting the discriminator's ability to locate spatial regions germane to specified attributes. Performance of CAFE-GAN was compared with the state-of-the-art methods via qualitative and quantitative study. The proposed approach in most of the target attributes demonstrated improved performances over the state-of-the-art methods, and in some attributes achieved significantly enhanced results.

Acknowledgment. Authors (Jeong gi Kwak and Hanseok Ko) of Korea University are supported by a National Research Foundation (NRF) grant funded by the MSIP of Korea (number 2019R1A2C2009480). David Han's contribution is supported by the US Army Research Laboratory.

References

1. Ak, K.E., Lim, J.H., Tham, J.Y., Kassim, A.A.: Attribute manipulation generative adversarial networks for fashion images. In: The IEEE International Conference on Computer Vision (ICCV) (2019)
2. Arjovsky, M., Chintala, S., Bottou, L.: Wasserstein GAN. arXiv:1701.07875 (2017)
3. Chang, H., Lu, J., Yu, F., Finkelstein, A.: PairedCycleGAN: asymmetric style transfer for applying and removing makeup. In: The IEEE Conference on Computer Vision and Pattern Recognition (CVPR) (2018)
4. Chen, Y.C., et al.: Facelet-bank for fast portrait manipulation. In: The IEEE Conference on Computer Vision and Pattern Recognition (CVPR) (2018)
5. Chen, Y.C., Shen, X., Lin, Z., Lu, X., Pao, I., Jia, J., et al.: Semantic component decomposition for face attribute manipulation. In: The IEEE Conference on Computer Vision and Pattern Recognition (CVPR) (2019)
6. Choi, Y., Choi, M., Kim, M., Ha, J.W., Kim, S., Choo, J.: StarGAN: unified generative adversarial networks for multi-domain image-to-image translation. In: The IEEE Conference on Computer Vision and Pattern Recognition (CVPR) (2018)
7. Fukui, H., Hirakawa, T., Yamashita, T., Fujiyoshi, H.: Attention branch network: learning of attention mechanism for visual explanation. In: The IEEE Conference on Computer Vision and Pattern Recognition (CVPR) (2019)
8. Goodfellow, I., et al.: Generative adversarial nets. In: Advances in Neural Information Processing Systems (NeurIPS) (2014)
9. Gulrajani, I., Ahmed, F., Arjovsky, M., Dumoulin, V., Courville, A.C.: Improved training of Wasserstein GANs. In: Advances in Neural Information Processing Systems (NeurIPS) (2017)
10. He, Z., Zuo, W., Kan, M., Shan, S., Chen, X.: AttGAN: facial attribute editing by only changing what you want. IEEE Trans. Image Process. (TIP) (2017)
11. Isola, P., Zhu, J.Y., Zhou, T., Efros, A.A.: Image-to-image translation with conditional adversarial networks. In: The IEEE Conference on Computer Vision and Pattern Recognition (CVPR) (2017)
12. Karras, T., Aila, T., Laine, S., Lehtinen, J.: Progressive growing of GANs for improved quality, stability, and variation (2017)
13. Karras, T., Laine, S., Aila, T.: A style-based generator architecture for generative adversarial networks. In: The IEEE Conference on Computer Vision and Pattern Recognition (CVPR) (2019)
14. Kim, T., Kim, B., Cha, M., Kim, J.: Unsupervised visual attribute transfer with reconfigurable generative adversarial networks (2017)
15. Kingma, D.P., Ba, J.: Adam: A method for stochastic optimization. arXiv:1412.6980 (2014)
16. Lai, W.S., Huang, J.B., Ahuja, N., Yang, M.H.: Deep laplacian pyramid networks for fast and accurate super-resolution. In: The IEEE Conference on Computer Vision and Pattern Recognition (CVPR) (2017)
17. Lample, G., Zeghidour, N., Usunier, N., Bordes, A., Denoyer, L., Ranzato, M.: Fader networks: manipulating images by sliding attributes. In: Advances in Neural Information Processing Systems (NeurIPS) (2017)

18. Ledig, C., et al.: Photo-realistic single image super-resolution using a generative adversarial network. In: The IEEE Conference on Computer Vision and Pattern Recognition (CVPR) (2017)
19. Li, M., Zuo, W., Zhang, D.: Deep identity-aware transfer of facial attributes (2016)
20. Lin, M., Chen, Q., Yan, S.: Network in network. In: The International Conference on Learning Representations (ICLR) (2014)
21. Liu, M., et al.: STGAN: a unified selective transfer network for arbitrary image attribute editing. In: The IEEE Conference on Computer Vision and Pattern Recognition (CVPR) (2019)
22. Liu, M.Y., Breuel, T., Kautz, J.: Unsupervised image-to-image translation networks. In: Advances in Neural Information Processing Systems (NeurIPS) (2017)
23. Liu, Z., Luo, P., Wang, X., Tang, X.: Deep learning face attributes in the wild. In: The IEEE Conference on Computer Vision and Pattern Recognition (CVPR) (2015)
24. Mao, X., Li, Q., Xie, H., Lau, R.Y., Wang, Z., Smolley, S.P.: Least squares generative adversarial networks. In: The IEEE International Conference on Computer Vision (ICCV) (2017)
25. Mirza, M., Osindero, S.: Conditional generative adversarial nets. arXiv:1411.1784 (2014)
26. Park, J., Han, D.K., Ko, H.: Fusion of heterogeneous adversarial networks for single image dehazing. IEEE Trans. Image Process. (TIP) **29**, 4721–4732 (2020)
27. Perarnau, G., Van De Weijer, J., Raducanu, B., Álvarez, J.M.: Invertible conditional GANs for image editing. arXiv:1611.06355 (2016)
28. Pumarola, A., Agudo, A., Martinez, A.M., Sanfeliu, A., Moreno-Noguer, F.: GANimation: anatomically-aware facial animation from a single image. In: Ferrari, V., Hebert, M., Sminchisescu, C., Weiss, Y. (eds.) ECCV 2018, Part X. LNCS, vol. 11214, pp. 835–851. Springer, Cham (2018). https://doi.org/10.1007/978-3-030-01249-6_50
29. Radford, A., Metz, L., Chintala, S.: Unsupervised representation learning with deep convolutional generative adversarial networks. arXiv:1511.06434 (2015)
30. Selvaraju, R.R., Cogswell, M., Das, A., Vedantam, R., Parikh, D., Batra, D.: Gradcam: visual explanations from deep networks via gradient-based localization. In: The IEEE International Conference on Computer Vision (ICCV) (2017)
31. Shen, W., Liu, R.: Learning residual images for face attribute manipulation (2017)
32. Simonyan, K., Zisserman, A.: Very deep convolutional networks for large-scale image recognition. arXiv:1409.1556 (2014)
33. Smilkov, D., Thorat, N., Kim, B., Viégas, F., Wattenberg, M.: SmoothGrad: removing noise by adding noise (2017)
34. Springenberg, J.T., Dosovitskiy, A., Brox, T., Riedmiller, M.: Striving for simplicity: The all convolutional net (2014)
35. Upchurch, P., et al.: Deep feature interpolation for image content changes. In: The IEEE Conference on Computer Vision and Pattern Recognition (CVPR) (2017)
36. Wu, P.W., Lin, Y.J., Chang, C.H., Chang, E.Y., Liao, S.W.: Relgan: multi-domain image-to-image translation via relative attributes. In: The IEEE International Conference on Computer Vision (ICCV) (2019)
37. Zhang, G., Kan, M., Shan, S., Chen, X.: Generative adversarial network with spatial attention for face attribute editing. In: Ferrari, V., Hebert, M., Sminchisescu, C., Weiss, Y. (eds.) ECCV 2018, Part VI. LNCS, vol. 11210, pp. 422–437. Springer, Cham (2018). https://doi.org/10.1007/978-3-030-01231-1_26

38. Zhao, B., Chang, B., Jie, Z., Sigal, L.: Modular generative adversarial networks. In: Ferrari, V., Hebert, M., Sminchisescu, C., Weiss, Y. (eds.) Computer Vision – ECCV 2018. LNCS, vol. 11218, pp. 157–173. Springer, Cham (2018). https://doi.org/10.1007/978-3-030-01264-9_10
39. Zhou, B., Khosla, A., Lapedriza, A., Oliva, A., Torralba, A.: Learning deep features for discriminative localization. In: The IEEE Conference on Computer Vision and Pattern Recognition (CVPR) (2016)
40. Zhu, J.Y., Park, T., Isola, P., Efros, A.A.: Unpaired image-to-image translation using cycle-consistent adversarial networks. In: The IEEE International Conference on Computer Vision (ICCV) (2017)

MimicDet: Bridging the Gap Between One-Stage and Two-Stage Object Detection

Xin Lu[1(✉)], Quanquan Li[1], Buyu Li[2], and Junjie Yan[1]

[1] SenseTime Research, Shenzhen, China
{luxin,liquanquan,yanjunjie}@sensetime.com
[2] The Chinese University of Hong Kong, Hong Kong, China
byli@ee.cuhk.edu.hk

Abstract. Modern object detection methods can be divided into one-stage approaches and two-stage ones. One-stage detectors are more efficient owing to straightforward architectures, but the two-stage detectors still take the lead in accuracy. Although recent work try to improve the one-stage detectors by imitating the structural design of the two-stage ones, the accuracy gap is still significant. In this paper, we propose MimicDet, a novel and efficient framework to train a one-stage detector by directly mimic the two-stage features, aiming to bridge the accuracy gap between one-stage and two-stage detectors. Unlike conventional mimic methods, MimicDet has a shared backbone for one-stage and two-stage detectors, then it branches into two heads which are well designed to have compatible features for mimicking. Thus MimicDet can be end-to-end trained without the pre-train of the teacher network. And the cost does not increase much, which makes it practical to adopt large networks as backbones. We also make several specialized designs such as dual-path mimicking and staggered feature pyramid to facilitate the mimicking process. Experiments on the challenging COCO detection benchmark demonstrate the effectiveness of MimicDet. It achieves 46.1 mAP with ResNeXt-101 backbone on the COCO test-dev set, which significantly surpasses current state-of-the-art methods.

Keywords: Object detection · Knowledge distillation

1 Introduction

In recent years, the computer vision community has witnessed significant progress in object detection with the development of deep convolutional neural networks (CNN). Current state-of-the-art detectors [2,8,10,18,19,26,33] show high performance on several very challenging benchmarks such as COCO [20] and Open Images [15]. These detectors can be divided into two categories, one-stage detectors and two-stage ones. Modern one-stage detectors [4,19,29,33,35] usually adopt a straightforward fully convolutional architecture, and the outputs

© Springer Nature Switzerland AG 2020
A. Vedaldi et al. (Eds.): ECCV 2020, LNCS 12359, pp. 541–557, 2020.
https://doi.org/10.1007/978-3-030-58568-6_32

of the network are classification probabilities and box offsets(w.r.t. pre-defined anchor box) at each spatial position. While two-stage detectors have a more complicated pipeline [10,18,22,26]. It first filters out the regions that have high probability to contain an object from the entire image with region proposal network (RPN) [26]. Then the proposals are fed into the region convolutional network (R-CNN) [8] and get their classification score and the spatial offsets. One-stage detectors are more efficient and elegant in design, but currently the two-stage detectors have domination in accuracy.

Compared to one-stage detectors, the two-stage ones have the following advantages: 1) By sampling a sparse set of region proposals, two-stage detectors filter out most of the negative proposals; while one-stage detectors directly face all the regions on the image and have a problem of class imbalance if no specialized design is introduced. 2) Since two-stage detectors only process a small number of proposals, the head of the network (for proposal classification and regression) can be larger than one-stage detectors, so that richer features will be extracted. 3) Two-stage detectors have high-quality features of sampled proposals by use of the RoIAlign [10] operation that extracts the location consistent feature of each proposal; but different region proposals can share the same feature in one-stage detectors and the coarse and spatially implicit representation of proposals may cause severe feature misalignment. 4) Two-stage detectors regress the object location twice (once on each stage) and the bounding boxes are better refined than one-stage methods.

We note that recent works try to overcome the drawbacks of one-stage detectors by imitating the structural design of the two-stage ones. For example, RefineDet [33] tries to imitate the two-stage detectors by introducing cascade detection flow into a one-stage detector. Based on RefineDet, AlignDet [4] proposes the RoIConv layer to solve the feature misalignment problem. The RoIConv layer imitates the RoIAlign [10] operation to match the feature and its corresponding anchor box with accurate and explicit location mapping. Although these works alleviate part of the disadvantages of traditional one-stage detectors, there still leaves a big gap to the two-stage detectors.

To further close the performance gap between one-stage and two-stage detectors, one natural idea is to imitate the two-stage approach not only in structure design, but also in feature level. An intuitive approach is to introduce the knowledge distillation(a.k.a mimic) workflow. However, unlike common mimic tasks, the teacher and student models are heterogeneous here. The features extracted from the backbone in teacher and student networks have different representing areas and also a big semantic gap. Thus it is inappropriate to directly apply existing mimic method in this heterogeneous situation. Moreover, conventional mimic workflow in object detection has many problems in practice. It needs to pre-train a teacher model with large backbone and then take the features from the teacher model as a supervision during the training of the student model. The entire workflow is complicated and the training efficiency is much lower. Besides, state-of-the-art object detectors usually adopt a relatively high resolution for input images, therefore forwarding both teacher and student networks simultaneously is hard in practice due to the large computation and GPU memory cost. In addition, traditional knowledge distillation methods usually use a powerful model as teacher to improve the performance of a student with small model, so it suffers scale-up limitation when the student model is comparatively large.

In this paper, we propose a novel training framework, named MimicDet, which can efficiently and significantly close the accuracy gap between the one-stage and two-stage detectors. Specifically, the network contains both one-stage detection head and two-stage detection head during training. The two-stage detection head, called T-head (teacher head), is a network branch with large amount of parameters, in order to extract high-quality features for sparsely sampled proposal boxes. The one-stage detection head, called S-head (student head), light-weighted branch for detecting all dense anchor boxes. Since the original teacher and student networks are heterogeneous, their feature maps have a spatial misalignment. To handle this problem, we exploit the guided deformable conv [31] layer where each convolution kernel can have an offset that computed by a lightweight neural network. Different from the AlignDet [4], the deformable offset can be optimized by mimicking the location consistent features generated by T-head. With the features matched, the similarity loss of the feature pairs will be optimized together with detection losses in MimicDet. Thus the features of S-head can acquire better properties from features of T-head. During inference, the T-head is discarded, which means a pure one-stage detector is implemented for object detection. This mechanism ensures that MimicDet inherits both high-efficiency and high-accuracy from the two architectures. Different from the conventional mimic methods in object detection [17,27], teacher and student share the same backbone in MimicDet and the mimic is between different detection heads instead of different backbones. Thus it does not need to pre-train a teacher model or require a stronger backbone model to serve as the teacher. These properties make MimicDet much more efficient and can be extended to larger models.

To further improve the performance of MimicDet, several specialized designs are also proposed. We design decomposed detection heads and conduct mimicking in classification and regression branches individually. The decomposed *dual-path mimicking* makes it easier to learn useful information from T-head. Furthermore, we propose the *staggered feature pyramid* from which one can extract a pair of features with different resolutions. For each anchor box, it obtains high-resolution features for T-head and low-resolution features for S-head from different layers of the feature pyramid. Thus MimicDet can benefit more from high-accuracy of T-head and high-efficiency of S-head without extra computation cost.

To demonstrate the effectiveness of MimicDet, we conduct experiments on the challenging COCO detection benchmark and its performance significantly surpasses other state-of-the-art one-stage detectors. Our best model, based on ResNeXt-101 backbone, achieves state-of-the-art 46.1 mAP on COCO *test-dev* benchmark.

2 Related Work

In this section we briefly review related works in the fields of object detection and network mimicking.

Two-Stage Detectors: For now, two-stage detectors take the lead in detection accuracy. In these detectors, sparse region proposals are generated in the first stage and then are further regressed and classified in the second stage. R-CNN [9] utilized low-level computer vision algorithms such as Selective Search [30] and Edge Boxes [36] to generate proposals, then adopt CNN to extract features for training SVM classifier and bounding box regressor. SPP-Net [11] and Fast R-CNN [8] proposed to extract features for each proposal on a shared feature map by spatial pyramid pooling. Faster R-CNN [26] integrated the region proposal process into the deep convnet and makes the entire detector an end-to-end trainable model. R-FCN [5] proposed a region-based fully convolutional network to generate region-sensitive features for detection. FPN [18] proposed a top-down architecture with lateral connection to generate a feature pyramid for multi-scale detection. In our work, instead of detecting object directly, the two-stage detection head performs as a teacher model in mimic mechanism and will not be used for inference.

One-stage Detectors: One-stage detectors perform classification and regression on dense anchor boxes without generating a sparse RoI set. YOLO [24] is an early exploration that directly detects objects on dense feature map. SSD [21] proposed to use multi-scale features for detecting variant scale objects. RetinaNet [19] proposed focal loss to address the extreme class imbalance problem in dense object detection. RefineDet [33] introduced anchor refinement module and the object detection module to imitate the cascade regression on dense anchor boxes. Guided Anchor [31] first used anchor-guided deformable convolutions to align features for RPN. AlignDet [4] designed RoI convolution, to imitate RoI Align operation in two-stage detectors and perform exact feature alignment for densely anchor box detection.

Network Mimicking: Typical network mimicking or knowledge distillation approaches utilize output logits or intermediate features of a well-trained large teacher model to supervise the training process of the efficient student model [1,14,27,28]. These approaches are widely used for model acceleration and compression in the classification task. Recently, [3,17] extend mimicking approaches to the object detection task and achieve satisfactory results.

Although we utilize mimicking as the critical method to improve the performance, MimicDet differs from the previous mimic-based object detection approaches in many aspects. First, the MimicDet use mimic between different types of detection heads instead of backbones. The teacher and student in MimicDet share the same backbone. Second, only an additional head branch as a teacher is introduced instead of an entire teacher model. Moreover, different from the pipeline in previous methods that train the teacher first and then mimic, teacher head and student head in MimicDet are trained jointly, which makes the training process much more efficient. Last, in conventional methods, student model usually needs a larger model to mimic, which limits the method to be applied on larger student model. MimicDet still works well in this situation because it does not rely on a stronger backbone to improve performance.

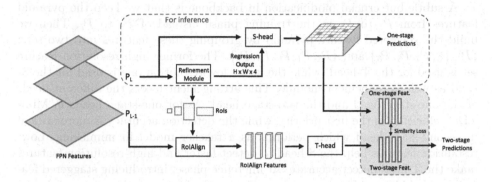

Fig. 1. Overview. In training phase, we first adopt the refinement module to classify and regress the pre-defined dense anchor boxes. Then these adjusted anchor boxes will be transferred to S-head and T-head for dense and sparse detection, respectively. S-head takes the regression output of the refinement module in to compute the offset for deformable conv layer. Then it performs dense detection for each anchor box. T-head performs two-stage detection for sampled anchor boxes. Features of these sampled anchor boxes in S-head and T-head will be extracted for mimicking. During the inference, only the one-stage part(in red dash line) will be used. (Color figure online)

3 MimicDet Framework

The overview of the MimicDet is shown in Fig. 1. During the inference phase, MimicDet consists of backbone, refinement module, and one-stage detection head called S-head. In the training phase, we add an extra two-stage detection head as teacher, called T-Head, to guide the S-Head and enhance its performance. We will expatiate on each component in following sections.

3.1 Backbone and Staggered Feature Pyramid

We build our model on a Feature Pyramid Network [18] backbone to efficiently extract multi-scale features for detecting objects that distributed in a wide range of scales. Generally, the FPN network adopts lateral connections to combine bottom-up feature with the top-down feature and generates a feature pyramid for further process. Here we follow the conventional denotation in the FPN, using C_l and P_l to denote the feature in ResNet [13] and FPN feature pyramid, respectively. Since the one-stage detection head is sensitive to the computation cost, we follow the modification in RetinaNet [19] and extend the original feature pyramid to P_7, which has a stride of 128 with respect to the input image. Specifically, P_6 is generated by adopting a 3×3 stride-2 convolution on C_5, P_7 is computed by applying ReLU [23] followed by a 3×3 stride-2 conv on P_6. The number of channels for all feature pyramid is 256.

A subtle but crucial modification in backbone is that we keep the pyramid features from P_2 to P_7 in the training phase instead of P_3 to P_7. Then we build the staggered feature pyramid by grouping these features into two sets: $\{P_2, P_3, P_4, P_5, P_6\}$ and $\{P_3, P_4, P_5, P_6, P_7\}$. The former high-resolution feature set is used for the T-head, while the latter low-resolution one is used for the S-head as well as refinement module. This arrangement meets the different needs of the one-stage head and the two-stage head. As for one-stage heads in MimicDet, efficiency is the first priority, while the detection accuracy is more critical in second-stage head which performs as a teacher model in mimicking. Low-resolution features help the S-head to detect faster and high-resolution features make the T-head detect accurate. At inference phase, introducing staggered feature pyramid is cost-free because we only use the low-resolution one and do not generate the P_2 for efficiency.

3.2   Refinement Module

After extracting features from the FPN, we adopt the refinement module to filter out easy negatives and adjust the location and size of pre-defined anchor boxes, which can mitigate the extreme class imbalance issue and provide better anchor initialization when training S-head and T-head. Specifically, the refinement module is formed by one 3×3 conv and two sibling 1×1 convs to perform class-agnostic binary classification and bounding box regression on top of the feature pyramid. Anchor boxes that adjusted by the refinement module will be transferred to T-head and S-head for sparse and dense detection, and only top-ranked boxes will contribute to the training process in T-head and S-head. In general, to MimicDet, the refinement module plays a similar role as the RPN in FPN [18](two-stage detector) and the ADM in RefineDet [33](one-stage detector).

We define anchors to have areas from 32^2 to 512^2 on feature pyramid P_3 to P_7, respectively. Unlike previous works which define multiple anchors on a certain position of the feature map, we only define one anchor on each position with 1:1 aspect ratio. We adopt this sparse anchor setting to avoid feature sharing because each anchor box in the one-stage head needs to have an exclusive and explicit feature for head mimicking.

Assigning label of the anchors in the refinement module is different from the conventional RoI based strategy because the pre-defined anchor boxes are much more sparse than those in RetinaNet. We assign the objects to feature pyramid from P_3 to P_7 according to their scale, and each feature pyramid learns to detect objects in a particular range of scales. Specifically, for pyramid P_l, the valid scale range of the target object is computed as $[S_l \times \eta_1, S_l \times \eta_2]$, where S_l is the basic scale for level l, and the η is set to control the valid scale range. We empirically set $S_l = 4 \times 2^l$, $\eta_1 = 1, \eta_2 = 2$. Any objects that smaller than $S_3 \times \eta_1$ or larger than $S_7 \times \eta_2$ will be assigned to P_3 or P_7, respectively. Given a ground-truth box B, we define its positive area as its central rectangle area that shrunk to 0.3 times of the original length and width. Suppose there exists a ground-truth box B lies in the valid scale range of an anchor box A, and the central point of A lies in the positive area of B. Only in this case, the A will be labeled as a positive sample, and the regression target is set to B. The regression target is kept the same as Faster R-CNN [26], and only positive anchors will contribute to the regression loss. We optimize the classification and regression branch by binary cross-entropy loss and $L1$ loss, respectively.

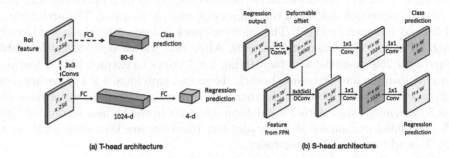

(a) T-head architecture   (b) S-head architecture

Fig. 2. Detection head architecture. Dash line denotes stacked operations. Green and orange features will be used for dual-path head mimicking, i.e. classification and regression mimicking. (Color figure online)

3.3 Detection Heads

T-head: T-head is a two-stage detection head that detects a sparse set of the anchor boxes adjusted by refinement module. The purpose of the T-head is to extract high-quality features and using these features to guide the training process of the S-head. We design the T-head as a heavy head with high resolution input for better features and the T-head will not be used during the inference.

As shown in Fig. 1, T-head access features from the high-resolution set of the staggered feature pyramid, i.e., one layer ahead of the original one. It first adopts the RoIAlign operation to generate location-sensitive features with 7×7 resolution for each anchor box. After that, the T-head will be split into two branches for classification and regression. In the classification branch, features of each anchor box are processed by two 1024-d fc layers, followed by one 81-d fc layer and a softmax layer to predict the classification probability. In the other branch, we adopt four consecutive 3×3 convs with 256 output channels and then flat the feature into a vector. Next, the flatten feature will be transformed into a 1024-d regression feature and a 4-d output sequentially by two fc layers. As shown in Fig. 2, the 81-d classification logits and 1024-d regression feature will be regarded as the mimicking target when training the S-head. The label assignment is based on the IoU criteria with 0.6 threshold. We optimize the classification and regression branch by cross-entropy loss and $L1$ loss respectively.

S-head: S-head is a one-stage detection head that directly performs dense detection on top of the feature pyramid without sampling. We design the S-head as a light-weight network that can overcome the feature misalignment and learning to extract high-quality features by mimicking the T-head. As we mentioned before, introducing the refinement module will break the location consistency between the anchor box and its corresponding features. The location inconsistency can lead to differences in representing area between S-head and T-head, which is harmful to head mimicking. So we use deformable convolution to capture the misaligned feature. The deformation offset is computed by a micro-network that takes the regression output of the refinement module as input. The architecture of S-head is shown in Fig. 2. The micro-network is formed by three 1×1 convs with 64 and 128 intermediate channels. After that we use one 5×5 deformable conv(with 256 channels) and two sibling 1×1 convs to extract 1024-d features for classification and regression branch. Then two individual 1×1 convs are used for generating predictions. To further reduce the computation, we replace the 5×5 deformable conv with 3×3 deformable conv in the highest resolution level P_3. The label assignment strategy and loss functions are kept same as those of the T-head for semantic consistency.

During the experiments we found that the proportion of positive samples of particular classes is too low, even though the refinement module has already rejected some easy negative samples. To deal with it, we use hard negative mining to mitigate the class imbalance problem, i.e., we always sample boxes with top classification loss to optimize the classification loss in S-head.

3.4 Head Mimicking

We use B_s to denote the set of all anchor boxes adjusted by the refinement module, and B_t to denote the sparse subset of the B_s that sampled for T-head. Based on B_t, we define B_m as a random sampled subset used for optimizing mimic loss. Given B_m, we can get its corresponding two-stage classification feature set F^{tc} and regression feature set F^{tr} by applying T-head on it. Similarly, classification and regression features of B_m in S-head can also be obtained and denoted as F^{sc} and F^{sr}. More specifically, in S-head, each pixel of its output feature map corresponds to an anchor box in B_s. To get the S-head feature of an adjusted anchor box, we trace back to its initial position and extract the pixel at that position in the feature map of S-head.

We define the mimic loss as follow:

$$
L_{mimic} = \frac{1}{N_{B_m}} \{ \sum_i [1 - cosine(F_i^{tr}, F_i^{sr})] \\
+ \sum_i [1 - cosine(F_i^{tc}, F_i^{sc})] \}
\tag{1}
$$

where i is the index of anchor boxes in B_m and N_{B_m} is the number of the anchor boxes in B_m. We adopt cosine similarity to evaluate the similarity loss on both classification and regression branches.

Finally, we formally define the multi-task training loss as: $L = L_R + L_S + L_T + L_{mimic}$, where L_R, L_S and L_T denotes loss of the refinement module, S-head and T-head respectively.

4   Implementation Details

4.1   Training

We adopt the depth 50 and 101 ResNet with FPN as our default backbone to generate the feature pyramid. Images are resized such that their shorter side is 800 pixels by convention. When training the refinement module, different from RPN, we utilize all anchor boxes with no sampling strategy involved. After that, we adopt NMS with 0.8 IoU threshold on anchor boxes that adjusted by refinement module and select top 2000 boxes as the proposal set. In T-head, 128 boxes are sampled from the proposal set with 1:3 ratio of positive to negative samples. In S-head, we sort boxes in the proposal set by their classification losses and select 128 boxes with top loss value for training the classification branch. As for the regression branch, we sample at most 64 positive boxes from the proposal set randomly. In addition to that, all 128 boxes that are sampled for T-head will be adopted to optimize the mimic loss.

We use SGD to optimize the training loss with 0.9 momentum and 0.0001 weight decay. The backbone parameters are initialized with the ImageNet-1k based pre-trained classification model. New parameters in FPN and T-head are initialized with He initialization [12], others newly added parameters are initialized by Gaussian initializer with $\sigma = 0.01$. By convention, we freeze the parameters in the first two conv layers and all batchnorm layers. No data augmentations except standard horizontal flipping are used. We set two default learning schedules named 1× and 2×, which has 13 and 24 epochs in total. The learning rate is set to 0.04 and then decreases by 10 at the 8th and 11th epoch in 1× schedule or 16th and 22nd epoch in 2× schedule. All models are trained on 16 GPUs with a total batch size of 32. We use warmup mechanisms in both baselines and our method to keep the multi-GPU training stable.

4.2   Inference

During inference, we discard the T-head and do not compute the P_2 feature map in FPN. The initial anchor boxes will be regressed twice by refinement module and S-head. To reduce computational consumption in post-processing, we filter out the top 1000 boxes according to their classification score predicted by refinement module. Then these boxes will be processed by NMS with 0.6 IoU threshold and 0.005 score threshold. Finally, the top 100 scoring boxes will be selected as the final detection result.

5 Experiments

We conduct experiments on the object detection track of the COCO benchmark and compare our method with other state-of-the-art detectors. All experiments below are trained on the union of 80k train images and 35k subset of val images, and the test is performed on the 5k subset of val (minival) and 20k test-dev.

5.1 Ablation Study

For simplicity, unless specified, all experiments in ablation study are conducted in 1× learning schedule, with 800 pixels input size and ResNet-50 backbone. Experiment results are evaluated on COCO *minival* set.

How Important is the Head Mimicking? To measure the impact of the head mimicking mechanism, here we design and train three models: (a) MimicDet without T-head and mimic loss; (b) MimicDet with T-head and without mimic loss; (c) MimicDet. Experiment results are listed in Table 1. Model (a) only achieves 36.0 AP, which is close to the RetinaNet baseline(we do not use focal loss in MimicDet). Model (b) achieves 36.4 AP and gains 0.4 AP comparing with model (a). Model (c) gets the best result with 39.0 AP, which achieves 2.6 AP improvement over model (b). These results demonstrate that adding a two-stage head hardly affects the performance of one-stage head, and the mimic mechanism is the principal factor of MimicDet's success.

In Fig. 3, we compare models that trained with and without mimic mechanism by visualizing the location consistency between their features and corresponding boxes. Although the plain guided deformable convolution has the potential for fixing the feature misalignment issue, results in practice are not that satisfactory. With the proposed mimic mechanism, guided deformable convolution learns to align the representing area between one-stage and two-stage branches, which helps to close the semantic gap between them. On the other hand, S-head can capture features precisely and consistently, which helps to mitigate the feature misalignment in one-stage detection.

Table 1. An ablation study to inspect the impact of introducing T-head and head-mimicking mechanism.

Method	AP	AP$_{.5}$	AP$_{.75}$
w/o T-head & w/o mimic	36.0	56.0	39.9
w T-head & w/o mimic	36.4	56.1	40.0
w T-head & w mimic	39.0	57.5	42.6

Design of S-head: In this section we explore the variant design of S-head. As discussed before, S-head should be designed as light as possible to ensure the computation efficiency. To this end, we first fix the architecture of S-head as a three-layer convolutional neural network. The first layer, called alignment layer, is adopted to capture the misaligned feature caused by cascade regression. It helps to align the representing area of S-head and T-head and close the semantic gap between them. The second convolutional layer project the aligned feature to a high dimensional feature space, and the third layer performs the final prediction. For simplicity, we keep second and third layer as 1×1 conv layer unchanged and explore the design of the alignment layer.

As shown in Table 2, we experiment on four typical variants denoted as (a) to (d). For (a), the alignment layer is a normal conv layer; For (b), it is a deformable conv layer; For (c), the original deformable conv is replaced by guided deformable conv layer. More specifically, the offsets in deformable conv layer is computed by the regression output of the refinement module. For (d), we replace 5×5 guided deformable conv with the 3×3 one for S-head on P_3 to reduce the computation cost.

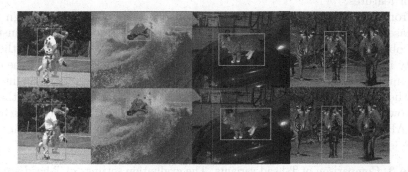

Fig. 3. MimicDet(top) *vs.* **MimicDet w/o head-mimicking mechanism (bottom).** Yellow box is the adjusted anchor box generated by refinement module. It's corresponding deformable convolution locations are denoted by red dots. (Colour figure online)

Experiments in Table 2 demonstrate that the alignment layer of S-head has marked impact on mimicking result. Without an efficient alignment layer, the representing areas in S-head and T-head will be different, and the performance of our mimic workflow deteriorates severely because of the large semantic gap between them. Model (b) and (c) yields 1.5 and 5.5 AP improvement over the model (a) respectively. Model (d) achieves 39.0 AP, which is comparable with model (c). For better speed-accuracy trade-off, we choose (d) as the default setting in MimicDet.

Design of T-head: In this section we inspect how the design of T-head impacts the mimicking result in MimicDet. We adopt a light-weight T-head as our baseline. In this baseline, T-head is formed by two consecutive 1024-d fully connected layers and two sibling output layers, which is a common baseline choice in many two-stage detectors. To further enhance the performance of T-head, we try to

Table 2. Comparison of S-head variants with different alignment layers. * indicates using 3×3 guided deformable conv for the S-head on P_3 layer.

Method	AP	AP$_{.5}$	AP$_{.75}$
(a) 5×5 normal conv	33.7	54.1	37.2
(b) 5×5 deform conv	35.2	55.3	39.2
(c) 5×5 guided deform conv	39.2	57.3	43.1
(d) 5×5 guided deform conv*	39.0	57.5	42.6

introduce a heavier architecture with conv layers. The heavy head consists of four conv layers followed by two 1024-d fully connected layers and two sibling output layers. Note that in all T-head variants above, features in both S-head and T-head are not split and the mimicking is performed on the last 1024-d vector features.

From the second row of Table 3, we observe that the heavy design does increase the AP of T-head as expected, but the AP of S-head deteriorates instead. This observation suggests that, as the complexity of T-head increases, the difficulty in mimicking also boosts simultaneously. In MimicDet, we propose dual-path mimicking to decompose the feature of classification and regression into two branches and conduct two mimicking processes individually. Comparing the second row and the third row of Table 3, the proposed dual-path mimicking achieves 39.0 AP in S-head, with a 1.9 AP gain from the naive heavy-head baseline.

Table 3. Comparison of T-head variants. The evaluation settings of T-head are same as those in S-head. Arrows denote the deterioration or amelioration in comparison with light head design.

Method	T-head AP	S-head AP
Light head	39.0	37.5
Heavy head	39.8 ↑	37.1 ↓
Dual-path mimicking	40.2 ↑	39.0 ↑

Staggered Feature Pyramid: In MimicDet we propose staggered feature pyramid to obtain features for T-head and S-head in different layers of the feature pyramid. In first row of Table 4, we use feature pyramid $\{P_3 - P_7\}$ for both S-head and T-head. In the second row, we use high-resolution feature pyramid $\{P_2 - P_6\}$ for T-head and low-resolution feature pyramid $\{P_3 - P_7\}$ for S-head. Results in Table 4 show that, by mimicking high-resolution features, MimicDet gains 0.6 AP improvement in ResNet-50 experiment.

Table 4. Ablation study for staggered feature pyramids. The second and third column denotes the layers in feature pyramid that will be used for T-head and S-head, respectively.

Method	T-head	S-head	AP
RetinaNet FPN [19]	$P_3 - P_7$	$P_3 - P_7$	38.4
Staggered feature pyramid	$P_2 - P_6$	$P_3 - P_7$	39.0

Shared Backbone versus Separated Backbone: The teacher and student in MimicDet share a same backbone. Here we compare MimicDet with its separated backbone counterpart. In separated backbone experiments, we will first train a two-stage detector as the teacher, then the pre-trained teacher model will be adopted to supervise the one-stage student model. Note that we inherit all the design in MimicDet except sharing the backbone. We conduct experiments on 16 Titan Xp GPUs. Results in Table 5 demonstrate that MimicDet achieves comparable accuracy with only about half of the training time in comparison with separated backbone counterpart.

Table 5. Training time in comparison with separated backbone counterpart. The learning rate schedule is set to 2×.

Backbone		AP	Training time
Res-50	shared	39.8	17.5 h
Res-50	separated	40.2	34 h
Res-101	shared	41.8	22 h
Res-101	separated	42.2	47 h

Speed/Accuracy Trade-off: Larger input size and backbone networks yield better accuracy, but also slower inference speed. we conduct experiments on variant input sizes and variant backbone networks to show the speed-accuracy trade-off, results are listed in Table 6. On single TITAN V GPU, MimicDet achieves 23 FPS with 38.7 mAP on Res-50 with 600 pixels input size and 13 FPS with 41.8 mAP on Res-101 with 800 pixels input size.

Table 6. The speed-accuracy trade-off by varying image sizes. We test the inference speed on single TITAN V GPU with batch size of 1. The learning rate schedule is set to 2×.

Size	Backbone	AP	AP$_{.5}$	AP$_{.75}$	AP$_S$	AP$_M$	AP$_L$	Inference time
400	ResNet-50	35.5	53.1	38.2	15.3	38.4	52.3	34 ms
600	ResNet-50	38.7	56.8	42.2	20.3	41.7	54.5	44 ms
800	ResNet-50	39.8	58.2	43.6	21.6	43.0	53.4	63 ms
400	ResNet-101	36.1	53.5	39.1	14.9	39.2	53.8	38 ms
600	ResNet-101	39.8	57.9	43.3	20.1	43.2	55.8	55 ms
800	ResNet-101	41.8	60.3	45.6	22.5	45.2	56.4	80 ms

5.2 Comparison with State-of-the-art Methods

We compare MimicDet to other state-of-the-art one-stage object detectors on COCO test-dev. We adopt ResNet-101 and ResNeXt-101-64 × 4d as our backbones. Follow the convention in RetinaNet, MimicDet is trained with scale jitter from 640 to 800 and 1.5× longer than the 2× training schedule. As shown in Table 7, MimicDet based on ResNet-101 and ResNeXt-101-64 × 4d achieves 44.4 AP and 46.1 AP respectively, which surpasses the current state-of-the-art one-stage methods by a large margin.

Table 7. Comparison with state-of-the-art one stage detectors on COCO *test-dev*. M means M40 or TitanX MaxWell, P means P100, Titan X(Pascal), Titan Xp or 1080Ti, V means TitanV. * indicates the results are based on flip test and soft NMS.

Method	Backbone	Input Size	Speed(ms)	AP	AP$_{.5}$	AP$_{.75}$	AP$_S$	AP$_M$	AP$_L$
SSD [21]	ResNet-101	513	125/M	31.2	50.4	33.3	10.2	34.5	49.8
DSSD [7]	ResNet-101	513	156/M	33.2	53.3	35.2	13.0	35.4	51.1
RefineDet [33]	ResNet101	512	-	36.4	57.5	39.5	16.6	39.9	51.4
YOLOv3 [25]	DarkNet-53	608	51/M	33.0	57.9	34.4	18.3	35.4	41.9
ExtremeNet* [34]	Hourglass-104	511	322/P	40.1	55.3	43.2	20.3	43.2	53.1
CornerNet* [16]	Hourglass-104	511	300/P	40.5	56.5	43.1	19.4	42.7	53.9
CenterNet* [6]	Hourglass-104	511	340/P	44.9	62.4	48.1	25.6	47.4	57.4
RetinaNet [19]	ResNet-101	800	104/P 91/V	39.1	59.1	42.3	21.8	42.7	50.2
FSAF [35]	ResNet-101	800	180/M	40.9	61.5	44.0	24.0	44.2	51.3
FCOS [29]	ResNet-101	800	86.2/V	41.5	60.7	45.0	24.4	44.1	51.0
RPDet [32]	ResNet-101	800	-	41.0	62.9	44.3	23.6	44.1	51.7
AlignDet [4]	ResNet-101	800	110/P	42.0	62.4	46.5	24.6	44.8	53.3
MimicDet(ours)	ResNet-101	800	80/V	**44.4**	63.1	48.8	25.8	47.7	55.8
RetinaNet [19]	ResNeXt-101-32 × 8d	800	177/P	40.8	61.1	44.1	24.1	44.2	51.2
FSAF [35]	ResNeXt-101-64 × 4d	800	362/M	42.9	63.8	46.3	26.6	46.2	52.7
FCOS [29]	ResNeXt-101-64 × 4d	800	143/V	43.2	62.8	46.6	26.5	46.2	53.3
MimicDet(ours)	ResNeXt-101-64 × 4d	800	132/V	**46.1**	65.1	50.8	28.6	49.2	57.5

6   Conclusion

In this paper, we propose a novel and efficient framework to train a one-stage detector called MimicDet. By introducing two-stage head with mimicking mechanism, MimicDet can obtain excellent properties of two-stage features and get rid of common issues that lie in the one-stage detectors. MimicDet inherits both high-efficiency of the one-stage approach and high-accuracy of the two-stage approach. It does not need to pre-train a teacher model or require a stronger backbone model to serve as the teacher. These properties make MimicDet much more efficient and can be easily extended to larger models in comparison with conventional mimic methods.

References

1. Buciluǎ, C., Caruana, R., Niculescu-Mizil, A.: Model compression. In: Proceedings of the 12th ACM SIGKDD International Conference on Knowledge Discovery and Data Mining, pp. 535–541. ACM (2006)
2. Cai, Z., Vasconcelos, N.: Cascade r-cnn: delving into high quality object detection. In: Proceedings of the IEEE Conference on Computer Vision and Pattern Recognition, pp. 6154–6162 (2018)
3. Chen, G., Choi, W., Yu, X., Han, T., Chandraker, M.: Learning efficient object detection models with knowledge distillation. In: Advances in Neural Information Processing Systems, pp. 742–751 (2017)
4. Chen, Y., Han, C., Wang, N., Zhang, Z.: Revisiting feature alignment for one-stage object detection. arXiv preprint arXiv:1908.01570 (2019)
5. Dai, J., Li, Y., He, K., Sun, J.: R-FCN: object detection via region-based fully convolutional networks. In: Advances in Neural Information Processing Systems, pp. 379–387 (2016)
6. Duan, K., Bai, S., Xie, L., Qi, H., Huang, Q., Tian, Q.: Centernet: keypoint triplets for object detection. In: Proceedings of the IEEE International Conference on Computer Vision, pp. 6569–6578 (2019)
7. Fu, C.Y., Liu, W., Ranga, A., Tyagi, A., Berg, A.C.: DSSD: deconvolutional single shot detector. arXiv preprint arXiv:1701.06659 (2017)
8. Girshick, R.: Fast R-CNN. In: Proceedings of the IEEE International Conference on Computer Vision, pp. 1440–1448 (2015)
9. Girshick, R., Donahue, J., Darrell, T., Malik, J.: Rich feature hierarchies for accurate object detection and semantic segmentation. In: Proceedings of the IEEE Conference on Computer Vision and Pattern Recognition, pp. 580–587 (2014)
10. He, K., Gkioxari, G., Dollár, P., Girshick, R.: Mask R-CNN. In: 2017 IEEE International Conference on Computer Vision (ICCV), pp. 2980–2988. IEEE (2017)
11. He, K., Zhang, X., Ren, S., Sun, J.: Spatial pyramid pooling in deep convolutional networks for visual recognition. In: Fleet, D., Pajdla, T., Schiele, B., Tuytelaars, T. (eds.) ECCV 2014. LNCS, vol. 8691, pp. 346–361. Springer, Cham (2014). https://doi.org/10.1007/978-3-319-10578-9_23
12. He, K., Zhang, X., Ren, S., Sun, J.: Delving deep into rectifiers: surpassing human-level performance on imagenet classification. In: Proceedings of the IEEE International Conference on Computer Vision, pp. 1026–1034 (2015)

13. He, K., Zhang, X., Ren, S., Sun, J.: Deep residual learning for image recognition. In: Proceedings of the IEEE Conference on Computer Vision and Pattern Recognition, pp. 770–778 (2016)
14. Hinton, G., Vinyals, O., Dean, J.: Distilling the knowledge in a neural network. arXiv preprint arXiv:1503.02531 (2015)
15. Kuznetsova, A., et al.: The open images dataset v4: unified image classification, object detection, and visual relationship detection at scale. arXiv preprint arXiv:1811.00982 (2018)
16. Law, H., Deng, J.: Cornernet: detecting objects as paired keypoints. In: Proceedings of the European Conference on Computer Vision (ECCV), pp. 734–750 (2018)
17. Li, Q., Jin, S., Yan, J.: Mimicking very efficient network for object detection. In: Proceedings of the IEEE Conference on Computer Vision and Pattern Recognition, pp. 6356–6364 (2017)
18. Lin, T.Y., Dollár, P., Girshick, R., He, K., Hariharan, B., Belongie, S.: Feature pyramid networks for object detection. In: 2017 IEEE Conference on Computer Vision and Pattern Recognition (CVPR), pp. 936–944. IEEE (2017)
19. Lin, T.Y., Goyal, P., Girshick, R., He, K., Dollár, P.: Focal loss for dense object detection. In: Proceedings of the IEEE International Conference on Computer Vision, pp. 2980–2988 (2017)
20. Lin, T.Y., et al.: Microsoft COCO: common objects in context. In: Fleet, D., Pajdla, T., Schiele, B., Tuytelaars, T. (eds.) ECCV 2014. LNCS, vol. 8693, pp. 740–755. Springer, Cham (2014). https://doi.org/10.1007/978-3-319-10602-1_48
21. Liu, W., et al.: SSD: single shot multibox detector. In: Leibe, B., Matas, J., Sebe, N., Welling, M. (eds.) ECCV 2016. LNCS, vol. 9905, pp. 21–37. Springer, Cham (2016). https://doi.org/10.1007/978-3-319-46448-0_2
22. Lu, X., Li, B., Yue, Y., Li, Q., Yan, J.: Grid R-CNN. In: Proceedings of the IEEE Conference on Computer Vision and Pattern Recognition, pp. 7363–7372 (2019)
23. Nair, V., Hinton, G.E.: Rectified linear units improve restricted boltzmann machines. In: Proceedings of the 27th International Conference on Machine Learning (ICML-10), pp. 807–814 (2010)
24. Redmon, J., Divvala, S., Girshick, R., Farhadi, A.: You only look once: unified, real-time object detection. In: Proceedings of the IEEE Conference on Computer Vision and Pattern Recognition, pp. 779–788 (2016)
25. Redmon, J., Farhadi, A.: Yolov3: an incremental improvement. arXiv preprint arXiv:1804.02767 (2018)
26. Ren, S., He, K., Girshick, R., Sun, J.: Faster R-CNN: towards real-time object detection with region proposal networks. In: Advances in Neural Information Processing Systems, pp. 91–99 (2015)
27. Romero, A., Ballas, N., Kahou, S.E., Chassang, A., Gatta, C., Bengio, Y.: Fitnets: hints for thin deep nets. arXiv preprint arXiv:1412.6550 (2014)
28. Sergey Zagoruyko, N.K.: Paying more attention to attention: improving the performance of convolutional neural networks via attention transfer. In: International Conference on Learning Representations (2017)
29. Tian, Z., Shen, C., Chen, H., He, T.: FCOS: fully convolutional one-stage object detection. arXiv preprint arXiv:1904.01355 (2019)
30. Uijlings, J.R., Van De Sande, K.E., Gevers, T., Smeulders, A.W.: Selective search for object recognition. Int. J. Comput. Vis. **104**(2), 154–171 (2013). https://doi.org/10.1007/s11263-013-0620-5
31. Wang, J., Chen, K., Yang, S., Loy, C.C., Lin, D.: Region proposal by guided anchoring. In: Proceedings of the IEEE Conference on Computer Vision and Pattern Recognition, pp. 2965–2974 (2019)

32. Yang, Z., Liu, S., Hu, H., Wang, L., Lin, S.: Reppoints: point set representation for object detection. In: Proceedings of the IEEE International Conference on Computer Vision, pp. 9657–9666 (2019)
33. Zhang, S., Wen, L., Bian, X., Lei, Z., Li, S.Z.: Single-shot refinement neural network for object detection. In: Proceedings of the IEEE Conference on Computer Vision and Pattern Recognition, pp. 4203–4212 (2018)
34. Zhou, X., Zhuo, J., Krahenbuhl, P.: Bottom-up object detection by grouping extreme and center points. In: Proceedings of the IEEE Conference on Computer Vision and Pattern Recognition, pp. 850–859 (2019)
35. Zhu, C., He, Y., Savvides, M.: Feature selective anchor-free module for single-shot object detection. arXiv preprint arXiv:1903.00621 (2019)
36. Zitnick, C.L., Dollár, P.: Edge boxes: locating object proposals from edges. In: Fleet, D., Pajdla, T., Schiele, B., Tuytelaars, T. (eds.) ECCV 2014. LNCS, vol. 8693, pp. 391–405. Springer, Cham (2014). https://doi.org/10.1007/978-3-319-10602-1_26

Latent Topic-Aware Multi-label Classification

Jianghong Ma[1(✉)] and Yang Liu[2]

[1] City University of Hong Kong, Hong Kong, China
jianghma2@cityu.edu.hk
[2] The Hong Kong University of Science and Technology, Hong Kong, China

Abstract. In real-world applications, data are often associated with different labels. Although most extant multi-label learning algorithms consider the label correlations, they rarely consider the topic information hidden in the labels, where each topic is a group of related labels and different topics have different groups of labels. In our study, we assume that there exists a common feature representation for labels in each topic. Then, feature-label correlation can be exploited in the latent topic space. This paper shows that the sample and feature exaction, which are two important procedures for removing noisy and redundant information encoded in training samples in both sample and feature perspectives, can be effectively and efficiently performed in the latent topic space by considering topic-based feature-label correlation. Empirical studies on several benchmarks demonstrate the effectiveness and efficiency of the proposed topic-aware framework.

Keywords: Multi-label learning · Sample and feature extraction · Feature-label correlation · Topic

1 Introduction

Multi-label classification (MLC) is a task of predicting labels of new samples based on training sample-label pairs [6,24]. Usually, the number of training samples is high and the number of features for each sample is also high. As involving irrelevant samples or features can negatively impact model performance, the academia has seen many efforts for extracting an informative subset of samples or features for classification. For sample extraction, some works select a subset of common training instances shared by all testing instances [5,25], while some works select a subset of different training instances for each testing instance [23,27]. For feature extraction, some works select the same features for all labels [20,29], while some works select different features for each label [9,10]. In our study, we assume each testing instances should have its own specific training instances. However, most of instance-specific sample extraction methods overlook the gap [1,27] between features and labels. The correlation between instances based on features cannot be assumed to be the same as that based on

© Springer Nature Switzerland AG 2020
A. Vedaldi et al. (Eds.): ECCV 2020, LNCS 12359, pp. 558–573, 2020.
https://doi.org/10.1007/978-3-030-58568-6_33

labels. Although the method in [23] has discovered this problem, it is still not clear why the input-output correlation can be well captured in the learned latent subspace. We also assume each label should have its own specific features. However, most of label-specific feature extraction methods overlook the relationship between features and labels. They often select discriminative features for each label based on label correlations rather than based on feature-label correlations.

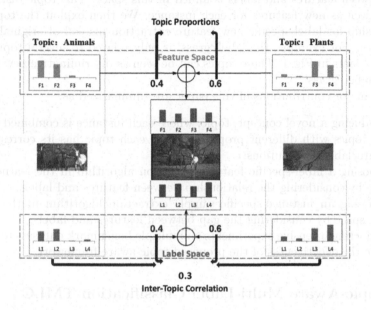

Fig. 1. A simple example of topic-aware data factorization

Based on the above discussion, this paper focuses on bridging the input-output gap and exploiting input-output correlation for sample and feature extraction respectively. Here, we propose a novel topic-aware framework by assuming each sample can be seen as a combination of topics with different proportions. The input and output share the same topic proportions, but they have different feature and label distributions for different topics. For showing a simple example, an image in the corel5k dataset [4] in Fig. 1 can be seen as a weighted combination of topic animals and plants with proportion coefficient being 0.4 and 0.6 respectively, where each topic has its own feature and label distributions. The important labels for topic animals are 'horses' and 'foals' while the important labels for topic plants are 'trees' and 'grass' in the given example. The important features for each topic should also be different. It should be noted that we assume some topics are correlated to each other. Two topics are assumed to be highly correlated to each other if they share similar label distributions or they often co-occur in samples. The topic proportions and feature/label proportions can be mined by non-negative matrix factorization on both input and output spaces. As features and labels share the same latent topic space, there is no gap between features and labels in this space. We can exploit the inter-instance relationship in the latent topic space. This kind of relationship can be directly applied to the output space. For example, if an image and the image in Fig. 1 share the similar topic distribution instead of feature distribution, we assume these two images have similar label sets. Because a shared structure between features and labels is extracted in the latent topic space, the correlation between features and labels is mined in this space. This topic distribution can be seen as new features for each instance. We then exploit the topic-label relationship for label-specific new feature extraction instead of original feature extraction. For example, the label 'grass' is only related to several topics, such as topic plants in Fig. 1. These topics can be seen as discriminative new features for the label 'grass'.

The major contributions of this paper are summarized as:

– Introducing a novel concept, topic, where each instance is combined by multiple topics with different proportions and each topic has its corresponding feature/label distributions;
– Proposing a label-specific feature extraction algorithm in the learned topic space by considering the relationship between features and labels;
– Proposing an instance-specific sample extraction algorithm in the learned topic space by considering the gap between features and labels;
– Conducting intensive experiments on multiple benchmark datasets to demonstrate the effectiveness of the proposed topic-aware framework.

2   Topic-Aware Multi-Label Classification-TMLC

In this section, we introduce the formulation of the proposed framework.

2.1   Preliminaries

Here, the matrix and the vector are denoted by the uppercase character (e.g., Γ) and lowercase character (e.g., γ) respectively. For matrix Γ, its $(i,j)^{th}$ entry is represented as $\Gamma_{i,j}$; its ith row and j^{th} column is represented as $\Gamma_{i,:}$ and $\Gamma_{:,j}$ respectively. The column vector e is a vector with all entries being 1.

Suppose in each multi-labeled dataset, the input data is represented by $X = [X^t, X^s] \in \mathbb{R}^{d \times n}$ with $X^t = [x_1^t, ..., x_{n_t}^t]$ and $X^s = [x_1^s, ..., x_{n_s}^s]$ as training and testing input matrices respectively; the output data is represented by $Y^t = [y_1^t, ..., y_{n_t}^t] \in \{0,1\}^{k \times n_t}$, where n_t, n_s, d and k is the number of training samples, testing samples, features and labels respectively.

2.2   The Overview of TMLC

In order to perform topic-aware multi-label classification, we consider two-level mapping. The commonly used mappings in the traditional MLC algorithms is illustrated in Fig. 2(a). First, the predictive model h of the mapping between X^t

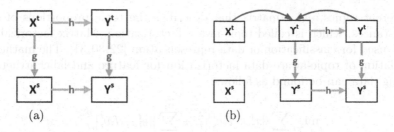

Fig. 2. (a) The mappings in the traditional MLC methods, (b) The mappings in the proposed MLC methods.

and Y^t in the training data can be applied to the testing data [17]. Second, the predictive model g of the mapping between X^t and X^s in the input data can be applied to the output data [14]. In fact, the first and second kind of mapping usually overlooks the input-output correlation and input-output gap respectively. In the proposed topic-aware framework shown in Fig. 2(b), we learn the mapping between Υ^t and Y^t which can be applied to the corresponding testing data in the latent topic space, where Υ^t encodes the input-output correlation by learning it as a shared structure between inputs and outputs. We also learn the mapping between Υ^t and Υ^s which can be applied to the output data, as there exists no gap between Υ^t and Y^t because inputs and outputs share the same Υ.

In our study, Υ^t can be learned from the original X^t and Y^t by assuming each instance can be seen as a combination of different topics where each topic has different feature and label distributions. The detailed exploration of Υ^t and Υ^s in the training and testing spaces respectively can be found in Sect. 2.3. We assume that some topics are often correlated to each other. The extraction of inter-topic correlation which can be used to guide the following topic-aware feature and sample exactions can be found in Sect. 2.4. For the first kind of mapping in Fig. 2(b), we perform label-specific new feature extraction by exploiting the topic-label relationship. The corresponding technical details can be found in Sect. 2.5. For the second kind of mapping in Fig. 2(b), we perform instance-specific sample with new representations extraction by exploiting the inter-instance relationship in the latent topic space. The corresponding technical details can be found in Sect. 2.5.

2.3   Topic-Aware Data Factorization

In our study, we assume each instance can be reconstructed by some topics with different weights. Each topic combines some related labels (objects), where these related labels share a common feature space. Then, the topic space can be seen as a common subspace shared by feature and labels. The common subspace establishes the correlations between the input and output. Specifically, X^t (Y^t) can be decomposed into a matrix that describes the feature (label) distributions of different topics and a matrix that describes the topic proportions of different instances, which is called topic-aware factorization. Matrix factorization is widely used for classification or data representation [22,30,31]. The mathematic formulation of topic-aware data factorization for feature and label structure in training data can be found as follows:

$$\min_{F,L,\Upsilon^t} \sum_i^{n_t} \left\|x_i^t - Fv_i^t\right\|_F^2 + \sum_i^{n_t} \left\|y_i^t - Lv_i^t\right\|_F^2$$
$$+ 2\lambda(\|F\|_1 + \|L\|_1), \quad s.t. \quad F, L, v_i^t \geq 0, \tag{1}$$

where $F = [f_1, ..., f_r] \in \mathbb{R}^{d \times r}$ and each f_i is a vector to indicate the feature distribution in topic i with r being the number of topics. The large (small) value in f_i denotes that the corresponding feature is highly (weakly) related to topic i. $L = [l_1, ..., l_r] \in \mathbb{R}^{k \times r}$ and each l_i is a vector to indicate the label distribution in topic i with r being the number of topics. The feature and label spaces share the same latent space $\Upsilon^t = [v_1^t, ..., v_{n_t}^t] \in \mathbb{R}^{r \times n_t}$ with each v_m^t is used to automatically weight different topics for training sample m, as features and labels are two parallel views to represent each topic but with different distributions. As each topic is only related to a few number of features (labels), the corresponding F (L) should be sparse. The parameter λ is used to control the sparsity.

It should be noted that the ith row of L ($L_{i,:}$) indicates the topic distribution for label i. Then, $L_{i,:}$ can be seen as the representation of label i in the topic space. If label i and label j are highly correlated, the corresponding $\|L_{i,:} - L_{j,:}\|_F^2$ should be small. Then, labels in the latent topic space can be jointly learned by involving label correlation as follows:

$$\min_L \sum_{i,j}^{k,k} C_{i,j} \|L_{i,:} - L_{j,:}\|_F^2, s.t. \quad L \geq 0, \tag{2}$$

where we utilize cosine similarity to calculate C with $C_{i,j} = \cos(Y_{i,:}^t, Y_{j,:}^t)$.

After obtaining the feature structure of topics in the training data (F), the topic proportions of testing samples can be found as follows:

$$\min_{\Upsilon^s} \sum_j^{n_s} \left\|x_j^s - Fv_j^s\right\|_F^2, \quad s.t. \quad v_j^s \geq 0, \tag{3}$$

where $\Upsilon^s = [v_1^s, ..., v_{n_s}^s] \in \mathbb{R}^{r \times n_s}$ and each v_j^s indicates the topic proportions of testing sample j.

The values in feature/label distributions in each topic and topic proportions in each sample are all assumed to be nonnegative, as nonnegativity is consistent with the biological modeling of data [8,13,18], especially for image data.

2.4   Inter-topic Correlation

In our study, we assume topics are correlated to each other, which is always neglected in many works. Because some topics may share similar label distributions and some topics may often co-occur in some samples, we can learn inter-topic correlation based on L which shows the label distributions of different topics and Υ^t which shows the topic combinations of different samples. As topic information can be represented by labels and labels can be inferred by each other, we can learn the inter-label correlation by considering the inter-topic correlation. Thus, the inter-topic correlation can be exploited by the following objective function

$$
\min_{\Xi} \left\| Y^t - L\Xi\Upsilon^t \right\|_F^2 + \left\| C - L\Xi L^T \right\|_F^2 ,
$$

$$
s.t. \ \Xi = \Xi^T, \Xi e = e, \Xi \geq 0, diag(\Xi) = 0,
\tag{4}
$$

where $\Xi \in \mathbb{R}^{r \times r}$ is the matrix to denote the relational coefficients between any pair of topics. In the next sections, we will show that inter-topic correlation can be directly used to guide feature and sample extractions in the topic space.

2.5   Topic-Aware Label-Specific Feature Extraction

One of the objective of our task is to learn the predictive mapping L from latent topics to labels. Each $L_{i,j}$ represents the selection weight for label i to topic j. The higher the relationship between label i and topic j, the higher weight should be given to $L_{i,j}$.

In order to learn the importance of each label in each topic, we add another term to Eq.(1) as follows:

$$
\min_{L,\Upsilon^t} \sum_{i,j}^{k,r} L_{i,j} \left\| Y_{i,:}^t - \Upsilon_{j,:}^t \right\|_F^2 , s.t. \ L, \Upsilon^t \geq 0,
\tag{5}
$$

where $Y_{i,:}^t$ indicates the distribution of label i in different samples and $\Upsilon_{j,:}^t$ indicates the distribution of topic j in different samples. The more similar $Y_{i,:}^t$ and $\Upsilon_{j,:}^t$, the more likely the topic j can be characterized by the label i.

After obtaining the predictive mapping L, the label prediction based on the label-specific feature extraction in the topic space is

$$
Y^s = L\Psi\Upsilon^s,
\tag{6}
$$

when considering inter-topic correlation, where

$$
\Psi = \iota I + (1 - \iota)\Xi,
\tag{7}
$$

where ι is a parameter to balance the weight between a topic and other topics.

2.6   Topic-Aware Instance-Specific Sample Extraction

One of the objective of our task is to learn the predictive mapping $\Theta \in \mathbb{R}^{n_t \times n_s}$ from training to testing samples in the latent topic space. Each $\Theta_{i,j}$ represents the selection weight for training sample i to testing sample j. The higher the relationship between two samples, the higher weight should be given to $\Theta_{i,j}$.

Here, we use Pearson correlation to measure the correlation between each training and testing samples in a topic level as follows:

$$\Phi_{i,j} = \frac{\sum_{p=1}^{r}(\Pi_{p,i}^{t} - \overline{\pi_i^t})(\Pi_{p,j}^{s} - \overline{\pi_j^s})}{(\sqrt{\sum_{p=1}^{r}(\Pi_{p,i}^{t} - \overline{\pi_i^t})^2})(\sqrt{\sum_{p=1}^{r}(\Pi_{p,j}^{s} - \overline{\pi_j^s})^2})}, \tag{8}$$

where

$$\Pi^t = \Psi\Upsilon^t, \Pi^s = \Psi\Upsilon^s, \tag{9}$$

by considering inter-topic correlation in each sample. $\overline{\pi_i^t}$ and $\overline{\pi_j^s}$ denotes the mean value of π_i^t and π_j^s, respectively.

The value of Φ, in the range of $[-1, 1]$, can show the positive and negative relationship between two samples. Then, each $\Theta_{i,j}$ can be learned by considering the sign consistency between Θ and Φ. The closer to 0 the absolute value of $\Phi_{i,j}$ is, the more independent of training sample i and testing sample j is, the less contribution of training sample i when predicting sample j, the smaller absolute value of the corresponding $\Theta_{i,j}$ should be. Then, each $\Theta_{i,j}$ can also be learned by considering the sparsity regularization between Θ and Φ. After combining the above analysis, Θ can be solved by the following objective function

$$\min_{\Theta} \left\| \Pi^t\Theta - \Pi^s \right\|_F^2 + \sum_{i,j}^{n_t, n_s} (-\alpha\Theta_{i,j}\Phi_{i,j} + \beta(1 - |\Phi_{i,j}|)|\Theta_{i,j}|), \tag{10}$$

where α and β are two parameters to balance above three terms.

After obtaining the predictive mapping Θ, the label prediction based on the instance-specific sample extraction is

$$Y^s = Y^t\Theta. \tag{11}$$

2.7   Optimization

Update F, L, Υ^t: Objective functions in Eq. (1), Eq. (2) and Eq. (5) can be combined to form an inequality and non-negative constrained quadratic optimization problem, which can be solved by introducing Lagrangian multipliers. The combined objective function can be extended to

$$\min_{F,L} \left\| X^t - F\Upsilon^t \right\|_F^2 + \left\| Y^t - L\Upsilon^t \right\|_F^2 + tr(Z^1F + Z^2L + Z^3\Upsilon^t)$$
$$+ \sigma tr((E^1A + AE^1 - 2LL^T)C) + \sigma tr((E^2B + SE^2 - 2\Upsilon^tY^{tT})L) \tag{12}$$
$$+ 2\lambda tr(E^1F + E^2L),$$

where σ is a parameter to balance terms of label-topic relationship and topic-based label-label relationship. E^i is an all-one matrix and Z^i is a Lagrange multiplier for nonnegative constraint. $A = diag(LL^T)$, $B = diag(Y^t Y^{tT})$ and $S = diag(\Upsilon^t \Upsilon^{tT})$, where $diag(A)$ indicates the diagonal entries in A. The KKT conditions of $Z_{p,q}^i H_{p,q} = 0$, where H is used to represent any variable from $\{F, L, \Upsilon^t\}$, to the derivative of the above function w.r.t. H can be applied to update one variable while fixing the other two variables, because the objective function is convex when any two variables are fixed. Specifically, each variable can be updated as follows:

$$F_{i,j} \leftarrow F_{i,j} \frac{(X^t \Upsilon^{tT})_{i,j}}{(F\Upsilon^t \Upsilon^{tT} + \lambda)_{i,j}}, \tag{13}$$

$$L_{i,j} \leftarrow L_{i,j} \frac{((1+\sigma)(Y^t \Upsilon^{tT}) + \sigma CL)_{i,j}}{(L\Upsilon^t \Upsilon^{tT} + \sigma(G + (E^2 C)^T \odot L) + \lambda)_{i,j}}, \tag{14}$$

$$\Upsilon_{i,j}^t \leftarrow \Upsilon_{i,j}^t \frac{(F^T X^t + (1+\sigma)L^T Y^t)_{i,j}}{((F^T F + L^T L)\Upsilon^t + (1/2)\sigma(E^3 L)^T \odot \Upsilon^t)_{i,j}}, \tag{15}$$

where $G = 1/2(BE^{2T} + E^{2T} S)$ and $\odot$ indicates the Hadamart product.

Update Υ^s: Similarly, the objective function in Eq. (3) in the testing phase can be extended to

$$\min_{\Upsilon^s} \|X^s - F\Upsilon^s\|_F^2 + Z^4 \Upsilon^s, \tag{16}$$

where ϱ_i is a Lagrange multiplier for sum-one constraint. Then, Υ^s can be updated as follows:

$$\Upsilon_{i,j}^s \leftarrow \Upsilon_{i,j}^s \frac{(F^T X^s)_{i,j}}{(F^T F \Upsilon^s)_{i,j}}. \tag{17}$$

Update Ξ: The problem (4) can be effectively solved by dividing into two subproblems [26] as follows:

$$\Xi = \arg\min_{\Xi} \|\Xi - \Xi_r^p\|_F^2, s.t. \ diag(\Xi) = 0, \Xi e = e, \Xi^T e = e, \tag{18}$$

and

$$\Xi = \arg\min_{\Xi} \|\Xi - \Xi_r^p\|_F^2, s.t. \ \Xi \geq 0, \tag{19}$$

where

$$\Xi_r^p = \frac{(\Xi^p - 1/\eta \nabla_{\Xi} \mathcal{L}(\Xi^p)) + (\Xi^p - 1/\eta \nabla_{\Xi} \mathcal{L}(\Xi^p))^T}{2}, \tag{20}$$

which is obtained from the problem (4) with its first order approximation at the previous point Ξ^p by considering the symmetric constraint $\Xi = \Xi^T$. The parameter η is the Lipschitz parameter, which is calculated according to $\nabla_{\Xi} \mathcal{L}(\Xi)$, is $r\sqrt{2(\sum_{i,j}^{r,r} \max^2((L^T L)_{:,i}(L^T L)_{j,:}) + \sum_{i,j}^{r,r} \max^2((L^T L)_{:,i}(\Upsilon^t \Upsilon^{tT})_{j,:}))}$.

By using the Lagrange multipliers for three constraints in Eq.(18), the first subproblem can be solved by

$$\Xi = \Xi_r^p - \frac{1}{r}tr(\Xi_r^p)I - \frac{r - e^T \Xi_r^p e + tr(\Xi_r^p)}{r - 1}I + R + R^T, \tag{21}$$

where $R = (\frac{I}{r} + \frac{2-r}{2r^2(r-1)}ee^T)(e - \Xi_r^p e + \frac{tr(\Xi_r^p)e}{r})e^T$.

The second subproblem can be solved by

$$\Xi = \lceil \Xi_r^p \rceil_{\geq 0}, \tag{22}$$

where $\lceil \Xi_r^p \rceil_{\geq 0}$ let all negative values in Ξ_r^p change to 0.

Thus, we solve the problem (4) by successively alternating between above two subproblems.

Update Θ: The predictive mapping Θ, which can be easily solved by using proximal gradient descend method. First, the gradient w.r.t. Θ in the problem (10) without considering the sparsity term is

$$\nabla_\Theta \mathcal{L} = \Pi^{tT}\Pi^t\Theta - \Pi^{tT}\Pi^s - \alpha\Phi. \tag{23}$$

The sparsity term can be solved by applying element-wise soft-thresholding operator [10]. Then, Θ can be updated as follows:

$$\Theta_{t+1}(i,j) \leftarrow \text{prox}_{\frac{\beta(1-|\Phi_{i,j}|)}{L_f}}(\Theta^t(i,j) - \frac{1}{L_f}\nabla_\Theta \mathcal{L}(\Theta^t)(i,j)), \tag{24}$$

where $\Theta^t(i,j) = \Theta_t(i,j) + \frac{b_{t-1}-1}{b_t}(\Theta_t(i,j) - \Theta_{t-1}(i,j))$. L_f is the Lipschitz parameter which is treated as the trace of the second differential of $F(\Theta)$ ($L_f = \|\Pi^t\Pi^{tT}\|_F$). The sequence b_t should satisfy the condition of $b_t^2 - b_t \leq b_{t-1}^2$. The prox is defined as

$$\text{prox}_\varepsilon(a) = \text{sign}(a)\max(|a| - \varepsilon, 0). \tag{25}$$

The procedure of the proposed TMLC is summarized in Algorithm 1. The update of Υ^t, L and F requires $O(n_t dr + n_t kr + r^2 d + r^2 k)$, $O(n_t kr + k^2 r + r^2 k)$ and $O(n_t dr)$ respectively. The update of Ξ requires $O(n_t kr + k^2 r + r^2 k + r^2 n_t)$. For each new instance, the update of Υ^s and Θ requires $O(r^2 d + r^2 + rd)$ and $O(n_t r^2 + n_t r)$ respectively. As the proposed method scales linearly with the number of instances, making it suitable for the large-scale datasets.

3   Relations to Previous Works and Discussions

Our work is related to sparse feature extraction methods, because Υ^t can be seen as new features in the proposed topic-aware framework and only several new features are extracted for each label. Feature extraction has been studied over decades, as its corresponding algorithms can be used to select the most informative features for enhancing the classification accuracy. Among these approaches, sparse learning strategies are widely used for their good performance. For example, some works focus on selecting features shared by all labels [3,7,12,20] by imposing $l_{2,1}$-norm regularizer. Although these works can deliver favorable results, some researchers prefer to use $l_{2,0}$-norm regularizer [2,21] to solve the original $l_{2,0}$-norm constrained feature selection problem. However, each label may be related to different features. Recently, certain works impose lasso to select label-specific features [9,10,15,19]. The above works are different from the proposed label-specific feature extraction in the latent topic space algorithm, as they target on selecting a subset of original features for each label whereas we focus on selecting a subset of new features for each label where the correlation between the original features and labels is exploited in the new feature space.

Algorithm 1. Topic-aware Multi-label Classification (TMLC)

Initialize: $f_i = 1/|\Omega_i| \sum_{x_m^t \in \Omega_i} x_m^t$, $l_j = 1/|\Omega_i| \sum_{y_m^t \in \Omega_i} y_m^t$, where $\{\Omega_1, ..., \Omega_r\}$ are r clusters generated from the training data [11], $|\Omega_i|$ is the number of samples in Ω_i, Υ^t and Υ^s as all one matrices, Θ as a random matrix.

1: Compute C with $C_{i,j} = \cos(Y_{i,:}^t, Y_{j,:}^t)$;
2: **Repeat**
3: update Υ^t according to Eq. (13);
4: update L according to Eq. (14);
5: update F according to Eq. (15);
6: **Until Convergence**;
7: **Repeat**
8: update $\varXi$ according to Eq.(21) an Eq.(22);
9: **Until Convergence**;
10: Compute $\varPsi$ according to Eq.(7);
11: **Repeat**
12: update Υ^s according to Eq. (17);
13: **Until Convergence**;
14: compute $\varPi^t$ and $\varPi^s$ according to Eq.(9);
15: compute $\varPhi$ according to Eq.(8);
16: **Repeat**
17: update Θ according to Eq. (24);
18: **Until Convergence**;
19: Predict $Y^s = L\varPsi\Upsilon^s + Y^t\Theta$.

Our work is also related to methods that use kNN technique between training and testing samples. The typical works are lazy kNN [27], CoE [23], LM-kNN [16] and SLEEC [1]. These works select highly correlated k training samples for each testing sample in the original or projected spaces. The labels of testing samples are selected from the labels of these k training samples. The above works are different from the proposed instance-specific sample extraction in the latent topic space algorithm, because they only consider the positive relationship between training sets and testing sets while we consider the positive and negative relationships between these two sets. Specifically, the positively (negatively) related training samples are given positive (negative) weights when predicting the corresponding testing sample. We deem that the classification accuracy can be improved by combining the positively and negatively related training samples, as certain testing samples may share the similar positively related training samples but have different negatively related training samples.

4   Experiments

In this section, results of intensive experiments on real-world multi-labeled datasets are used to demonstrate the effectiveness of the proposed TMLC.

4.1   Datasets

In our study, we conduct experiments on various benchmarks from different domains, where features of image datasets and other datasets are obtained from LEAR[1] and MULAN[2], respectively. Details of these datasets are listed in Table 1. All features of each dataset are normalized in the range of $[0, 1]$.

Table 1. Details of seven benchmarks

Dataset	# samples	# Features	# Labels	Domain
corel5k	4999	1000	260	image
pascal07	9963	1000	20	image
iaprtc12	19627	1000	291	image
espgame	20770	1000	268	image
mirflickr	25000	1000	457	image
yeast	2417	103	14	biology
cal500	502	68	174	music

4.2   Evaluation Metrics

In our experiments, three widely adopted F_1 measures including Macro-F_1, Micro-F_1 and Example-F_1 [28] are used to evaluate the multi-label classification performance.

---

[1] https://lear.inrialpes.fr/people/guillaumin/data.php.
[2] http://mulan.sourceforge.net/datasets-mlc.html.

4.3   Methods

We compared the following state-of-the-art related multi-label methods for classification in the experiments.

1. LM-kNN: it proposes a large margin distance metric learning with k nearest neighbors constraints for multi-label classification [16].
2. CoE: it conducts multi-label classification through the cross-view k nearest neighbor search among learned embeddings [23].
3. JFSC: it learns label-specific features and shared features for the discrimination of each label by exploiting two-order label correlations [10].
4. LLSFDL: it learns label-specific features and class-dependent labels for multi-label classification by mining high-order label correlations [9] .

The former two works are instance-specific sample extraction methods, while the latter two works are label-specific sample extraction methods.

4.4   Experimental Results

These are some parameters that need to be tuned for all the compared methods. In TMLC, the number of topics r is set to 50 and 100 for datasets with the number of instances smaller and larger than $15,000$ respectively. The parameter ι is set to 0.5 for equally weighting of a topic and other topics. The parameters σ and λ are selected from $\{10^{-4}, 10^{-3}, 10^{-2}, 10^{-1}\}$. The parameters α and β are selected from $\{10^{-3}, 10^{-2}, 10^{-1}, 1, 10^{1}, 10^{2}, 10^{3}\}$. These parameters are tuned by

Table 2. Performance in terms of Macro-F_1

Methods	Macro-F1				
	TMLC	CoE	LM-kNN	JFSC	LLSFDL
cal500	**0.240 ± 0.009**	0.111 ± 0.015	0.106 ± 0.004	0.123 ± 0.005	0.153 ± 0.020
corel5k	**0.083 ± 0.007**	0.075 ± 0.016	0.069 ± 0.003	0.044 ± 0.008	0.023 ± 0.000
yeast	**0.473 ± 0.014**	0.391 ± 0.002	0.372 ± 0.014	0.440 ± 0.005	0.434 ± 0.005
iaprtc12	0.135 ± 0.019	**0.150 ± 0.003**	0.132 ± 0.008	0.047 ± 0.002	0.020 ± 0.001
pascal07	**0.347 ± 0.004**	0.323 ± 0.007	0.289 ± 0.006	0.254 ± 0.008	0.215 ± 0.003
espgame	0.086 ± 0.012	**0.141 ± 0.012**	0.137 ± 0.005	0.055 ± 0.014	0.024 ± 0.001
mirflickr	**0.010 ± 0.016**	0.005 ± 0.020	0.002 ± 0.001	0.002 ± 0.019	0.001 ± 0.000

Table 3. Performance in terms of Micro-F_1

Methods	Micro-F1				
	TMLC	CoE	LM-kNN	JFSC	LLSFDL
cal500	**0.475 ± 0.009**	0.412 ± 0.007	0.369 ± 0.002	0.472 ± 0.010	0.468 ± 0.017
corel5k	**0.312 ± 0.013**	0.301 ± 0.001	0.287 ± 0.010	0.109 ± 0.003	0.124 ± 0.006
yeast	**0.651 ± 0.007**	0.620 ± 0.006	0.613 ± 0.011	0.487 ± 0.009	0.482 ± 0.006
iaprtc12	**0.323±0.005**	0.309 ± 0.009	0.296 ± 0.015	0.112 ± 0.003	0.096 ± 0.023
pascal07	0.451 ± 0.009	**0.454 ± 0.006**	0.450 ± 0.010	0.342 ± 0.012	0.335 ± 0.020
espgame	**0.218 ± 0.025**	0.215 ± 0.005	0.212 ± 0.003	0.064 ± 0.013	0.042 ± 0.005
mirflickr	**0.022 ± 0.008**	0.007 ± 0.000	0.005 ± 0.001	0.003 ± 0.010	0.002 ± 0.000

Table 4. Performance in terms of Example-F_1

Methods	Example-F1				
	TMLC	CoE	LM-kNN	JFSC	LLSFDL
cal500	**0.470 ± 0.017**	0.410 ± 0.010	0.361 ± 0.004	0.468 ± 0.009	0.464 ± 0.017
corel5k	**0.296 ± 0.012**	0.255 ± 0.020	0.245 ± 0.009	0.149 ± 0.006	0.112 ± 0.001
yeast	**0.638 ± 0.002**	0.598 ± 0.001	0.583 ± 0.010	0.473 ± 0.008	0.469 ± 0.007
iaprtc12	**0.276 ± 0.004**	0.257 ± 0.010	0.234 ± 0.013	0.093 ± 0.003	0.065 ± 0.005
pascal07	**0.413 ± 0.003**	0.403 ± 0.021	0.388 ± 0.010	0.322 ± 0.018	0.320 ± 0.011
espgame	**0.185 ± 0.018**	0.169 ± 0.016	0.153 ± 0.000	0.048 ± 0.013	0.033 ± 0.003
mirflickr	**0.011 ± 0.011**	0.004 ± 0.017	0.002 ± 0.001	0.001 ± 0.019	0.001 ± 0.000

5-cross validation on the training set. The parameters of other compared methods are tuned according to their corresponding papers.

In the experiment, each dataset is divided into 5 equal-sized subsets. In each run, one subset is used as the testing set and the remaining 4 sets are used as the training set. Each of these subsets is used in turn to be the testing set. Then, there are total 5 runs. Table 2, Table 3 and Table 4 illustrate the average results (mean ± std) of different multi-label algorithms in terms of three different kinds of F_1 measures. Based on the experimental results, the following observations can be made.

1. In the two instance-specific sample extraction algorithms (LM-kNN and CoE), the latter method always delivers the better results. It may due to the fact that like the proposed TMLC, the latter method also exploits the feature-label correlation. CoE mines the feature-label correlation in a cross-view perspective while TMLC mines the feature-label correlation in a topic-view perspective.
2. In the two label-specific feature extraction algorithms (JFSC and LLSFDL), the former method always delivers the better results. It may due to the fact that like the proposed TMLC, the former method also exploits the sample-label correlation. JFSC mines the sample-label correlation based on a discriminant model while TMLC mines the sample-label correlation based on a topic model.
3. Compared with other methods, TMLC always gets better performances, because the proposed method has considered both the gap and correlation between inputs and outputs.

4.5 Parameter Analysis

The proposed label-specific feature extraction has two parameters including σ and λ. To study how these parameters affect the classification results, the performance variances with different values of these two parameters on the corel5k dataset are illustrated in Fig. 3. During this process, the parameters α and β are set to their optimal values. Obviously, the performance is good when λ is larger compared with σ, it indicates that each topic is only related to several labels.

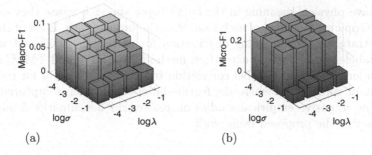

Fig. 3. The results with different σ and λ in the corel5k dataset based on (a) Macro-F_1; (b) Micro-F_1.

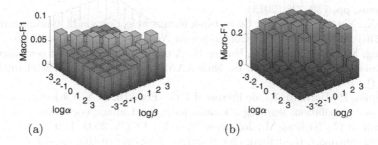

Fig. 4. The results with different α and β in the corel5k dataset based on (a) Macro-F_1; (b) Micro-F_1.

The proposed instance-specific sample extraction also has two parameters including α and β. To study how these parameters affect the classification results, the performance variances with different values of these two parameters on the corel5k dataset are illustrated in Fig. 4. During this process, the parameters σ and λ are set to their optimal values. Clearly, the classification results are always favorable when α and β have equal values. It indicates that sign consistency and sparsity regularization are equally important for instance-specific sample extraction by exploiting inter-sample correlation in the latent topic space.

5 Conclusion

In this study, we propose a novel topic-aware multi-label classification framework where the informative training samples are extracted for each testing sample in the latent topic space and the informative features are also extracted for each label in the latent topic space. Compared with the existing instance-specific sample extraction methods, the proposed TMLC method has bridged the feature-label gap by aligning features and labels in a latent topic space. The mapping between training and testing inputs in the latent topic space can be directly applied to the corresponding outputs. It is worth noting that the input and output have physical meaning in the latent topic space, because they can both illustrate topic proportions of each sample in the latent space. Each of them can also illustrate its corresponding distribution for each topic. Different with the current label-specific feature extraction methods, the proposed TMLC method has considered the feature-label correlation by selecting features for each label in the latent topic space where the feature-label correlation is captured in this space. The intensive empirical studies on real-world benchmarks demonstrate the efficacy of the proposed framework.

References

1. Bhatia, K., Jain, H., Kar, P., Varma, M., Jain, P.: Sparse local embeddings for extreme multi-label classification. In: Advances in Neural Information Processing Systems, pp. 730–738 (2015)
2. Cai, X., Nie, F., Huang, H.: Exact top-k feature selection via l2, 0-norm constraint. In: 23rd International Joint Conference on Artificial Intelligence (2013)
3. Chang, X., Nie, F., Yang, Y., Huang, H.: A convex formulation for semi-supervised multi-label feature selection. In: 28th AAAI Conference on Artificial Intelligence (2014)
4. Duygulu, P., Barnard, K., de Freitas, J.F.G., Forsyth, D.A.: Object recognition as machine translation: learning a lexicon for a fixed image vocabulary. In: Heyden, A., Sparr, G., Nielsen, M., Johansen, P. (eds.) ECCV 2002. LNCS, vol. 2353, pp. 97–112. Springer, Heidelberg (2002). https://doi.org/10.1007/3-540-47979-1_7
5. Elhamifar, E., Sapiro, G., Sastry, S.S.: Dissimilarity-based sparse subset selection. IEEE Trans. Pattern Anal. Mach. Intell. **38**(11), 2182–2197 (2016). https://doi.org/10.1109/TPAMI.2015.2511748
6. Gibaja, E., Ventura, S.: A tutorial on multilabel learning. ACM Comput. Surv. (CSUR) **47**(3), 52 (2015)
7. Guo, Y., Xue, W.: Probabilistic multi-label classification with sparse feature learning. In: IJCAI, pp. 1373–1379 (2013)
8. Hoyer, P.O.: Modeling receptive fields with non-negative sparse coding. Neurocomputing **52**, 547–552 (2003)
9. Huang, J., Li, G., Huang, Q., Wu, X.: Learning label-specific features and class-dependent labels for multi-label classification. IEEE Trans. Knowl. Data Eng. **28**(12), 3309–3323 (2016)
10. Huang, J., Li, G., Huang, Q., Wu, X.: Joint feature selection and classification for multilabel learning. IEEE Trans. Cybern. **48**(3), 876–889 (2017)
11. Huang, S.J., Zhou, Z.H.: Multi-label learning by exploiting label correlations locally. In: 26th AAAI Conference on Artificial Intelligence (2012)
12. Jian, L., Li, J., Shu, K., Liu, H.: Multi-label informed feature selection. In: IJCAI, pp. 1627–1633 (2016)
13. Lee, D.D., Seung, H.S.: Learning the parts of objects by non-negative matrix factorization. Nature **401**(6755), 788 (1999)
14. Liu, H., Li, X., Zhang, S.: Learning instance correlation functions for multilabel classification. IEEE Trans. Cybern. **47**(2), 499–510 (2017)
15. Liu, H., Zhang, S., Wu, X.: Mlslr: multilabel learning via sparse logistic regression. Inf. Sci. **281**, 310–320 (2014)
16. Liu, W., Tsang, I.W.: Large margin metric learning for multi-label prediction. In: 29th AAAI Conference on Artificial Intelligence (2015)

17. Ma, J., Zhang, H., Chow, T.W.S.: Multilabel classification with label-specific features and classifiers: a coarse- and fine-tuned framework. IEEE Trans. Cybern. 1–15 (2019). https://doi.org/10.1109/TCYB.2019.2932439
18. Ma, J., Chow, T.W.: Topic-based algorithm for multilabel learning with missing labels. IEEE Trans. Neural Netw. Learn. Syst. **30**(7), 2138–2152 (2018)
19. Ma, J., Chow, T.W., Zhang, H.: Semantic-gap-oriented feature selection and classifier construction in multilabel learning. IEEE Trans. Cybern. 1–15 (2020)
20. Nie, F., Huang, H., Cai, X., Ding, C.H.: Efficient and robust feature selection via joint $l_{2,1}$-norms minimization. In: Advances in Neural Information Processing Systems, pp. 1813–1821 (2010)
21. Pang, T., Nie, F., Han, J., Li, X.: Efficient feature selection via $l_{\{2,0\}}$-norm constrained sparse regression. IEEE Trans. Knowle. Data Eng. **31**(5), 880–893 (2018)
22. Ren, J., et al.: Learning hybrid representation by robust dictionary learning in factorized compressed space. IEEE Trans. Image Process. **29**, 3941–3956 (2020)
23. Shen, X., Liu, W., Tsang, I.W., Sun, Q.S., Ong, Y.S.: Multilabel prediction via cross-view search. IEEE Trans. Neural Netw. Learn. Syst. **29**(9), 4324–4338 (2017)
24. Tsoumakas, G., Katakis, I.: Multi-label classification: an overview. Int. J. Data Warehouse. Min. **3**(3), 1–13 (2006)
25. Wei, K., Iyer, R., Bilmes, J.: Submodularity in data subset selection and active learning. In: International Conference on Machine Learning, pp. 1954–1963 (2015)
26. Zhang, H., Sun, Y., Zhao, M., Chow, T.W., Wu, Q.J.: Bridging user interest to item content for recommender systems: an optimization model. IEEE Trans. Cybern. **50**, 4268–4280 (2019)
27. Zhang, M.L., Zhou, Z.H.: ML-KNN: a lazy learning approach to multi-label learning. Pattern Recogn. **40**(7), 2038–2048 (2007)
28. Zhang, M.L., Zhou, Z.H.: A review on multi-label learning algorithms. IEEE Trans. Knowl. Data Eng. **26**(8), 1819–1837 (2014)
29. Zhang, R., Nie, F., Li, X.: Self-weighted supervised discriminative feature selection. IEEE Trans. Neural Netw. Learn. Syst. **29**(8), 3913–3918 (2018)
30. Zhang, Z., et al.: Jointly learning structured analysis discriminative dictionary and analysis multiclass classifier. IEEE Trans. Neural Netw. Learn. Syst. **29**(8), 3798–3814 (2017)
31. Zhang, Z., Zhang, Y., Liu, G., Tang, J., Yan, S., Wang, M.: Joint label prediction based semi-supervised adaptive concept factorization for robust data representation. IEEE Trans. Knowl. Data Eng. **32**(5), 952–970 (2019)

Finding It at Another Side:
A Viewpoint-Adapted Matching Encoder
for Change Captioning

Xiangxi Shi[1], Xu Yang[1], Jiuxiang Gu[2], Shafiq Joty[1], and Jianfei Cai[1,3(✉)]

[1] Nanyang Technological University, 50 Nanyang Avenue, Singapore, Singapore
{xxshi,srjoty,asjfcai}@ntu.edu.sg, s170018@e.ntu.edu.sg
[2] Adobe Research, College Park, USA
jigu@adobe.com
[3] Monash University, Clayton, VIC 3800, Australia
jianfei.cai@monash.edu

Abstract. Change Captioning is a task that aims to describe the difference between images with natural language. Most existing methods treat this problem as a difference judgment without the existence of distractors, such as viewpoint changes. However, in practice, viewpoint changes happen often and can overwhelm the semantic difference to be described. In this paper, we propose a novel visual encoder to explicitly distinguish viewpoint changes from semantic changes in the change captioning task. Moreover, we further simulate the attention preference of humans and propose a novel reinforcement learning process to fine-tune the attention directly with language evaluation rewards. Extensive experimental results show that our method outperforms the state-of-the-art approaches by a large margin in both Spot-the-Diff and CLEVR-Change datasets.

Keywords: Image captioning · Change captioning · Attention · Reinforcement learning

1 Introduction

The real world is a complex dynamic system. Changes are ubiquitous and significant in our day-to-day life. As humans, we can infer the underlying information from the detected changes in the dynamic task environments. For instance, a well-trained medical doctor can judge the development of a patient's condition better by comparing the CT images captured in different time steps in addition to locating the lesion.

In recent years, a great deal of research has been devoted to detecting changes in various image content [1,5,7,27], which typically require a model to figure out the differences between the corresponding images. However, most of them focus on the changes in terms of object classes without paying much attention to various attribute changes (e.g.., color and size). Moreover, *change detection*

© Springer Nature Switzerland AG 2020
A. Vedaldi et al. (Eds.): ECCV 2020, LNCS 12359, pp. 574–590, 2020.
https://doi.org/10.1007/978-3-030-58568-6_34

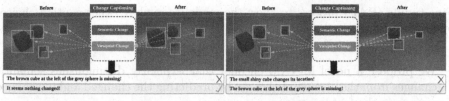

(a) Only viewpoint change happen   (b) Both object change and viewpoint change happen

Fig. 1. Current change captioning methods can be influenced by the viewpoint change between image pairs, which results in generating wrong captions. To address this problem, we propose a novel framework that aims at explicitly separating viewpoint changes from semantic changes.

usually requires pixel-level densely labeled ground-truth data for training the model. Acquiring such labeled data is too time-consuming and labor-intensive, some of which, such as in medical image applications even require human experts for generating annotations.

In addition, current change detection methods usually assume that the differences occur in an ideal situation. However, in practice, changes such as illumination and viewpoint changes occur quite frequently, which makes the change detection task even more challenging.

As humans, we can describe the differences between images accurately, since we can identify the most significant changes in the scene and filter out tiny differences and irrelevant noise. Inspired by the research development from dense object detection to image captioning, which generates natural language descriptions (instead of object labels) to describe the salient information in an image, the *change captioning task* has recently been proposed to describe the salient differences between images [14, 19, 20, 22]. Arguably, captions describing the changes are more accessible (hence preferred) for users, compared to the map-based labels (e.g.., pixel-level binary maps as change labels).

Despite the progress, the existing change captioning methods cannot handle the viewpoint change properly [22]. As shown in Fig. 1, the viewpoint change in the images can overwhelm the actual object change leading to incorrect captions. Handling viewpoint changes is more challenging as it requires the model to be agnostic of the changes in viewpoints from different angles, while being sensitive to other salient changes.

Therefore, in this paper, following the prevailing architecture of a visual encoder plus a sentence decoder, we propose a novel viewpoint-agnostic image encoder, called Mirrored Viewpoint-Adapted Matching (M-VAM) encoder, for the change captioning task. Our main idea is to exhaustively measure the feature similarity across different regions in the two images so as to accurately predict the changed and unchanged regions in the feature space. The changed and unchanged regions are formulated as probability maps, which are used to synthesize the changed and unchanged features. We further propose a Reinforcement Attention

Fine-tuning (RAF) process to allow the model to explore other caption choices by perturbing the probability maps.

Overall, the contributions of this work include:

- We propose a novel M-VAM encoder that explicitly distinguishes semantic changes from the viewpoint changes by predicting the changed and unchanged regions in the feature space.
- We propose a RAF module that helps the model focus on the semantic change regions so as to generate better change captions.
- Our model outperforms the state-of-the-art change captioning methods by a large margin in both Spot-the-Diff and CLEVR-Change datasets. Extensive experimental results also show that our method produces more robust prediction results for the cases with viewpoint changes.

2   Related Work

Image Captioning. Image captioning has been extensively studied in the past few years [3,9,11,12,26,28,30]. Farhadi *et al.* [9] proposed a template-based alignment method, which matches the visual triplets detected from the images with the language triplets extracted from the sentences. Unlike the template-based solutions, Vinyals *et al.* [26] proposed a Long Short-Term Memory (LSTM) network-based method. It takes the image features extracted from a pre-trained CNN as input and generates the image descriptions word by word. After that, Xu *et al.* [28] introduced an attention mechanism, which predicts a spatial mask over encoded features at each time step to select the relevant visual information automatically. Anderson *et al.* [3] attempted to mimic the human's visual system with a top-down attention structure. Yang *et al.* [29] introduced a method to match the scene graphs extracted from the captions and images using Graph Convolutional Networks.

Another theme of improvement is to use reinforcement learning (RL) to address the training and testing mismatch (aka. exposure bias) problem [23,31]. Rennie *et al.* [24] proposed the Self-Critical Sequence Training (SCST) method, which optimizes the captioning model with sentence-level rewards. Gu *et al.* [10] further improved the SCST-based method and proposed a stack captioning model to generate the image description from coarse to fine. Chen *et al.* [6], introduced Recurrent Neural Network based discriminator to evaluate the naturality of the generated sentences.

Captioning for Change Detection. Change detection tasks aim to identify the differences between a pair of images taken at different time steps to recognize the information behind them. It has a wide range of applications, such as scene understanding, biomedical, and aerial image analysis. Bruzzone *et al.* [5] proposed an unsupervised method to overcome the huge cost of human labeling. Wang *et al.* [27] used CNN to predict the changing regions with their proposed mixed-affinity matrix. Recently, Alcantarilla *et al.* [1] proposed a deconvolutional network combined with a multi-sensor fusion SLAM (Simultaneous Localization and Mapping) system to achieve street-view change detection in the real world.

Instead of detecting the difference between images, some studies have explored to describe the difference between images with natural language. Jhamtani et al. [14] first proposed the change captioning task and published a dataset extracted from the VIRAT dataset [18]. They also proposed a language decoding model with an attention mechanism to match the hidden state and features of annotated difference clusters. Oluwasanmi et al. [20] introduced a fully convolutional siamese network and an attention-based language decoder with a modified update function for the memory cell in LSTM. To address the viewpoint change problem in the change captioning task, Park et al. [22] built a synthetic dataset with a viewpoint change between each image pair based on the CLEVR engine. They proposed a Dual Dynamic Attention Model (DUDA) to attend to the images ("before" and "after") for a more accurate change object attention.

In contrast, in our work, we propose a M-VAM encoder to overcome the confusion raised by the viewpoint changes. Compared with Difference Description with Latent Alignment (DDLA) proposed by Jhamtani et al. [14], our method does not need ground-truth difference cluster masks to restrict the attention weights. Instead, our method can self-predict the difference (changed) region maps by feature-level matching without any explicit supervision. Both DUDA [22] and Fully Convolutional CaptionNet (FCC) [20] apply subtraction between image features at the same location in the given image pair, which has problems with large viewpoint changes and thus cannot reliably produce correct change captions. Different from them, our model overcomes the viewpoint change problem by matching objects in the given image pair without the restriction of being at the same feature location.

3   Method

Figure 2 gives an overview of our proposed change captioning framework. The entire framework basically follows the prevailing encoder-decoder architecture and consists of two main parts: a Mirrored Viewpoint-Adapted Matching encoder to distinguishing the semantic change from the viewpoint-change surrounding, and a top-down sentence decoder to generate the captions. During training, we use two types of supervisions: we first optimize our model with traditional cross-entropy (XE) loss with the model at the bottom branch. And then we finetune our model with reinforcement learning using both top and bottom branches, which share the parameters, to further explore the information of objects matched in the image pairs.

3.1   Mirrored Viewpoint-Adapted Matching Encoder

We first encode the given "before" image I_b and the "after" image I_a into the spatial image feature maps $f_b \in \mathbb{R}^{W \times H \times C}$ and $f_a \in \mathbb{R}^{W \times H \times C}$, respectively,

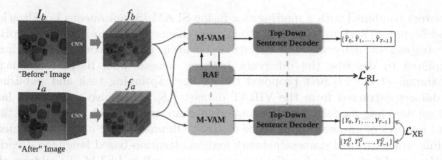

Fig. 2. Overview of the proposed change captioning framework. Our model mainly consists of one M-VAM encoder and one top-down sentence decoder. We use dashed lines to denote that the parameters of two models are shared with each other. Note that with a pre-trained CNN, the model first encodes the *before* and *after* images (I_b and I_a) into features (f_b and f_a) before feeding into M-VAM. The entire model is trained with a cross-entropy loss ($\mathcal{L}_{XE}$) first, followed by a reinforcement learning loss ($\mathcal{L}_{RL}$) based on our RAF process.

with a pretrainned CNN, where W, H, and C are the width, the height, and the channel size of the feature maps, respectively.

Figure 3 shows our proposed M-VAM encoder. It has of two VAM cells with shared parameters. Each VAM cell takes f_b and f_a as input and outputs the changed feature and the unchanged feature. Here, different orders of the inputs lead to different outputs. For example, when a VAM takes the input (f_b, f_a), it treats f_b as the reference and computes the changed (or difference) feature $f_{d_{b \to a}}$ and the unchanged (or background) feature f_{bb}. Similarly, for input order (f_a, f_b), the VAM computes the difference feature $f_{d_{a \to b}}$ and the background feature f_{ba}.

Viewpoint-Adapted Matching (VAM) Cell. Given the extracted image features f_a and f_b, we aim to detect where the changes are in the feature space. The traditional way is to directly compute the difference of the two feature maps [22], which can hardly represent the semantic changes of the two images in the existence of a viewpoint change. To solve this problem, we propose to identify the changed and the unchanged regions by comparing the similarity between different patches in the two feature maps.

In particular, for any cell location (x, y) in f_b and any cell location (i, j) in f_a, we compute their feature-level similarity as follows

$$E_{x,y}(i,j) = f_{b(x,y)}^T f_{a(i,j)} \tag{1}$$

With the obtained $E \in \mathbb{R}^{WH \times WH}$, we then generate a synthesized feature map f_{sb} for image I_b from the other image feature f_a as

$$f_{sb(x,y)} = \sum_{i,j} f_{a(i,j)} \cdot \text{softmax}(E_{x,y}(i,j)) \tag{2}$$

The softmax in Eq. (2) serves as a soft feature selection operation for $f_{a(i,j)}$ according to the similarity matrix E. When an object in $f_{b(x,y)}$ can be found at $f_{a(i,j)}$, the score of $E_{x,y}(i,j)$ should be higher than the score of other features. In this way, the most corresponding feature can be selected out to synthesize the feature of $f_{sb(x,y)}$, i.e., the objects in the *after* image can "move back" to the original position in the *before* image at the feature level.

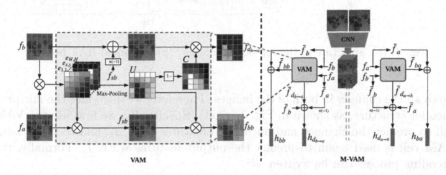

Fig. 3. Left: structure of a VAM Cell. Right: proposed M-VAM encoder, which consists of two VAM cells with shared parameters but with the inputs of f_a and f_b in different orders.

So far, our method just blindly searches for the most similar patch in f_a to reconstruct f_b without taking into account the fact that the most similar one in f_a might not be a correct match. Thus, we further compute an unchanged probability map $U_{x,y} \in [0,1]^{WH}$ and a changed probability map $C_{x,y} \in [0,1]^{WH}$ as follows.

$$U_{x,y} = \max_{i,j} \left(\sigma(a_u E_{x,y}(i,j) + b_u) \right) \tag{3}$$

$$C_{x,y} = 1 - U_{x,y} \tag{4}$$

where σ is the sigmoid function, a_u and b_u are the learned (scalar) parameters. Finally, as shown in Fig. 3 (left), with the obtained probability maps, the original and the synthesized feature maps f_b and f_{sb}, we compute the outputs of the VAM, i.e., the unchanged feature f_{bb} and the changed feature $f_{d_{b \to a}}$ as:

$$f_{bb} = f_{sb} \odot U \tag{5}$$

$$f_{d_{b \to a}} = (f_b - f_{sb}) \odot C \tag{6}$$

where $\odot$ is the element-wise product.

M-VAM. Using one VAM cell with the input order of (f_b, f_a), we can now predict the unchanged and changed features (f_{bb} and $f_{d_{b \to a}}$), but all of them are with respect to the "before" image, I_b. For an ideal change captioner, it should

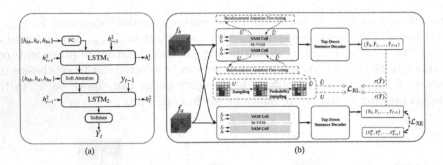

Fig. 4. (a): structure of the top-down decoder. (b): proposed RAF process.

catch all the changes in both of the images. Therefore, we introduce the mirrored encoder structure, as shown in Fig. 3 (right). Specifically, we first use the VAM cell to predict the changed and unchanged features w.r.t. I_b, and then the same VAM cell is used again to predict the output features w.r.t. I_a. Formally, the encoding process can be written as

$$f_{d_{b \to a}}, f_{bb} = \text{VAM}(f_b | f_a) \qquad (7)$$

$$f_{d_{a \to b}}, f_{ba} = \text{VAM}(f_a | f_b) \qquad (8)$$

Note that such a mirror structure also implicitly encourages the consistency of the matched patch pairs from the two opposite directions.

Finally, we treat the features obtained in Eq. (7) and Eq. (8) as the residual information and add them to the original image feature maps:

$$h_{bi} = \bar{f}_{bi} + \bar{f}_i, \qquad (9)$$

$$h_{d_{i \to j}} = \bar{f}_{d_{i \to j}} + \bar{f}_i - \bar{f}_j, \qquad (10)$$

where $i, j \in \{a, b\}$ and $i \neq j$, and $\bar{f}_*$ is the resultant feature vector after applying an average pooling on f_*.

3.2   Sentence Decoder

As shown in Fig. 4(a), we adopt a top-down structure [3] as our sentence decoder. It consists of two LSTM networks and a soft attention module. Since we have two changed features generated from the M-VAM encoder, to obtain a consistent changed feature, we first apply a fully connected layer to fuse the two features

$$h_d = \mathbf{W_1}[h_{d_{a \to b}}, h_{d_{b \to a}}] + \mathbf{b_1} \qquad (11)$$

where $\mathbf{W_1}$ and $\mathbf{b_1}$ are learned parameters.

With the set of the encoded changed and unchanged features $[h_d, h_{bb}, h_{ba}]$, the sentence decoder operates as follows:

$$h_t^1 = \text{LSTM}_1([\mathbf{W_2}[h_d, h_{bb}, h_{ba}] + \mathbf{b_2}, h_{t-1}^2]) \tag{12}$$

$$a_{i,t} = \text{softmax}(\mathbf{W_3} \tanh(h_i + h_t^1) + \mathbf{b_3}), \tag{13}$$

$$h_t^2 = \text{LSTM}_2([\sum a_{i,t} \times h_i, y_{t-1}]) \tag{14}$$

$$Y_t \sim \text{softmax}(\mathbf{W_4} h_t^2 + \mathbf{b_4}) \tag{15}$$

where h_t^1 and h_t^2 are the hidden states of the first LSTM and the second LSTM, respectively, $a_{i,t}$ is the soft attention weight for the visual feature h_i at time step t, $i \in \{d, bb, ba\}$ refers to different input features, y_{t-1} is the previous word embedding, and Y_t is the output word drawn from the dictionary at time step t according to the softmax probability.

3.3   Learning Process

Cross-Entropy Loss. We first train our model by minimizing the traditional cross-entropy loss. It aims to maximize the probability of the ground truth caption words given the two images:

$$\mathcal{L}_{\text{XE}} = -\sum_{t=0}^{T-1} \log(p_\theta(Y_t^G | Y_{0:t-1}^G, I_a, I_b)) \tag{16}$$

where θ is the parameter of our model, and Y_t^G is the t-th ground-truth word.

Reinforcement Attention Fine-Tuning (RAF). Training with the XE loss alone is usually insufficient, and the model suffers from the exposure bias problem [4]. The ground-truth captions of an image contain not only language style information but also information about preferential choices. When it comes to describing the change in a pair of images, a good captioning model needs to consider the relations between the objects and how the objects change. In other words, the model needs to learn what and when to describe. Besides, as humans, when we describe the changed objects, we often include some surrounding information. All these different expressions are related to different visual information in the images in a complicated way. To model such complicated relations, we propose to predict attention regions (i.e., focus areas to describe the changes) in a more human-like way that can help the model to generate more informative captions. Different from image captioning tasks, the images here are encoded only once to decode the sentence (i.e., before staring the decoding, our M-VAM encoder encodes an image pair into features, h_d, h_{bb}, h_{ba}, and keeps them unchanged during the entire decoding process as opposed to computing a different attention or context vector at each decoding step), for which the standard RL would have much larger influence on the decoder rather than on the encoder. Therefore, we propose a RAF process based on the attention perturbation to further optimize our encoder-decoder model.

As shown in Fig. 4(b), our RAF process generates two captions for the same image pair. The first caption $Y = \{Y_0, Y_1, \cdots, Y_{T-1}\}$ is generated in the usual way using our model as shown in the bottom branch of Fig. 4(b), while the second caption $\hat{Y} = \{\hat{Y}_0, \hat{Y}_1, \cdots, \hat{Y}_{T-1}\}$ is generated in a slightly different way according to a sampled probability map $\hat{U}$ (and $\hat{C}$), as shown in the top branch. Here, $\hat{U}$ is a slightly modified map of U by introducing some small noise, which will lead to the generation of a different caption. By comparing the two captions Y and $\hat{Y}$, and rewarding the better one with the reinforcement learning, our model can essentially *explore* for a better solution.

Specially, to generate $\hat{U}$, we first randomly select some elements to perturb from the unchanged probability map U with a sampling rate of γ. For the selected element $U_{i,j}$, we get its perturbed version by sampling from the following Gaussian distribution:

$$\hat{U}_{i,j} \sim \mathcal{N}(U_{i,j}, c) \tag{17}$$

where the standard deviation c is a small number (e.g.., 0.1 in our experiments) that is considered as a hyperparameter. For the elements that are not selected for perturbation, $\hat{U}_{i,j} = U_{i,j}$. With $\hat{U}$, we can then calculate the corresponding changed map $\hat{C}$ using Eq. (4) and generate the caption $\hat{Y}$ accordingly.

After that, we calculate the CIDEr scores for the two captions and treat them as rewards, i.e., $r(Y) = \text{CIDEr}(Y)$ and $r(\hat{Y}) = \text{CIDEr}(\hat{Y})$. If $r(\hat{Y})$ is higher than $r(Y)$, it means that $\hat{Y}$ describes the change of the objects better than Y (according to CIDEr). Consequently, the predicted probability map U should be close to the sampled probability map $\hat{U}$. On other hand, if $r(\hat{Y})$ is smaller than $r(Y)$, it suggests that the sampled probability map is worse than the original one, thus should not be used to guide the learning. Based on this intuition, we define our reinforcement loss as follows.

$$\mathcal{L}_{\text{RL}} = \mathbb{1}^+(r(\hat{Y}) - r(Y))||\hat{U} - U||_2^2, \tag{18}$$

where $\mathbb{1}^+(x)$ represents $\max(x, 0)$. The gradients from $\mathcal{L}_{\text{RL}}$ will backward to the M-VAM module. The overall training loss becomes:

$$\mathcal{L}_{\text{XE+RL}} = \lambda \mathcal{L}_{\text{RL}} + \mathcal{L}_{\text{XE}} \tag{19}$$

where λ is a hyper-parameter to control the relative weights of the two losses.

4   Experiment

4.1   Datasets and Metrics

We evaluate our method on two published datasets: Spot-the-Diff [14] and CLEVR-Change [22]. Spot-the-Diff contains 13,192 image pairs extracted from the VIRAT dataset [18], which were captured by stationary ground cameras. Each image pair can differ in multiple ways; on average, each image pair contains 1.86 reported differences. CLEVR-Change is a synthetic dataset with a set of basic geometry objects, which is generated based on the CLEVR engine. Different from Spot-the-Diff dataset, to address the viewpoint change problem, CLEVR-Change involves a viewpoint change between each image pair. The dataset contains 79,606 image pairs and 493,735 captions. The changes in the dataset can be categorized into six cases, which are Color, Texture, Movement, Add, Drop, and Distractor.

We evaluate our generated captions with five evaluation metrics: BLEU [21], METEOR [8], CIDEr [25], ROUGE_L [17], and SPICE [2]. For comparison in terms of BLEU scores, we mainly show the results for BLEU4 which reflects the matching ratio of 4-word sub-sequences between the generated caption and the ground truth caption.

4.2   Training Details

We use ResNet101 [13] trained on the ImageNet dataset [16] to extract the image features. The images are resized to 224×224 before extracting the features. The feature map size is $\mathbb{R}^{14 \times 14 \times 1024}$. During the training, we set the maximum length of captions as 20 words. The dimension of hidden states in the Top-Down decoder is set to 512. We apply a drop-out rate of 0.5 before the word probability prediction (Eq.(15)). We apply ADAM [15] optimizer for training.

When training with the cross-entropy loss, we train the model for 40 epochs with a learning rate of 0.0002. During the reinforcement attention fine-tuning (RAF) process, we set the learning rate to 0.00002. Due to the low loss value from Eq. (18), we set the hyperparameter λ in Eq. (19) as 1000.0 to put more weight to the reinforcement learning loss. In the RAF process, we set the random sampling rate γ as 0.025. As for the Gaussian distribution (Eq. (17)), we set c as 0.1.

4.3   Model Variations

To analyze the contribution of each component in our model, we consider the following two variants of our model.

1) **M-VAM**: This variant models the encoder-decoder captioning framework with the proposed M-VAM encoder introduced in Sect. 3.1 and the top-down decoder introduced in Sect. 3.2.
2) **M-VAM + RAF**: This variant fine-tunes M-VAM with the proposed RAF process.

4.4   Results

Quantitative Results on Spot-the-Diff. Quantitative results of different methods on Spot-the-Diff dataset are shown at Table 1. We compare with the four existing methods – FCC [20], SDCM [19], DDLA [14] and DUDA [22]. It can be seen that both of our models outperform existing methods in all evaluation measures. Compared with FCC, M-VAM provides improvements of 0.2, 1.4 and

1.3 in BLEU4, ROUGE_L and CIDEr, respectively. After using the RAF process, the scores are further improved from 10.1 to 11.1 in BLEU4, 31.3 to 33.2 in ROUGE_L, 38.1 to 42.5 in CIDEr, and 14.0 to 17.1 in SPICE. The big gains in CIDEr (+4.4) and SPICE (+3.1) for the RAF fine-tuning demonstrate its effectiveness in change captioning.

Quantitative Results on CLEVR-Change. In Table 2, we compare our results with DUDA [22] on the CLEVR-Change dataset. CLEVR-Change contains additional distractor image pairs, which may not be available in other scenarios. Therefore, we performed separate experiments with and without the distractors to show the efficacy of our method.

Table 1. Comparisons of the results of different methods on Spot-the-Diff dataset.

Model	BLEU4	ROUGE_L	METEOR	CIDEr	SPICE
DUDA [22]	8.1	28.3	11.5	34.0	-
FCC [20]	9.9	29.9	**12.9**	36.8	-
SDCM [19]	9.8	29.7	12.7	36.3	-
DDLA [14]	8.5	28.6	12.0	32.8	-
M-VAM	10.1	31.3	12.4	38.1	14.0
M-VAM + RAF	**11.1**	**33.2**	**12.9**	**42.5**	**17.1**

Table 2. Comparisons of the results of different methods on CLEVR-Change dataset.

Model	BLEU4	ROUGE_L	METEOR	CIDEr	SPICE
DUDA [22]	47.3	-	33.9	112.3	24.5
With Distractor					
M-VAM	50.3	69.7	37.0	114.9	30.5
M-VAM + RAF	**51.3**	70.4	**37.8**	115.8	30.7
Without Distractor					
M-VAM	45.5	69.5	35.5	117.4	29.4
M-VAM + RAF	50.1	**71.0**	36.9	**119.1**	**31.2**

We can see that our model tested on the entire dataset (with distractor) achieves state-of-the-art performance and outperforms DUDA by a wide margin. Our M-VAM pushes the BLEU4, METEOR, SPICE and CIDEr scores from 47.3, 33.9, 24.4 and 112.3 to 50.3, 69.7, 37.0, 30.7 and 114.9, respectively. The RAF process further improves the scores by about 1 point in all measures.

Our model also achieves impressive performance on the dataset without the distractors. The M-VAM achieves a CIDEr score of 117.4. RAF further improves the scores – from 45.5 to 50.1 in BLEU4 and from 117.4 to 119.1 in CIDEr. The gains are higher in this setting than the gains with the entire dataset. There is also a visible increase in SPICE compared with the models trained with the distractors (from 30.7 to 31.2). This is because, without the influence of the distractors, the model can distinguish the semantic change more accurately from the viewpoint change.

Table 3. The Detailed breakdown of evaluation on the CLEVR-Change dataset.

Model	Metrics	C	T	A	D	M	DI
DUDA [22]	CIDEr	120.4	86.7	108.2	103.0	56.4	110.8
M-VAM + RAF (w DI)	CIDEr	122.1	98.7	126.3	115.8	82.0	**122.6**
M-VAM + RAF (w/o DI)	CIDEr	**135.4**	**108.3**	**130.5**	**113.7**	**107.8**	-
DUDA [22]	METEOR	32.8	27.3	33.4	31.4	23.5	45.2
M-VAM + RAF(w DI)	METEOR	35.8	32.3	37.8	36.2	27.9	**66.4**
M-VAM + RAF (w/o DI)	METEOR	**38.3**	**36.0**	**38.1**	**37.1**	**36.2**	-
DUDA [22]	SPICE	21.2	18.3	22.4	22.2	15.4	28.7
M-VAM + RAF (w DI)	SPICE	28.0	26.7	30.8	32.3	22.5	**33.4**
M-VAM + RAF (w/o DI)	SPICE	**29.7**	**29.9**	**32.2**	**33.3**	**30.7**	-

Table 3 shows the detailed breakdown of the evaluation according to the categories defined in [22]. In the table, "Metrics" represent the metric used to evaluate the results for a certain model. "C" represents the changes related to "Color" attributes. Similarly, "T", "A", "D", "M" and "DI" represent "Texture", "Add", "Drop", "Move" and "Distractor", respectively.

Comparing our results with DUDA, we see that our method can generate more accurate captions for all kinds of changes. In particular, it obtains large improvements in CIDEr for "Move" and "Add" cases (from 108.2 to 126.3 and from 56.4 to 82.0, respectively). For SPICE, the "Drop" cases have the largest improvement (from 22.2 to 32.3) among the six cases. This indicates that our method can better distinguish the object movement from a viewpoint change scene, which is the most challenging case as reported in [22]. The model can also match the objects well when objects are missing from the 'before' image or added to the 'after' image. Referring to the METEOR score, we get the highest improvement for the "Distractor" cases. This means that our method can better discriminate the scene only with the viewpoint change from the other cases.

Comparing the model trained with and without the distractors, we can see that the performance is further boosted in "Color", "Texture" and "Move" cases. The SPICE scores also show an increase in the "Move" case (from 22.5 to 30.7). It means that such cases are easier to be influenced by the distractor data. This is because the movement of objects can be confused with the viewpoint movement. As for "Color" and "Texture", it may be caused by the surrounding description mismatching according to the observation. When the model is trained without the distractors, our RAF process, which is proposed to capture a better semantic surrounding information, is more efficient and thus improves the performance in "Color" and "Texture".

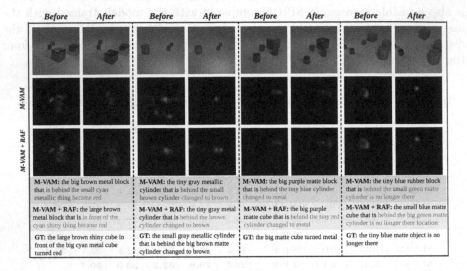

Fig. 5. Visualization of examples from the CLEVR-Change dataset. Red and blue color respectively denote the inconsistent and consistent words. The middle two rows show the changed probability maps of the two models. (Color figure online)

Qualitative Results. We show the visualization of some examples from the CLEVR-Change dataset in Fig. 5. The images at the top of the figure are the given image pairs. The images in the middle two rows are the changed probability maps of M-VAM and M-VAM+RAF, respectively. Comparing the probability maps, we can see that the maps after the RAF process are brighter and more concentrated at the changed regions than the ones from M-VAM, which indicates RAF can force the model to focus on the actual changed regions. Note that a rigorous overlap comparison is not meaningful here since the model does not perform an explicit change detection to find the objects in the pixel level (the size of the attention map is only 14 × 14).

In the first example, we can see that the attention map from M-VAM covers the changed object in the image only partially, while also attending to some irrelevant portions above the changed object. Thus, with the shifted attention region in the "after" image, the model mistakenly infers that the cube is behind (as opposed to in the front) the cyan cube. The RAF fine-tuning helps the model focus more appropriately and model the change correctly. In the second column, we observe that RAF can disregard small irrelevant attentions and keep concentration on the semantic change that matters (e.g.., the brown cylinder in the "after" image in the second column). In the third column, the example

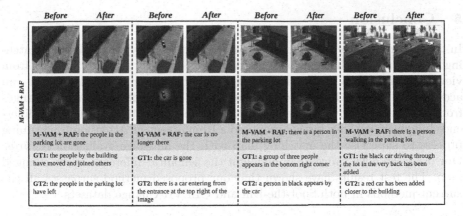

Fig. 6. Visualizations of the generated captions on Spot-the-Diff.

shows that our model can detect changes in *Texture* successfully, which has a low reported score in Park *et al.* [22]. In the last column, we can see that our model shows robustness to large viewpoint change; it can predict a clear and accurate changed probability map, filtering the distraction caused by the viewpoint shift. The small bright region in the "after" image represents a location where the blue cube is missing.

When summarizing an image with a short description like caption, humans tend to describe the surroundings and relationships with respect to the central entities. In the third column, the M-VAM picks "tiny blue cylinder" to describe the change of "big purple matte cube", whereas the M-VAM+RAF picks "tiny red cylinder" which is closer to the central entity.

Figure 6 shows examples from the Spot-the-Diff dataset. From the first to the third column, we can see that our model is also good at finding the difference in real-world scenes. In the first example, our model could successfully detect the people and their non-existence. In the second example, our model can detect the car and the fact that it is not in its position anymore. Although there is a mismatch at the changed probability map of the "after" image, the information from the 'before' image is strong enough to filter such noise and generate an accurate caption. Similarly, our model can successfully detect the changes between the images in the third example and focus on the most salient information (i.e., the appearance of a person near the car) to generate the caption somewhat correctly.

A negative sample is shown in the fourth column. We suspect this is because of the resize operation,due to which some objects become too small to be recognized. As a result, although the matching map is able to find all the changes, the semantic change information is overshadowed by the information in the surrounding, leading to a misunderstanding.

5   Conclusion

In this paper, we have introduced the novel Mirrored Viewpoint-Adapted Matching encoder for the change captioning task to distinguish semantic changes from viewpoint changes. Compared with other methods, our M-VAM encoder can accurately filter out the viewpoint influence and figure out the semantic changes from the images. To further utilize the language information to guide the difference judgment, we have further proposed a Reinforcement Attention Fine-tuning process to supervise the matching map with the language evaluation rewards. Our proposed RAF process can highlight the semantic difference in the changed map. The test results have shown that our proposed method surpasses all the current methods in both Spot-the-Diff and CLEVR-Change datasets.

Acknowledgement. This research is partially supported by the MOE Tier-1 research grants: RG28/18 (S) and the Monash University FIT Start-up Grant.

References

1. Alcantarilla, P.F., Stent, S., Ros, G., Arroyo, R., Gherardi, R.: Street-view change detection with deconvolutional networks. Auton. Robots **42**(7), 1301–1322 (2018)
2. Anderson, P., Fernando, B., Johnson, M., Gould, S.: SPICE: semantic propositional image caption evaluation. In: Leibe, B., Matas, J., Sebe, N., Welling, M. (eds.) ECCV 2016, Part V. LNCS, vol. 9909, pp. 382–398. Springer, Cham (2016). https://doi.org/10.1007/978-3-319-46454-1_24
3. Anderson, P., et al.: Bottom-up and top-down attention for image captioning and visual question answering. In: Proceedings of the IEEE Conference on Computer Vision and Pattern Recognition, pp. 6077–6086 (2018)
4. Bengio, S., Vinyals, O., Jaitly, N., Shazeer, N.: Scheduled sampling for sequence prediction with recurrent neural networks. In: Proceedings of the Advances in Neural Information Processing Systems, pp. 1171–1179 (2015)
5. Bruzzone, L., Prieto, D.F.: Automatic analysis of the difference image for unsupervised change detection. IEEE Trans. Geosci. Remote Sensi. **38**(3), 1171–1182 (2000)
6. Chen, C., Mu, S., Xiao, W., Ye, Z., Wu, L., Ju, Q.: Improving image captioning with conditional generative adversarial nets. In: Proceedings of the Thirty-Third AAAI Conference on Artificial Intelligence, vol. 33, pp. 8142–8150 (2019)
7. Daudt, R.C., Le Saux, B., Boulch, A.: Fully convolutional siamese networks for change detection. In: Proceedings of the 2018 25th IEEE International Conference on Image Processing, pp. 4063–4067. IEEE (2018)
8. Denkowski, M., Lavie, A.: Meteor universal: language specific translation evaluation for any target language. In: Proceedings of the Ninth workshop on Statistical Machine Translation, pp. 376–380 (2014)
9. Farhadi, A., et al.: Every picture tells a story: generating sentences from images. In: Daniilidis, K., Maragos, P., Paragios, N. (eds.) ECCV 2010, Part IV. LNCS, vol. 6314, pp. 15–29. Springer, Heidelberg (2010). https://doi.org/10.1007/978-3-642-15561-1_2
10. Gu, J., Cai, J., Wang, G., Chen, T.: Stack-captioning: Coarse-to-fine learning for image captioning. In: Proceedings of the Thirty-Second AAAI Conference on Artificial Intelligence (2018)

11. Gu, J., Joty, S., Cai, J., Wang, G.: Unpaired image captioning by language pivoting. In: Ferrari, V., Hebert, M., Sminchisescu, C., Weiss, Y. (eds.) ECCV 2018, Part I. LNCS, vol. 11205, pp. 519–535. Springer, Cham (2018). https://doi.org/10.1007/978-3-030-01246-5_31

12. Gu, J., Joty, S., Cai, J., Zhao, H., Yang, X., Wang, G.: Unpaired image captioning via scene graph alignments. In: Proceedings of the IEEE International Conference on Computer Vision (2019)

13. He, K., Zhang, X., Ren, S., Sun, J.: Deep residual learning for image recognition. In: Proceedings of the IEEE Conference on Computer Vision and Pattern Recognition, pp. 770–778 (2016)

14. Jhamtani, H., Berg-Kirkpatrick, T.: Learning to describe differences between pairs of similar images. In: arXiv preprint arXiv:1808.10584 (2018)

15. Kingma, D.P., Ba, J.: Adam: A method for stochastic optimization. arXiv preprint arXiv:1412.6980 (2014)

16. Krizhevsky, A., Sutskever, I., Hinton, G.E.: Imagenet classification with deep convolutional neural networks. In: Proceedings of the Advances in Neural Information Processing Systems, pp. 1097–1105 (2012)

17. Lin, C.Y.: Rouge: a package for automatic evaluation of summaries. In: Text summarization branches out, pp. 74–81 (2004)

18. Oh, S., et al.: A large-scale benchmark dataset for event recognition in surveillance video. In: Proceedings of the IEEE Conference on Computer Vision and Pattern Recognition, pp. 3153–3160. IEEE (2011)

19. Oluwasanmi, A., Aftab, M.U., Alabdulkreem, E., Kumeda, B., Baagyere, E.Y., Qin, Z.: Captionnet: automatic end-to-end siamese difference captioning model with attention. IEEE Access **7**, 106773–106783 (2019)

20. Oluwasanmi, A., Frimpong, E., Aftab, M.U., Baagyere, E.Y., Qin, Z., Ullah, K.: Fully convolutional captionnet: siamese difference captioning attention model. IEEE Access **7**, 175929–175939 (2019)

21. Papineni, K., Roukos, S., Ward, T., Zhu, W.J.: Bleu: a method for automatic evaluation of machine translation. In: Proceedings of the 40th Annual Meeting on Association for Computational Linguistics, pp. 311–318. Proceedings of Association for Computational Linguistics (2002)

22. Park, D.H., Darrell, T., Rohrbach, A.: Robust change captioning. In: Proceedings of the IEEE International Conference on Computer Vision, pp. 4624–4633 (2019)

23. Ranzato, M., Chopra, S., Auli, M., Zaremba, W.: Sequence level training with recurrent neural networks. arXiv preprint arXiv:1511.06732 (2015)

24. Rennie, S.J., Marcheret, E., Mroueh, Y., Ross, J., Goel, V.: Self-critical sequence training for image captioning. In: Proceedings of the IEEE Conference on Computer Vision and Pattern Recognition, pp. 7008–7024 (2017)

25. Vedantam, R., Zitnick, C.L., Parikh, D.: Cider: Consensus-based image description evaluation. In: Proceedings of the IEEE Conference on Computer Vision and Pattern Recognition (2015)

26. Vinyals, O., Toshev, A., Bengio, S., Erhan, D.: Show and tell: a neural image caption generator. In: Proceedings of the IEEE Conference on Computer Vision and Pattern Recognition, pp. 3156–3164 (2015)

27. Wang, Q., Yuan, Z., Du, Q., Li, X.: Getnet: A general end-to-end 2-D CNN framework for hyperspectral image change detection. IEEE Trans. Geosci. Remote Sens. **57**(1), 3–13 (2018)

28. Xu, K., et al.: Show, attend and tell: Neural image caption generation with visual attention. In: Proceedings of the International Conference on Machine Learning, pp. 2048–2057 (2015)

29. Yang, X., Tang, K., Zhang, H., Cai, J.: Auto-encoding scene graphs for image captioning. In: Proceedings of the IEEE Conference on Computer Vision and Pattern Recognition, June 2019
30. Yang, X., Zhang, H., Cai, J.: Deconfounded image captioning: A causal retrospect. arXiv preprint arXiv:2003.03923 (2020)
31. Yu, L., Zhang, W., Wang, J., Yu, Y.: Seqgan: sequence generative adversarial nets with policy gradient. In: Proceedings of the Thirty-First AAAI Conference on Artificial Intelligence (2017)

Attract, Perturb, and Explore: Learning a Feature Alignment Network for Semi-supervised Domain Adaptation

Taekyung Kim and Changick Kim[(✉)]

Korea Advanced Institute of Science and Technology, Daejeon, South Korea
{tkkim93,changick}@kaist.ac.kr

Abstract. Although unsupervised domain adaptation methods have been widely adopted across several computer vision tasks, it is more desirable if we can exploit a few labeled data from new domains encountered in a real application. The novel setting of the semi-supervised domain adaptation (SSDA) problem shares the challenges with the domain adaptation problem and the semi-supervised learning problem. However, a recent study shows that conventional domain adaptation and semi-supervised learning methods often result in less effective or negative transfer in the SSDA problem. In order to interpret the observation and address the SSDA problem, in this paper, we raise the intra-domain discrepancy issue within the target domain, which has never been discussed so far. Then, we demonstrate that addressing the intra-domain discrepancy leads to the ultimate goal of the SSDA problem. We propose an SSDA framework that aims to align features via alleviation of the intra-domain discrepancy. Our framework mainly consists of three schemes, i.e., attraction, perturbation, and exploration. First, the attraction scheme globally minimizes the intra-domain discrepancy within the target domain. Second, we demonstrate the incompatibility of the conventional adversarial perturbation methods with SSDA. Then, we present a domain adaptive adversarial perturbation scheme, which perturbs the given target samples in a way that reduces the intra-domain discrepancy. Finally, the exploration scheme locally aligns features in a class-wise manner complementary to the attraction scheme by selectively aligning unlabeled target features complementary to the perturbation scheme. We conduct extensive experiments on domain adaptation benchmark datasets such as DomainNet, Office-Home, and Office. Our method achieves state-of-the-art performances on all datasets.

Keywords: Domain adaptation · Semi-supervised learning

Electronic supplementary material The online version of this chapter (https://doi.org/10.1007/978-3-030-58568-6_35) contains supplementary material, which is available to authorized users.

1 Introduction

Despite the promising success of deep neural networks in several computer vision tasks, these networks often show performance degradation when tested beyond the training environment. One way to mitigate this problem is to collect large amounts of data from the new domain and train the network. Such heavy demands on data annotation cause great interest in domain adaptation and semi-supervised learning on deep neural networks. However, most recent studies on deep domain adaptation are focused on unsupervised approaches, and deep semi-supervised learning is still concentrated on addressing the identical domain problem. Though these methods can be directly applied to the semi-supervised domain adaptation (SSDA) problem only with an additional supervision on the extra labeled samples, a recent study [26] reveals that unsupervised domain adaptation (UDA) methods and semi-supervised learning (SSL) methods often show less effective or even worse performances than just training on the labeled source and target samples in the SSDA problem.

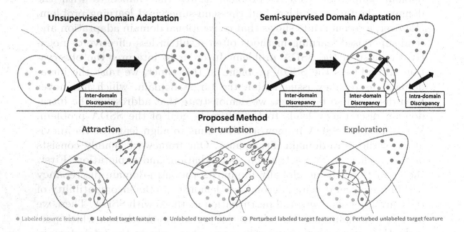

Fig. 1. Conceptual descriptions of the feature alignment approaches. The top row describes the different feature alignment behaviors between the UDA and SSDA problem. Supervision on labeled target samples attracts the corresponding features and their neighborhood toward the source feature cluster, which causes the intra-domain discrepancy. The bottom row describes the proposed attraction, perturbation, and exploration schemes, which are explained in Sect. 4 in detail.

In this paper, we introduce a new concept called *intra-domain discrepancy* to analyze the failure of the UDA and SSL methods and address the SSDA problems. Intra-domain discrepancy is a chronic issue in the SSDA problem that occurs during labeled sample supervision, but has never been discussed so far. In the UDA problem, supervision on the labeled source samples does not critically affect the target domain distribution in general but implicitly attracts some alignable target features similar to the source features. Thus, aligning the source and target domains by reducing their inter-domain discrepancy is reasonable. However, in the SSDA problem, supervision on the labeled target samples enforces the corresponding features and their neighborhood to be attracted toward source feature clusters, which guarantees partial alignment between two domain distributions. Besides, unlabeled target samples that less correlate with the labeled target samples are less affected by the supervision and eventually remain unaligned (Top row in Fig. 1). Thus, the target domain distribution is separated into an aligned target subdistribution and an unaligned target subdistribution, causing the intra-domain discrepancy within the target domain. The failure of the UDA and SSL methods will be discussed in Sect. 3 in detail.

Motivated by the insight, we propose an SSDA framework that aligns cross-domain features by addressing the intra-domain discrepancy within the target domain. Our framework focuses on enhancing the discriminability on the unaligned target samples and modulating the class prototypes, the representative features of each class. It consists of three schemes, i.e., attraction, perturbation, and exploration, as shown in Fig. 1. First, the attraction scheme aligns the unaligned target subdistribution to the aligned target subdistribution through the intra-domain discrepancy minimization. Second, we discuss why conventional adversarial perturbation methods are ineffective in the SSDA problem. Unlike these approaches, our perturbation scheme perturbs target subdistributions into their intermediate region to propagate labels to the unaligned target subdistribution. Note that our perturbation scheme does not ruin the already aligned target features since it additionally generates perturbed features temporarily for regularization. Finally, the exploration scheme locally modulates the prototypes in a class-aware manner complementary to the attraction and perturbation schemes. We perform extensive experiments to evaluate the proposed method on domain adaptation datasets such as DomainNet, Office-Home, Office, and achieved state-of-the-art performances. We also deeply analyze our methods in detail.

Our contributions can be summarized as follows:

- We introduce the intra-domain discrepancy issue within the target domain in the SSDA problem.
- We propose an SSDA framework that addresses the intra-domain discrepancy issues via three schemes, i.e., attraction, perturbation, and exploration.
  - The attraction scheme aligns the unaligned target subdistribution to the aligned target subdistribution through the intra-domain discrepancy minimization.
  - The perturbation scheme perturb target subdistributions into their intermediate region to propagate labels to the unaligned target subdistribution.
  - The exploration scheme locally modulate the prototypes in a class-aware manner complementary to the attraction and perturbation schemes.
- We conduct extensive experiments on DomainNet, Office-Home, and Office. We achieve state-of-the-art performances among various methods, including vanilla deep neural networks, UDA, SSL, and SSDA methods.

2   Related Work

2.1   Unsupervised Domain Adaptation

The recent success of deep learning-based approaches and the following enormous demand for massive amounts of data attract great interest in domain adaptation (DA). Even in the midst of significant interest, most recent works are focused on unsupervised domain adaptation (UDA). Recent UDA methods can be categorized into three approaches. The first approach is to reduce the cross-domain divergence. This can be achieved by minimizing the estimated domain divergence such as MMD [9] or assimilating feature distributions through adversarial confusion using a domain classifier [7,17–19]. The second approach is to translate the appearance of one domain into the opposite domain so that the translated data can be regarded as sampled from the opposite domain [10,12,13]. The last approach is to consider the source domain as partially labeled data and utilize the semi-supervised learning schemes. For example, Drop-to-Adapt [16] exploits a virtual adversarial perturbation scheme [21]. Recently, these approaches are widely adopted across several computer vision tasks beyond image classification such as object detection [4,14,28], semantic segmentation [11,30], person re-identification [35], and even in depth estimation [34].

2.2   Semi-supervised Learning

Similar to domain adaptation (DA), semi-supervised learning (SSL) has also attracted great attention as a way to overcome the shortages of the labeled data. The difference between DA and SSL is that domain adaptation assumes to deal with data sampled from two distributions with significant domain discrepancy, while SSL assumes to deal with the labeled and unlabeled data sampled from the identical distribution. With the rise of the deep learning approaches, several methods have been recently proposed for deep SSL. Some works add data augmentation and regularize the model by enforcing a consistency between the given and the augmented data [15,29]. Miyato *et al.* [21] extend this scheme by adversarially searching the bounded and small perturbation which leads the model to the most unstable state. Laine and Aila [15] ensemble the prediction of the model by averaging them throughout the training phase, while Targainen and Valpola [29] ensemble the parameter of the model itself. Other few works use self-training schemes with a memory module or a regularization through convergence speed [3,5]. Recently, Wang *et al.* [32] propose an augmentation distribution alignment approach to explicitly address the empirical distribution mismatch problem in semi-supervised learning.

2.3   Semi-supervised Domain Adaptation

Semi-supervised domain adaptation (SSDA) is an important task which bridges the well-organized source distribution toward target distribution via partially labeled target samples, while a few works have explored so far [1,6,26,33]. Donahue *et al.* [6] address the domain discrepancy by optimizing the auxiliary constrains on the labeled data. Yao *et al.* [33] learn a subspace that can reduce the data distribution mismatch. Ao *et al.* [1] estimate the soft label of the given labeled target sample with the source model and interpolated with the hard label for target model supervision. Saito *et al.* [26] minimize the distance between the unlabeled target samples and the class prototypes through minimax training on entropy. However, none of these methods discuss and mitigate the intra-domain discrepancy issue in the SSDA problem. Different from previous works, we address the SSDA problem with a new perspective of the intra-domain discrepancy.

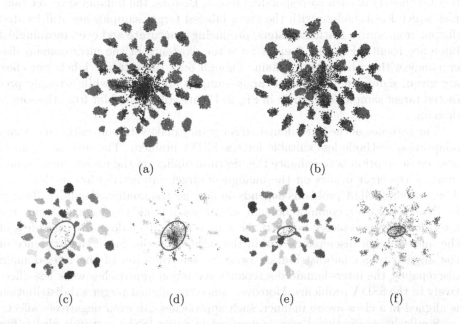

Fig. 2. (a)–(f) The t-SNE visualization of the source and target features in (a) the UDA problem and (b) the SSDA problem with three target labels for each class. We adopted the Real to Sketch scenario of the DomainNet dataset on the AlexNet backbone and visualized for partial classes. (c) and (e) visualize the source distribution of the UDA and SSDA problems and (d) and (f) visualize the target distribution of the UDA and SSDA problems, respectively. Even only three labeled target samples per class can attract their neighborhoods in this degree and separate the target domain into the aligned and unaligned subdistributions. (Color figure online)

3 Intra-domain Discrepancy

Intra-domain discrepancy of a domain is an internal distribution gap among subdistributions within the domain. Though we demonstrate the intra-domain discrepancy issue in the semi-supervised domain adaptation (SSDA) problem, such subdistributions can also appear in the unsupervised domain adaptation (UDA) problem since there usually exist target samples alignable to the source clusters. However, since each domain generally has a unique correlation among domain samples, the target domain distribution is not easily separated into distinctive subdistributions, which eventually causes the insufficient intra-domain discrepancy. Thus, the conventional inter-domain discrepancy minimization approaches have been effectively applied to the UDA problem. In contrast, in the SSDA problem, supervision on the labeled target samples enforces the target domain to be separated into the aligned subdistribution and the unaligned subdistribution deterministically. More specifically, as shown in the top row of Fig. 1, the presence of the label pulls the target samples and its neighborhoods toward source feature clusters of each corresponding labels. Besides, the unlabeled target samples which less correlate with the given labeled target samples are still located distant from source feature clusters, producing inaccurate and even meaningless inference results. Figure 2 demonstrates the existence of the intra-domain discrepancy within the target domain. Though only three target labels per class are given, significant number of target samples are aligned while wrongly predicted target samples (red circle in Fig. 2 (f)) are still located far from the source domain.

The presence of the intra-domain discrepancy makes the conventional domain adaptation methods less suitable for the SSDA problem. The ultimate goal of domain adaptation is to enhance the discriminability on the target domain, and most of the error occurs on the unaligned target subdistribution in this case. Thus, solving SSDA problems depends on how far the unaligned subdistribution is aligned. However, common domain adaptation methods focus on reducing the inter-domain discrepancy between the source and target domains regardless of the intra-domain discrepancy within the target domain. Since the existence of the aligned target subdistribution cause underestimation of the inter-domain discrepancy, the inter-domain discrepancy reduction approaches work less effectively in the SSDA problems. Moreover, since the aligned target subdistribution is aligned in a class-aware manner, such approaches can even negatively affect.

Similarly, conventional semi-supervised learning (SSL) methods also suffer from the intra-domain discrepancy issue in the SSDA problem. It stems from the different assumptions between the SSDA and SSL problems. Since the SSL problem assumes to sample labeled and unlabeled data from the identical distribution, SSL methods mainly focus on propagating the correct labels to their neighbors. In contrast, SSDA problems assume that there is a significant distribution divergence between the source and target domains, and that labeled samples are dominated by the source domain. Since correctly predicted target samples are highly aligned with the source distribution, whereas incorrectly predicted target samples are located far from them, we can no longer assume that these target samples share the same distribution. Thus, the SSL methods only propagate errors within the wrongly predicted subdistribution, and the propagation is also meaningless in the correctly predicted subdistribution due to the rich distribution of the source domain. Motivated by the interpretation, we propose a framework that addresses the intra-domain discrepancy.

4   Method

4.1   Problem Formulation

Let us denote the set of source domain samples by $\mathcal{D}_s = \{(\mathbf{x}_i^s, y_i^s)\}_{i=1}^{m_s}$. For the target domain, $\mathcal{D}_t = \{(\mathbf{x}_i^t, y_i^t)\}_{i=1}^{m_t}$ and $\mathcal{D}_u = \{(\mathbf{x}_i^u)\}_{i=1}^{m_u}$ denote the sets of labeled and unlabeled target samples, respectively. SSDA aims to enhance the target domain discriminability through training on $\mathcal{D}_s$, $\mathcal{D}_t$, and $\mathcal{D}_u$.

4.2   Spherical Feature Space with Prototypes

When aligning feature distributions, it is crucial to determine which feature space to adapt. Even if the same method is used, performance may not be improved depending on the feature space applied. Thus, we adopt similarity-based proto-typical classifier in [2] to prepare suitable feature space for better adaptation. Briefly, the prototypical classifier inputs normalized feature and compare the similarities among all class-wise prototypes, which reduces intra-class variations as results. For the classifier training, we use a cross-entropy loss as our classification loss to train an embedding function $f_\theta(\cdot)$ with parameters θ and the prototypes $\mathbf{p}_k$ $(k = 1, ..., K)$ on the source domain samples and the labeled target samples:

$$
\begin{aligned}
\mathcal{L}_{cls} &= \mathbb{E}_{(\mathbf{x},y)\in\mathcal{D}_s\cup\mathcal{D}_t}\left[-\log p(y|\mathbf{x},\mathbf{p})\right] \\
&= \mathbb{E}_{(\mathbf{x},y)\in\mathcal{D}_s\cup\mathcal{D}_t}\left[-\log\left(\frac{exp(\mathbf{p}_y\cdot f_\theta(\mathbf{x})/T)}{\Sigma_{i=1}^K exp(\mathbf{p}_i\cdot f_\theta(\mathbf{x})/T)}\right)\right].
\end{aligned}
\tag{1}
$$

While the prototypical classifier is trying to reduce the intra-class variation of the labeled sample features, the proposed schemes focus on aligning distributions of the normalized features on the spherical feature space.

4.3   Attraction Scheme

The attraction scheme aims to globally align the unaligned target subdistribution to the aligned target subdistribution in a subdistribution-level through the estimated intra-domain discrepancy minimization. The scheme measures the feature distribution divergence between the target subdistributions to estimate the intra-domain discrepancy within the target domain. However, the limited number of the labeled target samples would not be sufficient to represent the feature distribution of the aligned target subdistribution. Thus, motivated by the observation that the features of the aligned target subdistribution are highly aligned with that of the source domain in a class-aware manner, we instead use the complex distribution of the labeled source and target data. For the empirical estimation of the intra-domain discrepancy, we adopt Maximum Mean

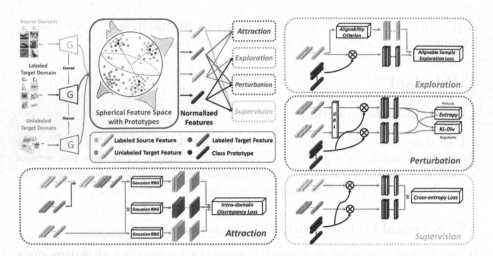

Fig. 3. An overall framework of the proposed method. Our framework consists of the feature extractor, trainable class prototypes, supervision module, and each module for the proposed schemes. The class prototypes and all normalized features of the samples are embedded in the same spherical feature space.

Discrepancy (MMD) [9], a kernel two-sample test that measures the distribution difference. We exploit a mixture $k(\cdot, \cdot)$ of Gaussian Radial Basis Function (RBF) kernels with multiple kernel widths σ_i (i = 1, ..., N). Thus, the estimated intra-domain discrepancy on the spherical feature space can be written as:

$$
\begin{aligned}
d(\mathcal{D}_s \cup \mathcal{D}_t, \mathcal{D}_u) = \; & \mathbb{E}_{(\mathbf{x},y),(\mathbf{x}',y') \in \mathcal{D}_s \cup \mathcal{D}_t}[k(f_\theta(\mathbf{x}), f_\theta(\mathbf{x}'))] \\
& + \mathbb{E}_{(\mathbf{z},w),(\mathbf{z}',w') \in \mathcal{D}_u}[k(f_\theta(\mathbf{z}), f_\theta(\mathbf{z}'))] \\
& - 2\mathbb{E}_{(\mathbf{x},y) \in \mathcal{D}_s \cup \mathcal{D}_t,(\mathbf{z},w) \in \mathcal{D}_u}[k(f_\theta(\mathbf{x}), f_\theta(\mathbf{z}))],
\end{aligned}
\tag{2}
$$

where $\mathbf{x}'$, $\mathbf{z}$, and $\mathbf{z}'$ represent samples, y', w, and w' represent the corresponding labels. Since our attraction scheme directly minimizes the intra-domain discrepancy, the attraction loss can be written by:

$$
L_a = d(\mathcal{D}_s \cup \mathcal{D}_t, \mathcal{D}_u).
\tag{3}
$$

4.4 Perturbation Scheme

Conventional adversarial perturbation, one of the semi-supervised learning (SSL) approaches, turns out to be ineffective or even cause negative transfer in the SSDA problem. In the same context as discussed in Sect. 3, the labeled target samples and its neighborhoods are aligned to the source domain separated from the inaccurate target samples, causing the intra-domain discrepancy. Then, the aligned features already guaranteed its confidence by rich information of the source domain, while the unaligned features can only propagate inaccurate predictions. Thus, the perturbation on both the aligned and unaligned target subdistribution are less meaningful.

Unlike the common adversarial perturbation approaches, our scheme perturbs target subdistributions toward their intermediate region for 1) accurate prediction propagation from the aligned subdistribution to the unaligned subdistribution and 2) class prototypes modulation toward the region. Such perturbation can be achieved by searching the direction of the anisotropically high entropy of the target features since the element-wise entropy increases as the feature move away from the prototype while the feature far from the prototypes can be attracted toward the prototypes. Note that the perturbation scheme does not ruin the already aligned subdistribution since it temporally generates additional perturbed features of the aligned feature for regularization. To achieve this, we first perturb the class prototypes in an entropy maximization direction. Then, we optimize a small and bounded perturbation toward the perturbed prototypes. Finally, we regularize the perturbed data and the given data through Kullback–Leibler divergence. To summarize, the perturbation loss can be formulated as follows:

$$H_{\mathbf{p}}(\mathbf{x}) = -\sum_{i=1}^{K} p(y = i|\mathbf{x}) \log p(y = i|\mathbf{x}, \mathbf{p})$$

$$r_{\mathbf{x}} = \operatorname*{argmin}_{\|r\| < \epsilon} \max_{\mathbf{p}} H_{\mathbf{p}}(\mathbf{x} + r)$$

$$\mathcal{L}_p = \mathbb{E}_{\mathbf{x} \in \mathcal{D}_u} \left[\sum_{i=1}^{K} D_{KL}[p(y = i|\mathbf{x}, \mathbf{p}), p(y = i|\mathbf{x} + \mathbf{r_x}, \mathbf{p})] \right] \tag{4}$$

$$+ \mathbb{E}_{(\mathbf{z},w) \in \mathcal{D}_t} \left[\sum_{i=1}^{K} D_{KL}[p(y = i|\mathbf{z}, \mathbf{p}), p(y = i|\mathbf{z} + \mathbf{r_z}, \mathbf{p})] \right].$$

where $H_{\mathbf{p}}(\cdot)$ is an element-wise entropy function defined upon similarities between the given feature and the prototypes, $\mathbf{x}$ and $\mathbf{z}$ represent samples, and y represent the corresponding label.

4.5 Exploration Scheme

The exploration scheme aims to locally modulate the prototypes in a class-aware manner complementary to the attraction scheme, while selectively aligns the unlabeled target features via suitable criteria complementary to the perturbation scheme. Though the attraction scheme globally aligns the target subdistributions on the feature space regardless of the prototypes, it does not explicitly enforce the prototypes to be modulated, which can be complemented by local and class-aware alignment. On the other hand, since the perturbation scheme regularizes the perturbed features of the anisotropically high entropy, the entropy of the perturbed feature and its neighborhood gradually became low. The exploration scheme aligns these features so that their entropy became isotropic, and thus the aligned features can be perturbed farther toward the unaligned subdistribution. To practically achieve this, we selectively collect unlabeled target data

Table 1. Classification accuracy (%) on the DomainNet dataset on the AlexNet and ResNet-34 backbone networks. The performance comparisons were done for seven scenarios with one or three labeled target samples for each class.

Net	Method	R to C		R to P		P to C		C to S		S to P		R to S		P to R		MEAN	
		1-shot	3-shot	1-shot	3-shot	1-shot	3-shot	1-shot	3-shot	1-shot	3-shot	1-shot	3-shot	1-shot	3-shot	1-shot	3-shot
AlexNet	S+T	43.3	47.1	42.4	45.0	40.1	44.9	33.6	36.4	35.7	38.4	29.1	33.3	55.8	58.7	40.0	43.4
	DANN	43.3	46.1	41.6	43.8	39.1	41.0	35.9	36.5	36.9	38.9	32.5	33.4	53.6	57.3	40.4	42.4
	ADR	43.1	46.2	41.4	44.4	39.3	43.6	32.8	36.4	33.1	38.9	29.1	32.4	55.9	57.3	39.2	42.7
	CDAN	46.3	46.8	45.7	45.0	38.3	42.3	27.5	29.5	30.2	33.7	28.8	31.3	56.7	58.7	39.1	41.0
	ENT	37.0	45.5	35.6	42.6	26.8	40.4	18.9	31.1	15.1	29.6	18.0	29.6	52.2	60.0	29.1	39.8
	MME	48.9	55.6	48.0	49.0	46.7	51.7	36.3	39.4	**39.4**	**43.0**	33.3	37.9	56.8	60.7	44.2	48.2
	SagNet	45.8	49.1	45.6	46.7	42.7	46.3	36.1	39.4	37.1	39.8	34.2	37.5	54.0	57.0	42.2	45.1
	Ours	47.7	54.6	**49.0**	**50.5**	**46.9**	**52.1**	**38.5**	**42.6**	38.5	42.2	**33.8**	**38.7**	**57.5**	**61.4**	**44.6**	**48.9**
ResNet	S+T	55.6	60.0	60.6	62.2	56.8	59.4	50.8	55.0	56.0	59.5	46.3	50.1	71.8	73.9	56.9	60.0
	DANN	58.2	59.8	61.4	62.8	56.3	59.6	52.8	55.4	57.4	59.9	52.2	54.9	70.3	72.2	58.4	60.7
	ADR	57.1	60.7	61.3	61.9	57.0	60.7	51.0	54.4	56.0	59.9	49.0	51.1	72.0	74.2	57.6	60.4
	CDAN	65.0	69.0	64.9	67.3	63.7	68.4	53.1	57.8	63.4	65.3	54.5	59.0	73.2	78.5	62.5	66.5
	ENT	65.2	71.0	65.9	69.2	65.4	71.1	54.6	60.0	59.7	62.1	52.1	61.1	75.0	78.6	62.6	67.6
	MME	70.0	72.2	67.7	69.7	69.0	71.7	56.3	61.8	**64.8**	**66.8**	61.0	61.9	76.1	78.5	66.4	68.9
	SagNet	59.4	62.0	61.9	62.9	59.1	61.5	54.0	57.1	56.6	59.0	49.7	54.4	72.2	73.4	59.0	61.5
	Ours	**70.4**	**76.6**	**70.8**	**72.1**	**72.9**	**76.7**	**56.7**	**63.1**	64.5	66.1	**63.0**	**67.8**	**76.6**	**79.4**	**67.6**	**71.7**

with its element-wise entropy less than a certain threshold, then apply a cross-entropy loss with the class of the nearest prototype. The objective function of the exploration scheme can be written as follows:

$$M_\epsilon = \{\mathbf{x} \in \mathcal{D}_u | H_\mathbf{p}(\mathbf{x}) < \epsilon\}$$
$$\hat{y}_\mathbf{x} = \underset{i \in \{1,\ldots,K\}}{\text{argmax}} \ p(y = i | \mathbf{x}, \mathbf{p}) \tag{5}$$
$$\mathcal{L}_e = \mathbb{E}_{\mathcal{D}_u}[-\mathbf{1}_{M_\epsilon}(\mathbf{x}) \log p(y = \hat{y}_\mathbf{x} | \mathbf{x}, \mathbf{p})].$$

where M_ϵ is a set of unlabeled target data with entropy value less than a hyperparameter ϵ, and $\mathbf{1}_{M_\epsilon}(\cdot)$ is an indicator function that filters out alignable samples from the given unlabeled target samples.

4.6 Overall Framework and Training Objective

The overall training objective of our method is the weighted sum of the supervision loss, the attraction loss, the perturbation loss, and the exploration loss. The optimization problem can be formulated as follows:

$$\min_{\mathbf{p},\theta} \mathcal{L}_{cls} + \alpha \mathcal{L}_a + \beta \mathcal{L}_e + \gamma \mathcal{L}_p. \tag{6}$$

We integrated all the schemes into one framework, as shown in the Fig. 3.

5 Experiments

5.1 Experimental Setup

Datasets. DomainNet [24] is a recently released large-scale domain adaptation benchmark dataset that contains six domains and approximately 0.6 million

Table 2. Classification accuracy (%) on the Office-Home dataset with the AlexNet and ResNet-34 backbone networks. The performance comparisons were done for a total of 12 scenarios on three-shot setting.

Net	Method	R to C	R to P	R to A	P to R	P to C	P to A	A to P	A to C	A to R	C to R	C to A	C to P	MEAN
AlexNet	S+T	44.6	66.7	47.7	57.8	44.4	36.1	57.6	38.8	57.0	54.3	37.5	57.9	50.0
	DANN	47.2	66.7	46.6	58.1	44.4	36.1	57.2	39.8	56.6	54.3	38.6	57.9	50.3
	ADR	45.0	66.2	46.9	57.3	38.9	36.3	57.5	40.0	57.8	53.4	37.3	57.7	49.5
	CDAN	41.8	69.9	43.2	53.6	35.8	32.0	56.3	34.5	53.5	49.3	27.9	56.2	46.2
	ENT	44.9	70.4	47.1	60.3	41.2	34.6	60.7	37.8	60.5	58.0	31.8	63.4	50.9
	MME	51.2	73.0	50.3	**61.6**	47.2	40.7	63.9	43.8	**61.4**	**59.9**	**44.7**	64.7	55.2
	Ours	**51.9**	**74.6**	**51.2**	**61.6**	**47.9**	**42.1**	**65.5**	**44.5**	60.9	58.1	44.3	**64.8**	**55.6**
ResNet	S+T	55.7	80.8	67.8	73.1	53.8	63.5	73.1	54.0	74.2	68.3	57.6	72.3	66.2
	DANN	57.3	75.5	65.2	69.2	51.8	56.6	68.3	54.7	73.8	67.1	55.1	67.5	63.5
	CDAN	61.4	80.7	67.1	76.8	58.1	61.4	74.1	59.2	74.1	70.7	60.5	74.5	68.2
	ENT	62.6	85.7	70.2	79.9	60.5	63.9	79.5	61.3	79.1	76.4	64.7	79.1	71.9
	MME	64.6	85.5	71.3	80.1	64.6	65.5	79.0	63.6	79.7	76.6	**67.2**	79.3	73.1
	Ours	**66.4**	**86.2**	**73.4**	**82.0**	**65.2**	**66.1**	**81.1**	**63.9**	**80.2**	**76.8**	66.6	**79.9**	**74.0**

images with 345 classes. **Office-Home** [31] and **Office** [25] are standard benchmarks for domain adaptation. Office-Home consists of Art, Clipart, Product, and Real-world domain with 65 classes. Office consists of Amazon, Webcam, and DSLR domains with 31 classes.

Evaluation Tasks. For a fair comparison with the state-of-the-art SSDA method [26], we performed experiments on 7 adaptation scenarios on the four domains (Real, Clipart, Painting, Sketch) with 126 classes for DomainNet, 12 adaptation scenarios on all the domains for Office-Home, and two challenging adaptation scenarios on Office. One or three labeled target samples are given for each class for these scenarios. Additionally, we compared the performances for 5, 10 and 20 labeled target samples for each class.

Implementation Details. We adopted AlexNet and ResNet-34 for the backbone network. Every mini-batch consists of the same number of labeled source and target samples with a doubled number of unlabeled target samples. We prepared 32 and 24 samples for each split of the mini-batch for AlexNet and Resnet-34, respectively. We used the Stochastic Gradient Descent (SGD) optimizer with an initial learning rate of 0.01, a momentum of 0.9, and a weight decay of 0.0005. All implementations were done in PyTorch [23] and on a single GeForce Titan XP GPU.

Baselines. We compared our method with the semi-supervised domain adaptation (SSDA), unsupervised domain adaptation (UDA), semi-supervised learning (SSL), and no adaptation methods. More specifically, the baselines consist of **MME** [26], **SagNet** [22], **DANN** [7], **ADR** [27], **CDAN** [18], **ENT** [8], and non-adapted model. For the UDA methods (DANN, ADR, and CDAN), the labeled target samples were supervised during the training process. S+T is a vanilla model trained on all labeled samples. DANN confuses the cross-domain distributions through adversarial learning. ADR adopts the dropout

Table 3. Classification accuracy (%) on the Office dataset with three-shot setting.

Net	Method	W to A	D to A	MEAN
AlexNet	S+T	61.2	62.4	61.8
	DANN	64.4	65.2	64.8
	ADR	61.2	61.4	61.3
	CDAN	60.3	61.4	60.8
	ENT	64.0	66.2	65.1
	MME	67.3	67.8	67.6
	Ours	**67.6**	**69.0**	**68.3**

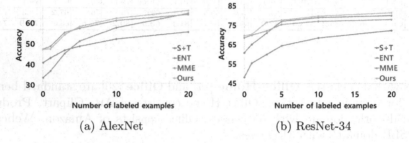

(a) AlexNet   (b) ResNet-34

Fig. 4. Trend in classification accuracy (%) with varying number of labeled target samples per class. The experiments are conducted on the Real to Clipart scenario of the DomainNet dataset.

scheme to modify the decision boundary for feature alignment. CDAN adversarially aligns the feature by fooling the conditional domain discriminator. ENT is an SSL method that minimizes the entropy of the unlabeled target data. For the fair comparison, all the methods have the same backbone architecture with our method.

5.2 Results

Performance Comparison on DomainNet. We summarized the classification accuracies of 7 scenarios on the DomainNet dataset in Table 1. On average, our method outperformed the best-performed baseline by 2.8% in the three-shot setting and 1.2% in the one-shot setting on ResNet-34, and by 0.7% in the three-shot setting and 0.4% in the one-shot setting on AlexNet. Moreover, our method outperformed most of the cases except for a few adaptation tasks. On the other hand, though UDA methods like DANN and ADR performed slightly better than S+T when only one labeled target per class is given, these methods become less effective or even cause negative transfer as the number of the labeled target samples increases. It verifies our statement that conventional domain adaptation methods are often less beneficial than the partial alignment effect from the given target labels. ENT showed significant improvement on ResNet-34, while it shows degenerative performance on AlexNet. Moreover, the performance enhancement gap increased as the number of labeled target samples increases, which will be discussed in more detail in Sect. 5.3.

Performance Comparison on Office-Home and Office. The comparison results of our method with the baselines on the Office-Home dataset are reported in Table 2. Our method outperformed all the baselines regardless of the backbone network on average. Similar to DomainNet, our method showed the best performance in most of the scenarios. While DANN performed at least similar to S+T on the AlexNet backbone, it showed degenerative performance on the ResNet-34 backbone. It demonstrates that capacity difference of the backbone network causes the difference in the degree of the target label exploitation, and DANN performed less effective than the exploitation of ResNet-34. The considerable results of ENT are reasonable since the three-shot setting provides approximately 5–10% of the target labels for training, and such ratio is quite rich in a perspective of the SSL problem. Table 3 showed the performance comparison on the Office dataset and our method also outperformed other baseline on the dataset.

Table 4. Ablation study results of the proposed schemes on the Real to Sketch task of DomainNet with three-shot setting.

Net	Method	Attract	Explore	Perturb	R to C	R to P	P to C	C to S	S to P	R to S	P to R	MEAN
AlexNet	S+T				47.1	45.0	44.9	36.4	38.4	33.3	58.7	43.4
	DANN				46.1	43.8	41.0	36.5	38.9	33.4	57.3	42.4
	MMD				47.9	45.5	44.6	38.1	38.4	35.5	56.6	43.8
	VAT				46.1	43.8	44.3	35.6	38.2	31.8	57.7	42.5
	Ours	✓			50.2	46.2	47.5	40.8	41.3	37.2	59.8	46.1
		✓	✓		53.9	49.8	50.5	42.0	41.9	38.0	60.7	48.3
				✓	57.2	47.5	54.1	38.8	39.7	38.5	59.2	47.9
		✓	✓	✓	54.6	50.5	52.1	42.6	42.2	38.7	61.4	48.9

5.3 Analysis

Performance Comparison with Varying Number of Target Labels. We compared the behavior of the methods by varying the number of labeled target samples from 0 to 20 for each class. As shown in Fig. 4, our methods showed superior performance for a large number of target labels even on the scenario where our method worked less effectively on a one-shot or three-shot setting. Moreover, it outperformed the other baselines throughout all the cases on the ResNet-34 backbone. On the other hand, ENT also significantly enhanced the accuracy for a large number of target labels, and it even outperformed the state-of-the-art SSDA methods when more than twenty and five target labels are given per class for the AlexNet and ResNet-34 backbone networks, respectively. It is reasonable since the increase of the labeled target sample ratio assimilates the SSDA problem to the SSL problem, which is suitable for SSL methods.

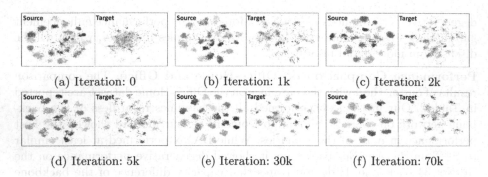

(a) Iteration: 0   (b) Iteration: 1k   (c) Iteration: 2k

(d) Iteration: 5k   (e) Iteration: 30k   (f) Iteration: 70k

Fig. 5. (a)–(f) The t-SNE visualization of the feature alignment progress through our method during the training phase.

Ablation Study on the Proposed Schemes. We conducted an ablation study on our schemes. To verify the effectiveness, we additionally evaluated DANN, MMD, and VAT [21] on the DomainNet dataset. As shown in Table 4, DANN and MMD rarely worked or even caused negative transfer while our attraction scheme showed meaningful improvement on average. It verifies that conventional UDA methods that focus on reducing the inter-domain discrepancy suffer from the intra-domain discrepancy issue, and it can be addressed by the intra-domain discrepancy minimization. Moreover, VAT also caused degenerative effect while our perturbation scheme significantly enhanced the performance, which demonstrates that conventional adversarial perturbation methods are not suitable for the SSDA problem, and our perturbation scheme can address it by modulating the perturbation direction toward the intermediate region of the target subdistributions. The exploration scheme also worked complementary to other schemes.

Convergence Analysis. To analyze the convergence of our method, we depicted the t-SNE visualization [20] of the cross-domain features over the training progress in Fig. 5. We conducted the experiment on the Real to Sketch scenario of the DomainNet. All 126 classes were used for the experiment, but we choose 20 classes for better visualization. Note that we did not specifically pick classes of top-20 accuracies. Figure 5 (a) clearly shows the initial domain divergence between the source and the target domain. Moreover, the feature depiction of the early stages often showed many unaligned source and target clusters. As the training goes on, our method aligned the corresponding source and target clusters and finally obtained well-accumulated target clusters, as shown in Fig. 5 (f).

6   Conclusions

In this work, we demonstrated the intra-domain discrepancy issue of the target domain in the SSDA problem. Motivated by this, we proposed an SSDA framework that aligns the cross-domain feature distributions by addressing the intra-domain discrepancy through the attraction, exploration, and perturbation schemes. The attraction scheme directly minimized the estimated intra-domain discrepancy within the target domain The perturbation scheme perturbed the well-aligned and unaligned target features into the intermediate region of the target subdistributions. The exploration scheme locally aligned features in a selective and class-wise manner complementary to the attraction and perturbation schemes. The experiments conducted on DomainNet, Office-Home, and Office datasets validate the effectiveness of our method, and it outperformed the conventional UDA and SSL methods on all the datasets.

References

1. Ao, S., Li, X., Ling, C.X.: Fast generalized distillation for semi-supervised domain adaptation. In: AAAI (2017)
2. Chen, W.Y., Liu, Y.C., Kira, Z., Wang, Y.C., Huang, J.B.: A closer look at few-shot classification. In: International Conference on Learning Representations (ICLR) (2019)
3. Chen, Y., Zhu, X., Gong, S.: Semi-supervised deep learning with memory. In: Proceedings of the European Conference on Computer Vision (ECCV) (2018)
4. Chen, Y., Li, W., Sakaridis, C., Dai, D., Van Gool, L.: Domain adaptive faster R-CNN for object detection in the wild. In: Proceedings of the IEEE Conference on Computer Vision and Pattern Recognition (CVPR) (2018)
5. Cicek, S., Fawzi, A., Soatto, S.: SaaS: speed as a supervisor for semi-supervised learning. In: Proceedings of the European Conference on Computer Vision (ECCV) (2018)
6. Donahue, J., Hoffman, J., Rodner, E., Saenko, K., Darrell, T.: Semi-supervised domain adaptation with instance constraints. In: Proceedings of the IEEE Conference on Computer Vision and Pattern Recognition (CVPR) (2013)
7. Ganin, Y., Lempitsky, V.: Unsupervised domain adaptation by backpropagation. In: Proceedings of the International Conference on Machine Learning (ICML) (2015)
8. Grandvalet, Y., Bengio, Y.: Semi-supervised learning by entropy minimization. In: Advances in Neural Information Processing Systems (NeurIPS) (2005)
9. Gretton, A., Borgwardt, K.M., Rasch, M.J., Schölkopf, B., Smola, A.: A kernel two-sample test. Journal of Machine Learning Research (JMLR) (2012)
10. Hoffman, J., Tzeng, E., Park, T., Zhu, J.Y., Isola, P., Saenko, K., Efros, A.A., Darrell, T.: Cycada: cycle-consistent adversarial domain adaptation. In: Proceedings of the International Conference on Machine Learning (ICML) (2018)
11. Hong, W., Wang, Z., Yang, M., Yuan, J.: Conditional generative adversarial network for structured domain adaptation. In: Proceedings of the IEEE Conference on Computer Vision and Pattern Recognition (CVPR) (2018)
12. Hu, L., Kan, M., Shan, S., Chen, X.: Duplex generative adversarial network for unsupervised domain adaptation. In: Proceedings of the IEEE Conference on Computer Vision and Pattern Recognition (CVPR) (2018)
13. Isola, P., Zhu, J.Y., Zhou, T., Efros, A.A.: Image-to-image translation with conditional adversarial networks. In: Proceedings of the IEEE Conference on Computer Vision and Pattern Recognition (CVPR) (2017)

14. Kim, T., Jeong, M., Kim, S., Choi, S., Kim, C.: Diversify and match: A domain adaptive representation learning paradigm for object detection. In: Proceedings of the IEEE/CVF Conference on Computer Vision and Pattern Recognition (CVPR) (2019)
15. Laine, S., Aila, T.: Temporal ensembling for semi-supervised learning. In: International Conference on Learning Representations (ICLR) (2017)
16. Lee, S., Kim, D., Kim, N., Jeong, S.G.: Drop to adapt: Learning discriminative features for unsupervised domain adaptation. In: Proceedings of the IEEE/CVF International Conference on Computer Vision (ICCV) (2019)
17. Long, M., Cao, Y., Wang, J., Jordan, M.: Learning transferable features with deep adaptation networks. In: Proceedings of the International Conference on Machine Learning (ICML) (2015)
18. Long, M., Cao, Z., Wang, J., Jordan, M.I.: Conditional adversarial domain adaptation. In: Advances in Neural Information Processing Systems (NeurIPS) (2018)
19. Long, M., Zhu, H., Wang, J., Jordan, M.I.: Unsupervised domain adaptation with residual transfer networks. In: Advances in Neural Information Processing Systems (NeurIPS) (2016)
20. van der Maaten, L., Hinton, G.: Visualizing data using t-SNE. Journal of Machine Learning Research (JMLR) (2008)
21. Miyato, T., Maeda, S.i., Ishii, S., Koyama, M.: Virtual adversarial training: a regularization method for supervised and semi-supervised learning. IEEE Transactions on Pattern Analysis and Machine Intelligence (TPAMI) (2018)
22. Nam, H., Lee, H., Park, J., Yoon, W., Yoo, D.: Reducing domain gap via style-agnostic networks (2019)
23. Paszke, A., Gross, S., Chintala, S., Chanan, G., Yang, E., DeVito, Z., Lin, Z., Desmaison, A., Antiga, L., Lerer, A.: Automatic differentiation in pytorch (2017)
24. Peng, X., Bai, Q., Xia, X., Huang, Z., Saenko, K., Wang, B.: Moment matching for multi-source domain adaptation. In: Proceedings of the IEEE/CVF International Conference on Computer Vision (ICCV) (2019)
25. Saenko, K., Kulis, B., Fritz, M., Darrell, T.: Adapting visual category models to new domains. In: Proceedings of the European Conference on Computer Vision (ECCV) (2010)
26. Saito, K., Kim, D., Sclaroff, S., Darrell, T., Saenko, K.: Semi-supervised domain adaptation via minimax entropy. In: Proceedings of the IEEE/CVF International Conference on Computer Vision (ICCV) (2019)
27. Saito, K., Ushiku, Y., Harada, T., Saenko, K.: Adversarial dropout regularization. In: Proceedings of the International Conference on Learning Representations (ICLR) (2018)
28. Saito, K., Ushiku, Y., Harada, T., Saenko, K.: Strong-weak distribution alignment for adaptive object detection. In: Proceedings of the IEEE/CVF Conference on Computer Vision and Pattern Recognition (CVPR) (2019)
29. Tarvainen, A., Valpola, H.: Mean teachers are better role models: Weight-averaged consistency targets improve semi-supervised deep learning results. In: Advances in neural information processing systems (NeurIPS) (2017)
30. Tsai, Y.H., Hung, W.C., Schulter, S., Sohn, K., Yang, M.H., Chandraker, M.: Learning to adapt structured output space for semantic segmentation. In: Proceedings of the IEEE Conference on Computer Vision and Pattern Recognition (CVPR) (2018)
31. Venkateswara, H., Eusebio, J., Chakraborty, S., Panchanathan, S.: Deep hashing network for unsupervised domain adaptation. In: Proceedings of the IEEE Conference on Computer Vision and Pattern Recognition (CVPR) (2017)

32. Wang, Q., Li, W., Gool, L.V.: Semi-supervised learning by augmented distribution alignment. In: Proceedings of the IEEE/CVF International Conference on Computer Vision (ICCV) (2019)
33. Yao, T., Yingwei Pan, Ngo, C., Houqiang Li, Tao Mei: Semi-supervised domain adaptation with subspace learning for visual recognition. In: Proceedings of the IEEE Conference on Computer Vision and Pattern Recognition (CVPR) (2015)
34. Zheng, C., Cham, T.J., Cai, J.: T2Net: Synthetic-to-realistic translation for solving single-image depth estimation tasks. In: Proceedings of the European Conference on Computer Vision (ECCV) (2018)
35. Zheng, Z., Yang, X., Yu, Z., Zheng, L., Yang, Y., Kautz, J.: Joint discriminative and generative learning for person re-identification. In: The IEEE Conference on Computer Vision and Pattern Recognition (CVPR) (2019)

Curriculum Manager for Source Selection in Multi-source Domain Adaptation

Luyu Yang[1]([✉]), Yogesh Balaji[1], Ser-Nam Lim[2], and Abhinav Shrivastava[1]

[1] University of Maryland, College Park, USA
{loyo,yogesh,abhinav}@cs.umd.edu
[2] Facebook AI, New York, USA
sernamlim@fb.com

Abstract. The performance of Multi-Source Unsupervised Domain Adaptation depends significantly on the effectiveness of transfer from labeled source domain samples. In this paper, we proposed an adversarial agent that learns a dynamic curriculum for source samples, called Curriculum Manager for Source Selection (CMSS). The Curriculum Manager, an independent network module, constantly updates the curriculum during training, and iteratively learns which domains or samples are best suited for aligning to the target. The intuition behind this is to force the Curriculum Manager to constantly re-measure the transferability of latent domains over time to adversarially raise the error rate of the domain discriminator. CMSS does not require any knowledge of the domain labels, yet it outperforms other methods on four well-known benchmarks by significant margins. We also provide interpretable results that shed light on the proposed method.

Keywords: Unsupervised domain adaptation · Multi-source · Curriculum learning · Adversarial training

1 Introduction

Training deep neural networks requires datasets with rich annotations that are often time-consuming to obtain. Previous proposals to mitigate this issue have ranged from unsupervised [8,18,21,29,30,42], self-supervised [17,35,36,41], to low shot learning [28,33,37,44]. Unsupervised Domain Adaptation (UDA), when first introduced in [15], sheds precious insights on how adversarial training can be utilized to get around the problem of expensive manual annotations. UDA aims to preserve the performance on an unlabeled dataset (target) using a model trained on a label-rich dataset (source) by making optimal use of the learned representations from the source.

Intuitively, one would expect that having more labeled samples in the source domain will be beneficial. However, having more labeled samples does not equal better transfer, since the source will inadvertently encompass a larger variety of domains. While the goal is to learn a common representation for both source

A. Vedaldi et al. (Eds.): ECCV 2020, LNCS 12359, pp. 608–624, 2020.
https://doi.org/10.1007/978-3-030-58568-6_36

and target in such a Multi-Source Unsupervised Domain Adaptation (MS-UDA) setting, enforcing each source domain distribution to exactly match the target may increase the training difficulty, and generate ambiguous representations near the decision boundary potentially resulting in negative transfer. Moreover, for practical purposes, we would expect the data source to be largely unconstrained, whereby neither the number of domains or domain labels are known. A good example here would be datasets collected from the Internet where images come from unknown but potentially a massive set of users.

To address the MS-UDA problem, we propose an adversarial agent that learns a dynamic curriculum [4] for multiple source domains, named Curriculum Manager for Source Selection (CMSS). More specifically, a constantly updated curriculum during training learns which domains or samples are best suited for aligning to the target distribution. The CMSS is an independent module from the feature network and is trained by maximizing the error of discriminator in order to weigh the gradient reversal back to the feature network. In our proposed adversarial interplay with the discriminator, the Curriculum Manager is forced to constantly re-measure the transferability of latent domains across time to achieve a higher error of the discriminator. Such a procedure of weighing the source data is modulated over the entire training. In effect, the latent domains with different transferability to the target distribution will gradually converge to different levels of importance without any need for additional domain partitioning prior or clustering.

We attribute the following contributions to this work:

- We propose a novel adversarial method during training towards the MS-UDA problem. Our method does not assume any knowledge of the domain labels or the number of domains.
- Our method achieves state-of-the-art in extensive experiments conducted on four well-known benchmarks, including the large-scale DomainNet ($\sim$ 0.6 million images).
- We obtain interpretable results that show how CMSS is in effect a form of curriculum learning that has great effect on MS-UDA when compared to the prior art. This positively differentiates our approach from previous state-of-the-art.

2   Related Work

UDA is an actively studied area of research in machine learning and computer vision. Since the seminal contribution of Ben-David et al. [1,2], several techniques have been proposed for learning representations invariant to domain shift [10,11, 23,25,45]. In this section, we review some recent methods that are most related to our work.

Multi-source Unsupervised Domain Adaptation. (MS-UDA) assumes that the source training examples are inherently multi-modal. The source

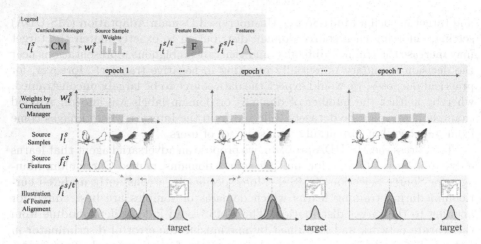

Fig. 1. Illustration of CMSS during training. All training samples are passed through the feature network F. CMSS prefers samples with better transferability to match the target, and re-measure the transferability at each iteration to keep up with the discriminator. At the end of training after the majority of samples are aligned, the CMSS weights tend to be similar among source samples.

domains contain labeled samples while the target domain contains unlabeled samples [15,22,27,32,46]. In [32], adaptation was performed by aligning the moments of feature distributions between each source-target pair. Deep Cocktail Network (DCTN) [40] considered the more realistic case of existence of category shift in addition to the domain shift, and proposes a k-way domain adversarial classifier and category classifier to generate a combined representation for the target.

Because domain labels are hard to obtain in the real world datasets, latent domain discovery [27] – a technique for alleviating the need for explicit domain label annotation has many practical applications. Xiong et al. [39] proposed to use square-loss mutual information based clustering with category distribution prior to infer the domain assignment for images. Mancini et al. [27] used a domain prediction branch to guide domain discovery using multiple batch-norm layers.

Domain-Adversarial Training has been widely used [7,9,31] since Domain-Adversarial Neural Network (DANN) [15] was proposed. The core idea is to train a discriminator network to discriminate source features from target, and train the feature network to fool the discriminator. Zhao et al. [46] first proposed to generalize DANN to the multi-source setting, and provides theoretical insights on the multi-domain adversarial bounds. Maximum Classifier Discrepancy (MCD) [33] is another powerful [19,24,32,38] technique for performing adaptation in an adversarial manner using two classifiers. The method first updates the classifiers to maximize the discrepancy between the classifiers' prediction on target samples, followed by minimizing the discrepancy while updating the feature generator.

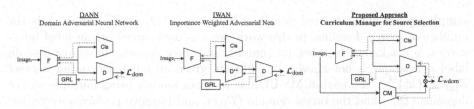

Fig. 2. Architecture comparison of *left*: DANN [15], *middle*: IWAN [43], and *right*: proposed method. Red dotted lines indicate backward passes. (*F*: feature extractor, *Cls*: classifier, *D*: domain discriminator, GRL: gradient reversal layer, CM: Curriculum Manager, $\mathcal{L}_{\text{dom}}$: Eq. 1 domain loss, $\mathcal{L}_{\text{wdom}}$: Eq. 3 weighted domain loss) (Color figure online)

Domain Selection and Weighting: Some previous methods that employed sample selection and sample weighing techniques for domain adaptation include [12–14]. Duan *et al.* [14] proposed using domain selection by leveraging a large number of loosely labeled web images from different sources. The authors of [14] adopted a set of base classifiers to predict labels for the target domain as well as a domain-dependent regularizer based on smoothness assumption. Bhatt *et al.* [5] proposed to adapt iteratively by selecting the best sources that learn shared representations faster. Chen *et al.* [9] used a hand-crafted re-weighting vector so that the source domain label distribution is similar to the unknown target label distribution. Mancini *et al.* [26] modeled the domain dependency using a graph and utilizes auxiliary metadata for predictive domain adaptation. Zhang *et al.* [43] employed an extra domain classifier that gives the probability of a sample coming from the source domain. The higher the confidence is from such an extra classifier, the more likely it can be discriminated from the target domain, in which case the importance of the said sample is reduced accordingly.

Curriculum for Domain Adaptation aims at an adaptive strategy over time in order to improve the effectiveness of domain transfer. The curriculum can be hand-crafted or learned. Shu *et al.* [34] designed the curriculum by combining the classification loss and discriminator's loss as a weighting strategy to eliminate the corrupted samples in the source domain. Another work with similar motivation is [8], in which Chen *et al.* proposed to use per-category prototype to measure the prediction confidence of target samples. A manually designed threshold τ is utilized to make a binary decision in selecting partial target samples for further alignment. Kurmi *et al.* [20] used a curriculum-based dropout discriminator to simulate the gradual increase of sample variance.

3 Preliminaries

Task Formulation: In multi-source unsupervised domain adaptation (MS-UDA), we are given an input dataset $\mathcal{D}_{\text{src}} = \{(\mathbf{x}_i^s, y_i^s)\}_{i=1}^{N_s}$ that contains samples from multiple domains. In this paper, we focus on classification problems, with the set of labels $y_i^s \in \{1, 2, \ldots, n_c\}$, where n_c is the number of classes. Each sample $\mathbf{x}_i^s$ has an associated domain label, $d_i^s \in \{1, 2, \ldots, S\}$, where S is the number of source domains. In this work, we assume source domain label information is not known *a priori*, i.e., number of source domains or source domain label per sample is not known. In addition, given an unlabeled target dataset $\mathcal{D}_{tgt} = \{\mathbf{x}_i^t\}_{i=1}^{N_t}$, the goal of MS-UDA is to train models using multiple source domains ($\mathcal{D}_{src}$) and the target domain ($\mathcal{D}_{tgt}$), and improve performance on the target test set.

Domain-Adversarial Training: First, we discuss the domain-adversarial training formulation [15]. The core idea of domain-adversarial training is to minimize the distributional distance between source and target feature distributions posed as an adversarial game. The model has a feature extractor, a classifier, and a domain discriminator. The classifier takes in feature from the feature extractor and classifies it in n_c classes. The discriminator is optimized to discriminate source features from target. The feature network, on the other hand, is trained to fool the discriminator while at the same time achieve good classification accuracy.

More formally, let $F_\theta : \mathbb{R}^{3 \times w \times h} \to \mathbb{R}^d$ denote the feature extraction network, $C_\phi : \mathbb{R}^d \to \mathbb{R}^{n_c}$ denote the classifier, and $D_\psi : \mathbb{R}^d \to \mathbb{R}^1$ denote the domain discriminator. Here, θ, ϕ and ψ are the parameters associated with the feature extractor, classifier, and domain discriminator respectively. The model is trained using the following objective function:

$$\max_{\psi} \min_{\theta, \phi} \quad \mathcal{L}_{cls} - \lambda \mathcal{L}_{dom} \tag{1}$$

$$\text{where} \quad \mathcal{L}_{cls} = -\frac{1}{N_s} \sum_{i=1}^{N_s} \tilde{\mathbf{y}}_i \log(C(F(\mathbf{x}_i^s)))$$

$$\mathcal{L}_{dom} = -\mathbb{E}_{\mathbf{x} \sim \mathcal{D}_{src}} \log(D(F(\mathbf{x}))) - \mathbb{E}_{\mathbf{x} \sim \mathcal{D}_{tgt}} \log(1 - D(F(\mathbf{x})))$$

$$= -\frac{1}{N_s} \sum_{i=1}^{N_s} \log(D(F(\mathbf{x}_i^s))) - \frac{1}{N_t} \sum_{i=1}^{N_t} \log(1 - D(F(\mathbf{x}_i^t)))$$

$\mathcal{L}_{cls}$ is is the cross-entropy loss in source domain (with $\tilde{\mathbf{y}}_i$ being the one-hot encoding of the label y_i), and $\mathcal{L}_{dom}$ is the discriminator loss that discriminates source samples from the target. Note that both these loss functions use samples from all source domains.

In principle, if domain labels are available, there are two possible choices for the domain discriminator: (1) k domain discriminators can be trained, each one discriminating one of the source domains from the target [15], or (2) a domain discriminator can be trained as a $(k + 1)$-way classifier to classify input samples as either one of the source domains or target [46]. However, in our setup, domain labels are unknown and, therefore, these formulations can not be used.

4   CMSS: Curriculum Manager for Source Selection

For the source domain that is inherently multi-modal, our goal is to learn a dynamic curriculum for selecting the best-suited samples for aligning to the target feature distribution. At the beginning of training, the Curriculum Manager is expected to prefer samples with higher *transferability* for aligning with the target, *i.e.*, source samples which have similar feature distributions to the target sample. Once the feature distributions of these samples are aligned, our Curriculum Manager is expected to prioritize the next round of source samples for alignment. As the training progresses, the Curriculum Manager can learn to focus on different aspects of the feature distribution as a proxy for better transferability. Since our approach learns a curriculum to prefer samples from different source domains, we refer to it is Curriculum Manager for Source Selection (CMSS).

Our approach builds on the domain-adversarial training framework (described in Sect. 3). In this framework, our hypothesis is that source samples that are hard for the domain discriminator to separate from the target samples are likely the ones that have similar feature distributions. Our CMSS leverages this and uses the discriminator loss to find source samples that should be aligned first. The preference for source samples is represented as per-sample weights predicted by CMSS. Since our approach is based on domain-adversarial training, weighing $\mathcal{L}_{\mathrm{dom}}$ using these weights will lead to the discriminator encouraging the feature network to bring the distributions of higher weighted source samples closer to the target samples. This signal between the discriminator and feature extractor is achieved using the gradient reversal layer (see [15] for details).

Therefore, our proposed CMSS is trained to predict weights for source samples at each iteration, which maximizes the error of the domain discriminator. Due to this adversarial interplay with the discriminator, the CMSS is forced to re-estimate the preference of source samples across training to keep up with the improving domain discriminator. The feature extractor, F, is optimized to learn features that are both good for classification and confuse the discriminator. To avoid any influence from the classification task in the curriculum design, our CMSS also has an independent feature extractor module that learns to predict weights per-sample given the source images and domain discriminator loss.

Training CMSS: The CMSS weight for every sample in the source domain, $\mathbf{x}_i^s$, is given by w_i^s. We represent this weighted distribution as $\tilde{\mathcal{D}}_{\mathrm{src}}$. The CMSS network is represented by $G_\rho : \mathbb{R}^{c \times w \times h} \to \mathbb{R}^1$ with parameters ρ. Given a batch of samples, $\mathbf{x}_1^s, \mathbf{x}_2^s, \ldots \mathbf{x}_b^s$, we first pass these samples to G_ρ to obtain an array of scores that are normalized using softmax function to obtain the resulting weight vector. During training, the CMSS optimization objective can be written as

$$\min_{\rho} \left[\frac{1}{N_s} \sum_{i=1}^{N_s} G_\rho(\mathbf{x}_i^s) \log(D(F(\mathbf{x}_i^s))) \right] \qquad (2)$$

With the source sample weights generated by CMSS, the loss function for domain discriminator can be written as

$$\mathcal{L}_{\text{wdom}} = -\frac{1}{N_s} \sum_{i=1}^{N_s} G_\rho(\mathbf{x}_i^s) \log(D(F(\mathbf{x}_i^s))) - \frac{1}{N_t} \sum_{i=1}^{N_t} \log(1 - D(F(\mathbf{x}_i^t)))$$

$$\text{s.t.} \sum_i G_\rho(\mathbf{x}_i^s) = N_s \tag{3}$$

The overall optimization objective can be written as

$$\max_{\psi} \min_{\theta,\phi,\rho} \mathcal{L}_{\text{cls}} - \lambda \mathcal{L}_{\text{wdom}} \tag{4}$$

where $\mathcal{L}_{\text{cls}}$ is the Cross-Entropy loss for source classification and $\mathcal{L}_{\text{wdom}}$ is the weighted domain discriminator loss from Eq. (3), with weights obtained by optimizing Eq. (2). λ is the hyperparameter in the gradient reversal layer. We follow [15] and set λ based on the following annealing schedule: $\lambda_p = \frac{2}{1+\exp(-\gamma \cdot p)} - 1$, where p is the current number of iterations divided by the total. γ is set to 10 in all experiments as in [15]. Details of training are provided in Algorithm 1.

4.1   CMSS: Theoretical Insights

We first state the classic generalization bound for domain adaptation [3,6]. Let $\mathcal{H}$ be a hypothesis space of VC-dimension d. For a given hypothsis class $\mathcal{H}$, define the symmetric difference operator as $\mathcal{H}\Delta\mathcal{H} = \{h(\mathbf{x}) \oplus h'(\mathbf{x}) | h, h' \in \mathcal{H}\}$. Let $\mathcal{D}_{\text{src}}$, $\mathcal{D}_{\text{tgt}}$ denote the source and target distributions respectively, and $\hat{\mathcal{D}}_{\text{src}}$, $\hat{\mathcal{D}}_{\text{tgt}}$ denote the empirical distribution induced by sample of size m drawn from $\mathcal{D}_{\text{src}}$, $\mathcal{D}_{\text{tgt}}$ respectively. Let ϵ_s (ϵ_t) denote the true risk on source (target) domain, and $\hat{\epsilon}_s$ ($\hat{\epsilon}_t$) denote the empirical risk on source (target) domain. Then, following Theorem 1 of [6], with probability of at least $1 - \delta$, $\forall h \in \mathcal{H}$,

$$\epsilon_t(h) \leq \hat{\epsilon}_s(h) + \frac{1}{2} d_{\mathcal{H}\Delta\mathcal{H}}(\hat{\mathcal{D}}_{\text{src}}, \hat{\mathcal{D}}_{\text{tgt}}) + C \tag{5}$$

Algorithm 1. Training CMSS (Curriculum Manager for Source Selection)

Require: N_{iter}: Total number of training iterations
Require: γ: For computing λ_p for $\mathcal{L}_{\text{wdom}}$
Require: N_b^s and N_b^t: Batch size for source and target domains
1: Shuffle the source domain samples
2: **for** t in $(1 : N_{\text{iter}})$ **do**
3:      Compute λ according to $2/(1 + \exp(-\gamma \cdot (t/N_{iter}))) - 1$
4:      Sample a training batch from source domains $\{(\mathbf{x}_i^s, y_i)\}_{i=1}^{N_b^s} \sim \mathcal{D}_{\text{src}}$ and from target domain $\{\mathbf{x}_i^t\}_{i=1}^{N_b^t} \sim \mathcal{D}_{\text{tgt}}$
5:      Update ρ by $\min_\rho -\lambda \mathcal{L}_{\text{wdom}}$
6:      Update ψ by $\min_\psi \lambda \mathcal{L}_{\text{dom}}$
7:      Update θ, ϕ by $\min_{\theta,\phi} \mathcal{L}_{\text{cls}} - \lambda \mathcal{L}_{\text{wdom}}$
8: **end for**

---

where C is a constant

$$C = \lambda + O\left(\sqrt{\frac{d\log(m/d) + \log(1/\delta)}{m}}\right)$$

Here, λ is the optimal combined risk (source + target risk) that can be achieved by hypothesis in $\mathcal{H}$. Let $\{x_i^s\}_{i=1}^m$, $\{x_i^t\}_{i=1}^m$ be the samples in the empirical distributions $\hat{\mathcal{D}}_{\text{src}}$ and $\hat{\mathcal{D}}_{\text{tgt}}$ respectively. Then, $P(x_i^s) = 1/m$ and $P(x_i^t) = 1/m$. The empirical source risk can be written as $\hat{\epsilon}_s(h) = 1/m \sum_i \hat{\epsilon}_{x_i^s}(h)$

Now consider a CMSS re-weighted source distribution $\hat{\mathcal{D}}_{\text{wsrc}}$, with $P(x_i^s) = w_i$. For $\hat{\mathcal{D}}_{\text{wsrc}}$ to be a valid probability mass function, $\sum_i w_i^s = 1$ and $w_i^s \geq 0$. Note that $\hat{\mathcal{D}}_{\text{src}}$ and $\hat{\mathcal{D}}_{\text{wsrc}}$ share the same samples, and only differ in weights. The generalization bound for this re-weighted distribution can be written as

$$\epsilon_t(h) \leq \sum_i w_i \hat{\epsilon}_{x_i^s}(h) + \frac{1}{2}d_{\mathcal{H}\Delta\mathcal{H}}(\hat{\mathcal{D}}_{\text{wsrc}}, \hat{\mathcal{D}}_{\text{tgt}}) + C$$

Since the bound holds for all weight arrays $\mathbf{w} = [w_1^s, w_2^s \ldots w_m^s]$ in a simplex, we can minimize the objective over $\mathbf{w}$ to get a tighter bound.

$$\epsilon_t(h) \leq \min_{\mathbf{w} \in \Delta^m} \sum_i w_i \hat{\epsilon}_{x_i^s}(h) + \frac{1}{2}d_{\mathcal{H}\Delta\mathcal{H}}(\hat{\mathcal{D}}_{\text{wsrc}}, \hat{\mathcal{D}}_{\text{tgt}}) + C \qquad (6)$$

The first term is the weighted risk, and the second term $d_{\mathcal{H}\Delta\mathcal{H}}(\hat{\mathcal{D}}_{\text{wsrc}}, \hat{\mathcal{D}}_{\text{tgt}})$ is the weighted symmetric divergence which can be realized using our weighted adversarial loss. Note that when $\mathbf{w} = [1/m, 1/m, \ldots 1/m]$, we get the original bound (5). Hence, the original bound is in the feasible set of this optimization.

Relaxations. In practice, deep neural networks are used to optimize the bounds presented above. Since the bound (6) is minimized over the weight vector $\mathbf{w}$, one trivial solution is to assign non-zero weights to only a few source samples. In this case, a neural network can overfit to these source samples, which could result in low training risk and low domain divergence. To avoid this trivial case, we present two relaxations:

- We use the unweighted loss for the source risk (first term in the bound (6)).
- For the divergence term, instead of minimizing $\mathbf{w}$ over all the samples, we optimize only over mini-batches. Hence, for every mini-batch, there is at least one w_i which is non-zero. Additionally, we make weights a function of input, i.e., $w_i = G_\rho(x_i^s)$, which is realized using a neural network. This will smooth the predictions of w_i, and make the weight network produce a soft-selection over source samples based on correlation with the target.

Note that the G_ρ network discussed in the previous section satisfies these criteria.

5   Experimental Results

In this section, we perform an extensive evaluation of the proposed method on the following tasks: digit classification(*MNIST, MNIST-M, SVHN, Synthetic Digits, USPS*), image recognition on the large-scale DomainNet dataset (*clipart, infograph, paiting, quickdraw, real, sketch*), PACS [22] (*art, cartoon, photo* and *sketch*) and Office-Caltech10 (*Amazon, Caltech, Dslr, Webcam*). We compare our method with the following contemporary approaches: Domain Adversarial Neural Network (**DANN**) [15], Multi-Domain Adversarial Neural Network (**MDAN**)[46] and two state-of-the-art discrepancy-based approaches: Maximum Classifier Discrepancy (**MCD**) [33] and Moment Matching for Multi-Source (M^3**SDA**) [32]. We follow the protocol used in other multi-source domain adaptation works [27,32], where each domain is selected as the target domain while the rest of domains are used as source domains. For **Source Only** and **DANN** experiments, all source domains are shuffled and treated as one domain. To guarantee fairness of comparison, we used the same model architectures, batch size and data pre-processing routines for all compared approaches. All our experiments are implemented in PyTorch.

Table 1. Results on Digits classification. The proposed CMSS achieves **90.8%** accuracy. Comparisons with MCD and M^3SDA are reprinted from [32]. All experiments are based on a 3-*conv*-layer backbone trained from scratch. (mt, mm, sv, sy, up: *MNIST, MNIST-M, SVHN, Synthetic Digits, UPSP*)

Models	mm, sv, sy, up $\rightarrow mt$	mt, sv, sy, up $\rightarrow mm$	mt, mm, sy, up $\rightarrow sv$	mt, mm, sv, up $\rightarrow sy$	mt, mm, sv, sy $\rightarrow up$	Avg
Source Only	92.3 ± 0.91	63.7 ± 0.83	71.5 ± 0.75	83.4 ± 0.79	90.7 ± 0.54	80.3 ± 0.76
DANN [15]	97.9 ± 0.83	70.8 ± 0.94	68.5 ± 0.85	87.3 ± 0.68	93.4 ± 0.79	83.6 ± 0.82
MDAN [46]	97.2 ± 0.98	**75.7 ± 0.83**	82.2 ± 0.82	85.2 ± 0.58	93.3 ± 0.48	86.7 ± 0.74
MCD [33]	96.2 ± 0.81	72.5 ± 0.67	78.8 ± 0.78	87.4 ± 0.65	95.3 ± 0.74	86.1 ± 0.64
M^3SDA [32]	98.4 ± 0.68	72.8 ± 1.13	81.3 ± 0.86	89.5 ± 0.56	96.1 ± 0.81	87.6 ± 0.75
CMSS	**99.0 ± 0.08**	75.3 ± 0.57	**88.4 ± 0.54**	**93.7 ± 0.21**	**97.7 ± 0.13**	**90.8 ± 0.31**

5.1   Experiments on Digit Recognition

Following DCTN [40] and M^3SDA [32], we sample 25000 images from training subset and 9000 from testing subset of *MNIST, MNIST-M, SVHN* and *Synthetic Digits*. The entire *USPS* is used since it contains only 9298 images in total.

In all the experiments, the feature extractor is composed of three *conv* layers and two *fc* layers. The entire network is trained from scratch with batch size equals 16. For each experiment, we run the same setting five times and report the mean and standard deviation. (See *Appendix* for more experiment details and analyses.) The results are shown in Table 1. The proposed method achieves an **90.8%** average accuracy, outperforming other baselines by a large margin ($\sim$ 3% improvement on the previous state-of-the-art approach).

5.2   Experiments on DomainNet

Next, we evaluate our method on **DomainNet** [32] – a large-scale benchmark dataset used for multi-domain adaptation. The DomainNet dataset contains samples from 6 domains: *Clipart, Infograph, Painting, Quickdraw, Real* and *Sketch*. Each domain has **345** categories, and the dataset has ∼**0.6 million** images in total, which is the largest existing domain adaptation dataset. We use ResNet-101 pretrained on ImageNet as the feature extractor for in all our experiments. For CMSS, we use a ResNet-18 pretrained on ImageNet. The batch size is fixed to 128. We conduct experiments over 5 random runs, and report mean and standard deviation over the 5 runs.

The results are shown in Table 2. CMSS achieves **46.5%** average accuracy, outperforming other baselines by a large margin. We also note that our approach achieves the best performance in each experimental setting. It is also worth mentioning that in the experiment when the target domain is *Quickdraw (q)*, our approach is the only one that outperforms Source Only baseline, while all other compared approaches result in negative transfer (lower performance than the source-only model). This is since *quickdraw* has a significant domain shift compared to all other domains. This shows that our approach can effectively alleviate negative transfer even in such challenging set-up.

Table 2. Results on the DomainNet dataset. CMSS achieves 46.5% average accuracy. When the target domain is *quickdraw q*, CMSS is the only one that outperforms Source Only which indicates *negative transfer* has been alleviated. *Source Only ** is reprinted from [32], *Source Only* is our implemented results. All experiments are based on ResNet-101 pre-trained on ImageNet. (*c: clipart, i: infograph, p: painting, q: quickdraw, r: real, s: sketch*)

Models	i, p, q $r, s \rightarrow c$	c, p, q $r, s \rightarrow i$	c, i, q $r, s \rightarrow p$	c, i, p $r, s \rightarrow q$	c, i, p $q, s \rightarrow r$	c, i, p $q, r \rightarrow s$	Avg
Source Only*	47.6±0.52	13.0±0.41	38.1±0.45	13.3±0.39	51.9±0.85	33.7±0.54	32.9±0.54
Source Only	52.1±0.51	23.4±0.28	47.7±0.96	13.0±0.72	60.7±0.32	46.5±0.56	40.6±0.56
DANN [15]	60.6±0.42	25.8±0.34	50.4±0.51	7.7±0.68	62.0±0.66	51.7±0.19	43.0±0.46
MDAN [46]	60.3±0.41	25.0±0.43	50.3±0.36	8.2±1.92	61.5±0.46	51.3±0.58	42.8±0.69
MCD [33]	54.3±0.64	22.1±0.70	45.7±0.63	7.6±0.49	58.4±0.65	43.5±0.57	38.5±0.61
M^3SDA [32]	58.6±0.53	26.0±0.89	52.3±0.55	6.3±0.58	62.7±0.51	49.5±0.76	42.6±0.64
CMSS	**64.2±0.18**	**28.0±0.20**	**53.6±0.39**	**16.0±0.12**	**63.4±0.21**	**53.8±0.35**	**46.5±0.24**

5.3   Experiments on PACS

PACS [22] is another popular benchmark for multi-source domain adaptation. It contains 4 domains: *art, cartoon, photo* and *sketch*. Images of 7 categories are collected for each domain. There are 9991 images in total. For all experiments, we used ResNet-18 pretrained on ImageNet as the feature extractor following [27]. For the Curriculum Manager, we use the same architecture as the feature extractor. Batch size of 32 is used. We conduct experiments over 5 random runs, and report mean and standard deviation over the runs. The results are shown in Table 3 (*a: art, c: cartoon, p: painting, s: sketch.*). CMSS achieves the state-of-the-art average accuracy of **89.5%**. On the most challenging *sketch* (s) domain, we obtain **82.0%**, outperforming other baselines by a large margin.

5.4   Experiments on Office-Caltech10

The office-Caltech10 [16] dataset has 10 object categories from 4 different domains: *Amazon, Caltech, DSLR,* and *Webcam.* For all the experiments, we use the same architecture (ResNet-101 pretrained on ImageNet) used in [32]. The experimental results are shown in Table 4 (A: *Amazon*, C: *Caltech*, D: *Dslr*, W: *Webcam*). CMSS achieves state-of-the-art average accuracy of **97.2%**.

Table 3. Results on PACS

Models	$c,p,s \to a$	$a,p,s \to c$	$a,c,s \to p$	$a,c,p \to s$	Avg
Source Only	74.9±0.88	72.1±0.75	94.5±0.58	64.7±1.53	76.6±0.93
DANN [15]	81.9±1.13	77.5±1.26	91.8±1.21	74.6±1.03	81.5±1.16
MDAN [46]	79.1±0.36	76.0±0.73	91.4±0.85	72.0±0.80	79.6±0.69
WBN [27]	**89.9±0.28**	89.7±0.56	**97.4±0.84**	58.0±1.51	83.8±0.80
MCD [33]	88.7±1.01	88.9±1.53	96.4±0.42	73.9±3.94	87.0±1.73
M^3SDA [32]	89.3±0.42	89.9±1.00	97.3±0.31	76.7±2.86	88.3±1.15
CMSS	88.6±0.36	**90.4±0.80**	96.9±0.27	**82.0±0.59**	**89.5±0.50**

Table 4. Results on Office-Caltech10

Models	A,C,D $\to W$	A,C,W $\to D$	A,D,W $\to C$	C,D,W $\to A$	Avg
Source Only	99.0	98.3	87.8	86.1	92.8
DANN [15]	99.3	98.2	89.7	94.8	95.5
MDAN [46]	98.9	98.6	91.8	95.4	96.1
MCD [33]	99.5	99.1	91.5	92.1	95.6
M^3SDA [32]	99.5	99.2	92.2	94.5	96.4
CMSS	**99.6**	**99.3**	**93.7**	**96.0**	**97.2**

Table 5. Comparing re-weighting methods

Models	i,p,q $r,s \to c$	c,p,q $r,s \to i$	c,i,q $r,s \to p$	c,i,p $q,s \to r$	c,i,p $q,r \to s$	c,i,p	Avg
DANN [15]	60.6	25.8	50.4	7.7	62.0	51.7	43.0
IWAN [43]	59.1	25.2	49.7	12.9	60.4	51.4	43.1
CMSS	**64.2**	**28.0**	**53.6**	**16.0**	**63.4**	**53.8**	**46.5**

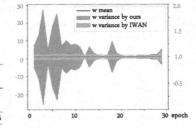

Fig. 3. Mean/var of weights over time.

5.5   Comparison with Other Re-weighting Methods

In this experiment, we compare CMSS with other weighing schemes proposed in the literature. We use IWAN [43] for this purpose. IWAN, originally proposed for partial domain adaption, reweights the samples in adversarial training using outputs of discriminator as sample weights (Refer to Fig. 2). CMSS, however, computes sample weights using a separate network G_ρ updated using an adversarial game. We adapt IWAN for multi-source setup and compare it against our approach. The results are shown in Table 5 (abbreviations of domains same as Table 2). IWAN obtained 43.1% average accuracy which is close to performance obtained using DANN with combined source domains. For further analysis, we plot how sample weights estimated by both approaches (plotted as mean ± variance) change as training progresses in Fig. 3. We observe that CMSS selects weights with larger variance which demonstrates its sample selection ability, while IWAN has weights all close to 1 (in which case, it becomes similar to DANN). This illustrates the superiority of our sample selection method. More discussions on sample selection can be found in Sect. 6.2. CMSS also achieves a faster and more stable convergence in test accuracy compared to DANN [15] where we assume a single source domain, which further supports the effectiveness of the learnt curriculum.

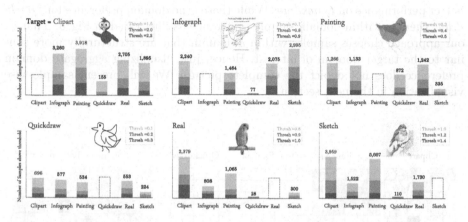

Fig. 4. Interpretation results of the sample selection on DomainNet dataset using the proposed method. In each plot, one domain is selected as the target. In each setting, predictions of CMSS are computed for each sample of the source domains. The bars indicate how many of these samples have weight prediction larger than a manually chosen threshold, with each bar denoting a single source domain. Maximum number of samples are highlighted in red. *Best viewed in color* (Color figure online)

6   Interpretations

In this section, we are interested in understanding and visualizing the source selection ability of our approach. We conduct two sets of experiments: (i) visualizations of the source selection curriculum over time, and (ii) comparison of our selection mechanism with other sample re-weighting methods.

6.1   Visualizations of Source Selection

Domain Preference. We first investigate if CMSS indeed exhibits domain preference over the course of training as claimed. For this experiment, we randomly select $m = 34000$ training samples from each source domain in DomainNet and obtain the raw weights (before softmax) generated by CMSS. Then, we calculate the number of samples in each domain passing a manually selected threshold τ. We use the number of samples passing this threshold in each domain to indicate the domain preference level. The larger the fraction, more weights are given to samples from the domains, hence, higher the domain preference. Figure 4 shows the visualization of domain preference for each target domain. We picked 3 different τ in each experiment for more precise observation. We observe that CMSS does display domain preference (*Clipart* - *Painting*, *Infograph* - *Sketch*, *Real* - *Clipart*) that is in fact correlated with the visual similarity of the domains. An exception is *Quickdraw*, where no domain preference is observed. We argue that this is because *Quickdraw* has significant domain shift compared to all other domains, hence no specific domain is preferred. However, CMSS still produces better performance on *Quickdraw*. While there is no domain preference for *Quickdraw*, there is within-domain sample preference as illustrated in Fig. 5. That is, our approach chooses samples within a domain that are structurally more similar to the target domain of interest. Hence, just visualizing aggregate domain preference does not depict the complete picture. We will present sample-wise visualization in the next section.

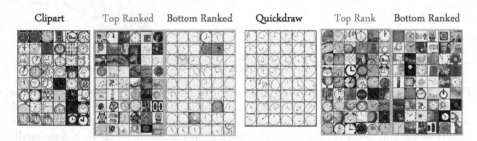

Fig. 5. Ranked source samples according to learnt weights (class "Clock" of Domain-Net dataset). *LHS*: Examples of unlabeled target domain *Clipart* and the Top/Bottom Ranked ~50 samples of the source domain composed of *Infograph, Painting, Quickdraw, Real* and *Sketch. RHS*: Examples of unlabeled target domain *Quickdraw* and the Ranked samples of source domain composed of *Clipart, Infograph, Painting, Real* and *Sketch*. Weights are obtained at inference time using CMSS trained after 5 epochs.

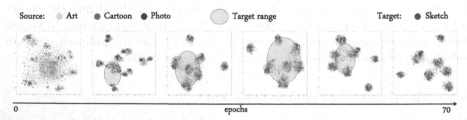

Fig. 6. t-SNE visualization of features at six different epochs during training. The shaded region is the migrated range of target features. Dateset used is PACS with *sketch* as the target domain. (Color figure online)

Beyond Domain Preference. In addition to domain preference, we are interested in taking a closer look at sample-wise source selection. To do this, we first obtain the weights generated by CMSS for all source samples and rank the source images according to their weights. An example is shown in Fig. 5. For better understanding, we visualize samples belonging to a fixed category ("Clock" in Fig. 5). See *Appendix* for more visualizations.

In Fig. 5, we find that notion of similarity discovered by CMSS is different for different domains. When the target domain is *Clipart* (left panel of Fig. 5), source samples with colors and cartoonish shapes are ranked at the top, while samples with white background and simplistic shapes are ranked at the bottom. When the target is *Quickdraw* (right panel of Fig. 5), one would think that CMSS will simply be selecting images with similar white background. Instead, it prefers samples which are structurally similar to the regular rounded clock shape (as most samples in *Quickdraw* are similar to these). It thus appears that structural similarity is favored in *Quickdraw*, whereas color information is preferred in *Clipart*. This provides support that CMSS selects samples according to ease of alignment to the target distribution, which is automatically discovered per domain. We argue that this property of CMSS has an advantage over approaches such as MDAN [46] which simply weighs manually partitioned domains.

6.2   Selection over Time

In this section, we discuss how source selection varies as training progresses. In Fig. 3, we plot mean and variance of weights (output of Curriculum Manager) over training iterations. We observe that the variance is high initially, which indicates many samples have weights away from the mean value of 1. Samples with higher weights are preferred, while those with low weights contribute less to the alignment. In the later stages, the variance is very low which indicates most of the weights are close to 1. Hence, our approach gradually adapts to increasingly many source samples over time, naturally learning a curriculum for adaptation. In Fig. 6, we plot a t-SNE visualization of features at different epochs. We observe that the target domain *sketch* (red) first adapts to *Art* (yellow), and then gradually aligns with *Cartoon* (green) and *Photo* (blue).

7   Conclusion

In this paper, we proposed Curriculum Manager for Source Selection (CMSS) that learns a curriculum for Multi-Source Unsupervised Domain Adaptation. A curriculum is learnt that iteratively favors source samples that align better with the target distribution over the entire training. The curriculum learning is achieved by an adversarial interplay with the discriminator, and achieves state-of-the-art on four benchmark datasets. We also shed light on the inner workings of CMSS, and we hope that will pave the way for further advances to be made in this research area.

Acknowledgement. This project was partially supported by Facebook AI and Defense Advanced Research Projects Agency (DARPA) via ARO contract number W911NF2020009. The views, opinions, and findings expressed are those of the authors and should not be interpreted as representing the official views or policies of the Department of Defense or the U.S. Government. There is no collaboration between Facebook and DARPA.

References

1. Ben-David, S., Blitzer, J., Crammer, K., Kulesza, A., Pereira, F., Vaughan, J.W.: A theory of learning from different domains. Machine Learn. 151–175 (2009). https://doi.org/10.1007/s10994-009-5152-4
2. Ben-David, S., Blitzer, J., Crammer, K., Pereira, F.: Analysis of representations for domain adaptation. In: Advances in Neural Information Processing Systems, pp. 137–144 (2007)
3. Ben-David, S., Blitzer, J., Crammer, K., Pereira, F.: Analysis of representations for domain adaptation. In: Schölkopf, B., Platt, J.C., Hoffman, T. (eds.) Advances in Neural Information Processing Systems, vol. 19, pp. 137–144. MIT Press (2007)
4. Bengio, Y., Louradour, J., Collobert, R., Weston, J.: Curriculum learning. In: Proceedings of the 26th Annual International Conference on Machine Learning. p. 41–48. ICML 2009, Association for Computing Machinery, New York, NY, USA (2009)
5. Bhatt, H.S., Rajkumar, A., Roy, S.: Multi-source iterative adaptation for cross-domain classification. In: IJCAI, pp. 3691–3697 (2016)
6. Blitzer, J., Crammer, K., Kulesza, A., Pereira, F., Wortman, J.: Learning bounds for domain adaptation. In: Platt, J.C., Koller, D., Singer, Y., Roweis, S.T. (eds.) Advances in Neural Information Processing Systems, vol. 20, pp. 129–136. Curran Associates, Inc. (2008)
7. Cao, Z., Long, M., Wang, J., Jordan, M.I.: Partial transfer learning with selective adversarial networks. In: Proceedings of the IEEE Conference on Computer Vision and Pattern Recognition, pp. 2724–2732 (2018)
8. Chen, C., et al.: Progressive feature alignment for unsupervised domain adaptation. In: Proceedings of the IEEE Conference on Computer Vision and Pattern Recognition, pp. 627–636 (2019)
9. Chen, Q., Liu, Y., Wang, Z., Wassell, I., Chetty, K.: Re-weighted adversarial adaptation network for unsupervised domain adaptation. In: Proceedings of the IEEE Conference on Computer Vision and Pattern Recognition, pp. 7976–7985 (2018)
10. Chen, Y., Li, W., Sakaridis, C., Dai, D., Van Gool, L.: Domain adaptive faster R-CNN for object detection in the wild. In: Proceedings of the IEEE Conference on Computer Vision and Pattern Recognition, pp. 3339–3348 (2018)
11. Ding, Z., Nasrabadi, N.M., Fu, Y.: Semi-supervised deep domain adaptation via coupled neural networks. IEEE Trans. Image Process. **27**(11), 5214–5224 (2018)
12. Duan, L., Tsang, I.W., Xu, D., Chua, T.S.: Domain adaptation from multiple sources via auxiliary classifiers. In: Proceedings of the 26th Annual International Conference on Machine Learning, pp. 289–296. ACM (2009)
13. Duan, L., Xu, D., Chang, S.F.: Exploiting web images for event recognition in consumer videos: a multiple source domain adaptation approach. In: 2012 IEEE Conference on Computer Vision and Pattern Recognition, pp. 1338–1345. IEEE (2012)

14. Duan, L., Xu, D., Tsang, I.W.H.: Domain adaptation from multiple sources: a domain-dependent regularization approach. IEEE Trans. Neural Netw. Learn. Syst. **23**(3), 504–518 (2012)
15. Ganin, Y., Lempitsky, V.: Unsupervised domain adaptation by backpropagation. arXiv preprint arXiv:1409.7495 (2014)
16. Gong, B., Shi, Y., Sha, F., Grauman, K.: Geodesic flow kernel for unsupervised domain adaptation. In: 2012 IEEE Conference on Computer Vision and Pattern Recognition, pp. 2066–2073. IEEE (2012)
17. Jeong, R., et al..: Self-supervised sim-to-real adaptation for visual robotic manipulation. arXiv preprint arXiv:1910.09470 (2019)
18. Kang, G., Jiang, L., Yang, Y., Hauptmann, A.G.: Contrastive adaptation network for unsupervised domain adaptation. In: Proceedings of the IEEE Conference on Computer Vision and Pattern Recognition, pp. 4893–4902 (2019)
19. Kumar, A., et al.: Co-regularized alignment for unsupervised domain adaptation. In: Advances in Neural Information Processing Systems, pp. 9345–9356 (2018)
20. Kurmi, V.K., Bajaj, V., Subramanian, V.K., Namboodiri, V.P.: Curriculum based dropout discriminator for domain adaptation. arXiv preprint arXiv:1907.10628 (2019)
21. Lee, S., Kim, D., Kim, N., Jeong, S.G.: Drop to adapt: Learning discriminative features for unsupervised domain adaptation. In: Proceedings of the IEEE International Conference on Computer Vision, pp. 91–100 (2019)
22. Li, D., Yang, Y., Song, Y.Z., Hospedales, T.M.: Deeper, broader and artier domain generalization. In: Proceedings of the IEEE International Conference on Computer Vision, pp. 5542–5550 (2017)
23. Li, Y., Tian, X., Gong, M., Liu, Y., Liu, T., Zhang, K., Tao, D.: Deep domain generalization via conditional invariant adversarial networks. In: Proceedings of the European Conference on Computer Vision, pp. 624–639 (2018)
24. Liu, H., Long, M., Wang, J., Jordan, M.: Transferable adversarial training: a general approach to adapting deep classifiers. In: International Conference on Machine Learning, pp. 4013–4022 (2019)
25. Luo, Y., Zheng, L., Guan, T., Yu, J., Yang, Y.: Taking a closer look at domain shift: category-level adversaries for semantics consistent domain adaptation. In: Proceedings of the IEEE Conference on Computer Vision and Pattern Recognition, pp. 2507–2516 (2019)
26. Mancini, M., Bulò, S.R., Caputo, B., Ricci, E.: Adagraph: unifying predictive and continuous domain adaptation through graphs. In: Proceedings of the IEEE Conference on Computer Vision and Pattern Recognition, pp. 6568–6577 (2019)
27. Mancini, M., Porzi, L., Rota Bulò, S., Caputo, B., Ricci, E.: Boosting domain adaptation by discovering latent domains. In: Proceedings of the IEEE Conference on Computer Vision and Pattern Recognition, pp. 3771–3780 (2018)
28. Motiian, S., Jones, Q., Iranmanesh, S., Doretto, G.: Few-shot adversarial domain adaptation. In: Advances in Neural Information Processing Systems, pp. 6670–6680 (2017)
29. Ouyang, C., Kamnitsas, K., Biffi, C., Duan, J., Rueckert, D.: Data efficient unsupervised domain adaptation for cross-modality image segmentation. In: Shen, D., Liu, T., Peters, T.M., Staib, L.H., Essert, C., Zhou, S., Yap, P.-T., Khan, A. (eds.) MICCAI 2019. LNCS, vol. 11765, pp. 669–677. Springer, Cham (2019). https://doi.org/10.1007/978-3-030-32245-8_74
30. Pan, Y., Yao, T., Li, Y., Wang, Y., Ngo, C.W., Mei, T.: Transferrable prototypical networks for unsupervised domain adaptation. In: Proceedings of the IEEE Conference on Computer Vision and Pattern Recognition, pp. 2239–2247 (2019)

31. Pei, Z., Cao, Z., Long, M., Wang, J.: Multi-adversarial domain adaptation. In: Thirty-Second AAAI Conference on Artificial Intelligence (2018)
32. Peng, X., Bai, Q., Xia, X., Huang, Z., Saenko, K., Wang, B.: Moment matching for multi-source domain adaptation. arXiv preprint arXiv:1812.01754 (2018)
33. Saito, K., Watanabe, K., Ushiku, Y., Harada, T.: Maximum classifier discrepancy for unsupervised domain adaptation. In: Proceedings of the IEEE Conference on Computer Vision and Pattern Recognition, pp. 3723–3732 (2018)
34. Shu, Y., Cao, Z., Long, M., Wang, J.: Transferable curriculum for weakly-supervised domain adaptation. Proc. AAAI Conf. Artif. Intell. **33**, 4951–4958 (2019)
35. Sun, Y., Tzeng, E., Darrell, T., Efros, A.A.: Unsupervised domain adaptation through self-supervision. arXiv preprint arXiv:1909.11825 (2019)
36. Valada, A., Mohan, R., Burgard, W.: Self-supervised model adaptation for multi-modal semantic segmentation. Int. J. Comput. Vis. **52**, 1–47 (2019)
37. Wang, T., Zhang, X., Yuan, L., Feng, J.: Few-shot adaptive faster r-cnn. In: Proceedings of the IEEE Conference on Computer Vision and Pattern Recognition, pp. 7173–7182 (2019)
38. Wu, Z., Wang, X., Gonzalez, J.E., Goldstein, T., Davis, L.S.: Ace: Adapting to changing environments for semantic segmentation. arXiv preprint arXiv:1904.06268 (2019)
39. Xiong, C., McCloskey, S., Hsieh, S.H., Corso, J.J.: Latent domains modeling for visual domain adaptation. In: Twenty-Eighth AAAI Conference on Artificial Intelligence (2014)
40. Xu, R., Chen, Z., Zuo, W., Yan, J., Lin, L.: Deep cocktail network: Multi-source unsupervised domain adaptation with category shift. In: Proceedings of the IEEE Conference on Computer Vision and Pattern Recognition, pp. 3964–3973 (2018)
41. Yoon, J.S., Shiratori, T., Yu, S.I., Park, H.S.: Self-supervised adaptation of high-fidelity face models for monocular performance tracking. In: Proceedings of the IEEE Conference on Computer Vision and Pattern Recognition, pp. 4601–4609 (2019)
42. You, K., Wang, X., Long, M., Jordan, M.: Towards accurate model selection in deep unsupervised domain adaptation. In: International Conference on Machine Learning, pp. 7124–7133 (2019)
43. Zhang, J., Ding, Z., Li, W., Ogunbona, P.: Importance weighted adversarial nets for partial domain adaptation. In: Proceedings of the IEEE Conference on Computer Vision and Pattern Recognition, pp. 8156–8164 (2018)
44. Zhang, J., Chen, Z., Huang, J., Lin, L., Zhang, D.: Few-shot structured domain adaptation for virtual-to-real scene parsing. In: Proceedings of the IEEE International Conference on Computer Vision Workshops (2019)
45. Zhao, H., Combes, R.T.d., Zhang, K., Gordon, G.J.: On learning invariant representation for domain adaptation. arXiv preprint arXiv:1901.09453 (2019)
46. Zhao, H., Zhang, S., Wu, G., Moura, J.M., Costeira, J.P., Gordon, G.J.: Adversarial multiple source domain adaptation. In: Advances in Neural Information Processing Systems, pp. 8559–8570 (2018)

Powering One-Shot Topological NAS with Stabilized Share-Parameter Proxy

Ronghao Guo[1], Chen Lin[2], Chuming Li[2], Keyu Tian[1], Ming Sun[2],
Lu Sheng[1(✉)], and Junjie Yan[2]

[1] College of Software, Beihang University, Beijing, China
{16211042,17375491,lsheng}@buaa.edu.cn
[2] SenseTime Research, Hong Kong, China
{linchen,lichuming,sunming1,yanjunjie}@sensetime.com

Abstract. One-shot NAS method has attracted much interest from the research community due to its remarkable training efficiency and capacity to discover high performance models. However, the search spaces of previous one-shot based works usually relied on hand-craft design and were short for flexibility on the network topology. In this work, we try to enhance the one-shot NAS by exploring high-performing network architectures in our large-scale Topology Augmented Search Space (*i.e*, over 3.4×10^{10} different topological structures). Specifically, the difficulties for architecture searching in such a complex space has been eliminated by the proposed stabilized share-parameter proxy, which employs Stochastic Gradient Langevin Dynamics to enable fast shared parameter sampling, so as to achieve stabilized measurement of architecture performance even in search space with complex topological structures. The proposed method, namely Stablized Topological Neural Architecture Search (ST-NAS), achieves state-of-the-art performance under Multiply-Adds (MAdds) constraint on ImageNet. Our lite model ST-NAS-A achieves 76.4% top-1 accuracy with only 326M MAdds. Our moderate model ST-NAS-B achieves 77.9% top-1 accuracy just required 503M MAdds. Both of our models offer superior performances in comparison to other concurrent works on one-shot NAS.

Keywords: Stablized one-shot NAS · Network topology

1 Introduction

Significant progress made by convolution neural networks (CNN) in challenging computer vision tasks has raised the demand to design powerful neural networks. Instead of manually design, Neural architecture search (NAS) has demonstrated great potentials in recent years. Early works of NAS by Real *et al.* [28,29] and Elsken *et al.* [11] achieved promising results but can only be applied to small datasets due to their large computation expenses. To this end, one-shot based

---

R. Guo and C. Lin—Contributed equally.

A. Vedaldi et al. (Eds.): ECCV 2020, LNCS 12359, pp. 625–641, 2020.
https://doi.org/10.1007/978-3-030-58568-6_37

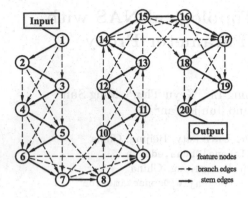

Fig. 1. An instance of our topology augmented search space. It contains over 3.4×10^{10} different network topologies which enables us to explore complex network typologies. Solid line denotes the chain-structured stem edges, and dotted line represents branch edges which connects feature maps with depth difference 2 or 3.

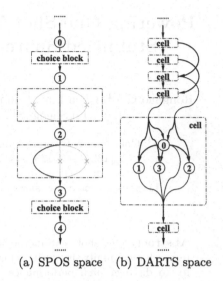

(a) SPOS space   (b) DARTS space

Fig. 2. Two typical search spaces in previous work. Figure (a) shows a chain-structured space while Fig. (b) shows a cell-based search space.

methods have drawn much interest thanks to its promising training efficiency and remarkable ability to discover high-performing models. One-shot method usually utilizes a hyper-network, which subsumes all architectures in the search space, and use shared weights to evaluate different architectures.

However, the search space of previous works (*e.g*, shown in Fig. 2) were usually carefully designed and did not enjoy too much flexibility on the network topology. For example, as one of mostly applied search spaces in the one-shot literature, the chain-structured search space [14] has sequentially connected intermediate feature maps, between which the edges are chosen from a set of computation operations. Networks with better operations can be discovered on this search space, but the network topology remains trivial. However, previous works [10,17] on network architecture design proved that complex topology will tremendously enhance the performance of deep learning models. We argue that complex topological structures added in search space will improve the performance of the searched network architectures as shown in Table 1.

In this work, we are interested in exploring complex network typologies with one-shot method. We propose a novel network architecture search space shown in Fig. 1 which contains over 3.4×10^{10} different network topologies, enabling the discovery of complex topology networks. The search space is obtained by introducing numerous computation modules as edges between nodes. A topology based architecture sampler is also introduced to sample architectures during one-shot training stage from the hyper-network. However, the great diversity introduced by topologies brings difficulties to the one-shot approach. Specifically, we observe high variance of performance estimation through the one-shot shared parameters in two cases: estimation through shared parameters at different epochs of a single run and estimation through shared parameters obtain in different runs. Zhang *et al.* [43] explore the reason behind the variance of ranking under weight sharing strategy. Thus the ranking ability of shared parameters is compromised.

To eliminate the interference of complex topologies, we estimate the expectation of architecture performance in additional training epochs of hyper-network via multiple samples of shared parameters. An fast weights sampling methods based on Stochastic Gradient Langevin Dynamics is developed to sample shared parameters efficiently.

The resulted Stabilized Topological Neural Architecture Search (ST-NAS) achieves compatible performance with the state-of-the-art NAS method. The resulted architecture ST-NAS-A obtains 76.4% top-1 accuracy with only 326M MAdds. A larger architecture ST-NAS-B obtains 77.9% top-1 accuracy with around 503M MAdds.

To summarize, our main contributions are as follows:

1. We introduce a topology augmented neural architecture search space that enables the discovery of efficient architectures with complex topology.
2. To relieve the complex topology's interference on model ranking, we modified model evaluation based on the expectation of the sharing parameters' performance.
3. We empirically demonstrate improvements on ImageNet classification under the same MAdds constraints compared with previous work, and show that the searched architectures transfer well to COCO object detection.

2   Related Work

Recently, auto machine learning methods have received a lot of attention due to its ability to design augmentation [9,22], loss function [19] and network architectures [4,13,14,20,21,25,28,44,45].

Early neural architecture search (NAS) works normally involves reinforcement learning [1,13,36,44–46] or evolution algorithm [26,29] to search for high-performing network architectures. However, these methods are usually computationally expensive which limits its uses in real scenarios.

Recent attentions have been focused on alleviating the computation cost via weight sharing method. This method usually contains a single training process of an over parameterized hyper-network which subsumes all candidate models, *i.e*, weights of the same operators are shared across different sub-models. Notably, Liu *et al.* [25] proposes continuous relaxations which enables optimizing network architectures with gradient decent, Cai *et al.* [5] proposes a proxy-less method to search on target datasets directly and Bender *et al.* [2] introduces one-shot method to decouple training and searching stages. Our NAS work take the use of the weight sharing hyper-network but relieve the variance during model training.

Early hand-craft neural networks [15,33,35] tend to stack repeated motifs. Works in [15–17,34] introduce different manual designed network topologies and result in performance gain.

Motivated by manual designed architectures [15,33,35], a widely used search space in works [13,25,26,44,45] are proposed to search for such motifs, dubbed cells or blocks, rather than all possible architectures. This search space is called cell-based space. Another widely used search space adopted in [5,14,36,42] is called chain-structured space. This space sequentially stacks several operation layers where each layer serves its output as next layer's input. NAS methods are adopted to search for operation layers in different position of this space. Work in [41] explores random wiring networks with less human prior and achieves comparable performance with manual designed networks.

3   Approach

Methods for NAS usually consist of three basic components: search space, performance estimation and search strategy. In this section, we first introduce our novel Topology Augmented Search Space and a new sampling strategy for hyper-network training in this particular space. Secondly, we provide new model performance estimation approach to relieve the variance of model ranks during the training of hyper-network. Finally the evolution algorithm for network search is described.

3.1   Topology Augmented Search Space

Motivation. To demonstrate the improvement of complex topology against a sequential structure, we take ResNet-18 as a baseline and shows a subtle change on the topology obtains obvious performance boost. We randomly add 4 residual blocks to connect the feature maps of blocks in ResNet-18's [15] chain structure with 3 random seeds, and rescale the width to keep the same FLOPs, the results are in Table 1. The 3 complex structures imply the great potential of topology-based structure search.

Table 1. The accuracy of ResNet-18 and three networks with 4 random skip residual blocks added on the baseline. The three networks are scaled to keep the FLOPs same with ResNet18. Obvious boosting is obtained via exploration in a more complex topology space.

Architecture	Res-18	Rand0	Rand1	Rand2
Accuracy (%)	70.2	71.5	72.0	69.6

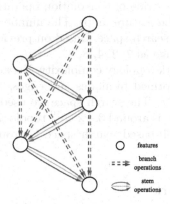

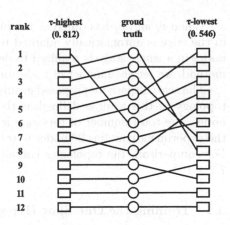

rank	τ-highest (0.812)	groud truth	τ-lowest (0.546)
1			
2			
3			
4			
5			
6			
7			
8			
9			
10			
11			
12			

Fig. 3. Illustration of candidates in stem edges and branch edges.

Fig. 4. Detailed ranking of the best and the worst among 20 runs. The ground truth ranking is provided in the middle. This figure shows that the quality of ranking exceedingly differs from.

Search Space. A neural network is denoted as a directed acyclic graph (DAG) defined by E, V, where the node $v_i \in V$ indicates the feature connected by edge $e_i \in E$ and edges represent CNN operators. The nodes $v_1, v_2, \ldots$ are indexed by the order of computation of their corresponding feature maps.

In our formulation, each e_i is a minimum search unit, also referred as a choice block, which contains a set of candidates computation blocks. A hyper-network is the network which subsumes all the sub-architectures in the search space. Following the previous works, we divide our search space into several sub DAGs (stages), each of which downsamples the input by a factor 2.

To enable the discovery of complex topology architectures, a novel topology augmented search space is proposed. In our search space, edges are divided into two categories, stem edges and branch edges, detailed in Fig. 3.

Stem Edges are non-removable edges which always appear in candidate architectures. The stem edges exist between all node pairs (v_i, v_j), where $|i-j| = 1$. Stem edges are chain-structured, which sequentially connect all consecutive nodes in each stages. We use the 9 kinds of linear bottlenecks (LB) [31] with SE module [16] as the candidate choices of stem edges. Further, on stem edges between feature maps with the same resolution, identity operation is added as an extra candidate to enhance topological diversity and depth flexibility. Therefore, there are 10 choices in the sequential structures.

Branch Edges are optional to contribute to topology diversity in the search space. The branch edges exist between all node pairs (v_i, v_j), where $|i - j| \in \{2, 3\}$. The candidate choices of branch edges are the same to stem edges. Differently, the branch edges could be abandoned flexibly.

When v_i and v_j has different resolution, the stride of convolution operation in the edge is automatically adapted to align the feature maps. The number of nodes in a single stage is required to define the search space. Based on previous method, we set the number of nodes in each stage as 2, 2, 4, 8, 4.

The search space we proposed ensures network topology complexity. Network topology in this work is defined as the DAG formed by nodes and edges. For nodes, the total number of topology is $2^{2(n-3)+1}$. The search space we used in the experiment contains 20 nodes in total, which is around 3.4×10^{10} topologies. For comparison, the topologies contained in cell-based search space is around 7.2×10^9.

3.2 Training the One-Shot Hyper-network

One-shot method uses the hyper-network to estimate the performance of architectures. Since huge amount of architectures exists in the hyper-network concurrently, training the hyper-network in whole will make the parameters of different architectures correlated with each other. To reduce the correlation, one-shot method samples a new network architectures m_k at each gradient step and update the only the activated part of the shared parameters.

$$\theta_T = \theta_{T-1} + \alpha \cdot \nabla L(\mathcal{F}(X, m_k, \theta_{T-1}), Y),$$
$$m_k \sim P_t(m_k). \tag{1}$$

$\mathcal{F}$ makes prediction of input X utilizing sampled model m_k. Thus the gradient of parameters unused by m_k remains zero. The architecture sampling distribution P_t is usually set to trivial uniform sampling [14] across the choice for each single edge.

Suppose there are I_{stem} choices for stem edges and I_{branch} for branch edges other than $none$, a simple uniform sampling strategy in our search space can be described as:

$$p_{stem}(o_i) = \frac{1}{I_{stem}}, \tag{2}$$

$$p_{branch}(o_i) = \frac{1}{I_{branch} + 1}. \tag{3}$$

However, the network sampled under this strategy in our space tends to sample architectures with high computational cost, because each of the large amount of branch edges has a low probability to be $none$. Consequently, the architecture with low computational cost in the hyper-network will under-fit, which would cause a bias in evaluation stage. Thus, the sampling strategy needs further consideration. The whole training process of hyper-network can be finded in Algorithm 1. Suppose that C_{target} is our target MAdds and C_{m_k} is the MAdds of architecture m_k, the sampling strategy should meet:

$$E_{m_k \sim P_t(m_k)}[C_{m_k}] = C_{target}. \tag{4}$$

To meet the constraints on expected computation, the sampling probability of *none* choice in branch edges, p_{drop}, is defined to adjust the expected computation cost of sampled networks:

$$p_{branch}(o_i) = \begin{cases} \frac{1-p_{drop}}{I_{branch}}, & o_i \neq none, \\ p_{drop}, & o_i = none. \end{cases} \tag{5}$$

Algorithm 1. Hyper-network Training

1: **Inputs:** D_{train}, T, B
2: $G(E, V) = $ InitializeHyperNetwork()
3: **for** $t = 1 : T$ **do**
4:    $p_{stem} = \frac{1}{I_{stem}}$
5:    $p_{branch} = \frac{1-p_{drop}}{I_{branch}}$ *if* $o_i \neq none$ *else* p_{drop}
6:    $E'_{stem} = $ Sample(E_{stem}, p_{stem})
7:    $E'_{branch} = $ Sample(E_{branch}, p_{branch})
8:    $m = G(E'_{stem} \cup E'_{branch}, V)$
9:    $D_{batch} = $ Sample$(D_{train}, uniform, B)$
10:    TrainForOneStep(m, D_{batch})
11: **end for**
12: **Outputs:** $G(E, V)$

3.3   Stabilizing Performance Estimation

In search stage, evaluating an architecture through the shared parameters is essential for exploring promising results. Previous work on one-shot method usually measure the network performance with fully trained hyper-network weights directly. In this section, we first demonstrate our observation on random shuffling of candidates architectures ranking in our search space. Then we introduce our approach to improve the ranking stability.

Instability of One-Shot NAS. Since the hyper-network is trained T iterations, the shared parameter obtained after training is denoted as θ_T. We define a accuracy function $Acc(m_k, \theta)$ which maps the model architecture m_k and hyper-network weights θ to the validation set accuracy. The value of Acc function can be estimated by simply loading the weight used by m_k and testing the model performance on validation set. The score function, denoted as $S(m_k)$, of previous approach is simply

$$S(m_k) = Acc(m_k, \theta_T). \tag{6}$$

However, the true score function should be the actual performance of the model m_k on validation set: $Acc(m_k, \theta_{m_k})$, where θ_{m_k} denotes the weight obtained by sampling and training m_k only. One-shot approach takes a approximation to

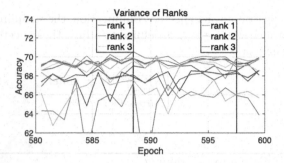

Fig. 5. The ranks of 10 random architectures during the last 20 training epochs, where drastic fluctuation can be observed. Ranks of three architectures at two time steps are shown in the figure, and each of them has different ranks at the two steps.

reuse the shared parameters for different architectures. Although this is empirically useful, we observe high variance of the model ranking in two cases: rankings at different epochs and rankings by different runs.

We randomly sample a set of architecture and obtain their independent weight. We rank their performance under shared parameters on validation set by different checkpoints at the last 20 epochs training of hyper-network. As shown in Fig. 5 , the rank of a single checkpoint fluctuates a lot during hyper-network training process and hardly distinguishes the performance of architectures. If we repeat the hyper-network training with different random seeds for 20 times and obtain shared parameters $\theta_T^i, i \in [1, 20]$. We quantify the correlation between rank of each θ_T^i and ground truth rank by Kendall's τ coefficient [18]. Here, we show the ranking performance of the best and worst runs in Fig. 4.

These two observations imply the necessity of a stabilized evaluation strategy. To present our strategy, formulation of the instability need to be introduced. In this paper, we model the performance estimation randomness as an unbiased noise. Since the shared parameters is fundamentally different from the parameters trained independently, we use a function ϕ to model the affect of weight sharing. General consensus has been reached: empirical $Acc(m_k, \theta_T)$ provides inaccurate but useful ranking, which demonstrates the desired rank preserving effect of ϕ. In summary, our model to describe the quantity relationship is:

$$Acc(m_k, \theta_T) = \phi(Acc(m_k, \theta_{m_k})) + v, \tag{7}$$
$$E[v] = 0. \tag{8}$$

It is obvious that the existence of the noise term v would hurt the model ranking. The most trivial approach to alleviate the negative effects of v is to train multiple hyper-networks, and eliminate the noise by taking expectation. However, this approach requires several times more computation resources for hyper-network training.

Algorithm 2. Shared Parameter Sampling by SGLD

1: **Inputs:** D_{train}, T_{SGLD}, T_{epoch}, B, $G(E,V)$, α
2: $\mathcal{G}_{sample} = \emptyset$
3: **for** $t = 1 : T_{SGLD}$ **do**
4:     $p_{stem} = \frac{1}{T_{stem}}$
5:     $p_{branch} = \frac{1-p_{drop}}{T_{branch}}$ if $o_i \neq none$ else p_{drop}
6:     $E'_{stem} = \text{Sample}(E_{stem}, p_{stem})$
7:     $E'_{branch} = \text{Sample}(E_{branch}, p_{branch})$
8:     $m = G(E'_{stem} \cup E'_{branch}, V)$
9:     $D_{batch} = \text{Sample}(D_{train}, uniform, B)$
10:    $\theta(m) = \theta(m) + \alpha\nabla L(\mathcal{F}(m, D_{batch})) + \sqrt{2\alpha}\,\epsilon$
11:    **if** $t \equiv 0 \bmod T_{epoch}$ **then**
12:      $\mathcal{G}_{sample}.\text{Append}(G(E,V))$
13:    **end if**
14: **end for**
15: **Outputs:** $\mathcal{G}_{sample}$

SG-MCMC Sampling. The sampling process is described in Algorithm 2. In order to obtain high-quality low correlation samples of optimized shared parameters efficiently, we investigate the rich literature of Markov Chain Monte Carlo (MCMC) sampling methods [3]. Recently, a few works demonstrate that constant learning rate stochastic gradient decent could be modified to Stochastic Gradient Langevin Dynamics (SGLD) to realize a Stochastic Gradient MCMC method under mild assumption [6,39]. Here, we apply SGLD [38,39] to approximate iid samples of share parameters posterior. The update rule we use, is simply

$$\Delta\theta_T = \alpha\nabla(\frac{1}{\mathcal{B}}\sum_i^{\mathcal{B}} L(\mathcal{F}(x_i, m_t, \theta_T), y_i)) + \sqrt{2\alpha}\,\epsilon, \tag{9}$$

$$m_t \sim P_t(m) \text{ and } \epsilon \sim \mathcal{N}(\mathbf{0}, I).$$

Here $\mathcal{B}$ is the number of data used to compute gradients (batch size). The step size α is set to the final learning rate of sub-net training. To ensure the independence, we generate each sample after SGLD update iterates for a data epoch.

To generate the iid samples of shared weights, we load the weights θ_T of the hyper-network after its training finishes, and set $\theta_T^0 = \theta_T$ as the initial sample. Then, for each θ_T^i, we apply SGLD to obtain the next sample of parameter posterior θ_T^{i+1} with the rule in Eq. (9). Thus we can obtain multiple samples of hyper-network parameters.

Average Accuracy and Parameter. Once we have K samples of $\{\theta_T^1, \theta_T^2, ..., \theta_T^K\}$ which approximates the parameters obtained by different run. To eliminate the effect of random noise v and stabilize the performance estimation, we propose two approaches: score expectation and parameter expectation.

Expectation over scores approach is to define the score $S(m)$ of each model m as the expectation of validation accuracy over K sampled shared weights.

$$S_s(m) = \frac{1}{K} \sum_i^K Acc(m, \theta_i). \tag{10}$$

Expectation over parameters approach is to take the average of sampled shared parameters and use average parameters to evaluate the performance of each model.

$$S_p(m) = Acc(m, \frac{1}{K} \sum_i^K \theta_i). \tag{11}$$

Independent Fine-Tuning. When evaluating the single architecture performance, loads the weights from the hyper-network and resuming training the architecture independently should be able to get more architecture-relevant weights. Thus we test this approach in our experiment.

3.4 Evolution Algorithm

Inspired by recent work [26,36], we apply evolution algorithm NSGA-II as the search agent. In this section, we first introduce some basic concept of NSGA-II. Next we discuss how we apply NSGA-II to our search space.

NSGA-II. We seek to obtain the model architecture with excellent performance under the constraint of computational expense. NSGA-II is the most popular choice among multi-objective evolutionary method. The core component of NSGA-II, is the Non Dominated Sorting which benefits the trade off between conflicting objectives. Since our optimization target is to minimize MAdds and maximize performance of architecture under different computational constraints.

Initialization. To reduce manual bias and explore the search space better, we use random initialization for all individuals of the first generation. More specifically, each architecture randomly select basic operators for each block in the search space.

Crossover and Mutation. Single-point crossover on random position is adopted in our evolution algorithm. For two certain individuals $m_1 = (x_1, x_2, ..., x_n)$ and $m_2 = (y_1, y_2, ..., y_n)$, a single-point crossover strategy on position p will result in a new individual $m_3 = (x_1, x_2, ..., x_p, y_{p+1}, y_{p+2}, ..., y_n)$.

We use random choice mutation to enhance generation diversity. When a mutation happens to an individual, a selected operation block in it is changed to another available choice randomly.

4   Experiments and Results

We verify the effectiveness of our method on a large classification benchmark, ImageNet [30]. In this section, we firstly describe our implementation details. Secondly, we present the performance of searching results on ImageNet as well as comparison with state of the art methods., Finally, we demonstrate the advantage of our designs via ablation study.

4.1   Experiments Settings

Datasets. We conduct experiments on the ImageNet, a standard benchmark for classification task. It has $1.28M$ training images and $50K$ validation images.

Train Details of Hyper-network. For the training of hyper-network, we adopt cosine learning rate scheduler with learning rate initialized as 0.1 and decaying to 2.5e–4 during 600 epochs. A L2 regularization is used and its weight is set to 1e–4. The optimizer is mini-batch stochastic gradient decent (SGD) with batch size 512 and we set momentum as 0 to decouple the gradients of architectures sampled in different batches. Hyper-parameter p_{drop} is set to 0.6. The hyper-network is trained on 32 GTX-1080Ti GPUs. We implement the stabilized evaluation of our method by saving 20 checkpoints at $600, 601, ..., 619$ epochs described in Sect. 3.3. The fine-tune strategy mentioned above is conducted with learning rate 2.5e–4.

Search Details. The evolution agent randomly generates 45 individuals for initialization. Then it repeats the exploitation and exploration loop where it generates 45 individuals via single-point crossover and random mutation. It conducts 22 loops and evaluates 990 models. At last, we choose the top ranked 2 models under different MAdds constraints.

Training Details of Resulted Architecture. For the independent training of resulted architectures, we use cosine learning rate scheduler with initial learning rate 0.8. We train the model for 300 epochs with batch size 2048 and adopt SGD optimizer with nesterov and momentum value 0.9. To prevent overfitting, we use L2 regularization with weight 1e–4 and standard augmentations including random crop and colorjitter.

4.2   Main Results

ST-NAS looks for models with objectives of low MAdds and high accuracy. We adopt expectation over parameters to select two resulted models separately under small and large MAdds constraints, namely, ST-NAS-A and ST-NAS-B. Architectures and performance of them compared with state-of-the-art methods are discussed in this subsection.

Performance on ImageNet. We compare ST-NAS method with efficient NAS methods, including DARTS, ProxyLessNAS and FBNet, in Table 2. Our model ST-NAS-A outperforms all of them while with the least MAdds and comparable parameter number.

Table 2. Performance comparison between ST-NAS and efficient NAS methods on ImageNet. Our model ST-NAS-A archieve best top 1 accuracy with the least MAdds.

Model	Search space	Params (M)	MAdds (M)	Top-1 acc (%)	Top-5 acc (%)
DARTS [25]	Cell-based	4.7	574	73.3	91.3
Proxyless-R [5]	Chain-structured	–	320	74.6	92.2
Single-path NAS [32]	Chain-structured	4.3	365	75.0	92.2
FairNAS-A [8]	Chain-structured	4.6	388	75.3	92.4
FBNet-C [40]	Chain-structured	5.5	375	74.9	–
SPOS [14]	Chain-structured	–	328	74.7	–
BetaNet-A [12]	Chain-structured	7.2	333	75.9	92.8
ST-NAS-A(ours)	**Topology augmented**	**5.2**	**326**	**76.4**	**93.1**

Table 3. Performance comparison among ST-NAS, manual designed networks and sample based NAS methods on ImageNet. Notably sample based methods with mark * takes much more computation resources. We show that the architecture discovered by ST-NAS perform better than both sample based NAS and manually designed architectures while maintaining least MAdds.

Model	Search space	Params (M)	MAdds (M)	Top-1 acc (%)	Top-5 acc (%)
*MNASNet-A1 [36]	Chain-structured	3.9	312	75.2	92.5
*MNASNet-A2 [36]	Chain-structured	4.8	340	75.6	92.7
*RCNet-B [42]	Chain-structured	4.7	471	74.7	92.0
*NASNet-B [46]	Cell-based	5.3	488	72.8	91.3
*EfficientNet-B0 [37])	Chain-structured	5.3	390	76.3	93.2
ST-NAS-A (ours)	**Topology augmented**	**5.2**	**326**	**76.4**	**93.1**
1.4-MobileNetV2 [31]	Chain-structured	6.9	585	74.7	92.5
2.0-ShuffleNetV2 [27]	Chain-structured	7.4	591	74.9	–
*NASNet-C [46]	Cell-based	4.9	558	72.5	91.0
*NASNet-A [46]	Cell-based	5.3	564	74.0	91.6
*1.4-MNASNet-A1 [36]	Chain-structured	–	600	77.2	93.7
*RENASNet [7]	Cell-based	5.4	580	75.7	92.6
*PNASNet [24]	Cell-based	5.1	588	74.2	91.9
ST-NAS-B (ours)	**Topology augmented**	**7.8**	**503**	**77.9**	**93.8**

For architectures resulted from high cost, *i.e*, manually designed networks and networks obtained by sample-based methods, we compare ST-NAS with them in two groups divided by MAdds, as shown in Table 3. At a much less search cost, our ST-NAS outperform all the methods in both MAdds groups.

Performance on COCO. Our implementation is based on feature pyramid network (FPN)[23]. Different models pretrained on ImageNet is utilized as feature extractor. All the models are trained for 13 epochs, known as 1× sched-

Table 4. Performance on COCO dataset. The channel number of ST-NAS-B is scaled to get ST-NAS-B*. ST-NAS-A outperforms MobileNetV2 by 1.5% COCO AP while maintaining same MAdds. ST-NAS-B* achieves 0.5% higher than ResNet50 but needs only a quarter MAdds.

Model	MAdds (backbone) (G)	mAP
MobileNetV2	0.33	31.7
ST-NAS-A	0.33	33.2
ResNet18	1.81	32.2
ST-NAS-B	0.50	35.3
ResNet50	4.09	36.9
ST-NAS-B*	1.03	37.7

ule. The results are shown in Table 4. Our ST-NAS-A backbone outperforms MobileNetV2. The ST-NAS-B performs comparably with ResNet50 with much less MAdds.

4.3 Ablation Studies

Improvement on Topology. We adopt our search method on chain structured space. The result on ImageNet is 77.0% with 500M MAdds. Compared with ST-NAS-B, searching for a topology obtain 0.9% improvement with similar computation cost.

Rank Fluctuation. To explain the importance of our stabilization mechanism, we randomly sample a set of architecture and rank their performance under shared parameters on validation set at the last 10 epochs training of hyper-network. As shown in Fig. 5, the rank of a single checkpoint fluctuates a lot during hyper-network training process and hardly distinguishes the performance of architectures, implying the necessity of a stabilized evaluation strategy (Fig. 6).

Table 5. τ of different rank stabilization approach we proposed. The original baseline is single checkpoint achieves 0.71 of Kendall τ which is 0.07 higher than the fine-tune approach. The parameter expectation and accuracy expectation method is tied and outperform the baseline with a margin.

Estimation approach	τ
Single checkpoint	0.71
Fine-tune	0.64
SGLD-param	0.84
SGLD-acc	0.81

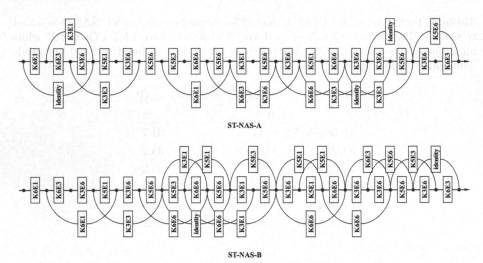

Fig. 6. Resulted architectures. Linear bottle-neck contains a $K \times K$ group-wise convolution layer between two 1×1 point-wise convolution layers. Expand ratio is defined as the ratio between group-wise convolution channels and point-wise convolution channels. We describe a linear bottle-neck with its expand ratio, *i.e*, the number after "E", and its group-wise convolution kernel size, *i.e*, the number after "K".

Ranking Verification. We further verify this reduction by quantifying the ranking ability of different evaluation strategies by correlation coefficient between ranks in hyper-network and the ground truth ranks. Kendall's tau coefficient is adopted as the metric in our verification. We randomly sample 12 networks and train them form scratch to obtain the ground truth rank. To compare with single checkpoint, we make use of checkpoints of 10 epochs at the 591-th, 592-th, ..., 600-th epoch to generate 10 ranks and get 10 correlation coefficients with the ground truth rank. The median of the 10 correlation coefficients is adopted to compare with other strategies. It is observed in Table 5 that SGLD consistently achieves higher correlation coefficients than fine-tune and single checkpoint, which verifies the effectiveness of SGLD in the reduction of parameter variance.

5   Conclusion

We proposed a topology-diverse search space and a novel search method, ST-NAS. In ST-NAS, we improve both the sampling strategy during hyper-network training and the architecture evaluation approach by rigorous theoretical analysis. Sound experiments demonstrate the effectiveness of our designs and achieve consistent improvements under different computation cost constraints.

Acknowledgement. This work was supported by the National Key Research and Development Project of China (No. 2018AAA0101900).

References

1. Baker, B., Gupta, O., Naik, N., Raskar, R.: Designing neural network architectures using reinforcement learning. arXiv preprint arXiv:1611.02167 (2016)
2. Bender, G., Kindermans, P.J., Zoph, B., Vasudevan, V., Le, Q.: Understanding and simplifying one-shot architecture search. In: International Conference on Machine Learning, pp. 549–558 (2018)
3. Bishop, C.M.: Pattern Recognition and Machine Learning. Springer, Heidelberg (2006). https://doi.org/10.1007/978-1-4615-7566-5
4. Brock, A., Lim, T., Ritchie, J.M., Weston, N.: Smash: one-shot model architecture search through hypernetworks. arXiv preprint arXiv:1708.05344 (2017)
5. Cai, H., Zhu, L., Han, S.: Proxylessnas: Direct neural architecture search on target task and hardware. arXiv preprint arXiv:1812.00332 (2018)
6. Chen, C., Carlson, D., Gan, Z., Li, C., Carin, L.: Bridging the gap between stochastic gradient MCMC and stochastic optimization. In: Artificial Intelligence and Statistics, pp. 1051–1060 (2016)
7. Chen, Y., et al.: Reinforced evolutionary neural architecture search. arXiv preprint arXiv:1808.00193 (2018)
8. Chu, X., Zhang, B., Xu, R., Li, J.: Fairnas: rethinking evaluation fairness of weight sharing neural architecture search. arXiv preprint arXiv:1907.01845 (2019)
9. Cubuk, E.D., Zoph, B., Mane, D., Vasudevan, V., Le, Q.V.: Autoaugment: learning augmentation policies from data. arXiv preprint arXiv:1805.09501 (2018)
10. Du, X., et al.: Spinenet: learning scale-permuted backbone for recognition and localization. arXiv preprint arXiv:1912.05027 (2019)
11. Elsken, T., Metzen, J.H., Hutter, F.: Simple and efficient architecture search for convolutional neural networks. arXiv preprint arXiv:1711.04528 (2017)
12. Fang, M., Wang, Q., Zhong, Z.: Betanas: balanced training and selective drop for neural architecture search. arXiv preprint arXiv:1912.11191 (2019)
13. Guo, M., Zhong, Z., Wu, W., Lin, D., Yan, J.: Irlas: inverse reinforcement learning for architecture search. In: Proceedings of the IEEE Conference on Computer Vision and Pattern Recognition, pp. 9021–9029 (2019)
14. Guo, Z., et al.: Single path one-shot neural architecture search with uniform sampling. arXiv preprint arXiv:1904.00420 (2019)
15. He, K., Zhang, X., Ren, S., Sun, J.: Deep residual learning for image recognition. In: Proceedings of the IEEE Conference on Computer Vision and Pattern Recognition, pp. 770–778 (2016)
16. Hu, J., Shen, L., Sun, G.: Squeeze-and-excitation networks. In: Proceedings of the IEEE Conference on Computer Vision and Pattern Recognition, pp. 7132–7141 (2018)
17. Huang, G., Liu, Z., Van Der Maaten, L., Weinberger, K.Q.: Densely connected convolutional networks. In: Proceedings of the IEEE Conference on Computer Vision and Pattern Recognition, pp. 4700–4708 (2017)
18. Kendall, M.G.: A new measure of rank correlation. Biometrika $30(1/2)$, 81–93 (1938)
19. Li, C., Yuan, X., Lin, C., Guo, M., Wu, W., Yan, J., Ouyang, W.: AM-LFS: automl for loss function search. In: Proceedings of the IEEE International Conference on Computer Vision, pp. 8410–8419 (2019)
20. Li, X., et al.: Improving one-shot nas by suppressing the posterior fading. arXiv preprint arXiv:1910.02543 (2019)

21. Liang, F., et al.: Computation reallocation for object detection. arXiv preprint arXiv:1912.11234 (2019)
22. Lin, C., et al.: Online hyper-parameter learning for auto-augmentation strategy. In: Proceedings of the IEEE International Conference on Computer Vision, pp. 6579–6588 (2019)
23. Lin, T.Y., Dollár, P., Girshick, R., He, K., Hariharan, B., Belongie, S.: Feature pyramid networks for object detection. In: Proceedings of the IEEE Conference on Computer Vision and Pattern Recognition, pp. 2117–2125 (2017)
24. Liu, C., et al.: Progressive neural architecture search. In: Proceedings of the European Conference on Computer Vision (ECCV), pp. 19–34 (2018)
25. Liu, H., Simonyan, K., Yang, Y.: Darts: differentiable architecture search. arXiv preprint arXiv:1806.09055 (2018)
26. Lu, Z., et al.: NSGA-net: a multi-objective genetic algorithm for neural architecture search. arXiv preprint arXiv:1810.03522 (2018)
27. Ma, N., Zhang, X., Zheng, H.T., Sun, J.: Shufflenet v2: practical guidelines for efficient cnn architecture design. In: Proceedings of the European Conference on Computer Vision (ECCV), pp. 116–131 (2018)
28. Real, E., Aggarwal, A., Huang, Y., Le, Q.V.: Regularized evolution for image classifier architecture search. In: Proceedings of the AAAI Conference on Artificial Intelligence, vol. 33, pp. 4780–4789 (2019)
29. Real, E., et al.: Large-scale evolution of image classifiers. In: Proceedings of the 34th International Conference on Machine Learning, vol. 70, pp. 2902–2911. JMLR. org (2017)
30. Russakovsky, O., et al.: Imagenet large scale visual recognition challenge. Int. J. Comput. Vis. **115**(3), 211–252 (2015)
31. Sandler, M., Howard, A., Zhu, M., Zhmoginov, A., Chen, L.C.: Mobilenetv 2: inverted residuals and linear bottlenecks. In: Proceedings of the IEEE Conference on Computer Vision and Pattern Recognition, pp. 4510–4520 (2018)
32. Stamoulis, D., et al.: Single-path nas: designing hardware-efficient convnets in less than 4 hours. arXiv preprint arXiv:1904.02877 (2019)
33. Szegedy, C., Ioffe, S., Vanhoucke, V., Alemi, A.A.: Inception-v4, inception-resnet and the impact of residual connections on learning. In: 31st AAAI Conference on Artificial Intelligence (2017)
34. Szegedy, C., et al.: Going deeper with convolutions. In: Proceedings of the IEEE Conference on Computer Vision and Pattern Recognition, pp. 1–9 (2015)
35. Szegedy, C., Vanhoucke, V., Ioffe, S., Shlens, J., Wojna, Z.: Rethinking the inception architecture for computer vision. In: Proceedings of the IEEE Conference on Computer Vision and Pattern Recognition, pp. 2818–2826 (2016)
36. Tan, M., et al.: MNASnet: Platform-aware neural architecture search for mobile. In: Proceedings of the IEEE Conference on Computer Vision and Pattern Recognition, pp. 2820–2828 (2019)
37. Tan, M., Le, Q.V.: Efficientnet: rethinking model scaling for convolutional neural networks. arXiv preprint arXiv:1905.11946 (2019)
38. Teh, Y.W., Thiery, A.H., Vollmer, S.J.: Consistency and fluctuations for stochastic gradient langevin dynamics. J. Mach. Learn. Res. **17**(1), 193–225 (2016)
39. Welling, M., Teh, Y.W.: Bayesian learning via stochastic gradient langevin dynamics. In: Proceedings of the 28th International Conference on Machine Learning (ICML-11), pp. 681–688 (2011)
40. Wu, B., et al.: FBNet: hardware-aware efficient convnet design via differentiable neural architecture search. In: Proceedings of the IEEE Conference on Computer Vision and Pattern Recognition, pp. 10734–10742 (2019)

41. Xie, S., Kirillov, A., Girshick, R., He, K.: Exploring randomly wired neural networks for image recognition. arXiv preprint arXiv:1904.01569 (2019)
42. Xiong, Y., Mehta, R., Singh, V.: Resource constrained neural network architecture search: will a submodularity assumption help? In: Proceedings of the IEEE International Conference on Computer Vision, pp. 1901–1910 (2019)
43. Zhang, Y., et al.: Deeper insights into weight sharing in neural architecture search. arXiv preprint arXiv:2001.01431 (2020)
44. Zhong, Z., et al.: Blockqnn: efficient block-wise neural network architecture generation. arXiv preprint arXiv:1808.05584 (2018)
45. Zoph, B., Le, Q.V.: Neural architecture search with reinforcement learning. arXiv preprint arXiv:1611.01578 (2016)
46. Zoph, B., Vasudevan, V., Shlens, J., Le, Q.V.: Learning transferable architectures for scalable image recognition. In: Proceedings of the IEEE Conference on Computer Vision and Pattern Recognition, pp. 8697–8710 (2018)

Classes Matter: A Fine-Grained Adversarial Approach to Cross-Domain Semantic Segmentation

Haoran Wang[1], Tong Shen[2], Wei Zhang[2(✉)], Ling-Yu Duan[3], and Tao Mei[2]

[1] ETH Zürich, Zurich, Switzerland
[2] JD AI Research, Nanjing, China
`wzhang.cu@gmail.com`
[3] Peking University, Beijing, China

Abstract. Despite great progress in supervised semantic segmentation, a large performance drop is usually observed when deploying the model in the wild. Domain adaptation methods tackle the issue by aligning the source domain and the target domain. However, most existing methods attempt to perform the alignment from a holistic view, ignoring the underlying class-level data structure in the target domain. To fully exploit the supervision in the source domain, we propose a fine-grained adversarial learning strategy for class-level feature alignment while preserving the internal structure of semantics across domains. We adopt a fine-grained domain discriminator that not only plays as a domain distinguisher, but also differentiates domains at class level. The traditional binary domain labels are also generalized to domain encodings as the supervision signal to guide the fine-grained feature alignment. An analysis with Class Center Distance (CCD) validates that our fine-grained adversarial strategy achieves better class-level alignment compared to other state-of-the-art methods. Our method is easy to implement and its effectiveness is evaluated on three classical domain adaptation tasks, i.e., GTA5→Cityscapes, SYNTHIA→Cityscapes and Cityscapes→Cross-City. Large performance gains show that our method outperforms other global feature alignment based and class-wise alignment based counterparts. The code is publicly available at https://github.com/JDAI-CV/FADA.

1 Introduction

The success of semantic segmentation [26] in recent years is mostly driven by a large amount of accessible labeled data. However, collecting massive densely

---

H. Wang and T. Shen—These authors contributed equally. This work was performed when Haoran Wang was visiting JD AI research as a research intern.

Electronic supplementary material The online version of this chapter (https://doi.org/10.1007/978-3-030-58568-6_38) contains supplementary material, which is available to authorized users.

A. Vedaldi et al. (Eds.): ECCV 2020, LNCS 12359, pp. 642–659, 2020.
https://doi.org/10.1007/978-3-030-58568-6_38

annotated data for training is usually a labor-intensive task [9]. Recent advances in computer graphics provide an alternative for replacing expensive human labor. Through physically based rendering, we can obtain photo-realistic images with the pixel-level ground-truth readily available in an effortless way [23,24].

However, performance drop is observed when the model trained with synthetic data (a source domain) is applied in realistic scenarios (a target domain), because the data from different domains usually share different distributions. This phenomenon is known as domain shift problem [27], which poses a challenge to cross-domain tasks [16].

Domain adaptation aims to alleviate the domain shift problem by aligning the feature distributions of the source and the target domain. A group of works focus on adopting an adversarial framework, where a domain discriminator is trained to distinguish the target samples from the source ones, while the feature network tries to fool the discriminator by generating domain-invariant features [8,15,16, 20,25,30,34,35,38].

Although impressive progress has been achieved in domain adaptive semantic segmentation, most of prior works strive to align global feature distributions without paying much attention to the underlying structures among classes. However, as discussed in recent works [3,17], matching the marginal feature distributions does not guarantee small expected error on the target domain. The class conditional distributions should also be aligned, meaning that class-level alignment also plays an important role. As illustrated in Fig. 1, the upper part shows the result of global feature alignment where the two domains are well-aligned but some samples are falsely mixed up. This motivates us to incorporate class information into the adversarial framework to enable fine-grained feature

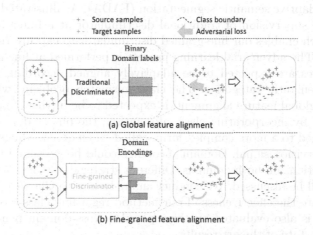

Fig. 1. Illustration of traditional and our fine-grained adversarial learning. Traditional adversarial learning pursues the marginal distribution alignment while ignoring the semantic structure inconsistency between domains. We propose to use a fine-grained discriminator to enable class-level alignment.

alignment. As illustrated in the bottom of Fig. 1, features are expected to be aligned according to specific classes.

There have been some pioneering works [7,20] trying to address this problem. Chen et al. [7] propose to use several independent discriminators to perform class-wise alignment, but independent discriminators might fail to capture the relationships between classes. Luo et al. [20] introduce an self-adaptive adversarial loss to apply different weights to each region. However, in fact, they do not explicitly incorporate class information in their methods, which might fail to promote class-level alignment.

Our motivation is to directly incorporate class information into the discriminator and encourage it to align features at a fine-grained level. Traditional adversarial training has been proven effective for aligning features by using a binary domain discriminator to model distribution $P(d|f)$ (d refers to domain and f is the feature extracted from input data). By confusing such a discriminator, expecting $P(d = 0|f) \approx P(d = 1|f)$ where 0 stands for the source domain and 1 for the target domain, the features become domain invariant and well aligned. To further take classes into account, we split the output into multiple channels according to $P(d|f) = \sum_{c=1}^{K} P(d, c|f)$ (where c refers to classes $\{1, \ldots, K\}$). We directly model the discriminator as $P(d, c|f)$ to formulate a fine-grained domain alignment task. Although in the setting of domain adaptation the category-level labels for target domain are inaccessible, we find that the model predictions on target domain also contain class information and prove that it is possible to supervise the discriminator with the predictions on both domains. In the adversarial learning process, class information is incorporated and the features are expected to be aligned according to specific classes.

In this paper, we propose such a fine-grained adversarial learning framework for domain adaptive semantic segmentation (FADA). As illustrated in Fig. 1, we represent the supervision of traditional discriminator at a fine-grained semantic level, which enables our fine-grained discriminator to capture rich class-level information. The adversarial learning process is performed at fine-grained level, so the features are expected to be adaptively aligned according to their corresponding semantic categories. The class mismatch problem, which broadly exists in the global feature alignment, is expected to be further suppressed. Correspondingly, by incorporating class information, the binary domain labels are also generalized to a more complex form, called "domain encodings" to serve as the new supervision signal. Domain encodings could be extracted from the network's predictions on both domains. Different strategies of constructing domain encodings will be discussed. We conduct an analysis with Class Center Distance to demonstrate the effectiveness of our method regarding class-level alignment. Our method is also evaluated on three popular cross-domain benchmarks and presents new state-of-the-art results.

The main contributions of this paper are summarized below.

- We propose a fine-grained adversarial learning framework for cross-domain semantic segmentation that explicitly incorporates class-level information.

- The fine-grained learning framework enables class-level feature alignment, which is further verified by analysis using Class Center Distance.
- We evaluate our methods with comprehensive experiments. Significant improvements compared to other state-of-the-art methods are achieved on popular domain adaptive segmentation tasks including GTA5 $\rightarrow$ Cityscapes, SYNTHIA $\rightarrow$ Cityscapes and Cityscapes $\rightarrow$ Cross-City.

2 Related Work

2.1 Semantic Segmentation

Semantic segmentation is a task of predicting unique semantic label for each pixel of the input image. With the advent of deep convolutional neural networks, the academia of computer vision witnesses a huge progress in this field. FCN [26] triggered the interests in introducing deep learning for this task. Many follow-up methods are proposed to enlarge the receptive fields to cover more context information [4–6,36]. Among all these works, the family of Deeplab [4–6] attracts a lot of attention and has been widely applied in many works for their simplicity and effectiveness.

2.2 Domain Adaptation

Domain adaptation strives to address the performance drop caused by the different distributions of training data and testing data. In the recent years, several works are proposed to approach this problem in image classification [3,25]. Inspired by the theoretical upper bound of risk in target domain [2], some pioneering works suggest to optimize some distance measurements between the two domains to align the features [18,29]. Recently, motivated by GAN [13], adversarial training becomes popular for its power to align features globally [7,25,30].

2.3 Domain Adaptive Semantic Segmentation

Unlike domain adaptation for image classification task, domain adaptive semantic segmentation receives less attention for its difficulty even though it supports many important applications including autonomous driving in the wild [8,16]. Based on the theoretical insight [2] on domain adaptive classification, most works follow the path of shortening the domain discrepancy between the two domains. Large progress is achieved through optimization by adversarial training or explicit domain discrepancy measures [15,16,30]. In the context of domain adaptive semantic segmentation task, AdaptSegnet [30] attempts to align the distribution in the output space. Inspired by CycleGAN [37], CyCADA [15] suggests to adapt the representation in pixel-level and feature-level. There are also many works focusing on aligning different properties between two domains such as entropy [32] and information [19].

Although huge progress has been made in this field, most of existing methods share a common limitation: Enforcing global feature alignment would inevitably

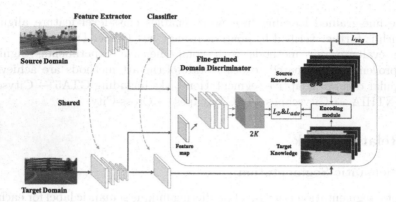

Fig. 2. Overview of the proposed fine-grained adversarial framework. Images from the source domain and target domain are randomly picked and fed to the feature extractor and the classifier. A segmentation loss is computed with the source predictions and the source annotations to help the segmentation network to generate discriminative features and learn task specific knowledge. The semantic features from both domains are fed to the convolutional fine-grained domain discriminator. The discriminator strives to distinguish the feature's domain information at a fine-grained class level using the domain encodings processed from the sample predictions.

mix samples with different semantic labels together when drawing two domains closer, which usually results in a mismatch of classes from different domains. CLAN [20] is a pioneer work to address category-level alignment. It suggests applying different adversarial weight to different regions, but it does not directly and explicitly incorporate the classes into the model.

3 Method

3.1 Revisit Traditional Feature Alignment

Semantic segmentation aims to predict per-pixel unique label for the input image [26]. In an unsupervised domain adaptation setting for semantic segmentation, we have access to a collection of labeled data $X_{\mathcal{S}} = \{(x_i^{(s)}, y_i^{(s)})\}_{i=1}^{n_s}$ in a source domain $\mathcal{S}$, and unlabeled data $X_{\mathcal{T}} = \{x_j^{(t)}\}_{j=1}^{n_t}$ in a target domain $\mathcal{T}$ where n_s and n_t are the numbers of samples from different domains. Domain $\mathcal{S}$ and domain $\mathcal{T}$ share the same K semantic class labels $\{1, \ldots, K\}$. The goal is to learn a segmentation model G which could achieve a low expected risk on the target domain. Generally, segmentation network G could be divided into a feature extractor F and a multi-class classifier C, where $G = C \circ F$.

Traditional feature-level adversarial training relies on a binary domain discriminator D to align the features extracted by F on both domains. Domain adaptation is tackled by alternatively optimizing G and D with two steps:

(1) D is trained to distinguish features from different domains. This process is usually achieved by fixing F and C and solving:

$$\min_{D} \mathcal{L}_D = -\sum_{i=1}^{n_s}(1-d)\log P(d=0|f_i) - \sum_{j=1}^{n_t} d\log P(d=1|f_j) \tag{1}$$

where f_i and f_j are the features extracted by F on source sample $x_i^{(s)}$ and target sample $x_j^{(t)}$; d refers to the domain variable where 0 refers to the source domain and 1 refers to the target domain. $P(d|f)$ is the probability output from the discriminator.

(2) G is trained with the task loss $\mathcal{L}_{seg}$ on the source domain and the adversarial loss $\mathcal{L}_{adv}$ on the target domain. This process requires fixing D and updating F and C:

$$\min_{F,C} \mathcal{L}_{seg} + \lambda_{adv}\mathcal{L}_{adv} \tag{2}$$

The cross-entropy loss $\mathcal{L}_{seg}$ on source domain minimizes the difference between the prediction and the ground truth, which helps G to learn the task specific knowledge.

$$\mathcal{L}_{seg} = -\sum_{i=1}^{n_s}\sum_{k=1}^{K} y_{ik}^{(s)}\log p_{ik}^{(s)}, \tag{3}$$

where $p_{ik}^{(s)}$ is the probability confidence of source sample $x_i^{(s)}$ belonging to semantic class k predicted by C, $y_{ik}^{(s)}$ is the entry for the one-hot label.

The adversarial loss $\mathcal{L}_{adv}$ is used to confuse the discriminator to encourage F to generate domain invariant features.

$$\mathcal{L}_{adv} = -\sum_{j=1}^{n_t}\log P(d=0|f_j) \tag{4}$$

3.2   Fine-Grained Adversarial Learning

To incorporate the class information into the adversarial learning framework, we propose a novel discriminator and enable a fine-grained adversarial learning process. The whole pipeline is illustrated in Fig. 2.

The traditional adversarial training strives to align the marginal distribution by confusing a binary discriminator. To make the discriminator not merely focus on distinguishing domains, we split each of the two output channels of the binary discriminator into K channels and encourage a fine-grained level adversarial learning. With this design, the predicted confidence for domains is represented as a confidence distribution over different classes, which enables the new fine-grained discriminator to model more complex underlying structures between classes, thus encouraging class-level alignment.

Correspondingly, the binary domain labels are also converted to a general form, namely domain encodings, to incorporate class information. Traditionally, the domain labels used for training the binary discriminator are [1, 0] and [0, 1] for the source and target domains respectively. The domain encodings are represented as a vector $[\mathbf{a}; \mathbf{0}]$ and $[\mathbf{0}; \mathbf{a}]$ for the two domains respectively, where $\mathbf{a}$ is the knowledge extracted from the classifier C represented by a K-dimensional vector; $\mathbf{0}$ is an all-zero K-dimensional vector. The choices of how to generate domain knowledge $\mathbf{a}$ will be discussed in Sect. 3.3.

During the training process, the discriminator not only tries to distinguish domains, but also learns to model class structures. The $\mathcal{L}_D$ in Eq. 1 becomes:

$$
\begin{aligned}
\mathcal{L}_D = &- \sum_{i=1}^{n_s} \sum_{k=1}^{K} a_{ik}^{(s)} \log P(d=0, c=k|f_i) \\
&- \sum_{j=1}^{n_t} \sum_{k=1}^{K} a_{jk}^{(t)} \log P(d=1, c=k|f_j)
\end{aligned}
\tag{5}
$$

where $a_{ik}^{(s)}$ and $a_{jk}^{(t)}$ are the kth entries of the class knowledge for the source sample i and target sample j. The adversarial loss $\mathcal{L}_{adv}$ used to confuse the discriminator and guide the generation of domain-invariant features in Eq. 4 becomes:

$$
\mathcal{L}_{adv} = - \sum_{j=1}^{n_t} \sum_{k=1}^{K} a_{jk}^{(t)} \log P(d=0, c=k|f_j),
\tag{6}
$$

$\mathcal{L}_{adv}$ is designed to maximize the probability of features from target domain being considered as the source features without hurting the relationship between features and classes.

The overall network in Fig. 2 is used in the training stage. During inference, the domain adaptation component is removed and one only needs to use the original segmentation network with the adapted weights.

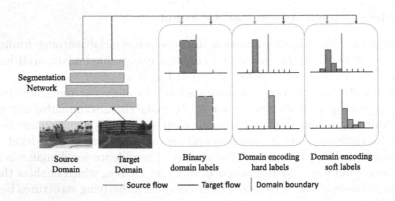

Fig. 3. Illustration of different strategies to generate domain encodings. Here we compare three different strategies to extract knowledge from segmentation network for constructing domain encodings: binary domain labels, one-hot hard labels and multi channel soft labels.

3.3 Extracting Class Knowledge for Domain Encodings

Now that we have a fine-grained domain discriminator, which could adaptively align features according to the class-level information contained in domain encodings, another challenge raises: how to get the class knowledge $a_{ik}^{(s)}$ and $a_{ik}^{(t)}$ in Eqs. 5 and 6 to construct domain encoding for each sample? Considering that in the unsupervised domain adaptive semantic segmentation task none of annotations in target domain is accessible, it seems contradictory to use the class knowledge on the target domain for guiding class-level alignment. However, during training, with ground-truth annotations from the source domain, the classifier C learns to map features into the semantic classes. Considering that the source domain and the target domain share the same semantic classes, it would be a natural choice to use the predictions of C as knowledge to supervise the discriminator.

As illustrated in Eqs. 5 and 6, the class knowledge for optimizing the fine-grained discriminator works as the supervision signal. The choices of $a_{ik}^{(s)}$ and $a_{jk}^{(t)}$ are open to many possibilities. For specific tasks, people could design different forms to produce class knowledge with prior knowledge. Here we discuss two general solutions to extract class knowledge from network predictions for constructing domain encodings. Because the class-level knowledge for different domains could be extracted in the same way, in the following discussion we would use a_k to represent kth entry for a single sample without differentiating the domain.

The one-hot hard labels could be a straightforward solution for generating knowledge, which could be denoted as:

$$a_k = \begin{cases} 1 \text{ if } k = \arg\max_k p_k \\ 0 \text{ otherwise} \end{cases} \tag{7}$$

where p_k is the softmax probability output of C for class k. In this way, only the most confident class is selected. In practice, in order to remove the impact of noisy samples, we can select samples whose confidence is higher than a certain threshold and ignore those with low confidence.

Another alternative is multi-channel soft labels, which has the following definition:

$$a_k = \frac{\exp\left(\frac{z_k}{T}\right)}{\sum_{j=1}^{K} \exp\left(\frac{z_j}{T}\right)} \tag{8}$$

where z_k is kth entry of logits and T is a temperature to encourage soft probability distribution over classes. Note that during training, an additional regularization could also be applied. For example, we practically find that clipping the values of the soft labels by a given threshold achieves more stable performance because it prevents from overfitting to certain classes.

An illustrative comparison of these two strategies with the traditional binary domain labels is presented in Fig. 3. We also conduct experiments in Sect. 4.6 to demonstrate the performance of different strategies.

Table 1. Experimental results for Cityscapes → Cross-City.

Cityscapes → Cross-City

City	Method	Road	Sidewalk	Building	Light	Sign	Veg	Sky	Person	Rider	Car	Bus	Mbike	Bike	mIoU
Rome	Source Dilation-Frontend	77.7	21.9	83.5	0.1	10.7	78.9	88.1	21.6	10.0	67.2	30.4	6.1	0.6	38.2
	Cross-City [7]	79.5	29.3	84.5	0.0	22.2	80.6	82.8	29.5	13.0	71.7	37.5	25.9	1.0	42.9
	Source DeepLab-v2	83.9	34.3	87.7	13.0	41.9	84.6	92.5	37.7	22.4	80.8	38.1	39.1	5.3	50.9
	AdaptSegNet [30]	83.9	34.2	**88.3**	18.8	**40.2**	**86.2**	**93.1**	47.8	21.7	80.9	**47.8**	48.3	8.6	53.8
	FADA	**84.9**	**35.8**	**88.3**	**20.5**	40.1	85.9	92.8	**56.2**	**23.2**	**83.6**	31.8	**53.2**	**14.6**	**54.7**
Rio	Source Dilation-Frontend	69.0	31.8	77.0	4.7	3.7	71.8	80.8	38.2	8.0	61.2	38.9	11.5	3.4	38.5
	Cross-City [7]	74.2	43.9	79.0	2.4	7.5	77.8	69.5	39.3	10.3	67.9	**41.2**	27.9	10.9	42.5
	Source DeepLab-v2	76.6	47.3	82.5	12.6	22.5	77.9	86.5	43.0	19.8	74.5	36.8	29.4	16.7	48.2
	AdaptSegNet [30]	76.2	44.7	**84.6**	9.3	**25.5**	**81.8**	87.3	55.3	**32.7**	74.3	28.9	43.0	27.6	51.6
	FADA	**80.6**	**53.4**	84.2	5.8	23.0	78.4	**87.7**	**60.2**	26.4	**77.1**	37.6	**53.7**	**42.3**	**54.7**
Tokyo	Source Dilation-Frontend	81.2	26.7	71.7	8.7	5.6	73.2	75.7	39.3	14.9	57.6	19.0	1.6	33.8	39.2
	Cross-City [7]	83.4	35.4	72.8	12.3	12.7	77.4	64.3	42.7	21.5	64.1	**20.8**	8.9	40.3	42.8
	Source DeepLab-v2	83.4	35.4	72.8	12.3	12.7	77.4	64.3	42.7	21.5	64.1	**20.8**	8.9	40.3	42.8
	AdaptSegNet [30]	81.5	26.0	77.8	**17.8**	**26.8**	82.7	90.9	55.8	**38.0**	**72.1**	4.2	24.5	50.8	49.9
	FADA	**85.8**	**39.5**	76.0	14.7	24.9	**84.6**	**91.7**	**62.2**	27.7	71.4	3.0	**29.3**	**56.3**	**51.3**
Taipei	Source Dilation-Frontend	77.2	20.9	76.0	5.9	4.3	60.3	81.4	10.9	11.0	54.9	32.6	15.3	5.2	35.1
	Cross-City [7]	78.6	28.6	80.0	13.1	7.6	68.2	82.1	16.8	9.4	60.4	34.0	26.5	9.9	39.6
	Source DeepLab-V2	78.6	28.6	80.0	13.1	7.6	68.2	82.1	16.8	9.4	60.4	34.0	26.5	9.9	39.6
	AdaptSegNet [30]	81.7	29.5	85.2	**26.4**	15.6	76.7	91.7	31.0	12.5	71.5	**41.1**	47.3	27.7	49.1
	FADA	**86.0**	**42.3**	**86.1**	6.2	**20.5**	**78.3**	**92.7**	**47.2**	**17.7**	**72.2**	37.2	**54.3**	**44.0**	**52.7**

4   Experiments

4.1   Datasets

We present a comprehensive evaluation of our proposed method on three popular unsupervised domain adaptive semantic segmentation benchmarks, e.g., Cityscapes → Cross-City, SYNTHIA → Cityscapes, and GTA5 → Cityscapes.

Cityscapes.   Cityscapes [9] is a real-world urban scene dataset consisting of a training set with 2,975 images, a validation set with 500 images and a testing set with 1,525 images. Following the standard protocols [15,16,30], we use the 2,975 images from Cityscapes training set as the unlabeled target domain training set and evaluate our adapted model on the 500 images from the validation set.

Cross-City.   Cross-City [7] is an urban scene dataset collected with Google Street View. It contains 3,200 unlabeled images and 100 annotated images of four different cities respectively. The annotations of Cross-City share 13 classes with Cityscapes.

SYNTHIA.   SYNTHIA [24] is a synthetic urban scene dataset. We pick its subset SYNTHIA-RAND-CITYSCAPES, which shares 16 semantic classes with

Cityscapes, as the source domain. In total, 9,400 images from SYNTHIA dataset are used as source domain training data for the task.

GTA5. GTA5 dataset [23] is another synthetic dataset sharing 19 semantic classes with Cityscapes. 24,966 urban scene images are collected from a physically-based rendered video game Grand Theft Auto V (GTAV) and are used as source training data.

Table 2. Experimental results for SYNTHIA → Cityscapes.

SYNTHIA → Cityscapes

Backbone	Method	Road	SW	Build	Wall	Fence	Pole	TL	TS	Veg	Sky	PR	Rider	Car	Bus	Motor	Bike	mIoU	mIoU*
VGG-16	FCNs in the wild [16]	11.5	19.6	30.8	4.4	0.0	20.3	0.1	11.7	42.3	68.7	51.2	3.8	54.0	3.2	0.2	0.6	20.2	22.9
	CDA [34]	65.2	26.1	74.9	0.1	0.5	10.7	3.5	3.0	76.1	70.6	47.1	8.2	43.2	20.7	0.7	13.1	29.0	34.8
	ST [38]	0.2	14.5	53.8	1.6	0.0	18.9	0.9	7.8	72.2	80.3	48.1	6.3	67.7	4.7	0.2	4.5	23.9	27.8
	CBST [38]	69.6	28.7	69.5	12.1	0.1	25.4	11.9	13.6	82.0	81.9	49.1	14.5	66.0	6.6	3.7	32.4	35.4	36.1
	AdaptSegNet [30]	78.9	29.2	75.5	-	-	-	0.1	4.8	72.6	76.7	43.4	8.8	71.1	16.0	3.6	8.4	-	37.6
	SIBAN [19]	70.1	25.7	80.9	-	-	-	3.8	7.2	72.3	80.5	43.3	5.0	73.3	16.0	1.7	3.6	-	37.2
	CLAN [20]	80.4	30.7	74.7	-	-	-	1.4	8.0	77.1	79.0	46.5	8.9	73.8	18.2	2.2	9.9	-	39.3
	AdaptPatch [31]	72.6	29.5	77.2	3.5	0.4	21.0	1.4	7.9	73.3	79.0	45.7	14.5	69.4	19.6	7.4	16.5	33.7	39.6
	ADVENT [32]	67.9	29.4	71.9	6.3	0.3	19.9	0.6	2.6	74.9	74.9	35.4	9.6	67.8	21.4	4.1	15.5	31.4	36.6
	Source only	10.0	14.7	52.4	4.2	0.1	20.9	3.5	6.5	74.3	77.5	44.9	4.9	64.0	21.6	4.2	6.4	25.6	29.6
	Baseline (feat. only) [30]	63.6	26.8	67.3	3.8	0.3	21.5	1.0	7.4	76.1	76.5	40.5	11.2	62.1	19.4	5.3	13.2	31.0	36.2
	FADA	**80.4**	**35.9**	80.9	2.5	0.3	**30.4**	7.9	**22.3**	81.8	**83.6**	48.9	**16.8**	**77.7**	**31.1**	**13.5**	17.9	**39.5**	**46.0**
ResNet-101	SIBAN [19]	82.5	24.0	79.4	-	-	-	16.5	12.7	79.2	82.8	**58.3**	18.0	79.3	25.3	17.6	25.9	-	46.3
	AdaptSegNet [30]	84.3	42.7	77.5	-	-	-	4.7	7.0	77.9	82.5	54.3	21.0	72.3	32.2	18.9	32.3	-	46.7
	CLAN [20]	81.3	37.0	80.1	-	-	-	16.1	13.7	78.2	81.5	53.4	21.2	73.0	32.9	22.6	30.7	-	47.8
	AdaptPatch [31]	82.4	38.0	78.6	8.7	0.6	26.0	3.9	11.1	75.5	**84.6**	53.5	21.6	71.4	32.6	19.3	31.7	40.0	46.5
	ADVENT [32]	**85.6**	**42.2**	79.7	8.7	0.4	25.9	5.4	8.1	80.4	84.1	57.9	**23.8**	73.3	36.4	14.2	**33.0**	41.2	48.0
	Source only	55.6	23.8	74.6	9.2	0.2	24.4	6.1	12.1	74.8	79.0	55.3	19.1	39.6	23.3	13.7	25.0	33.5	38.6
	Baseline (feat. only) [30]	62.4	21.9	76.3	**11.5**	0.1	24.9	11.7	11.4	75.3	80.9	53.7	18.5	59.7	13.7	20.6	24.0	35.4	40.8
	FADA	84.5	40.1	**83.1**	4.8	0.0	**34.3**	**20.1**	**27.2**	84.8	84.0	53.5	22.6	**85.4**	**43.7**	**26.8**	27.8	**45.2**	**52.5**

4.2 Evaluation Metrics

The metrics for evaluating our algorithm is consistent with the common semantic segmentation task. Specifically, we compute PSACAL VOC intersection-over-union (**IoU**) [11] of our prediction and the ground truth label. We have $\text{IoU} = \frac{TP}{TP+FP+FN}$, where TP, FP and FN are the numbers of true positive, false positive and false negative pixels respectively. In addition to the **IoU** for each class, a **mIoU** is also reported as the mean of **IoU**s over all classes.

Table 3. Experimental results for GTA5 → Cityscapes.

GTA5 → Cityscapes

Backbone	Method	Road	SW	Build	Wall	Fence	Pole	TL	TS	Veg	Terrain	Sky	PR	Rider	Car	Truck	Bus	Train	Motor	Bike	mIoU
VGG-16	FCNs in the wild [16]	70.4	32.4	62.1	14.9	5.4	10.9	14.2	2.7	79.2	21.3	64.6	44.1	4.2	70.4	8.0	7.3	0.0	3.5	0.0	27.1
	CDA [34]	74.9	22.0	71.7	6.0	11.9	8.4	16.3	11.1	75.7	13.3	66.5	38.0	9.3	55.2	18.8	18.9	0.0	16.8	14.6	28.9
	ST [38]	83.8	17.4	72.1	14.6	2.9	16.5	16.0	6.8	81.4	24.2	47.2	40.7	7.6	71.7	10.2	7.6	0.5	11.1	0.9	28.1
	CBST [38]	90.4	50.8	72.0	18.3	9.5	27.2	28.6	14.1	82.4	25.1	70.8	42.6	14.5	76.9	5.9	12.5	1.2	14.0	28.6	36.1
	CyCADA [15]	85.2	37.2	76.5	21.8	15.0	23.8	22.9	21.5	80.5	31.3	60.7	50.5	9.0	76.9	17.1	28.2	4.5	9.8	0.0	35.4
	AdaptSegNet [30]	87.3	29.8	78.6	21.1	18.2	22.5	21.5	11.0	79.7	29.6	71.3	46.8	6.5	80.1	23.0	26.9	0.0	10.6	0.3	35.0
	SIBAN [19]	83.4	13.0	77.8	20.4	17.5	24.6	22.8	9.6	81.3	29.6	77.3	42.7	10.9	76.0	22.8	17.9	5.7	14.2	2.0	34.2
	CLAN [20]	88.0	30.6	79.2	23.4	20.5	26.1	23.0	14.8	81.6	34.5	72.0	45.8	7.9	80.5	26.6	29.9	0.0	10.7	0.0	36.6
	AdaptPatch [31]	87.3	35.7	79.5	32.0	14.5	21.5	24.8	13.7	80.4	32.0	70.5	50.5	16.9	81.0	20.8	28.1	4.1	15.5	4.1	37.5
	ADVENT [32]	86.9	28.7	78.7	28.5	25.2	17.1	20.3	10.9	80.0	26.4	70.2	47.1	8.4	81.5	26.0	17.2	18.9	11.7	1.6	36.1
	Source only	35.4	13.2	72.1	16.7	11.6	20.7	22.5	13.1	76.0	7.6	66.1	41.1	19.0	69.8	15.2	16.3	0.0	16.2	4.7	28.3
	Baseline (feat. only) [30]	85.7	22.8	77.6	24.8	10.6	22.2	19.7	10.8	79.7	27.8	64.8	41.5	18.4	79.7	19.9	21.8	0.5	16.2	4.2	34.1
	FADA	92.3	51.1	83.7	33.1	29.1	28.5	28.0	21.0	82.6	32.6	85.3	55.2	28.8	83.5	24.4	37.4	0.0	21.1	15.2	43.8
ResNet-101	AdaptSegNet [30]	86.5	36.0	79.9	23.4	23.3	23.9	35.2	14.8	83.4	33.3	75.6	58.5	27.6	73.7	32.5	35.4	3.9	30.1	28.1	42.4
	SIBAN [19]	88.5	35.4	79.5	26.3	24.3	28.5	32.5	18.3	81.2	40.0	76.5	58.1	25.8	82.6	30.3	34.4	3.4	21.6	21.5	42.6
	CLAN [20]	87.0	27.1	79.6	27.3	23.3	28.3	35.5	24.2	83.6	27.4	74.2	58.6	28.0	76.2	33.1	36.7	6.7	31.9	31.4	43.2
	AdaptPatch [31]	92.3	51.9	82.1	29.2	25.1	24.5	33.8	33.0	82.4	32.8	82.2	58.6	27.2	84.3	33.4	46.3	2.2	29.5	32.3	46.5
	ADVENT [32]	89.4	33.1	81.0	26.6	26.8	27.2	33.5	24.7	83.9	36.7	78.8	58.7	30.5	84.8	38.5	44.5	1.7	31.6	32.4	45.5
	Source only	65.0	16.1	68.7	18.6	16.8	21.3	31.4	11.2	83.0	22.0	78.0	54.4	33.8	73.9	12.7	30.7	13.7	28.1	19.7	36.8
	Baseline (feat. only) [30]	83.7	27.6	75.5	20.3	19.9	27.4	28.3	27.4	79.0	28.4	70.1	55.1	20.2	72.9	22.5	35.7	8.3	20.6	23.0	39.3
	FADA	92.5	47.5	85.1	37.6	32.8	33.4	33.8	18.4	85.3	37.7	83.5	63.2	39.7	87.5	32.9	47.8	1.6	34.9	39.5	49.2
	FADA-MST	91.0	50.6	86.0	43.4	29.8	36.8	43.4	25.0	86.8	38.3	87.4	64.0	38.0	85.2	31.6	46.1	6.5	25.4	37.1	50.1

4.3 Implementation Details

Our pipeline is implemented by PyTorch [22]. For fair comparison, we employ DeeplabV2 [4] with VGG-16 [28] and ResNet-101 [14] as the segmentation base networks. All models are pre-trained on ImageNet [10]. For the fine-grained discriminator, we adopt a simple structure consisting of 3 convolution layers with channel numbers $\{256, 128, 2K\}$, 3×3 kernels, and stride of 1. Each convolution layer is followed by a Leaky-ReLU [21] parameterized by 0.2 except for the last layer.

To train the segmentation network, we use the Stochastic Gradient Descent (SGD) optimizer where the momentum is 0.9 and the weight decay is 10^{-4}. The learning rate is initially set to 2.5×10^{-4} and is decreased following a 'poly' learning rate policy with power of 0.9. For training the discriminator, we adopt the Adam optimizer with $\beta_1 = 0.9$, $\beta_2 = 0.99$ and the initial learning rate as 10^{-4}. The same 'poly' learning rate policy is used. λ_{adv} is constantly set to 0.001. Temperature T is set as 1.8 for all experiments.

Regarding the training procedure, the network is first trained on source data for 20k iterations and then fine-tuned using our framework for 40k iterations. The batch size is eight. Four are source images and the other four are target images. Some data augmentations are used including random flip and color jittering to prevent overfitting.

Although our model is already able to achieve new state-of-the-art results, we further boost the performance by using self distillation [1,12,33] and multi-scale testing. A detailed ablation study is conducted in Sect. 4.5 to reveal the effect of each component, which, we hope, could provide more insights into the topic.

4.4 Comparison with State-of-the-art Methods

Small Shift: Cross City Adaptation.   Adaptation between real images from different cities is a scenario with great potential for practical applications. Table 1 shows the results of domain adaptation on Cityscapes → Cross-City dataset. Our method has different performance gains for the four cities. On average over four cities, our FADA achieves 8.5% improvement compared with the source-only baselines, and 2.25% gain compared with the previous best method.

Large Shift: Synthetic to Real Adaptation.   Table 2 and 3 demonstrate the semantic segmentation performance on SYNTHIA → Cityscapes and GTA5 → Cityscapes tasks in comparison with existing state-of-the-art domain adaptation methods. We could observe that our FADA outperforms the existing methods by a large margin and obtain new state-of-the-art performance in terms of mIoU. Compared to the source model without any adaptation, a gain of 16.4% and 13.9% are achieved for VGG16 and ResNet101 respectively on SYNTHIA → Cityscapes. FADA also obtains 15.5% and 12.4% improvement on different baelines for GTA5 → Cityscapes task. Besides, compared to the state-of-the-art feature-level methods, a general improvement of over 4% is witnessed. Note that as mentioned in [34], the "train" images in Cityscapes are more visually similar to the "bus" in GTA5 instead of the "train" in GTA5, which is also a challenge

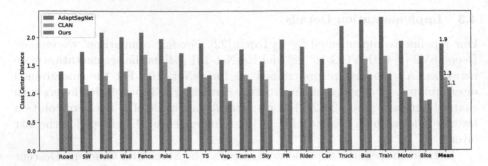

Fig. 4. Quantitative analysis of the feature joint distributions. For each class, we show the Class Center Distance as defined in Eq. 9. Our FADA shows a better aligned structure in class-level compared with other state-of-the-art methods.

to other methods. Qualitative results for GTA5 → Cityscapes task are presented at Fig. 5, reflecting that FADA also brings a significant visual improvement.

4.5   Feature Distribution

To verify whether our fine-grained adversarial framework aligns features on a class-level, we design an experiment to investigate to what degree the class-level features are aligned. Considering different networks map features to different feature spaces, it's necessarily to find a stable metric. CLAN [20] suggests to use a Cluster Center Distance, which is defined as the ratio of intra-class distance between the trained model and the initial model, to measure class-level alignment degree. To better evaluate the effectiveness of class-level feature alignment on the same scale, we propose to modify the Cluster Center Distance to the Class Center Distance (CCD) by taking inter-class distance into account. The CCD for class i is defined as follows:

$$CCD(i) = \frac{1}{K-1} \sum_{j=1,j \neq i}^{K} \frac{\frac{1}{|S_i|}\sum_{\mathbf{x} \in S_i} \|\mathbf{x} - \mu_i\|^2}{\|\mu_i - \mu_j\|^2} \tag{9}$$

where μ_i is the class center for class i, S_i is the set of all features belonging to class i. With CCD, we could measure the ratio of intra-class compactness over inter-class distance. A low CCD suggests the features of same class are clustered densely while the distance between different classes is relatively large. We randomly pick 2,000 source samples and 2,000 target samples respectively, and compare the CCD values with other state-of-the-art methods: AdaptSeg-Net for global alignment and CLAN for class-wise alignment without explicitly modeling the class relationship. As shown in the Fig. 4, FADA achieves a much lower CCD on most classes and get the lowest mean CCD value 1.1 compared to other algorithms. With FADA, we can achieve better class-level alignment and preserve consistent class structures between domains.

Table 4. Ablation studies of each component. F-Adv refers to fine-grained adversarial training; SD refers to self distillation; MST refers to multi-scale testing.

F-Adv	SD	MST	mIoU
			36.8
✓			46.9
✓	✓		49.2
✓	✓	✓	**50.1**

Table 5. Comparison of different strategies for extracting class-level knowledge on GTA5 → Cityscapes and SYNTHIA → Cityscapes tasks.

	GTA5	SYNTHIA
baseline [30]	39.4	35.4
hard labels	45.7	40.8
soft labels	**46.9**	**41.5**

4.6 Ablation Studies

Analysis of Different Components. Table 4 presents the impact of each component on DeeplabV2 with ResNet-101 on GTA5 → Cityscapes task. The fine-grained adversarial training brings an improvement of 10.1%, which already makes it the new state of the art. To further explore the potential of the model, the self distillation strategy leads to an improvement of 2.3% and multi-scale testing further boosts the performance by 0.7%.

Hard Labels vs. Soft Labels. As discussed in Sect. 3.3, the knowledge extracted from the classifier C could be produced from hard labels or soft labels. Here we compare these two forms of label on GTA5 → Cityscapes and SYN-THIA → Cityscapes tasks with DeeplabV2 ResNet-101. For soft labels, we use "confidence clipping "with threhold 0.9 as regularization. For hard labels, we only keep high-confidence samples, while ignoring the samples with confidence lower than 0.9. The results are reported in Table 5. Both choices give great boost to the baseline global feature alignment model. We observe that soft label is a more flexible choice and present more superior performance.

Impact of Confidence Clipping. In our experiments, we use "confidence clipping" as a regularizer to prevent overfitting on noisy soft labels. The values of the confidence are truncated by a given threshold, therefore the values are not encouraged to heavily fit to a certain class. We test several thresholds and the results are shown in Table 6. Note that when the threshold is 1.0, it means no regularization is used. We observe constant performance gain using the confidence clipping. The best result is found when the threshold is 0.9.

Table 6. Influence of threshold for confidence clipping.

GTA5 → Cityscapes				
threshold	0.7	0.8	0.9	1.0
mIoU	46.2	46.3	**46.9**	45.7

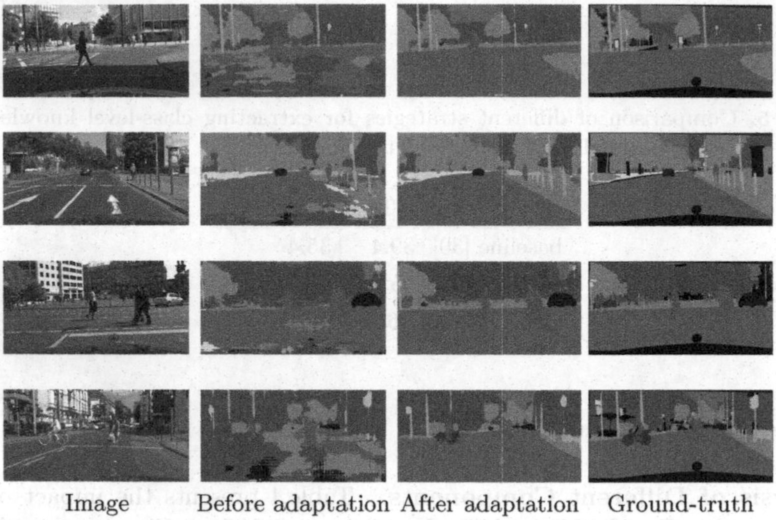

Image   Before adaptation After adaptation   Ground-truth

Fig. 5. Qualitative segmentation results for GTA5 → Cityscapes.

5   Conclusion

In this paper, we address the problem of domain adaptive semantic segmentation by proposing a fine-grained adversarial training framework. A novel fine-grained discriminator is designed to not only distinguish domains, but also capture category-level information to guide a fine-grained feature alignment. The binary domain labels used to supervise the discriminator are generalized to domain encodings correspondingly to incorporate class information. Comprehensive experiments and analysis validate the effectiveness of our method. Our method achieves new state-of-the-art results on three popular tasks, outperforming other methods by a large margin.

Acknowledgement. This work was partially supported by Beijing Academy of Artificial Intelligence (BAAI).

References

1. Bagherinezhad, H., Horton, M., Rastegari, M., Farhadi, A.: Label refinery: improving imagenet classification through label progression. CoRR abs/1805.02641 (2018), http://arxiv.org/abs/1805.02641

2. Ben-David, S., Blitzer, J., Crammer, K., Kulesza, A., Pereira, F., Vaughan, J.W.: A theory of learning from different domains. Machine Learn. **79**(1), 151–175 (2010). https://doi.org/10.1007/s10994-009-5152-4

3. Chen, C., et al.: Progressive feature alignment for unsupervised domain adaptation. In: The IEEE Conference on Computer Vision and Pattern Recognition (CVPR), June 2019

4. Chen, L., Papandreou, G., Kokkinos, I., Murphy, K., Yuille, A.L.: Deeplab: semantic image segmentation with deep convolutional nets, atrous convolution, and fully connected CRFS. IEEE Trans. Pattern Anal. Mach. Intell. **40**(4), 834–848 (2018). https://doi.org/10.1109/TPAMI.2017.2699184

5. Chen, L., Papandreou, G., Kokkinos, I., Murphy, K., Yuille, A.L.: Semantic image segmentation with deep convolutional nets and fully connected CRFS. In: 3rd International Conference on Learning Representations, ICLR 2015, San Diego, CA, USA, May 7–9, 2015, Conference Track Proceedings (2015). http://arxiv.org/abs/1412.7062

6. Chen, L.C., Zhu, Y., Papandreou, G., Schroff, F., Adam, H.: Encoder-decoder with atrous separable convolution for semantic image segmentation. In: ECCV (2018)

7. Chen, Y., Chen, W., Chen, Y., Tsai, B., Wang, Y.F., Sun, M.: No more discrimination: Cross city adaptation of road scene segmenters. In: IEEE International Conference on Computer Vision, ICCV 2017, Venice, Italy, October 22–29, 2017, pp. 2011–2020 (2017)

8. Chen, Y., Li, W., Sakaridis, C., Dai, D., Van Gool, L.: Domain adaptive faster r-cnn for object detection in the wild. In: Computer Vision and Pattern Recognition (CVPR) (2018)

9. Cordts, M., et al.: The cityscapes dataset for semantic urban scene understanding. In: Proceedings of the IEEE Conference on Computer Vision and Pattern Recognition (CVPR) (2016)

10. Deng, J., Dong, W., Socher, R., Li, L.J., Li, K., Fei-Fei, L.: ImageNet: a large-scale hierarchical image database. In: CVPR09 (2009)

11. Everingham, M., Eslami, S.M.A., Van Gool, L., Williams, C.K.I., Winn, J., Zisserman, A.: The pascal visual object classes challenge: a retrospective. Int. J. Comput. Vis. **111**(1), 98–136 (2015)

12. Furlanello, T., Lipton, Z.C., Tschannen, M., Itti, L., Anandkumar, A.: Born-again neural networks. In: Proceedings of the 35th International Conference on Machine Learning, ICML 2018, Stockholmsmässan, Stockholm, Sweden, July 10–15, 2018, pp. 1602–1611 (2018)

13. Goodfellow, I.J., et al.: Generative adversarial nets. In: Proceedings of the 27th International Conference on Neural Information Processing Systems - Volume 2, pp. 2672–2680. NIPS 2014, MIT Press, Cambridge, MA, USA (2014), http://dl.acm.org/citation.cfm?id=2969033.2969125

14. He, K., Zhang, X., Ren, S., Sun, J.: Deep residual learning for image recognition. arXiv preprint arXiv:1512.03385 (2015)

15. Hoffman, J., Tzeng, E., Park, T., Jun-Yan Zhu, A.P.I., Saenko, K., Efros, A.A., Darrell, T.: Cycada: Cycle consistent adversarial domain adaptation. In: International Conference on Machine Learning (ICML) (2018)

16. Hoffman, J., Wang, D., Yu, F., Darrell, T.: FCNS in the wild: Pixel-level adversarial and constraint-based adaptation (2016)

17. Kumar, A., et al.: Co-regularized alignment for unsupervised domain adaptation. In: Bengio, S., Wallach, H., Larochelle, H., Grauman, K., Cesa-Bianchi, N., Garnett, R. (eds.) Advances in Neural Information Processing Systems 31, pp. 9345–9356. Curran Associates, Inc. (2018), http://papers.nips.cc/paper/8146-co-regularized-alignment-for-unsupervised-domain-adaptation.pdf

18. Long, M., Cao, Y., Wang, J., Jordan, M.I.: Learning transferable features with deep adaptation networks. In: Proceedings of the 32nd International Conference on International Conference on Machine Learning, vol. 37. pp. 97–105. ICML 2015, JMLR.org (2015). http://dl.acm.org/citation.cfm?id=3045118.3045130

19. Luo, Y., Liu, P., Guan, T., Yu, J., Yang, Y.: Significance-aware information bottleneck for domain adaptive semantic segmentation. In: The IEEE International Conference on Computer Vision (ICCV), October 2019

20. Luo, Y., Zheng, L., Guan, T., Yu, J., Yang, Y.: Taking a closer look at domain shift: category-level adversaries for semantics consistent domain adaptation. In: The IEEE Conference on Computer Vision and Pattern Recognition (CVPR) (2019)

21. Maas, A., Hannun, A., Ng, A.: Rectifier nonlinearities improve neural network acoustic models. In: Proceedings of the International Conference on Machine Learning. Atlanta, Georgia (2013)

22. Paszke, A., et al.: Pytorch: An imperative style, high-performance deep learning library. In: Wallach, H., Larochelle, H., Beygelzimer, A., Alché-Buc, F., Fox, E., Garnett, R. (eds.) Advances in Neural Information Processing Systems, vol. 32, pp. 8024–8035. Curran Associates, Inc. (2019), http://papers.neurips.cc/paper/9015-pytorch-an-imperative-style-high-performance-deep-learning-library.pdf

23. Richter, S.R., Vineet, V., Roth, S., Koltun, V.: Playing for data: ground truth from computer games. In: Leibe, B., Matas, J., Sebe, N., Welling, M. (eds.) ECCV 2016. LNCS, vol. 9906, pp. 102–118. Springer, Cham (2016). https://doi.org/10.1007/978-3-319-46475-6_7

24. Ros, G., Sellart, L., Materzynska, J., Vazquez, D., Lopez, A.M.: The synthia dataset: a large collection of synthetic images for semantic segmentation of urban scenes. In: The IEEE Conference on Computer Vision and Pattern Recognition (CVPR), June 2016

25. Saito, K., Watanabe, K., Ushiku, Y., Harada, T.: Maximum classifier discrepancy for unsupervised domain adaptation. arXiv preprint arXiv:1712.02560 (2017)

26. Shelhamer, E., Long, J., Darrell, T.: Fully convolutional networks for semantic segmentation. IEEE Trans. Pattern Anal. Mach. Intell. **39**(4), 640–651 (2017). https://doi.org/10.1109/TPAMI.2016.2572683

27. Shimodaira, H.: Improving predictive inference under covariate shift by weighting the log-likelihood function. J. Stat. Plan. Inference **90**(2), 227–244 (2000)

28. Simonyan, K., Zisserman, A.: Very deep convolutional networks for large-scale image recognition. In: International Conference on Learning Representations, May 2015

29. Sun, B., Saenko, K.: Deep CORAL: correlation alignment for deep domain adaptation. In: Hua, G., Jégou, H. (eds.) ECCV 2016. LNCS, vol. 9915, pp. 443–450. Springer, Cham (2016). https://doi.org/10.1007/978-3-319-49409-8_35

30. Tsai, Y.H., Hung, W.C., Schulter, S., Sohn, K., Yang, M.H., Chandraker, M.: Learning to adapt structured output space for semantic segmentation. In: IEEE Conference on Computer Vision and Pattern Recognition (CVPR) (2018)

31. Tsai, Y.H., Sohn, K., Schulter, S., Chandraker, M.: Domain adaptation for structured output via discriminative patch representations. In: IEEE International Conference on Computer Vision (ICCV) (2019)

32. Vu, T.H., Jain, H., Bucher, M., Cord, M., Pérez, P.: Advent: Adversarial entropy minimization for domain adaptation in semantic segmentation. In: CVPR (2019)
33. Yim, J., Joo, D., Bae, J., Kim, J.: A gift from knowledge distillation: fast optimization, network minimization and transfer learning. In: 2017 IEEE Conference on Computer Vision and Pattern Recognition (CVPR), pp. 7130–7138, July 2017. https://doi.org/10.1109/CVPR.2017.754
34. Zhang, Y., David, P., Gong, B.: Curriculum domain adaptation for semantic segmentation of urban scenes. In: The IEEE International Conference on Computer Vision (ICCV), vol. 2, p. 6, October 2017
35. Zhang, Y., Qiu, Z., Yao, T., Liu, D., Mei, T.: Fully convolutional adaptation networks for semantic segmentation. CoRR abs/1804.08286 (2018)
36. Zhao, H., Shi, J., Qi, X., Wang, X., Jia, J.: Pyramid scene parsing network. In: CVPR (2017)
37. Zhu, J.Y., Park, T., Isola, P., Efros, A.A.: Unpaired image-to-image translation using cycle-consistent adversarial networkss. In: 2017 IEEE International Conference on Computer Vision (ICCV) (2017)
38. Zou, Y., Yu, Z., Kumar, B.V., Wang, J.: Unsupervised domain adaptation for semantic segmentation via class-balanced self-training. In: Proceedings of the European Conference on Computer Vision (ECCV), pp. 289–305 (2018)

Boundary-Preserving Mask R-CNN

Tianheng Cheng[1], Xinggang Wang[1(✉)], Lichao Huang[2], and Wenyu Liu[1]

[1] Huazhong University of Science and Technology, Wuhan, China
{thch,xgwang,liuwy}@hust.edu.cn
[2] Horizon Robotics Inc., Beijing, China
lichao.huang@horizon.ai

Abstract. Tremendous efforts have been made to improve mask localization accuracy in instance segmentation. Modern instance segmentation methods relying on fully convolutional networks perform pixel-wise classification, which ignores object boundaries and shap, leading coarse and indistinct mask prediction results and imprecise localization. To remedy these problems, we propose a conceptually simple yet effective Boundary-preserving Mask R-CNN (BMask R-CNN) to leverage object boundary information to improve mask localization accuracy. BMask R-CNN contains a boundary-preserving mask head in which object boundary and mask are mutually learned via feature fusion blocks. As a result, the predicted masks are better aligned with object boundaries. Without bells and whistles, BMask R-CNN outperforms Mask R-CNN by a considerable margin on the COCO dataset; in the Cityscapes dataset, there are more accurate boundary groundtruths available, so that BMask R-CNN obtains remarkable improvements over Mask R-CNN. Besides, it is not surprising to observe that BMask R-CNN obtains more obvious improvement when the evaluation criterion requires better localization (*e.g..*, AP_{75}) as shown in Fig. 1. Code and models are available at https://github.com/hustvl/BMaskR-CNN.

Keywords: Instance segmentation · Object detection · Boundary-preserving · Boundary detection

1 Introduction

Instance segmentation, a fundamental but challenging task in computer vision, aims to assign a pixel-level mask to localize and categorize each object in images, driving numerous vision applications such as autonomous driving, robotics and image editing. With the rapid development of deep convolutional neural networks (DCNN), various methods based on DCNN were proposed for instance segmentation. Prevalent methods for instance segmentation are based on object detection, which provides box-level localization information for instance-level segmentation, among which Mask R-CNN [21] is the most successful one. It extends Faster R-CNN [44] by adding a simple fully convolutional network (FCN) to predict the mask of each detected instance. Due to the great effectiveness and

© Springer Nature Switzerland AG 2020
A. Vedaldi et al. (Eds.): ECCV 2020, LNCS 12359, pp. 660–676, 2020.
https://doi.org/10.1007/978-3-030-58568-6_39

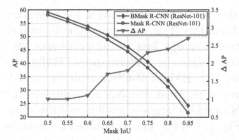

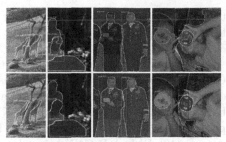

Fig. 1. AP curves of Mask R-CNN and BMask R-CNN under different mask IoU thresholds on the COCO *val2017* set. The blue line shows the AP gains of BMask R-CNN over Mask R-CNN. (Color figure online)

Fig. 2. **First row**: Selected cases of coarse boundaries appeared in the instance segmentation results of Mask R-CNN. **Second row**: Our proposed method can predict more precise boundaries.

flexibility, Mask R-CNN serves as a state-of-the-art baseline and has facilitated most recent instance segmentation research, such as [5,7,24,28,37].

In the Mask R-CNN framework, state-of-the-art instance segmentation networks [21,24,37] obtain instance masks by performing pixel-level classification via FCN. It treats all pixels in the proposal equally and ignores the object shape and boundary information. However, the pixels near boundaries are hard to be classified. Evidently, it is hard for pixel-level classifier to guarantee precise masks. We find that fine boundaries can provide better localization performance and make the object masks more distinct and clear. As illustrated in Fig. 2, Mask R-CNN (the first row) without consideration about boundaries is prone to output coarse and indistinct segmentation results with unreasonable overlaps between objects in comparison with the one that involves boundaries (the second row).

To address this issue, we leverage instance boundary information to enhance the mask prediction. Instance boundary is a dual representation of instance mask and it can guide the mask prediction network to output masks that are well-aligned with their groundtruths. Thus, the masks are more distinct and give more precise object location. Based on this motivation, we propose a conceptually simple and novel Boundary-preserving Mask R-CNN (BMask R-CNN) that unifies instance-level mask prediction and boundary prediction in one network.

Specifically, based on Mask R-CNN, we replace the original mask head with the proposed boundary-preserving mask head which contains two sub-networks for jointly learning object masks and boundaries. We insert two feature fusion blocks to strengthen the connection between boundary feature learning and mask feature learning. At last, mask prediction is guided by boundary features which contain abundant shape and localization information. The main purpose of learning boundaries is to capture features for precise object localization. Nevertheless, learning boundaries is non-trivial, because boundary groundtruths that generated from sparse annotated polygons (*i.e.*, in the COCO dataset [35]) are noisy and boundary classification has less training pixels than that for mask classification. To solve this problem, we further dive into the optimization for boundary learning by performing studies about boundary loss and exploit a boundary classification loss by combining binary cross-entropy loss and the dice loss [39].

We perform extensive experiments to evaluate the performance of BMask R-CNN. On the challenging COCO dataset [35], BMask R-CNN achieves considerably significant improvements compared with Mask R-CNN regardless of the backbones. Note that our BMask R-CNN provides larger gains if it requires more precise mask localization, as shown in Fig. 1. On the fine-annotated Cityscapes dataset [12], BMask R-CNN brings larger improvements with better mask annotations.

The main contributions of this paper can be summarized as follows.

- We present a novel Boundary-preserving Mask R-CNN (BMask R-CNN), which is the first work that explicitly exploits object boundary information to improve mask-level localization accuracy in the state-of-the-art Mask R-CNN framework.
- BMask R-CNN is conceptually simple yet effective. Without bells and whittles, BMask R-CNN outperforms Mask R-CNN by 1.7% AP and 2.2% AP on the COCO val set and the Cityscapes test set respectively. Further, BMask R-CNN obtains higher AP gains when the mask IoU threshold becomes higher, as shown in Fig. 1.
- We perform ablation studies on the components of BMask R-CNN, *e.g.*., feature fusion blocks, boundary features, boundary losses and the Sobel mask head, which are helpful to interpret how BMask R-CNN works and provide some thoughts for further research on instance segmentation.

2   Related Work

Instance Segmentation: Existing methods can be divided into two categories, i.e. detection-based methods and segmentation-based methods. Detection-based methods employ object detectors [15,18,34,44] to generate region proposals and then predict their masks after RoI pooling/align [18,21]. Based on CNN, [13,42,43] predict masks for object proposals. FCIS [33] extends Instance-FCN [13] by exploiting position-sensitive inside/outside score maps and fully convolutional networks for instance segmentation. BAIS [20] uses boundary-based distance transform to predict mask pixels that are beyond bounding boxes. Mask R-CNN [21] extends Faster R-CNN [44] by adding a mask prediction branch in parallel with the existing box regression and classification branches, demonstrating competitive performance on both object detection and instance segmentation. PANet [37] based on Mask R-CNN introduces the bottom-up path augmentation for FPN [34] to enhance information flow and adaptive feature pooling for better mask features. Mask scoring R-CNN [24] addresses the misalignment between mask quality and mask score in Mask R-CNN by explicitly learning the quality of predicted masks. [7] further improves Cascade Mask R-CNN [5] by interweaving box and mask branches in a multi-stage cascade manner and providing spatial context through semantic segmentation. Huang *et al.* apply a criss-cross attention module [25] to capture the full-image contextual information for instance segmentation. [30] draws on the idea of rendering and adaptively selects key points to recover fine details for high-quality image segmentation.

Segmentation-based methods first exploit pixel-level segmentation over the image and then group the pixels together for each object. InstanceCut [29] adopts boundaries to partition semantic segmentation into instance-level segmentation. SGN [36] groups pixels along rows and columns by line segments. [47] utilizes predicted instance centers and pixel-wise directions to group instances. Recently, several methods [4,17] take the advantage of deep metric learning to learnt the embedding to group pixels to for instance segmentation.

Boundary, Edge and Segmentation: Deep fully convolutional neural networks has achieved great progress in edge detection. Xie *et al.* propose the fully convolutional holistically-nested edge detector HED [49] which performs in an image-to-image manner and end-to-end training. CASENet [52] presents a novel challenging task semantic boundary detection, aiming to detect category-aware boundaries. [1,53] investigate the label misalignment problem caused by noisy labels in semantic boundary detection. [50] proposes geometric aware loss function for object skeleton detection in nature images. In semantic segmentation, Chen *et al.* [10] propose fully connected contditional random field (CRF) [32] to capture spatial details and refine boundaries. Recent semantic segmentation methods [3,8,23,45,51] leverage predicted boundaries or edges to facilitate semantic segmentation. [11,54] refine segmentation results with direction fields learned from predicted boundaries. Zimmermann *et al.* propose edge agreement head [55] to focus on boundaries of instances with an auxiliary edge loss. Different from these previous methods, BMask R-CNN explicitly predicts instance-level boundaries, from which we obtain instance shape information for better mask localization. Compared to semantic segmentation, boundaries in instance segmentation have dual relations to the masks. Therefore, we build fusion blocks to mutually learn boundary and mask features and improve the representations for mask localization and lead the mask prediction focus more on the boundaries.

3   Boundary-Preserving Mask R-CNN

3.1   Motivation

In Mask R-CNN, instance segmentation is performed based on pixel-level predictions. To learn a translation invariant predictor, predictions are made based on the local information. Though the local features extracted using deep network have large receptive fields, the shape information of object is ignored. Thus, the predicted masks often contain coarse and indistinct as well as some false positive predictions. For better understanding this problem, we analyze and visualize some raw mask prediction from Mask R-CNN with ResNet-50 [22] and FPN. As shown in Fig. 3, some mask predictions are rough and imprecise. Obviously, employing object boundaries will be helpful to address this issue by providing

Fig. 3. Visualization of some predicted masks (in the bottom row) of Mask R-CNN *vs.* their groundtruths (in the top row).

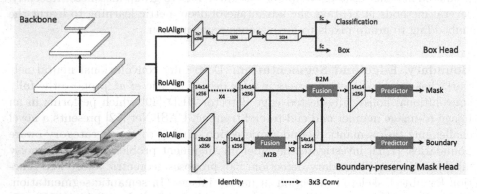

Fig. 4. The Overall architecture of **Boundary-preserving Mask R-CNN (BMask R-CNN)**. The dotted arrow denotes 3×3 convolution and the solid arrow denotes identity connection unless specified annotation in boundary-preserving mask head. "×4/×2" denotes a stack of four/two consecutive convs. The predictor contains a 2×2 deconvolution and a class-specific 1×1 convolution as the output layer for both boundary and mask prediction.

better localization and guidance. Therefore, we propose a Boundary-preserving Mask R-CNN to exploit boundary information to guide more precise mask prediction.

3.2   Boundary-Preserving Mask Head

BMask R-CNN improves the mask head in Mask R-CNN with boundary features and boundary prediction, as illustrated in Fig. 4. The new mask head termed as boundary-preserving mask head performs RoIAlign [21] to acquire RoI features for both boundary and mask prediction.

Boundary-preserving mask head jointly learns object boundaries and masks in an end-to-end manner. Note that object boundary and object mask have a close relation and we can easily convert either one to another. Features from the mask sub-network can provide high-level semantic information for learning boundaries. After obtaining boundaries, the shape information and localization information in boundary features can guide more precise mask predictions.

RoI Feature Extraction: We define $\mathcal{R}_m$ and $\mathcal{R}_b$ as Region of Interest (RoI) features for mask prediction and boundary prediction respectively. Following [34], $\mathcal{R}_m$ is extracted from the specific feature pyramid level ($P2 \sim P5$) according to the scale of the proposal, while $\mathcal{R}_b$ is obtained from the finest-resolution feature level $P2$, containing abundant spatial information. To preserve spatial information better for boundary prediction, the resolution of $\mathcal{R}_b$ is set to be larger than that of $\mathcal{R}_m$ when performing RoIAlgin. Then, it is downsampled by a strided 3×3 convolution and the output feature is denoted as $\widetilde{\mathcal{R}_b}$. $\widetilde{\mathcal{R}_b}$ has the same resolution as $\mathcal{R}_m$ and is used for feature fusion.

The feature fusion scheme in BMask R-CNN is illustrated in Fig. 4. Mask RoI features $\mathcal{R}_m$ is fed into 4 consecutive 3×3 convolutions and the output feature is denoted as $\mathcal{F}_m$. Boundary features $\widetilde{\mathcal{R}_b}$ is fused with $\mathcal{F}_m$ and then fed into two consecutive 3×3 convolutions.

Mask $\rightarrow$ Boundary (M2B) Fusion: Mask features $\mathcal{F}_m$ contain rich high-level information, i.e., the pixel-wise object category information, which is beneficial to predict object boundaries. Hence, we propose a simple fusion block to integrate boundary features and mask features for boundary prediction. The fusion block can be formulated as follows.

$$\mathcal{F}_b = f(\mathcal{F}_m) + \widetilde{\mathcal{R}_b}, \tag{1}$$

where $\mathcal{F}_b$ denotes the boundary features and f means a 1×1 convolution.

Boundary $\rightarrow$ Mask (B2M) Fusion: We fuse the final boundary features with mask features; thus, boundary information can be used to enrich mask features and guide precise mask prediction. The fusion block is the same as that of M2B.

3.3 Learning and Optimization

Following the common practice in edge detection [49,52], we regard boundary prediction as a pixel-level classification problem. The learned boundary features are fused with mask features to provide shape information for mask prediction.

Boundary Groundtruths: We use the Laplacian operator to generate soft boundaries from the binary mask groundtruths. The Laplacian operator is a second-order gradient operator and can produce thin boundaries. The produced boundaries are converted into binary maps by a threshold 0 as the final groundtruths.

Boundary Loss: Most boundary or edge detection methods [1,49,52] take the advantage of weighted cross-entropy to alleviate the class-imbalance problem in edge/boundary prediction. However, weighted binary cross-entropy leads to thick and coarse boundaries [16]. Following [16], we use dice loss [39] and binary cross-entropy to optimize the boundary learning. Dice loss measures the overlap between predictions and groundtruths and is insensitive to the number of foreground/background pixels, thus alleviating the class-imbalance problem. Our boundary loss $\mathcal{L}_b$ is formulated as follows.

$$\mathcal{L}_b(p_b, y_b) = \mathcal{L}_{Dice}(p_b, y_b) + \lambda \mathcal{L}_{BCE}(p_b, y_b), \tag{2}$$

where $p_b \in \mathbb{R}^{H \times W}$ denotes the predicted boundary for a particular category and $y_b \in \mathbb{R}^{H \times W}$ denotes the corresponding boundary groundtruth. H and W are height and width of the predicted boundary map respectively. λ is a hyperparameter to adjust the weight of dice loss (We set $\lambda = 1$ in all experiments). Dice loss is given as follows.

$$\mathcal{L}_{Dice}(p_b, y_b) = 1 - \frac{2 \sum_i^{H \times W} p_b^i y_b^i + \epsilon}{\sum_i^{H \times W} (p_b^i)^2 + \sum_i^{H \times W} (y_b^i)^2 + \epsilon}, \tag{3}$$

where i denotes the i-th pixel and ϵ is a smooth term to avoid zero division (We set $\epsilon = 1$.). In ablation experiments, we will analyze and evaluate different loss functions with quantitative results and qualitative results.

Multi-task Learning: Multi-task learning has been proved effective in many works [14,21,27,40,45], which achieves better performance for different tasks comparing with separate training. Since boundary and mask are crossed linked by two fusion blocks, jointly training can enhance the feature representation for both boundary and mask. We define a multi-task loss for each sample as follows.

$$\mathcal{L} = \mathcal{L}_{cls} + \mathcal{L}_{box} + \mathcal{L}_{mask} + \mathcal{L}_b, \tag{4}$$

where the classification loss $\mathcal{L}_{cls}$, regression loss $\mathcal{L}_{box}$, and Mask loss $\mathcal{L}_{mask}$ are inherited from Mask R-CNN. The boundary loss $\mathcal{L}_b$ has been introduced in detail in Eq. (2).

4 Experiments

We perform extensive experiments on the challenging COCO dataset [35] and the Cityscapes dataset [12] to demonstrate the effectiveness of Boundary-preserving Mask R-CNN. To better understanding each component of our method, we provide detailed ablation experiments on COCO.

Dataset and Metrics: COCO contains 115k images for training, 5k images for validation and 20k images for testing. Our models are trained on the training set (*train2017*). We report the results on the validation set (*val2017*) for ablation studies and the results on testing set (*test-dev2017*) to compare with other methods. The Cityscapes dataset is collected in urban scenes which contains 2975 training, 500 validation and 1525 testing images. As for instance segmentation,

Cityscapes involves 8 object categories and provides more precise instance-level segmentation annotations than COCO. We train our models on the training set and report our performance on the validation set and the testing set. For both COCO and Cityscapes, we use the same evaluation metric (i.e., COCO AP), which is the average precision over different IoU thresholds (from 0.5 to 0.95).

Implementation: We adopt Mask R-CNN [38] as our baseline and our method is developed based on it. All hyper-parameters are kept the same. Unless specified, we use ResNet-50 with FPN as our backbone network. We initialize our backbone networks with ImageNet pre-trained weights and freeze all BN [26] layers. The input images are resized such that the shorter side is 800 pixels and the longer is less than 1333 pixels. As for ablation experiments, we adopt 600 pixels for shorter side (the longer is less than 1000 pixels). Following the standard practice, we train all models on 4 NVIDIA GPUs using Synchronized SGD with initial learning rate 0.02 and 16 images per mini-batch for 90,000 iterations and reduce the learning rate by a factor of 0.1 and 0.01 after 60,000 and 80,000 iterations respectively. For larger backbones, we follow the linear scaling rule [19] to adjust the learning schedule when decreasing batch size.

4.1   Overall Results

We first evaluate our BMask R-CNN with different backbones on COCO and compare it with Mask R-CNN. As shown in Table 1, our method outperforms Mask R-CNN by remarkable APs in spite of different backbones. Compared with Mask R-CNN, BMask R-CNN significantly achieves 1.4, 1.7 and 1.5 AP improvements using ResNet-50-FPN, ResNet-101-FPN and HRNetV2-W32-FPN [48] respectively. Exploiting boundary information contributes to more precise mask localization due to the observation that our method yields noteworthy and stable improvements ($\approx$2.3 AP) on AP_{75}. AP^b shows AP for bounding box, on which BMask R-CNN very slightly improves over Mask R-CNN.

In Table 2, we compare BMask R-CNN with some state-of-the-art instance segmentation methods. All models are trained on COCO *train2017* and evaluated on COCO *test-dev2017*. Without bells and whistles, BMask R-CNN with ResNet-101-FPN can surpass these methods.

Figure 1 illustrates the AP curves of BMask R-CNN and Mask R-CNN under different IoU thresholds. Note that our method obtains larger gain when the IoU threshold increases, showing better localization performance of BMask R-CNN.

4.2   Ablation Experiments

In order to comprehend how BMask R-CNN works, we perform exhaustive experiments to analyze the components in BMask R-CNN. Table 3 shows the results of gradually adding components to the Mask R-CNN baseline. Each component of our proposed BMask R-CNN will be investigated in the following sections.

Table 1. Comparison with Mask R-CNN on COCO *val2017*

Method	Backbone	AP	AP_{50}	AP_{75}	AP^b	AP^b_{50}	AP^b_{75}
Mask R-CNN	ResNet-50-FPN	34.2	56.0	36.3	37.8	59.2	41.1
BMask R-CNN	ResNet-50-FPN	**35.6**	56.3	**38.4**	37.8	59.0	41.5
Mask R-CNN	ResNet-101-FPN	36.1	58.1	38.3	40.1	61.7	44.0
BMask R-CNN	ResNet-101-FPN	**37.8**	59.1	**40.6**	40.4	62.0	44.3
Mask R-CNN	HRNetV2-W32-FPN	36.6	58.7	38.9	40.8	61.9	44.9
BMask R-CNN	HRNetV2-W32-FPN	**38.1**	59.4	**40.7**	41.0	61.9	45.1

Table 2. Comparison with state-of-the-art methods for instance segmentation on COCO *test-dev2017* (* denotes our implementation)

Method	Backbone	AP	AP_{50}	AP_{75}	AP_S	AP_M	AP_L
MNC [14]	ResNet-101	24.6	44.3	24.8	4.7	25.9	43.6
FCIS+++ [33]	ResNet-101	33.6	54.5	-	-	-	-
Mask R-CNN [21]	ResNet-101-FPN	35.7	58.0	37.8	15.5	38.1	52.4
Mask R-CNN [21]	ResNeXt-101-FPN	37.1	60.0	39.4	16.9	39.9	53.5
MaskLab [9]	ResNet-101-FPN	35.4	57.4	37.4	16.9	38.3	49.2
MaskLab+ [9]	ResNet-101-FPN	37.3	59.8	39.6	19.1	40.5	50.6
Mask Scoring R-CNN [24]	ResNet-50-FPN	35.8	56.5	38.4	16.2	37.4	51.0
Mask Scoring R-CNN [24]	ResNet-101-FPN	37.5	58.7	40.2	17.2	39.5	53.0
CondInst [46]	ResNet-50-FPN	35.4	56.4	37.6	18.4	37.9	46.9
BlendMask [6]	ResNet-50-FPN	34.3	55.4	36.6	14.9	36.4	48.9
PointRend [30]	ResNet-50-FPN	36.3	-	-	-	-	-
Mask R-CNN*	ResNet-50-FPN	34.6	56.5	36.6	15.4	36.3	49.7
BMask R-CNN	ResNet-50-FPN	35.9	57.0	38.6	15.8	37.6	52.2
Mask R-CNN*	ResNet-101-FPN	36.2	58.6	38.4	16.4	38.4	52.1
BMask R-CNN	ResNet-101-FPN	37.7	59.3	40.6	16.8	39.9	54.6
BMask R-CNN w/ Mask Scoring	ResNet-101-FPN	38.7	59.1	41.9	17.4	40.7	55.5

Effects of Boundaries: To validate the effect of boundaries for mask prediction, we use mask targets to replace boundary targets and also evaluate the performance without bounadry supervision and loss with the architecture kept the same. Table 4 indicates that boundary supervision with our proposed boundary-preserving mask head improves mask results by 0.8 and 0.7 AP compared with mask supervision and no supervision respectively. Notably, using boundary can improve the mask localization performance (AP_{75}) by a significant margin.

RoI Feature Extraction: Compared with mask prediction, predicting boundaries requires more precise spatial information due to boundaries are spatially sparse. Therefore, we explore several strategies and present two considerations to extract better RoI features for boundaries. The first aspect is the source of RoI features. Lin *et al.* [34] propose that RoI features are extracted from the different levels ($P2 \sim P5$) in FPN depending on the scales of the corresponding proposals.

Table 3. Experiment results on COCO *val2017* of adding components to Mask R-CNN. We gradually add boundary supervision (BCE loss), fusions between mask and boundary features, BCE-Dice loss and our RoI feature extraction stategy for boundary features.

Boundary	Fusions	BCE-Dice	RoI Strategy	AP	AP_{50}	AP_{75}	AP^b
-	-	-	-	33.2	54.4	34.9	36.6
✓				33.9	54.8	35.8	36.7
✓	✓			34.2	55.4	36.4	36.8
✓	✓	✓		34.4	55.0	36.6	36.7
✓	✓	✓	✓	34.7	55.1	37.2	36.8

Table 4. Experiment results on COCO *val2017* of changing groundtruth of the boundary head. ✗ denotes no supervision on the boundary head.

Groundtruth	AP	AP_{50}	AP_{75}	AP^b
✗	34.0	55.0	36.0	36.3
Mask	33.9	54.3	35.9	36.0
Boundary	34.7	55.1	37.2	36.8

Features of high levels in FPN lacks spatial information which are inappropriate for boundaries. Figure 5 illustrates different sources for mask features $\mathcal{R}_m$ and boundary features $\mathcal{R}_b$. Figure 5(a) shows that boundary features are directly extracted from $P2$ while mask features are from from $P2 \sim P5$ according to the scale of the proposal. Figure 5(b) shows that both boundary features and mask features are extracted from the same feature level from $P2 \sim P5$. The other aspect is the feature resolution. Higher-resolution features preserve more spatial information which is beneficial to boundaries. Therefore, we explore the effects of 28×28 resolution and 14×14 resolution RoI features for learning boundaries. As Table 5 shows, directly extracting boundary RoI features from $P2$ is more effective with larger resolution. We employ boundary features extracted from $P2$ with 28×28 resolution in other experiments.

Feature Fusion: In Sect. 3.2, we have emphasized the relation between boundary features and mask features. Fusion blocks in our boundary-preserving mask head build explicit links to enrich both feature representation. Table 6 shows more results: if there is no fusion, it improves Mask R-CNN by 0.5 AP, which is the gain of multi-task learning; with both M2B and B2M fusion blocks, BMask R-CNN has 1.5 AP improvement over Mask R-CNN. We further investigate the influence of adding more subsequent fusion blocks. Keeping the overall computation cost substantially unchanged, adding more B2M or M2B fusions brings negligible improvements.

Table 5. Experiment results on COCO *val2017* for different RoI feature extraction strategies.

Source	Size	AP	AP_{50}	AP_{75}	AP^b
P2 ~ P5	14	34.4	55.1	36.5	36.8
P2	14	34.5	55.0	36.7	36.7
P2 ~ P5	28	34.4	55.1	36.8	36.6
P2	28	34.7	55.1	37.2	36.8

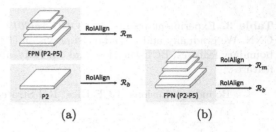

(a)   (b)

Fig. 5. Different RoI feature extraction strategies for Boundary-preserving Mask Head.

Table 6. Experiment results on COCO *val2017* for the impacts of fusion blocks, *i.e.*, mask → boundary (M2B) fusion and boundary → mask (B2M) fusion.

M2B Fusion	B2M Fusion	AP	AP_{50}	AP_{75}	AP^b
✗	✗	33.7	55.0	35.7	36.8
✓	✗	34.2	54.9	36.5	36.8
✗	✓	33.9	54.7	36.1	36.6
✓	✓	34.7	55.1	37.2	36.8

Loss Functions: We evaluate the impacts of different loss functions for optimizing boundary learning. Table 7 shows that the combination of BCE and Dice loss leads to better performance compared with individual BCE or Dice loss. Weighted BCE brings less gain than BCE in boundary prediction.

To investigate how the Dice-BCE combined loss provides such competitive improvements, we present detailed analysis on the visualization results of these experiments. As shown in Fig. 6, different loss functions have different impacts on learning boundaries. BCE loss provides considerably precisely-localized but unclear boundaries due to the class-imbalance problem. Weighted BCE solves this problem by applying balancing weights but this hard balancing leads to thick and coarse boundaries which exceed their corresponding masks. Dice loss also solves the class-imbalance problem without thick boundaries but lacks precise localization. Consequently, combining Dice loss and BCE can provide better-localized boundaries and avoid the class-imbalance problem.

Computation Cost: Compared with Mask R-CNN, our method involves four 3×3 and two 1×1 convolutional layers for boundary prediction and two fusion blocks which increase the computation cost. To clarify the improvements of BMask R-CNN are not from extra computation cost, we form a larger mask head by adding 4 more 3×3 convolutional layers as a comparison. Table 8 shows that BMask R-CNN still achieves a significant gain compared with Mask R-CNN with equal computation cost.

Table 7. Experiment results on COCO *val2017* for evaluating different loss functions.

Loss type	AP	AP_{50}	AP_{75}	AP^b
BCE	34.3	54.9	36.3	36.7
Weighted BCE	34.1	55.0	36.2	36.7
Dice	34.3	55.0	36.5	36.5
Dice-BCE	34.7	55.1	37.2	36.8

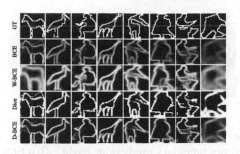

Fig. 6. Visualization results for analyzing the impacts of different loss functions. **GT** denotes groundtruth. **W-BCE** and **D-BCE** denote Weighted BCE and Dice-BCE respectively.

Table 8. Experiment results on COCO *val2017* for evaluating the impacts of computation cost. MRCNN and LMH denote Mask R-CNN and the larger mask head respectively. FLOPs are counted only for mask head without the final predictors. Inference time is tested on one NVIDIA RTX 2080Ti with the input size 600 * 1000

Method	FLOPs	Time(ms/img.)	AP	AP_{50}	AP_{75}	AP^b
MRCNN	0.46G	59.0	33.2	54.4	34.9	36.6
MRCNN w/ LMH	0.93G	61.3	33.7	54.6	35.8	36.7
BMask R-CNN	0.95G	63.7	34.7	55.1	37.2	36.8

Table 9. Experiment results on Cityscapes `val` (AP[`val`]) and `test`.(* denotes our implementation)

	AP [val]	AP_{75} [val]	AP	AP_{50}	Person	Rider	Car	Truck	Bus	Train	Mcycle	Bicycle
BAIS [20]	-	-	17.4	36.7	-	-	-	-	-	-	-	-
DIN [2]	-	-	20.0	38.8	16.5	16.7	25.7	20.6	30.0	23.4	17.1	10.1
SGN [36]	29.2	-	25.0	44.9	21.8	20.1	39.4	24.8	33.2	30.8	17.7	12.4
Mask R-CNN [21]	31.5	-	26.2	49.9	30.5	23.7	46.9	22.8	32.2	18.6	19.1	16.0
BshapeNet [28]	-	-	27.1	50.3	29.6	23.3	46.8	25.8	32.9	24.6	20.3	14.0
BshapeNet+ [28]	-	-	27.3	50.5	30.7	23.4	47.2	26.1	33.3	24.8	21.5	14.1
Neven *et al.* [41]	-	-	27.6	50.9	34.5	26.1	52.4	21.7	31.2	16.4	20.1	18.9
Mask R-CNN*	32.0	30.1	27.2	53.0	31.4	23.7	49.1	22.9	33.7	21.9	19.4	15.4
BMask R-CNN	35.0	33.6	29.4	54.7	34.3	25.6	52.6	24.2	35.1	24.5	21.4	17.1

4.3 Experiments on Cityscapes

To further explore the effects of BMask R-CNN on the fine-annotated Cityscapes dataset, we only use images with `fine` annotations to train and evaluate our models. For fair comparisons, we use ResNet-50-FPN as our backbone and resize images with shorter edge randomly selected from [800, 1024] for training. For inference, input images are kept the original size 1024 × 2048. Models are trained by SGD on 4 GPUs with mini-batch size 4 for 48,000 iterations. The learning

(a)   (b)

Fig. 7. (a): Qualitative comparison between COCO *val* annotations (left) and our instance boundary predictions (right). **(b): Sobel Mask Head:** We use Sobel operator to obtain 2-channel boundary features indicate X direction and Y direction and then apply two 3×3 convolutions to output boundary predictions.

rate is 0.005 at the beginning and reduced to 0.0005 after 36,000 iterations. Other settings are the same with experiments on COCO.

We report the results evaluated on Cityscapes `val` and `test` in Table 9. BMask R-CNN achieves 29.4 AP on `test` and obtains a remarkable 2.2 AP gain compared with the baseline Mask R-CNN. BMask R-CNN outperforms previous methods without extra data.

4.4   Discussions

Coarse Boundary Annotation *vs.* Precise Boundary Prediction: When datasets become larger and larger, obtaining precise mask annotations is unavoidably time-consuming. Though the COCO dataset provides abundant instance-level annotations, the mask and boundary annotations (represented by sparse polygons) are coarse, which limits the performance of our method BMask R-CNN. Nevertheless, BMask R-CNN can output more precise and smooth boundaries with fewer mask overlap between instances; some selected examples are shown in Fig. 7(a).

Sobel Mask Head: Instead of predicting boundaries using an extra branch, we also design a simple Sobel mask head to predict boundaries from masks, which is a improved version of [55]. As illustrated in Fig. 7(b), it has a Sobel operator [31] and two 3×3 convolutions following the mask predictions. We adopt the same Dice-BCE loss function for training. Using ResNet-50-FPN backbone and keep the rest settings the same, this Sobel mask head method obtains 34.0 AP which improves Mask R-CNN by 0.8 AP but is 0.7 AP worse than our main method.

4.5   Qualitative Results

We provide representative visualization results on COCO to compare our method with Mask R-CNN and further prove the effectiveness of our method. Figure 8(a) shows the qualitative results on COCO *val*. Mask R-CNN is more prone to generate masks with coarse boundaries which contain much background along with some false positive areas. Our proposed BMask R-CNN can alleviate this issue with the help of preserving boundaries. We further visualize our raw boundary and mask results in Fig. 8(b). It can be easily observed that predicted masks are more clear and highly coincident with their boundaries. Furthermore, utilizing predicted boundaries to refine masks brings minor improvement and the refinement is vulnerable to the noises.

(a)   (b)

Fig. 8. (a): Qualitative results on COCO dataset generated by Mask R-CNN and BMask R-CNN with ResNet-101-FPN. MRCNN and BMRCNN denotes Mask R-CNN and BMask R-CNN respectively. **(b):** Raw mask prediction and boundary prediction from boundary-preserving mask head. GT: the groundtruth segmentation. MRCNN: mask predicted by Mask R-CNN. BMRCNN: mask predicted by BMask R-CNN. Boundary: boundary predicted by BMask R-CNN. Results are obtained with ResNet-101-FPN backbone.

5 Conclusion

We address the issue that coarse boundaries and imprecise localization in instance segmentation and propose a novel Boundary-preserving Mask R-CNN. It incorporates boundary information to guide the mask learning for better boundaries and localization. Our experiments demonstrate that our method achieves remarkable and stable improvements on both COCO and Cityscapes especially in terms of localization performance. Extensive studies and visualization results provide a deep understanding of how our method BMask R-CNN works. Our method could also be plugged into Cascade Mask R-CNN and etc. for higher performance. We hope it can be a strong baseline and sheds light on this fundamental research topic.

Acknowledgements. This work was in part supported by NSFC (No. 61733007 and No. 61876212), Zhejiang Lab (No. 2019NB0AB02), and HUST-Horizon Computer Vision Research Center.

References

1. Acuna, D., Kar, A., Fidler, S.: Devil is in the edges: learning semantic boundaries from noisy annotations. In: CVPR, pp. 11075–11083 (2019)

2. Arnab, A., Torr, P.H.S.: Pixelwise instance segmentation with a dynamically instantiated network. In: CVPR, pp. 879–888 (2017)
3. Bertasius, G., Shi, J., Torresani, L.: Semantic segmentation with boundary neural fields. In: CVPR, pp. 3602–3610 (2016)
4. Brabandere, B.D., Neven, D., Gool, L.V.: Semantic instance segmentation with a discriminative loss function. CoRR abs/1708.02551 (2017)
5. Cai, Z., Vasconcelos, N.: Cascade R-CNN: high quality object detection and instance segmentation. CoRR abs/1906.09756 (2019)
6. Chen, H., Sun, K., Tian, Z., Shen, C., Huang, Y., Yan, Y.: Blendmask: top-down meets bottom-up for instance segmentation. In: CVPR (2020)
7. Chen, K., et al.: Hybrid task cascade for instance segmentation. In: CVPR, pp. 4974–4983 (2019)
8. Chen, L., Barron, J.T., Papandreou, G., Murphy, K., Yuille, A.L.: Semantic image segmentation with task-specific edge detection using CNNs and a discriminatively trained domain transform. In: CVPR, pp. 4545–4554 (2016)
9. Chen, L., Hermans, A., Papandreou, G., Schroff, F., Wang, P., Adam, H.: Masklab: instance segmentation by refining object detection with semantic and direction features. In: CVPR, pp. 4013–4022 (2018)
10. Chen, L., Papandreou, G., Kokkinos, I., Murphy, K., Yuille, A.L.: Deeplab: semantic image segmentation with deep convolutional nets, atrous convolution, and fully connected CRFs. IEEE Trans. Pattern Anal. Mach. Intell. **40**(4), 834–848 (2018)
11. Cheng, F., et al.: Learning directional feature maps for cardiac MRI segmentation. In: Martel, A.L., Abolmaesumi, P., Stoyanov, D., Mateus, D., Zuluaga, M.A., Zhou, S.K., Racoceanu, D., Joskowicz, L. (eds.) MICCAI 2020, Part IV. LNCS, vol. 12264, pp. 108–117. Springer, Cham (2020). https://doi.org/10.1007/978-3-030-59719-1_11
12. Cordts, M., et al.: The cityscapes dataset for semantic urban scene understanding. In: CVPR (2016)
13. Dai, J., He, K., Li, Y., Ren, S., Sun, J.: Instance-sensitive fully convolutional networks. In: Leibe, B., Matas, J., Sebe, N., Welling, M. (eds.) ECCV 2016. LNCS, vol. 9910, pp. 534–549. Springer, Cham (2016). https://doi.org/10.1007/978-3-319-46466-4_32
14. Dai, J., He, K., Sun, J.: Instance-aware semantic segmentation via multi-task network cascades. In: CVPR, pp. 3150–3158 (2016)
15. ai, J., Li, Y., He, K., Sun, J.: R-FCN: object detection via region-based fully convolutional networks. In: NIPS, pp. 379–387 (2016)
16. Deng, R., Shen, C., Liu, S., Wang, H., Liu, X.: Learning to predict crisp boundaries. In: Ferrari, V., Hebert, M., Sminchisescu, C., Weiss, Y. (eds.) ECCV 2018. LNCS, vol. 11210, pp. 570–586. Springer, Cham (2018). https://doi.org/10.1007/978-3-030-01231-1_35
17. Fathi, A., et al.: Semantic instance segmentation via deep metric learning. CoRR abs/1703.10277 (2017)
18. Girshick, R.B.: Fast R-CNN. In: ICCV, pp. 1440–1448 (2015)
19. Goyal, P., et al.: Accurate, large minibatch SGD: training imagenet in 1 hour. CoRR abs/1706.02677 (2017)
20. Hayder, Z., He, X., Salzmann, M.: Boundary-aware instance segmentation. In: CVPR, pp. 587–595 (2017)
21. He, K., Gkioxari, G., Dollár, P., Girshick, R.B.: Mask R-CNN. In: ICCV (2017)
22. He, K., Zhang, X., Ren, S., Sun, J.: Deep residual learning for image recognition. In: CVPR, pp. 770–778 (2016)

23. Huang, Q., Xia, C., Zheng, W., Song, Y., Xu, H., Jay Kuo, C.C.: Object boundary guided semantic segmentation. In: Lai, S.H., Lepetit, V., Nishino, K., Sato, Y. (eds.) ACCV 2016. LNCS, vol. 10111, pp. 197–212. Springer, Cham (2017). https://doi.org/10.1007/978-3-319-54181-5_13

24. Huang, Z., Huang, L., Gong, Y., Huang, C., Wang, X.: Mask scoring R-CNN. In: CVPR, pp. 6409–6418 (2019)

25. Huang, Z., et al.: Ccnet: criss-cross attention for semantic segmentation. IEEE Trans. Pattern Anal. Mach. Intell. (2020). https://doi.org/10.1109/TPAMI.2020.3007032

26. Ioffe, S., Szegedy, C.: Batch normalization: accelerating deep network training by reducing internal covariate shift. In: ICML, pp. 448–456 (2015)

27. Kendall, A., Gal, Y., Cipolla, R.: Multi-task learning using uncertainty to weigh losses for scene geometry and semantics. In: CVPR, pp. 7482–7491 (2018)

28. Kim, H.Y., Kang, B.R.: Instance segmentation and object detection with bounding shape masks. CoRR abs/1810.10327 (2018)

29. Kirillov, A., Levinkov, E., Andres, B., Savchynskyy, B., Rother, C.: Instancecut: from edges to instances with multicut. In: CVPR. pp. 7322–7331 (2017)

30. Kirillov, A., Wu, Y., He, K., Girshick, R.: Pointrend: Image segmentation as rendering. In: CVPR (2020)

31. Kittler, J.: On the accuracy of the sobel edge detector. Image Vis. Comput. 1(1), 37–42 (1983)

32. Krähenbühl, P., Koltun, V.: Efficient inference in fully connected CRFs with gaussian edge potentials. In: NIPS, pp. 109–117 (2011)

33. Li, Y., Qi, H., Dai, J., Ji, X., Wei, Y.: Fully convolutional instance-aware semantic segmentation. In: CVPR, pp. 4438–4446 (2017)

34. Lin, T., Dollár, P., Girshick, R.B., He, K., Hariharan, B., Belongie, S.J.: Feature pyramid networks for object detection. In: CVPR (2017)

35. Lin, T.Y., et al.: Microsoft COCO: common objects in context. In: Fleet, D., Pajdla, T., Schiele, B., Tuytelaars, T. (eds.) ECCV 2014. LNCS, vol. 8693, pp. 740–755. Springer, Cham (2014). https://doi.org/10.1007/978-3-319-10602-1_48

36. Liu, S., Jia, J., Fidler, S., Urtasun, R.: SGN: sequential grouping networks for instance segmentation. In: ICCV, pp. 3516–3524 (2017)

37. Liu, S., Qi, L., Qin, H., Shi, J., Jia, J.: Path aggregation network for instance segmentation. In: CVPR, pp. 8759–8768 (2018)

38. Massa, F., Girshick, R.: maskrcnn-benchmark: Fast, modular reference implementation of Instance Segmentation and Object Detection algorithms in PyTorch (2018). https://github.com/facebookresearch/maskrcnn-benchmark

39. Milletari, F., Navab, N., Ahmadi, S.: V-net: Fully convolutional neural networks for volumetric medical image segmentation. In: 3DV, pp. 565–571 (2016)

40. Misra, I., Shrivastava, A., Gupta, A., Hebert, M.: Cross-stitch networks for multi-task learning. In: CVPR, pp. 3994–4003 (2016)

41. Neven, D., Brabandere, B.D., Proesmans, M., Gool, L.V.: Instance segmentation by jointly optimizing spatial embeddings and clustering bandwidth. In: CVPR, pp. 8837–8845 (2019)

42. Pinheiro, P.H.O., Collobert, R., Dollár, P.: Learning to segment object candidates. In: NIPS, pp. 1990–1998 (2015)

43. Inheiro, P.O., Lin, T.Y., Collobert, R., Dollár, P.: Learning to refine object segments. In: Leibe, B., Matas, J., Sebe, N., Welling, M. (eds.) – ECCV 2016ECCV 2016. LNCS, vol. 9905, pp. 75–91. Springer, Cham (2016). https://doi.org/10.1007/978-3-319-46448-0_5

44. Ren, S., He, K., Girshick, R.B., Sun, J.: Faster R-CNN: towards real-time object detection with region proposal networks. IEEE Trans. Pattern Anal. Mach. Intell. **39**(6), 1137–1149 (2017)
45. Takikawa, T., Acuna, D., Jampani, V., Fidler, S.: Gated-SCNN: gated shape CNNs for semantic segmentation. In: ICCV (2019)
46. Tian, Z., Shen, C., Chen, H.: Conditional convolutions for instance segmentation. ArXiv abs/2003.05664 (2020)
47. Uhrig, J., Cordts, M., Franke, U., Brox, T.: Pixel-level encoding and depth layering for instance-level semantic labeling. In: Rosenhahn, B., Andres, B. (eds.) GCPR 2016. LNCS, vol. 9796, pp. 14–25. Springer, Cham (2016). https://doi.org/10.1007/978-3-319-45886-1_2
48. Wang, J., et al.: Deep high-resolution representation learning for visual recognition. CoRR abs/1908.07919 (2019)
49. Xie, S., Tu, Z.: Holistically-nested edge detection. Int. J. Comput. Vis. **125**(1–3), 3–18 (2017). https://doi.org/10.1007/s11263-017-1004-z
50. Xu, W., Parmar, G., Tu, Z.: Learning geometry-aware skeleton detection. In: BMVC (2019)
51. Yu, C., Wang, J., Peng, C., Gao, C., Yu, G., Sang, N.: Learning a discriminative feature network for semantic segmentation. In: CVPR, pp. 1857–1866 (2018)
52. Yu, Z., Feng, C., Liu, M., Ramalingam, S.: Casenet: Deep category-aware semantic edge detection. In: CVPR, pp. 1761–1770 (2017)
53. Yu, Z., et al.: Simultaneous edge alignment and learning. In: Ferrari, V., Hebert, M., Sminchisescu, C., Weiss, Y. (eds.) ECCV 2018. LNCS, vol. 11207, pp. 400–417. Springer, Cham (2018). https://doi.org/10.1007/978-3-030-01219-9_24
54. Yuan, Y., Xie, J., Chen, X., Wang, J.: SegFix: model-agnostic boundary refinement for segmentation. In: Vedaldi, A., Bischof, H., Brox, T., Frahm, J.M. (eds.) ECCV 2020. LNCS, vol. 12357, pp. 489–506. Springer, Cham (2020). https://doi.org/10.1007/978-3-030-58610-2_29
55. Zimmermann, R.S., Siems, J.N.: Faster training of mask R-CNN by focusing on instance boundaries. CoRR abs/1809.07069 (2018)

Self-supervised Single-View 3D Reconstruction via Semantic Consistency

Xueting Li[1]([✉]), Sifei Liu[2], Kihwan Kim[2], Shalini De Mello[2], Varun Jampani[2], Ming-Hsuan Yang[1], and Jan Kautz[2]

[1] University of California, Merced, Merced, USA
xli75@ucmerced.edu
[2] NVIDIA, Santa Clara, USA

Abstract. We learn a self-supervised, single-view 3D reconstruction model that predicts the 3D mesh shape, texture and camera pose of a target object with a collection of 2D images and silhouettes. The proposed method does not necessitate 3D supervision, manually annotated keypoints, multi-view images of an object or a prior 3D template. The key insight of our work is that objects can be represented as a collection of deformable parts, and each part is semantically coherent across different instances of the same category (e.g., wings on birds and wheels on cars). Therefore, by leveraging part segmentation of a large collection of category-specific images learned via self-supervision, we can effectively enforce semantic consistency between the reconstructed meshes and the original images. This significantly reduces ambiguities during joint prediction of shape and camera pose of an object, along with texture. To the best of our knowledge, we are the first to try and solve the single-view reconstruction problem without a category-specific template mesh or semantic keypoints. Thus our model can easily generalize to various object categories without such labels, e.g., horses, penguins, etc. Through a variety of experiments on several categories of deformable and rigid objects, we demonstrate that our unsupervised method performs comparably if not better than existing category-specific reconstruction methods learned with supervision. More details can be found at the project page https://sites.google.com/nvidia.com/unsup-mesh-2020.

Keywords: 3D from single images · Unsupervised learning

1 Introduction

Recovering both 3D shape and texture, and camera pose from 2D images is a highly ill-posed problem due to its inherent ambiguity. Existing methods resolve

X. Li—Work done during an internship at NVIDIA.

Electronic supplementary material The online version of this chapter (https://doi.org/10.1007/978-3-030-58568-6_40) contains supplementary material, which is available to authorized users.

A. Vedaldi et al. (Eds.): ECCV 2020, LNCS 12359, pp. 677–693, 2020.
https://doi.org/10.1007/978-3-030-58568-6_40

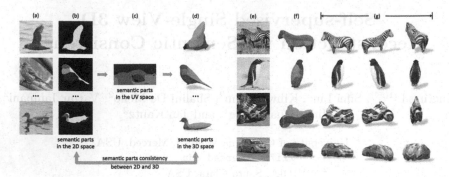

Fig. 1. Self-supervision with semantic part consistency (a–d): (a) Images of different objects in the same category (e.g., birds in this example). (b) Semantic part segmentation for each image learned via self-supervision. (c) Canonical semantic UV map for the category. (d) Semantic part segmentation on meshes. **Single-view 3D Mesh reconstruction (e–g):** Reconstruction (inference) of each single-view image (e) is demonstrated in (g), along with semantic labels of the mesh in (f). (Color figure online)

this task by utilizing various forms of supervision such as ground truth 3D shapes [3,33,34], 2D semantic keypoints [15], shading [11], category-level 3D templates [18] or multiple views of each object instance [17,27,35,40]. These types of supervision signals require tedious human effort, and hence make it challenging to generalize to many object categories that lack such annotations. On the other hand, learning to reconstruct by not using any 3D shapes, templates, or keypoint annotations, i.e., with only a collection of single-view images and silhouettes of object instances, remains challenging. This is because the reconstruction model learned without the aforementioned supervisory signals leads to erroneous 3D reconstructions. A typical failure case is caused by the "camera-shape ambiguity", wherein, incorrectly predicted camera pose and shape result in a rendering and object boundary that closely match the input 2D image and its silhouette, as shown in Fig. 2 (c) and (d).

Interestingly, humans, even infants who have never been taught about objects in a category, tend to mentally reconstruct objects in that category by perceiving them as a combination of several basic parts, e.g., a bird has two legs, two wings, and one head, etc., and use the parts to associate all the divergent instances of the category. By observing object parts, humans can also roughly infer the relative camera pose and 3D shape of any specific instance. In computer vision, a similar intuition is formulated by the deformable parts model, where objects are represented as a set of parts arranged in a deformable configuration [7,24].

Inspired by this intuition, we learn a single-view reconstruction model from a collection of images and silhouettes. We utilize the semantic parts in both the 2D and 3D space, along with their consistency to correctly estimate shape and camera pose. Specifically, we first leverage self-supervised co-part segmentation (SCOPS [14]) to decompose 2D images into a collection of semantic parts

(Fig. 1(b)). By exploiting the property of *semantic part invariance*, which states that the semantic part label of a point on the mesh surface does not change even when the mesh shape is deformed, we associate the semantic parts of *different* object instances with each other and build a category-level canonical semantic UV map (Fig. 1(c)). The semantic part label of each point on the reconstructed mesh surface (Fig. 1(d)) is then defined by this canonical semantic UV map. Finally, we resolve the aforementioned "camera-shape ambiguity" and learn the self-supervised reconstruction model by encouraging the consistency of semantic part labels in both the 2D and 3D space (Fig. 1, orange arrow). Furthermore, we train our model by iteratively learning (a) instance-level reconstruction and (b) a category-level template mesh from scratch. Thus, our model also does not require a pre-defined 3D template mesh or any other shape prior. Our main contribution is a 3D reconstruction model that is able to:

- Conduct single-view mesh reconstruction *without* any of the following forms of supervision: category-level 3D template prior, annotated keypoints, camera pose or multi-view images. In other words, the model can be generalized to other categories which do not have well-defined keypoints, e.g., penguin.
- Leverage the *semantic part invariance* property of object instances of a category as a deformable parts model.
- Learn a category-level 3D shape template from scratch via iterative learning.
- Perform comparably to the state-of-the-art supervised methods [15,18] trained with either pre-defined templates or annotated keypoints, while also improving the self-supervised semantic co-part segmentation model (SCOPS [14]).

2   Related Work

3D Shape Representation. Various representations have been explored for 3D processing tasks, including point clouds [6], implicit surfaces [21,22], triangular meshes [15–17,20,23,33,34] and voxel grids [3,8–10,31,35,40,44]. Among these, while both voxels and point clouds are more friendly to deep learning architectures (e.g., VON [36,43], PointNet [25,26], etc.), they suffer either from issues of memory inefficiency or are not amenable to differentiable rendering. Hence, in this work, we adopt triangular meshes [15–17,20,23,33,34] for 3D reconstruction.

Single-View 3D Reconstruction. Single-view 3D reconstruction [3,6,8,9,11, 31,35,40,44] aims to reconstruct a 3D shape given a single input image. One line of works have explored this ill-posed task with varying degree of supervision. Several methods [23,33,34] utilize image and ground truth 3D mesh pairs as supervision. This either requires significant manual annotation effort [38] or is restricted to synthetic data [1]. More recently, a few works [2,16,17,20] avoid 3D supervision by taking advantage of differentiable renderers [2,17,20] and the "analysis-by-synthesis" approach, with either multiple views, or known ground truth camera poses.

Fig. 2. Comparison with baselines. Each reconstructed mesh is rendered in the original view of the input image and the frontal view of the bird. (b) Shows the result from CMR with camera pose and template prior supervision. (c) Shows CMR with only template prior. (d) Shows CMR without both types of supervision where the model completely fails to learn the texture and shape. In contrast, our model in (e) reconstructs correctly even without supervision from camera pose or a template prior.

To further relax constraints on supervision, Kanazawa et al. [15] explored 3D reconstruction from a collection of images of different instances. However, their method still requires annotated 2D keypoints to infer camera pose correctly. It is also the first work to propose a learnable category-level 3D template shape, which, however, needs to be initialized from a keypoint-dependent 3D convex hull. Similar problem settings have also been explored in other methods [12,29, 37], but with object categories restricted to rigid or structured objects, such as cars or faces. Different from all these works, we target both rigid and non-rigid objects (e.g., birds, horses, penguins, motorbikes and cars shown in Fig. 1 (e)–(g)) and propose a method that jointly estimates a 3D mesh, texture, and camera pose from a single-view image, using only a collection of images with silhouettes as supervisions. In other words, we do not require 3D template priors, annotated keypoints, or multi-view images.

Self-supervised Correspondence Learning. Our work is also related to self-supervised cross-instance correspondence learning, via landmarks [13,28,30, 42], part segments [4,14], or canonical surface mapping [18]. We utilize self-supervised co-parts segmentation [14] to enforce semantic consistency, which was originally proposed purely for 2D images. The work of [18] learns a mapping function that maps pixels in 2D images to a predefined category-level template in a self-supervised manner. However, it dose not use the learned correspondence for 3D reconstruction. We show that our work, despite having a focus on 3D reconstruction, outperforms [18] at learning 2D to 3D correspondences as well.

3   Approach

To fully reconstruct the 3D mesh of an object instance from an image, a network should be able to jointly predict the shape and texture of the object, and the camera pose of the image. We start with the existing network from [15] (CMR) as the baseline reconstruction network. Given an input image, CMR extracts the image features using an encoder E and jointly predicts the mesh shape, camera pose and mesh texture by three decoders D_{shape}, D_{camera} and D_{texture}. The mesh shape V is reconstructed by predicting vertex offsets ΔV to a category-specific

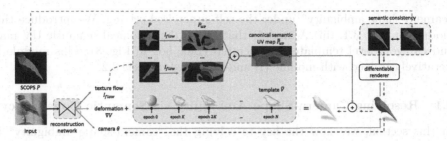

Fig. 3. Overview. (a) Green box: The reconstruction network. (b) Red box: Semantic part consistency constraint, see Sect. 3.1 for more details. (c) Blue box: Computing the canonical semantic UV map and the template shape using the reconstruction network, see Sect. 3.2. The red dashed arrows show that the gradients from the semantic part consistency constraint facilitate shape and viewpoint estimation. (Color figure online)

shape template $\bar{V}$, while the camera pose θ is represented by a weak perspective transformation. To reconstruct mesh textures, the texture decoder outputs a UV texture flow (I_{flow}) that maps pixels from the input image to the UV space. A pre-defined mapping function Φ further maps each pixel in the UV space to a point on the mesh surface.

One of the key elements for the CMR method to perform well is to exploit *manually annotated semantic keypoints* for (i) precisely pre-computing the ground truth camera pose for each instance, and (ii) estimating a category-level 3D template prior. However, annotating keypoints is tedious, not well-defined for most object categories in the world and impossible to generalize to new categories. Thus, we propose a method within a more scalable, but challenging self-supervised setting *without* using manually annotated keypoints to estimate camera pose or a template prior.

Not surprisingly, simply taking out the keypoints supervision, as well as all the related information (i.e., the camera pose and the template prior) from the CMR network makes it unable to predict camera pose and shape correctly, as shown in Fig. 2(c) and (d). This is due to the inherent ambiguity of hallucinating 3D meshes from only single-view 2D observations, where the model trivially picks a combination of camera pose and shape that yields the rendering that matches the given image and silhouette. Consider an extreme case, where the model predicts the front view for all instances, but is still able to match the image and silhouette observations by deforming each instance mesh accordingly.

In this work, we propose a framework (Fig. 3) designed for self-supervised mesh reconstruction learning, i.e., with only a collection of images and silhouettes as supervision. The framework consists of: (i) A reconstruction network (green box) that has the same architecture as [15] – it consists of an image encoder E and three decoders D_{shape}, D_{camera} and D_{texture} that jointly predict the mesh deformation ΔV, texture flow I_{flow} and camera pose θ for the instance in the image. (ii) A semantic consistency constraint (red box in Fig. 3) that regularizes the learning of module (i) and largely resolves the aforementioned "camera-shape ambiguity" under the self-supervised setting. We introduce this module in Sect. 3.1. (iii) A module that learns the canonical semantic UV map and category-level template from scratch (blue box in Fig. 3). This module is iteratively trained with module (i) and discussed in Sect. 3.2.

3.1   Resolving Camera-Shape Ambiguity via Semantic Consistency

In this section, we show the key to solving the "camera-shape ambiguity" is to make use of the semantic parts of object instances in both 3D and 2D. Specifically, we exploit the fact that (i) in the 2D space, self-supervised co-part segmentation [14] provides correct part segments for a majority of the object instances, even for those with large shape variations (see Fig. 1(b)); and (ii) in the 3D space, semantic parts are invariant to mesh deformations, i.e., the semantic part label of a specific point on the mesh surface is consistent across all reconstructed instances of a category. We demonstrate that this *semantic part invariance* allows us to build a category-level semantic UV map, namely the canonical semantic UV map, shared by all instances, which in turn allows us to assign semantic part labels to each point on the mesh. By enforcing consistency between the canonical semantic map and an instance's part segmentation in the 2D space, the camera-shape confusion can be largely resolved.

Part Segmentation in 2D via SCOPS. [14] SCOPS is a self-supervised method that learns semantic part segmentation from a collection of images of an object category (see Fig. 1(b)). The model leverages concentration and equivariance loss functions, as well as part basis discovery to output a probabilistic map w.r.t. the discovered parts that are semantically consistent across different object instances. We discuss in the supplementary as to how, besides generalizing SCOPS to reconstructing objects, our model also improves SCOPS in return.

Part Segmentation in 3D via Canonical Semantic UV Map. Given the semantic part segmentation of 2D images estimated by SCOPS, how can we obtain the semantic part labels for each point on the mesh surface? One intuitive way is to obtain a mapping from the 2D image space to the 3D shape space. Therefore, we propose to first utilize the learned texture flow I_{flow} by our reconstruction network that naturally forms a mapping from the 2D image space to the UV texture space, and then further map the semantic labels from the UV space to the mesh surface by the pre-defined mapping function Φ. We denote the semantic part segmentation of image i as $P^i \in \mathbb{R}^{H \times W \times N_p}$ (see Fig. 3 in the blue bbox), where H and W are the height and width of the image, respectively and N_p is the number of semantic parts. By mapping P^i from the 2D image space to the UV space using the learned texture flow, we obtain a "semantic UV map" denoted as $P^i_{\text{uv}} \in \mathbb{R}^{H_{\text{uv}} \times W_{\text{uv}} \times N_p}$, where H_{uv} and W_{uv} are the UV map's height and width, respectively.

Ideally, all instances should result in the same semantic UV map – the canonical semantic UV map for a category, regardless of shape differences of instances. This is because: (i) the *semantic part invariance* states that the semantic part

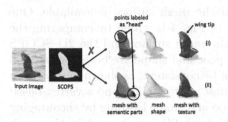

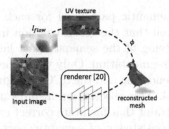

Fig. 4. Semantic part invariance.
(i) Incorrect reconstruction without semantic part consistency. (ii) Reconstruction with consistency. (Color figure online)

Fig. 5. The process of texture cycle consistency constraint computation. (Color figure online)

labels assigned to each point on the mesh surface are consistent across different instances; and (ii) the mapping function Φ that maps pixels from the UV space to the mesh surface is pre-defined and independent of deformations in the 3D space, such as face location or area changes. Thus, the semantic part labels of pixels in the UV map should also be consistent across different instances.

However, if we directly sample the individual P^i via the learned texture flow I_{flow}, the obtained semantic UV maps are indeed very different between instances, as shown in Fig. 3 (blue box). This is caused by the fact that (i) the part segmentation predictions produced by the self-supervised SCOPS method are noisy, and (ii) texture flow prediction is highly uncertain for the invisible faces of the reconstructed mesh. Therefore, we approximate the canonical semantic UV map, denoted as $\bar{P}_{\text{uv}}$ by aggregating the individual semantic UV maps:

$$\bar{P}_{\text{uv}} = \frac{1}{|\mathcal{U}|} \sum_{i \in \mathcal{U}} I_{\text{flow}}^i(P^i), \tag{1}$$

where $I_{\text{flow}}^i(P^i)$ indicates the sampling of P^i by I_{flow} and $\mathcal{U}$ is a subset of selected training samples with accurate texture flow prediction (details of the selection process can be found in the appendix). Through this aggregation process, $\bar{P}_{\text{uv}}$ produces a mean semantic UV map, which effectively eliminates outliers (i.e., instances with incorrect SCOPS), as well as the noisy pixel-level predictions.

Semantic Consistency Between 2D and 3D. As mentioned above, because our model learns via self-supervision and only relies on images and silhouettes that do not provide any semantic part information, it suffers from the "camera-shape ambiguity" introduced in Sect. 1. Take row (i) in Fig. 4 as an example. The model erroneously forms the wing's tip in the reconstructed bird by deforming the mesh faces assigned to the "head part" (colored in red). This incorrect shape reconstruction, associated with an incorrect camera pose, however, can yield a rendering that matches the observed image and its silhouette.

This ambiguity, although is not easy to spot by only comparing the rendering of the reconstruction with the input image, however, can be identified once the semantic part label for each point on the mesh surface is available. One can tell that the reconstruction in row (i) of Fig. 4 is wrong by comparing the rendering of the semantic part labels on the mesh surface and the 2D SCOPS part segmentation. Only when the camera pose and shape are both correct, will the rendering and the SCOPS segmentation be consistent, as shown in row (ii) in Fig. 4. This observation inspires us to propose a probability and a vertex-based constraint that facilitate correct camera pose and shape learning by encouraging the consistency of semantic part labels in both 2D images and in the mesh surface.

Probability-Based Constraint. For each reconstructed mesh instance i, we map the canonical semantic UV map $\bar{P}_{uv}$ onto its surface by the UV mapping Φ and render it using the predicted camera pose θ^i. We denote the projection from 3D to 2D as $\mathcal{R}$. We constrain the projected probability map to be close to the SCOPS part segmentation probability map P^i by computing the loss:

$$L_{sp} = \left\| P^i - \mathcal{R}(\Phi(\bar{P}_{uv}); \theta^i) \right\|^2. \tag{2}$$

We empirically found the mean squared error (MSE) metric to be more robust than the Kullback-Leibler divergence for comparing two probability maps.

Vertex-Based Constraint. We also propose a vertex-based constraint to enhance semantic part consistency (see Fig. 2 in the supplementary) by enforcing that 3D vertices assigned a part label p, after being projected to the 2D domain with the predicted camera pose θ^i, align with the area assigned to that part in the input image:

$$L_{sv} = \sum_{p=1}^{N_p} \frac{1}{|\bar{V}_p|} \text{Chamfer}(\mathcal{R}(\bar{V}_p; \theta^i), Y_p^i), \tag{3}$$

where $\bar{V}_p$ is the set of vertices on a learned category-level 3D *template* $\bar{V}$ (see Sect. 3.2) with the part label p, Y_p^i is the set of 2D pixels sampled from the part p in the original input image and N_p is the number of parts. Here we use the *Chamfer distance*, because the projected vertices and pixels with the same part label p in the input image do not have a strictly one-to-one correspondence.

Note that, $\bar{V}_p$ is a set of vertices on the category-level shape template $\bar{V}$ as opposed to each instance reconstruction V^i, since using V^i results in a degenerate solution where the network only alters 3D shape to satisfy this vertex-based constraint, rather than the camera pose. Instead, using $\bar{V}$ drives the network towards learning the correct camera pose, in addition to shape.

3.2   Progressive Training

We train the framework in Fig. 3 via progressive training based on two considerations: (a) building the canonical semantic UV map, introduced in Sect. 3.1, requires reliable texture flows to map the SCOPS from images to the UV space. Thus the canonical semantic UV map can only be obtained after the reconstruction network is able to predict texture flow reasonably well, and (b) a canonical

3D shape template [15, 18] is desirable, since it speeds up the convergence of the network [15] and also avoids degenerate solutions when applying the *vertex-based constrain* as introduced in Sect. 3.1. However, jointly learning the category-level 3D shape template and the instance-level reconstruction network leads to undesired trivial solutions. Thus, we propose an expectation-maximization (EM) style progressive training procedure below. In the E-step, we train the reconstruction network with the current template and canonical semantic UV map fixed, and in the M-step, we update the template and the canonical semantic UV map using the reconstruction network learned in the E-step.

E-Step: Learning Instance-Specific Reconstruction. In the E-step, we fix the canonical semantic UV map as well as the category-level template and train the reconstruction network mainly with the following objectives. (i) A negative IoU objective [16] between the rendered and the ground truth silhouettes for shape learning. (ii) A perceptual distance objective [15, 41] between the rendered and the input RGB images for texture learning. (iii) The probability and vertex-based constraints introduced in Sect. 3.1 to resolve the "camera-shape ambiguity" under the self-supervised setting. (iv) A texture consistency constraint to facilitate accurate texture flow learning that will be introduced in Sect. 3.3. Other constraints are included in the appendix. Note that in the first E-step, the template is a sphere and hence the probability and vertex-based constraints are not used.

M-Step: Canonical UV Map and Template Learning. In the M-step, we compute the canonical semantic UV map introduced in Sect. 3.1 and learn a category-level template from scratch, i.e., from a sphere primitive. As far as we know, we are the first method that learns a category-level template from scratch. This is in contrast to existing methods [18], where the template is either a readily available instance mesh from the category or is estimated from annotated keypoints [15]. Jointly learning the shape template along with the reconstruction network does not guarantee a meaningful "mean shape" which encapsulates the most representative characteristics of objects in a category. Instead, we propose a feed-forward template learning approach: the template starts out as a sphere and is updated every K training epochs by:

$$\bar{V}_t = \bar{V}_{t-1} + D_{\text{shape}}(\frac{1}{|\mathcal{Q}|} \sum_{i \in \mathcal{Q}} E(I^i)), \tag{4}$$

where $\bar{V}_t$ and $\bar{V}_{t-1}$ are the updated and current templates, respectively, I^i is the input image passed to the image encoder E and D_{shape} is the shape decoder (see the beginning of Sect. 3). $\mathcal{Q}$ is a set of selected training images with consistent mesh predictions and their selection procedure is discussed in the appendix. The template $\bar{V}_t$ is the mean shape of instances in a category for the current epoch, which enforces a meaningful shape (e.g., the template looks like a bird) rather than an arbitrary form for the category.

3.3   Texture Cycle Consistency Constraint

One issue with the learned texture flow is that the texture of 3D mesh faces with a similar color (e.g., black) can be incorrectly sampled from a single pixel location of the image (See Fig. 3 in the supplementary). Thus we introduce a texture cycle consistency objective to regularize the predicted texture flow (i.e., 2D→3D) to be consistent with the camera projection (i.e., 3D→2D). As shown in Fig. 5, considering the pixel marked with a yellow cross in the input image, it can be mapped to the mesh surface through the predicted texture flow I_{flow} along with the pre-defined mapping function Φ introduced in Sect. 3. Meanwhile, its mapping on the mesh surface can be re-projected back to the 2D image by the predicted camera pose, as shown by the green cross in Fig. 5. If the predicted texture flow conforms to the predicted camera pose, the yellow and green crosses would overlap, forming a $2D \to 3D \to 2D$ cycle.

Formally, given a triangle face j, we denote the set of input image pixels mapped to this face by texture flow as Ω_{in}^j. We further infer the set of pixels (denoted as Ω_{out}^j) projected from the triangle face j in the rendering operation by taking advantage of the probability map $\mathcal{W} \in \mathcal{R}^{|F| \times (H \times W)}$ in the differentiable renderer [20] where $|F|, H, W$ are the number of faces, height and width of the input image, respectively. Each entry in $\mathcal{W}_j^m$ indicates the probability of face j being projected onto the pixel m. We compute the geometric center of both sets (Ω_{in}^j and Ω_{out}^j), denoted by $\mathcal{C}_{\text{in}}^j$ and $\mathcal{C}_{\text{out}}^j$, respectively as:

$$\mathcal{C}_{\text{in}}^j = \frac{1}{N_c} \sum_{m=1}^{N_c} \Phi(I_{\text{flow}}(\mathcal{G}^m))_j; \quad \mathcal{C}_{\text{out}}^j = \frac{\sum_{m=1}^{H \times W} \mathcal{W}_j^m \times \mathcal{G}^m}{\sum_{m=1}^{H \times W} \mathcal{W}_j^m}, \tag{5}$$

where $\mathcal{G} \in \mathbb{R}^{(H \times W) \times 2}$ is a standard coordinate grid of the projected image (containing pixel location (u, v) values), and Φ is the fixed UV mapping that, along with the texture flow I_{flow} maps pixels from the 2D input image to a mesh face j, as discussed in the beginning of Sect. 3. N_c is the number of pixels in the input image mapped to each triangular face and $\times$ indicates multiplication between two scalars. We constrain the predicted texture flow to be consistent with the rendering operation by encouraging $\mathcal{C}_{\text{in}}^j$ to be close to $\mathcal{C}_{\text{out}}^j$:

$$L_{\text{tcyc}} = \frac{1}{|F|} \sum_{j=1}^{|F|} \left\| \mathcal{C}_{\text{in}}^j - \mathcal{C}_{\text{out}}^j \right\|_F^2. \tag{6}$$

We note that while not targeting 3D mesh reconstruction directly, a similar intuition, but with a different formulation was also introduced in [18].

4   Experimental Results

We first introduce our experimental settings in Sect. 4.1, and present qualitative evaluations for the bird, horse, motorbike and car categories in Sect. 4.2. Quantitative evaluations and ablation studies for the contribution of each proposed module are discussed in Sect. 4.3 and Sect. 4.4, respectively.

Fig. 6. Learned template and instance reconstructions from single-view images. (a) The learned template shape (first three columns) and semantic parts (last four columns). (b)–(d) 3D reconstruction from a single-view image. In each row from left to right, we show the input image, reconstruction rendered using the predicted camera view and from four other views. Please see the results for additional views in the appendix video. (Color figure online)

4.1 Experimental Settings

We validate our method on both rigid objects, i.e., *car* and *motorcycle* images from the PASCAL3D+ dataset [39], and non-rigid objects, i.e., *bird* images from the CUB-200-2011 dataset [32], *horse, zebra, cow* images from the ImageNet dataset [5] and *penguin* images from the OpenImages dataset [19]. We use progressive training (Sect. 3.2) to learn the model parameters. In each E-step, the reconstruction network is trained for 200 epochs and then used to update the template and the canonical semantic UV map in the M-step. The only exception is in the first round (a round consists of one E and M-step), where we train the reconstruction network without the semantic consistency constraint. This is because, at the beginning of training, I_{flow} is less reliable, which in turn makes the canonical semantic UV map less accurate.

4.2 Qualitative Results

Thanks to the self-supervised setting, our model is able to learn from a collection of images and silhouettes (e.g., horse and cow images [5] and penguin images [19]), which cannot be achieved by existing methods [15,17,33,40] that require extra supervisory signals.

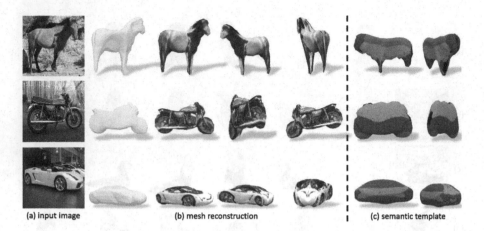

(a) input image   (b) mesh reconstruction   (c) semantic template

Fig. 7. More reconstruction results. Visualization of instance-level reconstructions and semantic templates for the *horse*, *motorbike* and *car* categories.

Template and Semantic Parts on 3D Meshes. We show the learned templates for the bird, horse, motorbike and car categories in Fig. 6 and Fig. 7, which capture the shape characteristics of each category, including the details such as the beak and feet of a bird, etc. We also visualize the canonical semantic UV map by showing the semantic part labels assigned to each point on the template's surface. For instance, bird meshes have four semantic parts – head (red), neck (green), belly (blue) and back (yellow) in Fig. 6, which are consistent with the part segmentation predicted by SCOPS [14].

Instance 3D Reconstruction. We show the results of 3D reconstruction from each single-view image in Fig. 6(b)–(d) and Fig. 7(b). Our model can reconstruct instances from an object category with highly divergent shapes, e.g., a thin bird in (b), a duck in (c) and a flying bird in (d). Our model also correctly maps the texture from each input image onto its 3D mesh, e.g., the eyes of each bird as well as fine textures on the back of the bird. Furthermore, the renderings of the reconstructed meshes under the predicted camera poses (2nd and 3rd columns in Fig. 6 and Fig. 7) match well with the input images in the first column, indicating that our model accurately predicts the original camera view.

4.3   Quantitative Evaluations

As a self-supervised approach, our model is more practically suited to reconstruct many non-rigid objects, e.g., animals captured in the wild that do not have 3D ground truth meshes available. Therefore, we treat the bird category [32] as the major one for qualitative evaluation, through the task of keypoint transfer following previous work [18]. Given a pair of source and target images of two different object instances from a category, we map a set of annotated keypoints

Table 1. Quantitative evaluation of mask IoU and keypoint transfer (KT) on the CUB dataset [32]. The comparisons are against the baseline supervised models [15,18].

(a) Metric	(b) CMR [15]	(c) CSM [18]	(d) Ours
Mask IoU ↑	0.706	-	0.734
KT (Camera) ↑	47.3	-	51.2
KT (Texture Flow) ↑	28.5	48.0	58.2

Table 2. Ablation studies of each proposed module by evaluating mask IoU and keypoint transfer (KT) on the CUB-200-2011 dataset [32].

(a) Metric	(b) Ours	(c) w/o L_{tcyc}	(d) w/o L_{sv} & L_{sp}	(e) with original [14]
Mask IoU ↑	0.734	0.731	0.744	0.731
KT (Camera) ↑	51.2	48.5	29.0	48.7
KT (Texture Flow) ↑	58.2	51.0	32.8	52.9

from the source image to the target image by first mapping them onto the learned shape template and then to the target image. Each mapping can be carried out by either the learned texture flow or the camera pose, as explained below.

To validate 3D reconstruction results, we also evaluate our model on rigid objects, e.g., cars [39], in terms of 3D IoU. However, we note that reconstruction of such rigid objects for which the ground truth 3D meshes/CAD models are easy to obtain, is not the major focus of this self-supervised method.

We first evaluate shape reconstruction on the bird category. Due to a lack of ground truth 3D shapes in the CUB-200-2011 dataset [32], we follow [15] and compute the mask reprojection accuracy – the intersection over union (IoU) between rendered and ground truth silhouettes. As shown in Table 1, our model is able to achieve comparable if not better mask reprojection accuracy compared to CMR [15], which unlike our method is learned with additional supervision from semantic keypoints. This indicates that our model is able to predict 3D mesh reconstructions and camera poses that are well matched to the 2D observations.

Next, we evaluate shape reconstruction on the car category. Although PAS-CAL3D+ [39] provides "ground truth" meshes (the most similar ones to the image in a mesh library), our reconstructed meshes are not aligned with these "ground truth" meshes since our self-suerpvised model is free to learn its own "canonical reference frame". Thus, to quantitatively evaluate the intersection over union (IoU) between the two meshes, following CMR [15], we exhaustively search a set of scale, translation and rotation parameters that best align to the "ground truth" meshes. Our method achieves an IoU (0.62) that is comparable to CMR [15] (0.64), even though the latter is trained with keypoints supervision.

Consider two different instances of a category as source and target images. To evaluate learned texture flow via keypoint transfer, given an annotated keypoint k^s in a source image (s), we map it to a triangle face (F_j) on the template using its learned flow I_{flow}^s. We then find all pixels (Ω_j) in the target image (t) that are mapped to the same triangle face F_j, by its texture flow I_{flow}^t and compute the geometric center of all pixels in Ω_j. We compare the location of the geometric center of Ω_j to the ground truth keypoint k^t and find the percentage of correct keypoints (PCK) as those that fall within a threshold distance $\alpha = 0.1$ of each other [18]. Figure 4 (a) in the appendix demonstrates qualitative visualizations of the keypoint transfer using texture flow and Table 1 shows that the texture flow learned by our method, even without supervision, outperforms the 2D→3D mappings learned by the supervised methods [15,18].

To evaluate the learned camera pose via keypoint transfer, we first find the 3D template's vertex v that corresponds to a source image's annotated 2D keypoint k^s by rendering all 3D vertices using its predicted pose θ^s. Then, v is the vertex whose 2D projection lies closest to the keypoint k^s. Next, we render the point v with a target image's predicted pose θ^t and compare it to its ground truth keypoint k^t to compute PCK. Figure 4 (b) in the supplementary demonstrates the keypoint transfer results by the predicted camera pose. Table 1 shows that our model achieves favourable performance against the baseline method [15].

4.4   Ablation Studies

In this section, we discuss the contribution of each proposed module: (i) The semantic consistency constraint discussed in Sect. 3.1. (ii) The texture cycle consistency introduced in Sect. 3.3. We evaluate on the CUB-200-2011 dataset [32] and use the mask reprojection accuracy as well as the keypoint transfer (via texture flow and via camera pose) accuracy discussed in Sect. 4.3 as our metrics.

As shown in Table 2 (b) *vs.* (d) our baseline model trained without the semantic consistency constraint performs much worse at the keypoint transfer task than our full model, indicating this baseline model predicts incorrect texture flow and camera views. We note that this baseline model achieves better mask IoU because the model trained without any constraint is more prone to overfit to the 2D silhouette observations.

Our model trained without the texture cycle consistency constraint achieves worse performance (Table 2 (b) *vs.* (c)) at transferring keypoints using the predicted texture flow. This proves the effectiveness of the texture cycle consistency constraint in encouraging the model to learn better texture flow.

5   Failure Case and Limitations

Our method performs sub-optimally for objects with large concavities and objects with a genus greater than 0, such as horses and chairs. It captures the major shape characteristics of each instance but ignores some details, e.g., the two wings of flying birds, and the legs of zebras or horses are not separated, as shown in Fig. 6 and Fig. 7. Moreover, our method utilizes the SCOPS method to provide semantic part segmentation, and so it suffers when the semantic part segmentation is not accurate, as shown in the first row of Fig. 8 in the supplementary or if the SCOPS model fails to discover meaningful parts for a certain category, such as airplanes, as shown in the supplementary document of [14]. We leave these failure cases and limitations to future works.

6 Conclusion

In this work, we learn a model to reconstruct 3D shape, texture and camera pose from single-view images, with only a category-specific collection of images and silhouettes as supervision. The self-supervised framework enforces semantic consistency between the reconstructed meshes and images and largely reduces ambiguities in the joint prediction of 3D shape and camera pose from 2D observations. It also creates a category-level template and a canonical semantic UV map, which capture the most representative shape characteristics and semantic parts of objects in each category, respectively. Experimental results demonstrate the efficacy of our proposed method in comparison to the state-of-the-art supervised category-specific reconstruction methods.

References

1. Chang, A.X., et al.: ShapeNet: An Information-Rich 3D Model Repository. arXiv preprint arXiv:1512.03012 (2015)
2. Chen, W., Gao, J., Ling, H., Smith, E., Lehtinen, J., Jacobson, A., Fidler, S.: Learning to predict 3d objects with an interpolation-based differentiable renderer. In: NeurIPS (2019)
3. Choy, C.B., Xu, D., Gwak, J., Chen, K., Savarese, S.: 3D–R2N2: a unified approach for single and multi-view 3D object reconstruction. In: Leibe, B., Matas, J., Sebe, N., Welling, M. (eds.) ECCV 2016. LNCS, vol. 9912, pp. 628–644. Springer, Cham (2016). https://doi.org/10.1007/978-3-319-46484-8_38
4. Collins, E., Achanta, R., Susstrunk, S.: Deep feature factorization for concept discovery. In: Ferrari, V., Hebert, M., Sminchisescu, C., Weiss, Y. (eds.) ECCV 2018. LNCS, vol. 11218. Springer, Cham (2016). https://doi.org/10.1007/978-3-030-01264-9_21
5. Deng, J., Dong, W., Socher, R., Li, L.J., Li, K., Fei-Fei, L.: Imagenet: a large-scale hierarchical image database. In: CVPR (2009)
6. Fan, H., Su, H., Guibas, L.J.: A point set generation network for 3D object reconstruction from a single image. In: CVPR (2017)
7. Felzenszwalb, P., Girshick, R., McAllester, D., Ramanan, D.: Object detection with discriminatively trained part-based models. TPAMI (2009)
8. Girdhar, R., Fouhey, D.F., Rodriguez, M., Gupta, A.: Learning a predictable and generative vector representation for objects. In: Leibe, B., Matas, J., Sebe, N., Welling, M. (eds.) ECCV 2016. LNCS, vol. 9910, pp. 484–499. Springer, Cham (2016). https://doi.org/10.1007/978-3-319-46466-4_29
9. Gwak, J., Choy, C.B., Chandraker, M., Garg, A., Savarese, S.: Weakly supervised 3d reconstruction with adversarial constraint. In: 3DV (2017)
10. Häne, C., Tulsiani, S., Malik, J.: Hierarchical surface prediction for 3D object reconstruction. In: 3DV (2017)
11. Henderson, P., Ferrari, V.: Learning to generate and reconstruct 3D meshes with only 2D supervision. In: BMVC (2018)

12. Henderson, P., Ferrari, V.: Learning single-image 3D reconstruction by generative modelling of shape, pose and shading. IJCV **128**, 835–854 (2019). https://doi.org/10.1007/s11263-019-01219-8
13. Honari, S., Molchanov, P., Tyree, S., Vincent, P., Pal, C., Kautz, J.: Improving landmark localization with semi-supervised learning. In: CVPR (2018)
14. Hung, W.C., Jampani, V., Liu, S., Molchanov, P., Yang, M.H., Kautz, J.: Scops: Self-supervised co-part segmentation. In: CVPR (2019)
15. Kanazawa, A., Tulsiani, S., Efros, A.A., Mali, J.: Learning category-specific mesh reconstruction from image collections. In: Ferrari, V., Hebert, M., Sminchisescu, C., Weiss, Y. (eds.) ECCV 2018. LNCS, vol. 11219. Springer, Cham (2018). https://doi.org/10.1007/978-3-030-01267-0_23
16. Kato, H., Harada, T.: Learning view priors for single-view 3D reconstruction. In: CVPR (2019)
17. Kato, H., Ushiku, Y., Harada, T.: Neural 3D mesh renderer. In: CVPR (2018)
18. Kulkarni, N., Gupta, A., Tulsiani, S.: Canonical surface mapping via geometric cycle consistency. In: ICCV (2019)
19. Kuznetsova, A., et al.: The open images dataset v4: Unified image classification, object detection, and visual relationship detection at scale. arXiv preprint arXiv:1811.00982 (2018)
20. Liu, S., Li, T., Chen, W., Li, H.: Soft rasterizer: a differentiable renderer for image-based 3D reasoning. In: ICCV (2019)
21. Liu, S., Saito, S., Chen, W., Li, H.: Learning to infer implicit surfaces without 3D supervision. In: NeurIPS (2019)
22. Mescheder, L., Oechsle, M., Niemeyer, M., Nowozin, S., Geiger, A.: Occupancy networks: learning 3D reconstruction in function space. In: CVPR (2019)
23. Pan, J., Han, X., Chen, W., Tang, J., Jia, K.: Deep mesh reconstruction from single rgb images via topology modification networks. In: ICCV (2019)
24. Pepik, B., Gehler, P., Stark, M., Schiele, B.: 3D2PM - 3D deformable part models. In: Fitzgibbon, A., Lazebnik, S., Perona, P., Sato, Y., Schmid, C. (eds.) ECCV 2012. LNCS, vol. 7577, pp. 356–370. Springer, Berlin, Heidelberg (2012). https://doi.org/10.1007/978-3-642-33783-3_26
25. Qi, C.R., Su, H., Mo, K., Guibas, L.J.: Pointnet: deep learning on point sets for 3D classification and segmentation. In: CVPR (2016)
26. Qi, C.R., Yi, L., Su, H., Guibas, L.J.: Pointnet++: deep hierarchical feature learning on point sets in a metric space. In: NeurIPS (2017)
27. Rezende, D.J., Eslami, S.A., Mohamed, S., Battaglia, P., Jaderberg, M., Heess, N.: Unsupervised learning of 3D structure from images. In: NeurIPS (2016)
28. Simon, T., Joo, H., Matthews, I., Sheikh, Y.: Hand keypoint detection in single images using multiview bootstrapping. In: CVPR (2017)
29. Szabó, A., Favaro, P.: Unsupervised 3d shape learning from image collections in the wild. arXiv preprint arXiv:1811.10519 (2018)
30. Thewlis, J., Bilen, H., Vedaldi, A.: Unsupervised learning of object landmarks by factorized spatial embeddings. In: ICCV (2017)
31. Tulsiani, S., Zhou, T., Efros, A.A., Malik, J.: Multi-view supervision for single-view reconstruction via differentiable ray consistency. In: CVPR (2017)
32. Wah, C., Branson, S., Welinder, P., Perona, P., Belongie, S.: The caltech-UCSD birds-200-2011 dataset (2011)
33. Wang, N., Zhang, Y., Li, Z., Fu, Y., Liu, W., Jiang, Y.G.: Pixel2Mesh: generating 3D mesh models from single RGB images. In: Ferrari, V., Hebert, M., Sminchisescu, C., Weiss, Y. (eds.) ECCV 2018. LNCS, vol. 11215, pp. 55–71. Springer, Cham (2018). https://doi.org/10.1007/978-3-030-01252-6_4

34. Wen, C., Zhang, Y., Li, Z., Fu, Y.: Pixel2mesh++: multi-view 3D mesh generation via deformation. In: ICCV (2019)
35. Wiles, O., Zisserman, A.: Silnet: Single-and multi-view reconstruction by learning from silhouettes. arXiv preprint arXiv:1711.07888 (2017)
36. Wu, J., Zhang, C., Xue, T., Freeman, W.T., Tenenbaum, J.B.: Learning a probabilistic latent space of object shapes via 3D generative-adversarial modeling. In: NeurIPS (2016)
37. Wu, S., Rupprecht, C., Vedaldi, A.: Photo-geometric autoencoding to learn 3d objects from unlabelled images. arXiv preprint arXiv:1906.01568 (2019)
38. Xiang, Y., et al.: ObjectNet3D: a large scale database for 3D object recognition. In: Leibe, B., Matas, J., Sebe, N., Welling, M. (eds.) ECCV 2016. LNCS, vol. 9912, pp. 160–176. Springer, Cham (2016). https://doi.org/10.1007/978-3-319-46484-8_10
39. Xiang, Y., Mottaghi, R., Savarese, S.: Beyond pascal: a benchmark for 3D object detection in the wild. In: WACV (2014)
40. Yan, X., Yang, J., Yumer, E., Guo, Y., Lee, H.: Perspective transformer nets: Learning single-view 3D object reconstruction without 3D supervision. In: NeurIPS (2016)
41. Zhang, R., Isola, P., Efros, A.A., Shechtman, E., Wang, O.: The unreasonable effectiveness of deep features as a perceptual metric. In: CVPR (2018)
42. Zhang, Y., Guo, Y., Jin, Y., Luo, Y., He, Z., Lee, H.: Unsupervised discovery of object landmarks as structural representations. In: CVPR (2018)
43. Zhu, J.Y., et al.: Visual object networks: Image generation with disentangled 3D representations. In: NeurIPS (2018)
44. Zhu, R., Kiani Galoogahi, H., Wang, C., Lucey, S.: Rethinking reprojection: closing the loop for pose-aware shape reconstruction from a single image. In: ICCV (2017)

MetaDistiller: Network Self-Boosting via Meta-Learned Top-Down Distillation

Benlin Liu[1](✉), Yongming Rao[2], Jiwen Lu[2], Jie Zhou[2], and Cho-Jui Hsieh[1]

[1] University of California, Los Angeles, Los Angeles, USA
{liubenlin,chohsieh}@cs.ucla.edu
[2] Tsinghua University, Beijing, China
raoyongming95@gmail.com, {lujiwen,jzhou}@tsinghua.edu.cn

Abstract. Knowledge Distillation (KD) has been one of the most popular methods to learn a compact model. However, it still suffers from high demand in time and computational resources caused by sequential training pipeline. Furthermore, the soft targets from deeper models do not often serve as good cues for the shallower models due to the gap of compatibility. In this work, we consider these two problems at the same time. Specifically, we propose that better soft targets with higher compatibility can be generated by using a label generator to fuse the feature maps from deeper stages in a top-down manner, and we can employ the meta-learning technique to optimize this label generator. Utilizing the soft targets learned from the intermediate feature maps of the model, we can achieve better self-boosting of the network in comparison with the state-of-the-art. The experiments are conducted on two standard classification benchmarks, namely CIFAR-100 and ILSVRC2012. We test various network architectures to show the generalizability of our MetaDistiller. The experiments results on two datasets strongly demonstrate the effectiveness of our method.

Keywords: Knowledge distillation · Meta learning

1 Introduction

Although deep neural networks have achieved astonishing performance in many important computer vision tasks, they usually require large number of model parameters (weights) which lead to extremely high time and space complexity in training and deployment. As a result, there exists a trade-off between accuracy and efficiency. To have a small and efficient model with similar performance as a deep one, many techniques have been proposed, including pruning [8,15,18], quantization [8,17], low-rank decomposition [3,27] and any others. Knowledge Distillation (KD) [10] is one of the most popular methods among them to train

Electronic supplementary material The online version of this chapter (https://doi.org/10.1007/978-3-030-58568-6_41) contains supplementary material, which is available to authorized users.

© Springer Nature Switzerland AG 2020
A. Vedaldi et al. (Eds.): ECCV 2020, LNCS 12359, pp. 694–709, 2020.
https://doi.org/10.1007/978-3-030-58568-6_41

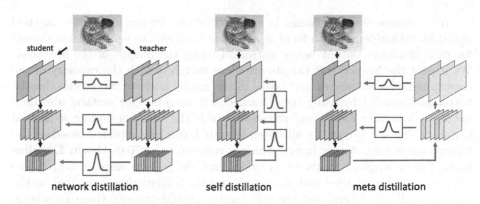

Fig. 1. The key idea of our approach. Traditional knowledge distillation methods often follow the time-consuming teacher-student pipeline as left. To overcome this problem, self-distillation (middle) abandons the large teacher model and uses the final output as soft teacher target for all intermediate outputs, which may be hampered by the capacity gap between deep layers and shallow layers [4]. Our MetaDistiller (right) proposes to generate more compatible soft target for each intermediate output respectively in a top-down manner. Best viewed in color. (Color figure online)

a compact network with high performance. Specifically, it first trains a teacher model, then uses the output of this teacher model as an additional soft target to transfer the knowledge learned by the teacher to the student model. This distillation process may go on for even more than one generation. Many variants [1,7,19,28,31] have been proposed subsequently to improve the guidance from teacher to student.

However, all these conventional distillation methods perform in a sequential way which is difficult to parallelize. This indicates that the distillation process may take a lot of time, especially for the learning of teacher model, while our goal is the compact student model. Hence, some works have explored to sidestep this sequential procedure. [26] proposes to use the output of previous iteration as the soft target for the current iteration, but this may lead to amplified errors, and it's rather hard to choose which previous iteration to use. Besides, either the generated soft targets of the whole dataset or the model at a certain iteration has to be kept in memory since they will serve as teachers. [30] proposes a self distillation method in the complete sense by extending [14], which adds classifiers to the intermediate hidden layers to directly supervise the learning of shallower layers. [30] further adopts the output of the final layer as teacher to transfer the knowledge from deep layers to the shallow ones. However, this design still suffers from the problem pointed out by [4] as conventional distillation methods. [4] finds that larger models do not often make better teachers due to the mismatched capacity between large models and smaller ones, which makes small students unable to mimic the large teachers. We can observe such problem in self-distillation given the capacity gap between the deepest model and the shallower ones.

To overcome these problems in self-distillation, we propose a new method called MetaDistiller where a label generator is employed to learn the soft targets for each shallower output layers instead of using the output of the last layer. We expect the learned soft targets are more matched with the capacity of the corresponding shallower output layers in the model and our model can thus be better self-boosted by using these learned soft targets. Our method also does not need a teacher model and therefore avoids the time-consuming sequential training. The soft targets for shallower output layers are obtained based on the feature maps from deeper layers, which is referred to as **Top-Down Distillation**. This is inspired by the observation that the feature maps in deep layers often encode higher-level and more task-specific information compared to the ones in shallower layers, and the soft targets should transfer these knowledge to the shallower layers for better training of the model. We argue this shares the same principle with the retrospection of human beings given that we can be aware of what we have missed in the previous stage as the learning proceeds. The top-down distillation aims at correcting this defect. Moreover, unlike [19], we learn the soft targets for the output probability distribution from intermediate layers rather than for the feature maps since the categorical information is shared across different stages while the contextual information needed for the reconstruction is not. To generate the soft targets that are more compatible with the intermediate outputs, the label generator is optimized by bi-level optimization which is commonly used in meta-learning, given we share the same motivation of 'learning to learn'. Though additional computational burden is added during training in comparison with self-distillation due to the existence of meta learning, it is relatively marginal when compared to traditional teacher-student distillation methods. Besides, neither time nor space complexity increases during inference time since label generator is then abandoned. Hence, such trade-off for the improvement in compatibility of soft targets is quite cost-effective. Figure 1 illustrates our motivation intuitively.

To show the effectiveness of our approach, we conduct experiments for various network architectures on two standard classification benchmarks, namely CIFAR-100 [13] and ILSVRC2012 [20]. The empirical study shows that we can consistently attain better results by employing our MetaDistiller compared to both self-distillation and traditional teacher-student distillation methods. We also find that the generated soft targets are more complementary to the one-hot ground-truth label compared to the final output. This is because the generated soft targets can better capture the information among negative classes, which is not covered by one-hot ground-truth, and thus can serve as a better regularization term to help the model generalize well to unseen data.

2   Related Work

Knowledge Distillation. Knowledge distillation is widely adopted to train a shallow network to achieve comparable or higher performance than a deep network. Hinton et al. [10] first proposes the idea of knowledge distillation where a well-optimized large model is employed as teacher to provide additional information not contained in the one-hot label to the compact student network. Fitnets [19] improves upon [10] by using the feature maps in the intermediate layers of teacher net as hints. [28] trains the student model to mimic the attention maps from the teacher model instead. [31] blurs the difference between teacher and student by exchanging their roles iteratively, which is similar to co-training. However, all these traditional distillation methods perform the time-consuming sequential training. Although [1,7] share the same architecture between teacher and student, they still require sequential training since each student is trained from scratch to match the teacher's soft labels.

To sidestep this resource-consuming pipeline, some works [26,30] explore to distill the network itself. [26] uses the outputs from previous iteration as soft targets, but this risks amplifying the error in the learning process, and decision on which iteration to adopt as teacher is rather hard. Strictly speaking, it is not self-boosting in a complete sense as we still need to keep the model of certain iteration as teacher. [30] improves upon [14] by further employing the final output as teacher to supervise the output from intermediate layers. Their self-distillation method totally abandons the need of keeping a big teacher model or the soft targets of the whole dataset produced by it to perform the distillation. Nevertheless, this simple formulation suffers from the problem as illustrated in [4] that shallower student is difficult to directly mimic the output from deep teacher model due to the gap between capacity. Our method overcomes this problem while still performs the distillation without resource-consuming teacher-student pipeline by generating more compatible soft targets through a meta-learned label generator to transfer the knowledge in a top-down manner.

Meta Learning. The core idea of meta learning is 'learning to learn', which means taking the optimization process of learning algorithm into the consideration of optimizing the learning algorithm itself. Bengio et al. [2] first applies meta learning to optimize hyper-parameters by differentiating through the preserved computational graph of training phase to obtain the gradients with regard to the hyper-parameters. Finn et al. [6] uses meta learning to learn better initialization for few-shot learning. [23] utilizes meta learning to augment training data for low-shot learning. [11] incorporates meta learning into traditional knowledge distillation. They automatically decide what knowledge in the teacher as well as its amount will be transferred to which part of the student based on meta gradients instead of hand-crafted rules. Sharing the similar motivation with the works mentioned above, in this paper, we adopt meta learning to learn the soft targets for self knowledge distillation, which can be subsequently used as supervision signals to boost the learning of the model.

3 Approach

In this section, we will introduce the proposed method, MetaDistiller. We first briefly review the formulation of knowledge distillation in Sect. 3.1, and then extend it to self-boosting in Sect. 3.2. In Sect. 3.3, we propose to perform the

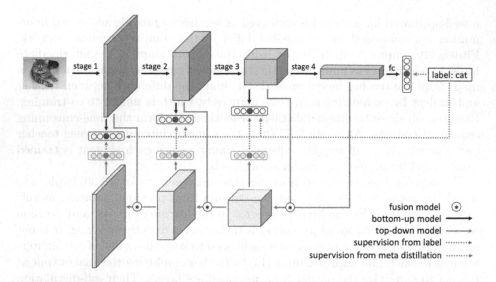

Fig. 2. The pipeline of our approach. The grey part and the yellow part correspond to main model and label generator respectively. We fuse the feature map produced in the feed forward process of main model through a top-down manner by adopting the operation defined in Eq. (5) iteratively to generate soft teacher targets (green). The intermediate outputs (blue) of the main model are supervised by the ground-truth label and generated soft teacher targets simultaneously. The label generator is trained by using meta learning. Best viewed in color. (Color figure online)

top-down distillation by incorporating feature maps from different stages progressively to generate soft targets. In Sect. 3.4, we then discuss how to apply meta learning to learn the soft target generator. Finally, we will present the whole training and inference pipeline of the proposed algorithm in Sect. 3.5.

3.1 Background: Knowledge Distillation

Knowledge Distillation (KD), proposed by Hinton et al. [10], aims at transferring knowledge from a deep ***teacher*** network, denoting as T, to a shallow ***student*** network, denoting as S. In order to transfer knowledge from teacher to student, the loss for training student network is modified by adding KL divergence between the teacher's and the student's output probability distributions. Formally, given labeled dataset $\mathcal{D}$ of N samples $\mathcal{D} = \{(x_1, y_1), \ldots, (x_N, y_N)\}$, we can write the loss function of student network as following,

$$\mathcal{L}_S\left(\mathcal{D}; \theta_S\right) = \frac{1}{N} \sum_{i=1}^{N} \alpha \mathcal{L}_{CE}\left(y_i, S(x_i; \theta_S)\right) + (1 - \alpha)\tau^2 \mathcal{KL}\left(T(x_i; \theta_T) \| S(x_i; \theta_S)\right),$$

(1)

where α is the hyper-parameter to control the relative importance of the two terms, τ is the temperature hyper-parameter, θ_T and θ_S are the parameters of teacher T and student S respectively. $\mathcal{L}_{CE}$ refers to the cross entropy loss and $\mathcal{KL}$ refers to the KL divergence which measures how one probability distribution is different from another.

By minimizing this modified loss function, the student network will try to match the one-hot ground-truth distribution (the first term) while lowering its discrepancy with output probability distribution from teacher $T(x_i; \theta_T)$ to transfer the knowledge of negative classes learned by teacher network. The probability distribution over classes can be obtained by employing softmax function, where the probability of sample s_i belonging to class j can be expressed as,

$$P\left(y_i = j | x_i\right) = \frac{e^{z_i^j / \tau}}{\sum_{c=1}^{\mathcal{C}} e^{z_i^c / \tau}} \tag{2}$$

where $\mathcal{C}$ is number of classes, z_i is the output logits of sample x_i, τ is the same temperature hyper-parameter as (1). The role of τ is to control the smoothness of the output probability distribution. The higher the temperature, the higher the entropy of distribution, i.e. the smoother distribution.

3.2   Network Self-Boosting

One of the major problems of the above teacher-student pipeline is the long training time caused by the two-stage sequential training procedure. To overcome such difficulty, a method called self-distillation [30] has been proposed by adopting deep-supervised net [14] to achieve self-boosting without first training a large teacher network. In the following, we will introduce a general self-boosting framework where [30] is a special case under our framework, and then we will show our general framework leads to improved algorithms in the next two subsections.

In the self-boosting framework introduced in [14], the deep network is divided into K stages and there is an exit branch at the end of each early stage. In addition to the loss at the final output, the one-hot ground-truth label is also used to supervise these early stages. Besides using the one-hot hard label as supervision signal to the intermediate outputs, we propose to add soft teacher targets to supervise these intermediate outputs. Specifically, (1) can be modified as following:

$$\mathcal{L}_S(\mathcal{D}; \theta_S, T_k) = \frac{1}{N} \sum_{i=1}^{N} \left(\mathcal{L}_{CE}\left(y_i, S_K\left(x_i; \theta_S\right)\right) + \sum_{k=1}^{K-1} \mathcal{L}_k(S_k\left(x_i; \theta_S\right), y_i; T_k) \right) \tag{3}$$

where S_k indicates the output of the k-th stage in the student network, and S_K is the final output of the model. T_k is called the **soft teacher target** for stage k to perform knowledge distillation, which is a function depends on stage $k+1, \ldots, K$ of the network. Unlike (1), there is no distillation term for the final output S_K since we do not have a teacher model to provide such supervision signal now. $\mathcal{L}_k$ refers to the loss function of outputs at earlier stages. For the first $K - 1$ stages, the supervision signal of stage k can be formally written as

$$\mathcal{L}_k(S_k(x_i; \theta_S), y_i; T_k) = \alpha \mathcal{L}_{CE}(y_i, S_k(x_i; \theta_S)) + (1 - \alpha)\tau^2 \mathcal{KL}(T_k \| S_k(x_i; \theta_S)) \tag{4}$$

where outputs from the intermediate layers are supervised by hard labels and soft teacher targets simultaneously.

In self-distillation [30], they directly utilize the output from the last layer $S_K(x_i; \theta_S)$ as T_k. However, we will argue that this is not the best way to assign T_k and will propose an automatic way to generate T_k that can significantly improve the performance.

3.3   Top-Down Distillation

As mentioned in the previous subsection, self-distillation [30] sets the soft teacher target T_k as the final layer softmax output for all the intermediate stages, and we argue this is not the best way for self-supervision. First, there exists "gap of capacity" problem, which means the capacity of earlier stages may not be enough to learn the final softmax output of a deep network. Furthermore, compared to the output probability distribution, feature maps in the deeper layers undoubtedly contain more information beyond the categorical information, which shall be useful for getting better soft teacher targets for earlier stages.

Based on these ideas, we propose to distill the knowledge from the feature maps of deep layers to better supervise the learning of shallower layers through a top-down feature fusion process. This fusion process can be recursively defined as,

$$\hat{\mathcal{F}}_k = \text{conv}_{3\times3}(\text{UpSampling}_{\times2}(\hat{\mathcal{F}}_{k+1}) + \text{conv}_{1\times1}(\mathcal{F}_k)) \tag{5}$$

where $\mathcal{F}_k$ indicates the original output feature map of the k-th stage in the main model and $\hat{\mathcal{F}}_k$ indicates the feature map produced by our label generator based on feature maps of stage $k + 1, \ldots, K$. For the final stage K, we slightly modify (5) by only keeping the 1×1 convolution operation since we do not have $\hat{\mathcal{F}}_{K+1}$. Upsampling can be simply achieved by using bilinear interpolation to match the spatial size of the feature map from two neighboring stages as each stage scales the input feature map by 0.5. 1×1 convolution is responsible for matching the channel size to perform element-wise addition for fusion. We finally use a 3×3 convolution layer to refine this result to obtain the generated feature map $\hat{\mathcal{F}}_k$ for the k-th stage. Note that the parameters for these convolution kernels are learnable which makes our approach able to automatically generate the best soft target feature for each particular task. We will discuss how to learn these parameters in the next subsection.

The process described above can be considered as the retrospection of human beings if we take the feed forward of neural network as the learning process of humans. Through rethinking what we have missed in previous learning stages, we could correct this for better learning if we could access the past states. And the top-down distillation aims at transferring such retrospection back to early stages for better learning. Compared to categorical probability distribution, feature maps contain more information, thus can serve a better indicator for different learning stages to achieve a better retrospection.

Though we can directly supervise the discrepancy between $\mathcal{F}_k$ and $\hat{\mathcal{F}}_k$ by using loss function like L_2 distance, we argue this is not a good choice based on the fact that feature also encodes contextual information other than categorical information. Unlike categorical information which should be invariant across different stages, the downsampling operation often causes a loss in the contextual information. As a result, the evolution of feature maps is not fully invertible. Hence, instead of adding supervision signal on the feature map, we prefer to only supervise categorical information as the deep layers can help on this. We therefore learn a normalized probability distribution from $\hat{\mathcal{F}}_k$ as our T_k by using an additional bottleneck module, which is also employed to produce the intermediate outputs, and use T_k as soft teacher label to perform the knowledge distillation. The whole pipeline is illustrated in Fig. 2

3.4   Meta-Learned Soft Teacher Label Generator

The soft teacher target generator defined in (5) is actually a fusion network that can subsequently help us to learn a better model, and therefore learning the parameters of this generator network shares the same "learning-to-learn" intuition with meta learning. In the following, we show how to learn the label generator based on the commonly used bi-level optimization framework in meta learning.

Formally, denoting the parameters of the main network as θ_s and the parameters of label generator as θ_g, learning the label generator can be written as a bi-level problem,

$$\min_{\theta_g} \; \mathcal{L}^{\text{val}}(X_{\text{val}}, Y_{\text{val}}; \theta_s^*)$$
$$s.t. \;\; \theta_s^* = \arg\min_{\theta_s} \; \mathcal{L}_S(\mathcal{D}_{\text{train}}; \theta_s.T_k(\theta_g)), \tag{6}$$

where $\mathcal{L}^{\text{val}}$ is the standard cross entropy loss and $\mathcal{L}^{\text{train}}(X_{\text{train}}, Y_{\text{train}}; \theta_s, \theta_g)$ is the same loss function as (3) only with T_k changing to $T_k(\theta_g)$ as the soft teacher targets in our framework is generated through a label generator parameterized by θ_g.

To conduct updates on θ_g, we need to compute $\nabla_{\theta_g} \mathcal{L}^{\text{val}}(X_{\text{val}}, Y_{\text{val}}; \theta_s^*)$. However this is nontrivial since θ_s^* is an implicit function of θ_g. This could be potentially done by the implicit function theory but the exact solution will lead to huge computational overhead. Therefore, following a standard approach used in meta learning (e.g., [6]), we only approximate the θ_s^* by the current one with an additional gradient update step. This approximation can be written as

$$\theta_s^+ := \theta_s - \zeta \nabla_{\theta_s} \mathcal{L}_S(\mathcal{D}_{\text{train}}; \theta_s.T_k(\theta_g)), \tag{7}$$

where ζ is the learning rate for this inner optimization step.

According to the above procedure, gradient with respect to θ_g can be expressed formally as

$$\nabla_g := \nabla_g \mathcal{L}^{\text{test}}(X_{\text{test}}, Y_{\text{test}}; \theta_s^+)$$
$$s.t. \quad \theta_s^+ = \text{opt}_{\theta_g}(\theta_s - \zeta \nabla_{\theta_s} \mathcal{L}_{\text{train}}(X_{\text{train}}, Y_{\text{train}}; \theta_s, \theta_g)), \tag{8}$$

where $\text{opt}_{\theta_g}(\cdot)$ indicates the term inside the parentheses is differentiable with regard to θ_g for optimization.

Algorithm 1. Training process for solving optimization problem in ((6)):

Input: labeled dataset $\mathcal{D} = \{(x_i, y_i)\}$, main model $S(\theta_s)$, label generator $G(\theta_g)$
Output: Model $S^*(\theta_s)$
1: **initialize:** main model $S(\theta_s)$, label generator $G(\theta_g)$
2: **while** not converged **do**
3:     // train S
4:     Sample a random mini-batch $\{(x_i, y_i)\}$ from $\mathcal{D}$
5:     Forward pass on S to obtain K outputs $\{S_1(x_i; \theta_s), \cdots, S_K(x_i; \theta_s)\}$
6:     Forward pass on G to generate $\{T_1, \cdots, T_{K-1}\}$
7:     Optimize S following (3) using generated soft targets $\{T_1, \cdots, T_{K-1}\}$
8:     // train G for every M epochs
9:     **if** epoch_num % M ==0 **then**
10:         Randomly split $\mathcal{D}$ in half to get two disjoint dataset $\mathcal{D}_{\text{train}}$ and $\mathcal{D}_{\text{test}}$
11:         //inner update, retaining the computational graph
12:         sample a random mini-batch X_{train} with labels Y_{train} from $\mathcal{D}_{\text{train}}$
13:         compute θ_s^+ by using a single step gradient update following (7)
14:         //meta gradient step
15:         sample a random mini-batch X_{test} with labels Y_{test} from $\mathcal{D}_{\text{test}}$
16:         Calculate $\mathcal{L}_{\text{test}}(\theta_s^+)$ using cross entropy loss
17:         Obtain ∇_g following (8)
18:         //update label generator
19:         Update θ_g based on ∇_g using gradient descent
20:     **end if**
21: **end while**

3.5 Training and Inference

The whole training pipeline of our method is described in Algorithm 1. More specifically, we alternatively update θ_s (model parameters) and θ_g (generator's parameters) by SGD, where the gradient with respect to generator is computed by (8). The generator is used to perform top-down distillation to generate the soft teacher targets to facilitate the self-boosting of the main model. The reason why we split the whole training dataset into two disjoint sets of equal size is that we expect the model supervised by generated soft targets can generalize well to unseen data, which can be taken as an indicator of higher compatibility. We set M = 5, i.e. we update the label generator every 5 epochs. This can

Table 1. Our method consistently outperforms the self-distillation on CIFAR-100. We present the results on CIFAR-100 by testing our MetaDistiller on 8 different networks. We report the result in [30] in the parentheses.

Model	Baseline	Final Output		Ensemble Output	
		SD	MD	SD	MD
ResNet18 [9]	77.31(77.09)	78.25(78.64)	**79.05**	79.32(79.67)	**80.05**
ResNet50 [9]	77.78(77.68)	79.71(80.56)	**80.85**	80.38(81.04)	**81.41**
ResNet101 [9]	78.92(77.98)	80.92(81.23)	**81.39**	81.55(82.03)	**81.98**
ResNeXT29-2 [24]	77.78	79.43	**80.38**	80.14	**80.89**
WideResNet16-1 [29]	67.58	69.42	**70.30**	70.17	**70.81**
VGG19(BN)[22]	71.61(64.47)	74.27(67.73)	**75.12**	75.01(68.54)	**75.93**
MobileNetv2 [21]	71.81	74.46	**75.37**	75.33	**75.96**
Shufflenetv2 [16]	68.12	69.89	**70.66**	70.71	**71.29**

reduce the training time while still keep the performance from degradation. For instance, on ResNet18 [9], the training time increases by 1.1 h compared to simply training the model without any types of distillation while it takes 7.9 h to first train a ResNet101 [9] as teacher model if we perform the conventional two-stage knowledge distillation. Moreover, the performance of the well-trained ResNet101 is almost the same as our self-boosted ResNet18. This shows that our proposed method can preserve the speed of self-distillation while improving its final test accuracy.

During the inference phase, we can abandon the label generator and the layers utilized to produce the intermediate results, so there will be no additional computational cost required for deployment compared to the model trained by simply using the cross entropy loss. Furthermore, if the computational resources allow, we can further perform ensemble by using the intermediate output. We report the results of both the final output and the ensemble output in experiments, and ensemble often improves the accuracy of final output by around 1%. and ensemble often improves the accuracy of final output by around 1%.

4 Experimental Results

4.1 Setups

We evaluate our method on two widely used image classification datasets, namely CIFAR-100 [13] and ILSVRC2012 [5]. To verify the generalizability of our method, we test with multiple different network architectures. As for hyperparameters, we set balance weight α to 0.5 and temperature τ simply to 1 for all the datasets constantly. For optimization, we use SGD and Adam [12] optimizer for main model S and label generator G respectively with initial learning rates both as 0.1. We train the network for 200 epochs in total, and multiply

Table 2. Our method consistently outperforms the self-distillation on ImageNet. We present the results on ImageNet by testing our MetaDistiller on 3 different networks.

Model	Baseline	Final Output		Ensemble Output	
		SD	MD	SD	MD
VGG19(BN)[22]	70.35	72.45	**73.40**	73.03	**73.98**
ResNet18 [9]	68.12	69.84	**70.52**	68.93	**69.82**
ResNet50 [9]	73.56	75.24	**76.11**	74.73	**75.59**

the learning rate by 0.1 at epoch 80 and 140. The learning rate of inner step gradient update ζ is kept the same with the learning rate of main model S. All the experiments are conducted with PyTorch.

4.2   CIFAR-100

We first conduct experiments on CIFAR-100 [13] for all 8 architectures mentioned in Sec 4.2. For all three different types of ResNets [9] and MobileNetv2 [21], we divide the network into four stages, and the output feature maps for each of the first three stages are taken as input to a bottleneck module, following self-distillation [30], to produce the intermediate output for that stage. For ResNeXt [24], WideResNet [29] and Shufflenetv2 [16], the network is divided into three stages and use the output feature maps from first two stages to produce intermediate output in the similar way. For VGG19 [22], we directly perform the average pooling and use a fully connected layer to produce the intermediate output for the feature maps before the second, third and fourth max-pooling operation. We implement the experiments on all other network architectures for self-distillation and use it as our baseline results. We also report the numbers in their original paper regarding three ResNet-type model and VGG. In Table 1, we denote self-distillation as SD and denote our MetaDistiller as MD.

The detailed experimental results are listed in Table 1. Besides the accuracy of final output, we also report the performance of the ensemble of all outputs. We can observe that our MetaDistiller consistently outperforms the baseline and self-distillation, indicating the generated soft teacher targets can better facilitate the training compared to the final output, and this generalize to different network architectures. The average improvement for three-stage model is slightly smaller due to less supervision is injected into the training of early stage.

4.3   ILSVRC2012

To verify the effectiveness of our method on the large-scale dataset, we also conduct experiments on ImageNet. We evaluate our method with three widely used networks including VGG19 [22], ResNet18 [9] and ResNet50 [9]. The results are presented in Table 2. We observe that our method consistently outperforms the

Table 3. Our method outperforms the traditional distillation methods We present the results on CIFAR-100 by testing our MetaDistiller with 5 two-stage distillation methods.

Teacher	Student	Baseline	KD [10]	FitNet [19]	AT [28]	DML [31]	MD
ResNet152 [9]	ResNet18 [9]	77.31	77.98	78.41	78.81	77.72	**79.05**
ResNet152 [9]	ResNet50 [9]	77.78	79.39	80.20	79.47	78.45	**80.85**

self-distillation baseline – our method improves the baseline by 0.95%, 0.68% and 0.87% on VGG19, ResNet18 and ResNet50, respectively. These results clearly demonstrate the effectiveness of our method. More notably, our method can significantly improve the original models without introducing extra computational cost. The enhanced VGG19, ResNet18 and ResNet50 outperform the original models by 3.05%, 2.40% and 2.03%, respectively. Since our method does not requires modifications on the original models during inference, MetaDistiller is very useful in boosting various prevalent CNN models for real-world applications.

4.4 Comparison with Traditional Distillation

We further compare our method with some widely-used traditional knowledge distillation methods, including KD [10], FitNet [19], AT [28] and DML [31]. All these methods follow the teacher-student pipeline. Concretely, they first train a ResNet of 152 layers (ResNet152) and then use the output from this large network as teacher to train a rather smaller network like ResNet18 and ResNet50. The experiments are conducted on CIFAR-100 dataset. We present the results in Table 3.

Our method can consistently improve the directly trained model by even larger margin compared to these five traditional distillation methods even though we do not have a large network to serve as a teacher. We outperform the runner-up by 0.44% and 0.78% on ResNet18 and ResNet50 respectively. This shows that our approach improves not only the training speed but also the accuracy in comparison with traditional distillation methods. We assume this is due to the output from large teacher network cannot adapt well to the small model. This also proves the necessity to solve the capacity discrepancy problem in knowledge distillation.

4.5 Ablation Study

There are major objectives for ablation study are two-folds. One is to show the improvement brought by our generated soft teacher targets is **much more** compared to self-distillation [30] upon the deeply-supervised network (DSN), which only uses the hard label to supervise the intermediate outputs; and the other is to show our generated soft target and hard label are complementary, i.e. they both contribute to the training. We denote DSN, self-distillation and our approach as 'Hard Only', 'Hard+Simple Soft' and 'Hard+Better Soft' respectively.

Table 4. Our method improves upon backbone by much larger margin compared to self-distillation. We test on CIFAR-100 for ResNet of three different depth

Model	Method	Final Output	Ensemble Output
ResNet18 [9]	DSN(Hard Only)	78.01	79.18
	SD(Hard+Simple Soft)	78.25	79.32
	MD(Hard+Better Soft)	**79.05**	**80.05**
ResNet50 [9]	DSN(Hard Only)	79.33	80.04
	SD(Hard+Simple Soft)	79.71	80.38
	MD(Hard+Better Soft)	**80.85**	**81.41**
ResNet101 [9]	DSN(Hard Only)	80.54	81.29
	SD(Hard+Simple Soft)	80.92	81.55
	MD(Hard+Better Soft)	**81.39**	**81.98**

We present all the results in Table 4. We conduct the experiments for three ResNet-type networks on CIFAR-100.

From Table 4, we can observe that both hard label and soft targets contribute to the improvement of accuracy. Yet, the simple soft teacher target usually only improves upon DSN by around 0.4%, which is relatively smaller compared to using our generated soft target labels. On the other hand, our method can improve the DSN by large margin, always more than 1%. This also indicates that our generated label is complementary to hard label given they can both contribute to boost the training, implying our generated soft teacher targets contain some helpful knowledge beyond the scope of ground-truth one-hot label. Notably, our generated soft targets bring more improvement than one-hot ground-truth label as well, which can be assumed that these hidden information is more helpful to the training of early stages compared to injecting more categorical information. These observations strongly demonstrate the effectiveness of the generated soft target, illustrating the benefits of using it to replace the final output as teacher label.

4.6   Visualization and Discussion

Why the generated soft teacher targets attain better results than using the final output? What 'dark knowledge' does the generated soft teacher target contain but final output does not? To answer this question, we visualize the normalized generated soft targets for each intermediate stage as well as the final output categorical distribution both from ResNet18 in Fig. 3. We can find that the generated soft targets are smoother while the final output often approaches the one-hot ground-truth label. Due to this fact, the final output cannot provide enough information for training that are not covered by the hard label. Moreover, we can see that the generated targets for shallower layers are even softer than that for deep layers. We argue this is because the feature maps in shallow layers

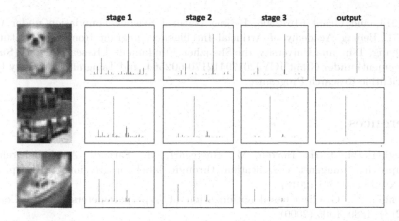

Fig. 3. Visualization. We exhibit the output from a well-optimized ResNet18 and the generated soft targets of three intermediate stages produced by its corresponding label generator. We can see that the final output is quite similar to the one-hot distribution and the soft targets differs from each other and the ones of shallower stages is softer.

often encode more generic information while the deep layers contain more specific information. Therefore, using the output from deepest layer often lead to the compatibility problem [4]. These observations imply that knowledge distillation is more like a kind of regularization, which avoids the model overfitting to the one-hot label on training data so that it can generalize well to unseen data.

The above phenomena are also observed by previous works [4,10,25] and they solve this by fine-tuning the temperature [10], early stopping the teacher training [4] and adding a regularization to the teacher to encourage it to be less strict [25]. Our method achieves the same to make the teacher label more tolerant yet through a fully automatic way. We do not need to carefully fine-tune the hyper-parameter like temperature given our generated teacher label can learn to adapt itself to be more compatible.

5   Conclusion

We proposed a new knowledge distillation method, namely MetaDistiller, to sidestep the time-consuming sequential training and improves the capability of the model at the same time. To achieve this, we generate more compatible soft teacher targets for the intermediate output layers by using a label generator to fuse the feature maps from deep layers of the main model in a top-down way. The experimental results on CIFAR and ImageNet prove the effectiveness of our method and the generalizability to different network architectures.

Acknowledgement. Cho-Jui Hsieh would like to acknowledge the support by NSF IIS-1719097, Intel, and Facebook. This work was also supported by the National Key Research and Development Program of China under Grant 2017YFA-0700802, National Natural Science Foundation of China under Grant 61822603, Grant U1813218, Grant

U1713214, and Grant 61672306, Beijing Natural Science Foundation under Grant L172051, Beijing Academy of Artificial Intelligence, a grant from the Institute for Guo Qiang, Tsinghua University, the Shenzhen Fundamental Research Fund (Subject Arrangement) under Grant JCYJ20170412170-602564, and Tsinghua University Initiative Scientific Research Program.

References

1. Bagherinezhad, H., Horton, M., Rastegari, M., Farhadi, A.: Label refinery: improving imagenet classification through label progression. arXiv preprint arXiv:1805.02641 (2018)
2. Bengio, Y.: Gradient-based optimization of hyperparameters. Neural Comput. **12**(8), 1889–1900 (2000)
3. Chen, P., Si, S., Li, Y., Chelba, C., Hsieh, C.J.: GroupReduce: block-wise low-rank approximation for neural language model shrinking. In: Advances in Neural Information Processing Systems, pp. 10988–10998 (2018)
4. Cho, J.H., Hariharan, B.: On the efficacy of knowledge distillation. In: Proceedings of the IEEE International Conference on Computer Vision, pp. 4794–4802 (2019)
5. Deng, J., Dong, W., Socher, R., Li, L.J., Li, K., Fei-Fei, L.: ImageNet: a large-scale hierarchical image database. In: 2009 IEEE Conference on Computer Vision and Pattern Recognition, pp. 248–255. IEEE (2009)
6. Finn, C., Abbeel, P., Levine, S.: Model-agnostic meta-learning for fast adaptation of deep networks. In: Proceedings of the 34th International Conference on Machine Learning, vol. 70, pp. 1126–1135. JMLR. org (2017)
7. Furlanello, T., Lipton, Z.C., Tschannen, M., Itti, L., Anandkumar, A.: Born again neural networks. arXiv preprint arXiv:1805.04770 (2018)
8. Han, S., Mao, H., Dally, W.J.: Deep compression: compressing deep neural networks with pruning, trained quantization and Huffman coding. arXiv preprint arXiv:1510.00149 (2015)
9. He, K., Zhang, X., Ren, S., Sun, J.: Deep residual learning for image recognition. In: Proceedings of the IEEE Conference on Computer Vision and Pattern Recognition, pp. 770–778 (2016)
10. Hinton, G., Vinyals, O., Dean, J.: Distilling the knowledge in a neural network. arXiv preprint arXiv:1503.02531 (2015)
11. Jang, Y., Lee, H., Hwang, S.J., Shin, J.: Learning what and where to transfer. arXiv preprint arXiv:1905.05901 (2019)
12. Kingma, D.P., Ba, J.: Adam: a method for stochastic optimization. arXiv preprint arXiv:1412.6980 (2014)
13. Krizhevsky, A., Hinton, G., et al.: Learning multiple layers of features from tiny images (2009)
14. Lee, C.Y., Xie, S., Gallagher, P., Zhang, Z., Tu, Z.: Deeply-supervised nets. In: Artificial Intelligence and Statistics, pp. 562–570 (2015)
15. Lin, J., Rao, Y., Lu, J., Zhou, J.: Runtime neural pruning. In: Advances in Neural Information Processing Systems, pp. 2181–2191 (2017)
16. Ma, N., Zhang, X., Zheng, H.T., Sun, J.: ShuffleNet V2: practical guidelines for efficient CNN architecture design. In: Proceedings of the European Conference on Computer Vision (ECCV), pp. 116–131 (2018)
17. Polino, A., Pascanu, R., Alistarh, D.: Model compression via distillation and quantization. arXiv preprint arXiv:1802.05668 (2018)

18. Rao, Y., Lu, J., Lin, J., Zhou, J.: Runtime network routing for efficient image classification. IEEE Trans. Pattern Anal. Mach. Intell. **41**(10), 2291–2304 (2018)
19. Romero, A., Ballas, N., Kahou, S.E., Chassang, A., Gatta, C., Bengio, Y.: FitNets: hints for thin deep nets. arXiv preprint arXiv:1412.6550 (2014)
20. Russakovsky, O., et al.: ImageNet large scale visual recognition challenge. Int. J. Comput. Vis. **115**(3), 211–252 (2015). https://doi.org/10.1007/s11263-015-0816-y
21. Sandler, M., Howard, A., Zhu, M., Zhmoginov, A., Chen, L.C.: MobileNetV2: inverted residuals and linear bottlenecks. In: Proceedings of the IEEE Conference on Computer Vision and Pattern Recognition, pp. 4510–4520 (2018)
22. Simonyan, K., Zisserman, A.: Very deep convolutional networks for large-scale image recognition. arXiv preprint arXiv:1409.1556 (2014)
23. Wang, Y.X., Girshick, R., Hebert, M., Hariharan, B.: Low-shot learning from imaginary data. In: Proceedings of the IEEE Conference on Computer Vision and Pattern Recognition, pp. 7278–7286 (2018)
24. Xie, S., Girshick, R., Dollár, P., Tu, Z., He, K.: Aggregated residual transformations for deep neural networks. In: Proceedings of the IEEE Conference on Computer Vision and Pattern Recognition, pp. 1492–1500 (2017)
25. Yang, C., Xie, L., Qiao, S., Yuille, A.: Knowledge distillation in generations: more tolerant teachers educate better students. arXiv preprint arXiv:1805.05551 (2018)
26. Yang, C., Xie, L., Su, C., Yuille, A.L.: Snapshot distillation: teacher-student optimization in one generation. In: Proceedings of the IEEE Conference on Computer Vision and Pattern Recognition, pp. 2859–2868 (2019)
27. Yu, X., Liu, T., Wang, X., Tao, D.: On compressing deep models by low rank and sparse decomposition. In: Proceedings of the IEEE Conference on Computer Vision and Pattern Recognition, pp. 7370–7379 (2017)
28. Zagoruyko, S., Komodakis, N.: Paying more attention to attention: improving the performance of convolutional neural networks via attention transfer. arXiv preprint arXiv:1612.03928 (2016)
29. Zagoruyko, S., Komodakis, N.: Wide residual networks. arXiv preprint arXiv:1605.07146 (2016)
30. Zhang, L., Song, J., Gao, A., Chen, J., Bao, C., Ma, K.: Be your own teacher: improve the performance of convolutional neural networks via self distillation. In: Proceedings of the IEEE International Conference on Computer Vision, pp. 3713–3722 (2019)
31. Zhang, Y., Xiang, T., Hospedales, T.M., Lu, H.: Deep mutual learning. In: Proceedings of the IEEE Conference on Computer Vision and Pattern Recognition, pp. 4320–4328 (2018)

Learning Monocular Visual Odometry via Self-Supervised Long-Term Modeling

Yuliang Zou[1]([✉])[iD], Pan Ji[2][iD], Quoc-Huy Tran[2][iD], Jia-Bin Huang[1][iD],
and Manmohan Chandraker[2,3][iD]

[1] Virginia Tech, Blacksburg, USA
ylzou@vt.edu
[2] NEC Labs America, Princeton, USA
[3] UCSD, San Diego, USA

Abstract. Monocular visual odometry (VO) suffers severely from error accumulation during frame-to-frame pose estimation. In this paper, we present a self-supervised learning method for VO with special consideration for consistency over longer sequences. To this end, we model the long-term dependency in pose prediction using a pose network that features a two-layer convolutional LSTM module. We train the networks with purely self-supervised losses, including a cycle consistency loss that mimics the loop closure module in geometric VO. Inspired by prior geometric systems, we allow the networks to see beyond a small temporal window during training, through a novel a loss that incorporates temporally distant (e.g., $O(100)$) frames. Given GPU memory constraints, we propose a stage-wise training mechanism, where the first stage operates in a local time window and the second stage refines the poses with a "global" loss given the first stage features. We demonstrate competitive results on several standard VO datasets, including KITTI and TUM RGB-D.

1 Introduction

Most existing VO systems are either *geometric* or *learning-based*. In this paper, we argue that a truly robust VO system should combine the best of both worlds (i.e., geometry and learning). In particular, we propose a self-supervised method to learn monocular VO with long-term modeling, where the training scheme is directly inspired by traditional geometric methods (see Fig. 1).

At the heart of the state-of-the-art VO systems [8–10, 24] is the incorporation of several long-studied geometric modules, including keypoint tracking, motion estimation, keyframe insertion, and bundle adjustment (BA) [38]. With all these modules, a key insight is to optimize the states (e.g., 6-DoF camera poses) over

Electronic supplementary material The online version of this chapter (https://doi.org/10.1007/978-3-030-58568-6_42) contains supplementary material, which is available to authorized users.

© Springer Nature Switzerland AG 2020
A. Vedaldi et al. (Eds.): ECCV 2020, LNCS 12359, pp. 710–727, 2020.
https://doi.org/10.1007/978-3-030-58568-6_42

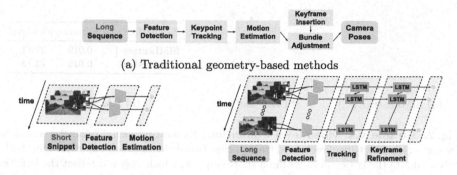

(a) Traditional geometry-based methods

(b) Existing self-supervised methods   (c) Our proposed method

Fig. 1. Learning monocular visual odometry with long-term modeling. Existing self-supervised methods only see *short* snippets during the *training* time, which makes it hard to learn to leverage temporal consistency over *long* sequences. Our method, in contrast, inspired by geometry-based visual odometry methods, combines the best of both the geometry and learning to aggregate long-term temporal information.

long-term observations such that the system suffers less from error accumulation [27]. While being robust in normal scenarios, monocular VO still suffers from the difficulty in initialization for slow motions [23], and the tracking tends to fail miserably in unconstrained environments with large texture-less regions, fast movements, or other adverse factors [48] such as rolling shutter effect [28,53] and unknown camera intrinsics [3,54].

In contrast, learning-based VO methods [44–47] have the potential of being more robust to the aforementioned challenges by harnessing the rich priors from data. However, training neural networks in a supervised way involves collecting large-scale, diverse datasets with ground truth annotations, which could be labor-extensive and time-consuming. Recently, self-supervised methods [13,20,26,42,50,52,55] have been proposed to tackle this task. Instead of supervising the networks with ground truth labels, the idea is to couple the depth and pose networks with photometric errors across adjacent frames and jointly train them in an end-to-end manner. Nonetheless, the performance of these methods still falls behind that of geometric methods [23] for general scenarios.

One of the potential reasons for their performance gap is that the pose networks do not exploit the temporal coherence over long sequences. During training, these networks receive *short* snippets (e.g., 3-frame or 5-frame) as input and predict the ego-motions that are optimized *locally* for the current snippet. When evaluating these methods in short snippets, they compare favorably even with the state-of-the-art geometric methods [23]. However, if we concatenate all the predictions to form the full trajectory, it is often the case that the learning-based methods generate much larger pose errors, as illustrated in Fig. 2.

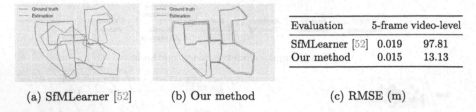

Evaluation	5-frame	video-level
SfMLearner [52]	0.019	97.81
Our method	0.015	13.13

(a) SfMLearner [52]   (b) Our method   (c) RMSE (m)

Fig. 2. 5-frame v.s. video-level evaluation. Evaluating visual odometry requires having a global picture of recovered trajectories. However, most self-supervised methods only evaluate the trajectories within a short snippet, which may not reflect the holistic performance. "5-frame" means that we evaluate the results using 5-frame snippets, and "video-level" means that we evaluate on the entire trajectory.

In this paper, we argue that learning VO requires explicit *long-term* modeling to infuse the insights from geometric methods [8–10,24]. To this end, we propose a novel self-supervised VO learning framework that draws inspiration from geometric modules. Specifically, we build our learning framework upon a depth network of an auto-encoder structure with skip-connections [13] and a pose network with a two-layer LSTM module [47]. In contrast to the supervised method by Xue et al. [47], our method incorporates extra depth information and uses a completely different training scheme, leading to a purely self-supervised learning framework. To mitigate error accumulation, we propose a cycle consistency constraint between the two-layer predictions, mimicking a mini *loop closure* module, which improves the pose consistency over the sequence. In order to model long-term dependency in VO, we propose a two-stage training strategy, which considers both short-term and long-term constraints. The proposed two training stages correspond to the *local* and *global* bundle adjustment modules in the geometric VO, allowing us to refine the poses within a large temporal range.

In summary, our contributions are:

- We propose a novel self-supervised VO learning framework that explicitly models long-term temporal dependency.
- We build connections between our method and key building blocks of geometric VO systems and demonstrate well-motivated designs.
- We evaluate the *full* pose trajectories by our method, against the state-of-the-art geometric and learning-based baselines, and achieve competitive results on standard VO datasets, including KITTI and TUM RGB-D.

To the best of our knowledge, our method is the first of the kind that is able to learn from "truly" long sequences (e.g., ~100 frames) in the training stage. Our experiments show that our proposed method gives rise to significant empirical benefits by explicitly considering long-term modeling.

2   Related Work

Geometric Methods. Visual odometry is a long-standing problem that estimates the ego-motion incrementally [25,27] using visual input. A conventional geometric VO system usually consists of the following components [27]: feature detection, feature matching (or tracking), motion estimation (e.g., triangulation [16]), and local optimization (e.g., bundle adjustment). A keyframe mechanism [18] is also adopted for improved robustness in motion estimation. Incorporated with a mapping system that reconstructs the 3D scene structures, a VO system turns to a system called Simultaneous Localization and Mapping (SLAM) [4]. The key to the robustness of the modern VO/SLAM systems [23,37] lies in their capability to extract reliable image measurements and optimize the states (e.g., 6-DoF camera poses) over a large number of frames. In this work, we leverage these geometric insights to design a robust learning-based VO system.

Fully-Supervised Methods. With the success of deep neural networks, end-to-end learning-based methods [44–47] have been proposed to tackle the visual odometry problem. These methods often rely on a supervised loss using the ground-truths to regress the 6-DoF camera relative pose from a pair of consecutive images. Recently, some methods [2,35,36,39,51] exploit CNNs to predict the scene depth and camera pose jointly, utilizing the geometric connection between the structure and the motion. This corresponds to learning Structure-from-Motion (SfM) in a supervised manner. Although the methods above achieve good performance, they require ground-truth annotations to train the networks. In contrast, our method is self-supervised, requiring nothing but the monocular video frames.

Self-Supervised Methods. To mitigate the requirement of data annotations, self-supervised methods [13,20,26,42,50,52] have been proposed to tackle the SfM task. The main supervisory signal of these methods comes from the photometric-consistency between corresponding pixels of neighboring frames. While they achieve good performance on single-view depth estimation, the performance of ego-motion estimation still lags behind the traditional SLAM/VO methods. Recently, Bian et al. [1] argue that the pose networks cannot provide full camera trajectories over long sequences due to the inconsistent scale of per-frame estimations and thus propose a geometry consistency constraint. However, their method only enforces the globally consistent trajectories by propagating the consistency on overlapping *short* snippets during training. In contrast, our method directly optimizes over *long* sequences via long-term modeling. Inspired by the keyframe mechanism in geometric methods, Sheng et al. [30] propose to jointly learn depth, ego-motion, and keyframe selection simultaneously in a self-supervised manner. Similarly, the training of this method considers only *short* snippets and thus is unable to model long-term dependency.

Sequential Modeling. Sequential modeling based on recurrent neural networks (RNNs) has been successfully applied to many applications, such as speech recognition [5], machine translation [14], and video prediction [33,40]. Aiming to esti-

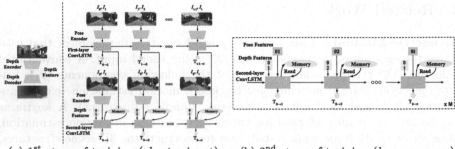

(a) 1st stage of training (*short* snippet)   (b) 2nd stage of training (*long* sequence)

Fig. 3. Overview. Our method adopts a stage-wise training strategy. (a) In the first stage of training, we jointly train all the components: the depth encoder, the depth decoder, the pose encoder, the first and second layer of ConvLSTM (Sect. 3.2); (b) In the second stage of training, we pre-extract features as input and fine-tune the second layer of the ConvLSTM module only (Sect. 3.3).

mate the full trajectory over a long sequence of frames, VO can be naturally formulated as a sequential learning problem and thus modeled with RNNs [42, 44–46]. Recently, Xue et al. [47] propose to use a two-layer LSTM network for pose estimation, where the first layer estimates the relative motion between consecutive frames, and the second layer estimates global absolute poses.

Despite using a similar pose network, our method differs from Xue et al. [47] in being self-supervised v.s. full-supervised and the associated training strategies. First of all, our method further incorporates depth information, while the method in [47] relies only on pose features. Apart from the photometric discrepancy as to the supervisory signal, we enforce a cycle consistency between the predictions from the two-layer LSTM modules, which serves as a mini "loop closure" module that mimics the geometry VO system. More importantly, we decouple our network training into two stages, allowing our method to optimize over *long* sequences (more than 90 frames) during training, whereas the method in [47] only trains with 11-frame snippets. To our knowledge, this is the *first* deep learning approach for visual odometry that takes *long* sequences as input in the training stage.

3   Method

Figure 3 provides a high-level overview of the proposed monocular VO system. Our system has two major components: a depth network and a pose network.[1] The single-image depth network employs an encoder-decoder structure with skip-connections [13]. The pose network consists of a FlowNet backbone [7], a two-layer LSTM module [47], and two pose prediction heads (with one after

---

[1] Since accurate pose prediction is the primary focus of this paper, we name our method as VO instead of SfM or SLAM.

each of the LSTM layers). In the two-layer recurrent architecture, the first-layer focuses on predicting consecutive frame motions, while the second-layer refines the estimations from the first-layer [47].

3.1 Background

We formulate the monocular visual odometry task as a view synthesis problem, by training the networks to predict a target image from the source image with the estimated depth and camera pose. Such a system typically consists of two components: a depth network which takes a single RGB image as input to predict the depth map, and a pose network which takes a concatenation of two consecutive frames as input to estimate the 6-DoF ego-motion.

Given two input images I_t and I_{t+1}, the estimated depth map $\hat{D}_t$ and camera pose $\hat{T}_{t \to t+1}$, we can then compute the per-pixel correspondence between the two input images. Assume a known camera intrinsic matrix K, and let p_t represent the 2D homogeneous coordinate of a pixel in I_t. We can find the corresponding point of p_t in I_{t+1} following the equation [52]:

$$p_{t+1} \sim K\hat{T}_{t \to t+1}\hat{D}_t(p_t)K^{-1}p_t . \tag{1}$$

Appearance Loss. For a self-supervised visual odometry system, the primary supervision comes from the appearance dissimilarity between the synthesis image and the target image. To effectively handle occlusion, we use three consecutive frames to compute the per-pixel minimum photometric reprojection loss [13], i.e.,

$$L_A = \frac{1}{N-2} \sum_{t=1}^{N-2} \min_{t' \in \{t-1, t+1\}} \rho(I_t, \hat{I}_{t' \to t}) , \tag{2}$$

where ρ is a weighted combination of the L2 loss and the Structured SIMilarity (SSIM) loss, $\hat{I}_{t' \to t}$ denotes the frame synthesized from $I_{t'}$ using Eq. (1). To handle static pixels, we adopt the auto-masking mechanism following Godard et al. [13].

Smoothness Loss. Since the appearance loss cannot provide meaningful supervision for texture-less or homogeneous regions of the scene, a smoothness prior of disparity is incorporated. We here use the edge-aware smoothness loss L_S [41].

Remark 1. The appearance loss in Eq. (2) corresponds to a local photometric bundle adjustment objective, which is also commonly used in the geometric direct VO/SLAM systems [8,9,43].

3.2 Cycle Consistency Within Memory-Aided Sequential Modeling

With the above setting, current state-of-the-art self-supervised methods estimate the ego-motion within a *local* range, discarding the sequential dependence and dynamics in the *long* sequences. Such information, however, is essential for a pose network to recover the entire trajectory in a consistent manner. We thus adopt a recurrent structure of our pose network to utilize the temporal information.

Sequential Modeling. To learn to utilize the temporal information, we adopt the recurrent network structure with a convolution LSTM (ConvLSTM) module [31]. Previously, the pose network takes the concatenation of two frames and outputs the 6-DoF camera pose directly. After incorporating the ConvLSTM module, the pose network also takes the previous estimation information into account when predicting the output. Formally, we have

$$F_t = \text{PEnc}(I_t, I_{t-1}) \,, \tag{3}$$

$$O_t, H_t = \text{ConvLSTM}(F_t, H_{t-1}) \,, \tag{4}$$

$$\hat{T}_{t-1 \to t} = g_1(O_t) \,, \tag{5}$$

where $\text{PEnc}(\cdot)$ is the pose encoder, O_t, H_t denotes the output and hidden state of ConvLSTM at time t, $g_1(\cdot)$ is a linear layer to predict the 6-DoF motion $\hat{T}_{t-1 \to t}$. By doing this, the network implicitly learns to aggregate temporal information and learns the motion pattern.

Memory Buffer and Refinement. In the sequential modeling setting above, the pose network estimates the relative pose for every two consecutive frames. However, the motions between consecutive frames are often tiny, which results in difficulties in extracting good features for relative pose estimation. Thus, predicting the camera pose from a non-adjacent "anchor" frame to the current frame could be a better option. Note that many traditional SLAM systems [23, 24] adopt a keyframe mechanism and always compute camera poses from the current frame to the most recent keyframe.

Inspired by the keyframe mechanism, we incorporate the second-layer ConvLSTM and adopt the memory module proposed by Xue et al. [47]. After each step in the first-layer ConvLSTM, we store the hidden state tensor in a memory buffer, whose length is set to the length of the input snippet. When we read out from the memory buffer, we compute the weighted average of all the memory slots in the memory buffer as in [47].

We also compute the depth and pose features for the first frame and the current frame as additional input to the second-layer ConvLSTM. This can be formally written as

$$E_t = \text{DEnc}(I_t) \,, \tag{6}$$

$$F_{t,abs} = \text{PEnc}(I_0, I_t) \,, \tag{7}$$

$$O_{t,abs}, H_{t,abs} = \text{ConvLSTM}(F_{t,abs}, E_0, E_t, M_t, H_{t-1,abs}) \,, \tag{8}$$

$$\hat{T}_{0 \to t} = g_2(O_{t,abs}) \,, \tag{9}$$

where $\text{DEnc}(\cdot)$ is the depth encoder, M_t is the read-out memory, $O_{t,abs}$, $H_{t,abs}$ denote the output and hidden state from the second-layer at time t, and $g_2(\cdot)$ is another linear layer predicting the absolute pose in the current snippet.

Remark 2. The ConvLSTMs explicitly model the sequential nature of the VO problem and meanwhile facilitate the implementation of a keyframe mechanism. Compared to the memory module by Xue et al. [47], which only considers pose features, our memory module accommodates both depth and pose features. As verified in our experiments [2], incorporating the additional depth information in memory improves the overall performance.

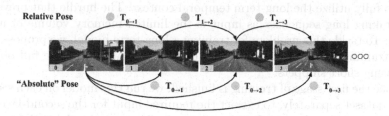

Fig. 4. Cycle-consistency over two-layer poses. In our model, the first layer ConvLSTM estimates the relative pose between consecutive frames, and the second layer ConvLSTM predicts the "absolute" pose within the current snippet. By exploiting the transitivity of camera poses, we incorporate a cycle-consistency constraint.

Cycle Consistency Over Two-Layer Poses. To train the second-layer ConvLSTM, we utilize the photometric error between the first frame and the other frames of the input snippet, i.e.,

$$L_{A,abs} = \frac{1}{N-1} \sum_{t=1}^{N-1} \rho(I_0, \hat{I}_{t \to 0}) , \tag{10}$$

where N is the number of frames for the input snippet, which is set to 7 in our model.

Also, according to the transitivity of the camera transformation, we have an additional constraint to ensure the consistency between the first and second layer ConvLSTM (as shown in Fig. 4), i.e.,

$$L_P = \frac{1}{N-1} \sum_{t=1}^{N-1} ||\hat{T}_{0 \to t} - \hat{T}_{t-1 \to t}\hat{T}_{0 \to t-1}||_2^2 . \tag{11}$$

Thus, the overall objective is

$$L_{\text{full}} = L_A + \lambda_1 L_S + L_{A,abs} + \lambda_2 L_P , \tag{12}$$

where λ_1, λ_2 are the hyper-parameters to balance the scale of different terms, which are empirically set to 0.001.

Remark 3. The loss in Eq. (11) can be thought of as a mini "loop closure" module that enforces the cycle-consistency between the outputs of two ConvLSTM layers. Note that our method is also compatible with the existing full loop closure techniques [19], which we will consider in the future work.

---

[2] Table 1 in the supplementary material.

3.3 Long-Range Constraints via Stage-Wise Training

Although we adopt a recurrent network structure to aggregate temporal information for better performance, the network has never seen *long* sequences but only *short* snippets during the training time. Thus, the network may not learn how to fully utilize the long-term temporal context. The hurdle that prevents us from taking long sequences as input is the limited memory volume of modern GPUs. To tackle this problem for training a long-term model, we propose a two-stage training strategy. We first train our whole model with the full objective L_{full} using short snippets.

Once the first stage of training is finished, we run this model on each sequence in the dataset separately, to extract the required input for the second-layer ConvLSTM and store them. After that, we only fine-tune the lightweight second-layer ConvLSTM, without the heavy feature extraction and depth networks, which saves us a lot of memories. By doing this, we can now feed long sequences into the network during the training time, allowing the network to better learn how to utilize the temporal context. Since only the second-layer ConvLSTM is optimized, our loss for the second stage of training is

$$L_{long} = \frac{1}{M} \sum_{m=0}^{M-1} \frac{1}{N-1} \sum_{t=m(N-1)+1}^{m(N-1)+N-1} \rho(I_{m(N-1)}, \hat{I}_{t \to m(N-1)}), \quad (13)$$

where N is the number of frames of each snippet, which is set to 7; M is the number of snippets in the input sequence, which is set to 16. Note that consecutive snippets have one frame in common, and thus the total number of frames in the input sequence is 97. The synthesized image $\hat{I}_{t \to m(N-1)}$ is a function of depth and pose, where pose encodes long-range constraints through hidden states of ConvLSTMs, yielding an effective window of 97 frames.

Remark 4. The second training stage can be viewed as a motion-only bundle adjustment module [23] that considers long-term modeling.

4   Experimental Results

4.1 Settings

Datasets. To evaluate our method, we conduct the main experiments on the KITTI dataset [11,12], which consists of urban and highway driving sequences for road scene understanding [6,32]. The odometry split of KITTI is a widely used benchmark for odometry/SLAM evaluation. It contains 22 sequences, among which Sequence 00-10 have ground truth trajectory labels, and the annotations of the remaining sequences are not publicly available. Following Zhou et al.[52], we use Sequence 00-08 as our training set and validate the models on Sequence 09 and 10. Besides, we select 18 more sequences from KITTI raw data, which have no overlaps with the odometry split, for further evaluation. Since the ground truth trajectories of Sequence 11-21 are not available, we run ORB-SLAM2 (stereo version) to get predictions as (pseudo) ground truth for evaluation. In addition to these outdoor scenes, we also train and evaluate our model on the TUM RGB-D dataset [34]. This dataset is collected by hand-held cameras in indoor environments with challenging conditions. We use the same train/test split as in Xue et al. [47].

Evaluation Metrics. For the KITTI dataset, we adopt the absolute trajectory RMSE and relative translation/rotation errors for all possible subsequences of length (100, 200, ..., 800 m). For the TUM RGB-D dataset, we use the translational RMSE as our evaluation metrics. For self-supervised monocular methods, since the absolute scale is unknown, we align the trajectory globally using the evo toolbox [15].

Implementation Details. We use ImageNet pretrained ResNet-18 as our depth encoder. Our depth decoder structure is the same as Godard et al. [13]. For the pose encoder, we take the encoder of FlowNet-S structure until the last layer, which is pre-trained on FlyingChairs [7] for optical flow estimation. We implement our system using the publicly available PyTorch framework and conduct all our experiments with a single TitanXP GPU. For both stages, we train the network with the Adam optimizer [17] for 20 epochs. The learning rate is set to 5e−5 for the first 15 epochs and drops to 5e−6 for the remaining epochs. The input size is 640×192, and the batch size is set to 2. In the first stage of training, the number of frames is set to 7, while the number of frames is set to 97 in the second stage. Note that the long-term optimization only happens in the training time. At the test time, our model runs at 14.3 frames per second.

4.2 Results

Ablation Study. To validate our design choice, we perform an ablation study on Sequence 09 and 10 of the KITTI Odometry dataset. We consider the following variants. 1) Baseline: the pose network takes as input the concatenation of two consecutive frames to generate pose estimation; 2) One-layer ConvLSTM: we incorporate a one-layer ConvLSTM for the pose network; 3) Two-layer ConvLSTM: we use the full model, but only conduct the first stage of training; 4) Two-layer ConvLSTM + Two-stage training: our final model with the two-stage training strategy.

As shown in Table 1, the performance gradually improves as we add more components. Specifically, adding a recurrent module improves the overall performance over the baseline; adding the second layer LSTM leads to further improvement, which validates the effectiveness of the second layer in the self-supervised learning setting; applying our second stage long-term training again boosts the performance, achieving a new state-of-the-art for self-supervised methods. We show a visual comparison in Fig. 5.

Table 1. Ablation study. We evaluate different variants of the proposed method on sequences 09 and 10 of the KITTI Odometry dataset [12]. The best performance is in bold and the second best is underlined.

Method	Seq. 09			Seq. 10		
	RMSE (m)	Rel. trans. (%)	Rel. rot. (deg/m)	RMSE (m)	Rel. trans. (%)	Rel. rot. (deg/m)
Baseline	22.71	7.55	0.028	17.87	10.43	0.046
One-layer ConvLSTM	23.45	5.59	0.016	11.93	7.23	0.023
Two-layer ConvLSTM	9.77	4.23	0.013	12.68	6.02	0.023
Two-layer ConvLSTM + Two-stage training	11.30	3.49	0.010	11.80	5.81	0.018

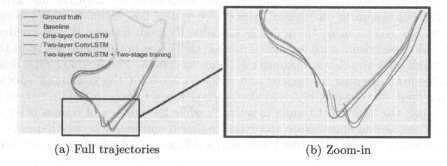

(a) Full trajectories   (b) Zoom-in

Fig. 5. Visual comparison on Sequence 09. We compare different variants of our method to validate the design choice. As we can see, adding each of the components gradually improves the overall performance. Best viewed in color. (Color figure online)

Comparison with the State-of-the-art Methods. For comparison, we select several state-of-the-art methods, including the monocular version of ORB-SLAM2 [24] (denoted as ORB-SLAM2-M) (with and without loop closure optimization), several supervised learning methods [36,44–47], and some other self-supervised methods [1,13,20–22,26,29,42,49,50,52]. Note that all supervised methods are trained on Sequence 00, 02, 08, 09 of the KITTI Odometry dataset [12], except DeepV2D [36], which is trained on the Eigen split of KITTI raw dataset [11]. As we can see in Table 2, our final model outperforms other self-supervised methods by a significant margin. In particular, our method outperforms the recent proposed SC-SfMLearner [1], which aims to reconstruct the scale-consistent camera trajectory. This indicates that the explicit long-term modeling used in our approach is more effective than propagating the geometric constraint among overlapping short snippets in SC-SfMLearner [1]. Our model also compares favorably with the geometric method and outperforms all supervised methods except Xue et al. [47] on Sequence 10.

Table 2. Comparison with the state-of-the-art. The results of ORB-SLAM2-M methods are the medians of 5 runs. '-' means the results are not available from that paper. For DeepV2D [36], SfMLearner [52], GeoNet [49], CC [26], DeepMatchVO [29], and MonoDepth2 [13], we take the pre-trained models and run on the sequences to get the results. The best performance of each block is in bold, and the second best is underlined.

Method		Seq. 09			Seq. 10		
		RMSE (m)	Rel. trans. (%)	Rel. rot. (deg/m)	RMSE (m)	Rel. trans. (%)	Rel. rot. (deg/m)
Geo.	ORB-SLAM2-M (w/o LC) [24]	<u>44.10</u>	<u>9.67</u>	0.003	6.43	4.04	0.003
	ORB-SLAM2-M [24]	8.84	3.22	<u>0.004</u>	<u>8.51</u>	<u>4.25</u>	0.003
Sup.	DeepVO [44]	-	-	-	-	8.11	0.088
	ESP-VO [45]	-	-	-	-	9.77	0.102
	GFS-VO [46]	-	-	-	-	<u>6.32</u>	<u>0.023</u>
	GFS-VO-RNN [46]	-	-	-	-	7.44	0.032
	BeyondTracking [47]	-	-	-	-	3.94	0.017
	DeepV2D [36]	79.06	8.71	0.037	48.49	12.81	0.083
Self-Sup.	SfMLearner [52]	24.31	8.28	0.031	20.87	12.20	<u>0.030</u>
	GeoNet [49]	158.45	28.72	0.098	43.04	23.90	0.090
	Depth-VO-Feat [50]	-	11.93	0.039	-	12.45	0.035
	vid2depth [22]	-	-	-	-	21.54	0.125
	UnDeepVO [20]	-	7.01	0.036	-	10.63	0.046
	Wang et al. [42]	-	9.88	0.034	-	12.24	0.052
	CC [26]	29.00	<u>6.92</u>	<u>0.018</u>	<u>13.77</u>	7.97	0.031
	DeepMatchVO [29]	<u>27.08</u>	9.91	0.038	24.44	12.18	0.059
	PoseGraph [21]	-	8.10	0.028	-	12.90	0.032
	MonoDepth2 [13]	55.47	11.47	0.032	20.46	<u>7.73</u>	0.034
	SC-SfMLearer [1]	-	11.2	0.034	-	10.1	0.050
	Ours	11.30	3.49	0.010	11.80	5.81	0.018

Results on Additional KITTI Sequences. From the raw data of the KITTI dataset, we select 18 short sequences that have no overlaps with either the training or test split of the KITTI Odometry dataset.[3] We then apply the same pre-trained models on these test sequences. As we can see in Table 3, our method outperforms other learning-based methods and even compares favorably with ORB-SLAM2-M methods in terms of RMSE and relative translation error.

Results on KITTI Test Sequences. Since the ground truth trajectories of Sequence 11-21 on the KITTI Odometry dataset are not available, we cannot directly recover the global scale using similarity transformation. Thus, we choose to run the stereo version of ORB-SLAM2 (denoted as ORB-SLAM2-S), which is one of the state-of-the-art methods on these sequences. To compare with other methods, we treat the estimations from ORB-SLAM2-S as pseudo ground truth. As we can see in Table 4, our method achieves state-of-the-art performance among the learning-based methods and even compares favorably with ORB-SLAM2-M in terms of the relative translation error. We visualize trajectories

---

[3] The sequence names are available in the supplementary material.

Table 3. Average results on 18 additional KITTI sequences. The results of ORB-SLAM2-M methods are the medians of 5 times. The best performance of each block is in **bold**, and the second best is <u>underlined</u>.

Method		RMSE (m)	Rel. trans (%)	Rel. rot. (deg/m)
Geo.	ORB-SLAM2-M (w/o LC) [24]	**7.17**	**9.41**	**0.008**
	ORB-SLAM2-M [24]	<u>8.12</u>	<u>10.64</u>	**0.008**
Sup.	DeepV2D [36]	10.94	11.81	0.028
Self-Sup.	SfMLearner [52]	10.79	13.82	0.041
	GeoNet [49]	14.41	18.99	0.076
	CC [26]	<u>7.51</u>	<u>10.49</u>	<u>0.024</u>
	DeepMatchVO [29]	8.53	12.76	0.033
	MonoDepth2 [13]	<u>7.51</u>	11.99	0.028
	Ours	**6.47**	**9.99**	**0.019**

Table 4. Results on KITTI Odometry official test split. Since ground truth trajectories are not publicly available, we use estimations from the stereo version of ORB-SLAM2 as pseudo ground truth. The best performance of each block is in **bold**, and the second best is <u>underlined</u>.

Method		RMSE (m)	Rel. trans (%)	Rel. rot. (deg/m)
Geo.	ORB-SLAM2-M (w/o LC) [24]	<u>81.20</u>	<u>19.60</u>	<u>0.009</u>
	ORB-SLAM2-M [24]	**44.09**	**12.96**	**0.007**
Sup.	DeepV2D [36]	221.33	24.61	0.041
Self-Sup.	SfMLearner [52]	75.00	26.54	0.045
	GeoNet [49]	94.98	29.11	0.062
	CC [26]	55.44	16.65	0.032
	DeepMatchVO [29]	95.79	17.31	0.038
	MonoDepth2 [13]	99.36	<u>12.28</u>	<u>0.031</u>
	Ours	<u>71.63</u>	**7.28**	**0.014**

of four sequences in Fig. 6. As we can see, our method better aligns with the reference trajectories from ORB-SLAM2-S. We also submit the globally aligned results to the KITTI evaluation server and get similar numbers, which indicates

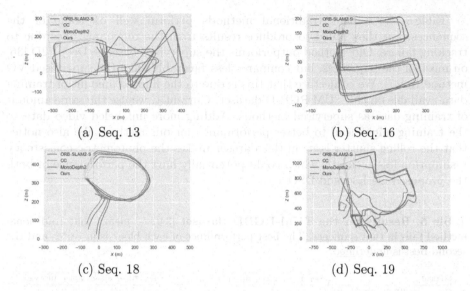

(a) Seq. 13   (b) Seq. 16

(c) Seq. 18   (d) Seq. 19

Fig. 6. Visual comparison on the KITTI Odometry test set. We show the trajectories of ORB-SLAM2-S, CC [26], MonoDepth2 [13] and our method. Our method aligns best with the reference ORB-SLAM2-S trajectories. Best viewed in color. (Color figure online)

Fig. 7. Qualitative Results on the TUM RGB-D dataset. We overlay the first frame of each sequence with the trajectory. Best viewed in color. (Color figure online)

using ORB-SLAM2-S as pseudo ground truth for evaluation is reasonable. Please refer to the supplementary material for more details (Fig. 7).

Results on the TUM RGB-D Dataset. The TUM RGB-D dataset was created to evaluate the performance of RGB-D SLAM and is thus very challenging for monocular methods. To test our model in indoor environments, we instead compare our model with several strong baselines. For traditional methods, we choose the monocular version of ORB-SLAM2 (denoted as ORB-SLAM2-M) and DSO [8]. For learning-based methods, we choose the BeyondTracking method from Xue et al. [47] and the recent DeepV2D [36]. For DeepV2D, since it is trained on another indoor dataset, in case it cannot generalize its scale to TUM RGB-D, we provide two options: with and without global scale alignment.

Table 5 shows that traditional methods perform well on some of the sequences, but they failed to produce results from the remaining ones due to tracking failure. Our method outperforms the supervised baseline DeepV2D [36] on most of the sequences, but compares less favorably than the supervised VO method in [47]. We conjecture that this is due to the limited amount of training data available on the TUM RGB-D dataset. Currently, we use the same amount of training data as supervised methods. Adding more unlabeled video data to the training might lead to better performance for our method. We also notice that the rolling shutter issue in this dataset makes the photometric consistency assumption less accurate, which could potentially hurt the performance of both the proposed method and DSO.

Table 5. Results on the TUM-RGBD dataset [34]. '-' means that traditional method fails in that sequence. The best performance of each block is in **bold**, and the second best is underlined.

Method		Seq. 1	Seq. 2	Seq. 3	Seq. 4	Seq. 5	Seq. 6	Seq. 7	Seq. 8	Seq. 9	Seq. 10	Avg
Geo.	ORB-SLAM2-M [24]	0.041	0.184	-	-	-	-	-	0.057	-	0.018	-
	DSO [8]	-	0.197	-	0.737	-	0.082	-	0.093	0.543	0.040	-
Sup.	BeyondTracking [47]	0.153	0.208	0.056	0.070	0.172	0.015	0.123	0.007	0.035	0.042	0.088
	DeepV2D [36]	0.232	0.651	0.186	0.167	0.171	0.029	0.435	0.106	0.085	0.082	0.214
	DeepV2D (aligned) [36]	0.087	0.300	0.114	0.106	0.181	0.013	0.380	0.110	0.094	0.098	0.148
Self-Sup.	Ours	0.192	0.190	0.083	0.122	0.177	0.016	0.219	0.102	0.179	0.107	0.139

Discussions. Our learning framework is motivated by geometric VO methods. The FlowNet backbone mimics the *tracking module* to extract pair-wise image features, and the LSTMs model the *sequential nature* of the VO problem. The design of the two-layer LSTM module also resembles the *keyframe mechanism* of geometric VO in the sense that the second LSTM predicts the motion between a keyframe and a non-keyframe, refining the initial consecutive estimations from the first LSTM. The cycle consistency constraint between the two-layer LSTM estimations serves as a mini *loop closure* to enforce the transitivity consistency of poses. The second stage of training allows our network to explicitly optimize over *long* sequences, which resembles the *motion-only bundle adjustment module*. We combine the best of both geometry and learning by building a self-supervised VO framework whose components (network, loss, training scheme) are fully inspired by the well-studied geometric modules. As verified in our experiments, these geometry inspired designs lead to significantly better results than the existing self-supervised baselines.

Limitations. Although the proposed system achieves a good camera pose estimation performance in terms of translation error, the improvement on the rotation prediction is not as substantial as the translation. We conjecture that the large rotation error may be due to the bias within the training data. Specifically, for driving scenarios, the translational motions occur much more frequently than the rotational ones. Training on more diverse video sequences or synthetic data could potentially alleviate the inherent bias in the existing datasets. Also, we observe that our system fails under the over-exposure scenarios since our method still relies on visual input to extract information.

5   Conclusions

In this work, we learn a monocular visual odometry system in a self-supervised manner to mimic critical modules in traditional geometric methods. We first adopt a two-layer convolutional LSTM module to model the long-term dependency in the pose estimation. To allow the network to see beyond *short* snippets (e.g., 3 or 5 frames) during the training time, we propose a stage-wise training strategy. Combining the recurrent architecture and the proposed decoupled training scheme, our system achieves state-of-the-art performance among self-supervised methods. In the current form, we do not have a mechanism to detect loops and perform full loop closure. In the future, we plan to study how to incorporate it into our learning framework.

Acknowledgment. This work was part of Y. Zou's internship at NEC Labs America, in San Jose. Y. Zou and J.-B. Huang were also supported in part by NSF under Grant No. (#1755785).

References

1. Bian, J.W., et al.: Unsupervised scale-consistent depth and ego-motion learning from monocular video. In: NeurIPS (2019)
2. Bloesch, M., Czarnowski, J., Clark, R., Leutenegger, S., Davison, A.J.: CodesLAM– learning a compact, optimisable representation for dense visual slam. In: CVPR (2018)
3. Bogdan, O., Eckstein, V., Rameau, F., Bazin, J.C.: DeepCalib: a deep learning approach for automatic intrinsic calibration of wide field-of-view cameras. In: CVMP (2018)
4. Cadena, C., et al.: Past, present, and future of simultaneous localization and mapping: toward the robust-perception age. IEEE Trans. Robot. **32**(6), 1309–1332 (2016)
5. Chorowski, J., Bahdanau, D., Cho, K., Bengio, Y.: End-to-end continuous speech recognition using attention-based recurrent NN: first results. arXiv preprint arXiv:1412.1602 (2014)
6. Dhiman, V., Tran, Q.H., Corso, J.J., Chandraker, M.: A continuous occlusion model for road scene understanding. In: CVPR (2016)
7. Dosovitskiy, A., et al.: FlowNet: learning optical flow with convolutional networks. In: ICCV (2015)
8. Engel, J., Koltun, V., Cremers, D.: Direct sparse odometry. TPAMI **40**(3), 611–625 (2017)
9. Engel, J., Schöps, T., Cremers, D.: LSD-SLAM: large-scale direct monocular SLAM. In: Fleet, D., Pajdla, T., Schiele, B., Tuytelaars, T. (eds.) ECCV 2014. LNCS, vol. 8690, pp. 834–849. Springer, Cham (2014). https://doi.org/10.1007/978-3-319-10605-2_54

10. Forster, C., Pizzoli, M., Scaramuzza, D.: SVO: fast semi-direct monocular visual odometry. In: ICRA (2014)
11. Geiger, A., Lenz, P., Stiller, C., Urtasun, R.: Vision meets robotics: the kitti dataset. IJRR **32**(11), 1231–1237 (2013)
12. Geiger, A., Lenz, P., Urtasun, R.: Are we ready for autonomous driving? The kitti vision benchmark suite. In: CVPR (2012)
13. Godard, C., Mac Aodha, O., Firman, M., Brostow, G.: Digging into self-supervised monocular depth estimation. In: ICCV (2019)
14. Graves, A., Jaitly, N.: Towards end-to-end speech recognition with recurrent neural networks. In: ICML (2014)
15. Grupp, M.: evo: Python package for the evaluation of odometry and SLAM (2017). https://github.com/MichaelGrupp/evo
16. Hartley, R.I., Sturm, P.: Triangulation. CVIU **68**(2), 146–157 (1997)
17. Kingma, D., Ba, J.: Adam: a method for stochastic optimization. In: ICLR (2014)
18. Klein, G., Murray, D.: Parallel tracking and mapping for small AR workspaces. In: ISMAR (2007)
19. Kümmerle, R., Grisetti, G., Strasdat, H., Konolige, K., Burgard, W.: g²o: a general framework for graph optimization. In: ICRA (2011)
20. Li, R., Wang, S., Long, Z., Gu, D.: UnDeepVO: monocular visual odometry through unsupervised deep learning. In: ICRA (2018)
21. Li, Y., Ushiku, Y., Harada, T.: Pose graph optimization for unsupervised monocular visual odometry. In: ICRA (2019)
22. Mahjourian, R., Wicke, M., Angelova, A.: Unsupervised learning of depth and ego-motion from monocular video using 3D geometric constraints. In: CVPR (2018)
23. Mur-Artal, R., Montiel, J.M.M., Tardos, J.D.: ORB-SLAM: a versatile and accurate monocular SLAM system. IEEE Trans. Robot. **31**(5), 1147–1163 (2015)
24. Mur-Artal, R., Tardós, J.D.: ORB-SLAM2: an open-source SLAM system for monocular, stereo and RGB-D cameras. IEEE Trans. Robot. **33**(5), 1255–1262 (2017)
25. Nistér, D., Naroditsky, O., Bergen, J.: Visual odometry. In: CVPR (2004)
26. Ranjan, A., et al.: Competitive collaboration: joint unsupervised learning of depth, camera motion, optical flow and motion segmentation. In: CVPR (2019)
27. Scaramuzza, D., Fraundorfer, F.: Visual odometry [Tutorial]. IEEE Robot. Autom. Mag. **18**(4), 80–92 (2011)
28. Schubert, D., Demmel, N., Usenko, V., Stuckler, J., Cremers, D.: Direct sparse odometry with rolling shutter. In: ECCV (2018)
29. Shen, T., et al.: Beyond photometric loss for self-supervised ego-motion estimation. In: ICRA (2019)
30. Sheng, L., Xu, D., Ouyang, W., Wang, X.: Unsupervised collaborative learning of keyframe detection and visual odometry towards monocular deep SLAM. In: ICCV (2019)
31. Shi, X., Chen, Z., Wang, H., Yeung, D.Y., Wong, W.K., Woo, W.C.: Convolutional LSTM network: a machine learning approach for precipitation nowcasting. In: NeurIPS (2015)
32. Song, S., Chandraker, M.: Robust scale estimation in real-time monocular SFM for autonomous driving. In: CVPR (2014)
33. Srivastava, N., Mansimov, E., Salakhudinov, R.: Unsupervised learning of video representations using LSTMs. In: ICML (2015)
34. Sturm, J., Engelhard, N., Endres, F., Burgard, W., Cremers, D.: A benchmark for the evaluation of RGB-D SLAM systems. In: IROS (2012)

35. Tang, C., Tan, P.: BA-Net: dense bundle adjustment network. In: ICLR (2019)
36. Teed, Z., Deng, J.: DeepV2D: video to depth with differentiable structure from motion. In: ICLR (2020)
37. Tiwari, L., Ji, P., Tran, Q.H., Zhuang, B., Anand, S., Chandraker, M.: Pseudo RGB-D for self-improving monocular SLAM and depth prediction. In: ECCV (2020)
38. Triggs, B., McLauchlan, P.F., Hartley, R.I., Fitzgibbon, A.W.: Bundle adjustment– a modern synthesis. In: International Workshop on Vision Algorithms (1999)
39. Ummenhofer, B., et al.: Demon: depth and motion network for learning monocular stereo. In: CVPR (2017)
40. Villegas, R., Yang, J., Zou, Y., Sohn, S., Lin, X., Lee, H.: Learning to generate long-term future via hierarchical prediction. In: ICML (2017)
41. Wang, C., Miguel Buenaposada, J., Zhu, R., Lucey, S.: Learning depth from monocular videos using direct methods. In: CVPR (2018)
42. Wang, R., Pizer, S.M., Frahm, J.M.: Recurrent neural network for (un-) supervised learning of monocular video visual odometry and depth. In: CVPR (2019)
43. Wang, R., Schworer, M., Cremers, D.: Stereo DSO: large-scale direct sparse visual odometry with stereo cameras. In: ICCV (2017)
44. Wang, S., Clark, R., Wen, H., Trigoni, N.: DeepVO: towards end-to-end visual odometry with deep recurrent convolutional neural networks. In: ICRA (2017)
45. Wang, S., Clark, R., Wen, H., Trigoni, N.: End-to-end, sequence-to-sequence probabilistic visual odometry through deep neural networks. IJRR **37**(4–5), 513–542 (2018)
46. Xue, F., Wang, Q., Wang, X., Dong, W., Wang, J., Zha, H.: Guided feature selection for deep visual odometry. In: Jawahar, C.V., Li, H., Mori, G., Schindler, K. (eds.) ACCV 2018. LNCS, vol. 11366, pp. 293–308. Springer, Cham (2019). https://doi.org/10.1007/978-3-030-20876-9_19
47. Xue, F., Wang, X., Li, S., Wang, Q., Wang, J., Zha, H.: Beyond tracking: selecting memory and refining poses for deep visual odometry. In: CVPR (2019)
48. Yang, N., Wang, R., Gao, X., Cremers, D.: Challenges in monocular visual odometry: photometric calibration, motion bias, and rolling shutter effect. RAL **3**(4), 2878–2885 (2018)
49. Yin, Z., Shi, J.: GeoNet: unsupervised learning of dense depth, optical flow and camera pose. In: CVPR (2018)
50. Zhan, H., Garg, R., Saroj Weerasekera, C., Li, K., Agarwal, H., Reid, I.: Unsupervised learning of monocular depth estimation and visual odometry with deep feature reconstruction. In: CVPR (2018)
51. Zhou, H., Ummenhofer, B., Brox, T.: DeepTAM: deep tracking and mapping. In: ECCV (2018)
52. Zhou, T., Brown, M., Snavely, N., Lowe, D.G.: Unsupervised learning of depth and ego-motion from video. In: CVPR (2017)
53. Zhuang, B., Tran, Q.H., Ji, P., Cheong, L.F., Chandraker, M.: Learning structure-and-motion-aware rolling shutter correction. In: CVPR (2019)
54. Zhuang, B., Tran, Q.H., Lee, G.H., Cheong, L.F., Chandraker, M.: Degeneracy in self-calibration revisited and a deep learning solution for uncalibrated SLAM. In: IROS (2019)
55. Zou, Y., Luo, Z., Huang, J.B.: DF-Net: unsupervised joint learning of depth and flow using cross-task consistency. In: ECCV (2018)

The Devil Is in Classification: A Simple Framework for Long-Tail Instance Segmentation

Tao Wang[1,4](✉) , Yu Li[2,4], Bingyi Kang[4], Junnan Li[3], Junhao Liew[4],
Sheng Tang[2], Steven Hoi[3], and Jiashi Feng[4]

[1] NGS, National University of Singapore, Singapore, Singapore
twangnh@gmail.com
[2] Institute of Computing Technology, Chinese Academy of Sciences, Beijing, China
liyu,ts@ict.ac.cn
[3] Salesforce Research Asia, Singapore, Singapore
{junnan.li,shoi}@salesforce.com
[4] ECE Department, National University of Singapore, Singapore, Singapore
{kang,liewjunhao}@u.nus.edu, elefjia@nus.edu.sg

Abstract. Most existing object instance detection and segmentation models only work well on fairly balanced benchmarks where per-category training sample numbers are comparable, such as COCO. They tend to suffer performance drop on realistic datasets that are usually long-tailed. This work aims to study and address such open challenges. Specifically, we systematically investigate performance drop of the state-of-the-art two-stage instance segmentation model Mask R-CNN on the recent long-tail LVIS dataset, and unveil that a major cause is the inaccurate classification of object proposals. Based on such an observation, we first consider various techniques for improving long-tail classification performance which indeed enhance instance segmentation results. We then propose a simple calibration framework to more effectively alleviate classification head bias with a bi-level class balanced sampling approach. Without bells and whistles, it significantly boosts the performance of instance segmentation for tail classes on the recent LVIS dataset and our sampled COCO-LT dataset. Our analysis provides useful insights for solving long-tail instance detection and segmentation problems, and the straightforward *SimCal* method can serve as a simple but strong baseline. With the method we have won the 2019 LVIS challenge. Codes and models are available at https://github.com/twangnh/SimCal.

Importantly, after the challenge submission [38], we find significant improvement can be further achieved by modifying the head from 2fc_rand to 3fc_ft (refer to Sect. 5.4 and Table 6 for details), which is expected to generates much higher test set result. We also encourage readers to read our following work [28] that calibrates the last classification layer with a specifically re-designed softmax classification module.

Electronic supplementary material The online version of this chapter (https://doi.org/10.1007/978-3-030-58568-6_43) contains supplementary material, which is available to authorized users.

© Springer Nature Switzerland AG 2020
A. Vedaldi et al. (Eds.): ECCV 2020, LNCS 12359, pp. 728–744, 2020.
https://doi.org/10.1007/978-3-030-58568-6_43

Keywords: Long-tail distribution · Instance segmentation · Object detection · Long-tail classification

1 Introduction

Object detection and instance segmentation aim to localize and segment individual object instances from an input image. The widely adopted solutions to such tasks are built on region-based two-stage frameworks, *e.g.*, Faster R-CNN [34] and Mask R-CNN [17]. Though these models have demonstrated remarkable performance on several class-balanced benchmarks, such as Pascal VOC [11], COCO [31] and OpenImage [1], they are seldom evaluated on datasets with long-tail distribution that is common in realistic scenarios [33] and dataset creation [11,26,31]. Recently, Gupta et al. [15] introduce the LVIS dataset for large vocabulary long-tail instance segmentation model development and evaluation. They observe the long-tail distribution can lead to severe performance drop of the state-of-the-art instance segmentation model [15]. However, the reason for such performance drop is not clear yet.

In this work, we carefully study why existing models are challenged by long-tailed distribution and develop solutions accordingly. Through extensive analysis on Mask R-CNN in Sect. 3, we show one major cause of performance drop is the inaccurate classification of object proposals, which is referred to the bias of classification head. Figure 1 shows a qualitative example. Due to long-tail distribution, under standard training schemes, object instances from the tail classes are exposed much less frequently to the classifier than the ones from head classes[1], leading to poor classification performance on tail classes.

To improve proposal classification, we first consider incorporating several common strategies developed for long-tail classification into current instance segmentation frameworks, including loss re-weighting [18,36], adaptive loss adjustment (focal loss [29], class-aware margin loss [7]), and data re-sampling [16,35]. We find such strategies indeed improve long-tail instance segmentation performance, but their improvement on tail classes is limited and facing the trade-off problem of largely sacrificing performance on head classes. We thus propose a simple and efficient framework after a thorough analysis of the above strategies. Our method, termed *SimCal*, aims to correct the bias in the classification head with a decoupled learning scheme. Specifically, after normal training of an instance segmentation model, it first collects class balanced proposal samples with a new bi-level sampling scheme that combines image-level and instance-level sampling, and then uses these collected proposals to calibrate the classification head. Thus performance on tail classes can be improved. *SimCal* also incorporates a simple dual head inference component that effectively mitigates performance drop on head classes after calibration.

Based on our preliminary findings, extensive experiments are conducted on LVIS [15] dataset to verify the effectiveness of our methods. We also validate the proposed method with SOTA multi-stage instance segmentation model HTC [9]

---

[1] we use head classes and many-shot classes interchangeably.

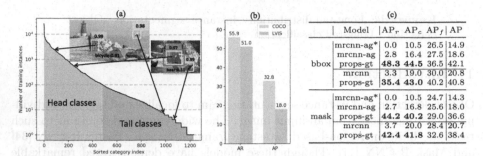

| | (a) | (b) | (c) |

	Model	AP_r	AP_c	AP_f	AP
	mrcnn-ag*	0.0	10.5	26.5	14.9
	mrcnn-ag	2.8	16.4	27.5	18.6
bbox	props-gt	**48.3**	**44.5**	36.5	42.1
	mrcnn	3.3	19.0	30.0	20.8
	props-gt	**35.4**	**43.0**	40.2	40.8
	mrcnn-ag*	0.0	10.5	24.7	14.3
	mrcnn-ag	2.7	16.8	25.6	18.0
mask	props-gt	**44.2**	**40.2**	29.0	36.6
	mrcnn	3.7	20.0	28.4	20.7
	props-gt	**42.4**	**41.8**	32.6	38.4

Fig. 1. (a) Examples of object proposal and instance segmentation results from ResNet50-FPN Mask R-CNN, trained on long-tail LVIS dataset. The RPN can generate high-quality object proposals (yellow bounding boxes with high confidence scores) even on long-tail distribution, *e.g.*, cargo ship (7 training instances) and vulture (4 training instances). However, they are missed in final detection and segmentation outputs (green bounding boxes and masks) due to poor proposal classification performance. Other proposal candidates and detection results are omitted from images for clarity. (b) Comparison of proposal recall (COCO style Average Recall) and AP between COCO and LVIS dataset with Mask R-CNN model. (c) Pilot experiment results on Mask R-CNN with class-agnostic and class-wise box and mask heads on ResNet50-FPN backbone, evaluated with LVIS v0.5 val set. *mrcnn-ag** denotes standard inference with 0.05 confidence threshold as in optimal settings of COCO, while *mrcnn-ag* means inference with threshold 0.0. Note for all later experiments we use 0.0 threshold. AP^{bb} denotes box AP. *props-gt* means testing with ground truth labels of the proposals (Color figure online)

and our sampled long-tail version of COCO dataset (COCO-LT). From our systematic study, we make the following intriguing observations:

- Classification is the primary obstacle preventing state-of-the-art region-based object instance detection and segmentation models from working well on long-tail data distribution. There is still a large room for improvement along this direction.
- By simply calibrating the classification head of a trained model with a bi-level class balanced sampling in the decoupled learning scheme, the performance for tail classes can be effectively improved.

2   Related Works

Object Detection and Segmentation. Following the success of R-CNN [13], Fast R-CNN [12] and Faster R-CNN [34] architectures, the two-stage pipeline has become prevailing for object detection. Based on Faster R-CNN, He et al. [17] propose Mask R-CNN that extends the framework to instance segmentation with a mask prediction head to predict region based mask segments. Lots of later works try to improve the two-stage framework for object detection and instance segmentation. For example, [20,21] add IOU prediction branch to improve confidence scoring for object detection and instance segmentation respectively. Feature augmentation and various training techniques are thoroughly examined by [32]. Recently, [6] and [9] further extend proposal based object detection and instance segmentation to multi-stage and achieve state-of-the-art performance. In this work, we study how to improve proposal-based instance segmentation models over long-tail distribution.

Long-Tailed Recognition. Recognition on long-tail distribution is an important research topic as imbalanced data form a common obstacle in real-world applications. Two major approaches tackling long-tail problems are sampling [4,5,16,35] and loss re-weighting [10,18,19]. Sampling methods over-sample minority classes or under-sample majority classes to achieve data balance to some degree. Loss re-weighting assigns different weights to different classes or training instances adaptively, e.g., by inverse class frequency. Recently, [10] proposes to re-weight loss by the number of inversed effective samples. [7] explores class aware margin for classifier loss calculation. In addition, [23] tries to examine the relation of feature and classifier learning in an imbalanced setting. [39] develops a meta learning framework that transfers knowledge from many-shot to few-shot classes. Existing works mainly focus on classification, while the crucial tasks of long-tail object detection and segmentation on remain largely unexplored.

3   Analysis: Performance Drop on Long-Tail Distribution

We investigate the performance decline phenomenon of popular two-stage frameworks for long-tail instance detection and segmentation.

Our analysis is based on experiments on LVIS v0.5 train and validation sets. The LVIS dataset [15] is divided into 3 sets: *rare, common, and frequent*, among which *rare* and *common* contain tail classes and *frequent* includes head classes. We report AP on each set, denoted as AP_r, AP_c, AP_f. For simplicity, we train a baseline Mask R-CNN with ResNet50-FPN backbone and *class agnostic box and mask prediction heads*. As shown in Fig. 1 (c), our baseline model (denoted as mrcnn-ag*) performs poorly, especially on tail categories (*rare* set, AP_r, AP_r^b), with 0 box and mask AP.

Usually, the confidence threshold is set to a small positive value (e.g. 0.05 for COCO) to filter out low-quality detections. Since LVIS contains 1,230 categories, the softmax activation gives much lower average confidence scores, thus we minish the threshold here. However, even lowering the threshold to 0 (mrcnn-ag), the performance remains very low for tail classes, and improvement on *rare* is much smaller than that of *common* (6.1 vs 2.7 for segmentation AP, 5.9 vs 2.8 for bbox AP). This reveals the Mask R-CNN model trained with the normal setting is heavily biased to the head classes.

We then calculate proposal recall of *mrcnn-ag* model and compare with the one trained on COCO dataset with the same setting. As shown in Fig. 1 (b), the same baseline model trained on LVIS only has a drop of 8.8% (55.9 to 51.0) in proposal recall compared with that on COCO, but notably, has a 45.1% (32.8 to 18.0) drop in overall mask AP. Since the box and mask heads are class agnostic, we hypothesize that the performance drop is mainly caused by the degradation of proposal classification.

To verify this, for the proposals generated by RPN [34], we assign their ground truth class labels to the second stage as its classification results. Then we evaluate the AP. As shown in Fig. 1 (c), the mask AP for tail classes is increased by a large margin, especially on *rare* and *common* sets. Such findings also hold for the box AP. Surprisingly, with normal class-wise box and mask heads (standard version of Mask R-CNN), performance on tail classes is also boosted significantly. This suggests the box and mask head learning are less sensitive to long-tail training data than classification.

The above observations indicate that the low performance of the model over tail classes is mainly caused by poor proposal classification on them. We refer to this issue as *classification head bias*. Addressing the bias is expected to effectively improve object detection and instance segmentation results.

4   Solutions: Alleviating Classification Bias

Based on the above analysis and findings, we first consider using several existing strategies of long-tail classification, and then present a new calibration framework to correct the classification bias for better detection and segmentation on long-tail distribution.

4.1   Using Existing Long-Tail Classification Approaches

We adapt some popular approaches of image classification to solving our long-tail instance detection and segmentation problem, as introduced below. We conduct experiments to see how our adapted methods work in Sect. 5.2. Given a sample x_i, the model outputs logits denoted as y_i, and p_i is probability prediction on the true label z.

Loss Re-weighting [10,18,25,36,37]. This line of works alleviate the bias by applying different weights to different samples or categories, such that tail classes or samples receive higher attention during training, thus improving the classification performance. For LVIS, we consider a simple and effective inverse class frequency re-weighting strategy adopted in [18,39]. Concretely, the training samples of each class are weighted by $w = N/N_j$ where N_j is the training instance number of class j. N is a hyperparameter. To handle noise, the weights are clamped to $[0.1, 10.0]$. The weight for the background is also a hyperparameter. During training, the second stage classification loss is weighted as $L = -w_i \log(p_i)$.

Focal Loss [29]. Focal loss can be regarded as loss re-weighting that adaptively assigns a weight to each sample by the prediction. It was originally developed for foreground-background class imbalance for one-stage detectors, and also applicable to alleviating the bias in long-tail problems since head-class samples tend to get smaller losses due to sufficient training, and the influence of tail-class samples would be again enlarged. Here we use the multi-class extension of Focal loss $L = -(1 - p_i)^\gamma \log(p_i)$.

Class-Aware Margin Loss [7]. This method assigns a class dependent margin to loss calculation. Specifically, a larger margin will be assigned for the tail classes, so they are expected to generalize better with limited training samples. We adopt the margin formulation $\Delta_j = C/N_j^{1/4}$ [7] where N_j is the training instance number N_j for class j as above and plug the margin into cross entropy loss $L = -\log e^{y_{iz}-\Delta_z}/(e^{y_{iz}-\Delta_z} + \sum_{c \neq z} e^{y_{ic}})$.

Repeat Sampling [16,35]. Repeat sampling directly over-samples data (images) with a class-dependent repeating factor, so that the tail classes can be more frequently involved in optimization. Consequently, the training steps for each epoch will be increased due to the over-sampled instances. However, this type of methods are not trivially applicable to detection frameworks since multiple instances from different classes frequently exist in one image. [15] developed a specific sampling strategy for LVIS dataset, calculating a per-image repeat factor based on a per-category repeat threshold and over-sampling each training image according to the repeat factor in each epoch. Note that box and mask learning will also be affected by this method.

We implement the adapted version of [7,10,15,30] for experiments. See Sect. 5.2 for details. From the results, we find the above approaches indeed bring some performance improvements over the baselines, which however are very limited. **Re-weighting methods** tend to complicate the optimization of deep models with extreme data imbalance [10], which is the case for object detection with long-tail distribution, leading to poor performance on head classes. **Focal loss** well addresses the imbalance between foreground and easy background samples, but has difficulty in tackling the imbalance between foreground object classes with more similarity and correlation. For **class-aware margin loss**, the prior margin enforced in loss calculation also complicates the optimization of a deep model, leading to larger drop of performance on head classes. The **repeat sampling** strategy suffers from overfitting since it repeatedly samples from tail classes. Also, it additionally samples more data during training, leading to increased computation cost. In general, the diverse object scale and surrounding context in object instance detection further complicate above-discussed limitations, making these methods hardly suitable for our detection tasks.

4.2   Proposed *SimCal*: Calibrating the Classifier

We find in Sect. 3 that significant performance gain on tail classes can be achieved with merely ground truth proposal labels, and as discussed in Sect. 4.1, exiting classification approaches are not very suitable for tackling our long-tail instance segmentation task. Here, we propose a new *SimCal* framework to calibrate the classification head by retraining it with a new bi-level sampling scheme while keeping the other parts frozen after standard training. This approach is very simple and incurs negligible additional computation cost since only the classification head requires gradient back-propagation. The details are given as follows.

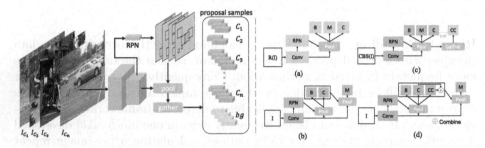

Fig. 2. Left: Illustration of proposed bi-level sampling scheme. Refer to Sect. 4.2 for more details. Right: Architecture of proposed method. I: training or test image sets; R: random sampling; CBS: class-balanced sampling; C: classification head; B: box regression head; M: mask prediction head; CC: calibrated classification head. *Blue* modules are in training mode while *grey* modules indicate frozen. (a) (b) show standard Mask R-CNN training and inference respectively. (c) (d) show proposed calibration and dual head inference respectively (Color figure online)

Calibration Training with Bi-level Sampling. As shown in Fig. 2, we propose a bi-level sampling scheme to collect training instances for calibrating the classification head through retraining. To create a batch of training data, first, n object classes (*i.e.*, c_1 to c_n) are sampled uniformly form all the classes (which share the same probability). Then, we randomly sample images that contain the categories respectively (*i.e.*, I_{c_1} to I_{c_n}), and feed them to the model. At the object level, we only collect proposals that belong to the sampled classes and background for training. Above, we only sample 1 image for each sampled class for simplicity, but note that the number of sampled images can also be larger. As shown in Fig. 2 right (a), after standard training, we freeze all the model parts (including backbone, RPN, box and mask heads) except for the classification head, and employ the bi-level sampling to retrain the classification head, which is initialized with the original head. Then, the classification head is fed with fairly balanced proposal instances, thus enabling the model to alleviate the bias. Different from conventional fine-tuning conducted on a small scale dataset after pretraining on a large one, our method only changes the data sample distribution. Refer to supplementary material for more implementation details, including foreground and background ratio and matching IOU threshold for proposals. Formally, the classification head is trained with loss:

$$L = \frac{1}{\sum_{i=0}^{N} n_i} \sum_{i=0}^{N} \sum_{j=1}^{n_i} L_{cls}(p_{ij}, p_{ij}^*) \tag{1}$$

where N is the number of sampled classes per batch, n_i is the number of proposal samples for class i, $i = 0$ is for background, L_{cls} is cross entropy loss, and p_{ij} and p_{ij}^* denotes model prediction and ground truth label.

Dual Head Inference. After the above calibration, the classification head is now balanced over classes and can perform better on tail classes. However, the performance on head classes drops. To achieve optimal overall performance, here we consider combining the new balanced head and the original one that have higher performance respectively on tail classes and on head classes. We thus propose a dual head inference architecture.

An effective combining scheme is to simply average the models' classification predictions [2,3,27], but we find this is not optimal as the original head is heavily biased to many-shot classes. Since the detection models adopt class-wise post-processing (i.e., NMS) and the prediction does not need to be normalized, we propose a new combining scheme that directly selects prediction from the two classifiers for the head and tail classes:

$$p[z] = \begin{cases} p^{cal}[z] & N_z \leq T \\ p^{orig}[z] & \text{otherwise,} \end{cases} \tag{2}$$

where $z \in [0, C]$ indexes the classes, C is the number of classes, $z = 0$ stands for background, p^{cal} and p^{orig} denote the $(C+1)$-dimensional predictions of calibrated and original heads respectively, p is the combined prediction, N_z is the training instance number of class z, and T is the threshold number controlling the boundary of head and tail classes. Other parts of inference remain the same (Fig. 2 (d)). Our dual head inference is with small overhead compared to the original model.

Bi-level Sampling vs. Image Level Repeat Sampling. Image level repeat sampling (e.g., [15]), which is traditionally adopted, balances the long-tail distribution at the image level, while our bi-level sampling alleviates the imbalance at the proposal level. Image level sampling approaches train the whole model directly, while we decouple feature and classification head learning, and adjust the classification head only with bi-level class-centric sampling and keep other parts the freezed after training under normal image-centric sampling. We also empirically find the best setting ($t = 0.001$) of IS [15] additionally samples about 23k training images (56k in total) per epoch, leading to more than 40% increase of training time. Comparatively, our method incurs less than 5% additional time and costs much less GPU memory since only a small part of the model needs backpropagation.

5   Experiments

In this section, we first report experiments of using exiting classification approaches to solve our long-tail instance segmentation problem. Then we evaluate our proposed solution, *i.e.* the *SimCal* framework, analyze its model designs and test its generalizality.

Our experiments are mainly conducted on LVIS dataset [15]. Besides, to check the generalizability of our method, we sample a new COCO-LT dataset from COCO [31]. We devise a complimentary instance-centric category division scheme that helps to more comprehensively analyze model performance. For each experiment, we report result with median overall AP over 3 runs.

5.1   Datasets and Metrics

Datasets. 1) LVIS [15]. It is a recent benchmark for large vocabulary long-tail instance segmentation [15]. The source images are from COCO dataset, while the annotation follows an iterative object spotting process that captures the long-tail category statistic naturally appearing in images. Current released version v0.5 contains 1,230 and 830 object classes respectively in its train and validation set, with test set unknown. Refer to Fig. 1

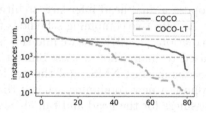

Fig. 3. Category distribution of COCO (2017) and sampled COCO-LT datasets. The categories are sorted in descending numbers of training instances

(a) for train set category distribution. The three sets contain about 50k, 5k and 20k images correspondingly. 2) COCO-LT. We sample it from COCO [31] by following an exponential distribution on training instance statistics to create a long-tail version. COCO-LT contains 80 classes and about 100k images. Figure 3 shows the category distribution. Due to space limitations, we defer details of sampling process to supplement.

Table 1. Different category division scheme, with LVIS v0.5 dataset [15]. The left part is division based on training image number as in [15], the right part is proposed scheme based on training intance number. Train-on-val means categories that appear in the validation set

Set	Total	Divided by #image			Divided by #instance			
		Rare	*Common*	*Frequent*	(0, 10)	[10, 100)	[100, 1000)	[1000, −]
Train	1230	454	461	315	294	453	302	181
Train-on-val	830	125	392	313	67	298	284	181

Metrics. We adopt AP as overall evaluation metric. Object categories in LVIS are divided into *rare, common, frequent* sets [15], respectively containing <10, 10–100, and $>=100$ training images. We show in Table 1 the category distribution of training and validation sets. Besides data splitting based on image number, we devise a complimentary instance-centric category division scheme, considering number of instances is a widely adopted measurement for detection in terms of benchmark creation, model evaluation [1,11,31]. In particular, we divide all the categories into four bins[2] based on the number of training instances, with #instances <10, 10–100, 100–1000, and $>=1000$, as shown in Table 1. Accordingly, we calculate AP on each bin as complementary metrics, denoted as AP_1, AP_2, AP_3, and AP_4. Such a division scheme offers a finer dissection of model performance. For example, AP_1 corresponds to the commonly referred few-shot object detection regime [8,22,24]. *rare* set (≤ 10 training images) contains categories that have up to 219 training instances ('chickpea'), so AP_r cannot well reflect model's few-shot learning capability. AP_4 reflects performance on classes with COCO level training data, while most classes in *frequent* set (>100 images) have much less than 1,000 training instances (*e.g.*, 'fire-alarm': 117). With the two division schemes, we can report AP on both image-centric (AP_r, AP_c, AP_f) and instance-centric (AP_1, AP_2, AP_3, AP_4) bins for LVIS. For COCO-LT, since the per-category training instance number varies in a much larger range, we divide the categories into four bins with <20, 20–400, 400–8000, and $>=8000$ training instances and report performance as AP_1, AP_2, AP_3, AP_4 on these bins. Unless specified, AP is evaluated with COCO style by mask AP.

5.2   Evaluating Adapted Existing Classification Methods

We apply adapted discussed methods in Sect. 4.1 to classification head of Mask R-CNN for long-tail instance segmentation, including [7,10,15,30]. Results are summarized in Table 2. We can see some improvements have been achieved on tail classes. For example, 6.0, 6.2, 7.7 absolute margins on AP_1 and 10.1, 8.7, 11.6 on AP_r for loss re-weighting (LR), focal loss (FL) and image level repeat sampling (IS) are observed, respectively. However, on the other hand, they inevitably lead to drop of performance on head classes, *e.g.*, more than 2.0 drop for all methods on AP_4 and AP_f. Performance drop on head classes is also observed in imbalanced classification [16,36]. Overall AP is improved by at most 2.5 in absolute value (*i.e.*, IS). Similar observation holds for box AP.

5.3   Evaluating Proposed *SimCal*

In this subsection, we report the results of our proposed method applied on mask R-CNN. We evaluate both class-wise and class-agnostic versions of the model. Here T for dual head inference is set to 300.

---

[2] Note we use "bin" and "set" interchangeably.

Table 2. Results on LVIS by adding common strategies in long-tail classification to Mask R-CNN in training. r50 means Mask R-CNN on ResNet50-FPN backbone with class-wise box and mask heads (standard version). CM, LR, FL and IS denote discussed class aware margin loss, loss re-weighting, Focal loss and image level repeat sampling respectively. AP^b denotes box AP. We report result with median overall AP over 3 runs

Model	AP_1	AP_2	AP_3	AP_4	AP_r	AP_c	AP_f	AP	AP_1^b	AP_2^b	AP_3^b	AP_4^b	AP_r^b	AP_c^b	AP_f^b	AP^b
r50	0.0	17.1	23.7	29.6	3.7	20.0	28.4	20.7	0.0	15.9	24.6	30.5	3.3	19.0	30.0	20.8
CM	2.6	21.0	21.8	26.6	8.4	21.2	25.5	21.0	2.8	20.0	22.0	26.6	6.8	20.5	26.4	20.7
LR	6.0	23.3	22.0	25.1	13.8	22.4	24.5	21.9	6.0	21.2	22.3	25.5	11.3	21.5	24.9	21.4
FL	6.2	21.0	22.0	27.0	12.4	20.9	25.9	21.5	5.8	20.5	22.7	28.0	10.5	21.0	27.0	21.7
IS	7.7	25.6	21.8	27.4	15.3	23.7	25.6	23.2	6.7	22.8	22.1	27.4	11.6	22.2	26.7	22.0

Table 3. Results on LVIS by applying *SimCal* to Mask R-CNN with ResNet50-FPN. r50-ag and r50 denote models with class-agnostic and class-wise heads (box/mask) respectively. cal and dual means calibration and dual head inference. Refer to supplementary file for an anlaysis on LVIS result mean and std

	Model	Cal	Dual	AP_1	AP_2	AP_3	AP_4	AP_r	AP_c	AP_f	AP
bbox	r50-ag			0.0	12.8	22.8	28.3	2.8	16.4	27.5	18.6
		✓		12.4	23.2	20.6	23.5	17.7	21.2	23.4	21.5
		✓	✓	**12.4**	**23.4**	21.4	27.3	**17.7**	**21.3**	26.4	**22.7**
	r50			0.0	15.9	24.6	30.5	3.3	19.0	30.0	20.8
		✓		8.1	21.0	22.4	25.5	13.4	20.6	25.7	21.4
		✓	✓	**8.2**	**21.3**	23.0	29.5	**13.7**	**20.6**	28.7	**22.6**
mask	r50-ag			0.0	13.3	21.4	27.0	2.7	16.8	25.6	18.0
		✓		13.2	23.1	20.0	23.0	18.2	21.4	22.2	21.2
		✓	✓	**13.3**	**23.2**	20.7	26.2	**18.2**	**21.5**	24.7	**22.2**
	r50			0.0	17.1	23.7	29.6	3.7	20.0	28.4	20.7
		✓		10.2	23.5	21.9	25.3	15.8	22.4	24.6	22.3
		✓	✓	**10.2**	**23.9**	22.5	28.7	**16.4**	**22.5**	27.2	**23.4**

Table 4. Results for augmenting discussed long-tail classification methods with proposed decoupled learning and dual head inference.

Model	AP_1	AP_2	AP_3	AP_4	AP_r	AP_c	AP_f	AP
r50	0.0	17.1	23.7	29.6	3.7	20.0	28.4	20.7
CM	4.6	21.0	22.3	28.4	10.0	21.1	27.0	21.6
LR	6.9	23.0	22.1	28.8	13.4	21.7	26.9	22.5
FL	7.1	21.0	22.1	28.4	13.1	21.5	26.5	22.2
IS	6.8	23.2	22.5	28.0	14.0	22.0	27.0	22.7
Ours	**10.2**	**23.9**	22.5	28.7	**16.4**	**22.5**	27.2	**23.4**

Table 5. Comparison between proposed combining scheme (sel) and averaging (avg).

	AP_1	AP_2	AP_3	AP_4	AP
Orig	0.0	13.3	21.4	27.0	18.0
Cal	8.5	20.8	17.6	19.3	18.4
Avg	8.5	20.9	19.6	24.6	20.3
Sel	**8.6**	**22.0**	19.6	26.6	**21.1**

Calibration Improves Tail Performance. From results in Table 3, we observe consistent improvements on tail classes for both class-agnostic and class-wise version of Mask R-CNN (more than 10 absolute mask and box AP improvement on tail bins). Overall mask and box AP are boosted by a large margin. But we also observe a significant drop of performance on head class bins, $e.g.$, 23.7 to 21.9 on AP_3 and 29.6 to 25.3 on AP_4 for the class-wise version of Mask R-CNN. With calibration, the classification head is effectively balanced.

Dual Head Inference Mitigates Performance Drop on Head Classes. The model has a minor performance drop on the head class bins but an enormous boost on the tail class bins. For instance, we observe 0.8 drop of AP_4 but 13.3 increase on AP_1 for r50-ag model. It can be seen that with the proposed combination method, the detection model can effectively gain the advantage of both calibrated and original classification heads.

Class-Wise Prediction Is Better for Head Classes While Class-Agnostic One Is Better for Tail Classes. We observe AP_1 of r50-ag with cal and dual is 3.1 higher (13.3 vs 10.2) than that of r50 while AP_4 is 2.5 lower (26.2 vs 28.7), which means class-agnostic heads (box/mask) have an advantage on tail classes, while class-wise heads perform better for many-shot classes. This phenomenon suggests that a further improvement can be achieved by using class-agnostic head for tail classes so they can benefit from other categories for box and mask prediction, and class-wise head for many-shot classes as they have abundant training data to learn class-wise prediction, which is left for future work.

Comparing with Adapted Existing Methods. For fair comparison, we also consider augmenting the discussed imbalance classification approaches with proposed decoupled learning framework. With the same baseline Mask R-CNN trained in the normal setting, we freeze other parts except for classification head, and use these methods to calibrate the head. After that, we apply the dual head inference for evaluation. As shown in Table 4, they have similar performance on head classes as dual head inference is used. They nearly all get improved on tail classes than the results in Table 2 ($e.g.$, 4.6 vs 2.6, 6.9 vs 6.0, and 7.1 vs 6.2 on AP_1 for CM, LR, and FL methods respectively), indicating the effectiveness of the decoupled learning scheme for recognition of tail classes. The image level repeat sampling (IS) gets worse performance than that in Table 2, suggesting box and mask learning also benefits a lot from the sampling. Our method achieves higher performance, $i.e.$, 10.2 and 23.9 for AP_1 and AP_2, which validates effectiveness of the proposed bi-level sampling scheme.

5.4   Model Design Analysis of *SimCal*

Calibration Dynamics. As shown in Fig. 4 (a), with the progress of calibration, model performance is progressively balanced over all the class bins. Increase

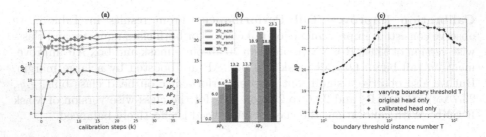

Fig. 4. (a) Model performance as a function of calibration steps. The result is obtained with r50-ag model (Table 3). (b) Effect of design choice of calibration head. Baseline: original model result; 2fc_ncm [14]: we have tried to adopt the deep nearest class mean classifier learned with 2fc representation. 2fc_rand: 2-layer fully connected head with 1024 hidden units, random initialized; 3fc-rand: 3-layer fully connected head with 1024 hidden units, random initialized. 3fc-ft: 3fc initialized from original head. (c) Effect of boundary number T (with r50-ag)

of AP on tail bins (*i.e.*, AP_1, AP_2) and decrease of AP on the head (*i.e.*, AP_3, AP_4) are observed. With about 10–20 steps, AP on all the bins and overall AP converge to a steady value.

Design Choice of Calibration Head. While the proposed calibration method tries to calibrate the original head, we can perform the calibration training on other head choices. As shown in Fig. 4 (b), we have tried different instantiations instead of the original head. It is interesting that with random initialization, 3-layer fully connected head performs worse than 2-layer head on AP_1 (*i.e.*, 2fc_rand vs 3fc-rand). But when it is initialized from the original 3-layer head, the performance is significantly boosted by 4.1 and 4.3 AP respectively on AP_1 and AP_2 (*i.e.*, 3fc_ft). This phenomenon indicates that training under random sampling can help the classification head learn general features and perform well when calibrating with balanced sampling. We only compare them on the tail class bins since they perform on par on head class bins with dual head inference.

Combining Scheme and Head/Tail Boundary for Dual Heads. As shown in Table 5. Our combination approach achieves much higher performance than simple averaging. Refer to supplementary material for more alternative combining choices. We also examine the effect of head/tail boundary as in Fig. 4 (c). For the same model, we vary the boundary threshold instance number T from 10 to 1000. The AP is very close to optimal ($T = 300$) in $T \in [90, 500]$. Thus dual head is insensitive to the exact value of hyperparameter T in a wide range.

5.5 Generalizability Test of *SimCal*

Performance on SOTA Models. We further apply the proposed method to state-of-the-art multi-stage cascaded instance segmentation model, Hybrid

Table 6. Results with Hybrid Task Cascade (HTC) on LVIS. With backbone of ResNeXt101-64x4d-FPN. best denotes best single model performance reported in [15]. The remaining rows are our experiment results with HTC. 2fc_rand and 3fc_ft are different design choices of classification head (Sect. 5.4). Only 2fc_rand is available on test set as the evaluation server is closed

Model	Val								Test			
	AP_1	AP_2	AP_3	AP_4	AP_r	AP_c	AP_f	AP	AP_r	AP_c	AP_f	AP
best [15]	–	–	–	–	15.6	27.5	31.4	27.1	9.8	21.1	30.0	20.5
htc-x101	5.6	33.0	33.7	37.0	13.7	34.0	36.6	31.9	5.9	25.7	35.3	22.9
IS	10.2	32.3	33.2	36.6	17.6	33.0	36.1	31.9	–	–	–	–
2fc_rand	12.9	32.2	33.5	37.1	18.5	33.3	36.1	32.1	10.3	25.3	35.1	23.9
3fc_ft	**18.8**	**34.9**	33.0	36.7	**24.7**	33.7	36.4	**33.4**	–	–	–	–

Table 7. Results on COCO-LT, evaluated on minival set. AP_1, AP_2, AP_3, AP_4 correspond to bins of [1, 20), [20, 400), [400, 8000), [8000, -) training instances

Model	Cal	Dual	AP_1	AP_2	AP_3	AP_4	AP	AP_1^b	AP_2^b	AP_3^b	AP_4^b	AP^b
r50-ag			0.0	8.2	24.4	26.0	18.7	0.0	9.5	27.5	30.3	21.4
r50-ag	✓		15.0	16.2	22.4	24.1	20.6	14.5	17.9	24.8	27.6	22.9
r50-ag	✓	✓	**15.0**	**16.2**	24.3	26.0	**21.8**	**14.5**	**18.0**	27.3	30.3	**24.6**

Task Cascade [9] (HTC), by calibrating classification heads at all the stages. As shown in Table 6, our method brings significant gain on tail classes and minor drop on many-shot classes. Notably, the proposed approach leads to much higher gain than the image level repeat sampling method (IS), (*i.e.*, 8.5 and 2.5 higher on AP_1 and AP_2 respectively). We achieve state-of-the-art single model performance on LVIS, which is 6.3 higher in absolute value than the best single model reported in [15] (33.4 vs 27.1). And with test set, a consistent gain is observed.

Performance on COCO-LT. As shown in Table 7, similar trend of performance boost as LVIS dataset is observed. On COCO-LT dual head inference can enjoy nearly full advantages of both the calibrated classifier on tail classes and the original one on many shot classes. But larger drop of performance on many-shot classes with LVIS is observed. It may be caused by the much stronger inter-class competition as LVIS has much larger vocabulary.

6 Conclusions

In this work, we carefully investigate two-stage instance segmentation model's performance drop with long-tail distribution data and reveal that the devil is in proposal classification. Based on this finding, we first try to adopt several common strategies in long-tail classification to improve the baseline model. We also propose a simple calibration approach, *SimCal*, for improving the second-stage classifier on tail classes. It is demonstrated that *SimCal* significantly enhances Mask R-CNN and SOTA multi-stage model HTC. A large room of improvement still exists along this direction. We hope our pilot experiments and in-depth analysis together with the simple method would benefit future research.

Acknowledgement. Jiashi Feng was partially supported by MOE Tier 2 MOE 2017-T2-2-151, NUS_ECRA_FY17_P08, AISG-100E-2019-035.

References

1. Openimage dataset. https://storage.googleapis.com/openimages/web/factsfigures.html
2. Alpaydin, E.: Multiple networks for function learning. In: IEEE International Conference on Neural Networks, pp. 9–14. IEEE (1993)
3. Breiman, L.: Bagging predictors. Mach. Learn. **24**(2), 123–140 (1996)
4. Buda, M., Maki, A., Mazurowski, M.A.: A systematic study of the class imbalance problem in convolutional neural networks. Neural Netw. **106**, 249–259 (2018)
5. Byrd, J., Lipton, Z.: What is the effect of importance weighting in deep learning? In: International Conference on Machine Learning, pp. 872–881 (2019)
6. Cai, Z., Vasconcelos, N.: Cascade R-CNN: delving into high quality object detection. In: Proceedings of the IEEE Conference on Computer Vision and Pattern Recognition, pp. 6154–6162 (2018)
7. Cao, K., Wei, C., Gaidon, A., Arechiga, N., Ma, T.: Learning imbalanced datasets with label-distribution-aware margin loss. arXiv preprint arXiv:1906.07413 (2019)
8. Chen, H., Wang, Y., Wang, G., Qiao, Y.: LSTD: a low-shot transfer detector for object detection. In: Thirty-Second AAAI Conference on Artificial Intelligence (2018)
9. Chen, K., et al.: Hybrid task cascade for instance segmentation. In: Proceedings of the IEEE Conference on Computer Vision and Pattern Recognition, pp. 4974–4983 (2019)
10. Cui, Y., Jia, M., Lin, T.Y., Song, Y., Belongie, S.: Class-balanced loss based on effective number of samples. In: Proceedings of the IEEE Conference on Computer Vision and Pattern Recognition, pp. 9268–9277 (2019)
11. Everingham, M., Van Gool, L., Williams, C.K., Winn, J., Zisserman, A.: The pascal visual object classes (voc) challenge. Int. J. Comput. Vis. **88**(2), 303–338 (2010)
12. Girshick, R.: Fast R-CNN. In: Proceedings of the IEEE International Conference on Computer Vision, pp. 1440–1448 (2015)
13. Girshick, R., Donahue, J., Darrell, T., Malik, J.: Rich feature hierarchies for accurate object detection and semantic segmentation. In: Proceedings of the IEEE Conference on Computer Vision and Pattern Recognition, pp. 580–587 (2014)
14. Guerriero, S., Caputo, B., Mensink, T.: DeepNCM: deep nearest class mean classifiers (2018)
15. Gupta, A., Dollar, P., Girshick, R.: LVIS: a dataset for large vocabulary instance segmentation. In: CVPR (2019)
16. He, H., Garcia, E.A.: Learning from imbalanced data. IEEE Trans. Knowl. Data Eng. **21**(9), 1263–1284 (2009)
17. He, K., Gkioxari, G., Dollár, P., Girshick, R.: Mask R-CNN. In: Proceedings of the IEEE International Conference on Computer Vision, pp. 2961–2969 (2017)

18. Huang, C., Li, Y., Change Loy, C., Tang, X.: Learning deep representation for imbalanced classification. In: Proceedings of the IEEE Conference on Computer Vision and Pattern Recognition, pp. 5375–5384 (2016)

19. Huang, C., Li, Y., Chen, C.L., Tang, X.: Deep imbalanced learning for face recognition and attribute prediction. IEEE Trans. Pattern Anal. Mach, Intell (2019)

20. Huang, Z., Huang, L., Gong, Y., Huang, C., Wang, X.: Mask scoring R-CNN. In: Proceedings of the IEEE Conference on Computer Vision and Pattern Recognition, pp. 6409–6418 (2019)

21. Jiang, B., Luo, R., Mao, J., Xiao, T., Jiang, Y.: Acquisition of Localization Confidence for Accurate Object Detection. In: Ferrari, V., Hebert, M., Sminchisescu, C., Weiss, Y. (eds.) Computer Vision – ECCV 2018. LNCS, vol. 11218, pp. 816–832. Springer, Cham (2018). https://doi.org/10.1007/978-3-030-01264-9_48

22. Kang, B., Liu, Z., Wang, X., Yu, F., Feng, J., Darrell, T.: Few-shot object detection via feature reweighting. In: Proceedings of the IEEE International Conference on Computer Vision, pp. 8420–8429 (2019)

23. Kang, B., et al.: Decoupling representation and classifier for long-tailed recognition. arXiv preprint arXiv:1910.09217 (2019)

24. Karlinsky, L., et al.: Repmet: representative-based metric learning for classification and few-shot object detection. In: Proceedings of the IEEE Conference on Computer Vision and Pattern Recognition, pp. 5197–5206 (2019)

25. Khan, S.H., Hayat, M., Bennamoun, M., Sohel, F.A., Togneri, R.: Cost-sensitive learning of deep feature representations from imbalanced data. IEEE Trans. Neural Netw. Learn. Syst. **29**(8), 3573–3587 (2017)

26. Krishna, R., et al.: Visual genome: connecting language and vision using crowd-sourced dense image annotations. Int. J. Comput. Vis. **123**(1), 32–73 (2017)

27. Krogh, A., Vedelsby, J.: Neural network ensembles, cross validation, and active learning. In: Advances in Neural Information Processing Systems, pp. 231–238 (1995)

28. Li, Y., et al.: Overcoming classifier imbalance for long-tail object detection with balanced group softmax. In: Proceedings of the IEEE/CVF Conference on Computer Vision and Pattern Recognition, pp. 10991–11000 (2020)

29. Lin, T.Y., Goyal, P., Girshick, R., He, K., Dollár, P.: Focal loss for dense object detection. In: Proceedings of the IEEE International Conference on Computer Vision, pp. 2980–2988 (2017)

30. Lin, T.Y., Goyal, P., Girshick, R., He, K., Dollár, P.: Focal loss for dense object detection. IEEE Trans. Pattern Anal. Mach. Intell. **39**, 2999–3007 (2018)

31. Lin, T.-Y., et al.: Microsoft COCO: common objects in context. In: Fleet, D., Pajdla, T., Schiele, B., Tuytelaars, T. (eds.) ECCV 2014. LNCS, vol. 8693, pp. 740–755. Springer, Cham (2014). https://doi.org/10.1007/978-3-319-10602-1_48

32. Liu, S., Qi, L., Qin, H., Shi, J., Jia, J.: Path aggregation network for instance segmentation. In: Proceedings of the IEEE Conference on Computer Vision and Pattern Recognition, pp. 8759–8768 (2018)

33. Reed, W.J.: The pareto, zipf and other power laws. Econ. Lett. **74**(1), 15–19 (2001)

34. Ren, S., He, K., Girshick, R., Sun, J.: Faster R-CNN: towards real-time object detection with region proposal networks. In: Advances in Neural Information Processing Systems, pp. 91–99 (2015)

35. Shen, L., Lin, Z., Huang, Q.: Relay Backpropagation for Effective Learning of Deep Convolutional Neural Networks. In: Leibe, B., Matas, J., Sebe, N., Welling, M. (eds.) ECCV 2016. LNCS, vol. 9911, pp. 467–482. Springer, Cham (2016). https://doi.org/10.1007/978-3-319-46478-7_29

744   T. Wang et al.

36. Tang, Y., Zhang, Y.Q., Chawla, N.V., Krasser, S.: SVMs modeling for highly imbalanced classification. IEEE Trans. Syst. Man Cybernet. Part B (Cybernet.) 39(1), 281–288 (2008)
37. Ting, K.M.: A comparative study of cost-sensitive boosting algorithms. In: Proceedings of the 17th International Conference on Machine Learning. Citeseer (2000)
38. Wang, T., et al.: Classification calibration for long-tail instance segmentation. arXiv preprint arXiv:1910.13081 (2019)
39. Wang, Y.X., Ramanan, D., Hebert, M.: Learning to model the tail. In: Advances in Neural Information Processing Systems. pp. 7029–7039 (2017)

What Is Learned in Deep Uncalibrated Photometric Stereo?

Guanying Chen[1]([✉]), Michael Waechter[2], Boxin Shi[3,4], Kwan-Yee K. Wong[1], and Yasuyuki Matsushita[2]

[1] The University of Hong Kong, Hong Kong, China
gychen@cs.hku.hk
[2] Osaka University, Suita, Japan
[3] Peking University, Beijing, China
[4] Peng Cheng Laboratory, Shenzhen, China

Abstract. This paper targets at discovering what a deep uncalibrated photometric stereo network learns to resolve the problem's inherent ambiguity, and designing an effective network architecture based on the new insight to improve the performance. The recently proposed deep uncalibrated photometric stereo method achieved promising results in estimating directional lightings. However, what specifically inside the network contributes to its success remains a mystery. In this paper, we analyze the features learned by this method and find that they strikingly resemble attached shadows, shadings, and specular highlights, which are known to provide useful clues in resolving the generalized bas-relief (GBR) ambiguity. Based on this insight, we propose a guided calibration network, named *GCNet*, that explicitly leverages object shape and shading information for improved lighting estimation. Experiments on synthetic and real datasets show that GCNet achieves improved results in lighting estimation for photometric stereo, which echoes the findings of our analysis. We further demonstrate that GCNet can be directly integrated with existing calibrated methods to achieve improved results on surface normal estimation. Our code and model can be found at https://guanyingc.github.io/UPS-GCNet.

Keywords: Uncalibrated photometric stereo · Generalized bas-relief ambiguity · Deep neural network

1 Introduction

Photometric stereo aims at recovering the surface normals of a scene from single-viewpoint imagery captured under varying light directions [47,50]. In contrast to multi-view stereo [41], photometric stereo works well for textureless surfaces and can recover highly detailed surface geometry.

Electronic supplementary material The online version of this chapter (https://doi.org/10.1007/978-3-030-58568-6_44) contains supplementary material, which is available to authorized users.

A. Vedaldi et al. (Eds.): ECCV 2020, LNCS 12359, pp. 745–762, 2020.
https://doi.org/10.1007/978-3-030-58568-6_44

Following the conventional assumption, this paper assumes the scene is illuminated by a single light direction in each image. Most existing photometric stereo methods [22,23,46] require *calibrated* light directions as input. *Uncalibrated* photometric stereo, on the other hand, simultaneously estimates light directions and surface normals. In multi-view stereo, this problem of auto-calibration (*i.e.*, calibration from images of the scene without the use of any explicit calibration targets) has been solved satisfactorily on "wild" imagery such as those from Internet photo sharing sites [3]. Auto-calibration for photometric stereo is without a doubt an important goal since it makes photometric stereo applicable to wild data [43] and useful for amateurs who know nothing about tedious calibration procedures [1,39,42]. Existing methods for uncalibrated photometric stereo [4,35] often assume a Lambertian reflectance model and their focus has been on resolving the generalized bas-relief (GBR) ambiguity [6]. Manifold embedding based methods [29,40] can deal with surfaces with general isotropic reflectances, but they rely on a roughly uniform lighting distribution which is usually not satisfied in real-world datasets.

Despite the impressive results on complex reflectances reported by recent deep learning methods for calibrated photometric stereo [10,22,27,38,49,53], not much work has been done on learning-based uncalibrated photometric stereo. Recently, Chen *et al.* [8] introduced a deep uncalibrated photometric stereo network, called Lighting Calibration Network (LCNet), to estimate light directions and intensities from input images, and a normal estimation network to predict surface normals. Compared with UPS-FCN [10] which directly estimates surface normals from images, Chen *et al.*'s two-stage approach achieves considerably better results. However, the features learned by LCNet to resolve the ambiguity in lighting estimation remain unknown.

This paper focuses on demystifying the problem of how deep uncalibrated photometric stereo learns to resolve the GBR ambiguity, and how to improve it for higher accuracy in lighting estimation. Our contributions are:

- We discuss the differences between the learning-based LCNet [8] and traditional uncalibrated methods, and analyze the features learned by LCNet to resolve the GBR ambiguity.
- We find that attached shadows, shadings, and specular highlights are key elements for lighting estimation, and that LCNet extracts features independently from each input image without exploiting any inter-image information ("inter-image" means information shared by all images).
- Based on our findings, we propose a guided calibration network (GCNet) that explicitly utilizes object shape and shading information as guidances for better lighting estimation.

2   Related Work

In this section, we briefly review recent deep learning methods for calibrated photometric stereo and existing methods for uncalibrated photometric stereo. Readers are referred to [2,20,45] for more comprehensive surveys of photometric stereo algorithms. In the rest of this paper, we will use "lighting" to refer to light direction and light intensity.

Deep Calibrated Photometric Stereo. Recently, deep learning methods have been proposed in the context of photometric stereo. Compared with traditional methods that often adopt a simplified reflectance model, learning-based methods can directly learn the mapping from observations to surface normals and achieve state-of-the-art results on a real-world benchmark [45] with complex reflectances. Santo et al. [38] first introduced a fully-connected deep photometric stereo network to estimate pixel-wise surface normals from a fixed number of observations. To handle a variable number of input images in an order-agnostic manner, Ikehata [22] proposed a fixed shape observation map representation, while Chen et al. [9] adopted an element-wise max-pooling operation to fuse features stemming from multiple inputs. Li et al. [27] and Zheng et al. [53] focused on reducing the number of required images while maintaining similar accuracy under the framework proposed by Ikehata [22]. Different from the above supervised methods that require a synthetic dataset for training, Taniai and Maehara [49] proposed an unsupervised framework to estimate surface normals via an on-line optimization process.

Uncalibrated Photometric Stereo. Most existing uncalibrated photometric stereo methods are based on matrix factorization (e.g., singular value decomposition) and assume a Lambertian reflectance model. A Lambertian surface's normals can be recovered up to a 3×3 linear ambiguity when light directions are unknown [19]. By considering the surface integrability constraint, this linear ambiguity can be reduced to a 3-parameter generalized bas-relief (GBR) ambiguity [6,15,26,52]. To further resolve the GBR ambiguity, many methods make use of additional clues like inter-reflections [7], specularities [12,13,16], albedo priors [4,44], isotropic reflectance symmetry [48,51], special light source distributions [54], or Lambertian diffuse reflectance maxima [35].

Manifold embedding methods [29,30,34,40] can handle surfaces with general isotropic reflectance based on the observation that the distance between two surface points' intensity profiles is closely related to their surface normals' angular difference. However, these methods often assume a uniform lighting distribution. Other methods related to uncalibrated photometric stereo include exemplar-based methods [21], regression-based methods [32], semi-calibrated photometric stereo [11], inaccurate lighting refinement [37], and photometric stereo under general lighting [5,18,33].

Recently, Chen et al. [8] introduced a Lighting Calibration Network (LCNet) to estimate lightings from images and then estimate surface normals based on the lightings. This two-stage method achieves considerably better results than the single-stage method [10]. It also has slightly better interpretability because the lightings estimated in the first stage can be visualized. However, the features learned by LCNet to estimate lightings remain unknown.

3   Learning for Lighting Calibration

In this section, we discuss the inherent ambiguity in uncalibrated photometric stereo of Lambertian surfaces, the fact that it can be resolved for non-Lambertian surfaces, and the features learned by LCNet [8] to resolve such ambiguity.

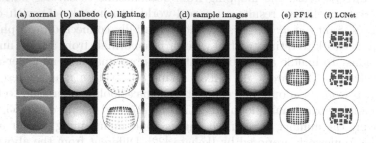

Fig. 1. Row 1 is the true shape of a *Sphere*, while rows 2 and 3 are shapes under two different GBR transformations. In column (c), the points' positions and colors indicate light direction and relative intensity, respectively. Columns (e) and (f) show the lightings estimated by PF14 [35] and LCNet [8].

Lambertian Surfaces and the GBR Ambiguity. When ignoring shadows (*i.e.*, attached and cast shadows) and inter-reflections, the image formation of a Lambertian surface with P pixels captured under F lightings can be written as

$$\mathbf{M} = \mathbf{N}^\top \mathbf{L}, \tag{1}$$

where $\mathbf{M} \in \mathbb{R}^{P \times F}$ is the measurement matrix. $\mathbf{N} \in \mathbb{R}^{3 \times P}$ is the surface normal matrix whose columns are albedo scaled normals $\mathbf{N}_{:,p} = \rho_p \boldsymbol{n}_p$, where ρ_p and $\boldsymbol{n}_p$ are the albedo and unit-length surface normal of pixel p. $\mathbf{L} \in \mathbb{R}^{3 \times F}$ is the lighting matrix whose columns are intensity scaled light directions $\mathbf{L}_{:,f} = e_f \boldsymbol{l}_f$, where e_f and $\boldsymbol{l}_f$ are the light intensity and unit-length light direction of image f.

By matrix factorization and applying the surface integrability constraint, $\mathbf{N}$ and $\mathbf{L}$ can be recovered up to an unknown 3-parameter GBR transformation $\mathbf{G}$ [6] such that $\mathbf{M} = (\mathbf{G}^{-\top}\mathbf{N})^\top(\mathbf{GL})$. This GBR ambiguity indicates that there are infinitely many combinations of albedo ρ, normal $\boldsymbol{n}$, light direction $\boldsymbol{l}$, and light intensity e that produce the same appearance $\mathbf{M}$ (see Fig. 1 (a)–(d)):

$$\hat{\rho} = \rho|\mathbf{G}^{-\top}\boldsymbol{n}|, \ \hat{\boldsymbol{n}} = \frac{\mathbf{G}^{-\top}\boldsymbol{n}}{|\mathbf{G}^{-\top}\boldsymbol{n}|}, \ \hat{\boldsymbol{l}} = \frac{\mathbf{G}\boldsymbol{l}}{|\mathbf{G}\boldsymbol{l}|}, \ \hat{e} = e|\mathbf{G}\boldsymbol{l}|. \tag{2}$$

Although the surface's appearance remains the same after GBR transformation (*i.e.*, $\hat{\rho}\hat{\boldsymbol{n}}^\top\hat{\boldsymbol{l}}\hat{e} = \rho\boldsymbol{n}^\top\boldsymbol{l}e$, see Fig. 1 (d)), a surface point's albedo will be scaled by $|\mathbf{G}^{-\top}\boldsymbol{n}|$. As a result, the albedo of an object will change gradually and become spatially-varying. Because this kind of spatially-varying albedo distribution resulting from GBR transformations rarely occurs on real world objects,

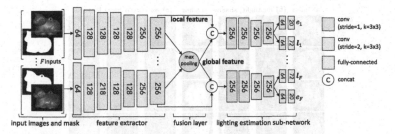

Fig. 2. Network architecture of LCNet [8]. Each layer's value indicates its output channel number. LCNet first extracts a local feature for each input with a shared-weight feature extractor. All local features are aggregated by element-wise max-pooling to produce the global feature, each local feature is concatenated with the global feature, and is fed into a shared-weight lighting estimation sub-network to estimate a 3D light direction l and a scalar light intensity e for each image.

Table 1. Light direction estimation results of PF14 [35] and LCNet [8] on a *Sphere* rendered with different BRDF types. Non-Lambertian BRDFs are taken from the MERL dataset [31].

model	Lambertian	fabric	plastic	phenolic	metallic	avg.
PF14	7.19	14.26	28.04	47.96	31.12	25.7
LCNet	5.38	4.07	3.08	3.05	4.09	3.93

Table 2. Light direction estimation results of LCNet [8] trained with different inputs. Values indicate mean angular error in degree.

Model input	Sphere	Bunny	Dragon	Armadillo
Images	3.03	4.88	6.30	6.37
(a) attached shadows	3.50	5.07	9.78	5.22
(b) specular component	2.53	6.18	7.33	4.08
(c) shading	**2.29**	**3.95**	**4.64**	**3.76**
(a) + (b) + (c)	1.87	2.06	2.34	2.12

some previous methods make explicit assumptions on the albedo distribution (*e.g.*, constant albedo [6,35] or low entropy [4]) to resolve the ambiguity.

PF14 [35], a state-of-the-art non-learning uncalibrated method [45], detects Lambertian diffuse reflectance maxima (*i.e.*, image points satisfying $n^{\top} l = 1$) to estimate **G**'s 3 parameters. We will later use it as a non-learning benchmark in our comparative experiments.

LCNet and the GBR Ambiguity. LCNet [8] is a state-of-the-art lighting calibration network for uncalibrated photometric stereo (see Fig. 2). Figure 1 (e)–(f) compare the results of LCNet and PF14 on surfaces that differ by GBR transformations. Since the input images are the same in all cases, LCNet estimates the same lightings in all cases, namely the most likely lightings for the input images. The same also applies to PF14. Although LCNet's result does not exactly equal the lightings for uniform albedo, we note that it learned from the training data that GBR-transformed surfaces are unlikely.

Although uncalibrated photometric stereo has an intrinsic GBR ambiguity for Lambertian surfaces, it was shown that GBR transformations do not preserve specularities [6,12,16]. Hence, specularities are helpful for ambiguity-free

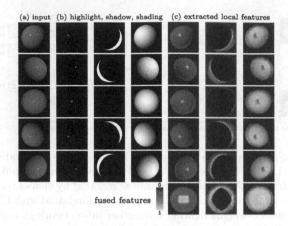

(a) input (b) highlight, shadow, shading   (c) extracted local features

fused features

Fig. 3. Feature visualization of LCNet on a non-Lambertian *Sphere*. Column 1: 5 of the 96 input images; Columns 2–4: Specular highlight centers, attached shadows, and shading rendered from ground truth; Columns 5–7: 3 of LCNet's 256 features maps. The last row shows the global features produced by fusing local features with max-pooling. All features are normalized to $[0, 1]$ and color coded.

lighting estimation. However, traditional methods often treat non-Lambertian observations as outliers, and thus fail to make full use of specularities for disambiguation [35]. In contrast, learning-based methods can learn the relation between specular highlights and light directions through end-to-end learning. As shown in Table 1, LCNet achieves good results for non-Lambertian surfaces while PF14 completely fails when non-Lambertian observations dominate.

Feature Analysis for LCNet. To analyze the features learned by LCNet, we first visualize the learned local and global features. Figure 3 shows 3 representative features selected from 256 feature maps extracted by LCNet from images of a non-Lambertian *Sphere* dataset[1]. Comparing Fig. 3's Column 2 with Column 5, Column 3 with Column 6, and Column 4 with Column 7, we can see that some feature maps are highly correlated with attached shadows (regions where the angle $\angle(n, l) \geq 90°$), shadings ($n^\top l$), and specular highlights (regions where n is close to the half angle $h = \frac{l+v}{|l+v|}$ of l and viewing direction v). As discussed earlier, these provide strong clues for resolving the ambiguity.

To further verify our observations, we did the following. We computed (a) attached shadows, (b) the "specular components" (with a bit of concept abuse, we denote $h^\top n$ as specular component), and (c) shadings for the publicly available synthetic Blobby and Sculpture datasets [10] from their ground-truth light directions and normals. We then trained 4 variants of the LCNet, taking (a), (b), (c), and (a) + (b) + (c), respectively, as input instead of regular images. We compared these 4 variant networks with LCNet (*i.e.*, the network trained with Blobby and Sculpture images) on a synthetic test dataset introduced in

---

[1] Please refer to our supplemental material for more visualizations.

Sect. 5.1. Similar to LCNet, the variant networks also took the object mask as input. Table 2 shows that the variant models achieve results comparable to or even better than the model trained on regular images.

We can see that shadings contribute more than attached shadows and specular components for lighting estimation. This is because shading information actually includes attached shadows (*i.e.*, pixels with a zero value in the shading for synthetic data), and can be considered as an image with a uniform albedo. The uniform albedo constraint is a well-known clue for resolving the GBR ambiguity [6,35]. In practice, attached shadows, shadings, and the specular components are not directly available as input, but this confirms our assumption that they provide strong clues for accurate lighting estimation.

4 Guided Calibration Network

In this section, we present the motivations for our guided calibration network (GCNet) and detail its structure.

4.1 Guided Feature Extraction

As we have seen, features like attached shadows, shadings, and specularities are important for lighting estimation, and a lighting estimation network may benefit greatly from being able to estimate them accurately. We further know that these features are completely determined by the light direction for each image as well as the inter-image shape information derived from the surface normal map. However, LCNet extracts features independently from each input image and thus cannot exploit any inter-image information during feature extraction. This observation also indicates that simply increasing the layer number of LCNet's shared-weight feature extractor cannot produce significant improvement.

Surface Normal as Inter-image Guidance. Intuitively, if we can provide such inter-image shape information as input to the network to guide the feature extraction process, it should be able to perform better. This, however, constitutes a chicken-and-egg problem where we require normals and lightings for accurate feature extraction but at the same time we require these features for estimating accurate lightings. We therefore suggest a cyclic network structure in which we first estimate initial lightings, and then use them to estimate normals as inter-image information to guide the extraction of local (*i.e.*, per-image) features to ultimately estimate final lightings. An alternative idea might be directly estimating surface normals from images. However, previous work (UPS-FCN [10]) shows that surface normals estimated directly from images are not accurate.

Shading as Intra-image Guidance. Another advantage of first estimating initial lighting and surface normals is that we can easily compute coarse attached shadows, shadings, or specular components as intra-image guidance for the feature extraction process (intra-image means the information is different for each image). As shading information already includes attached shadows, and not all

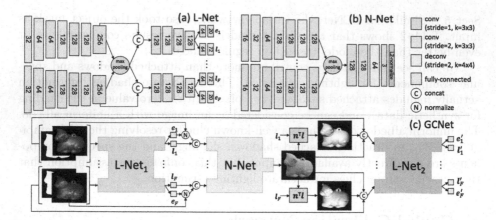

Fig. 4. Structure of (a) the lighting estimation sub-network L-Net, (b) the normal estimation sub-network N-Net, and (c) the entire GCNet. Values in layers indicate the output channel number.

materials exhibit specular highlights, we only compute the shading for each image as the dot-product of the estimated lighting with the surface normals, and use it as intra-image guidance. We experimentally verified that additionally including the specular component as network input does not improve results. The computed shading, on the other hand, does improve results and can assist the network to extract better features.

4.2 Network Architecture

As shown in Fig. 4, the proposed GCNet consists of two lighting estimation sub-networks (L-Net) and a normal estimation sub-network (N-Net). The first L-Net, "L-Net$_1$", estimates initial lightings given the input images and object masks. The N-Net then estimates surface normals from the lightings estimated by L-Net$_1$ and the input images. Finally, the second L-Net, "L-Net$_2$", estimates more accurate lightings based on the input images, object masks, the estimated normals, and the computed shadings.

L-Net. The L-Net is designed based on LCNet [8] but has less channels in the convolutional layers to reduce the model size (see Fig. 4 (a)). Compared to LCNet's 4.4×10^6 parameters, each L-Net has only 1.78×10^6 parameters.

Following LCNet, we discretize the lighting space and treat lighting estimation as a classification problem. Specifically, L-Net's output light direction and intensity are in the form of softmax probability vectors (a 32-vector for elevation, a 32-vector for azimuth, and a 20-vector for intensity). Given F images, the loss function for L-Net is

$$\mathcal{L}_{\text{light}} = \frac{1}{F} \sum_f (\mathcal{L}_{l_a}^f + \mathcal{L}_{l_e}^f + \mathcal{L}_e^f), \tag{3}$$

where $\mathcal{L}_{l_a}^f, \mathcal{L}_{l_e}^f$, and $\mathcal{L}_e^f$ are the cross-entropy loss for light azimuth, elevation, and intensity classifications for the fth input image, respectively. For example,

$$\mathcal{L}_{l_a}^f = -\sum_{i=1}^{32}\{y_i^f = 1\}\log(p_i^f), \tag{4}$$

where $\{\cdot\}$ is a binary indicator (0 or 1) function, y_i^f is the ground-truth label (0 or 1) and p_i^f is the predicted probability for bin i (32 bins in our case) for the fth image. The output probability vectors can be converted to a 3-vector light direction l_f and a scalar light intensity e_f by taking the probability vector's expectation, which is differentiable for later end-to-end fine-tuning.

L-Net$_1$ and L-Net$_2$ differ in that L-Net$_1$ has 4 input channels (3 for the image, 1 for the object mask) while L-Net$_2$ has 8 (3 additional channels for normals, 1 for shading).

N-Net. The N-Net is designed based on PS-FCN [10] but with less channels, resulting in 1.1×10^6 parameters compared to PS-FCN's 2.2×10^6 parameters (see Fig. 4 (b) for details). Following PS-FCN, the N-Net's loss function is

$$\mathcal{L}_{\text{normal}} = \frac{1}{P}\sum_p \left(1 - n_p^\top \tilde{n}_p\right), \tag{5}$$

where P is the number of pixels per image, and n_p and $\tilde{n}_p$ are the predicted and the ground-truth normal at pixel p, respectively.

End-to-End Fine-Tuning. We train L-Net$_1$, N-Net, and L-Net$_2$ one after another until convergence and then fine-tune the entire network end-to-end using the following loss

$$\mathcal{L}_{\text{fine-tune}} = \mathcal{L}_{\text{light}_1} + \mathcal{L}_{\text{normal}} + \mathcal{L}_{\text{shading}} + \mathcal{L}_{\text{light}_2}, \tag{6}$$

where $\mathcal{L}_{\text{light}_1}$ and $\mathcal{L}_{\text{light}_2}$ denote the lighting estimation loss for L-Net$_1$ and L-Net$_2$. The shading loss term $\mathcal{L}_{\text{shading}} = \frac{1}{FP}\sum_f\sum_p(n_p^\top l_f - \tilde{n}_p^\top \tilde{l}_f)^2$ is included to encourage better shading estimation, and l_f and $\tilde{l}_f$ denote the light direction predicted by L-Net$_1$ and ground-truth light direction for the fth image.

Training Details. Following LCNet [8], we trained the networks on the publicly available synthetic Blobby and Sculpture Dataset [10] which contains $85,212$ surfaces, each rendered under 64 random light directions.

First, we train L-Net$_1$ from scratch for 20 epochs, halving the learning rate every 5 epochs. Second, we train N-Net using ground-truth lightings and input images following PS-FCN's training procedure [10], and then retrain N-Net given the lightings estimated by L-Net$_1$ for 5 epochs, halving the learning rate every 2 epochs. Third, we train L-Net$_2$ given the input images, object masks, estimated normals, and computed shadings for 20 epochs, halving the learning rate every 5 epochs. The initial learning rate is 0.0005 for L-Net$_1$ and L-Net$_2$, and 0.0002 for retraining N-Net. End-to-end training is done for 20 epochs with an initial learning rate of 0.0001, halving it every 5 epochs.

We implemented our framework in PyTorch [36] and used the Adam optimizer [25] with default parameters. The full network has a total of 4.66×10^6 parameters which is comparable to LCNet (4.4×10^6). The batch size and the input image number for each object are fixed to 32 during training. The input images are resized to 128×128 at both training and test time.

5 Experimental Results

In this section, we evaluate our method on synthetic and real data. For measuring estimation accuracy, we used mean angular error (MAE) for light directions and surface normals, and scale-invariant relative error [8] for light intensities.

5.1 Evaluation on Synthetic Data

To quantitatively analyze the effects of object shapes, biased lighting distributions, and spatially-varying BRDFs (SVRBDFs) on the proposed method, we rendered a synthetic test dataset using the physically-based raytracer Mitsuba [24]. We rendered 4 different shapes (*Sphere, Bunny, Dragon* and *Armadillo*) using 100 MERL BRDFs [31], resulting in 400 test objects, each illuminated under 82 randomly sampled light directions. At test time, we randomly generated relative light intensities in the range $[0.2, 2.0]$ to scale the magnitude of the images (see Fig. 5).

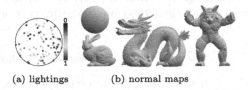

(a) lightings   (b) normal maps

Fig. 5. (a) Lighting distribution of the synthetic test dataset. (b) Normal maps of *Sphere, Bunny, Dragon* and *Armadillo*.

Ablation Study. To validate the design of the proposed network, we performed an ablation study and summarized the results in Table 3. The comparison between experiments with IDs 2–4 verifies that taking both the estimated normals and shading as input is beneficial for lighting estimation. The comparison between experiments with IDs 1 & 2 demonstrates that end-to-end fine-tuning further improves the performance. We can also see that L-Net$_1$ achieves results comparable to LCNet despite using only half of the network parameters, which indicates that simply increasing the channel number of the convolutional layers cannot guarantee better feature extraction. In the rest of the paper, we denote the results of "L-Net$_1$ + N-Net + L-Net$_2$ + Finetune" as "GCNet".

Table 3. Lighting estimation results on the synthetic test dataset. The results are averaged over 100 MERL BRDFs (bold fonts indicates the best).

ID	model	Sphere		Bunny		Dragon		Armadillo	
		Direction	Intensity	Direction	Intensity	Direction	Intensity	Direction	Intensity
0	LCNet [8]	3.03	0.064	4.88	0.066	6.30	0.072	6.37	0.074
1	L-Net$_1$ + N-Net + L-Net$_2$ + Finetune	**2.21**	**0.042**	**2.44**	**0.046**	**3.88**	**0.055**	**3.52**	0.060
2	L-Net$_1$ + N-Net + L-Net$_2$	2.52	0.052	2.90	0.054	4.20	0.061	3.92	0.060
3	L-Net$_1$ + N-Net + L-Net$_2^{(w/o\ normal)}$	2.45	0.050	3.35	0.051	5.82	0.070	5.25	**0.059**
4	L-Net$_1$	3.20	0.053	4.47	0.060	5.80	0.081	5.71	0.079

Table 4. Normal estimation results on the synthetic test dataset. The estimated normals are predicted by PS-FCN [10] given the lightings estimated by LCNet and GCNet.

Model	Sphere	Bunny	Dragon	Armadillo
LCNet [8] + PS-FCN [10]	2.98	4.06	5.59	6.73
GCNet + PS-FCN [10]	**2.93**	**3.68**	**4.85**	**5.01**

Table 5. Results on *Armadillo* under different lighting distributions. Direction, intensity, and normal are abbreviated to "dir.", "int.", and "norm.".

model	near uniform			narrow			upward-biased		
	dir.	int.	norm.	dir.	int.	norm.	dir.	int.	norm.
LCNet [8] + PS-FCN	6.09	0.072	6.49	5.92	0.059	8.44	7.10	0.065	8.80
GCNet + PS-FCN	**3.39**	**0.059**	**4.90**	**4.29**	**0.048**	**6.82**	**5.96**	**0.054**	**7.53**

Table 4 shows that, as expected, the calibrated photometric stereo method PS-FCN [10] can estimate more accurate normals given better estimated lighting.

Results on Different Lighting Distributions. To analyze the effect of biased lighting distributions on the proposed method, we evaluated GCNet on the *Armadillo* illuminated under three different lighting distributions: a near uniform, a narrow, and an upward-biased distribution. Table 5 shows that both GCNet and LCNet have decreased performance under biased lighting distributions (*e.g.*, the upward-biased distribution), but GCNet consistently outperforms LCNet.

Table 6. Lighting estimation results on *Bunny* rendered with SVBRDFs. (a) *Bunny* with uniform BRDF. (b) and (c) show the "ramp" and "irregular" material maps and two sample images of *Bunny* with the corresponding SVBRDFs.

model	uniform		ramp		irregular	
	direction	intensity	direction	intensity	direction	intensity
LCNet [8]	4.88	0.066	6.09	0.066	6.00	0.075
GCNet	**2.44**	**0.046**	**4.16**	**0.043**	**3.68**	**0.050**

(a) uniform   (b) ramp   (c) irregular

Table 7. Lighting estimation results on surface regions cropped from *Bunny*.

input	alumina-oxide				beige-fabric					
	LCNet		GCNet		LCNet		GCNet			
	dir.	int.	dir.	int.	dir.	int.	dir.	int.		
(a)	4.29	0.054	**1.35**	**0.025**	4.54	0.051	**2.29**	**0.026**		
(b)	3.83	0.050	**1.71**	**0.023**	4.45	0.044	**2.00**	**0.029**		
(c)	3.75	0.042	**2.46**	**0.024**	4.97	0.044	**3.13**	**0.025**		
(d)	4.04	0.047	**2.84**	**0.026**	4.55	0.051	**3.46**	**0.025**		

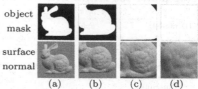

object mask

surface normal

(a)   (b)   (c)   (d)

Results on Surfaces with SVBRDFs. To analyze the effect of SVBRDFs, we used two different material maps to generate a synthetic dataset of surfaces with SVBRDFs following Goldman *et al.* [17]. Specifically, we rendered 100 test objects by randomly sampling two MERL BRDFs and blended the BRDFs for *Bunny* using "ramp" and "irregular" material maps shown in Table 6 (b) and (c). Table 6 shows that although both methods perform worse on surfaces with SVBRDFs compared to uniform BRDFs, our method is still reasonably good even though it was trained on surfaces with uniform BRDFs. This might be explained by that although SVBRDFs may affect the feature extraction of some important clues such as shading, others such as attached shadows and specular highlights are less affected and can still be extracted to estimate reliable lightings.

Effect of the Object Silhouette. Object silhouette can provide useful information for lighting calibration (*e.g.*, normals at the occluding contour are perpendicular to the viewing direction). To investigate the effect of the silhouette, we first rendered the *Bunny* using two different types of BRDFs (*alumina-oxide* and *beige-fabric*) under 100 lightings sampled randomly from the upper hemisphere, and then cropped surface regions with different sizes for testing. Table 7 shows that both LCNet and our method perform robustly for surface regions with or without silhouette, while our method consistently outperforms LCNet. This is because the training data for both methods was generated by randomly cropping image patches from the Blobby and Sculpture datasets [10], which contains surface regions without silhouette.

Runtime. The runtimes of LCNet and GCNet for processing an object (96 images in total) from the DiLiGenT benchmark are ~0.25 s and ~0.5 s including data loading and network feed-forward time, measured on a single 1080 Ti GPU. Even though LCNet runs slightly faster, both methods are very fast and run within a second.

5.2   Evaluation on Real Data

To demonstrate the proposed method's capability to handle real-world non-Lambertian objects, we evaluated our method on the challenging *DiLiGenT* benchmark [45] and the *Light Stage Data Gallery*[14].

Table 8. Lighting estimation results on DiLiGenT benchmark.

model	Ball dir.	int.	Cat dir.	int.	Pot1 dir.	int.	Bear dir.	int.	Pot2 dir.	int.	Buddha dir.	int.	Goblet dir.	int.	Reading dir.	int.	Cow dir.	int.	Harvest dir.	int.	average dir.	int.
PF14 [35]	4.90	0.036	5.31	**0.059**	2.43	**0.017**	5.24	0.098	13.52	**0.044**	9.76	0.053	33.22	0.223	21.77	0.122	16.34	0.074	24.99	0.156	13.75	0.088
LCNet [8]	3.27	0.039	**4.08**	0.095	5.44	0.058	3.47	**0.061**	2.87	0.048	4.34	0.048	10.36	0.067	**4.50**	0.105	4.52	0.073	6.32	0.082	4.92	0.068
GCNet	**1.75**	**0.027**	4.58	0.075	**1.41**	0.039	**2.44**	0.101	**2.81**	0.059	**2.86**	**0.032**	**2.98**	**0.042**	5.47	**0.048**	**3.15**	**0.031**	**5.74**	**0.065**	**3.32**	**0.052**

Table 9. Lighting estimation results on Light Stage Data Gallery.

model	Helmet Side dir.	int.	Plant dir.	int.	Fighting Knight dir.	int.	Kneeling Knight dir.	int.	Standing Knight dir.	int.	Helmet Front dir.	int.	average dir.	int.
PF14 [35]	25.40	0.576	20.56	0.227	69.50	1.137	46.69	9.805	33.81	1.311	81.60	**0.133**	46.26	2.198
LCNet [8]	6.57	0.212	16.06	0.170	15.95	0.214	19.84	0.199	11.60	0.286	11.62	0.248	13.61	0.221
GCNet	**5.33**	**0.096**	**10.49**	**0.154**	**13.42**	**0.168**	**14.41**	**0.181**	**5.31**	**0.198**	**6.22**	0.183	**9.20**	**0.163**

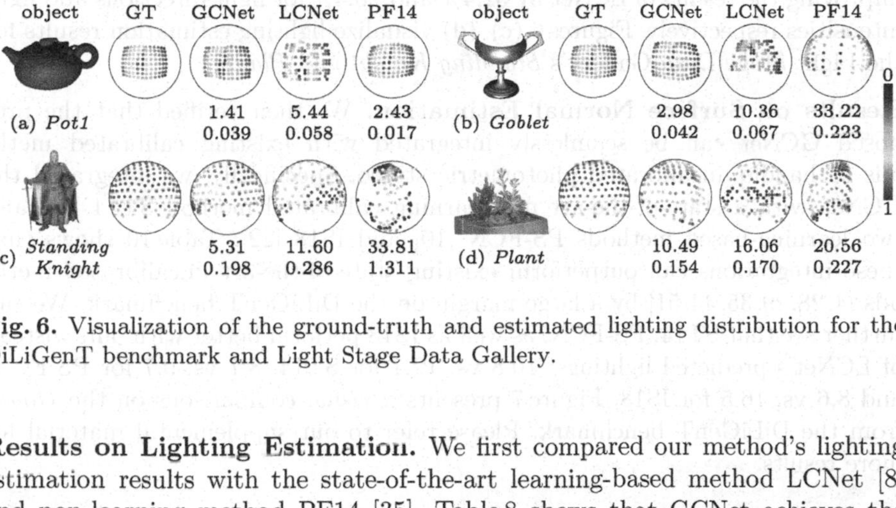

object	GT	GCNet	LCNet	PF14		object	GT	GCNet	LCNet	PF14	
(a) *Pot1*		1.41 0.039	5.44 0.058	2.43 0.017		(b) *Goblet*		2.98 0.042	10.36 0.067	33.22 0.223	0
(c) *Standing Knight*		5.31 0.198	11.60 0.286	33.81 1.311		(d) *Plant*		10.49 0.154	16.06 0.170	20.56 0.227	1

Fig. 6. Visualization of the ground-truth and estimated lighting distribution for the DiLiGenT benchmark and Light Stage Data Gallery.

Results on Lighting Estimation. We first compared our method's lighting estimation results with the state-of-the-art learning-based method LCNet [8] and non-learning method PF14 [35]. Table 8 shows that GCNet achieves the best average results on the DiLiGenT benchmark with an MAE of 3.32 for light directions and a relative error of 0.052 for light intensities. Although our method does not achieve the best results for all objects, it exhibits the most robust performance with a maximum MAE of 5.77 and a maximum relative error of 0.101 compared with LCNet (MAE: 10.36, relative error: 0.105) and PF14 (MAE: 33.22, relative error: 0.223). Figures 6 (a)–(b) visualize the lighting estimation results for the *Pot1* and the *Goblet*. The non-learning method PF14 works well for near-diffuse surfaces (*e.g.*, *Pot1*), but quickly degenerates on highly specular surfaces (*e.g.*, *Goblet*). Compared with LCNet, our method is more robust to surfaces with different reflectances and shapes.

Table 9 shows lighting estimation results on the Light Stage Data Gallery. Our method significantly outperforms LCNet and PF14, and achieves an average MAE of 9.20 for light directions and a relative error of 0.163 for light intensities,

Table 10. Normal estimation results on DiLiGenT benchmark. (* indicates the results of the calibrated method using ground-truth lightings as input.)

model	Ball	Cat	Pot1	Bear	Pot2	Buddha	Goblet	Reading	Cow	Harvest	average
AM07 [4]	7.3	31.5	18.4	16.8	49.2	32.8	46.5	53.7	54.7	61.7	37.3
SM10 [44]	8.9	19.8	16.7	12.0	50.7	15.5	48.8	26.9	22.7	73.9	29.6
WT13 [51]	4.4	36.6	9.4	6.4	14.5	13.2	20.6	59.0	19.8	55.5	23.9
LM13 [29]	22.4	25.0	32.8	15.4	20.6	25.8	29.2	48.2	22.5	34.5	27.6
PF14 [35]	4.8	9.5	9.5	9.1	15.9	14.9	29.9	24.2	19.5	29.2	16.7
LC18 [28]	9.3	12.6	12.4	10.9	15.7	19.0	18.3	22.3	15.0	28.0	16.3
LCNet + ST14	4.1	8.2	8.8	8.4	9.7	11.6	13.5	15.2	13.4	27.7	12.1
GCNet + ST14	2.0	7.7	7.5	5.7	9.3	10.9	10.0	14.8	13.5	26.9	10.8
ST14* [46]	1.7	6.1	6.5	6.1	8.8	10.6	10.1	13.6	13.9	25.4	10.3
LCNet + PS-FCN	3.2	7.6	8.4	11.4	7.0	8.3	11.6	14.6	7.8	17.5	9.7
GCNet + PS-FCN	2.5	7.9	7.2	5.6	7.1	8.6	9.6	14.9	7.8	16.2	8.7
PS-FCN* [10]	2.8	6.2	7.1	7.6	7.3	7.9	8.6	13.3	7.3	15.9	8.4
LCNet + IS18	6.4	15.6	10.6	8.5	12.2	13.9	18.5	23.8	29.3	25.7	16.5
GCNet + IS18	3.1	6.9	7.3	5.7	7.1	8.9	7.0	15.9	8.8	15.6	8.6
IS18* [22]	2.2	4.6	5.4	8.3	6.0	7.9	7.3	12.6	8.0	14.0	7.6

improving the results of LCNet by 32.4% and 26.4% for light directions and light intensities respectively. Figures 6 (c)–(d) visualize lighting estimation results for the Light Stage Data Gallery's *Standing Knight* and *Plant*.

Results on Surface Normal Estimation. We then verified that the proposed GCNet can be seamlessly integrated with existing calibrated methods to handle uncalibrated photometric stereo. Specifically, we integrated the GCNet with a state-of-the-art non-learning calibrated method ST14 [46] and two learning-based methods PS-FCN [10] and IS18 [22]. Table 10 shows that these integrations can outperform existing state-of-the-art uncalibrated methods [4,28,29,35,44,51] by a large margin on the DiLiGenT benchmark. We can further see that ST14, PS-FCN, as well as IS18 perform better with ours instead of LCNet's predicted lightings: 10.8 vs. 12.1 for ST14, 8.7 vs. 9.7 for PS-FCN, and 8.6 vs. 16.5 for IS18. Figure 7 presents a visual comparisons on the *Goblet* from the DiLiGenT benchmark. Please refer to our supplemental material for more results.

5.3 Failure Cases

As discussed in Sect. 3, LCNet [8] relies on features like attached shadows, shading, and specular highlights, which is also true for our method. For piecewise planar surfaces with a sparse normal distribution such as the one in Fig. 8 (a), few useful features can be extracted and as a result our method cannot predict reliable lightings for such surfaces. For highly-concave shapes under directional lightings, strong cast shadows largely affect the extraction of useful features. Figure 8 (b) shows that GCNet erroneously estimates a highly-concave bowl to be convex. Note that LCNet [8] and PF14 [35] also have similar problems.

GT / object GCNet + IS18 LCNet + IS18   PF14 [35]   GT / object GCNet + IS18 LCNet + IS18   PF14 [35]

(a) *Pot1*   MAE = 7.3 MAE = 10.6 MAE = 9.5   (b) *Goblet* MAE = 7.0 MAE = 18.5 MAE = 24.2

Fig. 7. Visual comparisons of normal estimation for *Pot1* and *Goblet* in the DiLi-GenT benchmark. We compared the normal estimation results of a calibrated method IS18 [22] given lightings estimated by our method and LCNet [8].

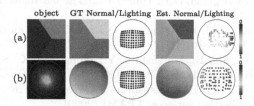

object   GT Normal/Lighting   Est. Normal/Lighting

Fig. 8. Failure cases. (a) Results on a piecewise planar surface with sparse normal distribution. (b) Results on a highly-concave bowl. The estimated normals are predicted by PS-FCN [8] given our method's estimated lightings.

6 Conclusions

This paper targeted discovering what is learned in deep uncalibrated photometric stereo to resolve the ambiguity. Specifically, we analyzed and discussed the behavior of the recent deep uncalibrated photometric stereo method LCNet. Based on our findings, we then introduced the guided calibration network (GCNet) that explicitly leverages inter-image information of object shape and intra-image information of shading to estimate more reliable lightings. Experiments on both synthetic and real datasets showed that our method significantly outperforms the state-of-the-art LCNet in lighting estimation, and demonstrated that our method can be integrated with existing calibrated photometric stereo methods to handle uncalibrated setups. Since strong cast shadows affect our method's feature extraction process and lead to unsatisfactory results, we will explore better methods to handle cast shadows in the future.

Acknowledgments. Michael Waechter was supported through a JSPS Postdoctoral Fellowship (JP17F17350). Boxin Shi is supported by the National Natural Science Foundation of China under Grant No. 61872012, National Key R&D Program of China (2019YFF0302902), and Beijing Academy of Artificial Intelligence (BAAI). Kwan-Yee K. Wong is supported by the Research Grant Council of Hong Kong (SAR), China, under the project HKU 17203119. Yasuyuki Matsushita is supported by JSPS KAK-ENHI Grant Number JP19H01123.

References

1. Ackermann, J., Fuhrmann, S., Goesele, M.: Geometric point light source calibration. In: Vision, Modeling & Visualization (2013)
2. Ackermann, J., Goesele, M.: A survey of photometric stereo techniques. Foundations and Trends in Computer Graphics and Vision (2015)
3. Agarwal, S., et al.: Building Rome in a day. Commun. ACM **54**, 105–112 (2011)
4. Alldrin, N.G., Mallick, S.P., Kriegman, D.J.: Resolving the generalized bas-relief ambiguity by entropy minimization. In: CVPR (2007)
5. Basri, R., Jacobs, D., Kemelmacher, I.: Photometric stereo with general, unknown lighting. IJCV **5**, 105–113 (2007)
6. Belhumeur, P.N., Kriegman, D.J., Yuille, A.L.: The bas-relief ambiguity. IJCV **35**, 33–44 (1999)
7. Chandraker, M.K., Kahl, F., Kriegman, D.J.: Reflections on the generalized bas-relief ambiguity. In: CVPR (2005)
8. Chen, G., Han, K., Shi, B., Matsushita, Y., Wong, K.Y.K.: Self-calibrating deep photometric stereo networks. In: CVPR (2019)
9. Chen, G., Han, K., Shi, B., Matsushita, Y., Wong, K.Y.K.: Deep photometric stereo for non-Lambertian surfaces. TPAMI (2020)
10. Chen, G., Han, K., Wong, K.-Y.K.: PS-FCN: a flexible learning framework for photometric stereo. In: Ferrari, V., Hebert, M., Sminchisescu, C., Weiss, Y. (eds.) ECCV 2018. LNCS, vol. 11213, pp. 3–19. Springer, Cham (2018). https://doi.org/10.1007/978-3-030-01240-3_1
11. Cho, D., Matsushita, Y., Tai, Y.W., Kweon, I.S.: Semi-calibrated photometric stereo. TPAMI (2018)
12. Drbohlav, O., Chantler, M.: Can two specular pixels calibrate photometric stereo? In: ICCV (2005)
13. Drbohlav, O., Šára, R.: Specularities reduce ambiguity of uncalibrated photometric stereo. In: Heyden, A., Sparr, G., Nielsen, M., Johansen, P. (eds.) ECCV 2002. LNCS, vol. 2351, pp. 46–60. Springer, Heidelberg (2002). https://doi.org/10.1007/3-540-47967-8_4
14. Einarsson, P., et al.: Relighting human locomotion with flowed reflectance fields. In: EGSR (2006)
15. Epstein, R., Yuille, A.L., Belhumeur, P.N.: Learning object representations from lighting variations. In: International Workshop on Object Representation in Computer Vision (1996)
16. Georghiades, A.S.: Incorporating the torrance and sparrow model of reflectance in uncalibrated photometric stereo. In: ICCV (2003)
17. Goldman, D.B., Curless, B., Hertzmann, A., Seitz, S.M.: Shape and spatially-varying BRDFs from photometric stereo. TPAMI (2010)
18. Haefner, B., Ye, Z., Gao, M., Wu, T., Quéau, Y., Cremers, D.: Variational uncalibrated photometric stereo under general lighting. In: ICCV (2019)
19. Hayakawa, H.: Photometric stereo under a light source with arbitrary motion. JOSA A **11**, 3079–3089 (1994)
20. Herbort, S., Wöhler, C.: An introduction to image-based 3D surface reconstruction and a survey of photometric stereo methods. 3D Research (2011)
21. Hertzmann, A., Seitz, S.M.: Example-based photometric stereo: shape reconstruction with general, varying BRDFs. TPAMI **27**, 1254–1264 (2005)

22. Ikehata, S.: CNN-PS: CNN-based photometric stereo for general non-convex surfaces. In: Ferrari, V., Hebert, M., Sminchisescu, C., Weiss, Y. (eds.) ECCV 2018. LNCS, vol. 11219, pp. 3–19. Springer, Cham (2018). https://doi.org/10.1007/978-3-030-01267-0_1
23. Ikehata, S., Aizawa, K.: Photometric stereo using constrained bivariate regression for general isotropic surfaces. In: CVPR (2014)
24. Jakob, W.: Mitsuba renderer (2010)
25. Kingma, D., Ba, J.: ADAM: a method for stochastic optimization. In: ICLR (2015)
26. Kriegman, D.J., Belhumeur, P.N.: What shadows reveal about object structure. JOSA A **18**, 1804–1813 (2001)
27. Li, J., Robles-Kelly, A., You, S., Matsushita, Y.: Learning to minify photometric stereo. In: CVPR (2019)
28. Lu, F., Chen, X., Sato, I., Sato, Y.: SymPS: BRDF symmetry guided photometric stereo for shape and light source estimation. TPAMI **40**, 221–234 (2018)
29. Lu, F., Matsushita, Y., Sato, I., Okabe, T., Sato, Y.: Uncalibrated photometric stereo for unknown isotropic reflectances. In: CVPR (2013)
30. Lu, F., Matsushita, Y., Sato, I., Okabe, T., Sato, Y.: From intensity profile to surface normal: Photometric stereo for unknown light sources and isotropic reflectances. TPAMI **37**, 1999–2012 (2015)
31. Matusik, W., Pfister, H., Brand, M., McMillan, L.: A data-driven reflectance model. In: SIGGRAPH (2003)
32. Midorikawa, K., Yamasaki, T., Aizawa, K.: Uncalibrated photometric stereo by stepwise optimization using principal components of isotropic BRDFs. In: CVPR (2016)
33. Mo, Z., Shi, B., Lu, F., Yeung, S.K., Matsushita, Y.: Uncalibrated photometric stereo under natural illumination. In: CVPR (2018)
34. Okabe, T., Sato, I., Sato, Y.: Attached shadow coding: estimating surface normals from shadows under unknown reflectance and lighting conditions. In: ICCV (2009)
35. Papadhimitri, T., Favaro, P.: A closed-form, consistent and robust solution to uncalibrated photometric stereo via local diffuse reflectance maxima. IJCV (2014)
36. Paszke, A., Gross, S., Chintala, S., Chanan, G.: PyTorch: tensors and dynamic neural networks in Python with strong GPU acceleration (2017)
37. Quéau, Y., Wu, T., Lauze, F., Durou, J.D., Cremers, D.: A non-convex variational approach to photometric stereo under inaccurate lighting. In: CVPR (2017)
38. Santo, H., Samejima, M., Sugano, Y., Shi, B., Matsushita, Y.: Deep photometric stereo network. In: ICCV Workshops (2017)
39. Santo, H., Waechter, M., Samejima, M., Sugano, Y., Matsushita, Y.: Light structure from pin motion: simple and accurate point light calibration for physics-based modeling. In: Ferrari, V., Hebert, M., Sminchisescu, C., Weiss, Y. (eds.) ECCV 2018. LNCS, vol. 11207, pp. 3–19. Springer, Cham (2018). https://doi.org/10.1007/978-3-030-01219-9_1
40. Sato, I., Okabe, T., Yu, Q., Sato, Y.: Shape reconstruction based on similarity in radiance changes under varying illumination. In: ICCV (2007)
41. Seitz, S.M., Curless, B., Diebel, J., Scharstein, D., Szeliski, R.: A comparison and evaluation of multi-view stereo reconstruction algorithms. In: CVPR (2006)
42. Shen, H.L., Cheng, Y.: Calibrating light sources by using a planar mirror. J. Electr. Imaging **20**(1), 013002-1–013002-6 (2011)
43. Shi, B., Inose, K., Matsushita, Y., Tan, P., Yeung, S.K., Ikeuchi, K.: Photometric stereo using internet images. In: 3DV (2014)
44. Shi, B., Matsushita, Y., Wei, Y., Xu, C., Tan, P.: Self-calibrating photometric stereo. In: CVPR (2010)

45. Shi, B., Mo, Z., Wu, Z., Duan, D., Yeung, S.K., Tan, P.: A benchmark dataset and evaluation for non-Lambertian and uncalibrated photometric stereo. TPAMI **41**, 271–284 (2019)
46. Shi, B., Tan, P., Matsushita, Y., Ikeuchi, K.: Bi-polynomial modeling of low-frequency reflectances. TPAMI **36**, 1078–1091 (2014)
47. Silver, W.M.: Determining shape and reflectance using multiple images. Ph.d. thesis, Massachusetts Institute of Technology (1980)
48. Tan, P., Mallick, S.P., Quan, L., Kriegman, D.J., Zickler, T.: Isotropy, reciprocity and the generalized bas-relief ambiguity. In: CVPR (2007)
49. Taniai, T., Maehara, T.: Neural inverse rendering for general reflectance photometric stereo. In: ICML (2018)
50. Woodham, R.J.: Photometric method for determining surface orientation from multiple images. Opt. Eng. (1980)
51. Wu, Z., Tan, P.: Calibrating photometric stereo by holistic reflectance symmetry analysis. In: CVPR (2013)
52. Yuille, A.L., Snow, D., Epstein, R., Belhumeur, P.N.: Determining generative models of objects under varying illumination: shape and albedo from multiple images using SVD and integrability. IJCV **35**, 203–222 (1999)
53. Zheng, Q., Jia, Y., Shi, B., Jiang, X., Duan, L.Y., Kot, A.C.: SPLINE-Net: sparse photometric stereo through lighting interpolation and normal estimation networks. In: ICCV (2019)
54. Zhou, Z., Tan, P.: Ring-light photometric stereo. In: Daniilidis, K., Maragos, P., Paragios, N. (eds.) ECCV 2010. LNCS, vol. 6312, pp. 265–279. Springer, Heidelberg (2010). https://doi.org/10.1007/978-3-642-15552-9_20

Prior-Based Domain Adaptive Object Detection for Hazy and Rainy Conditions

Vishwanath A. Sindagi$^{(\boxtimes)}$, Poojan Oza, Rajeev Yasarla, and Vishal M. Patel

Department of Electrical and Computer Engineering, Johns Hopkins University,
3400 N. Charles St, Baltimore, MD 21218, USA
{vishwanathsindagi,poza2,ryasarl1,vpatel36}@jhu.edu

Abstract. Adverse weather conditions such as haze and rain corrupt the quality of captured images, which cause detection networks trained on clean images to perform poorly on these corrupted images. To address this issue, we propose an unsupervised prior-based domain adversarial object detection framework for adapting the detectors to hazy and rainy conditions. In particular, we use weather-specific prior knowledge obtained using the principles of image formation to define a novel prior-adversarial loss. The prior-adversarial loss, which we use to supervise the adaptation process, aims to reduce the weather-specific information in the features, thereby mitigating the effects of weather on the detection performance. Additionally, we introduce a set of residual feature recovery blocks in the object detection pipeline to de-distort the feature space, resulting in further improvements. Evaluations performed on various datasets (Foggy-Cityscapes, Rainy-Cityscapes, RTTS and UFDD) for rainy and hazy conditions demonstrates the effectiveness of the proposed approach.

Keywords: Detection · Unsupervised domain adaptation · Adverse weather · Rain · Haze

1 Introduction

Object detection [12,16,17,30,39,48,53] is an extensively researched topic in the literature. Despite the success of deep learning based detectors on benchmark datasets [9,10,15,29], they have limited abilities in generalizing to several practical conditions such as adverse weather. This can be attributed mainly to the domain shift in the input images. One approach to solve this issue is to undo the effects of weather conditions by pre-processing the images using existing methods like image dehazing [11,19,60] and/or deraining [27,58,59]. However,

---

V. A. Sindagi and P. Oza—Equal contribution.

Electronic supplementary material The online version of this chapter (https://doi.org/10.1007/978-3-030-58568-6_45) contains supplementary material, which is available to authorized users.

© Springer Nature Switzerland AG 2020
A. Vedaldi et al. (Eds.): ECCV 2020, LNCS 12359, pp. 763–780, 2020.
https://doi.org/10.1007/978-3-030-58568-6_45

these approaches usually involve complex networks and need to be trained separately with pixel-level supervision. Moreover, as noted in [43], these methods additionally involve certain post-processing like gamma correction, which still results in a domain shift, thus prohibiting such approaches from achieving the optimal performance. Like [43], we observed minimal improvements in the detection performance when we used dehaze/derain methods as a pre-processing step before detection (see Sect. 4). Furthermore, this additional pre-processing results in increased computational overhead at inference, which is not preferable in resource-constrained/real-time applications. Another approach would be to retrain the detectors on datasets that include these adverse conditions. However, creating these datasets often involves high annotation/labeling cost [51].

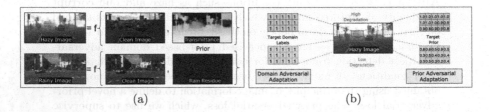

         (a)   (b)

Fig. 1. (a) Weather conditions can be modeled as function of clean image and the weather-specific prior, which we use to define a novel prior-adversarial loss. (b) Existing adaptation approaches use constant domain label for the entire. Our method uses spatially-varying priors that are directly correlated to the amount of degradations.

Recently, a few methods [6,41,45] have attempted to overcome this problem by viewing object detection in adverse weather conditions as an unsupervised domain adaptation task. These approaches consider that the images captured under adverse conditions (target images) suffer from a distribution shift [6,18] as compared to the images on which the detectors are trained (source images). It is assumed that the source images are fully annotated while the target images (with weather-based degradations) are not annotated. They propose different techniques to align the target features with the source features, while training on the source images. These methods are inherently limited in their approach since they employ only the principles of domain adaptation and neglect additional information that is readily available in the case of weather-based degradations.

We consider the following observations about weather-based degradations which have been ignored in the earlier work. *(i)* Images captured under weather conditions (such as haze and rain) can be mathematically modeled (see Fig. 1(a), Eqs. 8 and 9). For example, a hazy image is modeled by a superposition of a clean image (attenuated by transmission map) and atmospheric light [11,19]. Similarly, a rainy image is modeled as a superposition of a clean image and rain residue [27, 58,59] (see Fig. 1(a)). In other words, a weather-affected image contains weather specific information (which we refer to as prior) - transmission map in the case of hazy images and rain residue in the case of rainy images. These weather-specific information/priors cause degradations in the feature space resulting in poor detection performance. Hence, in order to reduce the degradations in the features, it is crucial to make the features weather-invariant by eliminating the weather-specific priors from the features. *(ii)* Further, it is important to note that the weather-based degradations are spatially varying and, hence do not affect the features equally at all spatial locations. Since, existing domain-adaptive detection approaches [6,41,45] label all the locations entirely either as target, they assume that the entire image has undergone constant degradation at all spatial locations (see Fig. 1(b)). This can potentially lead to incorrect alignment, especially in the regions of images where the degradations are minimal.

Motivated by these observations, we define a novel prior-adversarial loss that uses additional knowledge about the target domain (weather-affected images) for aligning the source and target features. Specifically, the proposed loss is used to train a prior estimation network to predict weather-specific prior from the features in the main branch, while simultaneously minimizing the weather-specific information present in the features. This results in weather-invariant features in the main branch, hence, mitigating the effects of weather. Additionally, the proposed use of prior information in the loss function results in spatially varying loss that is directly correlated to the amount of degradation (as shown in Fig. 1(b)). Hence, the use of prior can help avoid incorrect alignment.

Finally, considering that the weather-based degradations cause distortions in the feature space, we introduce a set of residual feature recovery blocks in the object detection pipeline to de-distort the features. These blocks, inspired by residual transfer framework proposed in [20], result in further improvements.

We perform extensive evaluations on different datasets such as Foggy-Cityscapes [43], RTTS [25] and UFDD [35]. Additionally, we create a Rainy-Cityscapes dataset for evaluating the performance different detection methods on rainy conditions. Various experiments demonstrate that the proposed method is able to outperform the existing methods on all the datasets.

2   Related Work

Object Detection: Object detection is one of the most researched topics in computer vision. Typical solutions for this problem have evolved from approaches involving sliding window based classification [8,53] to the latest anchor-based convolutional neural network approaches [30,38,39]. Ren *et al.* [39] pioneered the popular two stage Faster-RCNN approach. Several works have proposed single stage frameworks such as SSD [30], YOLO [38] *etc.*, that directly predict the object labels and bounding box co-ordinates. Following the previous work [6,23,24,41,45], we use Faster-RCNN as our base model.

Domain-Adaptive Object Detection in Adverse Conditions: Compared to the problem of general detection, detection in adverse weather conditions is relatively less explored. Existing methods, [6,24,41,45] have attempted to address this task from a domain adaptation perspective. Chen *et al.* [6] assumed that the adversarial weather conditions result in domain shift, and they overcome this by proposing a domain adaptive Faster-RCNN approach that tackles domain shift on image-level and instance-level. Following the similar argument of domain shift, Shan *et al.* [45] proposed to perform joint adaptation at image level using the Cycle-GAN framework [63] and at feature level using conventional domain adaptation losses. Saito *et al.* [41] proposed to perform strong alignment of the local features and weak alignment of the global features. Kim *et al.* [24] diversified the labeled data, followed by adversarial learning with the help of multi-domain discriminators. Cai *et al.* [5] addressed this problem in the semi-supervised setting using mean teacher framework. Zhu *et al.* [64] proposed region mining and region-level alignment in order to correctly align the source and target features. Roychowdhury *et al.* [40] adapted detectors to a new domain assuming availability of large number of video data from the target domain. These video data are used to generate pseudo-labels for the target set, which are further employed to train the network. Most recently, Khodabandeh *et al.* [23] formulated the domain adaptation training with noisy labels. Specifically, the model is trained on the target domain using a set of noisy bounding boxes that are obtained by a detection model trained only in the source domain.

3   Proposed Method

We assume that labeled clean data ($\{x_i^s, y_i^s\}_{i=1}^{n_s}$) from the source domain ($\mathcal{S}$) and unlabeled weather-affected data from the target domain ($\mathcal{T}$) are available. Here, y_i^s refers to all bounding box annotations and respective category label for the corresponding clean image x_i^s, x_i^t refers to the weather-affected image, n_s is the total number of samples in the source domain ($\mathcal{S}$) and n_t is the total number of samples in the target domain ($\mathcal{T}$). Our goal is to utilize the available information in both source and target domains to learn a network that lessens the effect of weather-based conditions on the detector. The proposed method contains three network modules – detection network, prior estimation network (PEN) and residual feature recovery block (RFRB). Figure 2 gives an overview of the proposed model. During source training, a source image (clean image) is passed to the detection network and the weights are learned by minimizing the detection loss, as shown in Fig. 2 with the source pipeline. For target training, a target image (weather-affected image) is forwarded through the network as shown in Fig. 2 by the target pipeline. As discussed earlier, weather-based degradations cause distortions in the feature space for the target images. In an attempt to de-distort these features, we introduce a set of residual feature recovery blocks in the target pipeline as shown in Fig. 2. This model is inspired from residual transfer framework proposed in [32] and is used to model residual features. The proposed PEN aids the detection network in adapting to the target domain by providing feedback through adversarial training using the proposed prior adversarial loss. In the following subsections, we briefly review the backbone network, followed by a discussion on the proposed prior-adversarial loss and residual feature recovery blocks.

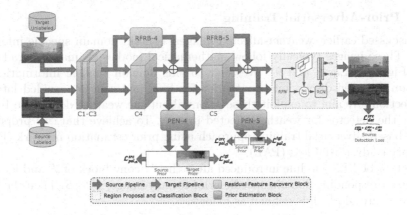

Fig. 2. Overview of the proposed adaptation method. We use prior adversarial loss to supervise the domain discriminators. For the source pipeline, additional supervision is provided by detection loss. For target pipeline, feed-forward through the detection network is modified by the residual feature recovery blocks.

3.1 Detection Network

Following the existing domain adaptive detection approaches [6,41,45], we base our method on the Faster-RCNN [39] framework. Faster-RCNN is among the first end-to-end CNN-based object detection methods and uses anchor-based strategy to perform detection and classification. For this paper we decompose the Faster-RCNN network into three network modules: feature extractor network ($\mathcal{F}$), region proposal network (RPN) stage and region classification network (RCN). The arrangement of these modules are shown in the Fig. 2 with VGG model architecture as base network. Here, the feature extractor network consists of first five conv blocks of VGG and region classification network module is composed of fully connected layers of VGG. The region proposal network uses output of feature extractor network to generate a set of candidate object regions in a class agnostic way. Features corresponding to these candidates are pooled from the feature extractor and are forwarded through the region classification network to get the object classifications and bounding box refinements. Since we have access to the source domain images and their corresponding ground truth, these networks are trained to perform detection on the source domain by minimizing the following loss function,

$$\min_{\mathcal{F},\,\mathcal{G}} \quad \mathcal{L}_{det}^{src}, \quad \text{where} \tag{1}$$

$$\mathcal{L}_{det}^{src} = \mathcal{L}_{rpn}^{src} + \mathcal{L}_{bbox}^{src} + \mathcal{L}_{rcn}^{src}. \tag{2}$$

Here, $\mathcal{G}$ represents both region proposal and region classification networks, $\mathcal{L}_{rpn}^{src}$ denotes the region proposal loss, $\mathcal{L}_{bbox}^{src}$ denotes the bounding-box regression loss and $\mathcal{L}_{rcn}^{src}$ denotes the region classification loss. The details of these individual loss components can be found in [39].

3.2 Prior-Adversarial Training

As discussed earlier, weather-affected images, contain domain specific information. These images typically follow mathematical models of image degradation (see Fig. 1(a), Eqs. 8 and 9). We refer to this domain specific information as a *prior*. Detailed discussion about prior for haze and rain is provided later in the section. We aim to exploit these priors about the weather domain to better adapt the detector for weather affected images. To achieve that, we propose a prior-based adversarial training approach using prior estimation network (PEN) and prior adversarial loss (PAL).

Let $\mathcal{P}_l$ be PEN module introduced after the lth conv block of $\mathcal{F}$ and let Z_{il}^{src} be the corresponding domain specific prior for any image, $x_i^s \in \mathcal{S}$. Then the PAL for the source domain is defined as follows,

$$\mathcal{L}_{pal_{cl}}^{src} = \frac{1}{n_s UV} \sum_{i=1}^{n_s} \sum_{j=1}^{U} \sum_{k=1}^{V} (Z_{il}^{src} - \mathcal{P}_l(\mathcal{F}_l(x_i^s)))_{jk}^2, \tag{3}$$

where, U and V are height and width of domain specific prior Z_{il}^{src} and output feature $\mathcal{F}_l(x_i^s)$. Z_{il}^{src} denotes the source image prior, scaled down from image-level prior to match the scale at l^{th} conv block. Similarly, PAL for the target domain images, $x_i^t \in \mathcal{T}$, with the corresponding prior Z_{il}^{tgt} can be defined as,

$$\mathcal{L}_{pal_{cl}}^{tgt} = \frac{1}{n_t UV} \sum_{i=1}^{n_t} \sum_{j=1}^{U} \sum_{k=1}^{V} (Z_{il}^{tgt} - \mathcal{P}_l(\mathcal{F}_l(x_i^t)))_{jk}^2, \tag{4}$$

where, we apply PAL after conv4 ($l = 4$) and conv5 ($l = 5$) block (as shown in Fig. 2). Hence, the final source and target adversarial losses can be given as,

$$\mathcal{L}_{pal}^{src} = \frac{1}{2}(\mathcal{L}_{pal_{c5}}^{src} + \mathcal{L}_{pal_{c4}}^{src}), \tag{5}$$

$$\mathcal{L}_{pal}^{tgt} = \frac{1}{2}(\mathcal{L}_{pal_{c5}}^{tgt} + \mathcal{L}_{pal_{c4}}^{tgt}). \tag{6}$$

The prior estimation networks ($\mathcal{P}_5$ and $\mathcal{P}_4$) predict the weather-specific prior from the features extracted from $\mathcal{F}$. However, the feature extractor network $\mathcal{F}$ is trained to fool the PEN modules by producing features that are weather-invariant (free from weather-specific priors) and prevents the PEN modules from correctly estimating the weather-specific prior. Since, this type of training includes prior prediction and is also reminiscent of the adversarial learning used in domain adaptation, we term this loss as prior-adversarial loss. At convergence, the feature extractor network $\mathcal{F}$ should have devoid itself from any weather-specific information and as a result both prior estimation networks $\mathcal{P}_5$ and $\mathcal{P}_4$ should not be able to correctly estimate the prior. *Note that our goal at convergence is not to estimate the correct prior, but rather to learn weather-invariant features so that the detection network is able to generalize well to the target domain.* This training procedure can be expressed as the following optimization,

$$\max_{\mathcal{F}} \min_{\mathcal{P}} \ \mathcal{L}_{pal}^{src} + \mathcal{L}_{pal}^{tgt}. \tag{7}$$

Furthermore, in the conventional domain adaptation, a single label is assigned for entire target image to train the domain discriminator (Fig. 1(c)). By doing this, it is assumed that the entire image has undergone a constant domain shift. However this is not true in the case of weather-affected images, where degradations vary spatially (Fig. 1(b)). In such cases, the assumption of constant domain shift leads to incorrect alignment especially in the regions of minimal degradations. Incorporating the weather-specific priors overcomes this issue as these priors are spatially varying and are directly correlated with the amount of degradations. Hence, utilizing the weather-specific prior results in better alignment.

Haze Prior. The effect of haze on images has been extensively studied in the literature [4,11,19,28,60–62]. Most existing image dehazing methods rely on the atmospheric scattering model for representing image degradations under hazy conditions and is defined as,

$$I(z) = J(z)t(z) + A(z)(1 - t(z)),   \tag{8}$$

where I is the observed hazy image, J is the true scene radiance, A is the global atmospheric light, indicating the intensity of the ambient light, t is the transmission map and z is the pixel location. The transmission map is a distance-dependent factor that affects the fraction of light that reaches the camera sensor. When the atmospheric light A is homogeneous, the transmission map can be expressed as $t(z) = e^{-\beta d(z)}$, where β represents the attenuation coefficient of the atmosphere and d is the scene depth.

Typically, existing dehazing methods first estimate the transmission map and the atmospheric light, which are then used in Eq. (8) to recover the observed radiance or clean image. The transmission map contains important information about the haze domain, specifically representing the light attenuation factor. We use this transmission as a domain prior for supervising the prior estimation (PEN) while adapting to hazy conditions. *Note that no additional human annotation efforts are required for obtaining the haze prior.*

Rain Prior. Similar to dehazing, image deraining methods [26,27,57–59] also assume a mathematical model to represent the degradation process and is defined as follows,

$$I(z) = J(z) + R(z),   \tag{9}$$

where I is the observed rainy image, J is the desired clean image, and R is the rain residue. This formulation models rainy image as a superposition of the clean background image with the rain residue. The rain residue contains domain specific information about the rain for a particular image and hence, can be used as a domain specific prior for supervising the prior estimation network (PEN) while adapting to rainy conditions. *Similar to the haze, we avoid the use of expensive human annotation efforts for obtaining the rain prior.*

In both cases discussed above (haze prior and rain prior), we do not use any ground-truth labels to estimate respective priors. Hence, our overall approach still falls into the category of unsupervised adaptation. Furthermore, these priors can be pre-computed for the training images to reduce the computational overhead during the learning process. Additionally, the prior computation is not required during inference and hence, the proposed adaptation method does not result in any computational overhead.

3.3   Residual Feature Recovery Block

As discussed earlier, weather-degradations introduce distortions in the feature space. In order to aid the de-distortion process, we introduce a set of residual feature recovery blocks (RFRBs) in the target feed-forward pipeline. This is inspired from the residual transfer network method proposed in [32]. Let $\Delta\mathcal{F}_l$ be the residual feature recovery block at the lth conv block. The target domain image feedforward is modified to include the residual feature recovery block. For $\Delta\mathcal{F}_l$ the feed-forward equation at the lth conv block can be written as,

$$\hat{\mathcal{F}}_l(x_i^t) \; = \; \mathcal{F}_l(x_i^t) \; + \; \Delta\mathcal{F}_l(\mathcal{F}_{l-1}(x_i^t)), \tag{10}$$

where, $\mathcal{F}_l(x_i^t)$ indicates the feature extracted from the lth conv block for any image x_i^t sampled from the target domain using the feature extractor network $\mathcal{F}$, $\Delta\mathcal{F}_l(\mathcal{F}_{l-1}(x_i^t))$ indicates the residual features extracted from the output $l - 1^{th}$ conv block, and $\hat{\mathcal{F}}_l(x_i^t)$ indicates the feature extracted from the lth conv block for any image $x_i^t \in \mathcal{T}$ with RFRB modified feedforward. The RFRB modules are also illustrated in Fig. 2, as shown in the target feedforward pipeline. It has no effect on source feedforward pipeline. In our case, we utilize RFRB at both conv4 ($\Delta\mathcal{F}_4$) and conv5 ($\Delta\mathcal{F}_5$) blocks. Additionally, the effect of residual feature is regularized by enforcing the norm constraints on the residual features. The regularization loss for RFRBs, $\Delta\mathcal{F}_4$ and $\Delta\mathcal{F}_5$ is defined as,

$$\mathcal{L}_{reg} = \frac{1}{n_t} \sum_{i=1}^{n_t} \sum_{l=4,5} \|\Delta\mathcal{F}_l(\mathcal{F}_{l-1}(x_i^t))\|_1, \tag{11}$$

3.4   Overall Loss

The overall loss for training the network is defined as,

$$\max_{\mathcal{P}} \min_{\mathcal{F},\Delta\mathcal{F},\mathcal{G}} \quad \mathcal{L}_{det}^{src} - \mathcal{L}_{adv} + \lambda\mathcal{L}_{reg}, \quad \text{where} \tag{12}$$

$$\mathcal{L}_{adv} = \frac{1}{2}(\mathcal{L}_{pal}^{src} + \mathcal{L}_{pal}^{tgt}). \tag{13}$$

Here, $\mathcal{F}$ represents the feature extractor network, $\mathcal{P}$ denotes both prior estimation network employed after conv4 and conv5 blocks, i.e., $\mathcal{P} = \{\mathcal{P}_5, \mathcal{P}_4\}$, and $\Delta\mathcal{F} = \{\Delta\mathcal{F}_4, \Delta\mathcal{F}_5\}$ represents RFRB at both conv4 and conv5 blocks. Also, $\mathcal{L}_{det}^{src}$ is the source detection loss, $\mathcal{L}_{reg}$ is the regularization loss, and $\mathcal{L}_{adv}$ is the overall adversarial loss used for prior-based adversarial training.

4   Experiments and Results

4.1   Implementation Details

We follow the training protocol of [6,41] for training the Faster-RCNN network. The backbone network for all experiments is VGG16 network [47]. We model the residuals using RFRB for the convolution blocks C4 and C5 of the VGG16 network. The PA loss is applied to only these conv blocks modeled with RFRBs. The PA loss is designed based on the adaptation setting (Haze or Rain). The parameters of the first two conv blocks are frozen similar to [6,41]. The detailed network architecture for RFRBs, PEN and the discriminator are provided in supplementary material. During training, we set shorter side of the image to 600 with ROI alignment. We train all networks 70K iterations. For the 50K iterations, the learning rate is set equal to 0.001 and for the 20K iterations it is set equal to 0.0001. We report the performance based on the trained model 70K iterations. We set λ equal to 0.1 for all experiments.

In addition to comparison with recent methods, we also perform an ablation study where we evaluate the following configurations to analyze the effectiveness of different components in the network. Note that we progressively add additional components which enables us to gauge the performance improvements obtained by each of them,

- **FRCNN:** Source only baseline experiment where Faster-RCNN is trained on the source dataset.
- **FRCNN + D_5:** Domain adaptation baseline experiment consisting of Faster-RCNN with domain discriminator after conv5 supervised by the domain adversarial loss.
- **FRCNN + D_5 + R_5:** Starting with FRCNN + D_5 as the base configuration, we add an RFRB block after conv4 in the Faster-RCNN. This experiment enables us to understand the contribution of the RFRB block.
- **FRCNN + P_5 + R_5:** We start with FRCNN + D_5 + R_5 configuration and replace domain discriminator and domain adversarial loss with prior estimation network (PEN) and prior adversarial loss (PAL). With this experiment, we show the importance of training with the proposed prior-adversarial loss.
- **FRCNN + P_{45} + R_{45}:** Finally, we perform the prior-based feature alignment at two scales: conv4 and conv5. Starting with FRCNN + P_5 + R_5 configuration, we add an RFRB block after conv3 and a PEN module after conv4. This experiment corresponds to the configuration depicted in Fig. 2. This experiment demonstrates the efficacy of the overall method in addition to establishing the importance of aligning features at multiple levels in the network.

Following the protocol set by the existing methods [6,41,45], we use mean average precision (mAP) scores for performance comparison.

4.2   Adaptation to Hazy Conditions

In this section, we present the results corresponding to adaptation to hazy conditions on the following datasets: (i) Cityscapes $\rightarrow$ Foggy-Cityscapes [43], (ii)

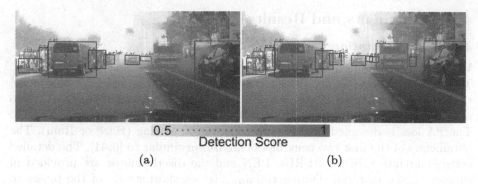

0.5 ······················· 1
Detection Score
(a)   (b)

Fig. 3. Detection results on Foggy-Cityscapes. (a) DA-Faster RCNN [6]. (b) Proposed method. The bounding boxes are colored based on the detector confidence. DA-Faster-RCNN produces detections with low confidence in addition to missing the truck class. Our method is able to output high confidence detections without missing any objects. (Color figure online)

Cityscapes → RTTS [25], and (iii) WIDER [55] → UFDD-Haze [35]. In the first two experiments, we consider Cityscapes [7] as the source domain. Note that the Cityscapes dataset contains images captured in clear weather conditions.

Cityscapes → Foggy-Cityscapes: In this experiment, we adapt from Cityscapes to Foggy-Cityscapes [43]. The Foggy-Cityscapes dataset was recently proposed in [43] to study the detection algorithms in the case of hazy weather conditions. Foggy-Cityscapes is derived from Cityscapes dataset by simulating fog on the clear weather images of Cityscapes. Both Cityscapes and Foggy-Cityscapes have the same number of categories which include, car, truck, motor-cycle/bike, train, bus, rider and person. Similar to [6,41], we utilize 2975 images of both Cityscapes and Foggy-Cityscapes for training. Note that we use annotations only from the source dataset (Cityscapes) for training the detection pipeline. For evaluation we consider a non overlapping validation set of 500 images provided by the Foggy-Cityscapes dataset.

We compare the proposed method with two categories of approaches: *(i) Dehaze + Detect:* Here, we employ dehazing network as pre-processing step and perform detection using Faster-RCNN trained on source (clean) images. For pre-processing, we chose two recent dehazing algorithms: DCPDN [60] and Grid-Dehaze [31]. *(i) DA-based methods:* Here, we compare with following recent domain-adaptive detection approaches: DA-Faster [6], SWDA [41], Diversify-Match [24], Mean Teacher with Object Relations (MTOR) [5], Selective Cross-Domain Alignment (SCDA) [64] and Noisy Labeling [23]. The corresponding results are presented in Table 1.

It can be observed from Table 1, that the performance of source-only training of Faster-RCNN is in general poor in the hazy conditions. Adding DCPDN and Gird-Dehaze as preprocessing step improves the performance by ~2% and ~4%, respectively. Compared to the domain-adaptive detection approaches, pre-processing + detection results in lower performance gains. This is because even

Table 1. Performance comparison for the Cityscapes → Foggy-Cityscapes experiment.

Method		prsn	rider	car	truc	bus	train	bike	bcycle	mAP
Baseline	FRCNN [39]	25.8	33.7	35.2	13.0	28.2	9.1	18.7	31.4	24.4
Dehaze	DCPDN [60]	27.9	36.2	35.2	16.0	28.3	10.2	24.6	32.5	26.4
	Grid-Dehaze [31]	29.7	40.4	40.3	21.3	30.0	9.1	25.6	36.7	29.2
DA-Methods	DAFaster [6]	25.0	31.0	40.5	22.1	35.3	20.2	20.0	27.1	27.6
	SCDA [64]	33.5	38.0	48.5	26.5	39.0	23.3	28.0	33.6	33.8
	SWDA [41]	29.9	42.3	43.5	24.5	36.2	32.6	30.0	35.3	34.3
	DM [24]	30.8	40.5	44.3	**27.2**	38.4	34.5	28.4	32.2	34.6
	MTOR [5]	30.6	41.4	44.0	21.9	38.6	**40.6**	28.3	35.6	35.1
	NL [23]	35.1	42.1	49.2	30.1	45.3	26.9	26.8	36.0	36.5
Ours	FRCNN + D_5	30.9	38.5	44.0	19.6	32.9	17.9	24.1	32.4	30.0
	FRCNN + D_5 + R_5	32.8	44.7	49.9	22.3	31.7	17.3	26.9	37.5	32.9
	FRCNN + P_5 + R_5	33.4	42.8	50.0	24.2	40.8	30.4	33.1	37.5	36.5
	FRCNN + P_{45} + R_{45}	**36.4**	**47.3**	**51.7**	22.8	**47.6**	34.1	**36.0**	**38.7**	**39.3**

after applying dehazing there still remains some domain shift as discussed in Sect. 1. Hence, using adaptation would be a better approach for mitigating the domain shift. Here, the use of simple domain adaptation [14] (FRCNN + D_5) improves the source-only performance. The addition of RFRB$_5$ (FRCNN + D_5 + R_5) results in further improvements, thus indicating the importance of RFRB blocks. However, the conventional domain adaptation loss assumes constant domain shift across the entire image, resulting in incorrect alignment. The use of prior-adversarial loss (FRCNN + P_5 + R_5) overcomes this issue. We achieved 3.6% improvement in overall mAP scores, thus demonstrating the effectiveness of the proposed prior-adversarial training. Note that, FRCNN + P_5 + R_5 baseline achieves comparable performance with state-of-the-art. Finally, by performing prior-adversarial adaptation at an additional scale (FRCNN + P_{45} + R_{45}), we achieve further improvements which surpasses the existing best approach [23] by 2.8%. Figure 3 shows sample qualitative detection results corresponding to the images from Foggy-Cityscapes. Results for the proposed method are compared with DA-Faster-RCNN [6]. It can be observed that the proposed method is able to generate comparatively high quality detections.

We summarize our observations as follows: (i) Using dehazing as a pre-processing step results in minimal improvements over the baseline Faster-RCNN. Domain adaptive approaches perform better in general. (ii) The proposed method outperforms other methods in the overall scores while achieving the best performance in most of the classes. See supplementary material for more ablations.

Cityscapes → RTTS: In this experiment, we adapt from Cityscapes to the RTTS dataset [25]. RTTS is a subset of a larger RESIDE dataset [25], and it contains 4,807 unannotated and 4,322 annotated real-world hazy images covering mostly traffic and driving scenarios. We use the unannotated 4,807 images for training the domain adaptation process. The evaluation is performed on the

774   V. A. Sindagi et al.

Table 2. Performance comparison for the Cityscapes → RTTS experiment.

Method		prsn	car	bus	bike	bcycle	mAP
Baseline	FRCNN [39]	46.6	39.8	11.7	19.0	37.0	30.9
Dehaze	DCPDN [60]	**48.7**	39.5	12.9	19.7	37.5	31.6
	Grid-Dehaze [31]	29.7	25.4	10.9	13.0	21.4	20.0
DA	DAFaster [6]	37.7	48.0	14.0	**27.9**	36.0	32.8
	SWDA [41]	42.0	46.9	15.8	25.3	37.8	33.5
Ours	Proposed	37.4	**54.7**	**17.2**	22.5	**38.5**	**34.1**

Table 3. Results (mAP) of the adaptation experiments from WIDER-Face to UFDD Haze and Rain.

Method	UFDD-Haze	UFDD-Rain
FRCNN [39]	46.4	54.8
DAFaster [6]	52.1	58.2
SWDA [41]	55.5	60.0
Proposed	**58.5**	**62.1**

annotated 4,322 images. RTTS has total five categories, namely motorcycle/bike, person, bicycle, bus and car. This dataset is the largest available dataset for object detection under real world hazy conditions.

In Table 2, the results of the proposed method are compared with Faster-RCNN [39], DA-Faster [6] and SWDA [41] and the dehaze+detection baseline as well. For RTTS dataset, the pre-processing with DCPDN improves the Faster-RCNN performance by ∼1%. Surprisingly, Grid-Dehaze does not help the Faster-RCNN baseline and results in even worse performance. Whereas, the proposed method achieves an improvement of 3.1% over the baseline Faster-RCNN (source-only training), while outperforming the other recent methods.

WIDER-Face → UFDD-Haze: Recently, Nada *et al.* [35] published a benchmark face detection dataset which consists of real-world images captured under different weather-based conditions such as haze and rain. Specifically, this dataset consists of 442 images under the haze category. Since, face detection is closely related to the task of object detection, we evaluate our framework by adapting from WIDER-Face [55] dataset to UFDD-Haze dataset. WIDER-Face is a large-scale face detection dataset with approximately 32,000 images 199K face annotations. The results corresponding to this adaptation experiment are shown in Table 3. It can be observed from this table that the proposed method achieves better performance as compared to the other methods.

Table 4. Performance comparison for the Cityscapes → Rainy-Cityscapes experiment.

Method		prsn	rider	car	truc	bus	train	bike	bcycle	mAP
Baseline	FRCNN	21.6	19.5	38.0	12.6	30.1	24.1	12.9	15.4	21.8
Derain	DDN [13]	27.1	30.3	50.7	23.1	39.4	18.5	21.2	24.0	29.3
	SPANet [54]	24.9	28.9	48.1	21.4	34.8	16.8	17.6	20.8	26.7
DA	DAFaster [6]	26.9	28.1	50.6	23.2	39.3	4.7	17.1	20.2	26.3
	SWDA [41]	29.6	**38.0**	52.1	27.9	**49.8**	28.7	24.1	25.4	34.5
Ours	FRCNN + D_5	29.1	34.8	52.0	22.0	41.8	20.4	18.1	23.3	30.2
	FRCNN + D_5 + R_5	28.8	33.1	51.7	22.3	41.8	24.9	22.2	24.6	31.2
	FRCNN + P_5 + R_5	29.7	34.3	52.5	23.6	47.9	32.5	24.0	25.5	33.8
	FRCNN + P_{45} + R_{45}	**31.3**	34.8	**57.8**	**29.3**	48.6	**34.4**	**25.4**	**27.3**	**36.1**

4.3 Adaptation to Rainy Conditions

In this section, we present the results of adaptation to rainy conditions. Due to lack of appropriate datasets for this particular setting, we create a new rainy dataset called Rainy-Cityscapes and it is derived from Cityscapes. It has the same number of images for training and validation as Foggy-Cityscapes. First, we discuss the simulation process used to create the dataset, followed by a discussion of the evaluation and comparison of the proposed method with other methods.

Rainy-Cityscapes: Similar to Foggy-Cityscapes, we use a subset of 3475 images from Cityscapes to create synthetic rain dataset. Using [1], several masks containing artificial rain streaks are synthesized. The rain streaks are created using different Gaussian noise levels and multiple rotation angles between 70° and 110°. Next, for every image in the subset of the Cityscapes dataset, we pick a random rain mask and blend it onto the image to generate the synthetic rainy image. More details and example images are provided in supplementary material.

Cityscapes → Rainy-Cityscapes: In this experiment, we adapt from Cityscapes to Rainy-Cityscapes. We compare the proposed method with recent methods such as DA-Faster [6] and SWDA [41]. Additionally, we also evaluate performance of two derain + detect baselines, where state of the art methods such as DDN [13] and SPANet [54] are used as a pre-processing step to the Faster-RCNN trained on source (clean) images. From the Table 4 we observe that such methods provide reasonable improvements over the Faster-RCNN baseline. However, the performance gains are much lesser as compared to adaptation methods, for the reasons discussed in the earlier sections (Sects. 1 and 4.2). Also, it can be observed from Table 4, that the proposed method outperforms the other methods by a significant margin. Additionally, we present the results of the ablation study consisting of the experiments listed in Sect. 4.1. The introduction of domain adaptation loss significantly improves the source only Faster-RCNN baseline, resulting in approximately 9% improvement for FRCNN + D_5 baseline in Table 4. This performance is further improved by 1% with the help of residual feature recovery blocks as shown in FRCNN + D_5 + R_5 baseline. When domain adversarial training is replaced with prior adversarial training

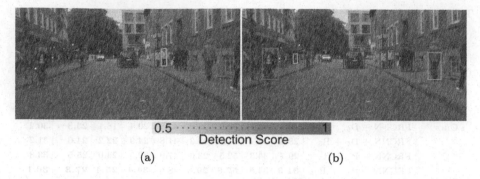

0.5
Detection Score
1

(a)   (b)

Fig. 4. Detection results on Rainy-Cityscapes. (a) DA-Faster RCNN [6]. (b) Proposed method. The bounding boxes are colored based on the detector confidence. DA-Faster-RCNN misses several objects. Our method is able to output high confidence detections without missing any objects. (Color figure online)

with PAL, i.e. FRCNN + P_5 + R_5 baseline, we observe 2.5% improvements, showing effectiveness of the proposed training methodology. Finally, by performing prior adversarial training at multiple scales, the proposed method FRCNN + P_{45} + R_{45} observes approximately 2% improvements and also outperforms the next best method SWDA [41] by 1.6%. Figure 4 illustrates sample detection results obtained using the proposed method as compared to a recent method [6]. The proposed method achieves superior quality detections.

WIDER-Face → UFDD-Rain: In this experiment, we adapt from WIDER-Face to UFDD-Rain [35]. The UFDD-Rain dataset consists of 628 images collected under rainy conditions. The results of the proposed method as compared to the other methods are shown in Table 3. It can be observed that the proposed method outperforms the source only training by 7.3%. We provide additional details about the proposed method including results and analysis in the y material.

5   Conclusions

We addressed the problem of adapting object detectors to hazy and rainy conditions. Based on the observation that these weather conditions cause degradations that can be mathematically modeled and cause spatially varying distortions in the feature space, we propose a novel prior-adversarial loss that aims at producing weather-invariant features. Additionally, a set of residual feature recovery blocks are introduced to learn residual features that can aid efficiently aid the adaptation process. The proposed framework is evaluated on several benchmark datasets such as Foggy-Cityscapes, RTTS and UFDD. Through extensive experiments, we showed that our method achieves significant gains over the recent methods in all the datasets.

Acknowledgement. This work was supported by the NSF gránt 1910141.

References

1. https://www.photoshopessentials.com/photo-effects/photoshop-weather-effects-rain/
2. Abavisani, M., Patel, V.M.: Domain adaptive subspace clustering. In: 27th British Machine Vision Conference, BMVC 2016 (2016)
3. Abavisani, M., Patel, V.M.: Adversarial domain adaptive subspace clustering. In: 2018 IEEE 4th International Conference on Identity, Security, and Behavior Analysis (ISBA), pp. 1–8. IEEE (2018)
4. Ancuti, C., Ancuti, C.O., Timofte, R.: NTIRE 2018 challenge on image dehazing: methods and results. In: Proceedings of the IEEE Conference on Computer Vision and Pattern Recognition Workshops, pp. 891–901 (2018)
5. Cai, Q., Pan, Y., Ngo, C.W., Tian, X., Duan, L., Yao, T.: Exploring object relation in mean teacher for cross-domain detection. In: Proceedings of the IEEE Conference on Computer Vision and Pattern Recognition, pp. 11457–11466 (2019)
6. Chen, Y., Li, W., Sakaridis, C., Dai, D., Gool, L.V.: Domain adaptive faster R-CNN for object detection in the wild. In: 2018 IEEE Conference on Computer Vision and Pattern Recognition, pp. 3339–3348 (2018)
7. Cordts, M., et al.: The cityscapes dataset for semantic urban scene understanding. In: Proceedings of the IEEE Conference on Computer Vision and Pattern Recognition, pp. 3213–3223 (2016)
8. Dalal, N., Triggs, B.: Histograms of oriented gradients for human detection. In: International Conference on Computer Vision and Pattern Recognition (CVPR 2005), vol. 1, pp. 886–893. IEEE Computer Society (2005)
9. Deng, J., Dong, W., Socher, R., Li, L.J., Li, K., Fei-Fei, L.: ImageNet: a large-scale hierarchical image database. In: 2009 IEEE Conference on Computer Vision and Pattern Recognition, pp. 248–255. IEEE (2009)
10. Everingham, M., Van Gool, L., Williams, C.K., Winn, J., Zisserman, A.: The PASCAL visual object classes (VOC) challenge. Int. J. Comput. Vis. 88(2), 303–338 (2010)
11. Fattal, R.: Single image dehazing. ACM Trans. Graph. (TOG) 27(3), 72 (2008)
12. Felzenszwalb, P.F., Girshick, R.B., McAllester, D., Ramanan, D.: Object detection with discriminatively trained part-based models. IEEE Trans. Pattern Anal. Mach. Intell. 32(9), 1627–1645 (2010)
13. Fu, X., Huang, J., Zeng, D., Huang, Y., Ding, X., Paisley, J.: Removing rain from single images via a deep detail network. In: Proceedings of the IEEE Conference on Computer Vision and Pattern Recognition, pp. 3855–3863 (2017)
14. Ganin, Y., Lempitsky, V.: Unsupervised domain adaptation by backpropagation. arXiv preprint arXiv:1409.7495 (2014)
15. Geiger, A., Lenz, P., Stiller, C., Urtasun, R.: Vision meets robotics: the KITTI dataset. Int. J. Robot. Res. 32(11), 1231–1237 (2013)
16. Girshick, R.: Fast R-CNN. In: Proceedings of the IEEE International Conference on Computer Vision, pp. 1440–1448 (2015)
17. Girshick, R., Donahue, J., Darrell, T., Malik, J.: Rich feature hierarchies for accurate object detection and semantic segmentation. In: Proceedings of the IEEE Conference on Computer Vision and Pattern Recognition, pp. 580–587 (2014)
18. Gopalan, R., Li, R., Chellappa, R.: Domain adaptation for object recognition: an unsupervised approach. In: 2011 International Conference on Computer Vision, pp. 999–1006. IEEE (2011)

19. He, K., Sun, J., Tang, X.: Single image haze removal using dark channel prior. IEEE Trans. Pattern Anal. Mach. Intell. **33**(12), 2341–2353 (2011)
20. He, K., Zhang, X., Ren, S., Sun, J.: Deep residual learning for image recognition. In: Proceedings of the IEEE Conference on Computer Vision and Pattern Recognition, pp. 770–778 (2016)
21. Hoffman, J., et al.: CyCADA: cycle-consistent adversarial domain adaptation. arXiv preprint arXiv:1711.03213 (2017)
22. Hu, L., Kan, M., Shan, S., Chen, X.: Duplex generative adversarial network for unsupervised domain adaptation. In: Proceedings of the IEEE Conference on Computer Vision and Pattern Recognition, pp. 1498–1507 (2018)
23. Khodabandeh, M., Vahdat, A., Ranjbar, M., Macready, W.G.: A robust learning approach to domain adaptive object detection. arXiv preprint arXiv:1904.02361 (2019)
24. Kim, T., Jeong, M., Kim, S., Choi, S., Kim, C.: Diversify and match: a domain adaptive representation learning paradigm for object detection. In: Proceedings of the IEEE Conference on Computer Vision and Pattern Recognition, pp. 12456–12465 (2019)
25. Li, B., et al.: Benchmarking single-image dehazing and beyond. IEEE Trans. Image Process. **28**(1), 492–505 (2019)
26. Li, S., Ren, W., Zhang, J., Yu, J., Guo, X.: Single image rain removal via a deep decomposition-composition network. Comput. Vis. Image Underst. **186**, 48–57 (2019)
27. Li, Y., Tan, R.T., Guo, X., Lu, J., Brown, M.S.: Rain streak removal using layer priors. In: IEEE Conference on Computer Vision and Pattern Recognition (CVPR), pp. 2736–2744 (2016)
28. Li, Y., You, S., Brown, M.S., Tan, R.T.: Haze visibility enhancement: a survey and quantitative benchmarking. Comput. Vis. Image Underst. **165**, 1–16 (2017)
29. Lin, T.-Y., et al.: Microsoft COCO: common objects in context. In: Fleet, D., Pajdla, T., Schiele, B., Tuytelaars, T. (eds.) ECCV 2014. LNCS, vol. 8693, pp. 740–755. Springer, Cham (2014). https://doi.org/10.1007/978-3-319-10602-1_48
30. Liu, W., et al.: SSD: single shot MultiBox detector. In: Leibe, B., Matas, J., Sebe, N., Welling, M. (eds.) ECCV 2016. LNCS, vol. 9905, pp. 21–37. Springer, Cham (2016). https://doi.org/10.1007/978-3-319-46448-0_2
31. Liu, X., Ma, Y., Shi, Z., Chen, J.: GridDehazeNet: attention-based multi-scale network for image dehazing. In: Proceedings of the IEEE International Conference on Computer Vision, pp. 7314–7323 (2019)
32. Long, M., Zhu, H., Wang, J., Jordan, M.I.: Unsupervised domain adaptation with residual transfer networks. In: Advances in Neural Information Processing Systems, pp. 136–144 (2016)
33. Long, M., Zhu, H., Wang, J., Jordan, M.I.: Deep transfer learning with joint adaptation networks. In: Proceedings of the 34th International Conference on Machine Learning, vol. 70, pp. 2208–2217. JMLR.org (2017)
34. Murez, Z., Kolouri, S., Kriegman, D., Ramamoorthi, R., Kim, K.: Image to image translation for domain adaptation. In: Proceedings of the IEEE Conference on Computer Vision and Pattern Recognition, pp. 4500–4509 (2018)
35. Nada, H., Sindagi, V.A., Zhang, H., Patel, V.M.: Pushing the limits of unconstrained face detection: a challenge dataset and baseline results. arXiv preprint arXiv:1804.10275 (2018)
36. Patel, V.M., Gopalan, R., Li, R., Chellappa, R.: Visual domain adaptation: a survey of recent advances. IEEE Signal Process. Mag. **32**(3), 53–69 (2015)

37. Perera, P., Abavisani, M., Patel, V.M.: In2I: unsupervised multi-image-to-image translation using generative adversarial networks. In: 2018 24th International Conference on Pattern Recognition (ICPR), pp. 140–146. IEEE (2018)
38. Redmon, J., Divvala, S., Girshick, R., Farhadi, A.: You only look once: unified, real-time object detection. In: Proceedings of the IEEE Conference on Computer Vision and Pattern Recognition, pp. 779–788 (2016)
39. Ren, S., He, K., Girshick, R., Sun, J.: Faster R-CNN: towards real-time object detection with region proposal networks. In: Advances in Neural Information Processing Systems, pp. 91–99 (2015)
40. RoyChowdhury, A., et al.: Automatic adaptation of object detectors to new domains using self-training. In: Proceedings of the IEEE Conference on Computer Vision and Pattern Recognition, pp. 780–790 (2019)
41. Saito, K., Ushiku, Y., Harada, T., Saenko, K.: Strong-weak distribution alignment for adaptive object detection. CoRR abs/1812.04798 (2018)
42. Saito, K., Watanabe, K., Ushiku, Y., Harada, T.: Maximum classifier discrepancy for unsupervised domain adaptation. In: Proceedings of the IEEE Conference on Computer Vision and Pattern Recognition, pp. 3723–3732 (2018)
43. Sakaridis, C., Dai, D., Gool, L.V.: Semantic foggy scene understanding with synthetic data. Int. J. Comput. Vis. **126**, 973–992 (2018)
44. Sankaranarayanan, S., Balaji, Y., Castillo, C.D., Chellappa, R.: Generate to adapt: aligning domains using generative adversarial networks. In: Proceedings of the IEEE Conference on Computer Vision and Pattern Recognition, pp. 8503–8512 (2018)
45. Shan, Y., Lu, W.F., Chew, C.M.: Pixel and feature level based domain adaption for object detection in autonomous driving (2018)
46. Shu, R., Bui, H.H., Narui, H., Ermon, S.: A DIRT-T approach to unsupervised domain adaptation. arXiv preprint arXiv:1802.08735 (2018)
47. Simonyan, K., Zisserman, A.: Very deep convolutional networks for large-scale image recognition. arXiv preprint arXiv:1409.1556 (2014)
48. Sindagi, V., Patel, V.: DAFE-FD: density aware feature enrichment for face detection. In: 2019 IEEE Winter Conference on Applications of Computer Vision (WACV), pp. 2185–2195. IEEE (2019)
49. Sindagi, V.A., Srivastava, S.: Domain adaptation for automatic OLED panel defect detection using adaptive support vector data description. Int. J. Comput. Vis. **122**(2), 193–211 (2017)
50. Sindagi, V.A., Yasarla, R., Babu, D.S., Babu, R.V., Patel, V.M.: Learning to count in the crowd from limited labeled data. arXiv preprint arXiv:2007.03195 (2020)
51. Sindagi, V.A., Yasarla, R., Patel, V.M.: Pushing the frontiers of unconstrained crowd counting: new dataset and benchmark method. In: Proceedings of the IEEE International Conference on Computer Vision, pp. 1221–1231 (2019)
52. Tzeng, E., Hoffman, J., Saenko, K., Darrell, T.: Adversarial discriminative domain adaptation. In: Proceedings of the IEEE Conference on Computer Vision and Pattern Recognition, pp. 7167–7176 (2017)
53. Viola, P., Jones, M., et al.: Rapid object detection using a boosted cascade of simple features. CVPR **1**(1), 511–518 (2001)
54. Wang, T., Yang, X., Xu, K., Chen, S., Zhang, Q., Lau, R.W.: Spatial attentive single-image deraining with a high quality real rain dataset. In: Proceedings of the IEEE Conference on Computer Vision and Pattern Recognition, pp. 12270–12279 (2019)

55. Yang, S., Luo, P., Loy, C.C., Tang, X.: WIDER FACE: a face detection benchmark. In: Proceedings of the IEEE Conference on Computer Vision and Pattern Recognition, pp. 5525–5533 (2016)

56. Yasarla, R., Sindagi, V.A., Patel, V.M.: Syn2Real transfer learning for image deraining using Gaussian processes. In: Proceedings of the IEEE/CVF Conference on Computer Vision and Pattern Recognition (CVPR) (2020)

57. You, S., Tan, R.T., Kawakami, R., Mukaigawa, Y., Ikeuchi, K.: Adherent raindrop modeling, detection and removal in video. IEEE Trans. Pattern Anal. Mach. Intell. **38**(9), 1721–1733 (2015)

58. Zhang, H., Patel, V.M.: Density-aware single image de-raining using a multi-stream dense network. In IEEE Conference on Computer Vision and Pattern Recognition (CVPR) (2018). abs/1802.07412

59. Zhang, H., Patel, V.M.: Image de-raining using a conditional generative adversarial network. arXiv preprint arXiv:1701.05957 (2017)

60. Zhang, H., Patel, V.M.: Densely connected pyramid dehazing network. In: Proceedings of the IEEE Conference on Computer Vision and Pattern Recognition, pp. 3194–3203 (2018)

61. Zhang, H., Sindagi, V., Patel, V.M.: Multi-scale single image dehazing using perceptual pyramid deep network. In: Proceedings of the IEEE Conference on Computer Vision and Pattern Recognition Workshops, pp. 902–911 (2018)

62. Zhang, H., Sindagi, V., Patel, V.M.: Joint transmission map estimation and dehazing using deep networks. IEEE Trans. Circ. Syst. Video Technol. **30**, 1975–1986 (2019)

63. Zhu, J.Y., Park, T., Isola, P., Efros, A.A.: Unpaired image-to-image translation using cycle-consistent adversarial networks. In: Proceedings of the IEEE International Conference on Computer Vision, pp. 2223–2232 (2017)

64. Zhu, X., Pang, J., Yang, C., Shi, J., Lin, D.: Adapting object detectors via selective cross-domain alignment. In: Proceedings of the IEEE Conference on Computer Vision and Pattern Recognition, pp. 687–696 (2019)

Adversarial Ranking Attack and Defense

Mo Zhou[1], Zhenxing Niu[2], Le Wang[1(✉)], Qilin Zhang[3], and Gang Hua[4]

[1] Xi'an Jiaotong University, Xi'an, China
lewang@xjtu.edu.cn
[2] Alibaba DAMO MIIL, Hangzhou, China
[3] HERE Technologies, Chicago, Illinois, USA
[4] Wormpex AI Research, Bellevue, USA

Abstract. Deep Neural Network (DNN) classifiers are vulnerable to adversarial attack, where an imperceptible perturbation could result in misclassification. However, the vulnerability of DNN-based image ranking systems remains under-explored. In this paper, we propose two attacks against deep ranking systems, *i.e.*, Candidate Attack and Query Attack, that can raise or lower the rank of chosen candidates by adversarial perturbations. Specifically, the expected ranking order is first represented as a set of inequalities, and then a triplet-like objective function is designed to obtain the optimal perturbation. Conversely, a defense method is also proposed to improve the ranking system robustness, which can mitigate all the proposed attacks simultaneously. Our adversarial ranking attacks and defense are evaluated on datasets including MNIST, Fashion-MNIST, and Stanford-Online-Products. Experimental results demonstrate that a typical deep ranking system can be effectively compromised by our attacks. Meanwhile, the system robustness can be moderately improved with our defense. Furthermore, the transferable and universal properties of our adversary illustrate the possibility of realistic black-box attack.

1 Introduction

Despite the successful application in computer vision tasks such as image classification [21,31], Deep Neural Networks (DNNs) have been found vulnerable to adversarial attacks. In particular, the DNN's prediction can be arbitrarily changed by just applying an imperceptible perturbation to the input image [17,69]. Moreover, such adversarial attacks can effectively compromise the state-of-the-art DNNs such as Inception [67,68] and ResNet [21]. This poses a serious security risk on many DNN-based applications such as face recognition, where recognition evasion or impersonation can be easily achieved [12,30,64,72].

Previous adversarial attacks primarily focus on *classification*, however, we speculate that DNN-based image ranking systems [3,6,15,29,35,52,70] also suffer from similar vulnerability. Taking the image-based product search as an

Electronic supplementary material The online version of this chapter (https://doi.org/10.1007/978-3-030-58568-6_46) contains supplementary material, which is available to authorized users.

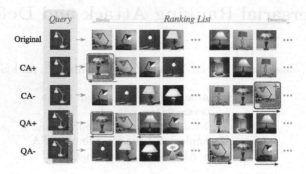

Fig. 1. Adversarial ranking attack that can *raise* or *lower* the rank of chosen candidates by adversarial perturbations. In Candidate Attack, adversarial perturbation is added to the candidate and its rank is *raised* (CA+) or *lowered* (CA−). In Query Attack, adversarial perturbation is added to the query image, and the ranks of chosen candidates are *raised* (QA+) or *lowered* (QA−).

example, a fair ranking system should rank the products according to their visual similarity to the query, as shown in Fig. 1 (row 1). Nevertheless, malicious sellers may attempt to raise the rank of their product by adding perturbation to the image (CA+, row 2), or lower the rank of his competitor's product (CA−, row 3); Besides, "man-in-the-middle" attackers (*e.g.*, a malicious advertising company) could hijack and imperceptibly perturb the query image in order to promote (QA+, row 4) or impede (QA−, row 5) the sales of specific products.

Unlike classification tasks where images are predicted independently, the rank of one candidate is related to the query as well as other candidates for image ranking. The relative relations among candidates and queries determine the final ranking order. Therefore, we argue that the existing adversarial classification attacks are incompatible with the ranking scenario. Thus, we need to thoroughly study the *adversarial ranking attack*.

In this paper, adversarial ranking attack aims to *raise* or *lower* the ranks of some chosen candidates $C = \{c_1, c_2, \ldots, c_m\}$ with respect to a specific query set $Q = \{q_1, q_2, \ldots, q_w\}$. This can be achieved by either Candidate Attack (CA) or Query Attack (QA). In particular, CA is defined as to raise (*abbr.* CA+) or lower (*abbr.* CA−) the rank of a single candidate c with respect to the query set Q by perturbing c itself; while QA is defined as to raise (*abbr.* QA+) or lower (*abbr.* QA−) the ranks of a candidate set C with respect to a single query q by perturbing q. Thus, adversarial ranking attack can be achieved by performing CA on each $c \in C$, or QA on each $q \in Q$. In practice, the choice of CA or QA depends on the accessibility to the candidate or query respectively, *i.e.*, CA is feasible for modifiable candidate, while QA is feasible for modifiable query.

An effective implementation of these attacks is proposed in this paper. As we know, a typical DNN-based ranking model maps objects (*i.e.*, queries and candidates) to a common embedding space, where the distances among them determine the final ranking order. Predictably, the object's position in the embedding space will be changed by adding a perturbation to it. Therefore, the essential of adversarial ranking attack is to find a proper perturbation, which could push the object to a desired position that leads to the expected ranking order. Specifically, we first represent the expected ranking order as a set of inequalities. Subsequently, a triplet-like objective function is designed according to those inequalities, and combined with Projected Gradient Descent (PGD) to efficiently obtain the desired adversarial perturbation.

Opposed to the proposed attacks, *adversarial ranking defense* is worth being investigated especially for security-sensitive deep ranking applications. Until now, the Madry defense [45] is regarded as the most effective method for classification defense. However, we empirically discovered a primary challenge of diverging training loss while directly adapting such mechanism for ranking defense, possibly due to the generated adversarial examples being too "strong". In addition, such defense mechanism needs to defend against distinct ranking attacks individually, but a *generic* defense method against all CA+, CA−, QA+ and QA− attacks is preferred.

To this end, a shift-distance based ranking defense is proposed, which could simultaneously defend against all attacks. Note that the position shift of objects in the embedding space is the key for all ranking attacks. Although different attacks prefer distinct shift directions (*e.g.*, CA+ and CA− often prefer opposed shifting directions), a large shift distance is their common preference. If we could reduce the shift distance of embeddings incurred by adversarial perturbation, all attacks can be simultaneously defensed. Specifically, we first propose a shift-distance based ranking attack, which aims to push the objects as far from their original positions as possible. And then, the adversarial examples generated from such attack is involved in the adversarial training. Experimental results manifest that our ranking defense can converge and moderately improve model robustness.

In addition, our ranking attacks have some good properties for realistic applications. First, our adversary is transferable, *i.e.*, the adversary obtained from a known DNN ranker can be directly used to attack an unknown DNN ranker (*i.e.*, the network architecture and parameters are unknown). Second, our attacks can be extended to *universal* ranking attacks with slight performance drop, *i.e.*, we could learn a *universal* perturbation to all candidates for CA, or a *universal* perturbation to all queries for QA. Such properties illustrate the possibility of practical black-box attack.

To the best of our knowledge, this is the first work that thoroughly studies the adversarial ranking attack and defense. In brief, our contributions are:

1. The adversarial ranking attack is defined and implemented, which can intentionally change the ranking results by perturbing the candidates or queries.
2. An adversarial ranking defense method is proposed to improve the ranking model robustness, and mitigate all the proposed attacks simultaneously.

2 Related Works

Adversarial Attacks. Szegedy et al. [69] claimed that DNN is susceptible to imperceptible adversarial perturbations added to inputs, due to the intriguing "blind spot" property, which was later ascribed to the local linearity [17] of neural networks. Following these findings, many white-box (model architecture and parameters are known to the adversary) attacking methods [5,7,8,10,16,32, 45,50,57,61,66,74] are proposed to effectively compromise the state-of-the-art DNN classifiers. Among them, PGD [45] is regarded as one of the most powerful attacks [1]. Notably, adversarial examples are discovered to be transferable [55, 56] among different neural network classifiers, which inspired a series of black-box attacks [11,24,41,65,73,76]. On the other hand, universal (*i.e.*, image-agnostic) adversarial perturbations are also discovered [37,49].

Deep Ranking. Different from the traditional "learning to rank" [27,38] methods, DNN-based ranking methods often embed data samples (including both queries and candidates) of all modalities into a common embedding space, and subsequently determine the ranking order based on distance. Such workflow has been adopted in distance metric learning [6,26,53,70], image retrieval [3], cross-modal retrieval [15,29,35,52], and face recognition [62].

Adversarial Attacks in Deep Ranking. For information retrieval and ranking systems, the risk of malicious users manipulating the ranking always exists [19,23]. However, only a few research efforts have been made in adversarial attacks in deep ranking. Liu et al. [42] proposed adversarial queries leading to incorrect retrieval results; while Li et al. [36] staged similar attack with universal perturbation that corrupts listwise ranking results. None of the aforementioned research efforts explore the *adversarial ranking attack*. Besides, adaptation of distance-based attacks (*e.g.* [61]) are unsuitable for our scenario.

Adversarial Defenses. Adversarial attacks and defenses are consistently engaged in an arms race [77]. Gradient masking-based defenses can be circumvented [2]. Defensive distillation [54,58] has been compromised by C&W [4,5]. As claimed in [22], ensemble of weak defenses are insufficient against adversarial examples. Notably, as an early defense method [69], adversarial training [13,17, 25,33,45,46,51,63,71,78] remains to be one of the most effective defenses. Other types of defenses include adversarial detection [43,48], input transformation/ reconstruction/replacement [14,20,44,47,60], randomization [39,40], network verification [18,28], *etc.* However, defense in deep ranking systems remains mostly uncharted.

3   Adversarial Ranking

Generally, a DNN-based ranking task could be formulated as a metric learning problem. Given the query q and candidate set $X = \{c_1, c_2, \ldots, c_n\}$, deep ranking is to learn a mapping f (usually implemented as a DNN) which maps all candidates and query into a common embedding space, such that the relative distances among the embedding vectors could satisfy the expected ranking order. For instance, if candidate c_i is more similar to the query q than candidate c_j, it is encouraged for the mapping f to satisfy the inequality

$\|f(q) - f(c_i)\| < \|f(q) - f(c_j)\|$[1], where $\| \cdot \|$ denotes ℓ_2 norm. For brevity, we denote $\|f(q) - f(c_i)\|$ as $d(q, c_i)$ in following text.

Therefore, adversarial ranking attack is to find a proper adversarial perturbation which leads the ranking order to be changed as expected. For example, if a less relevant c_j is expected to be ranked *ahead* of a relevant c_i, it is desired to find a proper perturbation r to perturb c_j, *i.e.* $\tilde{c}_j = c_j + r$, such that the inequality $d(q, c_i) < d(q, c_j)$ could be changed into $d(q, c_i) > d(q, \tilde{c}_j)$. In the next, we will describe Candidate Attack and Query Attack in detail.

3.1 Candidate Attack

Candidate Attack (**CA**) aims to raise (*abbr.* **CA+**) or lower (*abbr.* **CA−**) the rank of a *single* candidate c with respect to a set of queries $Q = \{q_1, q_2, \dots, q_w\}$ by adding perturbation r to the candidate itself, *i.e.* $\tilde{c} = c + r$.

Let $\text{Rank}_X(q, c)$ denote the rank of the candidate c with respect to the query q, where X indicates the set of all candidates, and a smaller rank value represents a higher ranking. Thus, the **CA+** that *raises* the rank of c with respect to every query $q \in Q$ by perturbation r could be formulated as the following problem,

$$r = \arg\min_{r \in \Gamma} \sum_{q \in Q} \text{Rank}_X(q, c + r), \tag{1}$$

$$\Gamma = \{r \,|\, \|r\|_\infty \leqslant \varepsilon; r, c + r \in [0, 1]^N\}, \tag{2}$$

where Γ is a ℓ_∞-bounded ε-neighbor of c, $\varepsilon \in [0, 1]$ is a predefined small positive constant, the constraint $\|r\|_\infty \leqslant \varepsilon$ limits the perturbation r to be "visually imperceptible", and $c + r \in [0, 1]^N$ ensures the adversarial example remains a valid input image. Although alternative "imperceptible" constraints exist (*e.g.*, ℓ_0 [9,66], ℓ_1 [8] and ℓ_2 [5,50] variants), we simply follow [17,32,45] and use the ℓ_∞ constraint.

However, the optimization problem Eq. (1)–(2) cannot be directly solved due to the discrete nature of the rank value $\text{Rank}_X(q, c)$. In order to solve the problem, a surrogate objective function is needed.

In metric learning, given two candidates $c_p, c_n \in X$ where c_p is ranked ahead of c_n, *i.e.* $\text{Rank}_X(q, c_p) < \text{Rank}_X(q, c_n)$, the ranking order is represented as an inequality $d(q, c_p) < d(q, c_n)$ and formulated in triplet loss:

$$L_{\text{triplet}}(q, c_p, c_n) = [\beta + d(q, c_p) - d(q, c_n)]_+, \tag{3}$$

where $[\cdot]_+$ denotes $\max(0, \cdot)$, and β is a manually defined constant margin. This function is known as the triplet (*i.e.* pairwise ranking) loss [6,62].

Similarly, the attacking goal of **CA+** in Eq. (1) can be readily converted into a series of inequalities, and subsequently turned into a sum of triplet losses,

$$L_{\text{CA+}}(c, Q; X) = \sum_{q \in Q} \sum_{x \in X} [d(q, c) - d(q, x)]_+. \tag{4}$$

---

[1] Sometimes cosine distance is used instead.

In this way, the original problem in Eq. (1)–(2) can be reformulated into the following constrained optimization problem:

$$r = \arg\min_{r \in \Gamma} L_{\mathrm{CA}+}(c + r, Q; X). \tag{5}$$

To solve the optimization problem, Projected Gradient Descent (PGD) method [32,45] (a.k.a the iterative version of FGSM [17]) can be used. Note that PGD is one of the most effective first-order gradient-based algorithms [1], popular among related works about adversarial attack.

Specifically, in order to find an adversarial perturbation r to create a desired adversarial candidate $\tilde{c} = c + r$, the PGD algorithm alternates two steps at every iteration $t = 1, 2, \ldots, \eta$. Step one updates $\tilde{c}$ according to the gradient of Eq. (4); while step two clips the result of step one to fit in the ε-neighboring region Γ:

$$\tilde{c}_{t+1} = \mathrm{Clip}_{c,\Gamma}\{\tilde{c}_t - \alpha \mathrm{sign}(\nabla_{\tilde{c}_t} L_{\mathrm{CA}+}(\tilde{c}_t, Q, X))\}, \tag{6}$$

where α is a constant hyper-parameter indicating the PGD step size, and $\tilde{c}_1$ is initialized as c. After η iterations, the desired adversarial candidate $\tilde{c}$ is obtained as $\tilde{c}_\eta$, which is optimized to satisfy as many inequalities as possible. Each inequality represents a pairwise ranking sub-problem, hence the adversarial candidate $\tilde{c}$ will be ranked ahead of other candidates with respect to every specified query $q \in Q$.

Likewise, the **CA−** that *lowers* the rank of a candidate c with respect to a set of queries Q can be obtained in similar way:

$$L_{\mathrm{CA}-}(c, Q; X) = \sum_{q \in Q} \sum_{x \in X} \left[-d(q, c) + d(q, x) \right]_+. \tag{7}$$

3.2 Query Attack

Query Attack (**QA**) is supposed to raise (*abbr.* **QA+**) or lower (*abbr.* **QA−**) the rank of a set of candidates $C = \{c_1, c_2, \ldots, c_m\}$ with respect to the query q, by adding adversarial perturbation r to the query $\tilde{q} = q + r$. Thus, **QA** and **CA** are two "symmetric" attacks. The **QA−** for *lowering* the rank could be formulated as follows:

$$r = \arg\max_{r \in \Gamma} \sum_{c \in C} \mathrm{Rank}_X(q + r, c), \tag{8}$$

where Γ is the ε-neighbor of q. Likewise, this attacking objective can also be transformed into the following constrained optimization problem:

$$L_{\mathrm{QA}-}(q, C; X) = \sum_{c \in C} \sum_{x \in X} \left[-d(q, c) + d(q, x) \right]_+, \tag{9}$$

$$r = \arg\min_{r \in \Gamma} L_{\mathrm{QA}-}(q + r, C; X), \tag{10}$$

and it can be solved with the PGD algorithm. Similarly, the **QA+** loss function $L_{\text{QA+}}$ for *raising* the rank of c is as follows:

$$L_{\text{QA+}}(q, C; X) = \sum_{c \in C} \sum_{x \in X} \left[d(q, c) - d(q, x) \right]_+ . \tag{11}$$

Unlike **CA**, **QA** perturbs the *query* image, and hence may drastically change its semantics, resulting in abnormal retrieval results. For instance, after perturbing a "lamp" query image, some unrelated candidates (*e.g.*, "shelf", "toaster", *etc.*) may appear in the top return list. Thus, an ideal query attack should preserve the query semantics, *i.e.*, the candidates in $X \backslash C^2$ should retain their original ranks if possible. Thus, we propose the Semantics-Preserving Query Attack (**SP-QA**) by adding the **SP** term to mitigate the semantic changes q, *e.g.*,

$$L_{\text{SP-QA-}}(q, C; X) = L_{\text{QA-}}(q, C; X) + \xi L_{\text{QA+}}(q, C_{\text{SP}}; X), \tag{12}$$

where $C_{\text{SP}} = \left\{ c \in X \backslash C | \text{Rank}_{X \backslash C}(q, c) \leqslant G \right\}$, *i.e.*, C_{SP} contains the top-G most-relevant candidates corresponding to q, and the $L_{\text{QA+}}(q, C_{\text{SP}}; X)$ term helps preserve the query semantics by retaining some C_{SP} candidates in the retrieved ranking list. Constant G is a predefined integer; and constant ξ is a hyper-parameter for balancing the attack effect and semantics preservation. Unless mentioned, in the following text **QA** means **SP-QA** by default.

3.3   Robustness and Defense

Adversarial defense for classification has been extensively explored, and many of them follows the adversarial training mechanism [25,33,45]. In particular, the adversarial counterparts of the original training samples are used to replace or augment the training samples. Until now, Madry defense [45] is regarded as the most effective [2,71] adversarial training method. However, when directly adapting such classification defense to improve ranking robustness, we empirically discovered a primary challenge of diverging training loss, possibly due to the generated adversarial examples being too "strong". Moreover, such defense mechanism needs to defend against distinct attacks individually. Therefore, a *generic* defense against all the proposed attacks is preferred.

Note that the underlying principle of adversarial ranking attack is to shift the embeddings of candidates/queries to a proper place, and a successful attack depends on a large shift distance as well as a correct shift direction. A large shift distance is an indispensable objective for all the **CA+**, **CA−**, **QA+** and **QA−** attacks. Predictably, a reduction in shift distance could improve model robustness against all attacks simultaneously.

To this end, we propose a "maximum-shift-distance" attack that pushes an embedding vector as far from its original position as possible (resembles Feature Adversary [61] for classification), $r = \arg\max_{r \in \Gamma} d(c + r, c)$. Then we use adversarial examples obtained from this attack to replace original training samples

---
² The complement of the set C.

Table 1. Adversarial ranking attack on vanilla model with MNIST. The "+" attacks (*i.e.* CA+ and QA+) raise the rank of chosen candidates towards 0 (%); while the "−" attacks (*i.e.* CA− and QA−) lower the ranks of chosen candidates towards 100 (%). Applying $\varepsilon = 0.01, 0.03, 0.1, 0.3$ QA+ attacks on the model, the SP term keeps the ranks of C_{SP} no larger than $3.6\%, 5.7\%, 7.7\%, 7.7\%$, respectively, regardless of m. With the QA− counterpart, the ranks of C_{SP} are kept no larger than $1.6\%, 1.6\%, 1.5\%, 1.5\%$, respectively, regardless of m. For all the numbers in the table, "%" is omitted.

ε	CA+				CA−				QA+				QA−			
	$w=1$	2	5	10	$w=1$	2	5	10	$m=1$	2	5	10	$m=1$	2	5	10
(CT) Cosine Distance, Triplet Loss (R@1=99.1%)																
0	50	50	50	50	2.1	2.1	2.1	2.1	50	50	50	50	0.5	0.5	0.5	0.5
0.01	44.6	45.4	47.4	47.9	3.4	3.2	3.1	3.1	45.2	46.3	47.7	48.5	0.9	0.7	0.6	0.6
0.03	33.4	37.3	41.9	43.9	6.3	5.9	5.7	5.6	35.6	39.2	43.4	45.8	1.9	1.4	1.1	1.1
0.1	12.7	17.4	24.4	30.0	15.4	14.9	14.8	14.7	14.4	21.0	30.6	37.2	5.6	4.4	3.7	3.5
0.3	2.1	9.1	13.0	17.9	93.9	93.2	93.0	92.9	6.3	11.2	22.5	32.1	8.6	6.6	5.3	4.8

for adversarial training, hence reduce the shift distance incurred by adversarial perturbations.

A ranking model can be normally trained with the defensive version of the triplet loss:

$$L_{\text{d-t}}(q, c_p, c_n) = L_{\text{triplet}}\Big(q + \arg\max_{r\in\Gamma} d(q+r,q), c_p + \arg\max_{r\in\Gamma} d(c_p+r,c_p),$$

$$c_n + \arg\max_{r\in\Gamma} d(c_n+r,c_n)\Big). \tag{13}$$

4 Experiments

To validate the proposed attacks and defense, we use three commonly used ranking datasets including MNIST [34], Fashion-MNIST [75], and Stanford Online Product (SOP) [53]. We respectively train models on these datasets with PyTorch [59], and conduct attacks[3] on their corresponding test sets (used as X).

Evaluation Metric. Adversarial ranking attack aims to change the ranks of candidates. For each candidate c, its *normalized* rank is calculated as $R(q, c) = \frac{\text{Rank}_X(q,c)}{|X|} \times 100\%$ where $c \in X$, and $|X|$ is the length of full ranking list. Thus, $R(q, c) \in [0, 1]$, and a top ranked c will have a small $R(q, c)$. The attack effectiveness can be measured by the magnitude of change in $R(q, c)$.

Performance of Attack. To measure the performance of a single CA attack, we average the rank of candidate c across every query $q \in Q$, i.e., $R_{\text{CA}}(c) = \sum_{q\in Q} R(q, c)/w$. Similarly, the performance of a single QA attack can be measured by the average rank across every candidate $c \in C$, i.e., $R_{\text{QA}}(q) = \sum_{c\in C} R(q, c)/m$. For the overall performance of an attack, we conduct T times of independent attacks and report the mean of $R_{\text{CA}}(c)$ or $R_{\text{QA}}(q)$, accordingly.

---

[3] Specifically, we use PGD without random starts [45].

Table 2. Adversarial ranking defense with MNIST. Applying $\varepsilon = 0.01, 0.03, 0.1, 0.3$ QA+ attacks on model, the ranks of candidates in C_{SP} are kept no larger than $0.5\%, 0.5\%, 0.5\%, 0.5\%$, respectively, regardless of m. With the QA− counterpart, the ranks of C_{SP} are kept less than $0.4\%, 0.4\%, 0.4\%, 0.4\%$, respectively, regardless of m.

ε	CA+				CA−				QA+				QA−			
	$w=1$	2	5	10	$w=1$	2	5	10	$m=1$	2	5	10	$m=1$	2	5	10
(CTD) Cosine Distance, Triplet Loss, Defensive (R@1=98.3%)																
0	50	50	50	50	2.0	2.0	2.0	2.0	50	50	50	50	0.5		0.5	0.5	0.5
0.01	48.9	49.3	49.4	49.5	2.2	2.2	2.2	2.1	49.9	49.5	49.5	49.7	0.5	0.5	0.5	0.5
0.03	47.4	48.4	48.6	48.9	2.5	2.5	2.4	2.4	48.0	48.5	49.2	49.5	0.6	0.6	0.5	0.5
0.1	42.4	44.2	45.9	46.7	3.8	3.6	3.5	3.4	43.2	45.0	47.4	48.2	1.0	0.8	0.7	0.7
0.3	30.7	34.5	38.7	40.7	7.0	6.7	6.5	6.5	33.2	37.2	42.3	45.1	2.4	1.9	1.6	1.5

CA+ and QA+. For CA+, the query set Q is randomly sampled from X. Likewise, for QA+, the candidate set C is from X. Without attack, both the $R_{\text{CA}}(c)$ and $R_{\text{QA}}(q)$ will approximate to 50%, and the attacks should significantly *decrease* the value.

CA− and QA−. In practice, the Q for CA− and the C for QA− cannot be randomly sampled, because the two attacks are often to lower some top ranked candidates. Thus, the two sets should be selected from the top ranked samples (top-1% in our experiments) in X. Formally, given the candidate c for CA−, we randomly sample the w queries from $\{q \in X | R(c, q) \leqslant 1\%\}$ as Q. Given the query q for QA−, m candidates are randomly sampled from $\{c \in X | R(q, c) \leqslant 1\%\}$ as C. Without attack, both the $R_{\text{CA}}(c)$ and $R_{\text{QA}}(q)$ will be close to 0%, and the attacks should significantly *increase* the value.

Hyper-parameters. We conduct CA with $w \in \{1, 2, 5, 10\}$ queries, and QA with $m \in \{1, 2, 5, 10\}$ candidates, respectively. In QA, we let $G = 5$. The SP balancing parameter ξ is set to 1 for QA+, and 10^2 for QA−. In addition, We investigate attacks of different strength ε, *i.e.* $0.01, 0.03, 0.1, 0.3$ on MNIST and Fashion-MNIST following [45], and $0.01, 0.03, 0.06$ on SOP following [33]. The PGD step size is empirically set to $\alpha = \min(\max(\frac{\varepsilon}{10}, \frac{1}{255}), 0.01)$, and the number of PGD iterations to $\eta = \min(\max(10, \frac{2\varepsilon}{\alpha}), 30)$. We perform $T = |X|$ times of attack to obtain the reported performance.

Adversarial Defense. Ranking models are trained using Eq. (13) with the strongest adversary following the procedure of Madry defense [45].

4.1 MNIST Dataset

Following conventional settings with the MNIST [34] dataset, we train a CNN ranking model comprising 2 convolutional layers and 1 fully-connected layer. This CNN architecture (denoted as C2F1) is identical to the one used in [45] except for the removal of the last fully-connected layer. Specifically, the ranking model is trained with cosine distance and triplet loss. The retrieval performance of the model is Recall@1 = 99.1% (R@1), as shown in Table 1 in grey highlight.

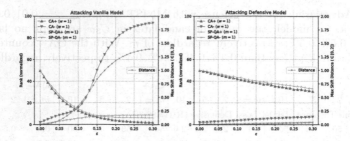

Fig. 2. Comparison of Attacks on vanilla and defensive models. Apart from the ranks of chosen candidates, We also measure the maximum shift distance of embedding vectors that adversarial perturbation could incur.

Attacking results against this vanilla model (*i.e.*, the ranking model which is not enhanced with our defense method) are presented in Table 1. For example, a strong **CA+** attack (*i.e.*, $\varepsilon = 0.3$) for $w = 1$ can raise the rank $R_{CA}(c)$ from 50% to 2.1%. Likewise, the rank of C can be raised to 9.1%, 13.0%, 17.9% for $w = 2, 5, 10$ chosen queries, respectively. On the other hand, a strong **CA−** attack for $w = 1$ can lower the rank $R_{CA}(c)$ from 2.1% to 93.9%. The results of strong **CA−** attacks for $w = 2, 5, 10$ are similar to the $w = 1$ case.

The results of **QA+** and **QA−** are also shown in Table 1. the rank changes with **QA** attacks are less dramatic (but still significant) than **CA**. This is due to the additional difficulty introduced by **SP** term in Eq. (12), and the **QA** attack effectiveness is inversely correlated with ξ. For instance, a strong **QA−** for $m = 1$ can only lower the rank $R_{QA}(q)$ from 0.5% to 8.6%, but the attacking effect can be further boosted by decreasing ξ. More experimental results are presented in following discussion. In brief, our proposed attacks against the vanilla ranking model is effective.

Next, we evaluate the performance of our defense method. Our defense should be able to enhance the robustness of a ranking model, which can be measured by the difference between the attack effectiveness with our defense and the attack effectiveness without our defense. As a common phenomenon of adversarial training, our defense mechanism leads to a slight retrieval performance degradation for unperturbed input (highlighted in blue in Table 2), but the attacking effectiveness is clearly mitigated by our defense. For instance, the same strong **CA+** attack for $w = 1$ on the defensive model (*i.e.*, the ranking model which is enhanced by our defense method) can only raise the rank $R_{CA}(c)$ from 50% to 30.7%, compared to its vanilla counterpart raising to 2.1%. Further analysis suggests that the weights in the first convolution layer of the defensive model are closer to 0 and have smaller variance than those of the vanilla model, which may help resist adversarial perturbation from changing the layer outputs into the local linear area of ReLU [17].

To visualize the effect of our attacks and defense, we track the attacking effect with ε varying from 0.0 to 0.3 on the vanilla and defensive models, as shown in Fig. 2. It is noted that our defense could significantly suppress the maximum

Table 3. Adversarial ranking attack and defense on Fashion-MNIST. The lowest ranks of C_{SP} are 3.0%, 5.2%, 7.8%, 8.3% in QA+, and 1.9%, 1.9%, 1.9%, 1.8% for QA+, respectively.

ε	CA+				CA−				QA+				QA−			
	$w=1$	2	5	10	$w=1$	2	5	10	$m=1$	2	5	10	$m=1$	2	5	10
(CT) Cosine Distance, Triplet Loss (R@1=88.8%)																
0	50	50	50	50	1.9	1.9	1.9	1.9	50	50	50	50	0.5	0.5	0.5	0.5
0.01	36.6	39.9	43.2	44.8	5.6	5.1	4.9	4.8	39.4	42.0	45.3	47.1	2.1	1.6	1.2	1.1
0.03	19.7	25.4	31.7	35.6	15.5	14.8	14.4	14.3	21.7	28.2	35.7	40.6	5.6	4.1	3.3	2.9
0.1	3.7	10.5	17.3	22.7	87.2	86.7	86.3	86.3	7.1	12.4	23.6	32.5	10.9	8.3	6.7	6.0
0.3	**1.3**	**9.4**	**16.0**	**21.5**	**100.0**	**100.0**	**100.0**	**100.0**	**6.3**	**10.8**	**21.8**	**31.7**	**12.6**	**9.4**	**7.5**	**6.6**
(CTD) Cosine Distance, Triplet Loss, Defensive (R@1=79.6%)																
0	50	50	50	50	1.2	1.2	1.2	1.2	50	50	50	50	0.5	0.5	0.5	0.5
0.01	48.9	48.9	49.3	49.3	1.4	1.4	1.4	1.4	49.4	49.9	49.9	50.0	0.5	0.5	0.5	0.5
0.03	47.1	47.9	48.3	48.3	2.0	1.9	1.8	1.8	48.3	49.1	49.5	49.8	0.7	0.6	0.6	0.6
0.1	42.4	43.5	44.5	44.8	4.6	4.2	4.0	3.9	45.4	47.2	48.7	49.2	1.4	1.2	1.1	1.1
0.3	**32.5**	**35.4**	**37.5**	**38.2**	**11.2**	**10.5**	**10.1**	**10.0**	**39.3**	**42.6**	**46.5**	**47.8**	**3.9**	**3.3**	**3.0**	**2.9**

embedding shift distance incurred by adversarial perturbation to nearly 0, but the defensive model is still not completely immune to attacks. We speculate the defensive model still has "blind spots" [69] in some local areas that could be exploited by the attacks.

In summary, these results and further experiments suggest that: (1) deep ranking models are vulnerable to adversarial ranking attacks, no matter what loss function or distance metric is selected; (2) vanilla models trained with contrastive loss are more robust than those trained with triplet loss. This is possibly due to contrastive loss explicitly reducing the intra-class embedding variation. Additionally, our defense method could consistently improve the robustness of all these models; (3) Euclidean distance-based models are harder to defend than cosine distance-based ones. Beyond these experiments, we also find that the margin hyper-parameter β of triplet loss and the dimensionality of the embedding space have marginal influences on model robustness.

4.2 Fashion-MNIST Dataset

Fashion-MNIST [75] is an MNIST-like but more difficult dataset, comprising $60,000$ training examples and $10,000$ test samples. The samples are 28×28 greyscale images covering 10 different fashion product classes, including "T-shirt" and "dress", *etc.* We train the vanilla and defensive models based on the cosine distance and triplet loss and conduct attack experiments.

The attack and defense results are available in Table 3. From the table, we note that our attacks could achieve better effect compared to experiments on MNIST. For example, in a strong **CA+** for $w=1$, the rank $R_{CA}(c)$ can be raised to 1.3%. On the other hand, despite the moderate improvement in robustness, the defensive model performs worse in unperturbed sample retrieval. The performance degradation is more pronounced on this dataset compared to MNIST. We speculate the differences are related to the increased dataset difficulty.

Table 4. Adversarial ranking attack and defense on SOP. With different ε, the worst ranks of C_{SP} in QA+ are $0.2\%, 0.7\%, 2.0\%, 3.3\%$, and those for QA− are $0.4\%, 0.7\%, 0.8\%, 1.0\%$, respectively.

ε	CA+				CA−				QA+				QA−			
	$w=1$	2	5	10	$w=1$	2	5	10	$m=1$	2	5	10	$m=1$	2	5	10
(ET) Euclidean Distance, Triplet Loss (R@1=63.1%)																
0	50	50	50	50	1.9	1.9	1.9	1.9	50	50	50	50	0.5	0.5	0.5	0.5
0.01	0.0	0.8	2.0	2.6	99.7	99.6	99.4	99.3	4.8	7.0	16.3	25.8	54.9	40.2	27.1	21.9
0.03	0.0	0.3	1.0	1.5	100.0	100.0	100.0	100.0	1.6	3.3	10.0	19.2	68.1	52.4	36.6	30.1
0.06	0.0	0.2	1.0	1.5	100.0	100.0	100.0	100.0	1.1	2.7	8.8	17.6	73.8	57.9	40.3	32.4
(ETD) Euclidean Distance, Triplet Loss, Defensive (R@1=46.4%)																
0	50	50	50	50	2.0	2.0	2.0	2.0	50	50	50	50	0.5	0.5	0.5	0.5
0.01	7.5	12.2	16.5	18.0	66.4	62.6	59.3	57.8	16.1	24.8	36.1	41.4	26.7	18.1	12.2	10.2
0.03	0.7	4.5	8.7	10.4	91.7	90.2	89.1	88.4	7.9	14.5	27.2	35.6	43.4	31.7	21.9	18.1
0.06	0.1	3.8	7.9	9.7	97.3	96.8	96.4	96.2	6.9	12.5	24.3	33.4	51.4	39.0	28.0	23.5

4.3 Stanford Online Products Dataset

Stanford Online Products (SOP) dataset [53] contains 120k images of 23k classes of real online products from eBay for metric learning. We use the same dataset split as used in the original work [53]. We also train the same vanilla ranking model using the same triplet ranking loss function with Euclidean distance, except that the GoogLeNet [67] is replaced with ResNet-18 [21]. The ResNet-18 achieves better retrieval performance.

Attack and defense results on SOP are present in Table 4. It is noted that our attacks are quite effective on this difficult large-scale dataset, as merely 1% perturbation ($\varepsilon = 0.01$) to any candidate image could make it ranked ahead or behind of nearly all the rest candidates (as shown by the CA+ and CA− results with $w = 1$). QA on this dataset is significantly effective as well. On the other hand, our defense method leads to decreased retrieval performance, $i.e.$ R@1 from 63.1% to 46.4%, which is expected on such a difficult dataset. Meanwhile, our defense could moderately improve the model robustness against relatively weaker adversarial examples ($e.g.\ \varepsilon = 0.01$), but improving model robustness on this dataset is more difficult, compared to experiments on other datasets.

By comparing the results among all the three datasets, we find ranking models trained on harder datasets more susceptible to adversarial attack, and more difficult to defend. Therefore, we speculate that models used in realistic applications could be easier to attack, because they are usually trained on larger-scale and more difficult datasets.

5 Discussions

White-box attacks are sometimes limited by data accessibility in practice, but it's possible to circumvent them with adversarial example transferability and universal perturbation, as will be discussed in this section. Such properties reveal the possibility of practical black-box attack.

Table 5. Transferring adversarial ranking examples generated from one model to another. We report the rank of the same c with respect to the same q across different models to illustrate the transfer attack effectiveness. Transferring adversarial examples to a model itself (the diagonal lines) is equivalent to white-box attack.

CA+ Transfer (Black Box), $w = 1$				QA+ Transfer (Black Box), $m = 1$			
From \ To	LeNet	C2F1	Res18	From \ To	LeNet	C2F1	Res18
LeNet	50→16.6	35.1	34.3	LeNet	50→20.5	43.0	45.8
C2F1	28.6	50→2.1	31.3	C2F1	43.5	50→6.3	45.4
Res18	24.4	27.0	50→2.2	Res18	41.4	40.4	50→14.1

CA- Transfer (Black Box), $w = 1$				QA- Transfer (Black Box), $m = 1$			
From \ To	LeNet	C2F1	Res18	From \ To	LeNet	C2F1	Res18
LeNet	2.5→63.7	2.1→10.0	2.1→9.1	LeNet	0.5→7.0	0.5→1.6	0.5→1.8
C2F1	2.5→9.1	2.1→93.9	2.1→9.3	C2F1	0.5→1.0	0.5→8.6	0.5→1.9
Res18	2.5→9.9	2.1→11.8	2.1→66.7	Res18	0.5→0.8	0.5→1.2	0.5→6.9

5.1 Adversarial Example Transferability

As demonstrated in the experiments, deep ranking models can be compromised by our white-box attacks. In realistic scenarios, the white-box attacks are not practical enough because the model to be attacked is often unknown (*i.e.*, the architecture and parameters are unknown). On the other hand, adversarial examples for classification have been found transferable [55,56] (*i.e.* model-agnostic) between different models with different network architectures. Typically, in this case, adversarial examples are generated from a replacement model [56] using a white-box attack, and are directly used to attack the black-box model.

Adversarial ranking attack could be more practical if the adversarial ranking examples have the similar transferability. Besides the C2F1 model, we train two vanilla models on the MNIST dataset: (1) LeNet [34], which has lower model capacity compared to C2F1; (2) ResNet-18 [21] (denoted as Res18), which has a better network architecture and higher model capacity.

The results are present in Table 5. For example, in the **CA+** transfer attack, we generate adversarial candidates from the C2F1 model and directly use them to attack the Res18 model (row 2, column 3, top-left table), and the ranks of the adversarial candidates with respect to the same query is still raised to 31.3%. We also find the **CA−** transfer attack is effective, where the ranks of our adversarial candidates are lowered, *e.g.* from 2.1% to 9.3% (row 2, column 3, bottom-left table). Similar results can be observed on the **QA** transfer experiments, and they show weaker effect due to the SP term.

Table 6. Universal adversarial perturbation for ranking on MNIST. Each pair of results presents the original rank of chosen candidates and that after adding adversarial perturbation. Both w, m are set to 1. Parameter ξ is set to 0 to reduce attack difficulty.

CA+	CA-	QA+	QA-
$50 \rightarrow 2.1$	$2.1 \rightarrow 93.9$	$50 \rightarrow 0.2$	$0.5 \rightarrow 94.1$
I-CA+	I-CA-	I-QA+	I-QA-
$50 \rightarrow 18.1$	$0.6 \rightarrow 9.5$	$50 \rightarrow 20.5$	$2.1 \rightarrow 7.6$
I-CA+ (unseen)	I-CA- (unseen)	I-QA+ (unseen)	I-QA- (unseen)
$50 \rightarrow 18.5$	$0.7 \rightarrow 9.4$	$50 \rightarrow 21.0$	$2.2 \rightarrow 7.4$

From these results, we find that: (1) CNN with better architecture and higher model capacity (*i.e.*, Res18), is less susceptible to adversarial ranking attack. This conclusion is consistent with one of Madry's [45], which claims that higher model capacity could help improve model robustness; (2) adversarial examples generated from the Res18 have the most significant effectiveness in transfer attack; (3) CNN of low model capacity (*i.e.*, LeNet), performs moderately in terms of both adversarial example transferability and model robustness. We speculate its robustness stems from a forced regularization effect due low model capacity. Beyond these, we also noted adversarial ranking examples are transferable disregarding the difference in loss function or distance metric.

Apart from transferability across different architectures, we also investigated the transferability between several independently trained C2F1 models. Results suggest similar transferability between them. Notably, when transferring adversarial examples to a defensive C2F1 model, the attacking effect is significantly mitigated. The result further demonstrates the effectiveness of our defense.

5.2   Universal Perturbation for Ranking

Recently, universal (*i.e.* image-agnostic) adversarial perturbation [49] for classification has been found possible, where a single perturbation may lead to misclassification when added to any image. Thus, we also investigate the existence of universal adversarial perturbation for adversarial ranking attack.

To this end, we follow [49] and formulate the image-agnostic CA+ (*abbr.* **I-CA+**). Given a set of candidates $C = \{c_1, c_2, \ldots, c_m\}$ and a set of queries $Q = \{q_1, q_2, \ldots, q_w\}$, **I-CA+** is to find a *single* universal adversarial perturbation r, so that the rank of every perturbed candidate $\tilde{c} = c + r$ ($c \in C$) with respect to Q can be raised. The corresponding optimization problem of **I-CA+** is:

$$r = \arg\min_{r \in \Gamma} \sum_{c \in C} L_{\text{CA+}}(c + r, Q; X). \tag{14}$$

When applied with such universal perturbation, the rank of any candidate $w.r.t$ Q is expected to be raised. The objective functions of **I-QA−**, **I-QA+** and **I-QA−** can be obtained in similar way. Note, unlike [36] which aims to find universal perturbation that can make image retrieval system return irrelevant results, our universal perturbations have distinct purposes.

We conduct experiment on the MNIST dataset. For **I-CA+** attack, we randomly sample 5% of X for generating the universal perturbation. Following [49], another non-overlapping 5% examples are randomly sampled from X to test whether the generated perturbation is generalizable on "unseen" (*i.e.*, not used for generating the perturbation) images. Experiments for the other image-agnostic attacks are conducted similarly. Note, we only report the **I-QA−** and **I-QA−** effectiveness on the 1% top ranked samples, similar to **CA−** and **QA−**.

As shown in Table 6, our **I-CA** can raise the ranks of C to 18.1%, or lower them to 9.5%. When added to "unseen" candidate images, the universal perturbation could retain nearly the same effectiveness, possibly due to low intra-class variance of the MNIST dataset.

6   Conclusion

Deep ranking models are vulnerable to adversarial perturbations that could intentionally change the ranking result. In this paper, the *adversarial ranking attack* that can compromise deep ranking models is defined and implemented. We also propose an *adversarial ranking defense* that can significantly suppress embedding shift distance and moderately improve the ranking model robustness. Moreover, the transferability of our adversarial examples and the existence of universal adversarial perturbations for ranking attack illustrate the possibility of practical black-box attack and potential risk of realistic ranking applications.

Acknowledgment. This work was supported partly by National Key R&D Program of China Grant 2018AAA0101400, NSFC Grants 61629301, 61773312, 61976171, and 61672402, China Postdoctoral Science Foundation Grant 2019M653642, Young Elite Scientists Sponsorship Program by CAST Grant 2018QNRC001, and Natural Science Foundation of Shaanxi Grant 2020JQ-069.

References

1. Athalye, A., Carlini, N.: On the robustness of the CVPR 2018 white-box adversarial example defenses. arXiv preprint arXiv:1804.03286 (2018)
2. Athalye, A., Carlini, N., Wagner, D.: Obfuscated gradients give a false sense of security: Circumventing defenses to adversarial examples. arXiv preprint arXiv:1802.00420 (2018)
3. Bui, T., Ribeiro, L., Ponti, M., Collomosse, J.: Compact descriptors for sketch-based image retrieval using a triplet loss convolutional neural network. CVIU **164**, 27–37 (2017)
4. Carlini, N., Wagner, D.: Defensive distillation is not robust to adversarial examples. arXiv preprint arXiv:1607.04311 (2016)
5. Carlini, N., Wagner, D.: Towards evaluating the robustness of neural networks. In: 2017 IEEE Symposium on Security and Privacy (SP), pp. 39–57. IEEE (2017)
6. Chechik, G., Sharma, V., Shalit, U., Bengio, S.: Large scale online learning of image similarity through ranking. JMLR **11**, 1109–1135 (2010)

7. Chen, J., Jordan, M.I.: Boundary attack++: Query-efficient decision-based adversarial attack. arXiv preprint arXiv:1904.02144 (2019)
8. Chen, P.Y., Sharma, Y., Zhang, H., Yi, J., Hsieh, C.J.: EAD: elastic-net attacks to deep neural networks via adversarial examples. In: AAAI (2018)
9. Croce, F., Hein, M.: Sparse and imperceivable adversarial attacks. In: ICCV, pp. 4724–4732 (2019)
10. Dong, Y., Liao, F., Pang, T., Su, H., Zhu, J., Hu, X., Li, J.: Boosting adversarial attacks with momentum. In: CVPR (June 2018)
11. Dong, Y., Pang, T., Su, H., Zhu, J.: Evading defenses to transferable adversarial examples by translation-invariant attacks. In: CVPR, pp. 4312–4321 (2019)
12. Dong, Y., et al.: Efficient decision-based black-box adversarial attacks on face recognition. In: CVPR, pp. 7714–7722 (2019)
13. Dong, Y., Su, H., Zhu, J., Bao, F.: Towards interpretable deep neural networks by leveraging adversarial examples. arXiv preprint arXiv:1708.05493 (2017)
14. Dubey, A., van der Maaten, L., Yalniz, Z., Li, Y., Mahajan, D.: Defense against adversarial images using web-scale nearest-neighbor search. In: CVPR, pp. 8767–8776 (2019)
15. Faghri, F., Fleet, D.J., Kiros, J.R., Fidler, S.: VSE++: Improved visual-semantic embeddings, vol. 2, no. 7, p. 8. arXiv preprint arXiv:1707.05612 (2017)
16. Ganeshan, A., Babu, R.V.: FDA: feature disruptive attack. In: ICCV, pp. 8069–8079 (2019)
17. Goodfellow, I.J., Shlens, J., Szegedy, C.: Explaining and harnessing adversarial examples. arXiv preprint arXiv:1412.6572 (2014)
18. Gopinath, D., Katz, G., Pasareanu, C.S., Barrett, C.: DeepSafe: A data-driven approach for checking adversarial robustness in neural networks. arXiv preprint arXiv:1710.00486 (2017)
19. Goren, G., Kurland, O., Tennenholtz, M., Raiber, F.: Ranking robustness under adversarial document manipulations. In: ACM SIGIR, pp. 395–404. ACM (2018)
20. Guo, C., Rana, M., Cisse, M., Van Der Maaten, L.: Countering adversarial images using input transformations. arXiv preprint arXiv:1711.00117 (2017)
21. He, K., Zhang, X., Ren, S., Sun, J.: Deep residual learning for image recognition. In: CVPR (June 2016)
22. He, W., Wei, J., Chen, X., Carlini, N., Song, D.: Adversarial example defense: ensembles of weak defenses are not strong. In: 11th USENIX Workshop on Offensive Technologies, WOOT 2017 (2017)
23. He, X., He, Z., Du, X., Chua, T.S.: Adversarial personalized ranking for recommendation. In: ACM SIGIR, pp. 355–364. ACM (2018)
24. Huang, Q., et al.: Intermediate level adversarial attack for enhanced transferability. arXiv preprint arXiv:1811.08458 (2018)
25. Huang, R., Xu, B., Schuurmans, D., Szepesvári, C.: Learning with a strong adversary. CoRR abs/1511.03034 (2015). http://arxiv.org/abs/1511.03034
26. Jacob, P., Picard, D., Histace, A., Klein, E.: Metric learning with horde: high-order regularizer for deep embeddings. In: ICCV, pp. 6539–6548 (2019)
27. Joachims, T.: Optimizing search engines using clickthrough data. In: ACM SIGKDD, pp. 133–142. ACM (2002)
28. Katz, G., Barrett, C., Dill, D.L., Julian, K., Kochenderfer, M.J.: Reluplex: an efficient SMT solver for verifying deep neural networks. In: Majumdar, R., Kunčak, V. (eds.) CAV 2017. LNCS, vol. 10426, pp. 97–117. Springer, Cham (2017). https://doi.org/10.1007/978-3-319-63387-9_5
29. Kiros, R., Salakhutdinov, R., Zemel, R.S.: Unifying visual-semantic embeddings with multimodal neural language models. arXiv preprint arXiv:1411.2539 (2014)

30. Komkov, S., Petiushko, A.: AdvHat: Real-world adversarial attack on ArcFace Face ID system. arXiv preprint arXiv:1908.08705 (2019)
31. Krizhevsky, A., Sutskever, I., Hinton, G.E.: ImageNet classification with deep convolutional neural networks. In: NeurIPS, pp. 1097–1105 (2012)
32. Kurakin, A., Goodfellow, I., Bengio, S.: Adversarial examples in the physical world. arXiv preprint arXiv:1607.02533 (2016)
33. Kurakin, A., Goodfellow, I., Bengio, S.: Adversarial machine learning at scale. arXiv preprint arXiv:1611.01236 (2016)
34. LeCun, Y., Bottou, L., Bengio, Y., Haffner, P., et al.: Gradient-based learning applied to document recognition. Proc. IEEE **86**(11), 2278–2324 (1998)
35. Lee, K.-H., Chen, X., Hua, G., Hu, H., He, X.: Stacked cross attention for image-text matching. In: Ferrari, V., Hebert, M., Sminchisescu, C., Weiss, Y. (eds.) ECCV 2018. LNCS, vol. 11208, pp. 212–228. Springer, Cham (2018). https://doi.org/10.1007/978-3-030-01225-0_13
36. Li, J., Ji, R., Liu, H., Hong, X., Gao, Y., Tian, Q.: Universal perturbation attack against image retrieval. In: ICCV, pp. 4899–4908 (2019)
37. Liu, H., et al.: Universal adversarial perturbation via prior driven uncertainty approximation. In: ICCV, pp. 2941–2949 (2019)
38. Liu, T.Y., et al.: Learning to rank for information retrieval. Found. Trends® Inf. Retr. **3**(3), 225–331 (2009)
39. Liu, X., Cheng, M., Zhang, H., Hsieh, C.-J.: Towards robust neural networks via random self-ensemble. In: Ferrari, V., Hebert, M., Sminchisescu, C., Weiss, Y. (eds.) ECCV 2018. LNCS, vol. 11211, pp. 381–397. Springer, Cham (2018). https://doi.org/10.1007/978-3-030-01234-2_23
40. Liu, X., Li, Y., Wu, C., Hsieh, C.J.: Adv-BNN: Improved adversarial defense through robust Bayesian neural network. arXiv preprint arXiv:1810.01279 (2018)
41. Liu, Y., Chen, X., Liu, C., Song, D.: Delving into transferable adversarial examples and black-box attacks. arXiv preprint arXiv:1611.02770 (2016)
42. Liu, Z., Zhao, Z., Larson, M.: Who's afraid of adversarial queries?: the impact of image modifications on content-based image retrieval. In: ICMR, pp. 306–314. ACM (2019)
43. Lu, J., Issaranon, T., Forsyth, D.: SafetyNet: detecting and rejecting adversarial examples robustly. In: ICCV, pp. 446–454 (2017)
44. Luo, Y., Boix, X., Roig, G., Poggio, T., Zhao, Q.: Foveation-based mechanisms alleviate adversarial examples. arXiv preprint arXiv:1511.06292 (2015)
45. Madry, A., Makelov, A., Schmidt, L., Tsipras, D., Vladu, A.: Towards deep learning models resistant to adversarial attacks. arXiv preprint arXiv:1706.06083 (2017)
46. Mao, C., Zhong, Z., Yang, J., Vondrick, C., Ray, B.: Metric learning for adversarial robustness. In: NeurIPS, pp. 478–489 (2019)
47. Meng, D., Chen, H.: MagNet: a two-pronged defense against adversarial examples. In: ACM SIGSAC, pp. 135–147. ACM (2017)
48. Metzen, J.H., Genewein, T., Fischer, V., Bischoff, B.: On detecting adversarial perturbations. arXiv preprint arXiv:1702.04267 (2017)
49. Moosavi-Dezfooli, S.M., Fawzi, A., Fawzi, O., Frossard, P.: Universal adversarial perturbations. In: CVPR, pp. 1765–1773 (2017)
50. Moosavi-Dezfooli, S.M., Fawzi, A., Frossard, P.: DeepFool: a simple and accurate method to fool deep neural networks. In: CVPR, pp. 2574–2582 (2016)
51. Mummadi, C.K., Brox, T., Metzen, J.H.: Defending against universal perturbations with shared adversarial training. In: ICCV, pp. 4928–4937 (2019)
52. Niu, Z., Zhou, M., Wang, L., Gao, X., Hua, G.: Hierarchical multimodal LSTM for dense visual-semantic embedding. In: ICCV, pp. 1881–1889 (2017)

53. Oh Song, H., Xiang, Y., Jegelka, S., Savarese, S.: Deep metric learning via lifted structured feature embedding. In: CVPR, pp. 4004–4012 (2016)
54. Papernot, N., McDaniel, P.: On the effectiveness of defensive distillation. arXiv preprint arXiv:1607.05113 (2016)
55. Papernot, N., McDaniel, P., Goodfellow, I.: Transferability in machine learning: from phenomena to black-box attacks using adversarial samples. arXiv preprint arXiv:1605.07277 (2016)
56. Papernot, N., McDaniel, P., Goodfellow, I., Jha, S., Celik, Z.B., Swami, A.: Practical black-box attacks against machine learning. In: Proceedings of the 2017 ACM on Asia Conference on Computer and Communications Security, pp. 506–519. ACM (2017)
57. Papernot, N., McDaniel, P., Jha, S., Fredrikson, M., Celik, Z.B., Swami, A.: The limitations of deep learning in adversarial settings. In: 2016 IEEE European Symposium on Security and Privacy (EuroS&P), pp. 372–387. IEEE (2016)
58. Papernot, N., McDaniel, P., Wu, X., Jha, S., Swami, A.: Distillation as a defense to adversarial perturbations against deep neural networks. In: 2016 IEEE Symposium on Security and Privacy (SP), pp. 582–597. IEEE (2016)
59. Paszke, A., et al.: Automatic differentiation in PyTorch (2017)
60. Prakash, A., Moran, N., Garber, S., DiLillo, A., Storer, J.: Deflecting adversarial attacks with pixel deflection. In: CVPR, pp. 8571–8580 (2018)
61. Sabour, S., Cao, Y., Faghri, F., Fleet, D.J.: Adversarial manipulation of deep representations. arXiv preprint arXiv:1511.05122 (2015)
62. Schroff, F., Kalenichenko, D., Philbin, J.: FaceNet: a unified embedding for face recognition and clustering. In: CVPR, pp. 815–823 (2015)
63. Shaham, U., Yamada, Y., Negahban, S.: Understanding adversarial training: increasing local stability of supervised models through robust optimization. Neurocomputing 307, 195–204 (2018)
64. Sharif, M., Bhagavatula, S., Bauer, L., Reiter, M.K.: Accessorize to a crime: real and stealthy attacks on state-of-the-art face recognition. In: ACM SIGSAC, pp. 1528–1540. ACM (2016)
65. Shi, Y., Wang, S., Han, Y.: Curls & Whey: boosting black-box adversarial attacks. arXiv preprint arXiv:1904.01160 (2019)
66. Su, J., Vargas, D.V., Sakurai, K.: One pixel attack for fooling deep neural networks. IEEE Trans. Evol. Comput. 23, 828–841 (2019)
67. Szegedy, C., et al.: Going deeper with convolutions. In: CVPR, pp. 1–9 (2015)
68. Szegedy, C., Vanhoucke, V., Ioffe, S., Shlens, J., Wojna, Z.: Rethinking the inception architecture for computer vision. In: CVPR, pp. 2818–2826 (2016)
69. Szegedy, C., et al.: Intriguing properties of neural networks. arXiv preprint arXiv:1312.6199 (2013)
70. Wang, J., et al.: Learning fine-grained image similarity with deep ranking. In: CVPR, pp. 1386–1393 (2014)
71. Wang, J., Zhang, H.: Bilateral adversarial training: towards fast training of more robust models against adversarial attacks. In: ICCV, pp. 6629–6638 (2019)
72. Wang, Z., Zheng, S., Song, M., Wang, Q., Rahimpour, A., Qi, H.: advPattern: physical-world attacks on deep person re-identification via adversarially transformable patterns. In: ICCV, pp. 8341–8350 (2019)
73. Wu, L., Zhu, Z., Tai, C., et al.: Understanding and enhancing the transferability of adversarial examples. arXiv preprint arXiv:1802.09707 (2018)
74. Xiao, C., Zhu, J.Y., Li, B., He, W., Liu, M., Song, D.: Spatially transformed adversarial examples. arXiv preprint arXiv:1801.02612 (2018)

75. Xiao, H., Rasul, K., Vollgraf, R.: Fashion-MNIST: a novel image dataset for benchmarking machine learning algorithms. arXiv preprint arXiv:1708.07747 (2017)
76. Xie, C., et al.: Improving transferability of adversarial examples with input diversity. In: CVPR, pp. 2730–2739 (2019)
77. Yuan, X., He, P., Zhu, Q., Li, X.: Adversarial examples: attacks and defenses for deep learning. IEEE TNNLS **30**, 2805–2824 (2019)
78. Zhong, Y., Deng, W.: Adversarial learning with margin-based triplet embedding regularization. In: ICCV, pp. 6549–6558 (2019)

Author Index

Printed in the United States
By Bookmasters